A Textbook of

Electrical Technology

VOLUME I

A Textbook of Electrical Technology

VOLUME I

BASIC ELECTRICAL ENGINEERING

IN

SI SYSTEM OF UNITS

(Including rationalized MKSA System)

For the Examinations of BE (Common Course to All Branches), B Tech., BSc (Engg), Sec. A & B of AMIE(I), AMIEE (London), IERE (London), Grade IETE, Diploma and other Competitive Examinations

BL THERAJA

AK THERAJA

S Chand And Company Limited

(ISO 9001 Certified Company)

S Chand And Company Limited

(ISO 9001 Certified Company)

Head Office: D-92, Sector–2, Noida – 201301, U.P. (India), Ph. 91-120-4682700

Registered Office: A-27, 2nd Floor, Mohan Co-operative Industrial Estate, New Delhi – 110 044, Phone: 011-49731800

www.**schandpublishing.com**; e-mail: **info@schandpublishing.com**

Marketing Offices:

Chennai	:	Ph: 23632120; chennai@schandpublishing.com
Guwahati	:	Ph: 2738811, 2735640; guwahati@schandpublishing.com
Hyderabad	:	Ph: 40186018; hyderabad@schandpublishing.com
Jalandhar	:	Ph: 4645630; jalandhar@schandpublishing.com
Kolkata	:	Ph: 23357458, 23353914; kolkata@schandpublishing.com
Lucknow	:	Ph: 4003633; lucknow@schandpublishing.com
Mumbai	:	Ph: 25000297; mumbai@schandpublishing.com
Patna	:	Ph: 4011400; patna@schandpublishing.com

S CHAND'S Seal of Trust

In our endeavour to protect you against counterfeit/fake books, we have pasted a hologram over the cover of this book. The hologram displays full visible effect, emboss effect, relief effect, mirror lens effect, pearl effect, motion effect, animated text, kinetic effect, concealed effect, micro structure, multicolour small text 'S CHAND', nanotext '50 micron' 'ORIGINAL' 'S CHAND', mirror strip '6.5 mm', mirror lens 3 mm with text 'SC', microtext 'OK', scratch strip '7 mm', color sparkling effect under scratch and QR code size 10 mm x 10 mm, etc.

A fake hologram does not display ALL these effects.

First Edition 1959
Subsequent Editions and Reprints 1960, 61, 62, 64, 65, 66, 67, 68, 69, 70, 71, 72, 73 (Twice), 74, 75, 76, 77, 78, 79, 80, 81, 82, 83, 84 (Twice), 85, 86, 89, 91, 93, 94, 95, 96, 97, 98, 99, 2001, 2002 (Twice), 2003, 2004; First Multicolour Edition 2005, Reprints 2006 (Twice), 2009 (Twice), 2010 (Twice), 2011 (Twice), 2012, 2013 (Twice), 2014, 2015, 2016 2018 (Twice), 2019 (Twice), 2020 (Twice), 2022 (Thrice), 2023 (Twice), 2024

Twenty Fourth Edition 2025; Reprint 2025
Reprint 2026

ISBN: 978-93-5501-826-7 **Product Code:** H3BEE52BEEL10ENAX25O

PRINTED IN INDIA

By Shri Shyamlal Printing Press Private Limited, Plot 20/4, Site-IV, Industrial Area Sahibabad, Ghaziabad – 201 010 and Published by S Chand And Company Limited, A-27, 2nd Floor, Mohan Co-operative Industrial Estate, New Delhi – 110 044.

Preface to the Twenty Fourth Edition

The field of Electrical Technology is vast and foundational to numerous engineering disciplines, making it essential for students, practitioners, and educators alike. "A Textbook of Electrical Technology: Volume I" is a comprehensive resource that aims to bridge the gap between theoretical concepts and practical applications. It covers the fundamental principles as well as advanced topics in electrical engineering, providing a solid foundation for all its readers.

This book is designed to cater to the needs of undergraduate students pursuing degrees in Electrical, Electronics and related engineering fields. It provides a solid grounding in essential topics such as Electric Current, Ohm's Law, DC Network Theorems, Electrostatics, Magnetism, AC Fundamentals, Electrochemical Power Sources, Electrical Instruments and Measurements, Polyphase Circuits, Fourier Series, Electrical Energy Generation and much more. The content is carefully curated to ensure that readers not only grasp the theoretical aspects but also gain insights into real-world applications and problem-solving techniques, preparing them for the challenges of the industry.

Each chapter in this book is structured to offer a clear and logical progression of topics. Beginning with basic concepts, the discussion gradually advances to more complex theories and applications. This approach ensures that readers develop a strong foundation before tackling intricate subjects. Numerous examples, diagrams and illustrations are included to facilitate understanding and to make learning more engaging.

Additionally, we have the following chapters available on our website for further gleaning of the subject available at https://www.schandpublishing.com/books/tech-professional/electrical-engineering-electronics/a-textbook-electrical-technology-volume-i-basic-electrical-engineering/9789355018267/

- Chapter 21: AC Bridges
- Chapter 22: AC Filter Network
- Chapter 23: Circle Diagram
- Chapter 24: Symmetrical Components

To aid in the retention of knowledge, the book features a variety of problems and exercises at the end of each chapter. These are designed to test comprehension, reinforce learning and encourage analytical thinking. Additionally, the text incorporates relevant mathematical tools and methodologies essential for solving electrical engineering problems.

Staying updated with the latest advancements is crucial in today's rapidly evolving technological landscape. This textbook integrates recent developments in the field, ensuring readers are well-equipped with the knowledge required to succeed in academic and professional settings. Whether you are preparing for examinations, conducting research or working in the industry, this book serves as a valuable reference.

Completing this textbook would not have been possible without the support and encouragement of many individuals. We extend our sincere thanks to our colleagues, students and professionals who provided valuable feedback and suggestions during the preparation of this text. Special gratitude is owed to the management and the editorial team of S Chand And Company Limited for their dedication and professionalism in bringing this book to fruition.

We hope this textbook will serve as a reliable and comprehensive guide for students and professionals alike. Our aim is to inspire a deep understanding of electrical technology, fostering innovation and excellence in this dynamic field. As you embark on your journey through the pages of this book, we encourage you to approach the material with curiosity and an eagerness to explore the boundless possibilities of electrical engineering.

Finally, we extend our best wishes to the readers of this book. May your journey through the world of electrical technology be enlightening and rewarding.

Authors

CONTENTS

On the Web:

For additional material, please go:

https://www.schandpublishing.com/books/tech-professional/electrical-engineering-electronics/a-textbook-electrical-technology-volume-i-basic-electrical-engineering/9789355018267/

Or

Scan for immediate access

CHAPTER

1

Electric Current and Ohm's Law

Chapter Outline

- ❖ The Idea of Electric Potential
- ❖ Resistance
- ❖ Unit of Resistance
- ❖ Law of Resistance
- ❖ Units of Resistivity
- ❖ Conductance and Conductivity
- ❖ Effect of Temperature on Resistance
- ❖ Temperature Coefficient of Resistance
- ❖ Ohm's Law
- ❖ Resistance in Series
- ❖ Voltage Divider Rule
- ❖ Resistance in Parallel
- ❖ Nonlinear Resistors
- ❖ Short and Open Circuits
- ❖ 'Shorts' in a Series Circuit
- ❖ 'Opens' in Series Circuit
- ❖ 'Open's in a Parallel Circuit
- ❖ 'Shorts' in Parallel Circuits
- ❖ Division of Current in Parallel Circuits
- ❖ Equivalent Resistance
- ❖ Duality Between Series and Parallel Circuits
- ❖ Relative Potential
- ❖ Voltage Divider Circuits

1.1 The Idea of Electric Potential

In Fig. 1.1, a simple voltaic cell is shown. It consists of copper plate (known as anode) and a zinc rod (*i.e.* cathode) immersed in dilute sulphuric acid (H_2SO_4) contained in a suitable vessel. The chemical action taking place within the cell causes the electrons to be removed from copper plate and to be deposited on the zinc rod at the same time. This transfer of electrons is accomplished through the agency of the diluted H_2SO_4 which is known as the electrolyte. The result is that zinc rod becomes negative due to the deposition of electrons on it and the copper plate becomes positive due to the removal of electrons from it. The large number of electrons collected on the zinc rod is being attracted by anode but is prevented from returning to it by the force set up by the chemical action within the cell.

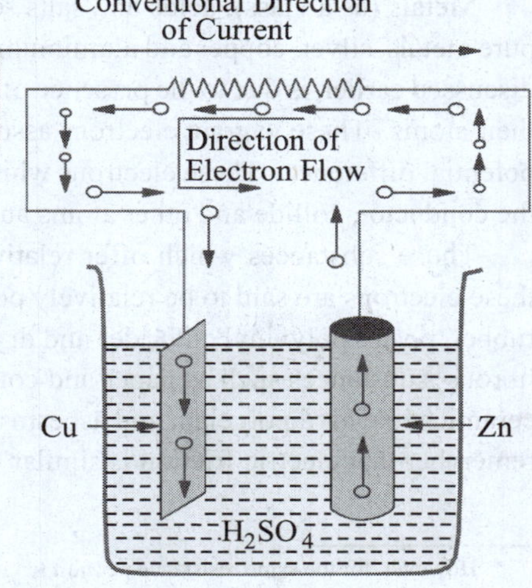

Fig. 1.1

But if the two electrodes are joined by a wire ***externally***, then electrons rush to the anode thereby equalizing the charges of the two electrodes. However, due to the continuity of chemical action, a continuous difference in the number of electrons on the two electrodes is maintained which keeps up a continuous flow of current through the external circuit. The action of an electric cell is similar to that of a water pump which, while working, maintains a continuous flow of water *i.e.,* water current through the pipe (Fig. 1.2).

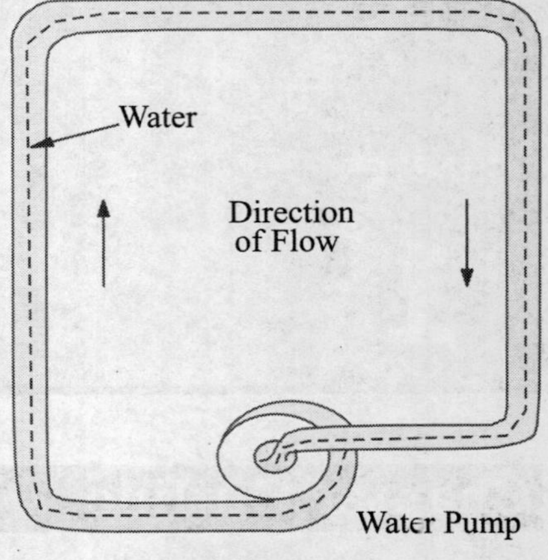

Fig. 1.2

It should be particularly noted that the direction of ***electronic*** current is from zinc to copper in the external circuit. However, the direction of ***conventional*** current (which is given by the direction of flow of positive charge) is from copper to zinc. In the present case, there is no flow of positive charge as such from one electrode to another. But we can look upon the arrival of electrons on copper plate (with subsequent decrease in its positive charge) as equivalent to an actual departure of positive charge from it.

When zinc is negatively charged, it is said to be at negative potential with respect to the electrolyte, whereas anode is said to be at positive potential relative to the electrolyte. Between themselves, copper plate is assumed to be at a higher potential than the zinc rod. The difference in potential is continuously maintained by the chemical action going on in the cell which supplies energy to establish this potential difference.

1.2 Resistance

It may be defined as the property of a substance due to which it opposes (or restricts) the flow of electricity (*i.e.*, electrons) through it.

Metals (as a class), acids and salts solutions are good conductors of electricity. Amongst pure metals, silver, copper and aluminium are very good conductors in the given order.* This, as discussed earlier, is due to the presence of a large number of free or loosely-attached electrons in their atoms. These vagrant electrons assume a directed motion on the application of an electric potential difference. These electrons while flowing pass ***through*** the molecules or the atoms of the conductor, collide and other atoms and electrons, thereby producing heat.

Those substances which offer relatively greater difficulty or hindrance to the passage of these electrons are said to be relatively poor conductors of electricity like bakelite, mica, glass, rubber, p.v.c. (polyvinyl chloride) and dry wood etc. Amongst good insulators can be included fibrous substances such as paper and cotton when dry, mineral oils free from acids and water, ceramics like hard porcelain and asbestos and many other plastics besides p.v.c. It is helpful to remember that electric friction is similar to friction in Mechanics.

---

* However, for the same resistance per unit length, cross-sectional area of aluminium conductor has to be 1.6 times that of the copper conductor but it weighs only half as much. Hence, it is used where economy of weight is more important than economy of space.

1.3 The Unit of Resistance

The practical unit of resistance is ohm.* A conductor is said to have a resistance of one ohm if it permits one ampere current to flow through it when one volt is impressed across its terminals.

For insulators whose resistances are very high, a much bigger unit is used *i.e.,* mega-ohm = 10^6 ohm (the prefix 'mega' or mego meaning a million) or kilo-ohm = 10^3 ohm (kilo means thousand). In the case of very small resistances, smaller units like milli-ohm = 10^{-3} ohm or micro-ohm = 10^{-6} ohm are used. The symbol for ohm is Ω.

Table 1.1. Multiples and Sub-multiples of Ohm

Prefix	*Its meaning*	*Abbreviation*	*Equal to*
Mega-	One million	M Ω	10^6 Ω
Kilo-	One thousand	k Ω	10^3 Ω
Centi-	One hundredth	c Ω	10^{-2} Ω
Milli-	One thousandth	m Ω	10^{-3} Ω
Micro-	One millionth	μ Ω	10^{-6} Ω

1.4 Laws of Resistance

The resistance R offered by a conductor depends on the following factors:

(*i*) It varies directly as its length, l.
(*ii*) It varies inversely as the cross-section A of the conductor.
(*iii*) It depends on the nature of the material.
(*iv*) It also depends on the temperature of the conductor.

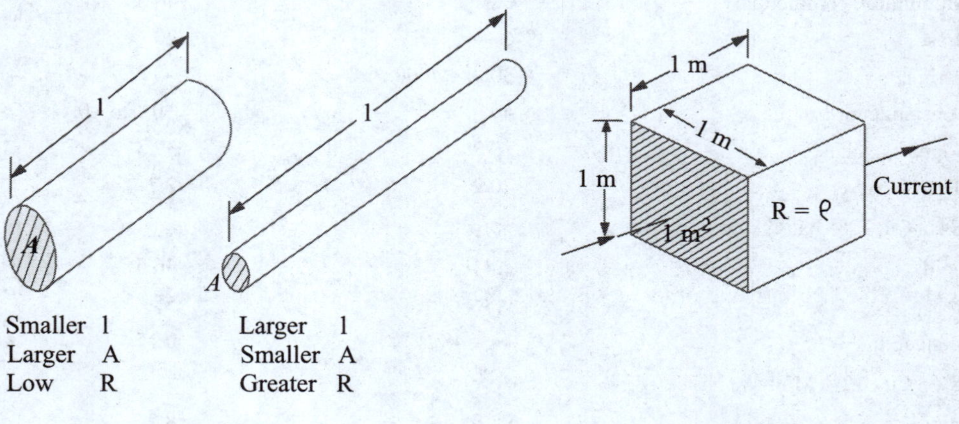

Fig. 1.3

Fig. 1.4

Neglecting the last factor for the time being, we can say that

$$R \propto \frac{l}{A} \quad \text{or} \quad R = \rho\frac{l}{A} \qquad ...(i)$$

* After George Simon Ohm (1787-1854), a German mathematician who in about 1827 formulated the law known after his name as Ohm's Law.

where ρ is a constant depending on the nature of the material of the conductor and is known as its ***specific resistance*** or ***resistivity***.

If in Eq. (*i*), we put

$$l = 1 \text{ metre} \quad \text{and} \quad A = 1 \text{ metre}^2, \text{ then } R = \rho \quad \text{(Fig. 1.4)}$$

Hence, specific resistance of a material may be defined as ***the resistance between the opposite faces of a metre cube of that material.***

1.5 Units of Resistivity

From Eq. (*i*), we have $\rho = \dfrac{AR}{l}$

In the S.I. system of units,

$$\rho = \frac{A \text{ metre}^2 \times R \text{ ohm}}{l \text{ metre}} = \frac{AR}{l} \text{ ohm-metre}$$

Hence, the unit of resistivity is ohm-metre (Ω-m).

It may, however, be noted that resistivity is sometimes expressed as so many ohm per m^3. Although, it is incorrect to say so but it means the same thing as ohm-metre.

If l is in centimetres and A in cm^2, then ρ is in ohm-centimetre (Ω-cm).

Values of resistivity and temperature coefficients for various materials are given in Table 1.2. The resistivities of commercial materials may differ by several per cent due to impurities etc.

Table 1.2. Resistivities and Temperature Coefficients

Material	*Resistivity in ohm-metre at 20°C ($\times 10^{-8}$)*	*Temperature coefficient at 20°C ($\times 10^{-4}$)*
Aluminium, commercial	2.8	40.3
Brass	6 – 8	20
Carbon	3000 – 7000	–5
Constantan or Eureka	49	+0.1 to –0.4
Copper (annealed)	1.72	39.3
German Silver	20.2	2.7
(84% Cu; 12% Ni; 4% Zn)		
Gold	2.44	36.5
Iron	9.8	65
Manganin	44 – 48	0.15
(84% Cu ; 12% Mn ; 4% Ni)		
Mercury	95.8	8.9
Nichrome	108.5	1.5
(60% Cu ; 25% Fe ; 15% Cr)		
Nickel	7.8	54
Platinum	9 – 15.5	36.7
Silver	1.64	38

Tungsten	5.5	47
Amber	5×10^{14}	
Bakelite	10^{10}	
Glass	$10^{10} - 10^{12}$	
Mica	10^{15}	
Rubber	10^{16}	
Shellac	10^{14}	
Sulphur	10^{15}	

Example 1.1. *A coil consists of 2000 turns of copper wire having a cross-sectional area of 0.8 mm². The mean length per turn is 80 cm and the resistivity of copper is 0.02 μΩ–m. Find the resistance of the coil and power absorbed by the coil when connected across 110 V d.c. supply.* **(F.Y. Engg. Pune Univ. May 1990)**

Solution. Length of the coil, $l = 0.8 \times 2000 = 1600$ m ; $A = 0.8$ mm² $= 0.8 \times 10^{-6}$ m².

$$R = \rho\frac{l}{A} = 0.02 \times 10^{-6} \times 1600/0.8 \times 10^{-6} = \mathbf{40\ \Omega}$$

Power absorbed $= V^2 / R = 110^2/40 = \mathbf{302.5\ W}$

Example 1.2. *An aluminium wire 7.5 m long is connected in 6 m long. When a current of 5 A is passed through the combinatic in the aluminium wire is 3 A. The diameter of the aluminium wire is 1 mm. Determine the diameter of the copper wire. Resistivity of copper is 0.017 μΩ-m ; that of the aluminium is 0.028 μΩ-m.* **(F.Y. Engg. Pune Univ. May 1991)**

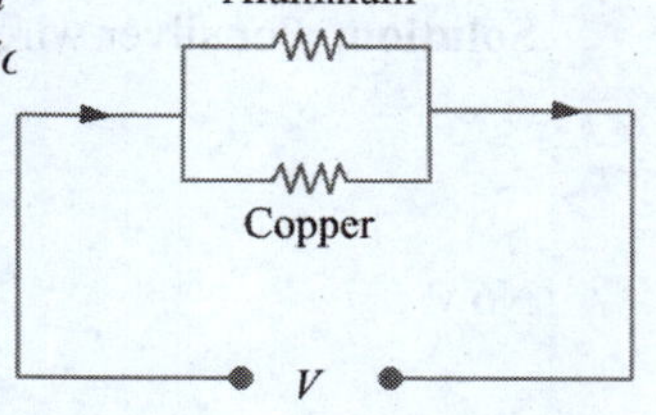

Solution. Let the subscript 1 represent aluminium and subscript 2 represent copper.

$$R_1 = \rho\frac{l_1}{a_1} \text{ and } R_2 = \rho_2\frac{l_2}{a_2} \qquad \therefore \quad \frac{R_2}{R_1} = \frac{\rho_2}{\rho_1}.\frac{l_2}{l_1}.\frac{a_1}{a_2}$$

$$\therefore \qquad a_2 = a_1.\frac{R_1}{R_2}.\frac{\rho_2}{\rho_1}.\frac{l_2}{l_1} \qquad ...(i)$$

Now $\qquad I_1 = 3$ A ; $I_2 = 5 - 3 = 2$ A.

If V is the common voltage across the parallel combination of aluminium and copper wires, then

$$V = I_1 R_1 = I_2 R_2$$

$$\therefore \qquad R_1/R_2 = I_2/I_1 = 2/3$$

$$a_1 = \frac{\pi d^2}{4} = \frac{\pi \times 1^2}{4} = \frac{\pi}{4}\text{ mm}^2$$

Substituting the given values in Eq. (i), we get

$$a_2 = \frac{\pi}{4} \times \frac{2}{3} \times \frac{0.017}{0.028} \times \frac{6}{7.5} = 0.2544 \text{ m}^2$$

$$\therefore \qquad \pi \times d_2^2/4 = 0.2544 \text{ or } d_2 = \mathbf{0.569\ mm}$$

Example 1.3. **(*a*)** *A rectangular carbon block has dimensions 1.0 cm × 1.0 cm × 50 cm.* **(*i*)** *What is the resistance measured between the two square ends ?* **(*ii*)** *between two opposing rectangular faces / Resistivity of carbon at 20°C is 3.5 ×* 10^{-5} *Ω-m.*

(*b*) *A current of 5 A exists in a 10-Ω resistance for 4 minutes* **(*i*)** *how many coulombs and* **(*ii*)** *how many electrons pass through any section of the resistor in this time ? Charge of the electron = 1.6 ×* 10^{-19} *C.* **(M.S. Univ. Baroda)**

Solution.

(*a*) (*i*) $R = \rho\, l/A$

Here, $A = 1 \times 1 = 1\ \text{cm}^2 = 10^{-4}\ \text{m}^2\ ;\ l = 0.5\ \text{m}$

$\therefore$ $R = 3.5 \times 10^{-5} \times 0.5/10^{-4} = \mathbf{0.175\ \Omega}$

(*ii*) Here, $l = 1\ \text{cm};\ A = 1 \times 50 = 50\ \text{cm}^2 = 5 \times 10^{-3}\ \text{m}^2$

$R = 3.5 \times 10^{-5} \times 10^{-2}/5 \times 10^{-3} = \mathbf{7 \times 10^{-5}\ \Omega}$

(*b*) (*i*) $Q = It = 5 \times (4 \times 60) = \mathbf{1200\ C}$

(*ii*) $n = \dfrac{Q}{e} = \dfrac{1200}{1.6 \times 10^{-19}} = \mathbf{75 \times 10^{20}}$ **electrons**

Example 1.4. *A piece of silver wire has a resistance of 1 Ω. What will be the resistance of manganin wire of one-third the length and one-third the diameter, if the specific resistance of manganin is 30 times that of silver.* **(Electrical Engineering-I, Delhi Univ.)**

Solution. For silver wire, $R_1 = \dfrac{l_1}{A_1}$; For manganin wire, $R = \rho_2 \dfrac{l_2}{A_2}$

$$\therefore \quad \frac{R_2}{R_1} = \frac{\rho_2}{\rho_1} \times \frac{l_2}{l_1} \times \frac{A_1}{A_2}$$

Now $A_1 = \pi d_1^2/4$ and $A_2 = \pi\, d_2^2/4 \quad \therefore \quad A_1/A_2 = d_1^2/d_2^2$

$$\therefore \quad \frac{R_2}{R_1} = \frac{\rho_2}{\rho_1} \times \frac{l_2}{l_1} \times \left(\frac{d_1}{d_2}\right)^2$$

$R_1 = 1\ \Omega;\ l_2/l_1 = 1/3,\ (d_1/d_2)^2 = (3/1)^2 = 9;\ \rho_2/\rho_1 = 30$

$\therefore$ $R_2 = 1 \times 30 \times (1/3) \times 9 = \mathbf{90\ \Omega}$

Tutorial Problem 1.1

1. Calculate the resistance of 100 m length of a wire having a uniform cross-sectional area of 0.1 mm² if the wire is made of manganin having a resistivity of 50×10^{-8} Ω-m.
   If the wire is drawn out to three times its original length, by how many times would you expect its resistance to be increased ? **[500 Ω; 9 times]**
2. A cube of a material of side 1 cm has a resistance of 0.001 Ω between its opposite faces. If the same volume of the material has a length of 8 cm and a uniform cross-section, what will be the resistance of this length ? **[0.064 Ω]**
3. A lead wire and an iron wire are connected in parallel. Their respective specific resistances are in the ratio 49: 24. The former carries 80 per cent more current than the latter and the latter is 47 per cent longer than the former. Determine the ratio of their cross-sectional area. **[2.5: 1]**
4. A rectangular metal strip has the following dimensions:
   $x = 10$ cm, $y = 0.5$ cm, $z = 0.2$ cm
   Determine the ratio of resistances R_x, R_y, and R_z between the respective pairs of opposite faces.
   [R_x: R_y: R_z: 10,000: 25: 4] **(*Elect. Engg. A.M.Ae. S.I.*)** **(RGPV, 2009)**

5. The resistance of a conductor 1 mm^2 in cross-section and 20 m long is 0.346 Ω. Determine the specific resistance of the conducting material.

[1.73×10^{-8} Ω-m] (*Elect. Circuits-1, Bangalore Univ. 1991*)

1.6 Conductance and Conductivity

Conductance (G) is reciprocal of resistance*. Whereas resistance of a conductor measures the ***opposition*** which it offers to the flow of current, the conductance measures the ***inducement*** which it offers to its flow.

From Eq. (*i*), $$R = \rho \frac{l}{A} \quad \text{or} \quad G = \frac{1}{\rho} \cdot \frac{A}{l} = \frac{\sigma A}{l}$$

where σ is called the ***conductivity* or *specific conductance*** of a conductor. The unit of conductance is siemens (S). Earlier, this unit was called mho; which is represented as ℧ or $\frac{1}{\Omega}$.

It is seen from the above equation that the conductivity of a material is given by

$$\sigma = G\frac{l}{A} = \frac{G \text{ siemens} \times l \text{ metre}}{A \text{ metre}^2} = G\frac{l}{A} \text{ siemens/metre}$$

Hence, the unit of conductivity is siemens/metre (S/m).

1.7 Effect of Temperature on Resistance

The effect of rise in temperature is:

(*i*) to ***increase*** the resistance of pure metals. The increase is large and fairly regular for normal ranges of temperature. The temperature/resistance graph is a straight line (Fig. 1.5). As would be presently clarified, metals have a positive temperature co-efficient of resistance.

(*ii*) to ***increase*** the resistance of alloys, though in their case, the increase is relatively small and irregular. For some high-resistance alloys like Eureka (60% Cu and 40% Ni) and manganin, the increase in resistance is (or can be made) negligible over a considerable range of temperature.

(*iii*) to ***decrease*** the resistance of electrolytes, insulators (such as paper, rubber, glass, mica etc.) and partial conductors such as carbon. Hence, insulators are said to possess a ***negative*** temperature-coefficient of resistance.

1.8 Temperature Coefficient of Resistance

Let a metallic conductor having a resistance of R_0 at 0°C be heated of t°C and let its resistance at this temperature be R_t. Then, considering normal ranges of temperature, it is found that the increase in resistance $\Delta R = R_t - R_0$ depends

(*i*) directly on its initial resistance

(*ii*) directly on the rise in temperature

(*iii*) on the nature of the material of the conductor.

or $$R_t - R_0 \propto R \times t \quad \text{or} \quad R_t - R_0 = \alpha R_0 t \qquad \text{...(}i\text{)}$$

where α (alpha) is a constant and is known as the *temperature coefficient of resistance* of the conductor.

---

* In a.c. circuits, it has a slightly different meaning.

Rearranging Eq. (*i*), we get $\alpha = \dfrac{R_t - R_0}{R_0 \times t} = \dfrac{\Delta R}{R_0 \times t}$

If $R_0 = 1\,\Omega, t = 1°\text{C}$, then $\alpha = \Delta R = R_t - R_0$

Hence, the temperature-coefficient of a material may be defined as:

the increase in resistance per ohm original resistance per °C rise in temperature.

From Eq. (*i*), we find that $R_t = R_0(1 + \alpha\, t)$ ...(*ii*)

It should be remembered that the above equation holds good for both rise as well as fall in temperature. As temperature of a conductor is decreased, its resistance is also decreased. In Fig. 1.5 is shown the temperature/resistance graph for copper and is practically a straight line. If this line is extended backwards, it would cut the temperature axis at a point where temperature is – 234.5°C (a number quite easy to remember). It means that theoretically, the resistance of copper conductor will become zero at this point though as shown by solid line, in practice, the curve departs from a straight line at very low temperatures. From the two similar triangles of Fig. 1.6 it is seen that:

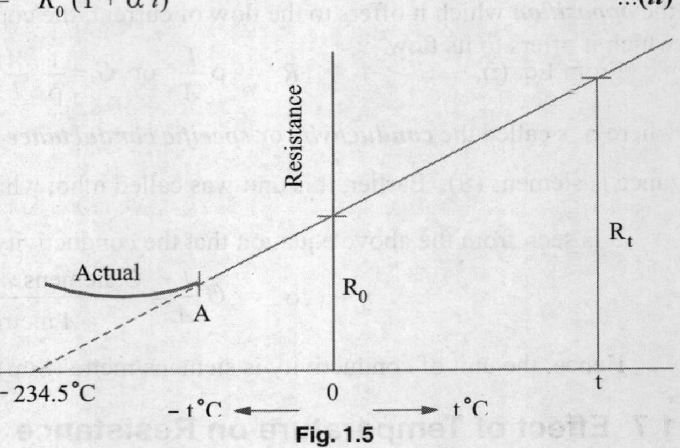

Fig. 1.5

$$\frac{R_t}{R_0} = \frac{t + 234.5}{234.5} = \left(1 + \frac{t}{234.5}\right)$$

$$\therefore \quad R_t = R_0\left(1 + \frac{t}{234.5}\right) \text{ or } R_t = R_0(1 + \alpha\, t) \text{ where } \alpha = 1/234.5 \text{ for copper.}$$

Example 1.5. *A copper conductor has its specific resistance of 1.6×10^{-6} ohm-cm at 0°C and a resistance temperature coefficient of 1/254.5 per °C at 20°C. Find **(i)** the specific resistance and **(ii)** the resistance - temperature coefficient at 60°C.* **(F.Y. Engg. Pune Univ. Nov.)**

Solution. $\alpha_{20} = \dfrac{\alpha_0}{1 + \alpha_0 \times 20}$ or $\dfrac{1}{254.5} = \dfrac{\alpha_0}{1 + \alpha_0 \times 20}$ $\therefore\ \alpha_0 = \dfrac{1}{234.5}$ per °C

(*i*) $\rho_{60} = \rho_0(1 + \alpha_0 \times 60) = 1.6 \times 10^{-6}(1 + 60/234.5) = \mathbf{2.01 \times 10^{-6}\ \Omega\text{-cm}}$

(*ii*) $\alpha_{60} = \dfrac{\alpha_0}{1 + \alpha_0 \times 60} = \dfrac{1/234.5}{1 + (60/234.5)} = \dfrac{\mathbf{1}}{\mathbf{294.5}}$ **per°C**

Example 1.6. *A platinum coil has a resistance of 3.146 Ω at 40°C and 3.767 Ω at 100°C. Find the resistance at 0°C and the temperature-coefficient of resistance at 40°C.*

(Electrical Science-II, Allahabad Univ.)

Solution. $R_{100} = R_0(1 + 100\,\alpha_0)$ ...(*i*)

$R_{40} = R_0(1 + 40\,\alpha_0)$ ...(*ii*)

$$\therefore \quad \frac{3.767}{3.146} = \frac{1 + 100\,\alpha_0}{1 + 40\,\alpha_0} \quad \text{or} \quad \alpha_0 = 0.00379 \quad \text{or} \quad 1/264 \text{ per°C}$$

From **(*i*)**, we have 3.767 = R_0 (1 + 100 × 0.00379) ∴ R_0 = **2.732 Ω**

Now, $\alpha_{40} = \dfrac{\alpha_0}{1+40\,\alpha_0} = \dfrac{0.00379}{1+40\times 0.00379} = \dfrac{\mathbf{1}}{\mathbf{304}}$ **per°C**

Example 1.7. *A potential difference of 250 V is applied to a field winding at 15°C and the current is 5 A. What will be the mean temperature of the winding when current has fallen to 3.91 A, applied voltage being constant. Assume* α_{15} *= 1/254.5.* **(Elect. Engg. Pune Univ.)**

Solution. Let R_1 = winding resistance at 15°C; R_2 = winding resistance at unknown mean temperature t_2°C.

∴ R_1 = 250/5 = 50 Ω; R_2 = 250/3.91 = 63.94 Ω.

Now $R_2 = R_1[1+\alpha_{15}(t_2-t_1)]$ ∴ $63.94 = 50\left[1+\dfrac{1}{254.5}(t_2-15)\right]$

∴ t_2 = **86°C**

Example 1.8. *A coil has a resistance of 18 Ω when its mean temperature is 20°C and of 20 Ω when its mean temperature is 50°C. Find its mean temperature rise when its resistance is 21 Ω and the surrounding temperature is 15° C.* **(Elect. Technology, Allahabad Univ.)**

Solution. Let R_0 be the resistance of the coil and α_0 its tempt. coefficient at 0°C.

Then, $18 = R_0(1+\alpha_0\times 20)$ and $20 = R_0(1+50\,\alpha_0)$

Dividing one by the other, we get

$$\frac{20}{18} = \frac{1+50\,\alpha_0}{1+20\,\alpha_0} \quad \therefore \ \alpha_0 = \frac{1}{250}\ \text{per°C}$$

If t°C is the temperature of the coil when its resistance is 21 Ω, then,

$$21 = R_0(1+t/250)$$

Dividing this equation by the above equation, we have

$$\frac{21}{18} = \frac{R_0\,(1+t/250)}{R_0\,(1+20\,\alpha_0)};\quad t = 65°\text{C; temp. rise} = 65-15 = \mathbf{50°C}$$

Example 1.9. *The coil of a relay takes a current of 0.12 A when it is at the room temperature of 15°C and connected across a 60-V supply. If the minimum operating current of the relay is 0.1 A, calculate the temperature above which the relay will fail to operate when connected to the same supply. Resistance-temperature coefficient of the coil material is 0.0043 per°C at 6°C.*

Solution. Resistance of the relay coil at 15°C is R_{15} = 60/0.12 = 500 Ω.

Let t°C be the temperature at which the minimum operating current of 0.1 A flows in the relay coil. Then, R_t = 60/0.1 = 600 Ω.

Now $R_{15} = R_0(1+15\,\alpha_0) = R_0(1+15\times 0.0043)$ and $R_t = R_0(1+0.0043\,t)$

$$\therefore \quad \frac{R_t}{R_{15}} = \frac{1+0.0043\,t}{1.0654} \ \text{or}\ \frac{600}{500} = \frac{1+0.0043\,t}{1.0645} \quad \therefore \ t = \mathbf{65.4°C}$$

If the temperature rises above this value, then due to increase in resistance, the relay coil will draw a current less than 0.1 A and, therefore, will fail to operate.

Tutorial Problem 1.2

1. It is found that the resistance of a coil of wire increases from 40 ohm at 15°C to 50 ohm at 60°C. Calculate the resistance temperature coefficient at 0°C of the conductor material. **[1/165 per °C]** ***(Elect. Technology, Indore Univ.)***
2. A tungsten lamp filament has a temperature of 2,050°C and a resistance of 500 Ω when taking normal working current. Calculate the resistance of the filament when it has a temperature of 25°C. Temperature coefficient at 0°C is 0.005/°C. **[50 Ω]** ***(Elect. Technology, Indore Univ.)***
3. An armature has a resistance of 0.2 Ω at 150°C and the armature Cu loss is to be limited to 600 watts with a temperature rise of 55°C. If α_0 for Cu is 0.0043/°C, what is the maximum current that can be passed through the armature ? **[50.8 A]**
4. A d.c. shunt motor after running for several hours on constant voltage mains of 400 V takes a field current of 1.6 A. If the temperature rise is known to be 40°C, what value of extra circuit resistance is required to adjust the field current to 1.6 A when starting from cold at 20°C ? Temperature coefficient = 0.0043/°C at 20°C. **[36.69 Ω]**
5. In a test to determine the resistance of a single-core cable, an applied voltage of 2.5 V was necessary to produce a current of 2 A in it at 15°C.

   (a) Calculate the cable resistance at 55°C if the temperature coefficient of resistance of copper at 0°C is 1/235 per°C.

   (b) If the cable under working conditions carries a current of 10 A at this temperature, calculate the power dissipated in the cable. **[*(a)* 1.45 Ω *(b)* 145 W]**
6. An electric radiator is required to dissipate 1 kW when connected to a 230 V supply. If the coils of the radiator are of wire 0.5 mm in diameter having resistivity of 60 μ Ω-cm, calculate the necessary length of the wire. **[1732 cm]**
7. An electric heating element to dissipate 450 watts on 250 V mains is to be made from nichrome ribbon of width 1 mm and thickness 0.05 mm. Calculate the length of the ribbon required (the resistivity of nichrome is 110×10^{-8} Ω-m). **[631 m]**
8. When burning normally, the temperature of the filament in a 230 V, 150 W gas-filled tungsten lamp is 2,750°C. Assuming a room temperature of 16°C, calculate (*a*) the normal current taken by the lamp (*b*) the current taken at the moment of switching on. Temperature coefficient of tungsten is 0.0047 Ω/Ω°C at 0°C. **[*(a)* 0.652 A *(b)* 8.45 A]** ***(Elect. Engg. Madras Univ.)***
9. An aluminium wire 5 m long and 2 mm diameter is connected in parallel with a wire 3 m long. The total current is 4 A and that in the aluminium wire is 2.5 A. Find the diameter of the copper wire. The respective resistivities of copper and aluminium are 1.7 and 2.6 μΩ-m. **[0.97 mm]**
10. The field winding of d.c. motor connected across 230 V supply takes 1.15 A at room temp. of 20°C. After working for some hours the current falls to 0.26 A, the supply voltage remaining constant. Calculate the final working temperature of field winding. Resistance temperature coefficient of copper at 20°C is 1/254.5. **[70.4°C]** ***(Elect. Engg. Pune Univ.)***

1.9 Ohm's Law

This law applies to electric to electric conduction through good conductors and may be stated as follows:

The ratio of potential difference (V) between any two points on a conductor to the current (I) flowing between them, is constant, provided the temperature of the conductor does not change.

In other words, $\dfrac{V}{I}$ = constant or $\dfrac{V}{I} = R$

where R is the resistance of the conductor between the two points considered.

Put in another way, it simply means that provided R is kept constant, current is directly proportional to the potential difference across the ends of a conductor. However, this linear relationship between V and I does not apply to all non-metallic conductors. For example, for silicon carbide, the relationship is given by $V = KI^m$ where K and m are constants and m is less than unity. It also does not apply to non-linear devices such as Zener diodes and voltage-regulator (VR) tubes.

Example 1.10. *A coil of copper wire has resistance of 90 Ω at 20°C and is connected to a 230-V supply. By how much must the voltage be increased in order to maintain the current consant if the temperature of the coil rises to 60°C ? Take the temperature coefficient of resistance of copper as 0.00428 from 0°C.*

Solution. As seen

$$\frac{R_{60}}{R_{20}} = \frac{1+60\times 0.00428}{1+20\times 0.00428} \qquad \therefore R_{60} = 90\times 1.2568/1.0856 = 104.2\Omega$$

Now, current at 20°C = 230/90 = 23/9 A

Since the wire resistance has become 104.2 Ω at 60°C, the new voltage required for keeping the current constant at its previous value = 104.2 × 23/9 = 266.3 V

∴ increase in voltage required = 266.3 – 230 = **36.3 V**

Example 1.11. *Three resistors are connected in series across a 12-V battery. The first resistor has a value of 1 Ω, second has a voltage drop of 4 V and the third has a power dissipation of 12 W. Calculate the value of the circuit current.*

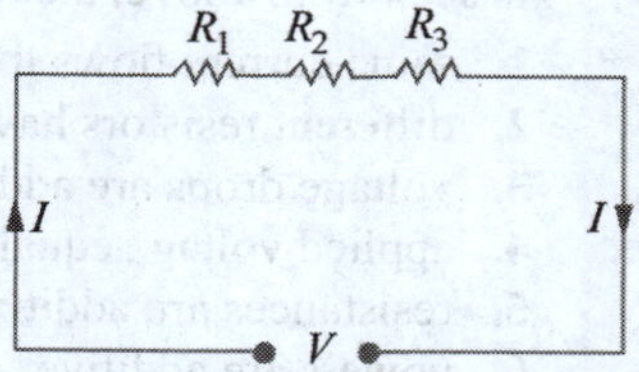

R_1 R_2 R_3 I I V

Fig. 1.6

Solution. Let the two unknown resistors be R_2 and R_3 and I the circuit current

$$\therefore \quad I^2 R_3 = 12 \quad \text{and} \quad IR_3 = 4$$

$$\therefore \quad R_3 = \frac{3}{4}R_2^2. \quad \text{Also,} \quad I = \frac{4}{R_2}$$

Now, $\quad I(1+R_2+R_3) = 12$

Substituting the values of I and R_3, we get

$$\frac{4}{R_2}\left(1+R_2+\frac{3}{4}R_2^2\right) = 12 \quad \text{or} \quad 3R_2^2 - 8R_2 + 4 = 0$$

$$\therefore \quad R_2 = \frac{8\pm\sqrt{64-48}}{6} \qquad \therefore R_2 = 2\,\Omega \quad \text{or} \quad \frac{2}{3}\Omega$$

$$\therefore \quad R_3 = \frac{3}{4}R_2^2 = \frac{3}{4}\times 2^2 = 3\,\Omega \quad \text{or} \quad \frac{3}{4}\left(\frac{2}{3}\right)^2 = \frac{1}{3}\Omega$$

$$\therefore \quad I = \frac{12}{1+2+3} = \mathbf{2\ A} \quad \text{or} \quad I = \frac{12}{1+(2/3)+(1/3)} = \mathbf{6\ A}$$

1.10 Resistances in Series

When some conductors having resistances R_1, R_2 and R_3 etc. are joined end-on-end as in Fig. 1.7, they are said to be connected in series. It can be proved that the equivalent resistance or total resistance between points A and D is equal to the sum of the three individual resistances. Being a series circuit, it should be remembered that ***(i)*** current is the same through all the three conductors ***(ii)*** but voltage drop across each is different due to its different resistance and is given by Ohm's Law and ***(iii)*** sum of the three voltage drops is equal to the voltage applied across the three conductors. There is a progressive fall in potential as we go from point A to D as shown in Fig. 1.8.

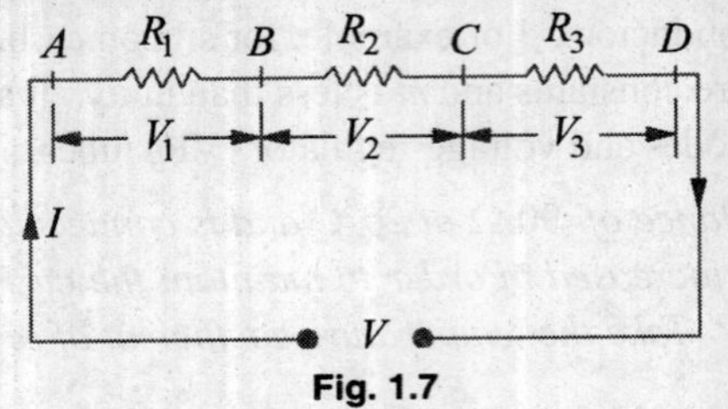

Fig. 1.7

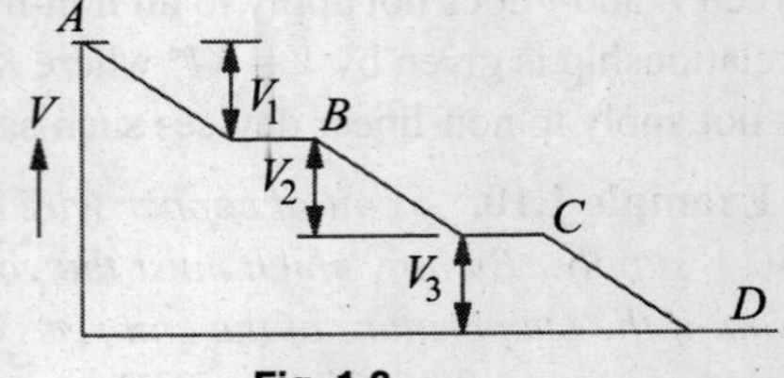

Fig. 1.8

$$\therefore \quad V = V_1 + V_2 + V_3 = IR_1 + IR_2 + IR_3 \qquad \text{—Ohm's Law}$$

But $\quad V = IR$

where R is the equivalent resistance of the series combination.

$$\therefore \quad IR = IR_1 + IR_2 + IR_3 \quad \text{or} \quad R = R_1 + R_2 + R_3$$

Also $$\frac{1}{G} = \frac{1}{G_1} + \frac{1}{G_2} + \frac{1}{G_3}$$

As seen from above, the main characteristics of a series circuit are:

1. same current flows through all parts of the circuit.
2. different resistors have their individual voltage drops.
3. voltage drops are additive.
4. applied voltage equals the sum of different voltage drops.
5. resistances are additive.
6. powers are additive.

1.11 Voltage Divider Rule

Since in a series circuit, same current flows through each of the given resistors, voltage drop varies directly with its resistance. In Fig. 1.9 is shown a 24-V battery connected across a series combination of three resistors.

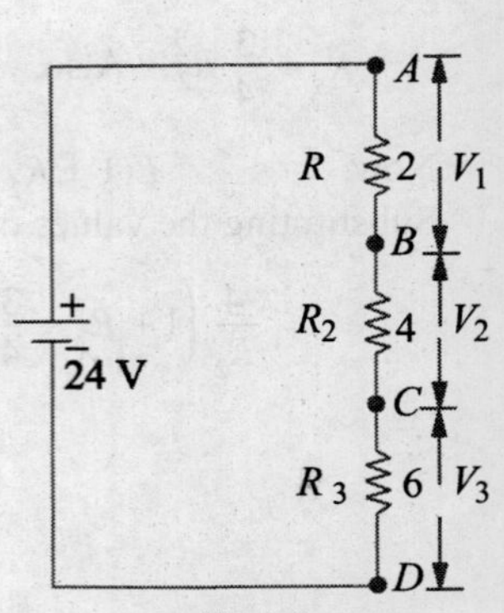

Fig.1.9

Total resistance $\quad R = R_1 + R_2 + R_3 = 12\ \Omega$

According to Voltage Divider Rule, various voltage drops are:

$$V_1 = V \cdot \frac{R_1}{R} = 24 \times \frac{2}{12} = 4\ \text{V}$$

$$V_2 = V \cdot \frac{R_2}{R} = 24 \times \frac{4}{12} = 8\ \text{V}$$

$$V_3 = V \cdot \frac{R_3}{R} = 24 \times \frac{6}{12} = 12\ \text{V}$$

1.12 Resistances in Parallel

Three resistances, as joined in Fig. 1.10 are said to be connected in parallel. In this case (*i*) p.d. across all resistances is the same (*ii*) current in each resistor is different and is given by Ohm's Law and (*iii*) the total current is the sum of the three separate currents.

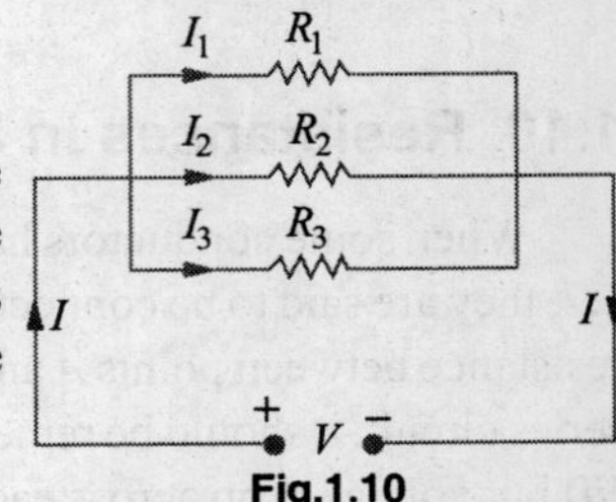

Fig.1.10

$$I = I_1 + I_2 + I_3 = \frac{V}{R_1} + \frac{V}{R_2} + \frac{V}{R_3}$$

Now, $\quad I = \dfrac{V}{R}$ where V is the applied voltage.

R = equivalent resistance of the parallel combination.

$$\therefore \quad \frac{V}{R} = \frac{V}{R_1} + \frac{V}{R_2} + \frac{V}{R_3} \quad \text{or} \quad \frac{1}{R} = \frac{1}{R_1} + \frac{1}{R_2} + \frac{1}{R_3}$$

Also $\quad G = G_1 + G_2 + G_3$

The main characteristics of a parallel circuit are:

1. same voltage acts across all parts of the circuit
2. different resistors have their individual current.
3. branch currents are additive.
4. conductances are additive.
5. powers are additive.

Example 1.12. *What is the value of the unknown resistor R in Fig. 1.12 if the voltage drop across the 500 Ω resistor is 2.5 volts ? All resistances are in ohm.*

(Elect. Technology, Indore Univ.)

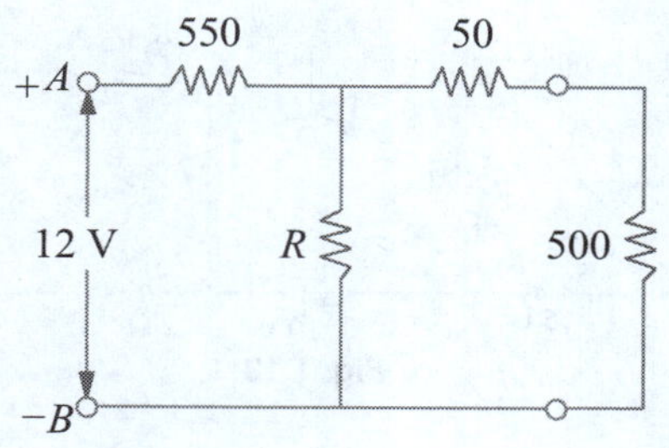

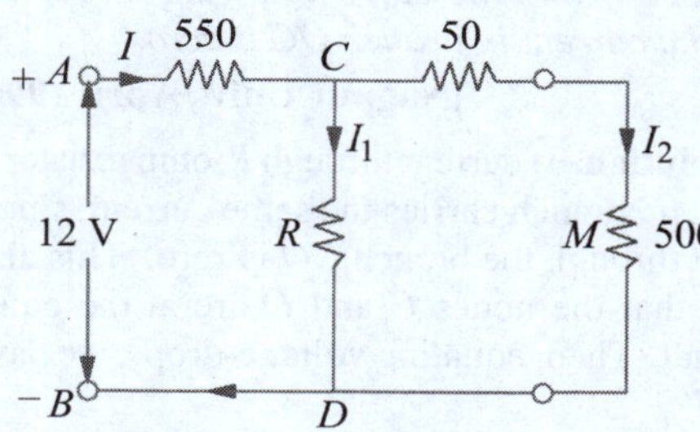

Fig. 1.11

Solution. By direct proportion, drop on 50 Ω resistance = 2.5 × 50/500 = 0.25 V

Drop across CMD or CD $\quad$ = 2.5 + 0.25 = 2.75 V

Drop across 550 Ω resistance $\quad$ = 12 − 2.75 = 9.25 V

I = 9.25/550 = 0.0168 A, I_2 = 2.5/500 = 0.005 A

I_1 = 0.0168 − 0.005 = 0.0118 A

∴ $\quad$ 0.0118 = 2.75/R; $\quad R$ = **233 Ω**

Example 1.13. *Calculate the effective resistance of the following combination of resistances and the voltage drop across each resistance when a P.D. of 60 V is applied between points A and B.*

Solution. Resistance between A and C (Fig. 1.12).

= 6 || 3 = 2 Ω

Resistance of branch ACD $\quad$ = 18 + 2 = 20 Ω

Now, there are two parallel paths between points A and D of resistances 20 Ω and 5 Ω.

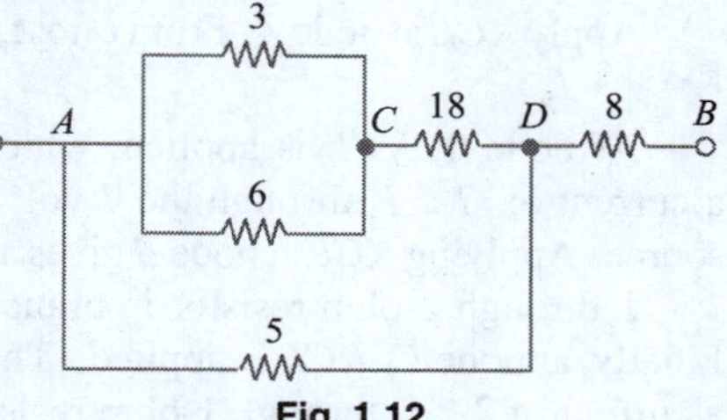

Fig. 1.12

Hence, resistance between A and D = 20 || 5 = 4 Ω

∴ Resistance between A and B = 4 + 8 = 12 Ω

Total circuit current = 60/12 = 5 A

Current through 5 Ω resistance = $5 \times \frac{20}{25} = 4$ A $\quad$ — **(Ref. Art. 1.19)**

Current in branch ACD $\quad = 5 \times \frac{5}{25} = 1$ A

$\therefore$ P.D. across 3 Ω and 6 Ω resistors = 1 × 2 = **2 V**

P.D. across 18 Ω resistors = 1 × 18 = **18 V**

P.D. across 5 Ω resistors = 4 × 5 = **20 V**

P.D. across 8 Ω resistors = 5 × 8 = **40 V**

Example 1.14. *A 50-ohm resistor is in parallel with 100-ohm resistor. Current in 50-ohm resistor is 7.2 A. How will you add a third resistor and what will be its value of the line-current is to be its value if the line-current is to be 12.1 amp ?* **[Nagpur Univ., Nov. 1997]**

Solution. Source voltage = 50 × 7.2 = 360 V, Current through 100–ohm resistor = 3.6 A

Total current through these two resistors in parallel = 10.8 A

For the total line current to be 12.1 A, third resistor must be connected in parallel, as the third branch, for carrying (12.1 – 10.8) = 1.3 A. If *R* is this resistor *R* = 360/1.3 = 277 ohms

Example 1.15. *In the circuit shown in Fig. 1.13, calculate the value of the unknown resistance R and the current flowing through it when the current in branch OC is zero.*

[Nagpur Univ., April 1996]

Fig. 1.13

Solution. If current through *R*-ohm resistor is *I* amp, *AO* branch carries the same current, since, current through the branch *CO* is zero. This also means that the nodes *C* and *O* are at the equal potential. Then, equating voltage-drops, we have $V_{AO} = V_{AC}$.

This means branch *AC* carries a current of 4*I*.

This is current of 4 *I* also flows through the branch *CB*. Equating the voltage-drops in branches *OB* and *CB*,

1.5 × 4 *I* = *R I*, giving *R* = 6 Ω

At node *A*, applying *KCL*, a current of 5 *I* flows through the branch *BA* from *B* to *A*. Applying *KVL* around the loop *BAOB*, *I* = 0.5 Amp.

Example 1.16. *For the given circuit find the current I_A and I_B.* **[Bombay Univ. 1991]**

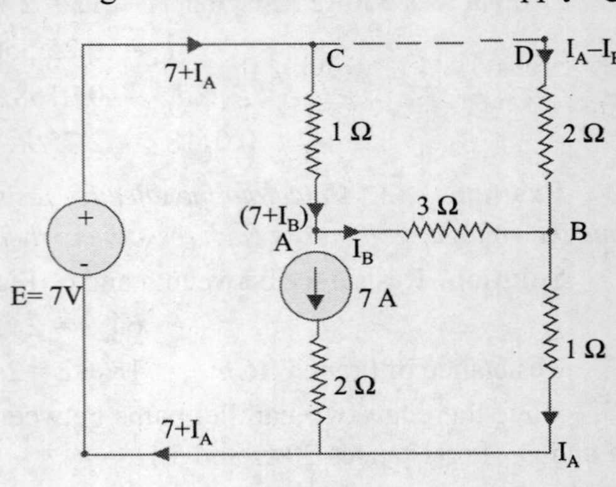

Fig. 1.14

Solution. Nodes *A, B, C, D* and reference node 0 are marked on the same diagram.

I_A and I_B are to be found.

Apply *KCL* at node *A*. From *C* to *A*, current E = 7 + I_B

At node 0, *KCL* is applied, which gives a current of 7 + I_A through the 7 volt voltage source. Applying *KCL* at node *B* gives a current $I_A - I_B$ through 2-ohm resistor in branch *CDA*. Finally, at node *C, KCL* is applied. This gives a current of 7 + I_B through 1-ohm resistor in branch *CA*.

KVL – Around the Loop *OECDBO*, $2\,(I_A - I_B) + 1.\,I_A = 7$

KVL – Around the Loop *CABDC*, $1\,(7 + I_B) + 3\,I_B - 2\,(I_A - I_B) = 0$

After rearranging the terms, $3\,I_A - 2\,I_B = 7,\ 2\,I_A + 6\,I_B = -7$

This gives I_A = 2 amp, I_B = – 0.5 amp.

This means that I_B is 0.5 amp from *B* to *A*.

Example 1.17. *Find R_{AB} in the circuit, given in Fig. 1.15 (a).* **[Bombay Univ. 2001]**

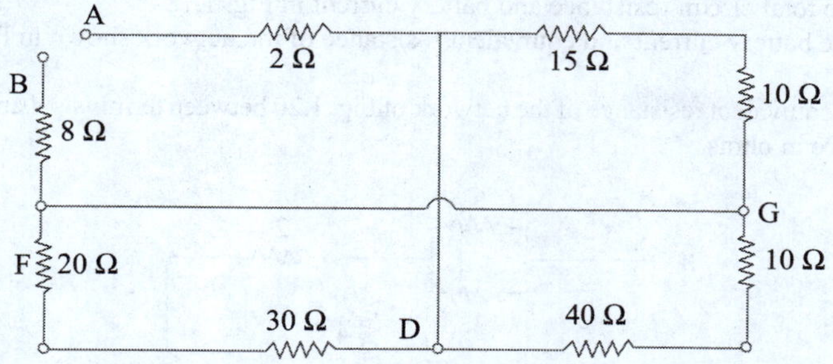

Fig. 1.15 (a)

Solution. Mark additional nodes on the diagram, *C, D, F, G*, as shown. Redraw the figure as in 1.15 (*b*), and simplify the circuit, to evaluate R_{AB}, which comes out to be 22.5 ohms.

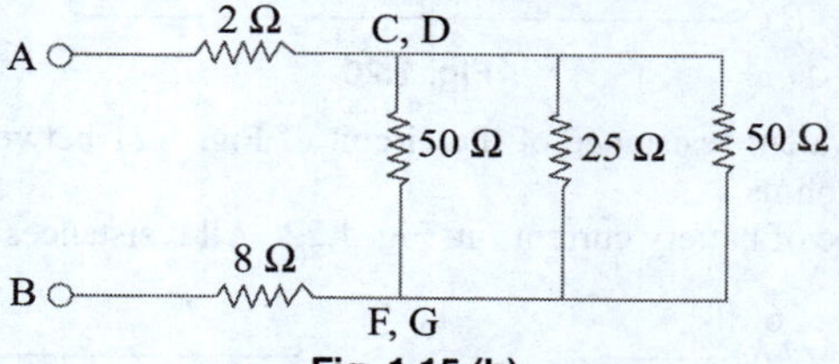

Fig. 1.15 (b)

Example 1.18. *Find current through 4 resistance.* **[Bombay Univ. 2001]**

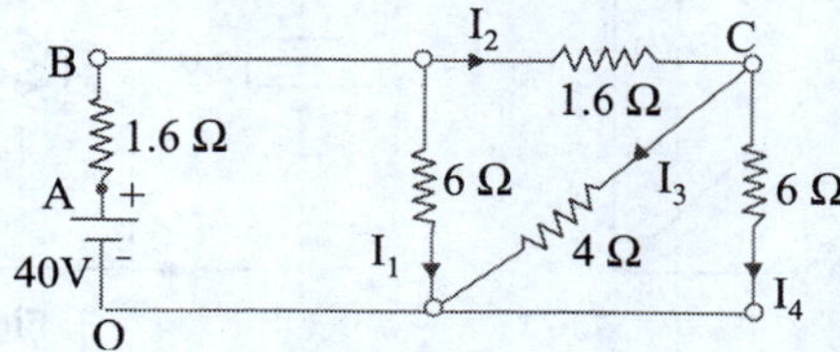

Fig. 1.16

Solution. Simplifying the series-parallel combinations, and solving the circuit, the source current is 10 amp. With respect to 0, $V_A = 40$, $V_B = 40 - 16 = 24$ volts.

$I_1 = 4$ amp, hence $I_2 = 6$ amp

$V_C = V_B - I_2 \times 1.6 = 24 - 9.6 = 14.4$ volts

$I_3 = 14.4/4 = 3.6$ amp, which is the required answer. Further $I_4 = 24$ amp.

Tutorial Problem 1.3

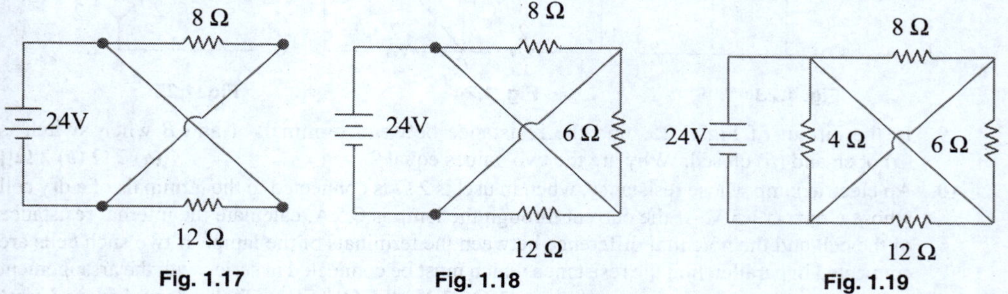

Fig. 1.17 **Fig. 1.18** **Fig. 1.19**

1. Find the current supplied by the battery in the circuit of Fig. 1.17 **[5 A]**
2. Compute total circuit resistance and battery current in Fig. 1.18 **[8/3 Ω, 9 A]**
3. Calculate battery current and equivalent resistance of the network shown in Fig. 1.19. **[15 A; 8/5 Ω]**
4. Find the equivalent resistance of the network of Fig. 1.20 between terminals *A* and *B*. All resistance values are in ohms. **[6 Ω]**

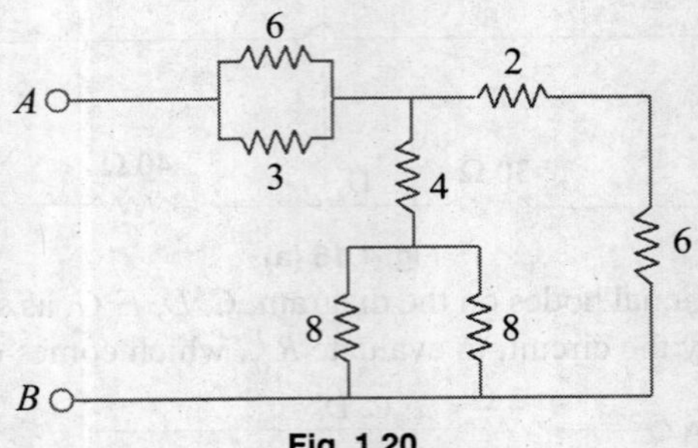

Fig. 1.20

5. What is the equivalent resistance of the circuit of Fig. 1.21 between terminals *A* and *B* ? All resistances are in ohms **[4 Ω]**
6. Compute the value of battery current *I* in Fig. 1.22. All resistances are in ohm. **[6 A]**

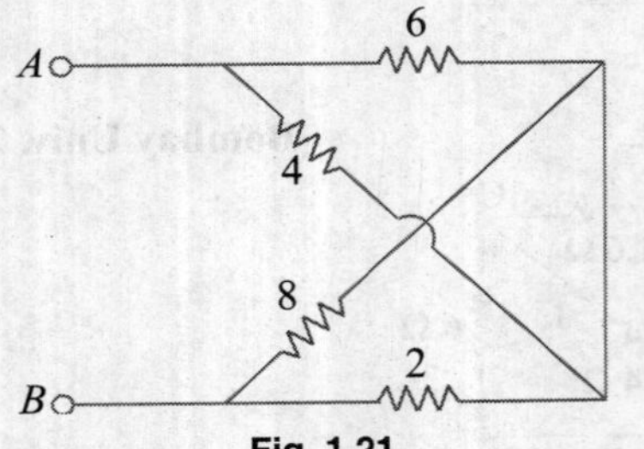

Fig. 1.21

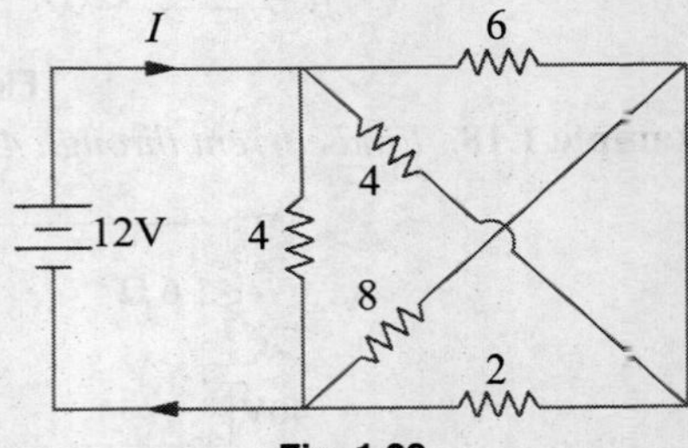

Fig. 1.22

7. Calculate the value of current *I* supplied by the voltage source in Fig. 1.23. All resistance values are in ohms. (**Hint:** Voltage across each resistor is 6 V) **[1 A]**
8. Compute the equivalent resistance of the circuit of Fig. 1.24 between points (*i*) *ab* (*ii*) *ac* and (*iii*) *bc*. All resistances values are in ohm. **[(*i*) 6 Ω, (*ii*) 4.5 Ω, (*iii*) 4.5 Ω]**

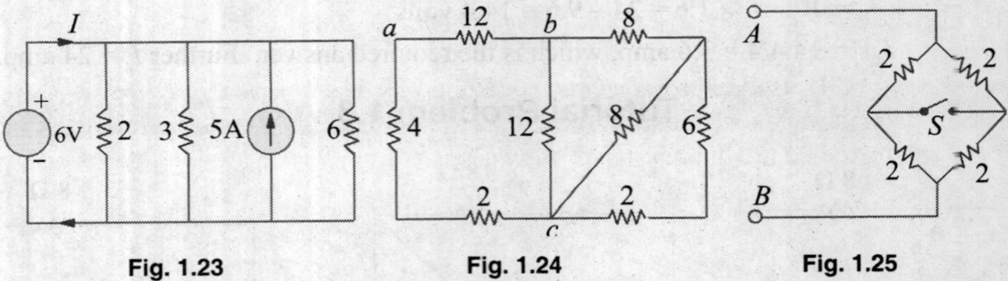

Fig. 1.23 **Fig. 1.24** **Fig. 1.25**

9. In the circuit of Fig. 1.25, find the resistance between terminals *A* and *B* when switch is (*a*) open and (*b*) closed. Why are the two values equal ? **[(*a*) 2 Ω (*b*) 2 Ω]**
10. An electric lamp whose resistance, when in use, is 2 Ω is connected to the terminals of a dry cell whose e.m.f. is 1.5 V. If the current through the lamp is 0.5 A, calculate the internal resistance of the cell and the potential difference between the terminals of the lamp. If two such cells are connected in parallel, find the resistance which must be connected in series with the arrangement to keep the current the same as before. **[1 Ω ; 1 V ; 0.5 Ω]** **(*Elect. Technology, Indore Univ.*)**

11. Determine the current by the source in the circuit shown below. ***(Bombay Univ. 2001)***

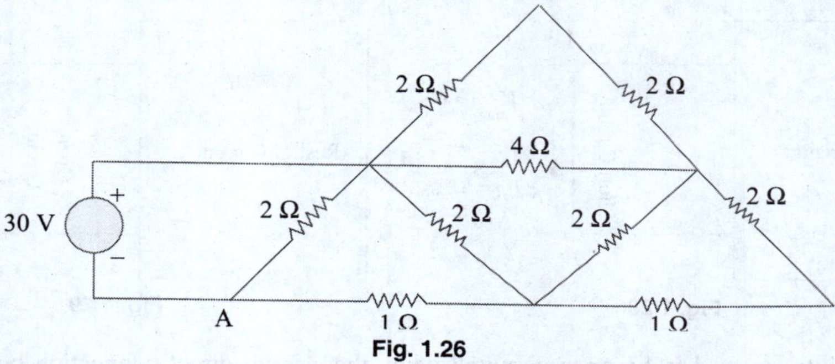

Fig. 1.26

Hint. Series-parallel combinations of resistors have to be dealt with. This leads to the source current of 28.463 amp.

1.13 Nonlinear Resistors

Those elements whose $V - I$ curves are not straight lines are called nonlinear elements because their resistances are nonlinear resistances. Their $V - I$ characteristics can be represented by a suitable equation.

Examples of nonlinear elements are filaments of incandescent lamps, diodes, thermistors and varistors. A varistor is a special resistor made of carborundum crystals held together by a binder. Fig. 1.27 (*a*) shows how current through a varistor increase rapidly when the applied voltage increases beyond a certain amount (nearly 100 V in the present case).

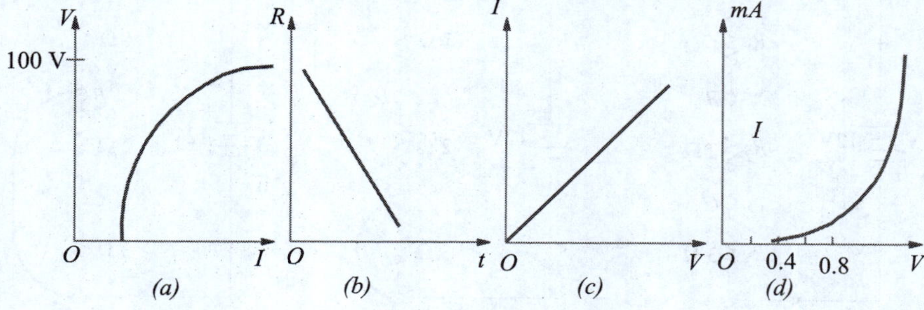

Fig. 1.27

There is a corresponding rapid decrease in resistance when the current increases. Hence, varistors are generally used to provide over-voltage protection in certain circuits.

A thermistor is made of metallic oxides in a suitable binder and has a large negative coefficient of resistance *i.e.,* its resistance decreases with increase in temperature as shown in Fig. 1.27 (*b*). Fig. 1.27 (*c*) shows how the resistance of an incandescent lamp increases with voltage whereas Fig. 1.27 (*d*) shows the *V-I* characteristics of a typical silicon diode. For a germanium diode, current is related to its voltage by the relation.

$$I = I_o (e^{V/0.026} - 1)$$

1.14 Short and Open Circuits

When two points of circuit are connected together by a thick metallic wire (Fig. 1.28), they are said to be *short-circuited.* Since 'short' has practically zero resistance, it gives rise to two important facts:

(*i*) no voltage can exist across it because $V = IR = I \times 0 = 0$

(*ii*) current through it (called short-circuit current) is very large (theoretically, infinity)

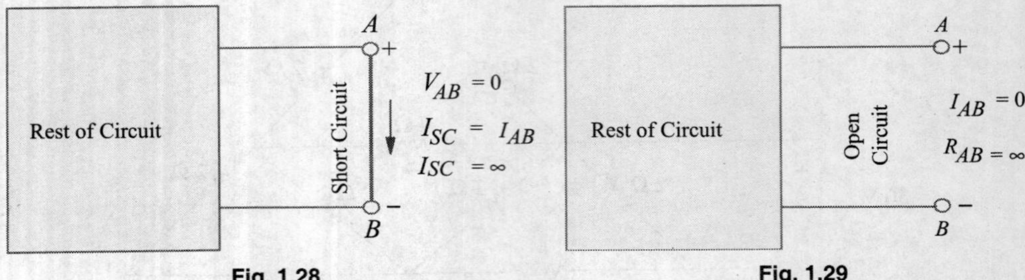

Fig. 1.28 Fig. 1.29

Two points are said to be open-circuited when there is no direct connection between them (Fig. 1.29). Obviously, an 'open' represents a break in the continuity of the circuit. Due to this break

(*i*) resistance between the two points is infinite.

(*ii*) there is no flow of current between the two points.

1.15 'Shorts' in a Series Circuit

Since a dead (or solid) short has almost zero resistance, it causes the problem of excessive current which, in turn, causes power dissipation to increase many times and circuit components to burn out.

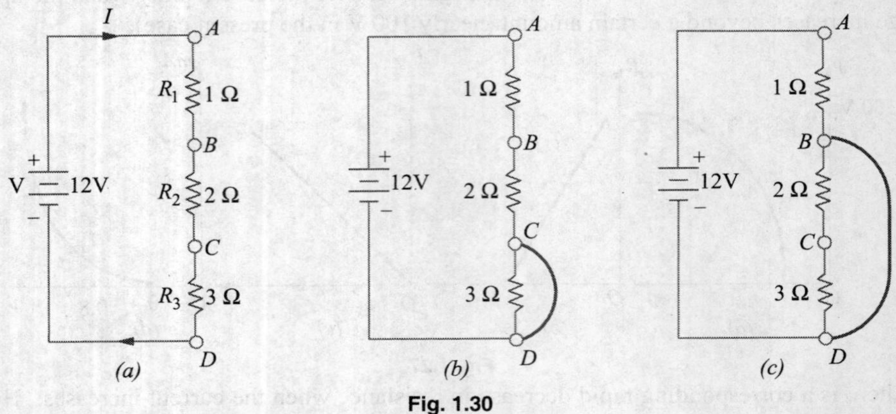

Fig. 1.30

In Fig. 1.30 (*a*) is shown a normal series circuit where

$$V = 12\text{ V}, R = R_1 + R_2 + R_3 = 6\ \Omega$$

$$I = V/R = 12/6 = 2\text{ A}, P = I^2R = 2^2 \times 6 = 24\text{ W}$$

In Fig. 1.30 (*b*), 3-Ω resistor has been shorted out by a resistanceless copper wire so that $R_{CD} = 0$. Now, total circuit resistance $R = 1 + 2 + 0 = 3\ \Omega$. Hence, $I = 12/3 = 4$ A and $P = 4^2 \times 3 = 48$ W.

Fig. 1.30 (*c*) shows the situation where both 2 Ω and 3 Ω resistors have been shorted out of the circuit. In this case,

$$R = 1\ \Omega, I = 12/1 = 12\text{ A} \quad \text{and} \quad P = 12^2 \times 1 = 144\text{ W}$$

Because of this excessive current (6 times the normal value), connecting wires and other circuit components can become hot enough to ignite and burn out.

1.16 'Opens' in a Series Circuit

In a normal series circuit like the one shown in Fig. 1.31 (*a*), there exists a current flow and the voltage drops across different resistors are proportional to their resistances. If the circuit becomes 'open' anywhere, following two effects are produced:

(i) since 'open' offers infinite resistance, circuit current becomes zero. Consequently, there is no voltage drop across R_1 and R_2.

***(ii) whole of the applied voltage (i.e. 100 V in this case) is felt across the 'open' i.e. across terminals A and B* [Fig. 1.31 *b*)].**

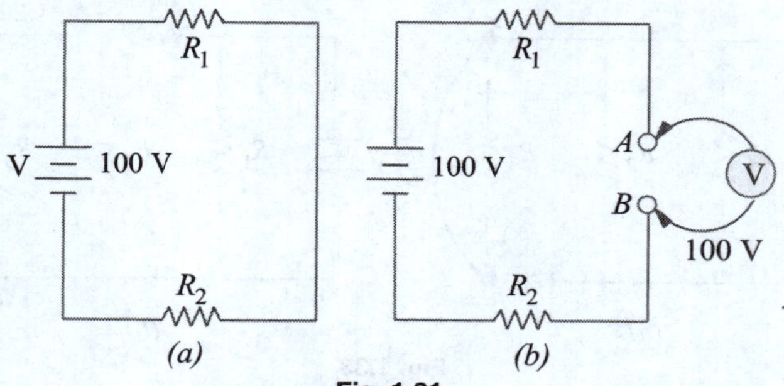

Fig. 1.31

The reason for this is that R_1 and R_2 become negligible as compared to the infinite resistance of the 'open' which has practicallly whole of the applied voltage dropped across it (as per Voltage Divider Rule of art. 1.11). Hence, voltmeter in Fig. 1.31 (*b*) will read nearly 100 *V i.e.* the supply voltage.

1.17 'Opens' in a Parallel Circuit

Since an 'open' offers infinite resistance, there would be no current in that part of the circuit where it occurs. In a parallel circuit, an 'open' can occur either in the main line or in any parallel branch.

As shown in Fig. 1.32 (*a*), an open in the main line prevents flow of current ***to all branches***. Hence, neither of the two bulbs glows. However, full applied voltage (*i.e.* 220 V in this case) ***is available across the open***.

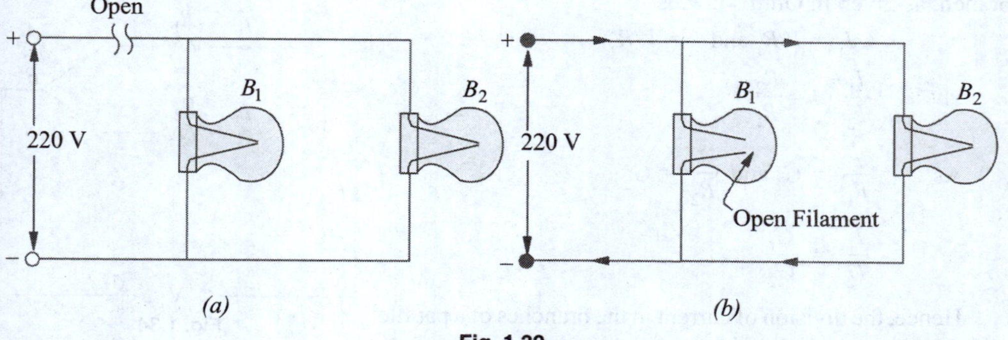

Fig. 1.32

In this Fig. 1.32 (*b*), 'open' has occurred in branch circuits of B_1. Since there is no current in this branch, B_1 will not glow. However, as the other bulb remains connected across the voltage supply, it would keep operating normality.

It may be noted that if a voltmeter is connected across the open bulb, it will read full supply voltage of 220 V.

1.18 'Shorts' in Parallel Circuits

Suppose a 'short' is placed across R_3 (Fig. 1.33). It becomes directly connected across the battery and draws almost infinite current because not only its own resistance but that of the connecting wires *AC* and *BD* is negligible. Due to this excessive current, the wires may get hot enough to burn out unless the circuit is protected by a fuse.

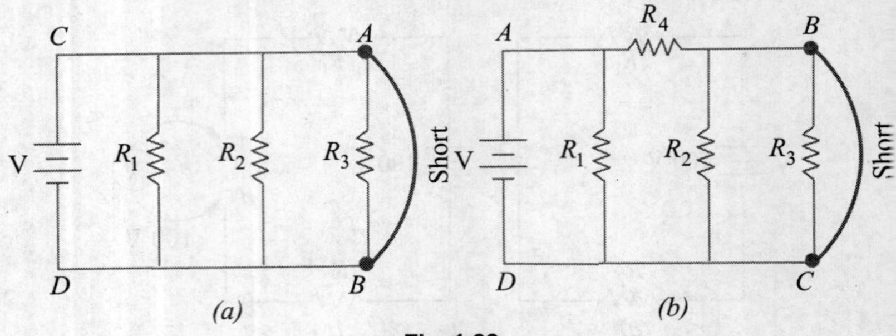

Fig. 1.33

Following points about the circuit of Fig. 1.33 (*a*) are worth noting.

1. not only is R_3 short-circuited but both R_1 and R_2 are also shorted out *i.e.* ***short across one branch means short across all branches.***
2. there is no current in shorted resistors. If there were three bulbs, they will not glow.
3. the shorted components are not damaged, For example, if we had three bulbs in Fig. 1.33 (*a*), they would glow again when circuit is restored to normal conditions by removing the short-circuited.

It may, however, be noted from Fig. 1.33 (*b*) that a short-circuit across R_3 may short out R_2 but not R_1 since it is protected by R_4.

1.19 Division of Current in Parallel Circuits

In Fig. 1.34, two resistances are joined in parallel across a voltage *V*. The current in each branch, as given in Ohm's law, is

$$I_1 = V/R_1 \text{ and } I_2 = V/R_2$$

$$\therefore \quad \frac{I_1}{I_2} = \frac{R_2}{R_1}$$

$$\text{As} \quad \frac{I}{R_1} = G_1 \text{ and } \frac{I}{R_2} = G_2$$

$$\therefore \quad \frac{I_1}{I_2} = \frac{G_1}{G_2}$$

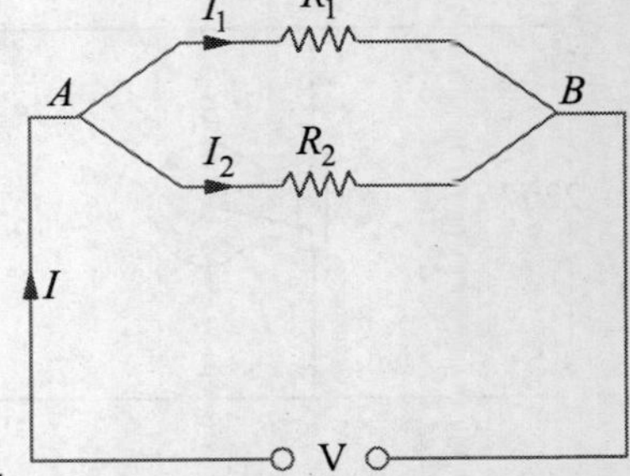

Fig. 1.34

Hence, the division of current in the branches of a parallel circuit is directly proportional to the conductance of the branches or inversely proportional to their resistances. We may also express the branch currents in terms of the total circuit current thus:

Now $I_1 + I_2 = I; \quad \therefore \quad I_2 = I - I_1 \quad \therefore \dfrac{I_1}{I - I_1} = \dfrac{R_2}{R_1}$ or $I_1R_1 = R_2\,(I - I_1)$

$$\therefore \quad I_1 = I\frac{R_2}{R_1 + R_2} = I\frac{G_1}{G_1 + G_2} \text{ and } I_2 = I\frac{R_1}{R_1 + R_2} = I\,.\frac{G_2}{G_1 + G_2}$$

This Current Divider Rule has direct application in solving electric circuits by Norton's theorem (Art. 2.23).

Take the case of three resistors in parallel connected across a voltage V (Fig. 1.34). Total current is $I = I_1 + I_2 + I_3$. Let the equivalent resistance be R. Then

$$V = IR$$

Also $\quad V = I_1 R_1 \quad \therefore \quad IR = I_1 R_1$

or $\quad \dfrac{I}{I_1} = \dfrac{R_1}{R}$ or $I_1 = IR/R_1$ ...**(i)**

Now $\quad \dfrac{I}{R} = \dfrac{I}{R_1} + \dfrac{I}{R_2} + \dfrac{I}{R_3}$

$$R = \frac{R_1R_2R_3}{R_1R_3 + R_2R_3 + R_1R_2}$$

From **(i)** above, $\quad I_1 = I\left(\dfrac{R_2R_3}{R_1R_2 + R_2R_3 + R_3R_1}\right) = I\,.\dfrac{G_1}{G_1 + G_2 + G_3}$

Similarly, $\quad I_2 = I\left(\dfrac{R_1R_3}{R_1R_2 + R_2R_3 + R_3R_1}\right) = I\,.\dfrac{G_2}{G_1 + G_2 + G_3}$

$$I_3 = I\left(\frac{R_1R_2}{R_1R_2 + R_2R_3 + R_3R_1}\right) = I\,.\frac{G_3}{G_1 + G_2 + G_3}$$

Example 1.19. *A resistance of 10 Ω is connected in series with two resistan es each of 15 Ω arranged in parallel. What resistance must be shunted across this pa allel combination so that the total current taken shall be 1.5 A with 20 V applied ?*

(Elements of Elect. Engg.-1; Banglore Univ.)

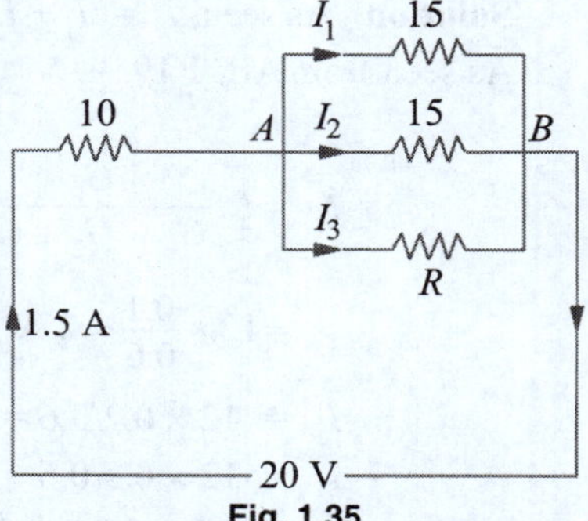

Fig. 1.35

Solution. The circuit connections are shown in Fig. 1.35.

Drop across 10-Ω resistor = 1.5 × 10 = 15 V

Drop across parallel combination, V_{AB} = 20 – 15 = 5 V

Hence, voltage across each parallel resistance is 5 V.

I_1 = 5/15 = 1/3 A, I_2 = 5/15 = 1/3 A

I_3 = 1.5 – (1/3 + 1/3) = 5/6 A

∴ $\quad I_3 R$ = 5 or (5/6) R = 5 or R = **6 Ω**

Example 1.20. *If 20 V be applied across AB shown in Fig. 1.36, calculate the total current, the power dissipated in each resistor and the value of the series resistance to have the total current.* **(Elect. Science-II, Allahabad Univ. 1992)**

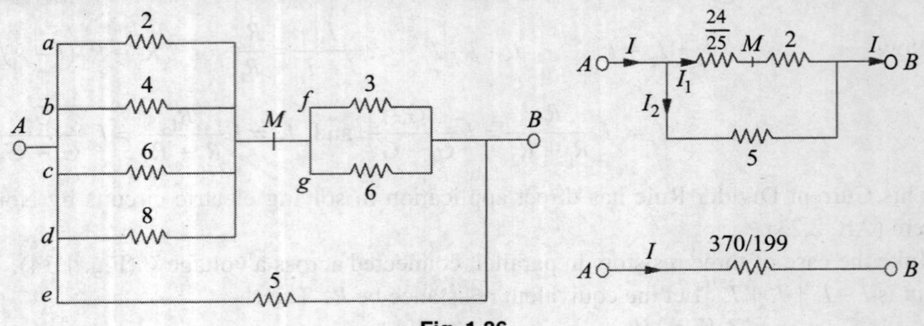

Fig. 1.36

Solution. As seen from Fig. 1.36. R_{AB} = 370/199 Ω.

Hence, total current $= 20 \div 370/199 = \mathbf{10.76\ A}$

$$I_1 = 10.76 \times 5(5 + 74.25) = 6.76\ A;\ I_2 = 10.76 - 6.76 = 4\ \text{A}$$

$$I_f = 6.76 \times 6/9 = 4.51\ \text{A};\ I_g = 6.76 - 4.51 = 2.25\ \text{A}$$

Voltag drop across *A* and *M*, $V_{AM} = 6.76 \times 24/25 = 6.48$ V

$$I_a = V_{AM}/2 = 6.48/2 = 3.24\ \text{A};\ I_b = 6.48/4 = 1.62\ \text{A};\ I_c = 6.48/6 = 1.08\ \text{A}$$

$$I_d = 6.48/8 = 0.81\ \text{A},\ I_e = 20/5 = 4\ \text{A}$$

Power Dissipation

$P_a = I_a^2 R_a = 3.24^2 \times 2 = \mathbf{21\ W}$, $P_b = 1.62^2 \times 4 = \mathbf{10.4\ W}$, $P_c = 1.08^2 \times 6 = \mathbf{7\ W}$

$P_d = 0.81^2 \times 8 = \mathbf{5.25\ W}$, $P_e = 4^2 \times 5 = \mathbf{80\ W}$, $P_f = 4.51^2 \times 3 = \mathbf{61\ W}$

$P_g = 2.25^2 \times 6 = \mathbf{30.4\ W}$

The series resistance required is **370/199 Ω**

Incidentally, total power dissipated = $I^2 R_{AB} = 10.76^2 \times 370/199 = 215.3$ W (as a check).

Example 1.21. *Calculate the values of different currents for the circuit shown in Fig. 1.37. What is the total circuit conductance ? and resistance ?*

Solution. As seen, $I = I_1 + I_2 + I_3$. The current division takes place at point *B*.

As seen from Art. 1.19.

$$I_1 = I \cdot \frac{G_1}{G_1 + G_2 + G_3}$$

$$= 12 \times \frac{0.1}{0.6} = \mathbf{2\ A}\ \mathbf{2\ A}$$

$$I_2 = 12 \times 0.2/0.6 = \mathbf{4\ A}$$

$$I_3 = 12 \times 0.3/0.6 = \mathbf{6\ A}$$

$$G_{BC} = 0.1 + 0.2 + 0.3 = 0.6\ \text{S}$$

$$\frac{1}{G_{AC}} = \frac{1}{G_{AB}} + \frac{1}{G_{BC}} = \frac{1}{0.4} + \frac{1}{0.6} = \frac{25}{6}\ \text{S}^{-1} \quad \therefore \quad R_{AC} = 1/G_{AC} = \mathbf{25/6\ \Omega}$$

Fig. 1.37

Example 1.22. *Compute the values of three branch currents for the circuits of Fig. 1.38 (a). What is the potential difference between points A and B ?*

Solution. The two given current sources may be combined together as shown in Fig. 1.38 (*b*). Net current = 25 – 6 = 19 A because the two currents flow in opposite directions.

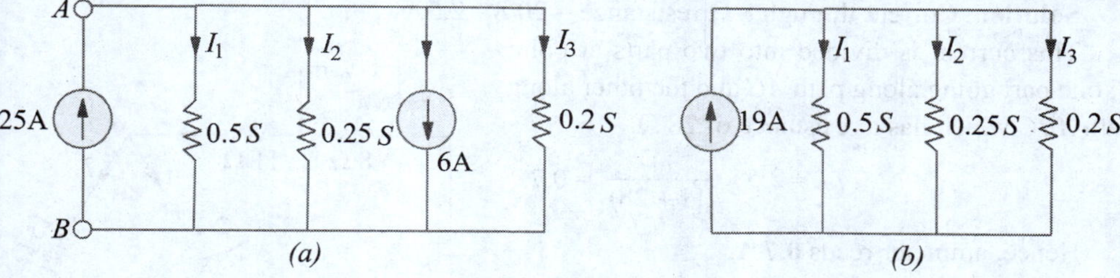

Fig. 1.38

Now, $G = 0.5 + 0.25 + 0.2 = 0.95$ S; $I_1 = I\dfrac{G_1}{G} = 19 \times \dfrac{0.5}{0.95} = \mathbf{10\ A}$

$$I_2 = I\frac{G_2}{G} = 19 \times \frac{0.5}{0.95} = \mathbf{5\ A};\ I_3 = I\frac{G_3}{G} = 19 \times \frac{0.2}{0.95} = \mathbf{4\ A}$$

$$V_{AB} = I_1 R_1 = \frac{I_1}{G_1} = \frac{I_2}{G_2} = \frac{I_3}{G_3} \therefore\ V_{AB} = \frac{10}{0.5} = \mathbf{20\ A}$$

The same voltage acts across the three conductances.

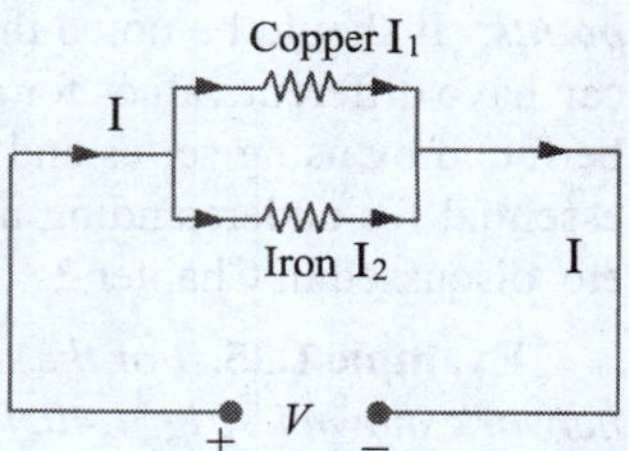

Example 1.23. *Two conductors, one of copper and the other of iron, are connected in parallel and at 20°C carry equal currents. What proportion of current will pass through each if the temperature is raised to 100°C ? Assume α for copper as 0.0042 and for iron as 0.006 per °C at 20°C. Find also the values of temperature coefficients at 100°C.*

(Electrical Engg. Madras Univ.)

Solution. Since they carry equal current at 20°C, the two conductors have the same resistance at 20°C *i.e.* R_{20}. As temperature is raised, their resistances increase through unequally.

For Cu, $R_{100} = R_{20}(1 + 80 \times 0.0042) = 1.336\ R_{20}$

For iron $R'_{100} = R_{20}(1 + 80 \times 0.006) = 1.48\ R_{20}$

As seen from Art. 1.19, current through Cu conductor is

$$I_1 = I \times \frac{R'_{100}}{R_{100} + R'_{100}} = I \times \frac{1.48\ R_{20}}{2.816\ R_{20}} = 0.5256\ I \text{ or } \mathbf{52.56\%\ of\ } \boldsymbol{I}$$

Hence, current through Cu conductor is 52.56 per cent of the total current. Obviously, the remaining current *i.e.* 47.44 per cent passes through iron.

Or current through iron conductor is

$$I_2 = I.\frac{R_{100}}{R_{100} + R'_{100}} = I \times \frac{1.336\ R_{20}}{2.816\ R_{20}} = 0.4744\ I \text{ or } \mathbf{47.44\%\ of\ } \boldsymbol{I}$$

For Cu, $$\alpha_{100} = \frac{1}{(1/0.0042) + 80} = \mathbf{0.00314°C^{-1}}$$

For iron, $$\alpha_{100} = \frac{1}{(1/0.006) + 80} = \mathbf{0.00406°C^{-1}}$$

Example 1.24. *A battery of unknown e.m.f. is connected across resistances as shown in Fig. 1.39. The voltage drop across the 8 Ω resistor is 20 V. What will be the current reading in the ammeter? What is the e.m.f. of the battery?* **(Basic Elect. Engg.; Bangladesh Univ., 1990)**

Solution. Current through 8 Ω resistance = 20/8 = 2.5 A

This current is divided into two parts at point *A*; one part going along path *AC* and the other along path *ABC* which has a resistance of 28 Ω.

$$I_2 = 2.5 \times \frac{11}{(11+28)} = 0.7$$

Hence, ammeter reads **0.7 A.**

Resistance between *A* and *C* = (28 × 11/39) ohm.

Total circuit resistance = 8 + 11 + (308/39) = 1049/39 Ω

∴ $E = 2.5 \times 1049/39 =$ **67.3 V**

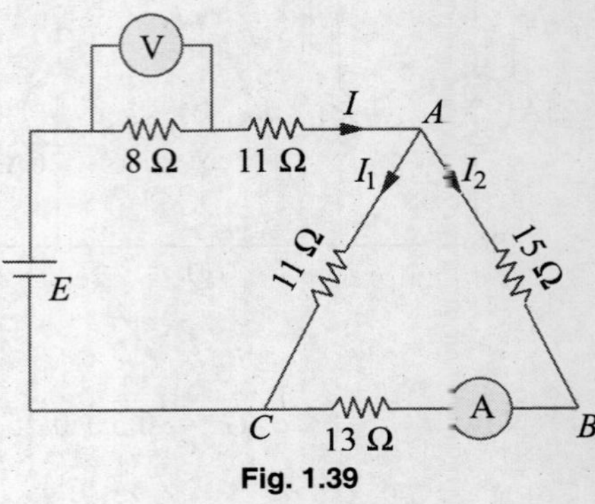

Fig. 1.39

1.20. Equivalent Resistance

The equivalent resistance of a circuit (or network) between its any two points (or terminals) is given by that ***single*** resistance which can replace the entire given circuit between ***these two points***. It should be noted that resistance is always between two ***given*** points of a circuit and can have different values for different point-pairs as illustrated by Example 1.42. it can usually be found by using series and parallel laws of resistances. Concept of equivalent resistance is essential for understanding network theorems like Thevenin's theorem and Norton's theorem etc. discussed in Chapter 2.

Example 1.25. *For the simple ladder network shown in Fig. 1.40, find the input voltage V_i which produces a current of 0.25 A in the 3 Ω resistor. All resistances are in ohm.*

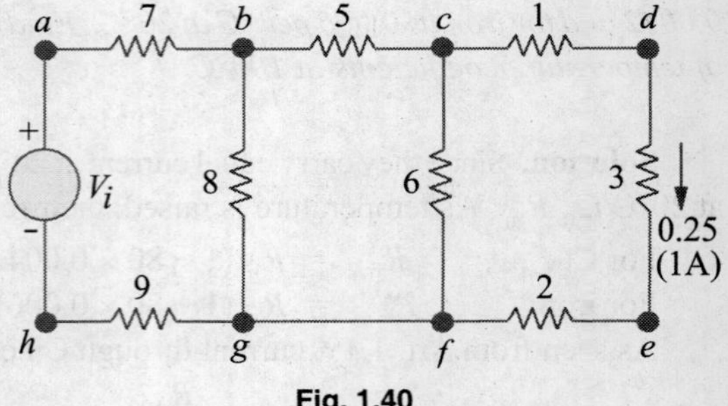

Fig. 1.40

Solution. We will assume a current of 1 A in the 3 Ω resistor. The voltage necessary to produce 1 A bears the same ratio to 1 A as V_i does to 0.25 A because of the linearity of the network. It is known as Current Assumption technique.

Since $R_{cdef} = R_{cf} = 6\ \Omega$

Hence, $I_{cf} = 1\text{ A}$

and $V_{cf} = V_{cdef} = 1 \times 6 = 6\text{ V}.$

Also, $I_{bc} = 1 + 1 = 2\text{ A}$

$V_{bg} = V_{bb} + V_{ef} = 2 \times 5 + 6 = 16\text{ V}$

$I_{bg} = 16/8 = 2\text{ A}$

$I_{ab} = I_{bc} + b_{bg} = 2 + 2 = 4\text{ A}$

$V_i = V_{ab} + V_{bg} + V_{gh}$

$= 4 \times 7 + 16 + 4 \times 9 = 80\text{ V}$

Taking the proportion, we get

$$\frac{80}{1} \quad \frac{V_i}{0.25} \qquad \therefore \quad V_i = 80 \times 0.25 = \mathbf{20\ V}$$

Example 1.26. *In this circuit of Fig. 1.41, find the value R_1 and R_2 so that $I_2 = I_1/n$ and the input resistance as seen from points A and B is R ohm.*

Solution. As seen, the current through R_2 is $(I_1 - I_2)$. Hence, p.d. across points C and D is $R_2(I_1 - I_2) = (R_1 + R) I_2$ or $R_2 I_1 = (R_1 + R_2 + R) I_2$

$$\frac{I_1}{I_2} \quad \frac{R_1 + R_2 + R}{R_2} = n \qquad ...(i)$$

The input resistance of the circuit as viewed from terminals A and B is required to be R.

$$\therefore \quad R = R_1 + R_2 \parallel (R_1 + R)$$

$$R_1 + \frac{R_1 + R}{n} \qquad \text{...using Eq. } (i)$$

$$R(n-1) = R_1(n+1)$$

$$\therefore \quad R_1 \frac{n-1}{n+1} R \text{ and } R_2 = \frac{R_1 + R}{(n-1)} \frac{2n}{n^2 - 1} R$$

1.21 Duality between Series and Parallel Circuits

There is a certain peculiar pattern of relationship between series and parallel circuits. For example, in a series circuit, current is the same whereas in a parallel circuit, voltage is the same. Also, in a series circuit, individual voltages are added and in a parallel circuit, individual currents are added. It is seen that while comparing series and parallel circuits, voltage takes the place of current and current takes the place of voltage. Such a pattern is known as "duality" and the two circuits are said to be duals of each other. Dual elements refer to same magnitude.

As arranged in Table 1.4 the equations involving voltage, current and resistance in a series circuit have a corresponding dual counterparts in terms of current, voltage and conductance for a parallel circuit.

Table 1.4

Series Circuit	Parallel Circuit
$I_1 = I_2 = I_3 = \ldots\ldots\ldots$	$V_1 = V_2 = V_3 = \ldots\ldots\ldots$
$V_T = V_1 + V_2 + V_3 + \ldots\ldots\ldots$	$I_T = I_1 + I_2 + I_3 + \ldots\ldots\ldots$
$R_T = R_1 + R_2 + R_3 + \ldots\ldots\ldots$	$G_T = G_1 + G_2 + G_3 + \ldots\ldots\ldots$
$I = \frac{V_1}{R_1} = \frac{V_2}{R_2} = \frac{V_3}{R_3} = \ldots\ldots$	$V = \frac{I_1}{G_1} = \frac{I_2}{G_2} = \frac{I_3}{G_3} = \ldots\ldots$
Voltage Divider Rule $V_1 = V_T \frac{R_1}{R_T}, V_2 = V_T \frac{R_2}{R_T}$	Current Divider Rule $I_1 = I_T \frac{G_1}{G_T}, I_2 = I_T \frac{G_2}{G_T}$

Tutorial Problem 1.4

1. Using the current-divider rule, find the ratio I_L/I_S in the circuit shown in Fig. 1.41. **[0.25]**

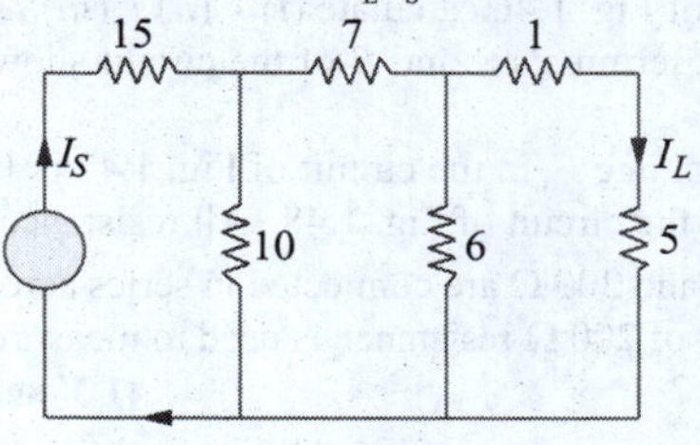

Fig. 1.41

2. Find the values of variables indicated in the circuit of Fig. 1.42. All resistances are in ohms. **[(*a*) 40 V (*b*) 21 V; 15 V (*c*) – 5 A; 3 A]**

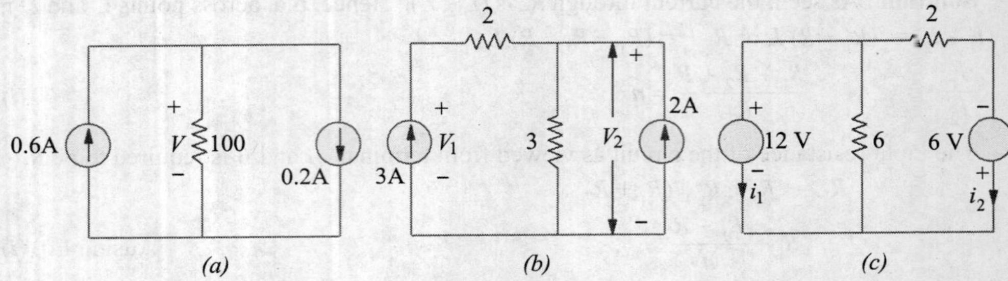

Fig. 1.42

3. An ohmeter is used for measuring the resistance of a circuit between its two terminals. What would be the reading of such an instrument used for the circuit of Fig. 1.43 at point (*a*) *AB* (*b*) *AC* and (*c*) *BC*? All resistances are in ohm. **[(*a*) 25 Ω (*b*) 24 Ω (*c*) 9 Ω]**
4. Find the current and power supplied by the battery to the circuit of Fig. 1.44 (*i*) under normal conditions and (*ii*) when a 'short' occurs across terminals *A* and *B*. All resistances are in kilo-ohm. **[(*i*) 2 mA; 24 m W; (*ii*) 34 mA; 36 mW]**

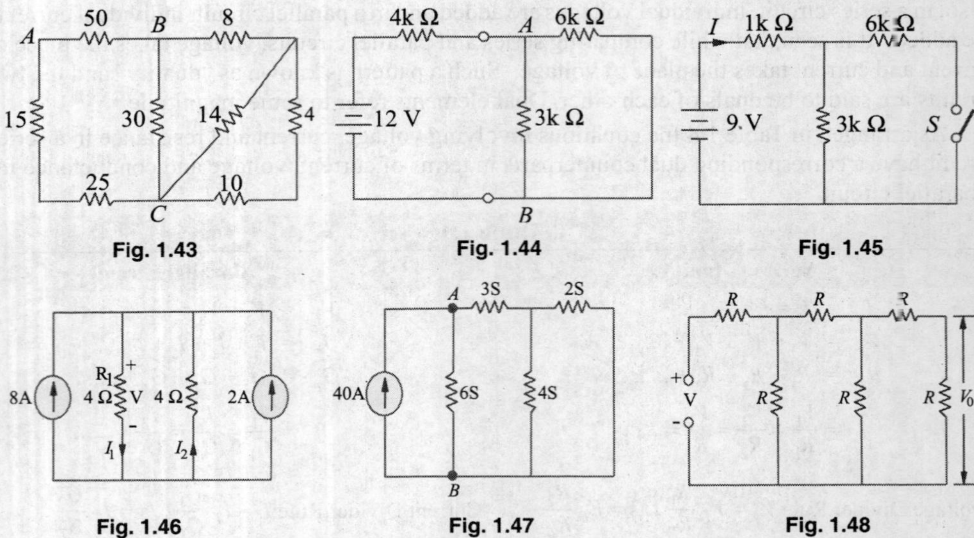

Fig. 1.43 **Fig. 1.44** **Fig. 1.45**

Fig. 1.46 **Fig. 1.47** **Fig. 1.48**

5. Compute the values of battery current *I* and voltage drop across 6 kΩ resistor of Fig. 1.45 when switch *S* is (*a*) closed and (*b*) open. All resistance values are in kilo-ohm. **[(*a*) 3mA; 6 V; (*b*) 2.25 mA; 0V]**
6. For the parallel circuit of Fig. 1.46 calculate (*i*) *V* (*ii*) I_1 (*iii*) I_2. **[(*i*) 20 V; (*ii*) 5 A; (*iii*) – 5 A]**
7. Find the voltage across terminals *A* and *B* of the circuit shown in Fig. 1.46. All conductances are in siemens (*S*). **[5 V]**
8. Prove that the output voltage V_0 in the circuit of Fig. 1.47 is *V*/13.
9. Deduce the current *I* in the circuit of Fig. 1.48. All resistances are in ohms. **[25 A]**
10. Two resistors of 100 Ω and 200 Ω are connected in series across a 4-V cell of negligible internal resistance. A voltmeter of 200 Ω resistance is used to measures P.D. across each. What will the voltage be in each case ? **[1 V across 100 Ω ; 2 V across 200 Ω]**

11. Using series–parallel combination laws, find the resistance between terminals *A* and *B* of the network shown in Fig. 1.49. **[3 *R*]**

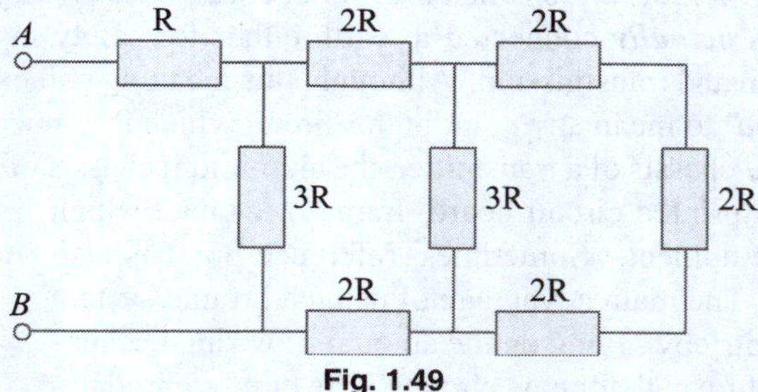

Fig. 1.49

12. A resistance coil *AB* of 100 Ω resistance is to be used as a potentiometer and is connected to a supply at 230 V. Find, by calculation, the position of a tapping point *C* between *A* and *B* such that a current of 2 A will flow in a resistance of 50 Ω connected across *A* and *C*.

[43.4 Ω from *A* to *C*] (*London Univ.*)

13. In the circuit shown in Fig. 1.50, calculate (*a*) current *I* (*b*) current I_1 and (*c*) V_{AB}. All resistances are in ohms. **[(*a*) 4 A (*b*) 0.25 *A* (*c*) 4 *V*]**

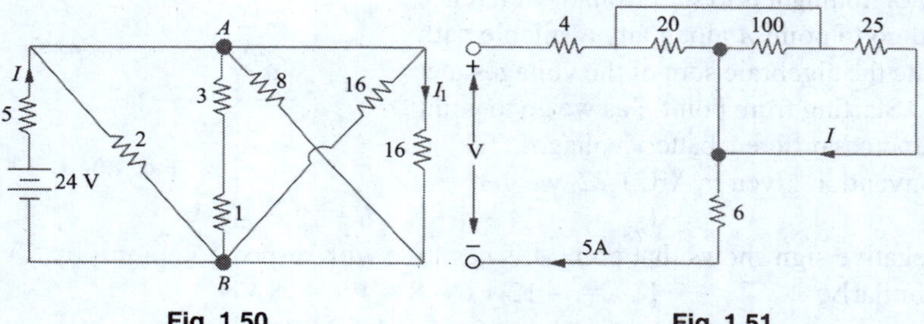

Fig. 1.50 **Fig. 1.51**

1.22 Relative Potential

It is the voltage of one point in a circuit with respect to that of another point (usually called the reference or common point).

Consider the circuit of Fig. 1.52 (*a*) where the most negative end-point *C* has been taken as the reference. With respect to point *C*, both points *A* and *B* are positive though *A* is more positive than *B*. The voltage of point *B* with respect to that of *C i.e.* V_{BC} = + 30 V.

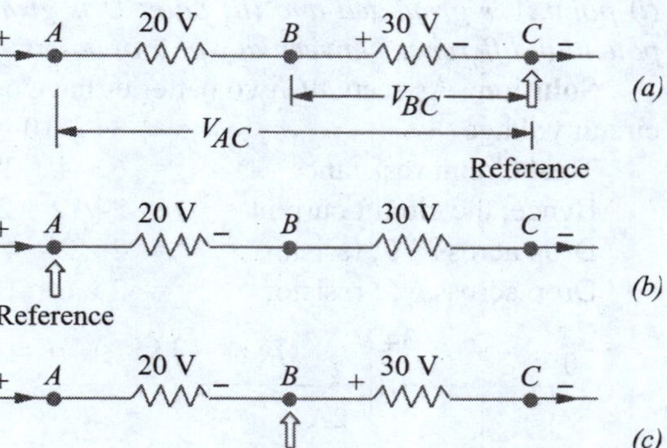

Fig. 1.52

Similarly, V_{AC} = + (20 + 30) = + 50 V.

In Fig. 1.52 (*b*), the most positive end point *A* has been taken as the reference point. With respect to *A*, both *B* and *C* are negative though *C* is more negative than *B*.

V_{BA} = –20 V, V_{CA} = –(20 + 30) = –50 V

In Fig. 1.52 (*c*), mid-point *B* has been taken as the reference point. With respect to *B, A* is at positive potential whereas *C* is at a negative potential.

Hence, V_{AB} = + 20 V and V_{CB} = – 30 V (of course, V_{BC} = + 30 V)

It may be noted that ***any point*** in the circuit can be chosen as the reference point to suit our requirements. This point is often called ***ground*** or *earth* because originally it meant a point in a circuit which was ***actually*** connected to earth either for safety in power systems or for efficient radio reception and transmission. Although, this meaning still exists, yet it has become usual today for ***'ground'*** to mean any point in the circuit which is connected to a large metallic object such as the metal chassis of a transmitter, the aluminium chassis of a receiver, a wide strip of copper plating on a printed circuit board, frame or cabinet which supports the whole equipment. Sometimes, reference point is also called ***common*** point. The main advantage of using a ground system is to simplify our circuitry by saving on the amount of wiring because ground is used as the return path for may circuits. The three commonly-used symbols for ***ground*** are shown in Fig. 1.53. Ground potential is assumed to be at zero volts.

Fig. 1.53

Example 1.27. *In Fig. 1.54, calculate the values of **(i)** V_{FA} **(ii)** V_{AE} and **(iii)** V_{BF}*

Solution. It should be noted that V_{FA} stands for the potential of point F with respect to point FA. The easiest way of finding it is to start from the reference point F and go to point A along any available path and calculate the algebraic sum of the voltages met on the way. Starting from point F as we go to point A, we come across different battery voltages. Taking the sign convention given in Art. 1.22, we get

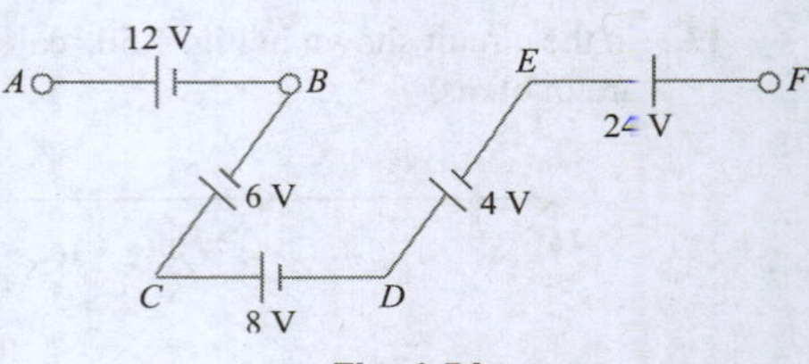

Fig. 1.54

(i) $V_{FA} = -24 + 4 + 8 - 6 + 12 = -6\ \text{V}$

The negative sign shows that point A is negative with respect to point F by 6 V.

(ii) Similarly, $V_{AE} = -12 + 6 - 8 - 4 = -18\ \text{V}$

(iii) Starting from point B, we get $V_{BF} = 6 - 8 - 4 + 24 = 18\ \text{V}$.

Since the result is positive it means that point B is at a higher potential than point F by 18 V.

Example 1.28. *In Fig. 1.55 compute the relative potentials of points A, B, C, D and E which **(i)** point A is grounded and **(ii)** point D is grounded. Does it affect the circuit operation or potential difference between any pair of points ?*

Solution. As seen, the two batteries have been connected in *series opposition.* Hence, net circuit voltage $= 34 - 10 = 24\ \text{V}$

Total circuit resistance $= 6 + 4 + 2 = 12\ \Omega$

Hence, the circuit current $= 24/12 = 2\ \text{A}$

Drop across 2 Ω resistor $= 2 \times 2 = 4\ \text{V}$, Drop across 4 Ω resistor $= 2 \times 4 = 8\ \text{V}$

Drop across 6 Ω resistor $= 2 \times 6 = 12\ \text{V}$

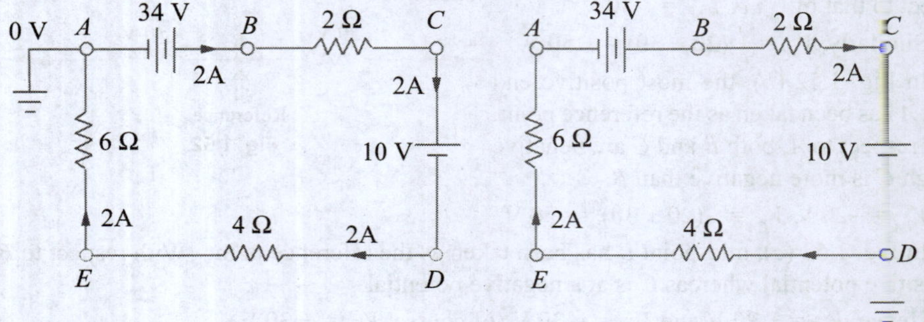

Fig. 1.55

Fig. 1.56

(i) Since point B is directly connected to the positive terminal of the battery whose negative terminal is earthed, hence $V_B = \mathbf{+34\ V.}$

Since there is a fall of 4 V across 2 Ω resistor, $V_C = 34 - 4 = \mathbf{30\ V}$

As we go from point C to D *i.e.* from positive terminal of 10-V battery to its negative terminal, there is a *decrease* in potential of 10 V. Hence, $V_D = 30 - 10 = \mathbf{20}$ *i.e.* point D is 20 V above the ground A.

Similarly, $\quad V_E = V_D$ – voltage fall across 4 Ω resistor = 20 – 8 = **+ 12 V**

Also $\quad V_A = V_E$ – fall across 6 Ω resistor = 12 (2 × 6) = **0 V**

(ii) In Fig. 1.57, point D has been taken as the ground. Starting from point D, as we go to E there is a fall of 8 V. Hence, $V_E = \mathbf{-8\ V.}$ Similarly, $V_A = -(8 + 12) = \mathbf{-20\ V.}$

As we go from A to B, there is a sudden increase of 34 V because we are going from negative terminal of the battery to its positive terminal.

$\therefore \quad V_B = -20 + 34 = \mathbf{+14\ V}$

$V_C = V_B$ – voltage fall across 2 Ω resistor = 14 – 4 = **+ 10 V.**

It should be so because C is connected directly to the positive terminal of the 10 V battery.

Choice of a reference point does not in any way affect the operation of a circuit. Moreover, it also does not change the voltage across any resistor or between any *pair* of points (as shown below) because the ground current $i_g = 0$.

Reference Point *A*

$$V_{CA} = V_C - V_A = 30 - 0 = +30\text{ V};\ V_{CE} = V_C - V_E = 30 - 12 = +18\text{ V}$$
$$V_{BD} = V_B - V_D = 34 - 20 = +14\text{ V}$$

Reference Point *D*

$$V_{CA} = V_C - V_A = 10 - (-20) = +30\text{ V};\ V_{CE} = V_C - V_E = 10 - (-8) = +18\text{ V}$$
$$V_{BD} = V_B - V_D = 14 - 0 = \mathbf{+14\ V}$$

Example 1.29. *Find the voltage V in Fig. 1.88 (a). All resistances are in ohms.*

Solution. The given circuit can be simplified to the final form shown in Fig. 1.57 (*d*). As seen, current supplied by the battery is 1 A. At point A in Fig. 1.57 (*b*), this current is divided into two equal parts of 0.5 A each.

Obviously, voltage V represents the potential of point B with respect to the negative terminal of the battery. Point B is above the ground by an amount equal to the voltage drop across the series combination of (40 + 50) = 90 Ω.

$$V = 0.5 \times 90 = 45\text{ V.}$$

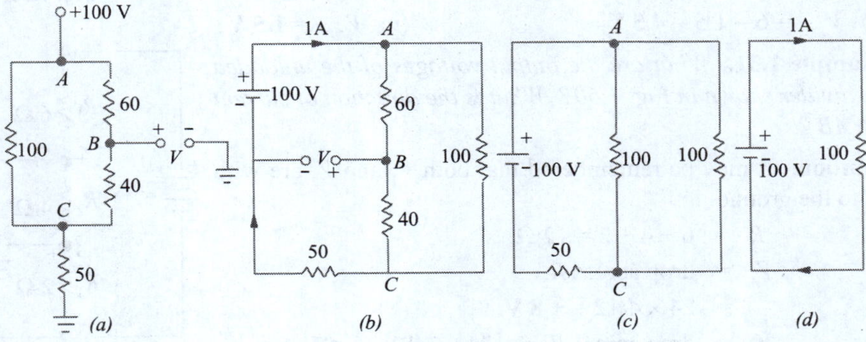

Fig. 1.57

1.23 Voltage Divider Circuit

A voltage divider circuit (also called potential divider) is a series network which is used to feed other networks with a number of different voltages and derived from a single input voltage source.

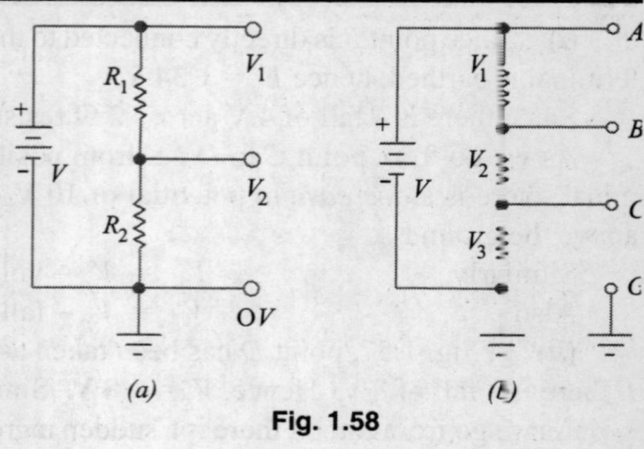

Fig. 1.58

Fig. 1.58 (*a*) shows a simple voltage divider circuit which provides two output voltages V_1 and V_2. Since no load is connected across the output terminals, it is called an ***unloaded*** voltage divider.

As seen from Art. 1.11.

$$V_1 = V\frac{R_1}{R_1 + R_2} \text{ and } V_2 = V \cdot \frac{R_2}{R_1 + R_2}$$

The ratio V_2/V is also known as voltage-ratio ***transfer function***.

As seen $\dfrac{V_2}{V} = \dfrac{R_2}{R_1 + R_2} = \dfrac{1}{1 + R_1/R_2}$

The voltage divider of Fig. 1.58 (*b*) can be used to get six different voltages:

$$V_{CG} = V_3, V_{BC} = V_2, V_{AB} = V_1, V_{BG} = (V_2 + V_3), V_{AC} = (V_1 + V_2) \text{ and } V_{AG} = V$$

Example 1.30. *Find the values of different voltages that can be obtained from a 12-V battery with the help of voltage divider circuit of Fig. 1.59.*

Solution. $R = R_1 + R_2 + R_3 = 4 + 3 + 1 = 8\ \Omega$

Drop across $R_1 = 12 \times 4/8 = 6$ V

$\therefore$ $V_B = 12 - 6 = 6$ V above ground

Drop across $R_2 = 12 \times 3/8 = 4.5$ V

$\therefore$ $V_C = V_B - 4.5 = 6 - 4.5 = 1.5$

Drop across $R_3 = 12 \times 1/8 = 1.5$ V

Different available load voltages are:

(*i*) $V_{AB} = V_A - V_B = 12 - 6 =$ **6 V**

(*ii*) $V_{AC} = 12 - 1.5 =$ **10.5 V** (*iii*) $V_{AD} =$ **12 V**

(*iv*) $V_{BC} = 6 - 1.5 =$ **4.5 V** (*v*) $V_{CD} =$ **1.5 V**

Fig. 1.59

Example 1.31. *What are the output voltages of the unloaded voltage divider shown in Fig. 1.60? What is the direction of current through AB?*

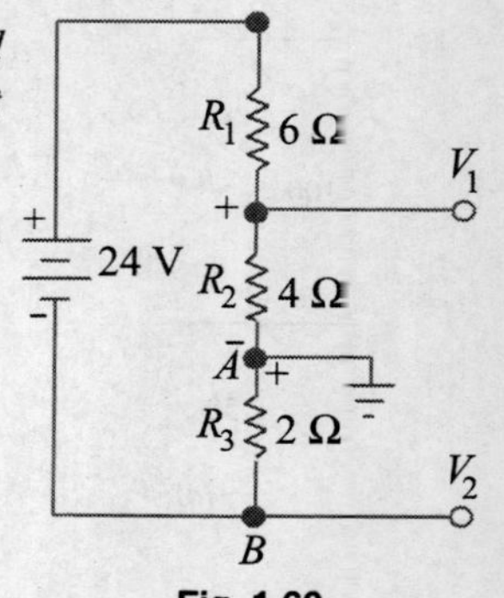

Fig. 1.60

Solution. It may be remembered that both V_1 and V_2 are with respect to the ground.

$$R = 6 + 4 + 2 = 12\ \Omega$$

$\therefore$ V_1 = drop across R_2

$= 24 \times 4/12 =$ **+ 8 V**

V_2 = drop across $R_3 = - 24 \times 2/12 =$ **– 4 V**

It should be noted that point B is at negative potential with respect to the ground.

Current flows from A to B *i.e.* from a point at a higher potential to a point at a lower potential.

Example 1.32. *Calculate the potentials of point A, B, C and D in Fig. 1.61. What would be the new potential values if connections of 6-V battery are reversed ? All resistances are in ohm.*

Solution. Since the two batteries are connected in additive series, total voltage around the circuit is = 12 + 6 = 18 V. The drops across the three resistors as found by the voltage divider rule as shown in Fig. 1.61 (*a*) which also indicates their proper polarities. The potential of any point in the circuit can be found by starting from the ground point G (assumed to be at 0V) and going to the point either in clockwise direction or counter-clockwise direction. While going around the circuit, the rise in potential would be taken as positive and the fall in potential as negative. Suppose we start from point G and proceed in the clockwise direction to point A. The only potential met on the way is the battery voltage which is taken as positive because there is a rise of potential since we are going from its negative to positive terminal. Hence, V_A is + 12 V.

$$V_B = 12 - 3 = 9\text{ V};\ V_C = 12 - 3 - 6 = 3\text{ V}$$

Similarly, $$V_D = 12 - 3 - 6 - 9 = -6\text{ V}.$$

It is also obvious that point D must be at – 6 V because it is directly connected to the negative terminal of the 6-V battery.

We would also find the potentials of various points by starting from point G and going in the counter-clockwise direction. For example, $V_B = -6 + 9 + 6 = 9$ V as before.

The connections of the 6 – V battery have been reversed in Fig. 1.61 (*b*). Now, the net voltage around the circuit is 12 – 6 = 6 V. The drop over the 1 Ω resistor is = 6 × 1/(1 + 2 + 3) = 1 V; Drop over 2 Ω resistor is = 6 × 2/6 = 2 V. Obviously, V_A = + 12 V, V_B = 12 – 1 = 11 V, V_C = 12 – 1 – 2 = 9 V. Similarly, V_D = 12 – 1 – 2 – 3 = + 6 V.

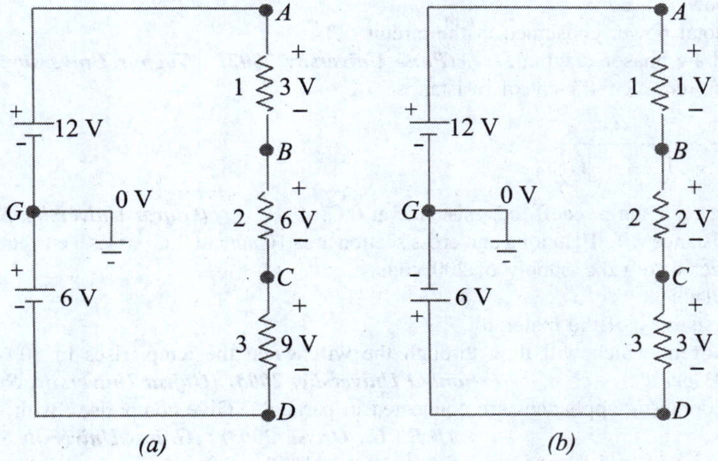

Fig. 1.61

Example 1.33. *Fig. 1.62 shows a transistor with proper voltages established across its base, collector are emitter for proper function. Assume that there is a vol age drop V_{BE} across the base-emitter junction of 0.6 V and collector current I_C is equal to collector current I_E. Calculate **(a)** V_1 **(b)** V_2 and V_B **(c)** V_4 and V_E **(d)** I_E and I_C **(e)** V_3 **(f)** V_C **(g)** V_{CE}. All resistances are given in kilo-ohm.*

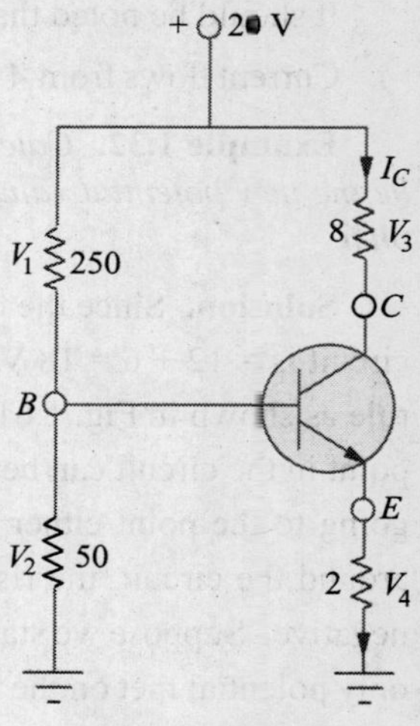

Fig. 1.62

Solution. **(*a*)** The 250 k and 50 k resistors form a voltage-divider bias network across 20 V supply.

∴ $V_1 = 20 \times 250/300 = \mathbf{16.7\ V}$

(*b*) $V_2 = 20 - 16.7 = \mathbf{3.3\ V}$

The voltage of point *B* with respect to ground is $V_2 = \mathbf{3.3\ V}$

(*c*) $V_E = V_2 - V_{BE} = 3.3 - 0.6 = 2.7$ V. Also $V_4 = \mathbf{2.7\ V}$

(*d*) $I_E = V_4/2 = 2.7\ V/2\ k = \mathbf{1.35\ mA}$. It also equals I_C.

(*e*) V_3 = drop across collector resistor
= 1.35 mA × 8 k = 10.8 V

(*f*) Potential of point *C* is $V_C = 20 - 10.8 = \mathbf{9.2\ V}$

(*g*) $V_{CE} = V_C - V_E = 9.2 - 2.7 = \mathbf{6.5\ V}$

Tutorial Problem 1.5

1. A direct - current circuit comprises two resistors, 'A' of value 25 ohms, and 'B' of unknown value, connected in parallel, together with a third resistor 'C' of value 5 ohms connected in series with the parallel group. The potential difference across C is found to 90 V. It the total power in the circuit is 4320 W, calculate:
(*i*) the value of resistor B, (*ii*) the voltage applied to the ends of the whole circuit,
(*iii*) the current in each resistor.
(Mumbai University 2002) (Nagpur University, Summer 2002)
2. A current of 5 A flows through a non inductive resistance connected in series with a choke coil when supplied at 250 V, 50 Hz. If voltage across resistance is 125 V and across coil is 200 V calculate:
(*i*) impedance, resistance and reactance of coil
(*ii*) power in coil
(*iii*) total power consumed in the circuit
(*iv*) draw phasor diagram. ***(Pune University 2002) (Nagpur University, Winter 2003)***
3. Define temp. coefficient of resistance.
Prove $\alpha_{t1} \dfrac{\alpha_0}{(1 + \alpha_0 t_1)}$
where α_0 = temp. coeff. of resistance at 0°C. ***(Gujrat University, Summer 2003)***
4. A resistance wire 10 m long and cross section area 10 mm^2 at 0°*C* passes a current of 10 A, when connected to a d.c. supply of 200 volts.
Calculate:
(*a*) resistivity of the material
(*b*) current which will flow through the wire when the temp. rises to 50°*C*. Given α_0 = 0.0003 *per°C*. ***(Mumbai University, 2003) (Gujrat University, Summer 2003)***
5. Why domestic appliances are connected in parallel ? Give comparison with series ckt.
(B.P.T.U., Orissa 2003) (Gujrat University,Summer 2003)
6. Two wires A and B made up of same material, wire B has twice the length of wire A and having half the diameter to that of A. Calculate the ratio R_B/R_A.
(Gujrat University,Summer 2003)

7. A resistor of 12 Ω is connected in series with a combination of 15 Ω and 20 Ω resistor in parallel. When a voltage of 120 V is applied across the whole circuit find the current taken from the supply. *(V.T.U., Belgaum, Karnataka University, Summer 2002)*
8. A network is arranged as shown in Fig. 1.63. Determine the value of currents in each resistor. *(V.T.U., Belgaum, Karnataka University, Summer 2002)*

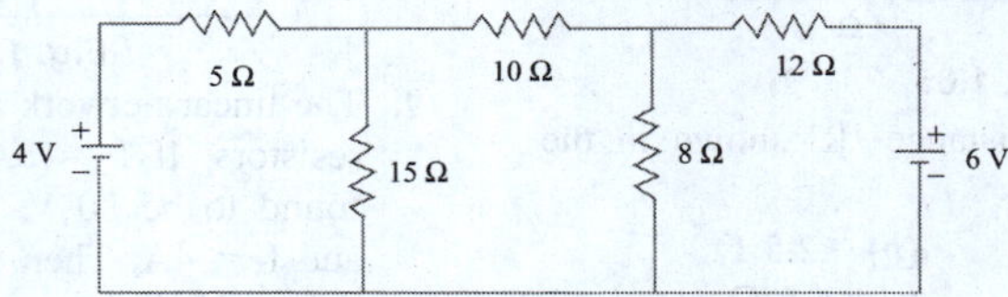

Fig. 1.63

9. A resistance of 100Ω is connected in series with 100μF capacitor across 200V, 60Hz supply. Find the impedance, current and power factor. *(V.T.U., Belgaum, Karnataka University, Summer 2002)*
10. An EMF whose instantaneous value is 100sin (314t – π/4) volts is applied to a circuit and the current flowing through it is 20sin (314t – 1.5708) Amperes. Find the frequency and the values of circuit elements, assuming a series combination of circuit elements. *(V.T.U.,Belgaum, Karnataka University, Wimter 2003, 2009)*
11. An inductive coil draws a current of 2A, when connected to a 230V, 50Hz supply. The power taken by the coil is 100 watts. Calculate the resistance and inductance of the coil. *(Pune University, 2003) (V.T.U.,Belgaum University, Winter 2003, 2007)*
12. Find the resistance between the terminals A and B for the network shown in Fig.1.64. *(Pune University, 2003) (V.T.U.,Belgaum University, Winter 2003)*

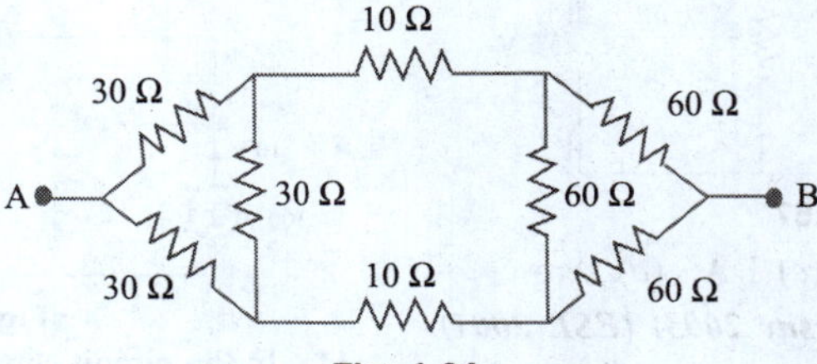

Fig. 1.64

OBJECTIVE TEST QUESTIONS

1. A 100 μA ammeter has an internal resistance of 100 Ω. For extending its range to measure 500 μA, the shunt required is of resistance (in Ω)
   (*a*) 20.0 (*b*) 22.22
   (*c*) 25.0 (*d*) 50.0
   (GATE 2001)
2. Resistances R_1 and R_2 have, respectively, nominal values of 10Ω and 5Ω, and tolerances of ± 5% and ± 10%. The range of values for the parallel combination of R_1 and R_2 is
   (*a*) 3.077 Ω to 3.636 Ω
   (*b*) 2.805 Ω to 3.371 Ω
   (*c*) 3.237 Ω to 3.678 Ω
   (*d*) 3.192 Ω to 3.435 Ω ***(GATE 2001)***
3. The open circuit impedance of a certain length of a loss-less line is 100 Ω. The short circuit impedance of the same line is also 100 Ω. The characteristic impedance of the line is
   (*a*) $100\sqrt{2}\,\Omega$ (*b*) 50 Ω
   (*c*) $\frac{100}{\sqrt{2}}\,\Omega$ (*d*) 100 Ω
   (ESE 2001)
4. The current in the given circuit with a dependent voltage source is
   (a) 10A (b) 12 A
   (c) 14 A (d) 16 A
   (ESE 2001)

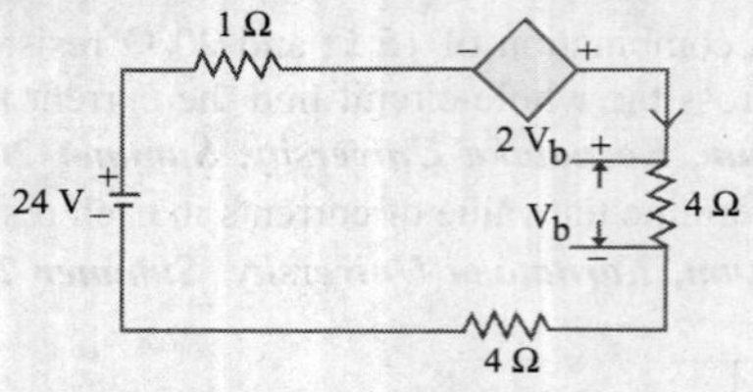

Fig. 1.65

5. The value of resistance 'R' shown in the given Fig. 1.66 is

(*a*) 3.5 Ω (*b*) 2.5 Ω
(*c*) 1 Ω (*d*) 4.5 Ω

(ESE 2001)

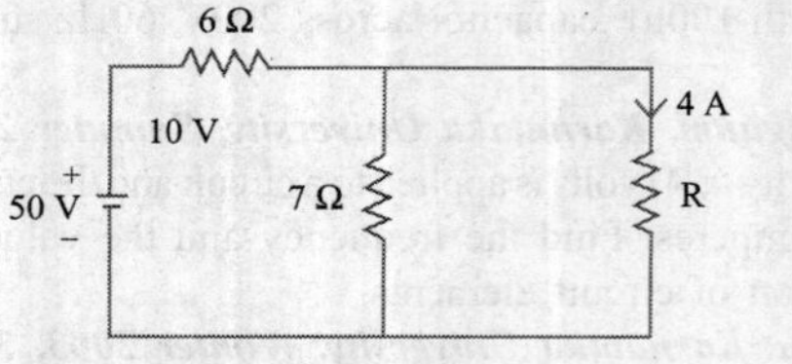

Fig. 1.66

6. For the circuit shown in the given Fig. 1.67 the current I is given by

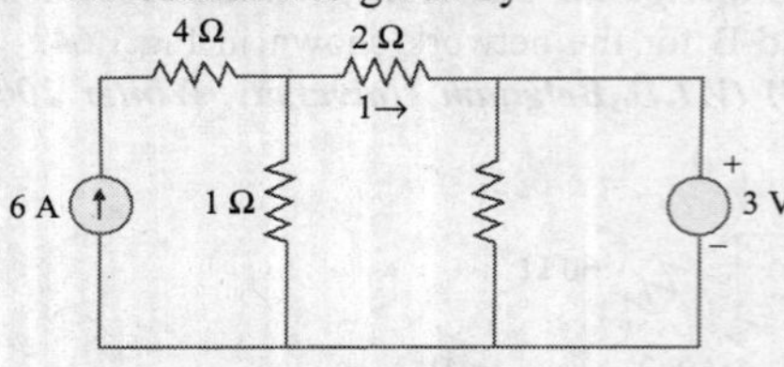

Fig. 1.67

(*a*) 3 A (*b*) 2 A (*c*) 1 A (*d*) zero

(Pune University 2003) (ESE 2001)

7. The value of V in the circuit shown in the given Fig. 1.68 is

(Mumbai University 2003)

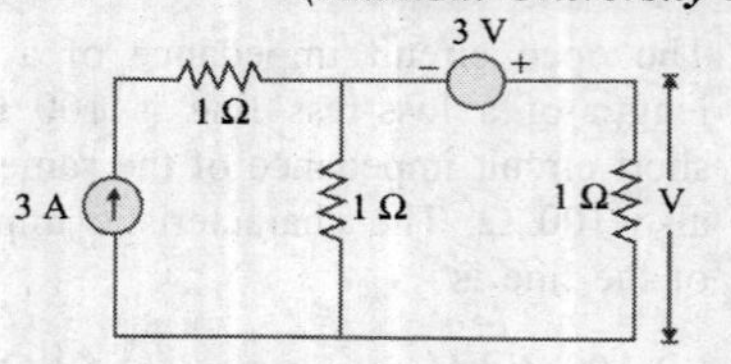

Fig. 1.68

(*a*) 1 V (*b*) 2 V
(*c*) 3 V (*d*) 4 V

(GATE 2003) (ESE 2001)

8. In the circuit shown in Fig. 1.69, the value of V_s is 0, when I = 4A. The value of I when V_s = 16V, is

(*a*) 6 A (*b*) 8 A (*c*) 10 A (*d*) 12 A

(GATE 2003) (ESE 2003)

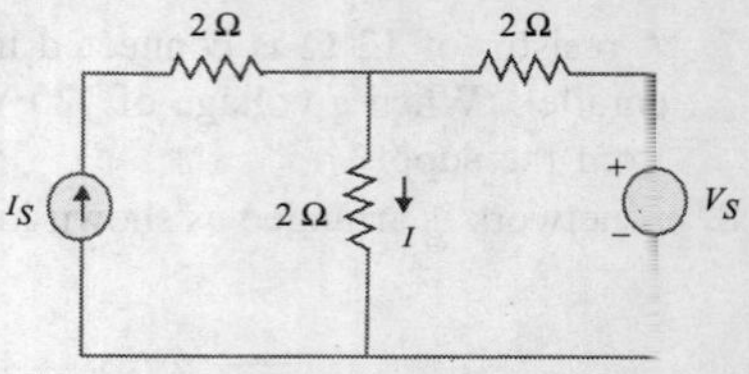

Fig. 1.69

9. The linear network as in Fig. 1.70 has only resistors. If I_1 = 8A and I_2 = 12 A; V is found to be 80 V. V = 0 when I_1 = –8A and I_2 = 4A. Then the value of V when I_1 = I_2 = 10 A, is

(*a*) 25 V (*b*) 50
(*c*) 75 V (*d*) 100 V

(GATE 2003) (ESE 2003)

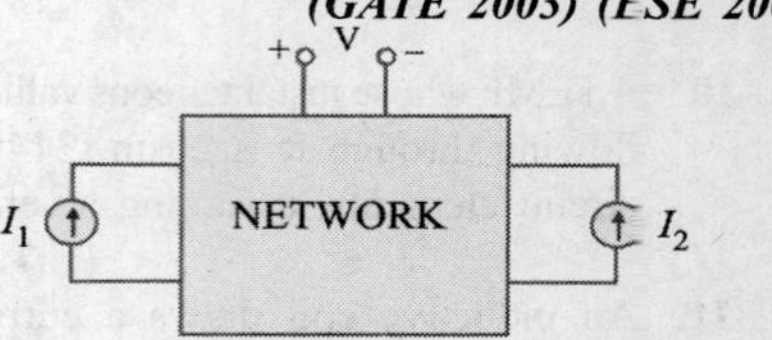

Fig. 1.70

10. In Fig. 1.71, the value of R is

(*a*) 10 Ω (*b*) 18 Ω
(*c*) 24 Ω (*d*) 12 Ω

(GATE 2003)

Fig. 1.71

11. In the circuit shown in Fig. 1.72, the switch S is closed at time t = 0. The voltage across the inductance at t = 0^+, is

(*a*) 2 V (*b*) 4 V
(*c*) –6 V (*d*) 8 V

(GATE 2003)

Fig. 1.72

12. The *rms* value of the resultant current in a wire which carries a dc current of 10 A and a sinusoidal alternating current of peak value 20 A is

(*a*) 14.1 A (*b*) 17.3 A
(*c*) 22.4 A (*d*) 30.0 A

(GATE 2004)

CHAPTER 2

DC Network Theorems

Chapter Outline

- ❖ Electric Circuits and Network Theorems
- ❖ Kirchhoff's Laws
- ❖ Determination of Voltage Sign
- ❖ Assumed Direction of Current
- ❖ Solving Simultaneous Equations
- ❖ Determinants
- ❖ Solving Equations with Two Unknowns
- ❖ Solving Equations With Three Unknowns
- ❖ Independent and Dependent Sources
- ❖ Maxwell's Loop Current Method
- ❖ Nodal Analysis with Voltage Sources
- ❖ Nodal Analysis with Current Sources
- ❖ Source Conversion
- ❖ Ideal Constant-Voltage Source
- ❖ Ideal Constant-Current Source
- ❖ Thevenin Theorem
- ❖ How to Thevenize a Given Circuit ?
- ❖ General Instructions for Finding Thevenin Equivalent Circuit
- ❖ Reciprocity Theorem
- ❖ Delta/Star Transformation
- ❖ Star/Delta Transformation
- ❖ Norton's Theorem
- ❖ How to Nortanize a Given Circuit ?
- ❖ General Instructions for Finding Norton Equivalent Circuit
- ❖ Millman's Theorem
- ❖ Generalised Form of Millman's Theorem
- ❖ Maximum Power Transfer Theorem

2.1 Electric Circuits and Network Theorems

There are certain theorems, which when applied to the solutions of electric networks, wither simplify the network itself or render their analytical solution very easy. These theorems can also be applied to an a.c. system, with the only difference that impedances replace the ohmic resistance of d.c. system. Different electric circuits (according to their properties) are defined below:

1. **Circuit.** A circuit is a closed conducting path through which an electric current either flows or is intended flow to flow. Circuit consists of active and passive elements in it.
2. **Parameters.** The various elements of an electric circuit are called its parameters like resistance, inductance and capacitance. These parameters may be ***lumped or distributed.***
3. **Liner Circuit.** A linear circuit is one whose parameters are constant with time also they do not change with voltage or current and circuit obeys Ohm's law.

4. **Non-linear Circuit.** It is that circuit whose parameters change with voltage or current.
5. **Bilateral Circuit.** A bilateral circuit is one whose properties or characteristics are the same in either direction. The usual transmission line is bilateral, because it can be made to perform its function equally well in either direction.
6. **Unilateral Circuit.** It is that circuit whose properties or characteristics change with the direction of its operation. A diode rectifier is a unilateral circuit, because it cannot perform rectification in both directions.
7. **Electric Network.** A combination of various electric elements, connected in any manner whatsoever, is called an electric network.
8. **Passive Network** is one which contains no source of e.m.f. in it.
9. **Active Network** is one which contains one or more than one source of e.m.f. along with passive elements.
10. **Node** is a point in a circuit where two or more circuit elements are connected together.
11. **Branch** is that part of a network which lies between two nodes.
12. **Loop.** It is a close path in a circuit in which no element or node is encountered more than once.
13. **Mesh.** It is a loop that contains no other loop within it. For example, the circuit of Fig. 2.1 (*a*) has six branches,five nodes, three loops and two meshes whereas the circuit of Fig. 2.1 (*b*) has four branches, two nodes, six loops and three meshes.

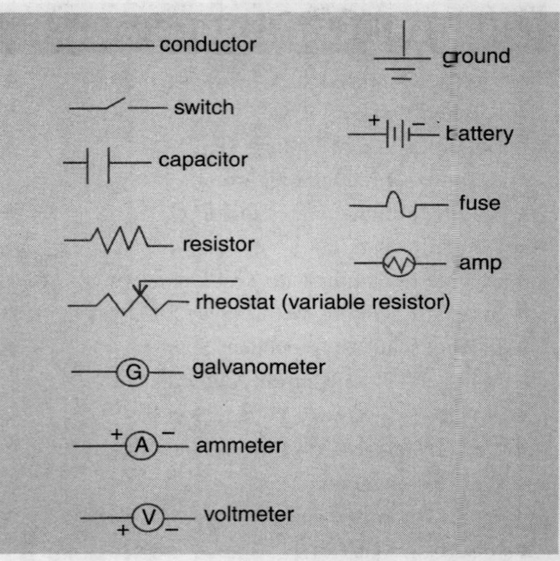

Standard symbols

It should be noted that, unless stated otherwise, an electric network would be assumed passive in the following treatment.

We will now discuss the various network theorems which are of great help in solving complicated networks. Incidentally, a network is said to be completely solved or analyzed when all voltages and all currents in its different elements are determined.

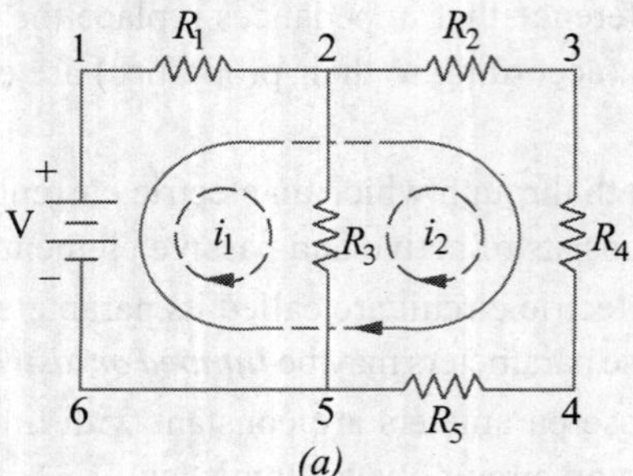

(*a*)

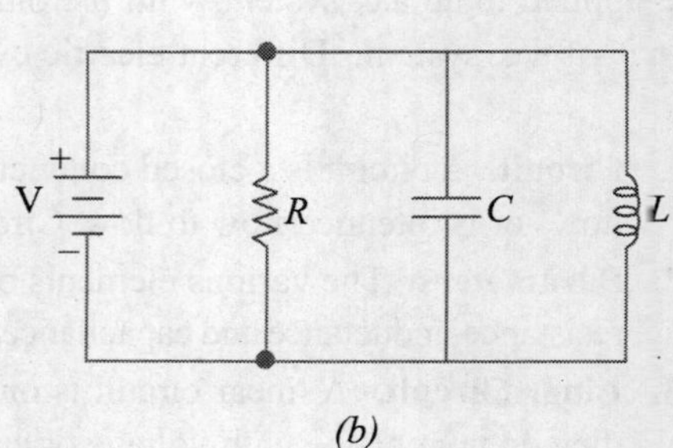

(*b*)

Fig. 2.1

There are two general approaches to network analysis:

(*i*) Direct Method

Here, the network is left in its original form while determining its different voltages and currents. Such methods are usually restricted to fairly simple circuits and include Kirchhoff's laws, Loop analysis, Nodal analysis, superposition theorem, Compensation theorem and Reciprocity theorem etc.

(*ii*) Network Reduction Method

Here, the original network is converted into a much simpler equivalent circuit for rapid calculation of different quantities. This method can be applied to simple as well as complicated networks. Examples of this method are: Delta/Star and Star/Delta conversions. Thevenin's theorem and Norton's Theorem etc.

2.2 Kirchhoff's Laws*

These laws are more comprehensive than Ohm's law and are used for solving electrical networks which may not be readily solved by the latter. Kirchhoff's laws, two in number, are particularly useful ***(a)*** in determining the equivalent resistance of a complicated network of conductors and ***(b)*** for calculating the currents flowing in the various conductors. The two-laws are:

1. Kirchhoff's Point Law or Current Law (KCL)

It states as follows:

In any electrical network, the algebraic sum of the currents meeting at a point (or junction) is zero.

Put in another way, it simply means that the total current *leaving* a junction is equal to the total current ***entering*** that junction. It is obviously true because there is no accumulation of charge at the junction of the network.

Consider the case of a few conductors meeting at a point *A* as in Fig. 2.2 (*a*). Some conductors have currents leading to point *A*, whereas some have currents leading away from point *A*. Assuming the incoming currents to be positive and the outgoing currents negative, we have

$$I_1 + (-I_2) + (-I_3) + (+I_4) + (-I_5) = 0$$

or $\quad I_1 + I_4 - I_2 - I_3 - I_5 = 0 \quad$ or $\quad I_1 + I_4 = I_2 + I_3 + I_5$

or $\quad$ **incoming currents = outgoing currents**

Similarly, in Fig. 2.2 (*b*) for node *A*

$$+I + (-I_1) + (-I_2) + (-I_3) + (-I_4) = 0 \quad \text{or} \quad I = I_1 + I_2 + I_3 + I_4$$

We can express the above conclusion thus: $\Sigma I = 0$ ....at a junction

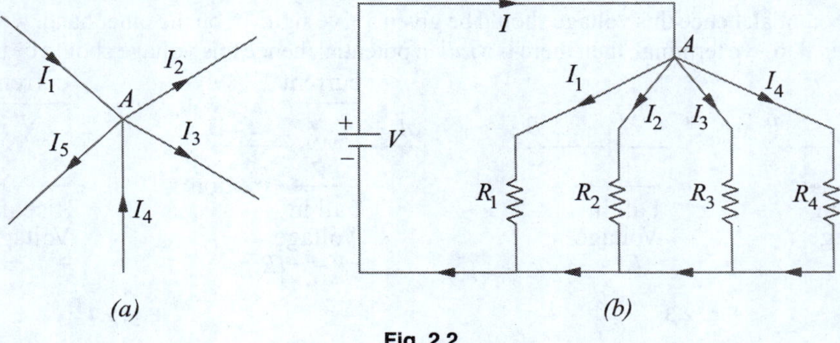

Fig. 2.2

* After Gustave Robert Kirchhoff (1824-1887), an outstanding German Physicist.

2. Kirchhoff's Mesh Law or Voltage Law (KVL)

It states as follows:

The algebraic sum of the products of currents and resistances in each of the conductors in any closed path (or mesh) in a network plus the algebraic sum of the e.m.fs. in that path is zero.

In other words, $\Sigma\, IR + \Sigma\, e.m.f. = 0$

...round a mesh

It should be noted that algebraic sum is the sum which takes into account the polarities of the voltage drops.

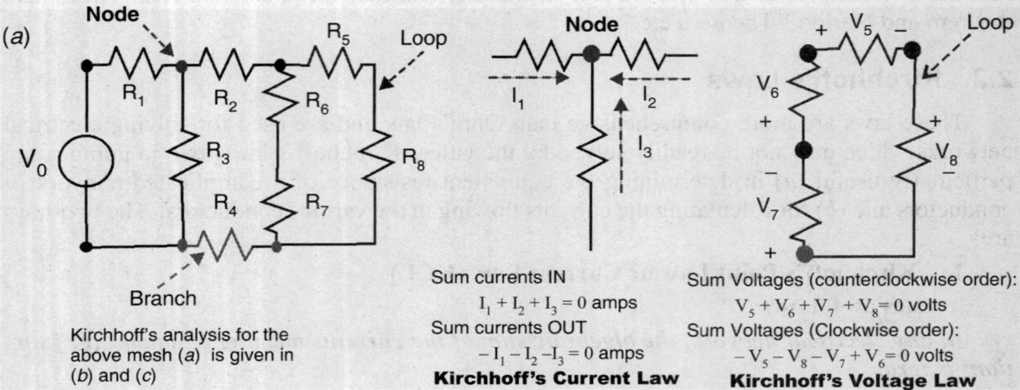

The basis of this law is this: If we start from a particular junction and go round the mesh till we come back to the starting point, then we must be at the same potential with which we started. Hence, it means that all the sources of e.m.f. met on the way must necessarily be equal to the voltage drops in the resistances, every voltage being given its proper sign, plus or minus.

2.3 Determination of Voltage Sign

In applying Kirchhoff's laws to specific problems, particular attention should be paid to the algebraic signs of voltage drops and e.m.fs., otherwise results will come out to be wrong. Following sign conventions is suggested:

(*a*) Sign of Battery E.M.F.

A *rise* in voltage should be given a + ve sign and a ***fall*** in voltage a –ve sign. Keeping this in mind, it is clear that as we go from the –ve terminal of a battery to its +ve terminal (Fig. 2.3), there is a *rise* in potential, hence this voltage should be given a + ve sign. If, on the other hand, we go from +ve terminal to –ve terminal, then there is a *fall* in potential, hence this voltage should be preceded

A E B — Rise in Voltage $+E$ A E B — Fall in Voltage $-E$

Fig. 2.3

current A + V – B R motion — Fall in Voltage $-V = -IR$ current A – V + B motion — Rise in Voltage $+V = +IR$

Fig. 2.4

by a –ve sign. ***It is important to note that the sign of the battery e.m.f. is independent of the direction of the current through that branch.***

(b) Sign of *IR* Drop

Now, take the case of a resistor (Fig. 2.4). If we go through a resistor in the *same* direction as the current, then there is a fall in potential because current flows from a higher to a lower potential. Hence, this voltage fall should be taken –ve. However, if we go in a direction opposite to that of the current, then there is a *rise* in voltage. Hence, this voltage rise should be given a positive sign.

It is clear that the sign of voltage drop across a resistor depends on the direction of current through that resistor but is independent of the polarity of any other source of e.m.f. in the circuit under consideration.

Consider the closed path *ABCDA* in Fig. 2.5. As we travel around the mesh in the clockwise direction, different voltage drops will have the following signs:

I_1R_1 is – ve (fall in potential)
I_2R_2 is – ve (fall in potential)
I_3R_3 is + ve (rise in potential)
I_4R_4 is – ve (fall in potential)
E_2 is – ve (fall in potential)
E_1 is + ve (rise in potential)

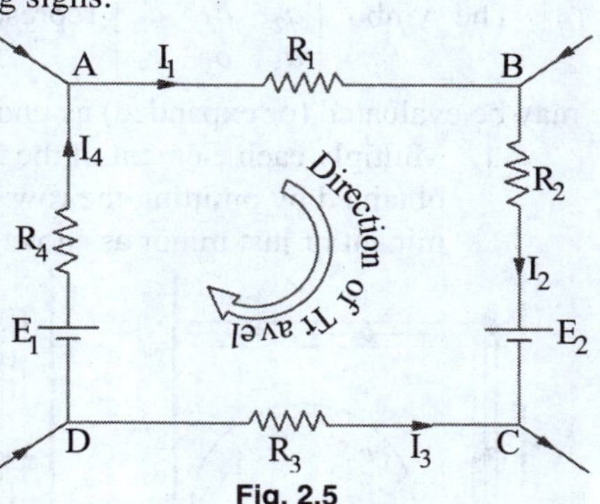

Fig. 2.5

Using Kirchhoff's voltage law, we get

$$-I_1R_1 - I_2R_2 - I_3R_3 - I_4R_4 - E_2 + E_1 = 0$$

or $$I_1R_1 + I_2R_2 - I_3R_3 + I_4R_4 = E_1 - E_2$$

2.4 Assumed Direction of Current

In applying Kirchhoff's laws to electrical networks, the question of assuming proper direction of current usually arises. The direction of current flow may be assumed either clockwise or anticlockwise. If the assumed direction of current is not the actual direction, then on solving the quesiton, this current will be found to have a minus sign. If the answer is positive, then assumed direction is the same as actual direction (Example 2.10). ***However, the important point is that once a particular direction has been assumed, the same should be used throughout the solution of the question.***

Note. It should be noted that Kirchhoff's laws are applicable both to d.c. and a.c. voltages and currents. However, in the case of alternating currents and voltages, any e.m.f. of self-inductance or that existing across a capacitor should be also taken into account (See Example 2.14).

2.5 Solving Simultaneous Equations

Electric circuit analysis with the help of Kirchhoff's laws usually involves solution of two or three simultaneous equations. These equations can be solved by a systematic elimination of the variables but the procedure is often lengthy and laborious and hence more liable to error. Determinants and Cramer's rule provide a simple and straight method for solving network equations through manipulation of their coefficients. Of course, if the number of simultaneous equations happens to be very large, use of a digital computer can make the task easy.

2.6 Determinants

The symbol $\begin{vmatrix} a & b \\ c & d \end{vmatrix}$ is called a determinant of the second order (or 2 × 2 determinant) because it contains two rows (*ab* and *cd*) and two columns (*ac* and *bd*). The numbers *a, b, c* and *d* are called the elements or constituents of the determinant. Their number in the present case is $2^2 = 4$.

The evaluation of such a determinant is accomplished by cross-multiplicaiton is illustrated below:

$$\Delta = \begin{vmatrix} a & b \\ c & d \end{vmatrix} = ad - bc$$

The above result for a second order determinant can be remembered as
upper left times lower right* minus *upper right times lower left

The symbol $\begin{vmatrix} a_1 & b_1 & c_1 \\ a_2 & b_2 & c_2 \\ a_3 & b_3 & c_3 \end{vmatrix}$ represents a third-order determinant having $3^2 = 9$ elements. It may be evaluated (or expanded) as under:

1. Multiply each element of the first row (or alternatively, first column) by a determinant obtained by omitting the row and column in which it occurs. (It is called minor determinant or just minor as shown in Fig. 2.6).

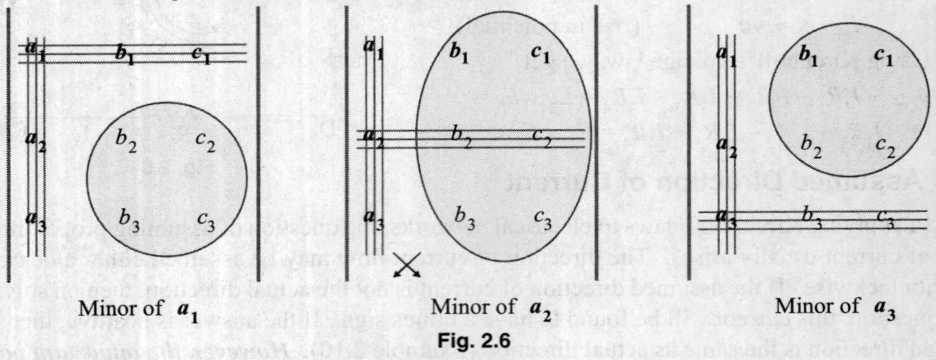

Fig. 2.6

2. Prefix + and – sing alternately to the terms so obtained.
3. Add up all these terms together to get the value of the given determinant.

Considering the first column, minors of various elements are as shown in Fig. 2.6.

Expanding in terms of first column, we get

$$\Delta = a_1 \begin{vmatrix} b_2 & c_2 \\ b_3 & c_3 \end{vmatrix} - a_2 \begin{vmatrix} b_1 & c_1 \\ b_3 & c_3 \end{vmatrix} + a_3 \begin{vmatrix} b_1 & c_1 \\ b_2 & c_2 \end{vmatrix}$$

$$= a_1 (b_2c_3 - b_3c_2) - a_2 (b_1c_3 - b_3c_1) + a_3 (b_1c_2 - b_2c_1) \quad \text{...(i)}$$

Expanding in terms of the first row, we get

$$\Delta = a_1 \begin{vmatrix} b_2 & c_2 \\ b_3 & c_3 \end{vmatrix} - b_1 \begin{vmatrix} a_2 & c_2 \\ a_3 & c_3 \end{vmatrix} + c_1 \begin{vmatrix} a_2 & b_2 \\ a_3 & b_3 \end{vmatrix}$$

$$= a_1 (b_2c_3 - b_3c_2) - b_1 (a_2c_3 - a_3c_2) + c_1 (a_2b_3 - a_3b_2)$$

which will be found to be the same as above.

Example 2.1. *Evaluate the determinant* $\begin{vmatrix} 7 & -3 & -4 \\ -3 & 6 & -2 \\ -4 & -2 & 11 \end{vmatrix}$

Solution. We will expand with the help of 1st column.

$$D = 7\begin{vmatrix} 6 & -2 \\ -2 & 11 \end{vmatrix} - (-3)\begin{vmatrix} -3 & -4 \\ -2 & 11 \end{vmatrix} + (-4)\begin{vmatrix} -3 & -4 \\ 6 & -2 \end{vmatrix}$$

$$= 7\,[(6 \times 11) - (-2 \times -2)] + 3\,[(-3 \times 11) - (-4 \times -2)] - 4\,[(-3 \times -2) - (-4 \times 6)]$$

$$= 7\,(66 - 4) + 3\,(-33 - 8) - 4\,(6 + 24) = \mathbf{191}$$

2.7 Solving Equations with Two Unknowns

Suppose the two given simultaneous equations are

$$ax + by = c$$
$$dx + ey = f$$

Here, the two unknown are x and y, a, b, d and e are coefficients of these unknowns whereas c and f are constants. The procedure for solving these equations by the method of determinants is as follows:

1. Write the two equations in the matrix form as $\begin{bmatrix} a & b \\ d & e \end{bmatrix}\begin{bmatrix} x \\ y \end{bmatrix} = \begin{bmatrix} c \\ f \end{bmatrix}$

2. The ***common*** determinant is given as $\Delta = \begin{bmatrix} a & b \\ d & e \end{bmatrix} = ae - bd$

3. For finding the determinant for x, replace the coefficients of x in the original matrix by the constants so that we get determinant Δ_1 given by $\Delta_1 = \begin{vmatrix} c & b \\ f & e \end{vmatrix} = (ce - bf)$

4. For finding the determinant for y, replace coefficients of y by the constants so that we get $\Delta_2 = \begin{vmatrix} a & c \\ d & f \end{vmatrix} = (af - cd)$

5. Apply Cramer's rule to get the value of x and y

$$x = \frac{\Delta_1}{\Delta} = \frac{ce - bf}{ae - bd} \text{ and } y = \frac{\Delta_2}{\Delta} = \frac{af - cd}{ae - bd}$$

Example 2.2. *Solve the following two simultaneous equations by the method of determinants:*

$$4i_1 - 3i_2 = 1$$
$$3i_1 - 5i_2 = 2$$

Solution. The matrix form of the equations is $\begin{bmatrix} 4 & -3 \\ 3 & -5 \end{bmatrix}\begin{bmatrix} i_1 \\ i_2 \end{bmatrix} = \begin{bmatrix} 1 \\ 2 \end{bmatrix}$

$$\Delta = \begin{vmatrix} 4 & -3 \\ 3 & -5 \end{vmatrix} = (4 \times -5) - (-3 \times 3) = -11$$

$$\Delta_1 = \begin{vmatrix} 1 & -3 \\ 2 & -5 \end{vmatrix} = (1 \times -5) - (-3 \times 2) = 1$$

$$\Delta_2 = \begin{vmatrix} 4 & 1 \\ 3 & 2 \end{vmatrix} = (4 \times 2) - (1 \times 3) = 5$$

$$\therefore \quad i_1 = \frac{\Delta_1}{\Delta} = \frac{1}{-11} = -\frac{1}{11}; \quad i_2 = \frac{\Delta_2}{\Delta} = -\frac{5}{11}$$

2.8 Solving Equations with Three Unknowns

Let the three simultaneous equations be as under:

$$ax + by + cz = d$$
$$ex + fy + gz = h$$
$$jx + ky + lz = m$$

The above equations can be put in the matrix form as under:

$$\begin{bmatrix} a & b & c \\ e & f & g \\ j & k & l \end{bmatrix} \begin{bmatrix} x \\ y \\ z \end{bmatrix} = \begin{bmatrix} d \\ h \\ m \end{bmatrix}$$

The value of common determinant is given by

$$\Delta = \begin{vmatrix} a & b & c \\ e & f & g \\ j & k & l \end{vmatrix} = a\,(fl - gk) - e(bl - ck) + j(bg - cf)$$

The determinant for x can be found by replacing coefficients of x in the original matrix by the constants.

$$\therefore \qquad \Delta_1 = \begin{vmatrix} d & b & c \\ h & f & g \\ m & k & l \end{vmatrix} = d\,(fl - gk) - h(bl - ck) + m(bg - cf)$$

Similarly, determinant for y is given by replacing coefficients of y with the three constants.

$$\Delta_2 = \begin{vmatrix} a & d & c \\ e & h & g \\ j & m & l \end{vmatrix} = a\,(hl - mg) - e(dl - mc) + j\,(dg - hc)$$

In the same way, determinant for z is given by

$$\Delta_3 = \begin{vmatrix} a & b & d \\ e & f & h \\ j & k & m \end{vmatrix} = a\,(fm - hk) - e(bm - dk) + j\,(bh - df)$$

As per Cramer's rule $\qquad x = \dfrac{\Delta_1}{\Delta},\ y = \dfrac{\Delta_2}{\Delta},\ z = \dfrac{\Delta_3}{\Delta}$

Example 2.3. *Solve the following three simultaneous equations by the use of determinants and Cramer's rule*

$$i_1 + 3i_2 + 4i_3 = 14$$
$$i_1 + 2i_2 + i_3 = 7$$
$$2i_1 + i_2 + 2i_3 = 2$$

Solution. As explained earlier, the above equations can be written in the form

$$\begin{bmatrix} 1 & 3 & 4 \\ 1 & 2 & 1 \\ 2 & 1 & 2 \end{bmatrix} \begin{bmatrix} i_1 \\ i_2 \\ i_3 \end{bmatrix} = \begin{bmatrix} 14 \\ 7 \\ 2 \end{bmatrix}$$

$$\Delta = \begin{bmatrix} 1 & 3 & 4 \\ 1 & 2 & 1 \\ 2 & 1 & 2 \end{bmatrix} = 1(4-1) - 1(6-4) + (3-8) = -9$$

$$\Delta_1 = \begin{bmatrix} 14 & 3 & 4 \\ 7 & 2 & 1 \\ 2 & 1 & 2 \end{bmatrix} = 14(4-1) - 7(6-4) + 2(3-8) = 18$$

$$\Delta_2 = \begin{bmatrix} 1 & 14 & 4 \\ 1 & 7 & 1 \\ 2 & 2 & 2 \end{bmatrix} = 1(14-2) - 1(28-8) + 2(14-28) = -36$$

$$\Delta_3 = \begin{bmatrix} 1 & 3 & 14 \\ 1 & 2 & 7 \\ 2 & 1 & 2 \end{bmatrix} = 1(4-7) - 1(6-14) + 2(21-28) = -9$$

According to Cramer's rule,

$$i_1 = \frac{\Delta_1}{\Delta} = \frac{18}{-9} = \mathbf{-2A};\ i_2 = \frac{\Delta_2}{\Delta} = \frac{-36}{-9} = \mathbf{-4A};\ i_3 = \frac{\Delta_3}{\Delta} = \frac{-9}{-9} = \mathbf{1A}$$

Example 2.4. *What is the voltage V_s across the open switch in the circuit of Fig. 2.7 ?*

Solution. We will apply *KVL* to find V_s. Starting from point *A* in the clockwise direction and using the sign convention, we have

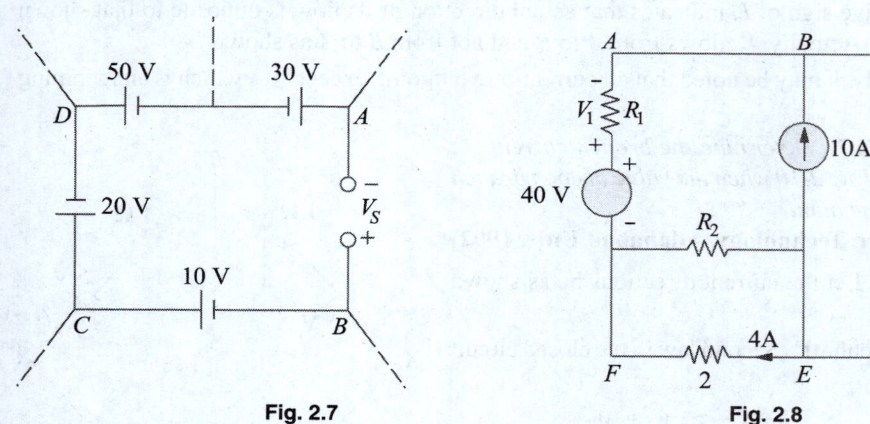

Fig. 2.7

Fig. 2.8

$+V_s + 10 - 20 - 50 + 30 = 0 \quad \therefore \quad V_s = \mathbf{30\ V}$

Example 2.5. *Find the unknown voltage V_1 in the circuit of Fig. 2.8.*

Solution. Initially, one may not be clear regarding the solution of this question. One may think of Kirchhoff's laws or mesh analysis etc. But a little thought will show that the question can be solved by the simple application of Kirchhoff's voltage law. Taking the outer closed loop ***ABCDEFA*** and applying ***KVL*** to it, we get

$-16 \times 3 - 4 \times 2 + 40 - V_1 = 0\ ; \quad \therefore \quad V_1 = -\mathbf{16\ V}$

The negative sign shows there is a fall in potential.

Example 2.6. *Using Kirchhoff's Current Law and Ohm's Law, find the magnitude and polarity of voltge V in Fig. 2.9 (a). Directions of the two current sources are as shown.*

Solution. Let us arbitrarily choose the directions of I_1, I_2 and I_3 and polarity of V as shown in Fig. 2.9.(*b*). We will use the sign convention for currents. Applying KCL to node *A*, we have

$$-I_1 + 30 + I_2 - I_3 - 8 = 0$$

or $$I_1 - I_2 + I_3 = 22 \qquad ...(i)$$

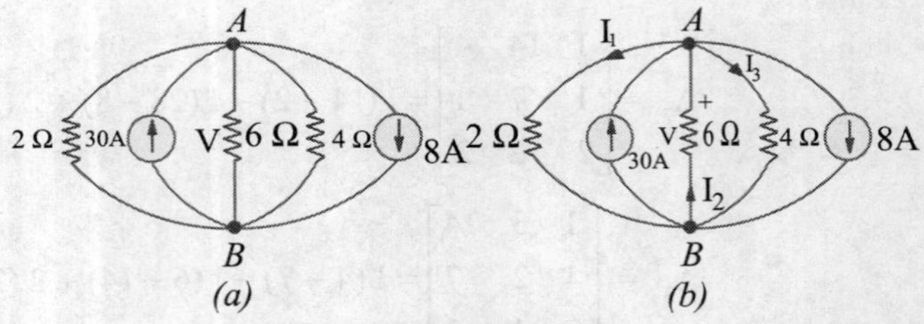

Fig. 2.9

Applying Ohm's law to the three resistive branches in Fig. 2.9 (*b*), we have

$$I_1 = \frac{V}{2}, I_3 = \frac{V}{4}, I_2 = -\frac{V}{6} \qquad \text{(Please note the – ve sign.)}$$

Substituting these values in (*i*) above, we get

$$\frac{V}{2} - \left(\frac{-V}{6}\right) + \frac{V}{4} = 22 \quad \text{or} \quad V = \mathbf{24\ V}$$

$$\therefore \qquad I_1 = V/2 = 24/2 = 12 \text{ A}, I_2 = -24/6 = -4 \text{ A}, I_3 = 24/4 = 6 \text{ A}$$

The negative sign of I_2 indicates that actual direction of its flow is opposite to that shown in Fig. 2.9 (*b*). Actually, I_2, flows from *A* to *B* and not from *B* to *A* as shown.

Incidentally, it may be noted that all currents are outgoing except 30A which is an incoming current.

Example 2.7. *Determine the branch currents in the network of Fig. 2.10 when the value of each branch resistance is one ohm.*

(Elect. Technology, Allahabad Univ. 1992)

Solution. Let the current directions be as shown in Fig. 2.10.

Apply Kirchhoff's Second law to the closed circuit *ABDA*, we get

$$5 - x - z + y = 0$$

or $$x - y + z = 5 \qquad ...(i)$$

Similarly, circuit *BCDB* gives

$$-(x - z) + 5 + (y + z) + z = 0$$

or $$x - y - 3z = 5 \qquad ...(ii)$$

Lastly, from circuit *ADCEA*, we get

$$-y - (y + z) + 10 - (x + y) = 0$$

or $$x + 3y + z = 10 \qquad ...(iii)$$

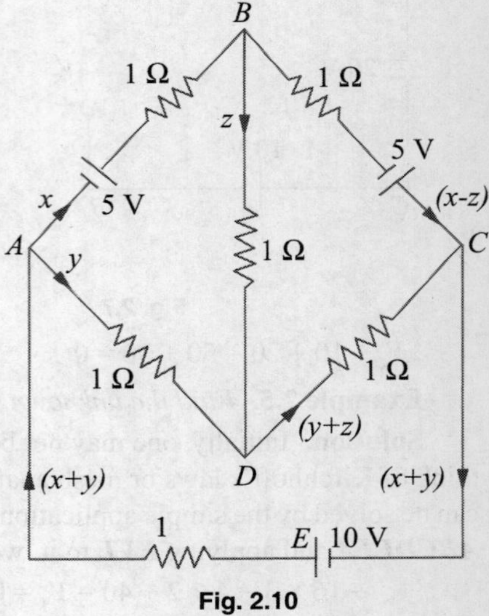

Fig. 2.10

From Eq. (*i*) and (*ii*), we get, $z = 0$

Substituting $z = 0$ either in Eq. (*i*) or (*ii*) and in Eq. (*iii*), we get

$$x - y = 5 \qquad ...(\textbf{iv})$$

$$x + 3y = 10 \qquad ...(\textbf{v})$$

Subtracting Eq. (**v**) from (**iv**), we get

$-4y = -5$ or $y = 5/4 = 1.24$ A

Eq. (*iv*) gives $x = 25/4$ A $= 6.25$ A

Current in branch *AB* = current in branch *BC* = **6.25 A**

Current in branch *BD* = 0; current in branch *AD* = current in branch *DC* = **1.25 A; current** in branch *CEA* = 6.25 + 1.25 = **7.5 A.**

Example 2.8. *State and explain Kirchhoff's laws. Determine the current supplied by the battery in the circuit shown in Fig. 2.11.* **(Elect. Engg. I, Bombay Univ.)**

Solution. Let the current distribution be as shown in the figure. Considering the close circuit *ABCA* and applying Kirchhoff's Second Law, we have

$$-100x - 300z + 500y = 0$$

or $\quad x - 5y + 3z = 0 \qquad ....(\textbf{i})$

Similarly, considering the closed loop *BCDB*, we have

$$-300z - 100(y + z) + 500(x - z) = 0$$

or $\quad 5x - y - 9z = 0 \quad ...(\textbf{ii})$

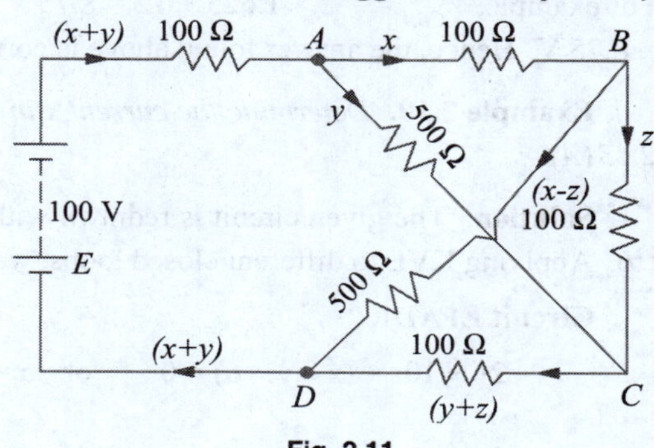

Fig. 2.11

Taking the circuit *ABDEA*, we get $-100x - 500(x - z) + 100 - 100(x + y) = 0$

or $\quad 7x + y - 5z = 1 \qquad ...(\textbf{iii})$

The value of *x*, *y* and *z* may be found by solving the above three simultaneous equations.

$$\therefore \qquad x = \frac{48}{240} = \frac{\mathbf{1}}{\mathbf{5}}\ \mathbf{A};\ y = \frac{24}{240} = \frac{\mathbf{1}}{\mathbf{10}}\ \mathbf{A};\ z = \frac{24}{240} = \frac{\mathbf{1}}{\mathbf{10}}\ \mathbf{A}$$

Current supplied by the battery is $x + y = 1/5 + 1/10 =$ **3/10 A.**

Example 2.9. *Two batteries A and B are connected in parallel and load of 10 Ω is connected across their terminals. A has an e.m.f. of 12 V and an internal resistance of 2 Ω ; B has an e.m.f. of 8 V and an internal resistance of 1 Ω. Use Kirchhoff's laws to determine the values and directions of the currents flowing in each of the batteries and in the external resistance. Also determine the potential difference across the external resistance.*

(F.Y. Engg. Pune Univ.)

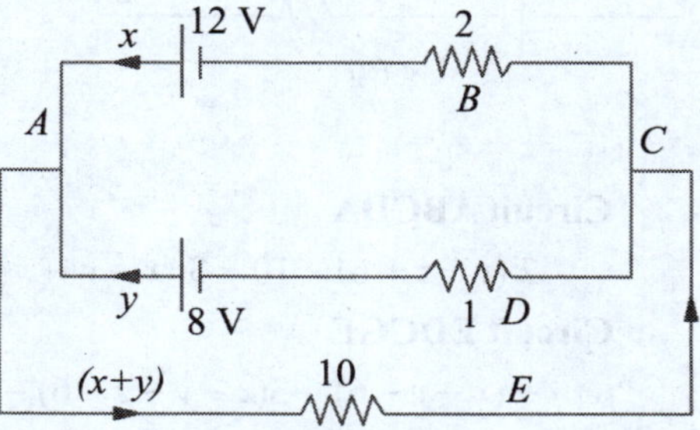

Fig. 2.12

Solution. Applying KVL to the closed circuit *ABCDA* of Fig. 2.12, we get

$$-12 + 2x - 1y + 8 = 0 \quad \text{or} \quad 2x - y = 4 \qquad ...(\textbf{i})$$

Similarly, from the closed circuit *ADCEA*, we get

$$-8+1y+10(x+y)=0 \quad \text{or} \quad 10x+11y=8 \qquad ...(ii)$$

From Eq. (*i*) and (*ii*), we get

$x =$ **1.625 A** and $y =$ **– 0.75 A**

The negative sign of y shows that the current is flowing into the 8-V battery and not out of it. In other words, it is a charging current and not a discharging current.

Current flowing in the external resistance $= x + y = 1.625 - 0.75 =$ **0.875 A**

P.D. across the external resistance $= 10 \times 0.875 =$ **8.75 V**

Note. To confirm the correctness of the answer, the simple check is to find the value of the external voltage available across point *A* and *C* with the help of the two parallel branches. If the value of the voltage comes out to be the same, then the answer is correct, otherwise it is wrong. For example, $V_{CBA} = -2 \times 1.625 + 12 = 8.75$ V. From the second branch $V_{CDA} = 1 \times 0.75 + 8 = 8.75$ V. Hence, the answer found above is correct.

Example 2.10. *Determine the current x in the 4-Ω resistance of the circuit shown in Fig. 2.13 (A).*

Solution. The given circuit is redrawn with assumed distribution of currents in Fig. 2.13 (*b*). Applying KVL to different closed loops, we get

Circuit EFADE

$$-2y+10z+(x-y-6)=0 \quad \text{or} \quad x-3y+10z=6 \qquad ...(i)$$

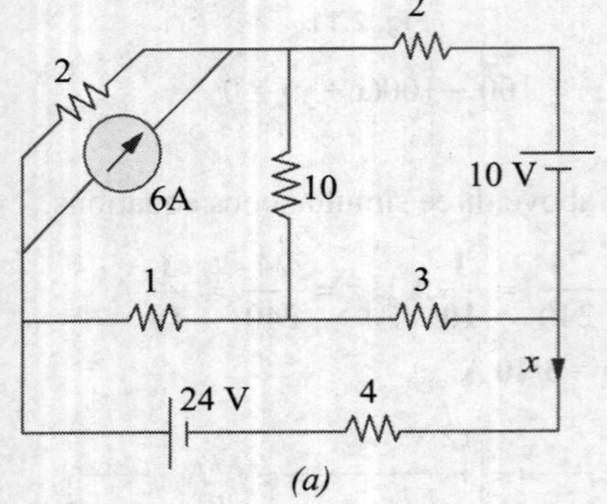

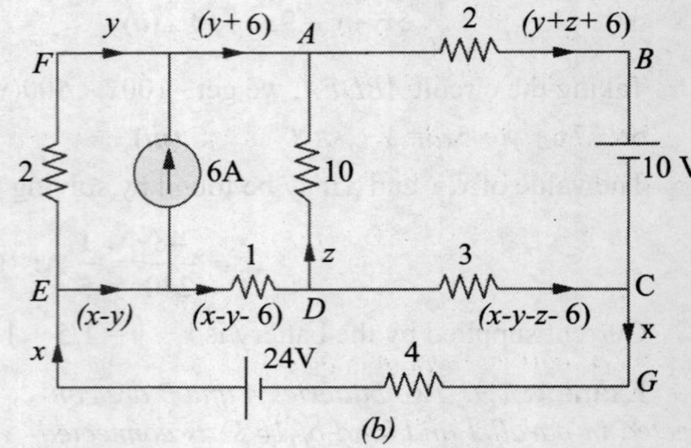

Fig. 2.13

Circuit ABCDA

$$2(y+z+6)-10+3(x-y-z-6)-10z=0 \quad \text{or} \quad 3x-5y-14z=40 \qquad ...(ii)$$

Circuit EDCGE

$$-(x-y-6)-3(x-y-z-6)-4x+24=0 \quad \text{or} \quad 8x-4y-3z=48 \qquad ...(iii)$$

From above equations we get $x =$ **4.1 A**

Example 2.11. *Applying Kirchhoff's laws to different loops in Fig. 2.14, find the values of* V_1 *and* V_2.

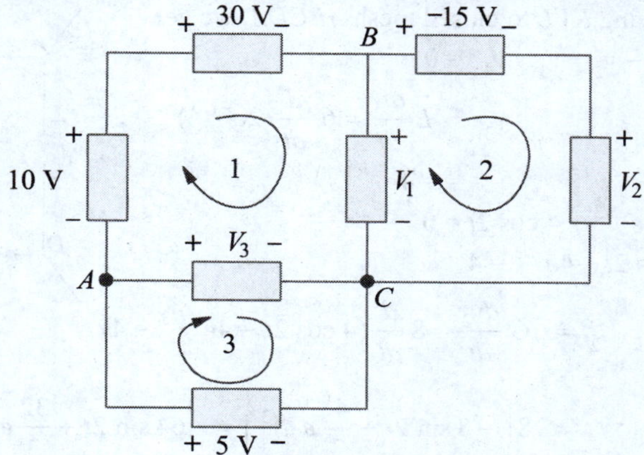

Fig. 2.14

Solution. Starting from point *A* and applying Kirchhoff's voltage law to loop No.3, we get

$$-V_3 + 5 = 0 \quad \text{or} \quad V_3 = 5\text{ V}$$

Starting from point *A* and applying Kirchhoff's voltage law to loop No. 1, we get

$10 - 30 - V_1 + 5 = 0$ or $V_1 = \mathbf{-15\ V}$

The negative sign of V_1 denotes that its polarity is opposite to that shown in the figure.

Starting from point *B* in loop No. 3, we get

$-(-15) - V_2 + (-15) = 0$ or $V_2 = \mathbf{0}$

Example 2.12. *In the network of Fig. 2.15, the different currents and voltages are as under:*

$$i_2 = 5e^{-2t},\ i_4 = 3\sin t \text{ and } v_3 = 4e^{-2t}$$

Using KCL, find voltage v_1.

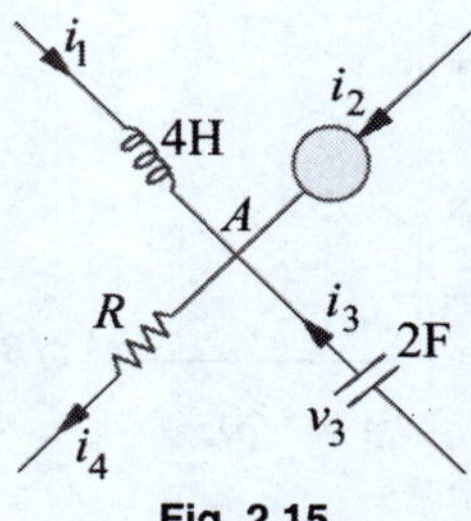

Fig. 2.15

Solution. According to *KCL*, the algebraic sum of the currents meeting at juncion *A* is zero *i.e.*

$$\vec{i}_1 + \vec{i}_2 + \vec{i}_3 + (-\vec{i}_4) = 0$$

$$\vec{i}_1 + \vec{i}_2 + \vec{i}_3 - i_4 = 0 \quad \text{...}(i)$$

Now, current through a capacitor is given by $i = C\,dv/dt$

$$\therefore \quad \vec{i}_3 = C\frac{dv_3}{dt} = \frac{2d(4e^{-2t})}{dt} = -16e^{-2t}$$

Substituting this value in Eq (*i*) above, we get

$$\vec{i}_1 + 5e^{-2t} - 16e^{-2t} - 3\sin t = 0$$

or

$$\vec{i}_1 = 3\sin t + 11e^{-2t}$$

The voltage v_1 developed across the coil is

$$v_1 = L\frac{di_1}{dt} = 4.\frac{d}{dt}(3\sin t + 11e^{-2t})$$

$$= 4(3\cos t - 22e^{-2t}) = 12\cos t - 88e^{-2t}$$

Example 2.13. *In the network shown in Fig. 2.16, $v_1 = 4V$, $v_4 = 4\cos 2t$ and $i_3 = 2e^{-t/3}$. Determine i_2.*

Solution. Applying *KVL* to closed mesh *ABCDA*, we get

$$-v_1 - v_2 + v_3 + v_4 = 0$$

Now $$v_3 = L\frac{di_3}{dt} = 6.\frac{d}{dt}(2e^{-t/3})$$

$$= -4e^{-t/3}$$

$\therefore$ $$-4 - v_2 - 4e^{-t/3} + 4\cos 2t = 0$$

or $$v_2 = 4\cos 2t - 4e^{-t/3} - 4$$

Fig. 2.16

Now, $$i_2 = C\frac{dv_2}{dt} = 8\frac{d}{dt}(4\cos 2t - 4e^{-t/3} - 4)$$

$\therefore$ $$i_2 = 8\left(-8\sin 2t + \frac{4}{3}e^{-t/3}\right) = -64\sin 2t + \frac{32}{3}e^{-t/3}$$

2.9 Independent and Dependent Sources

Those voltage or current sources, which do not depend on any other quantity in the circuit, are called independent sources. An independent d.c. voltage source is shown in Fig. 2.17 (*a*) whereas a time-varying voltage source is shown in Fig. 2.17 (*b*). The positive sign shows that terminal *A* is positive with respect to terminal *B*. In other words, potential of terminal *A* is *v* volts higher than that of terminal *B*.

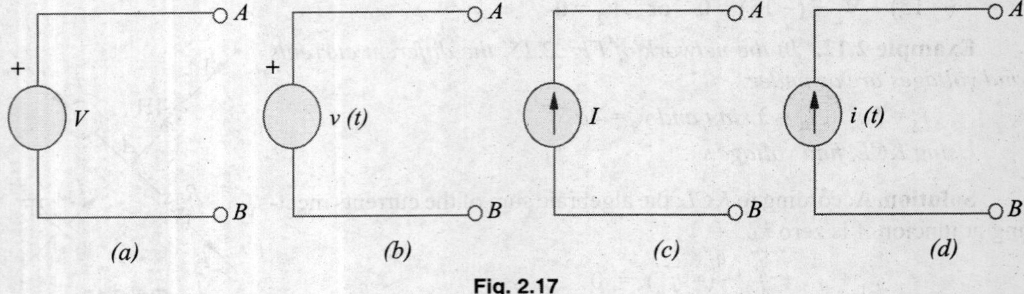

Fig. 2.17

Similarly, Fig. 2.17 (*c*) shows an ideal constant current source whereas Fig. 2.20 (*d*) depicts a time-varying current source. The arrow shows the direction of flow of the current at any moment under consideration.

A dependent voltage or current source is one which depends on some other quantity in the circuit which may be either a voltage or a current. Such a source is represented by a diamond-shaped symbol as shown in Fig. 2.18 so as not to confuse it with an independent source. There are four possible dependent sources:

1. Voltage-dependent voltage source [Fig. 2.18 (*a*)]
2. Current-dependent voltage source [Fig. 2.18 (*b*)]
3. Voltage-dependent current source [Fig. 2.18 (*c*)]
4. Current-dependent current source [Fig. 2.18 (*d*)]

Such sources can also be either constant sources or time-varying sources. Such sources are often met in electronic circuits. As seen above, the voltage or current source is dependent on the and is

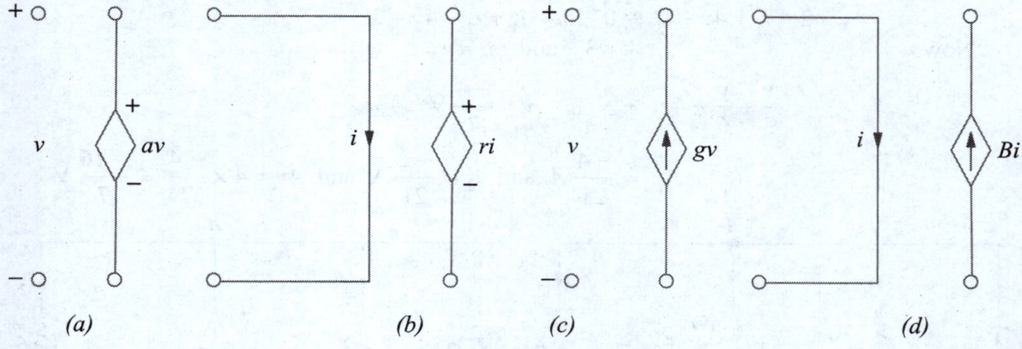

Fig. 2.18

proportional to another current or voltage. The constants of proportionality are written as *a, r, g* and β. The constants *a* and β have no unis, *r* has the unit of ohms and *g* has the unit of siemens.

Independent sources actually exist as physical entities such as a battery, a *d.c.* generator and an alternator etc. But dependent sources are parts of *models* that are used to represent electrical properties of electronic devices such as operational amplifiers and transistors etc.

Example 2.14. *Using Kirchhoff's current law, find the values of the currents i_1 and i_2 in the circuit of Fig. 2.19 (a) which contains a current-dependent current source. All resistances are in ohms.*

Solution. Applying *KCL* to node *A*, we get

$$2 - i_1 + 4\,i_1 - i_2 = 0 \quad \text{or} \quad -3i_1 + i_2 = 2$$

By Ohm's law, $i_1 = v/3$ and $i_2 = v/2$

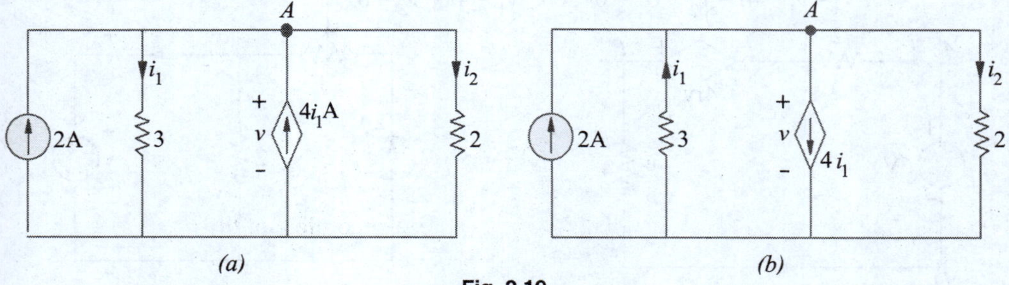

Fig. 2.19

Substituting these values above, we get

$$-3(v/3) + v/2 = 2 \quad \text{or} \quad v = -4 \text{ V}$$

$$\therefore \qquad i_1 = -4/3 \text{ A and } i_2 = -4/2 = -2 \text{ A}$$

The value of the dependent current source is $= 4i_1 = 4 \times (-4/3) =$ **− 16/3 A.**

Since i_1 and i_2 come out to be negative, it means that they flow upwards as shown in Fig. 2.19(*b*) and not downwards as presumed. Similarly, the current of the dependent source flows downwards as shown in Fig. 2.19(*b*). It may also be noted that the sum of the upwards currents equals that of the downward currents.

Example 2.15. *By applying Kirchhoff's current law, obtain the values of v, i_1 and i_2 in the circuit of Fig. 2.20 (a) which contains a voltage-dependent current source. Resistance values are in ohms.*

Solution. Applying *KCL* to node *A* of the circuit, we get

$$2 - i_1 + 4v - i_2 = 0 \quad \text{or} \quad i_1 + i_2 - 4v = 2$$

Now, $$i_1 = v/3 \quad \text{and} \quad i_2 = v/6$$

$$\therefore \quad \frac{v}{3} + \frac{v}{6} - 4v = 2 \quad \text{or} \quad v = \frac{-4}{7}\text{ V}$$

$$\therefore \quad i_1 = \frac{-4}{21} A \text{ and } i_2 = \frac{-2}{21}\text{ A and } 4v = 4 \times \frac{-4}{7} = \mathbf{\frac{-16}{7}\ V}$$

(*a*) (*b*)

Fig. 2.20

Since i_1 and i_2 come out to be negative and value of current source is also negative, their directions of flow are opposite to those presumed in Fig. 2.20(*a*). Actual current directions are shown in Fig. 2.20(*b*).

Example 2.16. *Apply Kirchhoff's voltage law, to find the values of current i and the voltage drops v_1 and v_2 in the circuit of Fig. 2.21 which contains a current-dependent voltage source. What is the voltage of the dependent source ? All resistance values are in ohms.*

Solution. Applying *KVL* to the circuit of Fig. 2.21 and starting from point *A*, we get

$$-v_1 + 4i - v_2 + 6 = 0 \quad \text{or} \quad v_1 - 4i + v_2 = 6$$

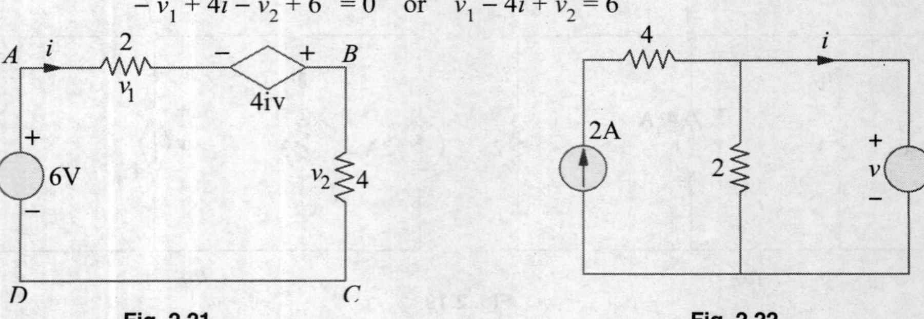

Fig. 2.21 **Fig. 2.22**

Now, $$v_1 = 2i \quad \text{and} \quad v_2 = 4i$$

$$\therefore \quad 2i - 4i + 4i = 6 \quad \text{or} \quad i = 3\text{A}$$

$$\therefore \quad v_1 = 2 \times 3 = 6\text{V} \quad \text{and} \quad v_2 = 4 \times 3 = \mathbf{12\ V}$$

Voltage of the dependent source = $4i = 4 \times 4 =$ **12 V**

Example 2.17. *In the circuit shown in Fig. 2.22, apply KCL to find the value of i for the case when* **(*a*)** *v = 2 V,* **(*b*)** *v = 4 V* **(*c*)** *v = 6 V. The resistor values are in ohms.*

Solution. (***a***) When $v = 2$ V, current through 2 Ω resistor which is connected in parallel with the 2 *v* source = 2/2 = 1A. Since the source current is 2 A, $i = 2 - 1 =$ **1 A.**

(***b***) When $v = 4$V, current through the 2Ω resistor = 4/2 = **2 A.** Hence $i = 2 - 2 =$ **0 A.**

(***c***) When $v = 6$ V, current through the 2Ω resistor = 6/2 = 3 A. Since current source can supply only 2 A, the balance of 1 A is supplied by the voltage source. Hence, $i = -1$ A *i.e.* it flows in a direction opposite to that shown in Fig. 2.22.

Example 2.18. *By using voltage-divider rule, calculate the voltages v_x and v_y in the network shown in Fig. 2.23.*

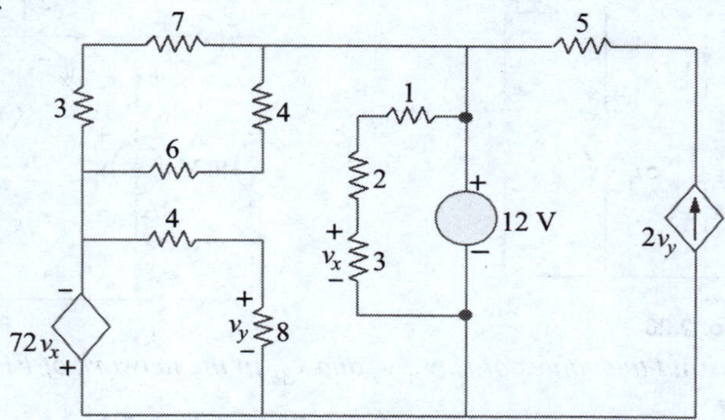

Fig. 2.23

Solution. As seen, 12 V drop in over the series combination of 1, 2 and 3 Ω resistors. As per voltage-divider rule v_x = drop over 3 Ω = 12 × 3/6 = 6 V.

The voltage of the dependent source = 12 × 6 = 72 V.

The voltage v_y equals the drop across 8 Ω resistor connected across the voltage source of 72 V.

Again using voltge-divider rule, drop over 8 Ω resistor = 72 × 8/12 = 48 V.

Hence, $v_y = -48$ V. The negative sign has been given because positive and negative signs of v_y are actually opposite to those shown in Fig. 2.23.

Example 2.19. *Use KCL to find the value of v in the circuit of Fig. 2.24.*

Solution. Let us start from ground and go to point *a* and find the value of voltage v_a. Obviously, $5 + v = v_a$ or $v = v_a - 5$. Applying *KCL* to point, we get

$6 - 2v + (5 - v_a)/1 = 0$ or $6 - 2(v_a - 5) + (5 - v_a) = 0$ or $v_a = 7$ V

Hence, $v = v_a - 5 = 7 - 5 = 2$ V. Since it turns out to be positive, its sign as indicated in the figure is correct.

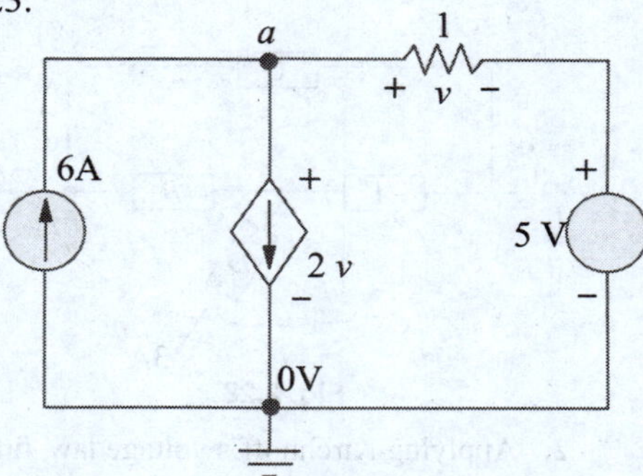

Fig. 2.24

Example 2.20. Basic Electric Circuits by Cunninghan. *Find the value of current i_2 supplied by the voltage-controlled current source (VCCS) shown in Fig. 2.25.*

Solution. Applying *KVL* to the closed circuit *ABCD*, we have $-4 + 8 - v_1 = 0 \therefore v_1 = 4$ V

The current supplied by *VCCS* is $10\,v_1 = 10 \times 4 = 40$ A. Since i_2 flows in an opposite direction to this current, hence $i_2 = -$ **40 A.**

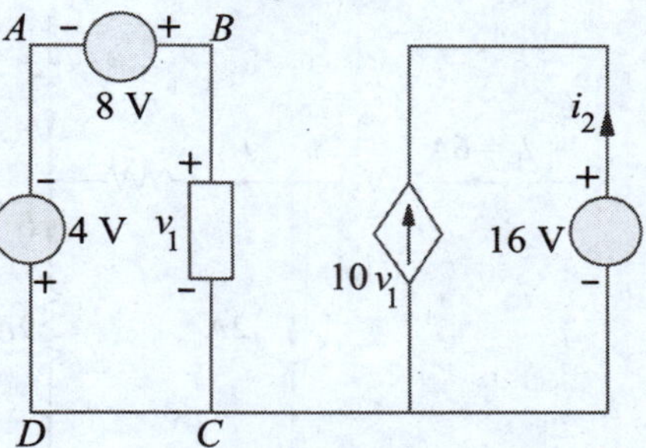

Fig. 2.25

Example 2.20. (*b*). *Find the voltage drop v_2 across the current-controlled voltage source (CCVS) shown in Fig. 2.26.*

Solution. Applying *KCL* to point *A*, we have $2 + 6 - i_1 = 0$ or $i_1 = 8$ A.

Application of *KVL* to the closed circuit on the right hand side gives $5\,i_1 - v_2 = 0$ or $v_2 = 5\,i_1 = 5 \times 8 = 40$ V.

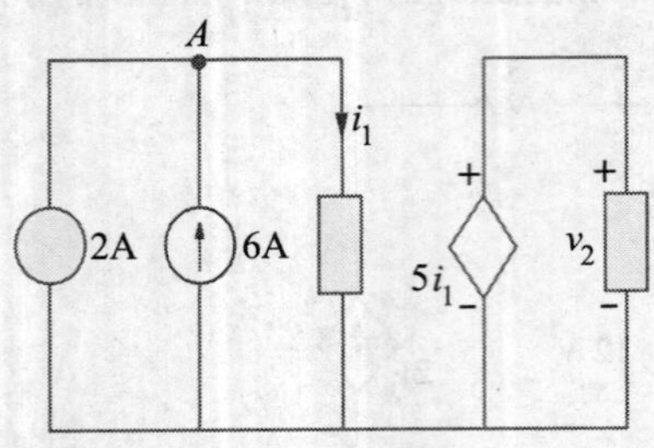

Fig. 2.26

Fig. 2.27

Example 2.21. *Find the values of i_1, v_1, v_x and v_{ab} in the network of Fig. 2.27 with its terminals a and b open.*

Solution. It is obvious that $i_1 = 4$ A. Applying *KVL* to the left-hand closed circuit, we get $-40 + 20 - v_1 = 0$ or $v_1 = -20$ V.

Similarly, applying *KVL* to the second closed loop, we get

$v_1 - v_x + 4v_1 - 50 = 0$ or $v_x = 5\,v_1 - 50 = -5 \times 20 - 50 = -150$ V

Again applying *KVL* to the right-hand side circuit containing v_{ab}, we get

$50 - 4v_1 - 10\,v_{ab} = 0$ or $v_{ab} = 50 - 4\,(-20) - 10 = 120$ V

Tutorial Problem 2.1

1. Apply *KCL* to find the value of *I* in Fig. 2.28. **[8 A]**

Fig. 2.28

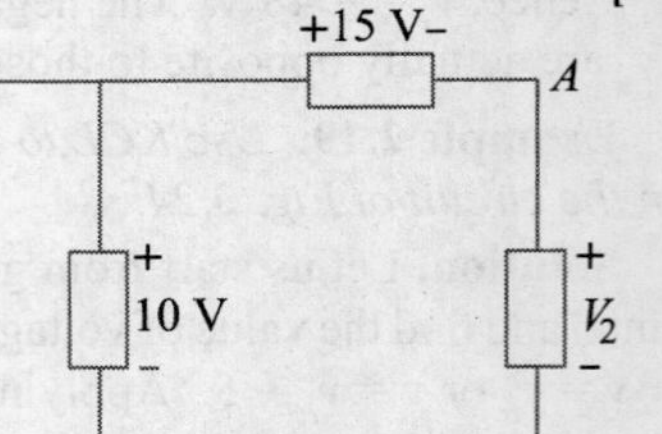

Fig. 2.29

2. Applying Kirchhoff's voltage law, find V_1 and V_2 in Fig. 2.29. **[$V_1 = 10$ V; $V_2 = 5$ V]**
3. Find the values of currents I_2 and I_4 in the network of Fig. 2.30. **[$I_2 = 4$ A ; $I_4 = 5$ A]**

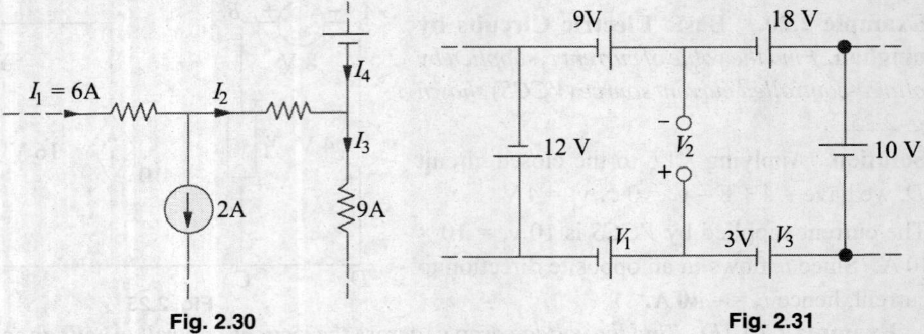

Fig. 2.30

Fig. 2.31

4. Use Kirchhoff's law, to find the values of voltages V_1 and V_2 in the network shown in Fig. 2.31.

[$V_1 = 2$ V ; $V_2 = 5$ V]

5. Find the unknown currents in the circuits shown in Fig. 2.32 (*a*). **[$I_1 = 2$ A ; $I_2 = 7$ A]**

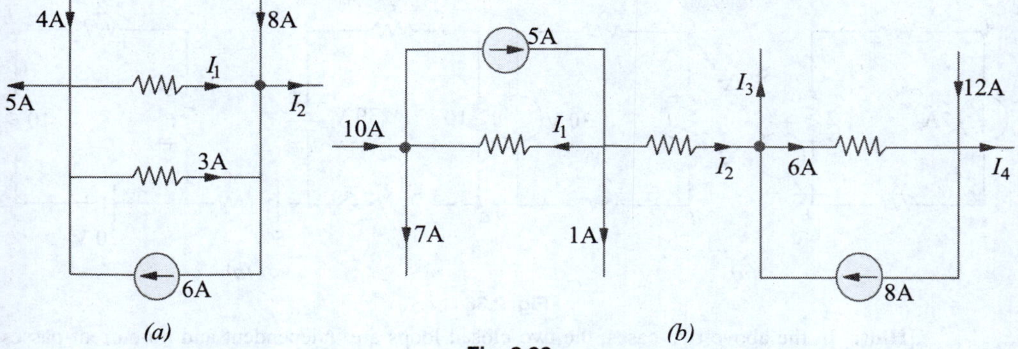

Fig. 2.32

6. Using Kirchhoff's current law, find the values of the unknown currents in Fig. 2.32 (*b*).

[$I_1 = 2$ A; $I_2 = 2$ A ; $I_3 = 4$ A ; $I_4 = 10$ A]

7. In Fig. 2.33, the potential of point *A* is – 30 V. Using Kirchhoff's voltage law, find (*a*) value of V and (*b*) power dissipated by 5 Ω resistance. All resistances are in ohms. **[100 V; 500 W]**

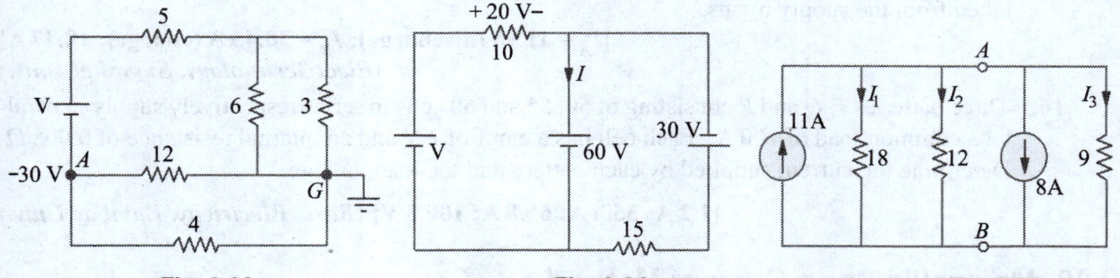

Fig. 2.33 Fig. 2.34 Fig. 2.35

8. Using *KVL* and *KCL*, find the values of *V* and *I* in Fig. 2.34. All resistances are in ohms.

[80 V; – 4 A]

9. Using *KCL*, find the values V_{AB}, I_1, I_2 and I_3 in the circuit of Fig. 2.35. All resistances are in ohms.

[$V_{AB} = 12$ V ; $I_1 = 2/3$ A; $I_2 = 1$ A; $I_3 = 4/3$ A]

10. A bridge network *ABCD* is arranged as follows:
Resistances between terminals *A–B*, *B–C*, *C–D*, *D–A*, and *B–D* are 10,20,15,5 and 40 ohms respectively. A 20 V battery of negligible internal resistance is connected between terminals *A* and *C*. Determine the current in each resistor.

[*AB* = 0.645 A; *BC* = 0.678 A; *AD* = 1.025 A; *DB* = 0.033 A; *DC* = 0.992 A]

11. Two batteries *A* and *B* are connected in parallel and a load of 10 Ω is connected across their terminals. A has an e.m.f. of 12 V and an internal resistance of 2 Ω ; *B* has an e.m.f. of 8 V and an internal resistance of 1 Ω. Use Kirchhoff's laws to determine the values and directions of the currents flowing in each of the batteries and in the external resistance. Also determine the p.d. across the external resistance.

[$I_A = 1.625$ A (discharge), $I_B = 0.75$ A (charge) ; 0.875 A; 8.75 V]

12. The four arms of a Wheatstone bridge have the following resistances ; *AB* = 100, *BC* = 10, *CD* = 4, *DA* = 50 ohms.
A galvanometer of 20 ohms resistance is connected across *BD*. Calculate the current through the galvanometer when a potential difference of 10 volts is maintained across *AC*.

[0.00513 A] [*Elect. Tech. Lond. Univ.*]

13. Find the voltage V_{da} in the network shown in Fig. 2.36 (*a*) if *R* is 10 Ω and (*b*) 20 Ω.

[(*a*) 5 V (*b*) 5 V]

14. In the network of Fig. 2.36 (*b*), calculate the voltage between points *a* and *b i.e.* V_{ab}.

[30 V] ***(Elect. Engg. I, Bombay Univ.)***

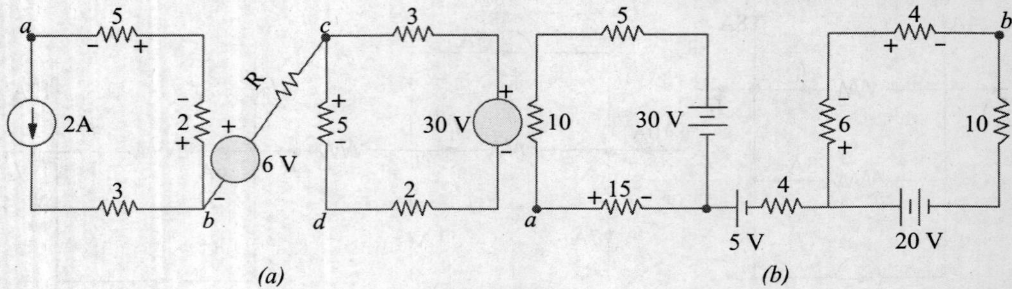

Fig. 2.36

[**Hint:** In the above two cases, the two closed loops are independent and no current passes between them].

15. A battery having an *E.M.F.* of 110 V and an internal resistance of 0.2 Ω is connected in parallel with another battery having an *E.M.F.* of 100 V and internal resistance 0.25 Ω. The two batteries in parallel are placed in series with a regulating resistance of 5 Ω and connected across 200 V mains. Calculate the magnitude and direction of the current in each battery and the total current taken from the supply mains.

[I_A = 11.96 (discharge); I_B = 30.43 A (charge): 18.47 A]

(Elect Technology, Sumbhal Univ.)

16. Three batteries *P, Q* and *R* consisting of 50, 55 and 60 cells in series respectively supply in parallel a common load of 100 A. Each cell has a e.m.f of 2 V and an internal resistance of 0.005 Ω. Determine the current supplied by each battery and the load voltage.

[1.2 A; 35.4 A: 65.8 A: 100.3 V] ***(Basic Electricity, Bombay Univ.)***

2.10 Maxwell's Loop Current Method

This method which is particularly well-suited to coupled circuit solutions employs a system of ***loop*** or ***mesh*** currents instead of ***branch*** currents (as in Kirchhoff's laws). Here, the currents in different meshes are assigned continuous paths so that they do not split at a junction into branch currents. This method eliminates a great deal of tedious work involved in the branch-current method and is best suited when energy sources are voltage sources rather than current sources. Basically, this method consists of writing loop voltage equations by Kirchhoff's voltage law in terms of unknown loop currents. As will be seen later, the number of independent equations to be solved reduces from *b* by Kirchhoff's laws to *b* – (*j* – 1) for the loop current method where *b* is the number of branches and *j* is the number of junctions in a given network.

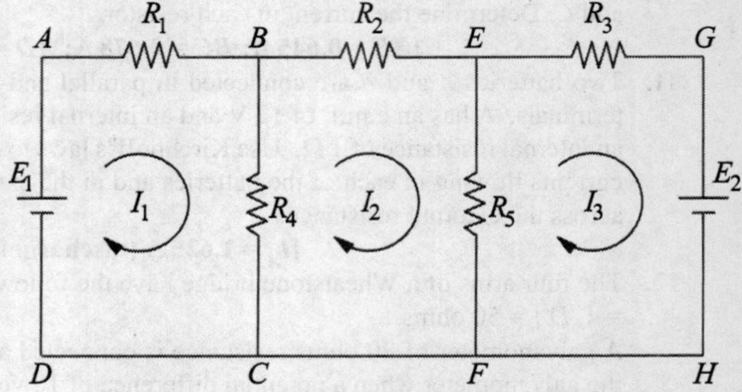

Fig. 2.37

Fig. 2.37 shows two batteries E_1 and E_2 connected in a network consisting of five resistors. Let the loop currents for the three meshes be I_1, I_2 and I_3. It is obvious that current through R_5 (when considered as a part of the first loop) is $(I_1 - I_2)$ and that through R_S is $(I_2 - I_3)$. However, when R_4 is considered part of the second loop, current through it is $(I_2 \rightarrow I_1)$. Similarly, when R_5 is considered part of the third loop, current through it is $(I_3 - I_2)$. Applying Kirchhoff's voltage law to the three loops, we get,

$$E_1 - I_1R_1 - R_4(I_1 - I_2) = 0 \text{ or } I_1(R_1 + R_4) - I_2R_4 - E_1 = 0 \quad \text{...loop 1}$$

Similarly, $-I_2R_2 - R_5(I_2 - I_3) - R_4(I_2 - I_1) = 0$

or $I_2R_4 - I_2(R_2 + R_4 + R_5) + I_3R_5 = 0$ **...loop 2**

Also $-I_3R_3 - E_2 - R_5(I_3 - I_2) = 0$ or $I_2R_5 - I_3(R_3 + R_5) - E_2 = 0$ **...loop 3**

The above three equations can be solved not only to find loop currents but branch currents as well.

Example 2.22. *Determine the current supplied by each battery in the circuit shown in Fig. 2.38.* **(Electrical Engg. Aligarh Univ.)**

Solution. Since there are three meshes, let the three mesh currents be shown in Fig. 2.38.

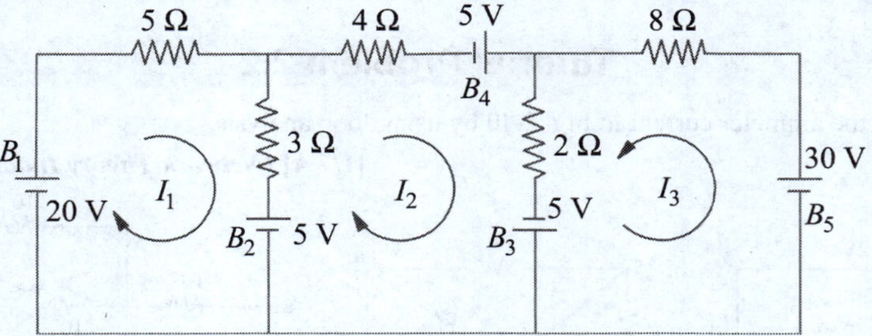

Fig. 2.38

For loop 1 we get

$$20 - 5I_1 - 3(I_1 - I_2) - 5 = 0 \quad \text{or} \quad 8I_1 - 3I_2 = 15 \quad \text{...}(i)$$

For loop 2 we have

$$-4I_2 + 5 - 2(I_2 - I_3) + 5 + 5 - 3(I_2 - I_1) = 0 \quad \text{or} \quad 3I_1 - 9I_2 + 2I_3 = -15 \quad \text{...}(ii)$$

Similarly, for loop 3, we get

$$-8I_3 - 30 - 5 - 2(I_3 - I_2) = 0 \quad \text{or} \quad 2I_2 - 10I_3 = 35 \quad \text{...}(iii)$$

Eliminating I_1 from **(*i*)** and **(*ii*)**, we get $63I_2 - 16I_3 = 165$ **...(*iv*)**

Similarly, for I_2 from **(*iii*)** and **(*iv*)**, we have $I_2 = 545/299$ A

From **(*iv*)**, $I_3 = -1875/598$ A

Substituting the value of I_2 in **(*i*)**, we get $I_1 = 765/299$ A

Since I_3 turns out to be negative, actual directions of flow of mesh currents are as shown in Fig. 2.39.

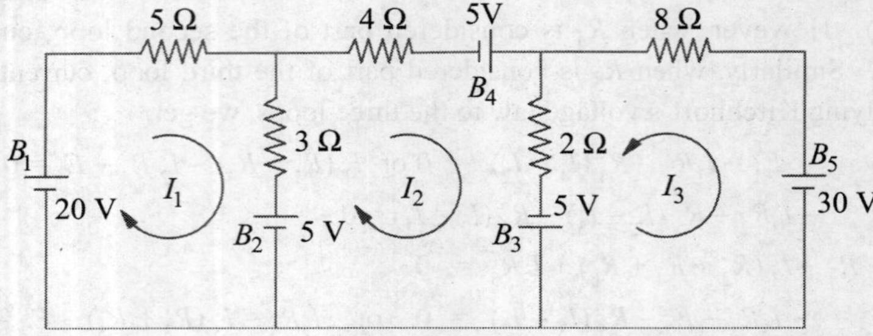

Fig. 2.39

Discharge current of $B_1 = \mathbf{765/299A}$

Charging current of $B_2 = I_1 - I_2 = \mathbf{220/299\ A}$

Discharge current of $B_3 = I_2 + I_3 = \mathbf{2965/598\ A}$

Discharge current of $B_4 = I_2 = \mathbf{545/299\ A}$; Discharge current of $B_5 = \mathbf{1875/598\ A}$

Tutorial Problem 2.2

1. Find the ammeter current in Fig. 2.40 by using loop analysis. **[1/7 A] (*Network Theory Indore Univ. 1981*)**

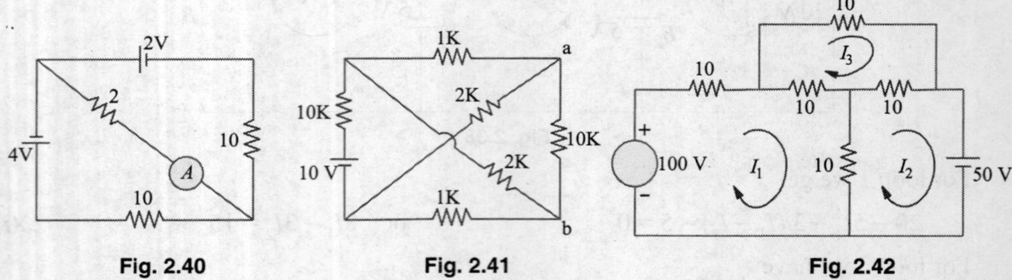

Fig. 2.40 Fig. 2.41 Fig. 2.42

2. Using mesh analysis, determine the voltage across the 10 kΩ resistor at terminals *a-b* of the circuit shown in Fig. 2.41. **[2.65 V] (*Elect. Technology, Indore Univ.*)**
3. Apply loop current method to find loop currents I_1, I_2 and I_3 in the circuit of Fig. 2.59. **[I_1 = 3.75 A, I_2 = 0, I_3 = 1.25 A]**

2.11 Nodal Analysis with Sources

The node-equation method is based directly on Kirchhoff's current law unlike loop-current method which is based on Kirchhoff's voltage law. However, like loop current method, nodal method also has the advantage that a minimum number of equations need be written to determine the unknown quantities. Moreover, it is particularly suited for networks having many parallel circuits with common ground connected such as electronic circuits.

For the application of this method, every junction in the network where two or more elements meet is regarded a node. One of these is regarded as the reference node or datum node or zero-potential node. Hence the number of simultaneous equations to be solved becomes $(n - 1)$ where n is the number of independent nodes. These node equations often become simplified if all voltage sources are converted into current sources.

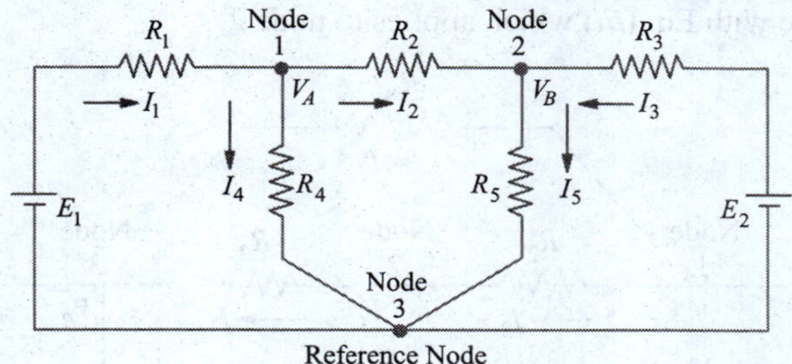

Fig. 2.43

***(i)* First Case**

Consider the circuit of Fig. 2.43 which has three nodes. One of these *i.e.* node 3 has been taken in as the reference node. V_A represents the potential of node 1 with reference to the datum node 3. Similarly, V_B is the potential difference between node 2 and node 3. Let the current directions which have been chosen arbitrary be as shown.

For node 1, the following current equation can be written with the help of *KCL*.

$$I_1 = I_4 + I_2$$

Now $\quad I_1R_1 = E_1 - V_A \quad \therefore \quad I_1 = (E_1 - V_A)/R_1$ ...***(i)***

Obviously, $\quad I_4 = V_A/R_4 \quad$ Also, $I_2R_2 = V_A - V_B \quad (\because V_A > V_B)$

$\therefore \quad I_2 = (V_A - V_B)/R_2$

Substituting these values in Eq. ***(i)*** above, we get,

$$\frac{E_1 - V_A}{R_1} = \frac{V_A}{R_4} + \frac{V_A - V_B}{R_2}$$

Simplifying the above, we have

$$V_A\left(\frac{1}{R_1} + \frac{1}{R_2} + \frac{1}{R_4}\right) - \frac{V_B}{R_2} - \frac{E_1}{R_1} = 0 \quad \text{...(ii)}$$

The current equation for node 2 is $I_5 = I_2 + I_3$

or
$$\frac{V_B}{R_5} = \frac{V_A - V_B}{R_2} + \frac{E_2 - V_B}{R_3} \quad \text{...(iii)}$$

or
$$V_B\left(\frac{1}{R_2} + \frac{1}{R_3} + \frac{1}{R_5}\right) - \frac{V_A}{R_2} - \frac{E_2}{R_3} = 0 \quad \text{...(iv)}$$

Though the above nodal equations ***(ii)*** and ***(iii)*** seem to be complicated, they employ a very simple and systematic arrangement of terms which can be written simply by inspection. Eq. ***(ii)*** at node 1 is represented by

1. The product of node potential V_A and $(1/R_1 + 1/R_2 + 1/R_4)$ *i.e.* the sum of the reciprocals of the branch resistance connected to this node.
2. ***Minus*** the ratio of adjacent potential V_B and the interconnecting resistance R_2.
3. ***Minus*** ratio of adjacent battery (or generator) voltage E_1 and interconnecting resistance R_1.
4. All the above set to zero.

Same is the case with Eq. (*iii*) which applies to node 2.

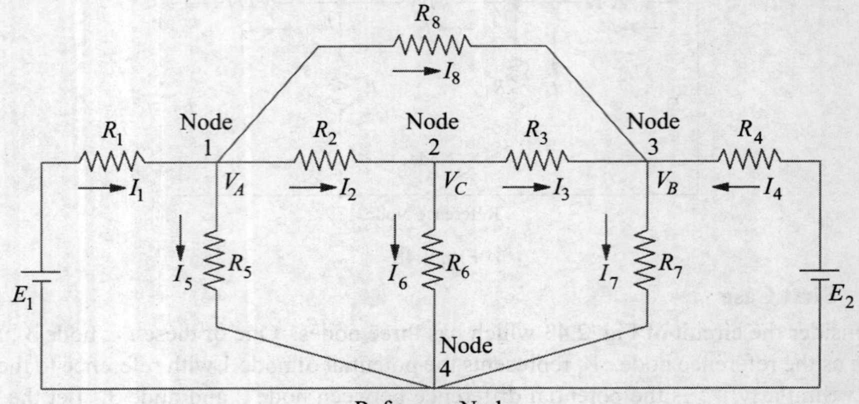

Fig. 2.44

Using conductances instead of resistances, the above two equations may be written as

$$V_A(G_1 + G_2 + G_4) - V_B G_2 - E_1 G_1 = 0 \quad \text{...}(iv)$$

$$V_B(G_2 + G_3 + G_5) - V_A G_2 - E_2 G_3 = 0 \quad \text{...}(v)$$

To emphasize the procedure given above, consider the circuit of Fig. 2.44.

The three node equations are $V_A\left(\dfrac{1}{R_1} + \dfrac{1}{R_2} + \dfrac{1}{R_5} + \dfrac{1}{R_8}\right) - \dfrac{V_C}{R_2} - \dfrac{V_B}{R_8} - \dfrac{E_1}{R_1} = 0$ (node 1)

$$V_C\left(\frac{1}{R_2} + \frac{1}{R_3} + \frac{1}{R_6}\right) - \frac{V_A}{R_2} - \frac{V_B}{R_3} = 0 \quad \text{(node 2)}$$

$$V_B\left(\frac{1}{R_3} + \frac{1}{R_4} + \frac{1}{R_7} + \frac{1}{R_8}\right) - \frac{V_C}{R_3} - \frac{V_A}{R_8} - \frac{E_4}{R_4} = 0 \quad \text{(node 3)}$$

After finding different node voltages, various currents can be calculated by using Ohm's law.

(*ii*) Second Case

Now, consider the case when a third battery of e.m.f. E_3 is connected between nodes 1 and 2 as shown in Fig. 2.45.

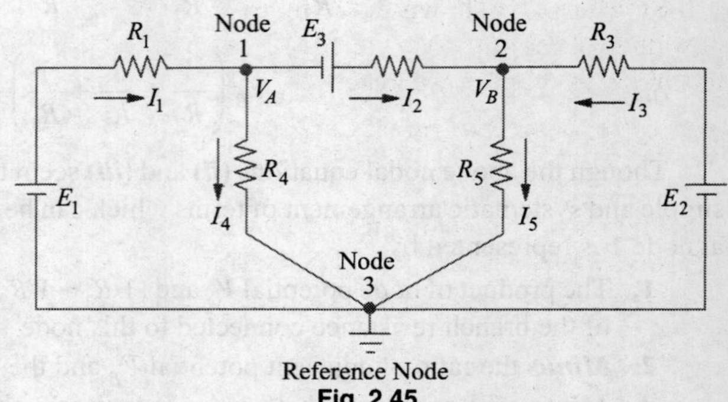

Fig. 2.45

It must be noted that as we travel from node 1 to node 2, we go from the −ve terminal of E_3 to its +ve terminal. Hence, according to the sign convention, E_3 must be taken as *positive.* However, if we travel from node 2 to node 1, we go from the +ve to the −ve terminal of E_3. Hence, ***when viewed from node* 2, E_3 is taken *negative*.**

For node 1

$$I_1 - I_4 - I_2 = 0 \text{ or } I_1 = I_4 + I_1 \text{ – as per KCL}$$

Now, $$I_1 = \frac{E_1 - V_A}{R_1}; I_2 = \frac{V_A + E_3 - V_B}{R_2}; I_4 = \frac{V_A}{R_4}$$

$\therefore$ $$\frac{E_1 - V_A}{R_1} = \frac{V_A}{R_4} + \frac{V_A + E_3 - V_B}{R_2}$$

or $$V_A\left(\frac{1}{R_1} + \frac{1}{R_2} + \frac{1}{R_3}\right) - \frac{E_1}{R_1} - \frac{V_B}{R_2} + \frac{E_2}{R_2} = 0 \quad \text{...(i)}$$

It is exactly the same expression as given under the First Case discussed above except for the additional term involving E_3. This additional term is taken as $+E_3/R_2$ (and not as $-E_3/R_2$) because this third battery is so connected that when viewed from mode 1, it represents a *rise* in voltage. Had it been connected the other way around, the additional term would have been taken as $-E_3/R_2$.

For node 2

$$I_2 + I_3 - I_5 = 0 \quad \text{or} \quad I_2 + I_3 = I_5 \quad \text{– as per } KCL$$

Now, as before, $$I_2 = \frac{V_A + E_3 - V_B}{R_2}, I_3 = \frac{E_2 - V_B}{R_3}, I_5 = \frac{V_B}{R_5}$$

$\therefore$ $$\frac{V_A + E_3 - V_B}{R_2} + \frac{E_2 - V_B}{R_3} = \frac{V_B}{R_5}$$

On simplifying, we get $$V_B\left(\frac{1}{R_2} + \frac{1}{R_3} + \frac{1}{R_5}\right) - \frac{E_2}{R_3} - \frac{V_A}{R_2} - \frac{E_3}{R_2} = 0 \quad \text{...(ii)}$$

As seen, the additional terms is $-E_3/R_2$ (and not $+ E_3/R_2$) because as viewed from this node, E_3 represents a ***fall*** in potential.

It is worth repeating that the additional term in the above Eq. (*i*) and (*ii*) can be either $+E_3/R_2$ or $-E_3/R_2$ depending on whether it represents a rise or fall of potential when viewed from ***the node under consideration.***

Example 2.23. *Using Node voltage method, find the current in the 3Ω resistance for the network shown in Fig. 2.46.* **(Elect. Tech. Osmania Univ.)**

Solution. As shown in the figure node 2 has been taken as the reference node. We will now find the value of node voltage V_1. Using the technique developed, we get

$$V_1\left(\frac{1}{5} + \frac{1}{2} + \frac{1}{2}\right) - \frac{4}{2} - \left(\frac{4+2}{5}\right) = 0$$

The reason for adding the two battery voltages of 2 V and 4 V is because they are connected in additive series. Simplifying above, we get $V_1 = 8/3$ V. The current flowing through the 3 Ω resistance towards node 1 is

$$= \frac{6 - (8/3)}{(3+2)} = \frac{2}{3} \text{ A}$$

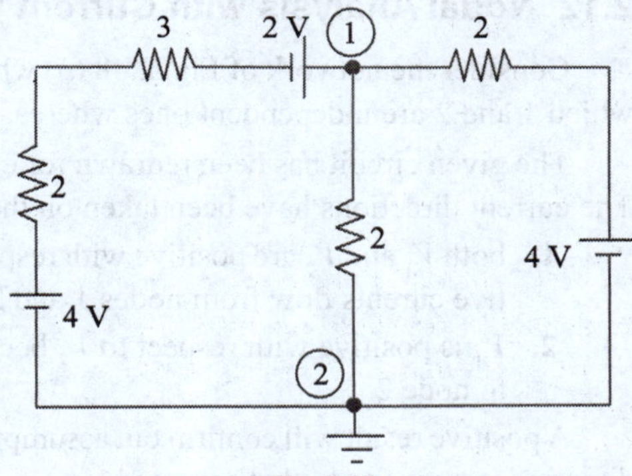

Fig. 2.46

Alternatively

$$\frac{6-V_1}{5}+\frac{4}{2}-\frac{V_1}{2}=0$$

$$12-2V_1+20-5V_1=0$$

$$7V_1=32$$

Also
$$\frac{6-V_1}{5}+\frac{4-V_1}{2}=\frac{V_1}{2}$$

$$12-2V_1+20-5V_1=5\,V_1$$

$$12V_1=32;\; V_1=8/3$$

Example 2.24. *Frame and solve the node equations of the network of Fig. 2.47. Hence, find the total power consumed by the passive elements of the network.*

(Elect. Circuits Nagpur Univ.)

Solution. The node equation for node 1 is

$$V_1\left(1+1+\frac{1}{0.5}\right)-\frac{V_2}{0.5}-\frac{15}{1}=0$$

or $4V_1-2V_2=15$ ...(*i*)

Similarly, for node 2, we have

$$V_1\left(1+\frac{1}{2}+\frac{1}{0.5}\right)-\frac{V_2}{0.5}-\frac{20}{1}=0$$

or $4V_1-7V_2=-40$ ...(*ii*)

$\therefore$ $V_2 = 11$ volt and $V_1 = 37/4$ volt

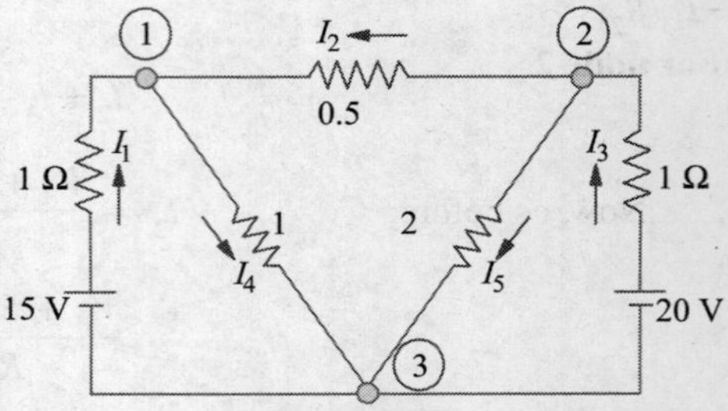

Datum Node

Fig. 2.47

Now,

$$I_1=\frac{15-37/4}{I}=\frac{23}{4}\text{ A}=5.75\text{ A};\; I_2=\frac{11-37/4}{0.5}=3.5\text{ A}$$

$$I_4=5.75+3.5=9.25\text{ A};\; I_3=\frac{20-11}{1}=9\text{ A};\; I_5=9-3.5=5.5\text{ A}$$

The passive elements of the network are its five resistances. Total power consumed by them is $= 5.75^2 \times 1 + 3.5^2 \times 0.5 + 9^2 \times 1 + 9.25^2 \times 1 + 5.5^2 \times 2 =$ **266.25**

2.12 Nodal Analysis with Current Sources

Consider the network of Fig. 2.48 (*a*) which has two current sources and three nodes out of which 1 and 2 are independent ones whereas No. 3 is the reference node.

The given circuit has been redrawn for ease of understanding and is shown in Fig. 2.48 (*b*). The current directions have been taken on the assumption that

1. both V_1 and V_2 are positive with respect to the reference node. That is why their respective curents flow from nodes 1 and 2 to node 3.
2. V_1 is positive with respect to V_2 because current has been shown flowing from node 1 to node 2.

A positive result will confirm out assumption whereas a negative one will indicate that actual direction is opposite to that assumed.

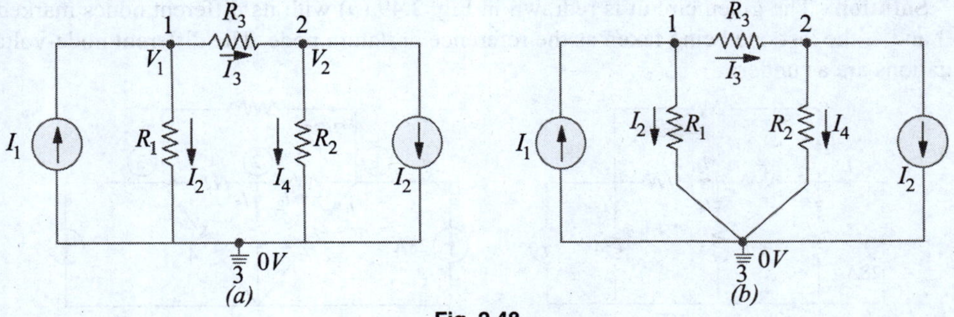

Fig. 2.48

We will now apply *KCL* to each node and use Ohm's law to express branch currents in terms of node voltages and resistances.

Node 1

$$I_1 - I_2 - I_3 = 0 \quad \text{or} \quad I_1 = I_2 + I_3$$

Now $$I_2 = \frac{V_1}{R_1} \quad \text{and} \quad I_3 = \frac{V_1 - V_2}{R_3}$$

$\therefore$ $$I_1 = \frac{V_1}{R_1} + \frac{V_1 - V_2}{R_3} \quad \text{or} \quad V_1\left(\frac{1}{R_1} + \frac{1}{R_3}\right) - \frac{V_2}{R_3} = I_1 \qquad ...(i)$$

Node 2

$$I_3 - I_2 - I_4 = 0 \quad \text{or} \quad I_3 = I_2 + I_4$$

Now, $$I_4 = \frac{V_2}{R_2} \quad \text{and} \quad I_3 = \frac{V_1 - V_2}{-R_3} \text{ – as before}$$

$\therefore$ $$\frac{V_1 - V_2}{R_3} = I_2 + \frac{V_2}{R_2} \quad \text{or} \quad V_2\left(\frac{1}{R_2} + \frac{1}{R_3}\right) - \frac{V_1}{R_3} = -I_1 \qquad ...(ii)$$

The above two equations can also be written by simple inspection. For example, Eq. (*i*) is represented by

1. ***product*** of potential V_1 and $(1/R_1 + 1/R_3)$ *i.e.* sum of the reciprocals of the branch resistances connected to this node.

2. ***minus*** the ratio of adjoining potential V_2 and the interconnecting resistance R_3.

3. all the above equated to the current supplied by the current source connected to this node.

This current is taken *positive* if flowing *into* the node and negative if flowing *out* of it. Same remarks apply to Eq. (*ii*) where I_2 has been taken negative because it flows *away* from node 2.

In terms of branch conductances, the above two equations can be put as

$$V_1(G_1 + G_3) - V_2 G_3 = I_1 \quad \text{and} \quad V_2(G_2 + G_3) - V_1 G_3 = -I_2$$

Example 2.25. *Use nodal analysis method to find currents in the various resistors of the circuit shown in Fig. 2.49 (a).*

Solution. The given circuit is redrawn in Fig. 2.49 (*b*) with its different nodes marked 1, 2, 3 and 4, the last one being taken as the reference or datum node. The different node-voltage equations are as under:

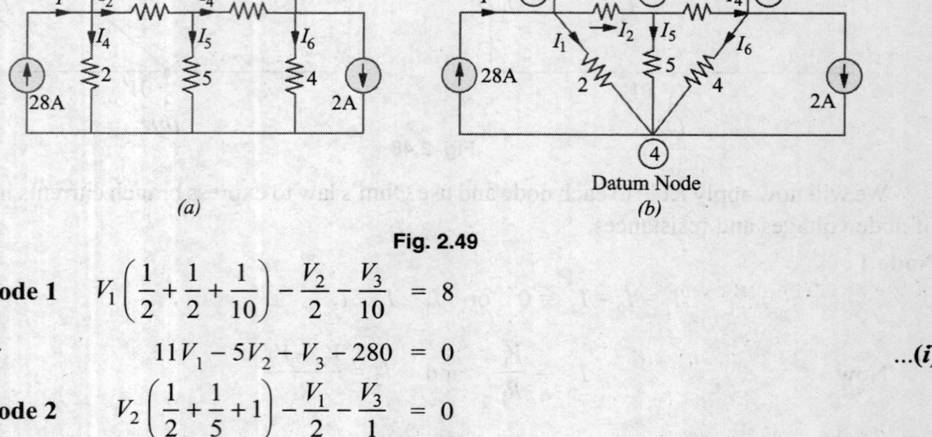

Fig. 2.49

Node 1 $\quad V_1\left(\frac{1}{2}+\frac{1}{2}+\frac{1}{10}\right)-\frac{V_2}{2}-\frac{V_3}{10} = 8$

or $\quad 11V_1 - 5V_2 - V_3 - 280 = 0$ ...(*i*)

Node 2 $\quad V_2\left(\frac{1}{2}+\frac{1}{5}+1\right)-\frac{V_1}{2}-\frac{V_3}{1} = 0$

or $\quad 5V_1 - 17\,V_2 + 10\,V_3 = 0$ ...(*ii*)

Node 3 $\quad V_3\left(\frac{1}{4}+1+\frac{1}{10}\right)-\frac{V_2}{1}-\frac{V_1}{10} = -2$

or $\quad V_1 + 10\,V_2 - 13.5\,V_3 - 20 = 0$ ...(*iii*)

$V_1 = 36$ V, $V_2 = 20$ V, $V_3 = 16$ V

It is obvious that all nodes are at a higher potential with respect to the datum node. The various currents shown in Fig. 2.69 (*b*) can now be found easily.

$$I_1 = V_1/2 = 36/2 = \mathbf{18\ A}$$
$$I_2 = (V_1 - V_2)/2 = (36-20)/2 = \mathbf{8\ A}$$
$$I_3 = (V_1 - V_3)/10 = (36-16)/10 = \mathbf{2\ A}$$

It is seen that total current, as expected, is 18 + 8 + 2 = **28 A**

$$I_4 = (V_2 - V_3)/1 = (20-16)/1 = \mathbf{4\ A}$$
$$I_5 = V_2/5 = 20/5 = \mathbf{4\ A},\ I_6 = V_3/4 = 16/4 = \mathbf{4\ A}$$

Example 2.26. *Using nodal analysis, find the different branch currents in the circuit of Fig. 2.50* (*a*)*. All branch conductances are in siemens* (*i.e. mho*).

Solution. Let the various branch currents be as shown in Fig. 2.50 (*b*). Using the procedure detailed, we have

First Node

$V_1\,(1+2) - V_2 \times 1 - V_3 \times 2 = -2$ or $3V_1 - V_2 - 2V_3 = -2$ ...(*i*)

Second Node

$V_2\,(1+4) - V_1 \times 1 = 5$ or $V_1 - 5V_2 = -5$ ...(*ii*)

Third Node

$V_3\,(2+3) - V_1 \times 2 = -5$ or $2V_1 - 5V_3 = 5$ ...(*iii*)

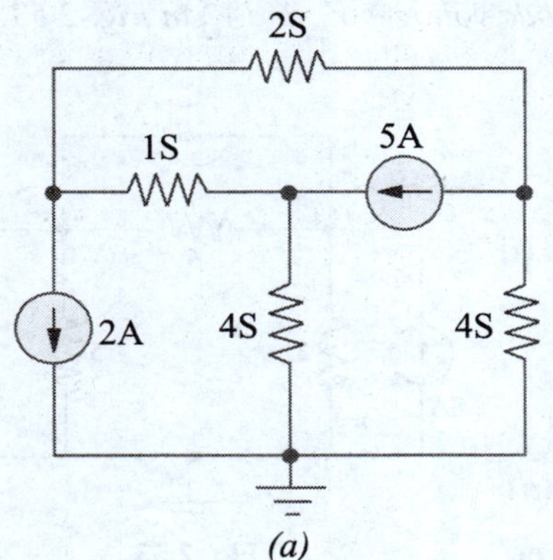

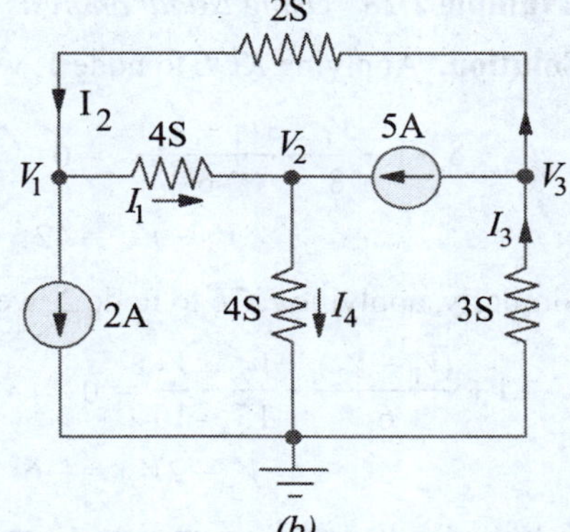

Fig. 2.50

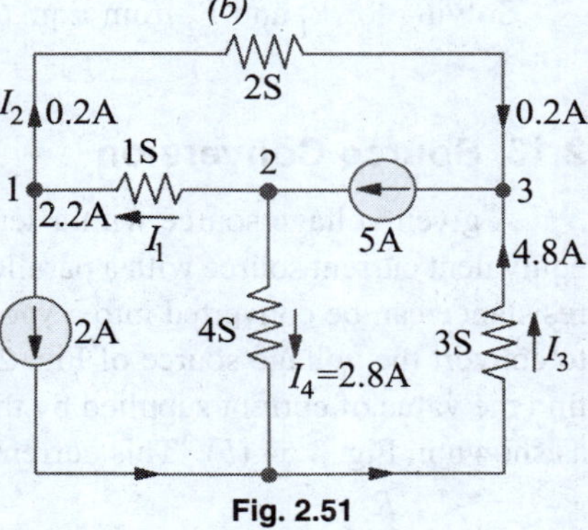

Fig. 2.51

Solving for the different voltages, we have

$$V_1 = -\frac{3}{2}\text{ V}, V_2 = \frac{7}{10}\text{ V and } V_3 = -\frac{8}{5}\text{ V}$$

$I_1 = (V_1 - V_2) \times 1 = (-1.5 - 0.7) \times 1 = \mathbf{-2.2\ A}$

$I_2 = (V_3 - V_1) \times 2 = [-1.6 - (-1.5)] \times 2 = \mathbf{-0.2\ A}$

$I_4 = V_2 \times 4 = 4 \times (7/10) = \mathbf{2.8\ A}$

$I_3 = 2 + 2.8 = \mathbf{4.8\ A}$

As seen, I_1 and I_2 flow in directions opposite to those originally assumed (Fig. 2.51).

Example 2.27. *Use Nodal analysis to determine the value of current i in the network of Fig. 2.52.*

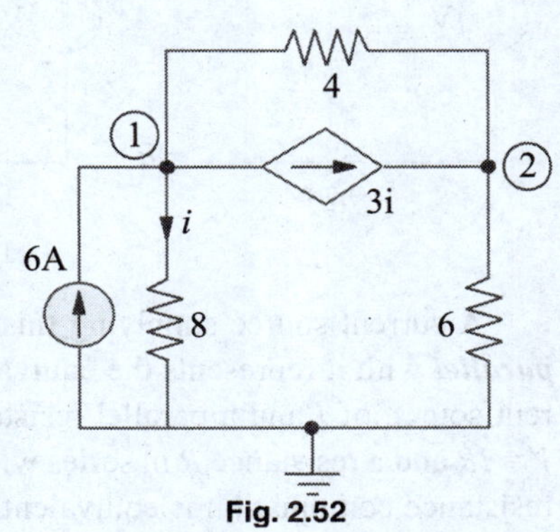

Fig. 2.52

Solution. We will apply *KCL* to the two nodes 1 and 2. Equating the incoming currents at node 1 to the outgoing currents, we have

$$6 = \frac{V_1 - V_2}{4} + \frac{V_1}{8} + 3\,i$$

As seen. $i = V_1/8$. Hence, the above equation becomes

$$6 = \frac{V_1 - V_2}{4} + \frac{V_1}{8} + 3\,\frac{V_1}{8}$$

or $3V_1 - V_2 = 24$

Similarly, applying *KCL* to node No. 2, we get

$$\frac{V_1 - V_2}{4} + 3\,i = \frac{V_2}{6} \quad\text{or}\quad \frac{V_1 - V_2}{4} + 3\,\frac{V_1}{8} = \frac{V_2}{6} \quad\text{or}\quad 3V_1 = 2V_2$$

From the above two equations, we get

$$V_1 = 16\text{ V} \quad \therefore \quad i = 16/8 = \mathbf{2\ A.}$$

Example 2.28. *Using Nodal analysis, find the node voltages V_1 and V_2 in Fig. 2.53.*

Solution. Applying *KCL* to node 1, we get

$$8 - 1 - \frac{V_1}{3} - \frac{(V_1 - V_2)}{6} = 0$$

or $$3V_1 - V_2 = 42 \qquad ...(i)$$

Similarly, applying *KCL* to node 2, we get

$$1 + \frac{(V_1 - V_2)}{6} - \frac{V_2}{15} - \frac{V_2}{10} = 0$$

or $$V_1 - 2V_2 = -6 \qquad ...(ii)$$

Fig. 2.53

Solving for V_1 and V_2 from Eqn. (*i*) and (*ii*), we get

$$V_1 = 18 \text{ V and } V_2 = 12 \text{ V.}$$

2.13 Source Conversion

A given voltage source with a series resistance can be converted into (or replaced by) an equivalent current source with a parallel resistance. Conversely, a current source with a parallel resistance can be converted into a vaoltage source with a series resistance. Suppose, we want to convert the voltage source of Fig. 2.54 (*a*) into an equivalent current source. First, we will find the value of current supplied by the source when a 'short' is put across in termials *A* and *B* as shown in Fig. 2.54 (*b*). This current is $I = V/R$.

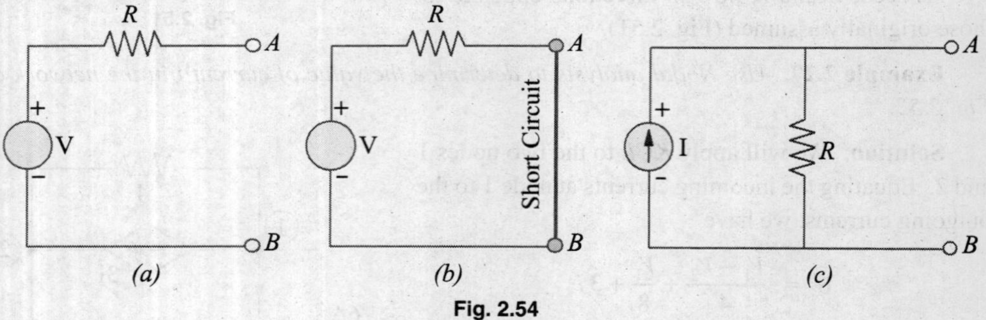

Fig. 2.54

A current source supplying this current *I* and having the same resistance *R* connected in ***parallel*** with it represents the equivalent source. It is shown in Fig. 2.54 (*c*). Similarly, a current source of *I* and a parallel resistance *R* can be converted into a voltage source of voltage $V = IR$ and a resistance *R* in series with it. It should be kept in mind that a voltage source-series resistance combination is equivalent to (or replaceable by) a current source-parallel resistance combination if, and only if their:

1. respective open-circuit voltages are equal, and
2. respective short-circuit currents are equal.
3. resistance remains same in both cases.

For example, in Fig. 2.54 (*a*), voltage across terminals *A* and *B* when they are open (*i.e.* open-circuit voltage V_{OC}) is *V* itself because there is no drop across *R*. Short-circuit current across $AB = I = V/R$.

Now, take the circuit of Fig. 2.54 (*c*). The open-circuit voltage across *AB* = drop across $R = IR = V$. If a short is placed across *AB*, whole of *I* passes through it because *R* is completely shorted out.

Example 2.29. *Convert the voltage source of Fig. 2.55 (a) into an equivalent current source.*

Solution. As shown in Fig 2.55 (*b*), current obtained by putting a short across terminals *A* and *B* is 10/5 = 2 A.

Hence, the equivalent current source is as shown in Fig. 2.55 (*c*).

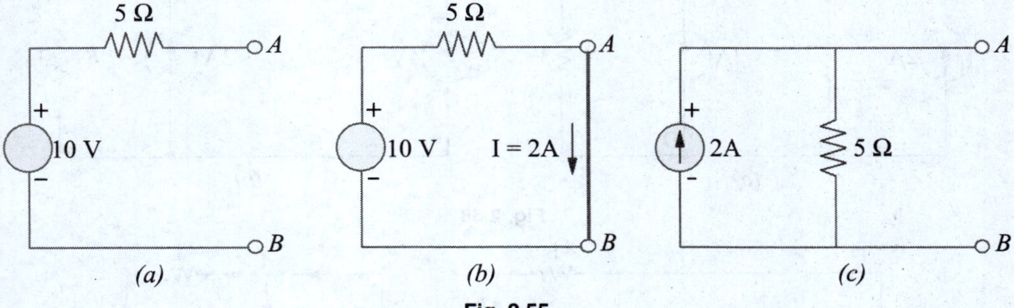

Fig. 2.55

Example 2.30. *Find the equivalent voltage source for the current source in Fig. 2.56 (a).*

Solution. The open-circuit voltage across terminals *A* and *B* in Fig. 2.56 (*a*) is

$$V_{OC} = \text{drop across } R$$
$$= 5 \times 2 = 10 \text{ V}$$

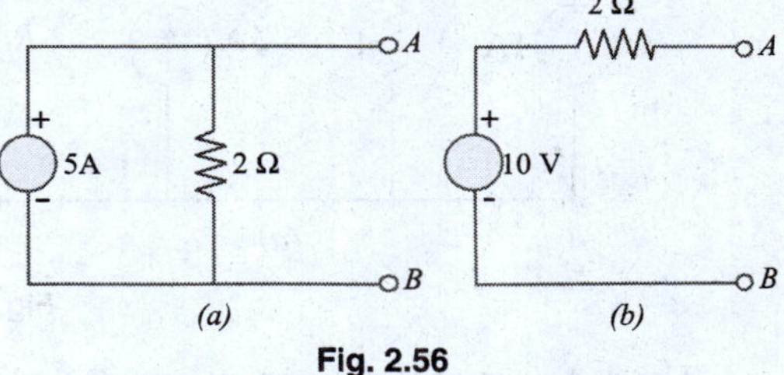

Fig. 2.56

Hence, voltage source has a voltage of **10 V** and the same resistance of 2 Ω through connected in series [Fig. 2.56 (*b*)].

Example 2.31. *Use Source Conversion technique to find the load current I_1 in the circuit of Fig. 2.57 (a).*

Solution. As shown in Fig. 2.57 (*b*). 6-V voltage source with a series resistance of 3 Ω has been converted into an equivalent 2 A current source with 3 Ω resistance in parallel.

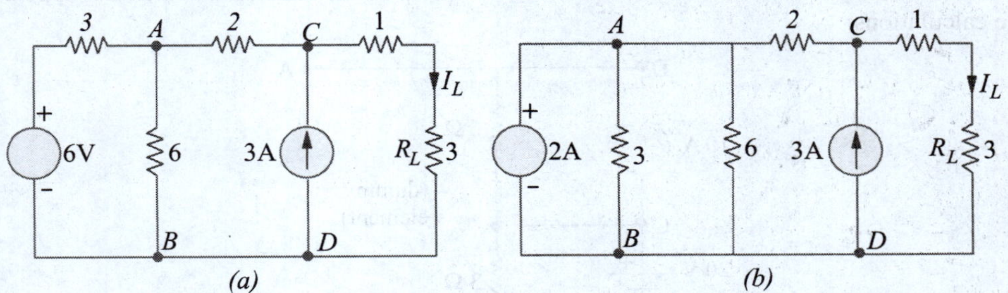

Fig. 2.57

The two parallel resistances of 3 Ω and 6 Ω can be combined into a single resistance of 2 Ω as shown in Fig. 2.58. (*a*)

The two current sources cannot be combined together because of the 2 Ω resistance present between points *A* and *C*. To remove this hurdle, we convert the 2 A current source into the equivalent 4 V voltage source as shown in Fig. 2.58 (*b*). Now, this 4 V voltage source with a series resistance of (2 + 2) = 4 Ω can again be converted into the equivalent current source as shown in Fig. 2.59 (*a*). Now, the two current sources can be combined into a single 4-A source as shown in Fig. 2.59 (*b*).

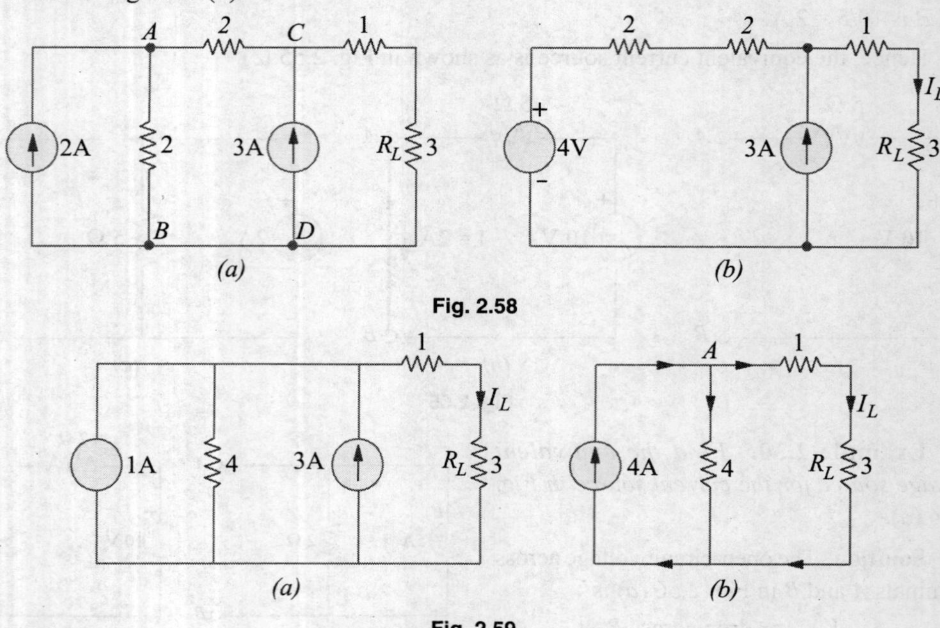

Fig. 2.58

Fig. 2.59

The 4-A current is divided into two equal parts at point *A* because each of the two parallel paths has a resistance of 4 Ω. Hence I_2 = **2 A.**

Example 2.32. *Replace the given network by a single current source in parallel with a resistance.* **[Bombay University 2001]**

Solution. The equivalence is expected for a load connected to the right-side of terminals *A* and *B*. In this case, the voltage-source has no resistive element in series. While handling such cases, the 3-ohm resistor has to be kept aside, treating it as an independent and separate loop. This voltage source will circulate a current of 20/3 amp in the resistor, and will not appear in the calculations.

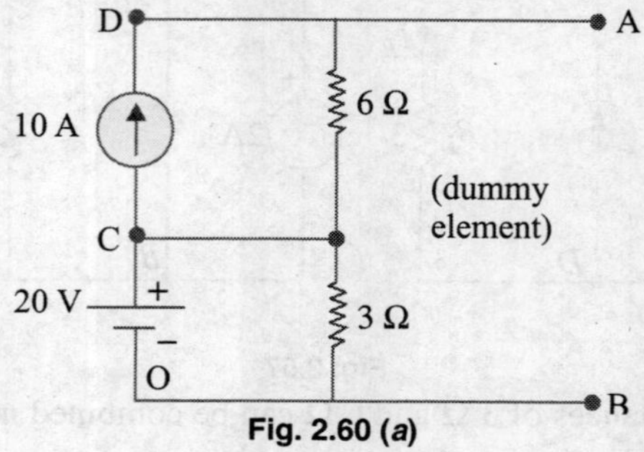

Fig. 2.60 (*a*)

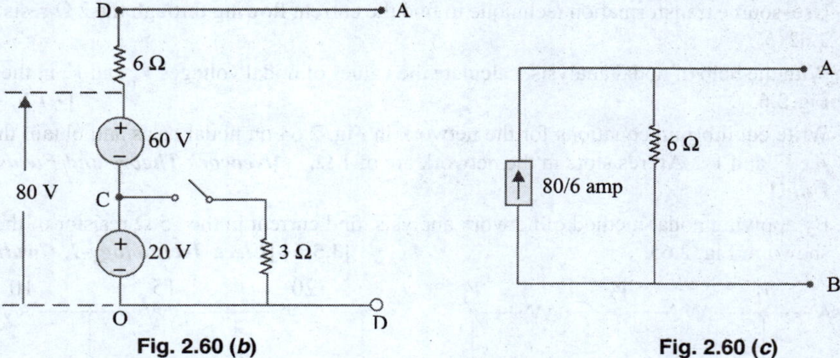

Fig. 2.60 (*b*) Fig. 2.60 (*c*)

This step does not affect the circuit connected to *A–B*.

Further steps are shown in Fig. 2.60 (*b*) and (*c*)

Tutorial Problem 2.3

1. Using Maxwell's loop current method, calculate the output voltage V_o for the circuits shown in Fig. 2.61. **[(*a*) 4 V (*b*) - 150/7 V (*c*) V_o = 0 (*d*) V_o = 0]**

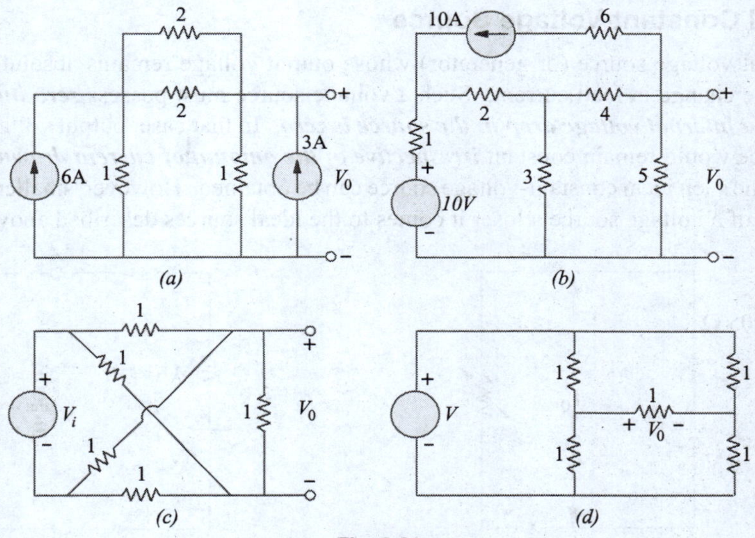

Fig. 2.61

2. Using nodal voltage method, find the magnitude and direction of current *I* in the network of Fig. 2.62.

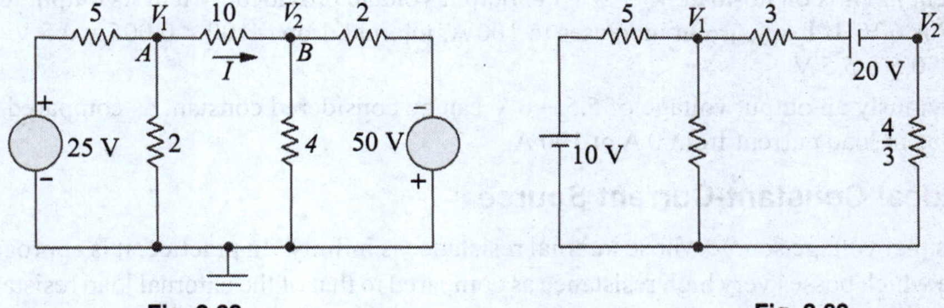

Fig. 2.62 Fig. 2.63

3. Use source transformation technique to find the current flowing through the 2 Ω resistor in Fig. 2.62 (*b*). **[10 A]**
4. With the help of nodal analysis, calculate the values of nodal voltages V_1 and V_2 in the circuit of Fig. 2.63. **[7.1 V; – 3.96 V]**
5. Write equilibrium equations for the network in Fig. 2.64 on nodal basis and obtain the voltage V_1, V_2 and V_3. All resistors in the network are of 1 Ω. ***[Network Theory and Fields, Madras Univ.]***
6. By applying nodal method of network analysis, find current in the 15 Ω resistor of the network shown in Fig. 2.65. **[3.5 A] *[Elect. Technology-1, Gwalior Univ.]***

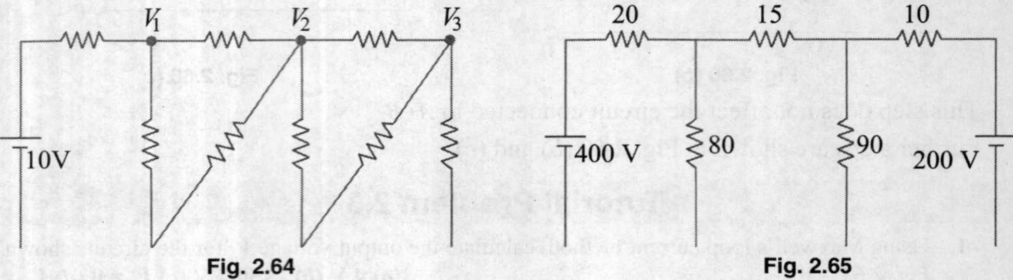

Fig. 2.64 **Fig. 2.65**

2.14 Ideal Constant-Voltage Source

It is that voltage source (or generator) whose output voltage remains absolutely constant whatever the change in load current. Such a voltage source must possess ***zero internal resistance so that internal voltage drop in the source is zero***. In that case, output voltage provided by the source would remain constant ***irrespective of the amount of current drawn from it***. In practice, none such ideal constant-voltage source can be obtained. However, smaller the internal resistance *r* of a voltage source, closer it comes to the ideal sources described above.

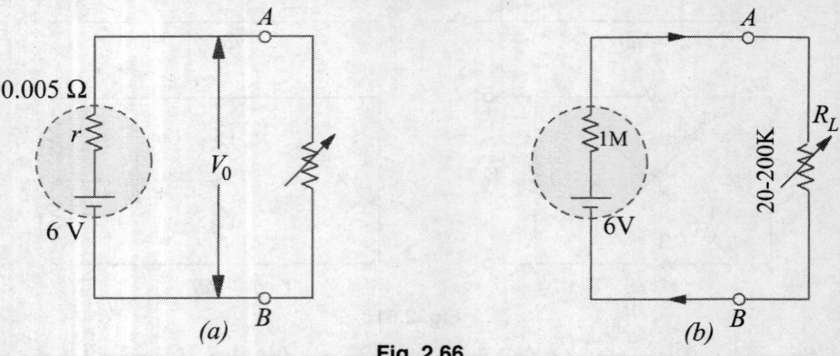

Fig. 2.66

Suppose, a 6-V battery has an internal resistance of 0.005 Ω [Fig. 2.66 (*a*)]. When it supplies no current *i.e.* it is on no-load, V_o = 6 V *i.e.* output voltage provided by it at its output terminals *A* and *B* is 6 V. If load current increases to 100 A, internal drop = 100 × 0.005 = 0.5 V. Hence, V_o = 6 – 0.5 = 5.5 V.

Obviously an output voltage of 5.5 – 6 V can be considered constant as compared to wide variations in load current from 0 A ot 100 A.

2.15 Ideal Constant-Current Source

It is that voltage source whose internal resistance is infinity. In practice, it is approached by a source which posses very high resistance as compared to that of the external load resistance. As shown in Fig. 2.67 (*b*), let the 6-V battery or voltage source have an internal resistance of 1 M Ω and let the load resistance vary from 20 K to 200 K. The current supplied by the source varies from 6.1/1.02 = 5.9 μ A to 6/1.2 = 5 μ A. As seen, even when load resistance increases 10 times, current decreases by 0.9 μA. Hence, the source can be considered, for all practical purposes, to be a constant-current source.

2.16 Superposition Theorem

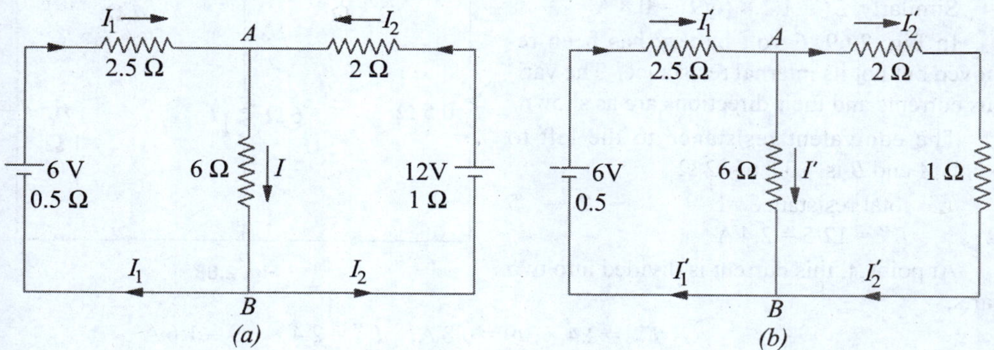

Fig. 2.67

According to this theorem, if there are a number of e.m.fs. acting simultaneously in any linear bilateral network, then each e.m.f. acts independently of the others *i.e.* as if the other e.m.fs. did not exist. The value of current in any conductor is the algebraic sum of the currents due to each e.m.f. Similarly, voltage across any conductor is the algebraic sum of the voltages which each e.m.f would have produced while acting singly. It is important to keep in mind that this theorem is applicable only to ***linear*** networks where current is ***linearly*** related to voltage as per Ohm's law.

Hence, this theorem may be stated as follows:

In a network of linear resistances containing more than one generator (or source of e.m.f.), the current which flows at any point is the sum of all the currents which would flow at that point if each generator where considered separately and all the other generators replaced for the time being by resistances equal to their internal resistances.

Explanation

In Fig. 2.67 (*a*) I_1, I_2 and I represent the values of currents which are due to the simultaneous action of the two sources of e.m.f. in the network. In Fig. 2.67 (*b*) are shown the current values which would have been obtained if left-hand side battery had acted alone. Similarly, Fig. 2.68 represents conditions obtained when right-hand side battery acts alone. By combining the current values of Fig. 2.68 (*b*) and 2.96 the actual values of Fig. 2.67 (*a*) can be obtained.

Obviously, $I_1 = I_1' - I_1''$, $I_2 = I_2'' - I_2'$, $I = I' + I''$.

Example 2.33. *In Fig. 2.67 (a) let battery e.m.fs. be 6 V and 12 V, their internal resistances 0.5 Ω and 1 Ω. The values of other resistances are as indicated. Find the different currents flowing in the branches and voltage across 60-ohm resistor.*

Solution. In Fig. 2.67 (*b*), 12-volt battery has been removed though its internal resistance of 1 Ω remains. The various currents can be found by applying Ohm's Law.

It is seen that there are two parallel paths between points A and B, having resistances of 6 Ω and (2 + 1) = 3 Ω.

∴ equivalent resistance $= 3 \parallel 6 = 2\ \Omega$

Total resistance $= 0.5 + 2.5 + 2 = 5\ \Omega$ ∴ $I_1' = 6/5 =$ **1.2 A.**

This current divides at point A inversely in the ratio of the resistances of the two parallel paths.

∴ $I' = 1.2 \times (3/9) = 0.4$ A.

Similarly, $I_2' = 1.2 \times (6/9) = 0.8$ A

In Fig. 2.69, 6 volt battery has been removed but not its internal resistance. The various currents and their directions are as shown.

The equivalent resistance to the left to points A and B is $= 3 \parallel 6 = 2\ \Omega$

∴ total resistance $= 1 + 2 + 2 = 5\ \Omega$ ∴ $I_2'' = 12/5 = 2.4$ A

At point A, this current is divided into two parts,

Fig. 2.68

$$I'' = 2.4 \times 3/9 = 0.8 \text{ A}, \quad I_1'' = 2.4 \times 6/9 = 1.6 \text{ A}$$

The actual current values of Fig. 2.95 (a) can be obtained by superposition of these two sets of current values.

∴ $I_1 = I_1' - I_1'' = 1.2 - 1.6 = \mathbf{-0.4\ A}$ (it is a charging current)

$I_2 = I_2'' - I_2' = 2.4 - 0.8 = \mathbf{1.6\ A}$

$I = I' + I'' = 0.4 + 0.8 = \mathbf{1.2\ A}$

Voltage drop across 6-ohm resistor $= 6 \times 1.2 =$ **7.2 V**

Example 2.34. *By using Superposition Theorem, find the current in resistance R shown in Fig. 2.69 (a)*

$$R_1 = 0.005\ \Omega,\ R_2 = 0.004\ \Omega,\ R = 1\ \Omega,\ E_1 = 2.05\text{ V},\ E_2 = 2.15\text{ V}$$

Internal resistances of cells are negligible.

(Electronic Circuits, Allahabad Univ. 1992)

Solution. In Fig. 2.69 (b), E_2 has been removed. Resistances of 1 Ω and 0.04 Ω are in parallel across poins A and C. $R_{AC} = 1 \parallel 0.04 = 1 \times 0.04/1.04 = 0.038\ \Omega$. This resistance is in series with 0.05 Ω. Hence, total resistance offered to battery $E_1 = 0.05 + 0.038 = 0.088\ \Omega$. $I = 2.05/0.088 = 23.3$ A. Current through 1-Ω resistance, $I_1 = 23.3 \times 0.04/1.04 = 0.896$ A from C to A.

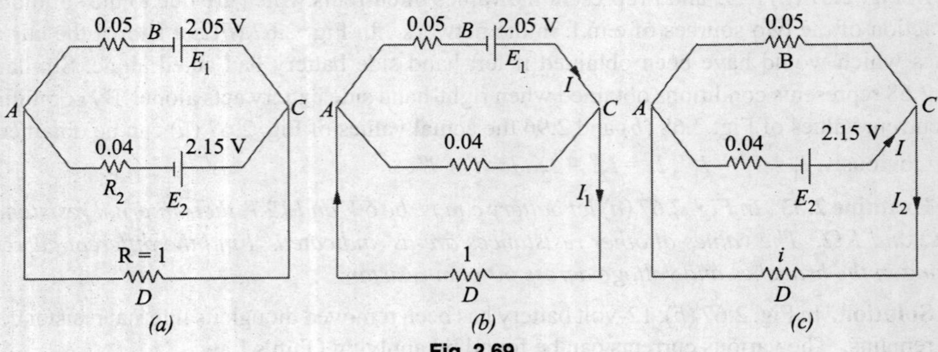

Fig. 2.69

When E_1 is removed, circuit becomes as shown in Fig. 2.69 (*c*). Combined resistance of paths *CBA* and *CDA* is = 1 || 0.05 = 1 × 0.05/1.05 = 0.048 Ω. Total resistance offered to E_2 is = 0.04 + 0.048 = 0.088 Ω. Current I = 2.15/0.088 = 24.4 A. Again, I_2 = 24.4 × 0.05/1.05 = 1.16 A.

To current through 1-Ω resistance when both batteries are present

$$= I_1 + I_2 = 0.896 + 1.16 = \mathbf{2.056\ A.}$$

Example 2.35. *Use Superposition theorem to find current I in the circuit shown in Fig. 2.70 (a). All resistances are in ohms.* **(Basic Circuit Analysis Osmania Univ. Jan/Feb 1992)**

Solution. In Fig. 2.70 (*b*), the voltage source has been replaced by a short and the 40 A current sources by an open. Using the current-divider rule, we get I_1 = 120 × 50/200 = 30 A.

In Fig. 2.70 (*c*), only 40 A current source has been considered. Again, using current-divider rule I_2 = 40 × 150/200 = 30 A.

In Fig. 2.70 (*d*), only voltage source has been considered. Using Ohm's law,

$$I_3 = 10/200 = 0.05\ A.$$

Since I_1 and I_2 cancel out, $I = I_3$ = 0.005 A.

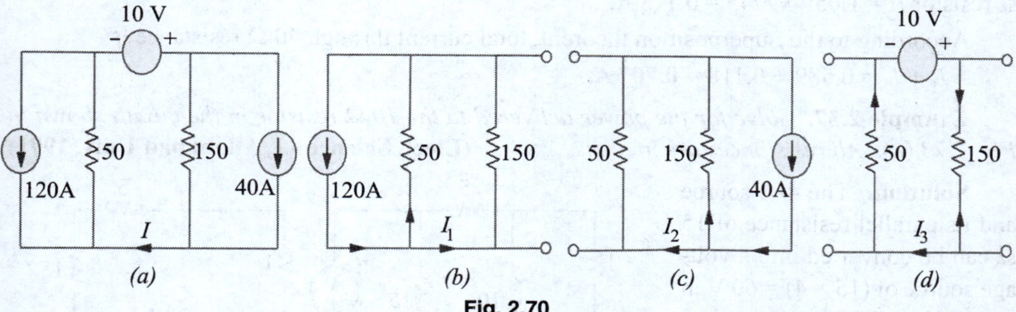

Fig. 2.70

Example 2.36. *Using Superposition theorem, find the current through the 40 Ω resistor of the circuit shown in Fig. 2.71 (a).* **(F.Y. Engg. Pune Univ. May 1990)**

Solution. We will first consider when 50 V battery acts alone and afterwards when 10-V battery is alone in the circuit. When 10-V battery is replaced by short-circuit, the circuit becomes as shown in Fig. 2.71 (*b*). It will be seen that the right-hand side 5 Ω resistor becomes connected in parallel with 40 Ω resistor giving a combined resistance of 5 || 40 = 4.44 Ω as shown in Fig. 2.71 (*c*). This 4.44 Ω resistance is in series with the left-hand side resistor of 5 Ω giving a total resistance of (5 + 4.44) = 9.44 Ω. As seen there are two resistances of 20 Ω and 9.44 Ω connected in parallel. In Fig. 2.71 (*c*) current I = 50/9.44 = 5.296 A.

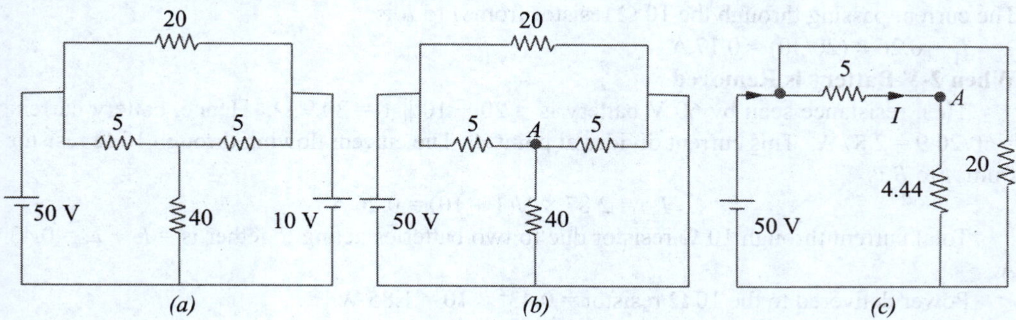

Fig. 2.71

At point A in Fig. 2.71 (b) there are two resistances of 5 Ω and 40 Ω connected in parallel, hence, current I divides between them as per the current-divider rule. If I_1 is the current flowing through the 40 Ω resistor, then

$$I_1 = I \times \frac{5}{5+40} = 5.296 \times \frac{5}{45} = \mathbf{0.589\ A.}$$

In Fig. 2.72 (a), 10 V battery acts alone because 50-V battery has been removed and replaced by a short-circuit.

As in the previous case, there are two parallel branches of resistances 20 Ω and 9.44 Ω across the 10-V battery. Current I through 9.44 Ω branch is I = 10/9.44 = 1.059 A. This current divides at point B between 5 Ω resistor and 40 Ω resistor. Current through 40 Ω resistor I_2 = 1.059 × 5/45 = 0.118 A.

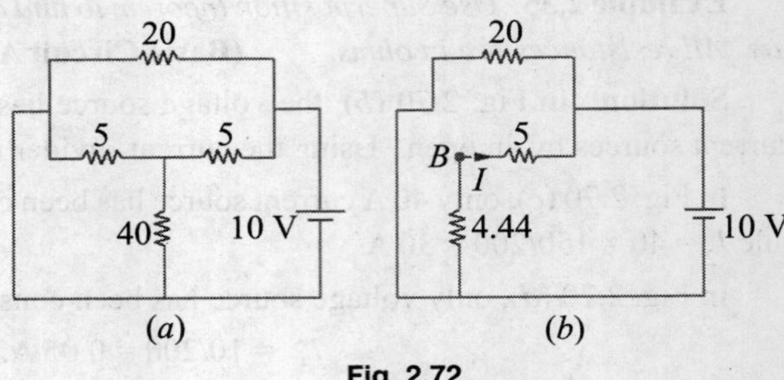

Fig. 2.72

According to the Superposition theorem, total current through 40 Ω resistance is $= I_1 + I_2 = 0.589 + 0.118 =$ **0.707 A.**

Example 2.37. *Solve for the power delivered to the 10 Ω resistor in the circuit shown in Fig. 2.73 (a). All resistances are in ohms.* **(Elect. Science - I, Allahabad Univ. 1991)**

Solution. The 4-A source and its parallel resistance of 15 Ω can be converted into a voltage source of (15 × 4) = 60 V in series with a 15 Ω resistances as shown in Fig. 2.73 (b).

Now, we will use Superposition theorem to find current through the 10 Ω resistances.

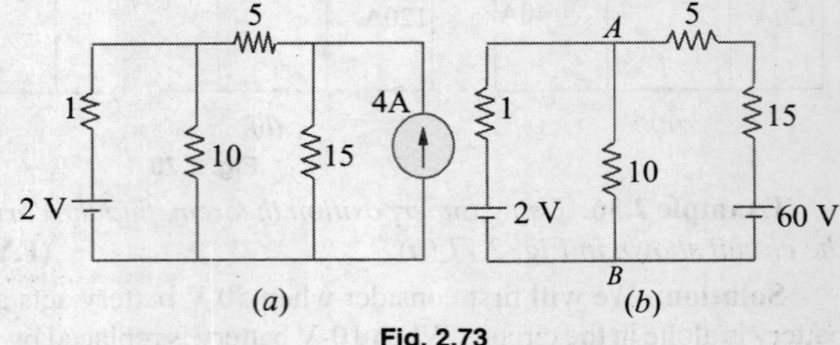

Fig. 2.73

When 60 – V Source is Removed

When 60 – V battery is removed the total resistance as seen by 2 V battery is = 1 + 10 || (15 + 5) = 7.67 Ω.

The battery current = 2/7.67 A = 0.26 A. At point A, this current is divided into two parts. The current passing through the 10 Ω resistor from A to B is

I_1 = 0.26 × (20/30) = 0.17 A

When 2-V Battery is Removed

Then resistance seen by 60 V battery is = 20 + 10 || 1 = 20.9 Ω. Hence, battery current = 60/20.9 = 2.87 A. This current divides at point A. The current flowing through 10 Ω resistor from A to B is

$$I_2 = 2.87 \times 1/(1 + 10) = 0.26\ \text{A}$$

Total current through 10 Ω resistor due to two batteries acting together is $= I_1 + I_2 = 0.43$ A.

Power delivered to the 10 Ω resistor = $0.43^2 \times 10$ = 1.85 W.

Example 2.38. *Apply Superposition theorem to the circuit of Fig. 2.74 (a) for finding the voltage drop V across the 5 Ω resistor.*

Solution. Fig. 2.74 (*b*) shows the redrawn circuit with the voltage source acting alone while the two current sources have been 'killed'. *i.e.* have been replaced by open circuits. Using voltage-divider principle, we get

$V_1 = 60 \times 5/(5 + 2 + 3) = 30$ V. It would be taken as positive, because current through the 5 Ω resistances flows from *A* to *B*, thereby making the upper end of the resistor positive and the lower end negative.

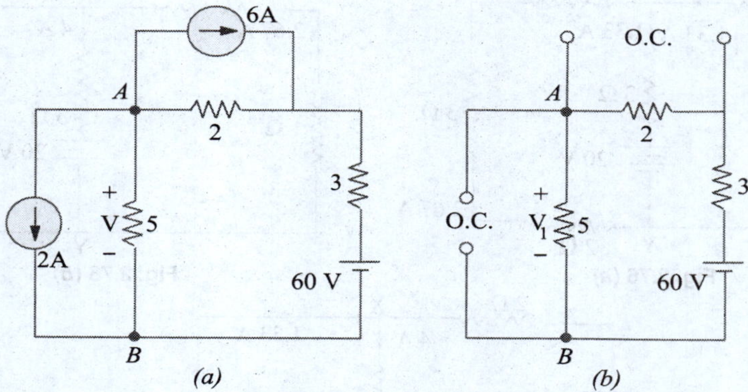

Fig. 2.74

Fig. 2.75 (*a*) shows the same circuit with the 6 A source acting alone while the two other sources have been 'killed'. It will be seen that 6 A source has to parallel circuits across it, one having a resistance of 2 Ω and the other (3 + 5) = 8 Ω. Using the current-divider rule, the current through the 5 Ω resistor = 6 × 2/(2 + 3 + 5) = 1.2 A.

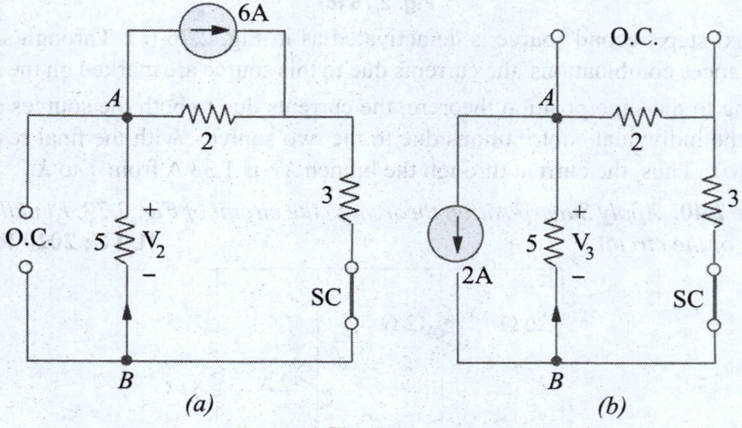

Fig. 2.75

∴ $V_2 = 1.2 \times 5 = 6$ V. It would be taken ***negative*** because current is flowing from *B* to *A*. *i.e.* point *B* is at a higher potential as compared to point *A*. Hence, $V_2 = -6$ V.

Fig. 2.75 (*b*) shows the case when 2-A source acts alone, while the other two sources are dead. As seen, this current divides equally at point *B*, because the two parallel paths have equal resistances of 5 Ω each. Hence, $V_3 = 5 \times 1 = 5$ V. It would also be taken as negative because current flows from *B* to *A*. Hence, $V_3 = -5$ V.

Using Superposition principle, we get

$$V = V_1 + V_2 + V_3 = 30 - 6 - 5 = \mathbf{19\ V}$$

Example 2.39. *Find the current flowing in the branch XY of the circuit shown in Fig. 2.76 (a) by superposition theorem.* **[Nagpur University, April 1996]**

Solution. As shown in Fig. 2.76 (*b*), one source is de-activated. Through series-parallel combinations of resistances, the currents due to this source are calculated. They are marked as on Fig. 2.76 (*b*).

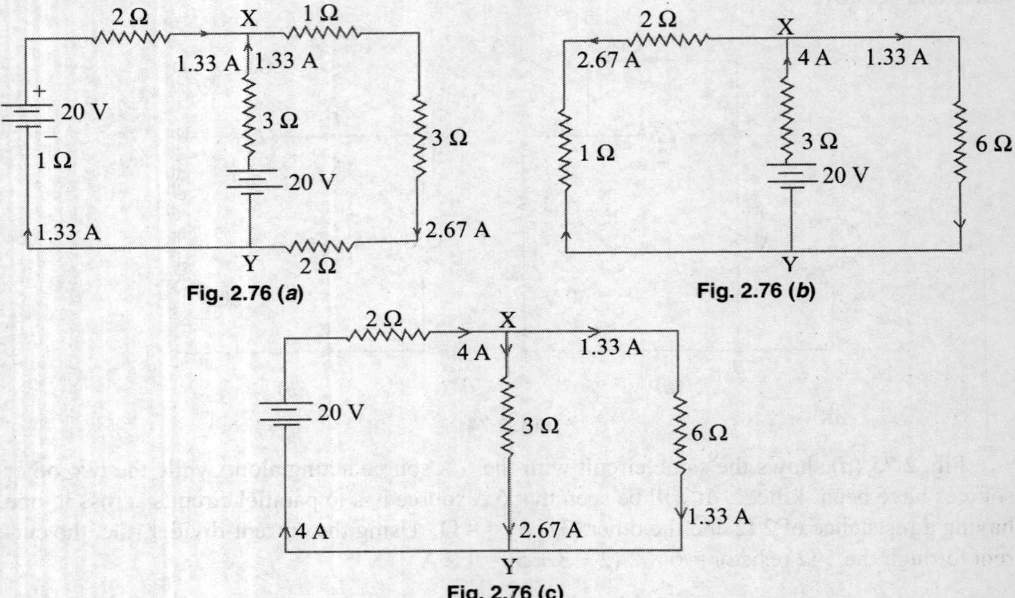

Fig. 2.76 (*a*)

Fig. 2.76 (*b*)

Fig. 2.76 (c)

In the next step, second source is de-activated as in Fig. 2.76 (*c*). Through simple series parallel resistances combinations, the currents due to this source are marked on the same figure.

According to the superposition theorem, the currents due to both the sources are obtained after adding the individual contributions due to the two sources, with the final results marked on Fig. 2.76 (*a*). Thus, the current through the branch *XY* is 1.33 A from *Y* to *X*.

Example 2.40. *Apply Superposition theorem to the circuit of Fig. 2.77. Find the current in 20 Ω resistor of the circuit.* **[UTU; 2020 RGPV 2016]**

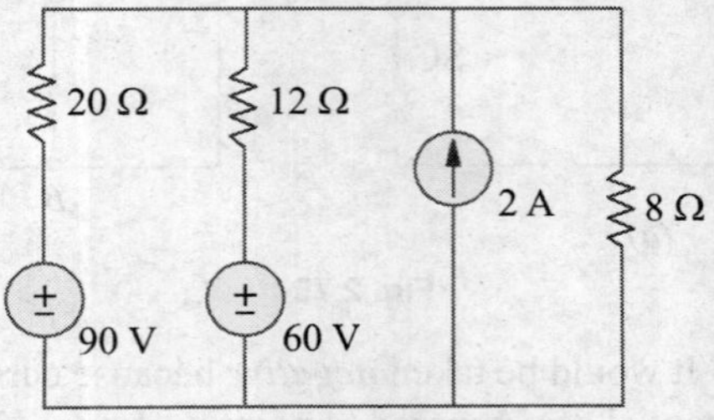

Fig. 2.77

Solution. Superposition theorem needs one source acting at a time.

Step I: Consider 90 V source alone, short 60 V source and open 2 A source as shown in the Fig. 2.77 (*a*)

Calculate the total resistance and total current in Fig. 2.78 (*a*) and (*b*).

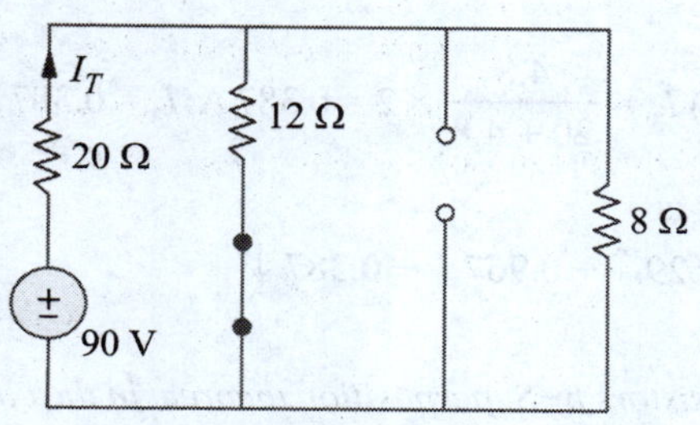

Fig. 2.78 (*a*)

Fig. 2.78 (*b*)

$$R_T = 20 + 12 \parallel 8 = 20 + 4.8$$

$$R_T = 24.8\ \Omega$$

$$I_T = \frac{V}{R_T} = \frac{90}{24.8} = 3.629\ \text{A}$$

$$I_T = 3.6629\ \text{A}$$

Current I_1 due to 90 V through 20 Ω is given by $I_T = I_1$

Step II: Now consider 60 V source alone. Short 90 V source and open 2 A current source as shown in Fig. 2.79 (*c*).

Calculate total resistane and total current from the Fig. (*c*)

$$R_T = 12 + 20 \parallel 8 = 17.71\ \Omega$$

$$R_T = 17.71\ \Omega$$

$$I_T = \frac{V}{R_T} = \frac{60}{17.71} = 3.387\ \text{A}$$

$$I_T = 3.387\ \text{A}$$

Fig. 2.79 (*c*)

Current I_2 due to 60 V through 20 Ω is given by current divide rule.

$$I_2 = \frac{8}{8+20} \times I_T = \frac{8}{28} \times 3.387 = 0.967\ \text{A} \downarrow$$

Step III: Consider 2A current source alone and short the two voltage sources as shown in Fig. 2.80 (*a*) and (*b*)

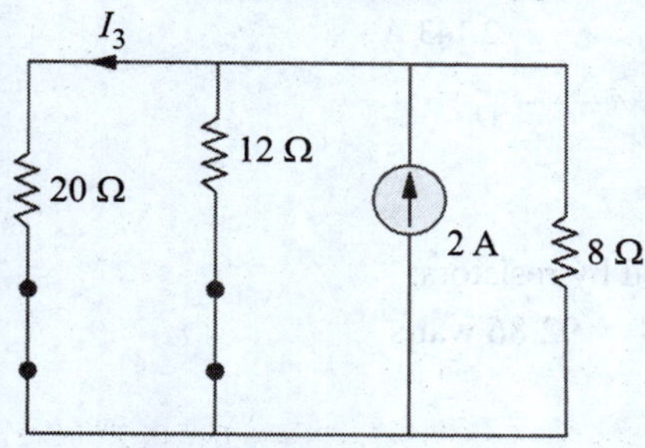

Fig. 2.80 (*a*)

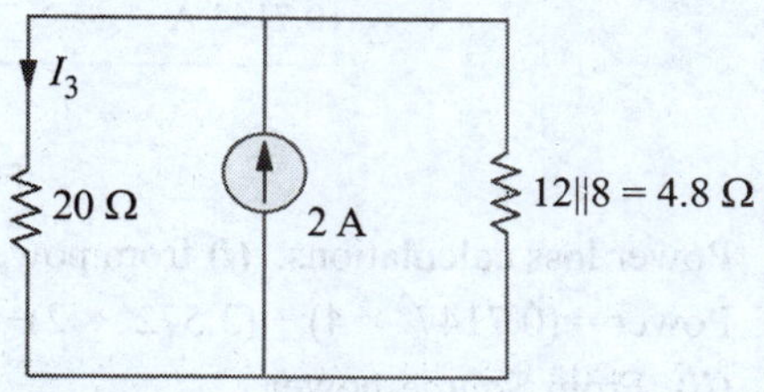

Fig. 2.80 (*b*)

$$R_T = 20 + 12 \parallel 8 = 20 + 4.8$$

$$R_T = 24.8\ \Omega$$

$$I_T = 2\text{A (Given)}$$

By current divided rule with the helf of Fig. (*e*) $I_3 = \dfrac{4.8}{20 + 4.8} \times 2 = 0.387\text{ A}$ $I_3 = 0.387\text{ A}\downarrow$

Step IV: Total current I through 20 Ω resistor is

$$I = I_1 + I_2 + I_3 = 3.629\uparrow + 0.967\downarrow + 0.387\downarrow$$

$$I = 2.274\text{ A}$$

Example 2.41. *Find the currents in all the resistors by Superposition theorem in the circuit shown in Fig. 2.81 (a). Calculate the power consumed.* **[Nagpur University, Nov. 1996]**

Solution. According to Superposition theorem, one source should be retained at a time, deactivating remaining sources. Contributions due to individual sources are finally algebraically added to get the answers required. Fig. 2.81 (*b*) shows only one source retained and the resultant currents in all branches/elements. In Fig. 2.81 (c), other source is shown to be in action, with concerned currents in all the elements marked.

To get the total current in any element, two component-currents in Fig. 2.81 (*b*) and Fig. 2.81 (*c*) for the element are to be algebraically added. The total currents are marked on Fig. 2.81 (*a*).

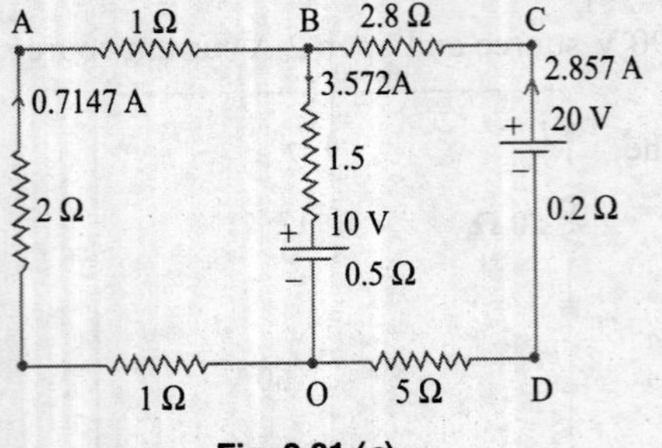

Fig. 2.81 (*a*)

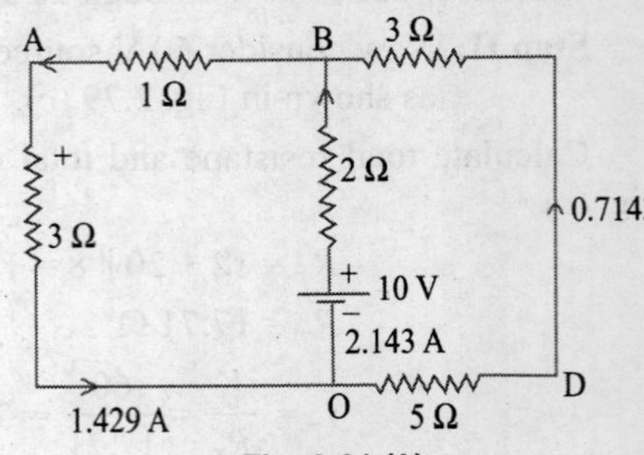

Fig. 2.81 (*b*)

All resistors are in ohms

Fig. 2.81 (*c*)

Power loss calculations. (*i*) from power consumed by resistors:

Power = $(0.7147^2 \times 4) + (3.572^2 \times 2) + (2.875^2 \times 8) = 92.86$ watts

(*ii*) From Source-power.

Power = $10 \times 3.572 + 20 \times 2.857 = 92.86$ watts

Tutorial Problem 2.4

1. Apply the principle of Superposition to the network shown in Fig. 2.82 to find out the current in the 10 Ω resistance. **[0.464 A]** ***(F.Y. Engg. Pune Univ.)***

2. Find the current through the 3 Ω resistance connected between *C* and *D* Fig. 2.83. **[1 A from C to D]** ***(F.Y. Engg. Pune Univ.)***

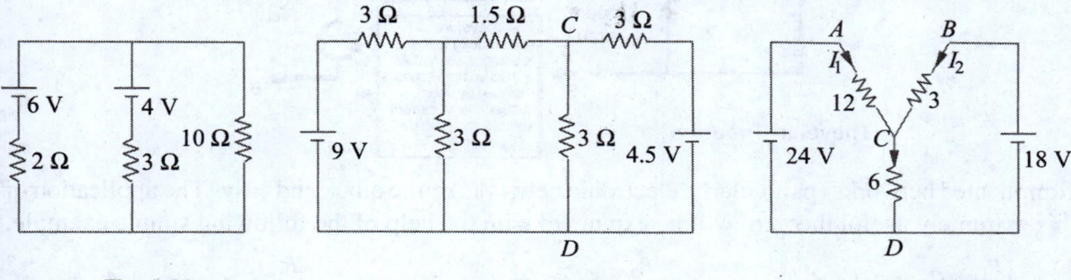

Fig. 2.82 **Fig. 2.83** **Fig. 2.84**

3. Using the Superposition theorem, calculate the magnitude and direction of the current through each resistor in the circuit of Fig. 2.85. **[I_1 = 6/7 A; I_2 = 10/7 A; I_3 = 16/7 A]**

4. For the circuit shown in Fig. 2.86 find the current in *R* = 8 Ω resistance in the branch *AB* using superposition theorem. **[0.875 A]** ***(F.Y. Engg. Pune Univ.)***

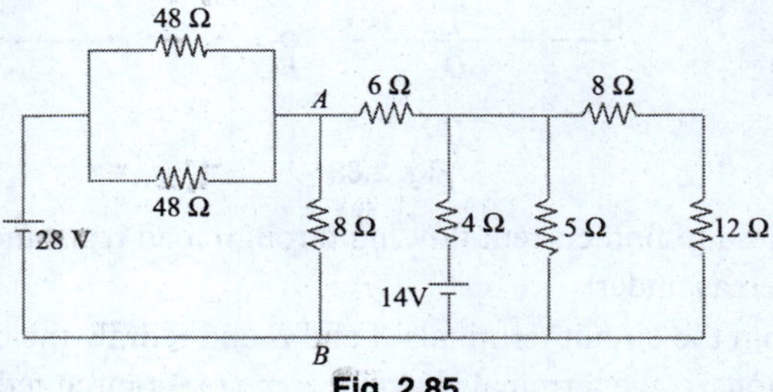

Fig. 2.85

5. Apply superposition principle to compute current in the 2-Ω resistor of Fig. 2.86. All resistors are in ohms. **[I_{ab} = 5 A]**

6. Use Superposition theorem to calculate the voltage drop across the 3 Ω resistor of Fig. 2.87. All resistance values are in ohms. **[18 V]**

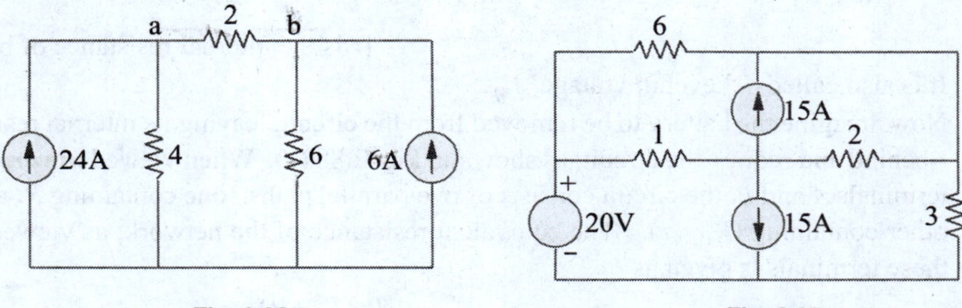

Fig. 2.86 **Fig. 2.87**

2.17 Thevenin Theorem

It provides a mathematical technique for replacing a given network, as viewed from two output terminals, by ***a single voltage source with a series resistance.*** It makes the solution of

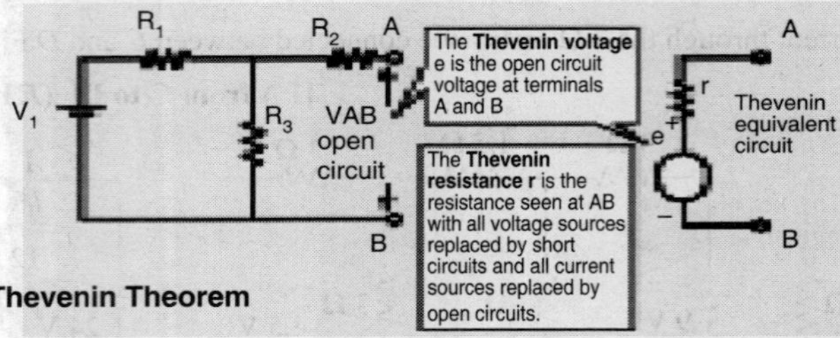

Thevenin Theorem

complicated networks (particularly, electronic networks) quite quick and easy. The application of this extremely useful theorem will be explained with the help of the following simple example.

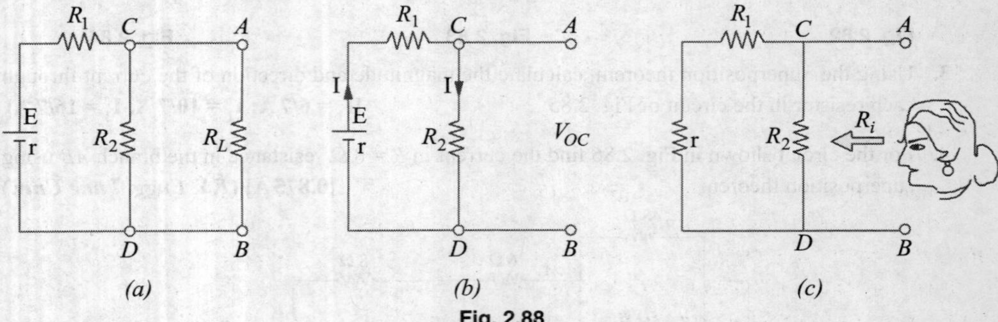

Fig. 2.88

Suppose, it is required to find current flowing through load resistance R_L, as shown in Fig. 2.88 (*a*). We will proceed as under:

1. Remove R_L from the circuit terminals *A* and *B* and redraw the circuit as shown in Fig. 2.89 (*b*). Obviously, the terminals have become open-circuited.

2. Calculate the open-circuit voltage V_{oc} which appears across terminals *A* and *B* when they are open *i.e.* when R_L is removed.

As seen, V_{oc} = drop across $R_2 = IR_2$ where I is the circuit current when *A* and *B* are open.

$$I = \frac{E}{R_1 + R_2 + r} \qquad \therefore \quad V_{oc} = IR_2 = \frac{ER_2}{R_1 + R_2 + r}$$

[*r* is the internal resistance of battery]

It is also called 'Thevenin voltage' V_{th}.

3. Now, imagine the battery to be removed from the circuit, leaving its internal resistance *r* behind and redraw the circuit, as shown in Fig. 2.88 (*c*). When viewed ***inwards*** from terminals *A* and *B*, the circuit consists of two parallel paths: one containing R_2 and the other containing $(R_1 + r)$. The equivalent resistance of the network, as viewed from these terminals is given as

$$R = R_2 \,\|\, (R_1 + r) = \frac{R_2(R_1 + r)}{R_2 + (R_1 + r)}$$

This resistance is also called,* Thevenin resistance R_{sh} (though, it is also sometimes written as R_i or R_0).

Consequently, as viewed from terminals A and B, the whole network (excluding R_1) can be reduced to a single source (called Thevenin's source) whose e.m.f. equals V_∞ (or V_{sh}) and whose internal resistance equals R_{sh} (or R_i) as shown in Fig. 2.89.

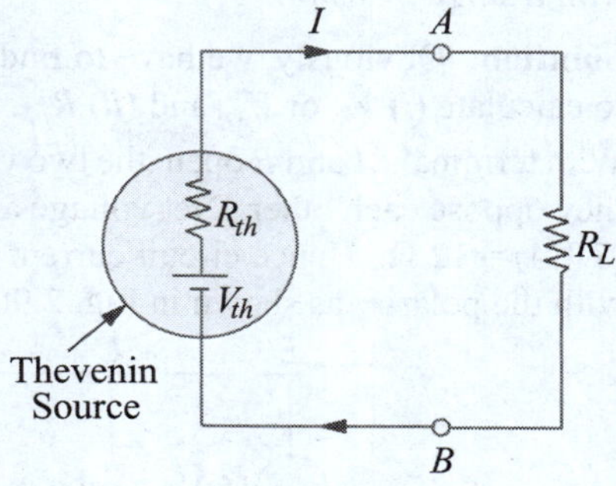

Fig. 2.89

4. R_L is now connected back across terminals A and B from where it was temporarily removed earlier. Current flowing through R_L is given by

$$I = \frac{V_{th}}{R_{th} + R_L}$$

It is clear from above that any network of resistors and voltage sources (and current sources as well) when viewed from any points A and B in the network, can be replaced by a single voltage source and a single resistance** in series with the voltage source.

After this replacement of the network by a single voltage source with a series resistance has been accomplished, it is easy to find current in any load resistance joined across terminals A and B. This theorem is valid even for those linear networks which have a nonlinear load.

Hence, Thevenin's theorem, as applied to d.c. circuits, may be stated as under:

The current flowing through a load resistance R_L connected across any two terminals A and B of a linear, active bilateral network is given by $V_{oc} \parallel (R_i + R_L)$ where V_{oc} is the open-circuit voltage (i.e. voltage across the two terminals when R_L is removed) and R_i is the internal resistance of the network as viewed back into the open-circuited network from terminals A and B with all voltage sources replaced by their internal resistance (if any) and current sources by infinite resistance.

2.18 How to Thevenize a Given Circuit ?

1. Temporarily remove the resistance (called load resistance R_L) whose current is required.
2. Find the open-circuit voltage V_{oc} which appears across the two terminals from where resistance has been removed. It is also called Thevenin voltage V_{th}.
3. Compute the resistance of the whole network as looked into from these two terminals after all voltage sources have been removed leaving behind their internal resistances (if any) and current sources have been replaced by open-circuit *i.e.* infinite resistance. It is also called Thevenin resistance R_{th} or T_t.
4. Replace the entire network by a single Thevenin source, whose voltage is V_{th} or V_{oc} and whose internal resistance is R_{th} or R_i.
5. Connect R_L back to its terminals from where it was previously removed.
6. Finally, calculate the current flowing through R_L by using the equation,

$$I = V_{th}/(R_{th} + R_L) \quad \text{or} \quad I = V_{oc}/(R_i + R_L)$$

* After the French engineer M.L. Thevenin (1857-1926) who while working in Telegraphic Department published a statement of the theorem in 1893.

** Or impedance in the case of a.c. circuits.

Example 2.42. *Convert the circuit shown in Fig. 2.90 (a), to a single voltage source in series with a single resistor.* **(AMIE Sec. B, Network Analysis Summer 1992)**

Solution. Obviously, we have to find equivalent Thevenin circuit. For this purpose, we have to calculate ***(i)*** V_{th} or V_{AB} and ***(ii)*** R_{th} or R_{AB}.

With terminals *A* and *B* open, the two voltage sources are connected in subtractive series because they oppose each other. Net voltage around the circuit is (15 − 10) = 5 V and total resistance is (8 + 4) = 12 Ω. Hence circuit current is = 5/12 A. Drop across 4 Ω resistor = 4 × 5/12 = 5/3 V with the polarity as shown in Fig. 2.90 (*a*).

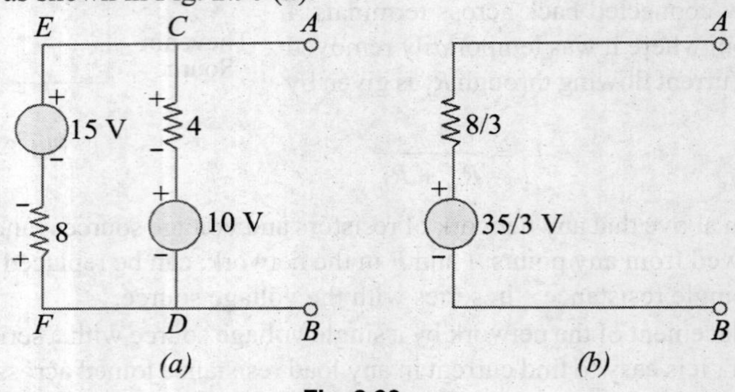

Fig. 2.90

$\therefore \quad V_{AB} = V_{th} = +10 + 5/3 = 35/3$ V.

Incidently, We could also find V_{AB} while going along the parallel route *BFEA*.

Drop across 8 Ω resistor = 8 × 5/12 = 10/3 V. V_{AB} equal the algebraic sum of voltages met on the way from *B* to *A*. Hence, $V_{AB} = (-10/3) + 15 = 35/3$ V.

As shown in Fig. 2.90 (*b*), the single voltage source has a voltage of 35/3 V.

For finding R_{th}, we will replace the two voltage sources by short-circuits. In that case, $R_{th} = R_{AB} = 4 \parallel 8 = 8/3$ Ω.

Example 2.43. *State Thevenin's theorem and give a proof. Apply this theorem to calculate the current through the 4 Ω resistor of the circuit of Fig. 2.91 (a).*

(A.M.I.E. Sec. B Network Analysis W.)

Solution. As shown in Fig. 2.91 (*b*), 4 Ω resistance has been removed there by open-circuiting the terminals *A* and *B*. We will now find V_{AB} and R_{AB} which will give us V_{th} and R_{th} respectively. The potential drop across 5 Ω resistor can be found with the help of voltage-divider rule. Its value is = 15 × 5/(5 + 10) = 5 V.

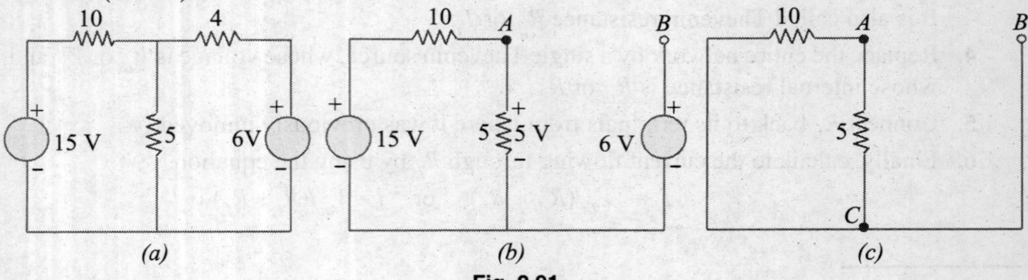

Fig. 2.91

For finding V_{AB}, we will go from point *B* to point *A* in the clockwise direction and find the algebraic sum of the voltages met on the way.

$\therefore \qquad V_{AB} = -6 + 5 = -1$ V.

It means that point *A* is negative with respect to point *E*, or point *B* is at a higher potential than point *A* by one volt.

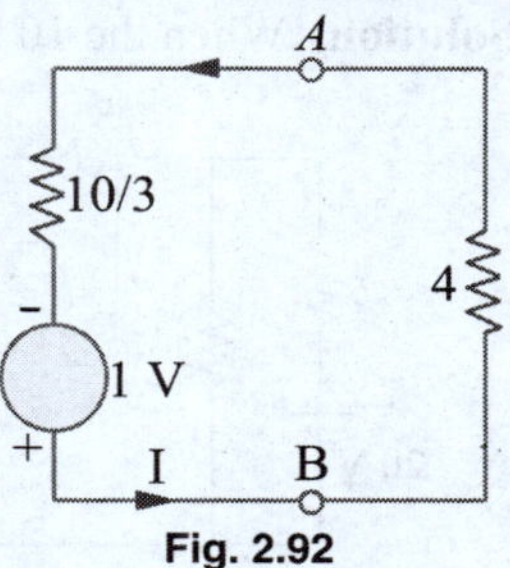

Fig. 2.92

In Fig. 2.91 (*c*), the two voltage source have been short-circuited. The resistance of the network as viewed from points *A* and *B* is the same as viewed from points *A* and *C*.

$\therefore \qquad R_{AB} = R_{AC} = 5 \parallel 10 = 10/3\ \Omega$

Thevenin's equivalent source is shown in Fig. 2.92 in which 4 Ω resistor has been joined back across terminals *A* and *B*. Polarity of the voltage source is worth nothing.

$\therefore \qquad I = \dfrac{1}{(10/3)+4} = \dfrac{3}{22} = 0.136$ A From *E* to *A*

Example 2.44. *With reference to the network of Fig. 2.93 (a), by applying Thevenin's theorem find the following:*

(*i*) *the equivalent e.m.f. of the network when viewed from terminals A and B.*

(*ii*) *the equivalent resistance of the network when looked into from terminals A and B.*

(*iii*) *current in the load resistance R_L of 15 Ω.* **(Basic Circuit Analysis, Nagpur Univ. 1993)**

Solution. (*i*) Current in the network before load resistance is connected [Fig. 2.94 (*a*)]

$$= 24/(12 + 3 + 1) = 1.5 \text{ A}$$

$\therefore$ voltage across terminals $AB = V_{oc} = V_{th} = 12 \times 1.5 =$ **18 V**

Hence, so far as terminals *A* and *B* are concerned, the network has an e.m.f. of 18 volt (and not 24 V).

(*ii*) There are two parallel paths between points *A* and *B*. Imagine that battery of 24 V is removed but not its internal resistance. Then, resistance of the circuit as looked into from point *A* and *B* is [Fig. 2.93 (*c*)]

$$R_i = R_{th} = 12 \times 4/(12 + 4) = 3\ \Omega$$

(*iii*) When load resistance of 15 Ω is connected across the terminals, the network is reduced to the structure shown in Fig. 2.93 (*d*).

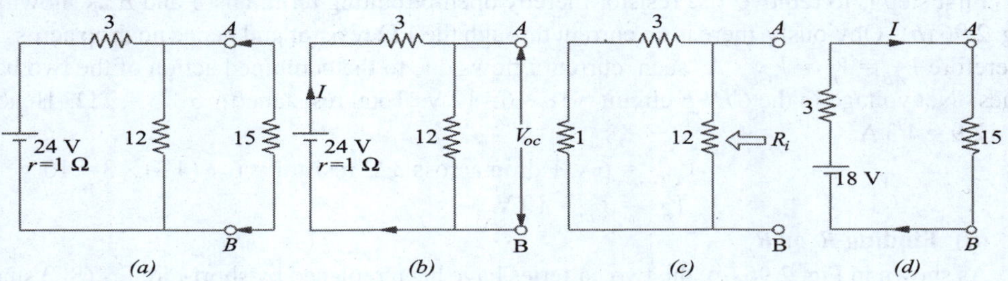

Fig. 2.93

$$I = V_{th}/(R_{th} + R_L) = 18/(15 + 3) = \mathbf{1\ A}$$

Example 2.45. *For the circuit shown in Fig. 2.94 (a), calculate the current in the 10 ohm resistance. Use Thevenin's theorem only.* **(Elect. Science-I Allahabad Univ. 1992)**

Solution. When the 10 Ω resistance is removed, the circuit becomes as shown in Fig. 2.94 (*b*).

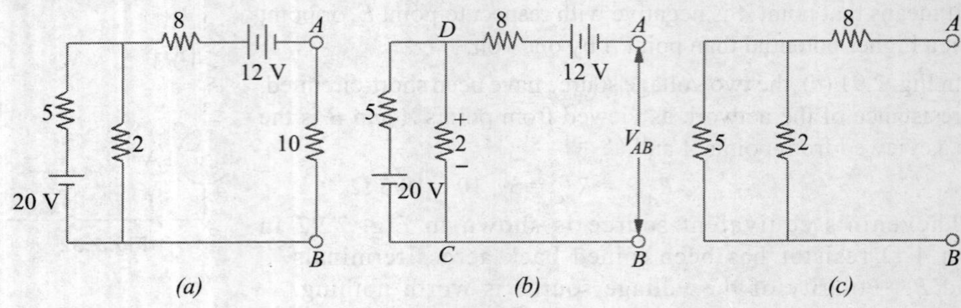

Fig. 2.94

Now, we will find the open-circuit voltage $V_{AB} = V_{th}$. For this purpose, we will go from point *B* to point *A* and find the algebraic sum of the voltages met on the way. It should be noted that with terminals *A* and *B* open, there is no voltage drop on the 8 Ω resistance. However the two resistances of 5 Ω and 2 Ω are connected in series across the 20-V battery. As per voltage-divider rule, drop on 2 Ω resistance = 20 × 2/(2 + 5) = 5.71 V with the polarity as shown in figure. As per the sign convention of Art.

$$V_{AB} = V_{th} = +5.71 - 12 = -6.29 \text{ V}$$

The negative sign shows that point *A* is negative with respect to point *B* or which is the same thing, point *B* is positive with respect to point *A*.

For finding $R_{AB} = R_{th}$, we replace the batteries by short-circuits as shown in Fig. 2.95.

$$\therefore \quad R_{AB} = R_{th} = 8 + 2 \parallel 5 = 9.43\ \Omega$$

Fig. 2.95

Hence, the equivalent Thevenin's source with respect to terminals *A* and *B* is as shown in Fig. 2.95. When 10 Ω resistance is reconnected across *A* and *B*, current through it is *I* = 6.24/(9.43 + 10) = **0.32 A.**

Example 2.46. *Using Thevenin's theorem, calculate the p.d. across terminals A and B in Fig. 2.96 (a).*

Solution. (*i*) Finding V_{oc}

First step is to remove 7 Ω resistor thereby open-circuiting terminals *A* and *B* as shown in Fig. 2.96 (*b*). Obviously, there is no current through the 1 Ω resistor and hence no drop across it. Therefore $V_{AB} = V_{oc} = V_{CD}$. As seen, current *I* flows due to the combined action of the two batteries. Net voltage in the *CDFE* circuit = 18 − 6= 12 V. Total resistance = 6 + 3 = 9 Ω. Hence, *I* = 12/9 = 4/3 A.

$$V_{CD} = 6 \text{ V} + \text{drop across } 3\ \Omega \text{ resistor} = 6 + (4/3) \times 3 = 10 \text{ V*}$$

$$\therefore \quad V_{oc} = V_{th} = 10 \text{ V.}$$

(*ii*) Finding R_i or R_{th}

As shown in Fig. 2.96 (*c*), the two batteries have been replaced by short-circuits (SC) since their internal resistances are zero. As seen, $R_i = R_{th} = 1 + 3 \parallel 6 = 3\ \Omega$. The Thevenin's equivalent circuit is as shown in Fig. 2.96 (*d*) where the 7 Ω resistance has been reconnected across terminals *A* and *B*. The p.d. across this resistor can be found with the help of Voltage Divider Rule.

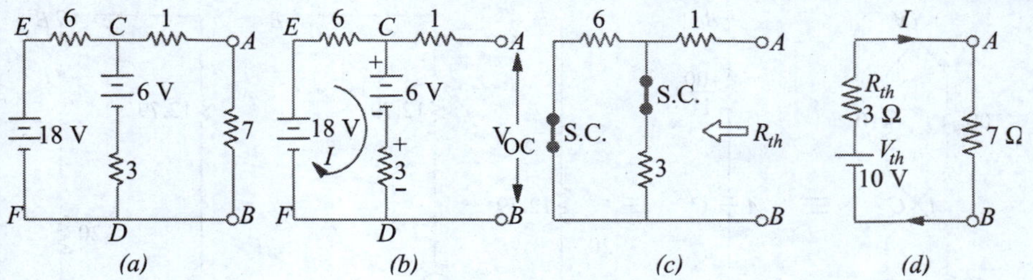

Fig. 2.96

Example 2.47. *The four arms of a Wheatstone bridge have the following resistances:*

AB = 100, BC = 10, CD = 4, DA = 50 Ω. A galvanometer of 20 Ω resistance is connected across BD. Use Thevenin's theorem to compute the current through the galvanometer when a p.d. of 10 V is maintained across AC. **(Elect. Technology, Vikram Univ. of Ujjain)**

Solution. **(*i*)** When galvanometer is removed from Fig. 2.97 (*a*), we get the circuit of Fig. 2.97 (*b*).

(*ii*) Let us next find the open-circuit voltage V_{oc} (also called Thevenin voltage V_{th}) between points *B* and *D*. Remembering that *ABC* (as well as *ADC*) is a potential divider on which a voltage drop of 10 V takes place, we get

Potential of *B* w.r.t. *C* = 10 × 10/110 = 10/11 = 0.909 V

Potential of *D* w.r.t. *C* = 10 × 4/54 = 20/27 = 0.741 V

∴ p.d. between *B* and *D* is V_{oc} or V_{th} = 0.909 − 0.741 = 0.168 V

(*iii*) Now, remove the 10-V battery retaining its internal resistance which, in this case, happens to be zero. Hence, it amounts to short-circuiting points *A* and *C* as shown in Fig. 2.97 (*d*).

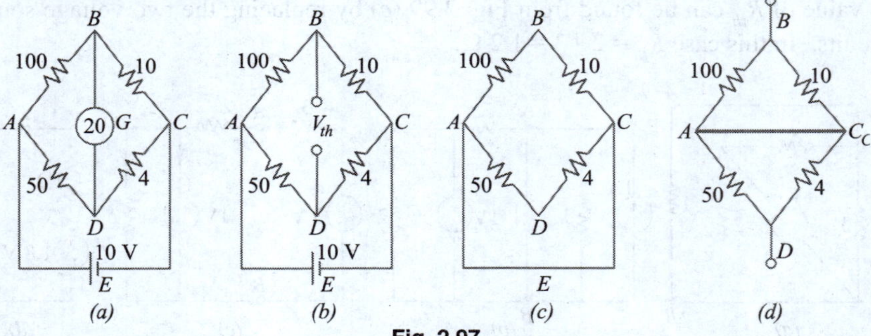

Fig. 2.97

(*iv*) Next, let us find the resistance of the whole network as viewed from points *B* and *D*. It may be easily found by noting that electrically speaking, points *A* and *C* have become one as shown in Fig. 2.98 (*a*). It is also seen that *BA* is in parallel with *BC* and *AD* is in parallel with *CD*. Hence, R_{BD} = 10 || 100 + 50 || 4 = 12.79 Ω.

(*v*) Now, so far as points *B* and *D* are connected, the network has a voltage source of 0.168 V and internal resistance R_i = 12.79 Ω. This Thevenin's source is shown in Fig. 2.98 (*c*).

(*vi*) Finally, let us connect the galvanometer (initially removed) to this Thevenin source and calculate the current *I* flowing through it. As seen from Fig. 2.98 (*d*).

$$I = 0.168/(12.79 + 20) = 0.005\text{ A} = \mathbf{5\text{ mA}}$$

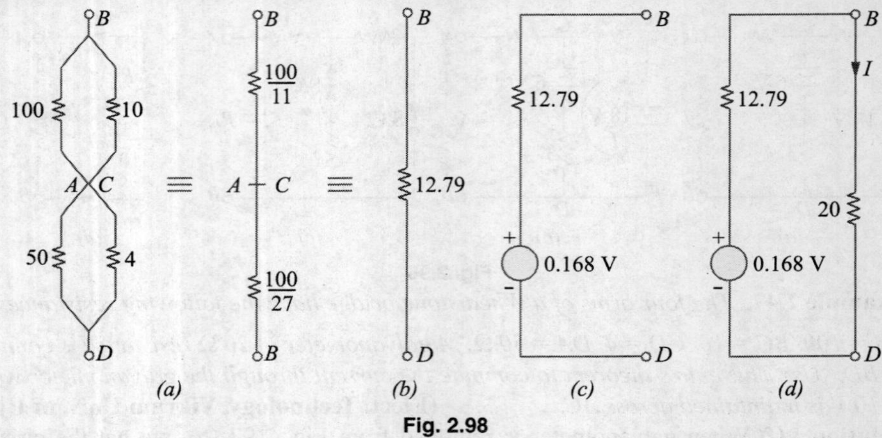

Fig. 2.98

Example 2.48. *Determine the current in the 1 Ω resistor across AB of network shown in Fig. 2.99 (a) using Thevenin's theorem.* **(Network Analysis, Nagpur Univ. 1993)**

Solution. The given circuit can be redrawn, as shown in Fig. 2.99 (*b*) with the 1 Ω resistor removed from terminals *A* and *B*. The current source has been converted into its equivalent voltage source as shown in Fig. 2.99 (*c*). For finding V_{th}, we will find the currents *x* and *y* in Fig. 2.99 (*c*). Applying *KVL* to the first loop, we get

$$3 - (3 + 2)\,x - 1 = 0 \quad \text{or} \quad x = 0.4 \text{ A}$$

$$\therefore \qquad V_{th} = V_{AB} = 3 - 3 \times 0.4 = 1.8 \text{ V}$$

The value of R_{th} can be found from Fig. 2.99 (*c*) by replacing the two voltage sources by short-circuits. In this case $R_{th} = 2 \parallel 3 = 1.2\ \Omega$.

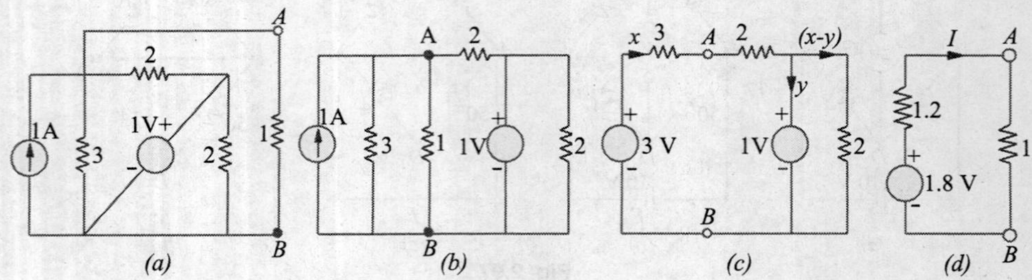

Fig. 2.99

Thevenin's equivalent circuit is shown in Fig. 2.99 (*d*). The current through the reconnected 1 Ω resistor is = 1.8/(12.1 + 1) = 0.82 A.

Example 2.49. *Find the current flowing through the 4 Ω resistor in Fig. 2.100 (a) when (i) E = 2 V and (ii) E = 12 V. All resistances are in series.*

Solution. When we remove *E* and 4 Ω resistor, the circuit becomes as shown in Fig. 2.100 (*b*). For finding R_{th} *i.e.* the circuit resistance as viewed from terminals *A* and *B*, the battery has been short-circuited, as shown. It is seen from Fig. 2.100 (*c*) that $R_{th} = R_{AB} = 15 \parallel 30 + 18 \parallel 9 = 16\ \Omega$.

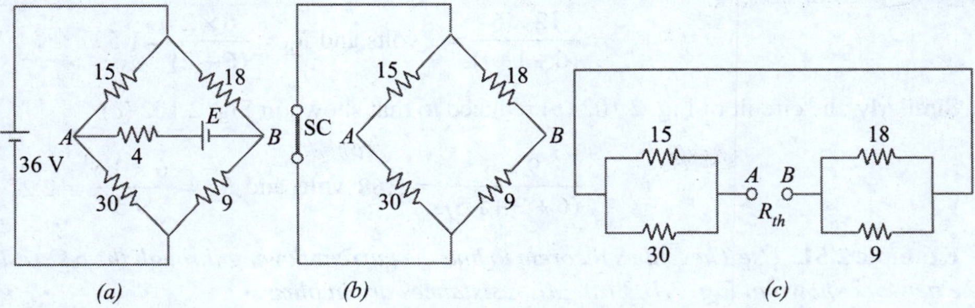

Fig. 2.100

We will find $V_{th} = V_{AB}$ with the help of Fig. 2.101 (*a*) which represents the original circuit, except with *E* and 4 Ω resistor removed. Here, the two circuits are connected in parallel across the 36 V battery. The potential of point *A* equals the drop on 30 Ω resistance, whereas potential of point *B* equals the drop across 9 Ω resistance. Using the voltage, divider rule, we have

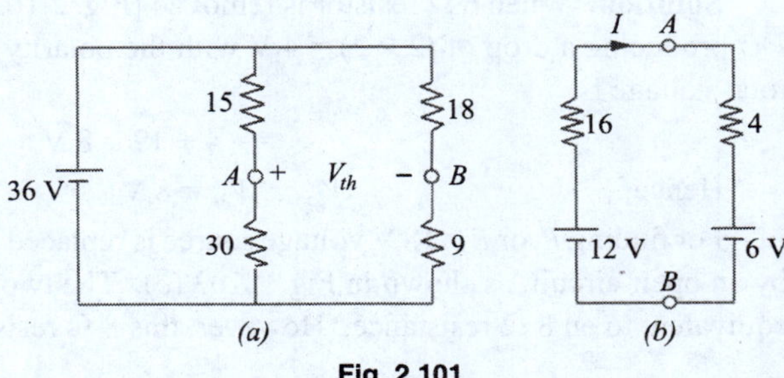

Fig. 2.101

$$V_A = 30 \times 30/45 = 24 \text{ V}$$

$$V_B = 36 \times 9/27 = 12 \text{ V}$$

$$\therefore \quad V_{AB} = V_A - V_B = 24 - 12 = 12 \text{ V}$$

In Fig. 2.101 (*b*), the series combination of *E* and 4 Ω resistors has been reconnected across terminals *A* and *B* of the Thevenin's equivalent circuit.

(*i*) $I = (12 - E)/20 = (12 - 2)/20 = 0.5$ A (*ii*) $I = (12 - 12)/20 = 0$

Example 2.50. *Determine Thevenin's equivalent circuit which may be used to represent the given network (Fig. 2.102) at the terminals AB.* **(Electrical Eng.; Calcutta Univ.)**

Solution. The given circuit of Fig. 2.102 (*a*) would be solved by applying Thevenin's theorem twice, first to the circuit to the left of point *C* and *D* and then to the left of points *A* and *B*. Using this technique, the network to the left of *CD* [Fig. 2.102 (*a*)] can be replaced by a source of voltage V_1 and series resistance R_{i1} as shown in Fig. 2.102 (*b*).

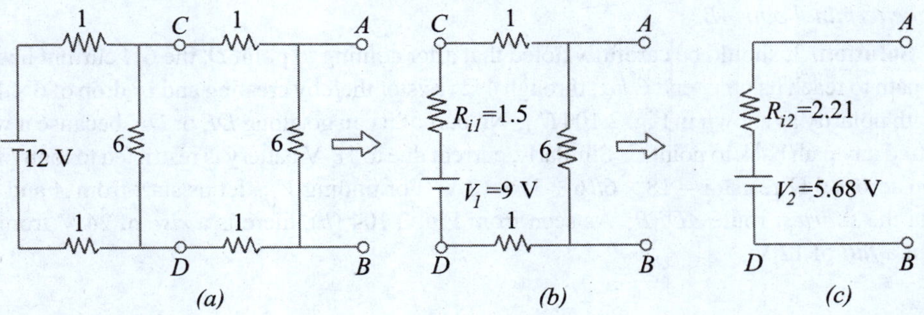

Fig. 2.102

$$V_1 = \frac{12\times 6}{(6+1+1)} = 9 \text{ volts and } R_{i1} = \frac{6\times 2}{(6+2)} = 1.5\ \Omega$$

Similarly, the circuit of Fig. 2.102 (*b*) reduced to that shown in Fig. 2.102 (*c*)

$$V_2 = \frac{9\times 6}{(6+2+1.5)} = \mathbf{5.68} \text{ volts and } R_{i2} = \frac{6\times 3.5}{9.5} = 2.21\ \Omega$$

Example 2.51. *Use Thevenin's theorem to find the current flowing through the 6 Ω resistor of the network shown in Fig. 2.103 (a). All resistances are in ohms.*

(Network Theory, Nagpur Univ. 1992)

Solution. When 6 Ω resistor is removed [Fig. 2.103 (*b*)], whole of 2 A current flows along *DC* producing a drop of (2 × 2) = 4 V with the polarity as shown. As we go along *BDCA*, the total voltage is

$= -4 + 12 = 8$ V —with *A* positive w.r.t. *B*.

Hence, $V_{oc} = V_{th} = 8$ V

For finding R_i or R_{th} 18 V voltage source is replaced by a short-circuit and the current source by an open-circuit, as shown in Fig. 2.103 (*c*). The two 4 Ω resistors are in series and are thus equivalent to an 8 Ω resistance. However, this 8 Ω resistor is in parallel with a short of 0 Ω.

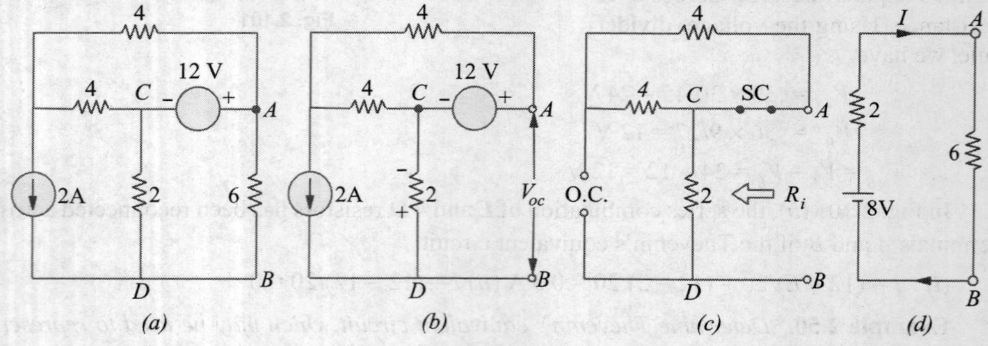

Fig. 2.103

Hence, their equivalent value is 0 Ω. Now this 0 Ω resistance is in series with the 2 Ω resistor. Hence, $R_i = 2 + 0 = 2$ Ω. The Thevenin's equivalent circuit is shown in Fig. 2.103 (*d*).

∴ $I = 8/(2 + 6) =$ **1 Amp** —from *A* to *B*

Example 2.52. *Find Thevenin's equivalent circuit for the network shown in Fig. 2.104 (a) for the terminal pair AB.*

Solution. It should be carefully noted that after coming to point *D*, the 6 *A* current has only one path to reach its other end *C i.e.,* through 4 Ω resistor thereby creating and *IR* drop of 6 × 4 = 24 V with polarity as shown in Fig. 2.104 (*b*). No part of it can go along *DE* or *DF* because it would not find any path back to point *C*. Similarly, current due to 18-V battery is restricted to loop *EDFE*. Drop across 6 Ω resistor = 18 × 6/(6 + 3) = 12 V. For finding V_{AB}, let us start from *A* and go to *B* via the *shortest* route *ADFB*. As seen from Fig. 2.104 (*b*), there is a *rise* of 24 V from *A* to *D* but a ***fall*** of 12 V.

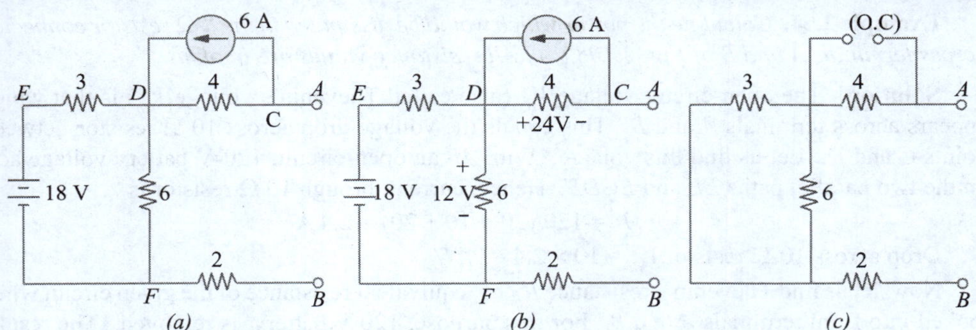

Fig. 2.104

from D to F. Hence, $V_{AB} = 24 - 12 = 12$ V with point A negative w.r.t. point B. Hence, $V_{th} = V_{AB} = -12$ V (or $V_{BA} = 12$ V).

For finding R_{th}, 18 V battery has been replaced by a short-circuit and 6 A current source by an open-circuit, as shown in Fig. 2.104 (*c*).

As seen,
$$R_{th} = 4 + 6 \,\|\, 3 + 2$$
$$= 4 + 2 + 2 = 8\ \Omega$$

Hence, Thevenin's equivalent circuit for terminals A and B is as shown in Fig. 2.105. It should be noted that if a load resistor is connected across AB, current through it will flow from B to A.

Fig. 2.105

Example 2.53. *The circuit shown in Fig. 2.106 (a) contains two voltage sources and two current sources. Calculate (a) V_{th} and (b) R_{th} between the open terminals A and B of the circuit. All resistance values are in ohms.*

Solution. It should be understood that since terminals A and B are open, 2 A current can flow only through 4 Ω and 10 Ω resistors, thus producing a drop of 20 V across the 10 Ω resistor, as shown in Fig. 2.106 (*b*). Similarly, 3 A current can flow through its own closed circuit between A and C thereby producing a drop of 24 V across 8 Ω resistor as shown in Fig. 2.106 (*b*). Also, there is no drop across 2 Ω resistor because no current flows through it.

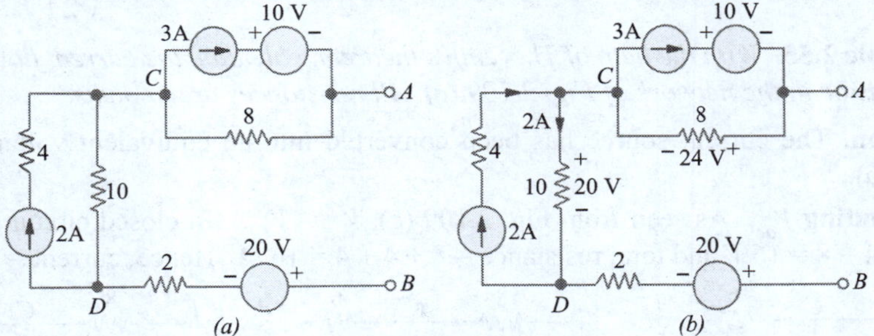

Fig. 2.106

Starting from point B and going to point A via points D and C, we get

$$V_{th} = -20 + 20 + 24 = 24 \text{ V} \qquad \text{—with point } A \text{ positive.}$$

For finding R_{th}, we will short-circuit the voltage sources and open-circuit the current sources, as shown in Fig. 2.107. As seen, $R_{th} = R_{AB} = 8 + 10 + 2 = 20\ \Omega$.

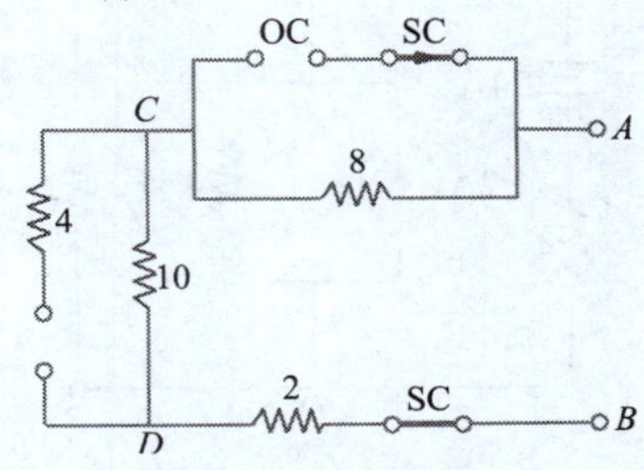

Fig. 2.107

Example 2.54. *Calculate the power which would be dissipated in the 8-Ω resistor connected across terminals A and B of Fig. 2.108 (a). All resistance values are in ohms.*

Solution. The open-circuit voltage V_{oc} (also called Thevenin's voltage V_{th}) is that which appears across terminals *A* and *B*. This equals the voltage drop across 10 Ω resistor between points *C* and *D*. Let us find this voltage. With *AB* an open-circuit, 120-V battery voltage acts on the two parallel paths *EF* and *ECDF*. Hence, current through 10 Ω resistor is

$$I = 120/(20 + 10 + 20) = 2.4 \text{ A}$$

Drop across 10-Ω resistor, $V_{th} = 10 \times 2.4 = 24$ V

Now, let us find Thevenin's resistance R_{th} *i.e.* equivalent resistance of the given circuit when looked into from terminals *A* and *B*. For this purpose, 120 V battery is removed. The results in shorting the 40-Ω resistance since internal resistance of the battery is zero as shown in Fig. 2.108 (*b*).

$$\therefore \quad R_i \text{ or } R_{th} = 16 + \frac{10 \times (20 + 20)}{10 + (20 + 20)} + 16 = 40\ \Omega$$

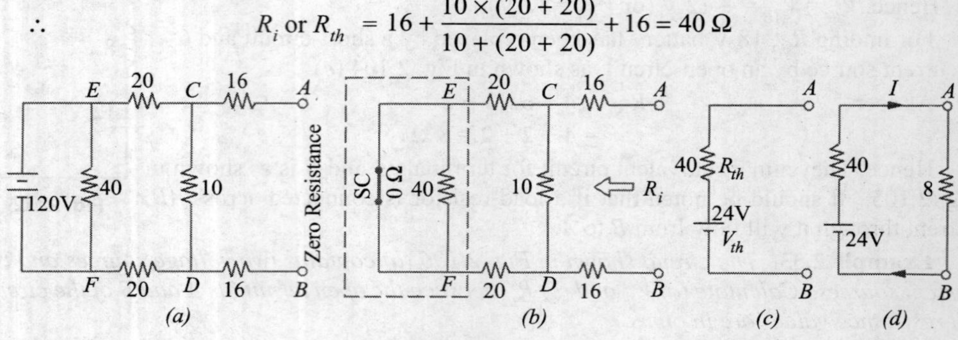

Fig. 2.108

Thevenin's equivalent circuit is shown in Fig. 2.108 (*c*). As shown in Fig. 2.108 (*d*), current through 8-Ω resistor is

$$I = 24/(40 + 8) = \frac{1}{2} \text{ A} \quad \therefore \quad P = I^2 R = \left(\frac{1}{2}\right)^2 \times 8 = \mathbf{2\ W}$$

Example 2.55. *With the help of Thevenin's theorem, calculate the current flowing through the 3-Ω resistor in the network of Fig. 2.109 (a). All resistances are in ohms.*

Solution. The current source has been converted into an equivalent voltage source in Fig. 2.109 (*b*).

(*i*) Finding V_{oc}. As seen from Fig. 2.109 (c), $V_{oc} = V_{CD}$. In closed circuit *CDFEC*, net voltage = 24 − 8 = 16 *V* and total resistance = 8 + 4 + 4 = 16 Ω. Hence, current = 16/16 = 1 A.

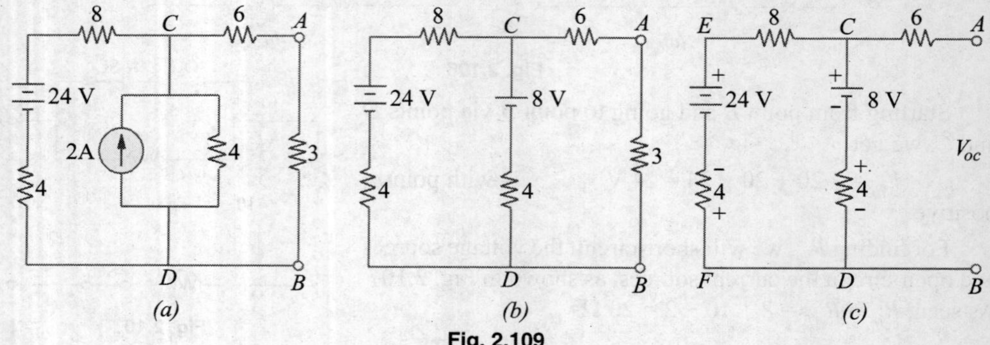

Fig. 2.109

Drop over the 4-Ω resistor in branch $CD = 4 \times 1 = 4$ V with a polarity which is in series addition with 8-V battery.

Hence, $V_{oc} = V_{th} = V_{CD} = 8 + 4 = 12$ V

***(ii)* Finding R_i or R_{th}.** In Fig. 2.110 (*a*), the two batteries have been replaced by short-circuits because they do not have any internal resistance.

As seen, $R_i = 6 + 4 \parallel (8 + 4) = 9\ \Omega$.

The Thevenin's equivalent circuit is as shown in Fig. 2.110 (*b*).

$I = 12/(9 + 3) = \mathbf{1\ A}$

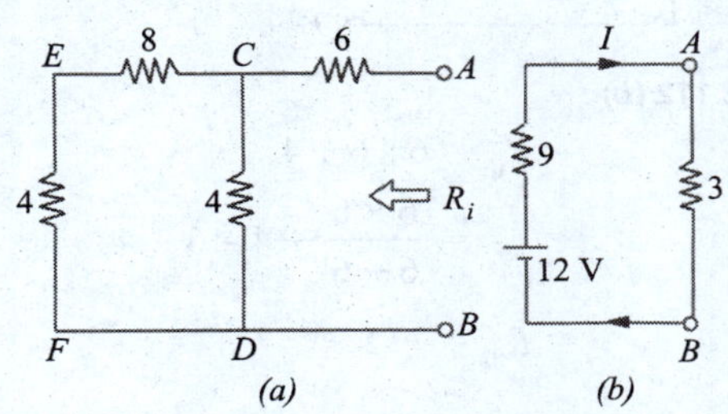

Fig. 2.110

Example 2.56. *Find V_{th} and R_{th} for a aircuit as shown in Fig 2.111. By theverin's theorem.*

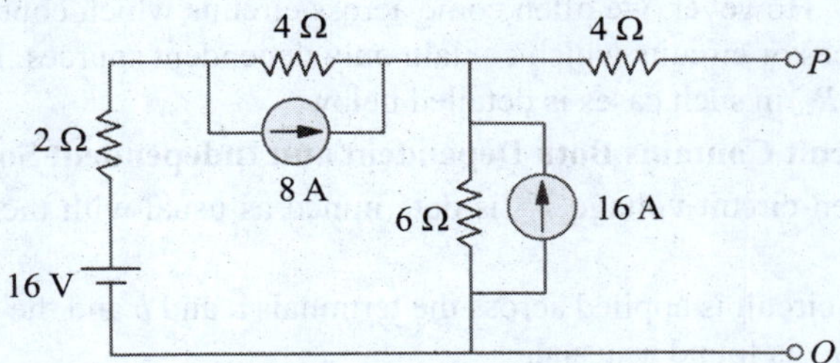

Fig. 2.111

Solution. Step I: Calculate the Vth across *P-Q* form the Fig. 2.112 (*a*) convert the current source into voltage source.

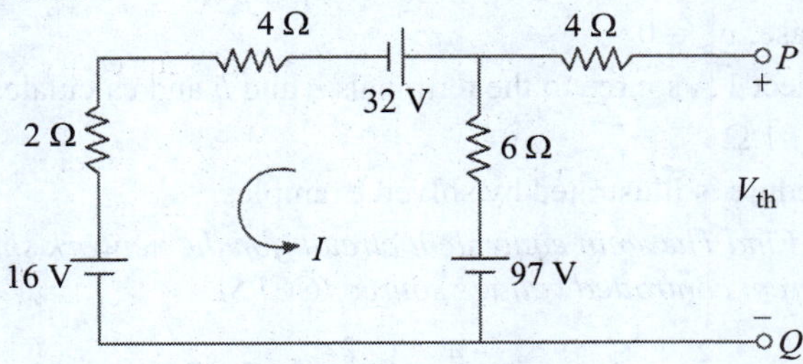

Fig. 2.112 (*a*)

$$V_{th} = V_{PQ}$$

Aplying KVL to the loop

$$96 - 6I - 32 - 4I - 2I - 16 = 0$$

$$12I = 48 \qquad I = 4\text{A}$$

Now the path from *P* to *Q*

$$V_{PQ} = 96 - 24 = 72\text{ V} = V_{th} \qquad V_{th} = 72\text{ V}$$

Stap II: Calculate the Thevenins Resistance R_{th} by shorting the voltage source and open the current source as shown in the Fig. 2.112 (*b*)

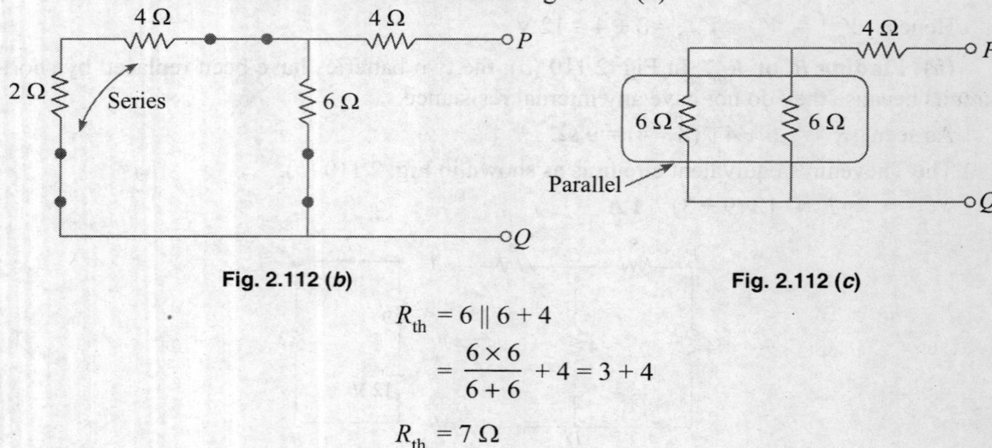

Fig. 2.112 (*b*) Fig. 2.112 (*c*)

$$R_{th} = 6 \parallel 6 + 4$$

$$= \frac{6 \times 6}{6 + 6} + 4 = 3 + 4$$

$$R_{th} = 7\ \Omega$$

2.19 General Instructions for Finding Thevenin Equivalent Circuit

So far, we have considered circuits which consisted of resistors and independent current or voltage sources only. However, we often come across circuits which contain both independent and dependent sources or circuits which contain only dependent sources. Procedure for finding the value of V_{th} and R_{th} in such cases is detailed below:

(*a*) When Circuit Contains Both Dependent and Independent Sources

(*i*) The open-circuit voltage V_{oc} is determined as usual with the sources activated or 'alive'.

(*ii*) A short-circuit is applied across the terminals *a* and *b* and the value of short-circuit current i_{sh} is found as usual.

(*iii*) Thevenin resistance $R_{th} = v_{oc}/i_{sh}$. It is the same procedure as adopted for Norton's theorem.

(*b*) When Circuit Contains Dependent Sources Only

(*i*) In this case, $v_{oc} = 0$

(*ii*) We connect 1 A source to the terminals *a* and *b* and calculate the value of v_{ab}.

(*iii*) $R_{th} = V_{ab}/ 1\ \Omega$

The above procedure is illustrated by solved examples.

Example 2.57. *Find Thevenin equivalent circuit for the network shown in Fig. 2.113 (a) which contains a current controlled voltage source (CCVS).*

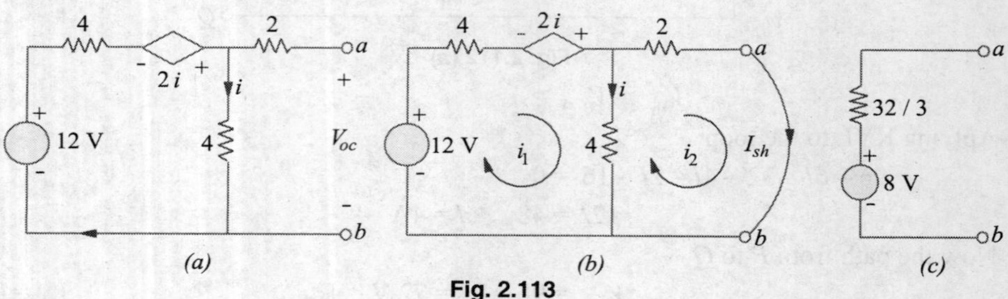

Fig. 2.113

Solution. For finding V_{oc} available across open-circuit terminals *a* and *b*, we will apply *KVL* to the closed loop.

$$\therefore \qquad 12 - 4\,i \times 2\,i - 4\,i \;= 0 \quad \therefore \quad i = 2\text{ A}$$

Hence, V_{oc} = drop across 4 Ω resistor = 4 × 2 = 8 V. It is so because there is no current through the 2 Ω resistor.

For finding R_{th}, we will put a short-circuit across terminals *a* and *b* and calculate I_{sh}, as shown in Fig. 2.114 (*b*). Using the two mesh currents, we have

$12 - 4\,i_1 + 2\,i - 4(i_1 - i_2) = 0$ and $-\,8\,i_2 - 4\,(i_2 - i_1) = 0$. Substituting $i = (i_1 - i_2)$ and Simplifying the above equations, we have

$$12 - 4\,i_1 + 2\,(i_1 - i_2) - 4\,(i_1 - i_2) = 0 \quad \text{or} \quad 3\,i_1 - i_2 = 6 \qquad \text{...}(i)$$

Similarly, from the second equation, we get $i_1 = 3\,i_2$. Hence, $i_2 = 3/4$ and $R_{th} = V_{oc}/I_{sh} = 8/(3/4)$ = 32/3 Ω. The Thevenin equivalent circuit is as shown in Fig. 2.113 (*c*).

Example 2.58. *Find the Thevenin equivalent circuit with respect to terminals a and b of the network shown in Fig. 2.114 (a).*

Solution. It will be seen that with terminals *a* and *b* open, current through the 8 Ω resistor is $v_{ab}/4$ and potential of point *A* is the same that of point a (because there is no current through 4 Ω resistor). Applying *KVL* to the closed loop of Fig. 2.114 (*a*), we get

$$6 + (8 \times v_{ab}/4) - v_{ab} \;= 0 \quad \text{or} \quad v_{ab} = 12\text{ V}$$

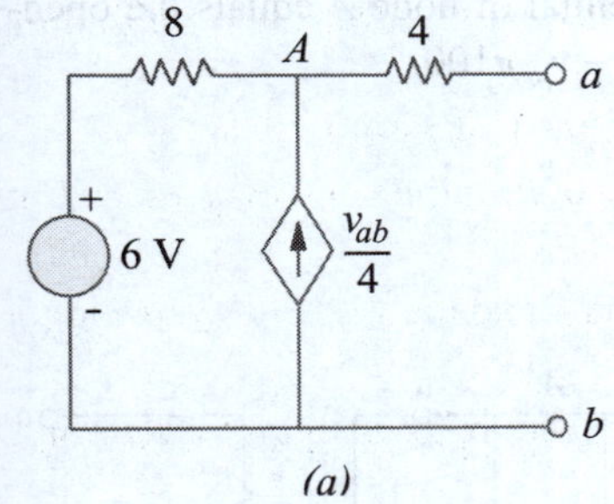

(*a*)

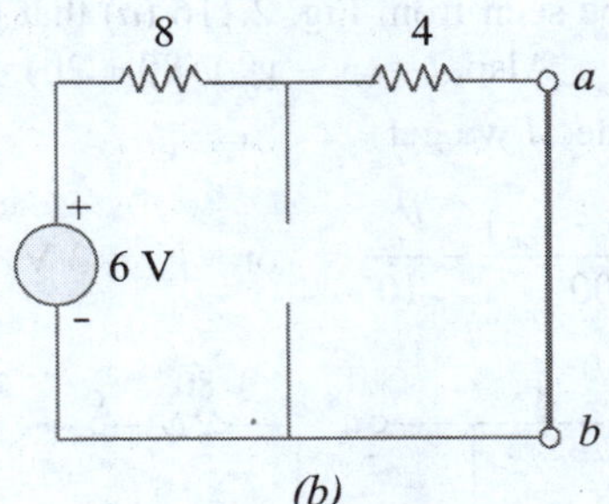

(*b*)

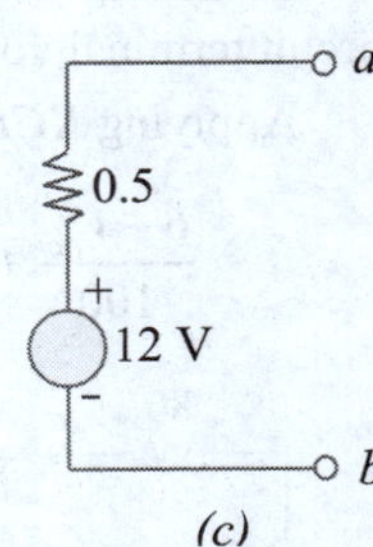

(*c*)

Fig. 2.114

It is also the value of the open-circuit voltage v_{oc}.

For finding short-circuit current i_{sh}, we short-circuit the terminals *a* and *b* as shown in Fig. 2.114 (*b*). Since with *a* and *b* short-circuited, $v_{ab} = 0$, the dependent current source also becomes zero. Hence, it is replaced by an open-circuit as shown. Going around the closed loop, we get

$$12 - i_{sh}\,(8 + 4) \;= 0 \quad \text{or} \quad i_{sh} = 6/12 = 0.5\text{ A}$$

Hence, the Thevenin equivalent is as shown in Fig. 2.114 (*c*).

Example 2.59. *Find the Thevenin equivalent circuit for the network shown in Fig. 2.115 (a) which contains only a dependent source.*

Solution. Since circuit contains no independent source, $i = 0$ when terminals *a* and *b* are open. Hence, $v_{oc} = 0$. Moreover, i_{sh} is zero since $v_{oc} = 0$.

Consequently, R_{sh} cannot be found from the relation $R_{th} = v_{oc}/i_{sh}$. Hence, we will connect a 1 A current source to terminals *a* and *b* as shown in Fig. 2.115 (*b*). Then by finding the value of v_{ab}, we will be able to calculate $R_{th} = v_{ab}/1$.

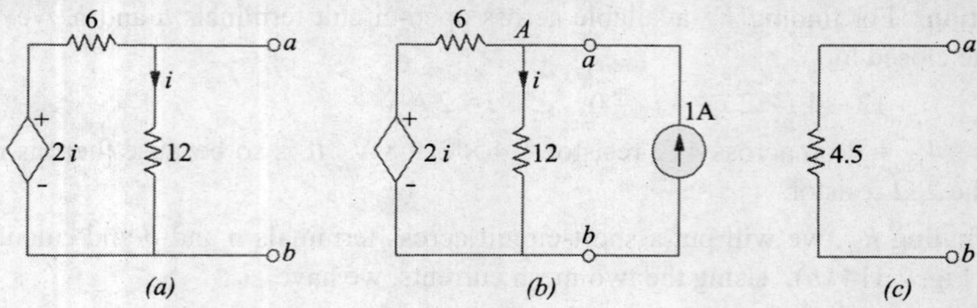

Fig. 2.115

It should be noted that potential of point A is the same as that of point a *i.e.* voltages across 12 Ω resistor is v_{ab}. Applying *KCL* to point A, we get

$$\frac{2i - v_{ab}}{6} - \frac{v_{ab}}{12} + 1 = 0 \quad \text{or} \quad 4i - 3v_{ab} = -12$$

Since $i = v_{ab}/12$, we have $4(v_{ab}/12) - 3v_{ab} = -12$ or $v_{ab} = 4.5$ V $\therefore R_{th} = v_{ab}/1 = 4.5/1 =$ **4.5 Ω.**

The Thevenin equivalent circuit is shown in Fig. 2.115 (*c*).

Example 2.60. *Determine the Thevenins equivalent circuit as viewed from the open-circuit terminals a and b of the network shown in Fig. 2.116 (a). All resistances are in ohms.*

Solution. It would be seen from Fig. 2.116 (*a*) that potential of node A equals the open-circuit terminal voltage v_{oc}. Also, $i = (v_s - v_{oc})/(80 + 20) = (6 - v_{oc})/100$.

Applying *KCL* to node, A we get

$$\frac{6 - V_{oc}}{100} + \frac{9 \times (6 - v_{oc})}{100} - \frac{V_{oc}}{10} = \quad \text{or} \quad V_{oc} = 3 \text{ V}$$

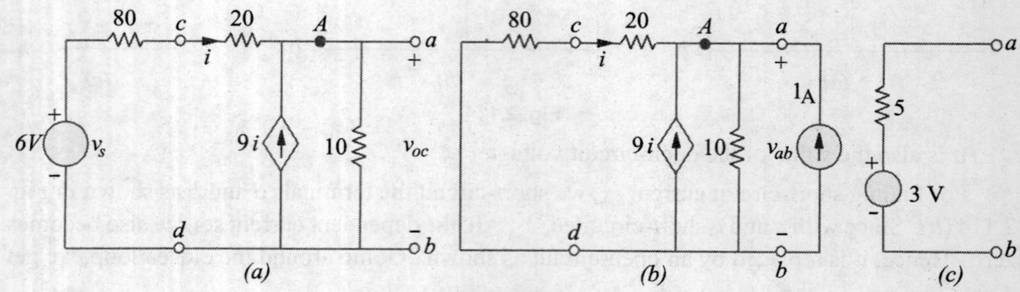

Fig. 2.116

For finding the Thevenin's resistance with respect to terminals a and b, we would first 'kill' the independent voltage source as shown in Fig. 2.116 (*b*). However, the dependent current source cannot be 'killed'. Next, we will connect a current source of 1 A at terminals a and b and find the value of v_{ab}. Then, Thevenin's resistance $R_{th} = v_{ab}/1$. It will be seen that current flowing away from node A *i.e.* from point c to d is $= v_{ab}/100$. Hence, $i = -v_{oc}/100$. Applying *KCL* to node A, we get

$$-\frac{v_{ab}}{100} + 9\left(-\frac{v_{ab}}{100}\right) - \frac{v_{ab}}{10} + 1 = 0 \quad \text{or} \quad v_{ab} = 5 \text{ V}$$

$\therefore \quad R_{th} = 5/1 = 5$ Ω. Hence, Thevenin's equivalent source is as shown in Fig. 2.116 (*c*).

2.20 Reciprocity Theorem

It can be stated in the following manner:

In any linear bilateral network, if a source of e.m.f. E in any branch produces a current I in any other branch, then the same e.m.f. E acting in the second branch would produce the same current I in the first branch.

In other words, it simply means that E and I are mutually transferrable. The ratio E/I is known as the ***transfer*** resistance (or impedance in a.c. systems). Another way of stating the above is that the receiving point and the sending point in a network are interchangebale. It also means that interchange of an *ideal* voltage sources and an ***ideal*** ammeter in any network will not change the ammeter reading. Same is the case with the interchange of an ideal current source and an ***ideal*** voltmeter.

Example 2.61. *In the netwrok of Fig. 2.117 (a), find* **(a)** *ammeter current when battery is at A and ammeter at B and* **(b)** *when battery is at B and ammeter at point A. Values of various resistances are as shown in diagram. Also, calculate the transfer resistance.*

Solution. **(*a*)** Equivalent resistance between points C and B in Fig. 2.117 (*a*) is

$= 12 \times 4/16 = 3\ \Omega$

∴ Total circuit reistance

$= 2 + 3 + 4 = 9\ \Omega$

∴ Battery current $= 36/9 = 4$ A

∴ Ammeter current

$= 4 \times 12/16 =$ **3 A.**

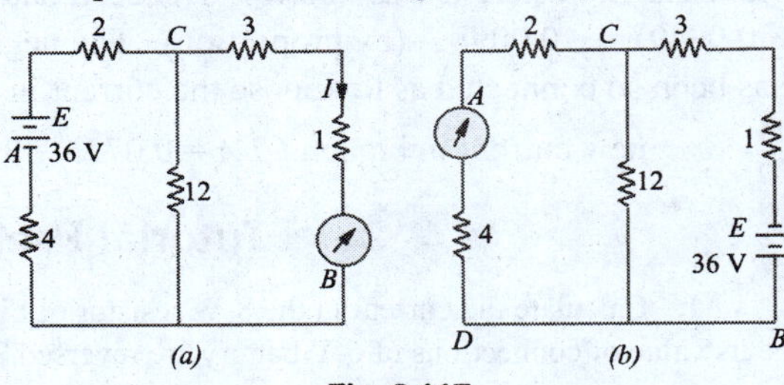

Fig. 2.117

(*b*) Equivalent resistance between points C and D in Fig. 2.117 (*b*) is

$= 12 \times 6/18 = 4\ \Omega$

Total circuit resistance $= 4 + 3 + 1 = 8\ \Omega$

Battery current $= 36/8 = 4.5$ A

∴ Ammeter current $= 4.5 \times 12/18 =$ **3 A**

Hence, ammeter current in both cases is the same.

Transfer resistance $= 36/3 =$ **12 Ω.**

Example 2.62. *Calculate the currents in the various branches of the network shown in Fig. 2.118 and then utilize the principle of Superposition and Reciprocity theorem together to find the value of the current in the 1-volt battery circuit when an e.m.f. of 2 votls is added in branch BD opposing the flow of original current in that branch.*

Solution. Let the currents in the various branches be as shown in the figure. Applying Kirchhoff's second law, we have

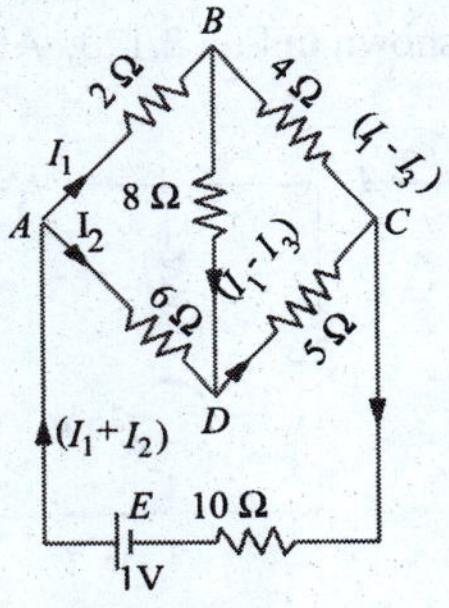

Fig. 2.118

Fig. 2.119

For loop *ABDA* ; $-2I_1 - 8I_3 + 6I_2 = 0$ or $I_1 - 3I_2 + 4I_3 = 0$ ...(*i*)

For loop *BCDB*, $-4(I_1 - I_3) + 5(I_2 + I_3) + 8I_3 = 0$ or $4I_1 - 5I_2 - 17I_3 = 0$ ...(*ii*)

For loop *ABCEA*, $-2I_1 - 4(I_1 - I_3) - 10(I_1 + I_2) + 1 = 0$ or $16I_1 + 10I_2 - 4I_3 = 1$ ...(*iii*)

Solving for I_1, I_2 and I_3, we get I_1 = **0.494 A;** I_2 = **0.0229 A;** I_3 = **0.0049 A**

∴ Current in the 1 volt battery circuit is $I_1 + I_2$ = **0.0723 A.**

The new circuit having 2 - V battery connected in the branch *BD* is shown in Fig. 2.119. According to the Principle of Superposition, the new current in the 1- volt battery circuit is due to the superposition of two currents; one due to 1 - volt battery and the other due to the 2 - volt battery when each acts independently.

The current in the external circuit due to 1 - volt battery when 2 - volt battery is not there, as found above, is 0.0723 A.

Now, according to Reciprocity theorem; if 1 - volt battery were tansferred to the branch *BD* (where it produced a current of 0.0049 A), then it would produce a current of 0.0049 A in the branch *CEA* (where it was before). Hence, a battery of 2 - V would produce a current of (– 2 × 0.0049) = – 0.0098 A (by proportion). The negative sign is used because the 2 - volt battery has been so connected as to oppose the current in branch *BD*,

∴ new current in branch *CEA* = 0.0723 – 0.0098 = **0.0625 A**

Tutorial Problem 2.5

1. Calculate the current in the 8-W resistor of Fig. 2.121 by using Thevenin's theorem. What will be its value of connections of 6-V battery are reversed ? **[0.8 A ; 0 A]**

2. Use Thevenin's theorem to calculate the p.d. across terminals *A* and *B* in Fig. 2.122. **[1.5 V]**

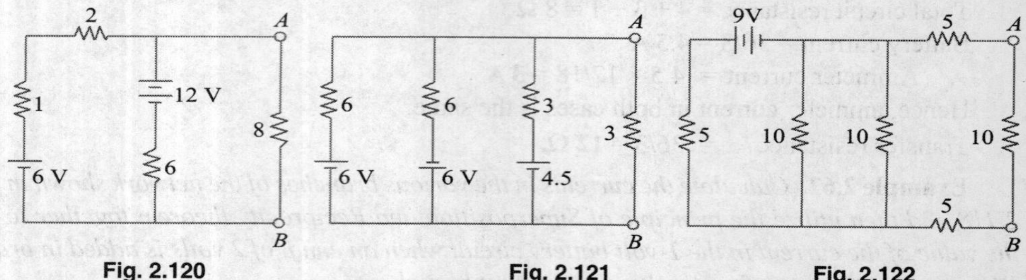

Fig. 2.120 Fig. 2.121 Fig. 2.122

3. Compute the current flowing through the load resistance of 10 Ω connected across terminals *A* and *B* in Fig. 2.122 by using Thevenin's theorem.

4. Find the equivalent Thevenin voltage and equivalent Thevenin resistance respectively as seen from open-circuited terminals *A* and *B* to the circuits shown in Fig. 2.123. All resistances are in ohms.

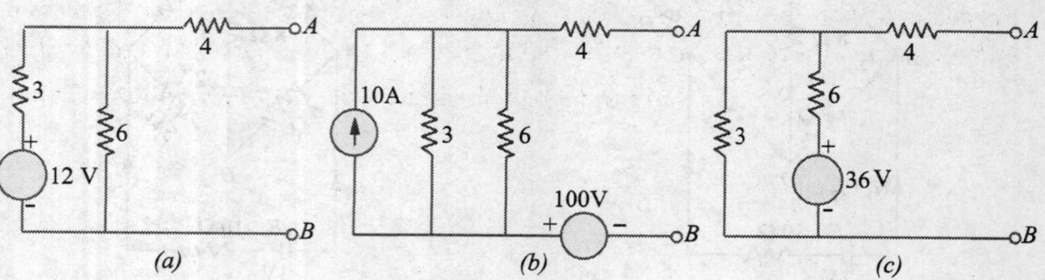

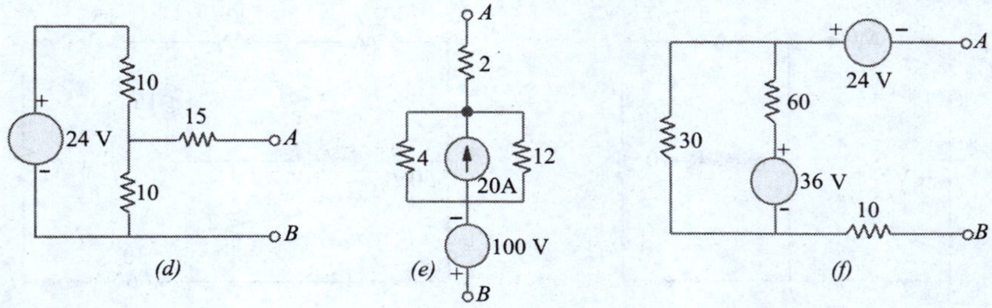

Fig. 2.123

[(a) 8 V, 6 Ω; (b) 120 V, 6 Ω; (c) 12 V, 6 Ω; (d) 12 V, 20 Ω; (e) – 40 V, 5 Ω; (f) – 12 V, 30 Ω]

5. Find Thevenin's equivalent of the circuits shown in Fig. 2.124 between terminals *A* and *B*.

[(a) $V_{th} = I\dfrac{R_1 R_2}{R_1 + R_2} + V\dfrac{R_2}{R_1 + R_2}$; $R_{th} = \dfrac{R_1 R_2}{R_1 + R_2}$ (b) $V_{th} = \dfrac{V_1 R_2 + V_2 R_1}{R_1 + R_2}$; $R_{th} = \dfrac{R_1 R_2}{R_1 + R_2}$

(c) $V_{th} = -IR$; $R_{th} = R_1$ (d) $V_{th} = -V_1 - IR$, $R_{th} = R$ (e) Not possible]

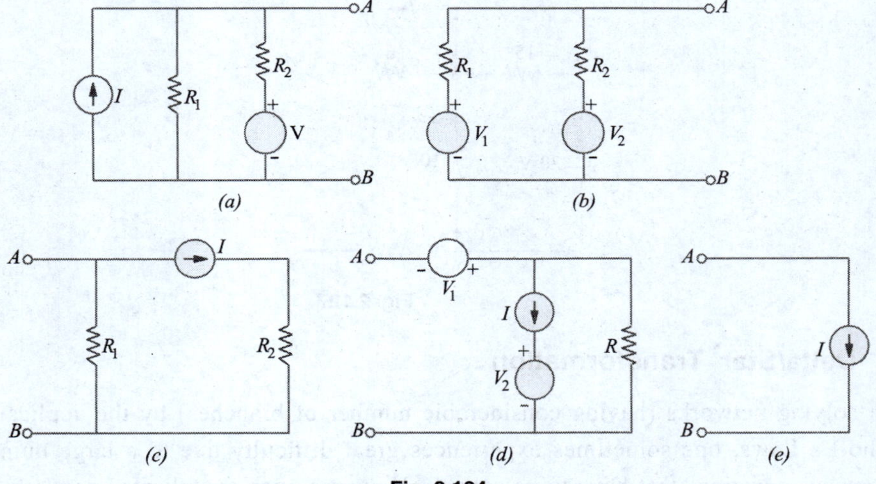

Fig. 2.124

6. The four arms of a Wheatstone bridge have the following resistances in ohms.

$$AB = 100, BC = 10, CD = 5, DA = 60$$

A galvanometer of 15 ohm resistance is connected across *BD*. Calculate the current through the galvanometer when a potential difference of 10 V is maintained across *AC*.

[*Elect. Engg. A.M.Ae. S.I.Dec. 1991*] **[4.88 mA]**

7. State and explain Thevenin's theorem. By applying Thevenin's theorem or otherwise, find the current through the resistance *R* and the voltage across it when connected as shown in Fig. 2.125.

[60.49 A, 600.49 V] (*Elect. and Mech. Technology, Osmania Univ.*)

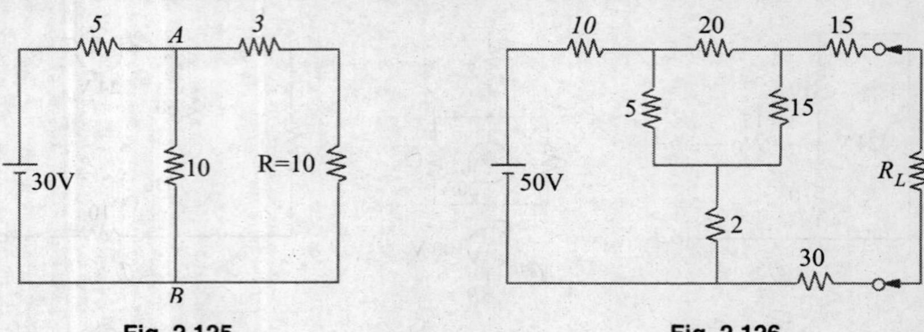

Fig. 2.125

Fig. 2.126

8. State and explain Thevenin's theorem.

For the circuit shown in Fig. 2.126, determine the current through R_L when its value is 50 Ω. Find the value of R_L for which the power drawn from the source is maximum.

(Elect. Technology I, Gwalior Univ.)

9. Find the Thevenin's equivalent circuit for terminal pair *AB* for the network shown in Fig. 2.127.

[V_{th} = – 16 V and R_{th} = 16 Ω]

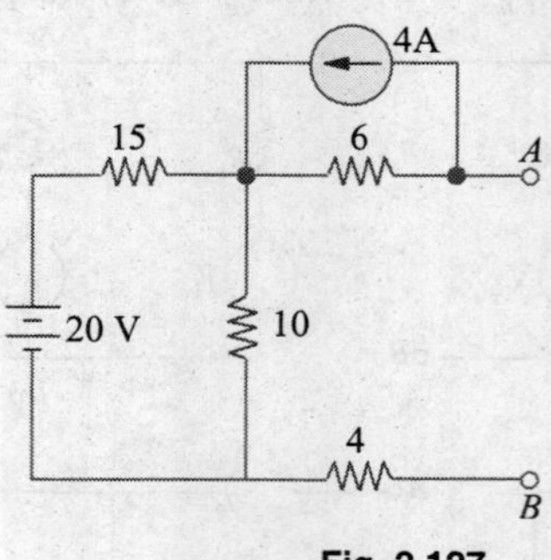

Fig. 2.127

2.21 Delta/Star* Transformation

In solving networks (having considerable number of branches) by the application of Kirchhoff's Laws, one sometimes experiences great difficulty due to a large number of simultaneous equations that have to be solved. However, such complicated network can be simplified by successively replacing delta meshes by equivalent star system and ***vice versa***.

Suppose we are given three resistances R_{12}, R_{23} and R_{31} connected in delta fashion between terminals 1, 2 and 3 as in Fig. 2.128 (*a*). So far as the respective terminals are concerned, these three given resistances can be replaced by the three resistances R_1, R_2 and R_3 connected in star as shown in Fig. 2.128 (*b*).

These two arrangements will be electrically equivalent if the resistance as measured between any pair of terminals is the same in both the arrangements. Let us find this condition.

* In Electronics, star and delta circuits are generally referred to as *T* and *π* circuits respectively.

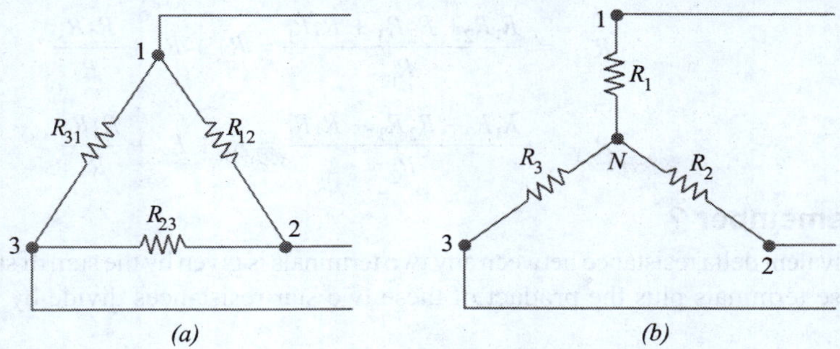

Fig. 2.128

First, take delta connection: Between terminals 1 and 2, there are two parallel paths; one having a resistance of R_{12} and the other having a resistance of $(R_{12} + R_{31})$.

$\therefore$ Resistance between terminals 1 and 2 is $= \dfrac{R_{12} \times (R_{23} + R_{31})}{R_{12} + (R_{23} + R_{31})}$

Now, take star connection: The resistance between the same terminals 1 and 2 is $(R_1 + R_2)$.

As terminal resistances have to be the same

$$\therefore \qquad R_1 + R_2 = \frac{R_{12} \times (R_{23} + R_{31})}{R_{12} + R_{23} + R_{31}} \qquad \text{...(i)}$$

Similarly, for terminals 2 and 3 and terminals 3 and 1, we get

$$R_2 + R_3 = \frac{R_{23} \times (R_{31} + R_{12})}{R_{12} + R_{23} + R_{31}} \qquad \text{...(ii)}$$

$$\text{and} \qquad R_3 + R_1 = \frac{R_{31} \times (R_{12} + R_{23})}{R_{12} + R_{23} + R_{31}} \qquad \text{...(iii)}$$

Now, subtracting **(*ii*)** from **(*i*)** and adding the result to **(*iii*)**, we get

$$R_1 = \frac{R_{12}\ R_{31}}{R_{12} + R_{23} + R_{31}};\ R_2 = \frac{R_{23}\ R_{12}}{R_{12} + R_{23} + R_{31}} \text{ and } R_3 = \frac{R_{31}\ R_{23}}{R_{12} + R_{23} + R_{31}}$$

How to Remember ?

It is seen from above that each numerator is the product of the two sides of the delta which meet at the point in star. Hence, it should be remembered that: ***resistance of each arm of the star is given by the product of the resistances of the two delta sides that meet at its end divided by the sum of the three delta resistances.***

2.22 Star/Delta Transformation

This tarnsformation can be easily done by using equations **(*i*)**, **(*ii*)** and **(*iii*)** given above. Multiplying **(*i*)** and **(*ii*)**, **(*ii*)** and **(*iii*)**, **(*iii*)** and **(*i*)** and adding them together and then simplifying them, we get

$$R_{12} = \frac{R_1 R_2 + R_2 R_3 + R_3 R_1}{R_3} = R_1 + R_2 + \frac{R_1 R_2}{R_3}$$

$$R_{23} = \frac{R_1R_2 + R_2R_3 + R_3R_1}{R_1} = R_2 + R_3 + \frac{R_2R_3}{R_1}$$

$$R_{31} = \frac{R_1R_2 + R_2R_3 + R_3R_1}{R_2} = R_1 + R_3 + \frac{R_3R_1}{R_2}$$

How to Remember ?

The equivalent delta resistance between any two terminals is given by the sum of star resistances between those terminals plus the product of these two star resistances divide by the third star resistances.

Example 2.63. *Find the input resistance of the circuit between the points A and B of Fig 2.129(a).* **(AMIE Sec. B Network Analysis Summer 1992)**

Solution. For finding R_{AB}, we will convert the delta *CDE* of Fig. 2.130 (*a*) into its equivalent star as shown in Fig. 2.129 (*b*).

$R_{CS} = 8 \times 4/18 = 16/9\ \Omega$; $R_{ES} = 8 \times 6/18 = 24/9\ \Omega$; $R_{DS} = 6 \times 4/18 = 12/9\ \Omega$.

The two parallel resistances between *S* and *B* can be reduced to a single resistance of 35/9 Ω.

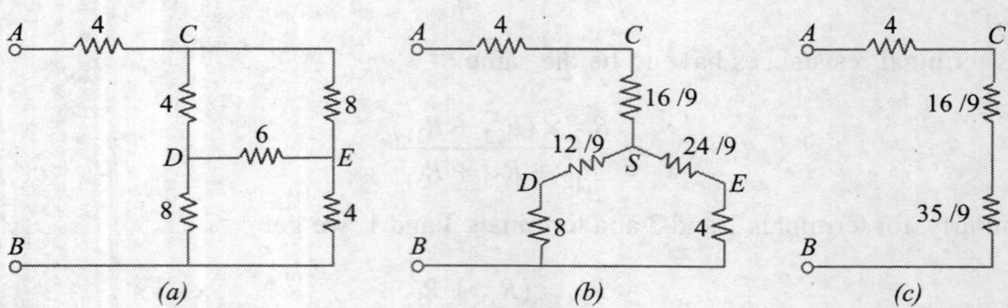

Fig 2.129

As seen from Fig. 2.129 (*c*), $R_{AB} = 4 + (16/9) + (35/9) = 87/9\ \Omega$.

Example 2.64. *Calculate the equivalent resistance between the terminals A and B in the network shown in Fig. 2.130 (a).*

Solution. The given circuit can be redrawn as shown in Fig. 2.130 (*b*). When the delta *BCD* is converted to its equivalent star, the circuit becomes as shown in Fig. 2.130 (*c*).

Each arm of the delta has a resistance of 10 Ω. Hence, each arm of the equivalent star has a resistance = 10 × 10/30 = 10/3 Ω. As seen, there are two parallel paths between points *A* and *N*, each having a resistance of (10 + 10/3) = 40/3 Ω. Their combined resistance is 20/3 Ω. Hence, $R_{AB} = (20/3) + 10/3 = 10\ \Omega$.

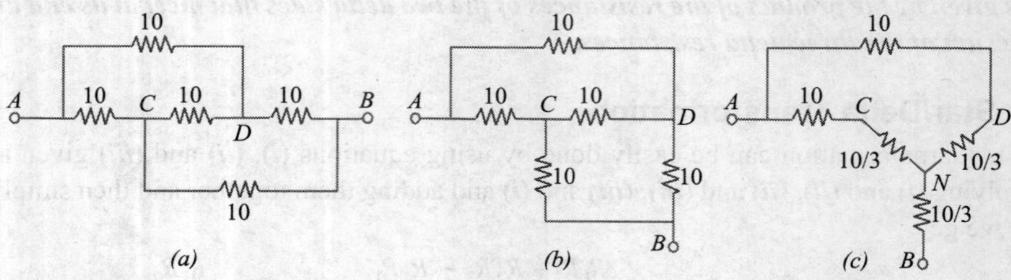

Fig. 2.130

Example 2.65. *Calculate the current flowing through the 10 Ω resistor of Fig. 2.131 (a) by using any method.* **(Network Theory, Nagpur Univ. 1993)**

Solution. It will be seen that there are two deltas in the circuit *i.e. ABC* and *DEF*. They have been converted into their equivalent stars as shown in Fig. 2.131 (*b*). Each arm of the delta *ABC* has a resistance of 12 Ω and each arm of the equivalent star has a resistance of 4 Ω. Similarly, each arm of the delta *DEF* has a resistance of 30 Ω and the equivalent star has a resistance of 10 Ω per arm.

The total circuit resistance between *A* and $F = 4 + 48 \parallel 24 + 10 = 30\ \Omega$. Hence $I = 180/30 = 6$ A.

Current through 10 Ω resistor as given by current-divider rule = 6 × 48/(48 + 24) = **4 A.**

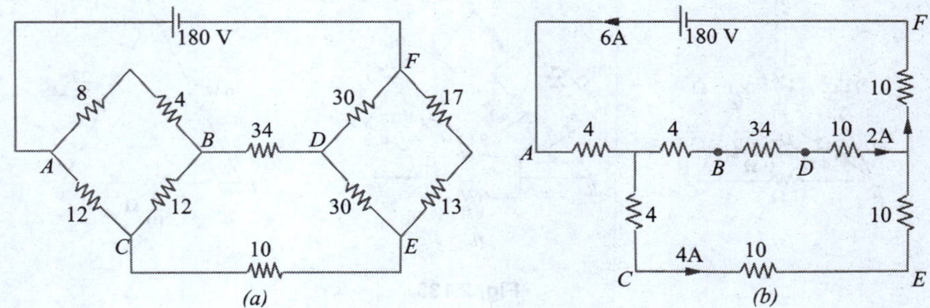

Fig. 2.131

Example 2.66. *A bridge network ABCD has arms AB, BC, CD and DA of resistances 1, 1, 2 and 1 ohm respectively. If the detector AC has a resistance of 1 ohm, determine by star/delta transformation, the network resistance as viewed from the battery terminals.*

(Basic Electricity, Bombay Univ.)

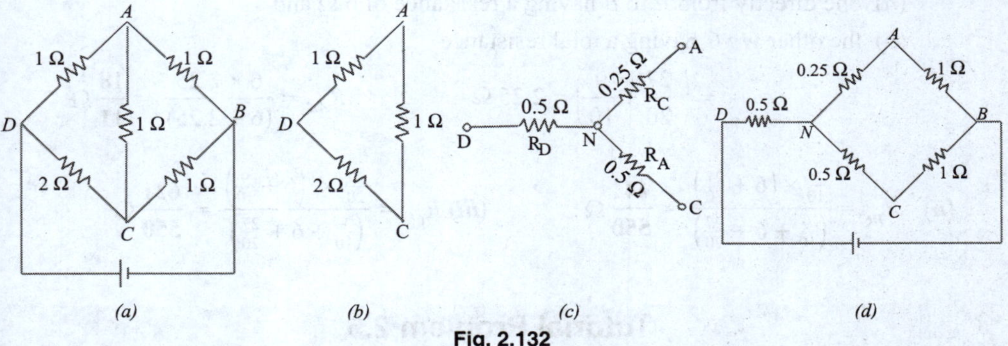

Fig. 2.132

Solution. As shown in Fig. 2.132 (*b*), delta *DAC* has been reduced to its equivalent star.

$$R_D = \frac{2 \times 1}{2 + 1 + 1} = 0.5\ \Omega,\ R_A = \frac{1}{4} = 0.25\ \Omega,\ R_C = \frac{2}{4} = 0.5\ \Omega$$

Hence, the original network of Fig. 2.132 (*a*) is reduced to the one shown in Fig. 2.132 (*d*). As seen, there are two parallel paths between points *N* and *B*, one of resistance 1.25 Ω and the other of resistance 1.5 Ω. Their combined resistance is

$$= \frac{1.25 \times 1.5}{1.25 + 1.5} = \frac{15}{22}\ \Omega$$

Total resistance of the network between points D and B is

$$= 0.5 + \frac{15}{22} = \mathbf{\frac{13}{11}}\ \Omega$$

Example 2.67. *A network of resistances is formed as follows as in Fig. 2.133 (a) AB = 9 Ω ; BC = 1 Ω; CA = 1.5 Ω forming a delta and AD = 6 Ω ; BD = 4 Ω and CD = 3 Ω forming a star. Compute the network resistance measured between* **(i)** *A and B* **(ii)** *B and C and* **(iii)** *C and A.* **(Basic Electricity, Bombay Univ. 1980)**

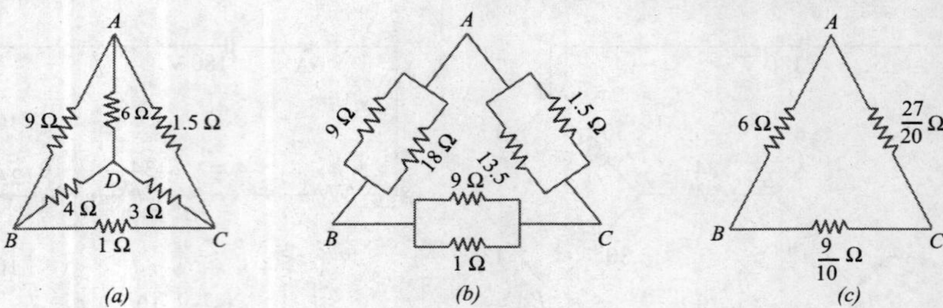

Fig. 2.133

Solution. The star of Fig. 2.133 (*a*) may be converted into the equivalent delta and combined in parallel with the given delta *ABC*. Using the rule the three equivalent delta resistance of the given star become as shown in Fig. 2.133 (*b*).

When combined together, the final circuit is as shown in Fig. 2.133 (*c*).

(*i*) As seen, there are two parallel paths across points *A* and *B*.

(*a*) one directly from *A* to *B* having a resistance of 6 Ω and

(*b*) the other *via C* having a total resistance

$$= \left(\frac{27}{20} + \frac{9}{10}\right) = 2.25\ \Omega \qquad \therefore \qquad R_{AB} = \frac{6 \times 2.25}{(6 + 2.25)} = \mathbf{\frac{18}{11}}\ \Omega$$

$$\textbf{(ii)}\ R_{BC} = \frac{\frac{9}{10} \times \left(6 + \frac{27}{20}\right)}{\left(\frac{9}{10} + 6 + \frac{27}{20}\right)} = \mathbf{\frac{441}{550}}\ \Omega \qquad \textbf{(iii)}\ R_{CA} = \frac{\frac{27}{20} \times \left(6 + \frac{9}{10}\right)}{\left(\frac{9}{10} + 6 + \frac{27}{20}\right)} = \mathbf{\frac{621}{550}}\ \Omega$$

Tutorial Problem 2.6

Delta/Star Conversion

1. Find the current in the 17 Ω resistor in the network shown in Fig. 2.134 (*a*) by using (*a*) star/delta conversion and (*b*) Thevenin's theorem. The numbers indicate the resistance of each member in ohms. **[10/3A]**

2. Convert the star circuit of Fig. 2.134 (*b*) into its equivalent delta circuit. Values shown are in ohms. Derive the formula used. ***(Elect. Technology, Indor Univ.)***

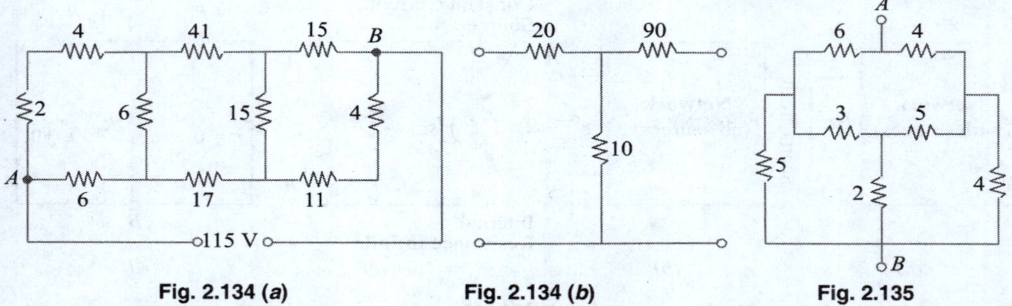

Fig. 2.134 (*a*) Fig. 2.134 (*b*) Fig. 2.135

3. Determine the resistance between points *A* and *B* in the network of Fig. 2.135.
[4.23 Ω] (*Elect. Technology, Indor Univ.*)
4. Three resistances of 20 Ω each are connected in star. Find the equivalent delta resistance. If the source of e.m.f. of 120 V is connected across any two terminals of the equivalent delta-connected resistances, find the current supplied by the source. **[60 Ω, 3A] (*Elect. Engg. Calcutta Univ.*)**

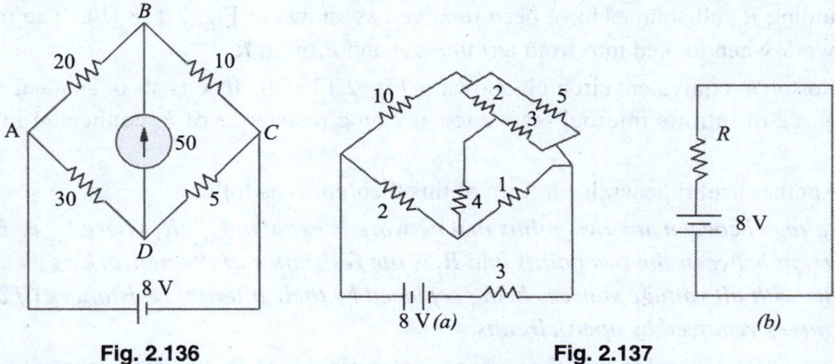

Fig. 2.136 Fig. 2.137

5. Using delta/star transformation determine the current through the galvanometer in the Wheatstone bridge of Fig. 2.136. **[0.025 A]**
6. With the aid of the delta star transformation reduce the network given in Fig. 2.137 (*a*) to the equivalent circuit shown at (*b*) **[R = 5.38 Ω]**

2.23 Norton's Theorem

This theorem is an alternative to the Thevenin's theorem. In fact, it is the dual of Thevenin's theorem. Whereas Thevenin's theorem reduces a two-terminal active network of linear resistances and generators to an equivalent constant-voltage source and series resistance, Norton's theorem replaces the network by an equivalent constant-current source and a parallel resistance.

This theorem may be stated as follows:

(i) Any two-terminal active network containing voltage sources and resistance when viewed from its output terminals , is equivalent to a constant-current source and a parallel resistance. The constant current is equal to the current which would flow in a short-circuit placed across the terminals and parallel resistance is the resistance of the network when viewed from these open-circuited terminals after all voltage and current sources have been removed and replaced by their internal resistances.

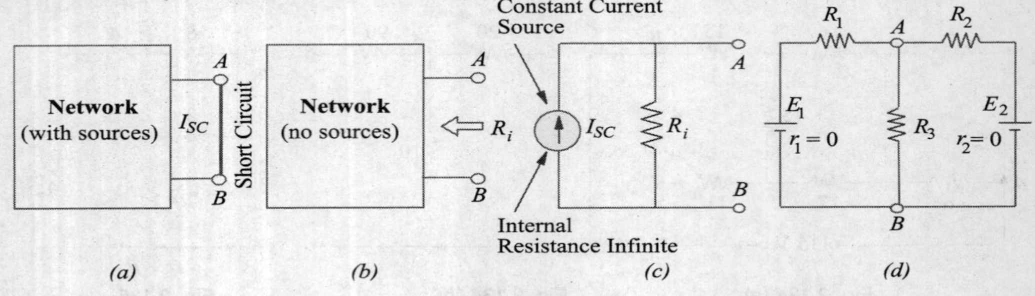

Fig. 2.138

Explanation

As seen from Fig. 2.138 (*a*), a short is placed across the terminals *A* and *B* of the network ***with all its energy sources present.*** The short-circuit current I_{SC} gives the value of constant-current source.

For finding R_i, all sources have been removed as shown in Fig. 2.138 (*b*). The resistance of the network when looked into from terminals *A* and *B* gives R_i.

The Norton's* equivalent circuit is shown in Fig. 2.138 (*c*). It consists of an ideal constant-current source of infinite internal resistance having a resistance of R_i connected in parallel with it.

(*ii*) Another useful generalized form of this theorem is as follows:

The voltage between any two points in a network is equal to I_{SC}. R_i where I_{SC} is the short-circuit current between the two points and R_i is the resistance of the network as viewed from these points with all voltage sources being replaced by their internal resistances (if any) and current sources replaced by open-circuits.

Suppose, it is required to find the voltage across resistance R_3 and hence current through it [Fig. 2.138 (*d*)]. If short-circuit is placed between *A* and *B*, then current in it due to battery of e.m.f. E_1 is E_1/R_1 and due to the other battery is E_2/R_2.

$$\therefore \quad I_{SC} = \frac{E_1}{R_1} + \frac{E_2}{R_2} = E_1 G_1 + E_2 G_2$$

where G_1 and G_2 are branch conductances.

Now, the internal resistance of the network as viewed from *A* and *B* simply consists of three resistances R_1, R_2 and R_3 connected in parallel between *A* and *B*. Please note that here load resistance R_3 ***has not been removed.*** In the first method given above, it has to be removed.

$$\frac{1}{R_i} = \frac{1}{R_1} + \frac{1}{R_2} + \frac{1}{R_3} = G_1 + G_2 + G_3$$

$$\therefore \quad R_i = \frac{1}{G_1 + G_2 + G_3} \quad \therefore \quad V_{AB} = I_{SC}.R_i = \frac{E_1 G_1 + E_2 G_2}{G_1 + G_2 + G_3}$$

Current through R_3 is $I_3 = V_{AB}/R_3$.

Solved example No. 2.96 illustrates this approach.

* After E/L/ Norton, formerely an engineer at Bell Telephone Laboratory. U.S.A.

2.24 How To Nortonize a Given Circuit ?

This procedure is based on the first statement of the theorem given above.

1. Remove the resistance (if any) across the two given terminals and put a short-circuit across them.
2. Compute the short-circuit current I_{SC}.
3. Remove all voltage sources but retain their internal resistances, if any. Similarly, remove all current sources and replace them by open-circuits *i.e.* by infinite resistance.
4. Next, find the resistance R_1 (also called R_N) of the network as looked into from the given terminals. It is exactly the same as R_{th}.
5. The current source (I_{SC}) joined in parallel across R_i between the two terminals gives Norton's equivalent circuit.

Example 2.68. *Determine the Thevenin and Norton equivalent circuits between terminals A and B for the voltage divider circuit of Fig. 2.139 (a).*

Solution. (*a*) **Thevenin Equivalent Circuit**

Obviosuly, V_{th} = drop across $R_2 = E\dfrac{R_2}{R_1 + R_2}$

When battery is replaced by a short-circuit.

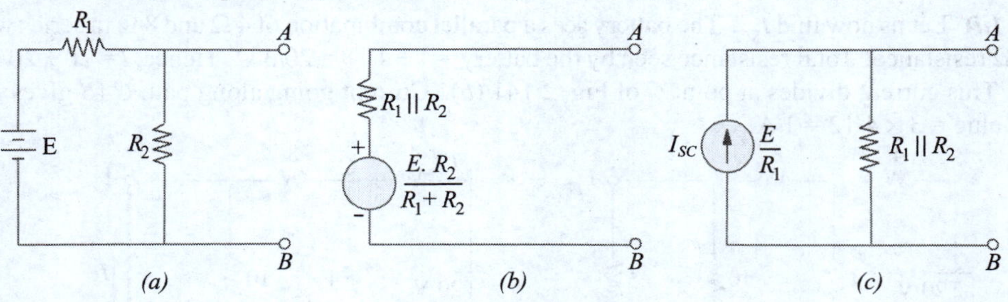

Fig. 2.139

$$R_i = R_1 \,||\, R_2 = R_1 R_2/(R_1 + R_2)$$

Hence, Thevenin equivalent circuit is as shown in Fig. 2.139 (*b*).

(*b*) Norton Equivalent Circuit

A short placed across terminals *A* and *B* will short out R_2 as well. Hence, $I_{SC} = E/R_1$. The Norton equivalent resistance is exactly the same as Thevenin resistance except that it is connected in parallel with the current source as shown in Fig. 2.139 (*c*)

Example 2.69. *Using Norton's theorem, find the constant-current equivalent of the circuit shown in Fig. 2.140 (a).*

Solution. When terminals *A* and *B* are short-circuited as shown in Fig. 2.140 (*b*), total resistance of the circuit, as seen by the battery, consists of a 10 Ω resistance in series with a parallel combination of 10 Ω and 15 Ω resistances.

$\therefore$ total resistance $= 10 + \dfrac{15 \times 10}{15 + 10} = 16\ \Omega$

$\therefore$ battery current $I = 100/16 = 6.25$ A

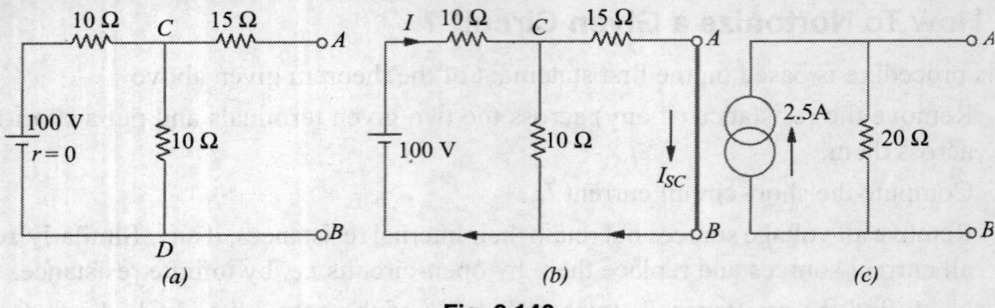

Fig. 2.140

This current is divided into two parts at point C of Fig. 2.140 (b).

Current through $A\ B$ is $I_{SC} = 6.25 \times 10/25 = 2.5$ A

Since the battery has no internal resistance, the input resistance of the network when viewed from A and B consists of a 15 Ω resistance in series with the parallel combination of 10 Ω and 10 Ω. Hence, $R_1 = 15 + (10/2) = 20$ Ω

Hence, the equivalent constant-current source is as shown in Fig. 2.140 (c).

Example 2.70. *Apply Norton's theorem to calculate current flowing through* 5 – Ω *resistor of Fig. 2.141 (a).*

Solution. (*i*) Remove 5 – Ω resistor and put a short across terminals A and B as shown in Fig. 2.141 (b). As seen, 10 – Ω resistor also becomes short-circuited.

(*ii*) Let us now find I_{SC}. The battery sees a parallel combination of 4 Ω and 8 Ω in series with a 4 Ω resistance. Total resistance seen by the battery = 4 + 4 || 8 = 20/3 Ω. Hence, $I = 20 + 20/3 = 3$ A. This current divides at point C of Fig. 2.141 (b). Current going along path CAB gives I_{SC}. Its value = 3 × 4/12 = 1 A.

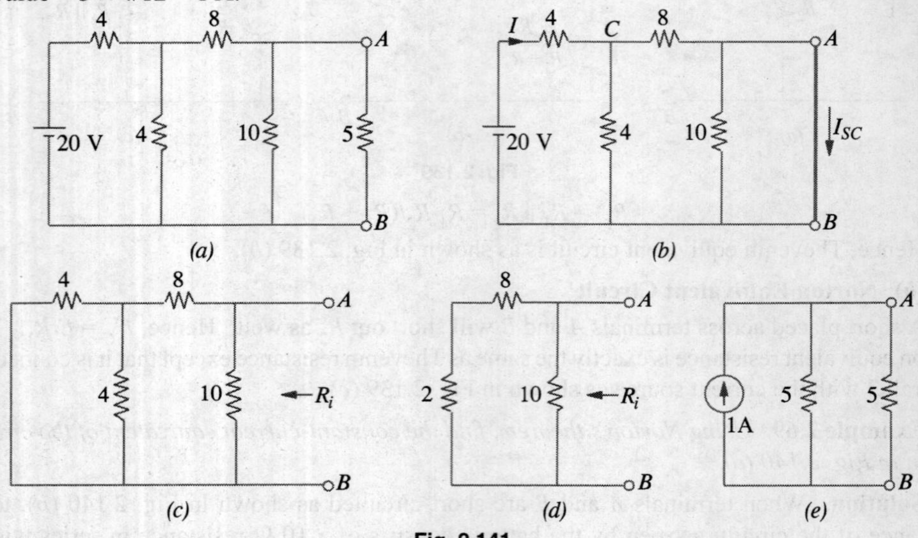

Fig. 2.141

(*iii*) In Fig. 2.141 (c), battery has been removed leaving behind its internal resistance which, in this case, is zero.

Resistance of the network looking into the terminals A and B in Fig. 2.141 (d) is

$$R_i = 10 \parallel 10 = 5\ \Omega$$

(*iv*) Hence, Fig. 2.141 (e), gives the Norton's equivalent circuit.

(v) Now, join the 5 – Ω resistance back across terminals *A* and *B*. The current flowing through it, obviously, is $I_{AB} = 1 \times 5/10 =$ **0.5 A.**

Example 2.71. *Using Norton's current-source equivalent circuit of the network shown in Fig. 2.142 (a), find the current that would flow through the resistor R_2 when it takes the values of 12, 24 and 36 Ω respectivley.* **[Elect. Circuits, South Gujarat Univ.]**

Solution. In Fig. 2.142 (*b*), terminals *A* and *B* have been short-circuited. Current in the shorted path due to E_1 is = 120/40 = 3 A from *A* to *B*. Current due to E_2 is 180/60 = 3 A from *A* to *B*. Hence I_{SC} = 6A. With batteries removed, the resistance of the network when viewed from open-circuited terminals is = 40 || 60 = 24 Ω.

(i) When R_L = 12 Ω $\qquad I_L = 6 \times 24\ (24 + 12) =$ **4 A**

(ii) When R_L = 24 Ω $\qquad I_L = 6/2 =$ **3 A.**

(iii) When R_L = 36 Ω $\qquad I_L = 6 \times 24/(24 + 36) =$ **2.4 A.**

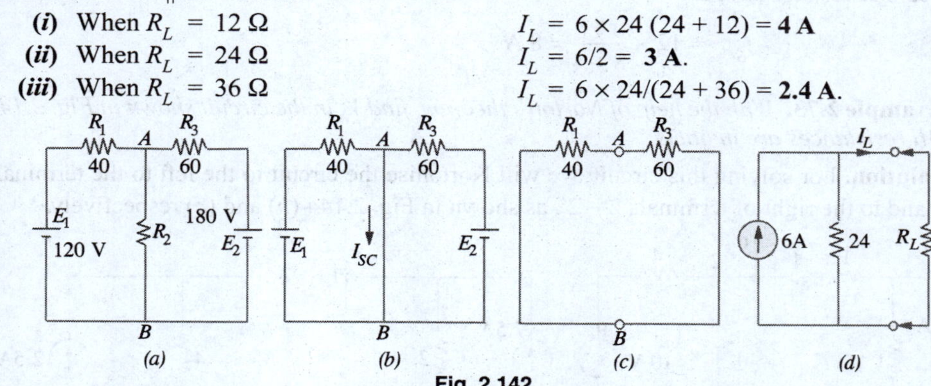

Fig. 2.142

Example 2.72. *Using Norton's theorem, calculate the current in the 6-Ω resistor in the network of Fig. 2.143 (a). All resistance are in ohms.*

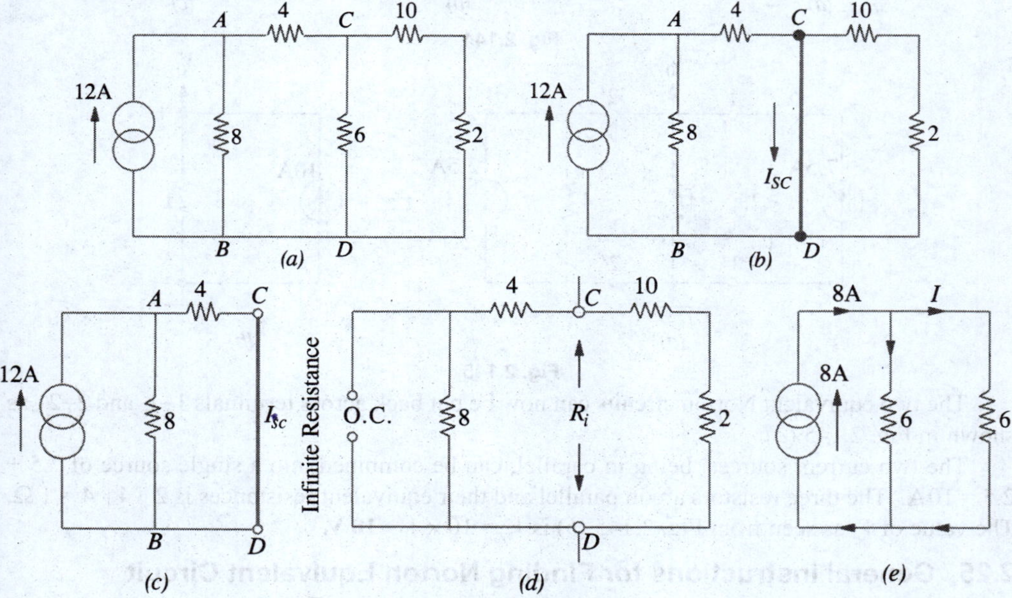

Fig. 2.143

Solution. When the branch containing 6 – Ω resistance is short-circuited, the given circuit is reduced to that shown in Fig. 2.143 (*b*) and finally to Fig. 2.143 (*c*). As seen, the 12 A current divides into two unequal parts at point *A*. The current passing through 4 Ω resistor forms the short-circuit current I_{SC}.

Resistance R_i between points C and D when they are open-circuited is

$$R_i = \frac{(4+8)\times(10+2)}{(4+8)+(10+2)} = 6\ \Omega$$

It is so because the constant-current source has *infinite* resistance *i.e.,* it behaves like an open circuit as shown in Fig. 2.143 (*d*).

Hence, Norton's equivalent circuit is as shown in Fig. 2.143 (*e*). As seen current of 8 A is divided equally between the two equal resistances of 6 Ω each. Hence, current through the required 6 Ω resistor is **4 A.**

$$I_{SC} = 12 \times \frac{8}{8+4} = 8\text{ A}$$

Example 2.73. *With the help of Norton's theorem, find V_o in the circuit shown in Fig. 2.144 (a). All resistances are in ohms.*

Solution. For solving this circuit, we will Nortonise the circuit to the left to the terminals 1 – 1′ and to the right of terminals 2 – 2′, as shown in Fig. 2.144 (*b*) and (*c*) respectively.

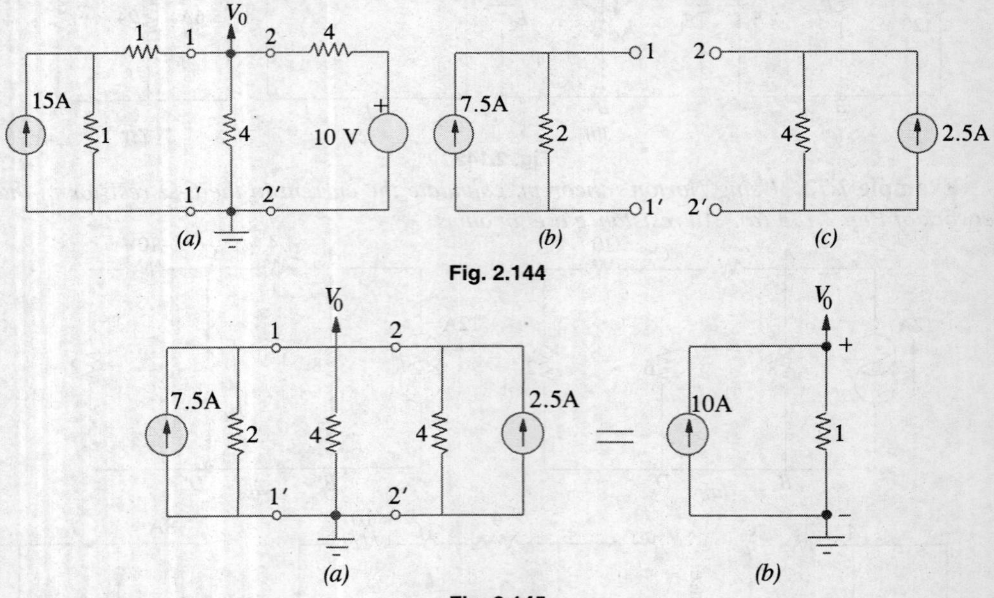

Fig. 2.144

Fig. 2.145

The two equivalent Norton circuits can now be put back across terminals 1–1′ and 2–2′, as shown in Fig. 2.145 (*a*).

The two current sources, being in parallel, can be combined into a single source of 7.5 + 2.5 = 10A. The three resistors are in parallel and their equivalent resistances is 2 || 4 || 4 = 1 Ω. The value of V_o as seen from Fig. 2.145 (*b*) is $V_o = 10 \times 1 =$ **10 V.**

2.25 General instructions for Finding Norton Equivalent Circuit

Procedure for finding Norton equivalent circuit of a given network has already been given in Art. That procedure applies to circuits which contain resistors and independent voltage or current sources. Similar procedures for circuits which contain both dependent and independent sources or only dependent sources are given below:

(a) Circuits Containing Both Dependent and Independent Sources

(i) Find the open-circuit voltage v_{∞} with all the sources activated or 'alive'.

(ii) Find short-circuit current i_{sh} by short-circuiting the terminals *a* and *b* but with all sources activated.

(iii) $R_N = V_{oc}/i_{sh}$

(b) Circuits Containing Dependent Sources Only

(i) $i_{sh} = 0.$

(ii) Connect 1 A source to the terminals *a* and *b* calculate v_{ab}.

(iii) $R_N = v_{ab}/1.$

Example 2.74. *Find the Norton equivalent for the transistor amplifier circuit shown is Fig. 2.146 (a). All resistances are in ohms.*

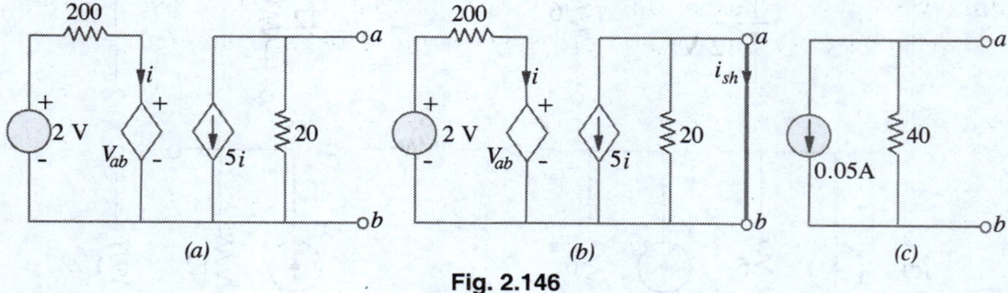

Fig. 2.146

Solution. We have to find the values of i_{sh} and R_N. It should be noted that when terminals *a* and *b* are short-circuited, $v_{ab} = 0$. Hence, in that case, we find from the left-hand portion of the circuit that $i = 2/200 = 1/100\text{A} = 0.01$ A. As seen from Fig. 2.146 (*b*), the short-circuit across terminals *a* and *b*, short circuits 20 Ω resistance also. Hence, $i_{sh} = -5\,i = -5 \times 0.01 = -0.05$ A.

Now, for finding R_N, we need $v_{oc} = v_{ab}$ from the left-hand portion of the Fig. 2.146 (*a*). Applying KVL to the closed circuit, we have

$$2 - 200\,i - v_{ab} = 0 \qquad \text{...}(i)$$

Now, from the right-hand portion of the circuit, we find v_{ab} = drop over 20 Ω resistance $= -20 \times 5i = -100\,i$. The negative sign is explained by the fact that currert flows from point *b* towards point *a*. Hence, $i = -v_b/100$. Substituting this value in Eqn. (*i*). above, we get

$$2 - 200\,(-v_b/100) - v_{ab} = 0 \quad \text{or} \quad v_{ab} = -2\text{ V}$$

$$\therefore \qquad R_N = v_{ab}/i_{sh} = -2/-0.05 = 40\ \Omega$$

Hence, the Norton equivalent circuit is as shown in Fig. 2.146 (*c*).

Example 2.75. *Using Norton's theorem, compute current through the 1-Ω resistor of Fig. 2.147.*

Solution. We will employ source conversion technique to simplify the given circuit. To begin with, we will convert the three voltge sources into their equivalent current sources as shown in Fig. 2.147 (*b*) and (*c*). We can combine together the two current sources on the left of *EF* but cannot combine the 2-A source across *CD* because of the 3-Ω resistance between *C* and *E*.

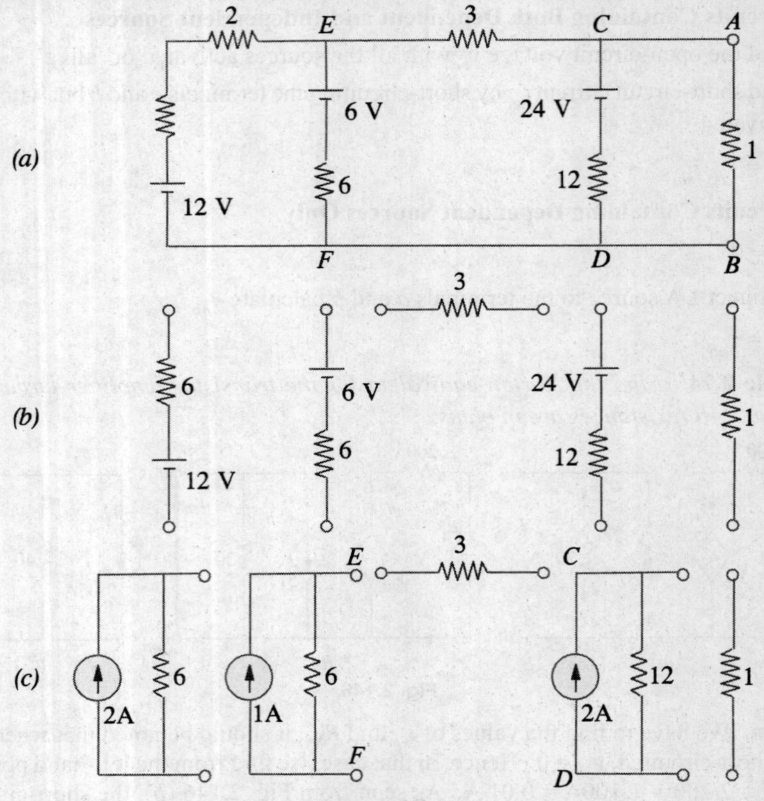

Fig. 2.147

In Fig. 2.148 (*b*), the two current sources at the left-hand side of 3 Ω resistor have been replaced by a single (2 A + 1 A) = 3 A current source having a single parallel resistance 6 || 6 = 3 Ω.

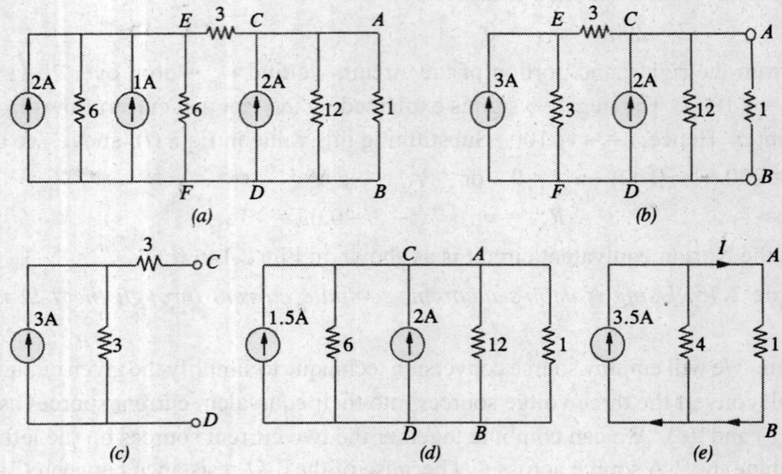

Fig. 2.148

We will now apply Norton's theorem to the circuit on the left-hand side of CD [Fig. 2.148 (*c*)] to convert it into a single current source with a single parallel resistor to replace the two 3 Ω resistors. As shown in Fig. 2.148 (*d*), it yields a 1.5 A current source in parallel with a 6 Ω resistor. This current source can now be combined with the one across *CD* as shown in Fig. 2.148 (*e*). The current through the 1-Ω resistor is

$$I = 3.5 \times 4/(4 + 1) = \mathbf{2.8\ A}$$

Example 2.76. *Obtain Thevenin's and Norton's equivalent circuits at AB shown in Fig. 2.149 (a).* **[Elect. Network, Analysis Nagpur Univ. 1993]**

Solution. ***Thevenin's Equivalent Circuit***

We will find the value of V_{th} by using two methods **(*i*)** *KVL* and **(*ii*)** mesh analysis.

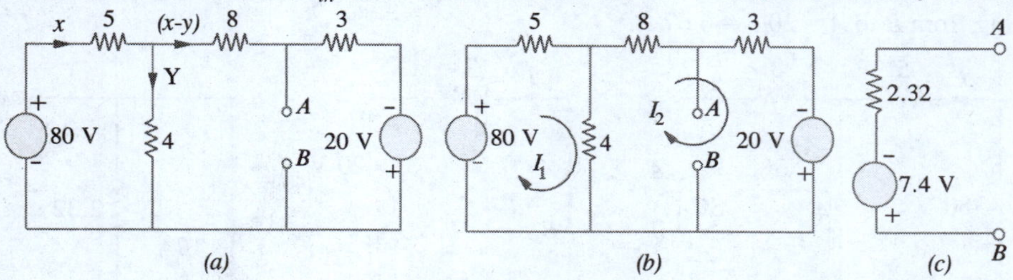

Fig. 2.149

(*a*) Using KVL

If we apply *KVL* to the first loop of Fig. 2.149 (*a*), we get

$$80 - 5x - 4y = 0 \quad \text{or} \quad 5x + 4y = 80 \qquad \text{...(i)}$$

From the second @ loop, we have

$$-11(x - y) + 20 + 4y = 0 \quad \text{or} \quad 11x - 15y = 20 \qquad \text{...(ii)}$$

From **(*i*)** and **(*ii*)**, we get $x = 10.75$ A; $y = 6.56$ A and $(x - y) = 4.2$ A.

Now, $V_{th} = V_{AB}$ *i.e.* voltage of point *A* with respect to point *B*. For finding its value, we start from point *B* and go to point *A* either *via* 3 Ω resistance or 4 Ω resistance or (5 + 8) = 13 Ω resistance and take the algebraic sum of the voltage met on the way. Taking the first route, we get

$$V_{AB} = -20 + 3(x - y) = -20 + 3 \times 4.2 = -7.4\ \text{V}$$

It shows that point *A* is negative with respect to point *B* or, which is the same thing, point *B* is positive with respect to point *A*.

(*b*) Mesh Analysis [Fig. 2.149 (*b*)]

Here, $R_{11} = 9$; $R_{22} = 15$; $R_{21} = -4$

$$\begin{vmatrix} 9 & -4 \\ -4 & 15 \end{vmatrix} \begin{vmatrix} I_1 \\ I_2 \end{vmatrix} = \begin{vmatrix} 80 \\ 20 \end{vmatrix}; \ \Delta = 135 - 16 = 119$$

$$\Delta_1 = \begin{vmatrix} 80 & -4 \\ 20 & 15 \end{vmatrix} = 1280; \ \Delta_2 = \begin{vmatrix} 9 & 80 \\ -4 & 20 \end{vmatrix} = 500$$

$$I_1 = 1280/119 = 10.75\ \text{A}; \ I_2 = 500/119 = 4.2\ \text{A}$$

Again $V_{AB} = -20 + 12.6 = -7.4$ V

Value of R_{th}

For finding R_{th}, we replace the two voltage sources by short-circuits.

$$\therefore \quad R_{th} = R_{AB} = 3 \parallel (8 + 4 \parallel 5) = 2.32\ \Omega$$

Norton's Equivalent Circuit

For this purpose, we will short-circuit the terminals A and B find the short-circuit currents produced by the two voltage sources. When viewed from the side of the 80-V source, a short across AB short-circuits everything on the right side of AB. Hence, the circuit becomes as shown in Fig. 2.150 (*a*). The short-circuit current I_1 can be found with the help of series-parallel circuit technique. The total resistance offered to the 80 – V source is $5 + 4 \parallel 8 = 23/3\ \Omega$.

$\therefore \quad I = 80 \times 3/23 = 10.43$ A; $\quad \therefore \quad I_1 = 10.43 \times 4/12 = 3.48$ A.

When viewed from the side of the 20-V source, a short across AB short-circuits everything beyond AB. In the case, the circuit becomes as shown in Fig. 2.150 (*b*). The short circuit current flowing from B to $A = 20/3 = 6.67$ A.

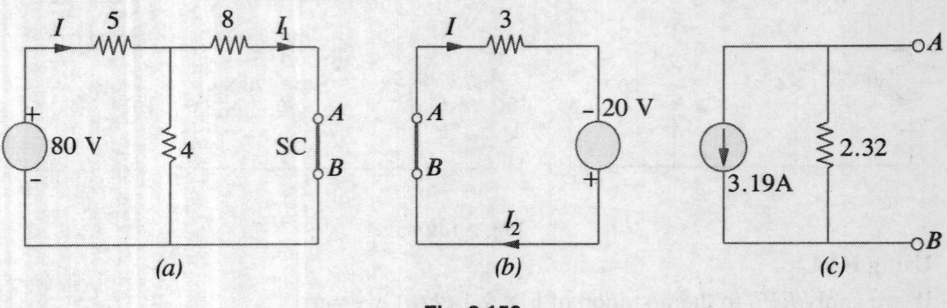

Fig. 2.150

Total short-circuit current $= 6.67 - 3.48 = 3.19$ A ... from B to A.

$$R_N = R_{th} = 3 \parallel (8 + 4 \parallel 5) = 2.32\ \Omega$$

Hence, the Norton's equivalent circuit becomes as shown in Fig. 2.150 (*c*).

2.26 Millman's Theorem

This theorem can be stated either in terms of voltage sources or current sources or both.

(*a*) As Applicable to Voltage Sources

This Theorem is a combination of Thevenin's and Norton's theorems. It is used for finding the common voltage across any network which contains a number of parallel voltage sources as shown in Fig. 2.151 (*a*). Then common voltage V_{AB} which appears across the output terminals A and B is affected by the voltage sources E_1, E_2 and E_3. The value of the voltage is given by

$$V_{AB} = \frac{E_1/R_1 + E_2/R_2 + E_3/R_3}{1/R_1 + 1/R_2 + 1/R_3} = \frac{I_1 + I_2 + I_3}{G_1 + G_2 + G_3} = \frac{\Sigma I}{\Sigma G}$$

This voltage represents the Thevenin's voltage V_{th}. The resistance R_{th} can be found, as usual, by replacing each voltage source by a short circuit. If there is a load resistance R_L across the terminals A and B, then load current I_L is given by

$$I_L = V_{th}/(R_{th} + R_L)$$

If as shown in Fig. 2.151 (*b*), a branch does not contain any voltage source, the same procedure is used except that the value of the voltage for that branch is equated to zero.

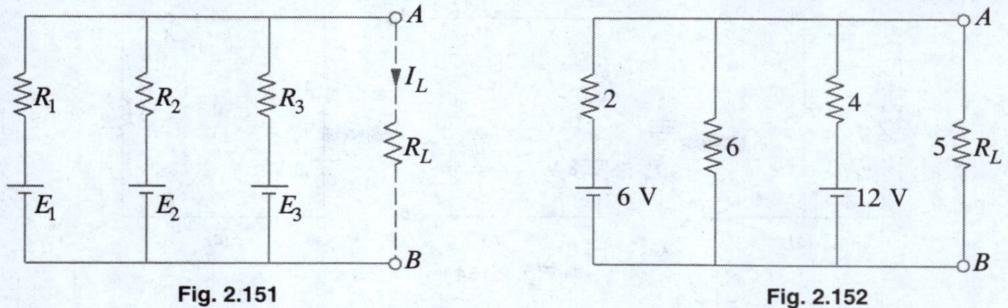

Fig. 2.151

Fig. 2.152

Example 2.77. *Use Millman's theorem, to find the common voltage across terminals A and B and the load current in the circuit of Fig. 2.152.*

Solution. As per Millman's Theorem,

$$V_{AB} = \frac{6/2 + 0/6 + 12/4}{1/2 + 1/6 + 1/4} = \frac{6}{11/12} = 6.55 \text{ V}$$

$$\therefore \quad V_{th} = 6.55 \text{ V}$$

$$R_{th} = 2 \parallel 6 \parallel 4 = 12/11 \ \Omega$$

$$I_L = \frac{V_{th}}{R_{th} + R_L} = \frac{6.55}{(12/11) + 5} = 1.05 \text{ A}$$

(*b*) As Applicable to Current Sources

This theorem is applicable to a mixture of parallel voltage and current sources that are reduced to a single final equivalent source which is either a constant current or a constant voltage source. This theorem can be stated as follows:

Any number of constant current sources which are directly connected in parallel can be converted into a single current source whose current is the algebraic sum of the individual source currents and whose total internal resistances equals the combined individual source resistances in parallel.

Example 2.78. *Use Millman's theorem, to find the voltage across and current through the load resistor R_L in the circuit of Fig. 2.153 (a).*

Solution. First thing to do is to convert the given voltage sources into equivalent current sources. It should be kept in mind that the two batteries are connected in opposite direction. Using source conversion technique we get the circuit of Fig. 2.153 (*b*).

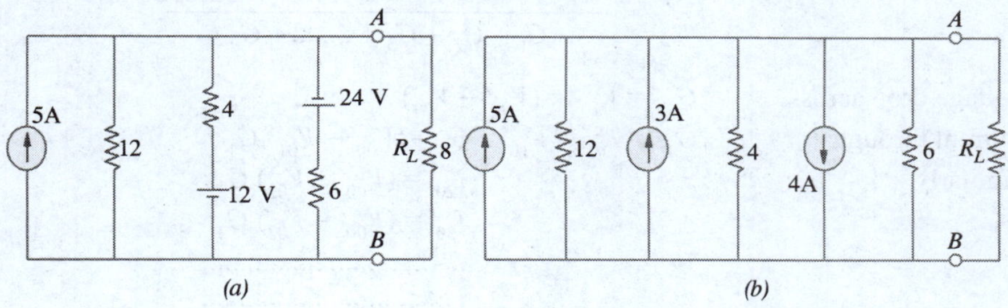

Fig. 2.153

The algebraic sum of the currents = 5 + 3 – 4 = 4 A. The combined resistance is = 12 || 4 || 6 = 2 Ω. The simplified circuit is shown in the current–source form in Fig. 2.154 (*a*) or voltage source form in Fig. 2.154 (*b*).

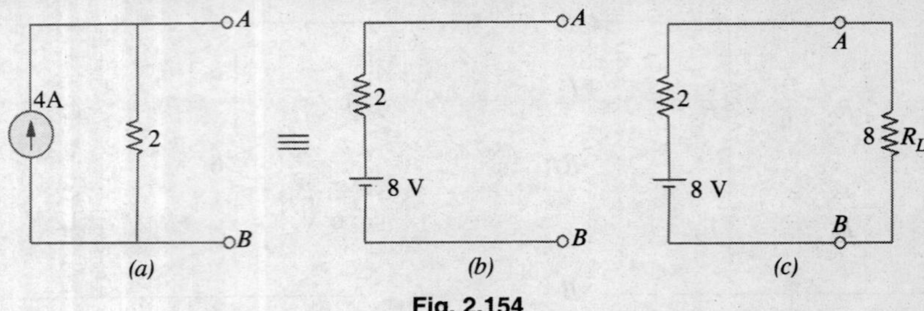

Fig. 2.154

As seen from Fig. 2.154 (*c*).

$$I_L = 8/(2+8) = 0.8 \text{ A}; \; V_L = 8 \times 0.8 = 64 \text{ V}$$

Alternatively, $V_L = 8 \times 8/(2+8) = 6.4 \text{ V}$

Following steps are necessary when using Millman's Theorem:

1. convert all voltage sources into their equivalent current sources.
2. calculate the algebraic sum of the individual dual source currents.
3. if found necessary, convert the final current source into its equivalent voltage source.

As pointed out earlier, this theorem can also be applied to voltage sources which must be initially converted into their constant current equivalents.

2.27 Generalised form of Millman's Theorem

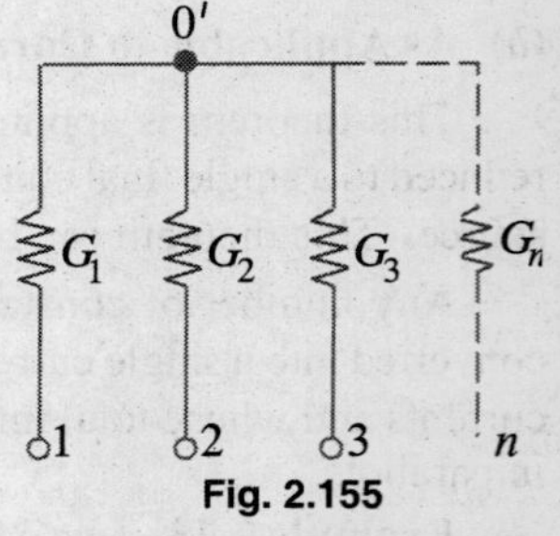

Fig. 2.155

This theorem is particularly useful for solving many circuits which are frequently encountered in both electronics and power applications.

Consider a number of admittances G_1, G_2, G_3... G_n which terminate at common point 0′ (Fig. 2.155). The other ends of the admittances are numbered as 1, 2, 3....*n*. Let *O* be any other point in the network. It should be clearly understood that it is not necessary to know anything about the inter-connection between point *O* and the end points 1, 2, 3...*n*. However, what is essential to know is the voltage drops from 0 to 1, 0 to 2, ... 0 to *n etc.*

According to this theorem, the voltage drop from 0 to 0′ (V_{oo}) is given by

$$V_{oo}' = \frac{V_{01}G_1 + V_{02}G_2 + V_{03}G_3 + ... + V_{0n}G_n}{G_1 + G_2 + G_3 + ... + G_n}$$

Proof

Voltage drop across $G_1 = V_{10}' = (V_{00}' - V_{01})$

Current through $G_1 = I_{10}' = V_{10}' \; G_1 = (V_{00}' - V_{01})\, G_1$

Similarly,

$$I_{20}' = (V_{00}' - V_{02})\, G_2$$

$$I_{30}' = (V_{00}' - V_{03})\, G_3$$

....

....

and $I_{n0}' = (V_{00}' - V_{0n})\, G_n$

By applying *KCL* to point 0′, we get

$$I_{10}' + I_{20}' + ... + I_{n0}' = 0$$

Substituting the values of these currents, we get

$$V_{00}' = \frac{V_{01}G_1 + V_{02}G_2 + V_{03}G_3 + ... + V_{0n}\,G_n}{G_1 + G_2 + G_3 + ... + G_n}$$

Precaution

It is worth repeating that only those resistances or admittances are taken into consideration which terminate at the common point. All those admittances are ignored which do not terminate at the common point even though they are connected in the circuit.

Example 2.79. *Use Millman's theorem to calculate the voltage developed across the 40 Ω resistor in the network of Fig. 2.156.*

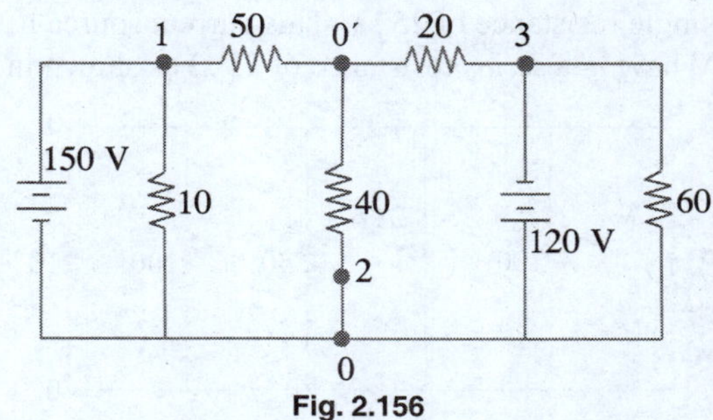

Fig. 2.156

Solution. Let the two ends of the 40 Ω resistor be marked as 0 and 0′. The end points of the three resistors terminating at the common point 0′ have been marked 1, 2 and 3.

The two resistors of values 10 Ω and 60 Ω will not come into the picture because they are not direclty connected to the common point 0′.

Here,

$$V_{01} = -150\text{ V};\quad V_{02} = 0;\quad V_{03} = 120\text{ V}$$

$$G_1 = 1/50\,;\quad G_2 = 1/40:\quad G_3 = 1/20$$

$$\therefore \qquad V_{00}' = \frac{(-150/50) + (0/40) + (120/20)}{(1/50) + (1/40) + (1/20)} = 31.6\text{ V}$$

It shows that point 0 is at a higher potential as compared to point 0′.

Example 2.80. *Calculate the voltage across the 10 Ω resistor in the network of Fig. 2.157 by using* ***(a)*** *Millman's theorem* ***(b)*** *any other method.*

Solution. (*a*) As shown in the Fig. 2.157 we are required to calculate voltage V_{00}'. The four resistances are connected to the common terminal 0′.

Let their other ends be marked as 1, 2, 3 and 4 as shown in Fig. 2.157. Now potential of point 0 with respect to point 1 is – 100 V becaus.

Fig. 2.157

$$\therefore \qquad V_{01} = -100\text{ V};\quad V_{02} = -100\text{ V};\quad V_{03} = 0\text{V};\quad V_{04} = 0\text{V}.$$

$$G_1 = 1/100 = 0.01\text{ Siemens};\qquad G_2 = 1/50 = 0.02\text{ Siemens};$$

$$G_3 = 1/100 = 0.01\text{ Siemens};\qquad G_4 = 1/10 = 0.1\text{ Siemens}$$

$$\therefore \qquad V_{00}' = \frac{V_{01}\,G_1 + V_{02}G_2 + V_{03}G_3 + V_{04}G_4}{G_1 + G_2 + G_3 + G_4}$$

$$= \frac{-100 \times 0.01 + (-100) \times 0.02 + 0. \times 0.01 + 0 \times 0.1}{0.01 + 0.02 + 0.01 + 0.1} = \frac{-3}{0.14}$$

$$= \mathbf{-21.4\ V}$$

Also, $V_{00}' = -V_{00}' = \mathbf{21.4\ V}$

(*b*) We could use the source conversion technique to solve this question. As shown in Fig. 2.158 (*a*), the two voltage sources and their series resistances have been converted into current sources with their parallel resistances. The two current sources have been combined into a single resistance current source of 3 A and the three parallel resistances have been combined into a single resistance of 25 Ω. This current source has been reconverted into a voltage source of 75 V having a series resistance of 25 Ω as shown in Fig. 2.158 (*c*).

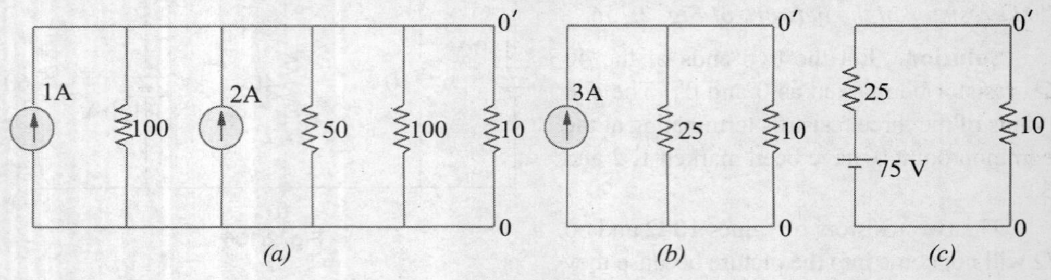

Fig. 2.158

Using the voltage divider formula the voltage drop across 10 Ω resistance is $V_{0'0} = 75 \times 10/(10 + 25) = 21.4$ V.

Example 2.81. *In the network shown in Fig. 2.159, using Millman's theorem, or otherwise find the voltage between A and B.* **(Elect. Engg. Paper-I Indian Engg. Services 1990)**

Solution. The end points of the different admittances which are connected directly to the common point *B* have been marked as 1, 2 and 3 as shown in the Fig. 2.159. Incidentally, 40 Ω resistance will not be taken into consideration because it is not directly connected to the common point B. Here $V_{01} = V_{A1} = -50$ V ; $V_{02} = V_{A2} =$ 100 V ; $V_{03} = V_{A3} = 0$ V.

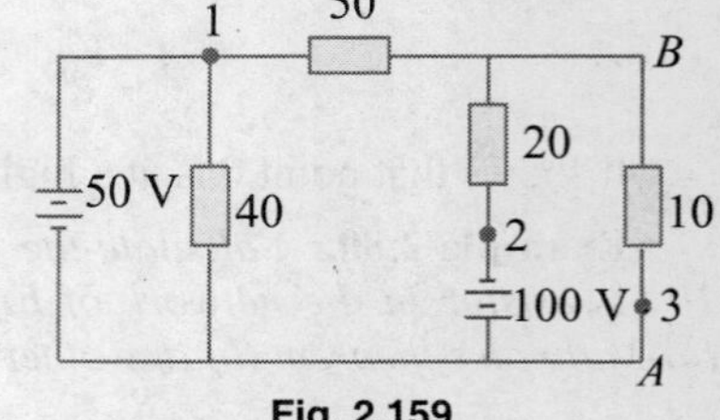

Fig. 2.159

$$\therefore V_{00}' = V_{AB} = \frac{(-50/50) + (100/20) + (0/10)}{(1/50) + (1/20) + (1/10)} = 23.5$$

V

Since the answer comes out to be positive, it means that point *A* is at a higher potential as compared to point *B*.

2.28 Maximum Power Transfer Theorem

Although applicable to all branches of electrical engineering, this theorem is particularly useful for analysing communication networks. The overall efficiency of a network supplying maximum power to any branch is 50 per cent. For this reason, the application of this theorem to power transmission and distribution networks is limited because, in their case, the goal is high efficiency and not maximum power transfer.

However, in the case of electronic and communication networks, very often, the goal is either to receive or transmit maximum power (through at reduced efficiency) specially when power involved is only a few milliwatts or microwatts. Frequently, the problem of maximum power transfer is of crucial significance in the operation of transmission lines and antennas.

As applied to d.c. networks, this theorem may be stated as follows:

A resistive load will abstract maximum power from a network when the load resistance is equal to the resistance of the network as viewed from the output terminals, with all energy sources removed leaving behind their internal resistances.

In Fig. 2.160 (*a*), a load resistance of R_L is connected across the terminals *A* and *B* of a network which consists of a generator of e.m.f. *E* and internal resistance R_g and a series resistance *R* which, in fact, represents the lumped resistance of the connecting wires. Let $R_i = R_g + R$ = internal resistance of the network as viewed from *A* and *B*.

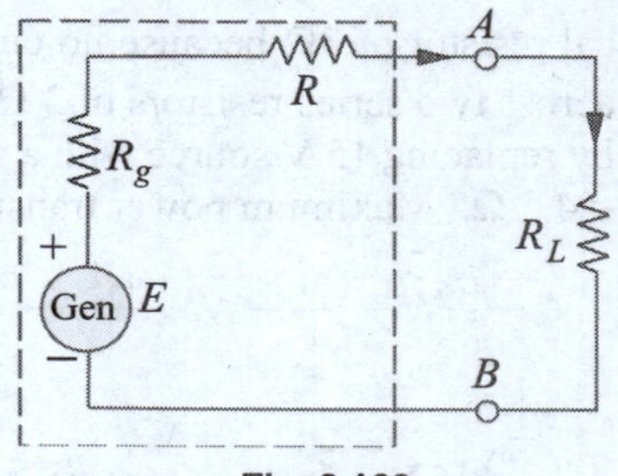

Fig. 2.160

According to this theorem, R_L will abstract maximum power from the network when $R_L = R_i$.

Proof. Circuit current $I = \dfrac{E}{R_L + R_i}$

Power consumed by the load is

$$P_L = I^2 R_L = \frac{E^2 R_L}{(R_L + R_i)^2} \qquad \text{...(}i\text{)}$$

For P_L to be maximu, $\dfrac{dP_L}{dR_L} = 0$.

Differentiating Eq. (*i*) above, we have

$$\frac{dP_L}{dR_L} = E^2 \left[\frac{1}{(R_L + R_i)^2} + R_L \left(\frac{-2}{(R_L + R_i)^3} \right) \right] = E^2 \left[\frac{1}{(R_L + R_i)^2} - \frac{2R_L}{(R_L + R_i)^3} \right]$$

$$\therefore \quad 0 = E^2 \left[\frac{1}{(R_L + R_i)^2} - \frac{2R_L}{(R_L + R_i)^3} \right] \quad \text{or} \quad 2R_L = R_L + R_i \quad \text{or} \quad R_L = R_i$$

It is worth noting that under these conditions, the voltage across the load is hold the open-circuit voltage at the terminals *A* and *B*.

$$\therefore \quad \text{Max. power is } P_{L\,max} = \frac{E^2 R_L}{4 R_L^2} = \frac{E^2}{4 R_L} = \frac{E^2}{4 R_i}$$

Let us consider an a.c. source of internal impedance $(R_1 + j X_1)$ supplying power to a load impedance $(R_L + jX_L)$. It can be proved that maximum power transfer will take place when the modules of the load impedance is equal to the modulus of the source impedance *i.e.* $|Z_L| = |Z_1|$

Where there is a completely free choice about the load, the maximum power transfer is obtained when load impedance is the complex conjugate of the source impedance. For example, if source impedance is $(R_1 + jX_1)$, then maximum transfer power occurs, when load impedance is $(R_1 - jX_1)$. It can be shown that under this condition, the load power is $= E^2/4R_1$.

Example 2.82. *In the network shown in Fig. 2.161 (a), find the value of R_L such that maximum possible power will be transferred to R_L. Find also the value of the maximum power and the power supplied by source under these conditions.*

(Elect. Engg. Paper I Indian Engg. Services)

Solution. We will remove R_L and find the equivalent Thevenin's source for the circuit to the left of terminals *A* and *B*. As seen from Fig. 2.161 (*b*) V_{th} equals the drop across the vertical resistor of 3Ω because no current flows through 2 Ω and 1 Ω resistors. Since 15 V drops across two series resistors of 3 Ω each, V_{th} = 15/2 = 7/5 V. Thevenin's resistance can be found by replacing 15 V source with a short-circuit. As seen from Fig. 2.161 (*b*), R_{th} = 2 + (3 || 3) + 1 = 4.5 Ω. Maximum power transfer to the load will take place when $R_L = R_{th}$ = 4.5 Ω.

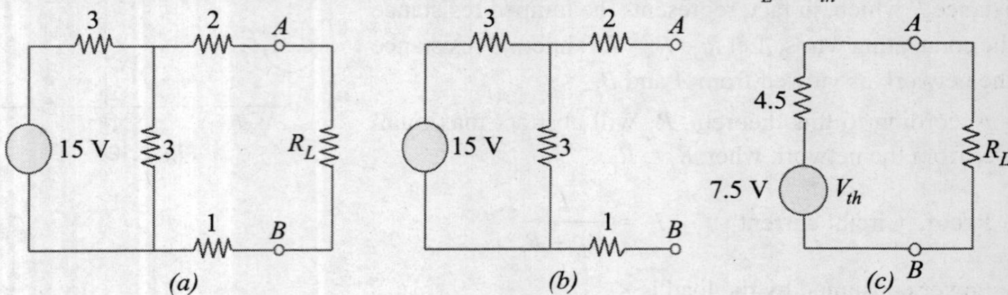

Fig. 2.161

Maximum power drawn by $R_L = V_{th}^2/4 \times R_L = 7.5^2/4 \times 4.5 = 3.125$ W.

Since same power in developed in R_{th}, power supplied by the source = 2 × 3.125 = 6.250 W.

Example 2.83. *In the circuit shown in Fig. 2.162 (a) obtain the condition from maximum power transfer to the load R_L. Hence determine the maximum power transferred.*

(Elect. Science-I Allahabad Univ. 1992)

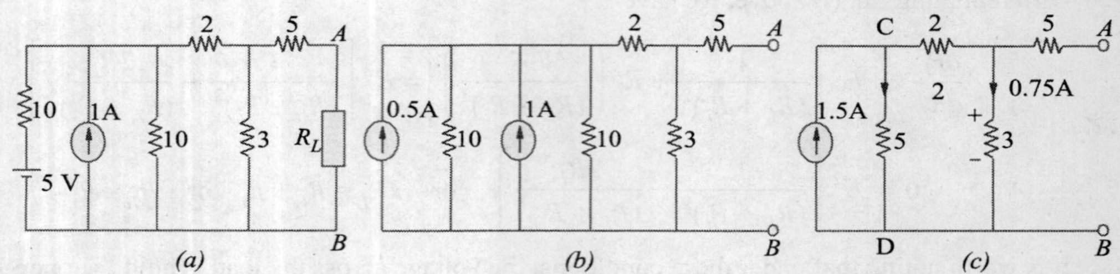

Fig. 2.162

Solution. We will find Thevenin's equivalent circuit to the left of trminals *A* and *B* for which purpose we will convert the battery source into a current source as shown in Fig. 2.162 (*b*). By combining the two current sources, we get the circuit of Fig. 2.162 (*c*). It would be seen that open circuit voltage V_{AB} equals the drop over 3 Ω resistance because there is no drop on the 5Ω resistance connected to terminal *A*. Now, there are two parallel path across the current source each of resistance 5 Ω. Hence, current through 3 Ω resistance equals 1.5/2 = 0.75 A. Therefore, $V_{AB} = V_{th} = 3 \times 0.75 = 2.25$ V with point *A* positive with respect to point *B*. For finding R_{AB}, current source is replaced by an infinite resistance.

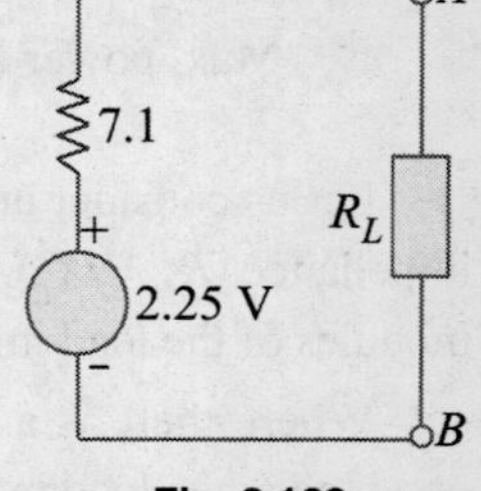

Fig. 2.163

$$\therefore \qquad R_{AB} = R_{th} = 5 + 3 \,||\, (2 + 5) = 7.1\ \Omega$$

The Thevenin's equivalent circuit alongwith R_L is shown in Fig. 2.163. As per the condition for *MPT* is that $R_L = 7.1\ \Omega$.

Maximum power transferred = $V_{th}^2 / 4R_L = 2.25^2/4 \times 7.1 = 0.178$ W = 178 mW.

Example 2.84. *Calculate the value of R which will absorb maximum power from the circuit of Fig. 2.164 (a). Also, compute the value of maximum power.*

Solution. For finding power, it is essential to know both *I* and *R*. Hence, it is essential to find an equation relating *I* to *R*.

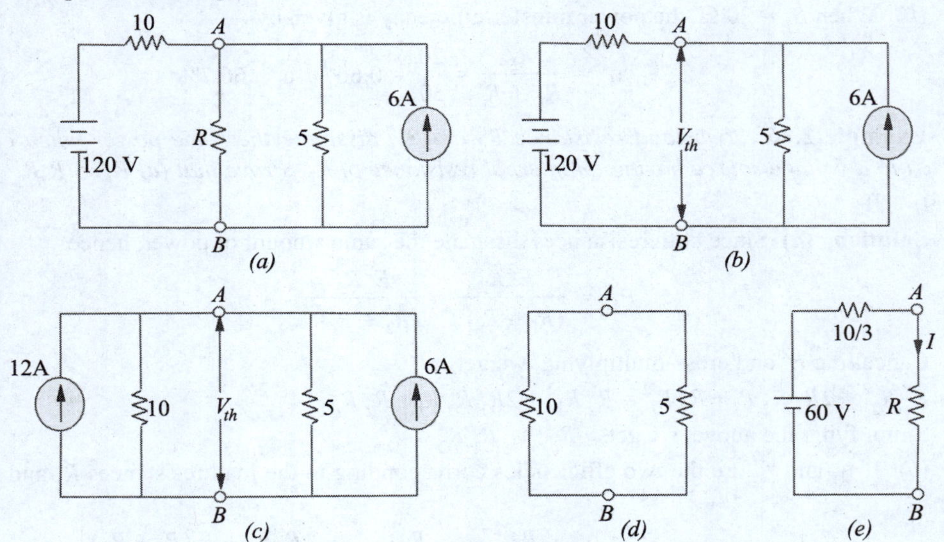

Fig. 2.164

Let us remove *R* and find Thevenin's voltage V_{th} across *A* and *B* as shown in Fig. 2.164 (*b*). It would be helpful to convert 120 V, 10-Ω source into a constant-current source as shown in Fig. 2.164 (*c*). Applying *KCL* to the circuit, we get

$$\frac{V_{th}}{10} + \frac{V_{th}}{5} = 12 + 6 \quad \text{or} \quad V_{th} = 60 \text{ V}$$

Now, for finding R_i and R_{th}, the two sources are reduced to zero. Voltage of the voltage-source is reduced to zero by short - circuiting it whereas current of the current source is reduced to zero by open-circuiting it. The circuit which results from such source suppression is shown in Fig. 2.164 (*d*). Hence, $R_i = R_{th} = 10 \parallel 5 = 10/3\ \Omega$. The Thevenin's equivalent circuit of the network is shown in Fig. 2.164 (*e*).

According to Maximum Power Transfer Theorem, *R* will absorb maximum power when it equals 10/3 Ω. In that case, $I = 60 \div 20/3 = 9$ A

$$P_{max} = I^2R = 9^2 \times 10/3 = \mathbf{270\ W}$$

Example 2.85. *A voltage source delivers 4 A when the load connected to it is 5 Ω and 2 A when the load becomes 20 Ω. Calculate*

(a) *maximum power which the source can supply* ***(b)*** *power transfer efficiency of the source with R_L of 20 Ω.*

Solution. We can find the values of E and R_i from the two given load conditions.

(*a*) When $R_L = 5\ \Omega$, $I = 4$ A and $V = IR_L = 4 \times 5 = 20$ V, then $20 = E - 4\,R_i$ ...(*i*)

When $R_L = 20\ \Omega$, $I = 2$ A and $V = IR_L = 2 \times 20 = 40$ V $\therefore$ $40 = E - 2\,R_i$ ...(*ii*)

From (*i*) and (*ii*), we get, $R_i = 10\ \Omega$ and $E = 60$ V

When $R_L = R_i = 10\ \Omega$

$$P_{L\,max} = \frac{E^2}{4R_i} = \frac{60 \times 60}{4 \times 10} = 90 \text{ W}$$

(*b*) When $R_L = 20\ \Omega$, the power transfer efficiency is given by

$$\eta = \frac{R_L}{R_L + R_i} = \frac{20}{30} = 0.667 \quad \text{or} \quad 66.7\%$$

Example 2.86. *Two load resistance R_1 and R_2 dissipate the same power when connected to a voltage source having an internal resistance of R_i. Prove that **(a)** $R_i^2 = R_1R_2$ and **(b)** $\eta_1 + \eta_2 = 1$.*

Solution. (*a*) Since both resistances dissipate the same amount of power, hence

$$P_L = \frac{E^2 R_1}{(R_1 + R_i)^2} = \frac{E^2 R_2}{(R_2 + R_i)^2}$$

Cancelling E^2 and cross-multiplying, we get

$R_1\,R_2^{\,2} + 2R_1\,R_2\,R_i + R_1\,R_i^2 = R_2\,R_1^{\,2} + 2R_1\,R_2\,R_i + R_2\,R_i^2$

Simplifying the above, we get, $R_i^2 = R_1\,R_2$

(*b*) If η_1 and η_2 are the two efficiencies corresponding to the load resistances R_1 and R_2, then

$$\eta_1 + \eta_2 = \frac{R_1}{R_1 + R_i} + \frac{R_2}{R_2 + R_i} = \frac{2\,R_1\,R_2 + R_i(R_1 + R_2)}{R_1R_2 + R_i^2 + R_i(R_1 + R_2)}$$

Substituting $R_i^2 = R_1\,R_2$, we get

$$\eta_1 + \eta_2 = \frac{2\,R_i^2 + R_i(R_1 + R_2)}{2R_i^2 + R_i(R_1 + R_2)} = 1$$

Example 2.87. *Determine the value of R_L for maximum power at the load. Determine maximum power also. The network is given in the Fig. 2.165 (a).* **[Bombay University 2001]**

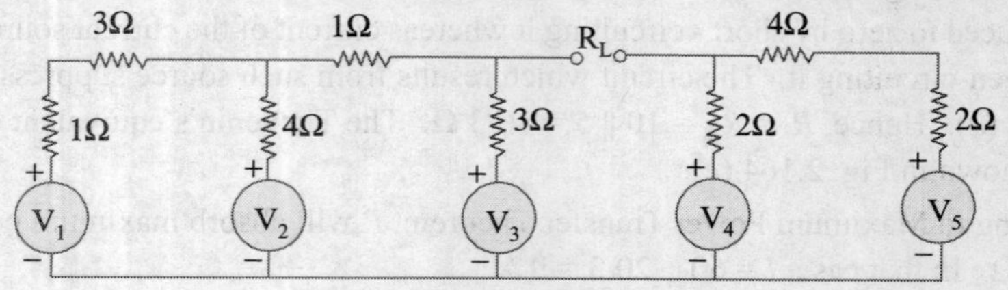

Fig. 2.165 (*a*)

Solution. This can be attempted by Thevenin's Theorem. As in the circuit, with terminals A and B kept open, from the right hand side, V_B (w.r. to reference node 0) can be calculated V_4 and V_5 will have a net voltage of 2 volts circulating a current of (2/8) = 0.25 amp in clockwise direction.

$$V_B = 10 - 0.25 \times 2 = 9.5 \text{ volts.}$$

On the Left-hand part of the circuit, two loops are there. V_A (w.r. to 0) has to be evaluated. Let the first loop (with V_1 and V_2 as the sources) carry a clockwise current of i_1 and the second loop (with V_2 and V_3 as the sources), a clockwise current of i_2. Writing the circuit equations.

$$8i - 4i_2 = +4$$
$$-4i + 8i_2 = +4$$

This gives $i_1 = 1$ amp, $i_2 = 1$ amp

Therefore, $V_A = 12 + 3 \times 1 = 15$ volts.

Thevenin – voltage, $V_{TH} = V_A - V_B = 15 - 9.5 = 5.5$ volts

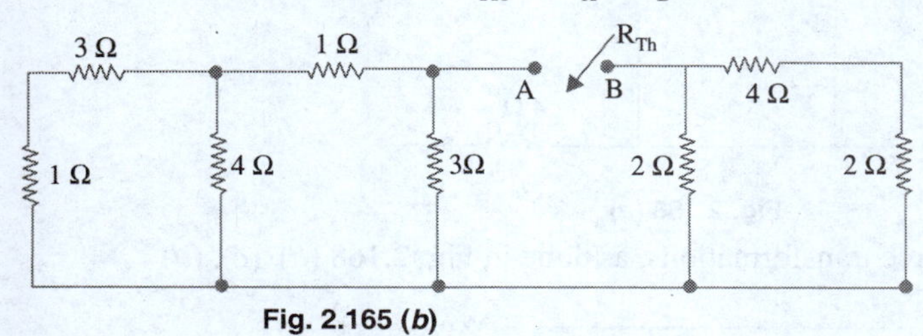

Fig. 2.165 (*b*)

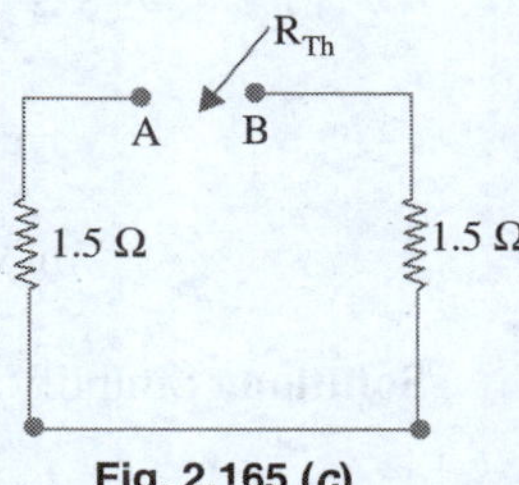

Fig. 2.165 (*c*)

Solving as shown in Fig. 2.165 (*b*) and (*c*).

$R_{TH} = 3$ ohms

For maximum power transfer, $R_L = 3$ ohms

Current = 5.5/6 = 0.9167 amp

Power transferred to load $0.9167^2 \times 3 = 2.52$ watts.

Example 2.88. *For the circuit shown below, what will be the value of R_L to get the maximum power ? What is the maximum power delivered to the load ?* **[Bombay University 2001]**

Solution. Detach R_L and apply Thevenin's Theorem.

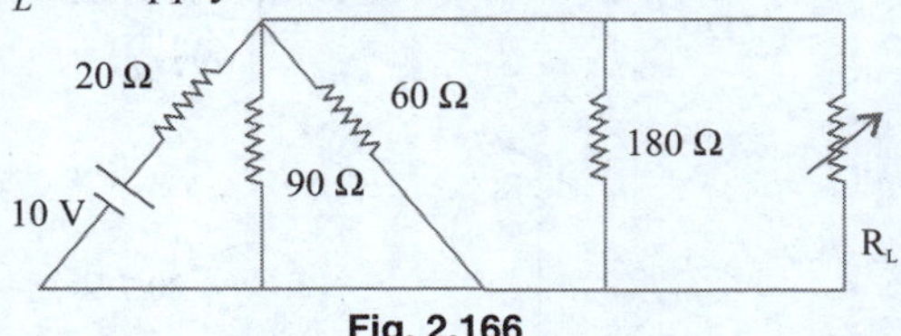

Fig. 2.166

$V_{TH} = 5.696$ volts, $R_{TH} = 11.39\ \Omega$

R_L must be 11.39 ohms for maximum power transfer.

$P_{\max} = 0.712$ watt.

Example 2.89. *Find the maximum power in 'R_L' which is variable in the circuit shown below in Fig. 2.167.* **[Bombay University, 2001]**

Solution. Apply Thevenin's theorem. For this R_L has to be detached from nodes *A* and *B*. Treat *O* as the reference node.

$V_A = 60$ V, $V_B = V_C + 2 = 50 + 2 = 52$ V

Thus, $V_{TH} = V_{AB} = 8$ volts, with A positive w.r. to *B*, $R_{TH} = (60//40) + (50//50) = 49$ ohms

Hence, for maximum power, $R_L = 49$ ohms

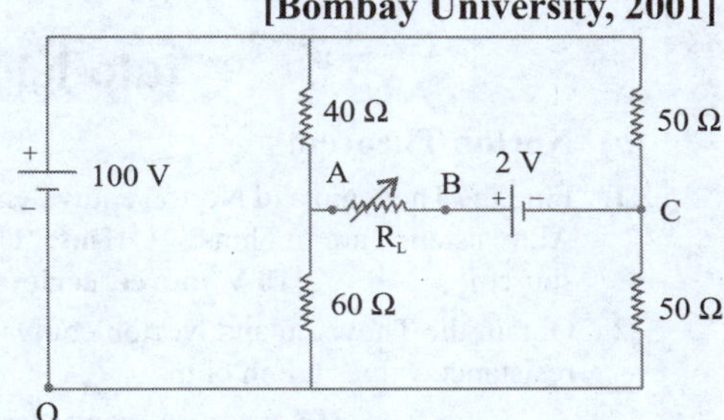

Fig. 2.167

With this R_L, Current = 8/98 amp = 0.08163 amp

Power to Load = $i^2 R_L$ = 0.3265 watt

Example 2.90. *Find the magnitude R_L for the maximum power transfer in the circuit shown in Fig. 2.168 (a). Also find out the maximum power.*

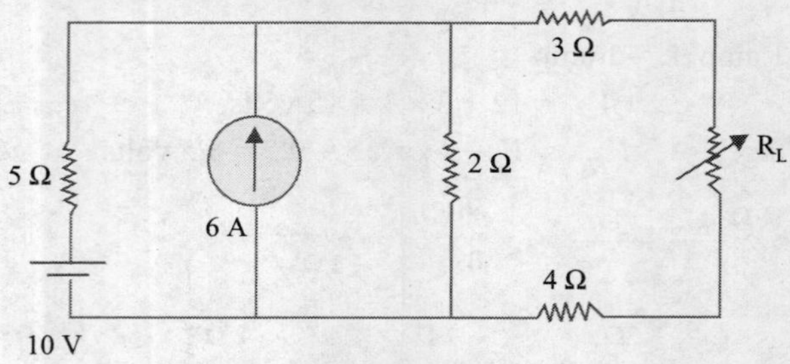

Fig. 2.168 (*a*)

Solution. Simplify by source transformations, as done in Fig. 2.168 (*b*), (*c*), (*d*)

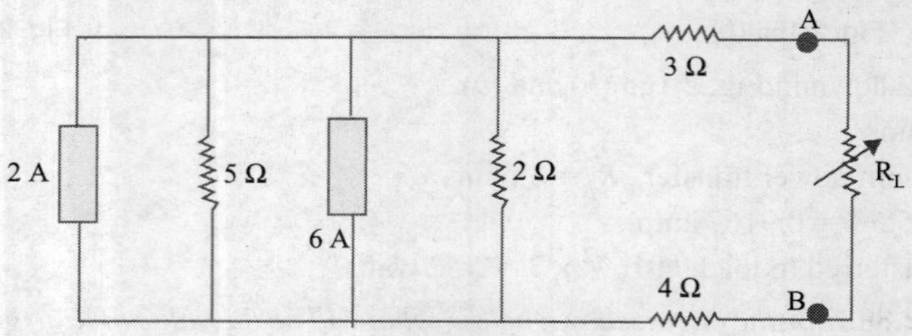

Fig. 2.168 (*b*)

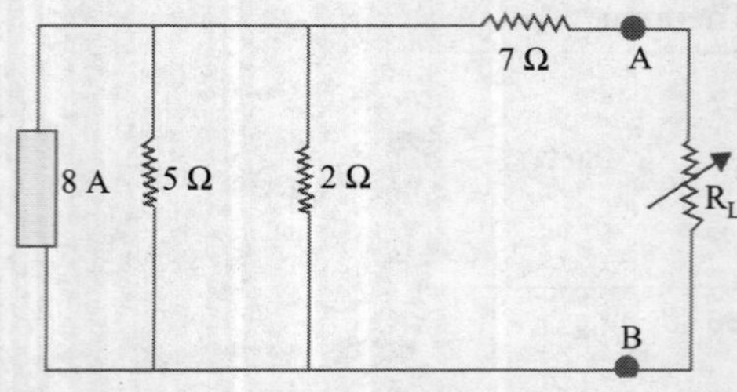

Fig. 2.168 (*c*)

Fig. 2.168 (*d*)

For maximum power, R_L = 7 + (10/7) = 8.43 Ω

Maximum power = $[(80/7)/16.68]^2 \times 8.43$ = 3.87 watts.

Tutorial Problem 2.7

(*a*) Norton Theorem

1. Find the Thevenin and Norton equivalent circuits for the active network shown in Fig. 2.169 (*a*). All resistance are in ohms. **[Hint:** Use Superposition principle to find contribution of each source] **[10 V source, series resistor = 5 Ω ; 2 A source, parallel resistance = 5 Ω]**
2. Obtain the Thevenin and Norton equivalent circuits for the circuit shown in Fig. 2.169 (b). All resistance values are in ohms.

   [15 V source, series resistance = 5 Ω ; 3 A source, parallel resistance = 5 Ω]

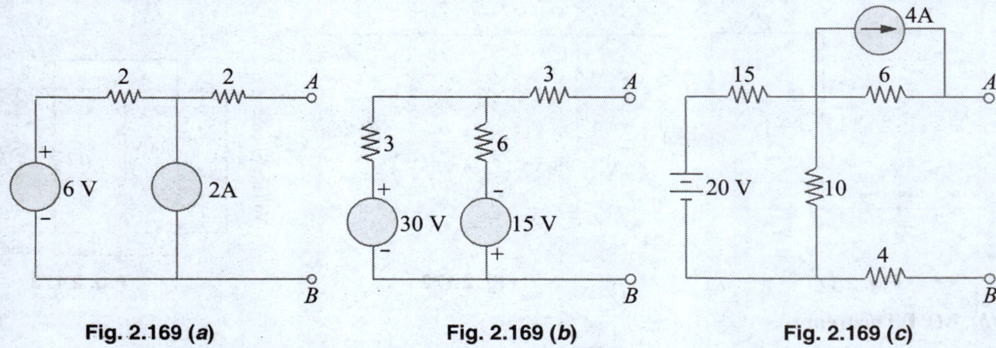

Fig. 2.169 (*a*) Fig. 2.169 (*b*) Fig. 2.169 (*c*)

3. Find the Norton equivalent circuit for the active linear network shown in Fig. 2.169 (*c*). All resistances are in ohms. **[Hint:** It would be easier to first find Thevenin's equivalent circuit].

[2 A source; parallel resistance = 16 Ω]

4. State the Tellegen's theorem and verify it by an illustration. Comment on the applicability of Tellegen's theorem on the types of networks.**(*Circuit and Field Theory, A.M.I.E. Sec. B, 1993*)**

Solution. Tellegen's Theorem can be stated as under:

For a network consisting of n elements if i_1, i_2.....i_n are the currents flowing through the elements satisfying Kirchhoff's current law and v_1, v_2......v_n are the voltages across these elements satisfying Kirchhoff's law, then

$$\sum_{k=1}^{n} v_k i_k = 0$$

where v_k is the voltage across and i_k is the current through the k_{th} element. In other words, according to Tellegen's Theorem, the sum of instantaneous powers for the n branches in a network is always zero.

This theorem has wide applications. It is valid for any lumped network that contains any elements linear or non-linear, passive or active, time-variant or time-invariant.

Explanation: This theorem will be explained with the help of the simple circuit shown in Fig. 2.171. The total resistance seen by the battery is = 8 + 4 || 4 = 10 Ω.

Battery current I = 100/10 = 10 A. This current divides equally at point B,

Drop over 8 Ω resistor = 8 × 10 = 80 V

Drop over 4 Ω resistor = 4 × 5 = 20 V

Drop over 1 Ω resistor = 1 × 5 = 5 V

Drop over 3 Ω resistor = 3 × 5 = 15 V

According to Tellegen's Theorem,

= 100 × 10 − 80 × 10 − 20 × 5 − 5 × 5 − 15 × 5 = 0

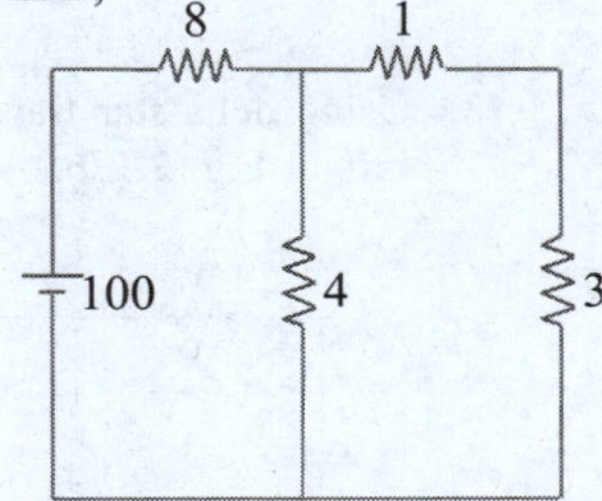

Fig. 2.170

(*b*) Millman's Theorem

5. Use Millman's theorem, to find the potential of point A with respect to the ground in Fig. 2.171. **[V_A = 8.18 V]**

6. Using Millman's theorem, find the value of output voltage V_0 in the circuit of Fig. 2.172. All resistances are in ohms. **[4 V]**

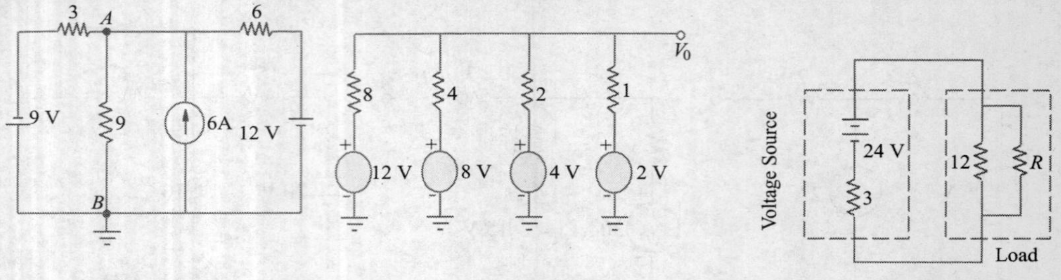

Fig. 2.171 Fig. 2.172 Fig. 2.173

(*b*) MPT Theorem

8. In Fig. 2.173 what value of R will allow maximum power transfer to the load ? Also calculate the maximum total load power. All resistances are in ohms. **[4 Ω ; 48 W]**

9. Find the total current flowing through the circuit shown in Fig. 2.174 using stat-delta transformation if the circuit is excited by 39 volts and the value of each resistor connected in circuit is 4 ohms.

(Ravishankar University, Raipur 2003) (Nagpur University, Summer 2003)

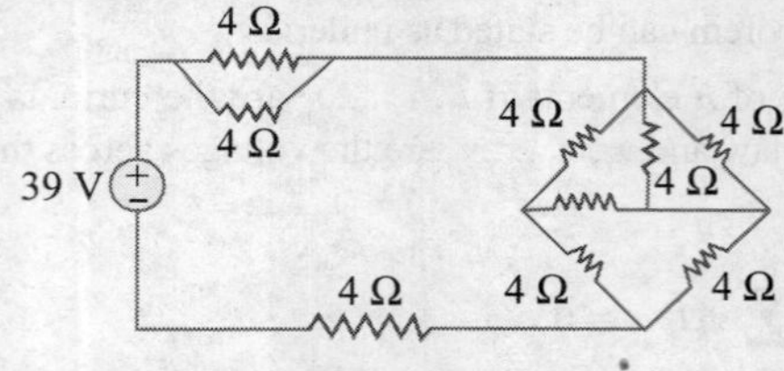

Fig. 2.174

10. State and explain Superposition Theorem.

(Pune University 2003) (Nagpur University, Summer 2004)

11. A cast iron ring of 40 cm diameter is wound with a coil. The coil carries a current of 3 amp and produces a flux of 3 mwb in the air gap. The length of air gap is 2 mm. The relative permeability of the cast iron is 800. The leakage coefficient is 1.2. Calculate no. of turns of the coil. ***(Nagpur University, Summer 2004)***

12. Using superposition theorem, calculate the current I_{AB} in the given circuit of Fig. 2.175.

(Gujrat University, Summer 2003)

13. Using delta-star transformation, determine the current drawn from the source in the given circuit Fig. 2.176. ***(Gujrat University,Summer 2003)***

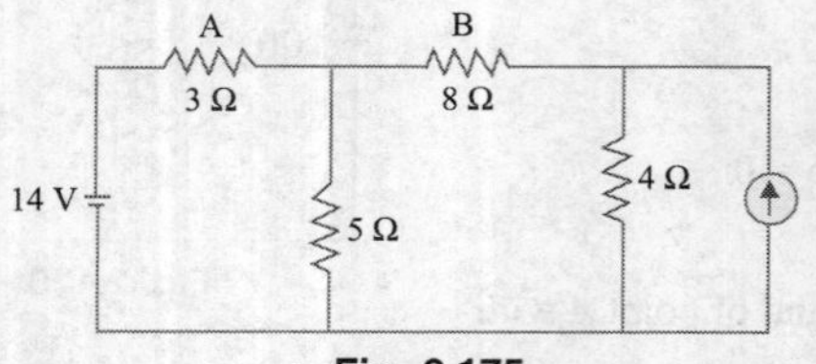

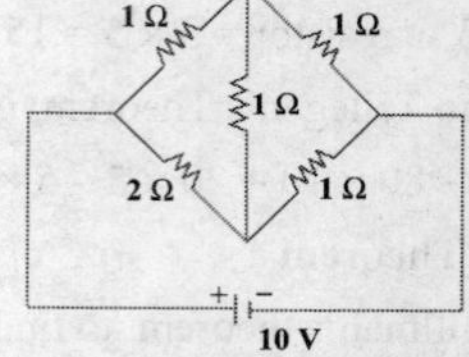

Fig. 2.175 Fig. 2.176

14. State and explain Kirchhoff's laws applied to electric circuit.

(Gujrat University, Summer2003)

15. State Kirchhoff's laws. ***(Madras University, April 2002)***

16. Three resistances R_{ab}, *Rbc* and *Rca* are connected in delta. Obtain expressions for their equivalent star resistances. ***(V.T.U., Belgaum Karnataka University, February 2002)***

OBJECTIVE TEST QUESTIONS

1. Kirchhoff's current law is applicable to only
   (*a*) closed loops in a network
   (*b*) electronic circuits
   (*c*) junctions in a network
   (*d*) electric circuits.
2. Kirchhoff's voltage law is concerned with
   (*a*) *IR* drops
   (*b*) battery e.m.fs.
   (*c*) junction voltages
   (*d*) both (*a*) and (*b*)
3. According to *KVL*, the *algebraic* sum of all *IR* drops and e.m.f.s in any closed loop of a network is always
   (*a*) zero
   (*b*) positive
   (*c*) negative
   (*d*) determined by battery e.m.fs.
4. The ***algebraic*** sign of an *IR* drop is primarily dependent upon the
   (*a*) amount of current flowing through it
   (*b*) value of *R*
   (*c*) direction of current flow
   (*d*) battery connection.
5. Maxwell's loop current method of solving electrical networks
   (*a*) uses branch currents
   (*b*) utilizes Kirchhoff's voltage law
   (*c*) is confined to single-loop circuits
   (*d*) is a network reduction method.
6. Point out of the WRONG statement. In the node-voltage technique of solving networks, choice of a reference node does not
   (*a*) affect the operation of the circuit
   (*b*) change the voltage across any element
   (*c*) alter the p.d. between any pair of nodes
   (*d*) affect the voltages of various nodes.
7. For the circuit shown in the given Fig. 2.177, when the voltage E is 10 V, the current i is 1 A. If the applied woltage across terminal C-D is 100 V, the short circuit current flowing through the terminal A-B will be

   (ESE 2001)

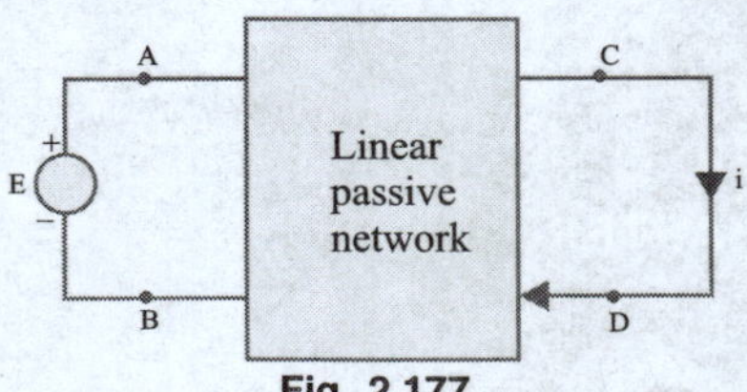

Fig. 2.177

   (*a*) 0.1 A (*b*) 1 A
   (*c*) 10 A (*d*) 100 A
8. The component inductance due to the internal flux-linkage of a non-magnetic straight solid circular conductor per metre length, has a constant value, and is independent of the conductor-diameter, because
   (*a*) All the internal flux due to a current remains concentrated on the peripheral region of the conductor.
   (*b*) The internal magnetic flux-density along the radial distance from the centre of the conductor increases proportionately to the current enclosed
   (*c*) The entire current is assumed to flow along the conductor-axis and the internal flux is distributed uniformly and concentrically
   (*d*) The current in the conductor is assumed to be uniformly distributed throughout the conductor cross-section

   (ESE 2003)
9. Two ac sources feed a common variable resistive load as shown n in Fig. 2.178. Under the maximum power transfer condition, the power absorbed by the load resistance R_L is

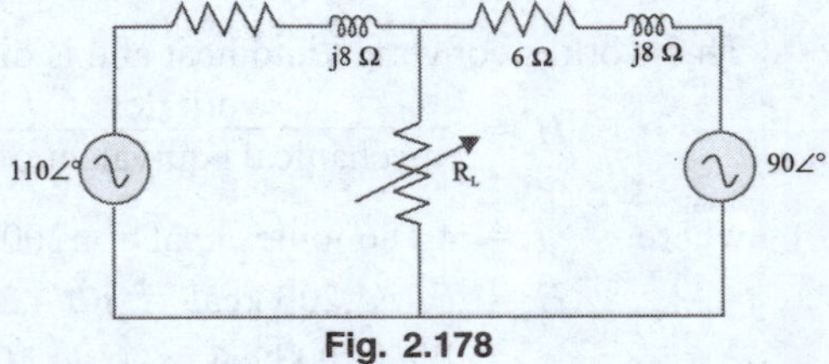

Fig. 2.178

   (*a*) 2200 W (*b*) 1250W
   (*c*) 1000 W (*d*) 625 W

   (GATE 2003)

CHAPTER

3

Work, Power and Energy

Chapter Outline

- Effect of Electric Current
- Joule's Law of Electric Heating
- Thermal Efficiency
- S-I. Units
- Calculation of Kilo-watt Power of a Hydroelectric Station

3.1 Effect of Electric Current

It is a matter of common experience that a conductor, when carrying current, becomes hot after some time. As explained earlier, an electric current is just a directed flow or drift of electrons through a substance. The moving electrons as they pass through molecules of atoms of that substance, collide with other electrons. This electronic collision results in the production of heat. This explains why passage of current is always accompanied by generation of heat.

3.2 Joule's Law of Electric Heating

The amount of work required to maintain a current of I amperes through a resistance of R ohm for t second is

$$\begin{aligned} \text{W.D.} &= I^2 Rt \text{ joules} \\ &= VIt \text{ joules} \quad (\because R = V/I) \\ &= Wt \text{ joules} \quad (\because W = VI) \\ &= V^2t/R \text{ joules} \quad (\because I = V/R) \end{aligned}$$

This work is converted into heat and is dissipated away. The amount of heat produced is

$$H = \frac{\text{work done}}{\text{mechanical equivalent of heat}} = \frac{W.D.}{J}$$

where J = 4,186 joules/kcal = 4,200 joules / kcal (approx)

$$\therefore \quad H = I^2Rt/4{,}200 \text{ kcal} = VIt/4{,}200 \text{ kcal}$$

$$= Wt/4{,}200 \text{ kcal} = V^2t/4{,}200\, R \text{ kcal}$$

3.3 Thermal Efficiency

It is defined as the ratio of the heat actually utilized to the total heat produced electrically. Consider the case of the electric kettle used for boiling water. Out of the total heat produced **(*i*)** some goes to heat the apparatus itself *i.e.* kettle **(*ii*)** some is lost by radiation and convection etc. and **(*iii*)** the rest is utilized for heating the water. Out of these, the heat utilized for useful purpose is that in **(*iii*)**. Hence, thermal efficiency of this electric apparatus is the ratio of the heat utilized for heating the water to the total heat produced.

Hence, the relation between heat produced electrically and heat absorbed usefully becomes

$$\frac{VIt}{J} \times \eta = ms\,(\theta_2 - \theta_1)$$

Example 3.1. *The heater element of an electric kettle has a constant resistance of 100 Ω and the applied voltage is 250 V. Calculate the time taken to raise the temperature of one litre of water from 15°C to 90°C assuming that 85% of the power input to the kettle is usefully employed. If the water equivalent of the kettle is 100 g, find how long will it take to raise a second litre of water through the same temperature range immediately after the first.*

(Electrical Engineering, Calcutta Univ.)

Solution. Mass of water = 1000 g = 1 kg ($\because$ 1 cm^3 weight 1 gram)

Heat taken by water = 1 × (90 – 15) = 75 kcal

Heat taken by the kettle = 0.1 × (90 – 15) = 7.5 kcal

Total heat taken = 75 + 7.5 = 82.5 kcal

Heat produced electrically $H = I^2Rt/J$ kcal

Now, I = 250/100 = 2/5 A, J = 4,200 J/kcal; $H = 2.5^2 \times 100 \times t/4200$ kcal

Heat actually utilized for heating one litre of water and kettle

= $0.85 \times 2.5^2 \times 100 \times t/4{,}200$ kcal

$$\therefore \quad \frac{0.85 \times 6.25 \times 100 \times t}{4{,}200} = 82.5 \quad \therefore \quad t = \mathbf{10\ min\ 52\ second}$$

In the second case, heat would be required only for heating the water because kettle would be already hot.

$$\therefore \quad 75 = \frac{0.85 \times 6.25 \times 100 \times t}{4{,}200} \quad \therefore \quad t = \mathbf{9\ min\ 53\ second}$$

Example 3.2. *Two heater A and B are in parallel across supply voltage V. Heater A produces 500 kcal in 200 min. and B produces 1000 kcal in 10 min. The resistance of A is 10 ohm. What is the resistance of B ? If the same heaters are connected in series across the voltage V, how much heat will be prduced in kcal in 5 min ?* **(Elect. Science - II, Allahabad Univ. 1992)**

Solution. Heat produced $= \dfrac{V^2 t}{JR}$ kcal

For heater A, $$500 = \frac{V^2 \times (20 \times 60)}{10 \times J} \quad \text{...(i)}$$

For heater B, $$1000 = \frac{V^2 \times (10 \times 60)}{R \times J} \quad \text{...(ii)}$$

From Eq. **(*i*)** and **(*ii*)**, we get, R = **2.5 Ω.**

When the two heaters are connected in series, let H be the amount of heat produced in kcal. Since combined resistance is (10 + 2.5) = 12.5 Ω, hence

$$H = \frac{V^2 \times (5 \times 60)}{12.5 \times J} \quad \text{...(iii)}$$

Dividing Eq. **(*iii*)** by Eq. **(*i*)**, we have H = **100 kcal.**

Example 3.3. *An electric kettle needs six minutes to boil 2 kg of water from the initial temperature of 20°C. The cost of electrical energy required for this operation is 12 paise, the rate being 40 paise per kWh. Find the kW-rating and the overall efficiency of the kettle.*

(F.Y. Engg. Pune Univ.)

Solution. Input energy to the kettle $= \dfrac{12 \text{ paise}}{40 \text{ paise/kWh}} = 0.3 \text{ kWh}$

$$\text{Input power} = \frac{\text{energy in kWh}}{\text{Time in hours}} = \frac{0.3}{(6/60)} = 3 \text{ kW}$$

Hence, the power rating of the electric kettle is 3 kW

Energy utilised in heating the water

$= mst = 2 \times 1 \times (100 - 20) = 160$ kcal $= 160/860$ kWh $= 0.186$ kWh.

Efficiency = output/input = 0.186/0.3 = 0.62 = **62%.**

3.4 S.I. Units

1. Mass. It is quantity of matter contained in a body.

Unit of mass is kilogram (kg). Other multiples commonly used are :

1 quintal = 100 kg, 1 tonne = 10 quintals = 1000 kg

2. Force. Unit of force is newton (N). Its definition may be obtained from Newton's Second Law of Motion *i.e.* $F = ma$.

If $m = 1$ kg ; $a = 1 \text{m/s}^2$, then $F = 1$ newton.

Hence, one newton is that force which can give an acceleration of 1 m/s^2 to a mass of 1 kg.

Gravitational unit of force is kilogram-weight (kg-wt). It may be defined as follows :

It is the force which can impart an acceleration of 9.8 m/s^2 to a mass of 1 kg.

or

It is the force which can impart an acceleration of 1 m/s^2 to a mass of 9.8 kg.

Obviously, 1 kg-wt. = 9.8 N

3. Weight. It is the force with which earth pulls a body downwards. Obviously, its units are the same as for force.

(*a*) Unit of weight is newton (N)

(*b*) Gravitational unit of weight is kg-wt.*

Note. If a body has a mass of m kg, then its weight, W = mg newtons = 9.8 newtons.

4. Work, If a force F moves a body through a distance S in its direction of application, then

Work done $W = F \times S$

(*a*) Unit of work is joule (J).

If, in the above equation, $F = 1$ N : $S = 1$ m ; then work done = 1 m.N or joule.

Hence, one joule is the work done when a force of 1 N moves a body through a distance of 1 m in the direction of its application.

(*b*) Gravitational unit of work is m-kg. wt or m-kg**.

* Often it is referred to as a force of 1 kg, the word 'wt' being omitted. To avoid confusion with mass of 1 kg, the force of 1 kg is written in engineering literature as kgf instead of kg. wt.

** Generally the work 'wt' is omitted and the unit is simply written as m-kg.

If $F = 1$ kg-wt; $S = 1$ m; then W.D. = 1 m-kg. Wt = 1 m-kg.

Hence, one m-kg is the work done by a force of one kg-wt when applied over a distance of one metre.

Obviously, 1 m-kg = 9.8 m-N or J.

5. Power. It is the rate of doing work. Its units is watt (W) which represents 1 joule per second.

$$1 \text{ W} = 1 \text{ J/s}$$

If a force of F newton moves a body with a velocity of v m./s then

$$\text{power} = F \times \text{v watt}$$

If the velocity v is in km/s, then

$$\text{power} = F \times \text{v kilowatt}$$

6. Kilowatt-hour (kWh) and kilocalorie (kcal)

$$1 \text{ kWh} = 1000 \times 1\frac{\text{J}}{\text{s}} \times 3600 \text{ s} = 36 \times 10^5 \text{ J}$$

$$1 \text{ kcal} = 4{,}186 \text{ J} \quad \therefore \; 1 \text{ kWh} = 36 \times 10^5/4,186 = 860 \text{ kcal}$$

7. Miscellaneous Units

(i) 1 watt hour (Wh) = $1\frac{\text{J}}{\text{s}} \times 3600$ s = 3600 J

(ii) 1 horse power (metric) = 75 m-kg/s = 75 × 9.8 = 735.5 J/s or watt

(iii) 1 kilowatt (kW) = 1000 W and 1 megawatt (MW) = 10^6 W

3.5 Calculation of Kilo-Watt Power of a Hydroelectric Station

Let Q = water discharge rate in cubic metres/second (m^3/s), H = net water head in metre (m). $g = 9.81$, η ; overall efficiency of the hydroelectric station expressed as a fraction.

Since 1 m^3 of water weighs 1000 kg., discharge rate is 1000 Q kg/s.

When this amount of water falls through a height of H metre, then energy or work available per second or available power is

$$= 1000\, QgH \text{ J/s} \quad \text{or} \quad \text{W} = QgH \text{ kW}$$

Since the overall station efficiency is η, power actually available is = 9.81 η QH kW.

Example 3.4. *A de-icing equipment fitted to a radio aerial consists of a length of a resistance wire so arranged that when a current is passed through it, parts of the aerial become warm. The resistance wire dissipates 1250 W when 50 V is maintained across its ends. It is connected to a d.c. supply by 100 metres of this copper wire, each conductor of which has resistance of 0.006 Ω/m.*

Calculate

(a) *the current in the resistance wire*

(b) *the power lost in the copper connecting wire*

(c) *the supply voltage required to maintain 50 V across the heater itself.*

Solution. ***(a)*** Current = wattage/voltage = 1250/50 = **25 A**

(b) Resistance of one copper conductor = 0.006 × 100 = 0.6 Ω

Resistance of both copper conductors = 0.6 × 2 = 1.2 Ω

Power loss = I^2R watts = 252 × 1.2 = **750 W**

(c) Voltage drop over connecting copper wire = IR volt = 25 × 1.2 = 30 V

∴ Supply voltage required = 50 + 30 = **80 V**

Example 3.5. *A factory has a 240-V supply from which the following loads are taken :*

Lighting : Three hundred 150-W, four hundred 100 W and five hundred 60-W lamps

Heating : 100 kW

Motors : A total of 44.76 kW (60 b.h.p.) with an average efficiency of 75 percent

Misc. : Various load taking a current of 40 A.

Assuming that the lighting load is on for a period of 4 hours/day, the heating for 10 hours per day and the remainder for 2 hours/day, calculate the weekly consumption of the factory in kWh when working on a 5-day week.

What current is taken when the lighting load only is switched on ?

Solution. The power consumed by each load can be tabulated as given below :

Power consumed

Lighting $300 \times 150 = 45{,}000 = 45$ kW

$400 \times 100 = 40{,}000 = 40$ kW

$500 \times 60 = 30{,}000 = \underline{30 \text{ kW}}$

Total = 115 kW

Heating = 100 kW

Motors = 44.76/0.75 = 59.7 kW

Misc. = $240 \times 40/1000 = 9.6$ kW

Similarly, the energy consumed/day can be tabulated as follows :

Energy consumed / day

Lighting	= 115 kW × 4 hr	= 460 kWh
Heating	= 100 kW × 10 hr	= 1,000 kWh
Motors	= 59.7 kW × 2 hr	= 119.4 kWh
Misc.	= 9.6 kW × 2 hr	= 19.2 kWh
Total daily consumption		= 1,598.6 kWh
Weekly consumption		= 1,598.6 × 5 = **7,993 kWh**
Current taken by the lighting load alone		= 115 × 1000/240 = **479 A**

Example 3.6. *A Diesel-electric generating set supplies an output of 25 kW. The calorific value of the fuel oil used is 12,500 kcal/kg. If the overall efficiency of the unit is 35%* ***(a)*** *calculate the mass of oil required per hour* ***(b)*** *the electric energy generated per tonne of the fuel.*

Solution. Output = 25 kW, Overall $\eta = 0.35$,

Input = 25/0.35 = 71.4 kW

$\therefore$ input per hour = 71.4 kWh = $71.4 \times 860 = 61{,}400$ kcal

Since 1 kg of fuel-oil produces 12,500 kcal

(*a*) $\therefore$ mass of oil required = 61,400/12,500 = **4.91 kg**

(*b*) 1 tonne of fuel = 1000 kg

Heat content = $1000 \times 12{,}500 = 12.5 \times 10^6$ kcal

= $12.5 \times 10^6/860 = 14{,}530$ kWh

Overall $\eta = 0.35\%$ $\therefore$ energy output = $14{,}530 \times 0.35 =$ **5,088 kWh**

Example 3.7. *A room measures 4 m × 7 m × 5 m and the air in it has to be always kept 15°C higher than that of the incoming air. The air inside has to be renewed every 35 minutes. Neglecting radiation loss, calculate the rating of the heater suitable for this purpose. Take specific heat of air as 0.24 and density as 1.27 kg/m³.*

Solution. Volume of air to be changed per second = $4 \times 7 \times 5/35 = 60 = 1/15$ m³

Mass of air to be changed/second = (1/15) × 1.27 kg

Heat required/second = mass/second × sp. heat × rise in temp.

= (1.27/15) × 0.24 × 15 kcal/s = 0.305 kcal/s

= 0.305 × 4186 J/s = **1277 watt.**

Example 3.8. *A motor is being self-started against a resisting torque of 60 N-m and at each start, the engine is cranked at 75 r.p.m. for 8 seconds. For each start, energy is drawn from a lead-acid battery. If the battery has the capacity of 100 Wh, calculate the number of starts that can be made with such a battery. Assume an overall efficiency of the motor and gears as 25%.*

(Principles of Elect. Engg.-I, Jadavpur Univ.)

Solution. Angular speed ω = 2π $N/60$ rad/s = 2π × 75/60 = 7.85 rad/s

Power required for rotating the engine at this angular speed is

P = torque × angular speed = ωT watt = 60 × 7.85 = **471 W**

Energy required per start is = power × time per start = 471 × 8 = 3,768 watt-s = 3,768 J

= 3,768/3600 = 1.047 Wh

Energy drawn from the battery taking into consideration the efficiency of the motor and gearing = 1.047/0.25 = 4.188 Wh

No. of start possible with a fully-charged battery = 100/4.188 = **24 (approx.)**

Example 3.9. *Find the amount of electrical energy expended in raising the temperature of 45 litres of water by 75ºC. To what height could a weight of 5 tonnes be raised with the expenditure of the same energy ? Assume efficiencies of the heating equipment and lifting equipment to be 90% and 70% respectively.* **(Elect. Engg. A.M. Ae. S.I.)**

Solution. Mass of water heated = 45 kg. Heat required = 45 × 75 = 3,375 kcal

Heat produced electrically = 3,375/0.9 = 3,750 kcal. Now, 1 kcal = 4,186 J

∴ electrical energy expended = 3,750 × 4,186 J

Energy available for lifting the load is = 0.7 × 3,750 × 4,186 J

If h metre is the height through which the load of 5 tonnes can be lifted, then potential energy of the load = mgh joules = 5 × 1000 × 9.81 h joules

∴ $5000 \times 9.81 \times h = 0.7 \times 3{,}750 \times 4{,}186$ ∴ h = **224 metres**

Example 3.10. *An hydro-electric station has a turbine of efficiency 86% and a generator of efficiency 92%. The effective head of water is 150 m. Calculate the volume of water used when delivering a load of 40 MW for 6 hours. Water weighs 1000 kg/m³.*

Solution. Energy output = 40 × 6 = 240 MWh

$= 240 \times 10^3 \times 36 \times 10^5 = 864 \times 10^9$ J

Overall $\eta = 0.86 \times 0.92$ ∴ Energy input = $\dfrac{864 \times 10^9}{0.86 \times 0.92} = 10.92 \times 10^{11}$ J

Since the head is 150 m and 1 m³ of water weighs 1000 kg, energy contributed by each m³ of water = 150 × 1000 m-kg (wt) = 150 × 1000 × 9.81 J = 147.2×10^4 J

∴ Volume of water for the required energy = $\dfrac{10.92 \times 10^{11}}{147.2 \times 10^4}$ = $\mathbf{74.18 \times 10^4\ m^3}$

Example 3.11. *An hydroelectric generating station is supplied form a reservoir of capacity 6 million m³ at a head of 170 m.*

(i) *What is the available energy in kWh if the hydraulic efficiency be 0.8 and the electrical efficiency 0.9 ?*

(ii) *Find the fall in reservoir level after a load of 12,000 kW has been supplied for 3 hours, the area of the reservoir is 2.5 km².*

(iii) If the reservoir is supplied by a river at the rate of 1.2 m³/s, what does this flow represent in kW and kWh/day ? Assume constant head and efficiency.

Water weighs 1 tonne/m³. **(Elect. Engineering-I, Osmania Univ.)**

Solution. (*i*) Wt. of water W = $6 \times 10^6 \times 1000$ kg wt = $6 \times 10^9 \times 9.81$ N

Water head = 170 m

Potential energy stored in this much water

= $Wh = 6 \times 10^9 \times 9.81 \times 170$ J = 10^{12} J

Overall efficiency of the station = $0.8 \times 0.9 = 0.71$

∴ Energy available = 0.72×10^{13} J = $72 \times 10^{11}/36 \times 10^5$

= **2×10^6 kWh**

(*ii*) Energy supplied = $12{,}000 \times 3 = 36{,}000$ kWh

Energy drawn from the reservoir after taking into consideration the overall efficiency of the station = $36{,}000/0.72 = 5 \times 10^4$ kWh

= $5 \times 10^4 \times 36 \times 10^5 = 18 \times 10^{10}$ J

If m kg is the mass of water used in two hours, then, since water head is 170 m

$$mgh = 18 \times 10^{10}$$

or $m \times 9.81 \times 170 = 18 \times 10^{10}$ ∴ $m = 1.08 \times 10^8$ kg

If h metre is the fall in water level, then

$$h \times \text{area} \times \text{density} = \text{mass of water}$$

∴ $h \times (2.5 \times 10^6) \times 1000 = 1.08 \times 10^8$ ∴ $h = 0.0432$ m = **4.32 cm**

(*iii*) Mass of water stored per second = $1.2 \times 1000 = 1200$ kg

Wt. of water stored per second = 1200×9.81 N

Power stored = $1200 \times 9.81 \times 170$ J/s = **2,000 kW**

Power actually available = $2{,}000 \times 0.72 = 1440$ kW

Energy delivered /day = 1440×24 = **34,560 kWh**

Example 3.12. *A large hydel power station has a head of 324 m and an average flow of 1370 cubic metres/sec. The reservoir is a lake covering an area of 6400 sq. km, Assuming an efficiency of 90% for the turbine and 95% for the generator, calculate*

(i) the available electric power ;

(ii) the number of days this power could be supplied for a drop in water level by 1 metre.

(AMIE Sec. B Power System I (E-6) Winter)

Solution. (*i*) Available power = $9.81\ \eta\ QH$ kW = $(0.9 \times 0.95) \times 1370 \times 324$ = 379, 524 kW = **379.52 MW.**

(*ii*) If A is the lake area in m^2 and h metre is the fall in water level, the volume of water used is = $A \times h = m^3$. The time required to discharge this water is Ah / Q second.

Now, $A = 6400 \times 10^6$ m² ; $h = 1$ m; $Q = 1370$ m³ /s.

∴ $t = 6400 \times 10^6 \times 1/1370 = 4.67 \times 10^6$ second = **540686 days.**

Example 3.13. *The reservoir area of a hydro-electric generating plant is spread over an area of 4 sq km with a storage capacity of 8 million cubic-metres. The net head of water available to the turbine is 70 metres. Assuming an efficiency of 0.87 and 0.93 for water turbine and generator respectively, calculate the electrical energy generated by the plant.*

Estimate the difference in water level if a load of 30 MW is continuously supplied by the generator for 6 hours. **(Power System I-AMIE Sec. B)**

Solution. Since 1 cubic metre of water weighs 1000 kg., the reservoir capacity = 8×10^6 m³ = $8 \times 10^6 \times 1000$ kg. = 8×10^9 kg.

Wt. of water, $W = 8 \times 10^9$ kg. Wt. $8 \times 10^9 \times 9.81 = 78.48 \times 10^9$ N. Net water head = 70 m.

Potential energy stored in this much water = Wh = $78.48 \times 10^9 \times 70 = 549.36 \times 10^{10}$J

Overall efficiency of the generating plant = $0.87 \times 0.93 = 0.809$

Energy available = $0.809 \times 549.36 \times 10^{10}$ J = 444.4×10^{10} J

= $444.4 \times 10^{10}/36 \times 10^5$ = $\mathbf{12.34 \times 10^5}$ **kWh**

Energy supplied in 6 hours = 30 MW × 6 h = 180 MWh

= 180,000 kWh

Energy drawn from the reservoir after taking into consideration, the overall efficiency of the station = 180,000/0.809 = 224,500 kWh = $224,500 \times 36 \times 10^5$

= 80.8×10^{10} J

If m kg. is the mass of water used in 6 hours, then since water head is 70 m,

$mgh = 80.8 \times 10^{10}$ or $m \times 9.81 \times 70 = 80.8 \times 10^{10}$ $\therefore$ $m = 1.176 \times 10^9$ kg.

If h is the fall in water level, then h × area × density = mass of water

$\therefore$ $h \times (4 \times 10^6) \times 1000 = 1.176 \times 10^9$ $\therefore$ $h = 0.294$ m = **29.4 cm.**

Example 3.14. *In a hydro-electric generating station, the mean head (i.e. the difference of height between the mean level of the water in the lake and the generating station) is 400 metres. If the overall efficiency of the generating stations is 70%, how many litres of water are required to generate 1 kWh of electrical energy ? Take one litre of water to have a mass of 1 kg.*

(F.Y. Engg. Pune Univ.)

Solution. Output energy = 1 kWh = 36×10^5 J

Input energy = $36 \times 10^5/0.7 = 5.14 \times 10^6$ J

If m kg. water is required, then

$mgh = 5.14 \times 10^6$ or $m \times 9.81 \times 400 = 5.14 \times 10^6$, $\therefore$ = **1310 kg.**

Example 3.15. *A pump driven by an electric motor lifts 1.5 m³ of water per minute to a height of 40 m. The pump has an efficiency of 90% and motor has an efficiency of 85%. Determine: **(a)** the power input to the motor. **(b)** The current taken from 480 V supply. **(c)** The electric energy consumed when motor runs at this load for 4 hours. Assume mass of 1 m³ of water to be 1000 kg.*

(Elect. Engg. Pune Univ.)

Solution. (*a*) Weight of the water lifted = 1.5 m³ = 1.5 × 1000 = 1500 kg. Wt = 1500 × 9.8 = 14700 N.

Height = 40 m; time taken = 1 min. = 60 s

$\therefore$ Motor output power = 14700 × 40/60 = 9800 W

Combined pump and motor efficiency = 0.9 × 0.85

$\therefore$ Motor power input = 9800/0.9 × 0.85 = 12810 W = **12.81 kW.**

(*b*) Current drawn by the motor = 12810/480 = **26.7 A**

Electrical energy consumed by the motor = 12.81 kW × 4 h = **51.2 kWh.**

Example 3.16. *An electric lift is required to raise a load of 5 tonne through a height of 30 m. One quarter of electrical energy supplied to the lift is lost in the motor and gearing. Calculate the energy in kWhr supplied. If the time required to raise the load is 27 minutes, find the kW rating of the motor and the current taken by the motor, the supply voltage being 230 V d.c. Assume the efficiency of the motor at 90%.* **(Elect. Engg. A.M. Ae. S.I. June)**

Solution. Work done by the lift = $Wh = mgh = (5 \times 1000) \times 9.8 \times 30 = 1.47 \times 10^6$ J

Since 25% of the electric current input is wasted, the energy supplied to the lift is 75% of the input.

$\therefore$ input energy to the lift = $1.47 \times 10^6/0.75 = 1.96 \times 10^6$ J

Now, 1 kWh = 26×10^5 J

$\therefore$ energy input to the lift = $1.96 \times 10^6/36 \times 10^5$ = 0.544 kWh

Motor energy output = 1.96×10^6 J; $\eta = 0.9$

Motor energy input = $1.96 \times 10^6/0.9 = 2.18 \times 10^6$ J : time taken = 27×60
= 1620 second

Power rating of the electric motor = work done/time taken
= $2.18 \times 10^6/1620 = 1.345 \times 10^3$ J/s = 1345 W

Current taken by the motor = 1345/230 = **5.85 A**

Example 3.17. *A current of 80 A flows for 1 hr, in a resistance across which there is a voltage of 2 V. Determine the velocity with which a weight of 1 tonne must move in order that its kinetic energy shall be equal to the energy dissipated in the resistance.*

(Elect. Engg. A.M.A.e. S.I.)

Solution. Energy dissipated in the resistance = $VIt = 2 \times 80 \times 3600 = 576{,}000$ J

A weight of one tonne represents a mass of one tonne *i.e.*, 1000 kg. Its kinetic energy is $= (1/2) \times 1000 \times v^2 = 500\, v^2$

$\therefore \quad 500\, v^2 = 576{,}000 \quad \therefore \quad v = 1152$ m/s.

Tutorial Problem 3.1

1. A heater is required to give 900 cal/min on a 100 V. d.c. circuit. What length of wire is required for this heater if its resistance is 3 Ω per metre ? **[53 metres]**
2. A coil of resistance 100 Ω is immersed in a vessel containing 500 gram of water of 16° C and is connected to a 220-V electric supply. Calculate the time required to boil away all the water (1kcal = 4200 joules, latent heat of steam = 536 keal/kg). **[44 min 50 second]**
3. A resistor, immersed in oil, has 62.5 Ω resistance and is connected to a 500-V d.c. supply. Calculate
   (*a*) the current taken
   (*b*) the power in watts which expresses the rate of transfer of energy to the oil.
   (*c*) the kilowatt-hours of energy taken into the oil in 48 minutes. **[8A ; 4000 W ; 3.2 kWh]**
4. An electric kettle is marked 500-W, 230 V and is found to take 15 minutes to raise 1 kg of water from 15° C to boiling point. Calculate the percentage of energy which is employed in heating the water. **[79%]**
5. A small electrically heated drying oven has two independent heating elements each of 1000 Ω in its heating unit. Switching is provided so that the oven temperature can be altered by rearranging the resistor connections. How many different heating positions can be obtained and what is the electrical power drawn in each arrangement from a 200 V battery of negligible resistance ? **[Three, 40, 20 and 80 W]**
6. Find the useful rating of a tin-smelting furnace in order to smelt 50 kg of tin per hour. Given: Smelting temperature of tin = 235°C, Specific heat of tin = 0.055 kcal/kg-K. Latent heat of liquefaction = 13.31 kcal per kg. Take initial temperature of metal as 15°C. **[1.5 kW]**
   (F.Y. Engg. Pune Univ.)
7. State the relation between
   (*i*) Kcal and kWh (*ii*) Horse power and watts
   (*iii*) kWh and joule (watt sec) (*iv*) K.E and joules.
   (Gujrat University, Summer 2003, AU 2010)
8. The electrical load in a small workshop consists of 14 lamps, each rated at 240 V, 60 W and 3 fans each rated at 240 V, 1 kW. What is the effective resistance of the total load, total current and energy utilised if run for 8 hrs.
   (Pune University 2003) (Gujrat University, Summer 2003)

OBJECTIVE TESTS – 3

1. If a 220 V heater is used on 110 V supply, heat produced by it will be —— as much.

   (*a*) one-half (*b*) twice

   (*c*) one-fourth (*d*) four times

2. For a given line voltage, four heating coils will produce maximum heat when connected

   (*a*) all in parallel (*b*) all in series

   (*c*) with two parallel pairs in series

   (*d*) one pair in parallel with the other two in series

3. The electric energy required to raise the temperature of a given amount of water is 1000 kWh. If heat losses are 25%, the total heating energy required is — kWh.

   (*a*) 1500 (*b*) 1250

   (*c*) 1333 (*d*) 1000

4. One kWh of energy equals nearly

   (*a*) 1000 W (*b*) 860 kcal

   (*c*) 4186 J (*d*) 735.5 W

5. One kWh of electric energy equals

   (*a*) 3600 J (*b*) 860 kcal

   (*c*) 3600 W (*d*) 4186 J

6. A force of 10,000 N accelerates a body to a velocity 0.1 km/s. This power developed is —— kW

   (*a*) 1,00,000 (*b*) 36,000

   (*c*) 3600 (*d*) 1000

7. A 100 W light bulb burns on an average of 10 hours a day for one week. The weekly consumption of energy will be —— unit/s

   (*a*) 7 (*b*) 70

   (*c*) 0.7 (*d*) 0.07

   (Principles of Elect. Engg. Delhi Univ.)

8. Two heaters, rated at 1000 W, 250 volts each, are connected in series across a 250 Volts 50 Hz A.C. mains. The total power drawn from the supply would be ——— watt.

   (*a*) 1000 (*b*) 500

   (*c*) 250 (*d*) 2000

   (Principles of Elect. Engg. Delhi Univ.)

ANSWERS

1. *c* **2.** *a* **3.** *c* **4.** *b* **5.** *b* **6.** *d*
7. *a* **8.** *b*

CHAPTER

4

Electrostatics

Chapter Outline

- Static Electricity
- Absolute and Relative Permittivity of a Medium
- Laws of Electrostatics
- Electric Field
- Electrostatic Induction
- Electric Flux and Faraday Tubes
- Field Strength or Field Intensity or Electric Intensity (E)
- Electric Flux Density or Electric Displacement D
- Gauss Law
- The Equations of Poisson and Laplace
- Electric Potential and Energy
- Potential and Potential Difference
- Potential at a Point
- Potential of a Charged Conducting Sphere
- Equipotential Surfaces
- Potential and Electric Intensity Inside a Conducting Sphere
- Potential Gradient
- Breakdown Voltage and Dielectric Strength
- Safety Factor of Dielectric
- Boundary Conditions

4.1 Static Electricity

In the previous chapters, we concerned ourselves exclusively with electric current *i.e.* electricity in motion. Now, we will discuss the behaviour of static electricity and the laws governing it. In fact, electrostatics is that branch of science which deals with the phenomena associated with electricity at rest.

It has been already discussed that generally an atom is electrically neutral *i.e.* in a normal atom the aggregate of positive charge of protons is exactly equal to the aggregate of negative charge of the electrons.

If, somehow, some electrons are removed from the atoms of a body, then it is left with a preponderance of positive charge. It is then said to be positively-charged. If, on the other hand, some electrons are added to it, negative charge out-balances the positive charge and the body is said to be negatively charged.

In brief, we can say that positive electrification of a body results from a deficiency of the electrons whereas negative electrification results from an excess of electrons.

The total deficiency or excess of electrons in a body is known as its charge.

4.2 Absolute and Relative Permittivity of a Medium

While discussing electrostatic phenomenon, a certain property of the medium called its ***permittivity*** plays an important role. Every medium is supposed to possess two permittivities:

(*i*) absolute permittivity (ε) and (*ii*) relative permittivity (ε_r).

For measuring relative permittivity, vacuum or free space is chosen as the reference medium. It has an absolute permittivity of 8.854×10^{-12} F/m.

Absolute permittivity $\varepsilon_0 = 8.854 \times 10^{-12}$ F/m

Relative permittivity, $\varepsilon_r = 1$

Being a ratio of two similar quantities, ε_r has no units.

Now, take any other medium. If its relative permittivity, as compared to vacuum is ε_r, then its absolute permittivity is $\varepsilon = \varepsilon_0\, \varepsilon_r$ F/m

If, for example, relative permittivity of mica is 5, then, its absolute permittivity is

$$\varepsilon = \varepsilon_0\, \varepsilon_r = 8.854 \times 10^{-12} \times 5 = 44.27 \times 10^{-12} \text{ F/m}$$

4.3 Laws of Electrostatics

First Law: It states that Like charges of electricity repel each other, whereas unlike charges attract each other.

Second Law: According to becond law, the force exerted between two *point* charges (*i*) is directly proportional to the product of their strengths (*ii*) is inversely proportional to the square of the distance between them.

This law is known as Coulomb's Law* and can be expressed mathematically as:

$$F \propto \frac{Q_1 Q_2}{d^2} \quad \text{or} \quad F = k\frac{Q_1 Q_2}{d^2}$$

In vector form, the Coulomb's law can be written as $\vec{F} = \frac{Q_1 Q_2}{d^2}\hat{d}$

$$= \frac{Q_1 Q_2}{d^2}\vec{d}$$

where $\vec{d}$ is the unit vector *i.e.* a vector of unit length in the direction of distance *d, i.e.* $d = \frac{\vec{d}}{\hat{d}}$ (where $\vec{d}$ is the vector notation for *d*, which is a scalar notation).

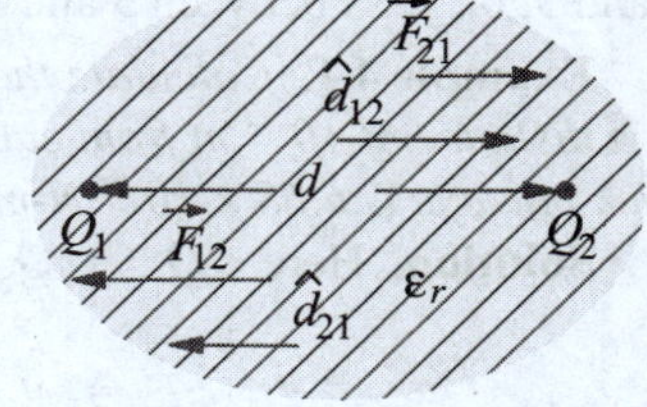

Fig. 4.1

Therefore, explicit forms of this law are:

$$\vec{F}_{21} = k\frac{Q_1 Q_2}{d_{12}^2}\hat{d}_{12} = k\frac{Q_1 Q_2}{d_{12}^3}\vec{d}_{12}$$

where $\vec{F}_{21}$ is the force on Q_2 due to Q_1 and $\hat{d}_{12}$ is the unit vector in direction from Q_1 to Q_2

and $\vec{F}_{12} = k\frac{Q_1 Q_2}{d_{21}^2}\hat{d}_{21} = k\frac{Q_1 Q_2}{d_{21}^2}\vec{d}_{21}$ where F_{12} is the force on Q_1 due to Q_2 and $\vec{d}_{21}$ is the unit vector in the direction from Q_2 to Q_1.

where *k* is the constant of proportionality, whose value depends on the system of units employed. In S.I. system, as well as M.K.S.A. system $k = 1/4\pi\varepsilon$. Hence, the above equation becomes.

$$F = \frac{Q_1 Q_2}{4\,\pi\varepsilon d^2} = \frac{Q_1 Q_2}{4\pi\varepsilon_0\varepsilon_r d^2}$$

* Coulomb is better known for his law which states that the force between two point charges is proportional to each charge and inversely proportional to the square of the distance between them.

If Q_1 and Q_2 are in colomb, d in metre and ε in fard/metre, then F is in newtons

Now $$\frac{1}{4\pi\varepsilon_0} = \frac{1}{4\pi \times 8.854 \times 10^{-12}} = 8.9878 \times 10^9 = 9 \times 10^9 \text{ (approx.)}$$

Hence, Coulomb's Law can be written as

$$F = 9 \times 10^9 \frac{Q_1 Q_2}{\varepsilon_r d^2} \quad \text{—in a medium}$$

$$= 9 \times 10^9 \frac{Q_1 Q_2}{d^2} \quad \text{—in air or vacuum ...(i)}$$

If in Eq. (*i*) above

$$Q_1 = Q_2 = Q \text{ (say)}, d = 1 \text{ metre}; F = 9 \times 10^9 \text{ N}$$

then $$Q^2 = 1 \quad \text{or} \quad Q = \pm 1 \text{ coulomb}$$

Hence, one coulomb of charge may be defined as ***that charge (or quantity of electricity) which when placed in air (strictly vacuum) from an equal and similar charge repels it with a force of 9×10^9 N.***

Unlike charges attract and like charges repel each other

Although coulomb is found to be a unit of convenient size in dealing with electric current, yet, from the standpoint of electrostatics, it is an enormous unit. Hence, its submultiples like micro-coulomb (μC) and micro-microcoulomb (μ μC) are generally used.

$$1 \ \mu\text{C} = 10^{-6} \text{ C}; \ 1 \ \mu\mu\text{C} = 10^{-12} \text{ C}$$

It may be noted here that relative permittivity of air is one, of water 81, of paper between 2 and 3, of glass between 5 and 10 and of mica between 2.5 and 6.

Example 4.1. *Calculate the electrostatic force of repulsion between two α-particles when at a distance of 10^{-13} m from each other. Charge of an α-particles is 3.2×10^{-12} C. If mass of each particle is 6.68×10^{-27} N-m²/kg².*

Solution. Here $Q_1 = Q_2 = 3.2 \times 10^{-19}$ C, $d = 10^{-13}$ m

$$F = 9 \times 10^9 \times \frac{3.2 \times 10^{-19} \times 3.2 \times 10^{-19}}{(10^{-13})^2} = \mathbf{9.2 \times 10^{-2} N}$$

The force of gravitational attraction between the two particles is given by

$$F = G \frac{m_1 m_2}{d^2} = \frac{6.67 \times 10^{-11} \times (6.68 \times 10^{-27})^2}{(10^{-13})^2} = \mathbf{2.97 \times 10^{-37} \ N}$$

Obviously, this force is negligible as compared to the electrostatic force between the two particles.

Example 4.2. *Calculate the distance of separation between two electrons (in vacuum) for which the electric force between them is equal to the gravitation force on one of them at the earth surface.*

mass of electron = 9.1×11^{-31} kg, charge of electron = 1.6×10^{-19} C.

Solution. Gravitational force on one electron = mg newton = $9.1 \times 10^{-31} \times 9.81$ N

Electrostatic force between the electrons

$$= 9 \times 10^9 \frac{Q^2}{d^2} = \frac{9 \times 10^9 \times (1.6 \times 10^{-19})^2}{d^2} N$$

Equating the two forces, we have

$$v \frac{9 \times 10^9 \times 2.56 \times 10^{-38}}{d^2} = 9.1 \times 10^{-31} \times 9.81 \quad \therefore \quad d = \mathbf{5.08\ m}$$

Example 4.3. *Determine resultant force on 3 μC charge due to – 4μC and 10 nC charges. All these three point charges are placed on the vertices of equilateral triangle ABC of side 50 cm.*

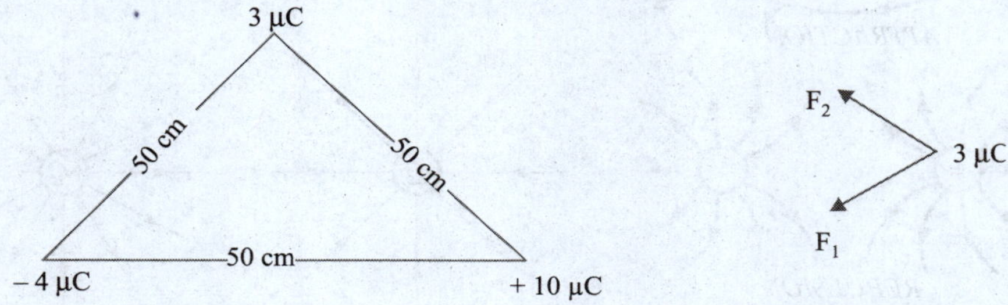

Fig. 4.2 (*a*) Fig. 4.2 (*b*)

Solution.
$$F_2 = \frac{Q_1 Q_2}{4\pi\varepsilon_0 d^2} = \frac{3 \times 10^{-6} \times 10 \times 10^{-9}}{4\pi \times 8.854 \times 10^{-12} \times 0.50 \times 0.50}$$

$= 1.08 \times 10^{-3}$ Newton, in the direction shown

Similarly, $F_1 = 0.432$ Newton, in the direction shown.

Resultant of F_1 and F_2 has to be found out.

Example 4.4. *A capacitor is composed of 2 plates separated by a sheet of insulating material 3 mm thick and of relative permitivity 4. The distance between the plates is increased to allow the insertion of a second sheet of 5 mm thick and of relative permitivity ε_r. If the equivalent capacitance is one third of the original capacitance. Find the value of ε_r.*

Solution.
$$C_1 \frac{\varepsilon_0 \varepsilon_r A}{d} = k(4/3), \quad \text{where} \quad k = \varepsilon_0 A \times 10^{+3}$$

The composite capacitor with one dielectric of $\varepsilon_{r1} = 4$ and other dielectric of ε_{r2} as relative permitivity has a capacitance of $C/3$. Two capacitors are effectively in series. Let the second dielectric contribute a capacitor of C_2.

$$K \,.\, (4/9) = \frac{C_1 C_2}{C_1 + C_2} = \frac{K \,.\, (4/3) \,.\, C_2}{K \,.\, (4/3) + C_2}$$

This gives $C_2 = 2/3\, K$

$$(2/3)\, K = \frac{\varepsilon_0 \varepsilon_{r2} A}{5 \times 10^{-3}}$$

$$E_{r2} = 10/3 \,.\, K\, 1/e_0\, A \times 10^{-3}$$

$$= 10/3\ e_0\, A \times 103\ 1/e_0\, A \times 10^{-3}$$

$$= 10/3 = \mathbf{3.33}$$

4.4 Electric Field

It is found that in the medium around a charge a force acts on a positive or negative charge when placed in that medium. If the charge is sufficiently large, then it may create such a huge stress as to cause the electrical rupture of the medium, followed by the passage of an arc discharge.

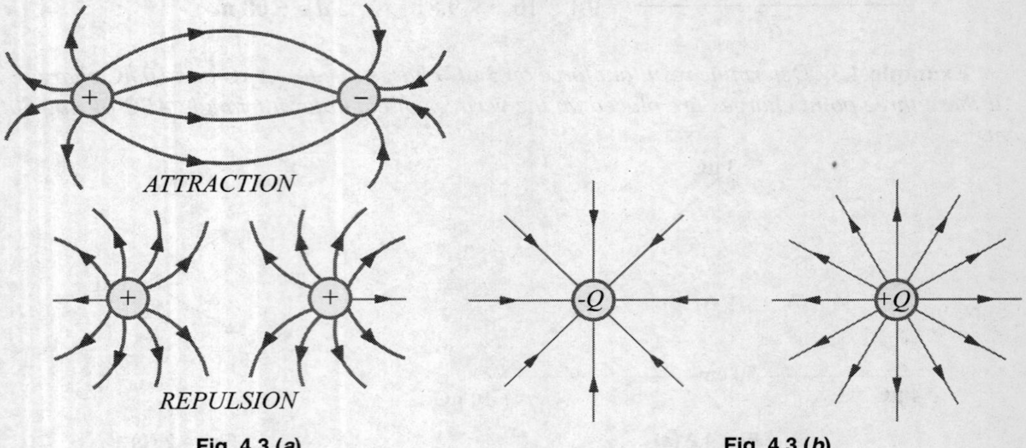

Fig. 4.3 (*a*) Fig. 4.3 (*b*)

The region in which the stress exists or in which electric forces act, is called an electric field or electrostatic field.

The stress is represented by imaginary lines of forces. The direction of the lines of force at any point is the direction along which a unit positive charge placed at that point would move if free to do so. It was suggested by Faraday that the electric field should be imagined to be divided into ***tubes of force*** containing a fixed number of ***lines*** of force. He assumed these tubes to the elastic and having the property of contracting longitudinally the repelling laterally. With the help of these properties, it becomes easy to explain ***(i)*** why unlike charges attract each other and try to come nearer to each other and ***(ii)*** why like charges repel each other [Fig. 4.3 (*a*)].

However, it is more common to use the term lines of force. These lines are supposed to emanate from a positive charge and end on a negative charge [Fig. 4.3 (*b*)]. These lines always leave or enter a conducting surface normally.

4.5 Electrostatic Induction

It is found that when an uncharged body is brought near a charged body, it acquires some charge. This phenomenon of an uncharged body getting charged merely by the nearness of a charged body is known as ***induction***. In Fig. 4.4, a positively-charged body *A* is brought close to a perfectly-insulated uncharged body *B*. It is found that the end of *B* nearer to *A* gets negatively charged whereas further end becomes positively charged. The negative and positive charges of *B* are known as ***induced charges***. The negative charge of *B* is called 'bound' charge because it must remain on *B* so long as positive charge of *A* remains there. However, the positive charge on the farther end of *B* is called free charge. In Fig. 4.5, the body *B* has been earthed by a wire. The positive charge flows to earth leaving negative charge behind. If next *A* is removed, then this negative charge will also go to earth, leaving *B* uncharged. It is found that:

(i) a positive charge induces a negative charge and ***vice-versa.***

(ii) each of the induced charges is equal to the inducing charge.

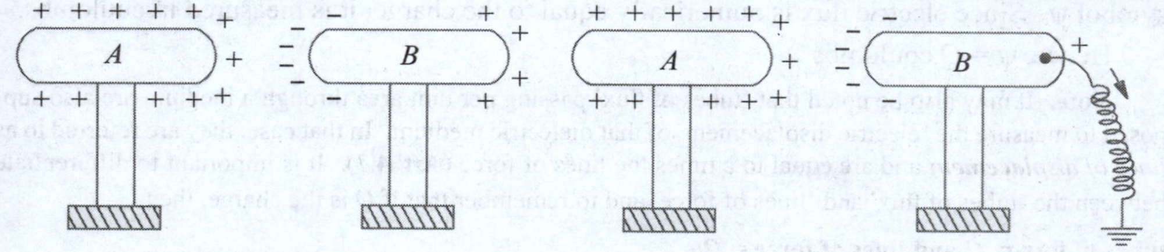

Fig. 4.4 Fig. 4.5

4.6 Electric Flux and Faraday Tubes

Consider a small closed curve in an electric field (Fig. 4.6). If we draw lines of force through each point of this closed curve, then we get a tube as shown in the figure. It is called the ***tube of the electric flux***. It may be defined as the region of space enclosed within the tubular surface formed by drawing lines of force through every point of a small closed curve in the electric field.

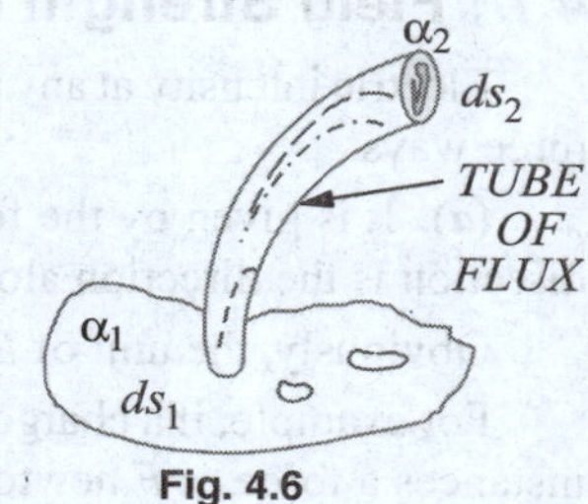

Fig. 4.6

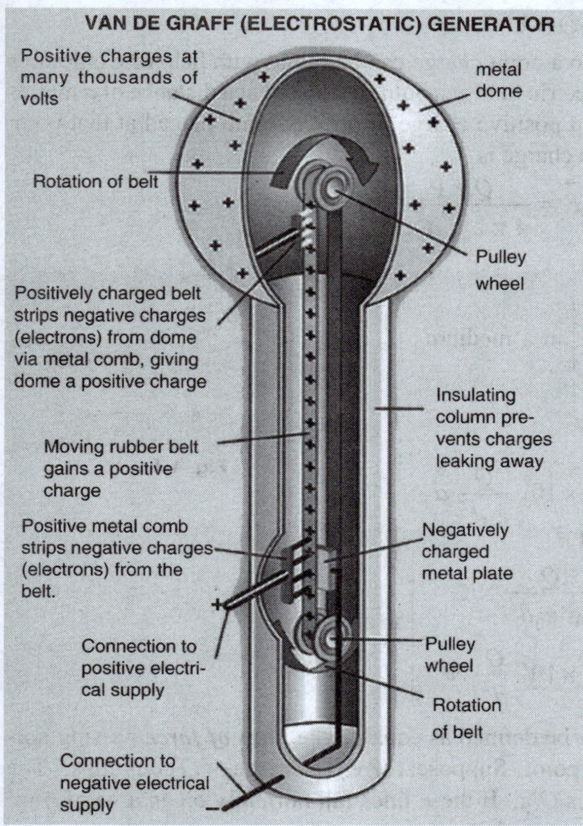

The Van de Graff generator is able to produce very high voltages, for example, up to 50 000 volts. When someone touches the dome of the generator, it will cause hair to stand on end (since like charges repel). Touching the VDG is not dangerous since the current is very small.

Since lines of force end on conductors, the two ends of a flux tube will consist of small area ds_1 and ds_2 on the conductor surfaces. If surface charge densities over these areas are σ_1 and $-\sigma_2$, then charges at the two ends of the flux tube will be $\sigma_1\, ds_1$ and $-\sigma_2\, ds_2$. These charges are assumed to be always equal but opposite to each other. The strength of a flux tube is represented by the charge at its ends.

A ***unit tube of flux*** is one in which the end charge is one unit of charge.

In the S.I. system of units, one such tube of flux is supposed to start from a positive charge of one coulomb and terminate on a negative charge of the same amount.

A ***unit tube of flux*** is known as Faraday tube. If the charge on a conductor is $\pm\, Q$ coulombs, then the ***number of Faraday tubes starting or terminating on it also Q.***

The number of Faraday tubes of flux passing through a surface in an electric field is called the *electric flux* (or dielectric flux) through that surface. Electric flux is represented by the symbol ψ. Since electric flux is numerically equal to the charge, it is measured in coulombs.

Hence, $\psi = Q$ coulombs

Note. It may also be noted that 'tubes of flux' passing per unit area through a medium are also supposed to measure the 'electric displacement' of that dielectric medium. In that case, they are referred to as *lines of* ***displacement*** and are equal to ε times the lines of force (Art. 4.7). It is important to differentiate between the 'tubes of flux' and 'lines of force' and to remember that if Q is the charge, then

tubes of flux = Q **and lines of force =** Q/ε

4.7 Field Strength or Field Intensity or Electric Intensity (*E*)

Electric intensity at any point within an electric field may be defined in either of the following three ways:

(a) It is given by the force experienced by a unit positive charge placed at that point. Its direction is the direction along which the force acts.

Obviously, the unit of E is newton/coulomb (N/C).

For example, if a charge of Q coulombs placed at a particular point P within an electric field instances a force of F newton, then electric field at that point is given by

$$E = F/Q \text{ N/C}$$

The value of E within the field due to a point charge can be found with help of Coulomb's laws. Suppose it is required to find the electric field at a point A situated at a distance of d metres from a charge of Q coulombs. Imagine a positive charge of one coulomb placed at that point (Fig. 4.7). The force experienced by this charge is

$$F = \frac{Q \times 1}{4\pi\varepsilon_0\varepsilon_r d^2} N \quad \text{or} \quad \vec{F} = \frac{Q \times 1}{4\pi\varepsilon_0\varepsilon_r d_{PA}^2} \hat{d}_{PA}$$

$$\left.\begin{aligned} \therefore \quad E &= \frac{Q \times 1}{4\pi\varepsilon_0\varepsilon_r d_{PA}^2} \text{ N/C} \\ &= 9 \times 10^9 \frac{Q}{\varepsilon_r d_{PA}^2} \text{ N/} \end{aligned}\right] \text{ in a medium}$$

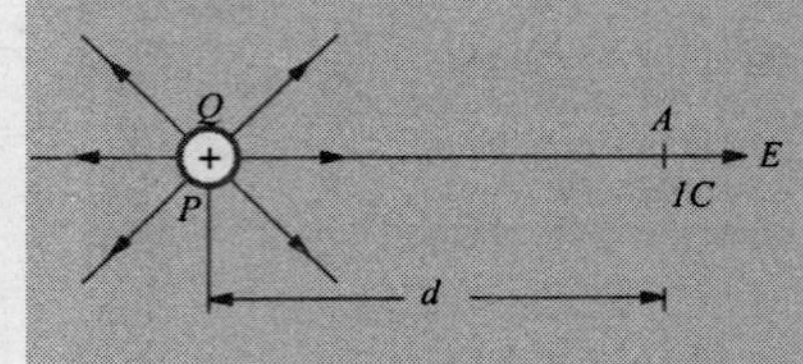

Fig. 4.7

or in vector notation,

$$\vec{E}(d) = 9 \times 10^9 \frac{Q}{\varepsilon_r d^2} \hat{d}$$

where $\vec{E}(d)$ denotes $\vec{E}$ as a function of d

$$\left.\begin{aligned} &= \frac{Q}{4\pi\varepsilon_0 d^2} \text{ N/C} \\ &= 9 \times 10^9 \frac{Q}{d^2} \text{ N/} \end{aligned}\right] \text{ in air}$$

(b) Electric intensity at a point may be defined as equal to the ***lines of force*** passing normally through a unit cross-section at that point. Suppose, there is a charge of Q coulombs. The number of lines of force produced by it is Q/ε. If these lines fall normally on an area of A m^2 surrounding the point, then electric intensity at that point is

$$E = \frac{Q/\varepsilon}{A} = \frac{Q}{\varepsilon A}$$

Now $\qquad Q/A = D \qquad$ —the flux density over the area

$$\therefore \quad E = \frac{D}{\varepsilon} = \frac{D}{\varepsilon_0 \varepsilon_r} \qquad \text{—in a medium}$$

$$= \frac{D}{\varepsilon_0} \qquad \text{—in air}$$

The unit of E is volt/metre.

(c) Electric intensity at any point in an electric field is ***equal to the potential gradient at that point.***

In other words, E is equal to the rate of fall of potential in the direction of the lines of force.

$$\therefore \quad E = \frac{-dV}{dx}$$

Obviously, the unit of E is volt/metre.

It may be noted that E and D are vector quantities having magnitude and direction.

$\therefore$ In vector notation, $\vec{D} = \varepsilon_0 \vec{E}$

Example 4.5. *Point charges in air are located as follows:*

$+ 5 \times 10^{-8}$ C at (0, 0) metres, $+ 4 \times 10^{-8}$ C at (3, 0) metres and $- 6 \times 10^{-8}$ C at (0, 4) metres. Find electric field intensity at (3, 4) metres.

Solution. Electric intensity at point D (3, 4) due to positive charge at point A is

$E_1 = 9 \times 10^9\, Q/d^2 = 9 \times 10^9 \times 5 \times 10^{-8}/5^2 = 18$ V/m

As shown in Fig. 4.8, it acts along AD.

Similarly, electric intensity at point D due to positive charge at point B is $E_2 = 9 \times 10^9 \times 4 \times 10^{-8}/4^2 = 22.5$ V/m. It acts along BD.

$E_1 = 9 \times 10^9 \times 6 \times 10^{-8}/3^2 = 60$ V/m. It acts along DC.

The resultant intensity may be found by resolving E_1, E_2 and E_3 into their X-and Y-components. Now, $\tan \theta = 4/3$; $\theta = 53°8'$.

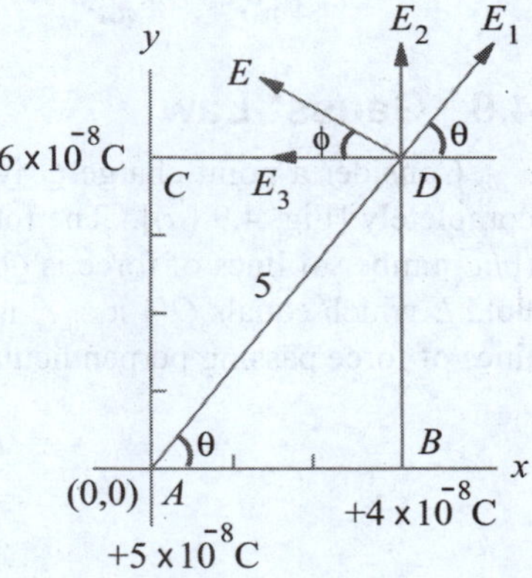

Fig. 4.8

X-component $= E_1 \cos \theta - E_2 = 18 \cos 53°8' - 60$

$= - 49.2$

Y-component $= E_1 \sin \theta + E_2 = 18 \sin 53°8' + 22.5$

$= 36.9$

$$\therefore \quad E = \sqrt{(-49.2)^2 + 36.9^2} = 61.5 \text{ V/m.}$$

It acts along DE such that $\tan \phi = 36.9/49.2 = 0.75$.

Hence $\quad \phi =$ **36.9°.**

4.8 Electric Flux Density or Electric Displacement

It is given by the normal flux per unit area.

If a flux of Ψ coulombs passes normally through an area of A m², then flux density is

$$D = \frac{\Psi}{A} \text{ C/m}^2$$

It is related to electric field intensity by the relation

$$D = \varepsilon_0 \varepsilon_r E \quad \text{...in a medium}$$
$$= \varepsilon_0 E \quad \text{...in free space}$$

In other words, the product of electric intensity E at any point within a dielectric medium and the absolute permittivity ε (= ε_0, ε_r) at the same point is called the ***displacement*** at that point.

Like electric intensity E, electric displacement D* is also a vector quantity (see 4.7) whose direction at every point is the same as that of E but whose magnitude is ε_0 ε_r times E. As E is represented by lines of force, similarly D may also be represented by lines called lines of electric ***displacement***. The tangent to these lines at any point gives the direction of D at that point and the number of lines per unit area perpendicular to their direction is numerically equal to the electric displacement at that point. Hence, the number of lines of electric displacement per unit area (D) is ε_0, ε_r times the number of lines of force per unit area at that point.

It should be noted that whereas the value of E depends on the permittivity of the surrounding medium, that of D is independent of it.

One useful property of D is that its surface integral over any closed surface equals the enclosed charge (Art. 4.9).

Let us find the value of D at a point distant r metres from a point charge of Q coulombs. Imagine a sphere of radius r metres surrounding the charge. Total flux = Q coulombs and it falls normally on a surface area of $4 \pi r^2$ metres. Hence, electric flux density.

$$\text{D} = \frac{\Psi}{4 \pi r^2} = \frac{Q}{4 \pi r^2} \text{ coulomb/metre}^2 \text{ or } \vec{D} = \frac{Q}{4 \pi r^2} \hat{r} = r \text{ (in vector notation)}$$

4.9 Gauss* Law

Consider a point charge Q lying at the centre of a sphere of radius r which surrounds it completely [Fig. 4.9 (*a*)]. The total number of ***tubes of flux*** originating from the charge is Q (but number of lines of force is Q/ε_0) and are normal to the surface of the sphere. The electric field E which equals $Q/4 \pi \varepsilon_0 r^2$ is also normal to the surface. As said earlier, total number of lines of force passing perpendicularly through the whole surface of the sphere is

$$= E \times \text{Area} = \frac{Q}{4\pi\varepsilon_0 r^2} \times 4 \pi r^2 = \frac{Q}{\varepsilon_0}$$

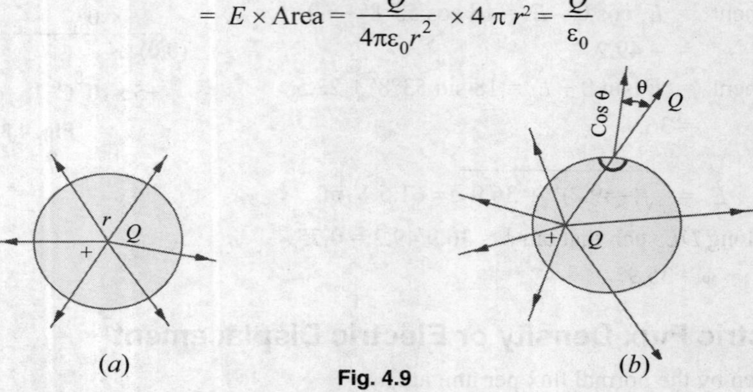

Fig. 4.9

* A more general definition of displacement D is that $D = \varepsilon_0 \varepsilon_r E + P$ where P is the polarisation of the dielectric and is equal to the dipole moment per unit volume.

* After the German mathematician and astronomer Karel Freidrich Gauss (1777-1855).

Now, suppose we draw another sphere surrounding the charge [Fig. 4.9 (*b*)] but whose centre does not lie at the charge but elsewhere. In this case also, the number of tubes of flux emanating from the charge is Q and lines of force is $Q.\varepsilon_0$ though they are not normal to the surface. These can, however, be split up into cos θ components and sin θ components. If we add up sin θ components all over the surface, they will be equal to zero. But if add up cos θ components over the whole surface of the sphere, the normal flux will again come out to be Q (or lines of force will come out to be Q/ε_0). Hence, it shows that irrespective of where the charge Q is placed within a closed surface completely surrounding it, the total normal flux is Q and the total number of lines of force passing out normally is Q/ε_0.

In fact, as shown in Fig. 4.10, if there are placed charges of value Q_1, Q_2, $-Q_3$ inside a closed surface, the total *i.e.* net charge enclosed by the surface is $(Q_1 + Q_2 - Q_3)/\varepsilon_0$ through the closed surface.

This is the meaning of Gauss's law which may be stated thus: the surface integral of the normal component of the electric intensity E over a closed surface is equal to $1/\varepsilon_0$ times the total charge inside it.

Mathematically, $\oint E_n ds = Q/\varepsilon_0$ (where the circle on the integral sign indicates that the surface of integration is a closed surface).

$$\text{or} \qquad \oint \varepsilon_n \varepsilon_0\, ds = Q,\ i.e.\ \oint D_n\, ds = Q \qquad [\therefore\ D_n = \varepsilon_0 E_n]$$

$$\text{or} \qquad \oint \varepsilon_0\, E \cos\theta\, ds = Q,\ i.e.\ \oint D \cos\theta\, ds = Q$$

$$\text{or} \qquad \oint \varepsilon_0\, E_{ds} \cos\theta = Q,\ i.e.,\ \oint D\, ds \cos\theta = Q$$

when E and D are not normal to the surface but make an angle θ with the normal (perpendicular) to the surface as shown in Fig. 4.11.

Proof. In Fig. 4.11, let a surface S completely surround a quantity of electricity or charge Q. Consider a small surface area ds subtending a small solid angle $d\omega$ at point charge Q. The field intensity at ds is $E = \dfrac{Q}{4\pi\varepsilon_0 d^2}$ where d is the distance between Q and ds.

In vector notation, $\oint \varepsilon_0\, \vec{E}.\vec{ds} = Q$ *i.e.* $\oint \vec{D}\,\vec{ds} = Q = \int_v \rho dv$ (where ρ is the volume density of charge in the volume enclosed by closed surface S).

Thus $\oint_s \vec{D}.\vec{ds} = \int_v \rho\, dv$ is the vector statement of Gauss Law* and its alternative statement is $\nabla \vec{D} = \rho$

* This results from the application of the Divergence theorem, also called the Gauss' Theorem, *viz.*, $\oint_S \nabla \vec{D} d\upsilon = \oint \vec{D}.\vec{ds}$ where vector operator called 'del' is defined as

$$\nabla = \frac{\partial}{\partial x}\hat{x} + \frac{\partial}{\partial y}\hat{y} + \frac{\partial}{\partial z}\hat{z}$$

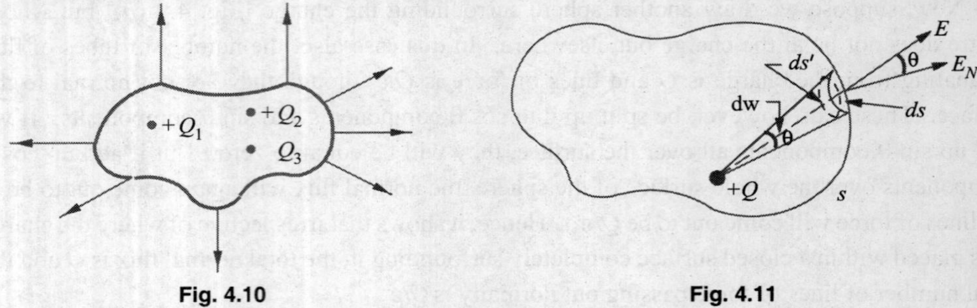

Fig. 4.10

Fig. 4.11

The normal component of the intensity $E_n = E \cos \theta$

$\therefore$ No. of lines of force passing normally through the area ds is

$$= E_n . d_s = E\, ds \cos \theta = \vec{E} . \vec{ds} \text{ in vector notation}$$

Now $\quad ds \cos \theta = ds' \quad \therefore \quad E.ds' = \dfrac{Q}{4 \pi \varepsilon_0 d^2} .ds'$

Now $\quad ds'/d^2 = d\omega$

Hence, the number of lines of force passing normally is $= \dfrac{Q}{4 \pi \varepsilon_0} d\omega$

Total number of lines of force over the whole surface

$$= \frac{Q}{4 \pi \varepsilon_0} \int_s d\omega = \frac{Q}{4 \pi \varepsilon_0} \times 4\pi = \frac{Q}{\varepsilon_0}$$

where sign $\oint$ denotes integration around the whole of the closed surface *i.e.* surface integral.

If the surface passes through a material medium, then the above law can be generalized to include the following:

the surface integral of the normal component of D over a closed surface equals the free charge enclosed by the surface.

As before $D = \dfrac{Q}{4 \pi \varepsilon d^2}$. The normal component $D_n = D \cos \theta = \dfrac{Q}{4 \pi d^2} \times \cos \theta$]

Hence, the normal electric flux from area ds is

$$d\psi = D_n \times ds = \frac{Q}{4 \pi d^2} . \cos \theta . ds = \frac{Q}{4 \pi d^2} . ds'$$

$$\therefore \quad d\psi = \frac{Q}{4\pi} \left(\frac{ds'}{d^2} \right) = \frac{Q}{4\pi} d\omega$$

or $$\psi = \int \frac{Q}{4\pi} . d\omega = \frac{Q}{4\pi} \int d\omega = \frac{Q}{4\pi} \times 4\pi = Q \qquad \therefore \quad \Psi = Q$$

which proves the statement made above.

Hence, we may state Gauss's law in two slightly different ways.

$$\oint_3 E_n . ds = \oint_3 E . \cos \theta . ds = Q / \varepsilon_0 \text{ or } \varepsilon_0 \oint_3 E_n . ds = Q$$

and $$\oint_3 D_n . ds = \oint_3 D_n . ds = Q \quad \text{(vector statement is given above)}$$

4.10 The Equations of Poisson and Laplace

These equations are useful in the solution of many problems concerning electrostatics especially the problem of space charge* present in an electronic valve. The two equations can be derived by applying Gauss's theorem. Consider the electric field set up between two charged plates *P* and *Q* [Fig. 4.12 (*a*)]. Suppose there is some electric charge present in the space between the two plates. It is, generally, known as the space charge. Let the space charge density be ρ coulomb/metre³. It will be assumed that the space charge density varies from one point of space to the another but is uniform throughout any thin layer taken parallel to the plates *P* and *Q*. If *X*-axis is taken perpendicular to the plates, then ρ is assumed to depend on the value of *x*. It will be seen from Fig. 4.12(*b*) that the value of electric intensity *E* increases with *x* because of the space charge.

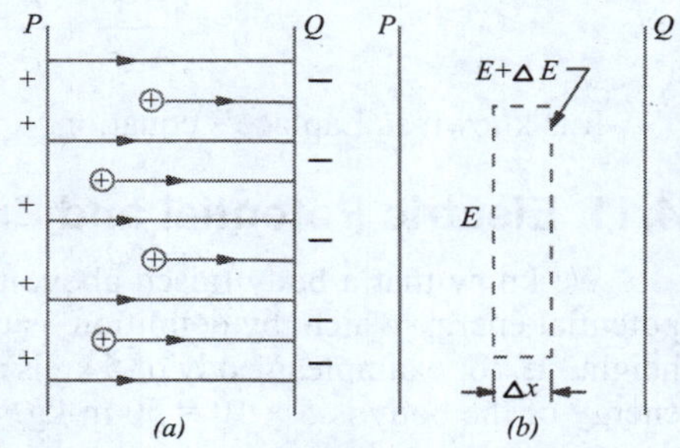

Fig. 4.12

Now, consider a thin volume element of cross-section *A* and thickness Δx as shown in Fig. 4.14(*b*). The values of electric intensity at the two opposite faces of this element are *E* and $(E + \Delta E)$. If dE/dx represents the rate of increases of electric intensity with distance, then

$$\Delta E = \frac{\partial E}{\partial x} \times \Delta x \quad \therefore \quad E + \Delta E = E + \frac{\partial E}{\partial x} \times \Delta x$$

The surface integral of electric intensity over the right-hand face of this element is

$$= \left(E + \frac{\partial E}{\partial x} . \Delta x \right) A$$

The surface integral over the left-hand face of the element is $= - E \times A$

The negative sign represents the fact that *E* is directed inwards over this face.

The surface integral over the entire surface, *i.e.,* the closed surface of the element is $= \left(E + \frac{\partial E}{\partial x} . \Delta x \right) A - E \times A = A . \Delta x . \frac{\partial E}{\partial x}$. From symmetry it is evident that along with *y* and *z* there is no field.

Now, according to Gauss's theorem (Art. 4.9), the surface integral of electric intensity over a closed surface is equal to $1/\varepsilon_0$ time the charge within that surface.

Volume of the element, $dV = A \times \Delta x$; charge $= \rho A . \Delta x$

$$\therefore \quad A . \Delta x . \frac{\partial E}{\partial x} = \rho A . \Delta x . \frac{1}{\varepsilon_0} \quad \text{or} \quad \frac{\partial E}{\partial x} = \frac{\rho}{\varepsilon_0}$$

$$\text{Now} \quad E = -\frac{\partial V}{\partial x} \quad \therefore \frac{\partial E}{\partial x} = \frac{\partial}{\partial x}\left(-\frac{dV}{dx} \right) = -\frac{\partial^2 V}{\partial x^2} \quad \therefore \quad \frac{\partial^2 V}{\partial x^2} = -\frac{\rho}{\varepsilon_0}$$

It is known as Poisson's equation in one dimension where potential varies with *x*.

When *V* varies with *x*, *y* and *z*, then $\frac{\partial^2 V}{\partial x^2} + \frac{\partial^2 V}{\partial y^2} + \frac{\partial^2 V}{\partial z^2} = -\frac{\rho}{\varepsilon_0} = \nabla^2 V$ in vector notation.

If, as a special case, where space charge density is zero, then obviously,

$$\partial^2 V / \partial x^2 = 0$$

In general, we have $\frac{\partial^2 V}{\partial x^2} + \frac{\partial^2 V}{\partial y^2} + \frac{\partial^2 V}{\partial z^2} = 0$ or $\nabla^2 V = 0$ in vector notation where ∇^2 is

---

* Such a space charge exists in the space between the cathode and anode of a vacuum tube.

defined (in cartesian co-ordinates) as the operation

$$\nabla^2 = \frac{\partial^2 V}{\partial x^2} + \frac{\partial^2 V}{\partial y^2} + \frac{\partial^2 V}{\partial z^2}$$

It is known as Laplace's equation.

4.11 Electric Potential and Energy

We know that a body raised above the ground level has a certain amount of mechanical potential energy which, by definition, is given by the amount of work done in raising it to that height. If, for example, a body of 5 kg is raised against gravity through 10 m, then the potential energy of the body is $5 \times 10 = 50$ m-kg. wt. $= 50 \times 9.8 = 490$ joules. The body falls because there is attraction due to gravity and always proceeds from a place of higher potential energy to one of lower potential energy. So, we speak of gravitational potential energy or briefly 'potential' at different points in the earth's gravitational field.

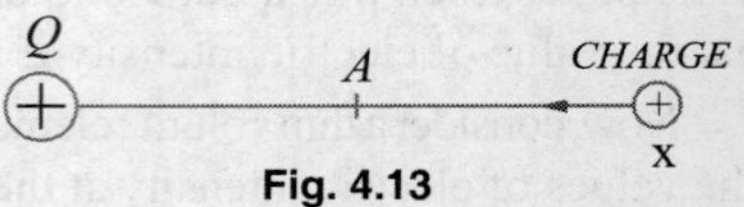

Fig. 4.13

Now, consider an electric field. Imagine an isolated positive charge Q placed in air (Fig. 4.13). Like earth's gravitational field, it has its own electrostatic field which theoretically extends upto infinity. If the charge X is very far away from Q, say, at infinity, then force on it is practically zero. As X is brought nearer to Q, a force of repulsion acts on it (as similar charges repel each other), hence work or energy is required to bring it to a point like A in the electric field. Hence, when at point A, charge X has some amount of electric potential energy. Similar other points in the field will also have some potential energy. In the gravitational field, usually 'sea level' is chosen as the place of 'zero' potential. In electric field infinity is chosen as the theoretical place of 'zero' potential although, in practice, earth is chosen as 'zero' potential, because earth is such a large conductor that its potential remains practically constant although it keeps on losing and gaining electric charge every day.

4.12 Potential and Potential Difference

As explained above, the force acting on a charge at infinity is zero, hence 'infinity' is chosen as the theoretical place of zero electric potential. Therefore, potential at any point in an electric field may be defined as

numerically equal to the work done in bringing a positive charge of one coulomb from infinity to that point against the electric field.

The unit of this potential will depend on the unit of charge taken and the work done.

If, in shifting one coulomb from infinity to a certain point in the electric field, the work done is one joule, then potential of that ponit is one volt.

Obviously, potential is work per unit charge,

$$\therefore \qquad 1 \text{ volt} = \frac{1 \text{ joule}}{1 \text{ coulomb}}$$

Similarly, potential difference (p.d.) of one volt exists between two points if one joule of work is done in shifting a charge of one coulomb from one point to the other.

4.13 Potential at a Point

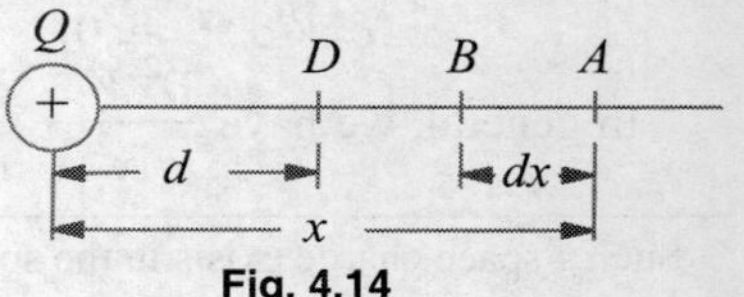

Fig. 4.14

Consider a positive point charge of Q coulombs placed in air. At a point x metres from it, the force on one coulomb positive charge is $Q/4\,\pi\varepsilon_0\, x^2$ (Fig. 4.14).

Suppose, this one coulomb charge is moved towards Q through a small distance dx. Then, work done is

$$dW = \frac{Q}{4\pi\varepsilon_0 r^2} \times (-dx)$$

The negative sign is taken because dx is considered along the negative direction of x.

The total work done in bringing this coulomb of positive charge from infinity to any point D which is d metres from Q is given by

$$W = -\int_{x=\infty}^{x=d} Q \cdot \frac{dx}{4\pi\varepsilon_0 x^2} = -\frac{Q}{4\pi\varepsilon_0}\int_{\infty}^{d} \frac{dx}{x^2}$$

$$= -\frac{Q}{4\pi\varepsilon_0}\left|-\frac{1}{x}\right|_{\infty}^{d} = -\frac{Q}{4\pi\varepsilon_0}\left[-\frac{1}{d} - \left(-\frac{1}{\infty}\right)\right] = \frac{Q}{4\pi\varepsilon_0 d} \text{ joules}$$

By definition, this work in joules in numerically equal to the potential of that point in volts.

$$\therefore \quad V = \frac{Q}{4\pi\varepsilon_0 d} = 9 \times 10^9 \frac{Q}{d} \text{ volt} \qquad \text{—in air}$$

$$\text{and} \quad V = \frac{Q}{4\pi\varepsilon_0\varepsilon_r d} = 9 \times 10^9 \frac{Q}{\varepsilon_r d} \text{ volt} \qquad \text{—in medium}$$

We find that as d increases, V decreases till it becomes zero at infinity.

4.14 Potential of a Charged Conducting Sphere

The above formula $V = Q/4\pi\varepsilon_0\varepsilon_r d$ applies only to a charge concentrated at a point. The problem of finding potential at a point outside a charged sphere sounds difficult, because the charge on the sphere is distributed over its entire surface and so, is not concentrated at a point. But the problem is easily solved by nothing that the lines of force of a charged sphere, like A in by noting that the lines of force of a charged sphere, like A in Fig. 4.15 spread out normally from its surface. If produced backwards, they meet at the centre of A. Hence for finding the potentials at points outside the sphere, we can imagine the charge on the sphere as concentrated at its centre O. If r is the radius of sphere in metres and Q its charge in coulomb then, potential of its surface is $Q/4\pi\,\varepsilon_0 r$ volt and electric intensity is $Q/4\pi\varepsilon_0 r^2$. At any other point 'd' metres from the centre of the sphere, the corresponding values are $Q/4\pi\,\varepsilon_0 d$ and $Q/4\pi\varepsilon_0 d^2$ respectively with $d > r$ as shown in Fig. 4.16 though its starting point is coincident with that of r. The variations of the potential and electric intensity with distance for a charged sphere are shown in Fig. 4.17.

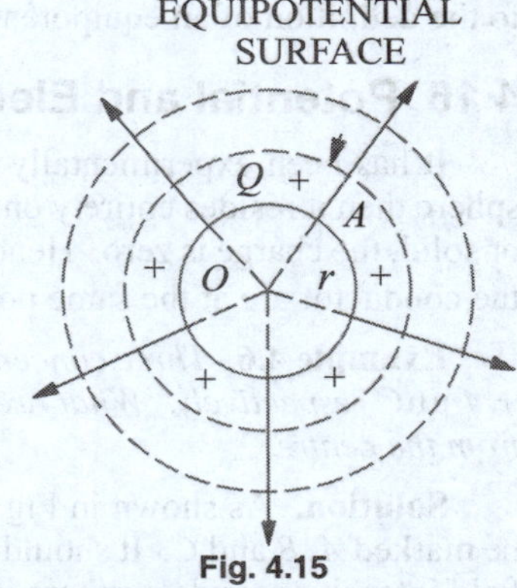

Fig. 4.15

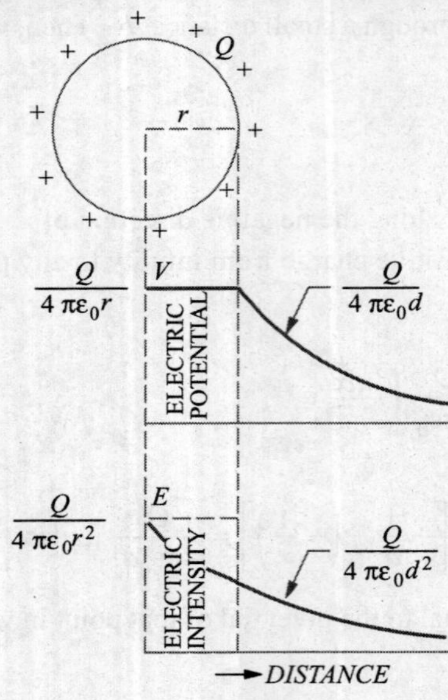

Fig. 4.16

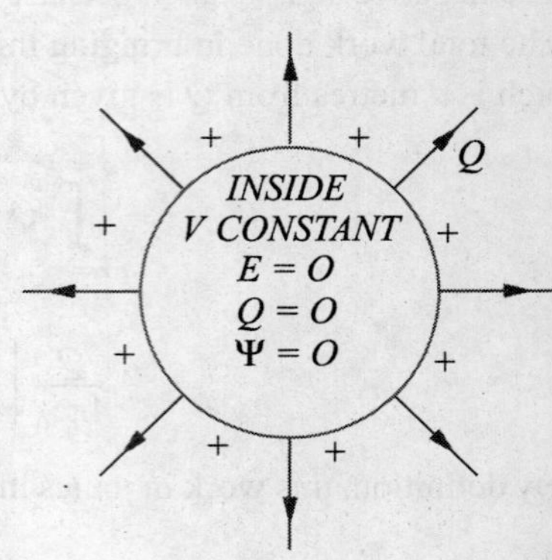

Fig. 4.17

4.15 Equipotential Surfaces

An equipotential surface is a surface in an electric field such that all points on it are at the same potential. For example, different spherical surfaces around a charged sphere are equipotential surfaces. One important property of an equipotential surface is that the direction of the electric field strength and flux density is always at right angles to the surface. Also, electric flux emerges out normal to such a surface. If, it is not so, then there would be some component of E along the surface resulting in potential difference between various points lying on it which is contrary to the definition of an equipotential surface.

4.16 Potential and Electric Intensity Inside a Conducting Sphere

It has been experimentally found that when charge is given to a conducting body say, a sphere then it resides entirely on its outer surface *i.e.*, within a conducting body whether hollow or solid, the charge is zero. Hence, **(*i*)** flux is zero **(*ii*)** field intensity is zero **(*iii*)** all points within the conductor are at the same potential as at its surface (Fig. 4.18).

Example 4.6. *Three concentric spheres of radii 4, 6 and 8 cm have charges of + 8, − 6 and + 4 μμC respectively. What are the potentials and field strengths at points, 2, 5, 7 and 10 cm from the centre.*

Solution. As shown in Fig. 4.18, let the three spheres be marked A, B and C. It should be remembered that **(*i*)** the field intensity outside a sphere is the same as that obtained by considering the charge at its centre **(*ii*)** inside the sphere, the field strength is zero **(*iii*)** potential anywhere inside a sphere is the same as at its surface.

(*i*) Consider point 'a' at a distance of 2 cm from the centre O. Since it is inside all the spheres, field strength at this point is zero.

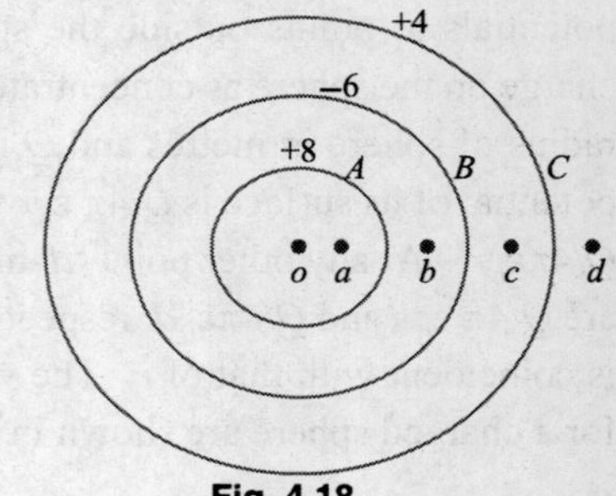

Fig. 4.18

Potential at '*a*'

$$= \sum \frac{Q}{4\pi\varepsilon_0 d} = 9 \times 10^9 \sum \frac{Q}{d}$$

$$= 9 \times 10^9 \left(\frac{8 \times 10^{-12}}{0.04} - \frac{6 \times 10^{-12}}{0.06} + \frac{4 \times 10^{-12}}{0.08} \right) = \mathbf{1.35\ V}$$

(ii) Since point '*b*' is outside sphere *A* but inside *B* and *C*.

$$\therefore \quad \text{Electrical field} = \frac{Q}{4\pi\varepsilon_0 d^2} = 9 \times 10^9 \frac{Q}{d}\ \text{N/C}$$

$$= 9 \times 10^9 \times \frac{8 \times 10^{-12}}{0.05^2} = \mathbf{28.8\ N/C}$$

$$\text{Potential at 'b'} = 9 \times 10^9 \times \left(\frac{8 \times 10^{-12}}{0.05} - \frac{6 \times 10^{-12}}{0.06} + \frac{4 \times 10^{-12}}{0.08} \right) = \mathbf{0.99\ V}$$

(iii) The field strength at point '*c*' distant 7 cm from centre *O*

$$= 9 \times 10^9 \times \left[\frac{8 \times 10^{-12}}{0.07^2} - \frac{6 \times 10^{-12}}{0.07^2} \right] = \mathbf{3.67\ N/C}$$

$$\text{Potential at 'c'} = 9 \times 10^9 \times \left[\frac{8 \times 10^{-12}}{0.07} - \frac{6 \times 10^{-12}}{0.07} + \frac{4 \times 10^{-12}}{0.08} \right] = \mathbf{0.71\ V}$$

(iv) Field strength at '*d*' distant 10 cm from point *O* is

$$= 9 \times 10^9 \times \left[\frac{8 \times 10^{-12}}{0.1^2} - \frac{6 \times 10^{-12}}{0.1^2} + \frac{4 \times 10^{-12}}{0.1^2} \right] = \mathbf{5.4\ N/C}$$

$$\text{Potential at 'd'} = 9 \times 10^9 \times \left[\frac{8 \times 10^{-12}}{0.1} - \frac{6 \times 10^{-12}}{0.1} + \frac{4 \times 10^{-12}}{0.1} \right] = \mathbf{0.54\ V}$$

Example 4.7. *Two positive point charges of 12×10^{-10} C and $8 \cdot \times 10^{-10}$ C are placed 10 cm apart. Find the work done in bringing the two charges 4 cm closer.*

Solution. Suppose the 12×10^{-10} C charge to be fixed. Now, the potential of a point 10 cm from this charge

$$= 9 \times 10^9 \frac{12 \times 10^{-10}}{0.1} = 108\ \text{V}$$

The potential of a point distant 6 cm from it

$$= 9 \times 10^9 \times \frac{12 \times 10^{-10}}{0.06} = 180\ \text{V}$$

$\therefore$ Potential difference $= 180 - 108 = 72$ V

Work done $=$ charge × p.d. $= 8 \times 10^{-10} \times 72$

$= \mathbf{5.76 \times 10^{-8}}$ **joule**

4.17 Potential Gradient

It is defined as the rate of change of potential with distance in the direction of electric force

i.e. $\frac{dV}{dx}$

Its unit is volt/metre although volt/cm is generally used in practice. Suppose in an electric field of strength E, there are two points dx metre apart. The p.d. between them is

$$dV = E\,.(-dx) = -E\,.dx \quad \therefore\; E = -\frac{dV}{dx} \qquad ...(i)$$

The –ve sign indicates that the electric field is directed outward, while the potential increases inward.

Hence, it means that electric intensity at a point is equal to the negative potential gradient at that point.

4.18 Breakdown Voltage and Dielectric Strength

An insulator or dielectric is a substance within which there are no mobile electrons necessary for electric conduction. However, when the voltage applied to such an insulator exceeds a certain value, then it breaks down and allows a heavy electric current (much larger than the usual leakage current) to flow through it. If the insulator is a solid medium, it gets punctured or cracked.

The disruptive or breakdown voltage of an insulator is the minimum voltage required to break it down.*

Dielectric strength of an insulator or dielectric medium is given by the ***maximum potential difference which a unit thickness of the medium can withstand without breaking down.***

In other words, the dielectric strength is given by the potential gradient necessary to cause breakdown of an insulator. Its unit is volt/metre (V/m) although it is usually expressed in kV/mm.

For example, when we say that the dielectric strength of air is 3 kV/mm, then it means that the maximum p.d. which one mm thickness of air can withstand across it without breaking down is 3 kV or 3000 volts. If the p.d. exceeds this value, then air insulation breaks down allowing large electric current to pass through.

Dielectric strength of various insulating materials is very important factor in the design of high-voltage generators, motors and transformers. Its value depends on the thickness of the insulator, temperature, moisture, content, shape and several other factors.

For example doubling the thickness of insulation does not double the safe working voltage in a machine.*

Note. It is obvious that the electric intensity E, potential gradient and dielectric strength are dimensionally equal.

4.19 Safety Factor of a Dielectric

It is given by the ratio of the dielectric strength of the insulator and the electric field intensity established in it. If we represent the dielectric strength by E_{bd} and the actual field intensity by E, then safety factor $\quad k = E_{bd}/E$

For example, for air $E_{bd} = 3 \times 10^6$ V/m. If we establish a field intensity of 3×10^5 V/m in it, then, $k = 3 \times 10^6/3 \cdot 10^5 = 10$.

4.20 Boundary Conditions

There are discontinuities in electric fields at the boundaries between conductors and dielectrics of different permittivities. The relationships existing between the electric field strengths and flux densities at the boundary are called the boundary conditions.

* Flashover is the disruptive discharge which taken places over the surface of an insulator and occurs when the air surrounding it breaks down. Disruptive conduction is luminous.

With reference to Fig. 4.19, first boundary conditions is that the normal component of flux density is continuous across a surface.

As shown, the electric flux approaches the boundary BB at an angle θ_1 and leaves it at θ_2. D_{1n} and D_{2n} are the normal components of D_1 and D_2. According to first boundary condition,

$$D_{1n} = D_{2n} \qquad ...(i)$$

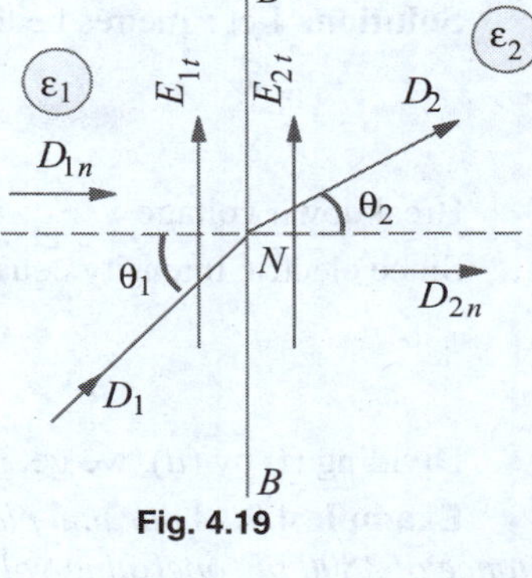

Fig. 4.19

The second boundary condition is that the tangential field strength is continuous across the boundary

$$\therefore \quad E_{1t} = E_{2t} \qquad ...(ii)$$

In Fig. 4.19, we see that

$$D_{1n} = D_1 \cos\theta_1 \quad \text{and} \quad D_{2n} = D_2 \cos\theta_2$$

Also $\quad E_1 = D_1/\varepsilon_1 \quad$ and $\quad E_{1t} = D_1 \sin\theta_2/\varepsilon_1$

Similarly, $\quad E_2 = D_2/\varepsilon_2 \quad$ and $\quad E_{2t} = D_2 \sin\theta_2/\varepsilon_2$

$$\therefore \quad \frac{D_{1n}}{E_1 t} = \frac{\varepsilon_1}{\tan\theta_1} \quad \text{and} \quad \frac{D_{2n}}{E_{2t}} = \frac{\varepsilon_2}{\tan\theta_2}$$

Since $\quad D_{1n} = D_{2n} \quad$ and $\quad E_{1t} = E_{2t} \quad \therefore \quad \dfrac{\tan\theta_1}{\tan\theta_2} = \dfrac{\varepsilon_1}{\varepsilon_2}$

This gives the law of electric flux refraction at a boundary.

It is seen that if $\varepsilon_1 > \varepsilon_2$, $\theta_1 > \theta_2$.

Table No. 4.1
Dielectric Constant and Strength
(*indicates average value)

Insulating material	*Dielectric constant or relative permittivity (e)*	*Dielectric Strength in kV/mm*
Air	1.0006	3.2
Asbestos*	2	2
Bakelite	5	15
Epoxy	3.3	20
Glass	5-12	12-100
Marble*	7	2
Mica	4-8	20-200
Micanite	4-5-6	25-35
Mineral Oil	2.2	10
Mylar	3	400
Nylon	4.1	16
Paper	1.8-2.6	18
Paraffin wax	1.7-2.3	30
Polyethylene	2.3	40
Polyurethane	3.6	35
Porcelain	5-6.7	15
PVC	3.7	50
Quartz	4.5-4.7	8
Rubber	2.5-4	12-20
Teflon	2	20
Vacuum	1	infinity
Wood	2.5-7	---

* The relation between the breakdown voltage V and the thickness of the dielectric is given approximately by the relation $V = At^{2/3}$
where A is a constant depending on the nature of the medium and also on the thickness t. The above statement is known as Baur's law.

Example 4.8. *Find the radius of an isolated sphere capable of being charged to 1 million volt potential before sparking into the air, given that breakdown voltage of air is 30,000 V/cm.*

Solution. Let r metres be the radius of the spheres, then

$$V = \frac{Q}{4\pi\varepsilon_0 r} = 10^6 \text{ V} \qquad \text{...}(i)$$

Breakdown voltage $= 30{,}000 \text{ V/cm} = 3 \times 10^6 \text{ V/m}$

Since electric intensity equals breakdown voltage

$$\therefore \quad E = \frac{Q}{4\pi\varepsilon_0 r^2} = 3 \times 10^6 \text{ V/m} \qquad \text{...}(ii)$$

Dividing (*i*) by (*ii*), we get $r = 1/3 =$ **0.33 metre**

Example 4.9. *A parallel plate capacitor having waxes paper as the insulator has a capacitance of 3800 pF, operating voltage of 600 V and safety factor of 2.5. The waxed paper has a relative permittivity of 4.3 and breakdown voltage of 15 · 10⁶ V/m. Find the spacing d between the two plates of the capacitor and the plate area.*

Solution. Breakdown voltage V_{bd} = operating voltage × safety factor = 600 · 2.5 = 1500 V

$V_{bd} = d \times E_{bd}$ or $d = 1500/15 \times 10^6 = 10^{-4}$ m = 0.1 mm

$C = \varepsilon_0 \varepsilon_r A/d$ or $A = Cd/\varepsilon_0\varepsilon_r = 3800 \times 10^{-9} \times 10^{-4}/8.854 \times 10^{-12} \cdot 4.3 =$ **0.01 m²**

Example 4.10. *Two brass plates are arranged horizontally, one 2 cm above the other and the lower plate is earthed. The plates are charged to a difference of potential of 6,000 volts. A drop of oil with an electric charge of 1.6 × 10^{-19} C is in equilibrium between the plates so that it neither rises nor falls. What is the mass of the drop ?*

Solution. The electric intensity is equal to the potential gradient between the plates.

$$g = 6{,}000/2 = 3{,}000 \text{ volt/cm} = 3 \times 10^5 \text{ V/m}$$

$$\therefore \quad E = 3 \times 10^5 \text{ V/m} \quad \text{or} \quad \text{N/C}$$

$$\therefore \quad \text{force on drop} = E \times Q = 3 \times 10^5 \times 1.6 \times 10^{-19} = 4.8 \times 10^{-14} \text{ N}$$

$$\text{Wt. of drop} = mg \text{ newton}$$

$$\therefore \quad m \times 9.81 = 4.8 \times 10^{-14} \quad \therefore \quad m = \mathbf{4.89 \times 10^{-15} \text{ kg}}$$

Tutorial Problem 4.1

1. Two parallel metal plates of large area are spaced at a distance of 1 cm from each other in air and a p.d. of 5,000 V is maintained between them. If a sheet of glass 0.5 cm thick and having a relative permittivity of 6 is introduced between the plates, what will be the maximum electric stress and where will it occur ? **[8.57 kV/cm; in air]**
2. A capacitor, formed by two parallel plates of large area, spaced 2 cm apart in air, is connected to a 10,000 V d.c. supply. Calculate the electric stress in the air when a flat sheet of glass of thickness 1.5 cm and relative permittivity 7 is introduced between the plates. **[1.4 × 10⁸ V/m]**
3. The capacitance of a capacitor formed by two parallel metal plates, each having an effective surface area of 50 cm² and separated by a dielectric 1 mm thick, is 0.0001 μF. The plates are charged to a p.d. of 200 V. Calculate (*a*) the charge stored (*b*) the electric flux density (*c*) the relative permittivity of the dielectric. **[(*a*) 0.02 μC (*b*) 4 μC/m² (*c*) 2.26]**
4. A capacitor is constructed from two parallel metallic circular plates separated by three layers of dielectric each 0.5 cm thick and having relative permittivity of 4, 6 and 8 respectively. If the metal discs are 15.25 cm in diameter, calculate the potential gradient in each dielectric when the applied voltage is 1,000 volts. ***(Elect. Engg.-I Delhi Univ.)***

5. A point electric charge of 8 μC is kept at a distance of 1 metre from another point charge of – 4 μC in free space. Determine the location of a point along the line joining two charges where in the electric field intensity is zero. ***(Elect. Engineering, Kerala Univ.)***

6. In a given R-L circuit, R = 35Ω and L = 0.1H. Find (*i*) current through the circuit (*ii*) power factor if a 50 Hz frequency, voltage V = 220∠30° is applied across the circuit.

(RGPV, Bhopal 2001)

7. Three voltage represented by e_1 = 20 sin ω *t*, e_2 = 30 sin (ω *t* = 45°) and e_3 = sin (ω t + 30°) are connected in series and then connected to a load of impedance (2 + *j* 3) Ω. Find the resultant current and power factor of the circuit. Draw the phasor diagram.

(B.P.T.U. Orissa 2003) (RGPV Bhopal 2001, 2000)

OBJECTIVE TESTS – 4

1. The unit of absolute permittivity of a medium is
(*a*) joule/coulomb
(*b*) newton-metre
(*c*) farad/metere
(*d*) farad/coulomb

2. If relative permittivity of mica is 5, its absolute permittivity is
(*a*) 5 ε_0 (*b*) $5/\varepsilon_0$
(*c*) $\varepsilon_0/5$ (*d*) 8.854×10^{-12}

3. Two similar electric charges of 1 C each are placed 1 m apart in air. Force of repulsion between them would be nearly...... newton
(*a*) 1 (*b*) 9×10^9
(*c*) 4 π (*d*) 8.854×10^{-12}

4. Electric flux emanating from an electric charge of + *Q* coulomb is
(*a*) Q/ε_0 (*b*) Q/ε_r
(*c*) $Q/\varepsilon_0\varepsilon_r$ (*d*) *Q*

5. The unit of electric intensity is
(*a*) joule/coulomb
(*b*) newton/coulomb
(*c*) volt/metre
(*d*) both (*b*) and (*c*)

6. If *D* is the electric flux density, then value of electric intensity in air is
(*a*) D/ε_0 (*b*) $D/\varepsilon_0\varepsilon_r$
(*c*) dV/dt (*d*) $Q/\varepsilon A$

7. For any medium, electric flux density *D* is related to electric intensity *E* by the equation
(*a*) $D = \varepsilon_0 E$ (*b*) $D = \varepsilon_0\varepsilon_r E$
(*c*) $D = E/\varepsilon_0\varepsilon_r$ (*d*) $D = \varepsilon_0 E/\varepsilon_r$

8. Inside a conducting sphere,...remains constant
(*a*) electric flux
(*b*) electric intensity
(*c*) charge
(*d*) potential

9. The SI unit of electric intensity is
(*a*) N/m (*b*) V/m
(*c*) N/C (*d*) either (*b*) or (*c*)

10. According to Gauss's theorem, the surface integral of the normal component of electric flux density *D* over a closed surface containing charge *Q* is
(*a*) *Q* (*b*) Q/ε_0
(*c*) $\varepsilon_0 Q$ (*d*) Q^2/ε_0

11. Which of the following is zero inside a charged conducting sphere ?
(*a*) potential
(*b*) electric intensity
(*c*) both (*a*) and (*b*)
(*d*) both (*b*) and (*c*)

12. In practice, earth is chosen as a place of zero electric potential because it
(*a*) is non-conducting
(*b*) is easily available
(*c*) keeps lossing and gaining electric charge every day
(*d*) has almost constant potential.

ANSWERS

1. *c*	**2.** *a*	**3.** *b*	**4.** *d*	**5.** *d*	**6.** *a*
7. *b*	**8.** *d*	**9.** *d*	**10.** *a*	**11.** *c*	**12.** *d*

CHAPTER

5 Capacitance

Chapter Outline

- Capacitor
- Capacitance
- Capacitance of an Isolated Sphere
- Spherical Capacitor
- Parallel-plate Capacitor
- Special Cases of Parallel-plate Capacitor
- Multiple and Variable Capacitors
- Capacitance between two Parallel Wires
- Capacitors in Series
- Capacitors in Parallel
- Insulation Resistance of a Cable Capacitor
- Energy Stored in a Capacitor
- Force of Attraction between Oppositely-charged Plates
- Current-Voltage Relationships in a Capacitor
- Charging of a Capacitor
- Time Constant
- Discharging of a Capacitor
- Transient Relations during Capacitor Charging Cycle
- Transient Relations during Capacitor Discharging Cycle
- Charging and Discharging of a Capacitor with Initial Charge

5.1. Capacitor

A capacitor essentially consists of two conducting surfaces separated by a layer of an insulating medium called ***dielectric***. The conducting surfaces may be in the form of either circular (or rectangular) plates or be of spherical or cylindrical shape. The purpose of a capacitor is to store electrical energy by electrostatic stress in the dielectric (the word 'condenser' is a misnomer since a capacitor does not 'condense' electricity as such, it merely stores it).

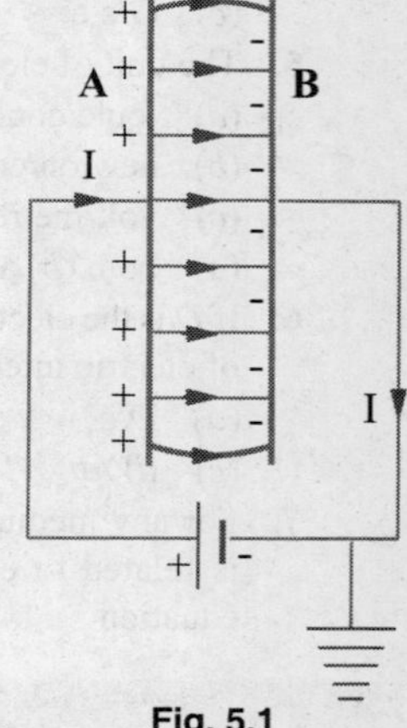

Fig. 5.1

A parallel-plate capacitor is shown in Fig. 5.1. One plate is joined to the positive end of the supply and the other to the negative end or is earthed. It is experimentally found that in the presence of an earthed plate B, plate A is capable of withholding more charge than when B is not there. When such a capacitor is put across a battery, there is a momentary flow of electrons from A to B. As negatively-charged electrons are withdrawn from A, it becomes positive and as these electrons collect on B, it becomes negative. Hence, a p.d. is established between plates A and B. The transient flow of electrons gives rise to charging current. The strength of the charging current is maximum when the two plates are uncharged but it then decreases and finally ceases when p.d. across the plates becomes slowly and slowly equal and opposite to the battery e.m.f.

5.2. Capacitance

The property of a capacitor to 'store electricity' may be called its capacitance.

As we may measure the capacity of a tank, not by the total mass or volume of water it can hold, but by the mass in kg of water required to raise its level by one metre, similarly, the capacitance of a capacitor is defined as ***"the amount of charge required to create a unit p.d. between its plates."***

Suppose we give Q coulomb of charge to one of the two plate of capacitor and if a p.d. of V volts is established between the two, then its capacitance is

$$C = \frac{Q}{V} = \frac{\text{charge}}{\text{potential differnce}}$$

Hence, capacitance is the ***charge required per unit potential difference.***

By definition, the unit of capacitance is coulomb/volt which is also called ***farad*** (in honour of Michael Faraday)

$$\therefore \qquad 1 \text{ farad} = 1 \text{ coulomb/volt}$$

One farad is defined as ***the capacitance of a capacitor which requires a charge of one coulomb to establish a p.d. of one volt between its plates.***

One farad is actually too large for practical purposes. Hence, much smaller units like microfarad (μF), nanofarad (nF) and micro-microfarad (μμF) or picofarad (pF) are generally employed.

$$1\ \mu\text{F} = 10^{-6}\ \text{F};\ 1\ \text{nF} = 10^{-9}\ \text{F};\ 1\ \mu\mu\text{F} \quad \text{or} \quad \text{pF} = 10^{-12}\text{F}$$

Incidentally, capacitance is that property of a capacitor which opposes any change of voltage across it.

5.3 Capacitance of an Isolated Sphere

Consider a charged sphere of radius r metres having a charge of Q coulomb placed in a medium of relative permittivity ε_r as shown in Fig. 5.2.

It has been proved that the free surface potential V of such a sphere with respect to infinity (in practice, earth) is given by

$$V = \frac{Q}{4\pi\varepsilon_0\varepsilon_r r} \quad \therefore \quad \frac{Q}{V} = 4\pi\varepsilon_0\varepsilon_r r$$

Fig. 5.2

By definition, $\quad Q/V =$ capacitance C

$\therefore \qquad = 4\pi\varepsilon_0\varepsilon_r r$ F — in a medium

$\qquad = 4\pi\varepsilon_0 r$ F — in air

Note: It is sometimes felt surprising that an isolated sphere can act as a capacitor because, at first sight, it appears to have one plate only. The question arises as to which is the second surface. But if we remember that the surface potential V is with reference to infinity (actually earth) then it is obvious that the other surface is earth. The capacitance $4\pi\varepsilon_0 r$ exists between the surface of the sphere and earth.

5.4 Spherical Capacitor

(*a*) When outer sphere is earthed

Consider a spherical capacitor consisting of two concentric spheres of radii 'a' and 'b' metres as shown in Fig. 5.3. Suppose, the inner sphere is given a charge of $+Q$ coulombs. It will induce

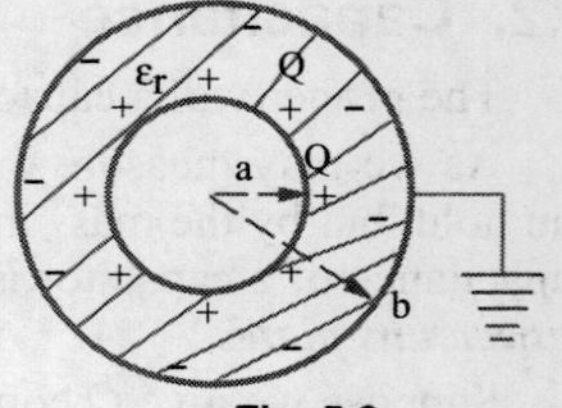

Fig. 5.3

a charge of $-Q$ coulombs on the inner surfaces which will go to earth. If the dielectric medium between the two spheres has a relative permittivity of ε_r, then the free surface potential of the inner sphere due to its own charge $Q/4\pi\varepsilon_0\varepsilon_r a$ volts. The potential of the inner sphere due to $-Q$ charge on the outer sphere is $-Q/4\pi\varepsilon_0\varepsilon_r b$ (remembering that potential anywhere inside a sphere is the same as that its surface).

$\therefore$ Total potential difference between two surfaces is

$$V = \frac{Q}{4\pi\varepsilon_0\varepsilon_r a} - \frac{Q}{4\pi\varepsilon_0\varepsilon_r b}$$

$$= \frac{Q}{4\pi\varepsilon_0\varepsilon_r}\left(\frac{1}{a} - \frac{1}{b}\right) = \frac{Q}{4\pi\varepsilon_0\varepsilon_r}\left(\frac{b-a}{ab}\right)$$

$$\frac{Q}{V} = \frac{4\pi\varepsilon_0\varepsilon_r ab}{b-a} \qquad \therefore \quad C = 4\pi\varepsilon_0\varepsilon_r\frac{ab}{b-a}\ \text{F}$$

(*b*) When inner sphere is earthed

Such a capacitor is shown in Fig. 5.4. If a charge of $+Q$ coulombs is given to the outer sphere *A*, it will distribute itself over both its inner and outer surfaces. Some charge Q_2 coulomb will remain on the outer surface of *A* because it is surrounded by earth all around. Also, some charge $+Q_1$ coulombs will shift to its inner side because there is an earthed sphere *B* inside *A*.

Obviously, $Q = Q_1 + Q_2$

The inner charge $+Q_1$ coulomb on *A* induces $-Q_1$ coulomb on *B* but the other induced charge of $+Q_1$ coulomb goes to earth.

Fig. 5.4

Now, there are two capacitors connected in parallel:

(*i*) One capacitor consists of the inner surface of *A* and the outer surface of *B*. Its capacitance, as found earlier, is

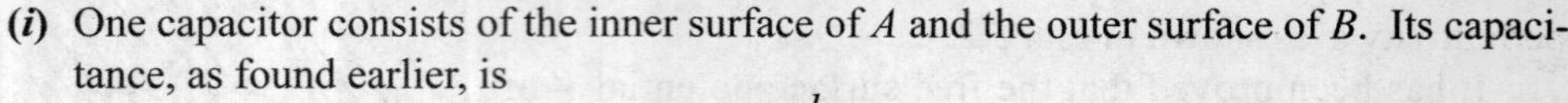

$$C_1 = 4\pi\varepsilon_0\varepsilon_r\frac{ab}{b-a}$$

(*ii*) The second capacitor consists of outer surfaces of *B* and earth. Its capacitance is $C_2 = 4\pi\varepsilon_0 b$ – if surrounding medium is air. Total capacitance $C = C_1 + C_2$.

5.5 Parallel-Plate Capacitor

(*i*) Uniform Dielectric-Medium

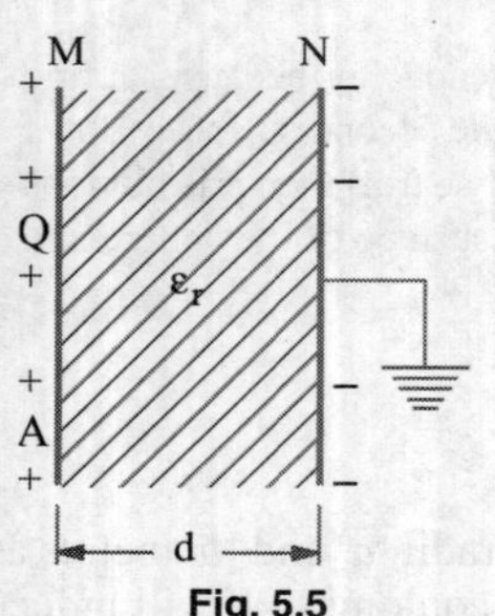

Fig. 5.5

A parallel-plate capacitor consisting of two plates *M* and *N* each of area *A* m² separated by a thickness *d* metres of a medium of relative permittivity ε_r is shown in Fig. 5.5. If a charge of $+Q$ coulomb is given to plate *M*, then flux passing through the medium is $\psi = Q$ coulomb. Flux density in the medium is

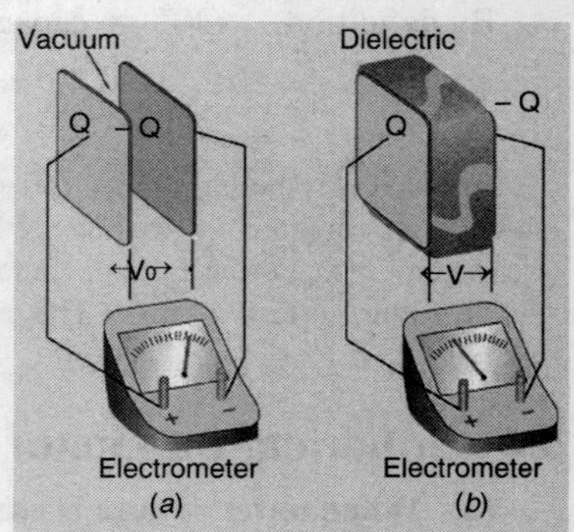

The figure shows how the capacitance changes when dielectric constant is changed

$$D = \frac{\Psi}{A} = \frac{Q}{A}$$

Electric intensity $E = V/d$ and

$$D = \varepsilon E$$

or $\quad \frac{Q}{A} = \varepsilon \frac{V}{d} \quad \therefore \quad \frac{Q}{V} = \frac{\varepsilon A}{d}$

$\therefore \quad C = \frac{\varepsilon_0 \varepsilon_r A}{d}$ farad – in a medium ...(*i*)

$= \frac{\varepsilon_0 A}{d}$ farad – with air as medium

(*ii*) Medium Partly Air

As shown in Fig. 5.6, the medium consists partly of air and partly of parallel-sided dielectric slab of thickness *t* and relative permittivity ε_r. The electric flux density $D = Q/A$ is the same in both media. But electric intensities are different.

$E_1 = \frac{D}{\varepsilon_0 \varepsilon_r}$ ... in the medium

$E_2 = \frac{D}{\varepsilon_0}$ ... in air

p.d. between plates, $V = E_1 . t + E_2 (d - t)$

$$= \frac{D}{\varepsilon_0 \varepsilon_r} t + \frac{D}{\varepsilon_0}(d - t) = \frac{D}{\varepsilon_0}\left(\frac{t}{\varepsilon_r} + d - t\right)$$

$$= \frac{Q}{\varepsilon_0 A}[d - (t - t/\varepsilon_r)]$$

Fig. 5.6

or $\quad \frac{Q}{V} = \frac{\varepsilon_0 A}{[d - (t - t/\varepsilon_r)]}$ or $C = \frac{\varepsilon_0 A}{[d - (t - t/\varepsilon_r)]}$ ...(*ii*)

If the medium were totally air, then capacitance would have been

$$C = \varepsilon_0 A/d$$

From (*ii*) and (*iii*), it is obvious that when a dielectric slab of thickness *t* and relative permittivity ε_r is introduced between the plates of an air capacitor, then its capacitance increases because as seen from (*ii*), the denominator decreases. The distance between the plates is effectively reduces by $(t - t/\varepsilon_r)$. To bring the capacitance back to its original value, the capacitor plates will have to be further separated by that much distance in air. Hence, the new separation between the two plates would be $= [d + (t - t/\varepsilon_r)]$

The expression given in (*i*) above can be written as $C = \frac{\varepsilon_0 A}{d/\varepsilon_r}$

If the space between the plates is filled with slabs of different thickness and relative permittivities, then the above expression can be generalized into $C = \frac{\varepsilon_0 A}{\Sigma d/\varepsilon_r}$

The capacitance of the capacitor shown in Fig. 5.7 can be written as

$$C = \frac{\varepsilon_0 A}{\left(\frac{d_1}{\varepsilon_{r1}} + \frac{d_2}{\varepsilon_{r2}} + \frac{d_3}{\varepsilon_{r3}}\right)}$$

(*iii*) Composite Medium

The above expression may be derived independently as given under:

If V is the total potential difference across the capacitor plates and V_1, V_2, V_3, the potential differences across the three dielectric slabs, then

$$V = V_1 + V_2 + V_3 = E_1 t_1 + E_2 t_2 + E_3 t_3$$

$$= \frac{D}{\varepsilon_0\,\varepsilon_{r1}}.t_1 + \frac{D}{\varepsilon_0\,\varepsilon_{r2}}.t_2 + \frac{D}{\varepsilon_0\,\varepsilon_{r3}}.t_3$$

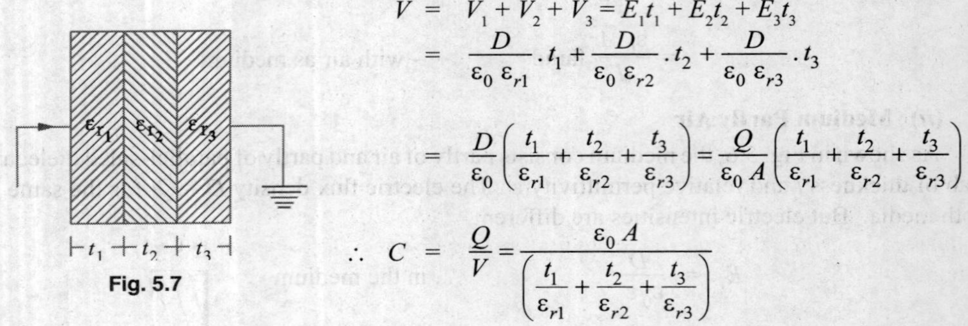

Fig. 5.7

$$= \frac{D}{\varepsilon_0}\left(\frac{t_1}{\varepsilon_{r1}} + \frac{t_2}{\varepsilon_{r2}} + \frac{t_3}{\varepsilon_{r3}}\right) = \frac{Q}{\varepsilon_0 A}\left(\frac{t_1}{\varepsilon_{r1}} + \frac{t_2}{\varepsilon_{r2}} + \frac{t_3}{\varepsilon_{r3}}\right)$$

$$\therefore \quad C = \frac{Q}{V} = \frac{\varepsilon_0 A}{\left(\frac{t_1}{\varepsilon_{r1}} + \frac{t_2}{\varepsilon_{r2}} + \frac{t_3}{\varepsilon_{r3}}\right)}$$

5.6 Special Cases of Parallel-Plate Capacitor

Consider the cases illustrated in Fig. 5.8.

(*i*) As shown in Fig. 5.8 (*a*), the dielectric is of thickness d but occupies only a part of the area. This arrangement is equal to two capacitors in parallel. Their capacitances are

$$C_1 = \frac{\varepsilon_0 A_1}{d} \quad \text{and} \quad C_2 = \frac{\varepsilon_0\,\varepsilon_r A_2}{d}$$

Total capacitance of the parallel-plate capacitor is

$$C = C_1 + C_2 = \frac{\varepsilon_0 A_1}{d} + \frac{\varepsilon_0\,\varepsilon_r A_2}{d}$$

(*ii*) The arrangement shown in Fig. 5.8 (*b*) consists of two capacitors connected in parallel.

(*a*) One capacitor having plate area A_1 and air as dielectric. Its capacitance is $C_1 = \frac{\varepsilon_0 A_1}{d}$

(*b*) The other capacitor has dielectric partly air and partly some other medium. Its capacitance is [Art 5.5 (*ii*)]. $C_2 = \frac{\varepsilon_0 A_2}{[d - (t - t/\varepsilon_r)]}$. Total capacitance is $C = C_1 + C_2$

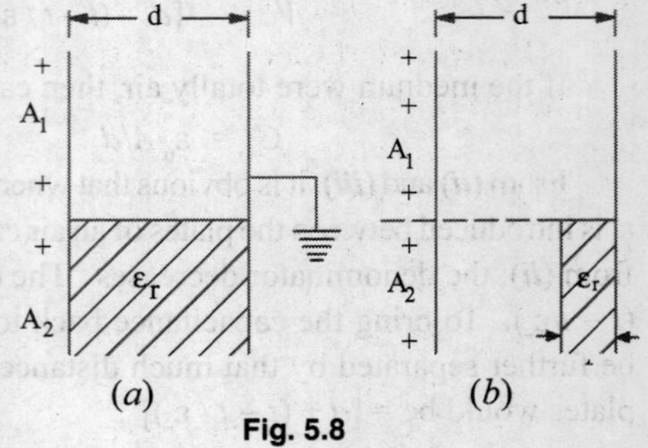

Fig. 5.8

5.7 Multiple and Variable Capacitors

The arrangement of Fig. 5.9. is equivalent to two capacitors joined in parallel. Hence, its capacitance is double that of a single capacitor. Similarly, the arrangement of Fig. 5.10 has four times the capacitance of single capacitor.

Multiple capacitors are shown in Fig. 5.9 and Fig. 5.10.

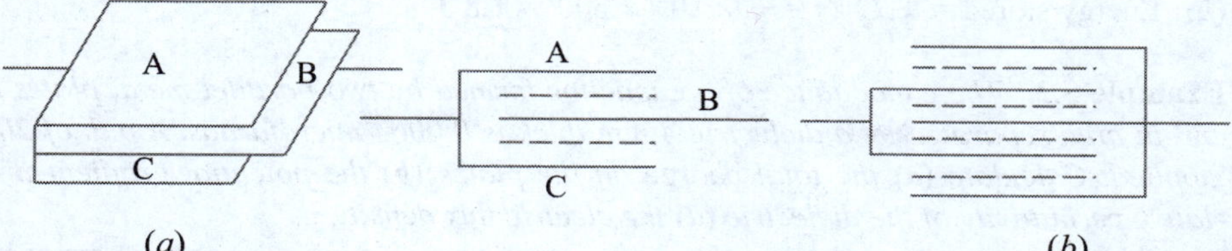

(*a*) Fig. 5.9 (*b*) Fig. 5.10

If one set of plates is fixed and the other is capable of rotation, then capacitance of such a multiplate capacitor can be varied. Such variable-capacitance air capacitors are widely used in radio work (Fig. 5.11). The set of fixed plates *F* is insulated from the other set *R* which can be rotated by turning the knob *K*. The common area between the two sets is varied by rotating *K*, hence the capacitance between the two is altered. Minimum capacitance is obtained when *R* is completely rotated out of *F* and maximum when *R* is completely rotated in *i.e.* when the two sets of plates completely overlap each other.

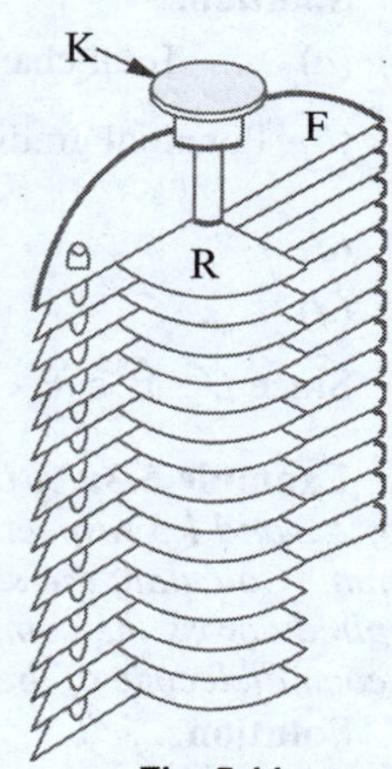

Fig. 5.11

The capacitance of such a capacitor is

$$= \frac{(n-1) \,.\, \varepsilon_0 \varepsilon_r \, A}{d}$$

where *n* is the number of plates which means that $(n-1)$ is the number of capacitors.

Example 5.1. *The voltage applied across a capacitor having a capacitance of 10 μ F is varied thus:*

The p.d. is increased uniformly from 0 to 600 V in 2 seconds. It is then maintained constant at 600 V for 1 second and subsequently decreased uniformly to zero in five seconds. Plot a graph showing the variation of current during these 8 seconds. Calculate **(a)** *the charge* **(b)** *the energy stored in the capacitor when the terminal voltage is 600.*

(Principles of Elect. Engg.-I, Jadavpur Univ.)

Solution. The variation of voltage across the capacitor is as shown in Fig. 5.12 (*a*).

The charging current is given by

$$i = \frac{dq}{dt} = \frac{d}{dt}(Cv) = C \,.\, \frac{dv}{dt}$$

Charging current during the first stage

$$= 10 \times 10^{-6} \times (600/2) = 3 \times 10^{-3} \text{ A} = 3 \text{ mA}$$

Charging current during the second stage is zero because $dv/dt = 0$ as the voltage remains constant.

Charging current through the third stage

$$= 10 \times 10^{-6} \times \left(\frac{0-600}{5}\right) = -1.2 \times 10^{-3} \text{ A} = -1.2 \text{ mA}$$

The waveform of the charging current or capacitor current is shown in Fig. 5.12 (*b*).

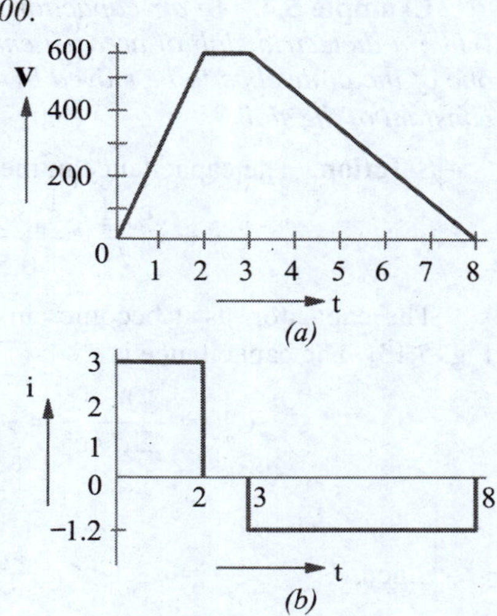

Fig. 5.12

(a) Charge when a steady voltage of 600 V is applied is $= 600 \times 10 \times 10^{-6} = \mathbf{6 \times 10^{-3}\ C}$

(b) Energy stored $= \frac{1}{2}\ C V^2 = \frac{1}{2} \times 10^{-5} \times 600^2 = \mathbf{1.8\ J}$

Example 5.2. *The capacitance of a capacitor formed by two parallel metal plates each 200 cm² in area separated by a dielectric 4 mm thick is 0.0004 microfarads. A p.d. of 20,000 V is applied. Calculate* **(a)** *the total charge on the plates* **(b)** *the potential gradient in V/m* **(c)** *relative permittivity of the dielectric* **(d)** *the electric flux density.*

(Elect. Engg. I Osmania Univ., RTU 2015)

Solution. $C = 4 \times 10^{-4}\ \mu F\ ;\ V = 2 \times 10^4\ V$

(a) ∴ Total charge $Q = CV = 4 \times 10^{-4} \times 2 \times 10^4\ \mu C = 8\ \mu C = \mathbf{8 \times 10^{-6}\ C}$

(b) Potential gradient $\dfrac{dV}{dx} = \dfrac{2 \times 10^4}{4 \times 10^{-3}} = \mathbf{5 \times 10^6\ V/m}$

(c) $D = Q/A = 8 \times 10^{-6}/200 \times 10^{-4} = \mathbf{4 \times 10^{-4}\ C/m^2}$

(d) $E = 5 \times 10^6\ V/m$

Since $D = \varepsilon_0\, \varepsilon_r\, E$ ∴ $\varepsilon_r = \dfrac{D}{\varepsilon_0 \times E} = \dfrac{4 \times 10^{-4}}{8.854 \times 10^{-12} \times 5 \times 10^6} = \mathbf{9}$

Example 5.3. *A parallel plate capacitor has 3 dielectrics with relative permittivities of 5.5, 2.2 and 1.5 respectively. The area of each plate is 100 cm² and thickness of each dielectric 1 mm. Calculate the stored charge in the capacitor when a potential difference of 5,000 V is applied across the composite capacitor so formed. Calculate the potential gradient developed in each dielectric of the capacitor.* **(Elect. Engg. A.M.Ae.S.I.)**

Solution.

$$C = \frac{\varepsilon_0 A}{\left(\dfrac{d_1}{\varepsilon_{r1}} + \dfrac{d_2}{\varepsilon_{r2}} + \dfrac{d_3}{\varepsilon_{r3}}\right)} = \frac{8.854 \times 10^{-12} \times (100 \times 10^{-4})}{\left(\dfrac{10^{-3}}{5.5} + \dfrac{10^{-3}}{2.2} + \dfrac{10^{-3}}{1.5}\right)} = \frac{8.854 \times 10^{-14}}{10^{-3} \times 0.303} = 292\ \text{pF}$$

$Q = CV = 292 \times 10^{-12} \times 5000 = 146 \times 10^{-8}$ coulomb

$D = Q/A = 146 \times 10^{-8}/(100 \times 10^{-4}) = 146 \times 10^{-6}\ C/m^2$

$g_1 = E_1 = D/\varepsilon_0\, \varepsilon_{r1} = 146 \times 10^{-6}/8.854 \times 10^{-12} \times 5.5 = \mathbf{3 \times 10^6\ V/m}$

$g_2 = E_2 = D/\varepsilon_0\, \varepsilon_{r2} = 7.5 \times 10^6\ V/m;\ g_3 = D/\varepsilon_0\, \varepsilon_{r3} = \mathbf{11 \times 10^6\ V/m}$

Example 5.4. *An air capacitor has two parallel plates 10 cm² in area and 0.5 cm apart. When a dielectric slab of area 10 cm² and thickness 0.4 cm was inserted between the plates, one of the plates has to be moved by 0.4 cm to restore the capacitance. What is the dielectric constant of the slab ?* **(Elect. Technology, Hyderabad Univ. 1992)**

Solution. The capacitance in the first case is

$$C_a = \frac{\varepsilon_0 A}{d} = \frac{\varepsilon_0 \times 10 \times 10^{-4}}{0.5 \times 10^{-2}} = \frac{\varepsilon_0}{5}$$

The capacitor, as it becomes in the second case, is shown in Fig. 5.13. The capacitance is

$$C_m = \frac{\varepsilon_0 A}{\Sigma d / \varepsilon_r} = \frac{\varepsilon_0 \times 10^{-3}}{\left(\dfrac{0.5 \times 10^{-3}}{\varepsilon_r}\right)} = \frac{\varepsilon_0}{\left(\dfrac{5}{\varepsilon_r} + 4\right)}$$

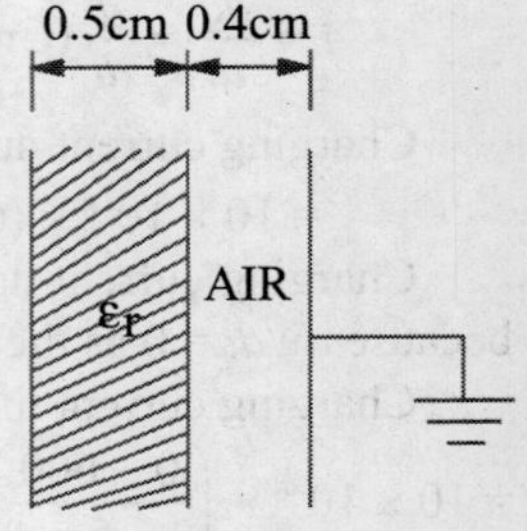

Fig. 5.13

Since, $C_a = C_m$ ∴ $\dfrac{\varepsilon_0}{5} = \dfrac{\varepsilon_0}{(5/\varepsilon_r + 4)}$ ∴ $\varepsilon_r = 5$

Note. Separation $= (t - t/\varepsilon_1)$ $\therefore$ $0.4 = (0.5 - 0.5/\varepsilon_r)$ or $\varepsilon_r = 5$

Tutorial Problem 5.1

1. Two parallel plate capacitors have plates of an equal area, dielectrics of relative permittivities ε_{r1} and ε_{r2} and plate spacing of d_1 and d_2. Find the ratio of their capacitances if $\varepsilon_{r1}/\varepsilon_{r2} = 2$ and $d_1/d_2 = 0.25$. **[$C_1/C_2 = 8$]**

2. A capacitor is made of two plates with an area of 11 cm² which are separated by a mica sheet 2 mm thick. If for mica $\varepsilon_r = 6$, find its capacitance. If, now, one plate of the capacitor is moved further to give an air gap 0.5 mm wide between the plates and mica, find the change in capacitance. **[29.19 pF, 11.6 pF]**

3. A parallel-plate capacitor is made of two plane circular plates separated by *d* cm of air. When a parallel-faced plane sheet of glass 2 mm thick is placed between the plates, the capacitance of the system is increased by 50% of its initial value. What is the distance between the plates if the dielectric constant of the glass is 6 ? **[0.5×10^{-3} m]**

4. A p.d. of 10 kV is applied to the terminals of a capacitor consisting of two circular plates, each having an area of 100 cm² separated by a dielectric 1 mm thick. If the capacitance is 3×10^{-4} μ F, calculate

(*a*) the total electric flux in coulomb

(*b*) the electric flux density and

(*c*) the relative permittivity of the dielectric. **[(*a*) 3×10^{-6}C (*b*) 3×10^{-4} μ C/m² (*c*) 3.39]**

5. Two slabs of material of dielectric strength 4 and 6 and of thickness 2 mm and 5 mm respectively are inserted between the plates of a parallel-plate capacitor. Find by how much the distance between the plates should be changed so as to restore the potential of the capacitor to its original value. **[5.67 mm]**

6. The oil dielectric to be used in a parallel-plate capacitor has a relative permittivity of 2.3 and the maximum working potential gradient in the oil is not to exceed 10^6 V/m. Calculate the approximate plate area required for a capacitance of 0.0003 μ F, the maximum working voltage being 10,000 V. **[147×10^{-3} m²]**

7. A parallel-plate capacitor has an effecting plate area of 100 cm² (each plate) separated by a dielectric 0.5 mm thick. Its capacitance is 442 μμ F and it is raised to a potential differences of 10 kV. Calculate from first principles:

(*a*) potential gradient in the dielectric

(*b*) electric flux density in the dielectric

(*c*) the relative permittivity of the dielectric material. **[(*a*) 20 kV/mm (*b*) 442 μC/m²(*c*) 2.5]**

5.8 Capacitance between Two Parallel Wires

This case is of practical importance in overhead transmission lines. The simplest system is 2-wire system (either *d.c.* or *a.c.*). In the case of *a.c.* system, if the transmission line is long and voltage high, the charging current drawn by the line due to the capacitance between conductors is appreciable and affects its performance considerably.

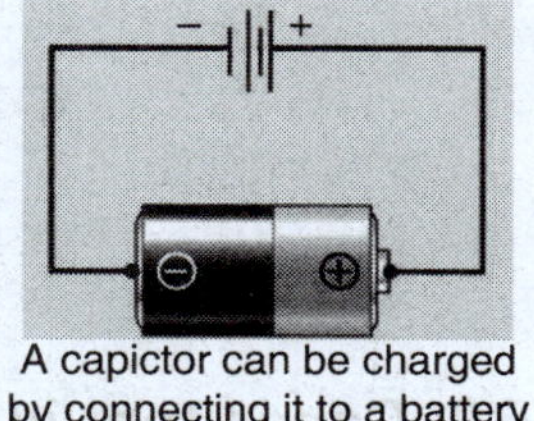
A capictor can be charged by connecting it to a battery

With reference to Fig. 5.14, let

d = distance between centres of the wires A and B

r = radius of each wire ($\leq d$)

Q = charge in coulomb/metre of each wire*

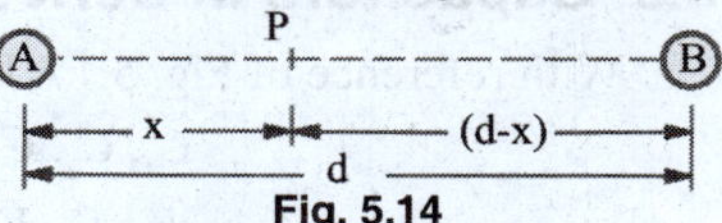

Fig. 5.14

* If charge on A is $+Q$, then on B will be $-Q$.

Now, let us consider electric intensity at any point P between conductors A and B.

Electric intensity at P* due to charge $+ Q$ coulomb/metre on A is

$$= \frac{Q}{2\pi\varepsilon_0\varepsilon_r x} \text{ V/m} \qquad \text{... towards } B.$$

Electric intesity at P due to charge $- Q$ coulomb/metre on B is

$$= \frac{Q}{2\pi\varepsilon_0\varepsilon_r (d-x)} \text{ V/m} \qquad \text{0... towards } B$$

$$\text{Total electric intensity at } P,\ E = \frac{Q}{2\pi\varepsilon_0\varepsilon_r}\left(\frac{1}{x}+\frac{1}{d-x}\right)$$

Hence, potential difference between the two wires is

$$V = \int_r^{d-r} E.dx = \frac{Q}{2\pi\varepsilon_0\varepsilon_r}\int_r^{d-r}\left(\frac{1}{x}+\frac{1}{d\ x}\right)dx$$

$$V = \frac{Q}{2\pi\varepsilon_0\varepsilon_r}\,|\log_e x - \log_e (d-x)|_r^{d-r} = \frac{Q}{\pi\varepsilon_0\varepsilon_r}\log_e\frac{d-r}{r}$$

$$\text{Now } C = Q/V \ \therefore\ C = \frac{\pi\varepsilon_0\varepsilon_r}{\log_e\frac{(d-r)}{r}} = \frac{\pi\varepsilon_0\varepsilon_r}{2.3\log_{10}\frac{(d-r)}{r}} = \frac{\pi\varepsilon_0\varepsilon_r}{2.3\log_{10}\left(\frac{d}{r}\right)} \text{ F/m (approx.)}$$

$$\text{The capacitance for a length of } l \text{ metres } C = \frac{\pi\varepsilon_0\varepsilon_r}{2.3\log_{10}\left(\frac{d}{r}\right)} \text{ F}$$

The capacitance per kilometre is

$$C = \frac{\pi\times 8.854\times10^{-12}\times\varepsilon_r\times100\times10^6}{2.3\log_{10}\left(\frac{d}{r}\right)} = \frac{0.0121\,\varepsilon_r}{\log_{10}\left(\frac{d}{r}\right)} = \mu\text{ F/km}$$

Example 5.5. *The conductors of a two-wire transmission line (4 km long) are spaced 45 cm between centre. If each conductor has a diameter of 1.5 cm, calculate the capacitance of the line.*

Solution. Formula used $\quad C = \dfrac{\pi\varepsilon_0\varepsilon_r}{2.3\log_{10}\left(\frac{d}{r}\right)}$ F

Here $l = 4000$ metres ; $r = 1.5/2$ cm ; $d = 45$ cm ; $\varepsilon_r = 1$ – for air $\quad\therefore\quad \dfrac{d}{r} = \dfrac{45\times2}{1.5} = 60$

$$C = \frac{\pi\times8.854\times10^{-12}\times4000}{2.3\log_{10}60} = \mathbf{0.0272\times10^{-6}\ F\ 0.0272\times10^{-6}\ F}$$

[or $\quad C = 4\times\dfrac{0.0121}{\log_{10}60} = \mathbf{0.0272\mu F}$]

5.9 Capacitors in Series

With reference of Fig. 5.17, let

C_1, C_2, C_3 = Capacitances of three capacitors

V_1, V_2, V_3 = p.ds. across three capacitors.

V = applied voltage across combination

C = combined or equivalent or joining capacitance.

In series combination, charge on all capacitors is the same but p.d. across each is different.

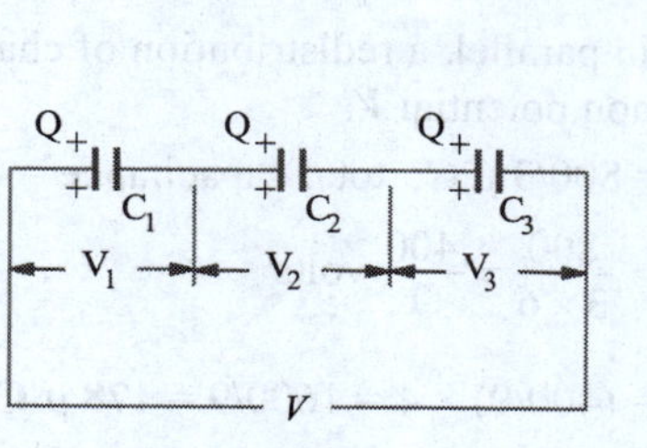

Fig. 5.15

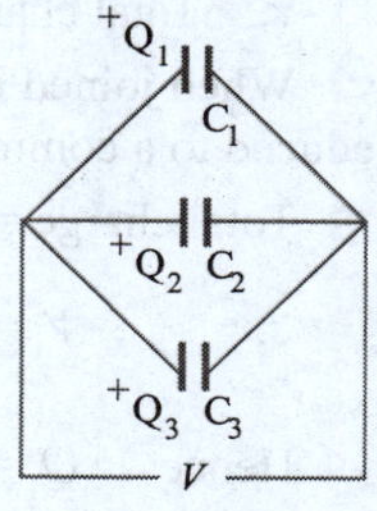

Fig. 5.16

$$\therefore \quad V = V_1 + V_2 + V_3$$

$$\text{or} \quad \frac{Q}{C} = \frac{Q}{C_1} + \frac{Q}{C_2} + \frac{Q}{C_3}$$

$$\text{or} \quad \frac{1}{C} = \frac{1}{C_1} + \frac{1}{C_2} + \frac{1}{C_3}$$

For a changing applied voltage,

$$\frac{dV}{dt} = \frac{dV_1}{dt} + \frac{dV_2}{dt} + \frac{dV_3}{dt}$$

We can also find values of V_1, V_2 and V_3 in terms of V. Now, $Q = C_1 V_1 = C_2 V_2 = C_3 V_3 = CV$

$$\text{where} \quad C = \frac{C_1 C_2 C_3}{C_1 C_2 + C_2 C_3 + C_3 C_1} = \frac{C_1 C_2 C_3}{\Sigma C_1 C_2}$$

$$\therefore \quad C_1 V_1 = C\,V \text{ or } V_1 = V\frac{C}{C_1} = V.\frac{C_2 C_3}{\Sigma C_1 C_2}$$

$$\text{Similarly,} \quad V_2 = V \cdot \frac{C_1 C_3}{\Sigma C_1 C_2} \text{ and } V_3 = V \cdot \frac{C_1 C_2}{\Sigma C_1 C_2}$$

5.10 Capacitors in Parallel

In this case, p.d. across each is the same but charge on each is different (Fig. 5.17).

$\therefore \quad Q = Q_1 + Q_2 + Q_3$ or $CV = C_1 V + C_2 V + C_3 V$ or $C = C_1 + C_2 + C_3$

For such a combination, dV/dt is the same for all capacitors.

Example 5.6. *Find the C_{eq} of the circuit shown in Fig. 5.17. All capacitances are in μ F.*

(Basic Circuit Analysis Osmania Univ. Jan./Feb. 1992)

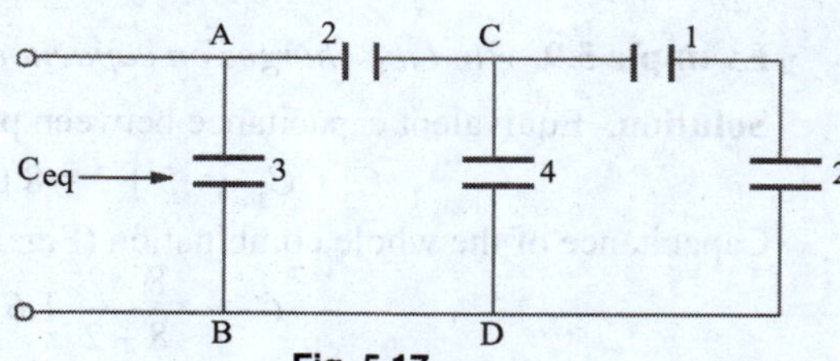

Fig. 5.17

Solution. Capacitance between C and $D = 4 + 1 \parallel 2 = 14/3$ μ F.

Capacitance between A and B *i.e.* $C_{eq} = 3 + 2 \parallel 14/3 = 4.4$ μ F

Example 5.7. *Two capacitors of a capacitance 4 μF and 2 μF respectively, are joined in series with a battery of e.m.f. 100 V. The connections are broken and the like terminals of the capacitors are then joined. Find the final charge on each capacitor.*

Solution. When joined in series, let V_1 and V_2 be the voltages across the capacitors. Then as charge across each is the same.

$\therefore \quad 4 \times V_1 = 2V_2 \qquad \therefore \quad V_2 = 2V_1$ Also $V_1 + V_2 = 100$

$\therefore \quad V_1 + 2V_1 = 100 \quad \therefore \quad V_1 = 100/3$ V and $V_2 = 200/3$ V

$\therefore \quad Q_1 = Q_2 = (200/3) \times 2 = (400/3)\ \mu C$

$\therefore$ Total charge on both capacitors = 800/3 μ C

When joined in parallel, a redistribution of charge takes place because both capacitors are reduced to a common potential V.

Total charge = 800/3 μ C; total capacitance = 4 + 2 = 6 μ F

$$\therefore \quad V = \frac{800}{3 \times 6} = \frac{400}{9} \text{ volts}$$

Hence $Q_1 = (400/9) \times 4 = 1600/9 = \mathbf{178\ \mu\ C}$

$Q_2 = (400/9) \times 2 = \mathbf{800/9 = 89\ \mu\ C\ (approx.)}$

Example 5.8. *Three capacitors A, B, C have capacitances 10, 50 and 25 μF respectively. Calculate* **(i)** *charge on each when connected in parallel to a 250 V supply* **(ii)** *total capacitance and* **(iii)** *p.d. across each when connected in series.* **(Elect. Technology, Gwalior Univ.)**

Solution. (*i*) Parallel connection is shown in Fig. 5.18 (*a*). Each capacitor has a p.d. of 250 V across it.

$Q_1 = C_1V = 10 \times 250 = \mathbf{2500\ \mu\ C}$

$Q_2 = 50 \times 250 = \mathbf{12{,}500\ \mu C}$

$Q_3 = 25 \times 250 = \mathbf{6{,}750\ \mu\ C.}$

(*ii*) $C = C_1 + C_2 + C_3 = 10 + 50 + 25 = \mathbf{85\ \mu F}$

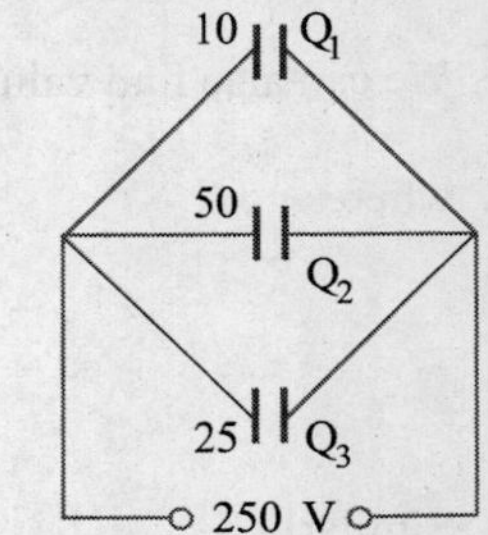

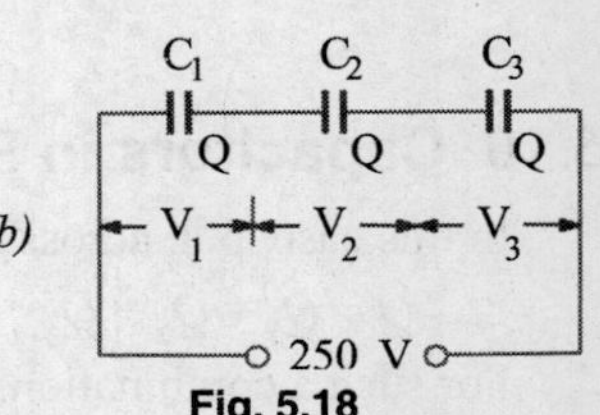

Fig. 5.18

(*iii*) Series connection is shown in Fig. 5.18 (*b*). Here charge on each capacitor is the same and is equal to that on the equivalent single capacitor.

$1/C = 1/C_1 + 1/C_2 + 1/C_3\ ;\ C = 25/4\ \mu$ F

$Q \quad = \quad CV = 25 \times 250/4 = 1562.5\ \mu$ F

$Q \quad = C_1V_1\ ;\ V_1 = 1562.5/10 = \mathbf{156.25\ V}$

$V_2 \quad = 1562.5/25 = \mathbf{62.5\ V;}\ V_3 = 1562.5/50 = \mathbf{31.25\ V.}$

Example 5.9. *Find the charges on capacitors in Fig. 5.19 and the p.d. across them.*

Solution. Equivalent capacitance between points *A* and *B* is

$$C_2 + C_3 = 5 + 3 = 8\ \mu\ F$$

Capacitance of the whole combination (Fig. 5.21)

$$C = \frac{8 \times 2}{8 + 2}\ 1.6\ \mu\ F$$

Fig. 5.19

Charge on the combination is

$Q_1 = CV = 100 \times 1.6 = \mathbf{160\ \mu C}$

$V_1 = \frac{Q_1}{C_1} = \frac{160}{2} = 80\ V\ ;\ V_2 = 100 - 80 = \mathbf{20\ V\ 20\ V}$

$Q_1 = C_2V_2 = 3 \times 10^{-6} \times 20 = \mathbf{60\ \mu C}$

$Q_3 = C_3V_2 = 5 \times 10^{-6} \times 20 = \mathbf{100\ \mu C}$

Example 5.10. *Three capacitors of 2 μ F, 5 μ F and 10 μ F have breakdown voltage of 200 V, 500 V and 100 V respectively. The capacitors are connected in series and the applied direct* *voltage to the circuit is gradually increased. Which capacitor will breakdown first ? Determine the total applied voltage and total energy stored at the point of breakdown.*

[Bombay Univeristy 2001]

Solution. C_1 of 2 μF, C_2 of 5 μ F, and C_3 of 10 μF are connected in series. If the equivalent single capacitor is C,

$1/C = 1/C_1 + 1/C_2 + 1/C_3$, which gives $C = 1.25$ μ F

If V is the applied voltage,

$V_1 = V \times C/C_1 = V \times (1.25 / 2)$

$= 62.5$ % of V

$V_2 = V \times (C/C_2) = C \times (1.25/5) = 25$ % of V

$V_3 = V \times (C/C_3) = V \times (1.25/10) = 12.5$ % of V

Fig. 5.20

If $V_1 = 200$ volts, $V = 320$ volts and $V_2 = 80$ volts, $V_3 = 40$ volts.

It means that, first capacitor C_1 will breakdown first.

Energy stored $= 1/2\ CV^2 = 1/2 \times 1.25 \times 10^{-6} \times 320 \times 320 =$ **0.064 Joule**

Example 5.11. (*a*) *A capacitor of 10 pF is connected to a voltage source of 100 V. If the distance between the capacitor plates is reduced to 50 % while it remains, connected to the 100 V supply. Find the new values of charge, energy stored and potential as well as potential gradient. Which of these quantities increased by reducing the distance and why ?*

[Bombay University 2000]

Solution.

(*i*) $C = 10$ pF	(*ii*) $C = 20$ pF, distance halved
Charge = 1000 p Coul	Charge = 2000 p-coul
Energy = $1/2\ CV^2 = 0.05$ μ J	Energy = 0.10 μ J

Potential gradient in the second case will be twice of earlier value.

Example 5.12 (*b*). *A capacitor 5 μ F charged to 10 V is connected with another capacitor of 10 μ F charged to 50 V, so that the capacitors have one and the same voltage after connection. What are the possible values of this common voltage ?* **[Bombay University 2000]**

Solution. The clearer procedure is discussed here.

Initial charges held by the capacitors are represented by equivalent voltage sources in Fig. 5.21 (*b*). The circuit is simplified to that in Fig. 5.21 (*c*). This is the case of C_1 and C_2 connected in series and excited by a 40-V source. If C is the equivalent capacitance of this series-combination,

$$1/C = 1/C_1 + C_2$$

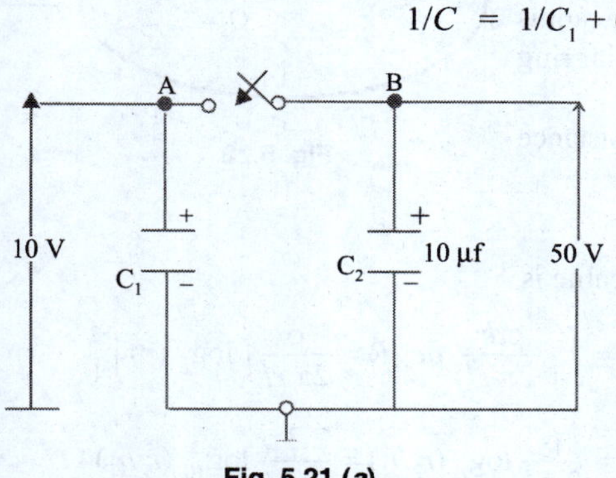

Fig. 5.21 (*a*)

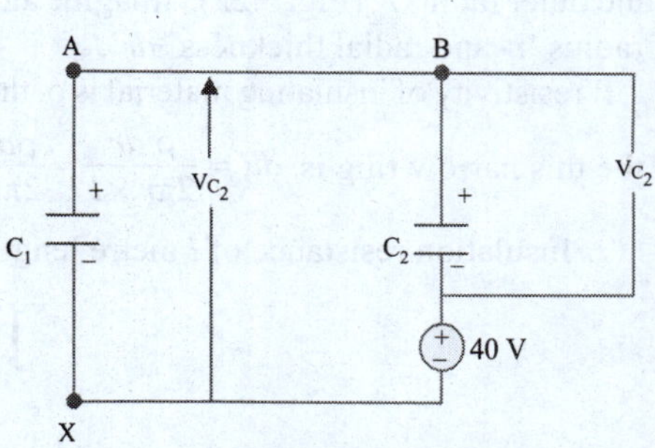

Fig. 5.21 (*b*) Simplification

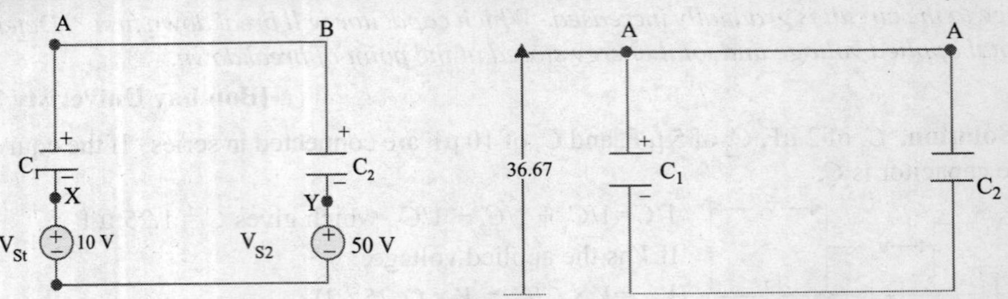

Fig. 5.21 (c). Initial charge represented by equiv-source

Fig. 5.21 (d). Final condition

This gives $C = 3.33\ \mu F$

In Fig. 5.21 (c), $V_{C1} = 40 \times C/C_1 = 40 \times 3.33/5 = 26.67$ volts

V_{S1} and V_{S2} are integral parts of C_1 and C_2 in Fig. 5.21 (c),

Voltage across $C_1 = 10 + 26.67 = 36.67$ (A w.r. to 0)

Voltage acorss $C_2 = 50 - 13.33 = 36.67$, (B w.r. to 0)

Thus, the final voltage across the capacitor is 36.67 volts.

Note: If one of the initial voltages on the capacitors happens to be the opposite to the single equivalent source voltage in Fig. 5.21 (c) will be 60 volts. Proceeding similarly, with proper care about signs, the final situation will be the common voltage will be 30 volts.

5.11 Insulation Resistance of a Cable Capacitor

In a cable capacitor, useful current flows along the axis of the core but there is always present some leakage of current. This leakage is radial *i.e.* at right angles to the flow of useful current. The resistance offered to this radial leakage of current is called ***insulation resistance*** of the cable. If cable length is greater, then leakage is also greater. It means that more current will leak. In other words, insulation resistance is decreased. Hence, we find that insulation resistance is inversely proportional to the cable length. This insulation resistance is not to be confused with conductor resistance which is directly proportional to the cable length.

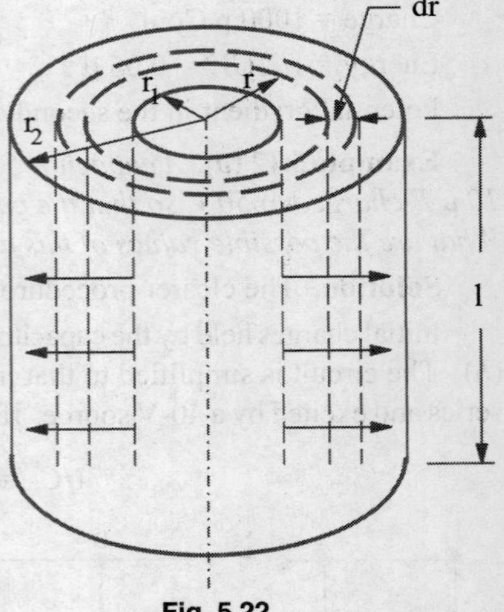

Fig. 5.22

Consider *i* metre of a single-core cable of inner-radius r_1 and outer radius r_2 (Fig. 5.22). Imagine an annular ring of radius '*r*' and radial thickness '*dr*'.

If resistivity of insulating material is ρ, then resistance of the this narrow ring is $dR = \dfrac{\rho\, dr}{2\pi r \times l} = \dfrac{\rho dr}{2\pi rl}$

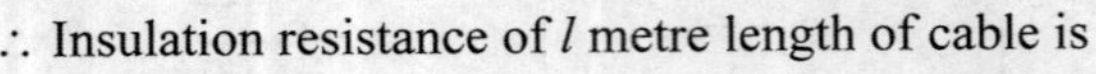

∴ Insulation resistance of *l* metre length of cable is

$$\int dR = \int_{r1}^{r2} \frac{\rho dr}{2\pi\, rl} \quad or \quad R = \frac{\rho}{2\pi\, rl} \Big|\, \log_e (r) \Big|_{r1}^{r2}$$

$$R = \frac{\rho}{2\,\pi\, l} \log_e (r_2/r_1) = \frac{2.3\,\rho}{2\,\pi\, l} \log_{10} (r_2/r_1)\ \Omega$$

It should be noted

(*i*) *that R is inversely proportional to the cable length*

(*ii*) *that R depends upon the ratio r_2/r_1 and NOT on the thickness of insulator itself.*

Example 5.13. *A liquid resistor consists of two concentric metal cylinders of diameters D = 35 cm and d = 20 cm respectively with water of specific resistance ρ = 8000 Ω cm between them. The length of both cylinders is 60 cm. Calculate the resistance of the liquid resistor.*

(Elect. Engg. Aligarh Univ.,)

Solution. $r_1 = 10$ cm ; $r_2 = 17.5$ cm; $\log_{10}(1.75) = 0.243$

$\rho = 8 \times 10^3\ \Omega - \text{cm};\ l = 60$ cm.

Resistance of the liquid resistor $R = \dfrac{2.3 \times 8 \times 10^3}{2\pi \times 60} \times 0.243 = 11.85\ \Omega.$

Example 5.14. *Two underground cables having conductor resistances of 0.7 Ω and 0.5 and insulation resistance of 300 M Ω respectively are joind* **(*i*)** *in series* **(*ii*)** *in parallel. Find the resultant conductor and insulation resistance.* **(Elect. Engineering, Calcutta Univ.)**

Solution. **(*i*)** The conductor resistance will add like resistances in series. However, the leakage resistances will decrease and would be given by the reciprocal relation.

Total conductor resistance = 0.7 + 0.5 = 1.2 Ω

If *R* is the combined leakage resistance, then

$$\frac{1}{R} = \frac{1}{300} + \frac{1}{600} \qquad \therefore\ R = 200\ M\Omega$$

(*ii*) In this case, conductor resistance is = 0.7 × 0.5/(0.7 + 0.5) = **0.3. Ω (approx)**

Insulation resistance = 300 + 600 = **900 M Ω**

Example 5.15. *The insulation resistance of a kilometre of the cable having a conductor diameter of 1.5 cm and an insulation thickness of 1.5 cm is 500 M Ω. What would be the insulation resistance if the thickness of the insulation were increased to 2.5 cm ?*

(Communication Systems, Hyderadad Univ. 1992)

Solution. The insulation resistance of a cable is

First Case $$R = \frac{2.3\,\rho}{2\pi l}\log_{10}(r_2/r_1)$$

$r_1 = 1.5/2 = 0.75$ cm ; $r_2 = 0.75 + 1.5 = 2.25$ cm

$\therefore\ r_2/r_1 = 2.25/0.75 = 3$; $\log_{10}(3) = 0.4771$ $\qquad \therefore\ 500 = \dfrac{2.3\,\rho}{2\,\pi\, l} \times 0.4771$ ...(*i*)

Second Case

$r_1 = 0.75$ cm – as before $r_2 = 0.75 + 2.5 = 3.25$ cm

$r_2/r_1 = 3.25/0.75 = 4.333$; $\log_{10}(4.333) = 0.6368 \qquad \therefore\ R = \dfrac{2.4\,\rho}{2\,\pi\, l} \times 0.6368$ ...(*ii*)

Dividing Eq. (*ii*) by Eq. (*i*), we get

$$\frac{R}{500} = \frac{0.6368}{0.4771};\ R = 500 \times 0.6368 / 0.4771 = 667.4\ \text{M}\ \Omega$$

5.12 Energy Stored in a Capacitor

Charging of a capacitor always involves some expenditure of energy by the charging agency. This energy is stored up in the electrostatic field set up in the dielectric medium. On discharging the capacitor, the field collapses and the stored energy is released.

To begin with, when the capacitor is uncharged, little work is done in transferring charge from one plate to another. But further instalments of charge have to be carried against the repulsive force due to the charge already collected on the capacitor plates. Let us find the energy spent in charging a capacitor of capacitance C to a voltage V.

Suppose at any stage of charging, the p.d. across the plates is v. By definition, it is equal to the work done in shifting one coulomb from one plate to another. If 'dq' is charge next transferred, the work done is

$$dW = v.dq$$

Now $q = Cv \quad \therefore \quad dq = C.dv \quad \therefore \quad dW = Cv.dv$

Total work done in giving V units of potential is

$$W = \int_0^V Cv.dv = C\left|\frac{v^2}{2}\right|_0^V \quad \therefore\ W = \frac{1}{2}CV^2$$

If C is in farads and V is in volts, then $W = \frac{1}{2}CV^2$ joules $= \frac{1}{2}QV$ joules $= = \frac{1}{2}QV$ joules

If Q is in coulombs and C is in farads, the energy stored is given in joules.

Note: As seen from above, energy stored in a capacitor is $E = \frac{1}{2}CV^2$

Now, for a capacitor of plate area A m^2 and dielectric of thickness d metre, energy per unit volume of dielectric medium.

$$= \frac{1}{2}\frac{CV^2}{Ad} = \frac{1}{2}\frac{\varepsilon A}{d}.\frac{V^2}{Ad} = \frac{1}{2}\varepsilon\left(\frac{V}{d}\right)^2 = \frac{1}{2}\varepsilon E^2 = \frac{1}{2}DE = D^2/2\varepsilon \text{ joules/m}^3 \, *$$

It will be noted that the formula $\frac{1}{2}DE$ is similar to the expression $\frac{1}{2}$ stress × strain which is used for calculating the mechanical energy stored per unit volume of a body subjected to elastic stress.

Example 5.16. *Since a capacitor can store charge just like a lead-acid battery, it can be used at least theoretically as an electrostatic battery. Calculate the capacitance of 12-V electrostatic battery which the same capacity as a 40 Ah, 12 V lead-acid battery.*

Solution. Capacity of the lead-acid battery = 40 Ah = 40 × 36 As = 144000 Coulomb

Energy stored in the battery = QV = 144000 × 12 = 1.728 × 10^6 J

Energy stored in an electrostatic battery = $\frac{1}{2}CV^2$

$$\therefore \quad \frac{1}{2}\times C\times 12^2 = 1.728\times 10^6 \quad \therefore\ C = 2.4\times 10^4 \text{ F} = 24 \text{ kF}$$

Example 5.17. *A capacitor-type stored-energy welder is to deliver the same heat to a single weld as a conventional welder that draws 20 kVA at 0.8 pf for 0.0625 second/weld. If C = 2000 μF, find the voltage to which it is charged.* **(Power Electronics, A.M.I.E. Sec B, 1993)**

Solution. The energy supplied per weld in a conventional welder is

$$W = VA \times \cos\phi \times \text{time} = 20{,}000 \times 0.8 \times 0.0625 = 1000 \text{ J}$$

Now, energy stored in a capacitor is (1/2) CV^2

$$\therefore \quad W = \frac{1}{2}CV^2 \text{ or } V = \sqrt{\frac{2W}{C}} = \sqrt{\frac{2\times 1000}{2000\times 10^{-6}}} = \mathbf{1000\ V\ 1000V}$$

* It is similar to the expression for the energy stored per unit volume of a magnetic field.

Example 5.18. *A parallel-plate capacitor is charged to 50 μC at 150 V. It is then connected to another capacitor of capacitance 4 times the capacitance of the first capacitor. Find the loss of energy.* **(Elect. Engg. Aligarh Univ.)**

Solution. $C_1 = 50/150 = 1/3\ \mu F$; $C_2 = 4 \times 1/3 = 4/3\ \mu F$

Before Joining

$$E_1 = \frac{1}{2} C_1 V_1^2 = \frac{1}{2} \times \left(\frac{1}{3}\right) 10^{-6} \times 150^2 = 37.5 \times 10^{-4}\ J;\ E_2 = 0$$

$$\text{Total energy} = 37.5 \times 10^{-4}\ J$$

After Joining

When the two capacitors are connected in parallel, the charge of 50 μ C gets redistributed and the two capacitors come to a common potential V.

$$V = \frac{\text{total charge}}{\text{total capacitance}} = \frac{50\ \mu C}{[(1/3) + (4/3)]\ \mu\ F} = 30V$$

$$E_1 = \frac{1}{2} \times (1/3) \times 10^{-6} \times 30^2 = 1.5 \times 10^{-4}\ J$$

$$E_2 = \frac{1}{2} \times (4/3) \times 10^{-6} \times 30^2 = 6.0 \times 10^{-4}\ J$$

Total energy $= 7.5 \times 10^{-4}$ J ; Loss of energy $= (37.5 - 7.5) \times 10^{-4} = \mathbf{3 \times 10^{-2}\ J}$

The energy is wasted away as heat in the conductor connecting the two capacitors.

Example 5.19. *An air-capacitor of capacitance 0.005 μ F is connected to a direct voltage of 500 V, is disconnected and then immersed in oil with a relative permittivity of 2.5. Find the energy stored in the capacitor before and after immersion.* **(Elect. Technology: London Univ.)**

Solution. Energy before immersion is

$$E_1 = \frac{1}{2} CV^2 = \frac{1}{2} \times 0.005 \times 10^{-6} \times 500^2 = \mathbf{625 \times 10^{-6}\ J}\ \mathbf{625 \times 10^{-6}\ J}$$

When immersed in oil, its capacitance is increased 2.5 times. Since charge is constant, voltage must become 2.5 times. Hence, new capacitances is $2.5 \times 0.005 = 0.0125\ \mu F$ and new voltage is $500/2.5 = 200$ V.

$$E_2 = \frac{1}{2} \times 0.0125 \times 10^{-6} \times (200)^2 = \mathbf{250 \times 10^{-6}\ J}\ \mathbf{250 \times 10^{-6}\ J}$$

Example 5.20. *When a capacitor C charges through a resistor R from a d.c. source voltage E, determine the energy appearing as heat.* **[Bombay University, 2000]**

Solution. *R-C* series ciruit switched on to a d.c. source of voltage *E*, at $t = 0$, results into a current $i\ (t)$, given by

$$i\ (t) = (E/R)\ e^{-t/\tau}$$

where $t = RC$

$$\Delta W_R = \text{Energy appearing as heat in time } \Delta t$$
$$= i^2 R\ \Delta t$$
$$\Delta W_R = \text{Energy appearing as heat in time } \Delta t$$
$$= i^2 R\ \Delta t$$
$$W_R = R \int_0^\infty i^2\ dt$$

$$= R\,(E/R)^2 \int_0^\infty (\varepsilon^{-t/\tau})^2 = \frac{1}{2} CE^2$$

Note: Energy stored by the capacitor at the end of charging process = 1/2 CE^2
Hence, energy received from the source = CF.

5.13 Force of Attraction between Oppositely-Charged Plates

In Fig. 5.23 are shown two parallel conducting plates A and B carrying constant charges of $+Q$ and $-Q$ coulombs respectively. Let the force of attraction between the two be F newtons. If one of the plates is pulled apart by distance dx, then work done is

$$= F \times dx \text{ joules} \qquad \text{...}(i)$$

Since the plate charges remain constant, no electrical energy comes into the arrangement during the movement dx.

$\therefore$ Work done = change in stored energy

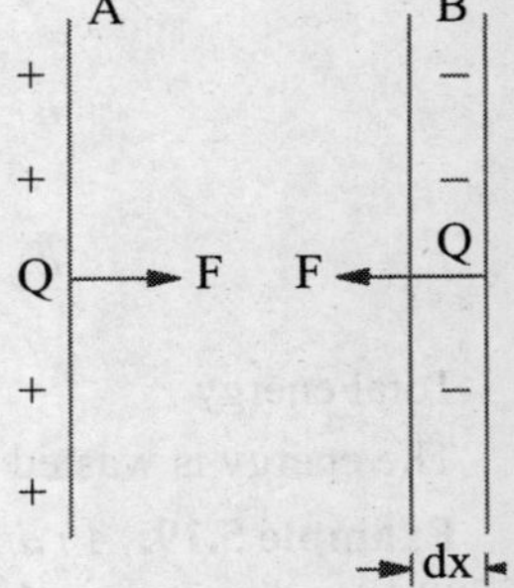

Fig. 5.23

Initial stored energy $= \frac{1}{2}\frac{Q_2}{C}$ joules

If capacitance becomes $(C - dC)$ due to the movement dx, then Final stored energy

$$= \frac{1}{2}\frac{Q^2}{(C - dC)} = \frac{1}{2}\cdot\frac{Q^2}{C}\,\frac{1}{\left(1 - \frac{dC}{C}\right)} = \frac{1}{2}\frac{Q^2}{C}\left(1 + \frac{dC}{C}\right) \text{ if } dC << C$$

$\therefore$ Change in stored energy

$$= \frac{1}{2}\frac{Q^2}{C}\left(1 + \frac{dC}{C}\right) - \frac{1}{2}\frac{Q^2}{C} = \frac{1}{2}\frac{Q^2}{C^2}.dC \quad \text{...}(ii)$$

Equating Eq. (*i*) and (*ii*), we have $F.dx = \frac{1}{2}\frac{Q^2}{C^2}.dC$

$$F = \frac{1}{2}\frac{Q^2}{C^2}.\frac{dC}{dx} = \frac{1}{2}V^2.\frac{dC}{dx} \qquad (\because V = Q/C)$$

Now $\qquad C = \frac{\varepsilon A}{x} \quad \therefore \quad \frac{dC}{dx} = -\frac{\varepsilon A}{x^2}$

$\therefore$
$$F = -\frac{1}{2}V^2.\frac{\varepsilon A}{x^2} = -\frac{1}{2}\varepsilon A\left(\frac{V}{x}\right)^2 \text{ newtons} = -\frac{1}{2}\varepsilon A E^2 \text{ newtons}$$

This represents the force between the plates of a parallel-plate capacitor charged to a p.d. of V volts. The negative sign shows that it is a force of attraction.

Example 5.21. *A parallel-plate capacitor is made of plates 1 m square and has a separation of 1 mm. The space between the plates is filled with dielectric of* $\varepsilon_r = 25.0$. *If 1 k V potential difference is applied to the plates, find the force squeezing the plates together.*

(Electromagnetic Theory, A.M.I.E. Sec B, 1993)

Solution. $F = -(1/2)\,\varepsilon_0\,\varepsilon_r\,AE^2$ newton

Now $E = V/d = 1000/1 \times 10^{-3} = 10^{-6}$ V/m

$$\therefore \quad F = -\frac{1}{2}\varepsilon_0\,\varepsilon_r\,AE^2 = -\frac{1}{2}\times 8.854\times 10^{-12}\times 25\times 1\times (10^6)^2 = -1.1\times 10^{-4}\text{ N}$$

Tutorial Problem 5.2

1. Find the capacitance per unit length of a cylindrical capacitor of which the two conductors have radii 2.5 and 4.5 cm and dielectric consists of two layers whose cylinder of contact is 3.5 cm in radius, the inner layer having a dielectric constant of 4 and the outer one of 6. **[440 pF/m]**
2. A parallel-plate capacitor, having plates 100 cm² area, has three dielectrics 1 mm each and of permittivities 3, 4 and 6. If a peak voltage of 2,000 V is applied to the plates, calculate:
   (*a*) potential gradient across each dielectric
   (*b*) energy stored in each dielectric.
   [8.89 kV/cm; 6.67 kV/cm ; 4.44 kV/cm ; 1047, 786, 524 × 10^{-7} joule]
3. The core and lead-sheath of a single-core cable are separated by a rubber covering. The cross-sectional area of the core is 16 mm². A voltage of 10 kV is applied to the cable. What must be the thickness of the rubber insulation if the electric field strength in it is not to exceed 6×10^6 V/m? **[2.5 mm (approx)]**
4. A p.d. of 10 kV is applied to the terminals of a capacitor of two circular plates each having an area of 100 sq. cm. separated by a dielectric 1 mm thick. If the capacitance is 3×10^{-4} microfarad, calculate the electric flux density and the relative permittivity of the dielectric. **[$D = 3 \times 10^{-4}$ C/m², $\varepsilon_r = 3.39$] (*City & Guilds, London*)**
5. Each electrode of a capacitor of the electrolytic type has an area of 0.02 sq. metre. The relative permittivity of the dielectric film is 2.8. If the capacitor has a capacitance of 10 μF, estimate the thickness of the dielectric film. **[4.95 × 10^{-8} m] (*I.E.E. London*)**

5.14 Current-Voltage Relationships in a Capacitor

The charge on a capacitor is given by the expression $Q = CV$. By differentiating this relation, we get

$$i = \frac{dQ}{dt} = \frac{d}{dt}(CV) = C\frac{dV}{dt}$$

Following important facts can be deduced from the above relations:

(*i*) since $Q = CV$, it means that the voltage across a capacitor is proportional to *charge,* not the *current.*

(*ii*) a capacitor has the ability to store charge and hence to provide a short of memory.

(*iii*) a capacitor can have a voltage across it even when there is *no current flowing.*

(*iv*) from $i = c\,dV/dt$, it is clear that current in the capacitor is present only when voltage on it changes with time. If $dV/dt = 0$ *i.e.* when its voltage is constant or for d.c. voltage, $i = 0$. Hence, the capacitor behaves like an *open circuit.*

(*v*) from $i = C\,dV/dt$, we have $dV/dt = i/C$. It shows that for a given value of (charge or discharge) current *i*, rate of change in voltage is inversely proportional to capacitance. Larger the value of *C*, slower the rate of change in capacitive voltage. Also, capacitor voltage ***cannot change instantaneously***.

(*vi*) the above equation can be put as $dv = \frac{i}{C}\,.\,dt$

Integrating the above, we get $\int dv = \frac{1}{C}\int i\,.\,dt$ or $dv = \frac{1}{C}\int_0^t i\,dt$

Example 5.22. *The voltage across a 5 μF capacitor changes uniformly from 10 to 70 V in 5 ms. Calculate* **(i)** *change in capacitor charge* **(ii)** *charging current.*

Solution. $Q = CV \quad \therefore \; dQ = C\,.\,dV$ and $i = C\, dV/dt$

(i) $dV = 70 - 10 = 60$ V, $\therefore \quad dQ = 5 \times 60 =$ **300 μ C.**

(ii) $i = C\,.\,dV/dt = 5 \times 60/5 =$ **60 mA**

Example 5.23. *An uncharged capacitor of 0.01 F is charged first by a current of 2 mA for 30 seconds and then by a current of 4 mA for 30 seconds. Find the final voltage in it.*

Solution. Since the capacitor is initially uncharged, we will use the principle of Superposition.

$$V_1 = \frac{1}{0.01}\int_0^{30} 2 \times 10^{-3}\,.\,dt = 100 \times 2 \times 10^{-3} \times 30 = 6 \text{ V}$$

$$V_2 = \frac{1}{0.01}\int_0^{30} 4 \times 10^{-3}\,.\,dt = 100 \times 4 \times 10^{-3} \times 30 = 12 \text{ V}\;;\; \therefore\; V = V_1 + V_2 = 6 + 12 = \mathbf{18\ V}$$

Example 5.24. *The voltage across two series-connected 10 μ F capacitors changes uniformly from 30 to 150 V in 1 ms. Calculate the rate of change of voltage for* **(i)** *each capacitor and* **(ii)** *combination.*

Solution. For series combination

$$V_1 = V\frac{C_2}{C_1 + C_2} = \frac{V}{3} \text{ and } V_2 = V\,.\,\frac{C_1}{C_1 + C_2} = \frac{2V}{3}$$

When $V = 30$ V $V_1 = V/3 = 30/3 = 10$ V ; $V_2 = 2V/3 = 2 \times 30/3 = 20$ V

When $V = 150$ V $V_1 = 150/3 = 50$ V and $V_2 = 2 \times 150/3 = 100$ V

(i) $\therefore \quad \dfrac{dV_1}{dt} = \dfrac{(50-10)}{1 \text{ ms}} = \mathbf{40\ kV/s}\;;\; \dfrac{dV_2}{dt} = \dfrac{(100-20)\text{ V}}{1\text{ ms}} = \mathbf{80\ kV/s}$

(ii) $\dfrac{dV}{dt} = \dfrac{(150-30)}{1\text{ ms}} =$ **120 kV/s**

It is seen that $dV/dt = dV_1/dt + dV_2/dt$.

5.15 Charging of a Capacitor

In Fig. 5.24. (*a*) is shown an arrangement by which a capacitor *C* may be charged through a high resistance *R* from a battery of *V* volts. The voltage across *C* can be measured by a suitable voltmeter. When switch *S* is connected to terminal (*a*), *C* is charged but when it is connected to *b*, *C* is short circuited through *R* and is thus discharged. As shown in Fig. 5.24. (*b*), switch *S* is

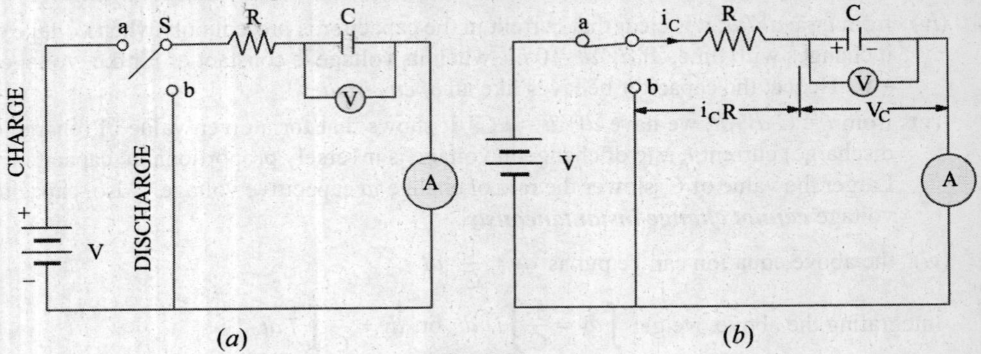

(*a*) (*b*)

Fig. 5.24

shifted to *a* for charging the capacitor for the battery. The voltage across *C* does not rise to *V* instantaneously but builds up slowly *i.e.* exponentially and not linearly. Charging current i_c is maximum at the start *i.e.* when *C* is uncharged, then it decreases exponentially and finally ceases when p.d. across capacitor plates becomes equal and opposite to the battery voltage *V*. At any instant during charging, let

v_c = p.d. across *C*; i_c = charging current

q = charge on capacitor plates

The applied voltage *V* is always equal to the sum of:

(*i*) resistive drop ($i_c R$) and (*ii*) voltage across capacitor (v_c)

$$\therefore \quad V = i_c R + v_c \qquad \text{...}(i)$$

Now
$$i_c = \frac{dq}{dt} = \frac{d}{dt}(Cv_c) = C\frac{dv_c}{dt} \quad \therefore \; V = v_c + CR\frac{dv_c}{dt} \qquad \text{...}(ii)$$

or
$$\frac{dv_c}{V - v_c} = -\frac{dt}{CR}$$

Integrating both sides, we get $\int \frac{-d\,V_c}{V - v_c} = -\frac{1}{CR}\int dt$; $\therefore$ $\log_c (V - v_c) = -\frac{t}{CR} + K$ ...(*iii*)

where *K* is the constant of integration whose value can be found from initial known conditions. We know that at the start of charging when $t = 0$, $v_c = 0$.

Substituting these values in (*iii*), we get $\log_c V = K$

Hence, Eq. (*iii*) becomes $\log_e (V - v_c) = \frac{-t}{CR} + \log_e V$

or
$$\log_c \frac{V - v_c}{V} = \frac{-t}{CR} = -\frac{1}{\lambda} \quad \text{where } \lambda = CR = \text{time constant}$$

$$\therefore \quad \frac{V - v_c}{V} = e^{-t/\lambda} \quad \text{or} \quad v_c = V(1 - e^{-t/\lambda}) \qquad \text{...}(iv)$$

This gives variation with time of voltage across the capacitor plates and is shown in Fig. 5.25 (*a*).

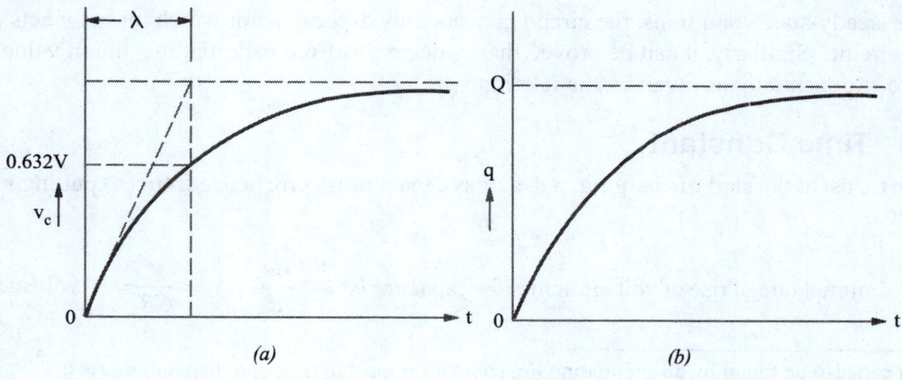

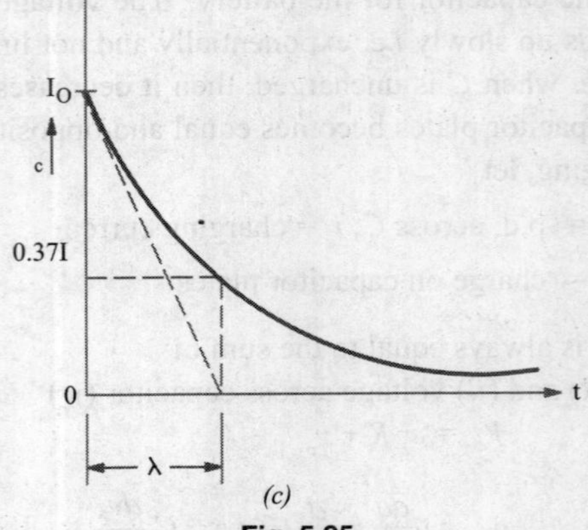

(c)

Fig. 5.25

Now $v_c = q/C$ and $V = Q/C$

Equation **(iv)** becomes $\frac{q}{c} = \frac{Q}{c}(1 - e^{-t/\lambda})\ \therefore\ q = Q(1 - e^{-t/\lambda})$ ...**(v)**

We find that increase of charge, like growth of potential, follows an exponential law in which the steady value is reached after infinite time (Fig. 5.25 *b*). Now, $i_c = dq/dt$.

Differentiating both sides of Eq. (*v*), we get

$$\frac{dq}{dt} = i_c = Q\frac{d}{dt}(1 - e^{-t/\lambda}) = Q\left(+\frac{1}{\lambda}e^{-t/\lambda}\right)$$

$$= \frac{Q}{\lambda}e^{-t/\lambda} = \frac{CV}{CR}e^{-t/\lambda} \qquad (\because Q = CV \text{ and } \lambda = CR)$$

$$\therefore \qquad i_c = \frac{V}{R}.e^{-t/\lambda} \text{ or } i_c = I_o\,e^{-t/\lambda} \qquad \text{...(vi)}$$

where I_0 = maximum current = V/R

Exponentially rising curves for v_c and q are shown in Fig. 5.25 (*a*) and (*b*) respectively. Fig. 5.25 (*c*) shows the curve for exponentially decreasing charging current. It should be particularly noted that i_c decreases in magnitude only but its direction of flow remains the same *i.e.* positive.

As charging continues, charging current decreases according to equation (*vi*) as shown in Fig. 5.25 (*c*). It becomes zero when $t = \infty$ (though it is almost zero in about 5 time constants). Under steady-state conditions, the circuit appears only as a capacitor which means it acts as an open-circuit. Similarly, it can be proved that v_R decreases from its initial maximum value of V to zero exponentially as given by the relation $v_R = V e^{-t/\lambda}$.

5.16 Time Constant

(***a***) Just at the start of charging, p.d. across capacitor is zero, hence from (*ii*) putting $v_c = 0$, we get

$$V = CR\frac{dv_c}{dt}$$

$\therefore$ initial rate of rise of voltage across the capacitor is* $= \left(\frac{dv_c}{dt}\right)_{t=0} = \frac{V}{CR} = \frac{V}{\lambda}$ volt/second

* It can also be found by differentiating Eq. (*iv*) with respect to time and then putting $t = 0$.

If this rate of rise were maintained, then time taken to reach voltage V would have been $V + V/CR = CR$. This time is known as ***time constant*** (λ) of the circuit.

Hence, time constant of an R-C circuit is defined as ***the time during which voltage across capacitor would have reached its maximum value V had it maintained its initial rate of rise.***

(*b*) In equation **(*iv*)** if $t = \lambda$, then

$$v_c = V\,(1 - e^{-t/\lambda}) = V\,(1 - e^{-t/\lambda}) = V(1 - e^{-1}) = V\left(1 - \frac{1}{e}\right) = V\left(1 - \frac{1}{2.718}\right) = \mathbf{0.632\ V}$$

Hence, time constant may be defined as ***the time during which capacitor voltage actually rises to 0.632 of its final steady value.***

(*c*) From equaiton **(*vi*)**, by putting $t = \lambda$, we get

$$i_c = I_0\,e^{-\lambda/\lambda} = I_0\,e^{-1} = I_0/2.718 \cong 0.37\,I_0$$

Hence, the constant of a circuit is also the ***time during which the charging current falls to 0.37 of its initial maximum value (or falls by 0.632 of its initial value).***

5.17 Discharging of a Capacitor

As shown in Fig. 5.26 (*a*), when S is shifted to b, C is discharged through R. It will be seen that the discharging current flows in a direction opposite to that the charging current as shown in Fig. 5.26 (*b*). Hence, if the direction of the charging current is taken positive, then that of the discharging current will be taken as negative. To begin with, the discharge current is maximum but then decreases exponentially till it ceases when capacitor is fully discharged.

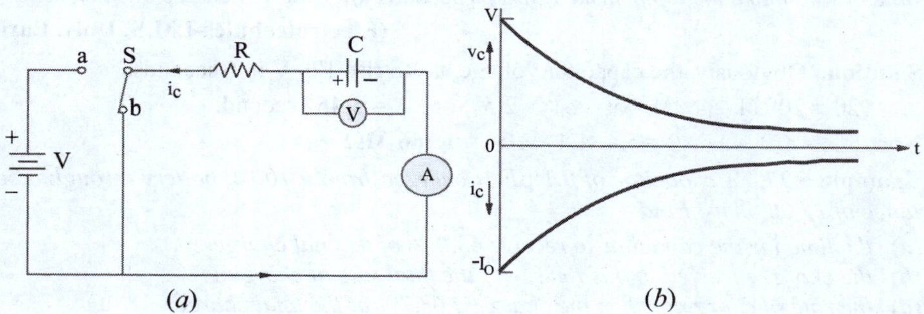

Fig. 5.26

Since battery is cut of the circuit, therefore, by putting $V = 0$ in equation (*ii*) of Art. 5.15, we get

$$0 = CR\frac{dv_c}{dt} + v_c \text{ or } v_c = -\,CR\frac{dv_c}{dt} \qquad \left(\because\ i_c = C\frac{dv_c}{dt}\right)$$

$$\therefore \qquad \frac{dv_c}{v_c} = \frac{dt}{CR} \text{ or } \int\frac{dv_c}{v_c} = -\frac{1}{CR}\int dt \ \therefore\ \log_e v_e = -\frac{t}{CR} + k$$

At the start of discharge, when $t = 0$, $v_c = V$ ∴ $\log_e V = 0 + K$; or $\log_e V = K$

Putting this value above, we get

$$\log_e v_c = -\frac{t}{\lambda} + \log_e V \text{ or } \log_e v_c/V = -\,t/\lambda$$

$$\text{or} \qquad \frac{v_c}{V} = e^{-t/\lambda} \text{ or } v_c = Ve^{-t/\lambda}$$

Similarly, $q = Q\,e^{-t/\lambda}$ and $i_c = -I_0\,e^{-t/\lambda}$

It can be proved that

$$v_R = -V\,e^{-t/\lambda}$$

The fall of capacitor potential and its discharging current are shown in Fig. 5.26 (*b*).

One practical application of the above charging and discharging of a capacitor is found in digital control circuits where a square-wave input is applied across an *R-C* circuit as shown in Fig. 5.26 (*a*). The different waveforms of the current and voltages are shown in Fig. 5.26 (*b*), (*c*), (*d*), (*e*). The sharp voltage pulses of V_R are used for control circuits.

Example 5.25. **(*a*)** *A capacitor is charged through a large non-reactive resistance by a battery of constant voltage V. Derive an expression for the instantaneous charge on the capacitor.*

(*b*) *For the above arrangement, if the capacitor has a capacitance of 10 μ F and the resistance is 1 M Ω, calculate the time taken for the capacitor to receive 90% of its final charge. Also, draw the charge/time curve.*

Solution. **(*a*)** For this part, please refer to Art. 5.15..

(*b*) $\lambda = CR = 10 \times 10^{-6} \times 1 \times 10^6 = 10\ s\ ; q = 0.9\ Q$

Now, $q = Q\,(1 - e^{-t/\lambda})$ $\therefore$ $0.9\,Q = Q\,(1 - e^{-t/10})$ or $e^{t/10} = 10$

$\therefore$ $0.1\,t\log_e e = \log_e 10$ or $0.1\,t = 2.3\log_{10} 10 = 2.3$ or $t =$ **23 s**

The charge/time curve is similar to that shown in Fig. 5.27 (*b*).

Example 5.26. *A resistance R and a 4 μF capacitor are connected in series across a 200 V. d.c. supply. Across the capacitor is a neon lamp that strikes (glows) at 120 V. Calculate the value of R to make the lamp strike (glow) 5 seconds after the switch has been closed.*

(Electrotechnics-I.M.S. Univ. Baroda)

Solution. Obviously, the capacitor voltage has to rise 120 V in 5 seconds.

$\therefore$ $120 = 200\,(1 - e^{-5/\lambda})$ or $e^{5/\lambda} = 2.5$ or $\lambda = 5.464$ second.

Now, $\lambda = CR$ $\therefore$ $R = 5.464/4 \times 10^{-6} =$ **1.366 MΩ**

Example 5.27. *A capacitor of 0.1 μF is charged from a 100-V battery through a series resistance of 1,000 ohms. Find*

(*a*) *the time for the capacitor to receive 63.2 % of its final charge.*

(*b*) *the charge received in this time* **(*c*)** *the final rate of charging.*

(*d*) *the rate of charging when the charge is 63.2% of the final charge.*

(Elect. Engineering, Bombay Univ., JNTU 2010)

Solution. **(*a*)** As seen from Art. 5.12, 63.2% of charge is received in a time equal to the time constant of the circuit.

Time required $= \lambda = CR = 0.1 \times 10^{-6} \times 1000 = 0.1 \times 10^{-3} =$ **10^{-4} second**

(*b*) Final charge, $Q = CV = 0.1 \times 100 = 10\ \mu C$

Charge received during this time is $= 0.632 \times 10 =$ **6.32 μ C**

(*c*) The rate of charging at any time is given by Eq. (*ii*) of Art. 5.15.

$$\frac{dv}{dt} = \frac{V - v}{CR}$$

Initially $v = 0$, Hence $\dfrac{dv}{dt} = \dfrac{V}{CR} = \dfrac{100}{0.1\times10^{-6}\times10^{3}} =$ **10^6 V/s**

(*d*) Here $v = 0.632\ V = 0.632 \times 100 = 63.2$ volts

$\therefore$ $\dfrac{dv}{dt} = \dfrac{100 - 63.2}{10^{-4}} =$ **368 kV/s**

Example 5.28. *A series combination having R = 2 M Ω and C = 0.01 μF is connected across a d.c. voltage source of 50 V. Determine*

(a) *capacitor voltage after 0.02 s, 0.04 s, 0.06 s and 1 hour*

(b) *charging current after 0.02 s, 0.04 s, 0.06 s and 0.1 s.* **(AKTU 2018)**

Solution. $\lambda = CR = 2 \times 10^6 \times 0.01 \times 10^{-6} = 0.02$ second

$I_m = V/R = 50/2 \times 10^6 = 25$ μA.

While solving this question, it should be remembered that **(*i*)** in each time constant, v_c increases further by 63.2% of its *balance* value and **(*ii*)** in each constant, i_c decreases to 37% its previous value.

(*a*) (*i*) *t* = 0.02 s

Since, initially at $t = 0$, $v_c = 0$ V and $V_e = 50$ V, hence, in one time constant

$v_c = 0.632\,(50 - 0) = \mathbf{31.6\ V}$

(*ii*) *t* = 0.04 *s*

This time equals two time-constants.

$\therefore \quad v_c = 31.6 + 0.632\,(50 - 31.6) = \mathbf{43.2\ V}$

(*iii*) *t* = 0.06 s

This time equals three time-constants.

$\therefore \quad v_c = 43.2 + 0.632\,(50 - 43.2) = \mathbf{47.\ 5\ V}$

Since in one hour, steady-state conditions would be established, v_c would have achieved its maximum possible value of **50 V**.

(*b*) **(*i*)** ***t* = 0.02 s,** $i_c = 0.37 \times 25 = \mathbf{9.25\ \mu A}$

(*ii*) ***t* = 0.4 s,** $i_c = 0.37 \times 9.25 = \mathbf{3.4\ \mu A}$

(*iii*) ***t* = 0.06 s,** $i_c = 0.37 \times 3.4 = \mathbf{1.26\ \mu A}$

(*iv*) ***t* = 0.1 s,** This time equals 5 time constants. In this time, current falls almost to **zero** value.

5.18 Transient Relations During Capacitor Charging Cycle

Whenever a circuit goes from one steady-state condition to another steady-state condition, it passes through a transient state which is of short duration. The first steady-state condition is called the ***initial condition*** and the second steady-state condition is called the ***final condition***. In fact, transient condition lies in between the initial and final conditions. For example, when switch *S* in Fig. 5.27 (*a*) is not connected either to *a* or *b*, the *RC* circuit is in its initial steady state with no current and hence no voltage drops. When *S* is shifted to point *a*, current starts flowing through *R* and hence, transient voltages are developed across *R* and *C* till they achieve their final steady values. The period during which current and voltage changes take place is called ***transient condition***.

The moment switch *S* is shifted to point '*a*' as shown in Fig. 5.27 (*b*), a charging current i_c is set up which starts charging *C* that is initially uncharged. At the beginning of the transient state, i_c is maximum because there is no potential across *C* to oppose the applied voltage *V*. It has maximum value = $V/R = I_0$. It produces maximum voltage drop across $R = i_c R = I_0 R$. Also, initially, $v_c = 0$, but as time passes, i_c decreases gradually so does v_R but v_c increases exponentially till it reaches the final steady value of *V*. Although *V* is constant, v_R and v_c are variable. However, at any time $V = v_R + v_c = i_c R + v_c$.

At the beginning of the transient state, $i_c = I_0$, $v_c = 0$ but $v_R = \text{V}$. At the end of the transient state, $i_c = 0$ hence, $v_R = 0$ but $v_c = \text{V}$.

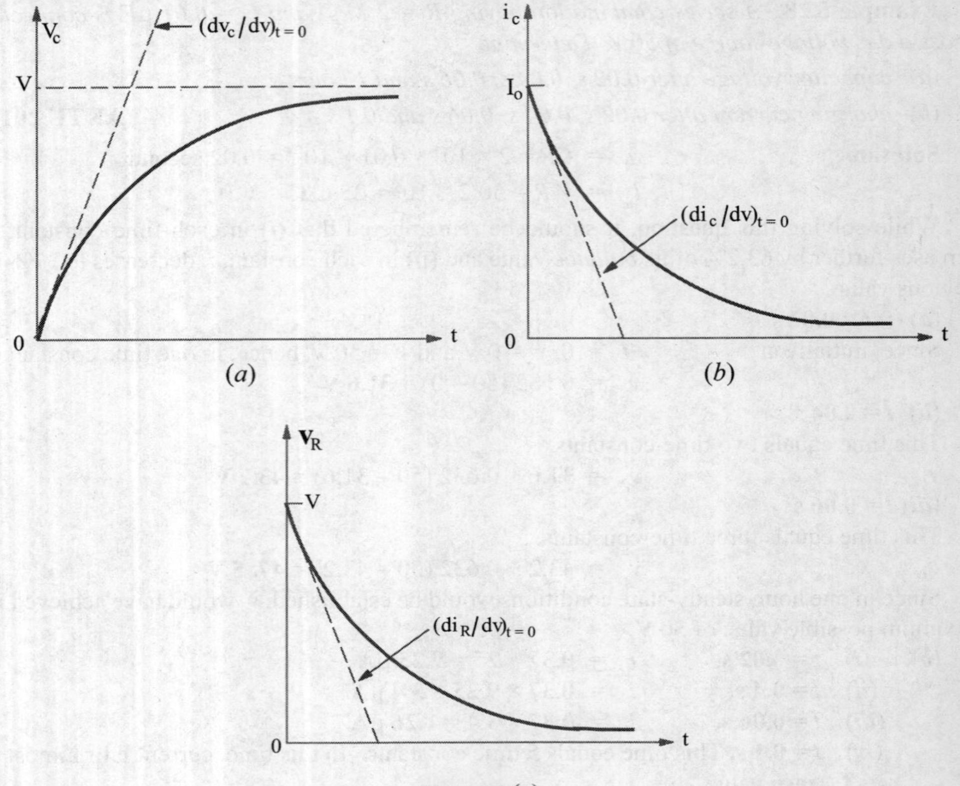

Fig. 5.27 (*c*)

The initial rates of change of v_c, v_R and i_c are given by

$$\left(\frac{dv_c}{dt}\right)_{t=0} = \frac{V}{\lambda} \text{ volt/second,}$$

$$\left(\frac{dv_R}{dt}\right)_{t=0} = \frac{I_0 R}{\lambda} = -\frac{V}{\lambda} \text{ volt/second}$$

$$\left(\frac{di_c}{dt}\right)_{t=0} = \frac{I_0}{\lambda} \text{ where } I_0 = \frac{V}{R}$$

These are the initial rates of change. However, their rate of change at any time during the charging transient are given as under:

$$\frac{dv_c}{dt} = \frac{V}{\lambda} e^{-t/\lambda}; \frac{di_c}{dt} = -\frac{dv_R}{dt} = -\frac{V}{\lambda} e^{-t/\lambda}$$

It is shown in Fig. 5.27 (*c*).

It should be clearly understood that a negative rate of change means a decreasing rate of change. It does not mean that the concerned quantity has reversed its direction.

5.19 Transient Relations During Capacitor Discharging Cycle

As shown in Fig. 5.28 (*b*), switch *S* has been shifted to *b*. Hence, the capacitor undergoes the discharge cycle. Just before the transient state starts, $i_c = 0$, $v_R = 0$ and $v_c = \text{V}$. The moment transient state begins, i_c has maximum value and decreases exponentially to zero at the end of the transient state. So does v_c. However, during discharge, all rates of change have polarity opposite to that during charge. For example, dv_c/dt has a positive rate of change during charging and negative rate of change during discharging.

Also, it should be noted that during discharge, v_c maintains its original polarity whereas i_c reverses its direction of flow. Consequently, during capacitor discharge, v_R also reverses its direction.

The various rates of change at *any time* during the discharge transients are as given

$$\frac{dv_c}{dt} = -\frac{V}{\lambda}e^{-t/\lambda};\ \frac{di_c}{dt} = \frac{I_0}{\lambda}e^{-t/\lambda};\ \frac{dv_R}{dt} = \frac{V}{\lambda}e^{-t/\lambda}$$

These are represented by the curves of Fig. 5.28.

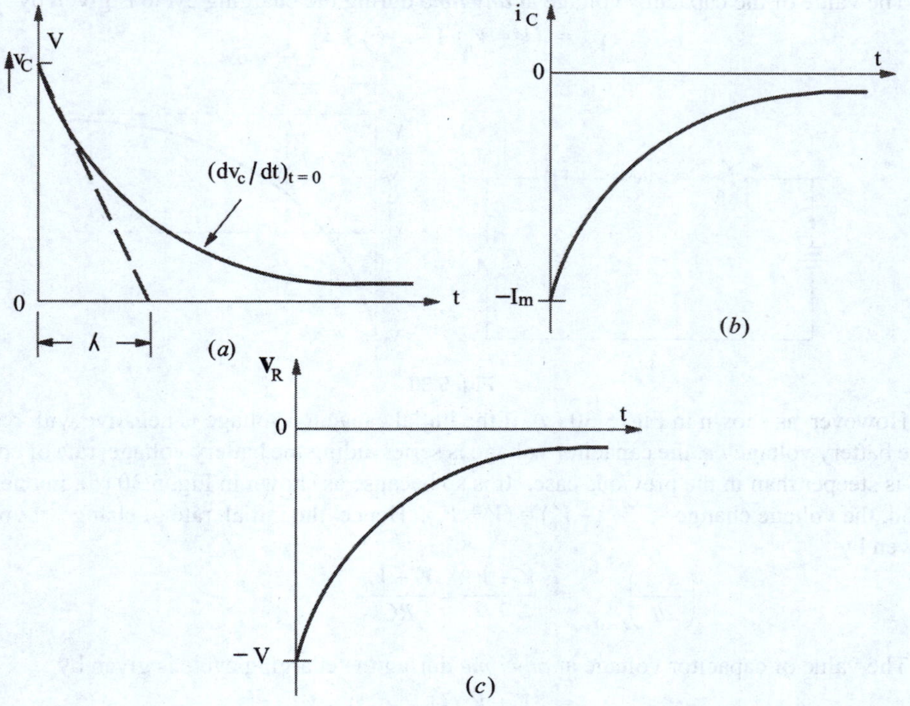

Fig. 5.28

5.20. Charging and Discharging of a Capacitor with Initial Charge

In Art. 5.15, we considered the case when the capacitor was initially uncharged and hence, had no voltage across it. Let us now consider the case, when the capacitor has an initial potential of V_0 (less than V) which opposes the applied battery voltage V as shown in Fig. 5.29 (*a*).

As seen from Fig. 5.29 (*b*), the initial rate of rise of v_c is now somewhat less than when the capacitor is initially uncharged. Since the capacitor voltage rises from an initial value of v_0 to the final value of V in one time constant, its initial rate of rise is given by

$$\left(\frac{dv_c}{dt}\right)_{t=0} = \frac{V - V_0}{\lambda} = \frac{V - V_0}{RC}$$

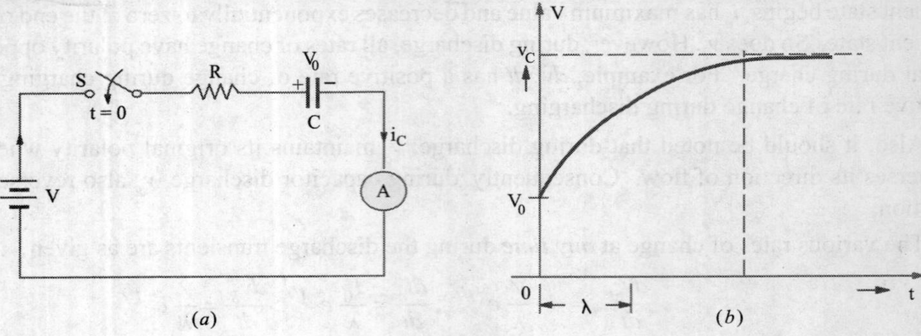

Fig. 5.29

The value of the capacitor voltage at *any time* during the charging cycle is given by

$$v_c = (V - V_0)(1 - e^{-t/\lambda}) + V_0$$

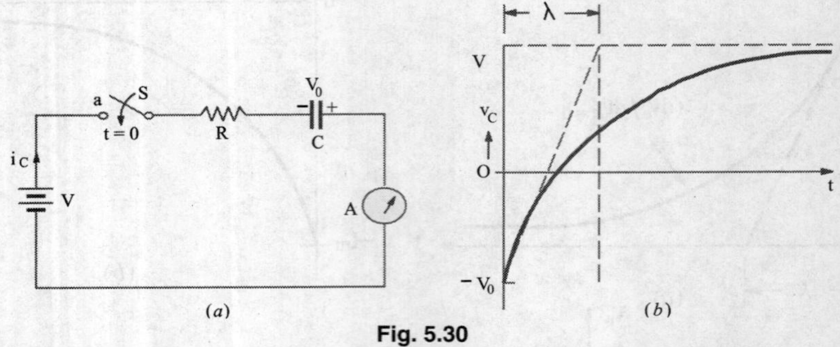

Fig. 5.30

However, as shown in Fig. 5.30 (*a*), if the initial capacitor voltage is negative with respect to the battery voltage *i.e.* the capacitor voltage is series aiding the battery voltage, rate of change of v_c is steeper than in the previous case. It is so because as shown in Fig. 5.30 (*b*), in one time period, the voltage change $= V - (-V_0) = (V + V_0)$. Hence, the initial rate of change of voltage is given by

$$\left(\frac{dv_c}{dt}\right)_{t=0} = \frac{V + V_0}{\lambda} = \frac{V + V_0}{RC}$$

The value of capacitor voltage at *any time* during the charging cycle is given by

$$v_c = (V + V_0)(1 - e^{-t/\lambda}) - V_0$$

The time required for the capacitor voltage to attain any value of v_c during the charging cycle is given by

$$t = \lambda \ln\left(\frac{V - V_0}{V - v_c}\right) = RC \ln\left(\frac{V - V_0}{V - v_c}\right) \quad \text{... when } V_0 \text{ is positive}$$

$$t = \lambda \ln\left(\frac{V + V_0}{V - v_c}\right) = RC \ln\left(\frac{V + V_0}{V - v_c}\right) \quad \text{... when } V_0 \text{ is negative}$$

Example 5.29. *In Fig. 5.31, the capacitor is initially uncharged and the switch S is then closed. Find the values of I, I_1, I_2 and the voltage at the point A at the start and finish of the transient state.*

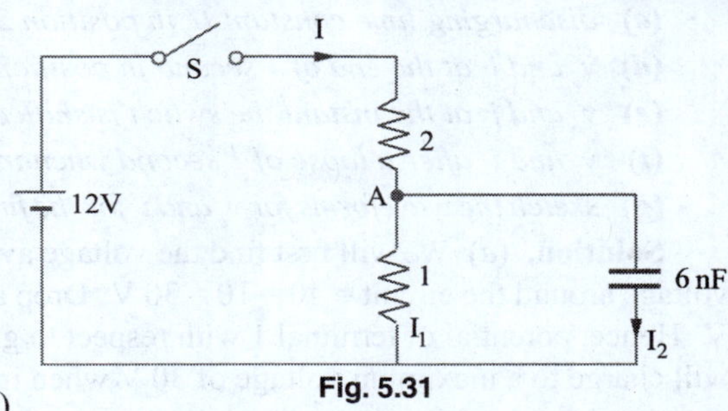

Fig. 5.31

Solution. At the moment of closing the switch *i.e.* at the start of the transient state, the capacitor acts as a short-circuit. Hence, there is only a resistance of 2 Ω in the circuit because 1 Ω resistance is shorted out thereby grounding point *A*. Hence, I_1 = **0;** $I = I_2 = 12/2 =$ **6A.** Obviously, V_A = **0 V.**

At the end of the transient state, the capacitor acts as an open-circuit. Hence,

$$I_2 = 0 \text{ and } I = I_1 = 12/(2+1)$$
$$= \mathbf{4\ A.}\ V_A = \mathbf{6\ V.}$$

Example 5.30. *Calculate the values of i_2, i_3, v_2, v_3, v_d, v_c and v_L of the network shown in Fig. 5.40 at the following times:*

(*i*) *At time, $t = 0^+$ immediately after the switch S is closed ;*

(*ii*) *At time, $t \to \infty$ i.e. in the steady state.* **(Network Analysis AMIE Sec. B Winter 1990)**

Solution. (*i*) In this case the coil acts as an open circuit, hence $i_2 = 0$; $v_2 = 0$ and $v_L = 20$ V.

Since a capacitor acts as a short circuit $i_3 = 20/(5 + 4) = 9 = 20/9$ A. Hence, $v_3 = (20/9) \times 4 = 80/9$ V and $v_c = 0$.

(*ii*) Under steady state conditions, capacitor acts as an open circuit and coil as a short circuit. Hence, $i_2 = 20/(5 + 7) = 20/12 = 5/3$ A; $v_2 = 7 \times 5/3 = 35/3$ V; $v_L = 0$. Also $i_3 = 0$, $v_3 = 0$ but v_c = **20 V.**

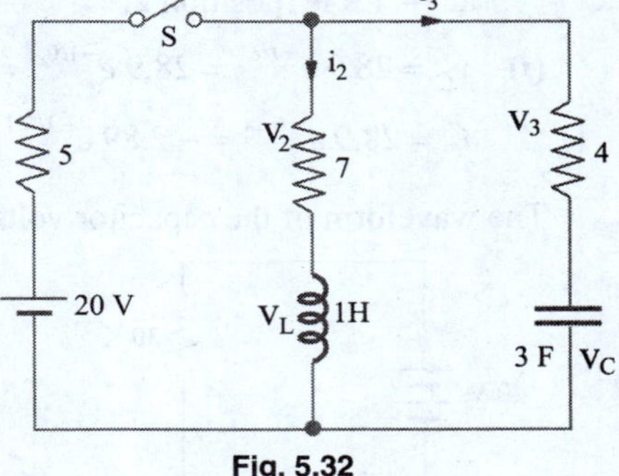

Fig. 5.32

Example 5.31. *If in the RC circuit of Fig. 5.32; R = 2 MΩ, C = 5 m F and V = 100 V, calculate:*

(*a*) *initial rate of change of capacitor voltage*

(*b*) *initial rate of change of capacitor current*

(*c*) *initial rate of change of voltage across the 2 MΩ resistor*

(*d*) *all of the above at t = 80 s.*

Solution. (*a*)
$$\left(\frac{dv_c}{dt}\right)_{t=0} = \frac{V}{\lambda} = \frac{100}{2\times10^6\times5\times10^{-6}} = \frac{100}{10} = \mathbf{10\ V/s}$$

(*b*)
$$\left(\frac{di_c}{dt}\right)_{t=0} = -\frac{I_0}{\lambda} = -\frac{V/R}{\lambda} = -\frac{100/2\times10^6}{10} = \mathbf{-5\mu A/s}$$

(*c*)
$$\left(\frac{dv_R}{dt}\right)_{t=0} = -\frac{V}{\lambda} = -\frac{100}{10} = \mathbf{-10\ V/s}$$

(*d*) All the above rates of change would be zero because the transient disappears after about $5\lambda = 5 \times 10 = 50$ s.

Example 5.32. *In Fig. 5.33 (a), the capacitor C is fully discharged, since the switch is in position 2. At time t = 0, the switch is shifted to position 1 for 2 seconds. It is then returned to position 2 where it remains indefinitely. Calculate*

(*a*) *the maximum voltage to which the capacitor is charged when in position 1.*

(*b*) *charging time constant λ_1 in position 1.*

(c) *discharging time constant λ_2 in position 2.*

(d) *v_c and i_c at the end of 1 second in position 1.*

(e) *v_c and i_c at the instant the switch is shifted to positon 2 at t = 1 second.*

(f) *v_c and i_c after a lapse of 1 second when in position 2.*

(g) *sketch the waveforms for v_c and i_c for the first 2 seconds of the above switching sequence.*

Solution. **(a)** We will first find the voltage available at terminal 1. As seen the net battery voltage around the circuit = 40 – 10 = 30 V. Drop across 30 K resistor = 30 × 30/(30 + 60) = 10 V. Hence, potential of terminal 1 with respect to ground G = 40 – 10 = 30 V. Hence, capacitor will charge to a maximum voltage of 30 V when in position 1.

(b) Total resistance, R = [(30 K ‖ 60 K) + 10 K] = 30 K

$\therefore\ \lambda_1 = RC = 30\text{ K} \times 10\ \mu\text{F} = 0.3\ s$

(c) $\lambda_2 = 10\text{ K} \times 10\ \mu\text{ F} = 0.1\text{ s}$

(d) $v_C = V(1 - e^{-t/\lambda 1}) = 30\,(1 - e^{-1/0.3}) = 28.9\text{ V}$

$$i_C = \frac{V}{R}e^{-t/\lambda_1} = \frac{30\text{ V}}{30\text{ K}}e^{-1/0.03} = 1 \times 0.0361 = 0.036\text{ mA}$$

(e) $v_C = 28.9$ V at $t = 1^+$S at position 2 but $i_C = -28.9$ V/10 K = – 2/89 mA at $t = 1^+$s in position 2.

(f) $v_C = 28.9\,e^{-t/\lambda_2} = 28.9\,e^{-1/0.1} = 0.0013\text{ V} = 0\text{ V}.$

$i_C = 28.9\,e^{-t/\lambda_2} = -\,2.89\,e^{-1/0.1} = 0.00013\text{ mA} \cong 0.$

The waveform of the capacitor voltage and charging current are sketched in Fig. 5.33 (*b*).

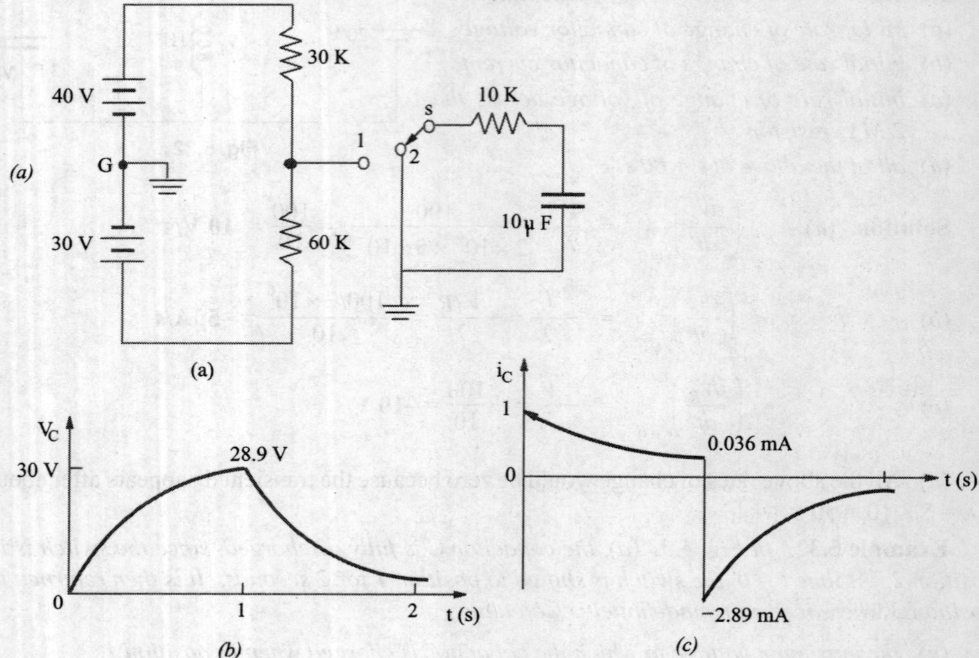

Fig. 5.33

Example 5.33. *In the RC circuit of Fig. 5.34, $R = 2\ M\Omega$ and $C = 5\ \mu F$, the capacitor is charged to an initial potential of 50 V. When the switch is closed at $t = 0^+$, calculate*

(*a*) *initial rate of change of capacitor voltage and*

(*b*) *capacitor voltage after a lapse of 5 times the time constant i.e. 5λ.*

If the polarity of capacitor voltage is reversed, calculate

(*c*) *the values of the above quantities and*

(*d*) *time for v_c to reach -10 V, 0 V and 95 V.*

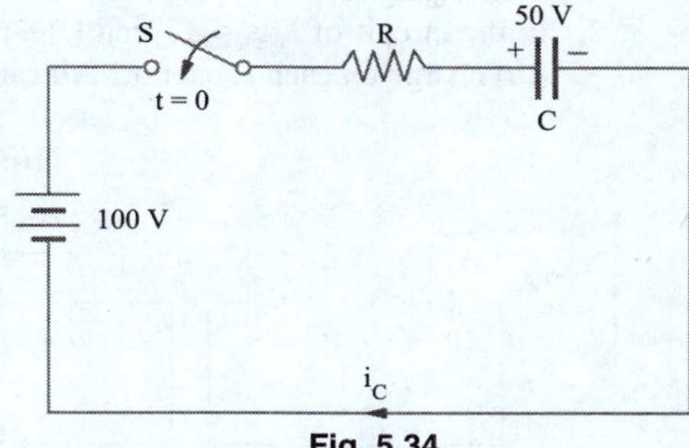

Fig. 5.34

Solution. (*a*) $\left(\dfrac{dv_c}{dt}\right)_{t=0} = \dfrac{V - V_0}{\lambda}$

$$= \frac{V - V_0}{RC} = \frac{100 - 50}{10} = \mathbf{5\ V/s}$$

(*b*) $v_C = (V - V_0)(1 - e^{-t/\lambda}) + V_0$

$$= (100 - 50)(1 - e^{-5\lambda/\lambda}) = 50 = 49.7 + 50 = \mathbf{99.7\ V}$$

(*c*) When $V_0 = -50\text{ V}, \left(\dfrac{dv_c}{dt}\right)_{t=0} = \dfrac{V - (-V_0)}{\lambda} = \dfrac{V + V_0}{\lambda} = \dfrac{150}{10} = \mathbf{15\ V/s}$

$$v_C = (V - V_0)(1 - e^{-t/\lambda}) + V_0 = [100 - (-50)](1 - e^{-5}) + (-50)$$

$$= 150(1 - e^{-5}) - 50 = \mathbf{99\ V.}$$

(*d*) $t = \lambda\, ln\left(\dfrac{V - V_0}{V - v_c}\right) = 10\, ln\left[\dfrac{100 - (-50)}{100 - (-10)}\right] = 10\, ln\left(\dfrac{150}{110}\right) = \mathbf{3.1\ s}$

$$t = 10\, ln\left[\frac{100 - (-50)}{100 - (0)}\right] = 10\, ln\left(\frac{150}{100}\right) = \mathbf{4.055\ s}$$

$$t = 10\, ln\left[\frac{100 - (-50)}{100 - 95}\right] = 10\, ln\left(\frac{150}{5}\right) = \mathbf{34\ s}$$

Example 5.34. *The uncharged capacitor, if it is initially switched to position 1 of the switch for 2 sec and then switched to position 2 for the next two seconds. What will be the voltage on the capacitor at the end of this period ? Sketch the variation of voltage across the capacitor.* **[Bombay University 2001]**

Solution. Uncharged capacitor is switched to position 1 for 2 seconds. It will be charged to 100 volts instantaneously since resistance is not present in the charging circuit. After 2 seconds, the capacitor charged to 100 volts will get discharged through *R-C* circuit with a time constant of

Fig. 5.35

$$\tau = RC = 1500 \times 10^{-3} = 1.5\text{ sec.}$$

Counting time from instant of switching over to position 2, the expression for voltage across the capacitor is $V(t) = 100 \exp(-t/\tau)$

After 2 seconds in this position,

$$v(t) = 100 \exp(-2/1.5) = 26/36\text{ Volts.}$$

Tutorial Problem 5.3

1. For the circuit shown in Fig. 5.36 calculate (*i*) equivalent capacitance and (*ii*) voltage drop across each capacitor. All capacitance values are in μF. **[(*i*) 6 μF (*ii*) V_{AB} = 50 V, V_{BC} = 40 V]**
2. In the circuit of Fig. 5.37 find (*i*) equivalent capacitance (*ii*) drop across each capacitor and (*iii*) charge on each capacitor. All capacitance values are in μF. **[(*i*) 1.82 μF (*ii*) V_1 = 50 V; V_2 = V_3 = 20 V; V_4 = 40 V (*iii*) Q_1 = 200 μC; Q_2 = 160 μC; Q_3 = 40 μC; Q_4 = 200 μC]**

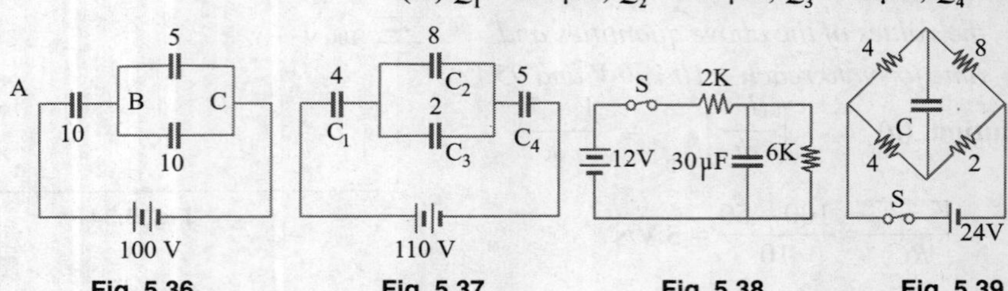

Fig. 5.36 Fig. 5.37 Fig. 5.38 Fig. 5.39

3. With switch in Fig. 5.38 closed and steady-state conditions established, calculate **(*i*)** steady-state current **(*ii*)** voltage and charge across capacitor **(*iii*)** what would be the discharge current at the instant of opening the switch ? **[(*i*) 1.5 mA (*ii*) 9V; 270 μC (*iii*) 1.5 mA]**
4. When the circuit of Fig. 5.39 is in steady state, what would be the p.d. across the capacitor ? Also, find the discharge current at the instant *S* is opened. **[8 V; 1.8 A]**
5. Find the time constant of the circuit shown in Fig. 5.40. **[200 μS]**
6. A capacitor of capacitance 0.01 μF is being charged by 1000 V d.c. supply through a resistor of 0.01 megaohm. Determine the voltage to which the capacitor has been charged when the charging current has decreased to 90 % of its initial value. Find also the time taken for the current to decrease to 90% of its initial value. **[100 V, 0.1056 ms]**

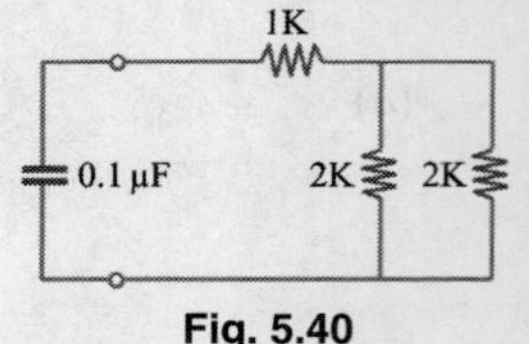

Fig. 5.40

7. An 8 μF capacitor is being charged by a 400 V supply through 0.1 mega-ohm resistor. How long will it take the capacitor to develop a p.d. of 300 V ? Also what fraction of the final energy is stored in the capacitor ? **[1.11 Second, 56.3% of full energy]**
8. An 10 μF capacitor is charged from a 200 V battery 250 times/second and completely discharged through a 5 Ω resistor during the interval between charges. Determine
   (*a*) the power taken from the battery.
   (*b*) the average value of the current in 5 Ω resistor. **[(*a*) 50 W (*b*) 0.5 A]**
9. When a capacitor, charged to a p.d. of 400 V, is connected to a voltmeter having a resistance of 25 MΩ, the voltmeter reading is observed to have fallen to 50 V at the end of an interval of 2 minutes. Find the capacitance of the capacitor. **[2.31 μF] (*App. Elect. London Univ.*)**
10. A capacitor and a resistor are connected in series with a d.c. source of V volts. Derive an expression for the voltage across the capacitor after '*t*' seconds during discharging. ***(Gujrat University, Summer 2003)***
11. Derive an expression for the equivalent capacitance of a group of capacitors when they are connected (*i*) in parallel (*ii*) in series. ***(Gujrat University, Summer 2003)***
12. The total capacitance of two capacitors is 0.03 μ*F* when joined in series and 0.16 μ*F* when connected in parallel. Calculate the capacitance of each capacitor. ***(Gujrat University, Summer 2003)***

13. In a capacitor with two plates separated by an insulator 3mm thick and of relative permittivity of 4, the distance between the plates is increased to allow the insertion of a second insulator 5mm thick and relative permittivity E. If the capacitance so formed is one third of the original capacitance, find E.***(V.TU., Belgaum Karnataka University, February 2002)***

14. Derive an expression for the capacitance of a parallel plate capacitor.

(V.TU., Belgaum Karnataka University, Summer 2002, RGPV 2016)

15. Three capacitors A, B and C are charged as follows

A = 10μF, 100 V B = 15μF, 150 V C = 25μF, 200 V

They are connected in parallel with terminals of like polarities together. Find the voltage across the combination. ***(V.TU., Belgaum Karnataka University, Summer 2002)***

16. Prove that average power consumed by a pure capacitance is zero.

(V.TU., Belgaum Karnataka University, Summer 2002)

17. Current drawn by a pure capacitor of 20μF is 1.382A from 220V AC supply. What is the supply frequency? ***(V.TU., Belgaum Karnataka University, Summer 2003)***

18. A capacitor consists of two plates of area 0.16m^2 spaced 6mm apart. This space is filled with a layer of 1mm thick paper of relative permittivity 2, and remaining space with glass of relative permittivity 5. A dc voltage of 10kV is applied between the plates. Determine the electric field strength in each dielectric.

(V.T.U. Belgaum Karnataka University, Wimter2004)

19. In a give R-L circuit, R = 35Ω and L = 0.1H. Find (*i*) current through the circuit (*ii*) power factor if a 50 Hz frequency, voltage V = 220∠30° is applied across the circuit.

(RGPV Bhopal 2001)

20. Three voltage represented by e_1 = 20 sin ω *t*, e_2 = 30 sin (ω *t* = 45°) and e_3 = sin (ω *t* + 30°) are connected in series and then connected to a load of impedance (2 + *j* 3) Ω. Find the resultant current and power factor of the circuit. Draw the phasor diagram.

(Mumbai University, 2002) (RGPV Bhopal 2001)

OBJECTIVE TESTS – 5

1. A capacitor consists of two
(*a*) insulation separated by a dielectric
(*b*) conductors separated by an insulator
(*c*) ceramic plates and one mica disc
(*d*) silver-coated insulators

2. The capacitance of a capacitor is NOT influenced by
(*a*) plate thickness
(*b*) plate area
(*c*) plate separation
(*d*) nature of the dielectric

3. A capacitor that stores a charge of 0.5 C at 10 volts has a capacitance of .....farad.
(*a*) 5 (*b*) 20
(*c*) 10 (*d*) 0.05

4. If dielectric slab of thickness 5 mm and ε_r = 6 is inserted between the plates of an air capacitor with plate separation of 8 mm, its capacitance is
(*a*) decreased (*b*) almost doubled
(*c*) almost halved (*d*) unaffected

5. For the circuit shown in the given figure, the current through *L* and the voltage across C_2 are respectively

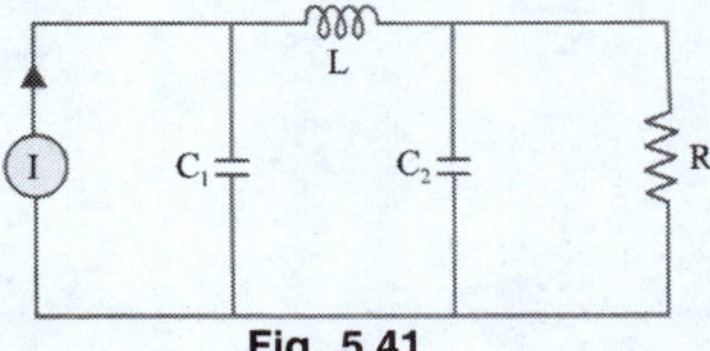

Fig. 5.41

(*a*) zero and RI
(*b*) I and zero
(*c*) zero and zero
(*d*) I and RI ***(ESE 2001)***

6. A parallel plate capacitor has an electrode area of 100 mm², with a spacing of 0.1 mm between the electrodes. The dielectric between the plates is air with a permittivity of 8.85×10^{-12} F/m. The charge on the capacitor is 100 V. the stored energy in the capacitor is

(*a*) 8.85 pJ (*b*) 440 pJ
(*c*) 22.1 nJ (*d*) 44.3 nJ

(GATE 2003)

7. A composite parallel plate capacitor is made up of two different dielectric materials with different thicknesses (t_1 and t_2) as shown in Fig. 5.42. The two different dielectric materials are separates by a conducting foil *F*. The voltage of the conducting foil is

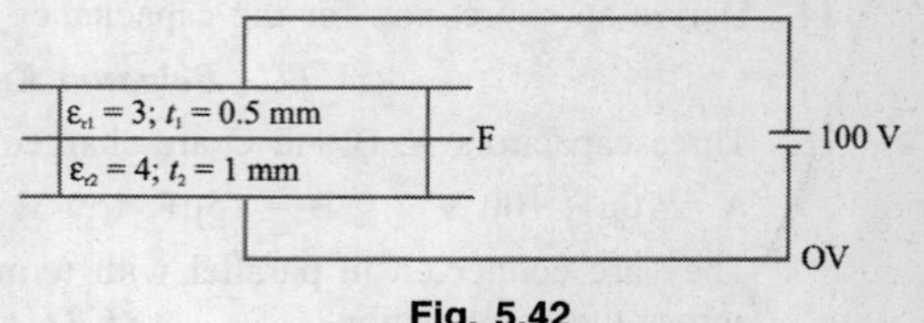

Fig. 5.42

(*a*) 52 V (*b*) 60 V
(*c*) 67 V (*d*) 33 V

(GATE 2003)

ANSWERS

1. *b* **2.** *a* **3.** *d* **4.** *b*

CHAPTER

6

Magnetism and Electromagnetism

Chapter Outline

6.1 Absolute and Relative Permeabilities of a Medium

The phenomena of magnetism and electromagnetism are dependent upon a certain property of the medium called its permeability. Every medium is supposed to possess two permeabilities: ***(i)*** absolute permeability (μ) and ***(ii)*** relative permeability (μ_r).

For measuring relative permeability, vacuum or free space is chosen as the reference medium. Permeability of Free Space or Vacuum is constant and denoted by μ_0 and is given by $\mu_0 = (4\pi \times 10^{-7})$ H/m Obviously, relative permeability of vacuum with reference to itself is unity. Hence, for free space,

absolute permeability $\mu = \mu_0 = 4\pi \times 10^{-7}$ H/m

relative permeability $\mu_r = 1.$

Now, take any medium other than vacuum. If its relative permeability, as compared to vacuum is μ_r, then its absolute permeability is $\mu = \mu_0\, \mu_r$

6.2 Laws of Magnetic Force

Coulomb was the first to determine experimentally the quantitative expression for the magnetic force between two *isolated* point poles. It may be noted here that, in view of the fact that magnetic poles always exist in pairs, it is impossible, in practice, to get an isolated pole. The concept of an isolated pole is purely theoretical. However, poles of a thin but long magnet may be assumed to be point poles for all practical purposes (Fig. 6.1). By using a torsion balance, he found that the force between two magnetic poles placed in a medium is

(*i*) directly proportional to their pole strengths
(*ii*) inversely proportional to the square of the distance between them and
(*iii*) inversely proportional to the absolute permeability of the surrounding medium.

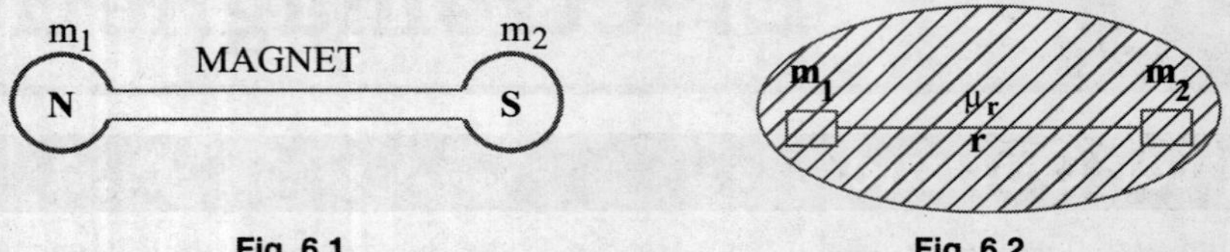

Fig. 6.1 Fig. 6.2

For example, if m_1 and m_2 represent the magnetic strength of the two poles (its unit as yet being undefined), r the distance between them (Fig. 6.2) and μ the absolute permeability of the surrounding medium, then the force F is given by

$$F \propto \frac{m_1 m_2}{\mu r^2} \quad \text{or} \quad F = k\frac{m_1 m_2}{\mu r^2} \quad \text{or} \quad \vec{F} = \frac{k\, m_1 m_2}{\mu r^2}\hat{r} \quad \text{in vector from}$$

where $\hat{r}$ is a unit vector to indicate direction of r.

$$\text{or} \qquad \vec{F} = k\frac{m_1\, m_2}{r^3}\vec{r} \quad \text{where } \vec{F} \text{ and } \vec{r} \text{ are vectors}$$

In the S.I. system of units, the value of the constant k is = $1/4\pi$.

$$F = \frac{m_1 m_2}{4\pi\mu r^2}\,\text{N} \quad \text{or} \quad F = \frac{m_1 m_2}{4\pi\mu_0 r^2}\,\text{N} \qquad \text{– in a medium}$$

$$\text{In vector form,} \qquad \vec{F} = \frac{m_1 m_2}{4\pi\mu r^3}\vec{r} = \frac{m_1 m_2}{4\pi\mu_0 r^2}\,\text{N}$$

If, in the above equation,

$$m_1 = m_2 = m \text{ (say)} \; ; r = 1 \text{ metre} \; ; F = \frac{1}{4\pi\,\mu_0}\,\text{N}$$

$$\text{Then} \qquad m^2 = 1 \quad \text{or} \quad m = \pm 1 \text{ weber*}$$

Hence, a unit magnetic pole may be defined as ***that pole which when placed in vacuum at a distance of one metre*** from a similar and equal pole repels it with a force of $1/4\pi\,\mu_0$ ***newtons.****

6.3 Magnetic Field Strength (H)

Magnetic field strength at any point within a magnetic field is numerically equally to the force experienced by a N-pole of one weber placed at that point. Hence, unit of H is N/Wb.

* To commemorate the memory of German physicist Wilhelm Edward Weber (1804 1891).

** A unit magnetic pole is also defined as that magnetic pole which when placed at a distance of one metre from a very long straight conductor carrying a current of one ampere experiences a force of $1/2\pi$ newtons (Art. 6.14)

Suppose, it is required to find the field intensity at a point A distant r metres from a pole of m webers. Imagine a similar pole of one weber placed at point A. The force experienced by this pole is

$$F = \frac{m \times 1}{4\pi\mu_0 r^2} \text{N} \qquad \therefore \quad H = \frac{m}{4\pi\mu_0 r^3} \text{N/Wb (or A/m)* or oersted.}$$

Also, if a pole of m Wb is placed in a uniform field of strength H N/Wb, then force experienced by the pole is = mH newtons.

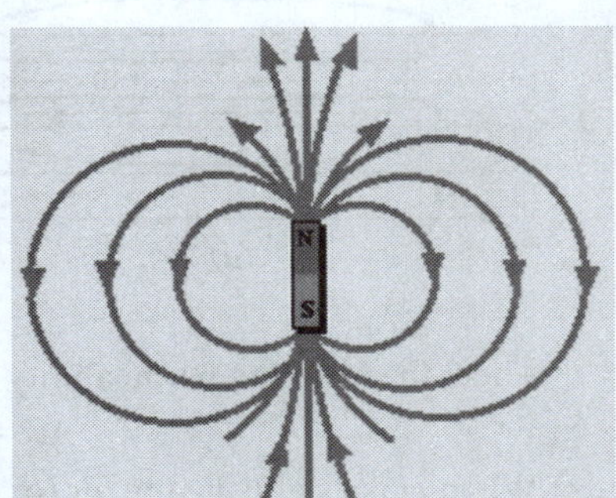

It should be noted that field strength is a vector quantity having both magnitude and direction

$$\therefore \qquad \vec{H} = \frac{m}{4\pi\mu_0 r^2} \qquad \hat{r} = \frac{m}{4\pi\mu_0 r^3} \vec{r}$$

It would be helpful to remember that following terms are sometimes interchangeably used with field intensity: Magnetising force, strength of field, magnetic intensity and intensity of magnetic field.

6.4 Magnetic Potential

The magnetic potential at any point within a magnetic field is measured by the work done in shifting a N-pole of one weber from infinity to that point against the force of the magnetic field. It is given by

$$M = \frac{m}{4\pi\mu_0 r} \text{J/Wb}$$

It is a scalar quantity.

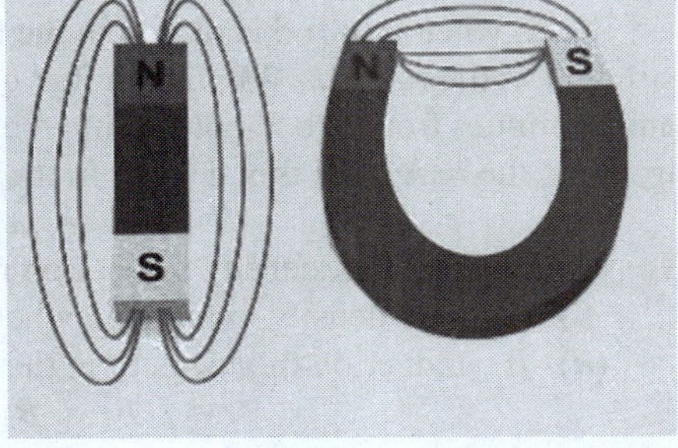

Magnetic lines of force

6.5 Flux Per Unit Pole

A unit N-pole is supposed to radiate out a flux of one weber. Its symbol is Φ. Therefore, the flux coming out of a N-pole of m weber is given by

$$\Phi = m \text{ Wb}$$

6.6 Flux Density (B)

It is given by the flux passing per unit area through a plane at right angles to the flux. It is usually designated by the capital letter B and is measured in weber/meter^2. It is a Vector Quantity.

It ΦWb is the total magnetic flux passing normally through an area of A m^2, then

$$B = \Phi/A \text{Wb/m}^2 \text{ or tesla (T)}$$

Note. Let us find an expression for the flux density at a point distant r metres from a unit N-pole (*i.e.* a pole of strength 1 Wb.) Imagine a sphere of radius r metres drawn round the unit pole. The flux of 1 Wb radiated out by the unit pole falls normally on a surface of $4\pi r^2.m^2$. Hence

$$B = \frac{\Phi}{A} = \frac{1}{4\pi r^2} \text{Wb/m}^2$$

* It should be noted that N/Wb is the same thing as ampere/metre (A/m) or just A/m cause 'turn' has no units

6.7 Absolute Permeability (μ) and Relative Permeability (μ_r)

In Fig. 6.3 is shown a bar of a magnetic material, say, iron placed in a uniform field of strength H N/Wb. Suppose, a flux density of B Wb/m² is developed in the rod.

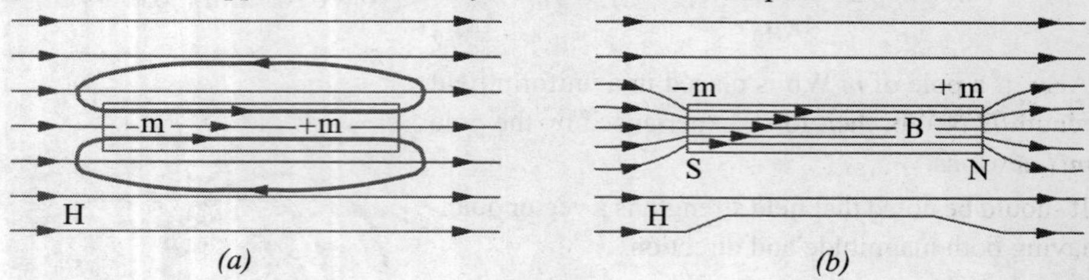

Fig. 6.3

Then, the absolute permeability of the material of the rod is defined as

$$\mu = B/H \text{ henry/metre or } B = \mu H = \mu_0 \mu_r H \text{ Wb/m}^2 \quad ...(i)$$

When H is established in air (or vacuum), then corresponding flux density developed in air is

$$B_0 = \mu_0 H$$

Now, when iron rod is placed in the field, it gets magnetised by induction. If induced pole strength in the rod is m Wb, then a flux of m Wb emanates from its N-pole, re-enters its S-pole and continues from S to N-pole within the magnet. If A is the face or pole area of the magentised iron bar, the *induction* flux density in the rod is

$$B_i = m/A \text{ Wb/m}^2$$

Hence, total flux density in the iron rod consists of two parts [Fig. 6.3 (*b*)].

(i) B_0 –flux density in air even when rod is not present

(ii) B_i –induction flux density in the rod

$$B = B_0 + B_i = \mu_0 H + m/\text{A}$$

Eq. (*i*) above may be written as $B = \mu_r \cdot \mu_0 H = \mu_r B_0$

$$\therefore \quad \mu_r = \frac{B}{B_0} = \frac{B\,(\text{material})}{B_0(\text{vacuum})} \quad \text{...for same } H$$

Hence, relative permeability of a material is equal to ***the ratio of the flux density produced in that material to the flux density produced in vacuum by the same magnetising force.***

6.8 Intensity of Magnetisation (I)

It may be defined as the induced pole strength developed per unit area of the bar. Also, it is the magnetic moment developed per unit volume of the bar.

Let m = pole strength induced in the bar in Wb

A = face or pole area of the bar in m²

Then $I = m/A$ Wb/m²

Hence, it is seen that intensity of magnetisation of a substance may be defined as ***the flux density produced in it due to its own induced magnetism.***

If l is the magnetic length of the bar, then the product ($m \times l$) is known as its magnetic moment M.

$$\therefore \quad I = \frac{m}{A} = \frac{m \times l}{A \times l} = \frac{M}{V} = \text{magnetic moment/volume}$$

6.9 Susceptibility (K)

Susceptibility is defined as ***the ratio of intensity of magnetisation I to the magnetising force H.***

$$\therefore \quad K = I/H \text{ henry/metre.}$$

6.10 Relation between B, H, I and K

It is obvious from the above discussion in Art. 6.7 that flux density B in a material is given by

$$B = B_0 + m/A = B_0 + I \qquad \therefore \quad B = \mu_0 H + I$$

Now absolute permeability is $\mu = \dfrac{B}{H} = \dfrac{\mu_0 H + I}{H} = \mu_0 + \dfrac{I}{H} \qquad \therefore \quad \mu = \mu_0 + K$

Also $\qquad \mu = \mu_0 \mu_r \quad \therefore \quad \mu_0 \mu_r = \mu_0 + K$ or $\mu_r = 1 + K/\mu_0$

For ferro-magnetic and para-magnetic substances, K is positive and for diamagnetic substances, it is negative. For ferro-magnetic substance (like iron, nickel, cobalt and alloys like nickel-iron and cobalt-iron) μ_r is much greater than unity whereas for para-magnetic substances (like aluminium), μ_r is slightly greater than unity. For diamagnetic materials (bismuth) $\mu_r < 1$.

Example 6.1. *The magnetic susceptibility of oxygen gas at 20ºC is 167×10^{-11} H/m. Calculate its absolute and relative permeabilities.*

Solution. $\qquad \mu_r = 1 + \dfrac{K}{\mu_0} = 1 + \dfrac{167 \times 10^{-11}}{4\pi \times 10^{-7}} = \mathbf{1.00133}$

Now, absolute permeability $\mu = \mu_0 \mu_r = 4\pi \times 10^{-7} \times 1.00133 = \mathbf{12.59 \times 10^{-7}}$ ***H/m***

ELECTROMAGNETISM

6.11 Force on a Current-Carrying Conductor Lying in a Magnetic Field

It is found that whenever a current-carrying conductor is placed in magnetic field, it experiences a force which acts in a direction perpendicular both to the direction of the current and the field. In Fig. 6.4 is shown a conductor *XY* lying at right angles to the uniform horizontal field of flux density B Wb/m^2 produced by two solenoids A and B. If l is the length of the conductor lying within this field and I ampere the current carried by it, then the magnitude of the force experienced by it is

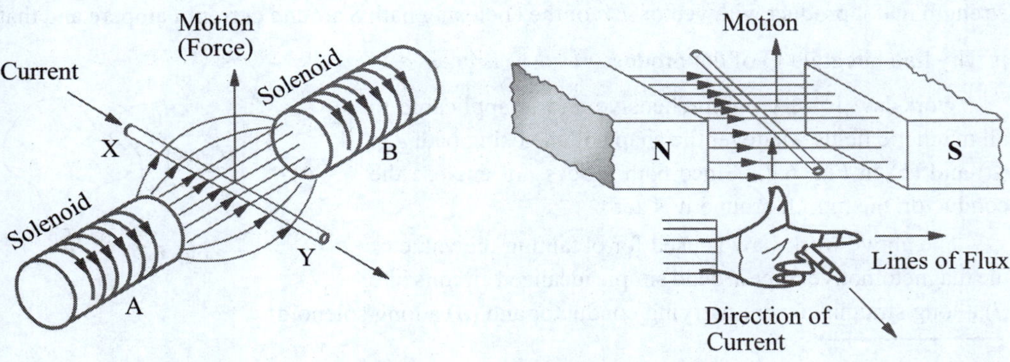

Fig. 6.4

Fig. 6.5

$$F = BIl = \mu_0 \mu_r HIl \text{ newton}$$

Using vector notation $\vec{F} = I\vec{l} \times \vec{B}$ and $F = IlB \sin\theta$ where θ is the angle between $\vec{l}$ and $\vec{B}$ which is 90º in the present case

or $$F = Il\,B \sin 90° = Il\,B \text{ newtons} \qquad (\because \sin 90° = 1)$$

The direction of this force may be easily found by Fleming's left-hand rule.

Hold out your left hand with forefinger, second finger and thumb at right angles to one another. If the forefinger represents the direction of the field and the second finger that of the current, then thumb gives the direction of the motion. It is illustrated in Fig. 6.5.

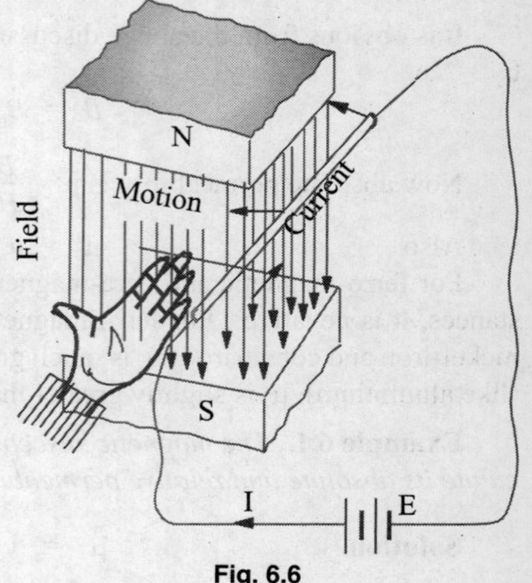

Fig. 6.6

Fig. 6.6 shows another method of finding the direction of force acting on a current carrying conductor. It is known as Flat Left Hand rule. The force acts in the direction of the thumb obviously, the direction of motor of the conductor is the same as that of the force.

It should be noted that no force is exerted on a conductor when it lies parallel to the magnetic field. In general, if the conductor lies at an angle θ with the direction of the field, then B can be resolved into two components, $B \cos\theta$ parallel to and $B \sin\theta$ perpendicular to the conductor. The former produces no effect whereas the latter is responsible for the motion observed. In that case,

$F = BIl \sin\theta$ newton, which has been expressed as cross product of vector above.*

6.12 Ampere's Work Law or Ampere's Circuital Law

The law states that m.m.f.** (magnetomotive force corresponding to e.m.f. *i.e.* electromotive force of electric field) around a closed path is equal to the current enclosed by the path. Mathematically, $\oint \vec{H} \cdot d\vec{s} = I$ amperes where $\vec{H}$ is the vector representing magnetic field strength in dot product with vector $d\vec{s}$ of the enclosing path S around current I ampere and that is why line integral ($\oint$) of dot product $\vec{H} \cdot d\vec{s}$ is taken.

Work law is very comprehensive and is applicable to all magnetic fields whatever the shape of enclosing path *e.g.* (*a*) and (*b*) in Fig. 6.7. Since path *c* does not enclose the conductor, the m.m.f. around it is zero.

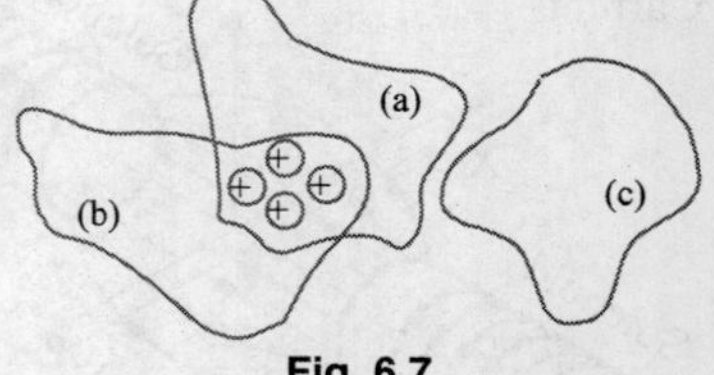

Fig. 6.7

The above work Law is used for obtaining the value of the magnetomotive force around simple idealized circuits like **(*i*)** a long straight current-carrying conductor and **(*ii*)** a long solenoid.

---

* It is simpler to find direction of Force (Motion) through cross product of given vectors $I\vec{l}$ and $\vec{B}$.

** M.M.F. is not a force, but is the work done.

(i) Magnetomotive Force around a Long Straight Conductor

In Fig. 6.8 is shown a straight conductor which is assumed to extend to infinity in either direction. Let it carry a current of I amperes upwards. The magnetic field consists of circular lines of force having their plane perpendicular to the conductor and their centres at the centre of the conductor.

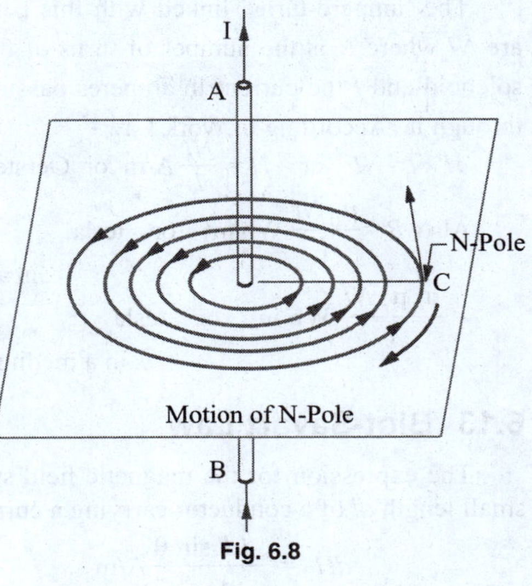

Fig. 6.8

Suppose that the field strength at point C distant r metres from the centre of the conductor is H. Then, it means that if a unit N-pole is placed at C, it will experience a force of H newtons. The direction of this force would be tangential to the circular line of force passing through C. If this unit N-pole is moved once round the conductor *against* this force, then work done *i.e.*

m.m.f. = force × distance = I

i.e. $I = H \times 2\pi r$ joules = Amperes

or $H = \dfrac{I}{2\pi r}$

$$= \oint \vec{H} \cdot d\vec{s} \text{ Joules = Amperes} = I$$

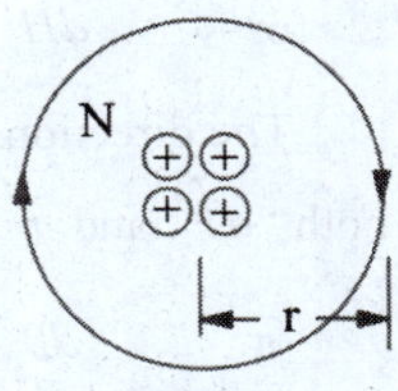

Fig. 6.9

Obviously, if there are N conductors (as shown in Fig. 6.9), then

$$H = \frac{NI}{2\pi r} \text{ A/m or Oersted}$$

and $B = \mu_0 \dfrac{NI}{2\pi r}$ Wb/m² tesla ...in air

$= \dfrac{\mu_0 \mu_r NI}{2\pi r}$ Wb/m² tesla ...in a medium

(ii) Magnetic Field Strength of a Long Solenoid

Let the Magnetic Field Strength along the axis of the solenoid be H. Let us assume that:

(i) the value of H remains constant throughout the length l of the solenoid and

(ii) the volume of H outside the solenoid is negligible.

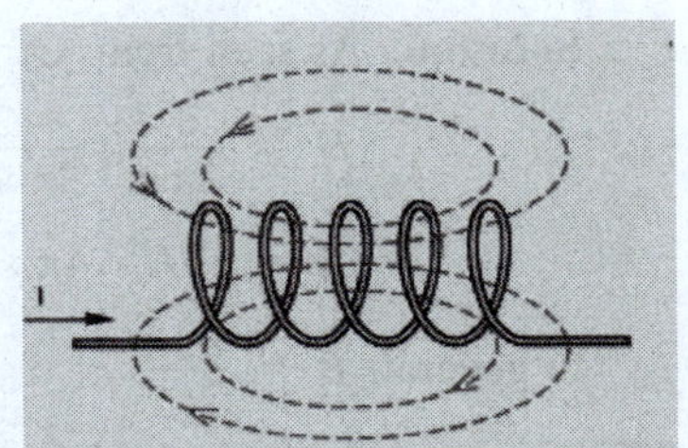

Magnetic field around a coil carrying electric current

Suppose, a unit N-pole is placed at point A outside the solenoid and is taken once round the completed path (shown dotted in Fig. 6.10) in a direction opposite to that of H. Remembering that the force of H newtons acts on the N-pole only over the length l (it being negligible elsewhere), the work done in one round is

$= H \times l$ joules = Amperes

The 'ampere-turns' linked with this path are NI where N is the number of turns of the solenoid and I the current in amperes passing through it. According to Work Law

$$H \times l = NI \quad \text{or} \quad H = \frac{NI}{l} \text{ A/m or Oersted}$$

Also $B = \dfrac{\mu_0 NI}{l}$ Wb/m² or tesla ...in air

$= \dfrac{\mu_0 \mu_r NI}{l}$ Wb/m² or tesla ...in a medium

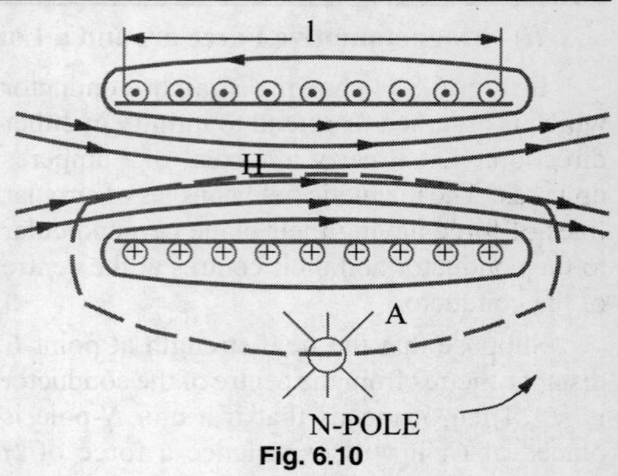

Fig. 6.10

6.13 Biot-Savart Law

The expression for the magnetic field strength dH produced at point P by a vanishingly small length dl of a conductor carrying a current of I amperes (Fig. 6.11) is given by

$$dH = \frac{Idl \sin\theta}{4\pi r^2} \text{ A/m}$$

or $$\vec{dH} = (Id\vec{l} \times \hat{r}) / 4\pi r^2 \text{ in vector form}$$

The direction of $\vec{dH}$ is perpendicular to the plane containing both '$\vec{dl}$' and $\vec{r}$ *i.e.* entering.

or $$dB_0 = \frac{\mu_0 Idl}{4\pi r^2} \sin\theta \text{ Wb/m}^2$$

and $$\vec{dB_0} = \frac{\mu_0 \vec{I}\, \vec{dl} \times \hat{r}}{4\pi r^2} \text{ in vector form}$$

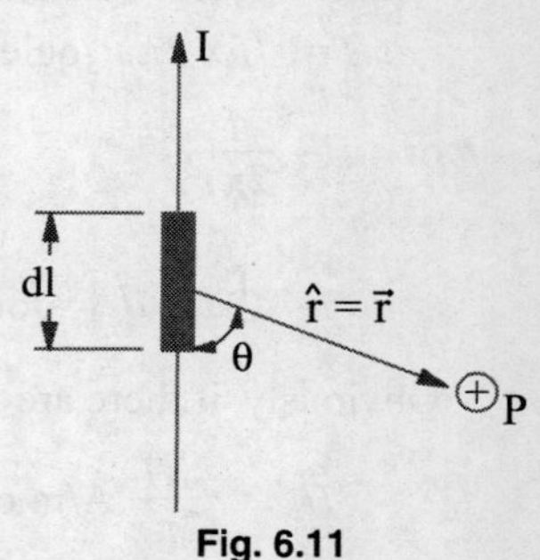

Fig. 6.11

Example 6.2. *Calculate the magnetising force and flux density at a distance of 5 cm from a long straight circular conductor carrying a current of 250 A and placed in air. Draw a curve showing the variation of B from the conductor surface outwards if its diameter is 2 mm.*

(AKTU 2022)

Solution. As seen from Art. 6.12 **(*i*)**,

$$H = \frac{I}{2\pi r} = \frac{250}{2\pi \times 0.05} = \mathbf{795.6\ AT/m}$$

$$B = \mu_0 H = 4\pi \times 10^{-7} \times 795.6 = \mathbf{10^{-3}\ Wb/m^2}$$

In general, $B = \dfrac{\mu_0 I}{2\pi r}$

Now, at the conductor surface, $r = 1$ mm $= 10^{-3}$ m

$$\therefore \quad B = \frac{4\pi \times 10^{-7} \times 250}{2\pi \times 10^{-3}} = \mathbf{0.05\ Wb/m^2}$$

The variation of B outside the conductor is shown in Fig. 6.12.

Fig. 6.12

6.14 Force between Two Parallel Conductors

(*i*) Currents in the same direction. In Fig. 6.13 are shown two parallel conductors P and Q carrying currents I_1 and I_2 amperes in the same direction *i.e.* upwards. The field strength in the space between the two conductors is decreased due to the two fields there being in opposition to each other. Hence, the resultant field is as shown in the figure. Obviously, the two conductors are attracted towards each other.

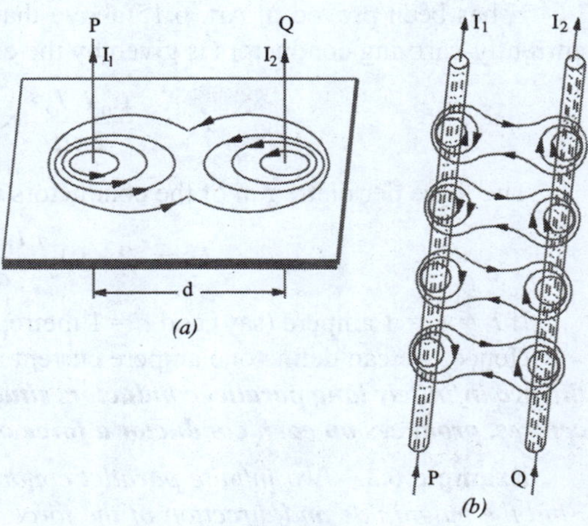

Fig. 6.13

(*ii*) Currents in opposite directions. If, as shown in Fig. 6.14, the parallel conductors carry currents in opposite directions, then field strength is increased in the space between the two conductors due to the two fields being in the same direction there. Because of the lateral repulsion of the lines of the force, the two conductors experience a mutual force of repulsion as shown separately in Fig. 6.14 (*b*).

6.15 Magnitude of Mutual Force

It is obvious that each of the two parallel conductors lies in the magnetic field of the other conductor. For example, conductor P lies in the magnetic field of Q and Q lies in the field of P. If 'd' metres is the distance between them, then flux density at Q due to P is [Art. 6.12 (b)]

$$B = \frac{\mu_0 I_1}{2\pi d} \text{ Wb/m}^2$$

If l is the length of conductor Q lying in this flux density, then force (either of attraction or repulsion).

$$F = BI_2 l \text{ newton} \quad \text{or} \quad F = \frac{\mu_0 I_1 I_2 l}{2\pi d} \text{ N}$$

Obviously, conductor P will experience an equal force in the opposite direction.

The above facts are known as Laws of Parallel Currents and may be stated as follows:

(*i*) Two parallel conductors attract each other if currents through them flow in the *same* direction and repel each other if the currents through them flow in the *opposite* directions.

(*ii*) The force between two such parallel conductors is proportional to the product of current strengths and to the length of the conductors considered and varies inversely as the distance between them.

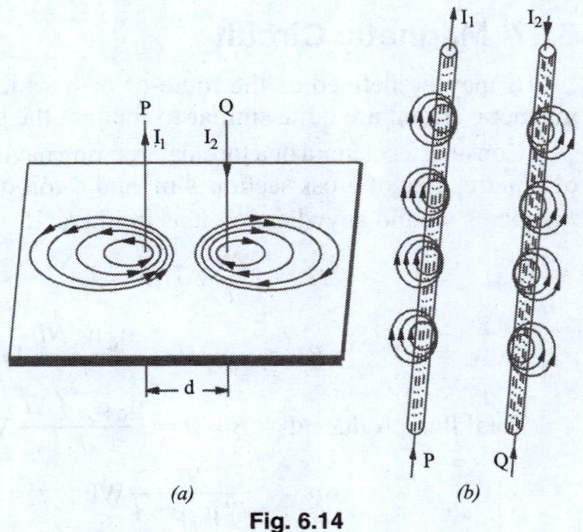

Fig. 6.14

6.16 Definition of Ampere

If has been proved in Art. 6.15 above that the force between two infinitely long parallel currently-carrying conductors is given by the expression

$$F = \frac{\mu_0 I_1 I_2 l}{2\pi d} \text{N} \quad \text{or} \quad F = \frac{4\pi \times 10^{-} \; I_1 I_2 l}{2\pi d} = 2 \times 10^{-7} \frac{I_1 I_2}{d} \text{N}$$

The force per metre run of the conductors is

$$F = 2 \times 10^{-7} \frac{I_1 I_2}{d} \text{N/m}$$

If $I_1 = I_2 = 1$ ampere (say) and $d = 1$ metre, then $F = 2 \times 10^{-7}$ N

Hence, we can define one ampere current as ***that current which when flowing in each of the two infinitely long parallel conductors situated in vacuum and separated 1 metre between centres, produces on each conductor a force of 2 × 10⁻⁷ N per metre length***.

Example 6.3. *Two infinite parallel conductors carry parallel currents of 10 amp. each. Find the magnitude and direction of the force between the conductors per metre length if the distance between them is 20 cm.* **(Elect. Engg. Material - II Punjab Univ. May 1990)**

Solution. $$F = 2 \times 10^{-7} \frac{10 \times 10 \times 1}{0.2} \text{N} = \mathbf{10^{-4}\ N}$$

The direction of force will depend on whether the two currents are flowing in the same direction or in the opposite direction. As per Art 6.15, it would be a force of attraction in the first case and that or repulsion in the second case.

Tutorial Problem 6.1

1. The force between two long parallel conductors is 15 kg/metre. The conductor spacing is 10 cm. If one conductor carries twice the current of the other, calculate the current in each conductor. **[6,060 A; 12,120 A]**
2. A wire is bent into a plane to form a square of 30 cm side and a current of 100 A is passed through it. Calculate the field strength set up at the centre of the square. **[300 AT/m]**
(Electrotechnics - I, M.S. Univ. Baroda)

MAGNETIC CIRCUIT

6.17 Magnetic Circuit

It may be defined as the route or path which is followed by magnetic flux. The law of magnetic circuit are quite similar to (but not the same as) those of the electric circuit.

Consider a solenoid or a toroidal iron ring having a magnetic path of l metre, area of cross section A m² and a coil of N turns carrying I amperes wound anywhere on it as in Fig. 6.15.

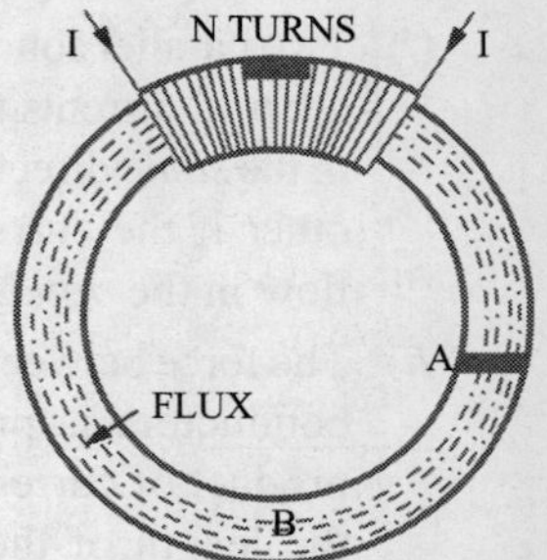

Fig. 6.15

$$H = \frac{NI}{l} \text{AT/m}$$

Now $$B = \mu_0\mu_r H = \frac{\mu_0 \mu_r NI}{l} \text{Wb/m}^2$$

Total flux produce $\Phi = B \times A = \dfrac{\mu_0 \mu_r ANI}{l}$ Wb

$$\therefore \quad \Phi = \frac{NI}{l/\mu_0 \mu_r A} \text{Wb}$$

The numerator 'NI' which produces magnetization in the magnetic circuit is known as magnetomotive force (m.m.f.). Obviously, its unit is ampere-turn (AT)*. It is analogous to e.m.f. in an electric circuit.

The denominator $\dfrac{l}{\mu\ \mu_r\ A}$ is called the ***reluctance*** of the circuit and is analogous to resistance in electric circuits.

$$\therefore \qquad \text{flux} = \frac{\text{m.m.f.}}{\text{reluctance}} \quad \text{or} \quad \Phi = \frac{m.mf.}{S}$$

Sometimes, the above equation is called the "Ohm's Law of Magnetic Circuit" because it resembles a similar expression in electric circuits *i.e.*

$$\text{current} = \frac{\text{e.m.f.}}{\text{resistance}} \quad \text{or} \quad I = \frac{V}{R}$$

6.18 Definitions Concerning Magnetic Circuit

1. Magnetomotive force (m.m.f.). It drives or tends to drive flux through a magnetic circuit and corresponds to electromotive force (e.m.f.) in an electric circuit.

M.M.F. is equal to the work done in joules in carrying a unit magnetic pole once through the entire magnetic circuit. It is measured in ampere-turns.

In fact, as p.d. between any two points is measured by the work done in carrying a unit charge from one points to another, similarly, m.m.f. between two points is measured by the work done in joules in carrying a unit magnetic pole from one point to another.

2. Ampere-turns (AT). It is the unit of magnetometre force (m.m.f.) and is given by the product of number of turns of a magnetic circuit and the current in amperes in those turns.

3. Reluctance. It is the name given to that property of a material which opposes the creation of magnetic flux in it. It, in fact, measures the opposition offered to the passage of magnetic flux through a material and is analogous to resistance in an electric circuit even in form. Its units is **AT/Wb.****

$$\text{reluctance} = \frac{l}{\mu A} = \frac{l}{\mu_0 \mu_r A}; \ \text{resistance} = \rho \frac{l}{A} = \frac{l}{\sigma A}$$

In other words, the reluctance of a magnetic circuit is the number of amp-turns required per weber of magnetic flux in the circuit. Since 1 AT/Wb = 1/henry, the unit of reluctance is "reciprocal henry."

4. Permeance. It is reciprocal of reluctance and implies the case or readiness with which magnetic flux is developed. It is analogous to conductance in electric circuits. It is measured in terms of Wb/AT or henry.

5. Reluctivity. It is specific reluctance and corresponds to resistivity which is 'specific resistance'.

6.19 Composite Series Magnetic Circuit

In Fig. 6.16 is shown a composite series magnetic circuit consisting of three different magnetic materials of different permeabilities and lengths and one air gap ($\mu_r = 1$). Each path will have its own reluctance. The total reluctance is the sum of individual reluctances as they are joined in series.

---

* Strictly speaking, it should be only 'ampere' because turns have no unit.

** From the ratio $\Phi = , \dfrac{\text{m.m.f.}}{\text{reluctance}}$ it is obvious that reluctance = m.m.f./Φ. Since m.m.f. is in ampere-turns and flux in webers, unit of reluctance is ampere-turn/weber (AT/Wb) or A/Wb.

$$\therefore \quad \text{total reluctance} = \Sigma \frac{l}{\mu_0 \mu_r A}$$

$$= \frac{l_1}{\mu_0 \mu_{r_1} A_1} + \frac{l_2}{\mu_0 \mu_{r_2} A_2} + \frac{l_3}{\mu_0 \mu_{r_3} A_3} + \frac{l_a}{\mu_0 A_g}$$

$$\therefore \quad \text{flux } \Phi = \frac{\text{m.m.f.}}{\dfrac{l}{\mu_0 \mu_r A}}$$

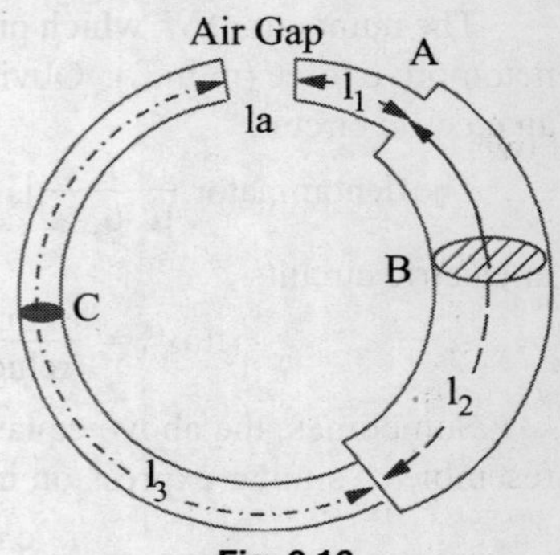

Fig. 6.16

6.20 How to Find Ampere-turns ?

It has been shown in that $H = NI/l$ AT/m or $NI = H \times l$

$\therefore$ ampere-turns AT $= H \times l$

Hence, following procedure should be adopted for calculating the total ampere turns of a composite magnetic path.

(i) Find H for each portion of the composite circuit. For air, $H = B/\mu_0$, otherwise $H = B/\mu_0\mu_r$.

(ii) Find ampere-turns for each path separately by using the relation AT $= H \times l$.

(iii) Add up these ampere-turns to get the total ampere-turns for the entire circuit.

6.21 Comparison between Magnetic and Electric Circuits

SIMILARITIES

	Magnetic Circuit	Electric Circuit
	MMF, I, I, FLUX Φ — Fig. 6.17	EMF, Current — Fig. 6.18
1.	Flux = $\dfrac{\text{m.m.f.}}{\text{reluctance}}$	Current = $\dfrac{\text{e.m.f.}}{\text{resistance}}$
2.	M.M.F. (ampere-turns)	E.M.F. (volts)
3.	Flux Φ (webers)	Current I (amperes)
4.	Flux density B (Wb/m^2)	Current density (A/m^2)
5.	Reluctance $S = \dfrac{l}{\mu A}\left(= \dfrac{l}{\mu_0 \mu_r A}\right)$	resistance $R = \rho \dfrac{l}{A} = \dfrac{l}{\rho A}$
6.	Permeance (= 1/reluctance)	Conductance (= 1/resistance)
7.	Reluctivity	Resistivity
8.	Permeability (= 1/reluctivity)	Conductivity (= 1/resistivity)
9.	Total m.m.f. $= \Phi S_1 + \Phi S_2 + \Phi S_3 + ...$	9. Total e.m.f. $= IR_1 + IR_2 + IR_3 + ...$

DIFFERENCES

1. Strictly speaking, flux does not actually 'flow' in the sense in which an electric current flows.

2. If temperature is kept constant, then resistance of an electric circuit is constant and is independent of the current strength (or current density). On the other hand, the reluctance of a magnetic circuit does depend on flux (and hence flux density) established in it. It is so because μ (which equals the slope of B/H curve) is not constant even for a given material as it depends on the flux density B. Value of μ is large for low value of B and ***vice-versa***. Hence, reluctance is small ($S = l/\mu A$) for small values of B and large for large values of B.

3. Flow of current in an electric circuit involves continuous expenditure of energy but in a magnetic circuit, energy is needed only creating the flux initially but not for maintaining it.

6.22 Parallel Magnetic Circuits

Fig. 6.19 (*a*) shown a parallel magnetic circuit consisting of two parallel magnetic paths *ACB* and *ADB* acted upon by the same *m.m.f.* Each magnetic path has an average length of $(2\,l_1 + l_2)$. The flux produced by the coil wound on the central core is divided equally at point *A* between the two outer parallel paths. The reluctance offered by the two parallel paths is = half the reluctance of each path.

Fig. 6.19 (*b*) shows the equivalent electrical circuit where resistance offered to the voltage source is $= R \| R = R/2$

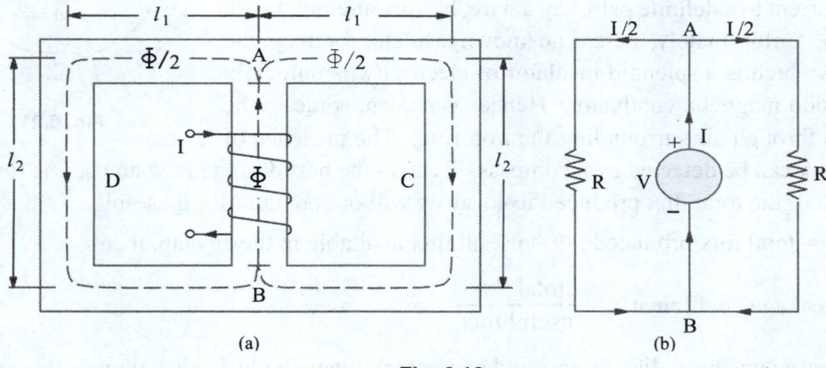

Fig. 6.19

It should be noted that reluctance offered by the central core AB has been neglected in the above treatment.

6.23 Series-Parallel Magnetic Circuits

Such a circuit is shown in Fig. 6.20 (*a*). It shows two parallel magnetic circuits *ACB* and *ACD* connected across the common magnetic path *AB* which contains an air-gap of length l_g. As usual, the flux Φ in the common core is divided equally at point *A* between the two parallel paths which have equal reluctance. The reluctance of the path *AB* consists of ***(i)*** air gap reluctance and ***(ii)*** the reluctance of the central core which comparatively negligible. Hence, the reluctance of the central core *AB* equals only the air-gap reluctance across which are connected two equal parallel reluctances. Hence, the *m.m.f.* required for this circuit would be the sum of ***(i)*** that required for the air-gap and ***(ii)*** that required for either of two paths (not both).

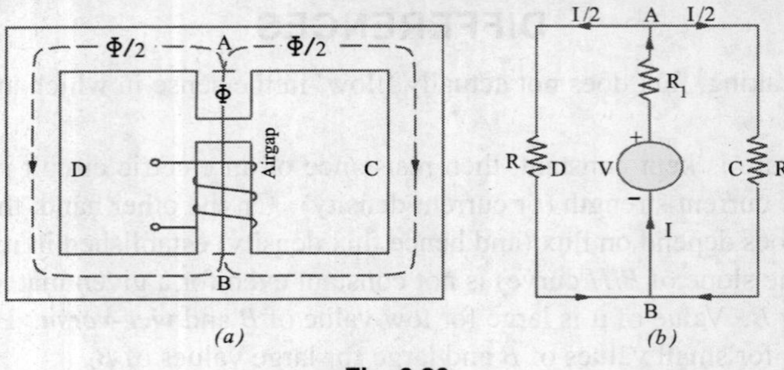

Fig. 6.20

The equivalent electrical circuit is shown in Fig. 6.20 (*b*) where the total resistance offered to the voltage source is $= R_1 + R \| R = R_1 + R/2$.

6.24 Leakage Flux and Hopkinson's Leakage Coefficient

Leakage flux is the flux which follows a path not intended for it. In Fig. 6.21 is shown an iron ring wound with a coil and having an air-gap. The flux in the air-gap is known as the useful flux because it is only this flux which can be utilized for various useful purposes.

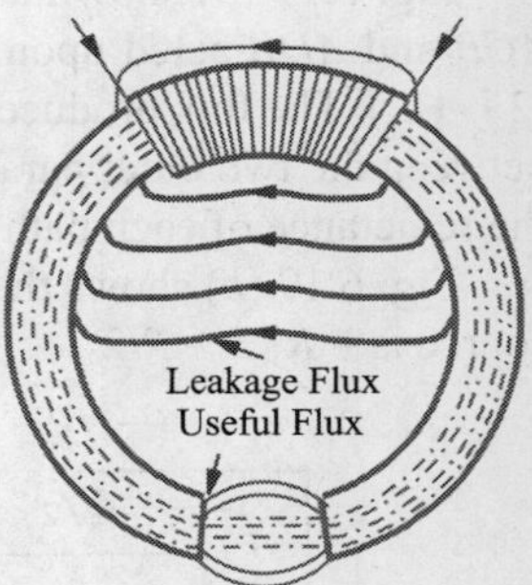

Fig. 6.21

It is found that it is impossible to confine all the flux to the iron path only, although it is usually possible to confine most of the electric current to a definite path, say a wire, by surrounding it with insulation. Unfortunately, there is no known insulator for magnetic flux. Air, which is a splendid insulator of electricity, is unluckily a fairly good magnetic conductor. Hence, as shown, some of the flux leaks through air surrounding the iron ring. The presence of leakage flux can be detected by a compass. Even in the best designed dynamos, it is found that 15 to 20% of the total flux produced leaks away without being utilised usefully.

If, Φ_t = total flux produced ; Φ = useful flux available in the air-gap, then

$$\text{leakage coefficient } \lambda = \frac{\text{total flux}}{\text{useful flux}} \quad \text{or} \quad \lambda = \frac{\Phi_t}{\Phi}$$

In electric machines like motors and generators, magnetic leakage is undesirable, because, although it does not lower their power efficiency, yet it leads to their increased weight and cost of manufacture. Magnetic leakage can be minimised by placing the exciting coils or windings as close as possible to the air-gap or to the points in the magnetic circuit where flux is to be utilized for useful purposes.

It is also seen from Fig. 6.21 that there is fringing or spreading of lines of flux at the edges of the air-gap. This fringing increases the effective area of the air-gap.

The value of λ for modern electric machines varies between 1.1 and 1.25.

6.25 Magnetisation Curves

The approximate magnetisation curves of a few magnetic materials are shown in Fig. 6.22. These curves can be determined by the following methods provided the materials are in the form of a ring:

(*a*) By means of a ballistic galvanometer and (*b*) By means of a fluxmeter.

6.26 Magnetisation Curves by Ballistic Galvanometer

In Fig. 6.22 shown the specimen ring of uniform cross-section wound uniformly with a coil *P* which is connected to a battery *B* through a reversing switch *RS*, a variable resistance R_1 and an ammeter. Another secondary coil *S* also wound over a small portion of the ring and is connected through a resistance *R* to a ballistic galvanometer *BG*.

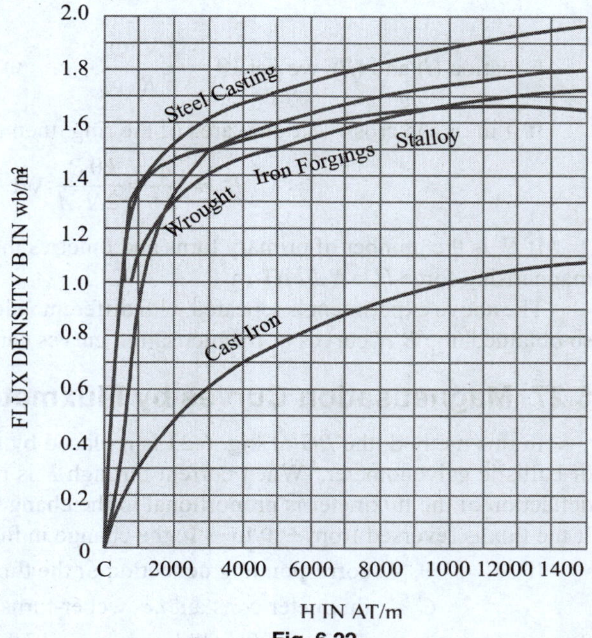

Fig. 6.22

The current through the primary *P* can be adjusted with the help of R_1. Suppose the primary current is *I*. When the primary current is reversed by means of *RS*, then flux is reversed through *S*, hence an induced e.m.f. is produced in it which sends a current through *BG*. This current is of very short duration. The first deflection or 'throw' of the *BG* is proportional to the quantity of electricity or charge passing through it so long as the time taken for this charge to flow is short as compared with the time of one oscillation.

If θ = first deflection or 'throw' of the galvanometer when primary current *I* is reversed.

k = ballistic constant of the galvanometer *i.e.* charge per unit deflection.

then, charge passing through *BG* is $= k\theta$ coulombs ...(*i*)

Let Φ = flux in Wb produced by primary current of *I* amperes ; t = time of reversal of flux;

then rate of change of flux $= \dfrac{2\Phi}{t}$ Wb/s

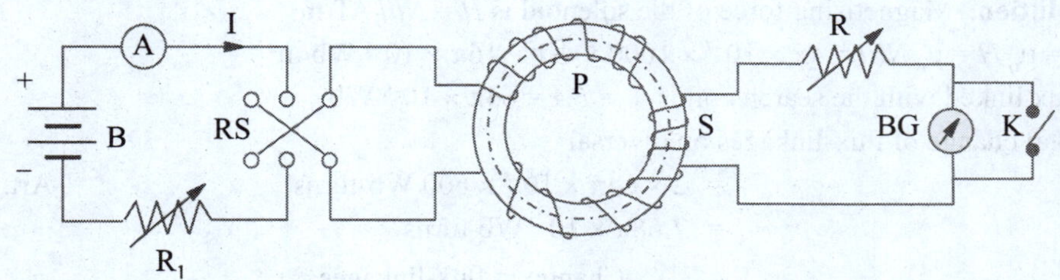

Fig. 6.23

If N_2 is the number of turns in secondary coil *S*, then average e.m.f. induces in it is

$$= N_2 \cdot \frac{2\Phi}{t} \text{ volt.}$$

Secondary current or current through $BG = \dfrac{2N_2\Phi}{R_s t}$ amperes

where R_s is the total resistance of the secondary circuit.

Charge flowing through BG = average current × time = $\dfrac{2N_2\Phi}{R_s t} \times t = \dfrac{2N_2\Phi}{R_s}$ coulomb ...(*ii*)

Equation (*i*) and (*ii*), we get $k\theta = \dfrac{2N_2\Phi}{R_s}$ $\quad \therefore \quad \Phi = \dfrac{k\theta R_s}{2N_2}$ Wb

If A m² is the cross-sectional area of the ring, then flux density is

$$B = \frac{\Phi}{A} = \frac{k\theta R_s}{2N_2 A} \text{ Wb/m}^2$$

If N_1 is the number of primary turns and l metres the mean circumference of the ring, then, magnetising force $H = N_1 I/l$ AT/m.

The above experiment is repeated with different values of primary current and form the data so obtained, the B/H curves or magnetisation curves can be drawn.

6.27 Magnetisation Curves by Fluxmeter

In this method, the BG of Fig. 6.23 is replaced by a fluxmeter which is just a special type of ballistic galvanometer. When current through P is reversed, the flux is also reversed. The deflection of the fluxmeter is proportional to the change in flux-linkages of the secondary coil. If the flux is reversed from $+\Phi$ to $-\Phi$, the change in flux-linkages in secondary S in $= 2\,\Phi\,N_2$.

If $\quad \theta$ = corresponding deflection of the fluxmeter

$\quad C$ = fluxmeter constant *i.e.* weber-turns per unit deflection.

then, $\quad$ change of flux-linkages in $S = C\,\theta$

$$\therefore \quad 2\Phi N_2 = C\theta \quad \text{or} \quad \Phi = \frac{C\theta}{N_2} \text{ Wb} \; ; \; B = \frac{\Phi}{A} = \frac{C\theta}{2N_2 A} \text{ Wb/m}^2$$

Example 6.4. *A fluxmeter is connected to a search-coil having 600 turns and mean area of 4 cm². The search coil is placed at the centre of an air-cored solenoid 1 metre long and wound with 1000 turns. When a current of 4 A is reversed, there is a deflection of 20 scale divisions on the fluxmeter. Calculate the calibration in Wb-turns per scale division.*

(Measurements-I, Nagpur Univ. 1991)

Solution. Magnetising force of the solenoid is $H = NI/l$ AT/m

$B = \mu_0 H = \mu_0 NI/l = 4\pi \times 10^{-7} \times 1000 \times 4/1 = 16\pi \times 10^{-4}$ Wb/m²

Flux linked with the search coil is $\Phi = BA = 64\pi \times 10^{-8}$ Wb

Total change of flux-linkages on reversal

$= 2 \times 64\pi \times 10^{-8} \times 600$ Wb-turns —Art. 6.25

$= 7.68\pi \times 10^{-4}$ Wb-turns

Fluxmeter constant C is given by = $\dfrac{\text{Change in flux-linkages}}{\text{deflection produced}}$

$= 7.68\pi \times 10^{-4}/20 =$ **1.206 × 10⁻⁴ Wb-turns/division**

Example 6.5. *A ballistic galvanometer, connected to a search coil for measuring flux density in a core, gives a throw of 100 scale divisions on reversal of flux. The galvanometer coil has a resistance of 180 ohm. The galvanometer constant is 100 μC per scale division. The search coil has an area of 50 cm², wound with 1000 turns having a resistance of 20 ohm. Calculate the flux density in the core.* **(Elect. Instru & Measu. Nagpur Univ. 1992)**

Solution. As seen from Art. 6.24.

$$k\theta = 2N_2\Phi/R_s \quad \text{or} \quad \Phi = k\theta R_s/2N_2 \text{ Wb}$$

$$\therefore \quad BA = k\theta R_s/2N_2 \quad \text{or} \quad B = k\theta R_s/2N_2A$$

Here

$$k = 100\ \mu\text{C/division} = 100 \times 10^{-6} = 10^{-4}\text{ C/division}$$

$$\theta = 100;\ A = 50\text{ cm}^2 = 5 \times 10^{-3}\text{ m}^2$$

$$R_s = 180 + 20 = 200\ \Omega$$

$$\therefore \quad B = 10^{-4} \times 100 \times 200/2 \times 1000 \times 5 \times 10^{-3} = \mathbf{0.2\ Wb/m^2}$$

Example 6.6. *A ring sample of iron, fitted with a primary and a secondary winding is to be tested by the method of reversals to obtain its B/H curve. Give a diagram of connections explain briefly how the test could be carried out.*

In such a test, the primary winding of 400 turns carries a current of 1.8 A. On reversal, a change of 8×10^{-3} Wb-turns is recorded in the secondary winding of 10 turns. The ring is made up of 50 laminations, each 0.5 mm thick with outer and inner diameters of 25 and 23 cm respectively. Assuming uniform flux distribution, determine the values of B, H and the permeability.

Solution. Here, change of flux-linkages = $2\Phi N_2 = 8 \times 10^{-3}$ Wb-turns

$$\therefore \quad 2\Phi \times 10 = 8 \times 10^{-3} \quad \text{or} \quad \Phi = 4 \times 10^{-4}\text{ Wb and } A = 2.5 \times 10^{-4}\text{ m}^2$$

$$\therefore \quad B = \frac{4 \times 10^{-4}}{2.5 \times 10^{-4}} = 1.6\text{ Wb/m}^2\ ;\ H = \frac{Nl}{l} = \frac{400 \times 1.8}{0.24\pi} = 955$$

AT/m

Now

$$\mu_0\,\mu_r = \frac{B}{H}\ ;\ \mu_r = \frac{B}{\mu_0 H} = \frac{1.6}{4\pi \times 10^{-7} \times 955} = \mathbf{1333}$$

Example 6.7. *An iron ring of 3.5 cm² cross-sectional area with a mean length of 100 cm is wound with a magnetising winding of 100 turns. A secondary coil of 200 turns of wire is connected to a ballistic galvanometer having a constant of 1 micro-coulomb per scale division, the total resistance of the secondary circuit being 2000 Ω. On reversing a current of 10 A in the magnetising coil, the galvanometer gave a throw of 200 scale divisions. Calculate the flux density in the specimen and the value of the permeability at this flux density.*

(Elect. Measure, A.M.I.E Sec.B. 1992)

Solution. Reference may please be made to Art. 6.24.

Here

$$N_1 = 100\ ;\ N_2 = 200\text{: } A = 3.5 \times 10^{-4}\text{ m}^2\ ;\ l = 100\text{ cm} = 1\text{m}$$

$$k = 10^{-6}\text{ C/division, } \theta = 100\text{ divisions; } R_s = 2000\ \Omega;\ I = 10\text{ A}$$

$$B = \frac{k\theta R_s}{2N_2A} = \frac{10^{-6} \times 100 \times 2000}{2 \times 200 \times 3.5 \times 10^{-4}} = 1.43\text{ Wb/m}^2$$

$$\text{Magnetising force } H = N_1 I/l = 100 \times 10/1 = 1000\text{ AT/m}$$

$$\mu = \frac{B}{H} = \frac{1.43}{1000} = \mathbf{1.43 \times 10^{-3}\ H/m}$$

Note. The relative permeability is given by $\mu_r = \mu / \mu_0 = 1.43 \times 10^{-3}/4\pi \times 10^{-7} =$ **1137.**

Example 6.8. *An iron ring has a mean diameter of 0.1 m and a cross-section of 33.5 × 10^{-6} m². It is wound with a magnetising winding of 320 turns and the secondary winding of 220 turns. On reversing a current of 10 A in the magnetising winding, a ballistic galvanometer gives a throw of 272 scale divisions, while a Hilbert Magnetic standard with 10 turns and a flux of 2.5 × 10^{-4} gives a reading of 102 scale divisions, other conditions remaining the same. Find the relative permeability of the specimen.* **(Elect. Measu. A.M.I.E. Sec B, 1991)**

Solution. Length of the magnetic path $l = \pi D = 0.1\ \pi$ m

Magnetising Force, $H = NI/l = 320 \times 10/0.1\ \pi = 10{,}186$ AT/m

Flux density $B = \mu_0 \mu_r \mathbf{H} = 4\pi \times 10^{-7} \times \mu_r \times 10{,}186 = 0.0128\ \mu_r$ ...(*i*)

Now, from Hilbert's Magnetic standard, we have

$$2.5 \times 10^{-4} \times 10 = K \times 102, K = 2.45 \times 10^{-5}$$

On reversing a current of 10 A in the magnetising winding, total change in Weber-turns is $2\Phi N_s = 2.45 \times 10^{-5} \times 272$ or $2 \times 220 \times \Phi = 2.45 \times 10^{-5} \times 272$ or $\Phi = 1.51 \times 10^{-5}$ Wb

$\therefore\ B = \Phi/A = 1.51 \times 10^{-5}/33.5 \times 10^{-6} = 0.45$ Wb/m²

Substituting this value in Eq. (*i*), we have $0.0128\ \mu_r = 0.45$, $\therefore\ \mu_r = \mathbf{35.1}$

Example 6.9. *A laminated soft iron ring of relative permeability 1000 has a mean circumference of 800 mm and a cross-sectional area 500 mm². A radial air-gap of 1 mm width is cut in the ring which is wound with 1000 turns. Calculate the current required to produce an air-gap flux of 0.5 mWb if leakage factor is 1.2 and stacking factor 0.9. Neglect fringing.*

Solution. Total AT reqd. $= \Phi_g S_g + \Phi_i S_i = \dfrac{\Phi_g l_g}{\mu_0 A_g} + \dfrac{\Phi_i l_i}{\mu_0 \mu_r A_i B}$

Now, air-gap flux $\Phi_s = 0.5$ mWb $= 0.5 \times 10^{-3}$ Wb, $l_g = 1$ mm $= 1 \times 10^{-3}$ m; $A_g = 500$ mm² $= 500 \times 10^{-6}$ m²

Flux in the iron ring, $\Phi_i = 1.2 \times 0.5 \times 10^{-3}$ Wb

Net cross-sectional area $= A_i \times$ stacking factor $= 500 \times 10^{-6} \times 0.9$ m²

$$\therefore\ \text{total AT reqd.} = \frac{0.5 \times 10^{-3} \times 1 \times 10^{-3}}{4\pi \times 10^{-7} \times 500 \times 10^{-6}} + \frac{1.2 \times 0.5 \times 10^{-3} \times 800 \times 10^{-3}}{4\pi \times 10^{-7} \times 1000 \times (0.9 \times 500 \times 10^{-6})} = 1644$$

$\therefore\ I = 1644/1000 = \mathbf{1.64\ A}$

Example 6.10. *A ring has a diameter of 21 cm and a cross-sectional area of 10 cm². The ring is made up of semicircular sections of cast iron and cast steel, with each joint having a reluctance equal to an air-gap of 0.2 mm. Find the ampere-turns required to produce a flux of 8 × 10⁻⁴ Wb. The relative permeabilities of cast steel and cast iron are 800 and 166 respectively. Neglect fringing and leakage effects.* **(Elect. Circuits, South Gujarat Univ.)**

Solution. $\Phi = 8 \times 10^{-4}$ Wb ; $A = 10$ cm² $= 10^{-3}$ m²; $B = 8 \times 10^{-4}/10^{-3} = 0.8$ Wb/m²

Air gap

$H = B/\mu_0 = 0.8/4\pi \times 10^{-7} = 6.366 \times 10^5$ AT/m

Total air-gap length $= 2 \times 0.2 = 0.4$ mm

$= 4 \times 10^{-4}$ m

$\therefore$ AT required $= H \times l = 6.366 \times 10^5 \times 4 \times 10^{-4} = 255$

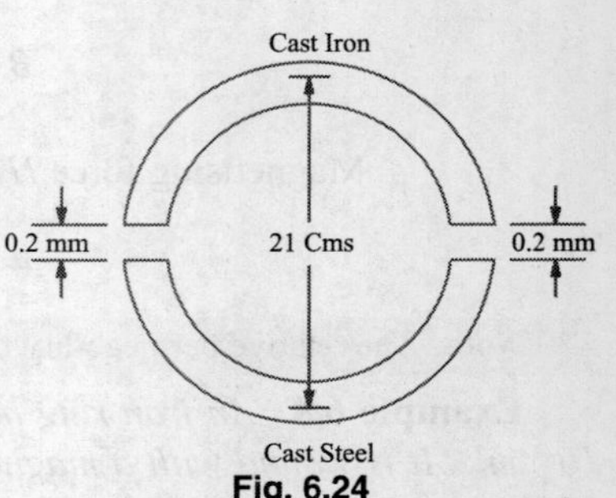

Fig. 6.24

Cast Steel Path (Fig. 6.24)

$H = B/\mu_0 \mu_r = 0.8/4\pi \times 10^{-7} \times 800 = 796$ AT/m

path $= \pi D/2 = 21\ \pi/2 = 33$ cm $= 0.33$ m

AT required $= H \times l = 796 \times 0.33 = 263$

Cast Iron Path

$H = 0.8/\pi \times 10^{-7} \times 166 = 3{,}835$ AT/m ; path $= 0.33$ m

AT required $= 3{,}835 \times 0.33 = 1265$

Total AT required $= 255 + 263 + 1265 = \mathbf{1783\ AT}$

Example 6.11. *A mild steel ring of 30 cm mean circumference has a cross-sectional area of 6 cm² and has a winding of 500 turns on it. The ring is cut through at a point so as to provide an air-gap of 1 mm in the magnetic circuit. It is found that a current of 4 A in the winding, produces a flux density of 1 T in the air-gap. Find **(i)** the relative permeability of the mild steel and **(ii)** inductance of the winding.* **(F.E. Engg. Pune Univ.)**

Solution. (*a*) Steel ring

$H \quad = B/\mu_0\,\mu_r = 1/4\pi \times 10^{-7} \times \mu_r$ AT/m $= 0.7957 \times 10^7/\mu_r$ AT/m

m.m.f. $\quad = H \times l = (0.7957 \times 10^7/\mu_r) \times 29.9 \times 10^{-2} = 0.2379 \times 10^6/\mu_r$ AT

(*b*) Air-gap

$H \quad = B/\mu_0 = 1/4\pi \times 10^{-7} = 0.7957 \times 10^6$ AT/m

m.m.f. reqd. $\quad = H \times l = 0.7957 \times 10^6 \times (1 \times 10^{-3}) = 795.7$ AT

Total m.m.f. $\quad = (0.2379 \times 10^6/\mu_r) + 795.7$

Total m.m.f. available $\quad = NI = 500 \times 4 = 2000$ AT

(*i*) $\therefore \quad 2000 = (0.2379 \times 10^6/\mu_r) + 795.7 \quad \therefore \quad \mu_r =$ **197.5**

(*ii*) Inductance of the winding $= \dfrac{N\Phi}{I} = \dfrac{NBA}{I} = \dfrac{500 \times 1 \times 6 \times 10^{-4}}{4} =$ **0.075 H**

Example 6.12. *An iron ring has a X-section of 3 cm² and a mean diameter of 25 cm. An air-gap of 0.4 mm has been cut across the section of the ring. The ring is wound with a coil of 200 turns through which a current of 2 A is passed. If the total magnetic flux is 0.24 mWb, find the relative permeability of iron, assuming no magnetic leakage.*

(Elect. Engg. A.M.Ae.S.I., June 1992)

Solution. $\Phi = 0.24$ mWb; $A = 3\text{ cm}^2 = 3 \times 10^{-4}\text{ m}^2$;

$B = \Phi/A = 0.24 \times 10^{-3}/3 \times 10^{-4} = 0.8\text{ Wb/m}^2$

AT for iron ring $= H \times l = (B/\mu_0\,\mu_r) \times l = (0.8/4\pi \times 10^{-7} \times \mu_r) \times 0.25 = 1.59 \times 10^{-5}/\mu_r$

AT for air-gap $= H \times l = (B/\mu_0) \times l = (0.8/4\pi \times 10^{-7}) \times 0.4 \times 10^{-3} = 255$

Total AT reqd. $= (1.59 \times 10^5/\mu_r) + 255$; total AT provided $= 200 \times 2 = 400$

$\therefore (1.59 \times 10^5/\mu_r) + 255 = 400 \quad$ or $\quad \mu_r =$ **1096.**

Example 6.13. *A rectangular iron core is shown in Fig. 6.25. It has a mean length of magnetic path of 100 cm, cross-section of (2 cm × 2 cm), relative permeability of 1400 and an air-gap of 5 mm cut in the core. The three coils carried by the core have number of turns N_a = 335, N_b = 600 and N_c = 600 ; and the respective currents are 1.6 A, 4 A and 3 A. The directions of the currents are as shown. Find the flux in the air-gap.*

(F.Y. Engg. Pune Univ.)

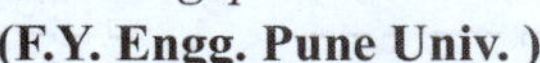

Fig. 6.25

Solution. By applying the Right-Hand Thumb rule, it is found that fluxes produced by the current I_a and I_b are directed in the clockwise direction through the iron core whereas that produced by current I_c is directed in the anticlockwise direction through the core.

$\therefore$ Total m.m.f. $= N_a I_a + N_b I_b - N_c I_c = 335 \times 1.6 + 600 \times 4 - 600 \times 3 = 1136$ AT

Reluctance of the air-gap $= \dfrac{l}{\mu_0\, A} = \dfrac{5 \times 10^3}{4\pi \times 10^{-7} \times 4 \times 10^{-4}} = 9.946 \times 10^6$ AT/Wb

Reluctance of the iron path $= \dfrac{l}{\mu_0\,\mu_r\, A} = \dfrac{100 - (0.5) \times 10^{-2}}{4\pi \times 10^{-7} \times 1400 \times 4 \times 10^{-4}} = 1.414 \times 10^6$ AT/Wb

Total reluctance = (9.946 + 1.414) × 10^6 = 11.36 × 10^6 AT/Wb

The flux in the air-gap is the same as in the iron core.

$$\text{Air-gap flux} = \frac{\text{m.m.f.}}{\text{reluctance}} = \frac{1136}{11.36 \times 10^6} = 100 \times 10^{-6}\ \text{Wb} = \mathbf{100\ \mu Wb}$$

Example 6.14. *A series magnetic circuit comprises of three sections* **(i)** *length of 80 mm with cross-sectional area 60 mm², **(ii)** length of 70 mm with cross-sectional area 80 mm² and **(iii)** and air-gap of length 0.5 mm with cross-sectional area of 60 mm². Sections **(i)** and **(ii)** are if a material having magnetic characteristics given by the following table.*

H (AT/m)	*100*	*210*	*340*	*500*	*800*	*1500*
B (Tesla)	*0.2*	*0.4*	*0.6*	*0.8*	*1.0*	*1.2*

Determine the current necessary in a coil of 4000 turns wound on section (ii) to produce a flux density of 0.7 Tesla in the air-gap. Neglect magnetic leakage. **(F.E. Pune Univ. May 1990)**

Solution. Section (*i*) It has the same cross-sectional area as the air-gap. Hence, it has the same flux density *i.e.* 0.7 Tsela as in the air-gap. The value of the magnetising force *H* corresponding to this flux density of 0.7 *T* as read from the *B-H* plot is 415 AT/m.

m.m.f. reqd = $H \times l$ = 415 × (80 × 10^{-3}) = 33.2 AT

Section (*ii*) Since its cross-sectional area is different from that of the air-gap, its flux density would also be different even though, being a series circuit, its flux would be the same.

Air-gap flux = $B \times L$ = 0 × (60 × 10^{-6}) = 42 × 10^{-6} Wb

Flux density in this section = 42 × 10^{-6}/80 × 10^{-6} = 0.525 T

The corresponding value of the *H* from the given garph is 285 AT/m

m.m.f. reqd, for this section = 285 × (70 × 10^{-3}) = 19.95 AT.

Air-gap

$H = B/\mu_0$ = 0.7/4π × 10^{-7} = 0.557 × 10^6 AT/m

∴ m.m.f. reqd. = 0.557 × 10^{-6} × (0.5 × 10^{-3}) = 278.5 AT

Total m.m.f. reqd. = 33.2 + 19.95 + 278.5 = 331.6

∴ NI = 331.6 or I = 331.6/4000 = **0.083 A**

Example 6.15. *A magnetic circuit made of mild steel is arranged as shown in Fig. 6.26. The central limb is wound with 500 turns and has a cross-sectional area of 800 mm². Each of the outer limbs has a cross-sectional area of 500 mm². The air-gap has a length of 1 mm. Calculate the current rquired to set up a flux of 1.3 mWb in the central limb assuming no magnetic leakage and fringing. Mild steel required 3800 AT/m to produce flux density of 1.625 T and 850 AT/m to produce flux density of 1.3 T.* **(F.Y. Engg. Pune Univ.)**

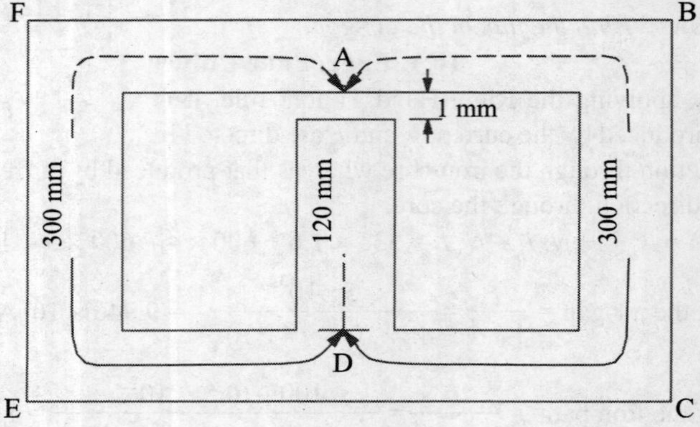

Fig. 6.26

Solution. Flux in the central limb is = 1.3 mWb = 1.3×10^{-3} Wb

Cross section $A = 800 \text{ mm}^2 = 800 \times 10^{-6} \text{ m}^2$

$\therefore \quad B = \Phi/A = 1.3 \times 10^{-6}/800 \times 10^{-6}$

$= 1.625$ T

Corresponding value of H for this flux density is given as 3800 AT/m.

Since the length of the central limb is 120 mm. m.m.f. required is

$= H \times l = 3800 \times (120 \times 10^{-3})$

$= 456$ AT/m.

Air-gap

Flux density in the air-gap is the same as that in the central limb.

H = B/μ0 = 1.625/4π × 10–7 = 0.1293 × 10–7 AT/m

Length of the air-gap = 1 mm = 10–3 m

m.m.f. reqd. for the air-gap = H × l = 0.1293 × 107 × 10–3 = 1293 AT.

The flux of the central limb divides equally at point A in figure along the two parallel path ABCD and AFED. We may consider either path, say ABCD and calculate the m.m.f. required for it. The same m.m.f. will also send the flux through the other parallel path *AFED*.

Flux through ABCD = 1.3 × 10–3/2 = 0.65 × 10–3 Wb

Flux density B = 0.65 × 10–3/500 × 10–6 = 1.3 T

The corresponding value of H for this value of B is given at 850 AT/m.

∴ m.m.f. reqd. for path ABCD = H × l = 850 × (300 × 10–3) = 255 AT

As said above, this, m.m.f. will also send the flux in the parallel path AFED.

Total m.m.f. reqd. = 456 + 1293 + 255 = 2004 AT

Since the number of turns is 500, I = 2004/500 = **4A.**

Example 6.16. *A cast steel d.c. electromagnet shown in Fig. 6.27 has a coil of 1000 turns on its central limb. Determine the current that the coil should carry to produce a flux of 2.5 mWb in the air-gap. Neglect leakage. Dimensions are given in cm. The magnetisation curve for cast steel is as under:*

Flux density (Wb/m^2):	*0.2*	*0.5*	*0.7*	*1.0*	*1.2*
Amp-turns/metre:	*300*	*540*	*650*	*900*	*1150*

(Electrotechnics-I, ; M.S. Univ. Baroda)

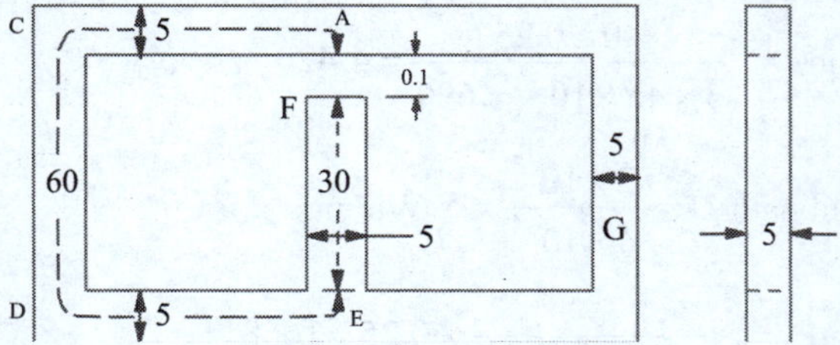

Fig. 6.27

Solution. Two points should be noted

(*i*) there are two (equal) parallel paths *ACDE* and *AGE* across the central path *AE*.

(*ii*) flux density in either paralle path is half of that in the central path because flux divides into two equal parts at point *A*.

Total m.m.f. required for the whole electromagnet is equal to the sum of the following three m.m.fs.

(*i*) that required for path *EF*
(*ii*) that required for air-gap
(*iii*) that required for either of the two parallel paths ; say, path $ACDE_2$

Flux density in the central limb and air gap is

$$= 2.5 \times 10^{-3}/(5 \times 5) \times 10^{-4} = 1 \text{ Wb/m}^2$$

Corresponding value of *H* as found from the given data is 900 AT/m.

$\therefore$ AT for central limb $= 900 \times 0.3 = 270$

H in air-gap $= B/\mu_0 = 1/4\pi \times 10^{-7} = 79.56 \times 10^4$ AT/m

AT required $= 79.56 \times 10^4 \times 10^{-3} = 795.6$

Flux density in path *ACDE* is 0.5 Wb/m² for which corresponding value of *H* is 540 AT/m.

$\therefore$ AT required for path $ACDE = 540 \times 0.6 = 324$

Total AT required = 270 + 795.6 + 324 = 1390 ;Current required = 1390/1000 = **1.39 A**

Example 6.17. *A cast steel magnetic structure made for a bar of section 8 cm × 2 cm is shown in Fig. 6.28. Determine the current that the 500 turn-magnetising coil on the left limb should carry so that a flux of 2 mWb is produced in the right limb. Take $\mu_r = 600$ and neglect leakage.* **(Elect. Technology Allahabad Univ. 1993)**

Solution. Since path *C* and *D* are in parallel with each other w.r.t. path *E* (Fig. 6.38), the m.m.f. across the two is the same.

$$\Phi_1 S_1 = \Phi_2 S_2$$

$$\therefore \quad \Phi_1 \times \frac{15}{\mu A} = 2 \times \frac{25}{\mu A}$$

$$\therefore \quad \Phi_1 = 10/3 \text{ mWb}$$

$$\therefore \quad \Phi = \Phi_1 + \Phi_2 = 16/3 \text{ mWb}$$

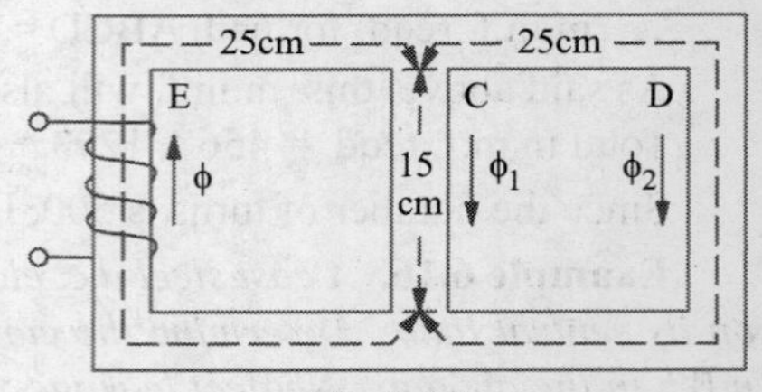

Fig. 6.28

Total AT required for the whole circuit is equal to the sum of

(*i*) that required for path *E* and (*ii*) that required for either of the two paths *C* or *D*.

$$\text{Flux density in path } E = \frac{16 \times 10^{-3}}{3 \times 4 \times 10^{-4}} = \frac{40}{3} \text{ Wb/m}^2$$

$$\text{AT reqd.} = \frac{40 \times 0.25}{3 \times 4\pi \times 10^{-7} \times 600} = 4,420$$

$$\text{Flux density in path } D = \frac{2 \times 10^{-3}}{4 \times 10^{-4}} = 5 \text{ Wb/m}^2$$

$$\text{AT reqd.} = \frac{5}{4\pi \times 10^{-7} \times 600} \times 0.25 = 1658$$

Total AT = 4,420 + 1,658 = 6,078 ;

Current needed = 6078/500 = **12.16 A**

Example 6.18. *Show how the ampere-turns per pole required to produce a given flux in d.c. generator are calculated.*

Find the amp-turns per pole required to produce a flux of 40 mWb per pole in a machine with a smooth core armature and having the following dimensions:

Length of air gap = 5 mm *Area of air-gap = 500 cm²*
Length of pole = 12 cm *Sectional area of pole core = 325 cm²*
Relative permeability of pole core = 1,500
Length of magnetic path in yoke betwen pole = 65 cm
Cross-sectional area of yoke = 450 cm² ; Relative permeability of yoke = 1,200
Leakage coefficient = 1.2
The ampere-turns for the armature core may be neglected.

Solution. Air-gap $\Phi = 40$ mWb $= 4 \times 10^{-2}$ Wb ; $A = 500 \times 10^{-4} = 5 \times 10^{-3}$ m²

$\therefore \quad B = 4 \times 10^{-2}/5 \times 10^{-2} = 0.8$ Wb/m² ; $H = B/\mu_0 = 0.8/4\pi \times 10^{-7} = 63.63 \times 10^{-4}$ AT/m

Air-gap length $= 5 \times 10^{-3}$ m ; AT reqd. $= 63.63 \times 10^4 \times 5 \times 10^{-3} = 3181.5$

Pole Core

$$\Phi = 1.2 \times 4 \times 10^{-2} = 4.8 \times 10^{-2} \text{ Wb}; \quad A = 325 \times 10^{-4} \text{ m}^2$$

$$B = 4.8 \times 10^{-2}/325 \times 10^{-4} = 1.477 \text{ Wb/m}^2$$

$$H = B/\mu_0 \mu_r = 1.477/4\pi \times 10^{-7} \times 1{,}500 = 783 \text{ AT/m}$$

$$\text{Pole length} = 0.12 \text{ m}; \text{ AT reqd.} = 783 \times 0.12 = 94$$

Yoke Path

$$\text{flux} = \text{half the pole flux} = 0.5 \times 4 \times 10^{-2} = 2 \times 10^{-2} \text{ Wb}$$

$$A = 450 \text{ cm}^2 = 45 \times 10^{-3} \text{ m}^2; B = 2 \times 10^{-2}/45 \times 10^{-3} = 4/9 \text{ Wb/m}^2$$

$$H = \frac{4/9}{4\pi \times 10^{-7} \times 1{,}200} = 294.5 \text{ AT/m Yoke length} = 0.65 \text{ m}$$

$$\text{At reqd} = 294.5 \times 0.65, \text{ Total AT/pole} = 3181.5 + 94 + 191.4 = \mathbf{3{,}467}$$

Example 6.19. *A shunt field coil is required to develop 1,500 AT with an applied voltage of 60 V. The rectangular coil is having a mean length of turn of 50 cm. Calculate the wire size. Resistivity of copper may be assumed to be 2 × 10⁻⁶ μΩ-cm at the operating temperature of the coil. Estimate also the number of turns if the coil is to be worked at a current density of 3 A/mm².*

(Basis Elect. Machines Nagpur Univ. 1992)

Solution. $NI = 1{,}500 \text{ (given) or } N.\dfrac{V}{R} = N.\dfrac{60}{R} = 1{,}500$

$\therefore \quad R = \dfrac{N}{25}$ ohm $\qquad$ Also $R = \rho.\dfrac{l}{A} = \dfrac{2 \times 10^{-6} \times 50\, N}{A}$

$\therefore \quad \dfrac{N}{25} = \dfrac{10^{-4} n}{A}$ $\qquad$ or $A = 25 \times 10^{-4}$ cm² or $A = 0.25$ mm²

$\therefore \quad \dfrac{\pi D^2}{4} = 0/25$ $\qquad$ or $D = \mathbf{0.568}$ **mm**

Current in the coil $= 3 \times 0.25 = 0.75$ A

Now, $NI = 1{,}500$; $\qquad \therefore \quad N = 1{,}500/0.75 = \mathbf{2{,}000}$

Example 6.20. *A wooden ring has a circular cross-section of 300 sq. mm and a mean diameter of the ring is 200 mm. It is uniformly wound with 800 turns.*

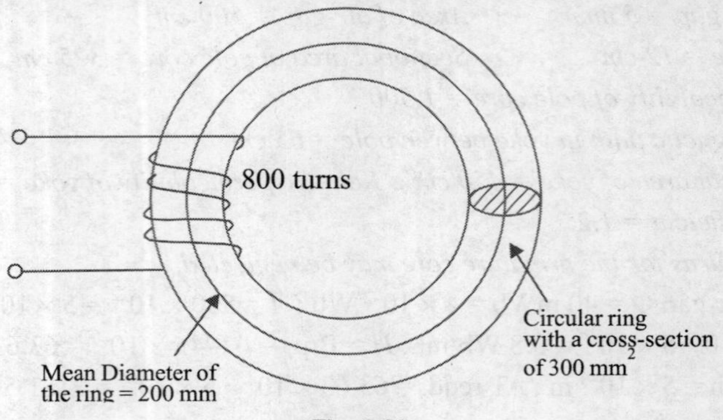

Fig. 6.29

Calculate:

(i) *the field strength produced in the coil by a current of 2 amperes: (assume = 1)*

(ii) *the magnetic flux density produced by this current and*

(iii) *the current required to produce a flux density of 0.02 wb/m².*

[Nagpur University (Summer 2000)]

Solution. The question assumes that the flux-path is through the ring, as shown by the dashed line, in figure, at the mean diameter, in Fig. 6.29.

With a current of 2 amp,

$$\text{Coil m.m.f.} = 800 \times 2 = 1600 \text{ AT}$$

$$\text{Mean length of path} = \pi \times 0.2 = 0.628 \text{ m}$$

(i) $$H = \frac{1600}{0.628} = 2548 \text{ amp-turns/meter}$$

(ii) $$B = \mu_0 \mu_r H = 4\pi \times 10^{-7} \times 1 \times 2548 = 3.20 \times 10^{-3} \text{ Wb/m}^2$$

This Flux density is produced by a coil current of 2-amp

(iii) For producing a flux of 0.02 Wb/m², the coil current required is

$$2 \times \frac{0.02}{0.0032} = \mathbf{12.5 \text{ amp}}$$

Example 6.21. *A magnetic core in the form of a closed circular ring has a mean length of 30 cm and a cross-sectional area of 1 cm². The relative permeability of iron is 2400. What direct-current will be needed in the coil of 2000 turns uniformly wound around the ring to create a flux of 0.20 mWb in iorn ? If an air-gap of 1 mm is cut through the core perpendicualr to the direction of this flux, what current will now be needed to maintain the same flux in the air gap?*

[Nagpur University Nov. 1999]

Solution.

$$\text{Reluctance of core} = \frac{1}{\mu_0 \mu_r} \frac{L}{a} = \frac{1}{10\pi \times 10^{-7} \times 2400} \times \frac{30 \times 10^2}{1 \times 10^{-4}}$$

$$= \frac{30 \times 10^{-9}}{4\pi \times 2400 \times 1} = 995223 \text{ MKS units}$$

$$\phi = 0.2 \times 10^{-3} \text{ Wb}$$

MMF required = $\phi \times$ Rel

= $0.2 \times 10^{-3} \times 995223 = 199$ amp-tunrs

Direct current required through the 2000 turn coil

$$= \frac{199}{2000} = 0.0995 \text{ amp}$$

Reluctance of 1 mm air gap

$$= \frac{1}{4\pi \times 10^{-7}} \times \frac{1 \times 10^{-3}}{1 \times 10^{-4}} = \frac{10^8}{4\pi} = 7961783 \text{ MKS units}$$

Addition of two reluctances

= 995223 + 7961783 = 8957006 MKS units

MMF required to establish the given flux

= $0.2 \times 10^{-3} \times 8957006 =$ **1791 amp turns**

Current required through the coil

$$= \frac{1791}{2000} = \mathbf{0.8955 \text{ amp}}$$

Note: Due to the high permeability of iron, which is given here as 2400, 1 mm of air-gap length is equivalent magnetically to 2400 mm of length through the core, for comparison of mmf required.

Example 6.22. *A steel-ring of 25 cm mean diameter and of circular section 3 cm in diameter has an air-gap of 1.5 mm length. It is wound uniformly with 700 turns of wire carrying a current of 2 amp. Calculate:* ***(i)*** *Magneto motive force* ***(ii)*** *Flux density* ***(iii)*** *Magnetic flux* ***(iv)*** *Relative permeability. Neglect magnetic leakage and assume that iron path takes 35 % of total magneto motive force.* **[Nagpur University, April 1996]**

Solution. From the given data, length of mean path in the ring ($= L_m$) is to be calculated. For a mean diameter of 25 cm, with 1.5 mm of air-gap length.

$$L_m = (\pi \times 0.25) - (1.5 \times 10^{-3}) = 0.7835 \text{ m}$$

Cross-sectional area of a 3 cm diameter ring = 7.065×10^{-4} sq.m.

Total m.m.f. due to coil = $700 \times 2 = 1400$ amp-turns

Since iron-path takes 355 of the total mmf, it is 490.

Remaining mmf of 910 is consumed by the air-gap.

Corresponding H for air-gap = $910/(1.5 \times 10^{-3}) = 606666$ amp-turns/m.

If Flux density is B_g, we have

$$B_g = \mu_0 H_g = 4\pi \times 10^{-7} \times 606666 = 0.762 \text{ Wb/m}^2$$

Iron-ring and air-gap are in series hence their flux is same. Since the two have some cross-sectional area, the flux density is also same. The ring has a mean length of 0.7835 m and needs an mmf of 490. For the ring.

$$H = 490/0.7845 = 625.4 \text{ amp-turns / m}$$

$$\mu_0 \mu_r = B/H = 0.752/625.4 = 1.218 \times 10^{-3}$$

$$\mu_r = (1.218 \times 10^{-3}) / (4\pi \times 10^{-7}) = 970$$

Flux = Flux density × Cross-sectional area = $0.762 \times 7.065 \times 10^{-4} = 0.538$ milli-webers

Check. μ_r of 970 means that 1.5 mm of air-gap length is equivalent to $(1.5 \times 10^{-3} \times 970)$ = 1.455 m of length through iron as a medium. With this equivalent.

$$\frac{\text{mmf of ring}}{\text{mmf for (ring + air-gap)}} = \frac{0.785}{0.785 + 1.455} = 0.35$$

which means that 35 % of total mmf is required for the ring

Example 23. *For the magnetic circuit shown in Fig. 6.30 the flux in the right limb is 0.24 mWb and the number of turns wound on the central-limb is 1000. Calculate* ***(i)*** *flux in the central limb* ***(ii)*** *the current required.*

The magnetization curve for the core is given as below:

H (AT/m):	*200*	*400*	*500*	*600*	*800*	*1060*	*1400*
B (Nb/m$_2$):	*0.4*	*0.8*	*1.0*	*1.1*	*1.2*	*1.3*	*1.4*

Neglect Leakage and fringing.

[Rajiv Gandhi Technical University, Bhopal, Summer 2001]

Solution. Area of cross-section of side-limbs = 2 × 3 = 6 sq.cm

Area of cross-section of core = 3 × 4 = 12 sq.cm

Flux in side Limbs = 0.24 mWb

Flux density in side Limbs = $(0.24 \times 10^{-3})/(6 \times 10^{-4}) = 0.4$ Wb/m²

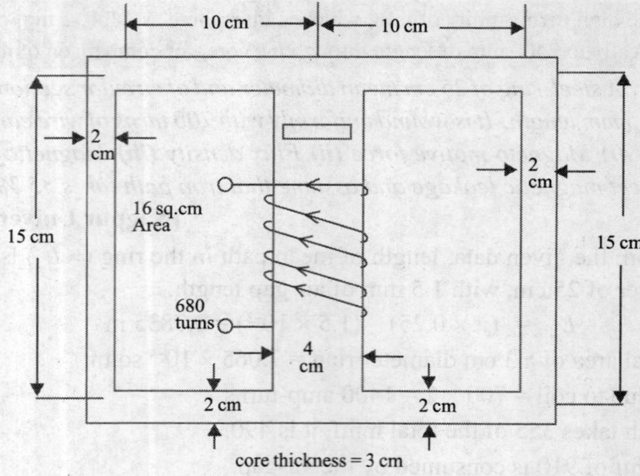

Fig. 6.30

Since the coil is wound on the central limb and the magnetic circuit is symmetrical, the flux in the central limb = 0.48 mWb. Flux density in the central limb = $(0.48 \times 10^{-3})/(12 \times 10^{-4}) = 0.4$ Wb/m²

For the flux density of 0.40 Wb/m², H = 200 AT/m

Central Limb has a path length of 15 cm.

Other part carrying 0.24 mWb has a total path length of 35 cm.

Total mmf required = (200 × 0.15) + (200 × 0.35) = 100 AT

Hence, coil current = 100/1000 = 0.1 Amp.

Example 6.24. *A ring composed of three sections. The cross-sectional area is 0.001 m² for each section. The mean arc length are l_a = 0.3 m, l_b = 0.2 m, l_c = 0.1 m. An air-gap length of 0.1 mm is cut in the ring. Mr for sections a, b, c are 5000, 1000, and 10,000 respectively. Flux in the air gap is 7.5×10^{-4} Wb. Find* ***(i)*** *mmf* ***(ii)*** *exciting current if the coil has 100 turns,* ***(iii)*** *reluctances of the sections.* **[Nagpur University April 1999]**

Solution. Area = 0.001 sq.m

$l_a = 0.3$ m, $l_b = 0.2$ m, $l_c = 0.1$ m, $l_g = 0.1 \times 10^{-3}$ m

$\mu_{ra} = 5000$, $\mu_{rb} = 1000$, $\mu_{rc} = 10{,}000$ $\mu_o = 4\pi \times 10^{-7}$

$\phi = 7.5 \times 10^{-4}$ Wb

(iii) Calculations of Reluctances of four parts of the magnetic circuit:

(a) Reluctance of air gap, $S_{ag} = \frac{1}{\mu_o} \times \frac{0.1 \times 10^{-3}}{0.001} = \frac{1000}{4\pi \times 0.001} = 79618$ AT/wb

(b) Reluctance of section '*a*' of ring

$$= S_a = \frac{1}{\mu_o \mu_{ra}} \times \frac{0.3}{0.001} = \frac{10^7 \times 0.3}{4\pi \times 47770 \times 5000 \times 0.001} = 47770 \text{ AT/wb}$$

(c) Reluctance of section '*b*' of the ring

$$= S_b = \frac{1}{\mu_o \mu_{rb}} \times \frac{0.20}{0.001} = \frac{10^7}{4\pi \times 1000} \times \frac{0.10}{0.001} = 15923.6 \text{ AT/wb}$$

(d) Reluctance of section '*c*' of the ring

$$= S_c = \frac{1}{\mu_o \mu_{rc}} \times \frac{0.10}{0.001} = \frac{10^7}{4\pi \times 1000} \times \frac{0.10}{0.001} = 7961 \text{ AT/wb}$$

Total Reluctance = $S_{ag} + S_a + S_b + S_c = 294585$ AT/wb

(i) Total mmf required = Flux × Reluctance

= $7.5 \times 10^{-4} \times 294585 = 221$ amp-turns

(ii) Current required = 221/100 = 2.21 amp

Tutorial Problem 6.2

1. An iron specimen in the form of a closed ring has a 350-turn magnetizing winding through which is passed a current of 4A. The mean length of the magnetic path is 75 cm and its cross-sectional area is 1.5 cm². Wound closely over the specimen is a secondary winding of 50 turns. This is connected to a ballistic glavanometer in series with the secondary coil of 9-mH mutual inductance and a limiting resistor. When the magnetising current is suddenly reversed, the galvanometer deflection is equal to that produced by the reversal of a current of 1.2 A in the primary coil of the mutual inductance. Calculate the *B* and *H* values for the iron under these conditions, deriving any formula used. **[1.44 Wb/m² ; 1865 AT/m]** ***(London Univ.)***
2. A moving-coil ballistic galvanometer of 150 Ω gives a throw of 75 divisions when the flux through a search coil, to which it is connected, is reversed.

   Find the flux density in which the reversal of the coil takes place, given that the galvanometer constant is 110 μ*C* per scale division and the search coil has 1400 turns, a mean are of 50 cm² and a resistance of 20 Ω. **[0.1 Wb/m²]** ***(Elect. Meas. & Measuring Inst. Gujarat Univ.)***
3. A fluxmeter is connected to a search coil having 500 turns and mean area of 5 cm². The search coil is placed at the centre of a solenoid one metre long wound with 800 turns. When a current of 5 A is reversed, there is a deflection of 25 scale divisions on the fluxmeter. Calculate the flux-meter constant.

   [10⁻⁴ Wb-turn/division] ***(Elect. Meas. & Measuring Inst., M.S. Univ. Baroda)***
4. An iron ring of mean length 50 cms has an air gap of 1 mm and a winding of 200 turns. If the permeability of iron is 300 when a current of 1 A flows through the coil, find the flux density.

   [94.2 mWb/m³] ***(Elect. Engg. A.M.Ae.S.I.)***

5. An iron ring of mean length 100 cm with an air gap of 2 mm has a winding of 500 turns. The relative permeability of iron is 600. When a current of 3 A flows in the winding, determine the flux density. Neglect fringing.

[0.523 Wb/m²] ***(Elect. Engg. & Electronic Bangalore Univ. 1990)***

6. A coil is wound uniformly with 300 turns over a steel ring of relative permeability 900, having a mean circumference of 40 mm and cross-sectional area of 50 mm². If a current of 25 amps is passed through the coil, find (*i*) m.m.f. (*ii*) reluctance of the ring and (*iii*) flux.

[(*i*) 7500 AT (*ii*) 0.7×10^6 AT/Wb (*iii*) 10.7 mWb]

(Elect. Engg. & Electronics Bangalore Univ.)

7. The length of the magnetic circuit of a relay is 25 cm and the cross-sectional area is 6.25 cm². The length of the air-gap in the operated position of the relay is 0.2 mm. Calculate the magnetomotive force required to produce a flux of 1.25 mWb in the air gap. The relative permeability of magnetic material at this flux density is 200. Calculate also the reluctance of the magnetic circuit when the relay is in the unoperated position, the air-gap then being 8 mm long (assume μ_r, remains const ant). **[2307 AT, 1.18×10^7 AT/Wb]**

8. In the magnetic circuit shown in Fig. 6.46 a coil of 500 turns is wound on the centre limb. The magnetic paths *A* to *B* by way of the outer limbs have a mean length of 100 cm each and an effective cross-sectional area of 2.5 cm². The centre limb is 25 cm long and 5cm² cross-sectional area. The air-gap is 0.8 cm long. A current of 9.2 A through the coil is found to produce a flux of 0.3 mWb.

9. The magnetic circuit of a choke is shown in Fig. 6.47. It is designed so that the flux in the central core is 0.003 Wb. The cross-section is square and a coil of 500 turns is wound on the central core. Calculate the exciting current. Neglect leakage and assume the flux to be uniformly distributed along the mean path shown dotted. Dimensions are in cm.

The characteristics of magnetic circuit are as given below:

B (Wb/m²):	0.38	0.67	1.07	1.2	1.26
H (AT/m):	100	200	600	1000	1400

(Elect. Technology I. Gwalior Univ.)

10. A magnetic circuit with a uniform cross-sectional area of 6 cm² consists of a steel ring with a mean magnetic length of 80 cm and an air gap of 2 mm. The magnetising winding has 540 ampere-turns. Estimate the magnetic flux produced in the gap. The relevant points on the magnetization curve of cast steel are:

B (Wb/m²):	0.12	0.14	0.16	0.18	0.20
H (AT/m):	200	230	260	290	320

[0.1128 m Wb] ***(City & Guilds, London)***

11. Explain the terms related to magnetic circuits:
(*i*) reluctance (*ii*) flux density (*iii*) magnetomotive force

(Nagpur University, Summer 2002)

12. A metal ring of mean diameter of 80 cm is made out of two semi-circular pieces of cast iron and cast steel separated at junctions by pieces of copper each of 1 mm thickness. If the ring is uniformly wound with 1000 turns, calculate the value of current required to produce a flux density of 0.85 wb/nV in the ring.
Given that relative permeability of cast iron as 200, that of cast steel is 1200 and for copper, $\mu_r = 1$. ***(Nagpur University, Summer 2002)***

13. A 1154 turns coil is wound on the central limb of the case steel frame shown in Fig. 6.50. A total flux of 1.6 mwb is required in the air gap. Find the current required through the gap. Assume that the gap density is uniform and there is no leakage. Frame dimensions are given in cm. Take permeability of cast steel as 1,200. ***(Nagpur University, Winter 2002)***

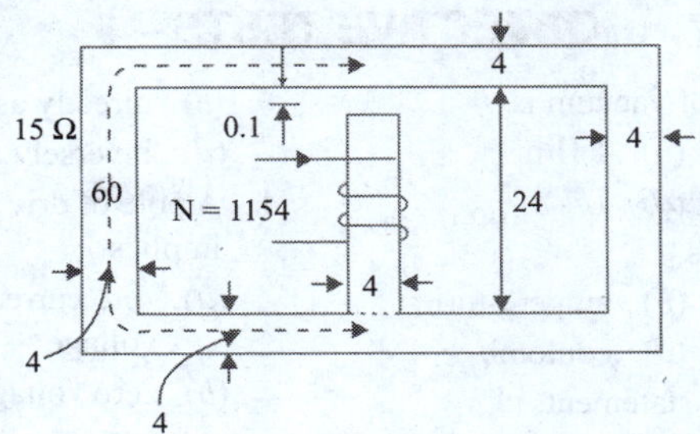

Fig. 6.31

14. Explain the terms related to magnetic circuits:
(*i*) Reluctance (*ii*) Flux density (*iii*) Coercive force (*iv*) Magnetomotive force (*v*) Residual flux. ***(Nagpur University, Summer 2003)***

15. Compare electric and magnetic circuit by their similarities and dissimilarities. ***(Nagpur University, Winter 2003)***

16. Compare electric and magnetic circuits with respect to their similarities and dissimilarities. ***(Nagpur University, Summer 2004)***

17. A steel wire of 25 cm mean diameter and circular cross section 3 cm in diameter has an airgap of 1 mm wide. It is wound with a coil of 700 turns carrying a current of 2 A.
Calculate: (*i*) m.m.f. (*ii*) Flux density (*iii*) Reluctance (*iv*) Relative permeability.
Assume that iron path take 30% of total m.m.f. ***(Gujrat University,Summer 2003)***

18. What is a search coil in magnetic measurements? ***(Anna University, April 2002)***

19. Name the magnestic squares used to find iron loss. ***(Anna University, April 2002)***

20. What is a magnetic circuit? A magnetic circuit is made up of 3 limbs A, B and C in prallel. The reluctances of the magnetic paths of A, B and C in AT/mWb are 312, 632.6 and 520 respectively. An exciting coil of 680 turns is wound on limb B. Find the exciting current to produce of flux of lmwb in the limb A.
(V.T.U., Belgaum Karnataka University, February 2002)

21. An iron ring of 300 cm mean circumference with a cross section of $5cm^2$ is wound uniformly with 350 turns of wire. Find the current required to produce a flux of 0.5 Mwb in iron. Take relative permeability of iron as 400.
(V.T.U. Belgaum Karnataka University, July/August 2002)

22. What is Biot-Savart law? Explain briefly. Find the magnetic field due to a small circular loop carrying current I at distances from loop that are large compared with its dimensions.
(Agra Univ. 1978 Supp.)

23. Magnetic potential ***(Mumbai University, 2002) (RGPV, Bhopal 2001)***

24. Flux density ***(Pune University,2002) (RGPV, Bhopal 2001)***

25. Susceptibility ***(Mumbai University, 2002) (RGPV, Bhopal 2001)***

26. Define mm *f*, flux, reluctance, absolute and relative permeabilities with reference to magnetic circuits. ***(U.P. Technical University 2003) (RGPV, Bhopal 2002)***

27. Discuss B-H curve of a ferro-magnetic material and explain the following.
(*i*) Magnetic saturation (*ii*) Hysteresis (*iii*) Residual magnetism (*iv*) Coercive force.
(RGPV, Bhopal 2002)

28. What is meant by leakage and fringing? Define leakage coefficient.
(RGPV, Bhopal 2002)

29. Define the following terms (any five):
(*i*) MMF (*ii*) Reluctance (*iii*) Permeance (*iv*) Magnetisation curve (*v*) flux density (*vi*) Magnetizing force (*vii*) Susceptibility (*viii*) Relative permeability (*ix*) Magnetic potential
(RGPV, Bhopal 2002)

30. Distinguish between leakage and fringing of flux. ***(RGPV, Bhopal 2002)***

31. Explain fringing of magnetic flux, magnetic leakage, staturation of ferowegnetic materials, B-H Curve, hysteresis and eddy current losses. ***(RGPV, Bhopal 2003)***

OBJECTIVE TESTS – 6

1. Relative permeability of vacuum is
   (*a*) $4\pi \times 10^{-7}$ H/m (*b*) 1 H/m
   (*c*) 1 (*d*) $1/4\ \pi$
2. Unit of magnetic flux is
   (*a*) weber (*b*) ampere-turn
   (*c*) tesla (*d*) coulomb
3. Point out the WRONG statement.
   The magnetising force at the centre of a circular coil varies.
   (*a*) directly as the number of its turns
   (*b*) directly as the current
   (*c*) directly as its radius
   (*d*) inversely as its radius
4. A pole of driving point admittance function implies
   (*a*) zero current for a finite value of driving voltage
   (*b*) zero voltage for a finite value of driving current
   (*c*) an open circuit condition
   (*d*) None of (*a*), (*b*) and (*c*) mentiboned in the question **(ESE 2001)**

ANSWERS

1. *c* **2.** *a* **3.** *a*

CHAPTER

7

Electromagnetic Induction

Chapter Outline

- Relation Between Magnetism and Electricity
- Production of Induced E.M.F. and Current
- Faraday's Laws of Electromagnetic Induction
- Direction of Induced E.M.F. and Current
- Lenz's Law
- Induced E.M.F.
- Dynamically-induced E.M.F.
- Statically-induced E.M.F.
- Self-Inductance
- Coefficient of Self-Inductance (L)
- Mutual Inductance
- Coefficient of Mutual Inductance (M)
- Coefficient of Coupling
- Inductances in Series
- Inductances in Parallel

7.1 Relation between Magnetism and Electricity

It is well known that whenever an electric current flows through a conductor, a magnetic field is immediately brought into existence in the space surrounding the conductor. It can be said that when electrons are in motion, they produce a magnetic field. The converse of this is also true *i.e.* when a magnetic field embracing a conductor moves *relative* to the conductor, it produces a flow of electrons in the conductor. This phenomenon whereby an e.m.f. and hence current (*i.e.* flow of electrons) is induced in any conductor which is cut across or is cut by a magnetic flux is known as ***electromagnetic induction***. The historical background of this phenomenon is this:

After the discovery **(by Oersted)** that electric current produces a magnetic field, scientists began to search for the converse phenomenon from about 1821 onwards. The problem they put to themselves was how to 'convert' magnetism into electricity. It is recorded that Michael Faraday* was in the habit of walking about with magnets in his pockets so as to constantly remind him of the problem. After nine years of continuous research and experimentation, he succeeded in producing electricity by 'converting magnetism'. In 1831, he formulated basic laws underlying the phenomenon of electromagnetic induction (known after his name), upon which is based the operation of most of the commercial apparatus like motors, generators and transformers etc.

7.2 Production of Induced E.M.F. and Current

In Fig. 7.1 is shown an insulated coil whose terminals are connected to a sensitive galvanometer *G*. It is placed close to a stationary bar magnet initially at position *AB* (shown dotted).

* Michael Faraday (1791-1867), an English physicist and chemist.

As seen, some flux from the *N*-pole of the magnet is linked with or threads through the coil but, as yet, there is no deflection of the galvanometer. Now, suppose that the magnet is ***suddenly*** brought closer to the coil in position *CD* (see figure). Then, it is found that there is a jerk or a

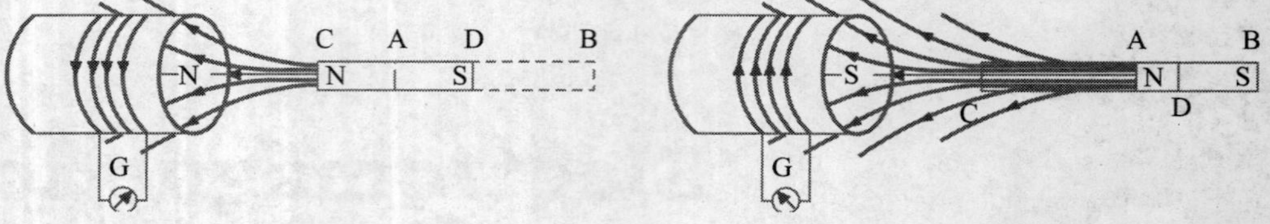

Fig. 7.1. **Fig. 7.2.**

sudden but a momentary deflection in the galvanometer and that this ***lasts so long as the magnet is in motion relative to the coil, not otherwise***. The deflection is reduced to zero when the magnet becomes again stationary at its new position *CD*. It should be noted that due to the approach of the magnet, flux linked with the coil is increased.

Next, the magnet is ***suddenly*** withdrawn away from the coil as in Fig. 7.2. It is found that again there is a ***momentary*** deflection in the galvanometer and it persists so long as the magnet is in motion, not when it becomes stationary. It is important to note that this deflection is in a direction opposite to that of Fig. 7.1. Obviously, due to the withdrawal of the magnet, flux linked with the coil is decreased.

The deflection of the galvanometer indicates the production of e.m.f. in the coil. The only cause of the production can be the sudden approach or withdrawal of the magnet from the coil. It is found that the actual cause of this e.m.f. is the change of flux linking with the coil. This e.m.f. exists so long as the change in flux exists. Stationary flux, however strong, will never induce any e.m.f. in a stationary conductor. In fact, the same results can be obtained by keeping the bar magnet stationary and moving the coil suddenly away or towards the magnet.

The direction of this electromagnetically-induced e.m.f. is as shown in the two figures given on back page.

The production of this electromagnetically-induced e.m.f. is further illustrated by considering a conductor *AB* lying within a magnetic field and connected to a galvanometer as shown in Fig. 7.3. It is found that whenever this conductor is moved up or down, *a* ***momentary*** deflection is produced in the galvanometer. It means that some transient e.m.f. is induced in *AB*. The magnitude of this induced e.m.f. (and hence the amount of deflection in the galvanometer) ***depends on the quickness of the movement of AB.***

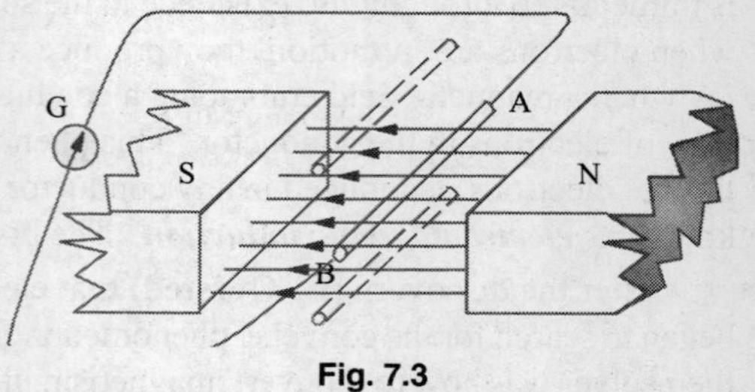

Fig. 7.3

From this experiment we conclude that whenever a conductor cuts or *shears* the magnetic flux, an e.m.f. is always induced in it.

It is also found that if the conductor is moved parallel to the direction of the flux so that it does not cut it, then no e.m.f. is induced in it.

7.3 Faraday's Laws of Electromagnetic Induction

Faraday summed up the above facts into two laws known as Faraday's Laws of Electromagnetic Induction.

First Law:

"Whenever the magnetic flux linked with a circuit changes, an e.m.f. is always induced in it."

or

"Whenever a conductor cuts magnetic flux, an e.m.f. is induced in that conductor."

Second Law:

"The magnitude of the induced e.m.f. is equal to the rate of change of flux-linkages."

Explanation: Suppose a coil has N turns and flux through it changes from an initial value of Φ_1 webers to the final value of Φ_2 webers in time t seconds. Then, remembering that by flux-linkages mean the product of number of turns and the flux linked with the coil, we have

Initial flux linkages = $N\Phi_1$, and Final flux linkages = $N\Phi_2$

$$\therefore \quad \text{Induced } e.m.f. \quad e = \frac{N\Phi_2 - N\Phi_1}{t} \text{ Wb/s or volt or } e = N\frac{\Phi_2 - \Phi_1}{t} \text{ volt}$$

Putting the above expression in its differential form, we get

$$e = \frac{d}{dt}(\text{N }\Phi) = N\frac{d\,\Phi}{dt} \text{ volt}$$

Usually, a minus sign is given to the right-hand side expression to signify the fact that the induced e.m.f. sets up current in such a direction that magnetic effect produced by it opposes the very cause producing it (Art. 7.5).

$$e = -N\frac{d\,\Phi}{dt} \text{ volt}$$

Example 7.1. *The field coils of a 6-pole d.c. generator each having 500 turns, are connected in series. When the field is excited, there is a magnetic flux of 0.02 Wb/pole. If the field circuit is opened in 0.02 second and residual magnetism is 0.002 Wb/pole, calculate the average voltage which is induced across the field terminals. In which direction is this voltage directed relative to the direction of the current.*

Solution. Total number of turns, $N = 6 \times 500 = 3000$

Total initial flux $= 6 \times 0.02 = 0.12$ Wb

Total residual flux $= 6 \times 0.002 = 0.012$ Wb

Change in flux, $d\Phi = 0.12 - 0.012 = 0.108$ Wb

Time of opening the circuit, $dt = 0.02$ second

$$\therefore \quad \text{Induced e.m.f.} = N\frac{d\,\Phi}{dt} \text{ volt} = 3000 \times \frac{0.108}{0.02} = \mathbf{16{,}200\ V}$$

The direction of this induced e.m.f. is the same as the initial direction of the exciting current.

Example 7.2. *A coil of resistance 100 Ω is placed in a magnetic field of 1 mWb. The coil has 100 turns and a galvanometer of 400 Ω resistance is connected in series with it. Find the average e.m.f. and the current if the coil is moved in 1/10th second from the given field to a field of 0.2 mWb.*

Solution. Induced e.m.f. $= N.\dfrac{d\,\Phi}{dt}$ volt

Here $d\,\Phi = 1 - 0.2 = 0.8 \text{ mWb} = 0.8 \times 10^{-3}$ Wb

$dt = 1/10 = 0.1$ second ; $N = 100$

$e = 100 \times 0.8 \times 10^{-3}/0.1 = \mathbf{0.8\ V}$

Total circuit resistance $= 100 + 400 = 500\ \Omega$

$\therefore$ Current induced $= 0.8/500 = 1.6 \times 10^{-3}$ A = **1.6 mA**

7.4 Direction of induced E.M.F. and Currents

There exists a definite relation between the direction of the induced current, the direction of the flux and the direction of motion of the conductor. The direction of the induced current may be found easily by applying either Fleming's Right-hand Rule or Flat-hand rule or Lenz's Law. Fleming's rule (Fig. 7.4) is used where induced e.m.f. is due to flux-cutting (*i.e.*, dynamically induced e.m.f.) and Lenz's when it is used to change by flux-linkages (*i.e.*, statically induced e.m.f.).

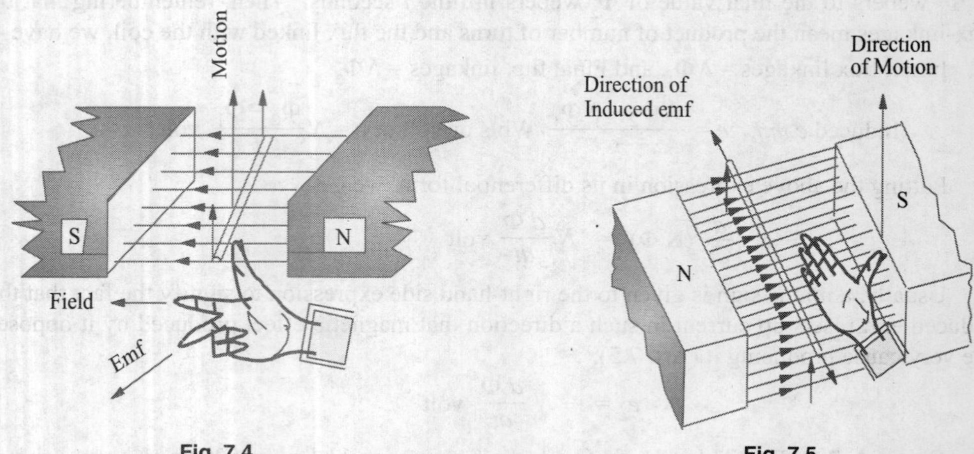

Fig. 7.4. Fig. 7.5.

Fig. 7.5 shows another way of finding the direction of the induced e.m.f. It is known as Right Flat-hand rule. Here, the front side of the hand is held perpendicular to the incident flux with the thumb pointing in the direction of the motion of the conductor. The direction of the fingers give the direction of the induced e.m.f. and current.

7.5 Lenz's Law

The direction of the induced current may also be found by this law which was formulated by Lenz* in 1835. This law states, in effect, that electromagnetically induced current always flows in such direction that the action of the magnetic field set up by it tends to oppose the very cause which produces it.

This statement will be clarified with reference to Fig. 7.1 and 7.2. It is found that when N-pole of the bar magnet approaches the coil, the induced current set up by induced e.m.f. flows in the ***anti-clockwise*** direction in the coil as seen from the magnet side. The result is that face of the coil becomes a N-pole and so tends to oppose the onward approach of the *N*-Pole of the magnet (like poles repel each other). The mechanical energy spent in overcoming this repulsive force is converted into electrical energy which appears in the coil.

When the magnet is withdrawn as in Fig. 7.2, the induced current flows in the clockwise direction thus making the face of the coil (facing the magnet) a *S*-pole. Therefore, the *N*-pole of the magnet has to withdrawn against this attractive force of the *S*-pole of coil. Again, the mechanical energy required to overcome this force of attraction is converted into electric energy.

It can be shown that Lenz's law is a direct consequence of Law of Conservation of Energy. Imagine for a moment that when *N*-pole of the magnet (Fig. 7.1) approaches the coil, induced current flows in such a direction as to make the coil face a *S*-pole. Then, due to inherent attrac-

* After the Russian born geologist and physicist Heinrich Friedrich Emil Lenz (1808 - 1865).

tion between unlike poles, the magnet would be automatically pulled towards the coil without the expenditure of any mechanical energy. It means that we would be able to create electric energy out of nothing, which is denied by the inviolable Law of Conservation of Energy. In fact, to maintain the sanctity of this law, it is imperative for the induced current to flow in such a direction that the magnetic effect produced by it tends to oppose the very cause which produces it. In the present case, it is relative motion of the magnet with magnet with respect to the coil which is the cause of the production of the induced current. Hence, the induced current always flows in such a direction to oppose this relative motion *i.e.*, the approach or withdrawal of the magnet.

7.6 Induced E.M.F.

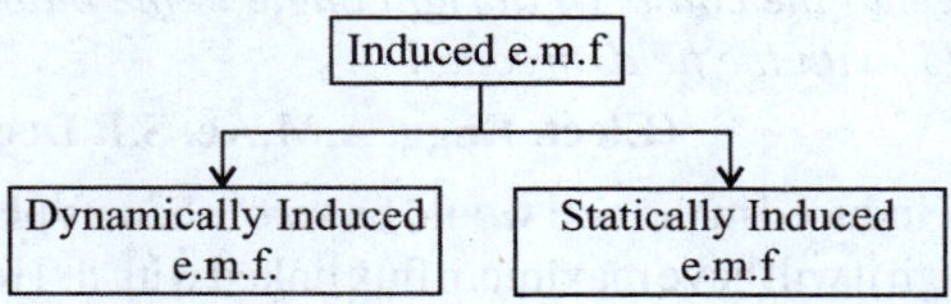

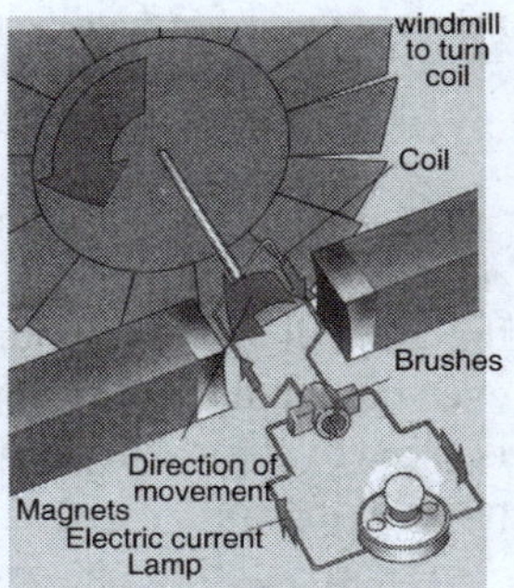

The principle of electric generation

Induced e.m.f. can be either ***(i)* dynamically induced** or ***(ii)* statically induced.** In the first case, usually the field is stationary and conductors cut across it (as in d.c. generators). But in the second case, usually the conductors or the coil remains stationary and flux linked with it is changed by simply increasing or decreasing the current producing this flux (as in transformers).

7.7 Dynamically induced E.M.F.

In Fig. 7.7. a conductor A is shown in cross-section, lying m^2 within a uniform magnetic field of flux density B Wb/m^2. The arrow attached to A shows its direction of motion. Consider the conditions shown in Fig. 7.6 (*a*) when A cuts across at right angles to the flux. Suppose '*l*' is its length lying within the field and let it move a distance dx in time dt. Then area swept by it is = ldx. Hence, flux cut = $l.dx \times B$ webers.

$$\text{Change in flux} = Bldx \text{ weber}$$
$$\text{Time taken} = dt \text{ second}$$

Hence, according to Faraday's Laws (Art. 7.3.) the e.m.f. induced in it (known as dynamically induced e.m.f.) is

$$\text{rate of change of flux linkages} = \frac{Bldx}{dt} = Bl\frac{dx}{dt} = Blv \text{ volt where } \frac{dx}{dt} = \text{velocity}$$

If the conductor A moves at an angle θ with the direction of flux [Fig. 7.6 (*b*)] then the induced e.m.f. is $e = Bl\upsilon \sin\theta$ volts $= \overrightarrow{l\upsilon} \times \vec{B}$ (*i.e.* as cross product vector $\vec{\upsilon}$ and $\vec{B}$).

The direction of the induced e.m.f. is given by Fleming's Right-hand rule (Art. 7.5) or Flat-hand rule and most easily by vector cross product given above.

It should be noted that generators work on the production of dynamically induced e.m.f. in the conductors housed in a revolving armature lying within a strong magnetic field.

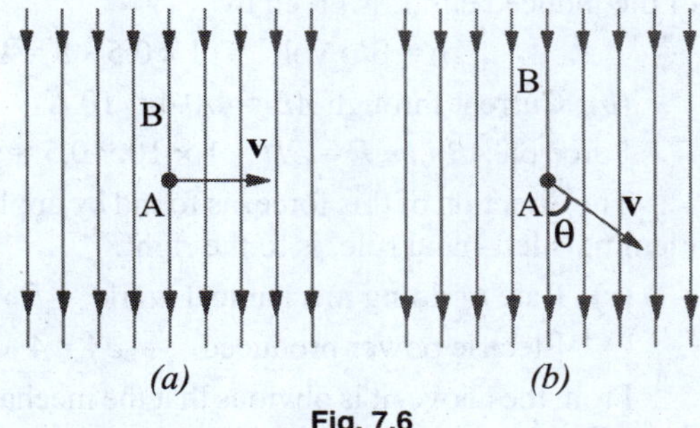

Fig. 7.6

Example 7.3. *A conductor of length 1 metre moves at right angles to a uniform magnetic field of flux density 1.5 Wb/m² with a velocity of 50 metre/second. Calculate the e.m.f. induced in it. Find also the value of induced e.m.f. when the conductor moves at an angle of 30° to the direction of the field.*

Solution. Here $B = 1.5 \text{ Wb/m}^2$ $l = 1$ m $\upsilon = 50$ m/s ; $e = ?$

Now $e = Bl\upsilon = 1.5 \times 1 \times 50 =$ **75 V.**

In the second case $\theta = 30°$ ∴ $\sin 30° = 0.5$ ∴ $e = 75 \times 0.5 =$ **37.5 V**

Example 7.4. *A square coil of 10 cm side and with 100 turns is rotated at a uniform speed of 500 rpm about an axis at right angle to a uniform field of 0.5 Wb/m². Calculate the instantaneous value of induced e.m.f. when the plane of the coil is* ***(i)*** *at right angle to the plane of the field.* ***(ii)*** *in the plane of the field.* ***(iii)*** *at 45° with the field direction.*

(Elect. Engg. A.M.Ae. S.I. Dec. 1991)

Solution. As seen from Art. 12.2, e.m.f. induced in the coil would be zero when its plane is at right angles to the plane of the field, even though it will have maximum flux linked with it. However, the coil will have maximum e.m.f. induced in it when its plane lies parallel to the plane of the field even though it will have minimum flux linked with it. In general, the value of the induced e.m.f. is given by $e = \omega N\Phi_m \sin\theta = E_m \sin\theta$ where θ is the angle between the axis of zero *e.m.f.* and the plane of the coil.

Here, $f = 500/60 = 25/3$ r.p.s ; $N = 100$; $B = 0.5$ Wb/ m² ; $A = (10 \times 10) \times 10^{-4} = 10^{-2}$ m².

∴ $E_m = 2\pi f NBA = 2\pi(25/3) \times 100 \times 0.5 \times 10^{-2} = 26.2$ V (*i*) since $\theta = 0$; $\sin\theta = 0$; therefore, $e = 0$. (*ii*) Here, $\theta = 90°$; $e = E_m \sin 90° = 26.2 \times 1 = 26.2$ V (*iii*) $\sin 45° = 1/\sqrt{2}$; $e = 26.2 \times 1/\sqrt{2} =$ **18.5 V**

Example 7.5. *A conducting rod AB (Fig. 7.7) makes contact with metal rails AD and BC which are 50 cm apart in a uniform magnetic field of B = 1.0 Wb/m² perpendicular to the plane ABCD. The total resistance (assumed constant) of the circuit ABCD is 0.4 Ω.*

(a) *What is the direction and magnitude of the e.m.f. induced in the rod when it is moved to the left with a velocity of 8 m/s ?*

(b) *What force is required to keep the rod in motion ?*

(c) *Compare the rate at which mechanical work is done by the force F with the rate of development of electric power in the circuit.*

Solution. **(*a*)** Since *AB* moves to the left, direction of the induced current, as found by applying Fleming's Right-hand rule is from *A* to *B*. Magnitude of the induced e.m.f. is given by

$$e = \beta l\upsilon \text{ volt } = 1 \times 0.5 \times 8 = \mathbf{4 \text{ volt}}$$

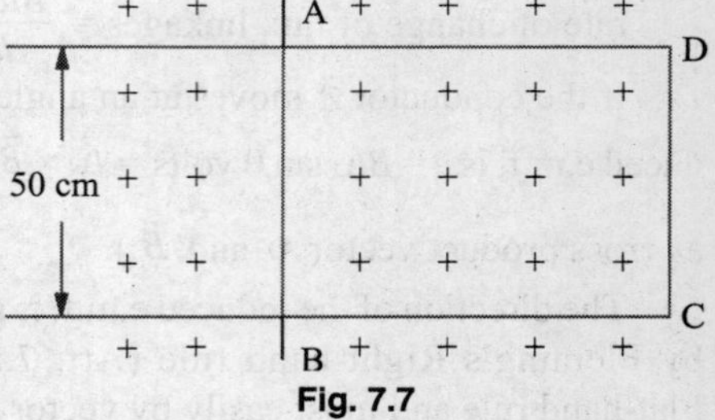

Fig. 7.7

(*b*) Current through $AB = 4/0.4 = 10$ A

Force on *AB i.e.* $F = BIl = 1 \times 10 \times 0.5 =$ **5 N**

The direction of this force, as found by applying Fleming's left-hand rule, is to the right.

(c) Rate of doing mechanical work $= F \times \upsilon = 5 \times 8 =$ **40 J/s or W**

Electric power produced $= e\,i = 4 \times 10 = 40$ W

From the above, it is obvious that the mechanical work done in moving the conductor against force *F* is converted into electric energy.

Example 7.6. *In a 4-pole dynamo, the flux/pole is 15 mWb. Calculate the average e.m.f. induced in one of the armature conductors, if armature is driven at 600 r.p.m.*

Solution. It should be noted that each time the conductor passes under a pole (whether *N* or *S*) it cuts a flux of 15 mWb. Hence, the flux cut in one revolution is 15 × 4 = 60 mWb. Since conductor is rotating at 600/60 = 10 r.p.s. time taken for one revolution is 1/10 = 0.1 second.

$$\therefore \quad \text{average e.m.f. generated} = N\frac{d\Phi}{dt} \text{ volt}$$

$$N = 1; \quad d\Phi = 60 \text{ mWb} = 6 \times 10^{-2} \text{ Wb} ; dt = 0.1 \text{ second}$$

$$\therefore \quad e = 1 \times 6 \times 10^{-2}/0.1 = \mathbf{0.6\ V}$$

Tutorial Problem 7.1

1. A conductor of active length 30 cm carries a current of 100 A and lies at right angles to a magnetic field of strength 0.4 Wb/m^2. Calculate the force in newtons exerted on it. If the force causes the conductor to move at a velocity of 10 m/s, calculate (*a*) the e.m.f. induced in it and (*b*) the power in watts developed by it. **[12 N; 1.2 V, 120 W]**
2. A straight horizontal wire carries a steady current of 150 A and is situated in a uniform magnetic field of 0.6 Wb/m^2 acting vertically downwards. Determine the magnitude of the force in kg/ metre length of conductor and the direction in which it works. **[9.175 kg/m horizontally]**
3. A conductor, 10 cm in length, moves with a uniform velocity of 2 m/s at right angles to itself and to a uniform magnetic field having a flux density of 1 Wb/m^2. Calculate the induced e.m.f. between the ends of the conductor. **[0.2 V]**

7.8 Statically Induced E.M.F.

It can be further sub-divided into two types as shown: **(*a*) *mutually induced e.m.f.*** **and (*b*) *self-induced e.m.f.***

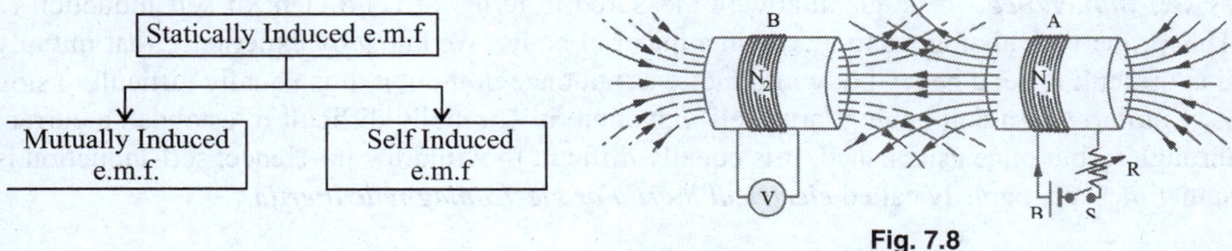

Fig. 7.8

(*a*) Mutually-induced e.m.f. Consider two coils A and B lying close to each other (Fig. 7.8).

Coil *A* is joined to a battery, a switch and a variable resistance *R* whereas coil *B* is connected to a sensitive voltmeter *V*. When current through *A* is established by closing the switch, its magnetic field is set up which partly links with or threads through the coil *B*. As current through *A* is changed, the flux linked with *B* is also changed. Hence, mutually induced e.m.f. is produced in *B* whose magnitude is given by Faraday's Laws (Art. 7.3) and direction by Lenz's Law (Art. 7.5).

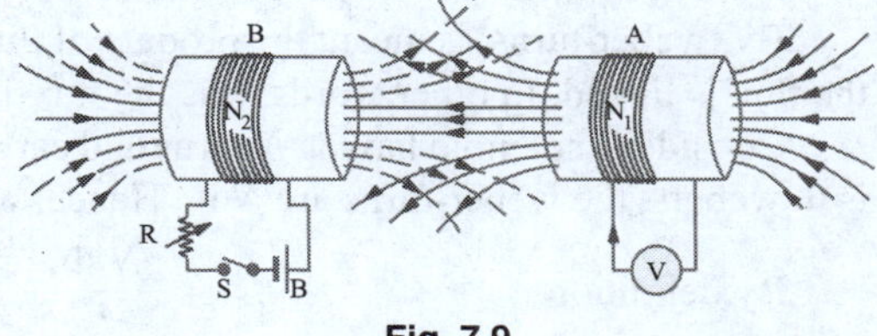

Fig. 7.9

If, now, battery is connected to *B* and the voltmeter across *A* (Fig. 7.9), then the situation is reversed and now a change of current in *B* will produce mutually-induced e.m.f. in *A*.

It is obvious that in the examples considered above, there is no movement of any conductor, the flux variations being brought about by variations in current strength only. Such an **e.m.f.** induced in one coil by the influence of the other coil is called (statically but) mutually induced e.m.f.

(*b*) Self-induced e.m.f. This is the e.m.f. induced in a coil due to ***the change of its own flux linked with it.*** If current through the coil (Fig. 7.10) is changed, then the flux linked with its own turns will also change, which will produce in it what is called *self-induced* e.m.f. The direction of this induced e.m.f. (as given by Lenz's law) would be such as to oppose any change of flux which is, in fact, the very cause of its production. Hence, it is also known as the opposing or counter e.m.f. of self-induction.

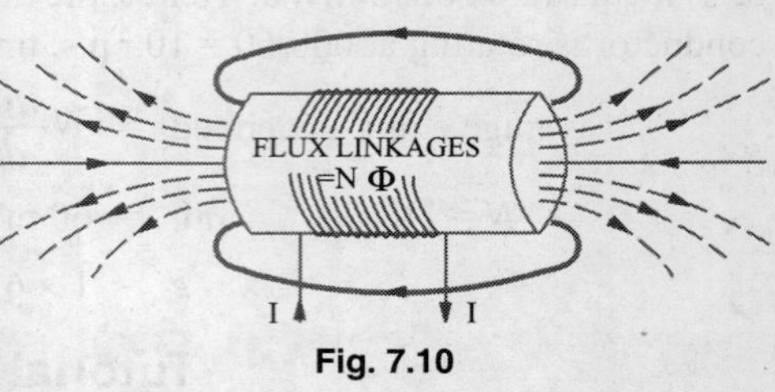

Fig. 7.10

7.9 Self-Inductance

Imagine a coil of wire similar to the one shown in Fig. 7.10 connected to a battery through a rheostat. It is found that whenever an effort is made to increase current (and hence flux) through it, it is always opposed by the instantaneous production of counter e.m.f. of self-induction. Energy required to overcome this opposition is supplied by the battery. As will be fully explained later on, this energy is stored in the additional flux produced.

If, now an effort is made to decrease the current (and hence the flux), then again it is delayed due to the production of self-induced e.m.f., this time in the opposite direction. This property of the coil due to which it opposes any increase or decrease or current of flux through it, is known as ***self-inductance***. It is quantitatively measured in terms of coefficient of self induction L. This property is analogous to inertia in a material body. We know by experience that initially it is difficult to set a heavy body into motion, but once in motion, it is equally difficult to stop it. Similarly, in a coil having large self-induction, it is initially difficult to establish a current through it, but once established, it is equally difficult to withdraw it. Hence, self-induction is sometimes analogously called ***electrical inertia or electromagnetic inertia.***

7.10 Coefficient of Self-Induction (L)

It may be defined in any one of the three ways given below:

(*i*) First Method for L

The coefficient of self-induction of a coil is defined as

the weber-turns per ampere in the coil

By 'weber-turns' is meant the product of flux in webers and the number of turns with which the flux is linked. In other words, it is the flux-linkages of the coil.

Consider a solenoid having N turns and carrying a current of I amperes. If the flux produced is Φ webers, the weber-turns are $N\Phi$. Hence, weber-turns per ampere are $N\,\Phi/I$.

By definition, $L = \dfrac{N\,\Phi}{I}$. The unit of self-induction is henry*.

If in the above relation, $N\Phi = 1$ Wb-turn, $I = 1$ ampere, then $L = 1$ henry (H)

* After the American scientist Joseph Henry (1797 - 1878), a company of Faraday.

Hence a coil is said to have a self-inductance of one henry if a current of 1 ampere when flowing through it produced flux-linkages of 1 Wb-turn in it.

Therefore, the above relation becomes $L = \dfrac{N\Phi}{I}$ henry

Example 7.7. *The field winding of a d.c. electromagnet is wound with 960 turns and has resistance of 50 Ω when the exciting voltages is 230 V, the magnetic flux linking the coil is 0.005 Wb. Calculate the self-inductance of the coil and the energy stored in the magnetic field.*

Solution. Formula used: $L = \dfrac{N\Phi}{L} H$

Current through coil = 230/50 = 4.6 A $\Phi = 0.005$ Wb ; $N = 960$

$L = \dfrac{960 \times 0.005}{4.6}$ = **1.0435 H**. Energy stored = $\dfrac{1}{2} L I^2 = 1\dfrac{1}{2} \times 1.0435 \times 4.6^2$ = **11.04 J**

Second Method for L

We have seen in Art. 6.16 that flux produced in a solenoid is

$$\Phi = \frac{NI}{l/\mu_0\mu_r A} \quad \therefore \frac{\Phi}{I} = \frac{N}{l/\mu_0\mu_r A} \quad \text{Now } L = N\frac{\Phi}{I} = N \cdot \frac{N}{l/\mu_0\mu_r A} H$$

$$\therefore \quad L = \frac{N^2}{l/\mu_0\mu_r A} = \frac{N^2}{S} \text{H} \quad \text{or} \quad L = \frac{\mu_0\mu_r AN^2}{l} \text{H}$$

It gives the value of self-induction in terms of the dimensions of the solenoid*.

Example 7.8. *An iron ring 30 cm mean diameter is made of square of iron of 2 cm × 2 cm cross-section and is uniformly wound with 400 turns of wire of 2 mm² cross-section. Calculate the value of the self-inductance of the coil. Assume $\mu_r = 800$.*

(Elect. Technology. I, Gwalior Univ.)

Solution. $L = \mu_0 \mu_r AN^2/l$. Here $N = 400$; $A = 2 \times 2 = 4$ cm² $= 4 \times 10^{-4}$ m² ; $l = 0.3\pi$ m ; $\mu_r = 800$

$$\therefore \quad L = 4\pi \times 10^{-7} \times 800 \times 4 \times 10^{-4} (400)^2/_{0.3\pi} = \textbf{68.3 mH}$$

Note. The cross-section of the wire is not relevant to the given question.

Third Method for L

It will be seen from Art. 7.10 (*i*) above that $L = \dfrac{N\Phi}{I}$ $\therefore$ $N\Phi = LI$ or $-N\Phi = -LI$

Differentiating both sides, we get $-\dfrac{d}{dt}(N\Phi) = -L \cdot \dfrac{dI}{dt}$ (assuming L to be constant);

$$-N \cdot \frac{d\Phi}{dt} = -L \cdot \frac{dI}{dt}$$

As seen from Art. 7.3, $-N \cdot \dfrac{d\Phi}{dt}$ = self-induced e.m.f. $\therefore e_L = -L\dfrac{dI}{dt}$

If $\dfrac{dI}{dt}$ = 1 ampere/second and $e_L = 1$ volt, then $L = 1\ H$

Hence, a coil has a self-inductance of one henry if one volt is induced in it when current through it changes at the rate of one ampere/second.

---

* In practice, the inductance of a short solenoid is given by L = Kμ0 μr.AN2/l where K is Nagaoka's constant.

Example 7.9. *If a coil of 150 turns is linked with a flux of 0.01 Wb when carrying current of 10 A, calculate the inductance of the coil. If this current is uniformly reversed in 0.01 second, calculating the induced electromotive force.*

Solution. $L = N\Phi/I = 150 \times 0.01/10 = \mathbf{0.15\ H}$

Now, $e_L = L\, dI/dt\ ;\ dI = -10 - (-10) = 20\ A$

$\therefore\quad e_L = 0.15 \times 20/0.01 = \mathbf{300\ V}$

Example 7.10. *An iron rod, 2 cm in diameter and 20 cm long is bent into a closed ring and is wound with 3000 turns of wire. It is found that when a current of 0.5 A is passed through this coil, the flux density in the coil is 0.5 Wb/m². Assuming that all the flux is linked with every turn of the coil, what is* ***(a)*** *the B/H ratio for the iron* ***(b)*** *the inductance of the coil ? What voltage would be developed across the coil if the current through the coil is interrupted and the flux in the iron falls to 10 % of its former value in 0.001 second ?*

(Principle of Elect. Engg. Jadavpur Univ.)

Solution. $H = NI/l = 3000 \times 0.2 = 7500\ \text{AT/m}\ B = 0.5\ \text{Wb/m}^2$

(a) Now, $\dfrac{B}{H} = \dfrac{0.5}{7500} = \mathbf{6.67 \times 10^{-5}\ H/m}$. Also $\mu_r = B/\mu_a H = 6.67 \times 10^{-5}/4\pi \times 10^{-7} = 53$

(b) $L = \dfrac{N\Phi}{I} = \dfrac{300 \times \pi \times (0.02)^2 \times 0.5}{4 \times 0.5} = \mathbf{0.94\ H}$

$e_L = N\dfrac{N\Phi}{dt}$ volt ; $d\Phi = 90\ \%$ of original flux $= \dfrac{0.9 \times \pi \times (0.02)^2 \times 0.5}{4} = 0.45\,\pi \times 10^{-4}$ Wb

$dt = 0.001$ second $\quad\therefore\quad e_L = 3000 \times 0.45\pi \times 10^{-4}/0.001 = \mathbf{424\ V}$

Example 7.11. *A circuit has 1000 turns enclosing a magnetic circuit 20 cm² in section. With 4 A, the flux density is 1.0 Wb/m² and with 9A, it is 1.4 Wbm². Find the mean value of the inductance between these current limits and the induced e.m.f. if the current falls from 9 A to 4 A in 0.05 seconds.* **(Elect. Engineering-1, Delhi Univ.)**

Solution. $L = N\dfrac{d\Phi}{dI} = N\dfrac{d}{dI}(BA) = NA\dfrac{dB}{dI}$ henry $= 1000 \times 20 \times 10^{-4}(1.4 - 1)/(9 - 4) = \mathbf{0.16\ H}$

Now, $e_L = L.dI/dt\ ;\ dI = (9 - 4) = 5\ A,\ dt = 0.05\ s \quad\therefore\quad e_L = 0.16 \times 5/0.05 = \mathbf{16\ V}$

7.11 Mutual Inductance

In Art. 7.8 (Fig. 7.8) we have that any change of current in coil *A* is always accompanied by the production of mutually-induced e.m.f. in coil *B*. Mutual inductance may, therefore, be defined as the ability of one coil (or circuit) to produce an e.m.f. in a nearby coil by induction when the current in the first coil changes. This action being reciprocal, the second coil can also induce an e.m.f. in the first when current in the second coil changes. This ability of reciprocal induction is measured in terms of the coefficient of mutual induction *M*.

Example 7.12. *A single element has the current and voltage functions graphed in figure 7.11. (a) and (b). Determine the element.* **[Bombay University 2001]**

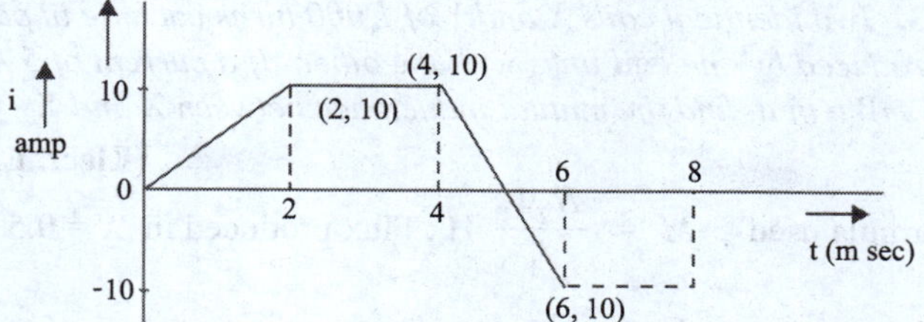

Fig. 7.11 (*a*)

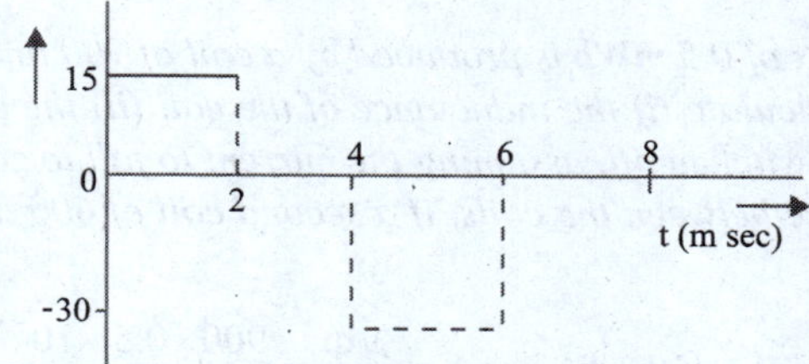

Fig. 7.11 (*b*)

Solution. Observations from the graph are tabulated below.

Sr. No.	Between time	*di*/*dt* amp/sec	*V*	*L*	
1	0 – 2 m Sec	5000	15	15/5000	= 3mH
2	2 – 4 m Sec	0	0	–	= 3 mH
3	4 – 6 m Sec	– 10,000	– 30	– 30 / (– 10,000)	
4	6 – 8 m Sec	0	0	–	

The element is a 3-mH inductor.

7.12 Coefficient of Mutual Inductance (M)

It can also be defined in three ways as given below:

***(i)* First Method for M**

Let there be two magnetically-coupled coils having N_1 and N_2 turns respectively (Fig. 7.9). Coefficient of mutual inductance between the two coils is defined as

the weber-turns in one coil due to one ampere current in the other.

Let a current I_1 ampere when flowing in the first coil produce a flux Φ_1 webers in it. ***It is supposed that whole of this flux links with the turns of the second coil*.*** Then, flux-linkages *i.e.*, webers-turns in the *second* coil for unit current in the *first* coil are $N_2\,\Phi_1/I_1$. Hence, by definition

$$M = \frac{N_2\,\Phi_1}{I_1}$$

If weber-turns in *second* coil due to one ampere current in the *first* coil *i.e.* $N_2\,\Phi_1/I_1 = 1$ then, as seen from above, $M = 1\text{H}$.

Hence, two coils are said to have a mutual inductance of 1 henry is one ampere current when flowing in one coil produces flux-linkages of one Wb-turn in the other.

---

* If whole of this flux does not link with turns of the second coil, then only that part of the flux which is actually linked is taken instead. (Ex. 7.13 and 7.17). In general, $M = N_2\Phi_2/I_1$.

Example 7.13. *Two identical coils X and Y of 1,000 turns each lie in parallel planes such that 80% of flux produced by one coil links with the other. If a current of 5 A flowing in X produces a flux of 0.5 mWb in it, find the mutual inductance between X and Y.*

(Elect. Engg. A,M.Ae.S.I.)

Solution. Formula used $M = \dfrac{N_2\Phi_1}{I_1}$ H ; Flux produced in $X = 0.5$ mWb $= 0.5 \times 10^{-3}$ Wb

Flux linked with $Y = 0.5 \times 10^{-3} \times 0.8 = 0.4 \times 10^{-3}$ Wb ; $M = \dfrac{1000 \times 0.4 \times 10^{-3}}{5} = \mathbf{0.08\ H}$

Example 7.14. *A flux of 0.5 mWb is produced by a coil of 900 turns wound on a ring with a current of 3 A in it. Calculate* **(i)** *the inductance of the coil* **(ii)** *the e.m.f. induced in the coil when a current of 5 A is switched off, assuming the current to fall to zero in 1 milli second and* **(iii)** *the mutual inductance between the coils, if a second coil of 600 turns is uniformly wound over the first coil.* **(F. E. Pune Univ.)**

Solution. (*i*) Inductance of the first coil $= \dfrac{N\Phi}{I} = \dfrac{900 \times 0.5 \times 10^{-3}}{3} = \mathbf{0.15\ H}$

(*ii*) e.m.f. induced $\quad e_1 = L\dfrac{di}{dt} = 0.15 \times \dfrac{(5-0)}{1 \times 10^{-3}} = \mathbf{750\ V}$

(*iii*) $\quad M \dfrac{N_2\Phi_1}{I_1} = \dfrac{600 \times 0.5 \times 10^{-3}}{3} = \mathbf{0.1\ H}$

(*ii*) Second Method for M

We will now deduce an expression for coefficient of mutual inductance in terms of the dimensions of the two coils.

Flux in the first coil $\quad \Phi_1 = \dfrac{N_2 I_1}{l/\mu_0\mu_r A}$ Wb ; Flux/ampere $= \dfrac{\Phi_1}{I_1} = \dfrac{N_1}{l/\mu_0\mu_r A}$

Assuming that whole of this flux (it usually is some percentage of it) is linked with the other coil having N_2 turns, the weber-turns in it due to the flux/ampere in the first coil is

$$M = \frac{N_2\Phi_1}{I_1} = \frac{N_2 N_1}{l/\mu_0\mu_r A} \quad \therefore \quad M = \frac{\mu_0\mu_r A\, N_1 N_2}{l}\ \text{H}$$

Also $\quad M = \dfrac{N_1 N_2}{l/\mu_0\mu_r A} = \dfrac{N_1 N_2}{\text{reluctance}} = \dfrac{N_1 N_2}{S}$ H

Example 7.15. *If a coil of 150 turns is linked with a flux of 0.01 Wb when carrying a current of 10 A ; calculate the inductance of the coil. If this current is uniformly reversed in 0.1 second, calculate the induced e.m.f. If a second coil of 100 turns is uniformly wound over the first coil, find the mutual inductance between the coils.* **(F. E. Pune Univ.)**

Solution. $\quad L_1 = N_1\Phi_1/I_1 = 150 \times 0.01/10 = \mathbf{0.15\ H}$

$e = L \times di/dt = 0.15 \times [10 - (-10)]/0.1 = 1 = \mathbf{30\ V}$

$M = N_2\Phi/I_1 = 100 \times 0.01/10 = \mathbf{0.1\ H}$

(*iii*) Third Method for M

As seen from Art. 7.12 (*i*) $\quad M = \dfrac{N_2\Phi_1}{I_1} \quad \therefore\ N_2\Phi_1 = MI_1$ or $-N_2\Phi_1 = -MI_1$

Differentiating both sides, we get: $-\frac{d}{dt}(N_2\Phi_1) = -M\frac{dI_1}{dt}$ (assuming M to be constant)

Now, $-\frac{d}{dt}(N_2\Phi_1)$ = mutually-induced e.m.f. in the second coil $= e_M \quad \therefore \; e_M = -M\frac{dI_1}{dt}$

If $dI_1/dt = 1$ A/s ; $e_M = 1$ volt, then $M =$ **1 H**.

Hence, two coils are said to have a mutual inductance of one henry if current changing at the rate of 1 ampere/second in one coil induces an e.m.f. of one volt in the other.

Example 7.16. *Two coils having 30 and 600 turns respectively are wound side-by-side on a closed iron circuit of area of cross-section 100 sq.cm. and mean length 200 cm. Estimate the mutual inductance between the coils if the relative permeability of the iron is 2000. If a current of zero ampere grows to 20 A in a time of 0.02 second in the first coil, find the e.m.f. induced in the second coil.* **(Elect. Engg. I, JNT Univ., Warangal)**

Solution. Formula used: $M = \frac{N_1N_2}{l/\mu_0\mu_r A}$ **H**, $N_1 = 30$; $N_2 = 600$; $A = 100 \times 10^{-4} = 10^{-2}\text{m}^2$, $l = 2\text{m}$

$$\therefore \quad M = \mu_0\mu_r A\, N_1N_2/l = 4\pi \times 10^{-7} \times 2000 \times 10^{-2} \times 30 \times 600/2 = \mathbf{0.226\ H}$$

$$dI_1 = 20 - 0 = 20\text{ A} ;\ dt = 0.02\text{ s} ;\ e_M = M\,dI_1/dt = 0.226 \times 20/0.2 = \mathbf{226\ V}$$

7.13 Coefficient of Coupling

Consider two magnetically-coupled coils A and B having N_1 and N_2 turns respectively. Their individual coefficients of self-induction are,

$$L_1 = \frac{N_1^2}{l/\mu_0\mu_r A} \quad \text{and} \quad L_2 = \frac{N_2^2}{l/\mu_0\mu_r A}$$

The flux Φ_1 produced in A due to a current I_1 ampere is $\Phi_1 = \frac{N_1I_1}{l/\mu_0\mu_r A}$

Suppose a fraction k_1 of this flux *i.e.* $k_1\Phi_1$ is linked with coil B.

Then $$M = \frac{k_1\Phi_1 \times N_2}{I_1} \quad \text{where } k_1 \leqslant 1.$$

Substituting the value of Φ_1, we have, $M = k_1 \times \frac{N_1N_2}{l/\mu_0\mu_r A}$ ...(*i*)

Similarly, the flux Φ_2 produced in B due to I_2 ampere in it is $\Phi_2 = \frac{N_2I_2}{l/\mu_0\mu_r A}$

Suppose a fraction k_2 of this flux *i.e.* $k_2\Phi_2$ is linked with A.

Then $$M = \frac{k_2\Phi_2 \times N_1}{I_2} = k_2\frac{N_1N_2}{l/\mu_0\mu_r A} \qquad \text{...(ii)}$$

Multiplying Eq. (*i*) and (*ii*), we get

$$M^2 = k_1k_2\frac{N_1^2}{l/\mu_0\mu_r A} \times \frac{N_2^2}{l/\mu_0\mu_r A} \quad \text{or} \quad M^2 = k_1k_2L_1L_2$$

Putting $\sqrt{k_1k_2} = k$, we have $M = k\sqrt{L_1L_2}$ or $k = \frac{M}{\sqrt{L_1L_2}}$

The constant k is called the ***coefficient of coupling*** and may be defined as the ratio of ***mutual inductance actually present between the two coils to the maximum possible value.*** If the flux due to one coil completely links with the other, then value of k is unity. If the flux of one coil does not at all link with the other, then $k = 0$. In the first case, when $k = 1$, coils are said to be tightly coupled and when $k = 0$, the coils are magnetically isolated from each other.

Example 7.17. *Two identical 750 turn coils A and B lie in parallel planes. A current changing at the rate of 1500 A/s in A induces an e.m.f. of 11.25 V in B. Calculate the mutual inductance of the arrangement. If the self-inductance of each coil is 15 mH, calculate the flux produced in coil A per ampere and the percentage of this flux which links the turns of B.*

Solution. Now, $e_M = M dI_1/dt$ ...Art. 7.12

$$M = \frac{e_M}{dI_1/dt} = \frac{11.25}{1500} = 7.5 \times 10^{-3}\text{ H} = \mathbf{7.5\ mH}$$

Now, $L_1 = \dfrac{N_1\Phi_1}{I_1} \quad \therefore \quad \dfrac{\Phi_1}{I_1} = \dfrac{L_1}{N_1} = \dfrac{15\times10^{-3}}{750} = \mathbf{2\times10^{-5}\ Wb/A}$ ...Art. 7.10

Now, $k = \dfrac{M}{\sqrt{L_1L_2}} = \dfrac{7.5\times10^{-3}}{\sqrt{L^2}} = \dfrac{7.5\times10^{-3}}{15\times10^{-3}} = 0.5 = \mathbf{50\%}$ $(\because L_1 = L_2 = L)$ ...Art. 7.13

Example 7.18. *Two coils, A and B, have self inductances of 120 μH and 300 μH respectively. A current of 1 A through coil A produces flux linkages of 100 μWb turns in coil B. Calculate* ***(i)*** *the mutual inductance between the coils* ***(ii)*** *the coupling coefficient and* ***(iii)*** *the average e.m.f. induced in coil B if a current of 1 A in coil A is reversed at a uniform rate in 0.1 sec.*

(F. E. Pune Univ.)

Solution. (*i*) $M = \dfrac{\text{flux-linkages of coil}}{\text{current in coil } A} = \dfrac{100\times10^{-6}}{1} = \mathbf{100\ \mu H}$

(ii) $M = k\sqrt{L_1L_2} \quad \therefore \quad k = \dfrac{M}{\sqrt{L_1L_2}} = \dfrac{100\times10^{-6}}{\sqrt{120\times10^{-6}\times300\times10^{-6}}} = \mathbf{0.527}$

(iii) $e_2 = M \times di/dt = (100 \times 10^{-6}) \times 2/0.1 = 0.002$ V or **2 mV.**

7.14 Inductances in Series

(*i*) Let the two coils be so joined in series that their fluxes (or m.m.fs) are additive *i.e.*, in the same direction (Fig. 7.12).

Let M = coefficient of mutual inductance

L_1 = coefficient of self-inductance of 1st coil

L_2 = coefficient of self-inductance of 2nd coil.

Then, self induced e.m.f. in A is $= e_1 = -L_1 . \dfrac{di}{dt}$

Mutually-induced e.m.f. in A due to change of current in B is $e' = -M . \dfrac{di}{dt}$

Self-induced e.m.f. in B is $= e_2 = -L_2 . \dfrac{di}{dt}$

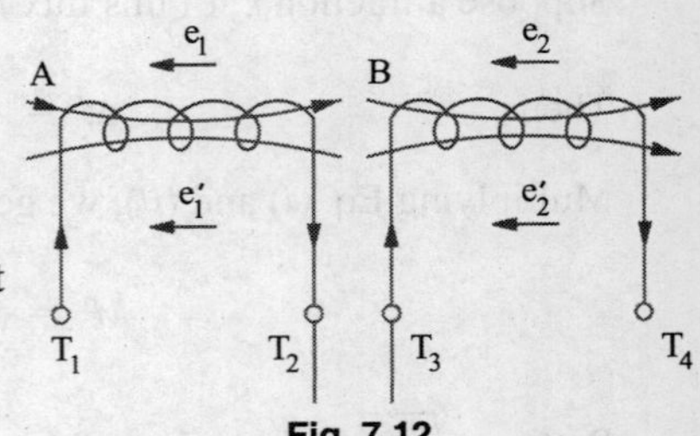

Fig. 7.12

Mutually-induced e.m.f. in B due to change of current in A is $= e_2' = -M.\dfrac{di}{dt}$

(All have –ve sign, because both self and mutally induced e.m.fs. are in opposition to the applied e.m.f.). Total induced e.m.f. in the combination $= -\dfrac{di}{dt}(L_1 + L_2 + 2M)$ ...(*i*)

If L is the equivalent inductance then total induced e.m.f. in that single coil would have been

$$= -L\frac{di}{dt} \qquad \text{...(ii)}$$

Equating (*i*) and (*ii*) above, we have $L = L_1 + L_2 + 2M$

(*ii*) When the coils are so joined that their fluxes are in opposite directions (Fig. 7.13).

As before $e_1 = -L_1\dfrac{di}{dt}$

$e_1' = +\text{M}.\dfrac{di}{dt}$ (mark this direction)

$e_2 = -L_2\dfrac{di}{dt}$ and $e_2' = +M.\dfrac{di}{dt}$

Total induced e.m.f. $= -\dfrac{di}{dt}(L_1 + L_2 - 2M)$

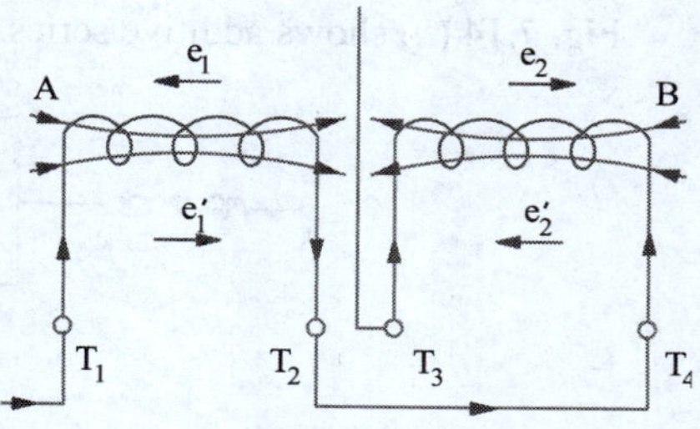

Fig. 7.13

∴ Equivalent inductance

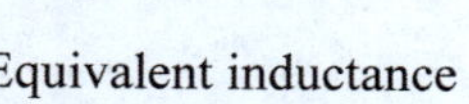

$$L = L_1 + L_2 - 2M$$

In general, we have: $L = L_1 + L_2 + 2M$ ... if m.m.fs are additive

and $L = L_1 + L_2 - 2M$ ... if m.m.fs. are subtractive

Example 7.19. *Two coils with a coefficient of coupling of 0.5 between them, are connected in series so as to magnetise **(a)** in the same direction **(b)** in the opposite direction. The corresponding values of total inductances are for **(a)** 1.9 H and for **(b)** 0.7 H. Find the self-inductances of the two coils and the mutual inductance between them.*

Solution. (*a*) $L = L_1 + L_2 + 2M$ or $1.9 = L_1 + L_2 + 2M$ ...(*i*)

(*b*) Here $L = L_1 + L_2 - 2M$ or $0.7 = L_1 + L_2 - 2M$ ...(*ii*)

Subtracting (*ii*) from (*i*), we get

$$1.2 = 4\text{M} \qquad \therefore \quad M = 0.3 \text{ H}$$

Putting this value in (*i*) above, we get $L_1 + L_2 = 1.3$ H ...(*iii*)

We know that, in general, $M = k\sqrt{L_1L_2}$

$$\therefore \quad \sqrt{L_1L_2} = \frac{M}{k} = \frac{0.3}{0.5} = 0.6 \qquad \therefore \quad L_1L_2 = \mathbf{0.36}$$

From (*iii*), we get $(L_1 + L_2)^2 - 4L_1L_2 = (L_1 - L_2)^2$

$$\therefore \quad (L_1 - L_2)^2 = 0.25 \quad \text{or} \quad L_1 - L_2 = 0.5 \qquad \text{...(iv)}$$

From (*iii*) and (*iv*), we get $L_1 = \mathbf{0.9\ H}$ and $L_2 = \mathbf{0.4\ H}$

Example 7.20. *The combined inductance of two coils connected in series is 0.6 H or 0.1 H depending on the relative directions of the currents in the coils. If one of the coils when isolated has a self-inductance of 0.2 H, calculate **(a)** mutual inductance and **(b)** coupling coefficient.*

(Elect. Technology, Univ. of Indore)

Solution. (*i*) $L = L_1 + L_2 + 2M$ or $0.6 = L_1 + L_2 + 2M$ ...(*i*)

and $0.1 = L_1 + L_2 - 2M$ ...(*ii*)

(*a*) From (*i*) and (*ii*) we get, $M = \mathbf{0.125\ H}$

Let $L_1 = 0.2$ H, then substituting this value in (*i*) above, we get $L_2 = 0.15$ H

(*b*) Coupling coefficient $k = M\sqrt{L_1L_2} = 0.125/\sqrt{0.2 \times 0.15} = \mathbf{0.72}$

Example 7.21. *Two coils with terminals T_1, T_2 and T_3, T_4 respectively are placed side by side. When measured separately, the inductance of the first coil is 1200 mH and that of the second is 800 mH. With T_2 joined to T_3, the inductance between T_1 and T_4 is 2500 mH. What is the mutual inductance between the two coils ? Also, determine the inductance between T_1 and T_3 when T_2 is joined to T_4.* **(Electrical Circuit, Nagpur Univ. 1991)**

Solution. $L_1 = 1200$ mH, $L_2 = 800$ mH

Fig. 7.14 (*a*) shows additive series.

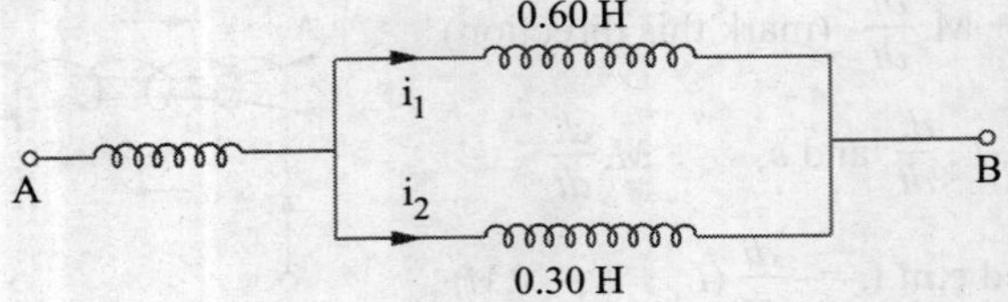

Fig. 7.14

$\therefore$ $L = L_1 + L_2 + 2M$

or $2500 = 1200 + 800 + 2M$; $M = \mathbf{250\ mH}$

Fig. 7.14 (*b*) shows the case of subtractive or opposing series.

Here, $L = L_1 + L_2 - 2M = 1200 + 800 - 2 \times 250 = \mathbf{1500\ mH}$

7.15. Inductance in Parallel

In Fig. 7.15, two inductances of values L_1 and L_2 henry are connected in parallel. Let the coefficient of mutual inductance between the two be M. Let i be the main supply current and i_1 and i_2 be the branch currents

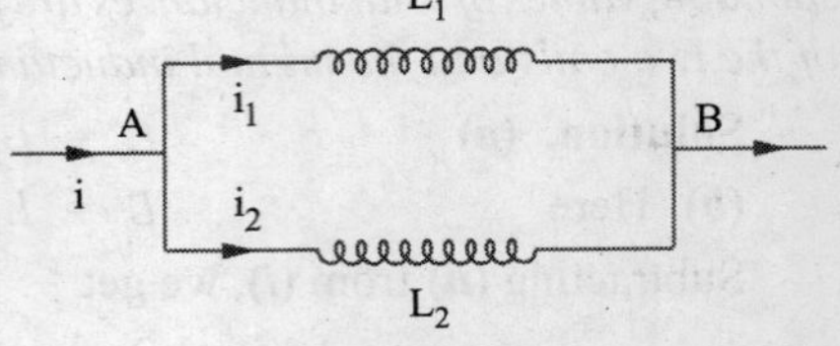

Fig. 7.15

Obviously, $i = i_1 + i_2$

$$\therefore \quad \frac{di}{dt} = \frac{di_1}{dt} + \frac{di_2}{dt} \qquad ...(i)$$

In each coil, both self and mutually induced e.m.fs. are produced. Since the coils are in parallel, these e.m.fs. are equal. For a case when self-induced e.m.f., we get

$$e = L_1\frac{di_1}{dt} + M\frac{di_2}{dt} = L_2\frac{di_2}{dt} + M\frac{di_1}{dt} \quad \therefore \quad L_1\frac{di_1}{dt} + M\frac{di_2}{dt} = L_2\frac{di_2}{dt} + M\frac{di_1}{dt}$$

$$\text{or} \quad \frac{di_1}{dt}(L_1 - M) = \frac{di_2}{dt}(L_2 - M) \quad \therefore \quad \frac{di_1}{dt} = \left(\frac{L_2 - M}{L_1 - M}\right)\frac{di_2}{dt} \qquad ...(ii)$$

Hence, (*i*) above becomes $$\frac{di}{dt} = \left[\left(\frac{L_2 - M}{L_1 - M}\right) + 1\right]\frac{di_2}{dt} \qquad ...(iii)$$

If L is the equivalent inductance, then $e = L.\dfrac{di}{dt}$ = induced e.m.f. in the parallel combination

$$= \text{induced e.m.f. in any one coil} = L_1.\frac{di_1}{dt} + M\frac{di_2}{dt}$$

$$\therefore \quad \frac{di}{dt} = \frac{1}{L}\left(L_1\frac{di_1}{dt} + M\frac{di_2}{dt}\right) \quad \text{...(iv)}$$

Substituting the value of di_1/dt from (*ii*) in (*iv*), we get $\dfrac{di}{dt} = \dfrac{1}{L}\left[L_1\left(\dfrac{L_2 - M}{L_1 - M}\right) + M\right]\dfrac{di_2}{dt}$ ...(v)

Hence, equating (*iii*) to (*iv*), we have $\dfrac{L_2 - M}{L_1 - M} + 1 = \dfrac{1}{L}\left[L_1\left(\dfrac{L_2 - M}{L_1 - M}\right) + M\right]$

or $$\frac{L_1 + L_2 - 2M}{L_1 - M} = \frac{1}{L}\left(\frac{L_1L_2 - M^2}{L_1 - M}\right)$$

$$\therefore \quad L = \frac{L_1L_2 - M^2}{L_1 + L_2 - 2M} \quad \text{when mutual field assists the separate fields.}$$

Similarly, $$L = \frac{L_1L_2 - M^2}{L_1 + L_2 + 2M} \quad \text{when the two fields oppose each other.}$$

Example 7.22. *Two coils of inductances 4 and 6 henry are connected in parallel. If their mutual inductance is 3 henry, calculate the equivalent inductance of the combination if (i) mutual inductance assists the self-inductance (ii) mutual inductance opposes the self-inductance.*

Solution. (*i*) $$L = \frac{L_1L_2 - M^2}{L_1 + L_2 - 2M} = \frac{4 \times 6 - 3^2}{4 + 6 - 2 \times 3} = \frac{15}{4} = \mathbf{3.75\ H}$$

(*ii*) $$L = \frac{L_1L_2 - M^2}{L_1 + L_2 + 2M} = \frac{24 - 9}{16} = \frac{15}{16} = \mathbf{0.94\ H\ (approx.)}$$

Tutorial Problem 7.2

1. Two coils are wound close together on the same paxolin tube. Current is passed through the first coil and is varied at a uniform rate of 500 mA per second, inducing an e.m.f. of 0.1 V in the second coil. The second coil has 100 turns. Calculate the number of turns in the first coil if its inductance is 0.4 H. **[200 turns]**
2. Two coils have 50 and 500 turns respectively are wound side by side on a closed iron circuit of section 50 cm² and mean length 120 cm. Estimate the mutual inductance between the coils if the permeability of iron is 1000. Also, find the self-inductance of each coil. If the current in one coil grows steadily from zero to 5A in 0.01 second, find the e.m.f. induced in the other coil. **[M = 0.131 H, L_1 = 0.0131 H, L_2 = 1.21 H, E = 65.4 V]**
3. An iron-cored choke is designed to have an inductance of 20 H when operating at a flux density of 1 Wb/m², the corresponding relative permeability of iron core is 4000. Determine the number of turns in the winding, given that the magnetic flux path has a mean length of 22 cm in the iron core and of 1 mm in air-gap that its cross-section is 10 cm². Neglect leakage and fringing. **[4100]**
4. A non-magnetic ring having a mean diameter of 30 cm and a cross-sectional area of 4 cm² is uniformly wound with two coils *A* and *B*, one over the other. *A* has 90 turns and *B* has 240 turns. Calculate from first principles the mutual inductance between the coils.
   Also, calculate the e.m.f. induced in *B* when a current of 6 A in *A* is reversed in 0.02 second. **[11.52 μH, 6.912 mV]**

5. Two coils *A* and *B*, of 600 and 100 turns respectively are wound uniformly around a wooden ring having a mean circumference of 30 cm. The cross-sectional area of the ring is 4 cm^2. Calculate (*a*) the mutual inductance of the coils and (*b*) the e.m.f. induced in coil *B* when a current of 2 A in coil *A* is reversed in 0.01 second. **[(*a*) 100.5 μH (*b*) 40.2 mV]**
6. A coil consists of 1,000 turns of wire uniformly wound on a non-magnetic ring of mean diameter 40 cm and cross-sectional area 20 cm^2.
Calculate (*a*) the inductance of the coil (*b*) the energy stored in the magnetic field when the coil is carrying a current of 15 A (*c*) the e.m.f. induced in the coil if this current is completely interrupted in 0.01 second. **[(*a*) 2mH (*b*) 0.225 joule (*c*) 3V]**
7. A coil of 50 turns having a mean diameter of 3 cm is placed co-axially at the centre of a solenoid 60 cm long, wound with 2,500 turns and carrying a current of 2 A. Determine mutual inductance of the arrangement. **[0.185 mH]**
8. A coil having a resistance of 2 Ω and an inductance of 0.5 H has a current passed through it which varies in the following manner ; (*a*) a uniform change from zero to 50 A in 1 second (*b*) constant at 50 A for 1 second (*c*) a uniform change from 50 A to zero in 2 seconds. Plot the current graph to a time base. Tabulate the p.d. applied to the coil during each of the above periods and plot the graph of p.d. to a time base.
[(*a*) 25 to 125 V (*b*) 100 V (*c*) 87.5 V to – 12.5 V]
9. State: (*i*) Flemming's right hand rule, and (*ii*) Fleming's left hand rule.
Mention their applications. ***(V.T.U., Belgaum Karnataka University, Winter 2003)***
10. Define: (*i*) Self inductance, and (*ii*) Mutual inductance.
Mention their units and formula to calculate each of them. Derive an expression for the energy stored in an inductor of self inductance 'L' henry carrying the current of 'I' amperes.
(V.T.U., Belgaum Karnataka University, Winter 2003)
11. State and explain Faraday's laws of electro magnetic induction, Lenz's Law. Fleming's right hand rule and Fleming's left hand rule.
(V.T.U., Belgaum Karnataka University, Summer 2003)
12. A coil of 300 turns wound on a core of non magnetic material has an inductance of 10mH. Calculate (*i*) the flux produced by a current of 5A (*ii*) the average value of the emf induced when a current of 5Amps is reversed in 8 mills seconds.
(V.T.U., Belgaum Karnataka University, Summer 2003)

OBJECTIVE TEST QUESTIONS

1. According to Faraday's Laws of Electromagnetic Induction, an e.m.f. is induced in a conductor whenever it
(*a*) lies in a magnetic field
(*b*) cuts magnetic flux
(*c*) moves parallel to the direction of the magnetic field
(*d*) lies perpendicular to the magnetic flux.
2. A pole of driving point admittance function implies
(*a*) zero current for a finite value of driving voltage
(*b*) zero voltage for a finite value of driving current
(*c*) an open circuit condition
(*d*) None of (*a*), (*b*) and (*c*) mentioned in the question ***(ESE 2001)***
3. The inductance of a long solenoid of length 1000 mm wound uniformly with 3000 turns on a cylindrical paper tube of 60 mm diameter is
(a) 3.2 μH (b) 3.2 mH
(c) 32.0 mH (d) 3.2 H
(GATE 2004)
4. A moving iron ammeter produced a full scale torque of 240 μNm with a deflection of 120^0 at a current of 10 A. The rate of change of self inductance (μH/radian) of the instrument at full scale is
(a) 2.0 μH/radian
(b) 4.8 μH/radian
(c) 12.0 μH/radian
(d) 114.6 μH/radian ***(GATE 2004)***
5. The self-inductance of a long cylindrical conductor due to its internal flux linkages is k H/m. If the diameter of the conductor is doubled, then the selfinductance of the conductor due its internal flux linkages would be
(*a*) 0.5 K H/m (*b*) K H/m
(*c*) 1.414 K H/m (*d*) 4 K H/m
(GATE)

CHAPTER

8 Magnetic Hysteresis

Chapter Outline

- ❖ Magnetic Hysteresis
- ❖ Area of Hysteresis Loop
- ❖ Properties and Application of Ferromagnetic Materials
- ❖ Permanent Magnet Materials
- ❖ Steinmetz Hysteresis Law
- ❖ Energy Stored in Magnetic Field
- ❖ Rate of Change of Stored Energy
- ❖ Energy Stored per Unit Volume
- ❖ Rise of Current in Inductive Circuit
- ❖ Decay of Current in Inductive Circuit

8.1 Magnetic Hysteresis

It may be defined as the lagging of magnetisation or induction flux density (B) behind the magnetising force (H). Alternatively, it may be defined as that quality of a magnetic substance, due to which energy is dissipated in it, on the reversal of its magnetism.

Let us take an unmagnetised bar of iron AB and magnetise it by placing it within the field of a solenoid (Fig. 8.1). The field H (= NI/l) produced by the solenoid is called the magnetising force. The value of H can be increased or decreased by increasing or decreasing current through the coil. Let H be increased in steps from zero up to a certain maximum value and the corresponding values of flux density (B) be noted. If we plot the relation between H and B, a curve like OA, as shown in Fig. 8.2, is obtained. The material becomes magnetically saturated for $H = OM$ and has at that time a maximum flux density of B_{max} established through it.

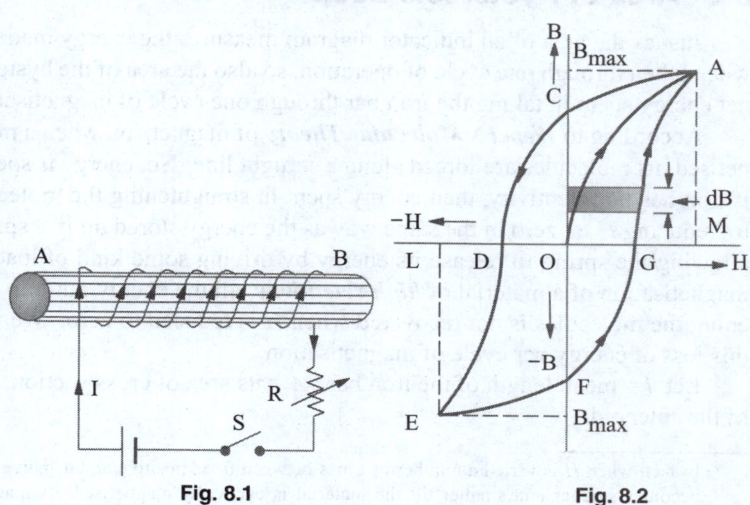

Fig. 8.1

Fig. 8.2

If H is now decreased gradually (by decreasing solenoid current), flux density B will not decrease along AO, as might be expected, but will decrease less rapidly along AC. When H is zero, B is not but has a definite value $B_r = OC$. It means that on removing the magnetising force H, the iron bar is not completely demagnetised. This value of B (= OC) measures the ***retentivity or remanence*** of the material and is called the remanent or residual flux density B_r.

To demagnetise the iron bar, we have to apply the magnetising force in the reverse direction. When H is reversed (by reversing current through the solenoid), then B is reduced to zero at point D where $H = OD$. This value of H required to wipe off residual magnetism is known as ***coercive force*** (H_c) and is a measure of the ***coercivity*** of the material *i.e.* its 'tenacity' with which it holds on to its magnetism.

If, after the magnetisation has been reduced to zero, value of H is further increased in the 'negative' *i.e.* reversed direction, the iron bar again reaches a state of magnetic saturation, represented by point L. By taking H back from its value corresponding to negative saturation, (= OL) to its value for positive saturation (= OM), a similar curve $EFGA$ is obtained. If we again start from G, the same curve $GACDEFG$ is obtained once again.*

It is seen that B always lag behind H. The two never attain zero value simultaneously. This lagging of B behind H is given the name 'hystereis' which literally means ***'to lag behind'***. The closed loop $ACDEFGA$ which is obtained when iron bar is taken through one complete cycle of magnetisation is known as 'hypothesis loop'.

By one cycle of magnetisation of a magnetic material is meant its being carried through one reversal of magnetisation, as shown in Fig. 8.3.

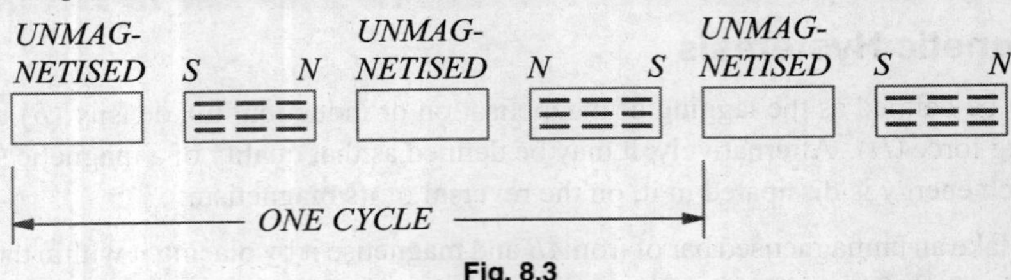

Fig. 8.3

8.2 Area of Hysteresis Loop

Just as the area of an indicator diagram measures the energy made available in a machine, when taken through one cycle of operation, so also the area of the hysteresis loop represents the net energy spent in taking the iron bar through one cycle of magnetisation.

According to ***Weber's Molecular Theory*** of magnetism, when a magnetic material is magnetised, its molecules are forced along a straight line. So, energy is spent in this process. Now, if iron has no retentivity, then energy spent in straightening the molecules could be recovered by reducing H to zero in the same way as the energy stored up in a spring can be recovered by allowing the spring to release its energy by driving some kind of load. Hence, in the case of magnetisation of a material of ***high retentivity***, all the energy put into it originally for straightening the molecules is not recovered when H is reduced to zero. We will now proceed to find this loss of energy per cycle of magnetisation.

Let l = mean length of the iron bar ; A = its area of cross-section; N = No. of turns of wire of the solenoid.

---

* In fact, when H is varied a number of times between fixed positive and negative maxima, the size of the loop becomes smaller and smaller till the material is cyclically magnetised. A material is said to be cyclically magnetised when for each increasing (or decreasing) value of H, B has the same value in successive cycles.

If B is the flux density at any instant, then $\Phi = BA$.

When current through the solenoid changes, then flux also changes and so produces an induced e.m.f. whose value is

$$e = N\frac{d\Phi}{dt} \text{ volt} = N\frac{d}{dt}(BA) = NA\frac{dB}{dt} \text{ volt}$$

(neglecting –ve sign)

Now $H = \frac{NI}{l}$ or $I = \frac{Hl}{N}$

The power or rate of expenditure of energy in maintaining the current 'I' against induced

e.m.f. 'e' is $= e\,I$ watt $= \frac{Hl}{N} \times NA\frac{dB}{dt} = AlH\frac{dB}{dt}$ watt

Energy spent in time 'dt' $= Al.H\frac{dB}{dt} \times dt = Al.H.dB$ joule

Total net work done for one cycle of magnetisation is $W = Al\oint H\,dB$ joule

where $\oint$ stands for integration over the whole cycle. Now, '$H\,dB$' represents the shaded area in Fig. 8.2. Hence, $\oint HdB$ = area of the loop *i.e.* the area between the B/H curve and the B-axis

$\therefore$ work done/cycle $= A_l \times$ (area of the loop) joule.

Now Al = volume of the material

$\therefore$ net work done/cycle/m^3 = (loop area) joule

or W_h = (Area of B/H loop) joule m^3/cycle

Precaution

Scale of B and H should be taken into consideration while calculating the actual loop area. For example, if the scales are, 1 cm = x AT/m –for H

and 1 cm = y Wb/m^2 –for B

then W_h = xy (area of B/H loop) joule/m^3/cycle

In the above expression, loop area has to be in cm^2.

As seen from above, hysteresis loop measures the energy dissipated due to hysteresis which appears in the form of heat and so raises the temperature of that portion of the magnetic circuit which is subjected to magnetic reversal. The shape of the hysteresis loop depends on the nature of the magnetic material (Fig. 8.4).

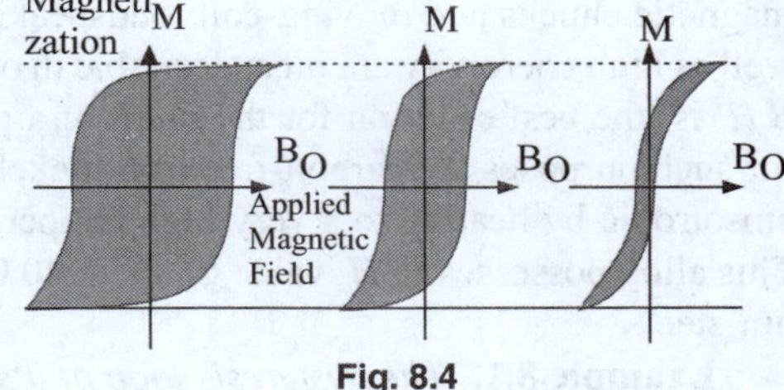

Fig. 8.4

Loop 1 is for hard steel. Due to its high retentivity and collectivity, it is well suited for making permanent magnets. But due to large hysteresis loss (as shown by large loop area) it is not suitable for rapid reversals of magne-tisation. Certain alloys of aluminium, nickel and steel called Alnico alloys have been found extremely suitable for making permanent magnets.

Loop 2 is for wrought iron and cast steel. It shows that these materials have high permeability and fairly good coercivity, hence making them suitable for cores of electromagnets.

Loop 3 is for alloyed sheet steel and it shows high permeability and low hysteresis loss. Hence, such materials are most suited for making armature and transformer cores which are subjected to rapid reversals of magnetisation.

8.3 Properties and Applications of Ferromagnetic Materials

Ferromagnetic materials having low retentivities are widely used in power and communication apparatus. Since silicon iron has high permeability and saturation flux density, it is extensively used in the magnetic circuits of electrical machines and heavy current apparatus where a high flux density is desirable in order to limit the cross-sectional area and, therefore, the weight and cost. Thin silicon-iron laminations (clamped together but insulated from each other by varnish, paper or their own surface scale) are used in the construction of transformer and armature cores where it is essential to minimize hysteresis and eddy-current losses.

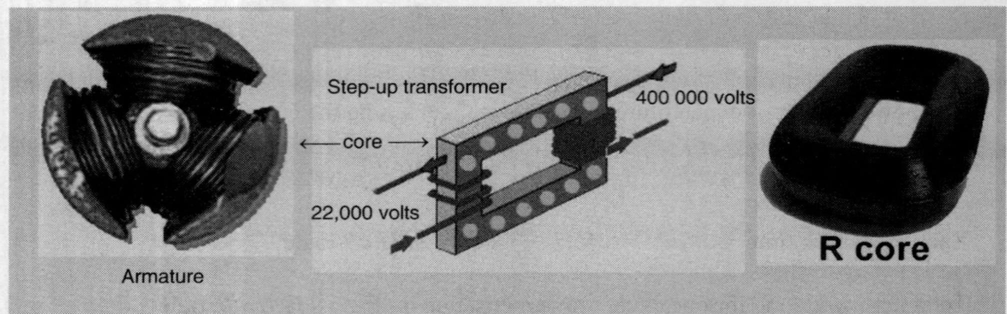

In field systems (where flux remains constant), a little residual magnetism is desirable. For such systems, high permeability and high saturation flux density are the only important requirements which are adequately met by fabricated rolled steel or cast or forged steel.

Frequencies used in line communication extend up to 10 MHz whereas those used in radio vary from about 100 kHz to 10 GHz. Hence, such material which have high permeability and low losses are very desirable. For these applications, nickel-iron alloys containing up to 80 per cent of nickel and a small percentage of molybdenum or copper, cold rolled and annealed are very suitable.

8.4 Permanent Magnet Materials

Permanent magnets find wide application in electrical measuring instruments, magnetos, magnetic chucks and moving-coil loudspeakers etc. In permanent magnets, high retentivity as well as high coercivity are most desirable in order to resist demagnetisation. In fact, the product $B_r H_c$ is the best criterion for the merit of a permanent magnet. The material commonly used for such purposes are carbon-free iron-nickel-aluminium copper-cobalt alloys which are made anisotropic by heating to a very high temperature and then cooling in a strong magnetic field. This alloy possesses $B_r H_c$ value of about 40,000 J/m^3 as compared with 2,500 J/m^3 for chromium-steel.

Example 8.1. *The hysteresis loop of a sample of sheet steel subjected to a maximum flux density of 1.3 Wb/m^2 has an area of 93 cm^2, the scales being 1 cm = 0.1 Wb/m^2 and 1 cm = 50 AT/m. Calculate the hysteresis loss in watts when 1500 cm^3 of the same material is subjected to an alternating flux density of 1.3 Wb/m^2 peak value of a frequency of 65 Hz.*

(Electromechanics, Allahabad Univ, 1992)

Solution. Loss $= xy$ (area of B/H loop) J/m^3/cycle

$= 0.1 \times 50 \times 93 = 465$ J/m^3/cycle

Volume $= 1500$ cm$^3 = 15 \times 10^{-4}$ m^3 ; No. of reversals/second = 65

$\therefore \quad W_h = 465 \times 15 \times 10^{-4} \times 65$ J/s = **45.3 W**

Note. The given value of $B_{max} = 1.3$ Wb/m^2 is not required for solution.

Example 8.2. *Calculate the hourly loss of energy in kWh in a specimen of iron, the hysteresis loop of which is equivalent in area to 250 J/m³. Frequency 50 Hz ; specific gravity of iron 7.5; weight of specimen 10 kg.* **(Electrical Engg. Materials, Nagpur Univ. 1991)**

Solution. Hysteresis loss = 250 J/m³/cycle, Mass of iron = 10 kg

Volume of iron specimen = $10/7.5 \times 10^3$ m³ = $10^{-2}/7.5$ m³

No. of cycles of reversals/hr = 60 × 50 = 3000

$\therefore$ loss/hour = $250 \times (10^{-2}/7.5) \times 3000 = 1000$ J = $1000/36 \times 10^5$ = $\mathbf{27.8 \times 10^{-5}}$ **kWh**

Example 8.3. *The hysteresis loop for a certain magnetic material is drawn to the following scales : 1 cm = 200 AT/m and 1 cm = 0.1 Wb/m². The area of the loop is 48 cm². Assuming the density of the material to be 7.8×10^3 kg/m³, calculate the hysteresis loss in watt/kg at 50 Hz.*

(Elect. Circuits & Fields, Gujarat Univ.)

Solution. Hysteresis loss = xy (area of B/H loop) J/m³/cycle

Now, 1 cm = 200 AT/m ; 1 cm = 0.1 Wb/m²

$\therefore$ x = 200, y = 0.1, area of loop = 48 cm²

$\therefore$ loss = $200 \times 0.1 \times 48 = 960$ J/m³/cycles, Density = 7.8×10^2 kg/m²

Volume of 1 kg of material = mass/density = $1/7.8 \times 10^3$ m³

$\therefore$ loss = $960 \times 1/7.8 \times 10^3$ J/cycle No. of reversals/second = 50

$\therefore$ loss = $960 \times 50 \times 10^{-3}/7.8 = 6.15$ J/s or watt

$\therefore$ hysteresis loss = **6.15 watt/kg.**

Example 8.4. *Determine the hysteresis loss in an iron core weighing 50 kg having a density of 7.8×10^3 kg/m³ when the area of the hysteresis loop is 150 cm², frequency is 50 Hz and scales on X and Y axes are : 1 cm = 30 AT/cm and 1 cm = 0.2 Wb/m² respectively.*

(Elements of Elect. Engg-1, Bangalore Univ.)

Solution. Hysteresis loss = xy (area of B/H loop) J/m³/cycle

1 cm = 30 AT/cm = 3000 AT/m ; 1 cm = 0.2 Wb/m²

x = 3000, y = 0.2, A = 150 cm²

$\therefore$ loss = $3000 \times 0.2 \times 150 = 90{,}000$ J/m³/cycle

Volume of 50 kg of iron = $m/\rho = 50/7.8 \times 10^{-3} = 6.4 \times 10^{-3}$ m³

$\therefore$ loss = $90{,}000 \times 6.4 \times 10^{-3} \times 50 = 28{,}800$ J/s or watts = **28.8 kW**

Example 8.5. *In a transformer core of volume 0.16 m³, the total iron loss was found to be 2,170 W at 50 Hz. The hysteresis loop of the core material, taken to the same maximum flux density, had an area of 9.0 cm² when drawn to scales of 1 cm = 0.1 Wb/m² and 1 cm = 250 AT/m. Calculate the total iron loss in the transformer core if it is energised to the same maximum flux density but at a frequency of 60 Hz.*

Solution. $W_h = xy \times$ (area of hysteresis loop) where x and y are the scale factors.

$W_h = 9 \times 0.1 \times 250 = 225$ J/m³/cycle

At 50 Hz

Hysteresis loss = $225 \times 0.16 \times 50 = 1{,}800$ W ; Eddy-current loss = $2{,}170 - 1800 = 370$ W

At 60 Hz

Hysteresis loss = $1800 \times 60/50 = 2{,}160$ W ; Eddy-current loss = $370 \times (60/50)^2 = 533$ W

Total iron loss = 2,160 + 533 = **2,693 W**

Tutorial Problem 8.1

1. The area of a hysteresis loop of a material is 30 cm². The scales of the co-ordinates are : 1 cm = 0.4 Wb/m² and 1 cm = 400 AT/m. Determine the hysteresis power loss if 1.2×10^{-3} m³ of the material is subjected to alternating flux density at 50 Hz. **[288 W] (*Elect. Engg., Aligarh Univ*)**

2. Calculate the loss of energy caused by hysteresis in one hour in 50 kg of iron when subjected to cyclic magnetic changes. The frequency is 25 Hz, the area of the hysteresis loop represents 240 joules/m^3 and the density of iron is 7800 kg/m^3.

[138,240] (*Principles of Elect. Engg. I, Jadvapur Univ.*)

3. The hysteresis loop of a specimen weighing 12 kg is equivalent to 300 joules/m^3. Find the loss of energy per hour at 50 Hz. Density of iron is 7500 kg/m^3.

[86,400] (*Electrotechnics – I, Gawahati Univ.*)

4. The area of the hysteresis loop for a steel specimen is 3.84 cm^2. If the ordinates are to the scales: 1 cm = 400 AT/m and 1 cm = 0.5 Wb/m^2, determine the power loss due to hysteresis in 1,200 cm^3 of the steel if it is magnetised from a supply having a frequency of 50 Hz. **[46.08 W]**
5. The armature of a 4-pole d.c. motor has a volume of 0.012 m^3. In a test on the steel iron used in the armature carried out to the same value of maximum flux density as exists in the armature, the area of the hysteresis loop obtained represented a loss of 200 J/m^3. Determine the hysteresis loss in watts when the armature rotates at a speed of 900 r.p.m. **[72 W]**
6. In a magnetisation test on a sample of iron, the following values were obtained.

H (AT/m)	1,900	2,000	3,000	4,000	4,500	3,000	1,000	0	−1,000	−1,900
B (Wb/m^2)	0	0.2	0.58	0.7	0.73	0.72	0.63	0.54	0.38	0

Draw the hysteresis loop and find the loss in watts if the volume of iron is 0.1 m^3 and frequency is 50 Hz. **[22 kW]**

8.5 Steinmetz Hysteresis Law

It was experimentally found by Steinmetz that hysteresis loss per m^3 per cycle of magnetisation of a magnetic meterial depends on **(*i*)** the maximum flux density established in it *i.e.* B_{max} and **(*ii*)** the magnetic quality of the material.

$$\therefore \quad \text{Hysteresis loss } W_h \propto B_{max}^{1.6} \text{ joule/m}^3\text{/cycle} = \eta\, B_{max}^{1.6} \text{ joule/m}^3 \text{ cycle}$$

where η is a constant depending on the nature of the magnetic material and is known as **Steinmetz hysteresis coefficient.** The index 1.6 is empirical and holds good if the value of B_{max} lies between 0.1 and 1.2 Wb/m^2. If B_{max} is either lesser than 0.1 Wb/m^2 or greater than 1.2 Wbm^2, the index is greater than 1.6.

$$\therefore \quad W_h = \eta B_{max}^{1.6}\, fV \text{ J/s or watt}$$

where f is frequency of reversals of magnetisation and V is the volume of the magnetic material.

The armatures of electric motors and generators and transformer cores etc. which are subjected to rapid reversals of magnetisation should, obviously, be made of substances having low hysteresis coefficient in order to reduce the hysteresis loss.

Example 8.6. *A cylinder of iron of volume 8 × 10^{-3} m^3 revolves for 20 minutes at a speed of 3,000 r.p.m in a two-pole field of flux density 0.8 Wb.m^2. If the hysteresis coefficient of iron is 753.6 joule/m^3, specific heat of iron is 0.11, the loss due to eddy current is equal to that due to hysteresis and 25% of the heat produced is lost by radiation, find the temperature rise of iron. Take density of iron as 7.8 × 10^3 kg/m^3.* **(Elect. Engineering-I, Osmania Univ.)**

Solution. An armature revolving in a multipolar field undergoes one magnetic reversal after passing under a pair of poles. In other words, number of magnetic reversals in the same as the number of *pair* of poles. If P is the number of poles, the magnetic reversals in one revolution are $P/2$. If speed of armature rotation is N r.p.m, then number of revolutions/second = $N/60$.

No. of reversals/second = reversals in one revolutions × No. of revolutions/second

$$= \frac{P}{2} \times \frac{N}{60} = \frac{PN}{120} \text{ reversals/second}$$

Here $N = 3{,}000 \text{r.p.m}; P = 2 \therefore f = \frac{3{,}000 \times 2}{120} = 50$ reversals/second

According to Steinmetz's hysteresis law, $W_h = \eta B_{max}^{1.6} f V$ watt

Note that f here stands for magnetic reversals/second and not for mechanical frequency armature rotation.

$$W_h = 753.6 \times (0.8)^{1.6} \times 50 \times 8 \times 10^{-3} = 211 \text{ J/s}$$

Loss in 20 minutes = 211 × 1,200 = 253.2 × 10^3 J

Eddy current loss = 253.2 × 10^3 J; Total loss = 506.4 × 10^3 J

Heat produced = 506.4 × 10^3/4200 = 120.57 kcal ; Heat utilized = 120.57 × 0.75 = 90.43 kcal

Heat absorbed by iron = (8 × 10^{-3} × 7.8 × 10^3) × 0.11 t kcal

$\therefore$ (8 × 10^{-3} × 7.8 × 10^3) × 0.11 × t = 90.43 $\therefore$ t = **13.17°C**

Example 8.7. *The area of the hysteresis loop obtained with a certain specimen of iron was 9.3 cm². The coordinates were such that 1 cm = 1,000 AT/m and 1 cm = 0.2 Wb/m². Calculate* **(a)** *the hysteresis loss per m³ per cycle and* **(b)** *the hysteresis loss per m³ at a frequency of 50 Hz if the maximum flux density were 1.5 Wb/m²* **(c)** *calculate the hysteresis loss per m³ for a maximum flux density of 1.2 Wb/m² and a frequency of 30 Hz, assuming the loss to be proportional to* $B_{max}^{1.6}$.

(Elect. Technology, Allahabad Univ. 1991)

Solution. (*a*) W_h = xy × (area of B/H loop)

= 1,000 × 0.2 × 9.3 = **1860 J/m²/cycle**

(*b*) W_h = 1,860 × 50 J/s/m³ = **93,000 W/m³**

(c) $W_h = \eta B_{max}^{1.8} f V W$ For a given specimen, $W_h \propto B_{max}^{1.8} f$

In (*b*) above, 93,000 α $1.5^{1.8}$ × 50 and W_h α $1.2^{1.8}$ × 30

$$\therefore \quad \frac{W_h}{93{,}000} = \left(\frac{1.2}{1.5}\right)^{1.8} \times \frac{30}{50}; \; W_h = 93{,}000 \times 0.669 \times 0.6 = \mathbf{37.360}$$

Example 8.8. *Calculate the loss of energy caused by hysteresis in one hour in 50 kg of iron if the peak density reached is 1.3 Wb/m² and the frequency is 25 Hz. Assume Steinmetz coefficient as 628 J/m³ and density of iron as 7.8 × 10³ kg/m³.*

What will be the area of B/H curve of this specimen if 1 cm = 12.4 AT/m and 1 cm = 0.1 Wb/m².

(Elect. Engg. ; Madras Univ.)

Solution. $W_h = \eta B_{max}^{1.6} f V$ watt ; volume $V = \frac{50}{7.8 \times 10^3} = 6.41 \times 10^{-3}$ m³

$\therefore$ W_h = 628 × $1.3^{1.6}$ × 25 × 6.41 × 10^{-3} = 152 J/s

Loss in one hour = 153 × 3,600 = 551,300 J

As per Steinmetz law, hysteresis loss = $\eta B_{max}^{1.6}$ J/m³/cycle

Also, hysteresis loss = xy (area of B/H loop)

Equating the two, we get

$$628 \times 1.3^{1.6} = 12.5 \times 0.1 \times \text{loop area}$$

$$\therefore \quad \text{loop area} = 628 \times 1.3^{1.6}/1.25 = \mathbf{764.3\ cm^2}$$

Tutorial Problem 8.2

1. In a certain transformer, the hysteresis loss is 300 W when the maximum flux density is 0.9 Wb/m^2 and the frequency 50 Hz. What would be the hysteresis loss if the maximum flux density were increased to 1.1 Wb/m^2 and the frequency reduced to 40 Hz. Assume the hysteresis loss over this range to be proportional to $B_{max}^{1.7}$ **[337 W]**
2. In a transformer, the hysteresis loss is 160 W when the value of B_{max} = 1.1 Wb/m^2 and when supply frequency is 60 Hz. What would be the loss when the value of B_{max} is reduced to 0.9 Wb/m^2 and the supply frequency is reduced to 50 Hz. **[97 W] (*Elect. Engg. II, Bangalore Univ.*)**

8.6 Energy Stored in a Magnetic Field

For establishing a magnetic field, energy must be spent, though no energy is required to ***maintain*** it. Take the example of the exciting coils of an electromagnet. The energy supplied to it is spent in two ways **(*i*)** part of it goes to meet I^2R loss and is lost once for all **(*ii*)** part of it goes to create flux and is stored in the magnetic field as potential energy and is similar to the potential energy of a raised weight. When a weight *W* is raised through a height of *h*, the potential energy stored in it is W_h. Work is done in ***raising*** this weight but once raised to a certain height, no further expenditure of energy is required to ***maintain*** it at that position. This mechanical potential energy can be recovered, so can be the electrical energy stored in the magnetic field.

When current through an inductive coil is gradually changed from zero to maximum value *I*, then every change of it is opposed by the self-induced e.m.f. produced due to this change. Energy is needed to overcome this opposition. This energy is stored in the magnetic field of the coil and is, later on, recovered when that field collapses. The value of this stored energy may be found in the following two ways:

(*i*) First Method. Let, at any instant,

i = instantaneous value of current ; e = induced e.m.f. at that instant = $L.di/dt$

Then, work done in time *dt* in overcoming this opposition is

$$dW = ei\ dt = L.\frac{di}{dt} \times i \times dt = Li\ di$$

Total work done in establishing the maximum steady current of *I* is

$$\int_0^W dW = \int_0^I L.i.di = LI^2 \text{ or } W = \frac{1}{2} L\,I^2$$

This work is stored as the energy of the magnetic field $\therefore\ E = \frac{1}{2} L\,I^2$ joules

(*ii*) Second Method

If current grows uniformly from zero value to its maximum steady value *I*, then average current is *I*/2. If *L* is the inductance of the circuit, then self-idcued e.m.f. is $e = LI/t$ where '*t*' is the time for the current change from zero to *I*.

$$\therefore \quad \text{Average power absorbed} = \text{induced e.m.f.} \times \text{average current}$$

$$= L\frac{1}{t} \times \frac{1}{2} I = \frac{1}{2} \frac{L\,I^2}{t}$$

$$\text{Total energy absorbed} = \text{power} \times \text{time} = \frac{1}{2}\frac{L I^2}{t} \times t = \frac{1}{2} L I^2$$

$$\therefore \quad \text{energy stored} \quad E = \frac{1}{2} L I^2 \text{ joule}$$

It may be noted that in the case of series-aiding coils, energy stored is

$$E = \frac{1}{2}(L_1 + L_2 + 2M) I^2 = \frac{1}{2} L_1 I^2 + \frac{1}{2} L_2 I^2 + M I^2$$

Similarly, for series-opposing coils, $E = \frac{1}{2} L_1 I^2 + \frac{1}{2} L_2 I^2 - M I^2$

Example 8.9. *Reluctance of a magnetic circuit is known to be 10^5 AT/Wb and excitation coil has 200 turns. Current in the coil is changing uniformly at 200 A/s. Calculate **(a)** inductance of the coil **(b)** voltage induced across the coil and **(c)** energy stored in the coil when instantaneous current at t = 1 second is 1 A. Neglect resistance of the coil.*

(Elect. Technology, Univ. of Indore, 1987)

Solution. (*a*) $L = N^2/S = 200^2/10^5 = \mathbf{0.4\ H}$

(*b*) $e_L = L\, dI/dt = 0.4 \times 200 = \mathbf{80\ V}$

(*c*) $E = \frac{1}{2} L I^2 = 0.5 \times 0.4 \times I^2 = \mathbf{0.2\ J}$

Example 8.10. *An iron ring of 20 cm mean diameter having a cross-section of 100 cm² is wound with 400 turns of wire. Calculate the exciting current required to establish a flux density of 1 Wb/m² if the relative permeability of iron is 1000. What is the value of energy stored ?*

(Elect. Engg-I, Nagpur Univ. 1992)

Solution. $B = \mu_0 \mu_r NI/l$ Wb/m²

$\therefore$ $1 = 4\pi \times 10^{-7} \times 1000 \times 400\, I/0.2\pi$ or $I = \mathbf{1.25\ A}$

Now, $L = \mu_0 \mu_r AN^2/l = 4\pi \times 10^{-7} \times 10^3 \times (100 \times 10^{-4}) \times (400)^2/0.2\pi = 32.H$

$$E = \frac{1}{2} LI^2 = \frac{1}{2} \times 3.2 \times 1.25^2 = \mathbf{2.5\ J}$$

8.7 Rate of Change of Stored Energy

As seen from Art. 8.6, $E = \frac{1}{2} L I^2$. The rate of change of energy can be found by differentiating the above equation

$$\frac{dE}{dt} = \frac{1}{2}\left[L.2.I.\frac{dI}{dt} + I^2\frac{dL}{dt}\right] = LI.\frac{dI}{dt} = \frac{1}{2} I^2 \frac{dL}{dt}$$

Example 8.11. *A relay (Fig. 8.5) has a coil of 1000 turns and an air-gap of area 10 cm² and length 1.0 mm. Calculate the rate of change of stored energy in the air-gap of the relay when*

(*i*) *armature is stationary at 1.0 mm from the core and current is 10 mA but is increasing at the rate of 25 A/s.*

(*ii*) *current is constant at 20 mA but inductance is changing at the rate of 100 H/s.*

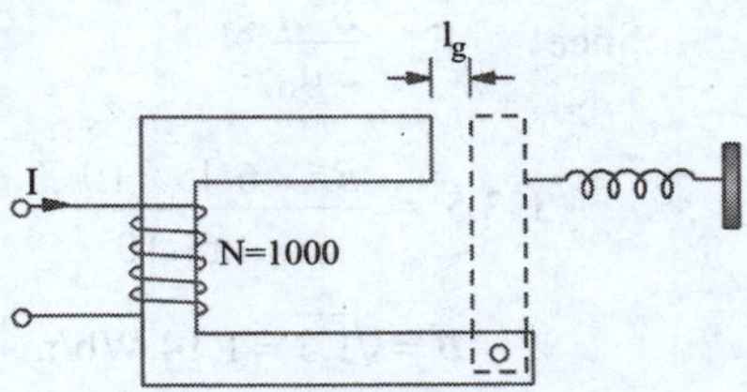

Fig. 8.5

Solution. $L = \frac{\mu_0 N^2 A}{l_g}$

$$= \frac{4\pi \times 10^{-7} \times (10^3)^2 \times 10 \times 10^{-4}}{1 \times 10^{-3}} = 1.26 \text{ H}$$

(*i*) Here, $dI/dt = 25$ A/s, $dL/dt = 0$ because armature is stationary.

$$\therefore \quad \frac{dE}{dt} = L\,I\,\frac{dI}{dt} = 1.26 \times 10 \times 10^{-3} \times 15 = \mathbf{0.315\ W}$$

(*ii*) Here, $dL/dt = 100$ H/s; $dI/dt = 0$ because current is constant.

$$\therefore \quad \frac{dE}{dt} = \frac{1}{2} I^2 \frac{dL}{dt} = \frac{1}{2}(20 \times 10^{-3})^2 \times 100 \ \mathbf{0.02\ W}$$

8.8 Energy Stored Per Unit Volume

It has already been shown that the energy stored in a magnetic field of length l metre and of cross-section A m² is $E = \frac{1}{2} L\,I^2$ joule or $E = \frac{1}{2} \times \frac{\mu_0 \mu_r AN^2}{l} . I^2$ joule

Now $H = \frac{NI}{l}$ $\therefore$ $E = \left(\frac{NI}{l}\right)^2 \times \frac{1}{2}\mu_0\mu_r\, Al = \frac{1}{2}\mu_0\,\mu_r\,H^2 \times Al$ joule

Now, Al = volume of the magnetic field in m³

$$\therefore \quad \text{energy stored/m}^3 = \frac{1}{2}\mu_0\mu_r H^2 = \frac{1}{2} BH \text{ joule} \qquad (\because \mu_0\,\mu_r\,H = B)$$

$$= \frac{B^2}{2\,\mu_0\,\mu_r} \text{ joule} \qquad \text{– in a medium}$$

or

$$= \frac{B^2}{2\,\mu_0} \text{ joule} \qquad \text{– in air}$$

Example 8.12. *A horse-shoe magnet is formed out of a bar of wrought iron 45.7 cm long, having a corss-section of 6.45 cm². Exciting coils of 500 turns are placed on each limb and connected in series. Find the exciting current necessary for the magnet to lift a load of 68 kg assuming that the load has negligible reluctance and makes close contact with the magnet. Relative permeability of iron = 700.* **(Elect. Engg. A.M.Ae. S.I., June, 1992)**

Solution. Horse-shoe magnet is shown in Fig. 8.6.

Force of attraction of each pole = 68/2 = 34 kg = 34 × 9.81 = 333.5 N

$A = 6.45 \text{ cm}^2 = 6.45 \times 10^{-4} \text{ m}^3$

Since $F = \frac{B^2 A}{2\,\mu_0}$ N

$$\therefore \quad 333.5 = \frac{B^2 \times 6.45 \times 10^{-4}}{2 \times 4\pi \times 10^{-7}}$$

$$\therefore \quad B = \sqrt{1.3} = 1.14 \text{ Wb/m}^2$$

and $H = B/\mu_0\,\mu_r = 1.14/4\pi \times 10^{-7} \times 700 = 1296$ AT/m

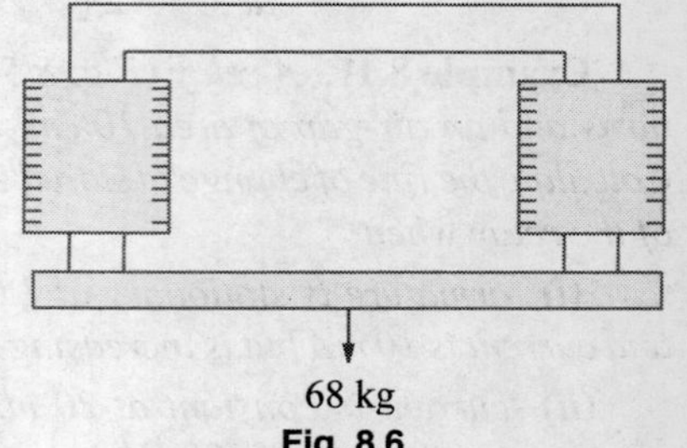

Fig. 8.6

Length of the plate = 45.7 cm = 0.457 m

$\therefore$ AT required = 1296 × 0.457 = 592.6

No. of turns = 500 × 2 = 1000 $\therefore$ current required = 592.6/1000 = **0.593 A**

Example 8.13. *The pole face area of an electromagnet is 0.5 m²/pole. It has to lift an iron ingot weighing 1000 kg. If the pole faces are parallel to the surface of the ingot at a distance of 1 millimetre, determine the coil m.m.f. required. Assume permeability of iron to be infinity at the permeability of free space is 4π × 10⁻⁷ H/m.* **(Elect. Technology, Univ. of Indore)**

Solution. Since iron has a permeability of infinity, it offers zero reluctance to the magnetic flux.

Force at two poles = $2 \times B^2A/2\mu_0 = B^2A/\mu_0$

$\therefore$ $B^2 \times 0.5/4\pi \times 10^{-7} = 1000 \times 9.8$, $B = 0.157$ Wb/m²

$\therefore$ $H = 0.157/4\pi \times 10^{-7} = 125 \times 10^3$ AT/m, $l = 2 \times 1 = 2$ mm $= 2 \times 10^{-3}$ m

$\therefore$ AT required = $125 \times 10^3 \times 2 \times 10^{-3}$ = **250.**

Example 8.14. *A soft iron ring having a mean circumference of 40 cm and cross-sectional area of 3 cm² has two radial saw cuts made at diametrically opposite points. A brass plate 0.5 mm thick is inserted in each gap. The ring is wound with 800 turns. Calculate the magnetic leakage and fringing. Assume the following data for soft iron :*

B (Wb/m²) :	*0.76*	*1.13*	*1.31*	*1.41*	*1.5*
H (AT/m) :	*50*	*100*	*150*	*200*	*250*

(Elect. Engineering-I, Delhi Univ.)

Solution. It should be noted that brass is a non-magnetic material.

Force at one separation = $B^2A/2\mu_0$ newton.

Force at both separations = B^2A/μ_0 newton.

Now $F = 12$ kg wt $= 12 \times 9.81 = 117.7$ *N*

$\therefore$ $117.7 = B^2 \times 3 \times 10^{-4}/4\pi \times 10^{-7}$; $B = 0.7$ Wb/m²

If B/H curve is drawn, it will be found that for $B = 0.7$ Wb/m², value of $H = 45$ AT/m.

Now, length of iron path = 40 cm = 0.4 m. AT required for iron path = 45 × 0.4 = 18

Value of H in the non-magnetic brass plates = $B/\mu_0 = 0.7/4\pi \times 10^{-7}$ = 557,042 AT/m

Total thickness of brass plates = 0.5 × 2 = 1 mm

AT required = $557{,}042 \times 1 \times 10^{-3}$ = 557, Total AT needed = 18 + 557 = 575

$\therefore$ magnetising current required = 557/800 = **0.72 A**

Example 8.15. *The following particulars are taken from the magnetic circuit of a relay; Mean length of iron circuit = 20 cm; length of air gap = 2 mm, number of turns on core = 8000, current through coil = 50 mA, relative permeability of iron = 500. Neglecting leakage, what is the flux density in the air-gap ? If the area of the core is 0.5 cm², what is the pull exerted on the armature ?*

Solution. Flux $\Phi = \dfrac{NI}{\Sigma l/\mu_0 \mu_r A}$.

Now, m.m.f $= NI = 8000 \times 50 \times 10^{-3} = 400$ AT

Total circuit reluctance $= \dfrac{l}{\Sigma \mu_0 \mu_r A}$ AT/Wb or H^{-1}

$$= \frac{0.2}{500 \times 4\pi \times 10^{-7} \times 0.5\ 10^{-4}} + \frac{2 \times 10^{-3}}{4\pi \times 10^{-7} \times 0.5 \times 10^{-4}}$$

$$\Phi = \frac{400}{12 \times 10^{7} / \phi} \text{ Wb ; Flux density } B = \frac{\Phi}{A} = \frac{400\pi}{12 \times 10^{7} \times 0.5 \times 10^{-4}} = \mathbf{0.21\ Wb/m^2}$$

The pull on the armature $= \dfrac{B^2 A}{2\,\mu_0} N = \dfrac{0.21^2 \times 0.5 \times 10^{-4}}{2 \times 4\pi \times 10^{-7}} = \mathbf{0.87\ N}$

Tutorial Problem 8.3

1. An air-cored solenoid has a length of 50 cm and a diameter of 2 cm. Calculate its inductance if it has 1,000 turns and also find the energy stored in it if the current rises from zero to 5 A.

[0.7 mH; 8.7 mJ] (*Elect. Engg. and Electronic Bangalore Univ. 1998*)

2. An air-cored solenoid 1 m in length and 10 cm in diameter has 5000 turns. Calculate (*i*) the self inductance (*ii*) the energy stored in the magnetic field when a current of 2 A flows in the solenoid.

[(*i*) 0.2468 H (*ii*) 0.4936 J] (*F.E. Pune Univ.*)

3. Determine the force required to separate two magnetic surfaces with contact area of 100 cm² if the magnetic flux density across the surface is 0.1 Wb/m². Derive formula used, if any.

[39.8 N] (*Elect. Engg. A.M.Ae.S.I.*)

4. In a telephone receiver, the size of each of the two poles is 1.2 cm × 0.2 cm and the flux between each pole and the diaphragm is 3×10^{-6} Wb; with what force is the diaphragm attracted to the poles ?

[0.125 N] (*Elect. Engg. A.M.Ae.S.I. June 1991*)

5. A lifting magnet is required to raise a load of 1,000 kg with a factor of safety of 1.5. If the flux density across the pole faces is 0.8 Wb/m², calculate the area of each pole. **[577 cm²]**

6. Magnetic material having a surface of 100 cm² are in contact with each other. They are in a magnetic circuit of flux. 0.01 Wb uniformly distributed across the surface. Calculate the force required to detach the two surfaces. **[3,978 N] (*Elect. Engg. Kerala Univ.*)**

7. A steel ring having a mean diameter of 35 cm and a cross-sectional area of 2.4 cm² is broken by a parallel-sided air-gap of length 1.2 cm. Short pole pieces of negligible reluctance extend the effective cross-sectional area of the air-gap to 12 cm². Taking the relative permeability of steel as 700 and neglecting leakage, determine (*a*) the current necessary in 300 turns of wire wound on the ring to produce a flux density in the air-gap of 0.25 Wb/m² (*b*) the tractive force between the poles. **[(*a*) 13.16 A (*b*) 29.9 N]**

8. A cast iron ring having a mean circumference of 40 cm and a cross-sectional area of 3 cm² has two radial saw-cuts at diametrically opposite points. A brass plate is inserted in each gap (thickness 0.5 mm). If the ring is wound with 800 turns, calculate the magnetising current to exert a total pull of 3 kg between the two halves. Neglect any magnetic leakage and fringing and assume the magnetic data for the cast iron to be:

B (Wb/m²) :	0.2	0.3	0.4	0.5
H (AT) :	850	1150	1500	2000

[1.04 A]

8.9 Rise of Current in an Inductive Circuit

In Fig. 8.7 is shown a resistance of R in series with a coil of self-inductance L henry, the two being put across a battery of V volt. The R-L combination becomes connected to battery when switch S is connected to terminal 'a' and is short-circuited when S is connected to 'b'. The inductive coil is assumed to be resistanceless, its actual small resistance being included in R.

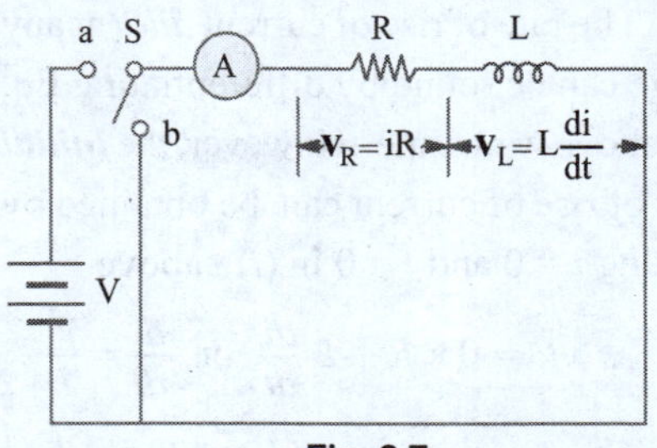

Fig. 8.7

When S is connected to 'a' the R-L combination is suddenly put across the voltage of V volt. Let us take the instant of closing S as the starting zero time. It is easily explained by recalling that the coil possesses electrical inertia *i.e.* self-inductance and hence, due to the production of the counter e.m.f. of self-inductance, delays the instantaneous full establishment of current through it.

We will now investigate the growth of current i through such an inductive circuit.

The applied voltage V must, at any instant, supply not only the ohmic drop iR over the resistance R but must also overcome the e.m.f. of self inductance *i.e.* $L di/dt$.

$$\therefore \quad V = v_R + v_L = iR + \frac{di}{dt}$$

$$\text{or} \quad (V - iR) = L\frac{di}{dt} \quad \therefore \quad \frac{di}{V - iR} = \frac{dt}{L}. \quad ...(i)$$

Multiplying both sides by $(-R)$, we get $(-R)\dfrac{di}{(V - iR)} = -\dfrac{R}{L}\,dt$

Integrating both sides, we get $\displaystyle\int \frac{(-R)\,di}{(V - iR)} = \int dt \;\therefore\; \log_e^{V - iR} = -\frac{R}{L}t + K \quad ...(ii)$

where e is the Napierian logarithmic base = 2.718 and K is constant of integration whose value can be found from the initial known conditions.

To begin with, when $t = 0$, $i = 0$, hence putting these values in **(*ii*)** above, we get

$$\log_e^V = K$$

Substituting this value of K in the above given equation, we have

$$\log_e^{V - iR} = \frac{R}{L}t + \log_e^V \text{ or } \log_e^{V - iR} - \log_e^V = -\frac{R}{L}t$$

$$\text{or} \quad \log_e \frac{V - iR}{V} = -\frac{R}{L}t = -\frac{1}{\lambda} \text{ where } L/R = \lambda \text{ 'time constant'}$$

$$\therefore \quad \frac{V - iR}{V} = e^{-t/\lambda} \text{ or } i = \frac{V}{R}(1 - e^{-t/\lambda})$$

Now, V/R represents the maximum steady value of current I_m that would eventually be established through the R-L circuit.

$$\therefore \quad i = I_m(1 - e^{-t/\lambda}) \quad ...(iii)$$

This is an exponential equation whose graph is shown in Fig. 8.8. It is seen from it that current rise is rapid at first and then decreases until at $t = \infty$, it becomes zero. Theoretically, current does not reach its maximum steady value I_m until infinite time. However, in practice, it reaches this value in a relatively short time of about 5λ.

The rate of rise of current di/dt at any stage can be found by differentiating Eq. (*ii*) above w.r.t. time. However, the ***initial*** rate of rise of current can be obtained by putting $t = 0$ and $i = 0$ in (*i*)* above.

$$\therefore \quad V = 0 \times R + L\frac{di}{dt} \quad \text{or} \quad \frac{di}{dt} = \frac{V}{L}$$

The constant $\lambda = L/R$ is known as the ***time-constant*** of the circuit. It can be variously defined as :

(*i*) It is the ***time*** during which current would have reached its maximum value of I_m (= V/R) had it maintained its initial rate of rise.

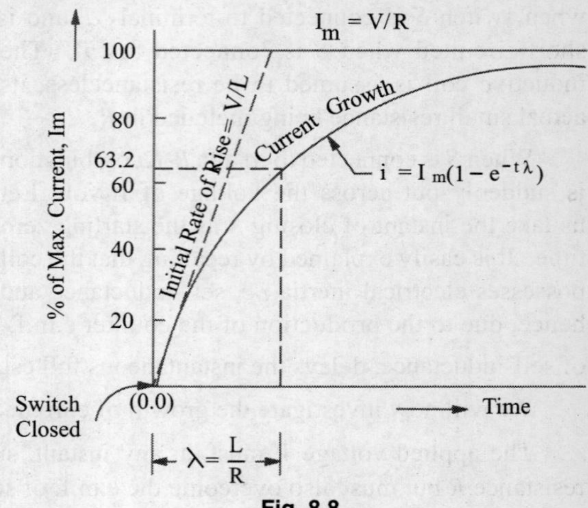

Fig. 8.8

$$\text{Time taken} = \frac{I_m}{\text{initial rate of rise}} = \frac{V/R}{V/L} = \frac{L}{R}$$

But actually the current takes more time because its rate of rise decreases gradually. In actual practice, in a time equal to the time constant, it merely reaches 0.632 of its maximum values as shown below:

Putting $t = L/R = \lambda$ in Eq. (*iii*) above, we get

$$i = I_m(1 - e^{-\lambda/\lambda}) = I_m\left(1 - \frac{1}{e}\right) = I_m\left(1 - \frac{1}{2.718}\right) = 0.632\, I_m$$

(*ii*) Hence, the time-constant λ of an *R-L* circuit may also be defined as the time during which the current ***actually*** rises to 0.632 of its maximum steady value (Fig. 8.8).

This delayed rise of current in an inductive circuit is utilized in providing time lag in the operation of electric relays and trip coils etc.

8.10 Decay of Current in an Inductive Circuit

When the switch *S* (Fig. 8.9) is connected to point '*b*', the *R-L* circuit is short-circuited. It is found that the current does not cease immediately, as it would do in a non-inductive circuit, but continues to flow and is reduced to zero only after an appreciable time has elapsed since the instant of short-circuit.

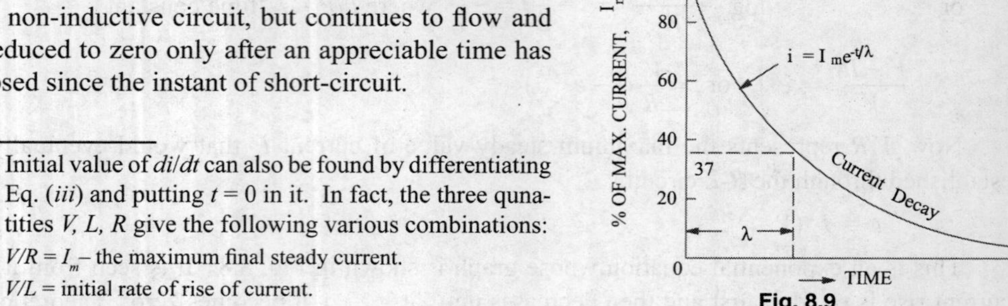

Fig. 8.9

* Initial value of di/dt can also be found by differentiating Eq. (*iii*) and putting $t = 0$ in it. In fact, the three qunatities *V, L, R* give the following various combinations:

$V/R = I_m$ – the maximum final steady current.

V/L = initial rate of rise of current.

L/R = time constant of the circuit.

The first rule of switching is that the current flowing through an inductance cannot change instantaneously. The second rule of switching is that the voltage across a capacitor cannot change instantaneously.

The equation for decay of current with time is found by putting $V = 0$.

$$0 = iR + L\frac{di}{dt} \text{ or } \frac{di}{i} = -\frac{R}{L}dt$$

Integrating both sides, we have

$$\int\frac{di}{i} = -\frac{R}{L}\int dt \therefore \log i = \frac{R}{L}t + K \qquad ...(i)$$

Now, at the instant of switching off current, $i = I_m$ and if time is counted from this instant, then $t = 0$

$$\therefore \qquad \log_e I_m = 0 + K$$

Putting the value of K in Eq (i) above, we get,

$$\log_e i = -\frac{t}{\lambda} = \log_e I_m$$

$$\therefore \qquad \log_e i/I_m = -\frac{t}{\lambda}$$

$$\therefore \qquad \frac{i}{I_m} = e^{-t/\lambda}$$

$$\text{or} \qquad i = I_m e^{-t/\lambda} \qquad ...(ii)$$

It is decaying exponential function and is plotted in Fig. 8.12. It can be shown again that theoretically, current should take infinite time to reach zero value although, in actual practice, it does so in a relatively short time of about 5λ.

Again, putting $t = \lambda$ in Eq. (ii) above, we get

$$i = \frac{I_m}{e} = \frac{I_m}{2.178} = 0.37\ I_m.$$

Hence, time constant (λ) of an R-L circuit may also be defined as the time during which current falls to 0.37 or 37% of its maximum steady value while decaying (Fig. 8.12).

Example 8.16. *A coil having an effective resistance of 20 Ω and an inductance of 5 H. is suddenly connected across a 50-V dc supply. What is the rate at which energy is stored in the field of the coil when current is **(a)** 0.5 A **(b)** 1.0 A and **(c)** steady ? Also find the induced EMF in the coil under the above conditions.*

Solution. (***a***) Power input = 50 × 0.5 = 25 W

Power wasted as heat = $i^2R = 0.5^2 \times 25 = 6.25$ W. Hence, rate of energy storage in the coil field is 25 – 6.25 = 18.75 W or J/s. (***b***) Power input = 50 × 1 = 50 W

Power lost as heat = $1^2 \times 25 = 25$ W. ∴ Rate of energy storage in field = 50 – 25 = 25 W or J/s.

(***c***) Steady value of current = 50/25 = 2 A. Power input = 50 × 2 = 100 W

Power lost as heat = $2^2 \times 25 = 100$ W

Rate of energy storage in field = 100 – 100 = 0; Now, $V = iR + e_L = V - iR$

(***a***) $e_L = 50 - 0.5 \times 25 = 37.5$ V (***b***) $e_L = 50 - 1 \times 25 = 25$ V (***c***) $e_L = 50 - 2 \times 25 = 0$ V.

Example 8.17. *A coil having a resistance of 10 Ω and an inductance of 4 H is switched across a 20-W dc source. Calculate **(a)** time required by the current to reach 50% of its final steady value and **(b)** value of the current after 0.5 second.*

Solution. The rise of current through an inductive circuit is given by the equation $i = I(1 - e^{-t/\lambda})$. It may be written as

$$e^{-t/\lambda} = \frac{I - i}{I} \quad \text{or} \quad \frac{I}{e^{t/h}} = \frac{I - i}{I} \quad \text{or} \quad e^{e/h} = \frac{I}{I - i}$$

Taking logs of both sides, we have

$$\frac{t}{\lambda}.\log^{e} = \log^{I/(I-i)} = I_n \frac{I}{(I-i)}$$

$$\therefore \quad \frac{Rt}{L} = \log^{I/(I-i)} = I_n \frac{I}{(I-i)}$$

(***a***) Now, $I = V/R = 20/10 = 2$ A

$$\therefore \quad t = \frac{4}{10} l_n \frac{2}{(2-1)} \aleph \frac{4}{10} \quad 0.693 \quad 0.2777 \text{ s.}$$

(***b***) $\lambda = L/R = 4/10 = 0.4$ s and $t = 0.5$ s

$$\therefore \quad i = 2(1 - e^{-t/0.4})$$

Example 8.18. *A coil has a time constant of 1 second and an inductance of 8 H. If the coil is connected to a 100 V d.c. source, determine:*

(*i*) *the rate of rise of current at the instant of switching* **(*ii*)** the steady value of the current and **(*iii*)** *the time taken by the current to reach 60% of the steady value of the current.*

(Electrotechnics-I, M.S. Univ. Baroda)

Solution. $\lambda = L/R$; $R = L/\lambda = 8/1 = 8$ ohm

(*i*) Initial $\quad di/dt = V/L = 100/8 =$ **12.5 A/s**

(*ii*) $\quad I_M = V/R = 100/8 =$ **12.5 A**

(***iii***) Here, $\quad i = 60\%$ of $12.5 = 7.5$ A

Now, $\quad i = I_m(1 - e^{-t/\lambda}) \quad \therefore \quad 7.5 = 12.5(1 - e^{-t/1})$; $\quad t =$ **0.915 second**

Example 8.19. *A d.c. voltage of 80 V is applied to a circuit containing a resistance of 80 Ω in series with an inductance of 20 H. Calculate the growth of current at the instant* **(*i*)** *of completing the circuit* **(*ii*)** *when the current is 0.5 A and* **(*iii*)** *when the current is 1 A.*

(Circuit Theory, Jadavpur Univ.)

Solution. The voltage equation for an *R-L* circuit is

$$V = iR + L\frac{di}{dt} \quad \text{or} \quad L\frac{di}{dt} = V - iR \quad \text{or} \quad \frac{di}{dt} = \frac{1}{L}(V - iR)$$

(***i***) when $\quad i = 0;\ \frac{di}{dt} = \frac{1}{L}(V - 0 \times R) = \frac{V}{L} = \frac{80}{20} = \mathbf{4\ A/s}$

(***ii***) when $\quad i = 0.5\ A;\ \frac{di}{dt} = \frac{80 - 0.5 \times 80}{20} = \mathbf{2\ A/s}$

(***iii***) when $\quad i = 1\ A;\ \frac{di}{dt} = \frac{80 - 80 \times 1}{20} = \mathbf{0.}$

In other words, the current has become steady at 1 ampere.

Example 8.20. *The two circuits of Fig. 8.10 have the same time constant of 0.005 second. With the same d.c. voltage applied to the two circuits, it is found that the steady state current of circuit (a) is 2000 times the initial current of circuit (b). Find R_1, L_1 and C.*

(Elect. Engg.-I, Bombay Univ.)

Solution. The time constant of circuit 8.10 (*a*) is $\lambda = L_1/R_1$ second, and that of circuit 8.10 (*b*) is $\lambda = CR_2$ second.

$$\therefore \quad L_1/R_1 = 0.005$$

$$C \times 2 \times 10^6 = 0.005, C = 0.0025 \times 10^{-6} = \mathbf{0.0025\ \mu F}$$

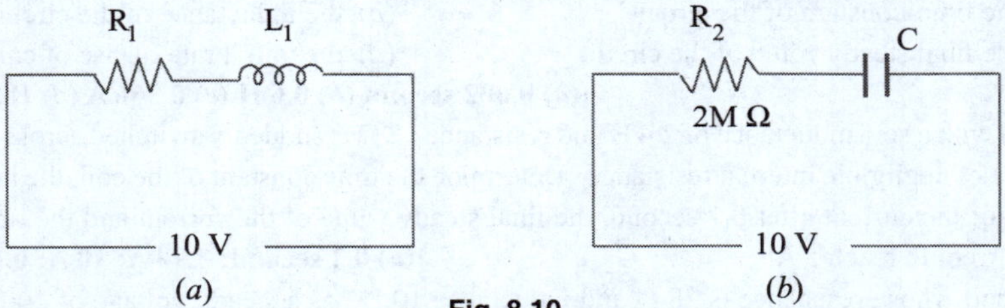

R1 L1 10 V (*a*) R2 C 2M Ω 10 V (*b*)

Fig. 8.10

Steady-state current of circuit 8.14 (*a*) is $= V/R_1 = 10/R_1$ amperes.

Initial current of circuit 8.14 (*b*) $= V/R_2 = 10/2 \times 10^6 = 5 \times 10^{-6} A$*

Now $\quad 10/R_1 = 2000 \times 5 \times 10^{-6} \therefore R_1 = 1000\ \Omega$.

Also $\quad L_1/R_1 = 0.005 \quad \therefore \quad L_1 = 1000 \times 0.005 = \mathbf{5\ H}$

Example 8.21. *A constant voltage is applied to a series R-L circuit at t = 0 by closing a switch. The voltage across L is 25 V at t = 0 and drops to 5 V at t = 0.025 second. If L = 2 H, what must be the value of R ?* **(Elect. Engg.-I Bombay Univ.)**

Solution. At $t = 0$, $i = 0$, hence there is no *iR* drop and the applied voltage must equal the back e.m.f. in the coil. Hence, the voltage across L at $t = 0$ represents the applied voltage.

At $t = 0.025$ second, voltage across L is 5 V, hence voltage across $R = 25 - 5 = 20$ V $\quad \therefore \quad iR = 20$ V – at $t = 0.025$ second. Now $i = I_m (I - e^{-t/\lambda})$

Here $\quad I_m = 25/R$ ampere, $t = 0.025$ second $\quad \therefore \quad i = \frac{25}{R}(1 - e^{-0.025/\lambda})$

$$R \times \frac{25}{R}(1 - e^{-0.025/\lambda}) = 20 \quad \text{or} \quad e^{0.025/\lambda} = 5 \quad \therefore \quad 0.025/\lambda = 2.3 \log_{10} 5 = 1.6077$$

$\therefore \quad \lambda = 0.025/1.6077 \quad$ Now $\quad \lambda = L/R = 2/R \quad \therefore \quad 2/R = 0.025/1.6077 \quad \therefore \quad \mathbf{R = 128.56\ \Omega}$

Example 8.22. *A circuit of resistance R ohms and inductance L henries has a direct voltage of 230 V applied to it. 0.3 second second after switching on, the current in the circuit was found to be 5 A. After the current had reached its final steady value, the circuit was suddenly short-circuited. The current was again found to be 5 A at 0.3 second second after short-circuiting the coil. Find the value of R and L.* **(Basic Electricity, Bombay Univ.)**

Solution. For growth ; $5 = I_m (1 - e^{-0.3/\lambda})$ ...(*i*)

For decay; $\quad 5 = I_m e^{-0.3/\lambda}$

Equating the two, we get, $I_m e^{-0.3/\lambda} = (1 - e^{-0.3/\lambda}) I_m$

or $\quad 2 e^{-0.3/\lambda} = 1 \quad \therefore \quad e^{-0.3/\lambda} = 0.5$ or $\lambda = 0.4328$

Putting this value in (*i*), we get,

$$5 = I_m = e^{0.3/0.4328} \quad \text{or} \quad I_m = 5 e^{+0.3/0.4328} = 5 \times 2 = 10 \text{ A.}$$

Now, $\quad I_m = V/R \quad \therefore \quad 10 = 230/R \quad$ or $\quad R = 230/10 = 23\ \Omega$ (approx.)

As $\quad \lambda = L/R = 0.4328$; $L = 0.4328 \times 23 = \mathbf{9.95\ H}$

* Because just at the time of starting the current, there is no potential drop across C so that the applied voltage is dropped across R_2. Hence, the initial charging current $= V/R_2$.

Tutorial Problem 8.4

1. A relay has a resistance of 300 Ω and is switched on to a 110 V d.c. supply. If the current reaches 63.2 per cent of its final steady value in 0.002 second, determine
   (*a*) the time-constant of the circuit (*b*) the inductance of the circuit
   (*c*) the final steady value of the circuit (*d*) the initial rate of rise of current.
   [(*a*) 0.002 second (*b*) 0.6 H (*c*) 0.366 A (*d*) 183 A/second]
2. A coil with a self-inductance of 2.4 H and resistance 12 Ω is suddenly switched across a 120-V d.c. supply of negligible internal resistance. Determine the time constant of the coil, the instantaneous value of the current after 0.1 second, the final steady value of the current and the time taken for the current to reach 5 A. **[(*a*) 0.2 second; 3.94 A; 10 A; 0.139 second]**
3. A circuit whose resistance is 20 Ω and inductance 10 H has a steady voltage of 100 V suddenly applied to it. For the instant 0.5 second after the voltage is applied, determine (*a*) the total power input to the circuit (*b*) the power dissipated in the resistance. Explain the reason for the difference between (*a*) and (*b*). **[(*a*) 316 W (*b*) 200 W]**
4. A lighting circuit is operated by a relay of which the coil has a resistance of 5 Ω and an inductance of 0.5 H. The relay coil is supplied from a 6-V d.c. source through a push-button switch. The relay operates when the current in the relay coil attains a value of 500 mA. Find the time interval between the pressing of the push-button and the closing of lighting circuit. **[53.8 ms]**
5. A 200-V d.c. supply is suddenly switched to a relay coil which has a time constant of 3 milli-second. If the current in the coil reaches 0.2 A after 3 milli-second, determine the final steady value of the current and the resistance of the coil. **[0.316 A; 632 Ω; 1.896 H]**
6. Explain the terms related to magnetic circuits:
   (i) coercive force (ii) residual flux. 1 ***(Nagpur University, Summer 2002)***
7. Derive an expression for the energy stored in the magnetic field of a coil of an inductance L henry. ***(Gujrat University, June/July 2003)***
8. What is the difference between B.H. curve and hysteresis loop? ***(Anna University, April 2002)***
9. Explain with neat diagram how can you obtain B.H. curve and hystersis loop of ring specimens. ***(Anna University, April 2002)***
10. Derive an expression for energy stored in an inductance. ***(V.T.U., Belgaum Karnataka University, Summer 2002)***
11. Derive an expression for the energy stored in a magnetic circuit. ***(V.T.U., Belgaum Karnataka University, January/February 2003)***
12. What is hysteresis and eddy current losses? What are the undesirable effects of eddy currents and hysteresis loss? How can they be minimised? Mention some applications of eddy currents. ***(U.P. Technical University, 2002) (RGPV Bhopal June 2002)***

OBJECTIVE TESTS – 8

1. Permanent magnets are normally made of
   (*a*) aluminium (*b*) wrought iron (*c*) cast iron (*d*) alnico alloys
2. A coil of 1000 turns is wound on a core. A current of 1 A flowing through the coil creates a core flux of 1 mWb. The energy stored in the magnetic field is
   (*a*) 0.25 J (*b*) 0.5 J (*c*) 1 J (*d*) 2 J
   (ESE 2003)

CHAPTER

9

Electrochemical Power Sources

Chapter Outline

- Faraday's Laws of electrolysis
- Value of Back e.m.f.
- Primary and Secondary Batteries
- Classification of Lead Storage Batteries
- Formation of Plates of Lead-acid Cells
- Plante Process
- Faure Process
- Positive Pasted Plates
- Negative Pasted Plates
- Comparison: Plante and Faure Plates
- Internal Resistance and Capacity of a Cell
- Electrical Characteristics of the Lead-acid Cell
- Battery Ratings
- Indications of a Fully-Charged Cell
- Voltage Regulators
- End-cell Control System
- Charging Systems
- Constant-current System-Constant-voltage System
- Trickle Charging
- Sulphation-Causes and Cure
- Mains operated Battery Chargers
- Car Battery Charger
- Automobile Battery Charger
- Static Uninterruptable Power Systems
- Alkaline Batteries
- Nickel-iron or Edison Batteries
- Nickel-Cadmium Batteries
- Silver-zinc Batteries
- High Temperature Batteries
- Secondary Hybrid Cells
- Fuel Cells

9.1. Faraday's Laws of Electrolysis

From his experiments, Faraday deduced two fund mental laws which govern the phenomenon of electrolysis. These are:

(i) **First Law.** The mass of ions liberated at an electrode is directly proportional to the quantity of electricity *i.e.* charge which passes through the electrolyte.

(ii) **Second Law.** The masses of ions of different substances liberated by the same quantity of electricity are proportional to their chemical equivalent weights.

Explanation of the First Law

If m = mass of the ions liberated, Q = quantity of electricity = $I \times t$ where I is the current and t is the time, then according to the first law $m \propto Q$ or $m = Z\,Q$ or $m = Z\,It$ where Z is a constant and is known as the electrochemical equivalent (*E.C.E.*) of the substance.

If Q = 1 coulomb *i.e.* I = 1 ampere and t = 1 second, then $m = Z$.

Hence, ***E.C.E.* of a substance is *equal to the mass of its ions liberated by the passage of one ampere current for one second through its electrolytic solution or by the passage of a charge of one coulomb.***

In fact, the constant Z is composite and it depends on the valency and atomic weight of the substance concerned. Its value is given by $Z = \left(\frac{1}{F} \cdot \frac{a}{v}\right)$ where a is the atomic weight, v the valency and F is Faraday's constant. It is so because m is proportional to atomic weight, since each ion carries a definite charge. Obviously, the charge carried by an ion is proportional to its valency. Now, consider the molecules of sulphuric acid and copper sulphate. The sulphion SO_4^- in the acid molecule is combined with two positive hydrogen ions, whereas in $CuSO_4$ molecule, it is combined only with one positive (bivalent) Cu^{++} ion. It is seen that a copper ion being bivalent carries twice the charge of a hydrogen ion which is univalent (monovalent). It means that in order to transfer a given quantity of electricity, only one-half as many bivalent copper ions as univalent hydrogen ions will be required. In other words, greater is the valency of an ion, smaller is the number of ions needed to carry a given quantity of electricity or charge which means that the mass of an ion liberated is inversely proportional to its valency.

$$\therefore \quad m = \left(\frac{1}{F} \cdot \frac{a}{v}\right) It = \left(\frac{1}{F} \cdot \frac{a}{v}\right) Q = \frac{E}{F} \cdot Q$$

where E is the chemical equivalent weight ($= a/v$).

The constant F is known as Faraday's constant. The value of Faraday's constant can be found thus. It is found that one coulomb liberates 0.001118 gram of silver. Moreover, silver is univalent and its atomic weight is 107.88. Hence, substituting these values above, we find that

$$0.001118 = \frac{1}{F} \cdot 107.88 \times 1$$

$\therefore$ $F = 107.88 / 0.001118 = 96{,}500$ coulomb, which corresponds to $96{,}500/3600 = 26.8$ Ah

Faraday's constant is defined as ***the charge required to liberate one gram-equivalent of any substance.***

For all substances, $\dfrac{\text{chemical equivalent } (E)}{\text{electrochemical equivalent } (Z)}$ = Faraday's constant (F) = 96,500 coulomb or $F = E/Z$

Explanation of the Second Law

Suppose an electric current is passed for the same time through acidulated water, solution of $CuSO_4$ and $AgNO_3$, then for every 1.0078 (or 1.008) gram of hydrogen evolved, 107.88 gram of silver and 31.54 gram of Cu are liberated. The values 107.88 and 31.54 represent the equivalent weights* of silver and copper respectively *i.e.* their atomic weights (as referred to hydrogen) divided by their respective valencies.

---

* The electro-chemical equivalents and chemical equivalents of different substance are inter-related thus:

$$\frac{E.C.E. \text{ of } A}{E.C.E. \text{ of } B} = \frac{\text{chemical equivalent of } A}{\text{chemical equivalent of } B}$$

Further, if m_1 and m_2 are masses of ions deposited at or liberated from an electrode, E_1 and E_2 their chemical equivalents and Z_1 and Z_2 their electrochemical equivalent weights, then

$$m_1/m_2 = E_1/E_2 = Z_1/Z_2$$

Example 9.1. *Calculate the time taken to deposit a coating of nickel 0.05 cm thick on a metal surface by means of a current of 8 A per cm^2 of surface. Nickel is a divalent metal of atomic weight 59 and of density 9 gram/cm^3. Silver has an atomic weight of 108 and an E.C.E. of 1.118 mg/C.*

Solution. Wt. of nickel to be deposited per cm^2 of surface = $1\times 0.05 \times 9 = 0.45$ g

$$\text{Now} \quad \frac{E.C.E. \text{ of Ni}}{E.C.E. \text{ of Ag}} = \frac{\text{chemical equivalent of Ni}}{\text{chemical equivalent of Ag}}$$

$$\therefore \quad E.C.E. \text{ of } Ni = 1.118 \times 10^{-3} \times \frac{(59/2)}{108} = 0.0003053 \text{ g/C}$$

(chemical equivalent = atomic wt./valency)

Now $m = ZIt$ $\therefore$ $0.45 = 0.0003053 \times 8 \times t$; $t = 184$ second = **3 min. 4 second**

Exmaple 9.2. *If 18.258 gm of nickel are deposited by 100 amp flowing for 10 minutes, how much copper would be deposited by 50 amp for 6 minutes: Atomic weight of nickel = 58.6 and that of copper 63.18. Valency of both is 2.* **(Electric Power AMIE Summer 1991)**

Solution. From Faraday's first law, we get $m = ZIt = m\left(\frac{1}{F}.\frac{a}{v}\right) It.$

If m_1 is the mass of nickel deposited and m_2 that of copper, then

$$m_1 = 18.258 = \left(\frac{1}{F}.\frac{58.6}{2}\right) \times 100\ (10 \times 60), \left(\frac{1}{F}.\frac{63.18}{2}\right) \times 50 \times (6 \times 60)$$

$$\therefore \quad \frac{m_2}{18.258} = \frac{31.59}{29.3} \times \frac{18{,}000}{60{,}000} \qquad \therefore\ m_2 = \mathbf{5.905\ gm}$$

Example 9.3. *The cylindrical surface of a shaft of diameter 12 cm and length 24 cm is to be repaired by electrodeposition of 0.1 cm thick nickel on it. Calculate the time taken if the current used is 100 A. The following data may be used:*

Specific gravity of nickel = 8.9 ; Atomic weight of nickel = 58.7 (divalent) ; E.C.E. of silver = 1.2 mg/C; Atomic weight of silver = 107.9. **(Elect. Engg. A.M.Ae. S.I. June, 1991)**

Solution. Curved surface of the salt = $\pi D \times l = \pi \times 12 \times 24$ cm^2

Thickness of nickel layer = 0.1 cm

Volume of nickel to be deposited = $12\pi \times 24 \times 0.1 = 90.5$ cm^3

Mass of nickel deposited = $90.5 \times 8.9 = 805.4$ g

$$\text{Chemical equivalent of } Ni = \frac{\text{atomic weight}}{\text{valency}} = \frac{58.7}{2} = 29.35$$

$$\text{Now} \quad \frac{E.C.E. \text{ of Ni}}{E.C.E. \text{ of Ag}} = \frac{\text{chemical equivalent of Ni}}{\text{chemical equivalent of Ag}}$$

$$\therefore \quad \frac{E.C.E. \text{ of Ni}}{1.12} = \frac{29.35}{107.9}$$

$$\therefore \quad E.C.E. \text{of Ni} = 1.12 \times 29.35/107.9 = 0.305 \text{ mg/C}$$

Now $m = ZIt$

$\therefore$ $805.4 = 0.305 \times 10^{-3} \times 100 \times t$ $\therefore$ $t = 26{,}406$ second or **7 hr, 20 min 7 s**

Example 9.4. *A sheet of iron having a total surface area of 0.36 m² is to be electroplated with copper to a thickness of 0.0254 mm. What quantity of electricity will be required ? The iron will be made the cathode and immersed, together with an anode of pure copper, in a solution of copper sulphate.*

(Assume the mass density of copper = 8.96×10^3 kg m^{-3}; E.C.E. of copper —32.9×10^{-8} kg C^1 Current density = 300 Am^{-2})

(AMIE Sec. B Utilisation of Electric Power Summer 1992)

Solution. Area over which copper is to be deposited = 0.36 m²

Thickness of the deposited copper = 0.0254×10^{-3} m

Volume of deposited copper $= 0.36 \times 0.0254 \times 10^{-3} = 9.144 \times 10^{-6}$ m³

Mass of copper deposited = volume × density

$= 9.144 \times 10^{-6} \times 8.96 \times 10^3 = 0.0819$ kg

Now, $m = Z\,Q$ $\quad\therefore\quad$ $Q = m/Z = 0.0819/32.9 \times 10^{-8} =$ **248936 C**

Tutorial Problem 9.1

1. A steady current was passed for 10 minutes through an ammeter in series with a silver voltameter and 3.489 grams of silver were deposited. The reading of the ammeter was 5A. Calculate the percentage error. Electrochemical equivalent of silver = 1.1138 mg/C.

   [3.85 %] ***(City and Guilds, London)***

2. Calculate the ampere-hours required to deposit a coating of silver 0.05 mm thick on a sphere of 5 cm radius. Assume electrochemical equivalent of silver = 0.001118 and density of silver to be 10.5 g cm³.

   [Utilization of Elect. Power, A.M.I.E. Summer]

9.2. Polarisation of Back E.M.F.

Let us consider the case of two platinum electrodes dipped in dilute sulphuric acid solution. When a small potential difference is applied across the electrodes, no current is found to flow. When, however, the applied voltage is increased, a time comes when a temporary flow of current takes place. The H^+ ions move towards the cathodes and O^{--} ions move towards the anode and are absorbed there. These absorbed ions have a tendency to go back into the electolytic solution, thereby leaving them as oppositely-charged electrodes. This tendency produces an e.m.f. that is in opposition to the applied voltage which is consequently reduced.

This opposing e.m.f. which is produced in an electrolyte due to the absorption of gaseous ions by the electrolyte from the two electrodes is known as the back e.m.f. of electolysis or polarisation.

The value of this back e.m.f. is different from different electrolytes. The minimum voltage required to decompose an electrolyte is called the *decomposition* voltage for that electrolyte.

9.3. Value of Back E.M.F.

For producing electrolysis, it is necessary that the applied voltage must be greater than the back e.m.f. of electrolysis for that electrolyte. The value of this back e.m.f. of electrolysis can be found thus:

Let us, for example, find the decomposition voltage of water. We will assume that the energy required to separate water into its constituents (*i.e.* oxygen and hydrogen) is equal to the energy liberated when hydrogen and oxygen combine to form water. Let *H* kcal be the amount of heat energy absorbed when 9 kg of water are decomposed into 1 kg of hydrogen and 8 kg of oxygen. If the electrochemical equivalent of hydrogen is *Z* kg/coulomb, then the passage of *q* coulomb liberates *Zq* kg of hydrogen. Now, *H* is the heat energy required to release 1 kg of hydrogen, hence for releasing *Zq* kg of hydrogen, heat energy required is *HZq* kcal to *JHZq* joules. If *E* is the decomposition voltage, then energy spent in circulating *q* coulomb of charge is *Eq* joule. Equating the two amounts of energies, we have

$$Eq = JHZq \text{ or } E = JHZ$$

where *J* is 4200 joule/kcal.

The e.m.f. of a cell can be calculated by determining the two electrode potentials. The electrode potential is calculated on the assumption that the electrical energy comes entirely from the heat of the reactions of the constituents. Let us take a zinc electrode. Suppose it is given that 1 kg of zinc when dissolved liberates 540 kcal of heat and that the electrochemical equivalent of zinc is 0.338×10^{-6} kg/coulomb. As calculated above,

$$E = JHZ = 4200 \times 540 \times 0.338 \times 10^{-6} = 0.76 \text{ volt}$$

The electrode potentials are usually referred to in terms of the potential of a standard hydrogen electrode *i.e.* an electrode of hydrogen gas at normal atmospheric pressure and in contact with a normal acid solution. In table No. 9.1 are given the eletrode potentials of various elements as referred to the standard hydrogen electrode. The elements are assumed to be in normal solution and at atmospheric pressure.

In the case of Daniel cell having copper and zinc electrodes, the copper electrode potential with respect to hydrogen ion is + 0.345 V and that of the zinc electrode is – 0.758 V. Hence, the cell e.m.f. is 0.345 – (– 0.758) = 1.103 volt. The e.m.f. of other primary cells can be found in a similar way.

Table No. 9.1

Electrode	*Potential (volt)*	*Electrode*	*Potential (volt)*
Cadmium	– 0.398	Mercury	+ 0.799
Copper	+ 0.345	Nickel	– 0.231
Hydrogen	0	Potassium	– 2.922
Iron	– 0.441	Silver	+ 0.80
Lead	– 0.122	Zinc	– 0.758

Example 9.5. *Calculate the weight of zinc and MnO_2 required to produce I ampere-hour in a leclanche cell.*

Atomic weights: Mn, 55 ; O, 16 ; Zn, 65. E.C.E. of hydrogen = 1.04×10^{-8} kg/C.

Solution. 1 ampere-hour = 3600 A-s = 3600 C

Wt. of hydrogen liberated = $Zq = 1.04 \times 10^{-8} \times 3600 = 37.44 \times 10^{-6}$ kg

Now, the chemical reactions in the cell are

$$Zn + 2NH_4Cl = ZnCl_2 + 2NH_3 + H_2$$

It is seen that 1 ***atom*** of zinc is used up in liberating ***two atoms*** of hydrogen. In other words, to produce 2 kg of hydrogen, 65 kg of zinc will have to go into chemical combination.

∴ Zinc required to produce 37.44×10^{-6} kg of hydrogen $= 37.44 \times 10^{-6} \times 65/2$

$$= 1.217 \times 10^{-3} \text{ kg}$$

The hydrogen liberated combines with manganese dioxide as under:

$$2MnO_2 + H_2 = H_2O + Mn_2O_3$$

Atomic weight of $2\ MnO_2 = 2(55 + 16 \times 2) = 174$

It is seen that 174 kg of MnO_2 combine with 2 kg of hydrogen, hence Wt. of MnO_2 needed to combine with 37.44×10^{-6} kg of hydrogen $= 37.44 \times 10^{-6} \times 174/2 = 3.258 \times 10^{-3}$ kg.

Hence, for 1 ampere-hour, $\mathbf{1.217 \times 10^{-3}}$ **kg** of zinc and $\mathbf{3.258 \times 10^{-3}}$ **kg** of MnO_2 are needed.

9.4. Primary and Secondary Batteries

An electric battery consists of a number of electrochemical cells, connected either in series or parallel. A cell, which is the basic unit of a battery, may be defined as a power generating device, which is capable of converting stored chemical energy into electrical energy. If the stored energy is inherently present in the chemical substances, it is called a primary cell or a non-rechargeable cell. Accordingly, the battery made of these cells is called primary battery. The examples of primary cells are Leclanche cell, zinc-chlorine cell, alkaline-manganese cell and metal air cells etc.

If, on the other hand, energy is induced in the chemical substances by applying an external source, it is called a secondary cell or rechargeable cell. A battery made out of these cells is called a secondary battery or storage battery or rechargeable battery. Examples of secondary cells are lead-acid cell, nickel-cadmium cell, nickel-iron cell, nickel-zinc cell, nickel-hydrogen cell, silver-zinc cell and high temperature cells like lithium-chlorine cell, lithium-sulphur cell, sodium-sulphur cell etc.

9.5. Classification of Secondary Batteries Based on Their Use

Various types of secondary batteries can be grouped in to the following categories as per their use :

1. ***Automotive Batteries or SLI Batteries or Portable Batteries***

These are used for starting, lighting and ignition (*SLI*) in internal-combustion-engined vehicles. Examples are; lead-acid batteries, nickel-cadmium batteries etc.

2. ***Vehicle Traction Batteries*** **or** ***Motive Power Batteries*** **or** ***Industrial Batteries***

These are used as a motive power source for a wide variety of vehicles. Lead-acid batteries, nickel-iron batteries, silver-zinc batteries have been used for this purpose. A number of advance batteries including high-temperature batteries are under development for electric vehicle (*EV*) use. These high-temperature batteries like sodium-sulphur and lithium-iron sulphide have energy densities in the range of 100-120 Wh/kg.

3. ***Stationary Batteries.***

These fall into two groups (***a***) standby power system which is used intermittently and (***b***) load-levelling system which stores energy when demand is low and, later on, uses it to meet peak demand.

9.6. Classification of Lead Storage Batteries

Lead storage batteries may be classified according to the service which they provide.

1. ***SLI Batteries***

The primary purpose of these batteries is to supply power for engine starting, lighting and ignition (*SLI*) of vehicles propelled by IC engines such as automobiles, buses, lorries and other heavy road vehicles and motor cycles etc. Usually, these batteries provide 12 V and consist of six series-connected lead-acid cells with capacity of the other of 100 Ah. Their present-day energy density is about 45 Wh/kg and 75 Wh/dm^3.

These days 'maintenance-free' (*MF*) *SLI* batteries have been designed, which do not require the addition of water throughout their normal service life of 2-5 years. *MF* versions of the *SLI* batteries are constructed of such material that no gassing occurs during charging. In *MF* batteries, the electrolyte is either absorbed within the microporous separators and the plates or is immobilized with suitable gelling agents.

These days the *SLI* battereis are charged from an alternator (*AC* generator) and not from dynamo (*DC* generator). The alternating current produced by the alternator is converted into direct current by a full-wave bridge rectifier, which uses semi-conductor diodes. In this arrangement, no cutout is needed and the transistorised voltage controller regulates the alternator output to suit the electrical load and the state of charge of the battery. The battery is charged under constant-voltage conditions.

2. *Vehicle Traction Batteries*

The recent universal concern over the levels of toxic gases (particularly in urban areas) emitted by the *IC* engines has revived interest in electric traction. There has been great development in the use of battery-powered vehicles, primarily industrial trucks and commercial road vehicles of various types like 'milk floats' (*i.e.* bottled-milk delivery trucks), fork lift trucks, mining, airport tractors, aircraft service vehicles, electric cars and, more recently, in robotics and guided vehicles.

Traction batteries are of higher quality than *SLI* batteries. They provide constant output voltage, high volumetric capacity, good resistance to vibration and a long service life. They can withstand prolonged and deep discharges followed by deep recharges usually on a daily basis. The voltage of traction batteries varies from 12 V to 240 V and they have a cycle life of 1000-1500 cycles.

A number of advanced batteries and under development for *EV* use ***(i)*** room temperature batteries like zinc-nickel oxide battery (75 Wh/kg) and zinc-chlorine hydrate battery (80 Wh/kg) and ***(ii)*** high-temperature batteries like sodium-sulphur battery (120 Wh/kg) and lithium-iron sulphide battery (100 Wh/kg).

3. *Stationary Batteries*

Their use falls into two groups:

(*a*) as standby power system and **(*b*)** as load-levelling system.

In the standby applications, the battery is used to power essential equipment or to provide alarms or emergency lighting, in case of break-down in the main power supply. Standby applications have increased in recent years with increasing demand for uninterruptable power systems (*UPS*) and a tremendous growth in new telecommunication networks. The *UPS* provides 'clean' a.c supply free of sage or surges in the line voltage, frequency variations, spikes and transients to modern computer and electronic equipment. Banks of sealed lead-acid (*SLA*) standby batteries have been recently used in telecommunication systems and for *UPS* applications.

Recently, advanced lead-acid batteries have been used for load-levelling purpose in the electric generating plants. A 100 M Wh lead-acid battery load-levelling system could occupy a building two and a half storey high and an area of about 250,000 m^2.

9.7. Parts of a Lead-Acid Battery

A battery consists of a number of cells and each cell of the battery-consists of (*a*) positive and negative plants **(*b*)** separators and **(*c*)** electrolyte, all contained in one of the many compartments of the battery container.* Different parts of a lead-acid battery are as under:

(*i*) *Plates.* A plate consists of a lattice type of grid of cast antimonial lead alloy which is covered with *active* material (Art. 9.8). The grid not only serves as a support for the fragile active material but also conducts electric current. Grid for the positive and negative plates are often of the same design although negative plate grids are made somewhat lighter. As discussed in Art. 9.10, positive plates are usually Plante plates whereas negative plates are generally of Faure or pasted type.

(*ii*) *Separators.* These are thin sheets of a porous material placed between the positive and negative plates for preventing contact between them and thus avoiding internal short-circuiting of the battery. A separator must, however, be sufficiently porous to allow diffusion or circulation of electrolyte between the plates. These are made of especially-treated cedar wood, glass wool mat, microporous rubber (mipor), microporous plastics (plastipore, miplast) and perforated p.v.c. as shown in Fig. 9.1. In addition to good porosity, a separator must possess high electrical resistance and mechanical strength.

(*iii*) *Electrolyte.* It is dilute sulphuric acid which fills the cell compartment to immerse the plates completely.

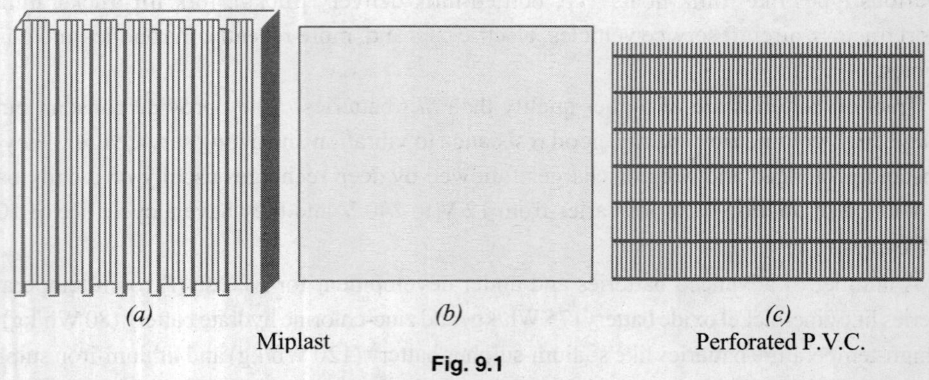

Fig. 9.1

(*iv*) *Container.* It may be made of vulcanised rubber or moulded hard rubber (ebonite), moulded plastic, ceramics, glass or celluloid. The vulcanised rubber containers are used for car service, while glass containers are superior for lighting plants and wireless sets. Celluloid containers are mostly used for portbable wireless set batteries. A single monoblock type container with 6 compartments generally used for starting batteries is shown in Fig. 9.2. Full details of a Russian 12-CAM-28 lead-acid battery parts are shown in Fig. 9.3. Details of some of these parts are as follows:

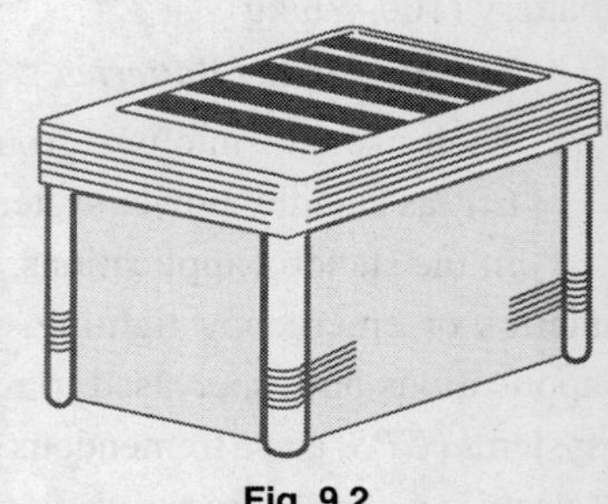

Fig. 9.2

(*a*) *Bottom Grooved Support Blocks.* These are raised ribs, either fitted in the bottom of the container or made with the container itself. Their function is to support the plates and hold

* The most common form of lead-acid cell used for marine applications is the tubular cell which consists of 'armoured' tubular positive plate of standard flat negative plate.

them in position and at the same time protect them from short-circuits that would otherwise occur as a result of fall of the active material from the plates onto the bottom of the container.

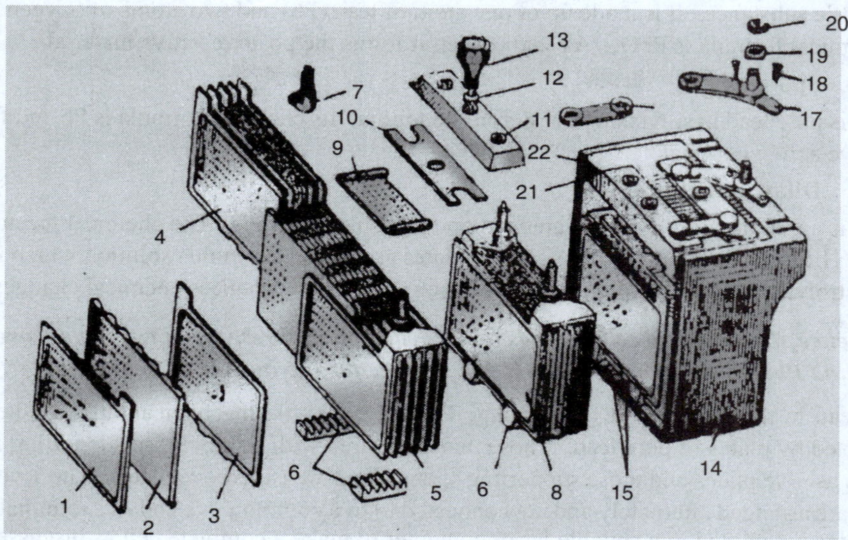

Fig. 9.3. (Courtesy MIR Publishers. Moscow)

1. – ve plate 2. separator 3. + ve plate. 4. + ve group 5. – ve group 6. – ve group grooved support block 7. lug 8. plate group 9. guard screen 10. guard plate 11. cell cover 12. plug washer 13. vent plug 14. monoblock jar 15. supporting prisms of + ve group 16. inter-cell connector 17. terminal lug 18. screw 19. washer 20. nut 21. rubber packing 22. sealing compound.

(*b*) *Connecting Bar.* It is the lead alloy link which joins the cells together in series connecting the positive pillar of one cell to the negative pillar of the next one.

(*c*) *Terminal Post* or *Pillar.* It is the upward extension from each connecting bar which passes through the cell cover for cable connections to the outside circuits. For easy indentification, the negative terminal post is smaller in diameter than the positive terminal post.

(*d*) *Vent Plugs* or *Filler Caps.* These are made of polystyrene or rubber and are usually screwed in the cover. Their function is to prevent escape of electrolyte but allow the free exit of the gas. These can be easily removed for topping up or taking hydrometer readings.

(*e*) *External Connecting Straps.* These are the antimonial lead alloy flat bars which connect the positive terminal post of one cell to the negative of the next across the top of the cover. These are of very solid construction especially in starting batteries because they have to carry very heavy currents.

9.8. Active materials of a Lead-Acid Cell

Those substances of the cell which take active part in chemical combination and hence absorb or porduce electricity during charging or discharging, are known as ***active materials*** of the cell.

The active materials of a lead-acid cell are:

1. ***Lead peroxide*** (PbO_2) for + ve plate **2.** ***Sponge Lead*** (Pb) for – ve plate **3.** ***Dilute Sulphuric*** *Acid* (H_2SO_4) as electrolyte.

1. Lead Peroxide

It is a combination of lead and oxygen, is dark chocolate brown in colour and is quite hard but brittle substance. It is made up of one atom of lead (Pb) and two atoms of oxygen (O_2) and its chemical formula is PbO_2. As said earlier, it forms the positive active material.

2. Sponge Lead

It is pure lead in soft sponge or porous condition. Its chemical formula is Pb and forms the negative active material.

3. Dilute Sulphuric Acid

It is approximately 3 parts water and one part sulphuric acid. The chemical formula of the acid is H_2SO_4. The positive and negative plates are immersed in this solution which is known as electrolyte. It is this medium through which the current produces chemical changes.

Hence, the lead-acid cell depends for its action on the presence of two plates covered with PbO_2 and Pb in a solution of dilute H_2SO_4 of specific gravity 1.21 or nearabout.

Lead in the form of PbO_2 or sponge Pb has very little mechanical strength, hence it is supported by plates of pure lead. Those plates covered with or otherwise supporting PbO_2 are known as + ve plates and those supporting sponge lead are called −ve plates. The + ve and −ve plates are arranged alternately and are connected to two common +ve and −ve terminals. These plates are assembled in a suitable jar or container to make a complete cell as discussed in Art. 9.4 above.

9.9. Internal Resistance and Capacity of a Cell

The secondary cell possesses internal resistance due to which some voltage is lost in the form of potential drop across it when current is flowing. Hence, the internal resistance of the cell has to be kept to the minimum.

One obvious way to lessen internal resistance is to increase the size of the plates. However, there is a limit to this because the cell will become too big to handle. Hence, in practice, it is usual to multiply the number of plate inside the cell and to join all the negative plates together and all the positives ones together as shown in Fig. 9.4.

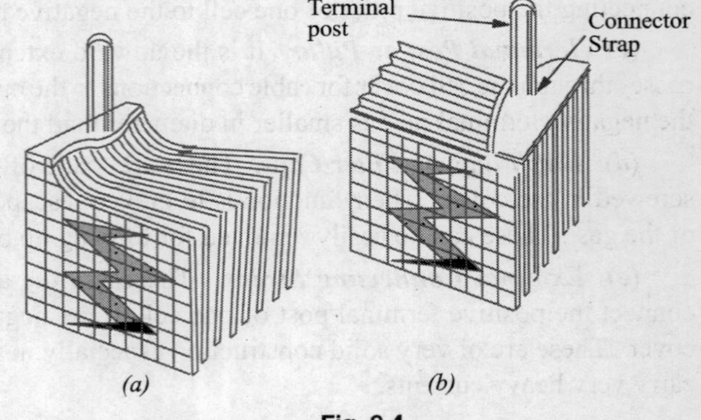

Fig. 9.4

The effect is equivalent to joining many cells in parallel. At the same time, the length of the electrolyte between the electrodes is decreased with a consequent reduction in the internal resistance.

The 'capacity' of a cell is given by the product of current in amperes and the time in hours during which the cell can supply current until its e.m.f. falls to 1.8 volt. It is expressed in ampere-hour (Ah).

The interlacing of plates not only decreases the internal resistance but additionally increases the capacity of the cell also. There is always one more negative plate than the positive paltes *i.e.* there is a negative plate at both ends. This gives not only more mechanical strength but also assures that both sides of a positive plate are used.

Since in this arrangement, the plates are quite close to each other, something must be done to make sure that a positive plate does not touch the negative plate otherwise an internal short-circuit will take place. The separation between the two plates is achieved by using separators which, in the case of small cells, are made of treated cedar wood, glass, wool mat, microporous rubber and mocroporous plastic and in the case of large stationary cells, they are in the form of glass rods.

9.10. Efficiencies of the Cell

The efficiency of a cell can be considered in two ways:

1. ***The quantity or ampere-hour (Ah) efficiency***
2. ***The energy or watt-hour (Wh) efficiency***

The Ah efficiency does not take into account the varying voltages of charge and discharge. The Wh efficiency does so and is always less than Ah efficiency because average p.d. during discharging is less than that during charging. Usually, during discharge the e.m.f. falls from about 2.1 V to 1.8 V whereas during charge it rises from 1.8 volt to about 2.5 V.

$$\text{Ah efficiency} = \frac{\text{amp-hour discharge}}{\text{amp-hour charge}}$$

The Ah efficiency of a lead-acid cell is normally between 90 to 95%, meaning that about 100 Ah must be put back into the cell for every 90-95 Ah taken out of it. Because of gassing which takes place during the charge, the Ah available for delivery from the battery decreases. It also decreases **(*i*)** due to self-discharge of the plates caused due to local reactions and **(*ii*)** due to leakage of current because of faulty insulation between the cells of the battery.

The Wh efficiency varies between 72-80%.

If Ah efficiency is given, Wh efficiency can be found from the following relation:

$$\text{Wh efficiency} = \text{Ah efficiency} \times \frac{\text{average volts on discharge}}{\text{average volts on charge}}$$

From the above, it is clear that anything that increases the charge volts or reduces the discharge volts will decrease Wh efficiency. Because high charge and discharge rates will do this, hence it is advisable to avoid these.

9.11. Electrical Characteristics of the Lead-Acid Cell

The three important features of an accumulator, of interests to an engineer, are **(1)** voltage **(2)** capacity and **(3)** efficiency.

1. Voltage

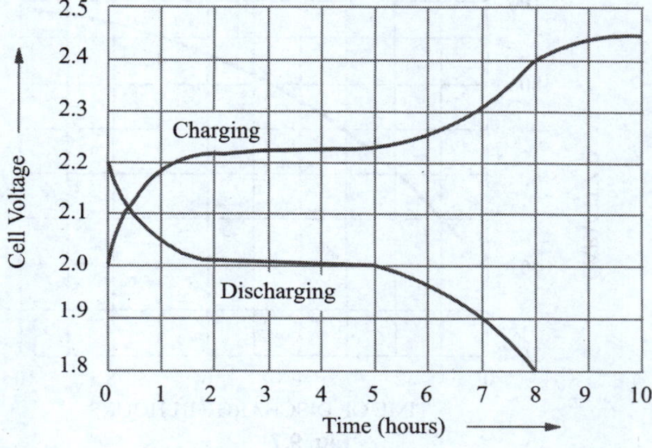

Fig. 9.5

The open-circuit voltage of a fully-charged cell is approximately 2.2 volt. This value is not fixed but depends on **(*a*)** length of time since it was last charged **(*b*)** specific gravity-voltage increasing with increase in sp. gravity and *vice versa*. If sp. gravity comes near to density of water *i.e.* 1.00 then voltage of the cell will disappear altogether **(*c*)** temperature-voltage increases, though not much, with increase in temperature.

The variations in the terminal p.d. of a cell on charge and discharge are shown in Fig. 9.5. The voltage-fall depends on the rate of discharge. Rates of discharge are generally specified by the number of hours during which the cell will sustain the rate in question before falling to 1.8 V. The voltage falls rapidly in the beginning (rate of fall depending on the rate of discharge), then vary slowly up to 1.85 and again suddenly to 1.8 V.

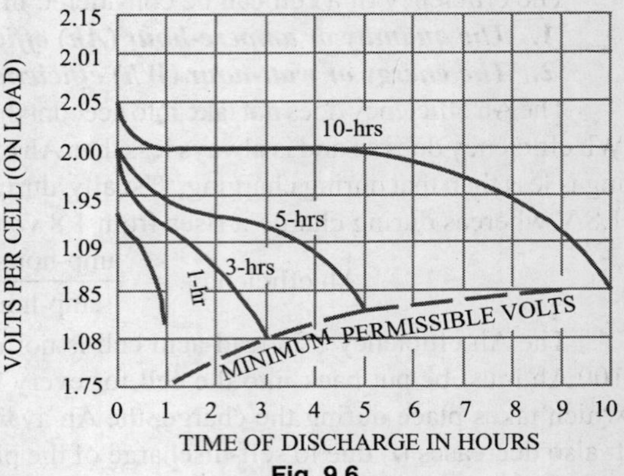

Fig. 9.6

The voltage should not be allowed to fall to lower than 1.8 V, otherwise hard insoluble lead sulphate is formed on the plate which increases the internal resistance of the cell.

The general form of the voltage-time curves corresponding to 1-, 3- 50 and 10-hour rates of corresponding to the steady currents which would discharge the cell in the above mentioned times (in hour). It will be seen that both the terminal voltage and the rate at which the voltage and the rate at which the voltage falls, depend on the rate of discharge. The more rapid fall in voltage at higher rates of discharge is due to the rapid increase in the internal resistance of the cell.

During charging, the p.d. increases (Fig. 9.5). The curve is similar to the discharge curve reversed but is everywhere higher due to the increased density of H_2SO_4 in the pores of the positive plate.

2. Capacity

It is measured in amp-hours (Ah). The capacity is always given at a specified rate of discharge (10-hour discharge in U.K., 8-hour discharge in U.S.A.). However, motor-cycle battery capacity is based on a 20-hour rate (at 30° C). The capacity depends upon the following:

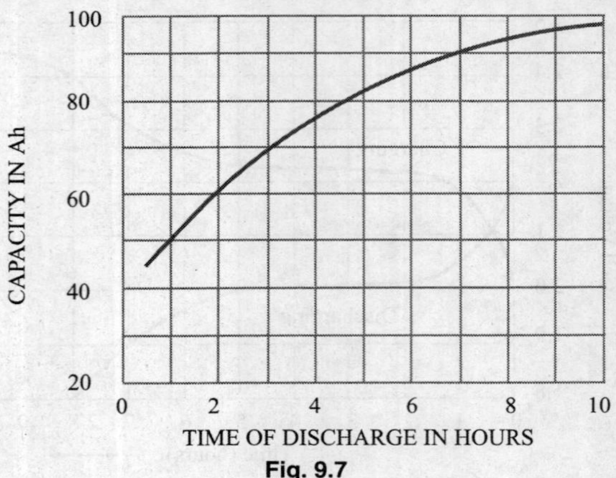

Fig. 9.7

(*a*) ***Rate of discharge.*** The capacity of a cell, as measured in Ah, depends on the discharge rate. It decreases with increased rate of discharge. Rapid rate of discharge means greater fall in p.d. of the cell due to internal resistance of the cell. Moreover, with rapid discharge the weakening of the acid in the pores of the plates is also greater. Hence, the chemical change produced at the plates by 1 ampere for 10 hours is not the same as produced by 2 A for 5 hours or 4 A for 2.5 hours. It is found that a cell having a 100 Ah capacity at 10 hour discharge rate, has its capacity reduced to 82.5 Ah at 5-hour rate and 50 Ah at 1-hour rate. The variation of capacity with discharge rate is shown in Fig. 9.7.

(*b*) ***Temperature.*** At high temperature,

(*i*) chemical reactions within the cell take place more vigorously.

(*ii*) the resistance of the acid is decreased and

(*iii*) there is a battery diffusion of the electrolyte.

Hence, high temperature increases the capacity of the lead-acid cell. Apparently, it is better to operate the battery at a high temperature. However, at high temperatures:

(*a*) the acid attacks the antimony-lead alloy grid, terminal posts and wooden separators.

(*b*) the paste is rapidly changed into lead sulphate. Sulphation is always accompanied by expansion of paste particularly at the positive plates and results in buckling and cracking of the grid.

Hence, it is not advisable to work batteries above 40° C.

As temperature is lowered, the speed of chemical reactions is decreased. Moreover, cell resistance also increases. Consequently, the capacity of the cell decreases with decrease in temperature till at freezing point the capacity is reduced to zero even though the battery otherwise be fully charged.

(*c*) ***Density of electrolyte.*** As the density of electrolyte affects the internal resistance and the vigour of chemical reaction, it has an important effect on the capacity. Capacity increases with the density.

(*d*) ***Quantity of active material.*** Since production of electricity depends on chemical action taking place within the cells, it is obvious that the capacity of the battery must depend directly upon the kind and amount of the active material employed. Consider the following calculations:

The gram-equivalent of lead is 103.6 gram and Faraday's constant is 96,500 coulombs which is = 96,500/3600 = 26.8 Ah. Hence, during the delivery of one Ah by the cell, the quantity of lead expended to form lead sulphate at the negative plate is 103.6/26.8 = 3.86 gram.

Similarly, it can be calculated that, at the same time, 4.46 gram of PbO_2 would be converted into lead sulphate at the positive plate while 3.66 gram of acid would be expended to form 0.672 gram of water. It is obvious that for obtaining a cell of a greater capacity, it is necessary to provide the plates with larger amounts of active material.

3. Efficiency

It has already been discussed in Art. 9.19

9.12. Battery Ratings

Following standards have been adopted, both by industry and government organisations to get a fair picture of battery quantity:

1. Ampere-hour Capacity

It is a function of the total plate area *i.e.* size of the individual plate multiplied by the number of plates. For measuring this capacity, the battery is discharged continuously for 20 hours and its current output supplied to a standard load is measured. Suppose that a battery delivers 4A current for 20 hours. Hence, its rating is 80 Ah which is stamped on the battery case.

2. Reserve Capacity

It is one of the newly-developed rating standards and is more realistic because it provides a double-check on the Ah figures. The capacity is given by the number of minutes a battery will tolerate a 25 A drain without dropping below 10.5 V. Higher this rating, better the battery.

3. Zero Cranking Power

It was the first cold weather rating and is applicable in relation to crafts which ply in freezing weather. This zero-degree performance test gives a valuable insight into battery quality. Large batteries are tested at a 300 A drain with battery chilled to – 18°C (0°F) whereas smaller sizes are tested at half this value.

This test consists of two part **(*a*)** the battery is first chilled to – 18°C (0°F) and the load applied for 5 second. Meanwhile, the voltage output is recorded. It is the first part of the zero-cranking-power rating. **(*b*)** The test is continued further till voltage drops to 5 V. The number of minutes it takes to reduce the voltage to 5 V forms the second half of the rating. Higher both the digits, better the battery quality.

4. Cold Cranking Power

This simple rating is applied to all 12-V storage batteries regardless of their size. The battery is loaded at – 18°C (0°F) till the total voltage drops to 7.2 V. The output current in amperes is measured for 30 seconds. Higher the output, better the battery.

Example 9.6. *An alkaline cell is discharged at a steady current of 4 A for 12 hours, the average terminals voltage being 1.2 V. To restore it to its original state of charge, a steady current of 3 A for 20 hours is required, the average terminal voltage being 1.44 V. Calculate the ampere-hour (Ah) efficiency and Wh efficiency in this particular case.*

(Principles of Elect. Engg.-I, Jadavpur Univ.)

Solution. As discussed in Art. 9.19

$$\text{Ah efficiency} = \frac{\text{Ah of discharge}}{\text{Ah of charge}} = \frac{12 \times 4}{20 \times 3} = \mathbf{0.8} \quad \textbf{or} \quad \mathbf{80\%}$$

$$\text{Wh efficiency} = \text{Ah effi.} \times \frac{\text{Av. volts on discharge}}{\text{Av. volts on charge}} = \frac{0.8 \times 1.2}{1.44} = \mathbf{0.667} \quad \textbf{or} \quad \mathbf{66.7\%}$$

Example 9.7. *A discharged battery is charged at 8 A for 2 hours after which it is discharged through a resistor of R Ω. If discharge period is 6 hours and the terminal voltage remains fixed at 12 V, find the value of R assuming the Ah efficiency of the battery as 80%.*

Solution. Input amp-hours $= 8 \times 2 = 16$

Efficiency $= 0.8 \quad \therefore \quad$ Ah output $16 \times 0.8 = 12.8$

Discharge current $= 12.8/6 \text{ A} \quad \therefore \quad R = \dfrac{12}{12.8/6} = \dfrac{6 \times 12}{12.8} = \mathbf{5.6\ \Omega}$

9.13. Indications of a Full-Charged Cell

The indications of a fully-charged cell are:

***(i)* gassing *(ii)* voltage *(iii)* specific gravity and *(iv)* colour of the plates.**

***(i)* Gassing**

When the cell is fully charged, it freely gives off hydrogen at cathode and oxygen at the anode, the process being known as "Gassing". Gassing at ***both*** plates indicates that the current is no longer doing any useful work and hence should be stopped. Moreover, when the cell is fully charged, the electrolyte assumes a milky appearance.

(*ii*) Voltage

The voltage ceases to rise when the cell becomes fully-charged. The value of the voltage of a fully-charged cell is a variable quantity being affected by the rate of charging, the temperature and specific gravity of the electrolyte etc. The approximate value of the e.m.f. is 2.1 V or so.

(*iii*) Specific Gravity of the Electrolyte

A third indication of the state of charge of a battery is given by the specific gravity of the electrolyte. We have seen from the chemical equations of Art. 9.9, that during discharging , the density of electrolyte decreases due to other production of water, whereas it increases during charging due to the absorption of water. The value of density when the cell is fully charged is 1.21 and 1.18 when discharged up to 1.8 V. Specific gravity can be measured with a suitable hydrometer which consists of a float, a chamber for the electrolyte and a squeeze bulb.

(*iv*) Colour

The colour plates, on full charge is deep chocolate brown for positive plate and clear slate gray for negative plate and the cell looks quite brisk and alive.

9.14. Applications of Lead-Acid Batteries

Storage batteries are these days used for a great variety and range of purposes, some of which are summarised below:

1. In Central Stations for supplying the whole load during light load periods, also to assist the generating plant during peak load periods, for providing reserve emergency supply during periods of plant breakdown and finally, to store energy at times when load is light for use at time when load is at its peak value.
2. In private generating plants both for industrial and domestic use, for much the same purpose as in Central Stations.
3. In sub-stations, they assist in maintaining the declared voltage by meeting a part of the demand and so reducing the load on and the voltage drop in, the feeder during peak-load periods.
4. As a power source for industrial and mining battery locomotives and for road vehicles like cars and trucks.
5. As a power source for submarines when submerged.
6. Marine applications include emergency or stand-by duties in case of failure of ship's electric supply, normal operations where batteries are subjected to regular cycles of charge and discharge and for supplying low-voltage current to bells, telephones, indicators and warning systems etc.
7. For petrol motor-car starting and ignition etc.
8. As a low voltage supply for operating purposes in many different ways such as high-tension switchgear, automatic telephone exchange and repeater stations, broadcasting stations and for wireless receiving sets.

9.15. Voltage Regulators

The voltage of a battery varies over a considerable range while under discharge. Hence, it is necessary to find some means to control the battery voltage upto the end so as to confine variations within reasonable limits – these limits being supplied by the battery.

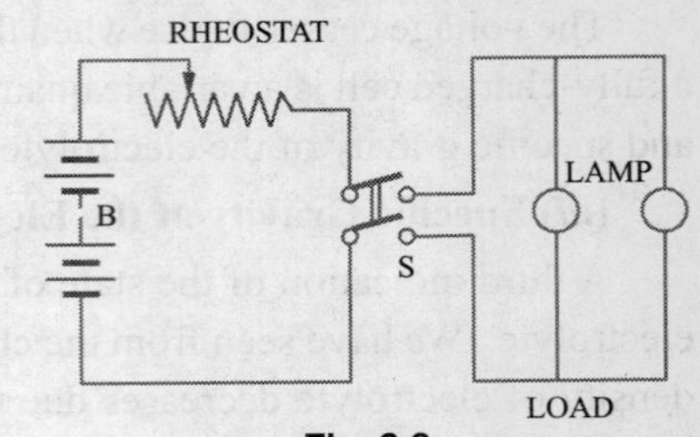

Fig. 9.8

The voltage control systems may be hand-operated or automatic. The simplest form of hand-operated control consists of a rheostat having a sufficient number of steps so that assistance can be inserted in the circuit when battery is fully charged and gradually cut out as the discharge continues, as shown in Fig. 9.8.

The above system can be designed for automatic operation as shown in Fig. 9.9. A rise in voltage results in the release of pressure on the carbon block rheostat, thereby increasing its resistance whereas a fall in voltage results in increasing the pressure on the block thereby decreasing its resistance. By this automatic variation of control resistance, variations in battery voltage are automatically controlled.

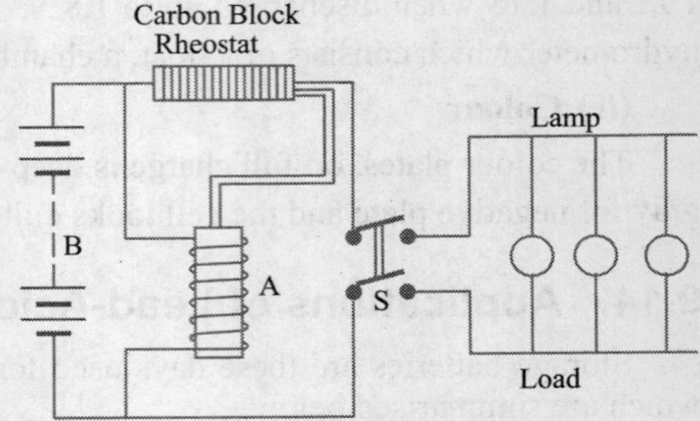

Fig. 9.9

9.16. End-Cell Control System

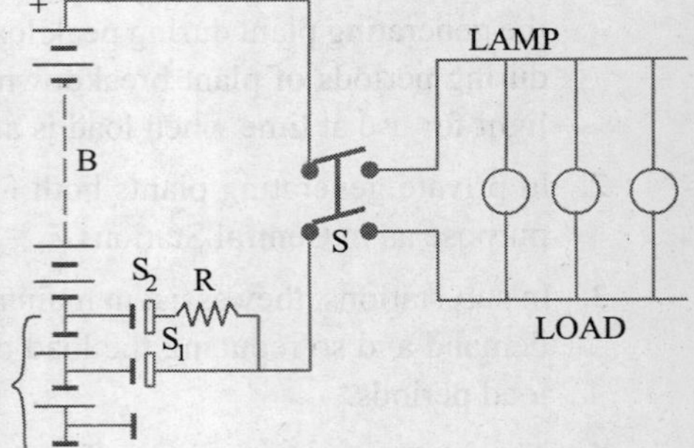

Fig. 9.10

The use of rheostat for controlling the battery voltage is objectionable especially in large-capacity installations where the I^2R loss would be considerable. Hence other more economical systems have been developed and put into use. One such system is the end-cell control system. It consists of suitable regulator switches which cut one or more of a selected number of cells out of the circuit when the battery is fully charged and into the circuit again as the discharge continues. To make the process of cutting cells in and out at the battery circuit simple, the group of cells selected for this control is situated at one end of the battery where from it derives the name ***end cell.*** By moving the contact arm of the switch to the left or right, cells are cut in or out of the discharge circuit and so the voltage is varied accordingly.

For making the end-cell switch operate without opening the circuit or shortcircuiting the cells during its passage from one cell to another, an auxiliary constraint S_2 is employed. S_2 prevents the circuit from being open entirely but has sufficient resistance R between it and the main contact arm S_1 to prevent any objectionably large current to flow on short-circuit. The above mechanism usually incorporates devices for preventing the stoppage of the switch in the short-circuit position.

9.17. Number of End-Cells

For maintaining a supply voltage of V volts from a battery of lead-acid cells when the latter are approaching their discharge voltage of 1.83 (depending on the discharge rate), the number of cells required is $V/1.83$. When the battery is fully charged with each cell having an e.m.f. of 2.1 V, then the number of cells required is $V/2.1$. Hence, the number of end-cells required is $(V/1.83 - V/2.1)$. These are connected to a regulating switch which adds them in series with the battery one or two at a time, as the discharge proceeds.

9.18. Charging Systems

In various installations, batteries are kept floating on the line and are so connected that they are being charged when load demands are light and automatically discharged during peak periods when load demands are heavy or when the usual power supply fails or is disconnected. In some other installations, the battery is connected to the feeder circuit as and when desired, allowed to discharge to a certain point, then removed and re-charged for further requirements.

For batteries other than the 'floating' and 'system-governed' type, following two general methods (though there are some variations of these) are employed.

(*i*) *The Constant-current System* and **(*ii*) *The Constant-voltage System.***

9.19. Constant-Current System

In this method, the charging current is kept constant by varying the supply voltage to overcome the increased back e.m.f. of cells. If a charging booster (which is just a shunt dynamo directly driven by a motor) is used, the current supplied by it can be kept constant by adjusting its excitations. If charged on a d.c. supply, the current is controlled by varying the rheostat connected in the circuit. The value of charging current should be so chosen that there would be no excessive gassing during final stages of charging and, also, the cell temperature does not exceed 45°C. This method takes a comparatively longer time.

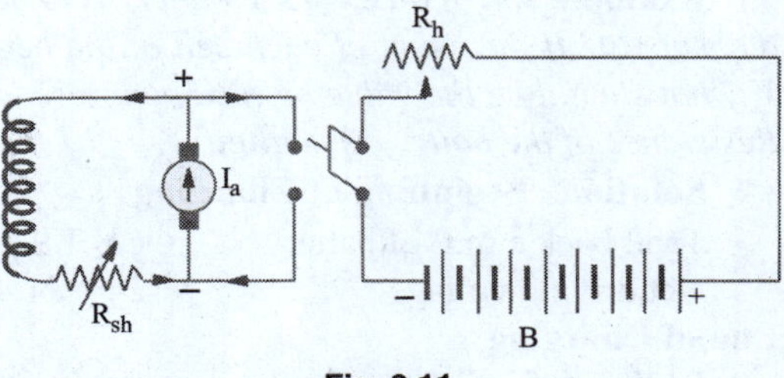

Fig. 9.11

9.20. Constant-Voltage System

In this method, the voltage is kept constant but it results in very large charging current in the beginning when the back e.m.f. of the cells is low and a small current when their back e.m.f. increases on being charged.

With this method, time of charging is almost reduced to half. It increases the capacity by approximately 20% but reduces the efficiency by 10% or so.

Calculations

When a secondary cell or a battery of such cells is being charged, then the emf of the cells acts in opposition to the applied voltage. If V is the supply voltage which sends a charging current of I against the back e.m.f. E_b, then input is VI but the power spent in overcoming the opposition (Fig. 9.12) is E_bI. This power E_bI is converted into the

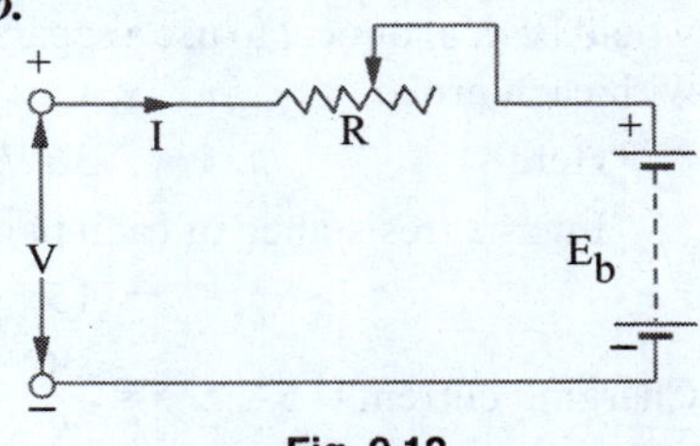

Fig. 9.12

chemical energy which is stored in the cell. The charging current can be found from the following equation:

$$I = \frac{V - E_b}{R}$$

where R = total circuit resistance including internal resistance of the battery
I = charging current

By varying R, the charging current can be kept constant throughout.

Example 9.8. *A battery of accumulators of e.m.f. 50 volt and internal resistance 2 Ω is charged on 100 volt direct means. What series resistance will be required to give a charging current of 2 A?*

If the price of energy is 50 paise per kWh, what will it cost to charge the battery for 8 hours and what percentage of energy supplied will be used in the form of heat ?

Solution. Applied voltage = 100 V; Back e.m.f. of the battery = 50 V
Net charging voltage = 100 – 50 = 50 V
Let R be the required resistance, then 2 = 50/(R + 2); R = 46/2 = 23 Ω.
Input for eight hours = 100 × 2 × 8 = 1600 Wh = 1.6 kWh
Cost = 50 × 1.5 = **80 paise;** Power wasted on total resistance = 25 × 2^2 = 100 W
Total input = 100 × 2 = 200 W: Percentage waste = 100 × 100/200 = **50 %**

Example 9.9. *A 6-cell, 12-V battery is to be charged at a constant rate of 10 A from a 24-V d.c. supply. If the e.m.f. of each cell at the beginning and end of the charge is 1.9 V and 2.4 V, what should be the value of maximum resistance to be connected in series with the battery. Resistance of the battery is negligible.*

Solution. Beginning of Charging
Total back e.m.f. of battery = 6 × 1.9 = 11.4 volt
Net driving voltage = 24 – 11.4 = 12.6 V ; R_{max} = 12.6/10 = **1.26 Ω**

End of Charging
Back e.m.f. of battery = 6 × 2.4 = 14.4 volt
Net driving voltage = 24 – 14.4 = 9.6 V; R_{min} = 9.6/10 = **0.96 Ω**

Example 9. 10. *Thirty accumulators have to be charged from their initial voltage of 1.8 V using a direct current supply of 36 volt. Each cell has an internal resistance of 0.02 Ω and can be charged at 5 amperes. Sketch a circuit by which this can be done, calculating the value of any resistance or resistances used. What will be the current taken from the mains towards the end of the charging period when the voltage has risen to 2.1 volt per cell ?1*

Solution. Since the supply voltage (36 V) is less than the back e.m.f. of the 30 cell battery (54 V), hence the cells are divided into two equal groups and placed in parallel across the supply for charging as shown in Fig. 9.13. It would be economical to use a separate resistance R in series with each group.

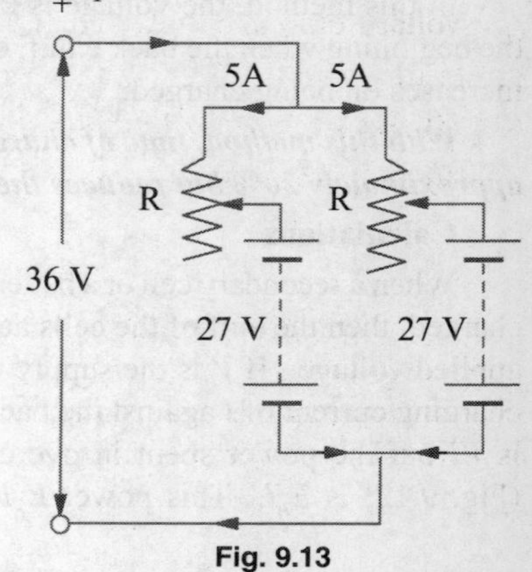

Fig. 9.13

Here V = 36 V, E_b = 15 × 1.8 = 27 V

Internal resistance of each parallel group

= 15 × 0.02 = 0.3 Ω

$$\text{Charging current} = 5\text{A} \therefore 5 = \frac{36 - 27}{R + 0.3}$$

R = **1.5 Ω**

Now, when the voltage per cell becomes 2.1 V, then back e.m.f. of each parallel group = 15 × 2.1 = 31.5 V

$$\therefore \quad \text{Charging current} = \frac{36 - 31.5}{1.5 + 0.3} = \mathbf{2.5\ A}$$

Example 9.11. *Two hundred and twenty lamps of 100 W each are to be run on a battery supply at 110 V. The cells of the battery when fully charged have an e.m.f. of 2.1 V each and when discharged 1.83 V each. If the internal resistance per cell is 0.00015 Ω* **(i)** *find the number of cells in the battery and* **(ii)** *the number of end cells. Take the resistance of the connecting wires as 0.005 Ω.*

Solution. Current drawn by lamps = 220 × 100/110 = 200 A

Voltage drop on the resistance of the connecting wires = 0.005 × 200 = 1.0 V

Battery supply voltage = 110 + 1 = 111 V

Terminal voltage/cell when fully charged and supplying the load

= 2.1 – (200 × 0.00015) = 2.08 V

Terminal voltage/cell when discharged

= 1.83 – (200 × 0.00015) = 1.8 V

(i) No. of cells in the battery = 111/2.08

= 53.4 say, **54**

(ii) No. of cells required when discharged

= 111/1.8 = 62

Hence, number of end cells = 62 – 54 = **8**

The connections are shown in Fig. 9.14.

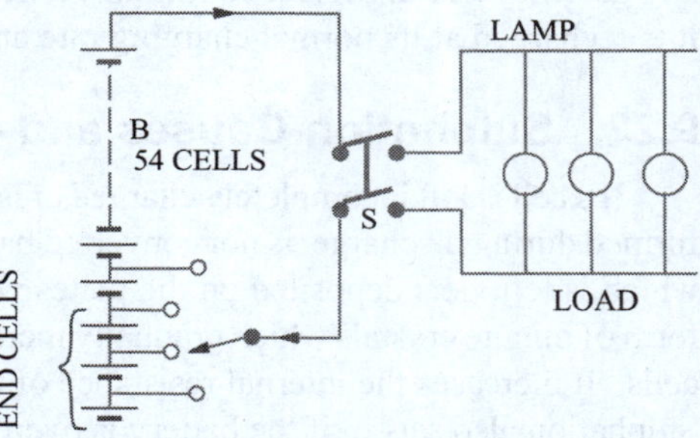

Fig. 9.14

Example 9.12. *A storage battery consists of 55 series-connected cells each of internal resistance 0.001 Ω and e.m.f. 2.1 V. Each cell consists of 21 plates, ten positive and eleven negative, each plate measuring 20 × 25 cm. If full-load current per cell is 0.01 A per cm² of positive plate surface, find* **(i)** *full-load terminal voltage of the battery and* **(ii)** *power wasted in the battery if the connectors have a total resistance of 0.025 Ω.*

Solution. Since both sides of a positive plate are utilized, the area of both sides will be taken into consideration.

Total area (both sides) of ten positive plates = 2 × 20 ×25 × 10 = 10,000 cm²

Full load current = 10,000 × 0.01 = 100 A

Voltage drop in battery and across connectors = 100 [(55 × 0.001) + 0.025] = 8 V

Battery e.m.f. = 55 × 1.2 = 115.5 V

(i) Battery terminal voltage on full-load = 115.5 – 8 = **107.5 V**

(ii) Total resistance = (55 × 0.001) + 0.025 = 0.08 Ω; Power loss = 100^2 × 0.08 = **800 W.**

Tutorial Problem 9.2

1. A 60-cell storage battery having a capacity of 360 Ah takes 8 hours when charged by a dc generator at a voltage of 220 V. Calculate the charging current and the range of the rheostat required to ensure a constant charging current. The emf of each cell is 1.8 V at the beginning of charging and 2.7 V at the end of the charging. Ignore the internal resistance of the cell.

[45 A; 2.45 to 1.29 Ω]

2. A storage battery consists of 55 series connected cells each of internal resistance 0.001 Ω and e.m.f. 2.1 V. Each cell consists of 21 plates, ten positive and eleven negative, each plate measuring 20 × 25 cm. If full-load current per cell is 0.01 A per cm^2 of positive plate surface, find (*i*) full-load terminal voltage of the battery and (*ii*) power wasted in the battery if the connectors have a total resistance of 0.025 Ω. **[(*i*) 107.5 V (*ii*) 800 W]**

9.21. Trickle Charging

When a storage battery is kept entirely as an emergency reserve, it is very essential that it should be found fully charged and ready for service when an emergency arises. Due to leakage action and other open-circuit losses, the battery deteriorates even when idle or on open-circuit. Hence, to keep it fresh, the battery is kept on a trickle charge. The rate of trickle charge is small and is just sufficient to balance the open-circuit losses. For example, a standby battery for station bus-bars capable of giving 2000 A for 1 hour or 400 Ah at the 10-hr rate, will be having a normal charging rate of 555 A, but a continuous 'trickle' charge of 1 A or so will keep the cells fully charged (without any gassing) and in perfect condition. When during an emergency, the battery gets discharged, it is recharged at its normal charging rate and then is kept on a continuous trickle charge.

9.22. Sulphation-Causes and Cure

If a cell is left incompletely charged or is not fully charged periodically, then the lead sulphate formed during discharge, is not converted back into PbO_2 and Pb. Some of the unreduced $PbSO_4$ which is left, gets deposited on the plates which are then said to be sulphated. $PbSO_4$ is in the form of minute crystals which gradually increase in size if not reduced by thoroughly charging the cells. It increases the internal resistance of the cell thereby reducing its efficiency and capacity. Sulphation also sets in if the battery is overcharged or left discharged for a long time.

Sulphated cells can be cured by giving them successive overcharges, for which purpose they are cut out of the battery during discharge, so that they can get two charges with no intervening discharge. The other method, in which sulphated cells need not be cut out of the battery, is to continue charging them with a '***milking booster***' even after the battery as a whole has been charged. A milking booster is a motor-driven low-voltage dynamo which can be connected directly across the terminals of the sulphated cells.

9.23. Maintenance of Lead-Acid Cells

The following important points should be kept in mind for keeping the battery in good condition:

1. Discharging should not be prolonged after the minimum value of the voltage for the particular rate of discharge is reached.
2. It should not be left in discharged condition for long.
3. The level of the electrolyte should always be 10 to 15 mm above the top of the plates which must not be left exposed to air. Evaporation of electrolyte should be made up by adding distilled water occasionally.
4. Since acid does not vaporise, none should be added.
5. Vent openings in the filling plug should be kept open to prevent gases formed within from building a high pressure.
6. The acid and corrosion on the battery top should be washed off with a cloth moistened with baking soda or ammonia and water.
7. The battery terminals and metal supports should be cleaned down to bare metal and covered with vaseline or petroleum jelly.

9.24. Main Operated Battery Chargers

A battery charger is an electrical device that is used for putting energy into a battery. The battery charger changes the a.c. from the power line into d.c. suitable for charger. However, d.c. generator and alternators are also used as charging sources for secondary batteries.

In general, a mains-operated battery charger consists of the following elements:

1. A step-down transformer for reducing the high a.c. mains voltage to a low a.c. voltage.
2. A half-wave or full-wave rectifier for converting alternating current into direct current.
3. A charger-current limiting element for preventing the flow of excessive charging current into the battery under charge.
4. A device for preventing the reversal of current *i.e.* discharging of the battery through the charging source when the source voltage happens to fall below the battery voltage.

In addition to the above, a battery charger may also have circuitry to monitor the battery voltage and automatically adjust the charging current. It may also terminate the charging process when the battery becomes fully charged. However, in many cases, the charging process is not totally terminated but only the charging rate is reduced so as to keep the battery on trickle charging. These requirements have been illustrated in Fig. 9.15.

Most of the modern battery chargers are fully protected against the following eventualities:

(*a*) They are able to operate into a short-circuit.
(*b*) They are not damaged by a reverse-connected battery.
(*c*) They are operated into a totally flat battery.
(*d*) They can be regulated both for current and voltage.

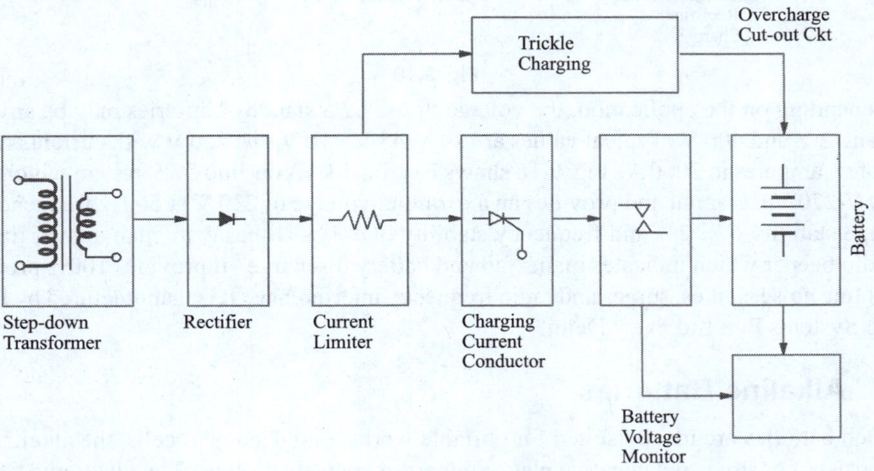

Fig. 9.15

9.25. Static Uninterruptable Power Systems (UPS)

The function of a *UPS* is to ensure absolute continuity of power to the computerised control systems thereby protecting critical equipment from electrical supply failure. A *UPS* makes it possible to provide a 'clean' reliable supply of alternating current free of sags of surges in the line voltage, frequency variation, spikes and transients. *UPS* systems achieve this by rectifying the standard mains supply, using the direct current to charge the standby battery and to produce 'clean' alternating current by passing through an inverter and filter system.

Components of a *UPS* System

The essential components of a *UPS* system as shown in Fig. 9.28 are as under:

1. A rectifier and thyristor-controlled battery charger which converts the *AC* input into regulated *DC* output and keeps the standby battery fully charged.
2. A standby battery which provides *DC* input power to inverter during voltage drops or on failure of the normal mains *AC* supply.
3. An inverter which converts *DC* to clean *AC* thus providing precisely regulated output voltage and frequency to the load as shown.

Working

As shown in Fig. 9.16 the main flow of energy is from the rectifier to the inverter with the standby battery kept on 'float'. If the supply voltage falls below a certain level or fails completely, the battery output to the inverter maintains a clean a.c. supply. When the mains power supply is resorted, the main energy flow against starts from the rectifier to the inverter but, in addition, the rectifier recharges the battery. When the standby battery gets fully charged, the charging current is automatically throttled back due to steep rise in the back e.m.f. of the battery . An automatic/ manual bypass switch is used to connect the load either directly to the mains a.c. supply or to the inverter a.c. supply.

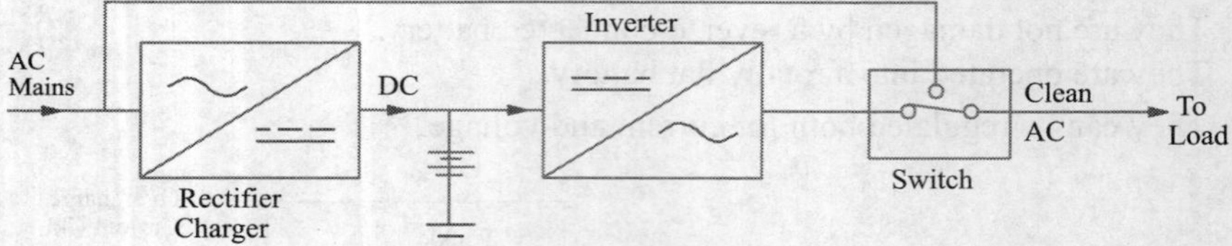

Fig. 9.16

Depending on the application, the voltage of the *UPS* standby battteries may be anywhere between 12 V and 400 V. Typical values are 24 V, 48 V, 110 V and 220 V with currents ranging from a few amperes to 2000 A. Fig. 9.16 shows Everon 4-kVA on-line *UPS* system which works on 170 V-270 V a.c. input and provides an a.c. output voltage of 230 V at 50 Hz frequency with a voltage stability of ± 2% and frequency stability of ± 1%. It has zero change over time and has audio beeper which indicates mains fail and battery discharge. It provides 100% protection against line noise, spikes, surges and radio frequency interference. It is manufactured by Everon Electro Systems Pvt. Ltd. New Delhi.

9.26. Alkaline Batteries

Such batteries are ideally suited for portable work. Like lead-acid cells, the alkaline cells also consist of positive and negative plates immersed in an electrolyte. The plates and the electrolyte are placed in a suitable-container.

The two types of alkaline batteries which are in general use are:

(*i*) nickel-iron type of Edison type.

(*ii*) nickel-cadmium type of Jungnor type which is commercially known as NIFE battery.

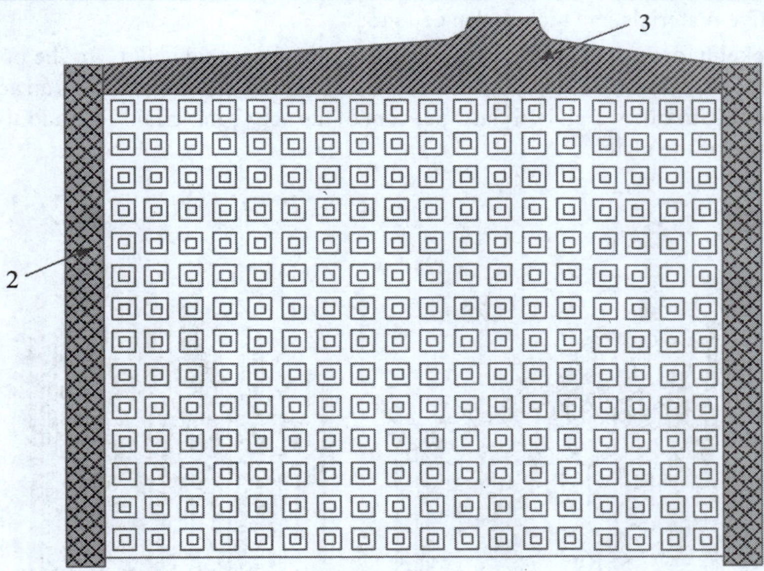

Fig. 9.17

Another alkaline battery which differs from the above only in the mechanical details of its plates is known as Alkum battery which uses nickel hydroxide and graphite in the positive plates and a powdered alloy of iron and chromium in the negative plates.

Silver-zinc type of alkaline batteries are also made whose active material for the positive plate is silver oxide (Ag_2O) and for negative plate is zinc oxide and zinc powder. The electrodes or plates of the alkaline cells are designed to be either of the enclosed-pocket type or open-pocket type. In the case of enclosed-pocket type plates, the active material is inside perforated metal envelopes whereas in the other type, the active material is outside directly in contact with the electrolyte. As shown in Fig. 9.17, the active material of the enclosed-pocket type plates is enclosed in nickel-plated perforated steel pockets or packs which are pressed into the steel frames of the plates.

The open-pocket type plates are made of the following three materials:

(i) metal-ceramic plate-the frame of the plate is a nickel-plated steel grid with the active material pressed in under a pressure of 800 to 1900 kg/cm^2.

(ii) foil plate-the base of such a plate is a thin nickel foil coated with a layer of nickel suspension deposited by a spray technique.

(iii) pressed plates–the base member of these plates is a nickel–plated pressed steel grid. The active material is pressed into them at a pressure of about 400 kg/cm^2.

9.27. Nickel–Iron and Edison Batteries

There is revived interest in the nickel-iron battery because it seems to be one of the few systems which may be developed into a high-energy density battery for electric vehicles. Since long the two main designs for this battery have been the tubular positive type and the flat pocket plate type although cells with sintered type negative are also being manufactured.

The active materials in a nickel-iron cell are:

(i) Nickel hydroxide $Ni(OH)_4$ or apple green nickel peroxide NiO_2 for the positive plate. About 17 per cent of graphite is added to increase conductivity. It also contains an activating additive barium hydroxide which is about 2 per cent of the active material. This additive increases the service life of the plates.

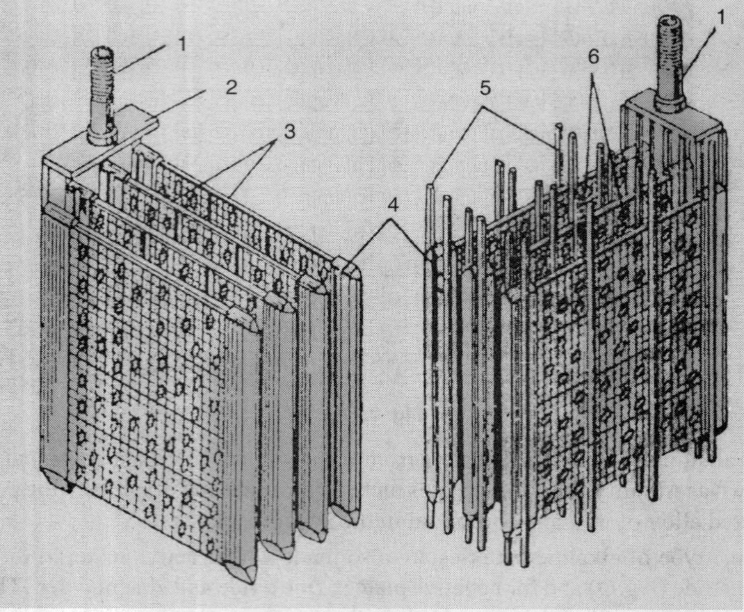

Fig. 9.18 Plates groups of an alkaline cell

(*a*) + ve group (*b*) – ve group, 1–terminal post 2–connecting strap, 3–plates. 4–plates side members. 5–ebonite spacer sticks. 6–pockets.

(ii) powdered iron and its oxides for the negative plate. Small quantities of nickel sulphate and ferrous sulphide are added to improve the performance of the coil.

(iii) the electrolyte is 21 per cent solution of caustic potash KOH (potassium hydrate) to which is added a small quantity of lithium hydrate LiOH for increasing the capacity of the cell.

As shown in Fig. 9.18, plates of the same polarity with their pockets filled, are assembled into cells groups for which purpose they are welded to a common strap having a threaded post.

The number of negative plates is one more than the positive plates. The extreme negative plates are electrically connected to the container. Ebonite separating sticks are placed between the positive and negative plates to prevent any short-circuiting.

The steel containers of the batteries are press-formed from steel and the joints are welded. The body and the cover are nickel-plated and have a dull finish. However, it should be kept in mind that since these containers are electrically alive, no loose wires should touch them owing to the danger of severe sparking from short-circuits.

9.28. Electrical Characteristics

The e.m.f. of an Edison cell, when fully charged, is nearly 1.4 V which decreases rapidly to 1.3 V and then try slowly on 1.1 or 1.0 V on discharge. The average discharge voltage for a 5-hour discharge rate is 1.2 V. Hence, for the same average value of the voltage, an alkali accumulator will consist of 1.6 to 1.7 times as many cells as in a lead-acid battery. Internal resistance of an alkali cell is nearly five times that of the lead-acid cell, hence there is a relatively greater difference between its terminal voltage when charging and discharging.

The average charging voltage for an alkali cell is about 1.7 V. The general shapes of the charge and discharge curves for such cells are, however, similar to those for lead-acid cells. The rated capacity of nickel accumulators usually refers to either 5-hour or 8-hour discharge rate unless stated otherwise.

The plates of such cells have greater mechanical strength because of all-steel construction. They are comparatively lighter because ***(i)*** their plates are lighter and ***(ii)*** they require less quantity of electrolyte. They can withstand heavy charge and discharge currents and do not deteriorate even if left discharged for long periods.

Due to its relatively higher internal resistance, the efficiencies of an Edison cell are power than those of the lead acid cell. On the average, its Ah efficiency is about 80% and Wh efficiency 60 or 50%. It has an average density of 50 Wh/kg.

With increase in temperature, e.m.f. is increased slightly but capacity increases by an appreciable amount. With decrease in temperature, the capacity decreases becoming practically zero at 4°C even through the cell is fully charged. This is serious drawback in the back in the case of electrically driven vehicles in cold weather and previous have to be taken to heat up the battery before starting, though, in practice, the $I^2 R$ loss in the internal resistance of the battery is sufficient to keep the battery cells warm when running.

The principal disadvantage of the Edison battery on nickel-iron battery is its high initial cost (which will probably be sufficiently reduced when patents expire). At present, an Edison battery costs approximately twice as much as a lead-acid battery designed for similar service. But since the alkaline battery outlasts an indeterminate number of lead-acid batteries, it is cheaper in the end.

Because of their lightness, compact construction, increased mechanical strength, ability to withstand rapid charging and discharging without injury and freedom from corrosive liquids and fumes, alkaline batteries are ideally suited for traction work such as propulsion of electric factory trucks, mine locomotives, miner's lamps, lighting and starting of public service vehicles and other services involving rough usage etc.

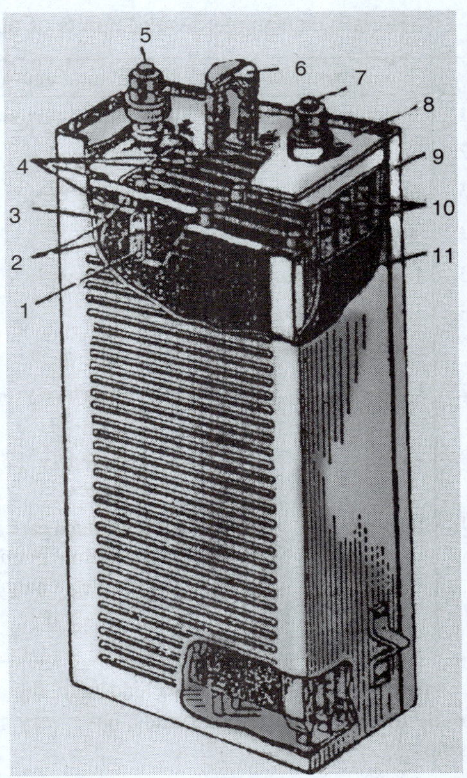

Fig. 9.19. Parts of Nickel-cadium alkaline cell. 1–active material 2–ebonite spacer sticks 3–pocket element 4–positive plates 5–positive terminal post 6–vent plug 7–negative terminal post 8–cover 9–container 10–negative plates 11–ebonite plate.

9.29. Nickel–Cadmium Batteries

The reactive materials in a nickel-cadmium cell (Fig. 9.19) are:

(i) $Ni(OH)_4$ for the positive plate exactly as in the nickel-iron cell.

(ii) a mixture of cadmium or cadmium oxide and iron mass to which is added about 3 per cent of solar oil for stabilizing the electrode capacity. The use of cadmium results in reduced internal resistance of the cell.

(iii) the electrolyte is the same as in the nickel-iron cell.

The cell grouping and plate arrangement is identical with nickel-iron batteries except that the number of positive plates is more than the negative plates. Such batteries are more suitable than nickel-iron batteries for floating duties in conjunction with a charging dynamo because, in their case, the difference between charging and discharging e.m.f.s is not as great as in nickel-iron batteries.

Nickel-cadmium sintered plate batteries were first manufactured by Germans for military aircrafts and rockets. Presently, they are available in a variety of designs and sizes and have energy density going upto 55 Wh/kg. Their capacity is less affected by high discharge rates and low operating temperature than any other rechargeable batteries. Since such batteries have very low open-circuit losses, they are well-suited for pleasure yatches and launches which may be laid up for long periods. They are also used in commercial airliners, military aeroplanes and helicopters for starting main engines or auxiliary turbines and for emergency power supply.

9.30. Comparison: Lead-Acid and Edison Cells

The relative strong and weak points of the cells have been summarised below:

	Particulars	*Lead-acid cell*	*Edison cell*
1.	Positive Plate	PbO_2 lead peroxide	Nickel hydroxide $Ni(OH)_4$ or NiO_2
2.	Negative Plate	Sponge lead	Iron
3.	Electrolyte	diluted H_2SO_4	KOH
4.	Average e.m.f.	2.0 V/cell	1.2 V/cell
5.	Internal resistance	Comparatively low	Comparatively higher
6.	Efficiency:	90–95%	nearly 80%
	amp-hour watt-hour	72–80%	about 60%
7.	Cost	Comparatively less than alkaline line cell	almost twice that of Pb-acid cell Easy maintenance
8.	Life	gives nearly 1250 charges and discharges	five years at least
9.	Strength	Needs much care and maintenance. Sulphation occurs often due to incomplete charge of discharge.	robust, mechanically strong, can withstand vibration, is light, unlimited rates of charge and discharge. Can be left discharged, free from corrosive liquids and fumes.

Moreover, as copmpared to lead-acid, the alkaline cells operate much better at low temperature, do not emit obnoxious fumes, have very small self-discharge and their plates do not buckle or smell.

9.31. Silver–Zinc Batteries

The active material of the positive plates is silver oxide which is pressed into the plate and then subjected to a heat treatment. The active material of the negative plates is a mixture of zinc powder and zinc oxide.

The chemical changes taking place within the cell can be represented by the following single equation:

$$\underset{\text{+ ve plate}}{Ag_2O} + \underset{\text{– ve plate}}{Zn} \underset{\text{Charge}}{\overset{\text{discharge}}{\rightleftharpoons}} \underset{\text{+ ve plate}}{2\,Ag} + \underset{\text{–ve plate}}{ZnO}$$

A silver-zinc cell has a specific capacity (*i.e.* capacity per unit weight) 4 to 5 times greater than that of other type of cells. Their ground applications are mainly military *i.e.* communications equipment, portable radar sets and night-vision equipment. Moreover, comparatively speaking, their efficiency is high and self-discharge is small. Silver-zinc batteries can withstand much heavier discharge currents than are permissible for other types and can operate over a temperature range of – 20° C to + 60°C. Hence, they are used in heavy – weight torpedoes and for submarine propulsion. It has energy density of 150 Wh/kg. Its life time in wet condition is 1-2 years and the dry storage life is upto 5 years. However, the only disadvantage of silver-zinc battery or cell is its higher cost.

9.32. High Temperature Batteries

It is a new group of source which requires operating temperatures above the embient. They possess the advantages of high specific energy and power coupled with low cost. They are particularly suitable for vehicle traction and load levelling purpose in the electric supply industry. We will beiefly describe the following cell from which high-temperature batteries are made.

1. ***Lithium/Chlorine Cell***

It has an emf of 3.5 V, a theoretical specific energy of 2200 Wh/kg at 614°C and operating temperature of 650°C.

2. ***Lithium/Sulphur Cell***

It has an emf of 2.25 V, specific energy of 2625 Wh/kg and an operating temperature of 365°C.

3. ***Lithium-Aluminium/Iron-Sulphide Cells***

The emf of these cells is 1.3 V and a theoretical specific energy of 450 Wh/kg.

4. ***Sodium/Sulphur Cells***

It utilises liquid sodium as negative electrode and sulphur as positive electrode and employs polycrystalline beta alumina as solid electrolyte. It was conceived in the 1960s by J.T. Kummer and N. Weber. The cell reaction can be written as $2\,Na + 3\,S = Na_2S_3$. The announcement of sodium/sulphur battery based on beta alumina was made by Ford Motor Company of USA in 1966. The open-circuit voltage of the cell is 2.1 V and it has a specific energy of 750 Wh/kg with an operating temperature of 350°C. The two unique features of this cell are (1) a Faradaic efficiency of 100% and an ampere-hour capacity which is invariant with discharge rate and (2) high self-life (which is critical for certain space applications).

9.33. Secondary Hybrid Cells

A hybrid cell may be defined as a galvanic electrotechnical generator in which one of the active reagents is in the gaseous state *i.e.* the oxygen of the air. Such cells take advantage of both battery and fuel cell technology. Examples of such cells are:

1. ***Metal-air cells such as iron oxygen and zinc oxygen cells.***

The Zn/O_2 cell has an open-circuit voltage of 1.65 V and a theoretical energy density of 1090 Wh/kg. The Fe/O_2 cell has an OCV of 1.27 V and energy density of 970 Wh/kg.

2. *Metal-halogen cells such as zinc-chlorine and zinc-bromine cells.*

The zinc-chlorine cell has an OCV of 2.12 V at 25°C and a theoretical energy density of 100 Wh/kg. Such batteries are being developed for EV and load levelling applications. The zinc-bromine cell has an OCV of 1.83 V at 25°C and energy density of 400 Wh/kg.

3. *Metal-hydrogen cells such as nickel-hydrogen cell.*

Such cells have an OCV of 1.4 V and a specific energy of about 65 Wh/kg. Nickel-hydrogen batteries have captured large share of the space battery market in recent years and are rapidly replacing Nickel/cadmium batteries as the energy storage system of choice. They are acceptable for geosynchronous orbit applications where not many cycles are required over the life of the system (1000 cycles, 10 years).

The impetus for research and development of metal-air cells has arisen from possible EV applications where energy density is a critical parameter. An interesting application suggested for a secondary zinc-oxygen battery is for energy storage on-board space craft where the cell could be installed inside one of the oxygen tanks thereby eliminating need for gas supply pipes and valves etc. These cells could be reacharged using solar converters.

Some of the likely future developments for nickel-hydrogen batteries are **(1)** increase in cycle life for low earth orbit applications upto 40,000 cycles (7 years) **(2)** increase in the specific energy upto 100 W/kg for geosynchronous orbit applications and **(3)** development of a bipolar nickel-hydrogen battery for high pulse power applications.

9.34. Fuel Cells

As discussed earlier, a secondary battery produces electric current by oxidation-reduction chemical reaction. Similar chemical reactions take place in fuel cells but there is a basic difference between the two. Whereas in secondary batteries the chemical energy is stored in the positive and negative electrodes, in fuel cells the oxidant and the fuel are stored outside the cells and must be fed to the electrodes continuously during the time the fuel cell supplies electric current. This gives an advantage to the fuel cells over the storage battery because fuels can be quickly replenished which is similar to filling up to the petrol tank of a car. Moreover, storage batteries when fully discharged take several hours to be recharged.

9.35. Hydrogen–Oxygen Fuel Cells

The first fuel battery was designed by F.T. Bacon in 1959. The construction of a simple fuel cell is shown in Fig. 9.20. The electrodes are made from sintered nickel plates having a coarse pore surface and a fine pore surface, the two surfaces being for gas and electrolyte respectively. The electrolyte used in KOH is of about 85 per cent concentration. The water vapour formed as a byproduct of the reaction is removed by condensation from the stream of hydrogen passing over the back of the fuel.

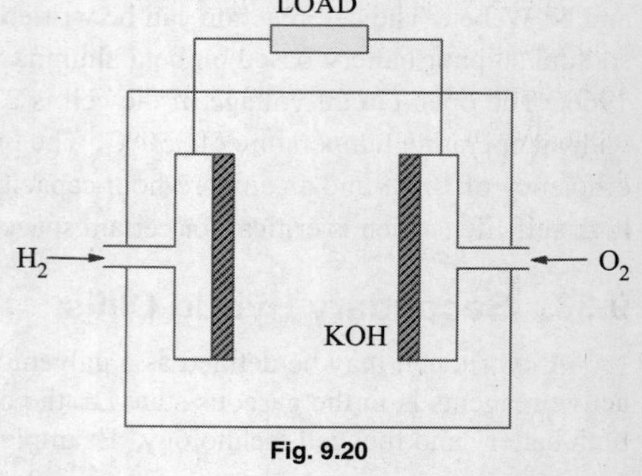

Fig. 9.20

The two electrodes of the fuel cell are fed with a continuous stream of hydrogen and oxygen (or air) as shown. The oxygen and hydrogen ions react with the potassium hydroxide electrolyte at the surface of the electrodes and produce water. The overall cell reaction is

$$2H_2 + O_2 = 2H_2O$$

The basic reaction taking place in the cells are shown in more details in the Fig. 9.20.

Fuel cell batteries have been used in the manned Apollo space mission for on-board power supply and also for power supply in unmanned satellites and space probes. These batteries have also been used for tractors, fork-lift trucks and golf carts etc. Research is being carried out to run these batteries with natural gas and alcohol. Fuel cell systems are particularly useful where electrical energy is required for long periods. Such applications include **(1)** road and rail traction **(2)** industrial trucks **(3)** naval craft and submarine **(4)** navigational aids and radio repeater stations etc.

9.36. Batteries for Aircraft

The on-board power requirements in aircraft have undergone many changes during the last three or four decades. The jet engines of the aircraft which require starting currents of about 1000 A, impose a heavy burden on the batteries. However, these days this load is provided by small turbogenerator sets and since batteries are needed only to start them, the power required is much less. These batteries possess good high-rate capabilities in order to supply emergency power for upto 1 h in the event of the generator failure. However, their main service is as a standby power for miscellaneous on-board equipment. Usually, batteries having 12 cells (of a nominal voltage of 24 V) with capacities of 18 and 34 Ah at the 10 h rate are used. In order to reduce weight, only light-weight high-impact polystyrene containers and covers are used and the cells are fitted with non-spill ventplugs to ensure complete unspillability in any aircraft position during aerobatics. Similarly, special plastic manifolds are moulded into the covers to provide outlet for gases evolved during cycling.

9.37. Batteries for Submarines

These batteries are the largest units in the traction service. In older types of submarines, the lead storage battery was the sole means of propulsion when the submarine was fully submerged and, additionally supplied the 'hotel load' power for lights, instruments and other electric equipment. When the introduction of the snorkel breathing tube made it possible to use diesel engines for propulsion, battery was kept in reserve for emergency use only. Even modern nuclear-powered submarines use storage batteries for this purpose. These lead-acid batteries may be flat, pasted plate or tubular positive plate type with 5 h capacities ranging from 10,000 to 12,000 Ah. One critical requirement for this service is that the rate of evolution of hydrogen gas on open-circuit should not exceed the specified low limit.

Double plate separation with the help of felted glass fibre mats and microscoporous separators is used in order to ensure durability, high performance and low standing losses.

OBJECTIVE TESTS – 9

1. Active materials of a lead-acid cell are:
   (*a*) lead peroxide (*b*) sponge lead
   (*c*) dilute sulphuric acid (*d*) all the above
2. During the charging of a lead-acid cell:
   (*a*) its cathode becomes dark chocolate brown in colour
   (*b*) its voltage increases
   (*c*) it gives out energy
   (*d*) specific gravity of H_2SO_4 is decreased
3. The ratio of Ah efficiency to Wh efficiency of a lead-acid cell is:
   (*a*) always less than one
   (*b*) just one
   (*c*) always greater than one
   (*d*) either (*a*) or (*b*)
4. The capacity of a cell is measured in:
   (*a*) watt-hours (*b*) watts
   (*c*) amperes (*d*) ampere-hours
5. The capacity of a lead-acid cell does NOT depend on its:
   (*a*) rate of charge
   (*b*) rate of discharge
   (*c*) temperature
   (*d*) quantity of active material
6. As compared to constant-current system, the constant-voltage system of charging a lead-acid cell has the advantage of:
   (*a*) avoiding excessive gassing
   (*b*) reducing time of charging
   (*c*) increasing cell capacity
   (*d*) both (*b*) and (*c*).
7. Sulphation in a lead-acid battery occurs due to:
   (*a*) trickle charging
   (*b*) incomplete charging
   (*c*) heavy discharging
   (*d*) fast charging
8. The active materials of a nickel-iron battery are.
   (*a*) nickel hydroxide
   (*b*) powdered iron and its oxides
   (*c*) 21% solution of caustic potash
   (*d*) all of the above.
9. During the charging and discharging of a nickel iron cell:
   (*a*) its e.m.f. remains constant
   (*b*) water is neither formed nor absorbed
   (*c*) corrosive fumes are produced
   (*d*) nickel hydroxide remains unsplit
10. As compared to a lead-acid cell, the efficiency of a nickel-iron cell is less due to its:
    (*a*) lower e.m.f.
    (*b*) smaller quantity of electrolyte used
    (*c*) higher internal resistance
    (*d*) compactness.
11. Trickle charging of a storage battery helps to:
    (*a*) prevent sulphation
    (*b*) keep it fresh and fully charged
    (*c*) maintain proper electrolyte level
    (*d*) increase its reserve capacity
12. A dead storage battery can be revived by:
    (*a*) a dose of H_2SO_4
    (*b*) adding so-called battery restorer
    (*c*) adding distilled water
    (*d*) none of the above
13. The sediment which accumulates at the bottom of a lead-acid battery consists largely of:
    (*a*) lead-peroxide (*b*) lead-sulphate
    (*c*) antimony-lead alloy (*d*) graphite
14. The reduction of battery capacity at high rates of discharge is primarily due to:
    (*a*) increase in its internal resistance
    (*b*) decrease in its terminal voltage
    (*c*) rapid formation of $PbSO_4$ on the plates
    (*d*) non-diffusion of acid to the inside active materials.
15. Floating battery systems are widely used for:
    (*a*) power stations
    (*b*) emergency lighting
    (*c*) telephone exchange installation
    (*d*) all of the above
16. Any charge given to the battery when taken off the vehicle is called:
    (*a*) bench charge (*b*) step charge
    (*c*) float charge (*d*) *trickle charge*

ANSWERS

1. *d* **2.** *b* **3.** *c* **4.** *d* **5.** *a* **6.** *d* **7.** *b* **8.** *d* **9.** *b* **10.** *c* **11.** *b* **12.** *d* **13.** *c* **14.** *c* **15.** *d* **16.** *a*

CHAPTER

10 Electrical Instruments and Measurements

Chapter Outline

- ❖ Absolute and Secondary Instruments
- ❖ Deflecting Torque
- ❖ Controlling Torque
- ❖ Damping Torque
- ❖ Moving-iron Ammeters and Voltmeters
- ❖ Moving-coil Instruments
- ❖ Permanent Magnet Type Instruments
- ❖ Voltmeter Sensitivity
- ❖ Electrodynamic or Dynamometer Type Instruments
- ❖ Hot-wire Instruments
- ❖ Megger
- ❖ Induction Voltmeter
- ❖ Wattmeters
- ❖ Energy Meters
- ❖ Electrolytic Meter
- ❖ Ampere-hour Mercury Motor Meter
- ❖ Friction Compensation
- ❖ Commutator Motor Meters
- ❖ Ballistic Galvanometer
- ❖ Vibratipon Galvanometer
- ❖ Vibrating-reed Frequency Meter
- ❖ Electrodynamic Frequency Meter
- ❖ Moving-iron Frequency Meter
- ❖ Electrodynamic Power Factor Meter
- ❖ Moving-iron Power Factor Meter
- ❖ Nalder-Lipman Moving-iron Power Factor Meter
- ❖ D.C. Potentiometer
- ❖ A.C. Potentiometers
- ❖ Instrument Transformers
- ❖ Potential Transformers.

10.1. Introduction

In the branch or Electrical Engineering, Electrical quantities like voltage, current and power are not visible by eyes. The instrument which are used to measure these quantities are called Electrical measuring instruments. The instrument which measures voltage, current and power are known as Voltmeter, Ammeter and Wattmeter respectively. The measuring instrument can be broadly divided into two category.

1. Analog Instrument

2. Digital Instrument

1. Analog Instrument. An analog instrument is one in which measured quantity is indicated by the movement of a pointer across the face of a scale.

2. Digital Instrument. An digital instrument is one in which measured quantity is indicated in the form of digital a series of number displayed on a screen.

10.2. Classification of Electrical Measuring Instrument

The electrical measuring instrument are mainly classified in three parts as shown in the block diagram.

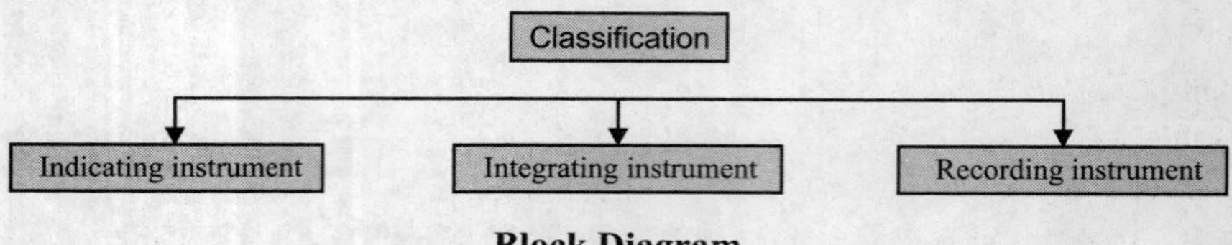

Block Diagram

1. Indicating instrument. Indicating instrument are those instruments which directly indicate the value of electrical quantity with the help of pointer.

Example: Ammeters, Voltmeters and Wattmeters

In such instrument, a pointer is moving over a calibrated scale and directly gives the value of electrical quantity being measured. **For example**, when an ammeter is connected in the circuit, the pointer of the ammeter directly indicates the value of current flowing in the circuit at that time.

2. Integrating Instrument. Integrating instrument are these instrument which measures the total quantity of electrical energy over a specified period of time. These instrument give the summation of electrical quantity being supplied for the given time only.

Example: Household Energy Meter, Amperhour meter

3. Recording Instrument. Recording instruments are these instruments which give is continuous record of the variations of the electrical quantity to be measured.

Example: The example of such instrument is Recording Voltmeters which is used in power stations to record the generated voltage.

10.3. Essentials of Indicating Instruments

As defined above, indicating instruments are those which indicate the value of the quantity that is being measured at the time at which it is measured. Such instruments consist essentially of a pointer which moves over a calibrated scale and which is attached to a moving system pivoted in jewelled bearings. The moving system is subjected to the following three torques:

1. A deflecting (or operating) torque

2. A controlling (or restoring) torque

3. A damping torque.

10.4. Deflecting Torque

The deflecting or operating torque (T_d) is produced by utilizing one or other effects *i.e.* magnetic, electrostatic, electrodynamic, thermal or inductive etc. The actual method of torque production depends on the type of instrument and will be discussed in the succeeding paragraphs. This deflecting torque causes the moving system (and hence the pointer attached to it) to move from its '***zero***' position *i.e.* its position when the instrument is disconnected from the supply.

10.5. Controlling Torque

The deflection of the moving system would be indefinite if there were no controlling or restoring torque. This torque oppose the deflecting torque and increases with the deflection of the moving system. The pointer is brought to rest at a position where the two opposing torques are equal. The deflecting torque ensures that currents of different magnitudes shall produce deflections of the moving system in proportion to their size. Without such at torque, the pointer would swing over to the maximum deflected position irrespective of the magnitude of the current to be measured. Moreover, in the absence of a restoring torque, the pointer once deflected, would not return to its zero position on removing the current. The controlling or restoring or balancing torque in indicating instruments is obtained either by a spring or by gravity as described below:

(*a*) Spring Control

A hair-spring, usually of phosphor-bronze, is attached to the moving system of the instrument as shown in Fig. 10.1 (*a*).

With the deflection of the pointer, the spring is twisted in the opposite direction. This twist in the spring produces restoring torque which is directly proportional to the angle of deflection of the moving system. The pointer comes to a position of rest (or equilibrium) when the deflecting torque (T_d) and controlling torque (T_c) are equal. For example, in permanent-magnet, moving-coil type of instruments, the deflecting torque is proportional to the current passing through them.

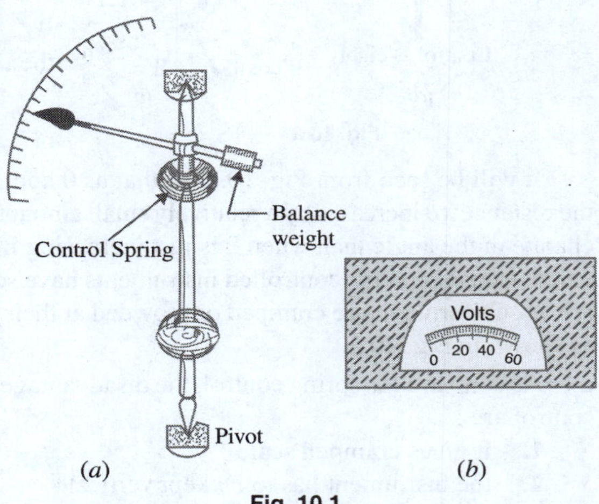

Fig. 10.1

$\therefore \quad T_d \propto I$

and for spring control $T_c \propto \theta$

As $T_c = T_d$

$\therefore \quad \theta \propto I$

Since deflection θ is directly proportional to current *I*, the spring-controlled instruments have a uniform or equally-spaced scales over the whole of their range as shown in Fig. 10.1 (*b*).

To ensure that controlling torque is proportional to the angle of deflection, the spring should have a fairly large number of turns so that angular deformation per unit length, on full-scale deflection, is small. Moreover, the stress in the spring should be restricted to such a value that it does not produce a permanent set in it.

Springs are made of such materials which

(*i*) are non-magnetic
(*ii*) are not subject to much fatigue
(*iii*) have low specific resistance-especially in cases where they are used for leading current in or out of the instrument
(*iv*) have low temperature-resistance coefficient.

The exact expression for controlling torque is $T_c = C\theta$ where *C* is *spring constant*. Its value is given by $C = \dfrac{Ebt^3}{L}$ N-m/rad. The angle θ is in radians.

(*b*) Gravity Control

Gravity control is obtained by attaching a small adjustable weight to some part of the moving system such that the two exert torques in the opposite directions. The usual arrangements is shown in Fig. 10. 2(*a*).

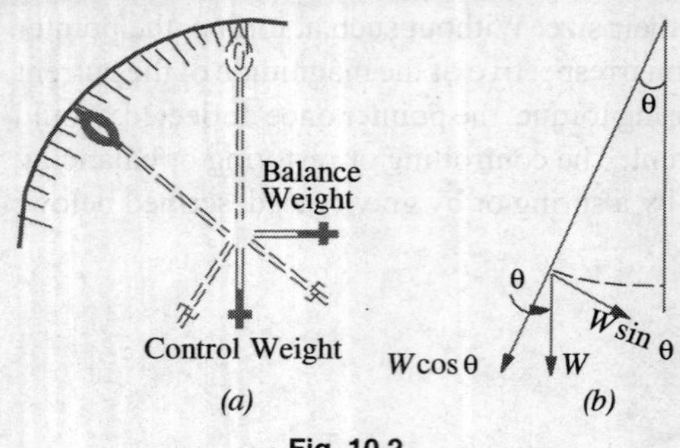

Fig. 10.2

It is seen from Fig. 10.2 (*b*) that the controlling or restoring torque is proportional to the sine of the angle of deflection *i.e.*

$$T_c \propto \sin\theta$$

The degree of control is adjusted by screwing the weight up or down the carrying system

It $\qquad T_d \propto I$

then for position of rest

$$T_d = T_c$$

or $\qquad I \propto \sin\theta \ (\textit{not}\ \theta)$

It will be seen from Fig. 10.2 (*b*) that as θ approaches 90º, the distance *AB* increases by a relatively small amount for a given change in the angle than when θ is just increasing from its zero value. Hence, gravity-controlled instruments have scales which are not uniform but are cramped or crowded at their lower ends as shown in Fig. 10.3.

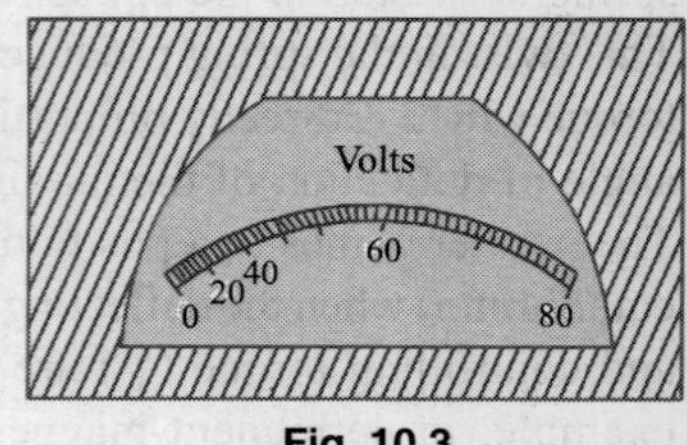

Fig. 10.3

As compared to spring control, the disadvantages of gravity control are :

1. it gives cramped scale
2. the instrument has to be kept vertical.

However, gravity control has the following advantages :

1. it is cheap
2. it is unaffected by temperature
3. it is not subjected to fatigue or deterioration with time.

Example 10.1 *The torque of an ammeter varies as the square of the current through it. If a current of 5 A produces a deflection of 90º, what deflection will occur for a current of 3 A when the instrument is* ***(i)*** *spring-controlled and* ***(ii)*** *gravity-controlled.* **(Elect. Meas. Inst and Meas. Jadavpur Univ.)**

Solution. Since deflecting torque varies as (current)², we have $T_d \propto I^2$

For spring control, $\quad T_c \propto \theta \quad \therefore\ \theta \propto I^2$

For gravity control, $T_c \propto \sin\theta \quad \therefore\ \sin\theta \propto I^2$

(i) **For spring control** $\quad 90^\circ \propto 5^2$ and $\theta \propto 3^2$; $\theta = 90^\circ \times 3^2/5^2 = \mathbf{32.4^\circ}$

(ii) **For gravity control** $\sin 90^\circ \propto 5^2$ and $\sin\theta \propto 3^2$

$$\sin\theta = 9/25 = 0.36\ ;\ \theta = \sin^{-1}(0.36) = \mathbf{21.1^\circ}.$$

10.6. Damping Torque

A damping force is one which acts on the moving system of the instrument ***only when it is moving*** and always opposes its motion. Such stabilizing or demping force is necessary to bring the pointer to rest ***quickly,*** otherwise due to inertia of the moving system, the pointer will oscillate about its final deflected position for quite some time before coming to rest in the steady position. The degree of damping should be adjusted to a value which is sufficient to enable the pointer to rise quickly to its deflected position without overshooting.

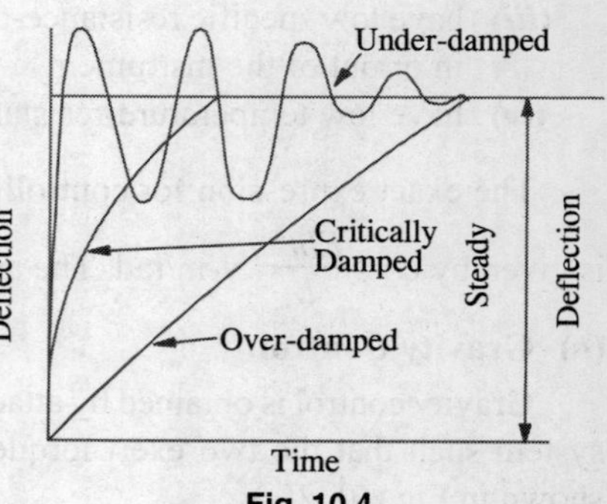

Fig. 10.4

In that case, the instrument is said to be ***dead-beat.*** Any increase of damping above this limit *i.e.* overdamping will make the instruments slow and lethargic. In Fig. 10.4 is shown the effect of damping on the variation of position with time of the moving system of an instrument.

The damping force can be produced by ***(i) air frictions (ii) eddy currents*** and ***(iii) fluid friction*** (used occasionally).

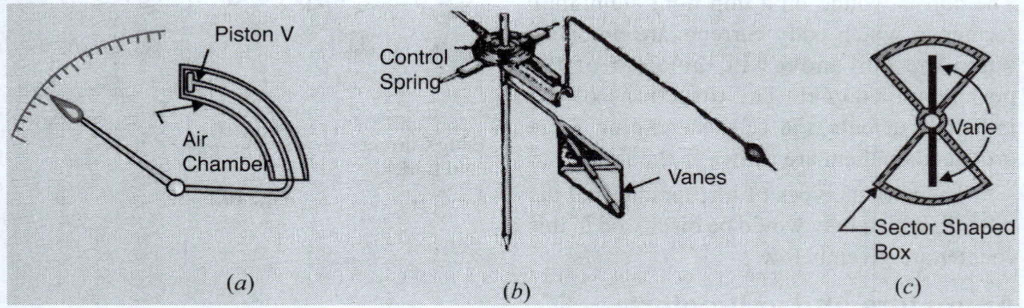

Fig. 10.5

Two methods of air-friction damping are shown in Fig. 10.5 (*a*) and 10.5 (*b*). In Fig. 10.5 (*a*), the light aluminium piston attached to the moving system of the instrument is arranged to travel with a very small clearance in a fixed air chamber closed at one end. The cross-section of the chamber is either circular or rectangular. Damping of the oscillation is affected by the compression and suction actions of the piston on the air enclosed in the chamber. Such a system of damping is not much favoured these days, those shown in Fig. 10.5 (*b*) and (*c*) being preferred. In the latter method, one or two light aluminium vanes are mounted on the spindle of the moving system which move in a closed sector-shaped box as shown.

Fluid-friction is similar in action to the air friction. Due to greater viscosity of oil, the damping is more effective. However, oil damping is not much used because of several disadvantages such as objectionable creeping of oil, the necessity of using the instrument always in the vertical position and its obvious unsuitability for use in portable instruments.

The eddy-current form of damping is the most efficient of the three. The two forms of such a damping are shown in Fig. 10.6 and 10.7. In Fig. 10.6 (*a*) is shown a thin disc of a conducting but ***non-magnetic*** material like copper or aluminium mounted on the spindle which carries the moving system and the pointer of the instrument. The disc is so positioned that its edges, when in rotation, cut the magnetic flux between the poles of a permanent magnet. Hence, eddy currents are produced in the disc which flow and so produce a damping force in such a direction as to oppose the very

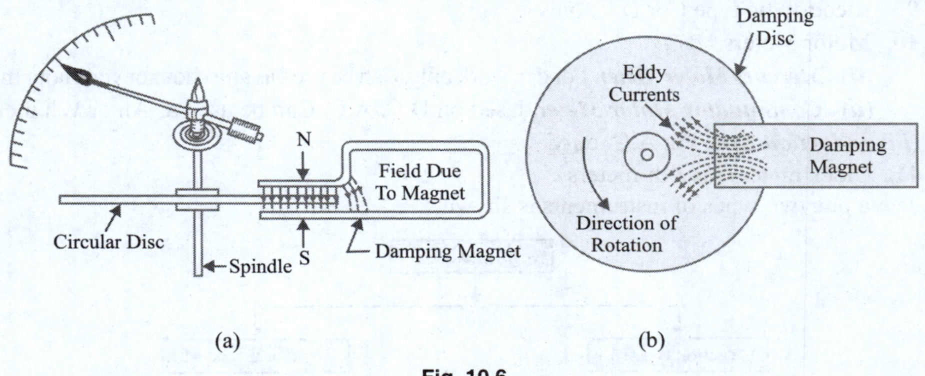

Fig. 10.6

cause producing them (Lenz's Law Art. 7.5). Since the cause producing them is the rotation of the disc, these eddy current retard the motion of the disc and the moving system as a whole.

In Fig. 10.7 is shown the second type of eddy-current damping generally employed in permanent-magnet moving coil instruments. The coil is wound on a thin light aluminium former in which eddy currents are produced when the coil moves in the field of the permanent magnet. The directions of the induced currents and of the damping force produced by them are shown in the figure.

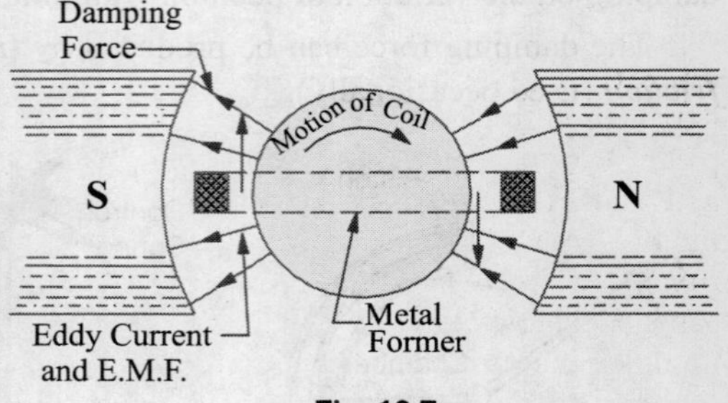

Fig. 10.7

The various types of instruments and the order in which they would be discussed in this chapter are given below.

Ammeters and voltmeters

1. Moving-iron type (both for D.C./A.C.)
   (a) ***the attraction type***
   (b) ***the repulsion type***
2. Moving-coil type
   (a) ***permanent-magnet type*** (for D.C. only)
   (b) ***electrodynamic or dynamometer type*** (*for D.C./A.C.*)
3. Hot-wire type (both for D.C./A.C.)
4. Induction type (for A.C. only)
   (a) ***Split-phase type***
   (b) ***Shaded-pole type***
5. Electrostatic type-for voltmeters only (for D.C./A.C.)

Wattmeter

6. Dynamometer type (both for D.C./A.C.),
7. Induction type (for A.C. only)
8. Electrostatic type (for D.C. only)

Energy Meters

9. Electrolytic type (for D.C. only)
10. Motor Meters
    (*i*) ***Mercury Motor Meter.*** For d.c. work only. Can be used as amp-hour or watt-hour meter.
    (*ii*) ***Commutator Motor Meter.*** Used on D.C./A.C. Can be used as Ah or Wh meter.
    (*iii*) ***Induction type.*** For A.C. only.
11. Clock meters (as Wh-meters).

There one two types of instruments is shown:

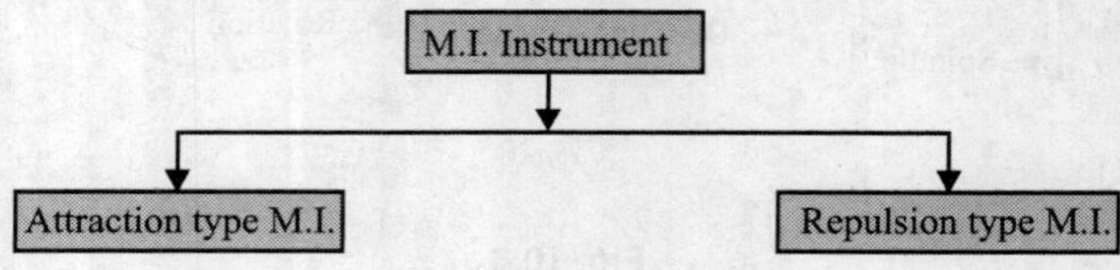

10.7. Moving-Iron Ammeters and Voltmeters

There are two basic forms of these instruments *i.e.* the ***attraction*** type and the ***repulsion*** type. The operation of the attraction type depends on the attraction of a single piece of soft iron into a magnetic field and that of repulsion type depends on the repulsion of two adjacent pieces of iron magnetised by the same magnetic field. For both types of these instruments, the necessary magnetic field is produced by the ampere-turns of a current-carrying coil. In case the instrument is to be used as an ammeter, the coil has comparatively fewer turns of thick wire so that the ammeter has low resistance because it is connected in series with the circuit. In case it is to be used as a voltmeter, the coil has high impedance so as to draw as small a current as possible since it is connected in parallel with the circuit. As the current through the coil is small, it has large number of turns in order to produce sufficient ampere-turns.

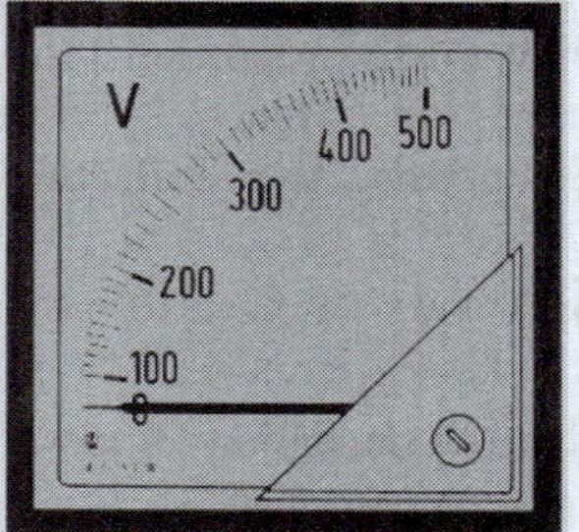

Moving iron voltmeter

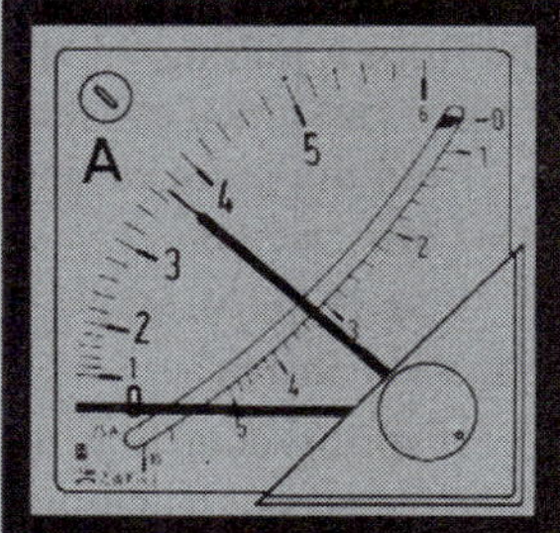

Moving iron ammeter

10.8. Attraction Type M.I. Insturments

The basic working principle of an attraction-type moving-iron instrument is illustrated in Fig. 10.8. It is well-known that if a piece of an unmagnetised soft iron is brought up near either of the two ends of a current-carrying coil, it would be attracted into the coil in the same way as it would be attracted by the pole of a bar magnet. Hence, if we pivot an oval-shaped disc of soft iron on a spindle between bearings near the coil (Fig. 10.8), the iron disc will swing into the coil when the latter has an electric current passing through it. As the field strength would be strongest at the centre of the coil, the ovalshaped iron disc is pivoted in such a way that the greatest bulk of iron moves into the centre of the coil. If a pointer is fixed to the spindle carrying the disc, then the passage of current through the coil will cause the pointer to deflect. The amount of deflection produced would be greater when the current producing the magnetic field is greater. Another point worth noting is that ***whatever the direction of current through the coil, the iron disc would always be magnetised in such a way that it is pulled inwards.*** Hence, such instruments can be used both for direct as well as alternating currents.

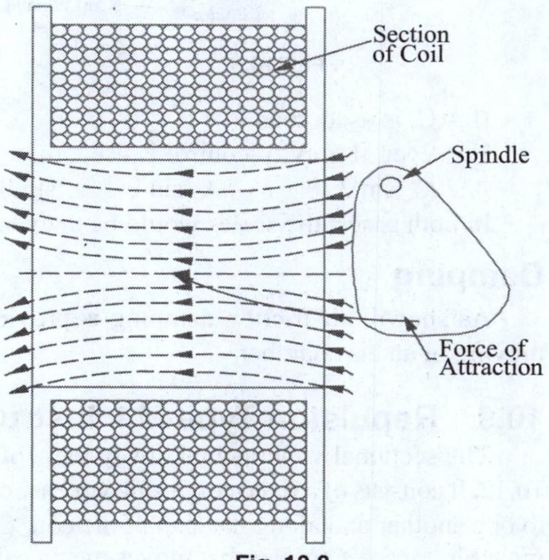

Fig. 10.8

A sectional view of the actual instrument is shown in Fig. 10.9. When the current to be measured is passed through the coil or solenoid, a magnetic field is produced, which attracts the eccentrically-mounted disc inwards, thereby deflecting the pointer, which moves over a calibrated scale.

Deflecting Torque

Let the axis of the iron disc, when in zero position, subtend an angle of ϕ with a direction perpendicular to the direction of the field H produced by the coil. Let the deflection produced be θ

corresponding to a current I through the coil. The magnetisation of iron disc is proportional to the component of H acting along the axis of the disc *i.e.* proportional to $H \cos [90 - (\phi + \theta)]$ or $H \sin (\theta + \phi)$. The force F pulling the disc inwards is proportional to MH or $H^2 \sin (\theta + \phi)$. If the permeability of iron is assumed constant, then, $H \propto I$. Hence, $F \propto I^2 \sin (\theta + \phi)$. If this force acted at a distance of l from the pivot of the rotating disc, then deflecting torque $T_d = Fl \cos (\theta + \phi)$. Putting the value of F, we get

$$T_d \propto I^2 \sin (\theta + \phi) \times l \cos (\theta + \phi) \propto I^2 \sin 2 (\theta + \phi) = KI^2 \sin 2 (\theta + \phi)$$

...sin l is constant

If spring-control is used, then controlling torque $T_c = K' \theta$

In the steady position of deflection, $T_d = T_c$

$\therefore \quad KI^2 \sin 2 (\theta + \phi) = K'\theta$; Hence $\theta \propto I^2$

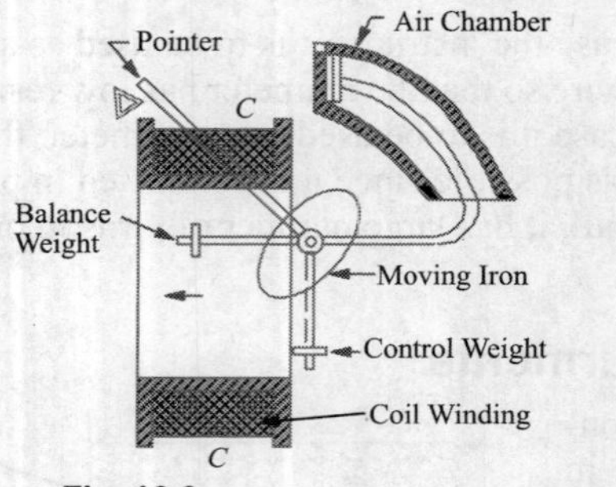

Fig. 10.9

Fig. 10.10

If A.C. is used, then $\theta \propto I^2_{r.m.s.}$

However, if gravity-control is used, then $T_c = K_l \sin \theta$

$\therefore \quad KI^2 \sin 2 (\theta + \phi) = K_l \sin \theta \qquad \therefore \sin \theta\ \alpha I^2 \sin 2 (\theta + \phi)$

In both cases, the scales would be uneven.

Damping

As shown, air-friction damping is provided, the actual arrangement being a light piston moving in an air-chamber.

10.9. Repulsion Type M.I. Instruments

The sectional view and cut-away view of such an instrument are shown in Fig. 10.11 and 10.12. It consists of a fixed coil inside which are placed two soft-iron rods (or bars) A and B parallel to one another and along the axis of the coil. One of them *i.e.* A is fixed and the other B which is movable carries a pointer that moves over a calibrated scale. When the current to be measured is passed through the fixed coil, it sets up its own magnetic field which magnetises the two rods

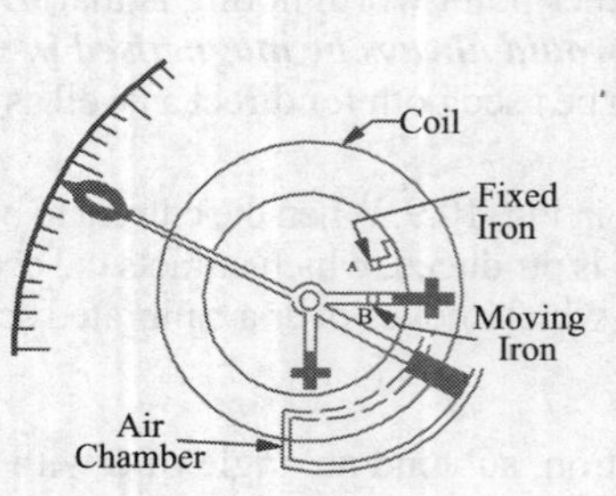

Fig. 10.11

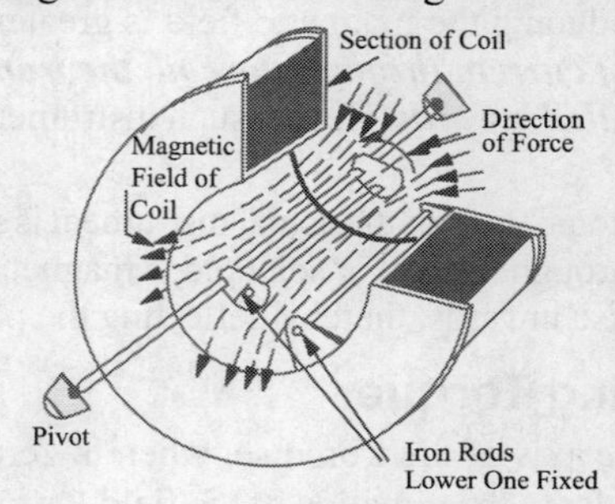

Fig. 10.12

similarly *i.e.* the adjacent points on the lengths of the rods will have the same magnetic polarity. Hence, they repel each other with the result that the pointer is deflected against the controlling torque of a spring or gravity. The force of repulsion is approximately proportional to the square of the current passing through the coil. Moreover, whatever may be the direction of the current through the coil, the two rods will be magnetised similaraly and hence will repel each other.

In order to achieve uniformity of scale, two tongue-shaped strips of iron are used instead of two rods. As shown in Fig. 10.13 (*a*), the fixed iron consists of a tongue-shaped sheet iron bent into a cylindrical form, the moving iron also consists of another sheet of iron and is so mounted as to move parallel to the fixed iron and towards its narrower end [Fig. 10.13 (*b*)].

Deflecting Torque

The deflecting torque is due to the repulsive force between the two similarly magnetised iron rods or sheets.

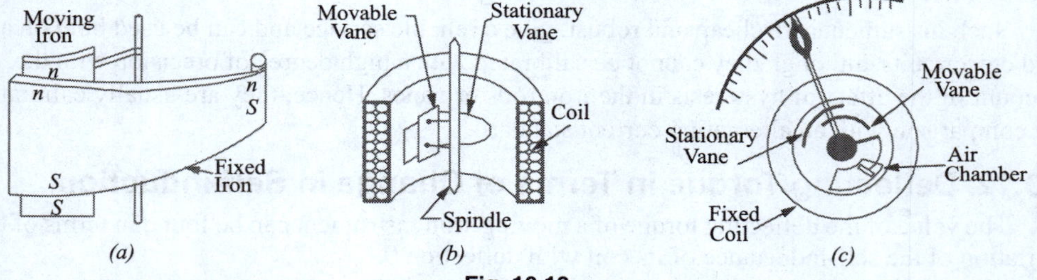

Fig. 10.13

Instantaneous torque $\propto$ repulsive force $\propto m_1 m_2$ ...product of pole strengths

Since pole strength are proportional to the magnetising force H of the coil,

$\therefore$ instantaneous torque $\propto H^2$

Since H itself is proportional to current (assuming ***constant*** permeability) passing through the coil,

$\therefore$ instantaneous torque $\propto I^2$

Hence, the deflecting torque, which is proportional to the mean torque is, in effect, proportional to the mean value of I^2. Therefore, when used on a.c. circuits, the instrument reads the r.m.s. value of current.

Scales of such instruments are uneven if rods are used and uniform if suitable-shaped pieces of iron sheet are used.

The instrument is either gravity-controlled or as in modern makes, is spring-controlled.

Damping is pneumatic, eddy current damping cannot be employed because the presence of a permanent magnet required for such a purpose would affect the deflection and hence, the reading of the instrument.

Since the polarity of both iron rods reverses simultaneously, the instrument can be used both for a.c. and d.c. circuits *i.e.* instrument belongs to the unpolarised class.

10.10. Sources of Error

There are two types of possible errors in such instruments, firstly, those which occur both in a.c. and d.c. work and secondly, those which occur in a.c. work alone.

(*a*) Errors with both d.c. and a.c. work

(i) Error due to hysteresis. Because of hysteresis in the iron parts of the moving system, readings are higher for descending values but lower for ascending values.

The hysteresis error is almost completely eliminated by using Mumetal or Perm-alloy, which have negligible hysteresis loss.

(ii) Error due to stray fields. Unless shielded effectively from the effects of stray external fields, it will give wrong readings. Magnetic shielding of the working parts is obtained by using a covering case of cast-rion.

(*b*) Errors with a.c. work only

Changes of frequency produce **(*i*)** change in the impedance of the coil and **(*ii*)** change in the magnitude of the eddy currents. The increase in impedance of the coil with increase in the frequency of the alternating current is of importance in voltmeters (Ex. 10.2). For frequencies higher than the one used for calibration, the instrument gives lower values. However, this error can be removed by connecting a capacitor of suitable value in parallel with the swamp resistance R of the instrument. It can be shown that the impedance of the whole circuit of the instrument becomes independent of frequency if $C = L/R^2$ where C is the capacitance of the capacitor.

10.11. Advantages and Disadvantages

Such instruments are cheap and robust, give a reliable service and can be used both on a.c. and d.c. circuits, although they cannot be calibrated with a high degree of precision with d.c. on account of the effect of hysteresis in the iron rods or vanes. Hence, they are usually calibrated by comparison with an alternating current standard.

10.12. Deflecting Torque in Terms of Change in Self-induction

The value of the deflecting torque of a moving-iron instrument can be found in terms of the variation of the self-inductance of its coil with deflection θ.

Suppose that when a direct current of I passes through the instrument, its deflection is θ and inductance L. Further suppose that when current changes from I to $(I + dI)$, deflection changes from θ to $(\theta + d\theta)$ and L changes to $(L + dL)$. Then, the increase in the energy stored in the magnetic field

is $$dE = d\left(\frac{1}{2}LI^2\right) = \frac{1}{2}L2I.dI + \frac{1}{2}I^2dL = LI.\,dI + \frac{1}{2}I^2.\,dL \text{ joule.}$$

If $T\frac{1}{2}$ N-m is the controlling torque for deflection θ, then extra energy stored in the control system is $T \times d\theta$ joules. Hence, the total increase in the stored energy of the system is

$$LI.dI + \frac{1}{2}I^2.\,dL + T \times d\theta \qquad \text{...}(i)$$

The e.m.f. induced in the coil of the instrument is $e = N.\dfrac{d\Phi}{dt}$ volt

where $d\phi$ = change in flux linked with the coil due to change in the position of the disc or the bars

dt = time taken for the above change ; N = No. of turns in the coil

Now $L = NF/I \quad \therefore \quad \Phi = LI/N \quad \therefore \quad \dfrac{d\Phi}{dt} = \dfrac{1}{N}.\dfrac{d}{dt}(LI)$

Induced e.m.f. $e = N.\dfrac{1}{N}.\dfrac{d}{dt}(LI) = \dfrac{d}{dt}(LI)$

The energy drawn from the supply to overcome this back e.m.f is

$$= e.Idt = \frac{d}{dt}(LI).Idt = I.d(LI) = I(L.dI + I.dL) = LI.dI + I^2.dL \qquad \text{...}(ii)$$

Equating **(*i*)** and **(*ii*)** above, we get $LI.dI + \frac{1}{2}I^2dL + T.d\theta = LI.dI + I^2.dL \quad \therefore \quad T = \frac{1}{2}I^2\frac{dL}{d\theta}$ N-m.

where $dL/d\theta$ is henry/radian and I in amperes.

10.13. Extension of Range by Shunts and Multipliers

(*i*) As Ammeter. The range of the moving-iron instrument, when used as an ammeter, can be extended by using a suitable shunt across its terminals. So far as the operation with direct current is concerned, there is no trouble, but with alternating current, the division of current between the instrument and shunt changes with the change in the applied frequency. For a.c. work, both the inductance and resistance of the instrument and shunt have to be taken into account.

Obviously, $\dfrac{\text{current through instruments, } i}{\text{current through shunt, } I_s} = \dfrac{R_s + j\omega L_s}{R + j\omega L} = \dfrac{Z_s}{Z}$

where R, L = resistance and inductance of the instrument

R_s, L_s = resistance and inductance of the shunt

It can be shown that above ratio *i.e.* the division of current between the instrument and shunt would be independent of frequency if the time-constants of the instrument coil and shunt are the same *i.e.* if $L/R = L_s/R_s$. The multiplying power (N) of the shunt is given by

$$N = \frac{I}{i} = 1 + \frac{R}{R_s}$$

where I = line current ; i = full-scale deflection current of the instrument.

***(ii)* As Voltmeter.** The range of this instrument, when used as a voltmeter, can be extended or multiplied by using a high non-inductive resistance R connected in series with it, as shown in Fig. 10.14. This series resistance is known as 'multiplier' when used on d.c. circuits. Suppose, the range of the instrument is to be extended from v to V. Then obviously, the excess voltage of $(V - \text{v})$ is to be dropped across R. If i is the full-scale deflection current of the instrument, then

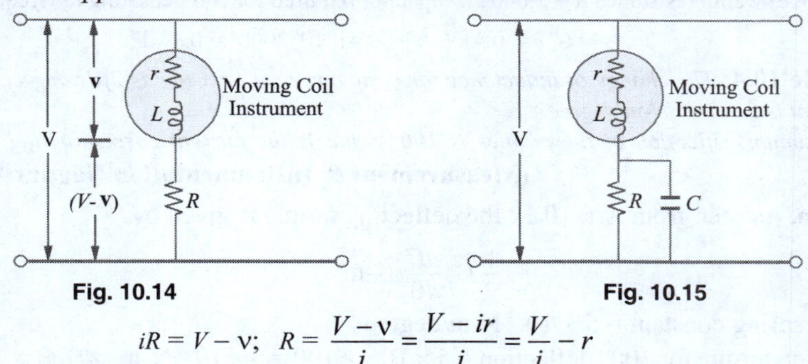

Fig. 10.14 Fig. 10.15

$$iR = V - \text{v}; \quad R = \frac{V - \text{v}}{i} = \frac{V - ir}{i} = \frac{V}{i} - r$$

Voltage magnification = V/v. Since $iR = V - \text{v}$; $\therefore \dfrac{iR}{\text{v}} = \dfrac{V}{\text{v}} - 1$

or $\dfrac{iR}{ir} = \dfrac{V}{\text{v}} - 1 \quad \therefore \quad \dfrac{V}{\text{v}} = \left(1 + \dfrac{R}{r}\right)$

Hence, greater the value of R, greater is the extension in the voltage range of the instrument.

For d.c. work, the principal requirement of R is that its value should remain constant *i.e.* it should have low temperature-coefficient. But for a.c. work it is essential that total impedance of the voltmeter and the series resistance R should remain as nearly constant as possible at different frequencies. That is why R is made as non-inductive as possible in order to keep the inductance of the whole circuit to the minimum. The frequency error introduced by the inductance of the instrument coil can be compensated by shunting R by a capacitor C as shown in Fig. 10.15. In case $r \ll R$, the impedance of the voltmeter circuit will remain practically constant (for frequencies upto 1000 Hz) provided.

$$C = \frac{L}{(1+\sqrt{2})R^2} = 0.41\frac{L}{R^2}$$

Example 10.2. *A 250-volt moving-iron voltmeter takes a current of 0.05 A when connected to a 250-volt d.c. supply. The coil has an inductance of 1 henry. Determine the reading on the meter when connected to a 250-volt, 100-Hz a.c. supply.* **(Elect. Engg., Kerala Univ.)**

Solution. When used on d.c. supply, the instrument offers ohmic resistance only. Hence, resistance of the instrument = 250/0.05 = 5000 Ω.

When used on a.c. supply, the instrument offers ***impedance*** instead of ohmic resistance.

$$\text{impedance at 100 Hz} = \sqrt{5000^2 + (2\pi \times 100 \times 1)^2} = 5039.3\ \Omega$$

$\therefore$ voltage of the instrument $= 250 \times 5000/5039.3 =$ **248 V**

Example 10.3. *A spring-controlled moving-iron voltmeter reads correctly on 250-V d.c. Calculate the scale reading when 250-V a.c. is applied at 50 Hz. The instrument coil has a resistance of 500 Ω and an inductance of 1 H and the series (non-reactive) resistance is 2000 Ω.*

(Elect. Instru. & Measure. Nagpur Univ. 1992)

Solution. Total circuit resistance of the voltmeter is

$$= (r + R) = 500 + 2{,}000 = 2{,}500\ \Omega$$

Since the voltmeter reads correctly on direct current supply, its full-scale deflection current is $= 250/2500 = 0.1$ A.

When used on a.c. supply, instrument offers an impedance

$$Z = \sqrt{2500^2 + (2\pi \times 50 \times 1)^2} = 2.520\ \Omega \quad \therefore \quad I = 0.099\ \text{A}$$

$\therefore$ Voltmeter reading on a.c. supply $= 250 \times 0.099/0.1 =$ **248 V***

Note. Since swamp resistance $R = 2{,}000\ \Omega$, capacitor required for compensating the frequency error is

$$C = 0.41\ L/R^2 = 0.41 \times 1/2000^2 = 0.1\ \mu\text{F}.$$

Example 10.4. *The change of inductance for a moving-iron ammeter is 2μH/degree. The control spring constant is 5 × 10⁻⁷ N-m/degree.*

The maximum deflection of the pointer is 100°, what is the current corresponding to maximum deflection? **(Measurement & Instrumentation Nagpur Univ. 1993)**

Solution. As seen from Art. 10.12 the deflecting torque is given by

$$T_d = \frac{1}{2} I^2 \frac{dL}{d\theta} \text{ N-m}$$

Control spring constant $= 5 \times 10^{-7}$ N-m/degree

Deflection torque for 100° deflection $= 5 \times 10^{-7} \times 100 = 5 \times 10^{-5}$ N-m ; $dL/d\theta = 2\ \mu$H/degree $= 2 \times 10^{-6}$ H/degree.

$$\therefore \quad 5 \times 10^{-5} = \frac{1}{2} I^2 \times 2 \times 10^{-6} \quad \therefore \quad I^2 = \mathbf{50} \text{ and } I = 7.07\ \text{A}$$

Example 10.5. *The inductance of attraction type instrument is given by $L = (10 + 5\theta - \theta^2)\mu H$ where θ is the deflection in radian from zero position. The spring constant is 12×10^{-6} N-m/rad. Find out the deflection for a current of 5 A.*

(Elect. and Electronics Measurements and Measuring Instruments Nagpur Univ. 1993)

Solution. $L = (10 + 5\,\theta - \theta^2) \times 10^{-6}$ H

$$\therefore \quad \frac{dL}{d\theta} = (0 + 5 - 2 \times \theta) \times 10^{-6} = (5 - 2\theta) \times 10^{-6} \text{ H/rad}$$

Let the deflection be θ radians for a current of 5A, then deflecting torque,

$$T_d = 12 \times 10^{-6} \times \theta \text{ N-m}$$

Also, $$T_d = \frac{1}{2} I^2 \frac{dL}{d\theta}$$

Equating the two torques, we get

$$12 \times 10^{-6} \times \theta = \frac{1}{2} \times 5^2 \times (5 - 2\theta) \times 10^{-6} \quad \therefore \quad \theta = \mathbf{1.689\ radian}$$

---

* or reading $= 250 \times 2500/2520 =$ **248 V.**

Tutorial Problem 10.1

1. Derive an expression for the torque of a moving-iron ammeter. The inductance of a certain moving-iron ammeter is $(8 + 4\theta - \frac{1}{2}\theta^2)$ μH where θ is the deflection in radians from the zero position. The control-spring torque is 12×10^{-6} N-m/rad. Calaculate the scale position in radians for a current of 3A. **[1.09 rad] (*I.E.E. London*)**

2. An a.c. voltmeter with a maximum scale reading of 50-V has a resistance of 500 Ω and an inductance of 0.09 henry, The magnetising coil is wound with 50 turns of copper wire and the remainder of the circuit is a non-inductive resistance in series with it. What additional apparatus is needed to make this instrument read correctly on both d.c. circuits or frequency 60 ? **[0.44 μF in parallel with series resistance]**

3. A 10-V moving-iron ammeter has a full-scale deflection of 40 mA on d.c. circuit. It reads 0.8% low on 50 Hz a.c. Hence, calculate the inductance of the ammeter. **[115.5 mH]**

4. It is proposed to use a non-inductive shunt to increase the range of a 10-A moving iron ammeter to 100 A. The resistance of the instrument, including the leads to the shunt, is 0.06 Ω and the inductance is 15 μH at full scale. If the combination is correct on a.d.c circuit, find the error at full scale on a 50 Hz a.c. circuit. **[3.5 %](*London Univ.*)**

10.14. Moving-Coil Instruments

There are two types of such instruments ***(i)* permanent-magnet type** which can be used for d.c. work only and ***(ii)* the *dynamometer type*** which can be used both for a.c. and d.c. work.

10.15. Permanent Magnet Type Instruments

The operation of a permanent-magnet moving-coil type instrument is be based upon the principle that when a current-carrying conductor is placed in a magnetic field, it is acted upon by a force which tends to move it to one side and out of the field.

Construction

As its name indicates, the instrument consists of a permanent magnet and a rectangular coil of many turns wound on a light aluminium or copper former inside which is an iron core as shown in Fig. 10.16. The powerfull U-shaped permanent magnet is made of Alnico and has soft-iron end-pole pieces which are bored out cylindrically. Between the magnetic poles is fixed a soft iron cylinder whose function is ***(i)*** to make the field radial and uniform and ***(ii)*** to decrease the reluctance of the air path between the poles and hence increase the magnetic flux. Surrounding the core is a rectangular coil of many turns wound on a light aluminium frame which is supported by delicate bearings and to which is attached a light pointer. The aluminium frame not only provides support for the coil but also provides damping by eddy currents induced in it. The sides of the coil are free to move in the two air-gaps between the poles and core as shown in Fig. 10.16 and Fig. 10.17. Control of the

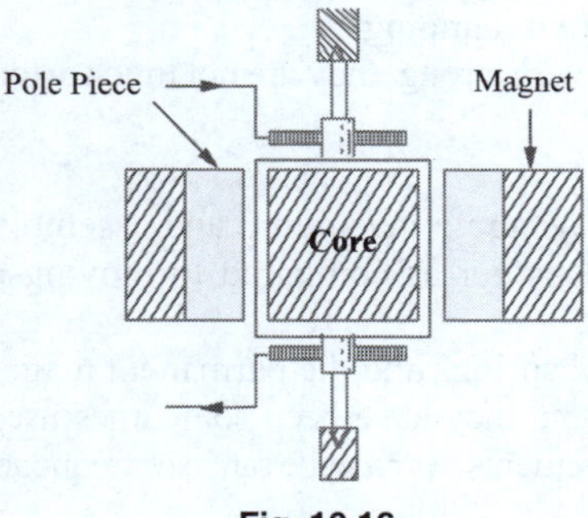

Fig. 10.16.

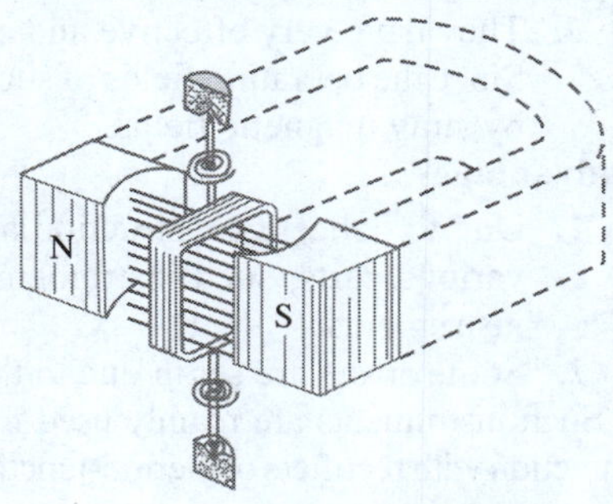

Fig. 10.17

coil movement is affected by two phosphor-bronze hair springs, one above and one below, which additionally serve the purpose of lending the current in and out of the coil. The two springs are spiralled in opposite directions in order to neutralize the effects of temperature changes.

Deflecting Torque

When current is passed through the coil, force acts upon its both sides which produce a deflecting torque as shown in Fig. 10.18. Let

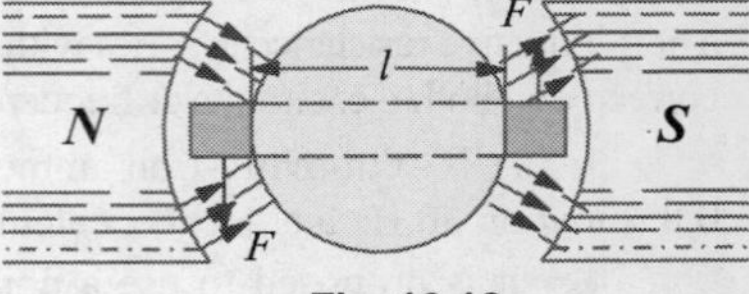

Fig. 10.18

B = flux density in Wb/m^2
l = length or depth of the coil in metre
b = breadth of coil in metre
N = number of turns in the coil

If I ampere is the current passing through the coil, then the magnitude of the force experienced by each of its sides is = BIl newton

For N turns, the force on each side of the coil is = $NBIl$ newton

$$\therefore \quad \text{deflecting torque } T_d = \text{force} \times \text{perpendicular distance}$$
$$= NBIl \times b = NBI(l \times b) = NBIA \text{ N-m}$$

where A is the face area of the coil.

It is seen that if B is constant, then T_d is proportional to the current passing through the coil *i.e.* $T_d \propto I$.

Such instruments are invariable spring-controlled so that $T_c \propto$ deflection θ.

Since at the final deflected position, $T_d = T_c \quad \therefore \theta \propto I$

Hence, such instruments have uniform scales. Damping is electromagnetic *i.e.* by eddy currents induced in the metal frame over which the coil is wound. Since the frame moves in an intense magnetic field, the induced eddy currents are large and damping is very effective.

10.16. Advantage and Disadvantages

The permanent-magnet moving-coil (PMMC) type instruments have the following advantages and disadvantages :

Advantages

1. They have low power consumption.
2. Their scales are uniform and can be designed to extend over an arc of 170° or so.
3. They possess high (torque/weight) ratio.
4. They can be modified with the help of shunts and resistances to cover a wide range of currents and voltages.
5. They have no hysteresis loss.
6. They have very effective and efficient eddy-current damping.
7. Since the operating fields of such instruments are very strong, they are not much affected by stray magnetic fields.

Disadvantages

1. Due to delicate construction and the necessary accurate machining and assembly of various parts, such instruments are somewhat costlier as compared to moving-iron instruments.
2. Some errors are set in due to the aging of control springs and the parmanent magnets.

Such instruments are mainly used for d.c. work only, but they have been sometimes used in conjunction with rectifiers or thermo-junctions for a.c. measurements over a wide range or frequencies.

Permanent-magnet moving-coil instruments can be used as ammeters (with the help of a low resistance shunt) or as voltmeters (with the help of a high series resistance).

The principle of permanent-magnet moving-coil type instruments has been utilized in the construction of the following:

1. For a.c. galvanometer which can be used for detecting extremely small d.c. currents. A galvanometer may be used either as an ammeter (with the help of a low resistance) or as a voltmeter (with the help of a high series resistance).
2. By eliminating the control springs, the instrument can be used for measuring the quantity of electricity passing through the coil. This method is used for *fluxmeters*.
3. If the control springs of such an instrument are purposely made of large moment of inertia, then it can be used as ballistic galvanometer.

10.17. Extension of Range

(*i*) As Ammeter

When such an instrument is used as an ammeter, its range can be extended with the help of a low-resistance shunt as shown in Fig. 10.19 (*a*). This shunt provides a bypath for extra current because it is connected across (*i.e.* in parallel with) the instrument. These shunted instruments can be made to record currents many times greater than their normal full-scale deflection currents. The ratio of maximum current (with shunt) to the full-scale deflection current (without shunt) is known as the 'multiplying power' or 'multiplying factor' of the shunt.

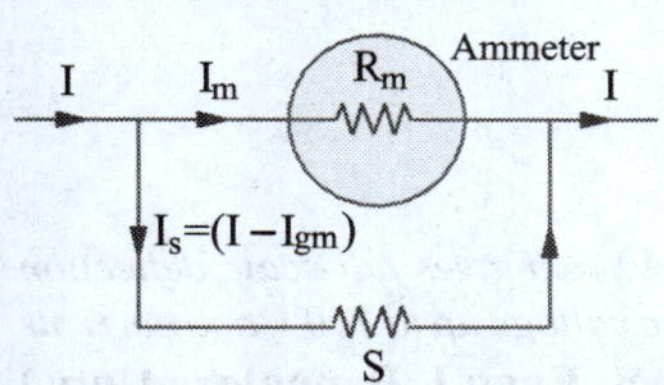

Fig. 10.19 (*a*)

Let R_m = instrument resistance

S = shunt resistance

I_m = full-scale deflection current of the instrument

I = line current to be measured

As seen from Fig. 10.20 (*a*), the voltage across the instrument coil and the shunt is the same since both are joined in parallel.

$$\therefore \quad I_m \times R_m = S I_s = S(I - I_m) \qquad \therefore \quad S = \frac{I_m R_m}{(I - I_m)}\text{; Also } \frac{I}{I_m} = \left(1 + \frac{R_m}{S}\right)$$

$$\therefore \quad \text{multiplying power} = \left(1 + \frac{R_m}{S}\right)$$

Obviously, lower the value of shunt resistance, greater its multiplying power.

(*ii*) As voltmeter

The range of this instrument when used as a voltmeter can be increased by using a high resistance in series with it [Fig. 10.19 (*b*)].

Let I_m = full-scale deflection current

R_m = instrument resistance

$\text{v} = R_m I_m$ = full-scale p.d. across it

V = voltage to be measured

R = series resistance required

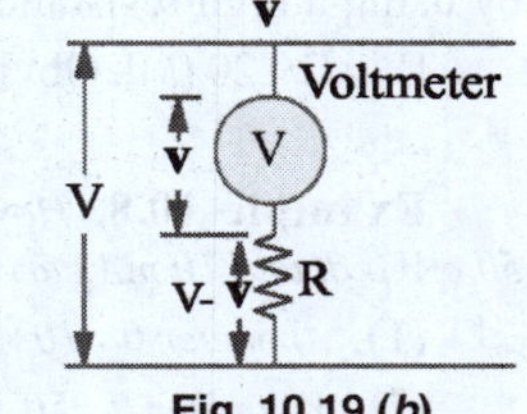

Fig. 10.19 (*b*)

Then it is seen that the voltage drop across R is $V - \text{v}$

$$\therefore \quad R = \frac{V - \text{v}}{I_m} \quad \text{or} \quad R \cdot I_m = V - \text{v}$$

Dividing both sides by v, we get

$$\frac{RI_m}{v} = \frac{V}{v} - 1 \quad \text{or} \quad \frac{R \cdot I_m}{I_m R_m} = \frac{V}{v} - 1 \quad \therefore \quad \frac{V}{v} = \left(1 + \frac{R}{R_m}\right)$$

$$\therefore \quad \text{voltage multiplication} = \left(1 + \frac{R}{R_m}\right)$$

Obviously, larger the value of R, greater the voltage multiplication or range. Fig. 10.20 (*b*) shown a voltmeter with a single multiplier resistor for one range. A multi-range voltmeter requires on multiplier resistor for each additional range.

Example 10.6. *A moving coil ammeter has a fixed shunt of 0.02 Ω with a coil circuit rtesistance of R = 1 kΩ and need potential difference of 0.5 V across it for full-scale deflection.*

(1) *To what total current does this correspond ?*

(2) *Calculate the value of shunt to give full scale deflection when the total current is 10 A and 75 A.* **(Measurement & Instrumentation Nagpur Univ. 1993)**

Solution. It should be noted that the shunt and the meter coil are in parallel and have a common p.d. of 0.5 V applied across them.

(1) $\therefore I_m = 0.5/1000 = 0.0005$ A; $I_s = 0.5./0.02 = 25$ A

$\therefore$ line current = **25.0005 A**

(2) When total current is 10 A, $I_s = (10 - 0.0005) = 9.9995$ A

$$\therefore \quad S = \frac{I_m R_m}{I_s} = \frac{0.0005 \times 1000}{9.9995} = 0.05\ \Omega$$

When total current is 75 A, $I_s = (75 – 0.0005) = 74.9995$ A

$\therefore \quad S = 0.0005 \times 1000/74.9995 =$ **0.00667 Ω**

Example 10.7. *A moving-coil instrument has a resistance of 10 Ω and gives full-scale deflection when carrying a current of 50 mA. Show how it can be adopted to measure voltage up to 750 V and currents upto 1000 A.* **(Elements of Elect. Engg.I, Bangalore Univ.)**

Solution. (*a*) As Ammeter. As discussed above, current range of the meter can be extended by using a shunt across it [Fig. 10.20 (*a*)].

Obviously,

$10 \times 0.05 = S \times 99.95$

$\therefore \quad S =$ **0.005 Ω**

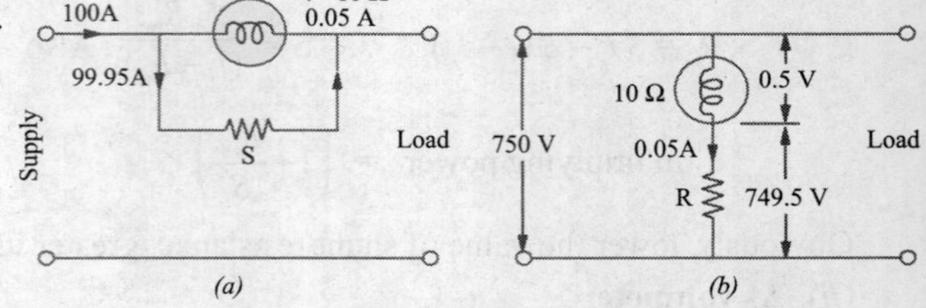

Fig. 10.20

(*b*) As Voltmeter. In this case, the range can be extended by using a high resistance R in series with it.

[Fig. 10.20 (*b*)]. Obviously, R must drop a voltage of (750–0.5) = 749.5 V while carrying 0.05 A.

$\therefore \quad 0.05\ R = 749.5 \quad \text{or} \quad R =$ **14.990 Ω**

Example 10.8. *How will you use a P.M.M.C. instrument which gives full scale deflection at 50 mV p.d. and 10 mA current as*

(1) *Ammeter 0 - 10 A range*

(2) *Voltmeter 0-250 V range* **(Elect. Instruments & Measurements Nagpur Univ. 1993)**

Solution. Resistance of the instrument R_m = 50 mV/10 mA = 5

(*i*) As Ammeter

full-scale meter current, I_m = 10 mA = 0.01 A

shunt current $I_s = I - I_m = 10 - 0.01 = 9.99$ A

Reqd. shunt resistance, $S = \frac{I_m R_m}{(I - I_m)} = \frac{0.01 \times 5}{9.99} = 0.0005\ \Omega$

***(ii)* As Voltmeter**

Full-scale deflection voltage, $v = 50$ mV $= 0.05$ V; $V = 250$ V

Reqd. series resistance, $R = \frac{V - v}{I_m} = \frac{250 - 0.05}{0.01} = 24{,}995\ \Omega$

Example 10.9. *A current galvanometer has the following parameters :*

$B = 10 \times 10^{-3}\ Wb/m^2$; $N = 200$ *turns*, $l = 16$ *mm*;

$d = 16$ *mm*; $k = 12 \times 10^{-9}$ *Nm/radian.*

Calculate the deflection of the galvanometer when a current of 1 μA flows through it.

(Elect. Measurement Nagpur Univ. 1993)

Solution. Deflecting torque $T_d = NBIA$ N–m $= 200 \times (10 \times 10^{-3}) \times (1 \times 10^{-6}) \times (1 \times 10^{-3}) \times (16 \times 10^{-3})$ N-m $= 512 \times 10^{-12}$ N–m

Controlling torque T_c = controlling spring constant × deflection $= 12 \times 10^{-9} \times \theta$ N–m

Equating the deflecting and controlling torques, we have $12 \times 110^{-9} \times \theta = 512 \times 10^{-12}$

$\therefore \qquad \theta = 0.0427$ radian $=$ **2.45°**

Example 10.10. *The coil of a moving coil permanent magnet voltmeter is 40 mm long and 30 mm wide and has 100 turns on it. The control spring exerts a torque of 120×10^{-6} N-m when the deflection is 100 divisions on full scale. If the flux density of the magnetic field in the air gap is 0.5 Wb/m², estimate the resistance that must be put in series with the coil to give one volt per division. The resistance of the voltmeter coil may be neglected.* **(Elect. Mesur. AMIE Sec. B Summer 1991)**

Solution. Let I be the current for full-scale deflection. Deflection torque $T_d = NBIA$ $= 100 \times 0.5 \times I \times (1200 \times 10^{-6}) = 0.06\,I$ N-m

Controlling torque $\qquad T_c = 120 \times 10^{-6}$ N-m

In the equilibrium position, the two torques are equal *i.e.* $T_d = T_c$.

$\therefore \quad 0.06\,I = 120 \times 10^{-6} \quad \therefore \quad I = 2 \times 10^{-3}$ A.

Since the instrument is meant to read 1 volt per division, its full-scale reading is 100 V.

Total resistance $= 100/2 \times 10^{-3} = 50{,}000\ \Omega$

Since voltmeter coil resistance is negligible, it represents the additional required resistance.

Example 10.11. *Show that the torque produced in a permanent-magnet moving-coil instrument is proportional to the area of the moving coil.*

A moving-coil voltmeter gives full-scale deflection with a current of 5 mA. The coil has 100 turns, effective depth of 3 cm and width of 2.5 cm. The controlling torque of the spring is 0.5 cm for full-scale deflection. Estimate the flux density in the gap. **(Elect. Meas, Marathwads Univ.)**

Solution. The full-scale deflecting torque is $T_d = NBIA$ N-m

where I is the full-scale deflection current ; $I = 5$ mA $= 0.005$ A

$T_d = 100 \times B \times 0.005 \times (3 \times 2.5 \times 10^{-4}) = 3.75 \times 10^{-4}\,B$ N-m

The controlling torque is

$$T_c = 0.5 \text{ g-cm} = 0.5 \text{ g. wt.cm} = 0.5 \times 10^{-3} \times 10^{-2} \text{ kg wt-m} = 0.5 \times 10^{-5} \times 9.8 = 4.9 \times 10^{-5} \text{ N-m}$$

For equilibrium, the two torques are equal and opposite.

$\therefore \qquad 4.9 \times 10^{-5} = 3.75 \times 10^{-4}$ B $\quad \therefore \quad B =$ **0.13 Wb/m²**

10.18. Voltmeter Sensitivity

It is defined in terms of resistance per volt (Ω/V). Suppose a meter movement of 1 kΩ internal resistance has s full-scale deflection current of 50 μ A. Obviously, full-scale voltage drop of the meter movement is = 50 μ A × 1000 Ω = 50 mV. When used as a voltmeter, its sensitivity would be 1000/50 × 10^{-3} = 20 kΩ/V. It should be clearly understood that a sensitivity of 20 kΩ/V means that the total resistance of the circuit in which the above movement is used should be 20 kΩ for a full-scale deflection of 1 V.

10.19. Multi-Range Voltmeter

It is a voltmeter which measures a number of voltage ranges with the help of different series resistances. The resistance required for each range can be easily calculated provided we remember one basic fact that the sensitivity of a meter movement is always the same regardless of the range selected. Moreover, the full-scale deflection current is the same in every range. For any range, the total circuit resistance is found by multiplying the sensitivity by the full-scale voltage for that range.

For example, in the case of the above-mentioned 50μA, 1 kΩ meter movement, total resistance required for 1 V full-scale deflection is 20 kΩ. It means that an additional series resistance of 19 kΩ is required for the purpose as shown in Fig. 10.21 (*a*).

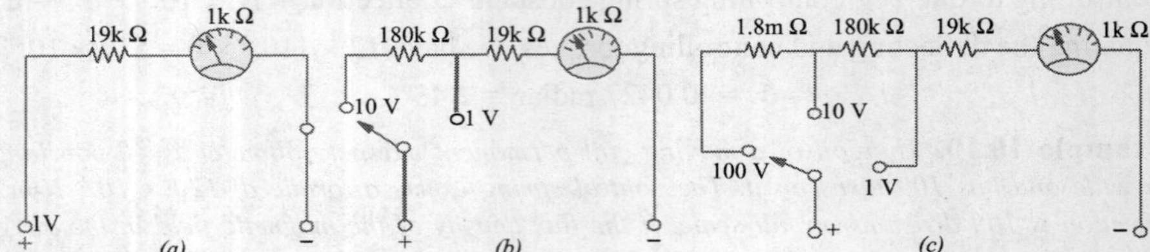

Fig. 10.21

For 10-V range, total circuit resistance must be (20 kΩ/V) (10 V) = 200 kΩ. Since total resistance for 1 V range is 20 kΩ, the series resistance *R* for 10-V range = 200 – 20 = 180 kΩ as shown in Fig. 10.21 (*b*).

For the range of 100 V, total resistance required is (20 kΩ/V) (100 V) = 2 MΩ. The additional resistance required can be found by subtracting the existing two-range resistance from the total resistance of 2 MΩ. Its value is

$$= 2 \text{ M}\Omega - 180 \text{ k}\Omega - 19 \text{ k}\Omega - 1 \text{ k}\Omega = 1.8 \text{ M}\Omega$$

It is shown in Fig. 10.21 (*c*).

Example 10.12. *A basic 1'Arsonval movement with internal resistance R_m = 100 Ω and full scale deflection current I_f = 1 mA is to be converted into a multirange d.c. voltmeter with voltage ranges of 0–10 V, 0–50 V, 0–250 V and 0–500 V. Draw the necessary circuit arrangement and find the values of suitable multipliers.* **(Instrumentation AMIE Sec. B Winter 1991)**

Solution. Full-scale voltage drop = (1 mA) (100 Ω) = 100 mV. Hence, sensitivity of this movement is 100/100 × 10^{-3} = 1 kΩ/V.

(i) 0–10 V range

Total resistance required = (1 kΩ/V) (10 V) = 10 kΩ. Since meter resistance is 1 kΩ additional series resistance required for this range R_1 = 10 – 1 = **9 kΩ**

(ii) 0-50 V range

R_T = (1 kΩ/V) (50 V) = 50 kΩ; R_2 = 50–9–1 = **40 kΩ**

(iii) 0–250 V range

R_T = (1 kΩ/V) (250 V) = 250 kΩ; R_3 = 250 – 50 = **200 kΩ**

(iv) 0–500 V range

$R_T = (1\ k\Omega/V)\ (500\ V) = 500\ k\Omega$; $R_4 = 500 - 250 = $ **250 kΩ**

The circuit arrangement is similar to the one shown in Fig. 10.21

Tutorial Problem 10.2

1. The flux density in the gap of a 1-mA (full scale) moving '-coil' ammeter is 0.1 Wb/m^2. The rectangular moving-coil is 8 mm wide by 1 cm deep and is wound with 50 turns. Calculate the full-scale torque which must be provided by the springs. **[4 × 10–7 N-m] (*App. Elec. London Univ.*)**

2. A moving-coil instrument has 100 turns of wire with a resistance of 10 Ω, an active length in the gap of 3 cm and width of 2 cm. A p.d. of 45 mV produces full-scale deflection. The control spring exerts a torque of 490.5 × 10^{-7} N-m at full-scale deflection. Calculate the flux density in the gap.

[0.1817 Wb/m2] (*I.E.E. London*)

3. A moving-coil instrument, which gives full-scale deflection with 0.015 A has a copper coil having a resistance of 1.5 Ω at 15°C and a temperature coefficient of 1/234.5 at 0°C in series with a swamp resistance of 3.5 Ω having a negligible temperature coefficient.

Determine (*a*) the resistance of shunt required for a full-scale deflection of 20 A and (*b*) the resistance required for a full-scale deflection of 250 V.

If the instrument reads correctly at 15°C, determine the percentage error in each case when the temperature is 25°C.

[(a) 0.00376 Ω ; 1.3 % (b) 16,662 Ω, negligible] (*App. Elect. London Univ.*)

4. A direct current ammeter and leads have a total resistance of 1.5 Ω. The instrument gives a full-scale deflection for a current of 50 mA. Calculate the resistance of the shunts necessary to give full-scale ranges of 2-5, 5.0 and 25.0 amperes **[0.0306 ; 1 0.01515 ; 0.00301 Ω] (*I.E.E. London*)**

5. The following data refer to a moving-coil voltmeter : resistance = 10,000 Ω, dimensions of coil = 3 cm × 3 cm ; number of turns on coil = 100, flux density in air-gap = 0.08 Wb/m^2, stiffness of springs = 3 × 10^{-6} N-m per degree. Find the deflection produced by 110 V. **[48.2°] (*London Univ.*)**

10.20. Electrodynamic or Dynamometer Type Instruments

An electrodynamic instrument is a moving-coil instrument in which the operating field is produced, not by a permanent magnet but by another fixed coil. This instrument can be used either as an ammeter or a voltmeter but is generally used as a wattmeter.

As shown in Fig. 10.22, the fixed coil is usually arranged in two equal sections *F* and *F* placed close together and parallel to each other. The two fixed coils are air-cored to avoid hysteresis effects when used on a.c. circuits. This has the effect of making the magnetic field in which moves the moving coil *M*, more uniform. The moving coil is spring-controlled and has a pointer attached to it as shown.

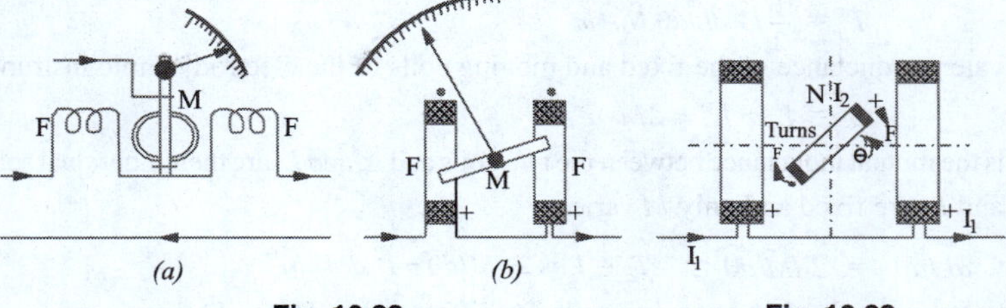

Fig. 10.22

Fig. 10.23

Deflecting Torque*

The production of the deflecting torque can be understood from Fig. 10.23. Let the current passing through the fixed coil be I_1 and that through the moving coil be I_2. Since there is no iron, the field strength and hence the flux density is proportional to I_1.

$\therefore \quad B = KI_1$ where K is a constant

Let us assume for simplicity that the moving coil is rectangular (it can be circular also) and of dimensions $l \times b$. Then, force on each side of the coil having N turns is (NBI_2l) newton.

The turning moment or deflecting torque on the coil is given by

$$T_d = NBI_2lb = NKI_1I_2lb \text{ N-m}$$

Now, putting $NKlb = K_1$, we have $T_d = K_1I_1I_2$ where K_1 is another constant.

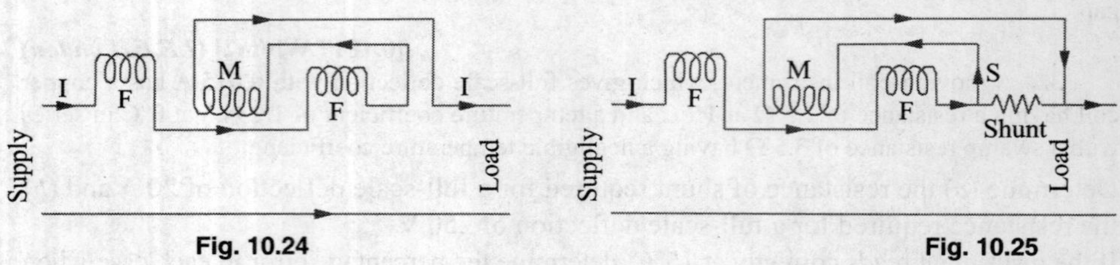

Fig. 10.24 Fig. 10.25

It shows that the deflecting torque is proportional to the product of the currents flowing in the fixed coils and the moving coil. Since the instrument tis spring-controlled, the restoring or control torque is proportional to the angular deflection θ.

i.e. $T_c \propto = K_2\theta \quad \therefore \quad K_1I_1I_2 = K_2\theta$ or $\theta \propto I_1I_2$

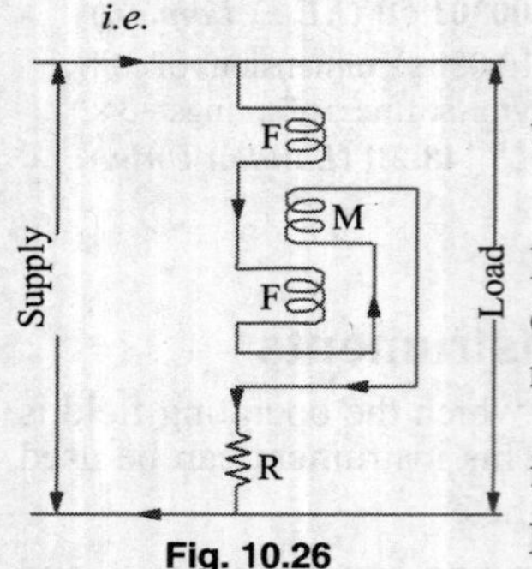

Fig. 10.26

When the instrument is used as an ammeter, the same current passes through both the fixed and the moving coils as shown in Fig. 10.24.

In that case $I_1 = I_2 = I$, hence $\theta \propto I^2$ or $I \propto \sqrt{\theta}$. The connections of Fig. 10.23 are used when small currents are to be measured. In the case of heavy currents, a shunt S is used to limit current through the moving coil as shown in Fig. 10.25.

When used as a voltmeter, the fixed and moving coils are joined in series along with a high resistance and connected as shown in Fig. 10.26. Here, again $I_1 = I_2 = I$, where $I = \frac{V}{R}$ in d.c. circuits and $I = V/Z$ in a.c. circuits.

$\therefore \quad \theta \propto V \times V$ or $\theta \propto V^2$ or $V \propto \sqrt{\theta}$

* As shown in Art. 10.12, the value of torque of a moving-coil instrument is

$$T_d = \frac{1}{2} I^2 \, dL/d\theta \; N-m$$

The equivalent inductance of the fixed and moving coils of the electrodynamic instrument is

$$L = L_1 + L_2 + 2M$$

where M is the mutual inductance between the two coils and L_1 and L_2 are their individual self-inductances. Since L_1 and L_2 are fixed and only M varies,

$$\therefore \quad dL/d\theta = 2dM/d\theta \quad \therefore \quad T_d \frac{1}{2} I^2 \times 2.dM/d\theta = I^2.dM/d\theta$$

If the currents in the fixed and moving coils are different, say I1 and I2 then

$$T_d = I_1 . I_2 . \; dM/d\theta \; N\text{-}m$$

Hence, it is found that whether the instrument is used as an ammeter or voltmeter, its scale is uneven throughout the whole of its range and is particularly cramped or crowded near the zero.

Damping is pneumatic, since owing to weak operating field, eddy current damping is inadmissible. Such instruments can be used for both a.c. and d.c. measurements. But it is more expensive and inferior to a moving-coil instrument for d.c. measurements.

As mentioned earlier, the most important application of electrodynamic principle is the wattmeter and is discussed in detail in Art. 10.34.

Errors

Since the coils are air-cored, the operating field produced is small. For producing an appreciable deflecting torque, a large number of turns is necessary for the moving coil. The magnitude of the current is also limited because two control springs are used both for leading in and for leading out the current. Both these factors lead to a heavy moving system resulting in frictional losses which are somewhat larger than in other types and so frictional errors tend to be relatively higher. The current in the field coils is limited for the fear of heating the coils which results in the increase of their resistance. A good amount of screening is necessary to avoid the influence of stray fields.

Advantages and Disadvantages

1. Such instruments are free from hysteresis and eddy-current errors.
2. Since (torque/weight) ratio is small, such instruments have low sensitivity.

Example 10.13 *The mutual inductance of a 25-A electrodynamic ammeter changes uniformly at a rate of 0.0035 μH/degree. The torsion constant of the controlling spring is 10^{-6} N-m per degree. Determine the angular deflection for full-scale.* **(Elect. Measurements, Poona Univ.)**

Solution. By torsion constant is meant the deflecting torque per degree of deflection. If full-scale deflecting is θ degree, then deflecting torque on full-scale is $10^{-6} \times \theta$ N-m.

Now, $T_d = I^2 dM/d\theta$ Also, $I = 25$ A

$dM/d\theta = 0.0035 \times 10^{-6}$ H/degree $= 0.0035 \times 10^{-6} \times 180/\pi$ H/radian

$10^{-6} \times \theta = 25^2 \times 0.0035 \times 10^{-6} \times 180/\pi \quad \therefore \quad \theta = \mathbf{125.4°}$

Example 10.14. *The spring constant of a 10-A dynamometer wattmeter is 10.5×10^{-6} N-m per radian. The variation of inductance with angular position of moving system is practically linear over the operating range, the rate of change being 0.078 mH per radian. If the full-scale deflection of the instrument is 83 degrees, calculate the current required in the voltage coil at full scale on d.c. circuit.*

(Elect. Inst. and Means. Nagpur Univ. 1991)

Solution. As seen from foot-note of Art. 10.20, $T_d = I_1 I_2 \, dM/d\theta$ N-m

Spring constant $= 10.5 \times 10^{-6}$ N/m/rad $= 10.5 \times 10^{-6} \times \pi/180$ N-m/degree

T_d = spring constant × deflection $= (10.5 \times 10^{-6} \times \pi/180) \times 83 = 15.2 \times 10^{-6}$ N-m

$\therefore \quad 15.2 \times 10^{-6} = 10 \times I_2 \times 0.078$; $I_2 = \mathbf{19.5\ \mu A}$.

10.21. Hot-Wire Instruments

The working parts of the instrument are shown in Fig. 10.27. It is based on the heating effect of current. It consists of platinum-iridium (It can withstand oxidation at high temperatures) wire *AB* stretched between a fixed end *B* and the tension-adjusting screw at *A*. When current is passed through *AB*, it expands according to I^2R formula. This sag in *AB* produces a slack in phosphor-bronze wire *CD* attached to the centre of *AB*. This slack in *CD* is taken up by the silk fibre which after passing round the pulley is attached to a spring *S*. As the silk thread is pulled by *S*, the pulley moves, thereby deflecting the pointer. It would be noted that even a small sag in *AB* is magnified (Art. 10.22) many times and is conveyed to the pointer. Expansion of *AB* is magnified by *CD* which is further magnified by the silk thread.

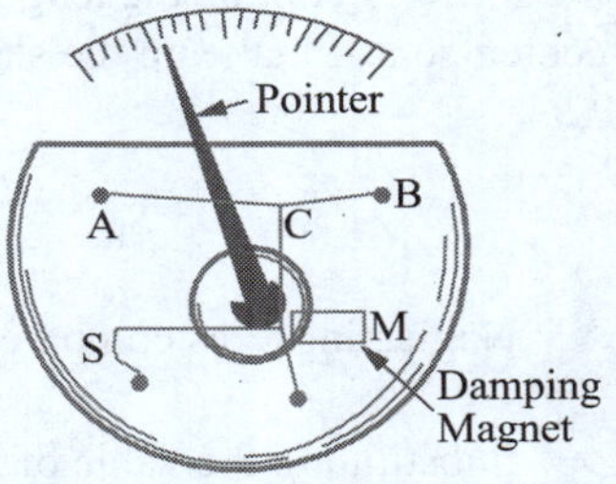

Fig. 10.27

It will be seen that the deflection of the pointer is proportional to the extension of *AB* which is itself proportional to I^2. Hence, deflection is $\propto I^2$. If spring control is used, then $T_c \propto \theta$.

Hence $\theta \propto I^2$

So, these instruments have a 'squre law' type scale. They read the r.m.s. value of current and their readings are independent of its form and frequency.

Damping

A thin light aluminium disc is attached to the pulley such that its edge moves between the poles of a permanent magnet *M*. Eddy currents produced in this disc give the necessary damping.

These instruments are primarily meant for being used as ammeters but can be adopted as voltmeters by connecting a high resistance in series with them. These instruments are suited both for a.c. and d.c. work.

Advantages of Hot-wire Instruments :

1. As their deflection depends on the r.m.s. value of the alternating current, they can be used on direct current also.

2. Their readings are independent of waveform and frequency.

3. They are unaffected by stray fields.

Disadvantages

1. They are sluggish owing to the time taken by the wire to heat up.

2. They have a high power consumption as compared to moving-coil instruments. Current consumption is 200 mA at full load.

3. Their zero position needs frequent adjustment.

4. They are fragile.

10.22. Magnification of the Expansion

As shown in Fig. 10.28 (*a*), let *L* be the length of the wire *AB* and *dL* its expansion after steady temperature is reached. The sag *S* produced in the wire as seen from Fig. 10.28 (*a*) is given by

$$S^2 = \left(\frac{L+dL}{2}\right)^2 - \left(\frac{L}{2}\right)^2 = \frac{2L.\,dL + (dL)^2}{4}$$

Neglecting $(dL)^2$, we have $S = \sqrt{L\,.\,dL/2}$

Magnification produced is $= \frac{S}{dL} = \frac{\sqrt{L\,.\,dL/2}}{dL} = \sqrt{\frac{L}{2.dL}}$

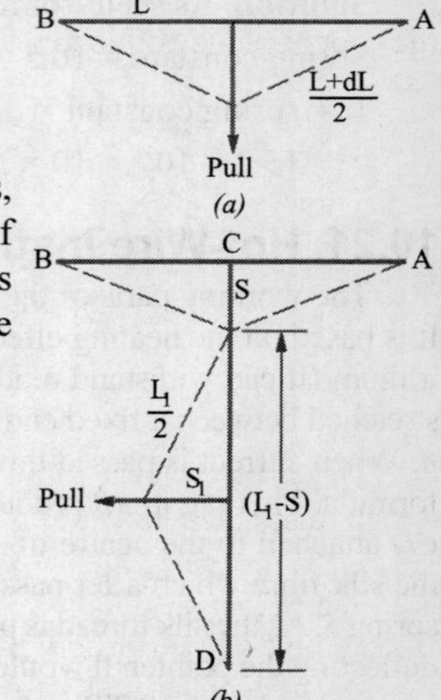

Fig. 10.28

As shown in Fig. 10.28 (*b*), in the case of double-sag instruments, this sag is picked up by wire *CD* which is under the constant pull of the spring. Let L_1 be the length of wire *CD* and let it be pulled at its center, so as to take up the slack produced by the sag *S* of the wire *AB*.

$$S_1^2 = \left(\frac{L_1}{2}\right)^2 - \left(\frac{L_1 - S}{2}\right)^2 = \frac{2L_1\,S - S^2}{4}$$

Neglecting S^2 as compared to $2L_1S$, we have $S_1 = \sqrt{L_1\,S/2}$

Substituting the value of *S*, we get $S_1 = \sqrt{\frac{L_1}{2}}\sqrt{\frac{L.dL}{2}}$

$$= \sqrt{\frac{L_1}{2\sqrt{2}}} \times \sqrt{L.dL}$$

Example 10.15. *The working wire of a single-sag hot wire instrument is 15 cm long and is made up of platinum-silver with a coefficient of linear expansion of* 16×10^{-6}. *The temperature rise of the wire is 85°C and the sag is taken up at the center. Find the magnification*

(*i*) *with no initial sag and* **(*ii*)** *with an initial sag of 1 mm.*

(Elect. Meas and Meas. Inst., Calcutta Univ.)

Solution. **(*i*)** Length of the wire at room temperature = 15 cm

Length when heated through 85°C is $= 15\,(1 + 16 \times 10^{-6} \times 85) = 15.02$ cm

Increase in length, $dL = 15.02 - 15 = 0.02$ cm

Magnification $= \sqrt{\dfrac{L}{2\,.\,dL}} = \sqrt{\dfrac{15}{0.04}} = \mathbf{19.36}$

(*ii*) When there is an initial sag of 1 mm, the wire is in the position *ACB* (Fig. 10.29). With rise in temperature, the new position becomes *ADB*. From the right-angled Δ *ADE*,

Fig. 10.29

we have $(S + 0.1)^2 = \left(\dfrac{L + dL}{2}\right)^2 - AE^2$

Now $AE^2 = AC^2 - EC^2 = \left(\dfrac{L}{2}\right)^2 - 0.1^2 = \dfrac{L^2}{4} - 0.1^2$

$$\therefore \quad (S + 0.1)^2 = \frac{L^2}{4} + \frac{(dL)^2}{4} + \frac{L\,.\,dL}{2} - \left[\frac{L^2}{4} - 0.1^2\right] = \frac{L^2}{4} + \frac{(dL)^2}{4} + \frac{L\,.\,dL}{2} - \frac{L^2}{4} + 0.01$$

$$= \frac{L\,.\,dL}{2} + 0.01 \qquad \text{...neglecting } (dL)^2/4$$

$$= \frac{15 \times 0.02}{2} + 0.01 = 0.16 \;\therefore\; (S + 0.1) = 0.4 \;\therefore\; S = 0.3 \text{ cm}$$

Magnification $= S/dL = 0.3/0.02 = \mathbf{15}$

10.23. Thermocouple Ammeter

The working principle of this ammeter is based on the Seebeck effect, which was discovered in 1821. A thermocouple, made of two dissimilar metals (usually bismuth and antimony) is used in the construction of this ammeter. The hot junction of the thermocouple is welded to a heater wire *AB*, both of which are kept in vacuum as shown in Fig. 10.30 (*a*). The cold junction of the thermocouple is connected to a moving-coil ammeter.

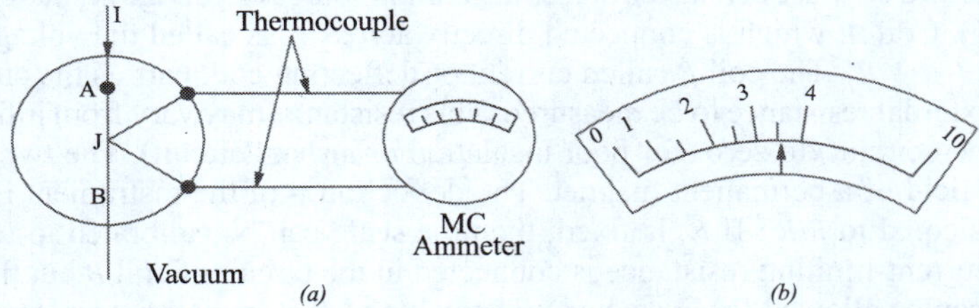

Fig. 10.30

When the current to be measured is passed through the heater wire *AB*, heat is generated, which raises the temperature of the thermocouple junction *J*. As the junction temperature rises, the generated thermoelectric EMF increases and derives a greater current through the moving-coil ammeter. The amount of deflection on the *MC* ammeter scale depend on the heating effect, since the amount of heat produced is directly proportional to the square of the current. The ammeter scale is non-linear so that, it is cramped at the low end and open at the high end as shown in Fig. 10.30 (*b*). This type of "current-squared: ammeter is suitable for reading both direct and alternating currents. It is particularly suitable for measuring radio-frequency currents such as those which occur in antenna system of broadcast transmitters. Once calibrated properly, the calibration of this ammeter remains accurate from dc upto very high frequency currents.

10.24. Megger

It is a portable instrument used for testing the insulation resistance of a circuit and for measuring resistances of the order of megaohms which are connected across the outside terminals XY in Fig. 10.31 (*b*).

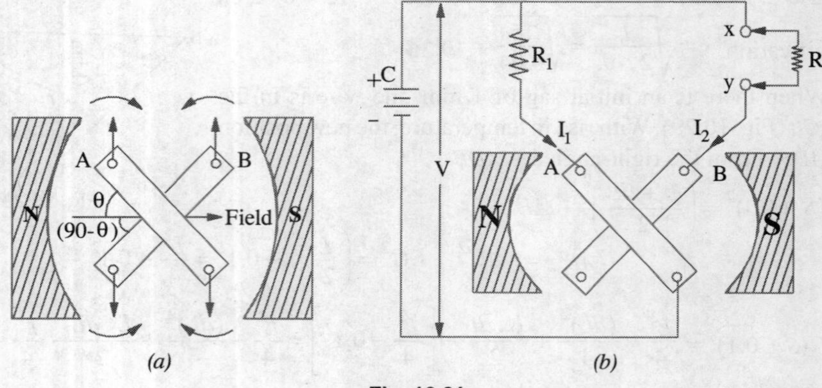

Fig. 10.31

1. Working Principle

The working principle of a 'corss-coil' type megger may be understood from Fig. 10.31 (*a*) which shows two coils *A* and *B* mounted rigidly at right angles to each other on a common axis and free to rotate in a magnetic field. When currents are passed through them, the two coils are acted upon by torques which are in opposite directions. The torque of coil *A* is proportional to $I_1 \cos\theta$ and that of *B* is proportional to $I_2 \cos(90 - \theta)$ or $I_2 \sin\theta$. The two coils come to a position of equilibrium where the two torques are equal and opposite *i.e.* where

$$I_1 \cos\theta = I_2 \sin\theta \quad \text{or} \quad \tan\theta = I_1/I_2$$

In practice, however, by modifying the shape of pole faces and the angle between the two coils, the ratio I_1/I_2 is made proportional to θ instead of tan θ in order to achieve a linear scale.

Suppose the two coils are connected across a common source of voltage *i.e.* battery *C*, as shown in Fig. 10.31 (*b*). Coil *A*, which is connected directly across *V*, is called the voltage (or control) coil. Its current $I_1 = V/R_1$. The coil *B* called current or deflecting coil, carries the current $I_2 = V/R$, where *R* is the external resistance to be measured. This resistance may vary from infinity (for good insulation or open circuit) to zero (for poor insulation or a short-circuit). The two coils are free to rotate in the field of a permanent magnet. The deflection θ of the instrument is proportional to I_1/I_2 which is equal to R/R_1. If R_1 is fixed, then the scale can be calibrated to read *R* directly (in practice, a current-limiting resistance is connected in the circuit of coil *B* but the presence of this resistance can be allowed for in scaling). The value of *V* is immaterial so long as it remains constant and is large enough to give suitable currents with the high resistance to be measured.

2. Construction

The essential parts of a megger are shown in Fig. 10.32. Instead of battery *C* of Fig. 10.31 (*b*), there is a hand-driven d.c. generator. The crank turns the generator armature through a clutch mechanism which is designed to slip at a pre-determined speed. In this way, the generator speed and voltage are kept constant and at their correct values when testing.

The generator voltage is applied across the voltage coil *A* through a fixed resistance R_1 and across deflecting coil *B* through a current-limiting resistance R' and the external resistance is connected across testing terminal *XY*. The two coils, in fact, constitute a moving-coil voltmeter and an ammeter combined into one instrument.

(i) Suppose the terminals *XY* are open-circuited. Now, when crank is operated, the generator voltage so produced is applied across coil *A* and current I_1 flows through it but no current flows through coil *B*. The torque so produced rotates the moving element of the megger until the scale points to 'infinity', thus indicating that the resistance of the external circuit is too large for the instrument to measure.

(ii) When the testing terminals *XY* are closed through a low resistance or are short-circuited, then a large current (limited only by *R*′) passes through the deflecting coil *B*. The deflecting torque produced by coil *B* overcomes the small opposing torque of coil *A* and rotates the moving element until the needle points to 'zero', thus shown that the external resistance is too small for the instrument to measure.

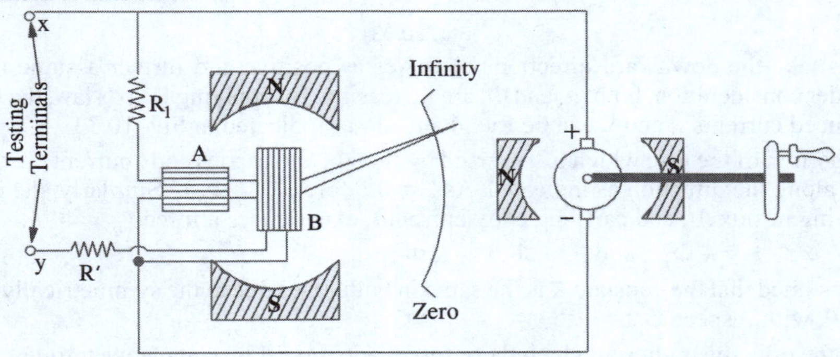

Fig. 10.32

Although, a megger can measure all resistance lying between zero and infinity, essentially it is a high-resistance measuring device. Usually, zero is the first mark and 10 kΩ is the second mark on its scale, so one can appreciate that it is impossible to accurately measure small resistances with the help of a megger.

The instrument described above is simple to operate, portable, very robust and independent of the external supplies.

10.25. Induction type Voltmeters and Ammeters

Induction type instruments are used only for a.c. measurements and can be used either as ammeter, voltmeter or wattmeter. However, the induction principle finds its widest application as a watt-hour or energy meter. In such instruments, the deflecting torque is produced due to the reaction between the flux of an a.c. magnet and the eddy currents induced by this flux. Before discussing the two types of most commonly-used induction instruments, we will first discuss the underlying principle of their operation.

Principle

The operation of all induction instruments depends on the production of torque due to the reaction between a flux Φ_1 (whose magnitude depends on the current or voltage to be measured) and eddy currents induced in a metal disc or drum by another flux Φ_2 (whose magnitude also depends on the current or voltage to be measured). Since the magnitude of eddy currents also depend on the flux producing them, the *instantaneous* value of torque is proportional to the square of current or voltage under measurement and the value of *mean* torque is proportional to the mean square value of this current or voltage.

Consider a thin aluminium or Cu disc *D* free to rotate about an axis passing through its centre as shown in Fig. 10.33. Two a.c. magnetic poles P_1 and P_2 produce alternating fluxes Φ_1 and Φ_2 respectively which cut this disc. Consider any annular portion of the disc around P_1 with center on the axis of P_1. This portion will be linked by flux Φ_1 and so an alternating e.m.f. e_1 be induced in it. This e.m.f. will circulate an eddy current i_1 which, as shown in Fig. 10.33, will pass under P_2. Similarly, Φ_2 will induce an e.m.f. e_2 which will further induce an eddy current i_2 in an annular portion of the disc around P_2. This eddy current i_2 flows under pole P_1.

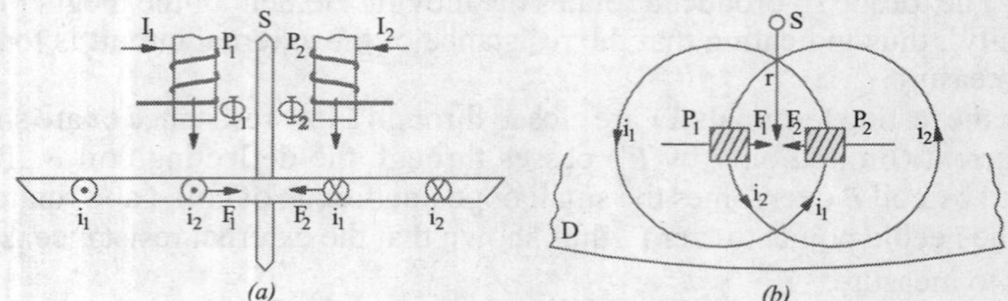

Fig. 10.33

Let us take the downward directions of fluxes as positive and further assume that at the instant under consideration, both Φ_1 and Φ_2 are increasing. By applying Lenz's law, the directions of the induced currents i_1 and i_2 can be found and are as indicated in Fig. 10.33.

The portion of the disc which is traversed by flux Φ_1 and carries eddy current i_2 experiences a force F_1 along the direction as indicated. As $F = Bil$, force $F_1 \propto \Phi_1 i_2$. Similarly, the portion of the disc lying in flux Φ_2 and carrying eddy current i_1 experiences a force $F_2 \propto \Phi_2 i_1$.

$\therefore \quad F_1 \propto \Phi_1 i_2 = K \Phi_1 i_2$ and $F_2 \propto \Phi_2 i_1 = K \Phi_2 i_1$.

It is assumed that the constant K is the same in both cases due to the symmetrically positions of P_1 and P_2 with respect to the disc.

If r is the effective radius at which these forces act, the net instantaneous torque T acting on the disc begin equal to the difference of the two torques, is given by

$$T = r(K\Phi_1 i_2 - K\Phi_2 i_1) = K_1(\Phi_1 i_2 - \Phi_2 i_1) \qquad ...(i)$$

Let the alternating flux Φ_1 be given by $\Phi_1 = \Phi_{1m} \sin \omega t$. The flux Φ_2 which is assumed to lag Φ_1 by an angle α radian is given by $\Phi_2 = \Phi_{2m} \sin(\omega t - \alpha)$

Induced e.m.f. $\quad e_1 = \dfrac{d\Phi}{dt} = \dfrac{d}{dt}(\Phi_{1m} \sin \omega t) = \omega \Phi_{1m} \cos \omega t$

Assuming the eddy current path to be purely resistive and of value R*, the value of eddy current is

$$i_1 = \frac{e_1}{R} = \frac{\omega\Phi_{1m}}{R} \cos \omega t \text{ Similarly } e_2 = \omega \Phi_{2m} (\omega t - \alpha) \text{ and } i_2 = \frac{\omega \Phi_{2m}}{R} \cos(\omega t - \alpha)^{**}$$

Substituting these values of i_1 and i_2 in Eq. (*i*) above, we get

$$T = \frac{K_1 \omega}{R}[\Phi_{1m} \sin \omega t. \Phi_{2m} \cos(\omega t - \alpha) - \Phi_{2m} \sin(\omega t - \alpha) \Phi_{1m} \cos \omega t]$$

$$= \frac{K_1 \omega}{R}.\Phi_{1m} \Phi_{2m} [\sin \omega t. \cos(\omega t - \alpha) - \cos \omega t. \sin(\omega t - \alpha)]$$

$$= \frac{K_1 \omega}{R}.\Phi_{1m} \Phi_{2m} \sin \alpha = k_2 \omega \Phi_{1m} \Phi_{2m} \sin \alpha \text{ (putting } K_1/R = K_2)$$

It is obvious that

(*i*) if $\alpha = 0$ *i.e.* if two fluxes are in phase, then *net* torque is zero. If on the other hand, $\alpha = 90°$, the net torque is maximum for given values of Φ_{1m} and Φ_{2m}.

(*ii*) the net torque is in such a direction as to rotate the disc from the pole with leading flux towards the pole with lagging flux.

---

* If it has a reactance of X, then impedance Z should be taken, whose value is given by $Z = \sqrt{R^2 + X^2}$.

** It being assumed that both paths have the same resistance.

(iii) since the expression for torque does not involve '*t*', it is independent of time *i.e.* it has a steady value at all time.

(iv) the torque T is inversely proportional to R-the resistance of the eddy current path. Hence, for large torques, the disc material should have low resistivity. Usually, it is made of Cu or, more often, of aluminium.

10.26. Induction Ammeters

It has been shown in Art.10.22 above that the net torque acting on the disc is

$$T = K_2 \omega \Phi_{1m} \Phi_{2m} \sin \alpha$$

Obviously, if both fluxes are produced by the same alternating current (of maximum value I_m) to be measured, then $T = K_3 \omega I_m^2 \sin \alpha$

Hence, for a given frequency ω and angle α, the torque is proportional to the square of the current. If the disc has spring control, it will take up a steady deflected position where controlling torque becomes equal to the deflecting torque. By attaching a suitable pointer to the disc, the apparatus can be used as an ammeter.

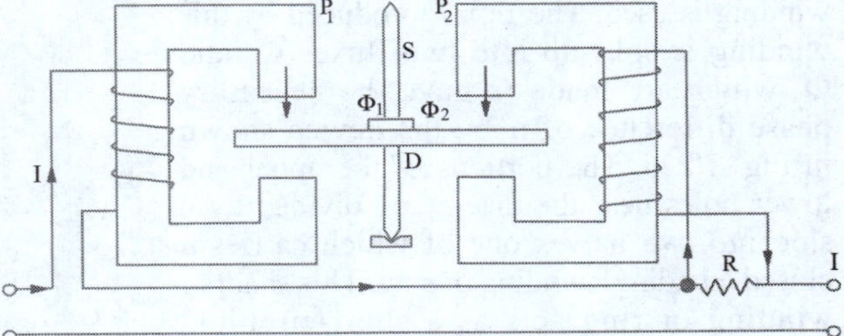

Fig. 10.34

There are three different possible arrangements by which the operational requirements of induction ammeters can be met as discussed below.

(i) **Disc Instrument with Split-phase Winding**

In this arrangement, the windings on the two laminated a.c. magnets P_1 and P_2 are connected in series (Fig. 10.34). But, the winding of P_2 is shunted by a resistance R with the result that the current in this winding lags with respect to the total line current. In this way, the necessary phase angle α is produced between two fluxes Φ_1 and Φ_2 produced by P_1 and P_2 respectively. This angle is of the order of 60°. If the hysteresis effects etc. are neglected, then each flux would be proportional to the current to be measured *i.e.* line current I

$$T_d \propto \Phi_{1m} \Phi_{2m} \sin \alpha$$

or $T_d \propto I^2$ where I is the r.m.s. value.

If spring control is used, then $T_c \propto \theta$

In the final deflected position, $T_c = T_d \quad \therefore \quad \theta \propto I^2$

Eddy current damping is employed in this instrument. When the disc rotates, it cuts the flux in the air-gap of the magnet and has eddy currents induced in it which provide efficient damping.

(ii) **Cylindrical type with Split-phase Winding**

The operating principle of this instrument is the same as that of the above instrument except that instead of a rotating disc, it employs a hollow aluminium drum as shown in Fig. 10.35. The poles P_1 produce the alternating flux Φ_1 which produces eddy current i_1 in those portions of the drum that lie under poles P_2. Similarly, flux F_2 due to poles P_2 produces eddy current i_2 in those parts of the drum that lie under poles P_1. The force F_1 which is $\propto \Phi_1 i_2$ and F_2 which is $\propto \Phi_2 i_1$ are tangential to the surface of the drum and the resulting torque tends to rotate the drum about its own axis. Again, the winding of P_2 is shunted by resistance R which helps to introduce the necessary phase difference α between F_1 and F_2.

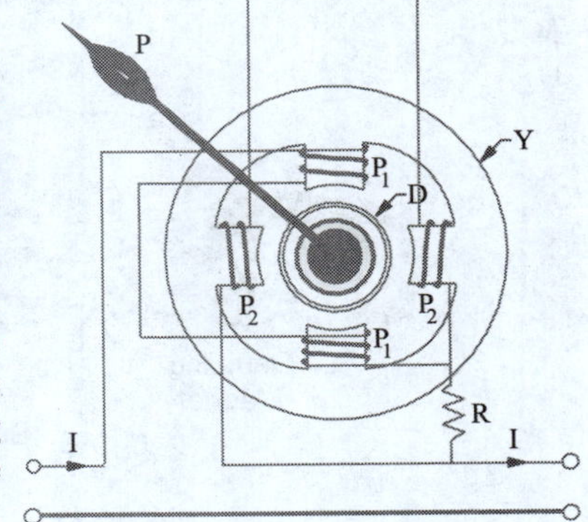

Fig. 10.35

The spiral control springs (not shown in the figure) prevent any continuous rotation of the drum and ultimately bring it to rest at a position where the deflecting torque becomes equal to the controlling torque of the springs. The drum has a pointer attached to it and is itself carried by a spindle whose two ends fit in jewelled bearings. There is a cylindrical laminated core inside the hollow drum whole function is to strengthen the flux cutting the drurm. The poles are laminated and magnetic circuits are completed by the yoke *Y* and the core.

Damping is by eddy currents induced in a separate aluminium disc (not shown in the figure) carried by the spindle when it moves in the air-gap flux of a horse-shoe magnet (also not shown in the figure).

(*iii*) Shaded-pole Induction Ammeter

In the shaded-pole disc type induction ammeter (Fig. 10.36) only single flux-producing winding is used. The flux *F* produced by this winding is split up into two fluxes Φ_1 and Φ_2 which are made to have the necessary phase difference of α by the device shown in Fig. 10.36. The portions of the upper and lower poles near the disc *D* are divided by a slot into two halves one of which carries a closed 'shading' winding or ring. This shading winding or ring acts as a short-circuited secondary and the main winding as a primary. The current induced in the ring by transformer action retards the phase of flux Φ_2 with respect to that of Φ_1 by about 50° or so. The two fluxes Φ_1 and Φ_2 passing through the unshaded and shaded parts respectively, react with eddy currents i_2 and i_1 respectively and so produce the net driving torque whose value is

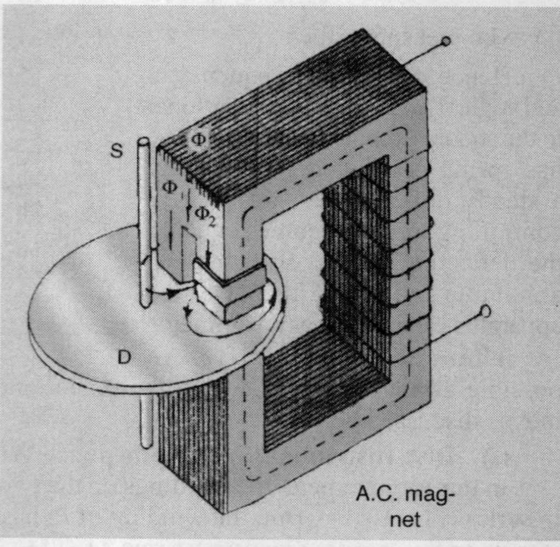

Fig. 10.36

$$T_d \propto \Phi_{1m} \, \Phi_{2m} \sin \alpha$$

Assuming that both Φ_1 and Φ_2 are proportional to the current *I*, we have

$$T_d \propto I^2$$

This torque is balanced by the controlling torque provided by the spiral springs.

The actual shaded-pole type induction instruments is shown in Fig. 10.37. It consists of a suitably-shaped aluminium or copper disc mounted on a spindly which is supported by jewelled bearings. The spindle carries a pointer and has a control spring attached to it. The edge or periphery of the disc moves in the air-gap of a laminated a.c. electromagnet which is energised either by the current to be measured (as ammeter) or by a current proportional to the voltage to be measured (as a voltmeter). Damping is by eddy currents induced by a permanent magnet embracing another portion of the ***same*** disc. As seen, the disc serves both for damping as well

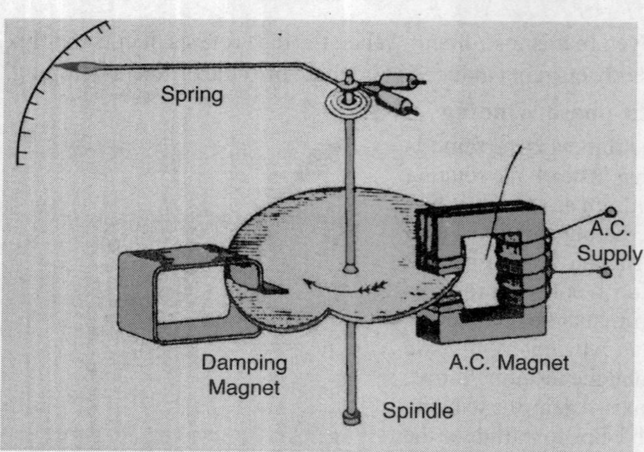

Fig. 10.37

as operating purposes. The main flux is split into two components fluxes by shading one-half of each pole. These two fluxes have a phase difference of 40° to 50° between them and they induce two eddy currents in the disc. Each eddy current has a component in phase with the ***other*** flux, so that two torques are produced which are oppositely directed. The resultant torque is equal to the difference between the two. This torque ***deflects*** the disc-continuous rotation being prevented by the control spring and the deflection produced is proportional to the square of the current or voltage being measured.

As seen, for a given frequency, $T_d \propto I^2 = KI^2$

For spring control $T_c \propto \theta$ or $T_c = K_1 \theta$

For steady deflection, we have $T_c = T_d$ or $\theta \propto I^2$

Hence, such instruments have uneven scales *i.e.* scales which are cramped at their lower ends. A more even scale can, however, be obtained by using a cam-shaped disc as shown in Fig. 10.37.

10.27. Induction Voltmeter

Its construction is similar to that of an induction ammeter except for the difference that its winding is wound with a large number of turns of fine wire. Since it is connected across the lines and carries very small current (5 – 10mA), the number of turns of its wire has to be large in order to produce an adequate amount of m.m.f. Split phase windings are obtained by connecting a high resistance R in series with the winding of one magnet and an inductive coil in series with the winding of the other magnet as shown in Fig. 10.38.

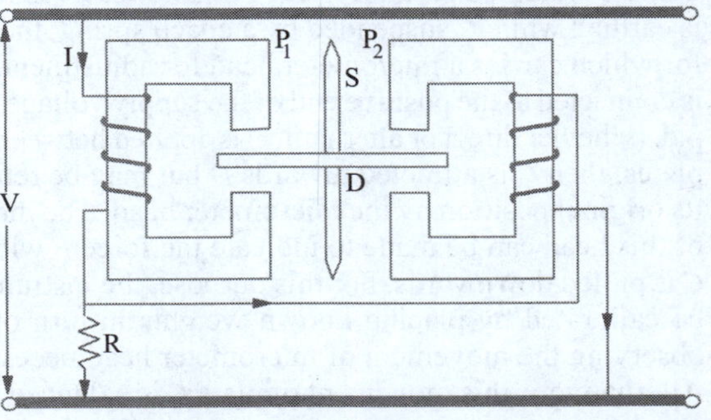

Fig. 10.38

10.28. Errors in Induction Instruments

There are two types of errors ***(i) frequency*** error and ***(ii) temperature*** error.

1. Since deflecting torque depends on frequency, hence unless the alternating current to be measured has same frequency with which the instrument was calibrated, there will be large error in its readings. Frequency errors can be compensated for by the use of a non-inductive shunt in the case of ammeters. In voltmeters, such errors are not large and, to a great extent, are self-compensating.
2. Serious errors may occur due to the variation of temperature because the resistances of eddy current paths depends on the temperature. Such errors can, however, be compensated for by hunting in the case of ammeters and by combination of shunt and swamping resistances in the case of voltmeters.

10.29. Advantages and Disadvantages

1. A full-scale deflection of over 200° can be obtained with such instruments. Hence, they have long open scales.
2. Damping is very efficient.
3. They are not much affected by external stray fields.
4. Their power consumption is fairly large and cost relatively high.
5. They can be used for a.c. measurements only.
6. Unless compensated for frequency and temperature variations, serious errors may be introduced.

10.30. Electrostatic Voltmeters

Electrostatic instruments are almost always used as voltmeters and that too more as a laboratory rather than as industrial instruments. The underlying principle of their operation is the force of attraction between electric charges on neighboring plates between which a p.d. is maintained. This force gives rise to a deflecting torque. Unless the p.d. is sufficiently large, the force is small. Hence, such instruments are used for the measurement of very high voltages.

There are two general types of such instruments :

(*i*) ***the quadrant type***-used upto 20 kV. (***ii***) ***the attracted disc type*** – used upto 500 kV.

10.31. Attracted-Disc Type Voltmeter

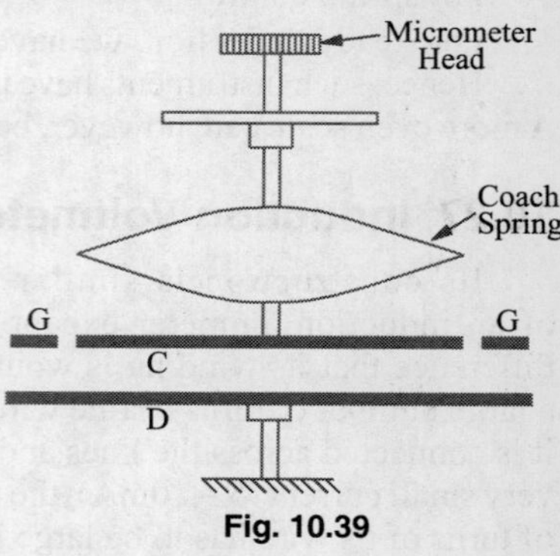

Fig. 10.39

As shown in Fig. 10.39, it consists of two-discs or plates *C* and *D* mounted parallel to each other. Plate *D* is fixed and is earthed while *C* suspended by a coach spring, the support for which carries a micrometer head for adjustment. Plate *C* is connected to the positive end of the supply voltage. When a p.d. (whether direct or alternating) is applied between the two plates, then *C* is attracted towards *D* but may be returned to its original position by the micrometer head. The movement of this head can be made to indicate the force *F* with which *C* is pulled downwards. For this purpose, the instrument can be calibrated by placing known weights in turn on *C* and observing the movement of micrometer head necessary to bring *C* back to its original position. Alternatively, this movement of plate *C* is balanced by a control device which actuates a pointer attached to it that sweeps over a calibrated scale.

There is a guard ring *G* surrounding the plate *C* and separated from it by a small air-gap. The ring is connected electrically to plate *C* and helps to make the field uniform between the two plates. The effective area of plate *C*, in that case, becomes equal to its actual area plus half the area of the air-gap.

Theory

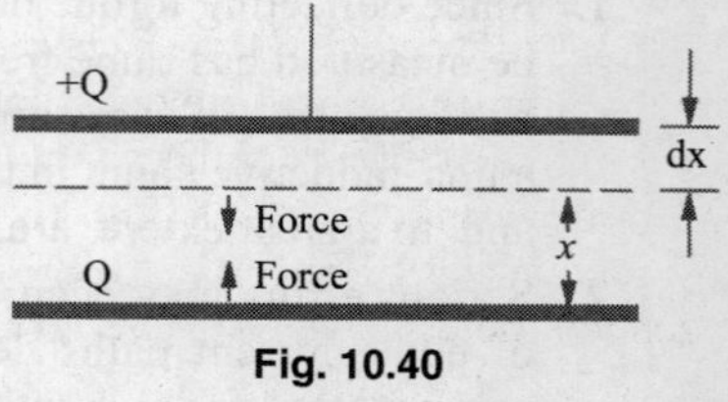

Fig. 10.40

In Fig. 10.40 are shown two parallel plates separated by a distance of *x* meters. Suppose the lower plate is fixed and carries a charge of – *Q* coulomb whereas the upper plate is movable and carries a charge of + *Q* coulomb. Let the mutual force of attraction between the two plates be *F* newtons. Suppose the upper plates is moved apart by a distance *dx*. Then mechanical work done during this movement is *F* × *dx* joule. Since charge on the plate is constant, no electrical energy can move into the system from outside. This work is done at the case of the energy stored in the parallel-plate capacitor formed by the two plates.

Before movement, let the capacitance of the capacitor be *C* farad. Then,

$$\text{Initial energy stored} = \frac{1}{2} \cdot \frac{Q^2}{C}$$

If the capacitance changes to (*C* + *dC*) because of the movement of plate, then

$$\text{Final energy stored} = \frac{1}{2}\frac{Q^2}{(C+dC)} = \frac{1}{2}\frac{Q^2}{C} \cdot \frac{1}{\left(1+\frac{dC}{C}\right)} = \frac{1}{2}\frac{Q^2}{C} \cdot \left(1+\frac{dC}{C}\right)^{-1}$$

$$= \frac{1}{2}\frac{Q^2}{C}\left(1-\frac{dC}{C}\right) \text{ if } dC \le C$$

$$\text{Change in stored energy} = \frac{1}{2}\frac{Q^2}{C} - \frac{1}{2}\frac{Q^2}{C}\left(1 - \frac{dC}{C}\right) = \frac{1}{2}\frac{Q^2}{C} \cdot \frac{dC}{C}$$

$$\therefore \qquad F \times dx = \frac{1}{2}\frac{Q^2}{2C} \cdot \frac{dC}{2C} \text{ or } F = \frac{1}{2}\frac{Q^2}{C^2} \cdot \frac{dC}{dx} = \frac{1}{2}V^2\frac{dC}{dx}$$

Now, $$C = \frac{\varepsilon_0 A}{x} \therefore \frac{dC}{dx} = -\frac{\varepsilon_0 A}{x^2} \therefore F = -\frac{1}{2}V^2\frac{\varepsilon_0 A}{x^2} \text{ N}$$

Hence, we find that force is directly proportional to the square of the voltage to be measured. The negative sign merely shows that it is a force of attraction.

10.32. Quadrant Type Voltmeters

The working principle and basic construction of such instruments can be understood from Fig. 10.41. A light aluminium vane C is mounted on a spindle S and is situated partially within a hollow metal quadrant B. Alternatively, the vane be suspended in the quadrant. When the vane and the quadrant are oppositely charged by the voltage under measurement, the vane is further attracted inwards into the quadrant thereby causing the spindle and hence the pointer to rotate. The amount of rotation and hence the deflecting torque is found proportional to V^2. The deflecting torque in the case of arrangement shown in Fig. 10.41 is very small unless V is extremely large.

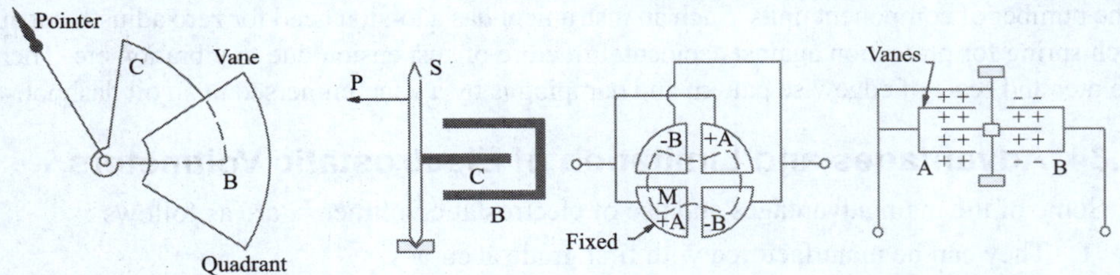

Fig. 10.41 Fig. 10.42

The force on the vane may be increased by using a larger number of quadrants and a double-ended vane. In Fig. 10.42 are shown four fixed metallic double quadrants arranged so as to form a circular box with short air-gaps between the quadrants in which is suspended or pivoted as aluminium vane. Opposite quadrants AA and BB are joined together and each pair is connected to on terminal of the a.c. or d.c. supply and at the same time, one pair is connected to the moving vane M. Under these conditions [Fig. 10.42.] the moving vane is recalled by quadrants AA and attracted by quadrants BB. Hence, a deflecting torque is produced which is proportional to (p.d.)^2. Therefore, such voltmeters have an uneven scale. Controlling torque is produced by torsion of the suspension spring or by the spring (used in pivoted type voltmeters). Damping is by a disc or vane immersed in oil in the case of suspended type or by air friction in the case of pivoted type instruments.

Theory

With reference to Fig 10.41, suppose the quadrant and vane are connected across a source of V volts and let the resulting deflection be θ. If C is a capacitance between the quadrant and vane in the deflected position, then the charge on the instrument will be CV coulomb. Suppose that the voltage is charged from V to $(V + dV)$, then as a result, let θ, C and Q charge to $(\theta + d\theta)$, $(C + dC)$ and $(Q + dQ)$ respectively. Then, the energy stored in the electrostatic field is increased by

$$dE = d\left(\frac{1}{2}CV^2\right) = \frac{1}{2}V^2 . dC + C.V\ dV \text{ joule}$$

If T is the value of controlling torque corresponding to a deflection of θ, then the additional energy stored in the control will be $T \times d\theta$ joule.

$$\text{Total increase in stored energy} = T \times d\theta + \frac{1}{2}V^2\ dC + CVdV \text{ joule}$$

It is seen that during this charge, the source supplies a change dQ at potential V. Hence, the value of energy supplied is

$$= V \times dQ = V \times d\,(CV) = V^2 \times dC + CV.\,dV$$

Since the energy supplied by the source must be equal to the extra energy stored in the field and the control

$$\therefore \qquad T \times d\theta + \frac{1}{2}V^2\ dC + CV.\,dV = V^2.\,dC + CVdV$$

or

$$T \times d\theta = \frac{1}{2}V^2.\,dC \ \therefore\ T = \frac{1}{2}V^2\frac{dC}{d\theta}\ \text{N-m}$$

The torque is found to be proportional to the square of the voltage to be measured whether that voltage is alternating or direct. However, in alternating circuits the scale will read r.m.s. values.

10.33. Kelvin's Multicellular Voltmeter

As shown in Fig. 10.43, it is essentially a quadrant type instrument, as described above, but with the difference that instead of four quadrants and one vane, it has a large number of fixed quadrants and vanes mounted on the same spindle. In this way, the deflecting torque for a given voltage is increased many times. Such voltmeters can be used to measure voltages as low as 30 V. As said above, this reduction in the minimum limit of voltage is due to the increasing operating force in proportion to the number of component units. Such an instrument has a torsion head for zero adjustment and a coach spring for protection against accidental fracture of suspension due to vibration etc. There is a pointer and scale of edgewise pattern and damping is by a vane immersed in an oil dashpot.

10.34. Advantages and Limitation of Electrostatic Voltmeters

Some of the main advantages and use of electrostatic voltmeters are as follows :

1. They can be manufactured with first grade accuracy.
2. They give correct reading both on d.c. and a.c. circuits. On a.c. circuits, the scale will, however, read r.m.s. values whatever the wave-form.
3. Since no iron is used in their construction, such instruments are free from hysteresis and eddy current losses and temperature errors.
4. They do not draw any continuous current on d.c. circuits and that drawn on a.c. circuits (due to the capacitance of the instrument) is extremely small. Hence, such voltmeters do not cause any disturbance to the circuits to which they are connected.
5. Their power loss is negligibly small.
6. They are unaffected by stray magnetic fields although they have to be guarded against any stray electrostatic field.
7. They can be used upto 1000 kHz without any serious loss of accuracy.

However, their main limitations are :

1. Low-voltage voltmeters (like Kelvin's Multicellular voltmeter) are liable to friction errors.
2. Since torque is proportional to the square of the voltage, their scales are not uniform although some uniformity can be obtained by suitably shaping the quadrants of the voltmeters.
3. They are expensive and cannot be made robust.

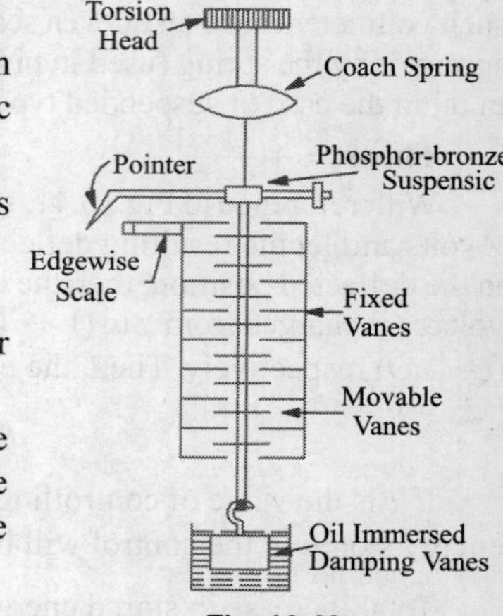

Fig. 10.43

10.35. Range Extension of Electrostatic Voltmeters

The range of such voltmeters can be extended by the use of multipliers which are in the form of a resistance potential divider or capacitance potential divider. The former method can be used both for direct and alternating voltages whereas the latter method is useful only for alternating voltages.

(*i*) Resistance Potential Divider

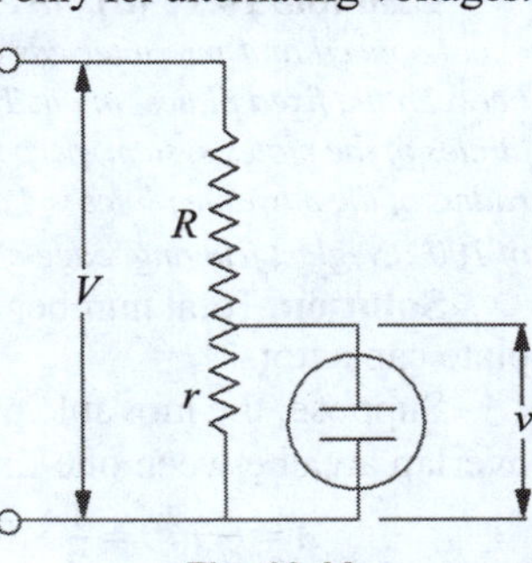

Fig. 10.44

This divider consists of a high non-inductive resistance across a small portion which is attached to the electrostatic voltmeter as shown in Fig. 10.44. Let R be the resistance of the whole of the potential divider across which is applied the voltage V under measurement. Suppose V is the maximum value of the voltage which the voltmeter can measure without the multiplier. If r is the resistance of the portion of the divider across which voltmeter is connected, then the multiplying factor is given by

$$\frac{V}{v} = \frac{R}{r}$$

The above expression is true for d.c. circuits but for a.c. circuits, the capacitance of the voltmeter (which is in parallel with r) has to be taken into account. Since this capacitance is variable, it is advisable to calibrate the voltmeter along with its multiplier.

(*ii*) Capacitance Potential Divider

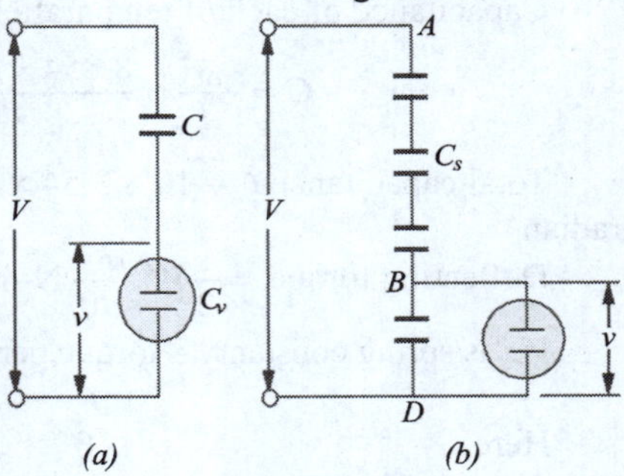

Fig. 10.45

In this method, the voltmeter may be connected in series with a single capacity C and put across the voltage V which is to be measured [Fig. 10.45 (*a*)] or a number of capacitors may be joined in series to form the potential divider and the voltmeter may be connected across one of the capacitors as shown in Fig. 10.45 (*b*).

Consider the connection shown in Fig. 10.45 (*a*). It is seen that the multiplying factor is given by

$$\frac{V}{v} = \frac{\text{reactance of total circuit}}{\text{reactance of voltmeter}}$$

Now, capacitance of the total circuit is $\dfrac{CC_v}{C + C_v}$ and its reactance is

$$= \frac{1}{\omega \times \text{capacitance}} = \frac{C + C_v}{\omega\, C\, C_v}$$

$$\text{Reactance of the voltmeter} = \frac{1}{\omega\, C_v}$$

$$\therefore \frac{V}{v} = \frac{(C + C_v)/\omega\, C\, C_v}{1/\omega C_v} = \frac{C + C_v}{C} \quad \therefore \text{ Multiplying factor} = \frac{C + C_v}{C} = 1 + \frac{C_v}{C}.$$

Example 10.16. *The reading '100' of a 120-V electrosatatic voltmeter is to* represent 10,000 volts when its range is extended by the use of a capacitor in series. If the capacitance of the voltmeter at the above reading is 70 μμF, find the capacitance of the capacitor multiplier required.*

Solution. Multiplying factor $= \dfrac{V}{v} = 1 + \dfrac{C_v}{C}$

Here, $V = 10{,}000$ volt, $v = 100$ volts ; C_v = capacitance of the voltmeter = 70μμ F

* It is helpful to compare it with a similar expression in Art. 10.17 for permanent magnet moving-coil instruments.

C = capacitance of the multiplier $\therefore \quad \frac{10{,}000}{100} = 1 + \frac{70}{C}$

or $70/C = 99 \quad \therefore \quad C = 70/99\ \mu\mu\ F =$ **0.707 μμ F (approx)**

Example 10.17 (*a*). *An electrostatic voltmeter is constructed with 6 parallel, semicircular fixed plates equal-spaced at 4 mm intervals and 5 interleaved semi-circular movable plates that move in planes midway between the fixed plates, in air. The movement of the movable plates is about an axis through the center of the circles of the plates system, perpendicular to the planes of the plates. The instrument is spring-controlled. If the radius of the movable plates is 4 cm, calculate the spring constant if 10 kV corresponds to a full-scale deflection of 100°. Neglect fringing, edge effects and plate thickness.* **(Elect. Measurements, Bombay Univ.)**

Solution. Total number of plates (both fixed and movable) is 11, hence there are 10 parallel plate capacitors.

Suppose, the movable plates are rotated into the fixed plates by an angle of θ radian. Then, overlap area between one fixed and one movable semi-circular plate is

$$A = \frac{1}{2} r^2 \theta = \frac{1}{2} \times 0.04^2 \times \theta = 8 \times 10^{-4}\ \theta\ \text{m}^2;\ d = 4/2 = 2\ \text{mm} = 2 \times 10^{-3}\ \text{m}$$

Capacitance of each of ten parallel-plate capacitors is

$$C = \frac{\varepsilon_0 A}{d} = \frac{8.854 \times 10^{-12} \times 8 \times 10^{-4}\ \theta}{2 \times 10^{-3}} = 3.54 \times 10^{-12}\ \theta\ \text{F}$$

Total capacitance $C = 10 \times 3.54 \times 10^{-12}\ \theta = 35.4 \times 10^{-12}\ \theta$ F $\therefore\ dC/d\theta = 35.4 \times 10^{-12}$ farad/radian

$$\text{Deflecting torque } = \frac{1}{2} V^2 \frac{dC}{d\theta}\ \text{N-m} = \frac{1}{2} \times (10{,}000)^2 \times 35.4 \times 10^{-12} = 17.7 \times 10^{-4}\ \text{N-m}$$

If S is spring constant *i.e.* torque per radian and θ is the plate deflection, then control torque is

$$T_c = S\theta$$

Here, $\theta = 100° = 100 \times \pi/180 = 5\pi/9$ radian

$\therefore \quad S \times 5\pi/9 = 17.7 \times 10^{-4} \quad \therefore \quad$ **S = 10.1 × 10– 4 N-m/rad.**

Example 10.18 (*b*). *A capacitance transducer of two parallel plates of overlapping area of 5×10^{-4} m² is immersed in water. The capacitance 'C' has been found to be 9.50 pF. Calculate the separation 'd' between the plates and the sensitivity, $S = \partial C/\partial d$, of this transducer, given : ε_r water = 81 ; ε_0 = 8.854 pF/m.* **(Elect. Measuer. A.M.I.E. Sec. B, 1992)**

Solution. Since $C = \varepsilon_0 \varepsilon_r A/d$, $d = 3\ \varepsilon_0 \varepsilon_r A/C$.

Substituting the given values we get, $d = 37.7 \times 10^{-3}$ m

$$\text{Sensitivity } \frac{\partial C}{\partial d} = \frac{\partial}{d}\left(\frac{\varepsilon_0 \varepsilon_r A}{d}\right) = -\frac{8.854 \times 10^{-12} \times 81 \times 5 \times 10^{-4}}{(37.7 \times 10^{-3})^2} = -0.025 \times 10^{-8}\ \text{F/m}$$

10.36. Wattmeters

We will discuss the two main types of wattmeters in general use, that is, ***(i)*** the dynamometer or electrodynamic type and ***(ii)*** the induction type.

10.37. Dynamometer Wattmeter

The basic principle of dynamometer instrument has already been explained in detail in Art. 10.20. The connections of a dynamometer type wattmeter are shown in Fig. 10.46. The fixed circular coil which carries the main circuit current I_1 is wound in two halves positioned parallel to each other. The distance between the two halves can be adjusted to give a uniform magnetic field. The moving coil which is pivoted centrally carries a current I_2 which is proportional to the voltage V. Current I_2 is led into the moving coil by two springs which also supply the necessary controlling torque. The equivalent diagrammatic view is shown in Fig. 10.47.

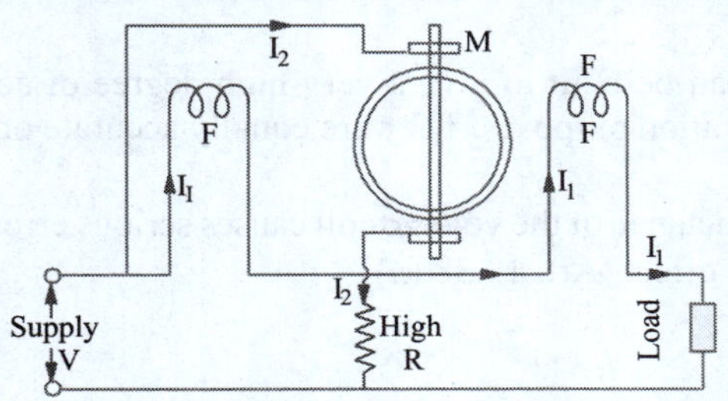

Fig. 10.46

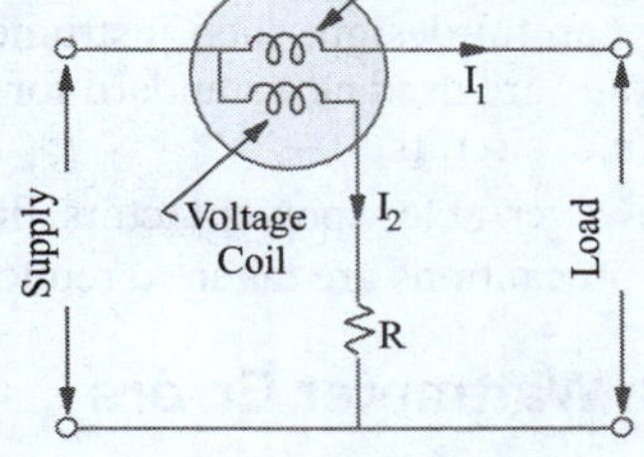

Fig. 10.47

Deflecting Torque

Since coils are air-cored, the flux density produced is directly proportional to the current I_1.

$\therefore \quad B \propto I_1$ or $B = K_1 I_1$; current $I_2 \propto V$ or $I_2 = K_2 V$

Now $\quad T_d \propto BI_2 \propto I_1 V \quad \therefore \quad T_d = KVI_1 = K \times$ power

In d.c. circuits, power is given by the product of voltage and current in amperes, hence torque is directly proportional to the power.

Let us see how this instrument indicates true power on a.c. circuits.

For a.c. supply, the value of instantaneous torque is given by $T_{inst} \propto vi = K\,vi$

where $\quad v$ = instantaneous value of voltage across the moving coil

$\quad i$ = instantaneous value of current through the *fixed* coils.

However, owing to the large inertia of the moving system, the instrument indicates the mean or average power.

$\therefore$ Mean deflecting torque $T_m \propto$ average value of vi

Let $v = V_{max} \sin\theta$ and $i = I_{max} \sin(\theta - \phi)$ $\therefore$ $T_m \propto \frac{1}{2\pi}\int_0^{2\pi} V_{max} \sin\theta \times I_{max} \sin(\theta - \phi)\, d\theta$

$$\propto \frac{V_{max} I_{max}}{2\pi}\int_0^{2\pi} \sin\theta \sin(\theta - \phi)\, d\theta \propto \frac{V_{max} I_{max}}{2\pi}\int_0^{2\pi} \frac{\cos\phi - \cos(2\theta - \phi)}{2}\, d\theta$$

$$\propto \frac{V_{max} I_{max}}{4\pi}\left[\theta\cos\phi - \frac{\sin(2\theta - \phi)}{2}\right]_0^{2\pi} \propto \frac{V_{max}}{\sqrt{2}} \cdot \frac{I_{max}}{\sqrt{2}} \cdot \cos\phi \propto V\,I\cos\phi$$

where V and I are the r.m.s. values. $\quad \therefore \quad T_m \propto VI\cos\phi \propto$ true power.

Hence, we find that in the case of a.c. supply also, the deflection is proportional to the true power in the circuit.

Scales of dynamometer wattmeters are more or less uniform because the deflection is proportional to the average power and for spring control, controlling torque is proportional to the deflection. Hence $\theta \propto$ power. Damping is pneumatic with the help of a piston moving in an air chamber as shown in Fig. 10.48.

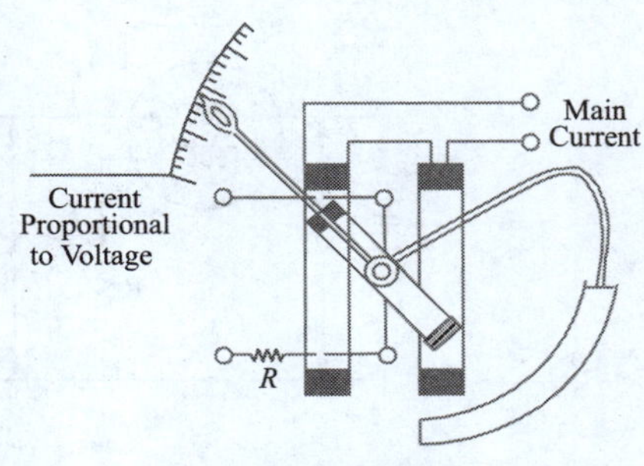

Fig. 10.48

Errors

The inductance of the moving or voltage coil is liable to cause error but the high non-inductive resistance connected in series with the coil swamps, to a great extent, the phasing effect of the voltage-coil inductance.

Another possible error in the indicated power may be due to (*i*) some voltage drop in the circuit or (*ii*) the current taken by the voltage coil. In standard wattmeters, this defect is overcome by having an additional compensating winding which is connected in series with the voltage coil but is so placed that is produces a field in opposite direction to that of the fixed or current coils.

Advantages and Disadvantages

By careful design, such instruments can be built to give a very high degree of accuracy. Hence they are used as a standard for calibration purposes. They are equally accurate on d.c. as well as a.c. circuits.

However, at low power factors, the inductance of the voltage coil causes serious error unless special precautions are taken to reduce this effect [Art. 10.38 (*ii*)].

10.38. Wattmeter Errors

(*i*) Error Due to Different Connections

Two possible ways of connecting a wattmeter in a single-phase a.c. circuit are shown in Fig. 10.49 along with their phasor diagrams. In Fig. 10.49 (*a*), the pressure or voltage-coil current does not pass through the current coil of the wattmeter whereas in the connection of Fig. 10.49 (*b*) is passes. A wattmeter is supposed to indicate the power consumed by the load but its actual reading is slightly higher due to power losses in the instrument circuits. The amount of error introduced depends on the connection.

(*a*) Consider the connection of Fig. 10.49 (*a*). If cos ϕ is the power factor of the load, then power in the load is $VI \cos \theta$.

Now, voltage across the pressure-coil of the wattmeter is V_1 which is the phasor sum of the load voltage V and p.d. across current-coil of the instrument *i.e.* V' (= Ir where r is the resistance of the current coil).

Hence, power reading as indicated by the wattmeter is = $V_1 I \cos \theta$

where θ = phase difference between V_1 and I as shown in the phasor diagram of Fig. 10.50 (*a*).

As seen from the phasor diagram, $V_1 \cos \theta = (V \cos \phi + V')$

$\therefore$ wattmeter reading $= V_1 \cos \theta \,.\, I = (V \cos \phi + V')\, I$

$= VI \cos \phi + V' I = VI \cos \phi + I^2 r$ = power in load + power in current coil.

(*b*) Next, consider the connection of Fig. 10.49 (*b*). The current through the current-coil of the wattmeter is the phasor sum of load current I and voltage-coil current $I' = V/R$. The power reading indicated by the wattmeter is = $VI_1 \cos \theta$.

As seen from the phasor diagram of Fig. 10.49 (*b*), $I_1 \cos \theta = (I \cos \phi + I')$

$\therefore$ wattmeter reading $= V(I \cos \phi + I') = VI \cos \phi + VI' = VI \cos \phi + V^2/R$

= power in load + power in pressure-coil circuit

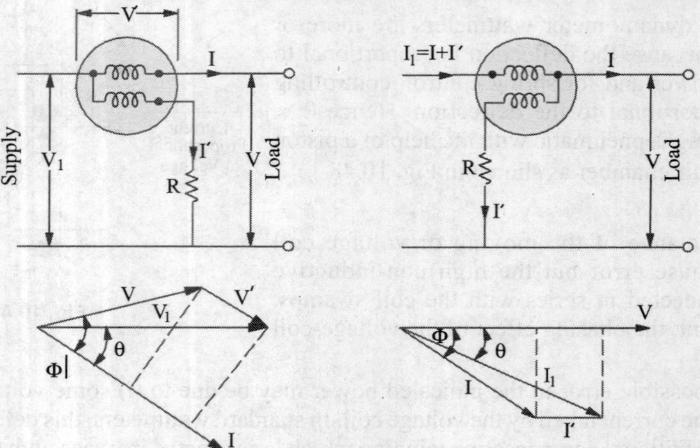

Fig. 10.49

***(ii)* Error Due to Voltage-coil Inductance**

While developing the theory of electrodynamic instruments, it was assumed that pressure-coil does not posses any inductance (and hence reactance) so that current drawn by it was = V/R. The wattmeter reading is proportional to the mean deflecting torque, which is itself proportional to $I_1 I_2 \cos\theta$, where θ is the angle between two currents (Fig. 10.51).

In case the inductance of the voltage-coil is neglected.

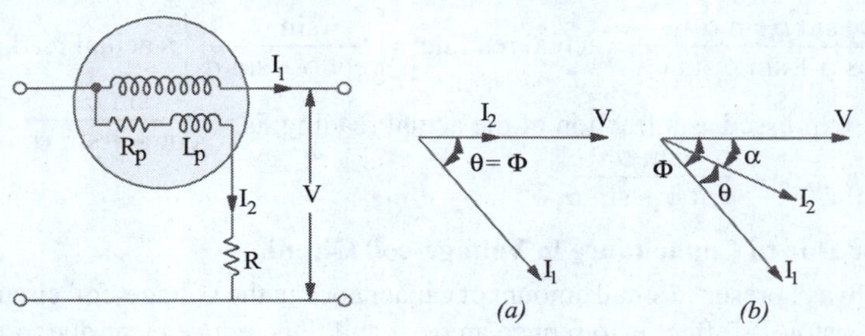

Fig. 10.50 **Fig. 10.51**

$$I_2 = V/(R + R_p) = V/R \text{ approximately}$$

and $$\theta = \phi \text{ as shown in the phasor diagram of 10.51 } (a)$$

$$\therefore \quad \text{wattmeter reading} \propto \frac{I_1 V}{R}\cos\phi \qquad \text{...}(i)$$

In case, inductance of the voltage coil is taken into consideration, then

$$I_2 = \frac{V}{\sqrt{(R_p + R)^2 + X_L^2}} = \frac{V}{\sqrt{R^2 + X_L^2}} = \frac{V}{Z_p}$$

It lags behind V by an angle α [Fig. 10.51 (*b*)] such that

$$\tan\alpha = X_L/(R_p + R) = X_L/R \text{ (approx.)} = \omega L_p/R$$

$$\therefore \quad \text{wattmeter reading} \propto \frac{I_1 V}{Z_p}\cos\theta \propto \frac{I_1 V}{Z_p}\cos(\phi - \theta)$$

Now $$\cos\alpha = \frac{R_p + R}{Z_p} = \frac{R}{Z_p} \quad \therefore \quad Z_p = \frac{R}{\cos\alpha}$$

$$\therefore \quad \text{wattmeter reading in this case is} \propto I_1 \frac{V}{R}\cos\alpha\cos(\phi - \alpha) \qquad \text{...}(ii)$$

Eq. (*i*) above, gives wattmeter reading when inductance of the voltage coil is neglected and Eq. (*ii*) gives the reading when it is taken into account.

The correction factor which is given by the ratio of the true reading (W_t) and the actual or indicated reading (W_a) of the wattmeter is

$$\frac{W_t}{W_a} = \frac{\frac{VI_1}{R_1}\cos\phi}{\frac{VI_1}{R_1}\cos\alpha\cos(\phi - \alpha)} = \frac{\cos\phi}{\cos\alpha\cos(\phi - \alpha)}$$

Since, in practice, α is very small, $\cos\alpha = 1$. Hence the correction factor becomes $= \dfrac{\cos\phi}{\cos(\phi - \alpha)}$

$$\therefore \quad \text{True reading} = \frac{\cos\phi}{\cos\alpha\cos(\phi - \alpha)} \times \text{actual reading} \cong \frac{\cos\phi}{\cos(\phi - \alpha)} \times \text{actual reading}$$

The error in terms of the actual wattmeter reading can be found as follows :

Actual reading–true reading

$$= \text{actual reading} - \frac{\cos\phi}{\cos\alpha\cos(\phi-\alpha)} \times \text{actual reading}$$

$$= \left[1 - \frac{\cos\phi}{\cos(\phi-\alpha)}\right] \times \text{actual reading} = \left[1 - \frac{\cos\phi}{\cos\phi + \sin\phi\sin\alpha}\right] \times \text{actual reading}$$

$$= \left[\frac{\sin\phi.\sin\alpha}{\cos\phi + \sin\phi.\sin\alpha}\right] \times \text{actual reading} = \left[\frac{\sin\alpha}{\cot\phi + \sin\alpha}\right] \times \text{actual reading}$$

The error, expressed as a fraction of the actual reading, is $= \dfrac{\sin\alpha}{\cot\phi + \sin\alpha}$

Percentage error $= \dfrac{\sin\alpha}{\cot\phi + \sin\alpha} \times 100$

(*iii*) Error Due to Capacitance in Voltage-coil Circuit

There is always present a small amount of capacitance in the voltage-coil circuit, particularly in the series resistor. Its effect is to reduce angle α and thus reduce error due to the inductance of the voltage coil circuit. In fact, in some wattmeters, a small capacitor is purposely connected in parallel with the series resistor for obtaining practically non-inductive voltage-coil circuit. Obviously, over-compensation will make resultant reactance capacitive thus making α negative in the above expressions.

(*iv*) Error Due to Stray Fields

Since operating field of such an instrument is small, it is very liable to stray field errors. Hence, it should be kept as far away as possible from stray fields. However, errors due to stray fields are, in general, negligible in a properly-constructed instrument.

(*v*) Error Due to Eddy Currents

The eddy current produced in the solid metallic parts of the instrument by the alternating field of the current coil changes the magnitude and strength of this operating field thus producing an error in the reading of the wattmeter. This error is not easily calculable although it can be serious if care is not taken to remove away solid masses of metal from the proximity of the current coil.

Example 10.19. *A dynamometer type wattmeter with its voltage coil connected across the load side of the instrument reads 250 W. If the load voltage be 200 V, what power is being taken by load? The voltage coil branch has a resistance of 2,000 Ω.* **(Elect. Engineering, Madras Univ.)**

Solution. Since voltage coil is connected across the load side of the wattmeter (Fig. 10.52), the power consumed by it is also included in the meter reading.

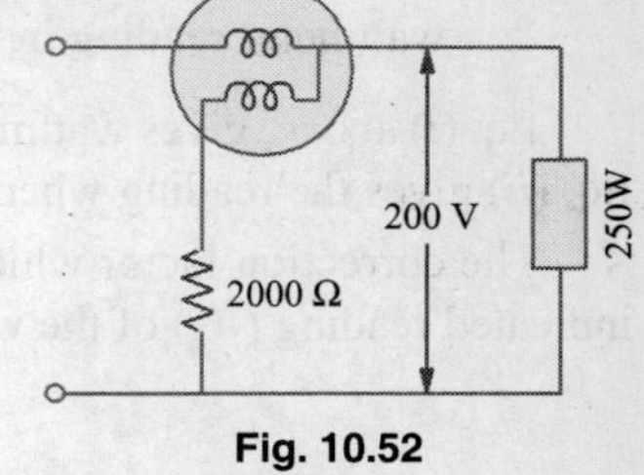

Fig. 10.52

Power consumed by voltage coil is

$= V^2/R = 200^2/2{,}000 = 20$ W

∴ Power being taken by load = 250 – 20

= 230 W

Example 10.20. *A 250-V, 10-A dynamometer type wattmeter has resistance of current and potential coils of 0.5 and 12,500 ohms respectively. Find the percentage error due to each of the two methods of connection when unity p.f. loads at 250 volts are of* ***(a)*** *4A* ***(b)*** *12 A.*

Neglect the error due to the inductance of pressure coil. **(Elect. Measurements, Pune. Univ.)**

Solution. (*a*) When I = 4 A

(*i*) Consider the type of connection shown in Fig. 10.50 (*a*)

Power loss in current coil of wattmeter = $I^2r = 4^2 \times 0.5 =$ **8 W**

Load power $= 250 \times 4 \times 1 = 1000$ W ; Wattmeter reading = 1008 W

∴ percentage error $= (8/1008) \times 100 =$ **0.794%**

(ii) Power loss in pressure coil resistance $= V^2/R = 250^2/12{,}500 = 5$ W

∴ Percentage error $= 5 \times 100/1005 =$ **0.497 %**

(*b*) When I = 12 A

(*i*) Power loss in current coil $= 12^2 \times 0.5 = 72$ W

Load power $= 250 \times 12 \times 1 = 3000$ W ; wattmeter reading = 3072 W

∴ percentage error $= 72 \times 100/3072 =$ **2.34 %**

(ii) Power loss in the resistance of pressure coil is $250^2/12{,}500 = 5$ W

∴ percentage error $= 5 \times 100/3005 =$ **0.166 %**

Example 10.21. *An electrodynamic wattmeter has a voltage circuit of resistance of 8000 Ω and inductance of 63.6 mH which is connected directly across a load carrying a current of 8A at a 50-Hz voltage of 240-V and p.f. of 0.1 lagging. Estimate the percentage error in the wattmeter reading caused by the loading and inductance of the voltage circuit.* **(Elect & Electronic Measu. & Instru. Nagpur, Univ. 1992)**

Solution. The circuit connections are shown in Fig. 10.53.

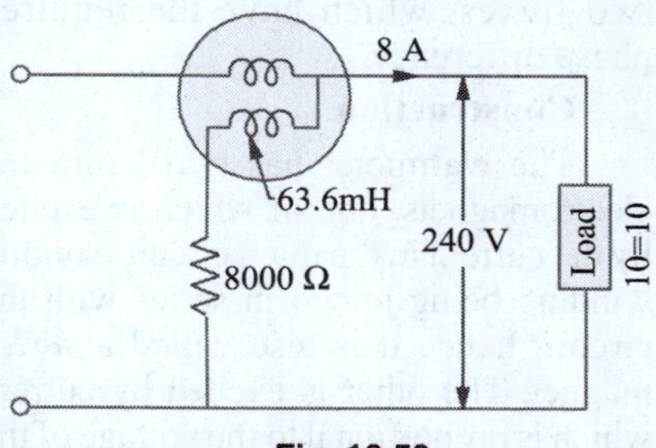

Fig. 10.53

Load power $= 240 \times 8 \times 0.1 = 192$ W

$\cos\phi = 0.1, \phi = \cos^{-1}(0.1) = 84^\circ\ 16'$

Power loss in voltage coil circuit is $= V^2/R$

$= 240^2/8000 = 7.2$ W

Neglecting the inductance of the voltage coil, the wattmeter reading would be

$= 192 + 7.2 =$ **199.2 W**

Now, $X_p = 2\pi \times 50 \times 63.3 \times 10^{-3} = 20\ \Omega$

$\alpha = \tan^{-1}(20/8000) = \tan^{-1}(0.0025) = 0^\circ\ 9'$

Error factor due to inductance of the voltage coil $= \dfrac{\cos(\phi - \alpha)}{\cos\phi} = \dfrac{\cos 84^\circ 7'}{\cos 84^\circ 16'} = 1.026$

Wattmeter reading $= 1.026 \times 199.2 = 204.4$ W

Percentage error $= \left(\dfrac{204.4 - 199.2}{199.2}\right) \times 100 =$ **2.6 %**

Example 10.22. *The inductive reactance of the pressure-coil circuit of a dynamometer wattmeter is 0.4 % of its resistance at normal frequency and the capacitance is negligible.*

Calculate the percentage error and correction factor due to reactance for load at ***(i)*** *0.707 p.f. lagging and* ***(ii)*** *0.5 p.f. lagging.* **(Elect. Measurement. Bombay Univ.)**

Solution. It is given that $X_p/R = 0.4\ R = 0.004$

$\tan\alpha = X_p/R = 0.004 \quad \therefore \quad \alpha = 0^\circ\ 14'$ and $\sin\alpha = 0.004$

(*i*) When p.f. = 0.707 (*i.e.* $\phi = 45^\circ$)

$$\text{Correction factor} = \frac{\cos\phi}{\cos(\phi - \alpha)} = \frac{\cos 45^\circ}{\cos 44^\circ\ 46'} = \mathbf{0.996}$$

$$\text{Percentage error} = \frac{\sin\alpha}{\cot\phi + \sin\alpha} \times 100 = \frac{\sin 0^\circ\ 14'}{\cot 45^\circ + \sin 0^\circ\ 14'} \times 100$$

$$= \frac{0.004 \times 100}{1 + 0.004} = \frac{0.4}{1.004} = \mathbf{0.4\ (approx)}$$

(*ii*) **When p.f.** $= \mathbf{0.5}$ (*i.e.,* $\phi = 60°$)

$$\text{Correction factor} = \frac{\cos 60°}{\cos 59° 46'} = \mathbf{0.993}$$

$$\text{Percentage error} = \frac{\sin 0° 14'}{\cot 60° + \sin 0° 14'} \times 100 = \frac{0.004 \times 100}{0.577 + 0.004} = \frac{0.4}{0.581} = \mathbf{0.7}$$

10.39. Induction Wattmeters

Principle of induction wattmeters is the same as that of induction ammeters and voltmeters. They can be used on a.c. supply only in contrast with dynamometer wattmeters, which can be used both on d.c. and a.c. supply. Induction wattmeters are useful only when the frequency and supply voltage are constant.

Since, both a current and a pressure element are required in such instrument, it is not essential to use the shaded-pole principle. Instead of this, two separate a.c. magnets are used, which produce two fluxes, which have the required phase difference.

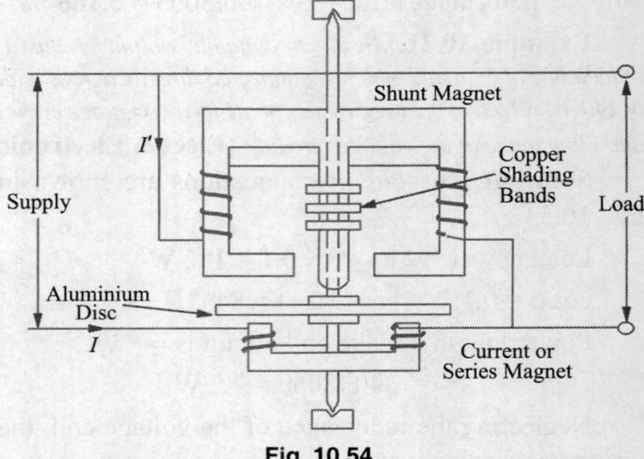

Fig. 10.54

Construction

The wattmeter has two laminated electromagnets, one of which is excited by the current in the main circuit-exciting winding being joined in series with the circuit, hence it is also called a ***series*** magnet. The other is excited by current which is proportional to the voltage of the circuit. Its exciting coil is joined in parallel with the circuit, hence this magnet is sometimes referred to as ***shunt*** magnet.

A thin aluminium disc is so mounted that it cuts the fluxes of both magnets. Hence, two eddy currents are produced in the disc. The deflection torque is produced due to the interaction of these eddy current and the inducing fluxes. Two or three copper rings are fitted on the central limb of the shunt magnet and can be so adjusted as to make the resultant flux in the shunt magnet lag behind the applied voltage by 90°.

Two most common forms of the electromagnets are shown in Fig. 10.54 and 10.55. It is seen that in both cases, one magnet is placed above and the other below the disc. The magnets are so positioned and shaped that their fluxes are cut by the disc.

In Fig. 10.54, the two pressure coils are joined in series and are so wound that both send the flux through the central limb in the same direction. The series magnet carries two coils joined in series and so wound that they magnetise their respective cores in the same direction. Correct phase displacement between the shunt and series magnet fluxes can be obtained by adjusting the position of the copper shading bands as shown.

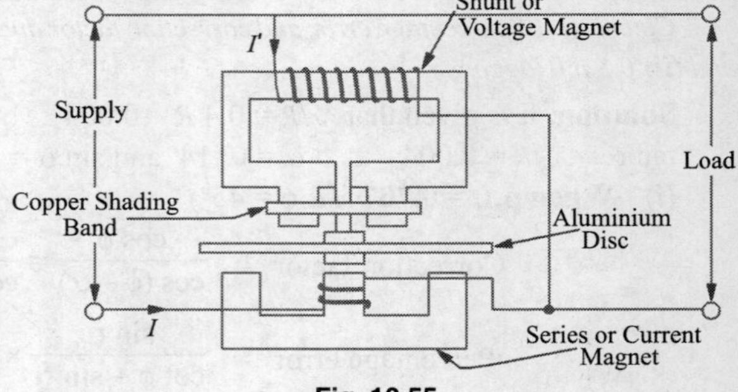

Fig. 10.55

In the type of instrument shown in Fig. 10.56, there is only one pressure winding and one current winding. The two projecting poles of the shunt magnet are surrounded by a copper shading band whose position can be adjusted for correcting the phase of the flux of this magnet with respect to the voltage.

Both types of induction wattmeters shown above, are spring-controlled, the spring being fitted to the spindle of the moving system which also carries the pointer. The scale is uniformly even and extends over 300º.

Currents upto 100 A can be handled by such wattmeters directly but for currents greater than this value, they are used in conjunction with current transformers. The pressure coil is purposely made as much inductive as possible in order that the flux through it should lag behind the voltage by 90º.

Theory

The winding of one magnet carries line current I so that $\Phi_1 \propto I$ and is in phase with I (Fig. 10.56). The other coil *i.e.*, pressure or voltage coil is made highly inductive having an inductance of L and and negligible resistance. This is connected across the supply voltage V. The current in the pressure coil is therefore, equal to $V/\omega L$. Hence, $\Phi_2 \propto V/\omega L$ and lags behind the voltage by 90º. Let the load current I lag behind V by ϕ *i.e.*, let the load power factor angle be ϕ. As shown in Fig. 10.55, the phase angle between Φ_1 and Φ_2 is $\alpha = (90 - \phi)$.

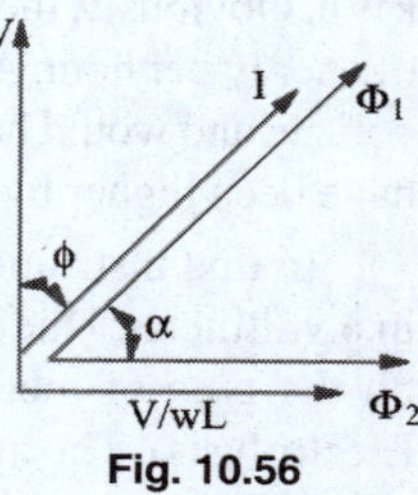

Fig. 10.56

The value of the torque acting on the disc is given by

$$T = k\omega\, \Phi_{1m}\, \Phi_{2m} \sin \alpha \qquad \text{– Art. 10.25}$$

or $$T \propto 2\, \omega . I . \frac{V}{\omega L} . \sin (90 - \phi) \propto VI \cos \phi \propto \text{power}$$

Hence, the torque is proportional to the power in the load circuit. For spring control, the controlling torque $T_c \propto \theta$.

$\therefore \quad \theta \propto$ power. Hence, the scale is even.

10.40. Advantage and Limitations of Induction Wattmeters

These wattmeters possess the advantages of fairly long scales (extending over 300º), are free from the effects of stray fields and have good damping. They are practically free from frequency errors. However, they are subject to (sometimes) serious temperature errors because the main effect of temperature is on the resistance of the eddy current paths.

10.41. Energy Meters

Energy meters are integrating instruments, used to measure quantity of electric energy supplied to a circuit in a given time. They give no direct indication of power *i.e.*, as to the rate at which energy is being supplied because their registrations are independent of the rate at which a given quantity of electric energy is being consumed. Supply or energy meters are generally of the following types:

(i) ***Electrolytic meters*** - their operation depends on electrolytic action.

(ii) ***Motor meters*** - they are really small electric motors.

(iii) ***Clock meters*** - they function as clock mechanisms.

10.42. Electrolytic Meter

It is used on d.c. circuits* only and is essentially an ampere-hour meter and not a true watt-hour meter. However, its registrations are converted into watt-hour by multiplying them by the voltage (assumed constant) of the circuits in which it is used. Such instruments are usually calibrated to read kWh directly at the declared voltage. Their readings would obviously be incorrect when

* Recently such instruments have been marketed for measurement of kilovoltampere-hours on a.c. supply, using a small rectifier unit, which consists of a current transformer and full-wave copper oxide rectifier.

used on any other voltage. ***Because of the question of power factor, such instrument cannot be used on a.c. circuits.***

The advantages of simplicity, cheapness and of low power consumption of ampere-hour meters are, to a large extent, discounted by the fact that variations in supply voltage are not taken into account by them. As an example suppose that the voltage of a supply whose nominal value is 220 V, has an average value of 216 volts in one hour during which a consumer draws a current of 100 A. Quantity of electricity as measured by the instrument which is calibrated on 220 V, is 220 × 100/1000 = 22 kWh. Actually, the energy consumed by the customer is only 216 × 100/1000 = 21.6 kWh. Obviously, the consumer is being overcharged to the extent of the cost of 22 − 21.6 = 0.4 kWh of energy per hour. A true watthour-meter would have taken into account the decrease in the supply voltage and would have, therefore, resulted in a saving to the consumer. If the supply voltage would have been higher by that amount, then the supply company would have been the loser (Ex. 10.27).

In this instrument, the operating current is passed through a suitable electrolyte contained in a voltmeter. Due to electrolysis, a deposit of mercury is given or a gas is liberated (depending on the type of meter) in proportion to the quantity of electricity passed (Faraday's Laws of Electrolysis). The quantity of electricity passed is indicated by the level of mercury in a graduated tube. Hence, such instruments are calibrated in amp-hour or if constancy of supply voltage is assumed, are calibrated in watt-hour or kWh.

Such instruments are cheap, simple and are accurate even at very small loads. They are not affected by stray magnetic fields and due to the absence of any moving parts are free from friction errors.

10.43. Motor Meters

Most commonly-used instruments of this type are :

(*i*) ***Mercury motor*** meters **(*ii*)** ***Commutator motor*** meters and **(*ii*)** ***Induction motor*** meters.

Of these, mercury motor meter is normally used on d.c. circuits whereas the induction type instrument is used only on a.c. circuits. However, the commutator type meter can be used both for d.c. as well as a.c. work.

Instruments used for d.c. work can be either in the form of a amp-hour meters or watt-hour meters. In both cases, the moving system is allowed to revolve continuously instead of being merely allowed to deflect or rotate through a fraction of a revolution as in indicating instruments. The speed of rotation is directly proportional to the current in the case of amp-hour meter and to power in the case of watt-hour meter. Hence the number of revolutions made in a given time is proportional, in the case of amp-hour meter, to the quantity of electricity ($Q = i \times t$) and in the case of Wh meter, to the quantity of energy supplied to the circuit. The number of revolutions made are registered by a counting mechanism consisting of a train of gear wheels and dials.

The control of speed of the rotating system is brought about by a permanent magnet (known as braking magnet) which is so placed as to set up eddy currents in some parts of the rotating system. These eddy currents produce a retarding torque which is proportional to their magnitude-their magnitude itself depending on the speed of rotation of the rotating system. The rotating system attains a *steady speed* when the braking torque exactly balances the driving torque which is produced either by the current or power in the circuit.

The essential parts of motor meters are :

1. An operating system which produces an operating torque proportional to the current or power in the circuit and which causes the rotation of the rotating system.

2. A retarding or braking device, usually a permanent magnet, which produces a braking torque is proportional to the speed of rotation. Steady speed of rotation is achieved when braking torque becomes equal to the operating torque.

3. A registering mechanism for the revolutions of the rotating system. Usually, it consists of a train of wheels driven by the spindle of the rotating system. A worm which is cut on the spindle engages a pinion and so driven a wheel-train.

10.44. Errors in Motor Meters

The two main errors in such instruments are : ***(i)*** friction error and ***(ii)*** braking error. Friction error is of much more importance in their case than the corresponding error in indicating instruments because ***(a)*** it operates continuously and ***(b)*** it affects the speed of the rotor. The braking action in such meters corresponding to damping in indicating instruments. The braking torque directly affects the speed for a given driving torque and also the number of revolution made in a given time.

Friction torque can be compensated for providing a small constant driving torque which is applied to the moving system independent of the load.

As said earlier, steady speed of such instruments is reached when driving torque is euqal to the braking torque. The braking torque is proportional to the flux of the braking magnet and the eddy current induced in the moving system due to its rotation in the field of the braking magnet

$$\therefore \qquad T_B \propto \Phi_i$$

where Φ is the flux of the braking magnet and i the induced current. Now $i = e/R$ where e is the induced e.m.f. and R the resistance of the eddy current path. Also $e \propto \Phi\, n$ where n is the speed of the moving part of the instrument.

$$\therefore \qquad T_b \;\alpha\; \phi \times \frac{\phi_n}{R} \propto \frac{\Phi^2 n}{R}$$

The torque T_B' at the steady speed of N is given $T_B' \propto \Phi^2 N/R$

Now $\qquad T_B' = T_D \qquad$ – the driving torque

$$\therefore \qquad T_D \propto \Phi^2 N/R \quad \text{or} \quad N \propto T_D R/\Phi^2$$

Hence for a given driving torque, the steady speed is direcly proportional to the resistance of the eddy current of path and inversely to the square of the flux.

Obviously, it is very important that the strength of the field of the brake magnet should be constant throughout the time the meter is in service. The constancy of field strengths can be assured by careful design treatment during the manufacturing of the brake magnet. Variations in temperature will affect the braking torque since the resistance of the eddy current path will change. This error is difficult to fully compensate for.

10.45. Quantity or Ampere-Hour Meters

The use of such meters is mostly confined to d.c. circuits. Their operation depends on the production of two torques ***(i)*** a driving torque which is proportional to the current I in the circuit and ***(ii)*** a braking torque which is proportional to the speed n of the spindle. This speed attains a steady value N when these two torques become numerically equal. In that case, speed becomes proportional to current *i.e.*, $N \propto I$. Over a certain period of time, the total number of revolutions $\int N dt$ will be proportional to the quantity of electricity $\int I dt$ passing through the meter. A worm cut in the spindle as its top engages gear wheels of the recording mechanism which has suitably marked dials reading directly in ampere hours. Since electric supply charges are based on watt-hour rather than ampere-hours, the dials of ampere-hour meters are frequently marked in corresponding watt-hour at the normal supply voltage. Hence, their indications of watt-hours are correct only when the supply voltages remains constant, otherwise reading will be wrong.

10.46. Ampere-Hour Mercury Motor Meter

It is one of the best and most popular forms of mercury Ah meter used for d.c. work.

Construction

It consists of a thin Cu disc D mounted at the base of a spindle, working in jewelled cup bearings and revolving between a pair of ***permanent*** magnets M_1 and M_2. One of the two magnets *i.e.*, M_2 is used for driving purposes whereas M_1 is used for braking. In between the poles of M_1 and M_2 is a hollow circular box B in which rotates the Cu disc and the rest of the space is filled up with mercury which exerts considerable upward thrust on the disc, thereby reducing the pressure on the bearings. The spindle is so weighted that it just sinks in the mercury bath. A worm cut in the spindle at its top engages the gear wheels of the recording mechanism as shown in Fig. 10.57.

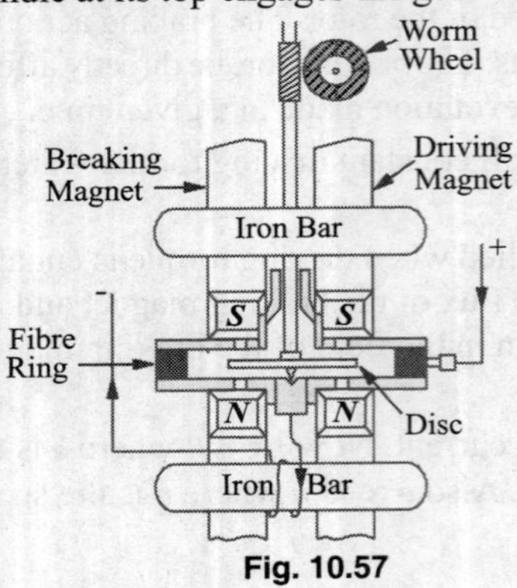

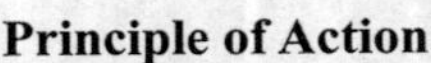
Fig. 10.57

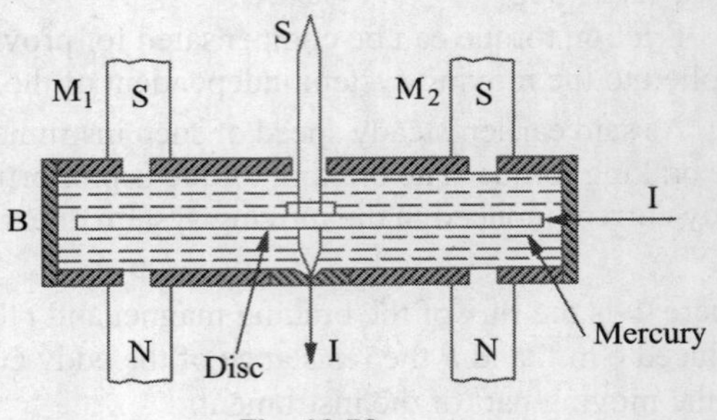

Fig. 10.58

Principle of Action

Its principle of action can be understood from Fig. 10.59 which shows a separate line drawing of the motor element.

The current to be measured is led into the disc through the mercury at a point at its circumference on the right-hand side. As shown by arrows, it flows radially to the centre of the disc where it passes out to the external circuit through the spindle and its bearings. It is worth noting ***that current flows takes place only under the right-hand side magnet M₂*** and not under the left-hand side magnet M_1. The field of M_2 will, therefore, exert a force on the right-side portion of the disc which carries the current (motor action). The direction of the force, as found by Fleming's Left-hand rule, is as shown by the arrow. The magnitude of the force depends on the flux density and current ($\because F = BIl$). The driving or motoring torque T_d so produced is given by the product of the force and the distance from the spindle at which this force acts. When the disc rotates under the influence of this torque, it cuts through the field of left-hand side magnet M_1 and hence eddy currents are produced in it which results in the production of braking torque. The magnitude of the retarding or braking torque is proportional to the speed of rotation of the disc.

Cu Disc
Direction of Induced Current
Driving Magnet
S
S
Current In
Direction of Braking Action
Direction of Disc Rotation
N
N
Braking Magnet
Current Out

Fig. 10.59

Theory

Driving torque $T_d \propto$ force on the disc $\times B\,I$

If the flux density of M_2 remains constant, then $T_d \propto I$.

The braking torque T_B is proportional to the flux Φ of braking magnet M_1 and eddy current i induced in the disc due to its rotation in the field of M_1.

$\therefore \quad T_B \propto \Phi_i$

Now $i = e/R$ where e is the induced e.m.f. and R the resistance of eddy current path.

Also $e \propto \Phi_n$ - where n is the speed of the disc $\therefore\ T_B \propto \Phi \times \frac{\phi_n}{R} \propto \frac{\Phi^2 n}{R}$

The speed of the disc will attain a steady value N when the driving and braking torques becomes equal. In that case, $T_B \propto \Phi^2 N/R$.

If Φ and R are constant, then $I \propto N$

The total number of revolution in any given time t *i.e.*, $\int_0^t N.dt$ will become proportional to $\int_0^t I.dt$ *i.e.*, to the total quantity of electricity passed through the meter.

10.47. Friction Compensation

There are two types of frictions in this ampere-hour meter.

(i) **Bearing Friction.** The effect of this friction is normally negligible because the disc and spindle float in mercury. Due to the upward thrust, the pressure on bearings is considerably reduced which results in freedom from wear as well as freat reduction in the bearing friction.

(ii) **Mercury Friction.** Since the disc revolves in mercury, there is friction between mercury and the disc, which gives rise to a torque, approximately proportional to the square of the speed of rotation. Hence, this friction causes the meter to run shown on heavy loads. It can be compensated for in the following two ways :

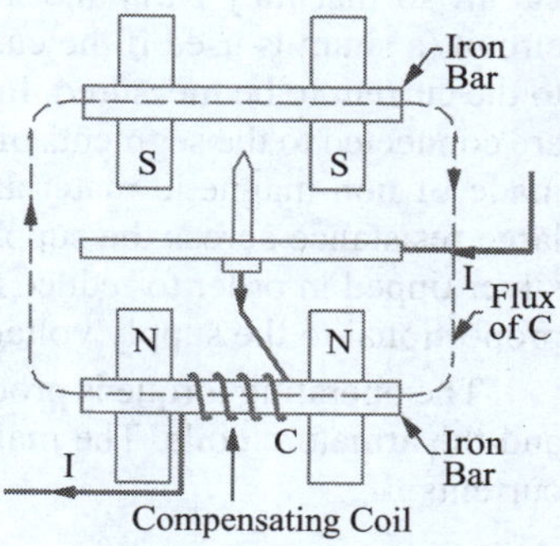

Fig. 10.60

(*a*) a coil of few turns is wound on one of the poles of the driving magnet M_2 and the meter current is passed through it in a suitable direction so as to increase the strength of M_2. The additional driving torque so produced can be made just sufficient to compensate for the mercury friction.

(*b*) in the other method, two iron bars are placed across the permanent magnets, one above and other one below the mercury chamber as shown in Fig. 10.60. The lower bar carries a small compensating coil through which is passed the load current. The local magnetic field set up by this coil strengthens the field of driving magnet M_2 and weakens that of the braking magnet M_1, thereby compensating for mercury friction.

10.48. Mercury Meter Modified as Watt-Hour Meter

If the permanent magnet M_2 of the amp-hour meter, used for producing the driving torque, is replaced by a wound electromagnet connected across the supply, the result is a watt-hour meter. The exciting current of this electromagnet is proportional to the voltage of the supply. The driving torque is exerted on the aluminium disc immersed in the mercury chamber below which is placed this electromagnet. The aluminimum disc has radial slots cut in it for ensuring the radial flow of current through it the current being led into and out of this disc through mercury contacts situated at diameterically opposite points. These radial slots, moreover, prevent the same disc being used for braking purposes. Braking is by a separate aluminium disc mounted on the same spindle and revolving in the air-gap of a separate braking magnet.

10.49. Commutator Motor Meters

These meters may be either ampere-hour or true watt-hour meters. In Fig. 10.61 is shown the principle of a common type of watt-hour meter known as Elihu-Thomson meter. It is based on the dynamometer principle (Art. 10.20) and is essentially an ironless motor with a wound armature having a commutator.

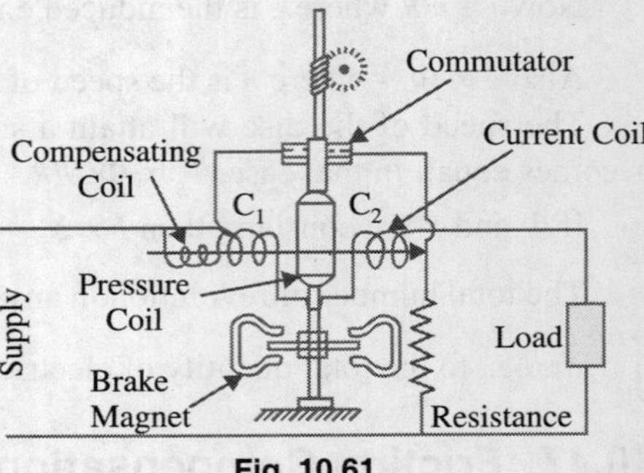

Fig. 10.61

Construction

There are two fixed coils C_1 and C_2 each consisting of a few turns of heavy copper strip and joined in series with each other and with the supply circuit so that they carry the main current in the circuit (a shunt is used if the current is too heavy). The field produced by them is proportional to the current to be measured. In this field rotates an armature carrying a number of coils which are connected to the segments of a small commutator. The armature coils are wound on a former made of non-magnetic material and are connected through the brushes and in series with a large resistance across the supply lines. The commutator is made of silver and the brushes are silver tipped in order to reduce friction. Obviously, the current passing through the armature is proportional to the supply voltage.

The operating torque is produced due to the reaction of the field produced by the fixed coils and the armature coils. The magnitude of this torque is proportional to the product of the two currents *i.e.*,

$$T_d \propto \phi I_1 \quad \text{or} \quad T_d \propto I_1 \times I \qquad (\because \Phi \propto I)$$

where I = main circuit current

I_1 = current in armature coils.

since $I_1 \propto V \quad \therefore \quad T_d \propto VI$ -power

Brake torque is due to the eddy currents induced in an aluminium disc mounted on the *same* spindle and running in the air-gaps of two permanent magnets. As shown in Art, 10.44, this braking torque is proportional to the speed of the disc if the flux of the braking magnet and the resistance of the eddy current paths are assumed constant.

When steady speed of rotation is reached, then

$$T_B = T_d \quad \therefore \quad N \propto VI \propto \text{power } W$$

Hence, steady number of revolutions in a given time is proportional to $\int Wt$ = the energy in the circuit.

The friction effect is compensated for by means of a small compensating coil placed coaxially with the two currents coils and connected in series with the armature such that it strengthens the field of current coil. But its position is so adjusted that with zero line current the armature just fails to rotate. Such meters are now employed mainly for switchboard use, house service meters being invariably of the mercury ampere-hour type.

10.50. Induction Type Single-Phase Watthour Meter

Induction type meters are, by far, the most common form of a.c. meters met with in every day domestic and industrial installations. These meters measure electric energy in kilo-watthours. The principle of these meters is practically the same as that of the induction wattmeters. Constructionally, the two are similar that the control spring and the pointer of the watt-meter are replaced, in the case of watthour meter, by a brake magnet and by a spindle of the meter.

The brake magnet induces eddy currents in the disc which revolves continuously instead of rotating through only a fraction of a revolution as in the case of wattmeters.

Construction

The meter consists of two a.c. electromagnets as shown in Fig. 10.62. (*a*), one of which *i.e.*, M_1 is excited by the line current and is known as ***series*** magnet. The alternating flux Φ_1 produced by it is proportional to and in phase with the line current (provided effects of hysteresis and iron saturation are neglected). The winding of the other magnet M_2 called ***shunt*** magnet, is connected across the supply line and carries current proportional to the supply voltage V. The flux Φ_2 produced by it is proportional to supply voltage V and lags behind it by 90º. This phase displacement of exact 90º is achieved by adjustment of the copper shading band C (also known as power factor compensator) on the shunt magnet M_2. Major portion of Φ_2 crosses the narrow gap between the centre and side limbs of M_2 but a small amount, which is the useful flux, passes through the disc D. The two fluxes Φ_1 and Φ_2 induce e.m.f.s in the disc which further produce the circulatory eddy currents. The reaction between these fluxes and eddy currents produces the driving torque on the disc in a manner similar to that explained in Art. 10.39. The braking torque is produced by a pair of magnets [Fig. 10.62 (*b*)] which are mounted diametrically opposite to the magnets M_1 and M_2. The arrangements minimizes the interaction between the fluxes of M_1 and M_2. This arrangements minimizes the interaction between the fluxes of M_1 and M_2 and that of the braking magnet. When the peripheral portion of the rotating disc passes through the air-gap of the braking magnet, the eddy currents are induced in it which give rise to the necessary torque. The braking torque $T_B \propto \Phi^2 N/R$ where Φ is the flux of braking magnet, N the speed of the rotating disc and R the resistance of the eddy current path. If Φ and R are constant, then $T_B \propto N$.

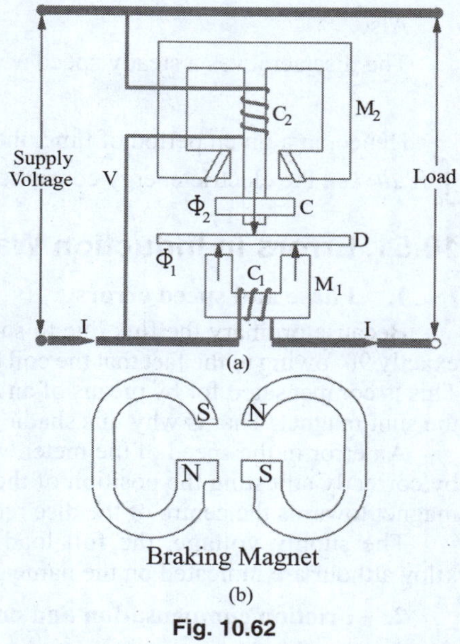

Fig. 10.62

The register mechanism is either of pointer type of cyclometer type. In the former type, the pinion on the rotor shaft drives, with the help of a suitable train of reduction gears, a series of five or six pointers rotating on dials marked with ten equal divisions. The gearing between different pointers is such that each pointer advances by 1/10th of a revolution for a complete revolution of the adjacent pointer on the main rotor disc in the train of gearing as shown in Fig. 10.63.

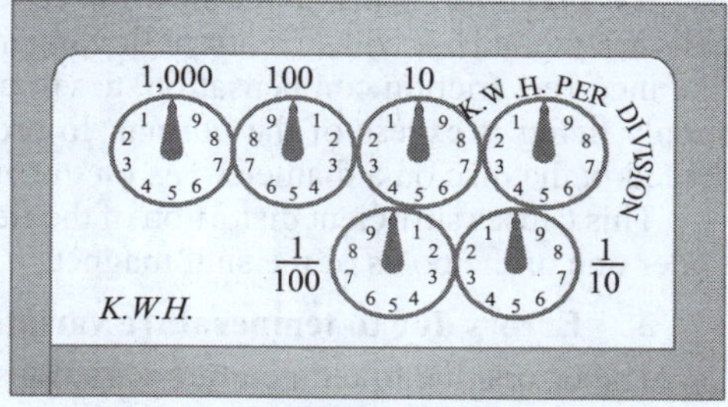

Fig. 10.63

Theory

As shown in Art. 10.39. and with reference to Fig. 10.56, the driving torque is given by $T_d \propto \omega\Phi_{1m}$ and Φ_{2m} where $\sin \propto, \Phi_{1m}$ and Φ_{2m} are the maximum fluxes produced by magnets M_1 and α the angle between these fluxes. Assuming that fluxes are proportional to the current, we have

Current through the windings of $M_1 = I$ –the line current

Current through the winding of $M_2 = V/\omega L$

$$\alpha = 90 - \phi \text{ where } \phi \text{ is the load p.f. angle}$$

$$T_d \propto \omega . \frac{V}{\omega L} . I \cos (90 - \phi) \propto VI \cos \phi \propto \text{power}$$

Also, $$T_b \propto N$$

The disc achieves a steady speed N when the two torques are equal *i.e.*, when

$$T_d = T_B \quad \therefore \quad N \propto \text{power } W$$

Hence, in a given period of time, the total number of revolution $\int_0^t N.dt$ is proportional to $\int_0^t W.dt$ *i.e.*, the electric energy consumed.

10.51. Errors in Induction Watthour Meters

1. Phase and speed errors

Because ordinary the flux due to shunt magnet does not lag behind the supply voltage by exactly 90° owing to the fact that the coil has some resistance, the torque is not zero power factor. This is compensated for by means of an adjustable shading ring placed over the central limb of the sunt magnet. That is why this shading ring is known as ***power factor compensator.***

An error in the speed of the meter, when tested on a non-inductive load, can be eliminated by correctly adjusting the position of the brake magnet. Movement of the poles of the braking magnet towards the centre of the disc reduces the braking torque and ***vice-versa***.

The supply voltage, the full load current and the correct number of revolutions per kilowatthour are indicated on the name plate of the meter.

2. Friction compensation and creeping error

Frictional forces at the rotor bearings and in the register mechanism gives rise to unwanted braking torque on the disc rotor. This can be reduced to an unimportant level by making the ratio of the shunt magnet flux Φ_2 and series magnet flux Φ_1 large with the help of two shading bands. These bands embrace the flux contained in the two outer limbs of the shunt magnet and so eddy currents are induced in them which cause phase displacement between the enclosed flux and the main-gap flux. As a result of this, a small driving torque is exerted on the disc rotor solely by the pressure coil and independent of the main driving torque. The amount of this corrective torque is adjusted by the variation of the position of the two bands, so as to exactly compensate for firctional torque in the instrument. Correctness of friction compensation is achieved when the rotor does not run on no-load ***with only the supply voltage connected.***

By **'*creeping*'** is meant the slow but continuous rotation of the rotor when only the pressure coils are excited but with no current flowing in the circuit. It may be caused due to various factors like incorrect friction compensation, to vibration, to stray magnetic fields or due to the voltage supply being is excess of the normal. In order to prevent creeping on no-load, two holes are drilled in the disc on a diameter *i.e.*, on the opposite sides of the spindle.

This causes sufficient distortion of the field to prevent rotation when one of the holes comes under one of the poles of the sunt magnet.

3. Errors due to temperature variations

The errors due to temperature variations of the instruments are usually small, because the various effects produced tend to neutralise one another.

Example 10.23. *An ampere-hour meter, calibrated at 210 V, is used on 230 V circuit and indicates a consumption of 730 units in a certain period. What is the actual energy supplied ?*

If this period is reckoned as 200 hours, what is the average value of the current ?

(Elect. Technology, Utkal Univ.)

Solution. As explained in Art. 10.42, ampere-hour meters are calibrated to read directly in mWh at the *declared voltage*. Obviously, their readings would be incorrect when used on any other voltage.

Reading on 210 volt = 730 kWh
Reading on 230 volt = 730 × 230/210 = **800 kWh (approx.)**
Average current = 800,000/230 × 200 = **17.4 A**

Example 10.24. *In a test run of 30 min. duration with a constant current of 5 A, mercury-motor amp-hour meter, was found to register 0.51 kWh. If the meter is to be used in a 200-V circuit, find its error and state whether it is running fast or slow. How can the instrument be adjusted to read correctly?*

(Elect. Mean Inst. and Mean., Jadavpur Univ.)

Solution. Ah passed in 30 minutes = 5 × 1/2 = 2.5
Assumed voltage = 0.51 × 1000/2.5 = 204 V

When used on 200-V supply, it would obviously show higher values because actual voltage is less than the assumed voltage. It would be fast by 4 × 100/200 = **2%**

Example 10.25. *An amp-hour meter is calibrated to read kWh on a 220-V supply. In one part of the gear train from the rotor to the first counting dial, there is a pinion driving a 75-tooth wheel. Calculate the number of teeth on a wheel which is required to replace 75-tooth wheel, in order to render the meter suitable for operation on 250-V supply.*

Solution. An amp-hour meter, which is calibrated on 220-V supply would run fast when operated on 250-V supply in the ratio 250/220 or 25/22. Hence, to neutralize the effect of increased voltage, the number of teeth in the wheel should be reduced by the same ratio.

∴ Teeth on the new wheel = 75 × 22/25 = **66**

Example 10.26. *A meter, whose constant is 600 revolutions per kWh, makes five revolution in 20 seconds. Calculate the load in kW.* **(Elect. Meas. and Meas. Inst. Gujarat Univ.)**

Solution. Time taken to make 600 revolution is = 600 × 20/5 = 2,400 second

During this time, the load consumes 1 kWh of energy. If W is load in kW, then

$$W \times 2400(60) \times 60 = 1 \quad \text{or} \quad W = \mathbf{1.5\ kW}$$

Example 10.27. *A current of 6 A flows for 20 minutes through a 220-V ampere hour meter. If during a test the initial and final readings on the meter are 3.53 and 4.00 kWh respectively, calculate the meter error as a percentage of the meter readings.*

If during the test, the spindle makes 480 revolutions, calculate the testing constant in coulomb/rev and rev/kWh.

Solution. Energy actually consumed $= 6 \times \left(\frac{20}{60}\right) \times \frac{200}{1000} = 0.44$ kWh

Energy as registered by meter = 4.00 − 3.53 = 0.47 kWh

Error = 0.47 − 0.44 = 0.03 kWh ; % error = 0.03 × 100/0.47 = 6.38 %

No. of coulombs passed through in 20 minutes = 6 × 20 × 7,2000 coulomb

Testing for 480 revolutions, only 0.44 kWh are consumed, hence testing constant = 480/0.44 = **1091 rev/kWh.**

Example 10.28. *A 230-V, single-phase domestic energy meter has a constant load of 4 A passing through it for 6 hours at unity power factor. If the meter disc makes 2208 revolutions during this period, what is the meter are 1472 when operating at 230 V and 5 V for 4 hours.* **(Elect, Measure, A.M.I.E. Sec B, 1991)**

Solution. Energy consumption in 6 hr = 230 × 4 × 1 × 6 = 5520 W = 5.52 kW

Meter constant = 2208/5.52 = 400 rev/kWh.

Now, 1472 revolution represents energy consumption of 1472/400 = 3.68 kWh

∴ $VI \cos\phi \times \text{hours} = 3.68 \times 10^{-10}$ or $230 \times 5 \times \cos\phi \times 4 = 3680$, ∴ $\cos\phi = \mathbf{0.8}$

Example 10.29. *A 230 V, single-phase domestic energy meter has a constant load of 4 A passing through it for 6 hours at unity power factor. If the motor disc makes 2208 revolution during this period, what is the constant in rev kWh ? Calculate the power factor of the load if the No. of rev made by the meter are 1472 when operating at 230 V and 5 A for 4 hours.* **(Elect. Measuring. AMIE Sec. Winter 1991)**

Solution. Energy supplied at unity p.f. = 230 × 4 × 6 × 1/1000 = 5.52

∴ 230 × 5 × 4 × cos ϕ/2000 = 3.68 ∴ cos ϕ = **0.8.**

Example 10.30. *A single phase kWhr meter makes 500 revolutions per kWh. It is found on testing as making 40 revolutions in 58.1 seconds at 5 kW full load. Find out the percentage error.*

(Elect. Measurement & Measuring Instrument Nagpur Univ. 1993)

Solution. The number of revolutions the meter will make in one hour on testing = 40 × 3600/58.1 = 2478.5

These revolutions correspond to an energy of 5 × 5 kWh

∴ No. of revolutions kWh = 2478.5/5 = 495.7

Percentage error = (500 – 495.7) × 100/500 = **0.86 %**

Example 10.31. *An energy meter is designed to make 100 revolution of the disc for one unit of energy. Calculate the number of revolutions made by it when connected to a load carrying 40 A at 230-V and 0.4 power factor for an hour. If it actually makes 360 revolutions, find the percentage error.*

(Elect. Engg - I Nagpur Univ. 1993)

Solution. Energy consumed in one hour = 230 × 40 × 0.4 × 1/1000 = 3.68 kWh

No. of revolutions the meter should make if it is correct = 3.68 × 100 = 368

No. of revolutions actually made = 360

∴ Percentage error = (369 – 360) × 100/368 = **2.17 %**

Example 10.32. *The disc of an energy meter makes 600 revolutions per unit of energy. When a 1000 watt load is connected, the disc rotates at 10.2 r.p.s if the load is on for 12 hours, how many units are recorded as error?* **(Measurs, Instru. Allahabad Univ. 1992)**

Solution. Since load power is one kWh, energy actually consumed is

= 1 × 12 = 12 kWh

Total number of revolutions made by the disc during the period of 12 hours

= 10.2 × 60 × 12 = 7,344

since 600 revolutions record one kWh, energy recorded by the meter is

= 7,344/600 = 12.24 kWh

Hence, **0.24** unit is recorded extra.

Example 10.33. *A d.c. ampere-hour meter is rated at 5-A, 250-V. The declared constant is 5 A-s/rev. Express this constant in rev/kWh. Also calculate the full-load speed of the meter.*

(Elect. Meas. Inst. and Meas., Jadavpur Univ.)

Solution. Meter constant = 5 A-s/rev

Now, 1 kWh = 10^3 Wh = 10^3 × 3600 volt-second

= 36 × 10^5 volt × amp × second

Hence, on a 250-V circuit, this corresponds to 36 × 10^5/250 = 14,400 A-s

Since for every 5 A-s, there is one revolution, the number of revolution is one kWh is

= 14,400/5 = 2,880 revolutions

∴ Meter constant = **2,800 rev/kWh**

Since full-load meter current is 5 A and its constant is 5 A-s/rev, it is obvious that it makes one revolution every second.

∴ full load speed = 60 r.p.m.

Example 10.34. (*a*). *A single-phase energy meter of the induction type is rated 230-V; 10-A, 50-Hz and has a meter constant of 600 rev/kWh when correctly adjusted. If 'quadrature' adjustment is slightly disturbed so that the lag is 85°, calculate the percentage error at full-load 0.8 p.f. lag.*

(Measu & Instru. Nagpur Univ 1991)

Solution. As seen from Art. 10.50, the driving torque. T_d depends among other factors, on sin α where α is the angle between the two alternating fluxes. ∴ $T_d \propto \sin\alpha$

If the voltage flux-lagging adjustment is disturbed so that the phase angle between the voltage flux and the voltage is less than 90° (instead of being exactly 90°) the error is introduced.

Now $\cos\phi = 0.8, \quad \phi = \cos^{-1}(0.8) = 36°52'$

$\therefore \quad \alpha = 85° - 36°52' = 48°8'$ where it should be $= 90 - 36°52' = 53°8'$

$$\therefore \quad \text{error} = \frac{\sin 53°8' - \sin 48°8'}{\sin 53°8'} \times 100 = \mathbf{7\%}$$

Example 10.35 (*b*). *A 50-A, 230-V energy meter is a full-load test makes 61 revolutions in 37 seconds if the normal speed of the disc is 520 revolutions/kWh, compute the percentage error.*

[Nagpur University November 1999]

Solution. Unity power-factor is assumed.

Energy consumed, in kWh, in 37 seconds

$$= \frac{50 \times 230}{1000} \times \frac{37}{3600} = 0.1182 \text{ kWh}$$

Number of revolutions corresponding to this energy = 520 × 0.1182 = 61.464

The meter makes 61 revolutions

$$\therefore \quad \%\text{ Error} = \frac{61 - 61.464}{61.464} \times 100\,\% = \mathbf{-\,0.755\,\%}$$

10.52. Ballistic Galvanometer

It is used principally for measuring small electric charges such as those obtained in magnetic flux measurements. Constructionally, it is similar to a moving-coil galvanometer except that **(*i*)** it has extremely small electromagnetic damping and **(*ii*)** has long period of undamped oscillation (several seconds). These conditions are necessary if the galvanometer is to measure electric charge. In fact, the moment of inertia of the coil is made so large that whole of the charge passes through the galvanometer before its coil has had time to move sufficiently. In that case, the first swing of the coil is proportional to the charge passing through the galvanometer. After this swing has been observed, the oscillating coil may be rapidly brought to rest by using eddy-current damping.

As explained above, the coil moves after the charge to be measured has passed through it. Obviously, during the movement of the coil, there is no current flowing through it. Hence, the equation of its motion is

$$J\frac{d^2\phi}{dt^2} + D\frac{d\theta}{dt} + C\theta = 0$$

where J is the moment of inertia, D is damping constant and C is restoring constant.

Since damping is extremely small, the approximate solution of the above equation is

$$0 = U_e^{-(D/2J)t} \sin(\omega_0 t + \phi)$$

At the start of motion, where

$r = 0, \theta = 0,$ hence $\phi = 0$

$\therefore \quad \theta = Ue^{-(D/2J)t} \sin \omega_0 t$ ...(*i*)

During the passage of charge, at any instant, there will be a deflecting torque of Gi acting on the coil. If t is the time taken by the whole charge to pass through, the torque impulse due to this charge is

$$\int_0^t Gidt \text{ . Now } \int_0^t i\,dt = Q$$

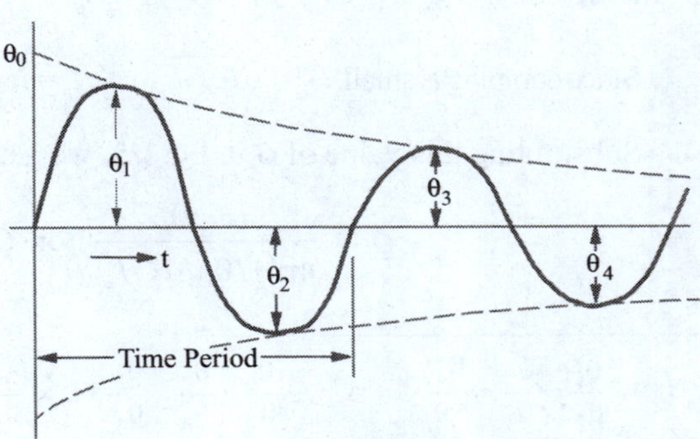

Fig. 10.65

Hence, torque impulse = GQ. This must be equal to the change of angular momentum produced *i.e.*, $j\alpha$ where α is the angular velocity of the coil at the end of the impulse period.

$$\therefore \quad GQ = J\alpha$$

or $$\alpha = GQ/J$$

Differentiating Eq. (*i*) above, we get

$$\frac{d\theta}{dt} = U\left(e^{-(D/2J)^t}\,\omega_0 t - \frac{D}{2}\,J\,e^{-(D/2J)r}\,\sin\omega_0 t\right)$$

Since duration of the passage of charge is very small, at the end of the passage, $t \cong 0$, so that from above, $dq/dt = U\omega_0$.

However, at this time, $\dfrac{d\theta}{dt} = \alpha = \dfrac{GQ}{J} \quad \therefore \quad \dfrac{GQ}{J} = U\,\omega_0 \quad$ or $\quad Q = \dfrac{J\omega_0}{G}U$

Now, U being the amplitude which the oscillations would have if the damping were zero, it may be called undamped swing θ_0.

$$\therefore \quad Q = \frac{J\omega_0}{G}\theta_0 \quad \text{or} \quad Q \propto \theta_0 \qquad \text{...(ii)}$$

However, in practice, due to the presence of small amount of damping, the successive oscillations diminish exponentially (Fig. 10.64). Even the first swing θ_1 is much less than θ_0. Hence, it becomes necessary to obtain the value of θ_0 from the observed value of first maximum swing θ_1.

As seen from Fig. 10.64, the successive peak value θ_1, θ_2, θ_3 etc. are ϕ radian apart or ϕ/ω_0 second apart. The ratio of the amplitude of any two successive peaks is

$$= \frac{\theta_0 e^{-(D/2J)^t}\,\sin\omega_0 t}{\theta_0 e^{-(D/2J)t + (\pi/\omega_0)}\,\sin(\omega_0 t + \pi)} = \frac{e^{-(D/2J)^t}\,\sin\omega_0 t}{e^{-(D/2J)^t}\,e^{-(D/2J)(\pi/\omega_0)}\,(-\sin\omega_0 t)} = -e^{(D/2J)(\pi/\omega_0)}$$

$$\therefore \quad \frac{\theta_1}{\theta_2} = \frac{\theta_2}{\theta_3} = \frac{\theta_3}{\theta_4} = \ldots.. = e^{(D/2J)(\pi/\theta_0)}$$

Let $e^{-(D/2J)(\pi/\omega 0)} = \Delta^2$ where Δ is called the damping factor*.

The time period of oscillation $T_0 = 2\pi/\omega_0$. If damping is very small $\theta_0 = \theta_1$, $t = T_0/4 = \pi/2\omega_0$ as a very close approximation.

Hence, from Eq. (*i*) above, putting $t = \pi/2\omega_0$, we have

$$\theta = \theta_0^{(D/2J)(\pi/2\omega 0)}\,\sin\pi/2 = \theta_0\Delta - 1 \quad \therefore \quad \theta_0 = \Delta\,\theta_1 \qquad \text{...(iii)}$$

or undamped swing = damping factor × 1st swing

Suppose, a steady current of I_s flowing through the galvanometer produces a steady deflection ϕ_s, then $\quad C\theta_s = GI_s \quad$ or $\quad G = C\theta_s/I_s$

Since damping is small, $\omega_0 = \sqrt{C/J} \quad \therefore \quad C = J\omega_0^2 = J \times \left(\dfrac{2\pi}{T_0}\right)^2 = 4\pi^2\dfrac{J^2}{T_0^2} \quad \therefore \quad G = \dfrac{4\pi^2\,J\theta s}{T_0^2\,I_s}$

Substituting this value of G in Eq. (*ii*), we get

$$Q = \frac{J\omega_0\,\theta_0}{4\pi^2\,J\theta_s/T_0^2\,I_s^2} \quad \text{or} \quad Q_0 = \frac{T_0}{2\pi}\cdot\frac{I_s}{\theta_s}\cdot\theta_0 = \frac{T_0}{2\pi}\cdot\frac{I_s}{2\pi}\cdot\Delta.\theta_1 \qquad \text{...(iv)}$$

---

* Now, $\dfrac{\theta_1}{\theta_2} = \dfrac{\theta_2}{\theta_3} = \dfrac{\theta_{n-1}}{\theta_n} = \Delta^3 \quad \therefore \quad \dfrac{\theta_1}{\theta_2}\times\dfrac{\theta_2}{\theta_3}\times\dfrac{\theta_3}{\theta_4}\times\ldots\times\dfrac{\theta_{n-1}}{\theta_n} = (\Delta^2)^{n-1}\;\dfrac{\theta_1}{\theta_2} = (\Delta^2)^{n-1}$ or $\therefore \left(\dfrac{\theta_1}{\theta_2}\right)^{1/2(n-1)}$

Hence, Δ may be obtained by observing the first and nth swing.

Alternatively, let quantity $(D/2J)\,(\pi/\omega_0)$ be called the logarithmic decrement λ. Since, $\Delta^2 = e^{\lambda}$, we have

$$\Delta = e^{\lambda/2} = 1 + (\lambda/2) + \frac{(\lambda/2)^2}{2!} + \ldots \cong \left(1 + \frac{\lambda}{2}\right) \text{ when } \lambda \text{ is small}^*$$

Hence, from Eq. **(*iv*)** above, we have $Q = \frac{T_0}{2\pi}, \frac{I_s}{\theta_s}\left(1+\frac{\lambda}{2}\right)\theta_1$ ...**(*v*)**

In general, Eq. **(*iv*)** may be put as $Q = k\theta_1$

Example. 10.36. *A ballistic galvanometer has a free period of 10 seconds and gives a steady deflection of 200 divisions with a steady current of 0.1 µA. A charge of 121 µC is instantaneously discharged through the galvanometer giving rise to a first maximum deflection of 100 divisions . Calculate the 'decrement' of the resulting oscillations.* **(Electrical Measurements, Bombay Univ.)**

Solution. From Eq. **(*v*)** of Art 10.52, we have $Q = \frac{T_0}{2\pi}\cdot\frac{I_s}{\theta_s}\left(1+\frac{\lambda}{2}\right)\theta_1$

Here, $Q = 121\ \mu C = 121 \times 10^{-6}\ C$; $T_0 = 10s$; $I_s = 0.1$ mA $= 10^{-4}$ A ; $\theta_s = 200$; $\theta_1 = 100$

$$\therefore \quad 121 \times 10^{-6} = \frac{10}{2\pi} \times \frac{10^{-4}}{200}\left(1+\frac{\lambda}{2}\right) \times 100; \quad \therefore\ \lambda = \mathbf{1.04}$$

10.53. Vibration Galvanometer

Such galvanometers are widely used as null-point detectors in a.c. bridges.

Construction

As shown in Fig. 10.65. (*a*), it consists of a moving coil suspended between poles of a strong permanent magnet. The natural frequency of oscillation of the coil is very high, this being achieved

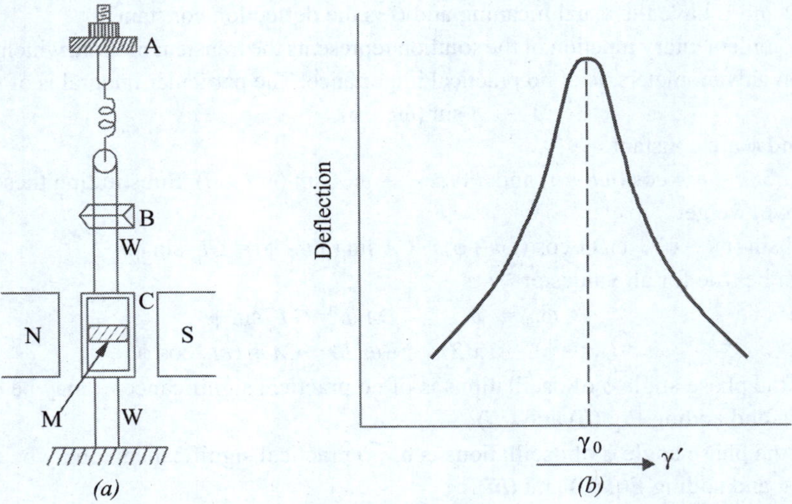

Fig. 10.65

* Since $\Delta^2 = e^{\lambda}$, taking logs, we have $2 \log_c \Delta = \lambda \log_e e = \lambda$

$$\frac{\lambda}{2} = \log_c \Delta = \log_e \left(\frac{\theta_1}{\theta_2}\right)^{1/2(n-1)} = \frac{1}{2\,(n-1)} \log_e \theta_1/\theta_n \;\therefore\; \lambda = \frac{1}{(n-1)} \log_e \frac{\theta_1}{\theta_n}$$

by the use of a large value of control constant and a moving system of very small inertia. The suspension (which provides control) is either a phosphor-bronze strip or is a bifilar suspension in which case the two suspension wires carry the coil and a small piece of mirror (or in some cases the two suspension wires themselves from the coil.)

As seen, *W* is the suspension, *C* is the moving coil and *M* the mirror on which as cast a beam of light. From mirror *M*, this beam is deflected on to a scale. When alternating current is passed through *C*, an alternating torque is applied to it so that the reflected spot of light on the scale is drawn out in the form of a band of light. The length of this band of light is maximum if the natural frequency of oscillation of *C* coincides with the the supply frequency due to resonance. The turning of *C* may be done in the following two ways :

(*i*) by changing the length of suspension *W*. This is achieved by raising or lowering bridge piece *B* against which the bifilar loop presses.

(*ii*) by adjusting tension in the suspension. This is achieved by turning the knurled knob *A*.

By making the damping very small, the resonance curve of the galvanometer can be made sharply-peaked [Fig. 10.65 (*b*)]. In that case, the instrument discriminates sharply against frequencies other than its own natural frequency. In other words, its deflection becomes very small even when the frequency of the applied current differs by a very small amount from its resonance frequency.

Theory

If the equation of the current passing through the galvanometer is $i = I_m \sin \omega_t$, then the equation of motion of the coil is :

$$J\frac{d^2\theta}{dt^2} + D\frac{d\theta}{dt} + C\theta = Gi = GI_m \sin \omega t \qquad \text{...}(i)$$

where *J*, *D* and *C* have the usual meaning and *G* is the deflection constant.

The complementary function of the solution represents the transient motion, which in the case of vibration galvanometers, is of no practical importance. The particular integral is of the form

$$\theta = A \sin (\omega t - \phi)$$

where *A* and ϕ are constant.

Now, $d\theta/dt = \omega A \cos (\omega t - \phi)$ and $d^2\theta/dt^2 = -\omega^2 A \sin (\omega t - \theta)$. Substituting these values in Eq. **(*i*)** above, we get

$$-\omega^2 JA \sin (\omega t - \phi) + \omega DA \cos (\omega t - \phi) + CA \sin (\omega t - \phi) = GI_m \sin \omega t$$

It must be true for all values or *i*

When $\quad \omega t = \phi \qquad DA\,\omega = G I_m \sin \phi \qquad$ **...(*ii*)**

When $\quad (wt - \phi) = \pi/2 \qquad -\omega^2 JA + CA = G I_m \cos \phi \qquad$ **...(*iii*)**

Since the phase angle ϕ of oscillations is of no practical significance, it may be eliminated by squaring and adding Eq. **(*ii*)** and **(*iii*)**.

Since the phase angle ϕ of oscillations is of no practical significance, it may be eliminated by squaring and adding Eq. **(*ii*)** and **(*iii*)**.

$$\therefore \quad \omega^2 D^2 A^2 + A^2 (C - \omega^2 J)^2 = G^2 I_m^2 \quad \text{or} \quad A = \frac{G I_m}{\sqrt{[D^2\omega^2 + (C - \omega^2 J)^2]}} \qquad \text{...}(iv)$$

This represents the amplitude *A* of the resulting oscillation for a sinusoidally alternating current of peak value I_m flowing through the moving coil of the galvanometer.

10.54. The Vibrating-Reed Frequency Meter

1. Working Principle

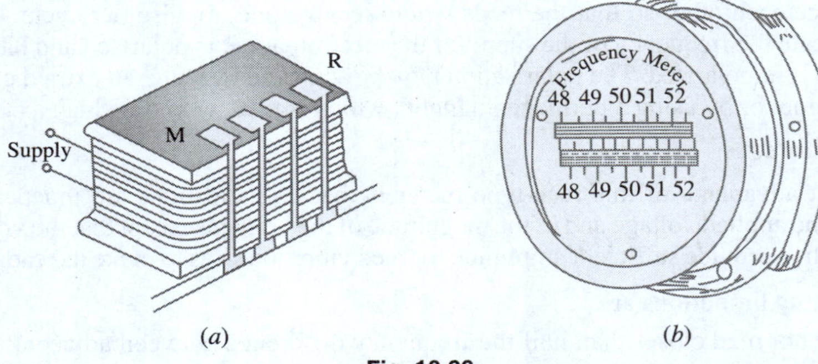

Fig. 10.66

The meter depends for its indication on the mechanical resonance of thin flat steel reeds arranged alongside and, close to, an electromagnet as shown in Fig. 10.66.

2. Construction

The electromagnet has a laminated armature and its winding, in series with a resistance, is connected across a.c. supply whose frequency is required. In that respect, the external connection of this meter is the same as that of a voltmeter.

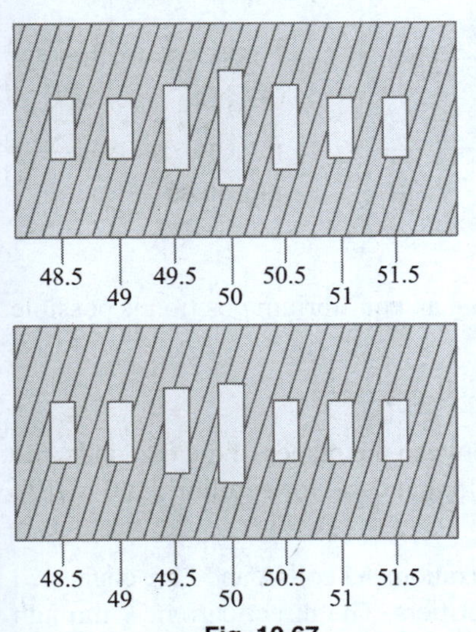

Fig. 10.67

The metallic reeds (about 4 mm wide and 0.5 mm thick) are arranged in a row and are mounted side by side on a common and slightly flexible base which also carries the armature of the electromagnet. The upper free ends of the reeds are bent over a right angles so as to serves as flags or targets and enamelled white for better visibility. The successive reeds are not exactly similar, their natural frequencies of vibration differing by $\frac{1}{2}$ cycle. The reeds are arranged in ascending order of natural frequency.

3. Working

When the electromagnet is connected across the supply whose frequency is to be measured, its magnetism alternates with the same frequency. Hence the electromagnet exerts attracting force on each reed once every half cycle. All reeds tend to vibrate but only that whose natural frequency is exactly double the supply frequency vibrates with maximum amplitude due to mechanical resonance [Fig. 10.67 (*a*)]. The supply frequency is read directly by noting the scale mark opposite the white painted flag which is vibrating the most ($f = 50$ Hz). The vibrations of other reeds would be so small as to be almost unobservable. For a frequency exactly midway between the natural frequencies of the two reeds ($f = 49.75$ Hz), both will vibrate with amplitudes which are equal but much less than when the supply frequency exactly coincides with that of the reeds.

4. Range

Such meters have a small range usually from 47 to 53 Hz or from 57 to 63 Hz etc.

The frequency range of a given set of reeds may be doubled by polarising the electromagnets as explained below. As seen from above description, each reeds is attracted twice per cycle of the supply *i.e.*, once every half-cycle and the reeds whose natural frequency is twice that of the current is of the one which responds most. Suppose the electromagnet carries an additional winding carrying direct current whose steady flux is equal in magnitude to the alternating flux of the a.c. winding. The resultant flux would be zero in one half-cycle and double in the other half-cycle when the two fluxes reinforce each other so that the reeds would receive one impulse per cycle. Obviously, a reed will indicate the frequency of the supply if the electromagnet is polarised and half the supply frequency if it is unpolarised. The polarisation may be achieved by using an extra d.c. winding on the electromagnet or by using a permanent tangent which is then wound with an a.c. winding.

5. Advantages

One great advantage of this reed-type meter is that its indications are independent of the waveform of the applied voltage and of the magnitude of the voltage, except that the voltage should be high enough to provide sufficient amplitude for reed vibration so as to make its readings reliable.

However, its limitations are :

(*a*) it cannot read closer than half the frequency difference between adjacent reeds.

(*b*) its error is dependent upon the accuracy with which reeds can be turned to a given frequency.

10.55. Electrodynamic Frequency Meter

It is also referred to as moving-coil frequency meter and is a ratiometer type of instrument.

1. Working Principle

The working principle may be understood from Fig. 10.68 which shows two moving coils rigidly fixed together with their planes at right angles to each other and mounted on the shaft or spindle situated in the field of a permanent magnet. There is no mechanical control torque acting on the two coils. If G_1 and G_2 are displacement constants of the two coils and I_1 and I_2 are the two currents, then their respective torques are $T_1 = G_1 I_1 \cos\theta$, $T_2 = G_2 I_2 \sin\theta$. These torques act in the opposite directions.

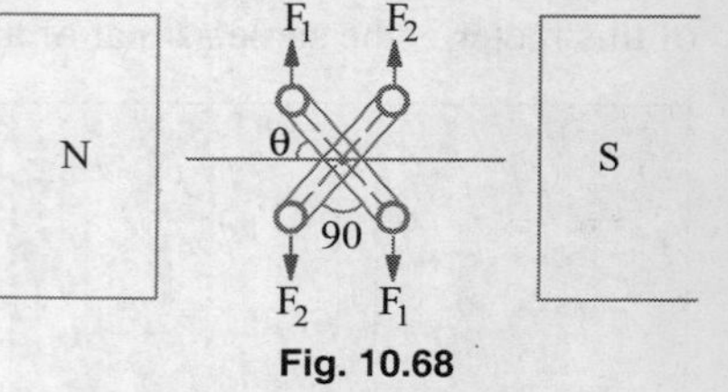

Fig. 10.68

Obviously, T_1 decreases with θ where as T_2 increases but an equilibrium position is possible for same angle θ for which

$$G_1 I_1 \cos\theta = G_2 I_2 \sin\theta \text{ or } \tan\theta = \frac{G_1}{G_2}\cdot\frac{I_1}{I_2}$$

By modifying the shape of pole faces and the angle between the planes of the two coils, the ratio I_1/I_2 is made proportional to angle θ instead of tan θ. In that case, for equilibrium $\theta \propto I_1/I_2$.

2. Construction

The circuit connections are shown in Fig. 10.69. The two ratiometer coils *X* and *Y* are connected across the supply lines through their respective bridge rectifiers. The direct current I_1 through

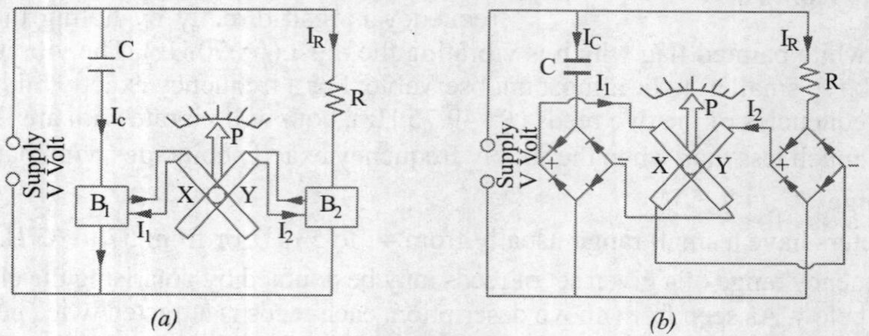

Fig. 10.69

coil X represents the R.M.S. value of capacitor current I_C as rectified by B_1. Similarly, direct current I_2 flowing through Y is the rectified current I_R passing through series resistance R.

3. Working

When the meter is connected across supply lines, rectified currents I_1 and I_2 pass through coils X and Y they come to rest at an angular position where their torques are equal but opposite. This angular position is dependent on the supply frequency which is read by a pointer attached to the coil.

As proved above, $\theta \propto I_1/I_2$

Assuming sinusoidal waveform, mean values of I_1 and I_2 are proportional to the R.M.S. values of I_C and I_R respectively.

$$\therefore \qquad \theta \propto \frac{I_1}{I_2} \propto \frac{I_C}{I_R} \qquad \text{Also } I_C \propto V_m\,\omega\,C \text{ and } I_R \propto V_m/R$$

where V_m is the maximum value of the supply voltage whose equation is assumed as $v = V_m \sin \omega t$.

$$\therefore \qquad \theta \propto \frac{V_m\,\omega\,C}{V_m / R} \propto \omega\,CR \propto \omega \qquad \therefore \quad \theta \propto f \qquad (\because \omega = 2\pi f)$$

Obviously, such meters have linear frequency scales. Moreover, since their readings are independent of voltage, they can be used over a fairly wide range of voltage although at too low voltages, the distortions introduced by rectifier prevent an accurate indication of frequency.

It will be seen that the range of frequency covered by the meter depends on the value of R and C and these may be chosen to get ranges of 40–60 Hz, 1200–2000 Hz or 8000–12,000 Hz.

10.56. Moving-Iron Frequency Meter

1. Working Principle

The action of this meter depends on the variation in current drawn by two parallel circuits – one inductive and the other non-inductive—when the frequency changes.

2. Construction

The construction and internal connections are shown in Fig. 10.70. The two coils A and B are so fixed that their magnetic axes are perpendicular to each other. At their centres is pivoted a long and thin soft-iron needle which aligns itself along the resultant magnetic field of the two coils. There is no control device used in the instrument.

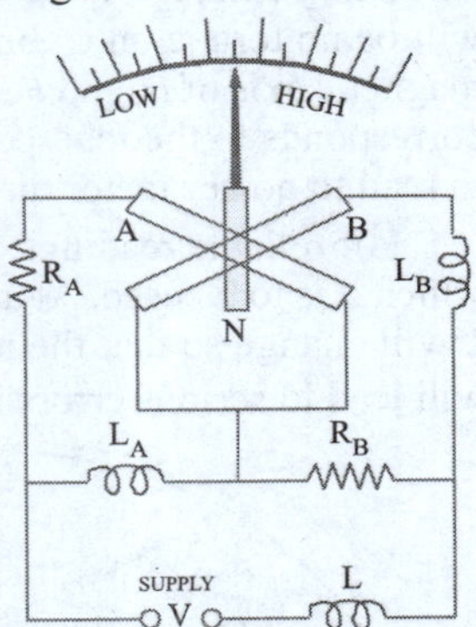

Fig. 10.70

It will be noted that the various circuit elements constitute a Wheatstone bridge which becomes balanced at the supply frequency. Coil A has a resistance R_A in series with it and a reactance L_A in parallel. Similarly R_B is in series with coil B and L_B is in parallel. The series inductance L helps to suppress higher harmonics in the current waveform and hence, tends to minimize the waveform errors in the indication of the instrument.

3. Working

On connecting the instrument across the supply, currents pass through coils A and B and produce opposing torques. When supply frequency is high, currents through coil A is more whereas that through coil B is less due to the increase in the reactance offered by L_B. Hence, magnetic field of coil A is stronger than that of coil B. Consequently, the iron needle lies more nearly to the magnetic axis of coil A than that of B. For low frequencies, coil B draws more current than coil A and, hence, the needle lies more nearly parallel to the magnetic axis of B than to that of coil A. The variations of frequency are followed by the needle as explained above.

The instrument can be designed to cover a broad or narrow range of frequencies determined by the parameters of the circuit.

10.57. Electrodynamic Power Factor Meter

1. Working Principle

The instrument is based on the dynamometer principle with spring control removed.

2. Construction

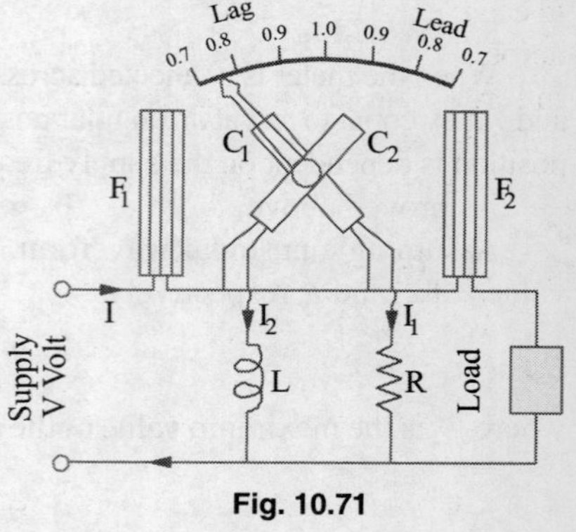

Fig. 10.71

As shown in Fig. 10.71 and 10.72, the instrument has a stationary coil which is divided into two sections F_1 and F_2. Being connected in series with the supply line, it carries the load current. Obviously, the uniform field produced by F_1 and F_2 is proportional to the line current. In this field are situated two moving coils C_1 and C_2 rigidly attached to each other and mounted on the same shaft or spindle. The two moving coils are 'voltage' coils but C_1 has a series resistance R whereas C_2 has a series inductance L. The values of R and L as well as turns on C_1 and C_2 are so adjusted that the ampere-turns of C_1 and C_2 are exactly equal. However, I_1 is in phase with the supply voltage V whereas I_2 lags behind V by nearly 90°. As mentioned earlier, there is no control torque acting on C_1 and C_2 – the currents being led into them by fine ligaments which exerts no control torque.

3. Working

Consider the case when load power factor is unity *i.e.* I is in phase with V. Then I_1 is in phase with I whereas I_2 lags behind by 90°. Consequently, a torque will act on C_1 which will set its plane perpendicular to the common magnetic axis of coils F_1 and F_2 *i.e.* corresponding to the pointer position of unity p.f. However, there will be no torque acting on coil C_2.

Now, consider the case when load power factor is zero *i.e.* I lags behind V by 90° (like current I_2). In that case, I_2 will be in phase with I where as I_1 will be 90° out of phase. As a result, there will be no toruqe on C_1 but that acting on C_2 will bring its plane perpendicular to the common magnetic axis of F_1 and F_2. For intermediate values of power factor, the deflection of the pointer corresponds to the load power factor angle ϕ or to cos ϕ, if the instrument has been calibrated to read to power factor direcly.

For reliable readings, the instrument has to be calibrated at the frequency of the supply on which it is to be used. At any other frequency (or when harmonics are present), the reactance of L will change so that the magnitude and phase of current through C_2 will be incorrect and that will lead to serious errors in the instrument readings.

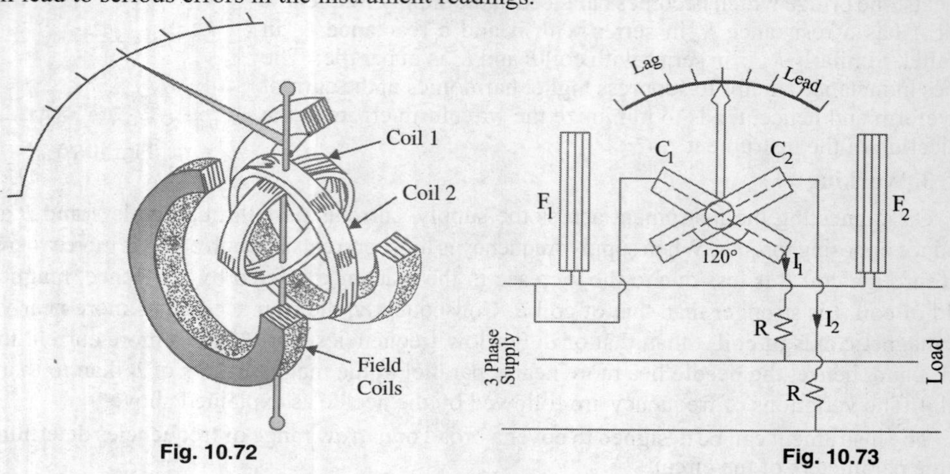

Fig. 10.72

Fig. 10.73

For use on balanced 3-phase load, the instrument is modified, so as to have C_1 and C_2 at 120° to each other, instead of 90°, as in 1-phase supply. As shown in Fig. 10.74, C_1 and C_2 are connected across two different phase of the supply circuit, the stationary coils F_1 and F_2 being connected in series with the third phase (so that it carries the line current). Since there is no need of phase splitting between the currents of C_1 and C_2, I_1 and I_2 are not determined by the phase-splitting circuit and consequently, the instrument is not affected by variations in frequency or waveform.

10.58. Moving-Iron Power Factor Meter

1. Construction

One type of power factor meter suitable for 3-phase balanced circuits is shown in Fig. 10.74. It consists of three fixed coils *R*, *Y* and *B* with axes mutually at 120° and intersecting on the centre line of the instrument. These coils are connected respectively in *R*, *Y* and *B* lines of the 3-phase supply through current transformers. When so energised, the three coils produce a synchronously rotating flux.

There is a fixed coil *B* at the centre of three fixed coils and is connected in series with a high resistance across one of the pair of lines, say, across *R* and *Y* lines as shown. Coil *B* is threaded by the instrument spindle which carries an iron cylinder *C* [Fig. 10.74 (*b*)] to which are fixed sector-shaped iron vanes V_1 and V_2. The same spindle also carries damping vanes and pointer (not shown in the figure) but ***there are no control springs.*** The moving system is shown separately in Fig. 10.74 (*b*).

2. Working

The alternating flux produced by coil *B* interacts with the fluxes produced by the three current coils and causes the moving system to take up a position determined by the power factor angle of the load. However, the instrument is calibrated to read the power factor cos ϕ directly instead of ϕ. In other words, the angular deflection ϕ of the iron vanes from the line *MN* in Fig. 10.74 (*a*) is equal to the phase angle ϕ.

Because of the rotating field produced by coils *R*, *Y* and *B*, there is a slight induction-motor action which tends to continuously turn the moving iron in the direction of the rotating flux. Hence, it becomes essential to design the moving iron as to make this torque negligibly small *i.e.* by using high-resistance metal for the moving iron in order to reduce eddy currents in it.

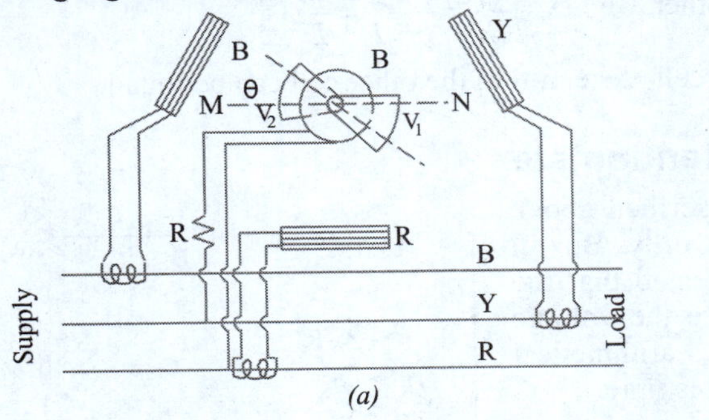

(*a*)

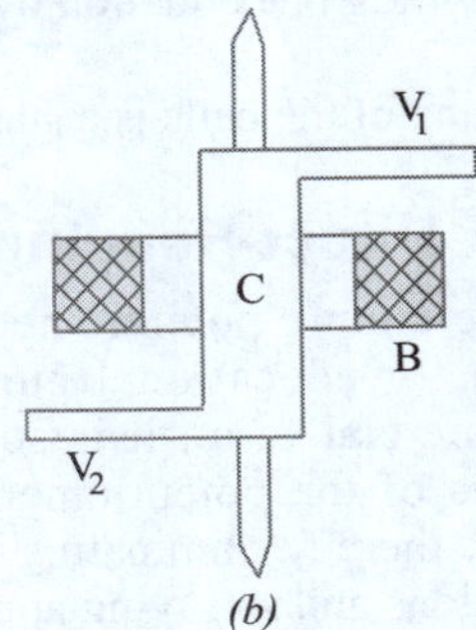

(*b*)

Fig. 10.74

3. Merits and Demerits

Moving iron p.f. meters are more commonly used as compared to the electrodynamic type because

(*i*) they are robust and comparatively cheap **(*ii*)** they have scales upto 360° and *(ii)* in their case, all coils being fixed, there are no electrical connections to the moving parts.

On the other hand, they are not as accurate as the electrodynamic type of instruments and, moreover, suffer from erros introduced by the hysteresis and eddy-current losses in the iron parts–these losses varying with load and frequency.

10.59. D.C. Potentiometer

A potentiometer is used for measuring and comparing the e.m.fs. of different cells and for calibrating and standardizing voltmeters, ammeters etc. In its simplest from, it consists of a German silver or manganin wire usually one meter long and stretched between two terminals as shown in Fig. 10.75.

This wire is connected in series with a suitable rheostat and battery *B* which sends a steady current through the resistance wire *AC.* As the wire is of uniform cross-section throughout, the fall in potential across it is uniform and the drop between any two points is proportional to the distance between them. As seen, the battery voltage is spread over the rheostat and the resistance wire *AC.* As we go along *AC,* there is a progressive fall of potential. If ρ is the resistance/cm of this wire, *L* its length, then for a current of *I* amperes, the fall of potential over the whole length of the wire is ρ *LI* volts.

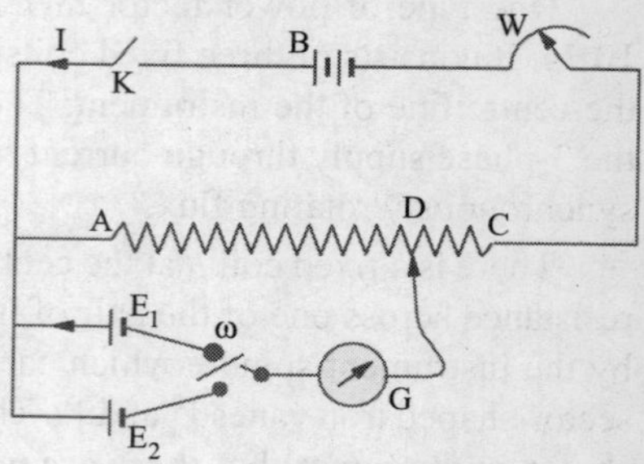

Fig. 10.75

The two cells whose e.m.fs are to be compared are joined as shown in Fig. 10.76, always remembering that ***positive terminals of the cells and the battery must be joined together.*** The cells can be joined with the galvanometer in turn through a two-way key. The other end of the galvanometer is connected to a movable contact on *AC.* By this movable contact, a point like *D* is found when there is no current in and hence no deflection of *G.* Then, it means that the e.m.f. of the cell just balances the potential fall on *AD* due to the battery current passing through it.

Suppose that the balance or null point for first cell of e.m.f. E_1 occurs at a length L_1 as measured from point *A*. The $E_1 = \rho\, L_1 I$.

Similarly, if the balance point is at L_2 for th other cell, then $E_2 = \rho\, L_2 I$.

Dividing one equation by the other, we have $\dfrac{E_1}{E_2} = \dfrac{\rho L_1 I}{\rho L_2 I} = \dfrac{L_1}{L_2}$

If one of the cells is a standard cell, the e.m.f. of the other cell can be found.

10.60. Direct-Reading Potentiometer

The simple potentiometer described above is used for educational purposes only. But in it commercial form, it is so calibrated that the readings of the potentiometer give the voltage directly, thereby eliminating tedious arithmetical calculations and so saving appreciable time.

Such a direct-reading potentiometer is shown in Fig. 10.76. The resistance *R* consists of 14 equal resistances joined in series, the resistance of each unit being equal to that of the whole slide were *S* (which is divided into 100 equal parts). The battery current is controlled by slide wire resistance *W*.

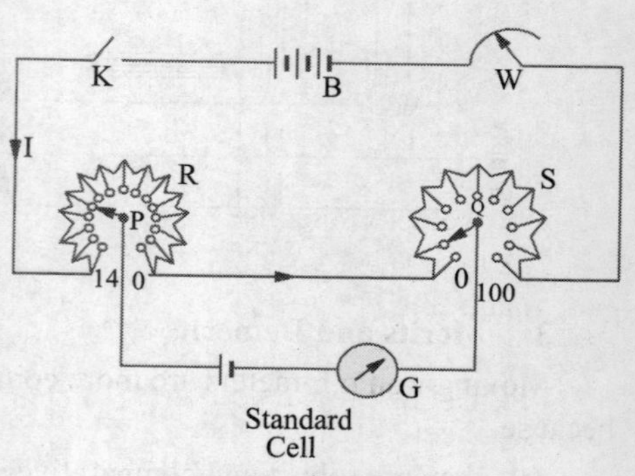

Fig. 10.76

10.61. Standardizing the Potentiometer

A standard cell *i.e.* Weston cadmium cell of e.m.f. 1.0183 V is connected to sliding contacts *P* and *Q* through a sensitive galvanometer *G*. First, *P* is put on stud No. 10 and *Q* on 18.3 division on *S* and then *W* is adjusted for zero deflection on *G*. In that case, potential difference between *P* and *Q* is equal to cell voltage *i.e.* 1.0183 V so that potential drop on each resistance of *R* is 1/10 = 0.1 V and every division of *S* represents 0.1/100 = 0.001 V. After standardizing this way, the ***position of W is not to be changed in any case*** otherwise the whole adjustment would go wrong. After this, the instrument becomes direct reading. Suppose in a subsequent experiment, for balance, *P* is moved to stud No. 7 and *Q* to 84 division, then voltage would be = $(7 \times 0.1) + (84 \times 0.001) = 0.784$ V.

It should be noted that since most potentiometers have fourteen steps on *R*, it is usually not possible to measure p.ds. exceeding 1.5 V. For measuring higher voltges, it is necessary to use a volt box.

10.62. Calibration of Ammeters

The ammeter to be calibrated is connected in series with a variable resistance and a standard resistance *F*, say, of 0.1 Ω across battery B_1 of ample current capacity as shown in Fig. 10.77. Obviously, the resistance of *F* should be such that with maximum current flowing through the ammeter *A*, the potential drop across *F* should not exceed 1.5 V. Some convenient current, say 6 amperes (as indicated by *A*) is passed through the circuit by adjusting the rheostat *RH*.

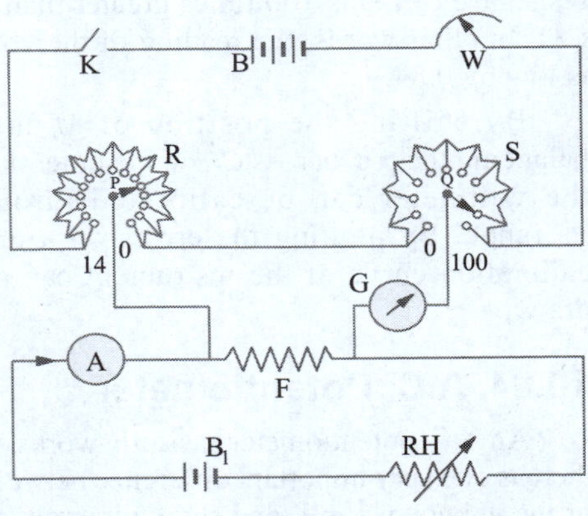

Fig. 10.77

The potential drop across *F* is applied between *P* and *Q* as shown. Next the sliding contacts *P* and *Q* are adjusted for zero deflection on *G*. Suppose *P* reads 5 and *Q* reads 86.7. Then it means that p.d. across *F* is 0.5867 V and since *F* is of 0.1 Ω, hence ***true*** value of current through *F* is 0.5867/0.1 = 5.867 amperes. Hence, the ammeter reads high by (6 – 5.867) = 0.133 A. The test is repeated for various values of current over the entire ranges of the ammeter.

10.63. Calibration of Voltmeters

As pointed out in Art. 10.62, a voltage higher than 1.5 cannot be measured by the potentiometer directly, the limit being set by the standard cell and the type of the potentiometer (since it has only 14 resistances on *R* as in Fig. 10.78). However, with the help of a volt-box which is nothing else but a voltage reducer, measurements of voltage up to 150 V or 300 V can be made, the upper limit of voltage depending on the design of the volt-box.

The diagram of connections for calibration of voltmeters is shown in Fig. 10.78. By calibration is meant the determination of the extent of error in the reading of the voltmeter throughout its range. A high value resistor *AB* is connected across the supply terminals of high voltage battery B_1 so that it acts as a voltage divider. The volt-box consists of a high resistance *CD* with tapings at accurately determined points like *E* and *F* etc. The resistance *CD* is usually 15,000 to 300 Ω. The two tappings *E* and *F* are such that the resistances of portions *CE* and *CF* are 1/100th and 1/10th the resistance *CD*. Obviously, whatever the potential drop across *CD*, the corresponding potential drop across *CE* is 1/100th and that across *CF*, 1/10th of that across *CD*.

If supply voltage is 150 V, then p.d. across AB is also 150 V and if M coincides with B, then p.d. across CD is also 150 V, so across CF is 15 volts and across CE is 1.5 V. Then p.d. across CE can be balanced over the potentiometer as shown in Fig. 10.78. Various voltages can be applied across the voltmeter by moving the contact point M on the resistance AB.

Suppose that M is so placed by voltmeter V reads 70 V and p.d. across CE is balanced by adjusting P and Q. If the readings on P and Q to give balance are 7 and 8.4 respectively, then p.d. across CE is 0.7084 V.

Hence, the true p.d. across AM or CD or voltmeter is 0.7048 × 100 = 70.84 V (because resistance of CD is 100 times greater than that of CE). In other words, the reading of the voltmeter is low by 0.84 V.

By shifting the position of M and then balancing the p.d. across CE on the potentiometer, the voltmeter can be calibrated throughout its range. By plotting the errors on a graph, a calibration curve of the instrument can also be drawn.

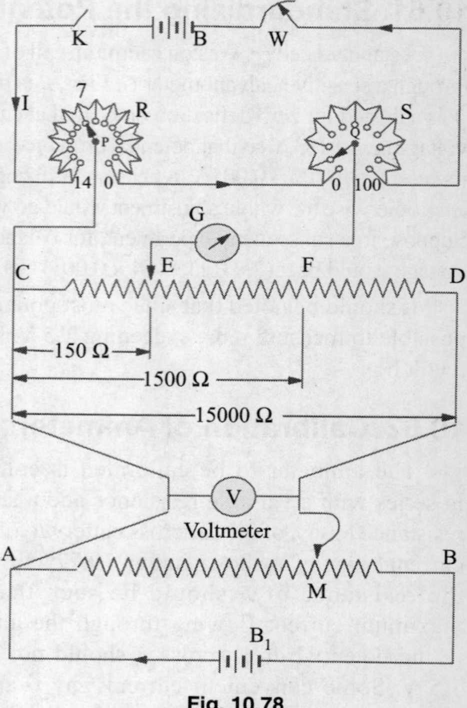

Fig. 10.78

10.64. A.C. Potentiometer

An A.C. potentiometer basically works on the same principle as a d.c. potentiometer. However, there is one very important difference between the two. In d.c. potentiometer, only the ***magnitudes*** of the unknown e.m.f. and slide-wire voltage drop are made equal for obtaining balance. But in an a.c. potentiometer, not only the magnitudes but *phases* as well have to be equal for obtaining balance. Moreover, to avoid frequency and waveform errors, the a.c. supply for slide-wire must be taken from same source as the voltage or current to be measured.

A.C. potentiometers are of two general types differing in the manner in which the value of the unknown voltage is presented by the instrument dials or scales. The two types are :

(i) Polar potentiometers in which the unknown voltage is mesured in polar form *i.e.* in terms of magnitude and relative phase.

(ii) Co-ordinate potentiometers which measure the rectangular co-ordinates of the voltage under test.

The two products are illustrated in Fig. 10.80. In Fig. 10.79 (*a*), vector OQ denotes the test voltage whose magnitude and phase are to be imitated. In polar potentiometer, the length r of the vector OP can be varied with the help of a sliding contact on the slidewire while its phase ϕ is varied independently with the help of a phase-shifter. Drysdale potentiometer is of this type.

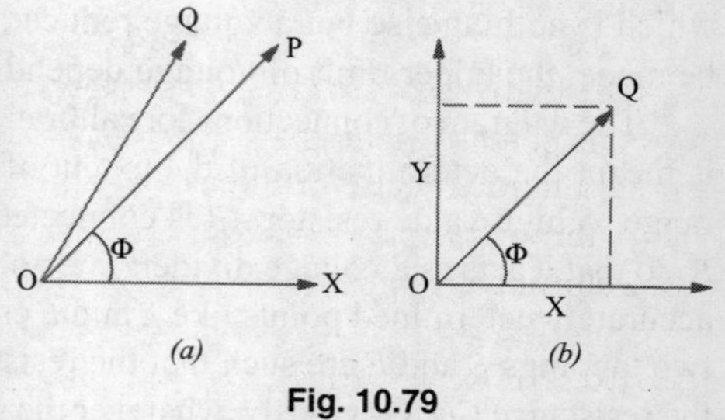

Fig. 10.79

In co-ordinate type potentiometers, the unknown voltage vector OQ is copied by the adjustment of 'in phase' and 'quadrature' components X and Y. Their values are read from two scales of the potentiometer. The magnitude of the required vector is = $\sqrt{X^2 + Y^2}$ and its phase is given by $\phi = \tan^{-1}(X/Y)$. Examples of this type are **(i)** Gall potentiometer and **(ii)** Campbell-Larsen potentiometer.

10.65. Drysdale Polar Potentiometer

As shown in Fig. 10.80 for a.c. measurements, the slide-wire *MN* is supplied from a phase shifting circuit so arranged that magnitude of the voltage supplied by it remains constant while its

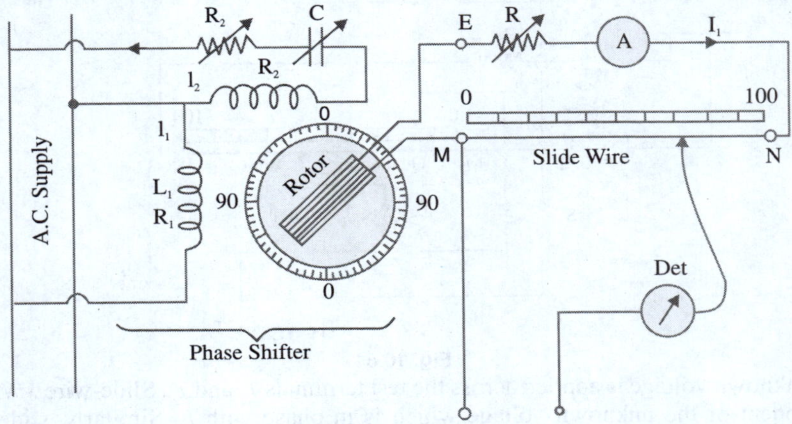

Fig. 10.80

phase can be varied through 360°. Consequently, slide-wire current *I* can be maintained constant in magnitude but varied in phase. The phase-shifting circuit consists of :

(i) Two stator coils supplied from the same source in parallel. Their currents I_1 and I_2 are made to differ by 90° by using well-known phase-splitting technique.

(ii) The two windings produce a rotating flux which induces a secondary e.m.f. in the rotor winding which is of constant magnitude but the phase of which can be varied by rotating the rotor in any position either manually or otherwise. The phase of the rotor e.m.f. is read from the circular graduated dial provided for the purpose.

The ammeter *A* in the slide-wire circuit is of electrodynamic of thermal type. Before using it for a.c. measurements, the potentiometer is first calibrated by using d.c. supply for slidwire and a standard cell for test terminals T_1 and T_2.

The unknown alternating voltage to be measured is applied across test terminals I_2 and T_2 balance is effected by the alternate adjustment of the slide-wire contact and the position of phase-shifting rotor. The slide-wire reading represents the magnitude of the test voltage phase-shifter reading gives its phase with reference to an arbitrary reference vector.

10.66. Gall Co-Ordinate Potentiometer

This potentiometer uses two slide-wires *CD* and *MN* with their currents I_1 and I_2 (Fig. 10.81) having a mutual phase difference of 90°. The two currents are obtained from the single phase supply through isolating transformers, the circuit for 'quadrature' slidewire *MN* incorporating a phase shifting arrangement.

Before use, then current I_1 is first standardised as described for Drysdale potentiometer (Art. 10.66). Next, current I_2 is standardised with the help of the mutually induced e.m.f. *E* in inductometer secondary. This e.m.f. $E = \omega MI_1$ and is in quadrature phase with I_1. Now, *E* is balanced against the voltage drop on slide-wire *MN*. This balance will be obtained only when I_2

is of correct magnitude and is in exact quadrature with I_1. Balance is achieved with the help of the phase-shifter and rheostat R_2.

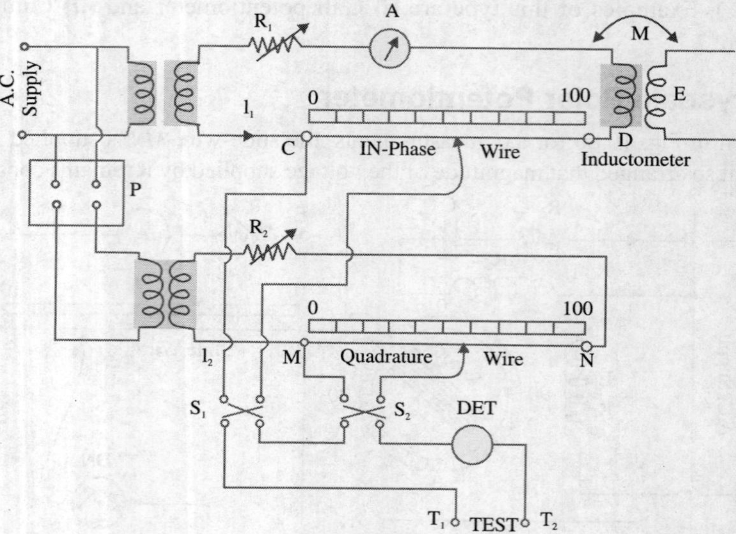

Fig. 10.81

The unknown voltage is applied across the test terminals T_1 and T_2. Slide-wire *MN* measures that component of the unknown voltage which is in phase with I_2. Similarly, slide-wire *CD* measures that component of the unknown voltage which is in phase with I_1. Since I_1 and I_2 are in quadrature, the two measured values are quadrature components of the unknown voltage. If V_1 and V_2 are these values, then

$$V = \sqrt{V_1^2 + V_2^2} \quad \text{and} \quad \phi = \tan^{-1}(V_2/V_1) \qquad \text{—with respect to } I_1$$

Reversing switches S_1 and S_2 are used for measuring both positive and negative in-phase and quadrature components of the unknown-voltage.

10.67. Instrument Transformers

The d.c. circuits when large currents are to be measured, it is usual to use low-range ammeters with suitable shunts. For measuring high voltages, low-range voltmeters are used with high resistances connected in series with them. But it is neither convenient nor practical to use this method with alternating current and voltage instruments. For this purpose, specially constructed accurate-ratio instrument transformers are employed in conjunction with standard low-range a.c. instruments. Their purpose is to reduce the line current or supply voltage to a value small enough to be easily measured with meters of moderates size and capacity. In other words, they are used for extending the range of a.c. ammeters and voltmeters. Instruments transformers are of two types:

(i) current transformers (*CT*) —for measuring large alternating currents.

(ii) potential transformers (*VT*) —for measuring high alternating voltages.

Advantages of using instrument transformers for range extension of a.c. meters are as follows:

(1) the instrument is insulated from the line voltage, hence it can be grounded. **(2)** the cost of the instrument (or meter) together with the instrument transformer is less than that of the instrument alone if it were to be insulated for high voltages. **(3)** it is possible to achieve standardisation of instruments and meters at secondary ratings of 100–120 volts and 5 or 1 amperes **(4)** if necessary, several instruments can be operated from a single transformer and 5 power consumed in the measuring circuits is low.

In using instrument transformers for current (or voltage) measurements, we must know the ratio of primary current (or voltage) to the secondary current (or voltage). These ratios give us the multiplying factor for finding the primary values from the instrument readings on the secondary side.

However, for energy or power measurements, it is essential to know not only the transformation ratio but also the phase angle between the primary and secondary currents (or voltages) because it necessitates further correction to the meter reading.

For range extension on a.c. circuits, instrument transformers are more desirable than shunts (for current) and multipliers (for voltage measurements) for the following reasons :

1. time constant of the shunt must closely match the time constant of the instrument. Hence, a different shunt is needed for each instrument.

2. range extension is limited by the current-carrying capacity of the shunt *i.e.* upto a few hundred amperes at the most.

3. if current is at high voltage, instrument insulation becomes a very difficult problem.

4. use of multipliers above 1000 becomes almost impracticable.

5. insulation of multipliers against leakage current and reduction of their distributed capacitance becomes not only more difficult but expensive above a few thousand volts.

10.68. Ratio and Phase-angle Errors

For satisfactory and accurate performance, it is necessary that the ratio of transformation of the instrument transformer should be constant within close limits. However, in practice, it is found that neither current transformation ratio I_1/I_2 (in the case of current transformers) nor voltage transformation ratio V_1/V_2 (in the case of potential transformers) remains constant. The transformation ratio is found to depend on the exciting current as well as the current and the power factor of the secondary circuit. This fact leads to an error called ratio error of the transformer which depends on the working component of primary.

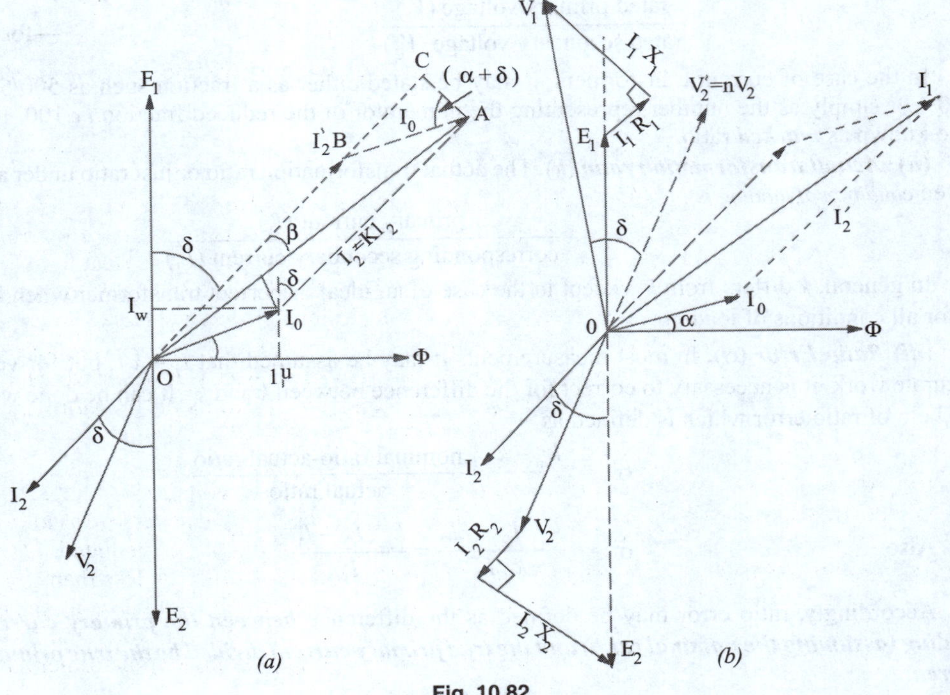

Fig. 10.82

It is seen from Fig. 10.82 (*a*) that the phase angle between the primary and secondary currents is not exactly 180° but slightly less than this value. This difference angle β may be found by reversing vector I_2. The angular displacement between I_1 and I_2 reversed is called the phase angle error of the current transformer. This angle is reckoned positive if the reversed secondary current ***leads*** the primary current. However, on very low power factors, the phase angle may be negative. Similarly, there is an angle of γ between the primary voltage V_1 and secondary voltage reversed-this angle represents the phase angle error of a voltage transformer. In either case, the phase error depends on the magnetising component $I\mu$ of the primary current. It may be noted that ratio error is primarily due to the reason that the ***terminal*** voltage transformation ratio of a transformer is not exactly equal to its turn ratio. The divergence between the two depends on the resistance and reactance of the transformer windings as well as upon the value of the exciting current of the transformer. Accuracy of voltage ratio is of utmost importance in a voltage transformer although phase angle error does not matter if it is to be merely connected *to **a voltmeter.*** Phase-angle error becomes important only when voltage transformer supplies the voltage coil of a wattmeter *i.e.* in power measurement. In that case, phase angle error causes the wattmeter to indicate on a wrong power factor.

In the case of current transformers, constancy of current ratio is of paramount importance. Again, phase angle error is of no significance if the current transformer is merely feeding an ammeter but it assumes importance when feeding the current coil of a wattmeter. While discussing errors, it is worthwhile to define the following terms :

(i) Nominal transformation ratio (k_n). It is the ratio of the ***rated*** primary to the ***rated*** secondary current (or voltage).

$$k_n = \frac{\text{rated primary current } (I_1)}{\text{rated secondary current } (I_2)} \qquad \text{—for } CT$$

$$= \frac{\text{rated primary voltage } (V_1)}{\text{rated secondary voltage } (V_2)} \qquad \text{—for } VT$$

In the case of current transformers, it may be stated either as a fraction such as 500/5 or 100/1 or simply as the number representing the numerator of the reduced fraction *i.e.* 100. It is also known as ***marked*** ratio.

(ii) Actual transformation ratio (k). The actual transformation ratio or just ratio under any given *condition of loading is*

$$k = \frac{\text{primary current } (I_1)}{\text{corresponding secondary current } (I_2)}$$

In general, k differs from k_n except in the case of an ideal or perfect transformer when $k = k_n$ for all conditions of loading.

(iii) Ratio Error (σ). In most measurements it may be assumed that $I_1 = k_n I_2$ but for very accurate work, it is necessary to correct for the difference between k and k_n. It can be done with the help of ratio error which is defined as

$$\sigma = \frac{k_n - k}{k} = \frac{\text{nominal ratio-actual ratio}}{\text{actual ratio}}$$

Also,
$$\sigma = \frac{k_n \,.\, I_2 - kI_2}{k \,.\, I_2} = \frac{k_n \,.\, I_2 - I_1}{I_1}$$

Accordingly, ratio error may be defined as the difference ***between the primary current reading (assuming the nominal ratio) and the true primary current divided by the true primary current.***

(iv) Ratio Correction Factor (R.C.F.). It is given by

$$R.C.F. = \frac{\text{actual ratio}}{\text{nominal ratio}} = \frac{k}{k_n}$$

10.69. Current Transformer

A current transformer takes the place of shunt in d.c. measurements and enables heavy alternating current to be measured with the help of a standard 5-A range a.c. ammeter.

As shown in Fig. 10.83, the current - or series-transformer has a primary winding of one or more turns of thick wire connected in series with the line carrying the current to be measured. The secondary consist of a large number of turns of fine wire and feeds a standard 5-A ammeter (Fig. 10.83) or the current coil of a watt-meter or watthour-meter (Fig. 10.84).

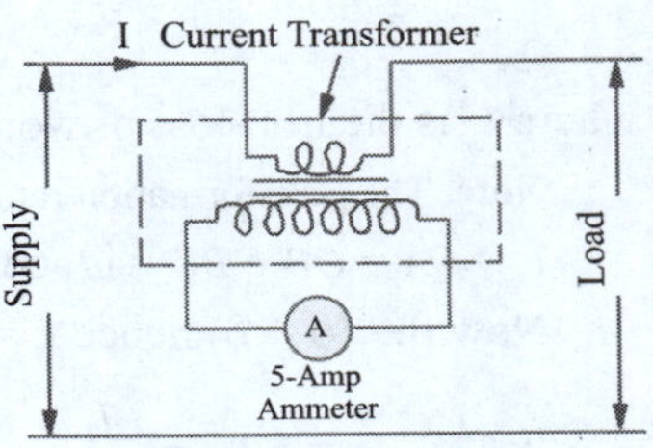

Fig. 10.83

For example, a 1,000/5A current transformer with in single-turn primary will have 200 secondary turns. Obviously, it steps down the current in the 200 : 1 ratio whereas it steps up the voltage drop across the single-turn primary (an extremely small quantity) in the ratio 1 : 200. Hence if we know the current ratio of the transformer and the reading of the a.c. ammeter, the line current can be calculated.

It is worth noting that ammeter resistance being extremely low, a current transformer operates with its secondary under nearly short-circuit conditions. Should it be necessasry to remove the ammeter of the current coils of the wattmeter or a relay, the seondary winding must, first of all, be short-circuited *before* the instrument is disconnected.

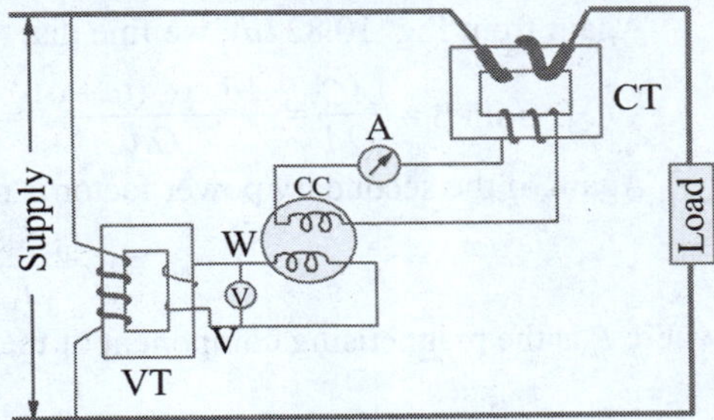

Fig. 10.84

If it is not done then due to the absence of counter ampere-turns of the secondary, the unopposed primary m.m.f. will set up an abnormally high flux in the core which will produce excessive core loss with subsequent heating of and damage of the transformer insulation and a high voltage across the secondary terminals. This is not the case with the ordinary constant-potential transformers because their primary current is determined by the load on their secondary whereas in a current transformer, primary current is determined entirely by the load on the system and not by the load on its own secondary. Hence, the secondary of a current transformer should ***never be left open under any circumstances.***

10.70. Theory of Current Transformer

Fig. 10.82 (*b*) represents the general phase diagram for a current transformer. Current I_0 has been exaggerated for clarity.

(a) Ratio Error. For obtaining an expression for the ratio error, it will be assumed that the turn ratio *n* (= secondary turns, N_2/primary turns N_1) is made equal to the nominal current ratio *i.e.* $n = k_n$. In other words, it will be assumed that $I_1/I_2 = n$ although actually $n = I_1/I_2'$. As seen from Art. 10.63.

$$\sigma = \frac{n\,I_2 - I_1}{I_1} = \frac{I_2' - I_1}{I_1} = \frac{OB - OA}{OA} \qquad \text{— } [\because n = k_n]$$

$$\equiv \frac{OB - OC}{OA} \qquad (\because \beta \text{ is very small angle})$$

$$= -\frac{BC}{OA} = -\frac{AB\sin(\alpha+\delta)}{OA} = -\frac{I_0\sin(\alpha+\delta)}{I_1} = -\frac{I_0\sin(\alpha+\delta)}{nI_2}$$

For most instrument transformers, the power factor of the secondary burden is nearly unity so that δ is very small. Hence, very approximately.

$$\sigma = \frac{I_0\sin\alpha}{I_1} = \frac{I_\omega}{I_1}$$

where I_ω is the iron-loss or working or wattful component of the exciting current I_0

Note. The transformation ratio R may be found from Fig. 10.82 (a) as under :

$I_1 = OA = OB + BC = nI_2\cos\beta + I_0\cos[90-(\delta+\beta+\alpha)] = nI_2\cos\beta + I_0\sin(\delta+\beta+\alpha)$

Now $\beta = (\alpha+\delta)$ hence $I_1 = nI_2 + I_0\sin(\alpha+\delta)$ where n is the turn ratio of the transformer.

$$\therefore \quad \text{ratio } R = \frac{I_1}{I_2} = \frac{nI_2 + I_0\sin(\alpha+\delta)}{I_2} \text{ or } R = n + \frac{I_0\sin(\alpha+\delta)}{I_2} \quad \text{...}(i)$$

If δ is negligible small, then $R = n + \frac{I_0\sin\alpha}{I_2} = n + \frac{I_\omega}{I_2}$

It is obvious from **(*i*)** above that ratio error can be eliminated if secondary turn are reduced by a number

$$= I_0\sin(\alpha+\delta)/I_2$$

(*b*) Phase angle (β)

Again from Fig. 10.82 (a), we fine that

$$\beta \cong \sin\beta = \frac{AC}{OA} = \frac{AB\cos(\alpha+\delta)}{OA} = \frac{I_0\cos(\alpha+\delta)}{I_1} = \frac{I_0\cos(\alpha+\delta)}{nI_2}$$

Again, if the secondary power factor is nearly unity, then δ is very small, hence

$$\beta \cong \frac{I_0\cos\alpha}{I_1} = \frac{I_\mu}{I_1} \text{ or } \frac{I_\mu}{nI_2}$$

where I_μ is the magnetising component of the exciting current I_0.

$$\therefore \quad \beta = \frac{I_\mu}{I_1} \text{ —in radian; } = \frac{180}{\pi}\times\frac{I_\mu}{I_1} \text{ —in degrees}$$

Note. As found above, $\beta = \frac{I_0\cos(\alpha+\delta)}{I_1} = \frac{I_0(\cos\alpha\cos\delta - \sin\alpha\sin\delta)}{I_1}$

$$= \frac{I_\mu\cos\delta - I_\omega\sin\delta}{I_1} = \frac{I_\mu\cos\delta - I_\omega\sin\delta}{nI_2} \text{ radian}$$

$$\therefore \quad \beta = \frac{180}{\pi}\times\frac{I_\mu\cos\delta - I_\omega\sin\delta}{nI_2} \text{ degrees.}$$

Dependence of ratio error on working component of I_0 and that of phase angle on the magnetising component is obvious. If R is to come closer to k and β is to become negligible small, then I_μ and I_ω and hence I_0 should be very small.

10.71. Clip-on Type Current Transformer

It has a laminated core which is so arranged that it can be opened out at a hinged section by merely pressing a trigger-like projection (Fig. 10.85). When the core is thus opened, it permits the admission of very heavy current-carrying bus-bars or feeders whereupon the trigger is released and the core is tightly closed by a spring. The current-carrying conductor of feeder acts as a single-turn primary whereas the secondary is connected across the standard ammeter conveniently mounted in the handle itself.

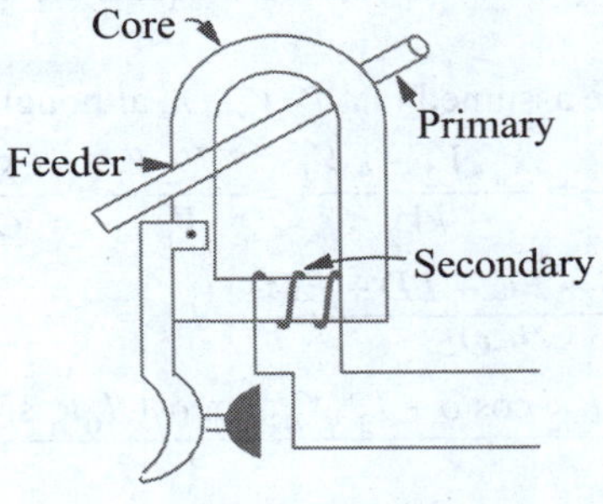

Fig. 10.85

10.72. Potential Transformers

These transformers are extremely accurate-ratio stepdown transformers and are used in conjunction with standard low-range voltmeters (100-120 V) whose deflection when divided by transformation ratio, gives the true voltage on the primary or high-voltage side. In general, they are of the shell type and do not differ much from the ordinary two-winding transformers except that their power rating is extremely small. Sine their secondary windings are required to operate instruments or relays or pilot lights, their ratings are usually of 40 to 100W. For safety, the secondary is completely insulated from the high voltage primary and is, in addition, grounded for affording protection to the operator. Fig. 10.86 shows the connection of such a transformer.

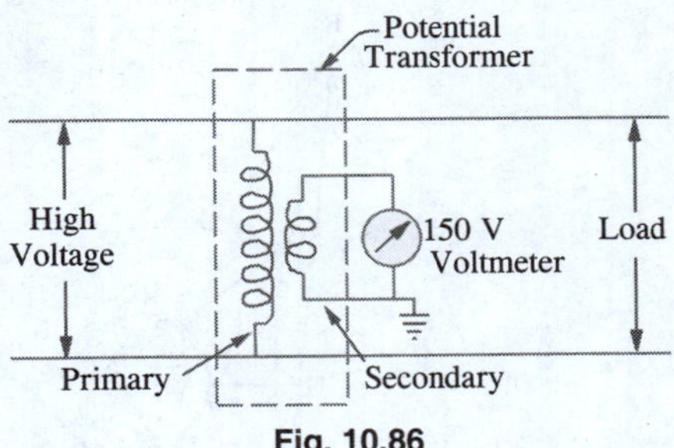

Fig. 10.86

10.73. Ratio and Phase-Angle Errors

In the case of a potential transformer, we are interested in the ratio of the primary to the secondary terminal voltage and in the phase angle γ between the primary and reversed secondary terminal voltage V_2'.

The general theory of voltage transformer is the same as for the power transformers except that, as the current in the secondary burden is very small, the total primary current I_1 is not much greater than I_0.

In the phasor diagram of Fig. 10.87, vectors *AB, BC, CD* and *DE* represent small voltage drops due to resistances and reactances of the transformer winding (they have been exaggerated for the sake of clarity). Since the drops as well as the phase angle γ are small, the top portion of diagram 10.87 (*a*) can be drawn with negligible loss of accuracy as in Fig. 10.87 (*b*) where V_2' vector has been drawn parallel to the vector for V_1.

In these diagrams, V_2' is the secondary terminal voltage as referred to primary assuming transformation without voltage drops. All actual voltage drops have been referred to the primary. Vector *AB* represents total resistive drop as referred to primary *i.e.* $I_2' R_{01}$. Similarly, *BC* represents total reactive drop as referred to primary *i.e.* $I_2' X_{01}$.

In a voltage transformer, the relatively large no-load current produces appreciable resistive drops which have been represented by vectors *CD* and *DE* respectively. Their values are I_0R_1 and I_0X_1 respectively.

(a) Ratio Error

In the following theory, *n* would be taken to represent the ratio of ***primary turns to secondary turns***. Further, it would be assumed, as before, that *n* equals the nominal transformation ratio *i.e.* $n = k_n$.

In other words, it would be assumed that $V_1/V_2 = n$, although, actually, $V_1/V_2' = n$.

Then $$\sigma = \frac{k_n - k}{k} = \frac{k_n . V_2 - k\,V_2}{kV_2} = \frac{V_2' - V_1}{V_1} = -\frac{EN}{OE} \quad \text{...Fig. 10.88 (a)}$$

$$= -\frac{AG + FC + LD + EM}{OE} \quad \text{...Fig. 10.88 (b)}$$

$$= -\frac{I_2'\,R_{02}\,\cos\delta + I_2'\,X_{02}\,\sin\delta + I_0 R_1\,\sin\alpha + I_0 X_1\,\cos\alpha}{V_1}$$

$$= -\frac{I_2'\,R_{02}\,\cos\delta + I_2'\,X_{02}\,\sin\delta + I_\mu R_1 + I_\mu X_1}{V_1}$$

where I_ω and I_μ are the iron-loss and magnetising components of the no-load primary current I_0.

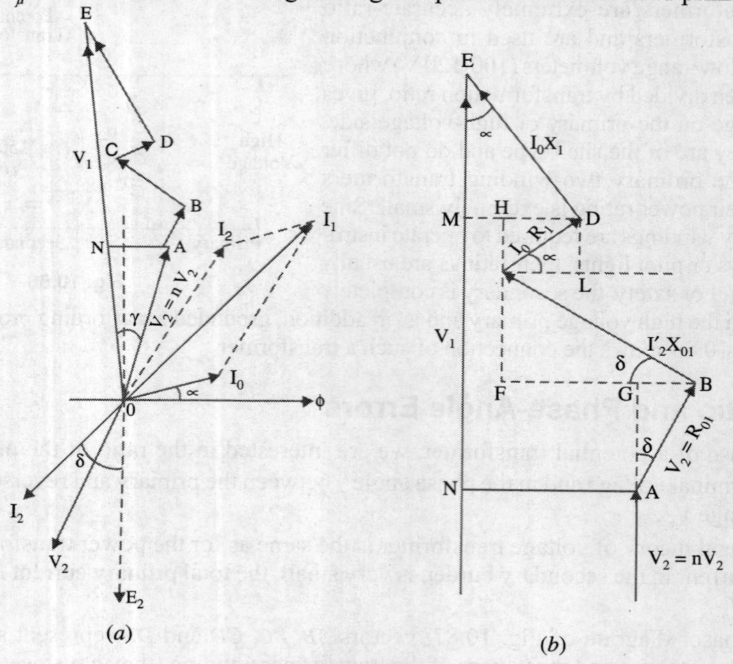

Fig. 10.87

(b) Phase Angle (γ)

To a very close approximation, value of γ is given by $\gamma = AN/OA$ —in radian

Now, $OA \cong OE$ provided ratio error is neglected. In that case,

$$\gamma = \frac{AN}{OE} = -\frac{GF + HM}{OE} \quad \text{...Fig. 10.90 (b)}$$

$$= -\frac{(BE - BG) + (DM - DH)}{OE}$$

$$= -\frac{I_2'\,X_{01}\,\cos\delta - I_2'\,R_{01}\,\sin\delta + I_0 X_1 \sin\alpha - I_0 R_1\,\cos\alpha}{V_1}$$

$$= -\frac{I_2' X_{01} \cos\delta - I_2' R_{01} \sin\delta + I_\omega X_1 - I_\mu R_1}{V_1}$$

The negative sign has been given because reversed secondary voltage *i.e.* V_2' lags behind V_1.

Example 10.37. *A relay current-transformer has a bar primary and 200 secondary turns. The secondary burden is an ammeter of resistance 1.2 Ω and resistance of 0.5 Ω and the secondary winding has a resistance of 0.2 Ω and reactance of 0.3 Ω. The core requires the equivalent of 100 AT for magnetisation and 50 AT for core losses.*

(*i*) *Find the primary current and the ratio error when the secondary ammeter indicates 5.0 A.*

(*ii*) *By how many turns should the secondary winding be reduced to eliminate the ratio error for this condition?* **(Electrical Measurements, Bombay Univ.)**

Solution. Total secondary impedance is

$$\mathbf{Z}_2 = 1.4 + j\,0.8 = 1.612 \angle 29°45' \qquad \therefore\ \delta = 29°45'$$

$$\mathbf{I0} = 100 + j50 = 111.8 \angle 26°34' \qquad \therefore\ \alpha = 26°34'$$

Turn ratio, $n = 200/I = 200$; Transformation ratio $R = n + \dfrac{I_0 \sin(\alpha+\delta)}{I_2}$

$$\therefore \quad R = 200 + \frac{111.8 \sin 56°19'}{5} = 218.6$$

(*i*) Primary current $= 5 \times 218.6 = \mathbf{1093\ A}$

Ratio error $\sigma = -\dfrac{I_0 \sin(\alpha+\delta)}{nI_2} = \dfrac{111.8 \times 0.8321}{200 \times 5} = \mathbf{-0.093\ or\ -9.3\%}$

(*ii*) No. of secondary turns to be reduced $= I_0 \sin(\alpha+\delta)/I_2 = 93/5 =$ **19 (approx).**

Example 10.38. *A current transformer has 3 primary turns and 300 secondary turns. The total impedance of the secondary is (0.583 + j 0.25) ohm. The secondary current is 5 A. The ampere-turns required to supply excitation and iron losses are respectively 10 and 5 per volt induced in the secondary.*

Determine the primary current and phase angle of the transformer.

(Elect Meas; M.S. Univ. Baroda)

Solution. $\mathbf{Z}_2 = 0.583 = j\,0.25 = 0.6343 \angle 23°\,10'$ $\therefore E_2 = I_2Z_2 = 5 \times 0.6343 = \mathbf{3.17\ V}$

Now, there are 10 magnetising *AT* per *secondary* volt induced in secondary.

$\therefore$ total magnetising $AT = 3.17 \times 10 = 31.7$; Similarly, iron-loss $AT = 3.17 \times 5 = 15.85$

Remembering that there are 3 primary turns, the magnetising and iron-loss components of primary current are as under :

Magnetising current, $I_\mu = 31.7/3 = 10.6$ A; iron-loss current $I_\omega = 15.85/3 = 5.28$ A

$$I_0 = \sqrt{10.6^2 + 5.28^2} = 11.84\ \text{A}$$

Now, $$R = n + \frac{I_0 \sin(\alpha+\delta)}{I_2}$$

Here, $n = 300/3 = 100$; $\alpha = \tan^{-1}(I_\omega/I_\mu) = \tan^{-1}(5.28/10.6)$

$= \tan^{-1}(0.498) = 26°30'$

$\delta =$ secondary load angle $= 23°10'$ —found earlier

$$\therefore \quad R = 100 + \frac{11.86}{5}(\sin 49°40') = 100 + 1.81 = 101.81$$

$$I_1 = R \times I_2 = 101.81 \times 5 = \mathbf{509.05\ A}$$

$$\beta = \frac{180}{\pi} \times \frac{I_\mu \cos\delta - I_\omega \sin\delta}{nI_2}$$

$$= \frac{180}{\pi} \times \frac{10.6 \cos 23°10' - 5.28 \sin 23°10'}{100 \times 5} = \mathbf{0.88°}$$

Example 10.39. *A current transformer with a bar primary has 300 turns in its secondary winding. The resistance and reactance of the secondary circuit are 1.5 Ω and 1.0 Ω respectively including the transformer winding. With 5 A flowing in the secondary circuit the magnetising ampere-turns required are 100 and iron loss is 1.2 W. Determine the ratio error at this condition.*

(Elect. Measure, A.M.I.E. Sec. B, 1992)

Solution. Turn ratio $n = 300/1 = 300$

Secondary impedance is $Z_2 = 1.5 + j\,1.0 = 1.8 \angle 33°42'$

Secondary induced e.m.f. $E_2 = I_2 Z_2 = 5 \times 1.8 = 9$ V

$E_1 = E_2/n = 9/300 = 0.03$ V

Let us now find the magnetising and working components of primary no-load current I_0

Magnetising $AT = 100$. Since threre is one primary turn, ∴ $I_\mu = 100/1 = 100$ A

Now, $E_1 I_\omega = 1.2 \quad \therefore \quad I_\omega = 1.2/0.03 = 40$ A

$$I_0 = 100 + j\,40 = 107.7 \angle 21°48'; \; \sigma = -\frac{I_0 \sin(\alpha + \delta)}{nI_2}$$

Now $\alpha = 21°48'$ and $\delta = 33°\,42'$

$$\therefore \qquad \sigma = -\frac{107.7 \sin 55°30'}{300 \times 5} = \mathbf{-0.0592 \text{ or } -5.92\%}$$

Phase angle
$$\beta = \frac{I_0 \cos(\alpha + \delta)}{nI_2} = \frac{107.7 \times \cos 55°30'}{1500}$$

$$= \frac{180}{\pi} \times \frac{107.7 \times 0.5664}{1500} = \mathbf{2°20'}$$

Tutorial Problem 10.3

1. A current transformer with 5 primary turns has a secondary burden consisting of a resistance of 0.16 Ω and an inductive reactance of 0.12Ω. When primary current is 200 A, the magnetising current is 1.5 A and the iron-loss component is 0.4 A. Determine the number of secondary turns needed to make the current ratio 100/1 and also the phase angle under these conditions. **[407 : 0.275°]**

2. A current transformer having a 1-turn primary is rated at 500/5 A, 50 Hz, with an output of 1.5 VA. At rated load with the non-inductive burden, the in-phase and quadrature components (referred to the flux) of the exciting ampere-turns are 8 and 10 respectively. The number of turns in the secondary is 98 and and the resistance and leakage reactance of the secondary winding are 0.35 Ω and 0.3 Ω respectively. Calculate the current ratio and the phase angle error. **[501.95/5; 0.533°]**

   (Elect. Inst. and Meas, M.S. Univ. Baroda)

3. A potential transformer, ratio 1000/100-V, has the following constants :

   primary resistance = 94.5 Ω ; secondary resistance = 0.86 Ω
   primary reactance = 66.2 Ω ; equivalent reactance = 66.2 Ω
   magnetising current = 0.02 A at 0.4 p.f.

   Calculate (*i*) the phase angle at no-load between primary and secondary voltages (*ii*) the load in VA at u.p.f. at which the phase angle would be zero. **[(i) 0°4′ (ii) 18.1 VA]**

4. PMMC instrument has FSD current of 50 milliampere and 2 ohm resistance. How the instrument can be converted to

   (*i*) 0.5 A range Ammeter (*ii*) 0.100 V range Voltmeter? ***(Nagpur University, Summer 2002)***

5. What are the essential torques of an indicating instruments? Justify their necessity.
(Nagpur University, Winter 2002)

6. Discuss the necessity of damping in Indicating instrument and explain eddy current damping.
(U.P. Technical University 2002) (Nagpur University, Summer 2003)

7. PMMC instrument has FSD current of 50 milliampere and 2 ohm resistance. How the instrument can be converted to (*i*) 0-5 A range Ammeter (*ii*) 0-100 V range Voltmeter?
(Nagpur University, Summer 2003)

8. What are the essential requirements of indicating type instruments? Explain each of them.
(U.P. TechnicalUniversity 2002) (Nagpur University, Winter 2003)

9. What are the different operating systems required in an instrument? Explain damping system in detail. ***(U.P. TechnicalUniversity 2002) (Nagpur University, Summer 2004)***

10. If a shunt for a moving coil instrument is a have a multiplicationfactor m, shown that its shunt resistance is given by

$$\mathbf{R}_{sh} = \frac{R_m}{m-1}$$

where $R_m \rightarrow$ resistance of the meter. ***(Nagpur University, Summer 2004)***

11. Find the value of a series resistance to be connected to a basic d'Arsonval movement with internal resistance R_m = 100 ohm and full scale deflection current is 1 mA, for conversion into 0–500 volt.
(Nagpur University, Summer 2004)

12. What is the basic principle of Induction type instruments? ***(Anna University, April 2002)***

13. What is damping torque in instruments? ***(Anna University, April 2002)***

14. Why does MI instruments have non-linear scale? ***(Anna University, April 2002)***

15. Explain how can you measure power by the Ammeters. ***(Anna University, April 2002)***

16. What is creeping in energy meter? ***(Anna University, April 2002)***

17. What is the advantage of Induction type meters? ***(Anna University, April 2002)***

18. What is a trivector meter? ***(Anna University, April 2002)***

19. What is the difference between PMMC instrument and ballistic galvanometer?
(Anna University, April 2002)

20. What is the basic principle of operation of reed type frequency meters?
(Anna University, April 2002)

21. Why can't you measure low resistance with meggar? ***(Anna University, April 2002)***

22. What are the advantages of electronic meters? ***(Anna University, April 2002)***

23. Explain the principle of operation, construction and the expression for deflection of a moving iron instrument. ***(Anna University, April 2002)***

24. Draw and explain the working of 2 element wattmeter. Explain how does the 2 element wattmeter reads total power in a 3 ϕ circuit. ***(Anna University, April 2002)***

25. Write the equations for wattmeter readings W_1 and W_2 in 3 phase power measurement and therefrom for power factor. ***(Anna University, Wimter 2002)***

26. Explain how power can be measured in a three phase circuit with the help of two wattmeters, for a balanced star connected load. Draw the phasor diagram. ***(Anna University Winter 2002)***

27. State the working principle of a dynamometer type wattmeter and show its connections.
(V.T.U., Belgaum Karnataka University, February 2002)

28. A.D.C. milliammeter having a resistance of 2Ω gives full scale deflection when the current is 50mA. How can it be used to measure (*i*) a current of 5A (*ii*) a voltage of 500 V.
(V.T.U., Belgaum Karnataka University, February 2002)

29. With the help of a neat diagram, explain the construction and principle of operation of single phase energy meter. ***(V.T.U., Belgaum Karnataka University, Wimter 2003)***
30. With a neat sketch explain the construction and working of a single phase induction type energy-meter. ***(V.T.U., Belgaum Karnataka University, Summer 2003)***

OBJECTIVE TESTS—10

1. The kWh meter can be classified as a/an-instrument :
(*a*) deflecting (*b*) digital
(*c*) recording (*d*) indicating
2. The moving system of an indicating type of electrical instrument is subjected to :
(*a*) a deflecting torque
(*b*) a controlling torque
(*c*) a damping torque
(*d*) all of the above
3. The damping force acts on the moving system of an indicating instrument only when it is :
(*a*) moving
(*b*) stationary
(*c*) near its full deflection
(*d*) just starting to move.
4. The most efficient form of damping employed in electrical instruments is :
(*a*) air friction
(*b*) fluid friction
(*c*) eddy currents
(*d*) none of the above.
5. Moving iron instruments can be used for measuring :
(*a*) direct currents and voltages
(*b*) alternating current and voltages
(*c*) radio frequency currents
(*d*) both (*a*) and (*b*).
6. Permanent-magnet moving-coil ammetere have uniform scales because :
(*a*) of eddy current damping
(*b*) they are spring-controlled
(*c*) their deflecting torque varies directly as current
(*d*) both (*b*) and (*c*).
7. The meter that is suitable for *only* direct current measurements is :
(*a*) moving-iron type
(*b*) permanent-magnet type
(*c*) electrodynamic type
(*d*) hot-wire type.
8. A moving coil voltmeters measures—
(*a*) only a.c. voltages
(*b*) only d.c. voltages
(*c*) both a.c. and d.c. voltages
(Principles of Elect. Engg. Delhi Univ.)
9. The reading of the voltmeter in Fig. 10.88 would be nearest to–volt :

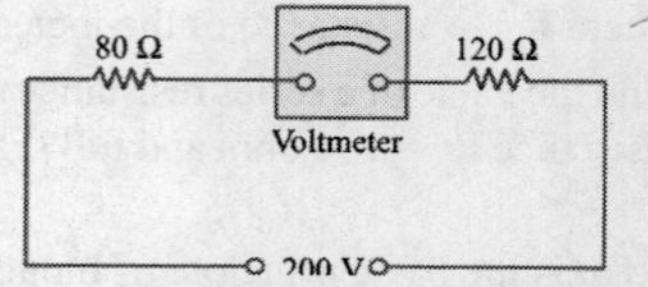

Fig. 10.88

(*a*) 80 (*b*) 120
(*c*) 200 (*d*) 0
10. The hot-wire ammeter:

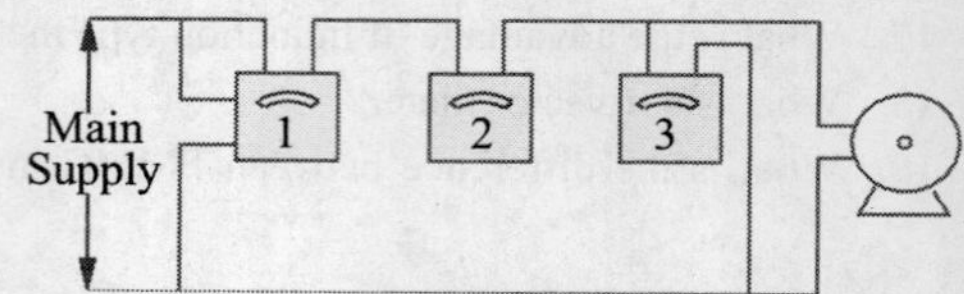

Fig. 10.89

(*a*) is used only for d.c. circuits
(*b*) is a high precision instrument
(*c*) is used only for a.c. circuits
(*d*) reads equally well on d.c. and/or a.c. circuits.
11. If an energy meter disc makes 10 revolutions in 100 seconds when a load of 450 w is connected to it, the meter constant (in rev/k Wh) is
(*a*) 1000 (*b*) 500
(*c*) 1600 (*d*) 800
(GATE 2001)
12. The inductance of a certain moving-iron ammeter is expressed as $L = 10 + 3\theta - \frac{\theta^2}{4}\,\mu H$ µH, where θ is the deflection in radians from the zero position. The control spring torque

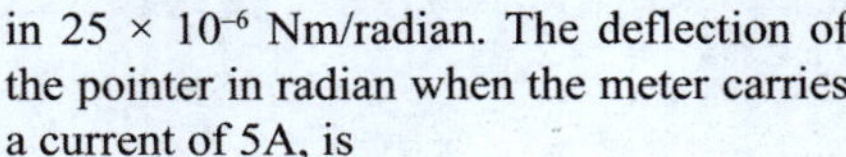

in 25×10^{-6} Nm/radian. The deflection of the pointer in radian when the meter carries a current of 5A, is

(*a*) 2.4 (*b*) 2.0

(*c*) 1.2 (*d*) 1.0

(GATE 2003)

13. A dc potentiometer is designed to measure up to about 2 V with a slide wire of 800 mm. A standard cell of emf 1.18 V obtains balance at 600 mm. A test cell is seen to obtain balance at 680 mm. The emf of the test cell is

(*a*) 1.00 V (*b*) 1.34 V

(*c*) 1.50 V (*d*) 1.70 V

(GATE 2004)

14. A galvanometer with a full scale current of 10 mA has a resistance of 1000 Ω. The multiplying power (the ratio of measured current to galvanometer current) of a 100 Ω shunt with this galvanometer is

(*a*) 110 (*b*) 100

(*c*) 11 (*d*) 10

(GATE 2004)

15. A moving coil of a meter has 100 turns, and a length and depth of 10 mm and 20 mm respectively. It is positioned in a uniform radial flux density of 200mT. The coil carries a current of 50 mA. The torque on the coil is

(*a*) 200 μNm

(*b*) 100 μNm

(*c*) 2 μNm

(*d*) 1 μNm ***(GATE 2004)***

16. A dc A-h meter is rated for 15 A, 250 V. The meter constant is 14.4 A-sec/rev. The meter constant at rated voltage may be expressed as

(*a*) 3750 rev/kWh

(*b*) 3600 rev/kWh

(*c*) 1000 rev/kWh

(d) 960 rev/kWh

(GATE 2004)

ANSWERS

1. *c* **2.** *d* **3.** *a* **4.** *c* **5.** *d* **6.** *d* **7.** *b* **8.** *b* **9.** *c* **10.** *d*

CHAPTER

11 A.C. Fundamentals

Chapter Outline

11.1. Generation of Alternating Voltages and Currents

Alternating voltage may be generated by rotating a coil in a magnetic field, as shown in Fig. 11.1 (*a*) or by rotating a magnetic field within a stationary coil, as shown in Fig. 11.1 (*b*).

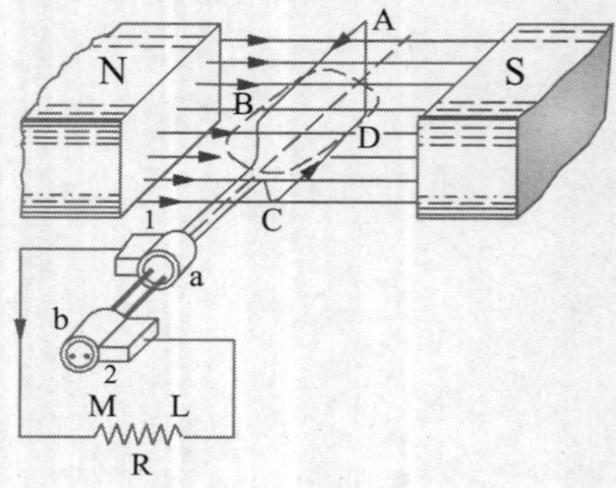

(*a*)

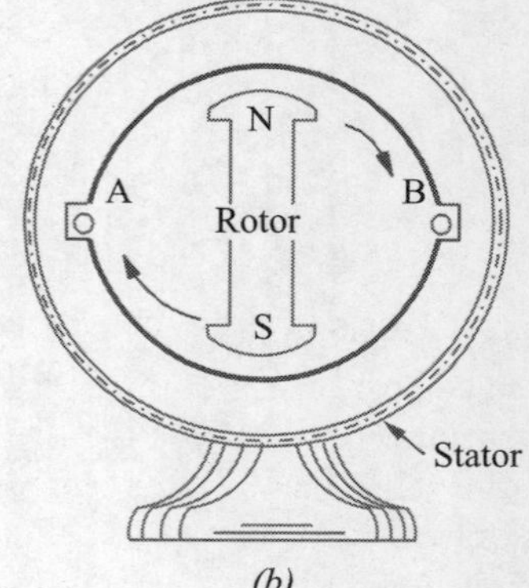

(*b*)

Fig. 11.1

The value of the voltage generated depends, in each case, upon the number of turns in the coil, strength of the field and the speed at which the coil or magnetic field rotates. Alternating voltage may be generated in either of the two ways shown above, but rotating-field method is the one which is mostly used in practice.

11.2. Equations of the Alternating Voltages and Currents

Consider a rectangular coil, having N turns and rotating in a uniform magnetic field, with an angular velocity of ω radian/second, as shown in Fig. 11.2. Let time be measured from the X-axis. Maximum flux Φ_m is linked with the coil, when its plane coincides with the X-axis. In time t seconds, this coil rotates through an angle $\theta = \omega t$. In this deflected position, the component of the flux which is perpendicular to the plane of the coil, is $\Phi = \Phi_m \cos \omega t$. Hence, ***flux linkages*** of the coil at any time are $N\Phi = N\Phi_m \cos \omega t$.

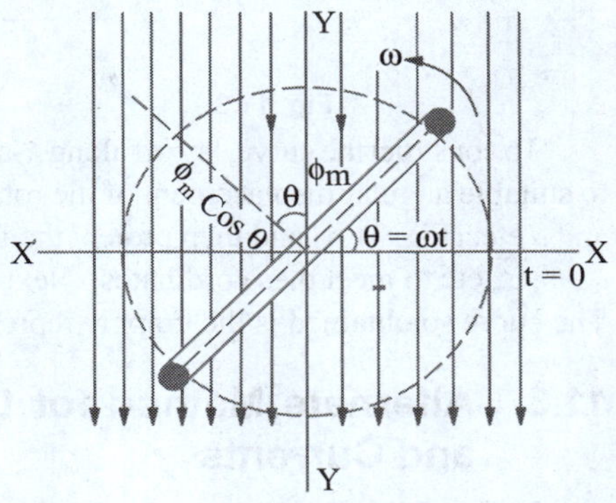

Fig. 11.2

According to Faraday's Laws of Electromagnetic Induction, the e.m.f. induced in the coil is given by the rate of change of flux-linkages of the coil. Hence, the value of the induced e.m.f. at this instant (*i.e.* when $\theta = \omega t$) or the instantaneous value of the induced e.m.f. is

$$e = -\frac{d}{dt}(N\,\Phi) \text{ volt} = -N.\frac{d}{dt}(\Phi_m \cos \omega t) \text{ volt} = -N\Phi_m\,\omega\,(-\sin \omega t) \text{ volt}$$

$$= \omega N \Phi_m \sin \omega t \text{ volt} = \omega N \Phi_m \sin \theta \text{ volt} \quad ...(i)$$

When the coil has turned through 90° *i.e.* when $\theta = 90°$, then $\sin \theta = 1$, hence e has maximum value, say E_m. Therefore, from Eq. **(*i*)** we get

$$E_m = \omega N \Phi_m = \omega N B_m A = 2\pi f N B_m A \text{ volt} \quad ...(ii)$$

where B_m = maximum flux density in Wb/m² ; A = area of the coil in m²

f = frequency of rotation of the coil in rev/second

Substituting this value of E_m in Eq. **(*i*)**, we get $e = E_m \sin \theta = E_m \sin \omega t$ **...(*iii*)**

Similarly, the equation of induced alternating current is $i = I_m \sin \omega t$ **...(*iv*)**

provided the coil circuit has been closed through a resistive load.

Since $\omega = 2\pi f$, where f is the frequency of rotation of the coil, the above equations of the voltage and current can be written as

$$e = E_m \sin 2\pi f t = E_m \sin\left(\frac{2\pi}{T}\right)t \text{ and } i = I_m \sin 2\pi f t = I_m \sin\left(\frac{2\pi}{T}\right)t$$

where T = time-period of the alternating voltage or current = $1/f$

It is seen that the induced e.m.f. varies as sine function of the time angle ωt and when e.m.f. is plotted against time, a curve similar to the one shown in Fig. 11.3 is obtained. This curve is known as sine curve and the e.m.f. which varies in this manner is known as ***sinusoidal*** e.m.f. Such a sine curve can be conveniently drawn, as shown in Fig. 11.4. A vector, equal in length to E_m is drawn. It rotates in the counter-clockwise direction with a velocity of ω radian/second, making one revolution while the generated e.m.f. makes two loops or one cycle. The projection of this vector on Y-axis gives the instantaneous value e of the induced e.m.f. *i.e.* $E_m \sin \omega t$.

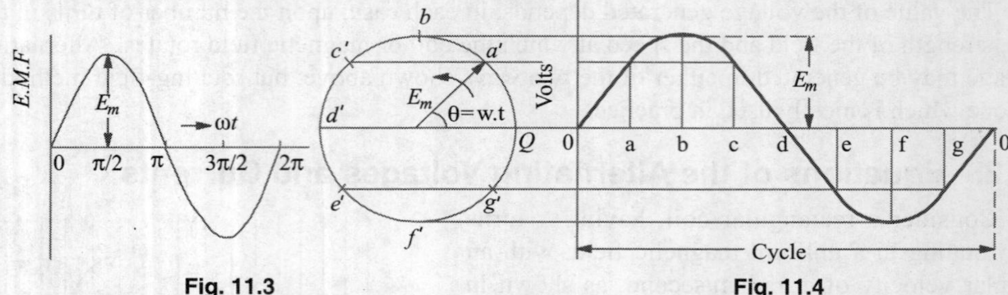

Fig. 11.3 Fig. 11.4

To construct the curve, lay off along *X*-axis equal angular distance *oa*, *ab*, *bc*, *cd* etc. corresponding to suitable angular displacement of the rotating vector. Now, erect coordinates at the points *a*, *b*, *c* and *d* etc. (Fig. 11.4) and then project the free ends of the vector E_m at the corresponding positions a', b', c', etc to meet these ordinates. Next draw a curve passing through these intersecting points. The curve so obtained is the graphic representation of equation ***(iii)*** above.

11.3. Alternate Method for the Equations of Alternating Voltages and Currents

In Fig. 11.5 is shown a rectangular coil *AC* having *N* turns and rotating in a magnetic field of flux density *B* Wb./m². Let the length of each of its sides *A* and *C* be *l* meters and their peripheral velocity *v* metre/second. Let angle be measured from the horizontal position *i.e.* from the *X*-axis. When in horizontal position, the two sides *A* and *C* move parallel to the lines of the magnetic flux. Hence, no flux is cut and so no e.m.f. is generated in the coil.

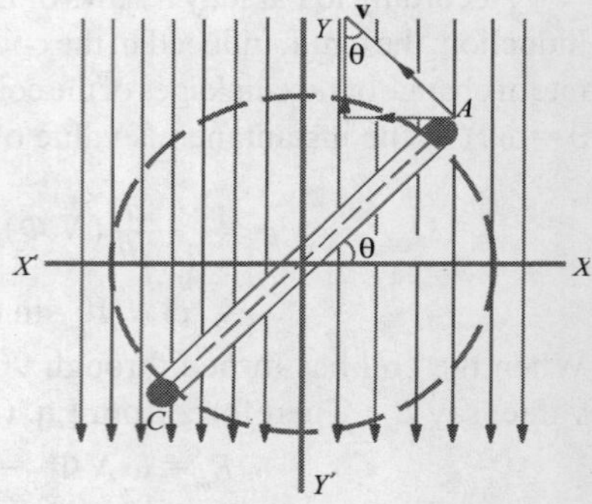

Fig. 11.5

When the coil has turned through angle θ, its velocity can be resolved into two mutually perpendicular components ***(i)*** $v \cos \theta$ component-parallel to the direction of the magnetic flux and ***(ii)*** $v \sin \theta$ component-perpendicular to the direction of the magnetic flux. The e.m.f. is generated due entirely to the perpendicular component *i.e.* $v \sin \theta$.

Hence, the e.m.f. generated in one side of the coil which contains *N* conductors, as seen from Art. 7.7, is given by, $e = N \times Bl\, v \sin \theta$.

Total e.m.f. generated in both sides of the coil is

$$e = 2BNl\, v \sin \theta \text{ volt} \qquad \text{...}(i)$$

Now, *e* has maximum value of E_m (say) when θ = 90°. Hence, from Eq. (*i*) above, we get, $E_m = 2\,B\,N\,l\,v$ volt. Therefore Eq. ***(i)*** can be rewritten as $e = E_m \sin \theta$ ...as before

If *b* = width of the coil in meters ; *f* = frequency of rotation of coil in Hz, then $v = \pi\, bf$

$$\therefore \quad E_m = 2\,B\,N\,l \times \pi\, b f = 2\,\pi\, f N B A \text{ volts} \qquad \text{...as before}$$

Example 11.1. *A square coil of 10 cm side and 100 turns is rotated at a uniform speed of 1000 revolutions per minute, about an axis at right angles to a uniform magnetic field of 0.5 Wb/m². Calculate the instantaneous value of the induced electromotive force, when the plane of the coil is* ***(i)*** *at right angles to the field* ***(ii)*** *in the plane of the field.* **(Electromagnetic Theory, A.M.I.E. Sec B, 1992)**

Solution. Let the magnetic field lie in the vertical plane and the coil in the horizontal plane. Also, let the angle θ be measured from *X*-axis.

Maximum value of the induced e.m.f., $E_m = 2\,\pi\, f N B_m A$ volt.

Instantaneous value of the induced e.m.f. $e = E_m \sin \theta$

Now $f = 100/60 = (50/3)$ rps, $N = 100$, $B_m = 0.5$ Wb/m², $A = 10^{-2}$ m²

(i) In this case, $\theta = 0°$

$\therefore \quad e = 0$

(ii) Here $\theta = 90°$, $\therefore \quad e = E_m \sin 90° = E_m$

Substituting the given values, we get

$e \;= 2\pi \times (50/3) \times 100 \times 0.5 \times 10^{-2} =$ **52.3 V**

11.4. Simple Waveforms

The shape of the curve obtained by plotting the instantaneous values of voltage or current as the ordinate against time as a abscissa is called its waveform or wave-shape.

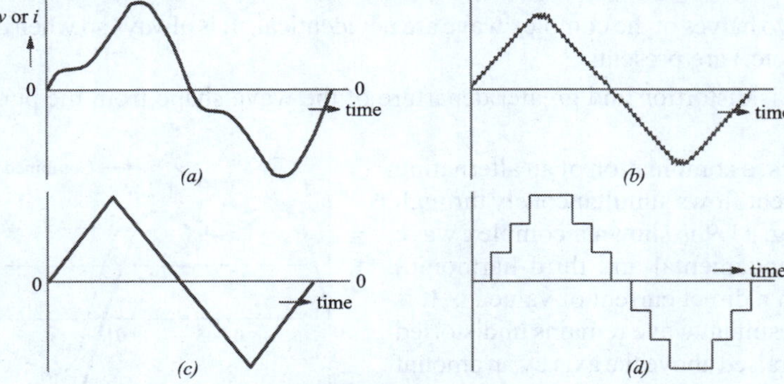

Fig. 11.6

An alternating voltage or current may not always take the form of a systematical or smooth wave such as that shown in Fig. 11.3. Thus, Fig. 11.6 also represents alternating waves. But while it is scarcely possible for the manufacturers to produce sine-wave generators or alternators, yet sine wave is the ideal form sought by the designers and is the accepted standard. The waves deviating from the standard sine wave are termed as distorted waves.

In general, however, ***an alternating current or voltage is one the circuit direction of which reverses at regularly recurring intervals.***

11.5. Complex Waveforms

Complex waves are those which depart from the ideal sinusoidal form of Fig. 11.4. All alternating complex waves, which are periodic and have equal positive and negative half cycles can be shown to be made up of a number of pure sine waves, having different frequencies but all these frequencies are integral multiples of that of the lowest alternating wave, called the ***fundamental*** (or first harmonic). These waves of higher frequencies are called ***harmonics.*** If the fundamental

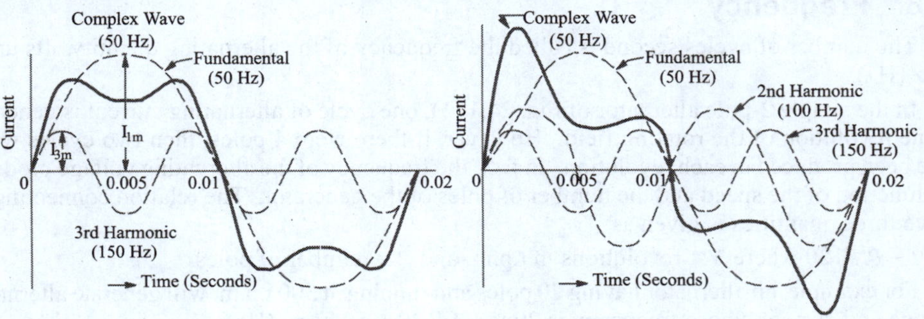

Fig. 11.7 **Fig. 11.8**

frequency is 50 Hz, then the frequency of the second ***harmonic*** is 100 Hz and of the third is 150 Hz and so on. The complex wave may be composed of the fundamental wave (or first harmonic) and any number of other harmonics.

In Fig. 11.7 is shown a complex wave which is made up of a fundamental sine wave of frequency of 50 Hz and third harmonic of frequency of 150 Hz. It is seen that

(i) the two halves of the complex wave are identical in shape. In other words, there is no distortion. This is always the case when only ***odd*** harmonic (3rd, 5th, 7th, 9th etc.) are present.

(ii) frequency of the complex wave is 50 Hz *i.e.* the same as that of the fundamental sine wave.

In Fig. 11.8 is shown a complex wave which is a combination of fundamental sine wave of frequency 50 Hz and 2nd harmonic of frequency 100 Hz and 3rd harmonic of frequency 150 Hz. It is seen that although the frequency of the complex wave even now remains 50 Hz, yet :

(i) the two halves of the complex wave are not identical. It is always so when *even* harmonics (2nd, 4th, 6th etc.) are present.

(ii) there is distortion and greater departure of the wave shape from the purely sinusoidal shape.

Sometimes, a combination of an alternating and direct current flows simultaneously through a circuit In Fig. 11.9 is shown a complex wave (containing fundamental and third harmonic) combined with a direct current of value I_D. It is seen that the resultant wave remains undistorted in shape but is raised above the axis by an amount I_D. It is worth noting that with reference to the original axis, the two halves of the combined wave are not equal in area.

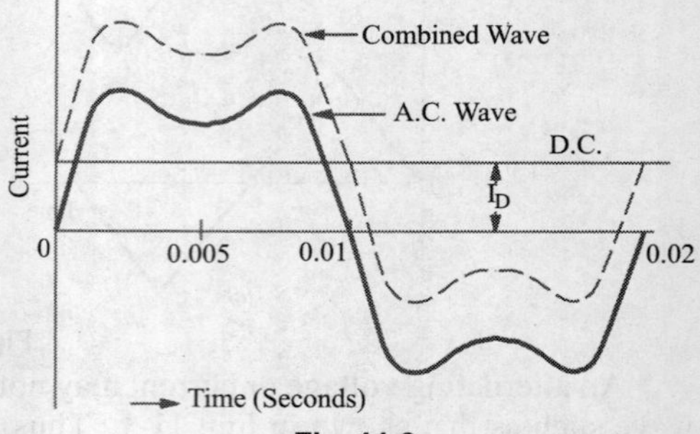

Fig. 11.9

11.6. Cycle

One complete set of positive and negative values of alternating quantity is known as cycle. Hence, each diagram of Fig. 11.6 represents one complete cycle.

A cycle may also be sometimes specified in terms of angular measure. In that case, one complete cycle is said to spread over 360° or 2π radians.

11.7. Time Period

The time taken by an alternating quantity to complete one cycle is called its time period *T*. For example, a 50-Hz alternating current has a time period of 1/50 second.

11.8. Frequency

The number of cycles/second is called the frequency of the alternating quantity. Its unit is hertz (Hz).

In the simple 2-pole alternator of Fig. 24.1 (*b*), one cycle of alternating current is generated in one revolution of the rotating field. However, if there were 4 poles, then two cycles would have been produced in each revolution. In fact, the frequency of the alternating voltage produced is a function of the speed and the number of poles of the generator. The relation connecting the above three quantities is given as

$f = PN/120$ where N = revolutions in r.p.m. and P = number of poles

For example, an alternator having 20 poles and running at 300 r.p.m. will generate alternating voltage and current whose frequency is 20 × 300/120 = 50 hertz (Hz).

It may be noted that the frequency is given by the reciprocal of the time period of the alternating quantity.

$$\therefore \qquad f = 1/T \quad \text{or} \quad T = 1/f$$

11.9. Amplitude

The maximum value, positive or negative, of an alternating quantity is known as its amplitude.

11.10. Different Forms of E.M.F. Equation

The standard form of an alternating voltage, as already given in Art. 11.2, is

$$e = E_m \sin\theta = E_m \sin\omega t = E_m \sin 2\pi f t = E_m \sin \frac{2\pi}{T} t$$

By closely looking at the above equations, we find that

(*i*) the maximum value or peak value or amplitude of an ***alternating voltage is given by the coefficient of the sine of the time angle.***

(*ii*) ***the frequency f is given by the coefficient of time divided by 2π..***

For example, if the equation of an alternating voltage is given by $e = 50 \sin 314t$ then its maximum value of 50 V and its frequency is $f = 314/2\pi = 50$ Hz.

Similarly, if the equation is of the form $e = I_m\sqrt{(R^2 + 4\omega^2 L^2)} \sin 2\omega t$, then its maximum value is $E_m = I_m\sqrt{(R^2 + 4\omega^2 L^2)}$ and the frequency is $2\omega/2\pi$ or ω/π Hz.

Example 11.2. *The maximum values of the alternating voltage and current are 400 V and 20 A respectively in a circuit connected to 50 Hz supply and these quantities are sinusoidal. The instantaneous values of the voltage and current are 283 V and 10 A respectively at t = 0 both increasing positively.*

(i) *Write down the expression for voltage and current at time t.*

(ii) *Determine the power consumed in the circuit.* **(Elect. Engg. Pune Univ. UPTU 2004)**

Solution. (*i*) In general, the expression for an a.c. voltage is $v = V_m \sin(\omega t + \phi)$ where ϕ is the phase difference with respect to the point where $t = 0$.

Now, $v = 283$ V ; $V_m = 400$ V. Substituting $t = 0$ in the above equation, we get

$283 = 400 (\sin\omega \times 0 + \phi)$ $\therefore$ $\sin\phi = 283/400 = 0.707$; $\therefore$ $\phi = 45°$ or $\pi/4$ radian.

Hence, general expression for voltage is

$v = 400 (\sin 2\pi \times 50 \times t + \pi/4)$

$= 400 \sin(100\pi t + \pi/4)$

Similarly, at $t = 0$, $10 = 20 \sin(\omega \times 0 + \phi)$

$\therefore$ $\sin\phi = 0.5$ $\therefore$ $\phi = 30°$ or $\pi/6$ radian

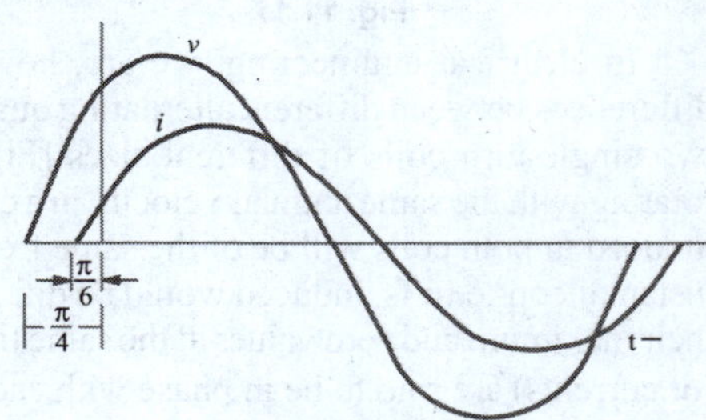

Fig. 11.10

Hence, the general expression for the current is

$i = 20 (\sin 100\pi t + 30°) = 20 \sin(100\pi t + \pi/6)$

(*ii*) $P = VI \cos\theta$ where V and I are rms values and θ is the phase difference between the voltage and current.

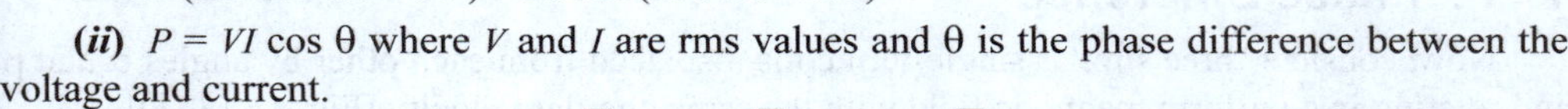

Now, $V = V_m/\sqrt{2} = 400/\sqrt{2}$; $I = 20/\sqrt{2}$; $\theta = 45° - 30° = 15°$ (See Fig. 11.10)

$\therefore$ $P = (400/\sqrt{2}) \times (20/\sqrt{2}) \times \cos 15° = \mathbf{3864\ W}$

Example 11.3. *An alternating current of frequency 60 Hz has a maximum value of 120 A. Write down the equation for its instantaneous value. Reckoning time from the instant the current is zero and is* *becoming positive, find* ***(a)*** *the instantaneous value after 1/360 second and* ***(b)*** *the time taken to reach 96 A for the first time.*

Solution. The instantaneous current equation is

$$i = 120 \sin 2\pi ft = 120 \sin 120\pi t$$

Now when $t = 1/360$ second, then

(a) $i = 120 \sin (120 \times \pi \times 1/360)$ ...angle in radians

$= 120 \sin (120 \times 180 \times 1/360)$ ...angle in degree

$= 120 \sin 60° = 103.9$ A

(b) $96 = 120 \times \sin 2 \times 180 \times 60 \times t$ ...angle in degree

or $\sin (360 \times 60 \times t) = 96/120 = 0.8 \quad \therefore \quad 360 \times 60 \times t = \sin^{-1} 0.8 = 53°$ (approx)

$\therefore \quad t = \theta/2\pi f = 53/360 \times 60 =$ **0.00245 second.**

11.11. Phase

By phase of an alternating current is meant the fraction of the time period of that alternating current which has elapsed since the current last passed through the zero position of reference. For example, the phase of current at point *A* is *T*/4 second, where *T* is time period or expressed in terms of angle, it is $\pi/2$ radians (Fig. 11.11). Similarly, the phase of the rotating coil at the instant shown in Fig. 11.1 is ωt which is, therefore, called its phase angle.

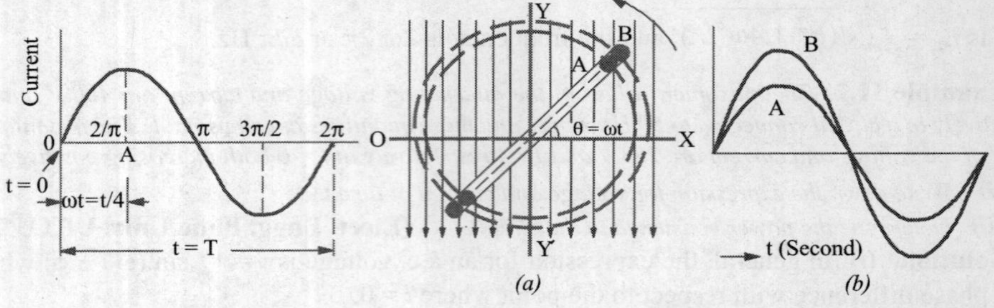

Fig. 11.11 **Fig. 11.12**

In electrical engineering, we are, however, more concerned with relative phases or phase differences between different alternating quantities, rather than with their absolute phases. Consider two single-turn coils of different sizes [Fig. 11.12 (*a*)] arranged radially in the same plane and rotating with the same angular velocity in a common magnetic field of uniform intensity. The e.m.fs. induced in both coils will be of the same frequency and of sinusoidal shape, although the values of instantaneous e.m.fs. induced would be different. However, the two alternating e.m.fs. would reach their maximum and zero values at the same time as shown in Fig. 11.12 (*b*). Such alternating voltages (or currents) are said to be in phase with each other. The two voltages will have the equations

$$e_1 = E_{m1} \sin \omega t \quad \text{and} \quad e_2 = E_{m2} \sin \omega t$$

11.12. Phase Difference

Now, consider three similar single-turn coils displaced from each other by angles α and β and rotating in a uniform magnetic field with the same angular velocity [Fig. 11.13 (*a*)].

In this case, the value of induced e.m.fs. in the three coils are the same, but there is one important difference. The e.m.fs. in these coils do not reach their maximum or zero values simultaneously but one after another. The three sinusoidal waves are shown in Fig. 11.13 (*b*). It is seen that curves *B* and *C* are displaced from curve *A* and angles β and $(\alpha + \beta)$ respectively. Hence, it means that phase difference between *A* and *B* is β and between *B* and *C* is α but between

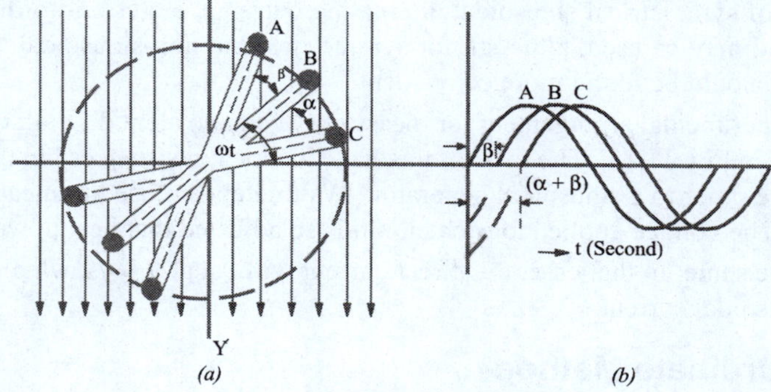

(a) (b)

Fig. 11.13

A and *C* is (α + β). The statement, however, does not give indication as to which e.m.f. reaches its maximum value first. This deficiency is supplied by using the terms ***'lag'*** or ***'lead'***.

A leading alternating quantity is one which reaches its maximum (or zero) value earlier as compared to the other quantity.

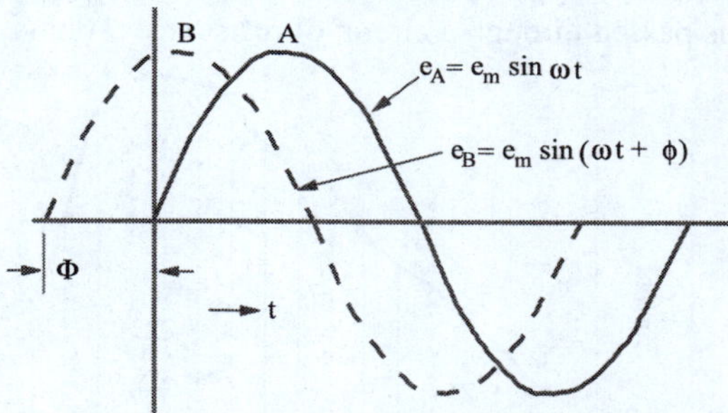

Fig. 11.14

Similarly, a lagging alternating quantity is one which reaches its maximum or zero value later than the other quantity. For example, in Fig. 11.13 (*b*), *B* lags behind *A* by β and *C* lags behind *A* by (α + β) because they reach their maximum values later.

The three equations for the instantaneous induced e.m.fs. are (Fig. 11.14)

$$e_A = E_m \sin \omega t \text{ ...reference quantity}$$

$$e_B = E_m \sin (\omega t - \beta)$$

$$e_C = E_m \sin [\omega t - (\alpha + \beta)]$$

In Fig. 11.14, quantity *B* leads *A* by an angle ϕ. Hence, their equations are

$$e_A = E_m \sin \omega t \text{ ...reference quantity}$$

$$e_B = E_m \sin (\omega t + \phi)$$

A plus (+) sign when used in connection with phase difference denotes 'lead' whereas a minus (–) sign denotes 'lag'.

11.13. Root-Mean-Square (R.M.S.) Value

The r.m.s. value of an alternating current ***is given by that steady (d.c.) current which when flowing through a given circuit for a given time produces the same*** **heat** ***as produced by the alternating current when flowing through the same circuit for the same time.***

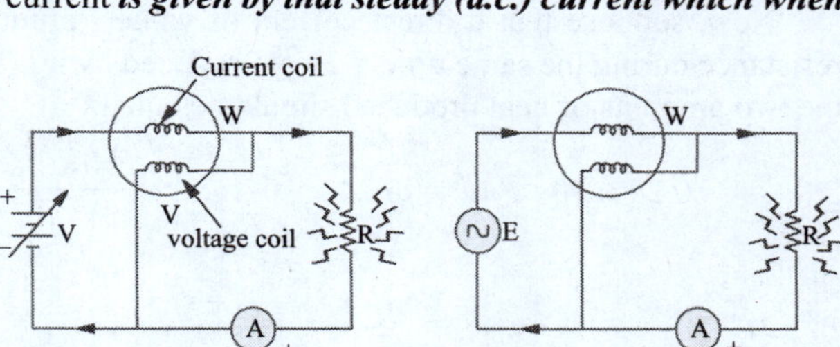

Fig. 11.15

It is also known as the ***effective*** or ***virtual*** value of the alternating current, the former term being used more extensively. For computing the r.m.s. value of symmetrical sinusoidal alternating currents, either mid-ordinate method or analytical method may be used, although for symmetrical but non-sinusoidal waves, the mid-ordinate method would be found more convenient.

A simple experimental arrangement for measuring the equivalent d.c. value of a sinusoidal current is shown in Fig. 11.15. The two circuits have identical resistances but one is connected to battery and the other to a sinusoidal generator. Wattmeters are used to measure heat power in each circuit. The voltage applied to each circuit is so adjusted that heat power production in each circuit is the same. In that case, the direct current will equal $I_m/\sqrt{2}$ which is called r.m.s. value of the sinusoidal current.

11.14. Mid-Ordinate Method

In Fig. 11.16 are shown the positive half cycles for both symmetrical sinusoidal and non-sinusoidal alternating currents. Divide time base '*t*' into *n* equal intervals of time each of duration *t*/*n* seconds. Let the average values of instantaneous currents during these intervals be respectively $i_1, i_2, i_3 .... i_n$ (*i.e.* mid-ordinates in Fig. 11.16). Suppose that this alternating current is passed through a circuit of resistance *R* ohms. Then,

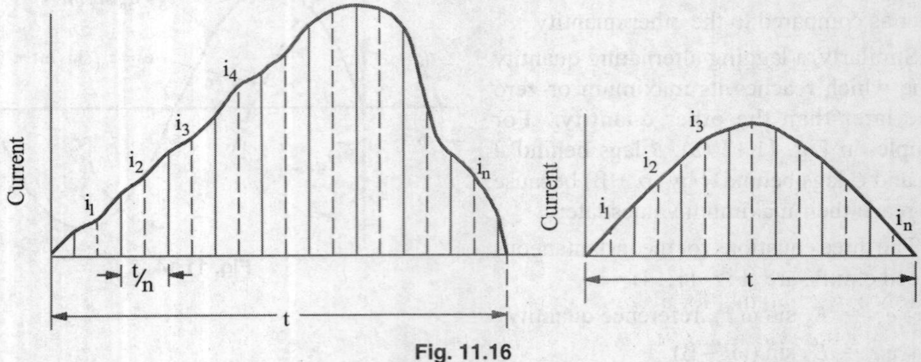

Fig. 11.16

Heat produced in 1st interval $= 0.24 \times 10^{-3}\, i_1^2\, Rt/n$ kcal ($\because 1/J = 1/4200 = 0.24 \times 10^{-3}$)

Heat produced in 2nd interval $= 0.24 \times 10^{-3}\, i_2^2\, Rt/n$ kcal

: : : : :

: : : : :

: : : : :

Heat produced in *n*th interval $= 0.24 \times 10^{-3}\, i_n^2\, Rt/n$ kcal

Total heat produced in *t* seconds is $= 0.24 \times 10^{-3}\, Rt \left(\dfrac{i_1^2 + i_2^2 + ... + i_n^2}{n}\right)$ kcal

Now, suppose that a direct current of value *I* produces the same heat through the same resistance during the same time *t*. Heat produced by it is $= 0.24 \times 10^{-3}\, I^2Rt$ kcal. By definition, the two amounts of heat produced should be equal.

$$\therefore \quad 0.24 \times 10^{-3} I^2 Rt = 0.24 \times 10^{-3}\, Rt \left(\frac{i_1^2 \aleph i_2^2 \quad ... \quad i_n^2}{n}\right)$$

$$\therefore \quad I^2 = \frac{i_1^2 + i_2^2 + ... + i_n^2}{n} \qquad \therefore \quad I = \sqrt{\left(\frac{i_1^2 + i_2^2 + ... + i_n^2}{n}\right)}$$

= square root of the mean of the squares of the instantaneous currents

Similarly, the r.m.s. value of alternating voltage is given by the expression

$$V = \sqrt{\left(\frac{v_1^2 + v_2^2 + ... + v_n^2}{n}\right)}$$

11.15. Analytical Method

The standard form of a sinusoidal alternating current is $i = I_m \sin \omega t = I_m \sin \theta$.

The mean of the squares of the instantaneous values of current over one complete cycle is (even the value over half a cycle will do).

$$= \int_0^{2\pi} \frac{i^2 \, d\theta}{(2\pi - 0)}$$

The square root of this value is $= \sqrt{\left(\int_0^{2\pi} \frac{i^2 \, d\theta}{2\pi}\right)}$

Hence, the r.m.s. value of the alternating current is

$$I = \sqrt{\left(\int_0^{2\pi} \frac{i^2 d\theta}{2\pi}\right)} = \sqrt{\left(\frac{I_m^2}{2\pi}\int_0^{2\pi} \sin^2 \theta \, d\theta\right)} \qquad \text{(put } i = I_m \sin \theta\text{)}$$

Now, $\cos 2\theta = 1 - 2\sin^2 \theta \quad \therefore \quad \sin^2 \theta = \frac{1 - \cos 2\theta}{2}$

$$\therefore \qquad I = \sqrt{\left(\frac{I_m^2}{4\pi}\int_0^{2\pi} (1 - \cos 2\theta)\, d\theta\right)} = \sqrt{\left(\frac{I_m^2}{4\pi}\left|\theta - \frac{\sin 2\theta}{2}\right|_0^{2\pi}\right)}$$

$$= \sqrt{\left(\frac{I_m^2}{4\pi} \times 2\pi\right)} = \sqrt{\left(\frac{I_m^2}{2}\right)} \qquad \therefore \; I = \frac{I_m}{\sqrt{2}} = 0.707\, I_m$$

Hence, we find that for a symmetrical sinusoidal current

r.m.s. value of current = 0.707 × max. value of current

The r.m.s. value of an alternating current is of considerable importance in practice, because the ammeters and voltmeters record the r.m.s. value of alternating current and voltage respectively. In electrical engineering work, ***unless indicated otherwise, the values of the given current and voltage are always the r.m.s. values.***

It should be noted that the average heating effect produced during one cycle is

$$= I^2R = (I_m/\sqrt{2})^2 R = \frac{1}{2} I_m^2 R$$

11.16. R.M.S. Value of a Complex Wave

In their case also, either the mid-ordinate method (when equation of the wave is not known) or analytical method (when equation of the wave is known) may be used. Suppose a current having the equation $i = 12 \sin \omega t + 6 \sin (3\omega t - \pi/6) + 4 \sin (5\omega t + \pi/3)$ flows through a resistor of R ohm. Then, in the time period T second of the wave, the effect due to each component is as follows :

Fundamental .... $(12/\sqrt{2})^2 \; RT$ watt

3rd harmonic .... $(6/\sqrt{2})^2 \; RT$ watt

5th harmonic .... $(4/\sqrt{2})^2 \; RT$ watt

$\therefore$ Total heating effect $= RT\,[(12/\sqrt{2})^2 + (6/\sqrt{2})^2 + (4/\sqrt{2})^2]$

If I is the r.m.s. value of the complex wave, then equivalent heating effect is I^2RT

$$\therefore \qquad I^2RT = RT\,[(12/\sqrt{2})^2 + (6/\sqrt{2})^2 + (4/\sqrt{2})^2]$$

$$\therefore \qquad I = \sqrt{[(12/\sqrt{2})^2 + (6/\sqrt{2})^2 + (4/\sqrt{2})^2]} = 9.74\text{ A}$$

Had there been a direct current of (say) 5 amperes flowing in the circuit also*, then the r.m.s. value would have been

$$= \sqrt{(12/\sqrt{2})^2 + (6/\sqrt{2})^2 + (4/\sqrt{2})^2 + 5^2} = 10.93\text{ A}$$

Hence, for complex waves the rule is as follows : ***The r.m.s. value of a complex current wave is equal to the square root of the sum of the squares of the r.m.s. values of its individual components.***

11.17. Average Value

The average value I_a of an alternating current is expressed ***by that steady current which transfers across any circuit the same charge as is transferred by that alternating current during the same time.***

In the case of a symmetrical alternating current (*i.e.* one whose two half-cycles are exactly similar, whether sinusoidal or non-sinusoidal), the average value over a complete cycle is zero. Hence, in their case, the average value is obtained by adding or integrating the instantaneous values of current over one half-cycle only. ***But in the case of an unsymmetrical alternating current (like half-wave rectified current) the average value must always be taken over the whole cycle.***

(*i*) Mid-ordinate Method

With reference to Fig. 11.16, $\quad I_{av} = \dfrac{i_1 + i_2 + \ldots + i_n}{n}$

This method may be used both for sinusoidal and non-sinusoidal waves, although it is specially convenient for the latter.

(*ii*) Analytical Method

The standard equation of an alternating current is, $i = I_m \sin\theta$

$$I_{av} = \int_0^{\pi} \frac{i\,d\theta}{(\pi - 0)} = \frac{I_m}{\pi}\int_0^{\pi} \sin\theta\,d\theta \qquad \text{(putting value of } i\text{)}$$

$$= \frac{I_m}{\pi}\left|-\cos\theta\right|_0^{\pi} = \frac{I_m}{\pi}\left|+1-(-1)\right| = \frac{2I_m}{\pi} = \frac{I_m}{\pi/2} = \frac{\text{twice the maximum current}}{\pi}$$

$$\therefore \qquad I_{av} = I_m/\tfrac{1}{2}\pi = 0.637\,I_m \quad \therefore \textbf{ average value of current} = \textbf{0.637} \times \textbf{maximum value}$$

Note. R.M.S. value is always greater than average value except in the case of a rectangular wave when both are equal.

11.18. Form Factor

It is defined as the ratio, $K_f = \dfrac{\text{r.m.s. value}}{\text{average value}} = \dfrac{0.707\,I_m}{0.637\,I_m} = 1.1.$ (for sinusoidal alternating currents only)

In the case of sinusoidal alternating voltage also, $K_f = \dfrac{0.707\,E_m}{0.637\,E_m} = 1.11$

As is clear, the knowledge of form factor will enable the r.m.s. value to be found from the arithmetic mean value and ***vice-versa.***

---

* The equation of the complex wave, in that case, would be,
i = 5 + 12 sin wt + 6 sin (3wt - p/6) + 4 sin (5wt + p/3)

11.19. Crest or Peak or Amplitude Factor

It is defined as the ratio $K_a = \dfrac{\text{maximum value}}{\text{r.m.s. value}} = \dfrac{I_m}{I_m/\sqrt{2}} = \sqrt{2} = 1.414$ (for sinusoidal a.c. only)

For sinusoidal alternating voltage also, $K_a = \dfrac{E_m}{E_m/\sqrt{2}} = 1.414$

Knowledge of this factor is of importance in dielectric insulation testing, because the dielectric stress to which the insulation is subjected, is proportional to the maximum or peak value of the applied voltage. The knowledge is also necessary when measuring iron losses, because the iron loss depends on the value of maximum flux.

Example 11.4. *An alternating current varying sinusoidally with a frequency of 50 Hz has an RMS value of 20 A. Write down the equation for the instantaneous value and find this value* ***(a)*** *0.0025 second* ***(b)*** *0.0125 second after passing through a positive maximum value. At what time, measured from a positive maximum value, will the instantaneous current be 14.14 A ?*

(RTU 2006, Elect. Science-I Allahabad Univ. 1992)

Solution. $I_m = 20\sqrt{2} = 28.2$ A, $\omega = 2\pi \times 50 = 100\,\pi$ rad/s.

The equation of the sinusoidal current wave with reference to point O (Fig. 11.17) as zero time point is $\quad i = \mathbf{28.2 \sin 100\,\pi t}$ **ampere**

Since time values are given from point A where voltage has positive and maximum value, the equation may itself be referred to point A. In the case, the equation becomes :

$$i = 28.2 \cos 100\,\pi t$$

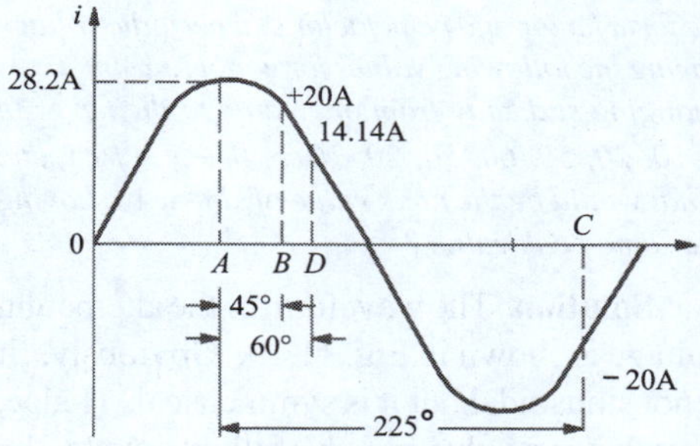

Fig. 11.17

(i) **When $t = 0.0025$ second**

$i = 28.2 \cos 100\pi \times 0.0025$ ...angle in radian

$= 28.2 \cos 100 \times 180 \times 0.0025$ ...angle in degrees

$= 28.2 \cos 45° = 20$ A ...point B

(ii) **When $t = 0.0125$ second**

$i = 28.2 \cos 100 \times 180 \times 0.0125$

$= 28.2 \cos 225° = 28.2 \times (-1/\sqrt{2})$

$= -20$ A ...point C

(iii) **Here $i = 14.14$ A**

$\therefore \quad 14.14 = 28.2 \cos 100 \times 180\, t \quad \therefore \quad \cos 100 \times 180\, t = \dfrac{1}{2}$

or $\quad 100 \times 180\, t = \cos^{-1}(0.5) = 60°$, $t = \mathbf{1/300}$ **second** ...point D

Example 11.5. *An alternating current of frequency 50 Hz has a maximum value of 100 A. Calculate* ***(a)*** *its value 1/600 second after the instant the current is zero and its value decreasing thereafter* ***(b)*** *how many seconds after the instant the current is zero (increasing thereafter wards) will the current attain the value of 86.6 A ?* **(Elect. Technology. Allahabad Univ. 1991)**

Solution. The equation of the alternating current (assumed sinusoidal) with respect to the origin O (Fig. 11.18) is

$$i = 100 \sin 2\pi \times 50t = 100 \sin 100\,\pi t$$

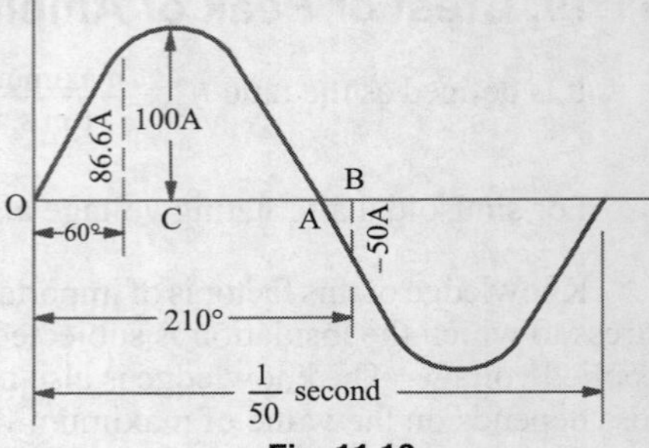

Fig. 11.18

(a) It should be noted that, in this case, time is being measured from point *A* (where current is zero and decreasing thereafter) and not from point *O*.

If the above equation is to be utilized, then, this time must be referred to point *O*. For this purpose, half time-period *i.e.* 1/100 second has to be added to 1/600 second. The given time as referred to point *O* becomes

$$= \frac{1}{100} + \frac{1}{600} = \frac{7}{600} \text{ second}$$

$\therefore \quad i = 100 \sin 100 \times 180 \times 7/600 = 100 \sin 210°$

$= 100 \times -1/2 = -$ **50 A** ...point *B*

(b) In this case, the reference point is *O*.

$\therefore \quad 86.6 = 100 \sin 100 \times 180\, t \quad$ or $\quad \sin 18{,}000\, t = 0.866$

or $\quad 18{,}000\, t = \sin^{-1}(0.866) = 60° \quad \therefore \quad t = 60/18{,}000 =$ **1/300 second**

Example 11.6. *Calculate the r.m.s. value, the form factor and peak factor of a periodic voltage having the following values for equal time intervals changing suddenly from one value to the next : 0, 5, 10, 20, 50, 60, 50, 20, 10, 5, 0, −5, −10 V etc. What would be the r.m.s value of sine wave having the same peak value ?*

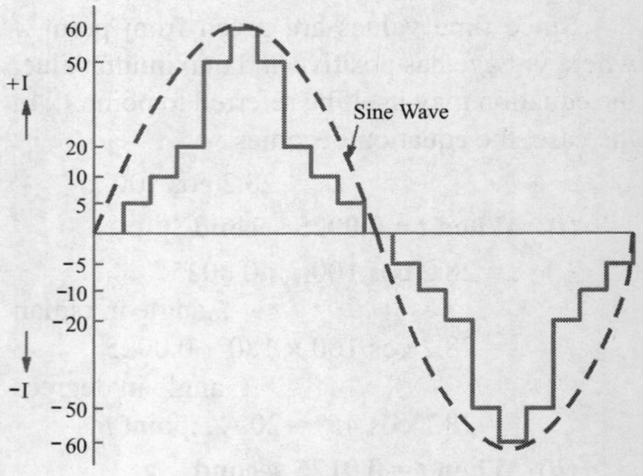

Fig. 11.19

Solution. The waveform of the alternating voltage is shown in Fig. 11.19. Obviously, it is not sinusoidal but it is symmetrical. Hence, though r.m.s value may be full one cycle, the average value has necessarily to be considered for half-cycle only, otherwise the symmetrical negative and positive half-cycles will cancel each other out.

$$\text{Mean value of } v^2 = \frac{0^2 + 5^2 + 10^2 + 20^2 + 50^2 + 60^2 + 50^2 + 20^2 + 10^2 + 5^2}{10} = \mathbf{965\ V}$$

$\therefore \quad$ r.m.s. value $= \sqrt{965} =$ **31 V (approx.)**

$$\text{Average value (half-cycle)} = \frac{0 + 5 + 10 + 20 + 50 + 60 + 50 + 20 + 10 + 5}{10} = \mathbf{23\ V}$$

$$\text{Form factor} = \frac{\text{r.m.s. value}}{\text{average value}} = \frac{31}{23} = \mathbf{1.35.}$$ Peak factor = 60/31 = **2 (approx.)**

R.M.S. value of a sine wave of the same peak value = 0.707 × 60 = **42.2 V.**

Alternative Solution

If '*t*' be the regular time interval, then area of the half-cycle is

$= (5t + 10t + 20t + 50t)\, 2 + 60t = 230t$, Base $= 10t \quad \therefore \quad$ Mean value $= 230t/10t = 23$ V.

Area when ordinates are squared $= (25t + 100t + 400t + 2500t)\, 2 + 3600t = 9650t$, Base $= 10t$

∴ Mean height of the squared curve = 9650t/10 t = 965

∴ r.m.s. value = $\sqrt{965}$ = 31 V

Further solution is as before.

11.20. R.M.S. Value of H.W. Rectified Alternating Current

Half-wave (H.W.) rectified alternating current is one whose one half-cycle has been suppressed *i.e.* one which flows for half the time during one cycle. It is shown in Fig. 11.20 where suppressed half-cycle is shown dotted.

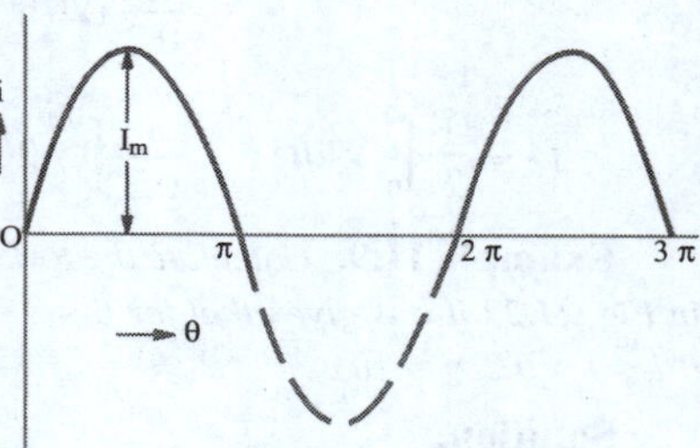

Fig. 11.20

As said earlier, for finding r.m.s. value of such an alternating current, summation would be carried over the period for which current ***actually*** flows *i.e.* from 0 to π, though it would be averaged for the whole cycle *i.e.* from 0 to 2π.

∴ R.M.S. current

$$I = \sqrt{\left(\int_0^{\pi} \frac{i^2 d\theta}{2\pi}\right)} = \sqrt{\left(\frac{I_m^2}{2\pi}\int_0^{\pi} \sin^2\theta \, d\theta\right)}$$

$$= \sqrt{\frac{I_m^2}{4\pi}\int_0^{\pi}(1-\cos 2\theta)\,d\theta}$$

$$= \sqrt{\left(\frac{I_m^2}{4\pi}\left|\theta - \frac{\sin 2\theta}{2}\right|_0^{\pi}\right)} = \sqrt{\left(\frac{I_m^2}{4\pi}\times\pi\right)} = \sqrt{\left(\frac{I_m^2}{4}\right)} \quad \therefore \quad I = \frac{I_m}{2} = \mathbf{0.5I_m}$$

11.21. Average Value of H.W. Rectified Alternating Current

For the same reasons as given in Art. 11.20, integration would be carried over from 0 – π

$$\therefore \quad I_{av} = \int_0^{\pi}\frac{i d\theta}{2\pi} = \frac{I_m}{2\pi}\int_0^{\pi}\sin\theta\, d\theta \qquad (\because \; i = I_m \sin\theta)$$

$$= \frac{I_m}{2\pi}\left|-\cos\theta\right|_0^{\pi} = \frac{I_m}{2\pi}\times 2 = \frac{I_m}{\pi}$$

11.22. Form Factor of H.W. Rectified Alternating Current

$$\text{Form factor} = \frac{\text{r.m.s. value}}{\text{average value}} = \frac{I_m/2}{I_m/\pi} = \frac{\pi}{2} = 1.57$$

Example 11.7. *An alternating voltage e = 200 sin 314t is applied to a device which offers an ohmic resistance of 20 Ω to the flow of current in one direction, while preventing the flow of current in opposite direction. Calculate RMS value, average value and form factor for the current over one cycle.*

(Elect. Engg. Nagpur Univ. 1992)

Solution. Comparing the given voltage equation with the standard form of alternating voltage equation, we find that V_m = 200 V, R = 20 Ω, I_m = 200/20 = 10 A. For such a half-wave rectified current, RMS value = I_m/2 = 10/2 = **5A.**

Average current = I_m/π = 10/π = **3.18 A** ; Form factor = 5/3.18 = **1.57**

Example 11.8. *Compute the average and effective values of the square voltage wave shown in Fig. 11.21.*

Solution. As seen, for $0 < t < 0.1$ *i.e.* for the time interval 0 to 0.1 second, $v = 20$ V. Similarly, for $0.1 < t < 0.3$, $v = 0$. Also time-period of the voltage wave is 0.3 second.

$$\therefore \quad V_{av} = \frac{1}{T}\int_0^T v\, dt = \frac{1}{0.3}\int_0^{0.1} 20\, dt$$

$$= \frac{1}{0.3}(20 \times 0.1) = \mathbf{6.67\ V}$$

$$V^2 = \frac{1}{T}\int_0^T v^2 dt = \frac{1}{0.3}\int_0^{0.1} 20^2\, dt = \frac{1}{0.3}(400 \times 0.1) = 133.3;\ V = \mathbf{11.5\ V}$$

Fig. 11.21

Example 11.9. *Calculate the RMS value of the function shown in Fig. 11.22 if it is given that for $0 < t < 0.1$, $y = 10(1 - e^{-100t})$ and $0.1 < t < 0.2$, $y = 10\, e^{-50(t-0.1)}$*

Solution.

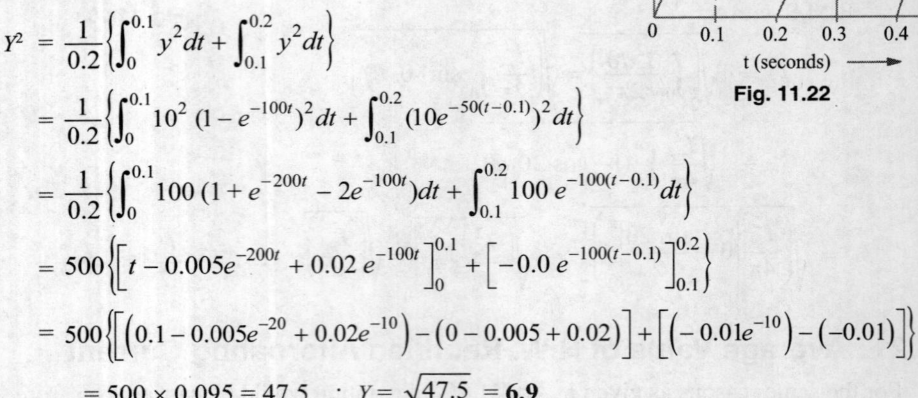

Fig. 11.22

$$Y^2 = \frac{1}{0.2}\left\{\int_0^{0.1} y^2 dt + \int_{0.1}^{0.2} y^2 dt\right\}$$

$$= \frac{1}{0.2}\left\{\int_0^{0.1} 10^2\,(1 - e^{-100t})^2 dt + \int_{0.1}^{0.2} (10e^{-50(t-0.1)})^2 dt\right\}$$

$$= \frac{1}{0.2}\left\{\int_0^{0.1} 100\,(1 + e^{-200t} - 2e^{-100t})dt + \int_{0.1}^{0.2} 100\, e^{-100(t-0.1)} dt\right\}$$

$$= 500\left\{\left[t - 0.005e^{-200t} + 0.02\, e^{-100t}\right]_0^{0.1} + \left[-0.0\, e^{-100(t-0.1)}\right]_{0.1}^{0.2}\right\}$$

$$= 500\left\{\left[\left(0.1 - 0.005e^{-20} + 0.02e^{-10}\right) - (0 - 0.005 + 0.02)\right] + \left[\left(-0.01e^{-10}\right) - (-0.01)\right]\right\}$$

$$= 500 \times 0.095 = 47.5 \quad \therefore\ Y = \sqrt{47.5} = \mathbf{6.9}$$

Example 11.10. *The half cycle of an alternating signal is as follows : It increases uniformly from zero at 0° to F_m at α°, remains constant from α° (180 – α)°, decreases uniformaly from F_m at (180 – α)° to zero at 180°. Calculate the average and effective values of the signal.*

(Elect. Science-I, Allahabad Univ. 1992)

Solution. For finding the average value, we would find the total area of the trapezium and divide it by π (Fig. 11.23).

Area = 2 × Δ *OAE* + rectangle *ABDE* = 2 × (1/2) × $F_m \alpha + (\pi - 2\alpha) F_m = (\pi - \alpha) F_m$

$\therefore$ average value $= (\pi - \alpha) F_m/\pi$

RMS Value From similar triangles, we get

$$\frac{y}{\theta} = \frac{F_m}{\alpha} \text{ or } y^2 = \frac{Fm^2}{\alpha^2}\theta^2$$

Fig. 11.23

This gives the equation of the signal over the two triangles *OAE* and *DBC*. The signal remains constant over the angle α to (π – α) *i.e.* over an angular distance of (π – α) – α = (π – 2α)

$$\text{Sum of the squares} = \frac{2F_m^2}{\alpha^2}\int_\theta^\alpha \theta^2\, d\theta + F_m^2(\pi - 2\alpha) = F_m^2(\pi - 4\alpha/3).$$

The mean value of the squares is $= \dfrac{1}{\pi}F_m^2\left(\pi - \dfrac{4\alpha}{3}\right) = F_m^2\left(1 - \dfrac{4\alpha}{3\pi}\right)$

$$\text{r.m.s. value} = F_m\sqrt{\left(1 - \frac{4\alpha}{3\pi}\right)}$$

Example 11.11. *Find the average and r.m.s values of the a.c. voltage whose waveform is given in Fig. 11.24 (a)*

Solution. It is seen [(Fig. 11.24 (a)] that the time period of the waveform is 5s. For finding the average value of the waveform, we will calculate the net area of the waveform over one period and then find its average value for one cycle.

$A_1 = 20 \times 1 = 20\ V-s, A_2 = -5 \times 2 = -10\ \text{V} - \text{s}$

Net area over the full cycle $= A_1 + A_2 = 20 - 10 = 10\ \text{V} - \text{s}$.

Average value $= 10\ V\text{–s}/5\text{s} = 2\ \text{V}$.

Fig. 11.24 (*b*) shows a graph of $v^2(t)$. Since the negative voltage is also squared, it becomes positive.

Average value of the area $= 400\ V^2 \times 1\ \text{s} + 25\ V^2 \times 2\ \text{s} = 450\ V^2 - \text{s}$. The average value of the sum of the square $= 450\ V^2/\text{- s}/5\text{s} = 90V^2$ rms value $= \sqrt{90\ V^2} = 9.49$ V.

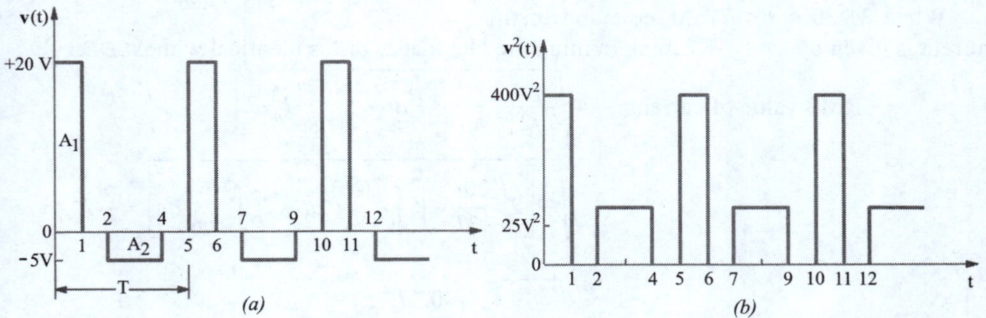

Fig. 11.24

Example 11.12. *What is the significance of the r.m.s and average values of a wave ? Determine the r.m.s. and average value of the waveform shown in Fig. 11.25*

(Elect. Technology, Indore Univ.)

Solution. The slope of the curve *AB* is *BC*/*AC* = 20/*T*. Next, consider the function *y* at any time *t*. It is seen that *DE*/*AE* = *BC*/*AC* = 10/*T*

or $(y - 10)/t = 10/T$

or $y = 10 + (10/T)t$

This gives us the equation for the function for one cycle.

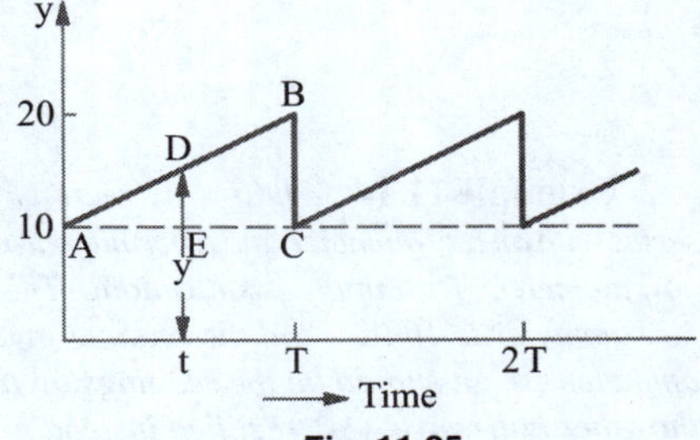

Fig. 11.25

$$Y_{av} = \frac{1}{T}\int_0^T y\,dt = \frac{1}{T}\int_0^T\left(10 + \frac{10}{T}t\right)dt$$

$$= \frac{1}{T}\int_0^T\left[10.dt + \frac{10}{T}.t.\,dt\right] = \frac{1}{T}\left|10t + \frac{5t^2}{T}\right|_0^T = \mathbf{15}$$

$$\text{Mean square value} = \frac{1}{T}\int_0^T y^2 dt = \int_0^T \left(10 + \frac{10}{T}t\right)^2 dt$$

$$= \frac{1}{T}\int_0^T \left(100 + \frac{100}{T^2}t^2 + \frac{200}{T}t\right) dt$$

$$= \frac{1}{T}\left|100t + \frac{100t^3}{3T^2} + \frac{100t^2}{T}\right|_0^T = \frac{700}{3}$$

or RMS value $= 10\sqrt{7/3}$ = **15.2**

Example 11.13. *For the trapezoidal current wave-form of Fig. 11.26, determine the effective value.* **(Elect. Technology, Vikram Univ. Ujjain , Similar Example, Nagpur Univ. 1999)**

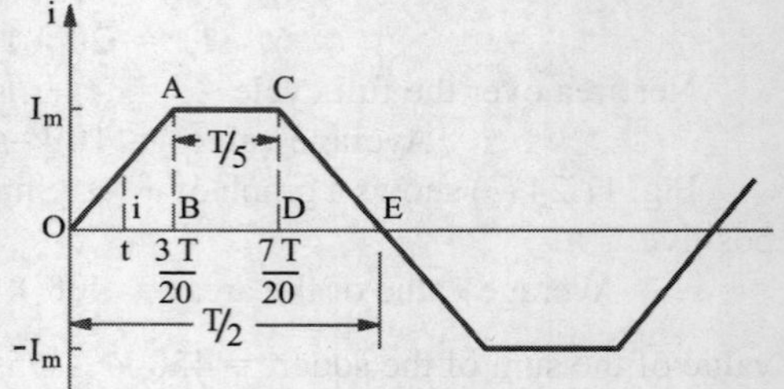

Fig. 11.26

Solution. For $0 < t < 3T/20$, equation of the current can be found from the relation

$$\frac{i}{t} = \frac{I_m}{3T/20} \quad \text{or } i = \frac{20I_m}{3T}.t$$

When $3T/20 < t < 7T/20$, equation of the current is given by $i = I_m$. Keeping in mind the fact that ΔOAB is identical with ΔCDE,

$$\text{RMS value of current} = \sqrt{\frac{1}{T/2}\left[2\int_0^{3T/20} i^2 dt + \int_{3T/20}^{7T/20} I_m^2 dt\right]}$$

$$= \sqrt{\frac{2}{T}\left[2\left(\frac{20I_m}{3T}\right)^2 \int_0^{3T/20} t^2 dt + I_m^2 \int_{3T/20}^{7T/20} dt\right]} = \frac{3}{5}I_m^2$$

$$\therefore \qquad I = \sqrt{(3/5)}.I_m = \mathbf{0.775\ I_m}$$

Incidentally, the average value is given by

$$I_{ac} = \frac{2}{T}\left\{2\int_0^{3T/20} idt + \int_{3T/20}^{7T/20} I_m dt\right\} = \frac{2}{T}\left\{2\int_0^{3T/20}\left(\frac{20I_m}{3T}\right)tdt + I_m \int_{3T/20}^{7T/20} dt\right\}$$

$$= \frac{2}{T}\left\{2\left(\frac{20I_m}{3T}\right)\left|\frac{t^2}{2}\right|_0^{3T/20} + I_m \left| t \right|_{3T/20}^{7T/20}\right\} = \frac{\mathbf{7}}{\mathbf{10}}.\mathbf{I_m}$$

Example 11.14. *A half-wave rectifier which prevents current flowing in one direction is connected in series with an a.c. ammeter and a permanent-magnet moving-coil ammeter. The supply is sinusoidal. The reading on the a.c. ammeter is 10 A. Find the reading given by the other ammeter. What should be the readings on the ammeters, if the other half-wave were rectified instead of being cut off ?*

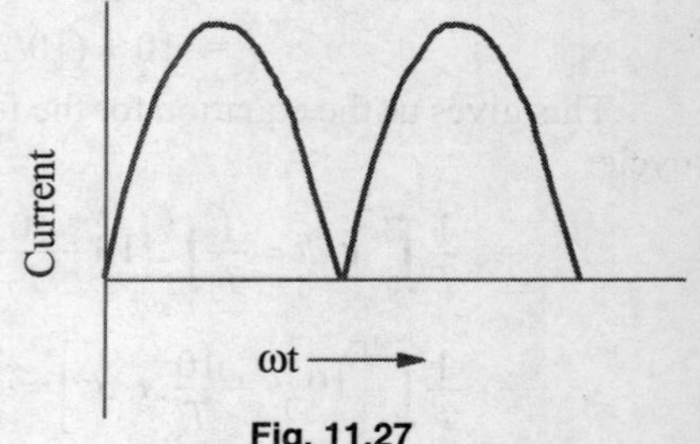

Fig. 11.27

Solution. It should be noted that an a.c. ammeter reads r.m.s. value whereas the d.c. ammeter reads the average value of the rectified current.

As shown in Art. 11.20 from H.W. rectified alternating current, $I = I_m/2$ and $I_{av} = I_m/\pi$

As a.c. ammeter reads 10 A, hence r.m.s. value of the current is 10 A.

$\therefore \quad 10 = I_m/2 \quad$ or $\quad I_m = 20$ A

$\therefore \quad I_{av} = 20/\pi =$ **6.365 A** – reading of d.c. ammeter.

The full-wave rectified current wave is shown in Fig. 11.28. In this case mean value of i^2 over a complete cycle is given as

$$= 2\int_0^\pi \frac{i^2\, d\theta}{2\pi - 0} = \frac{1}{\pi}\int_0^\pi I_m^2\; 2\sin^2\theta\, d\theta$$

$$= \frac{I_m^2}{2\pi}\int_0^\pi (1-\cos 2\theta)d\theta = \frac{I_m^2}{2\pi}\left|\theta - \frac{\sin 2\theta}{2}\right|_0^\pi = \frac{I_m^2}{2}$$

$\therefore \quad I = I_m/\sqrt{2} = 20/\sqrt{2} = 14.14$ A $\therefore$ a.c. ammeter will read **14.14 A**

Now, mean value of i over a complete cycle

$$= \frac{2\int_0^\pi I_m \sin\theta\, d\theta}{2\pi} = \frac{I_m}{\pi}\int_0^\pi \sin\theta d\theta = \frac{I_m}{\pi}\left|-\cos\theta\right|_0^\pi = \frac{2I_m}{\pi} = \frac{2\times 20}{\pi} = \mathbf{12.73\ A}$$

This value, as might have been expected, is twice the value obtained in the previous case.

$\therefore \quad$ d.c. ammeter will read **12.73 A.**

Example 11.15. *A full-wave rectified sinusoidal voltage is clipped at $1/\sqrt{2}$ of its maximum value. Calculate the average and RMS values of such a voltage.*

Solution. As seen from Fig. 11.28, the rectified voltage has a period of π and is represented by the following equations during the different intervals.

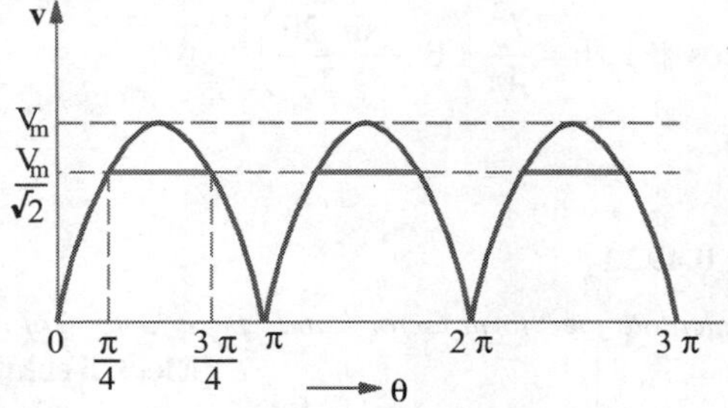

Fig. 11.28

$0 < \theta < \pi/4$; $v = V_m \sin\theta$

$\pi/4 < \theta < 3\pi/4$; $v = V_m/\sqrt{2} = 0.707\, V_m$

$3\pi/4 < \theta < \pi$; $v = V_m \sin\theta$

$$\therefore \quad V_{av} = \frac{1}{\pi}\left\{\int_0^{\pi/4} v d\theta + \int_{\pi/4}^{3\pi/4} v\, d\theta + \int_{3\pi/4}^{\pi} v\, d\theta\right\}$$

$$= \frac{1}{\pi}\left\{\int_0^{\pi/4} V_m \sin\theta\, d\theta + \int_{\pi/4}^{3\pi/4} 0.707\, V_m d\theta + \int_{3\pi/4}^{\pi} V_m \sin\theta\, d\theta\right\}$$

$$= \frac{V_m}{\pi}\left\{\left|-\cos\theta\right|_0^{\pi/4} + 0.707\left|\theta\right|_{\pi/4}^{3\pi/4} + \left|-\cos\theta\right|_{3\pi/4}^{\pi}\right\} = \frac{V_m}{\pi}(0.293 + 1.111 + 0.293) = \mathbf{0.54\ V_m}$$

$$V^2 = \frac{1}{\pi}\left\{\int_0^{\pi/4} V_m^2 \sin^2\theta\, d\theta + \int_{\pi/4}^{3\pi/4} (0.707\, V_m)^2 d\theta + \int_{3\pi/4}^{\pi} V_m^2 \sin^2\theta\, d\theta\right\} = 0.341\, V_m^2$$

$$\therefore \quad V = \mathbf{0.584\ V_m}$$

Example 11.16. *A delayed full-wave rectified sinusoidal current has an average value equal to half its maximum value. Find the delay angle θ.*

(Basic Circuit Analysis, Nagpur 1992)

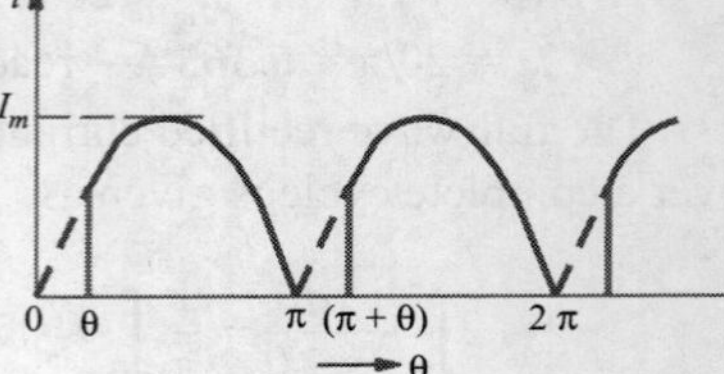

Fig. 11.29

Solution. The current waveform is shown in Fig. 11.29.

$$I_{av} = \frac{1}{\pi}\int_{\theta}^{\pi} I_m \sin\theta\, d\theta = \frac{I_m}{\pi}(-\cos\pi + \cos\theta)$$

Now, $I_{av} = I_m/2$

$$\therefore \quad \frac{I_m}{\pi}(-\cos\pi + \cos\theta) = \frac{I_m}{2}$$

$$\therefore \quad \cos\theta = 0.57,\ \theta = \cos^{-1}(0.57) = \mathbf{55.25^{\circ}}$$

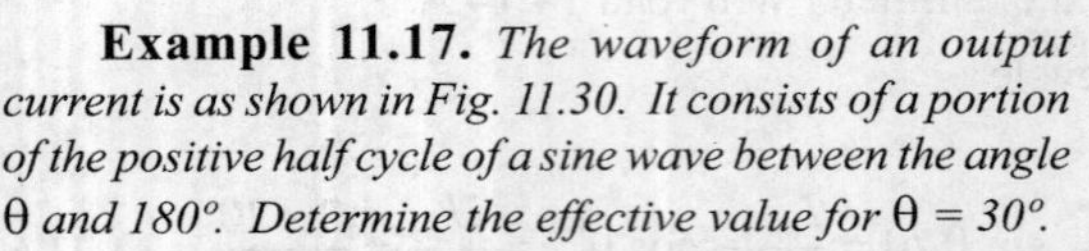

Example 11.17. *The waveform of an output current is as shown in Fig. 11.30. It consists of a portion of the positive half cycle of a sine wave between the angle θ and 180°. Determine the effective value for θ = 30°.*

(Elect. Technology, Vikram Univ. 1984)

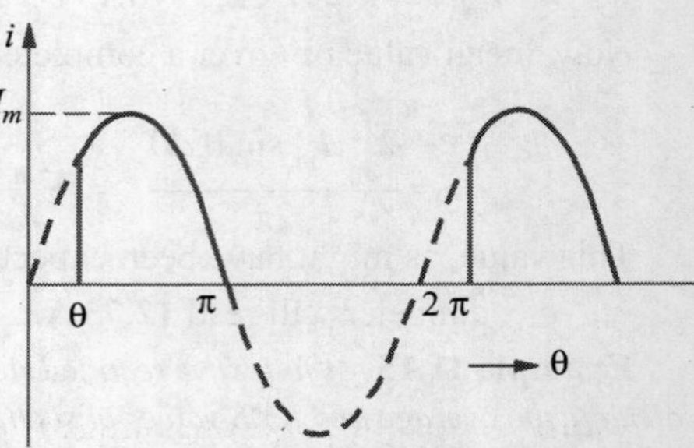

Fig. 11.30

Solution. The equation of the given delayed half-wave rectified sine wave is $i = I_m \sin \omega t = I_m \sin \theta$. The effective value is given by

$$I = \sqrt{\frac{1}{2\pi}\int_{\pi/6}^{\pi} i^2 d\theta} \text{ or } I^2 = \frac{1}{2\pi}\int_{\pi/6}^{\pi} I_m^2 \sin^2\theta . d\theta$$

$$= \frac{I_m^2}{4\pi}\int_{\pi/6}^{\pi}(1 - \cos 2\theta)\, d\theta = \frac{I_m^2}{4\pi}\left|\left(\theta - \frac{\sin 2\theta}{2}\right)_{\pi/6}^{\pi}\right|$$

$$= 0.242\, I_m^2$$

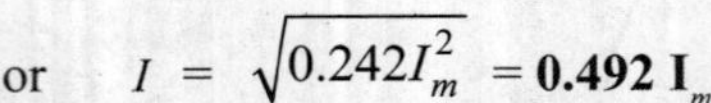

or $\quad I = \sqrt{0.242 I_m^2} = \mathbf{0.492\ I_m}$

Example 11.18. *Calculate the "form factor" and "peak factor" of the sine wave shown in Fig. 11.31.* **(Elect. Technology-I, Gwalior Univ.)**

Solution. For $0 < \theta < \pi$, $i = 100 \sin\theta$ and for $\pi < \theta < 2\pi$, $i = 0$. The period is 2π.

$$\therefore \quad I_{av} = \frac{1}{2\pi}\left\{\int_0^{\pi} i d\theta + \int_{\pi}^{2\pi} 0 d\theta\right\}$$

$$= \frac{1}{2\pi}\left\{100\int_0^{\pi}\sin\theta\, d\theta\right\} = 31.8 \text{ A}$$

i
i = 100 sin 314t
0 2π 3π
θ

Fig. 11.31

$$I^2 = \frac{1}{2\pi}\int_0^{\pi} i^2 d\theta = \frac{100^2}{2\pi}\int_0^{\pi}\sin^2\theta\, d\theta = \frac{100^2}{4} = 2500\ ; I = 50\text{A}$$

$\therefore$ form factor = 50/31.8 = **1.57** ; peak factor = 100/50 = **2**

Example 11.19. *Find the average and effective values of voltage of sinusoidal waveform shown in Fig. 11.32.* **(Elect. Science-I Allahabad Univ. 1991)**

Solution. Although, the given waveform would be integrated from π/4 to π, it would be averaged over the whole cycle because it is unsymmetrical. The equation of the given sinusoidal waveform is $v = 100 \sin\theta$.

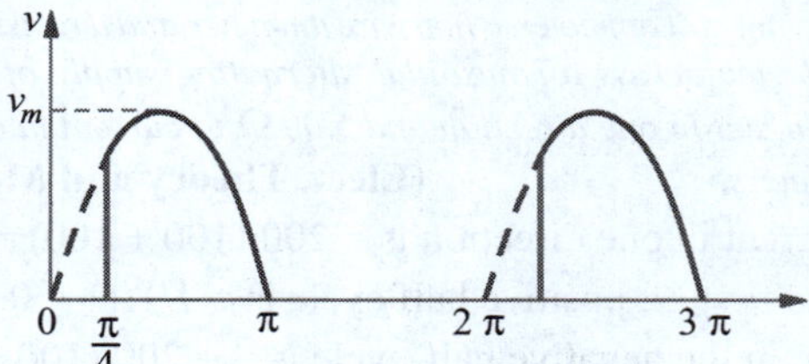

Fig. 11.32

$$\therefore \quad V_{av} = \frac{1}{2\pi}\int_{\pi/4}^{\pi} 100\sin\theta\, d\theta = \frac{100}{2\pi}\left| -\cos\theta \right|_{\pi/4}^{\pi} = 27.2 \text{ V}$$

$$V^2 = \frac{1}{2\pi}\int_{\pi/4}^{\pi} 100^2 \sin^2\theta\, d\theta = \frac{100^2}{4\pi}\frac{\pi}{4}(1-\cos 2\theta)\, d\theta = \frac{100^2}{4\pi}\left|\theta - \frac{\sin 2\theta}{2}\right|_{\pi/4}^{\pi} = \frac{100^2}{4\pi}\left(\pi - \frac{\pi}{4} + \frac{1}{2}\right)$$

$\therefore V$ = **47.7 V**

Example 11.20. *A circuit offers a resistance of 20 Ω in one direction and 100 Ω in the reverse direction. A sinusoidal voltage of maximum value 200 V is applied to the above circuit in series with*

(a) *a moving-iron ammeter* ***(b)*** *a moving-coil ammeter*

(c) *a moving-coil instrument with a full-wave rectifier* ***(d)*** *a moving-coil ammeter.*

Calculate the reading of each instrument.

Solution. (*a*) The deflecting torque of an *MI* instrument is proportional to (current)^2. Hence, its reading will be proportional to the average value of i^2 over the whole cycle. Therefore, the reading of such an instrument :

$$= \sqrt{\left[\frac{1}{2\pi}\left(\int_0^{\pi} 10^2 \sin^2\theta d\theta + \int_{\pi}^{2\pi} 2^2 \sin^2\theta d\theta\right)\right]}$$

$$= \sqrt{\left[\frac{1}{2\pi}\left(\frac{100}{2}\left|\theta - \frac{\sin 2\theta}{2}\right|_0^{\pi} + \frac{4}{2}\left|\theta - \frac{\sin 2\theta}{2}\right|_{\pi}^{2\pi}\right)\right]} = \sqrt{26} = \mathbf{5.1\ A}$$

(*b*) An MC ammeter reads the average current over the whole cycle.

Average current over positive half-cycle is = 10 × 0.637 = 6.37 A

Average current over positive half-cycle is = – 2 × 0.637 = – 1.27 A

∴ Average value over the whole cycle is = (6.37 – 1.27)/2 = **2.55 A**

(*c*) In this case, due to the full-wave rectifier, the current passing through the operating coil of the instrument would flow in the positive direction during both the positive and negative half cycles.

∴ reading = (6.37 + 1.27)/2 = **3.82 A**

(*d*) Average heating effect over the positive half-cycle is = $\frac{1}{2}I_{m_1}^2 R$

Average heating effect over the negative half-cycle is = $\frac{1}{2}I_{m_2}^2 R$

where I_{m1} = 200/20 = 10 A; I_{m2} = 200/100 = 2 *A*

Average heating effect over the whole cycle is = $\left(\frac{1}{2}\times 10^2 R + \frac{1}{2}\times 2^2 \times R\right)/2 = 26\ R$

If *I* is the direct current which produces the same heating effect, then

$I^2 R = 26\ R$ $\therefore\ I = \sqrt{26}$ = **5.1 A**

Example 11.21. *A moving coil ammeter, a hot-wire ammeter and a resistance of 100 Ω are connected in series with a rectifying device across a sinusoidal alternating supply of 200 V. If the device has a resistance of 100 Ω to the current in one direction and 5.00 Ω to current in opposite direction, calculate the readings of the two ammeters.* **(Elect. Theory and Meas. Madras University)**

Solution. R.M.S. current in one direction is = 200/(100 + 100) = 1 A

Average current in the first *i.e.* positive half cycle is = 1/1.11 = 0.9 A

Similarly, r.m.s. value in the negative half-cycle is = – 200/(100 + 500) = – 1/3 A

Average value = (–1/3)/1.11 = –0.3 A

Average value over the whole cycle is = (0.9 – 0.3)/2 = 0.3 A

Hence, M/C ammeter reads **0.3 A**

Average heating effect during the +ve half cycle = $I_{rms}^2 \times R = I^2 \times R = R$

Similarly, average heating effect during the –ve half-cycle is = $(-1/3)^2 \times R = R/9$

Here, R is the resistance of the hot-wire ammeter.

Average heating effect over the whole cycle is = $\frac{1}{2}\left(R + \frac{R}{9}\right) = \frac{5R}{9}$

If I is the direct current which produces the same heating effect, then

$$I^2 R = 5R/9 \qquad \therefore \; I = \sqrt{5/3} = 0.745 \text{ A}$$

Hence, hot-wire ammeter indicates **0.745 A**

Example 11.22. *Determine the r.m.s. value of a semi-circular current wave which has a maximum value of a.*

Solution. The equation of a semi-circular wave (shown in Fig. 11.33) is

$$x^2 + y^2 = a^2 \quad \text{or} \quad y^2 = a^2 - x^2$$

$$\therefore I_{rms} = \sqrt{\frac{1}{2a}\int_{-a}^{+a} y^2 dx} \quad \text{or} \quad I_{rms}^2 = \frac{1}{2a}\int_{-a}^{+a} (a^2 - x^2)\, dx$$

$$= \frac{1}{2a}\int_{-a}^{+a} (a^2\, dx - x^2\, dx)$$

$$= \frac{1}{2a}\left| a^2 x - \frac{x^3}{3} \right|_{-a}^{+a} = \frac{1}{2a}\left(a^3 - \frac{a^3}{3} + a^3 - \frac{a^3}{3}\right) = \frac{2a^2}{3}$$

$$\therefore \qquad I_{rms} = \sqrt{2a^2/3} = \mathbf{0.816\ a}$$

Fig. 11.33

Example 11.23. *Calculate the r.m.s. and average value of the voltage wave shown in Fig. 11.34.*

Solution. In such cases, it is difficult to develop a single equation. Hence, it is usual to consider two equations, one applicable from 0 to 1 and an other form 1 to 2 millisecond.

For t lying between 0 and 1 ms, $v_1 = 4$, For t lying between 1 and 2 ms, $v_2 = -4t + 4$

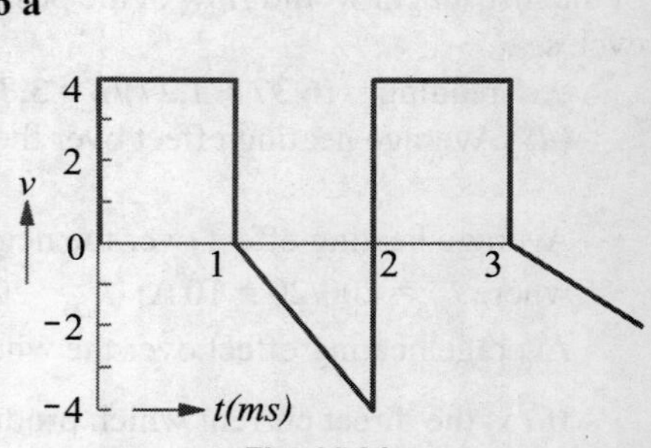

Fig. 11.34

$$\therefore \quad v_{rms} = \sqrt{\frac{1}{2}\left(\int_0^1 v_1^2\, dt + \int_1^2 v_2^2\, dt\right)}$$

$$V^2{}_{rms} = \frac{1}{2}\left[\int_0^1 4^2\, dt + \int_1^2 (-4t+4)^2\, dt\right]$$

$$= \frac{1}{2}\left[\left|16\,t\right|_0^1 + \left|\frac{16\,t^3}{3}\right|_1^2 + \left|16\,t\right|_1^2 - \left|\frac{32\,t^2}{2}\right|_1^2\right]$$

$$= \frac{1}{2}\left[16 + \frac{16\times 8}{3} - \frac{16}{3} + 16\times 2 - 16\times 1 - \frac{32\times 4}{2} + \frac{32\times 1}{2}\right] = \frac{32}{3} \quad \therefore \; V_{rms} = \sqrt{32/3} = \mathbf{3.265\ volt}$$

$$V_{av} = \frac{1}{2}\left[\int_0^1 v_1 dt + \int_1^2 v_2\, dt\right] = \frac{1}{2}\left[\int_0^1 4dt + \int_1^2 (-4t+4)\, dt\right] = \frac{1}{2}\left[\left|4t\right|_0^1 + \left|-\frac{4t^2}{2} + 4t\right|_1^2\right] = \mathbf{1\ volt}$$

Tutorial Problem 11.1

1. Calculate the maximum value of the e.m.f. generated in a coil which is rotating at 50 rev/s in a uniform magnetic field of 0.8 Wb/m^3. The coil is wound on a square former having sides 5 cm in length and is wound with 300 turns. **[188.5 V]**

2. (*a*) What is the peak value of a sinusoidal alternating current of 4.78 r.m.s. amperes ?

(*b*) What is the r.m.s. value of a rectangular voltage wave with an amplitude of 9.87 V ?

(*c*) What is the average value of a sinusoidal alternating current of 31 A maximum value ?

(*d*) An alternating current has a periodic time of 0.03 second. What is its frequency ?

(*e*) An alternating current is represented by i = 70.7 sin 520 *t*. Determine (*i*) the frequency (*ii*) the current 0.0015 second after passing through zero, increasing positively.

[6.76 A ; 9.87 V ; 19.75 A ; 33.3 Hz ; 82.8 Hz ; 49.7 A]

3. A sinusoidal alternating voltage has an r.m.s. value of 200 V and a frequency of 50 Hz. It crosses the zero axis in a positive direction when $t = 0$. Determine (*i*) the time when voltage first reaches the instantaneous value of 200 V and (*ii*) the time when voltage after passing through its maximum positive value reaches the value of 141.4 V. **[(*i*) (0.0025 second (*ii*) 1/300 second)]**

4. Find the form factor and peak factor of the triangular wave shown in Fig. 11.35 **[1.155; 1.732]**

5. An alternating voltage of 200 sin 471 *t* is applied to a h.w. rectifier which is in series with a resistance of 40 Ω. If the resistance of the rectifier is infinite in one direction and zero in the other, find the r.m.s. value of the current drawn from the supply source. **[2.5 A]**

100 V

0 $\frac{\pi}{2}$ → θ

Fig. 11.35

6. A sinusoidally varying alternating current has an average value of 127.4 A. When its value is zero, then its rate change is 62,800 A/s. Find an analytical expression for the sine wave. **[*i* = 200 sin 100 π*t*]**

7. A resistor carries two alternating currents having the same frequency and phase and having the same value of maximum current *i.e.* 10 A. One is sinusoidal and the other is rectangular in waveform. Find the r.m.s. value of the resultant current. **[12.24 A]**

8. A copper-oxide rectifier and a non-inductive resistance of 20 Ω are connected in series across a sinusoidal a.c. supply of 230 V (r.m.s.). The resistance of the rectifier is 2.5 Ω in forward direction and 3,000 Ω in the reverse direction. Calculate the r.m.s. and average values of the current.

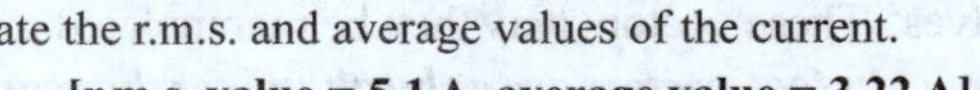

[r.m.s. value = 5.1 A, average value = 3.22 A]

9. Find the average and effective values for the waveshape shown in Fig. 11.36 if the curves are parts of a sine wave. **[27.2 V, 47.7V]** ***(Elect. Technology, Indore Univ.)***

10. Find the effective value of the resultant current in a wire which carries simultaneously a direct current of 10 A and a sinusoidal alternating current with a peak value of 15 A.

[14.58A] ***(Elect. Technology, Vikram Univ. Ujjain)***

11. Determine the r.m.s. value of the voltage defined by $e = 5 + 5 \sin (314\, t + \pi/6)$

[6.12 V] ***(Elect. Technology, Indore Univ.)***

12. Find the r.m.s. value of the resultant current in a wire which carries simultaneously a direct current of 10 A and a sinusoidal alternating current with a peak value of 10 A.

[12.25 A] ***(Elect. Technology-I; Delhi Univ.)***

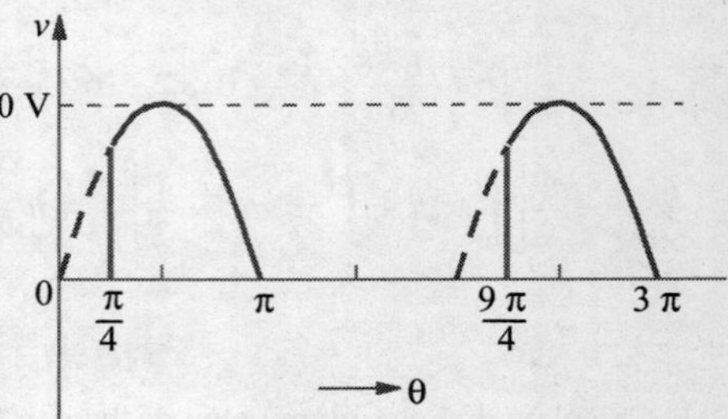

Fig. 11.36

13. An alternating voltage given by $e = 150 \sin 100\pi t$ is applied to a circuit which offers a resistance of 50 ohms to the current in one direction and completely prevents the flow of current in the opposite direction. Find the r.m.s. and average values of this current and its form factor.

[1.5 A, 0.95 A, 1.57] ***(Elect. Technology, Indore Univ.)***

14. Find the relative heating effects of three current waves of equal maximum value, one rectangular, the second semi-circular and the third sinusoidal in waveform.

[1: 2/2, 1/2] ***(Sheffield Univ. U.K.)***

15. Calculate the average and root mean-square value, the form factor and peak factor of a periodic current wave have the following values for equal time intervals over half-cycle, changing suddenly from one value of the next. **[0, 40, 60, 80, 100, 80, 60, 40, 0]** ***(A.M.I.E.)***

16. A sinusoidal alternating voltage of amplitude 100 V is applied across a circuit containing a rectifying device which entirely prevents current flowing in one direction and offers a resistance of 10 ohm to the flow of current in the other direction. A hot wire ammeter is used for measuring the current. Find the reading of instrument. ***(Elect. Technology. Punjab Univ.)***

11.23. Representation of Alternating Quantities

It has already been pointed out that an attempt is made to obtain alternating voltages and currents having sine waveform. In any case, a.c. computations are based on the assumption of sinusoidal voltages and currents. It is, however, cumbersome to continuously handle the instantaneous values in the form of equations of waves like $e = E_m \sin \omega t$ etc. A conventional method is to employ vector method of representing these sine waves. These vectors may then be manipulated instead of the sine functions to achieve the desired result. ***In fact, vectors are a shorthand for the represen-tation of alternating voltages and currents and their use greatly simplifies the problems in a.c. work.***

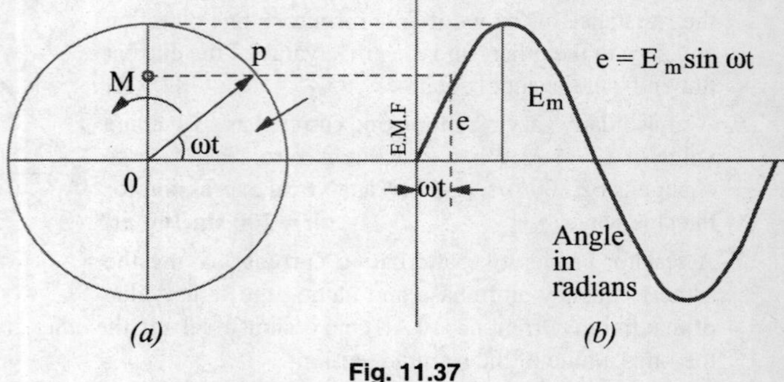

Fig. 11.37

A vector is a physical quantity which has magnitude as well as direction. Such vector quantities are completely known when particulars of their magnitude, direction and the sense in which they act, are given. They are graphically represented by straight lines called vectors. The length of the line represents the magnitude of the alternating quantity, the inclination of the line with respect to some axis of reference gives the direction of that quantity and an arrow-head placed at one end indicates the direction in which that quantity acts.

The alternating voltages and currents are represented by such vectors rotating counter-clockwise with the same frequency as that of the alternating quantity. In Fig. 11.37 (*a*), *OP* is such a vector which represents the maximum value of the alternating current and its angle with *X* axis gives its phase. Let the alternating current be represented by the equation $e = E_m \sin \omega t$. It will be seen that the projection of *OP* and *Y*-axis at any instant gives the instantaneous value of that alternating current.

$$\therefore \qquad OM = OP \sin \omega t \text{ or } e = OP \sin \omega t = E_m \sin \omega t$$

It should be noted that a vector like *OP* can be made to represent an alternating voltage of current if it satisfies the following conditions :

(i) Its length should be equal to the peak or maximum value of the sinusoidal alternating current to a suitable scale. ***(ii)*** It should be in the horizontal position at the same instant as the alternating quantity is zero and increasing. ***(iii)*** Its angular velocity should be such that it completes one revolution in the same time as taken by the alternating quantity to complete one cycle.

11.24. Vector Diagram using R.M.S. Values

Instead of using maximum values as above, it is very common practice to draw vector diagrams using r.m.s. values of alternating quantities. But it should be understood that in that case, the projection of the rotating vector on the *Y*-axis does not give the instantaneous value of that alternating quantity.

11.25. Vector Diagrams of Sine Waves of Same Frequency

Two or more sine waves of the same frequency can be shown on the same vector diagram because the various vectors representing different waves all rotate counter-clockwise at the same frequency and maintain a fixed position relative to each other. This is illustrated in Fig. 11.38 where a voltage *e* and current *i* of the same frequency are shown. The current wave is supposed to pass upward through zero at the instant when $t = 0$ while at the same time the voltage wave has already advanced an angle α from its zero value. Hence, their equations can be written as

$$i = I_m \sin \omega t$$

and
$$e = E_m \sin (\omega t + \alpha)$$

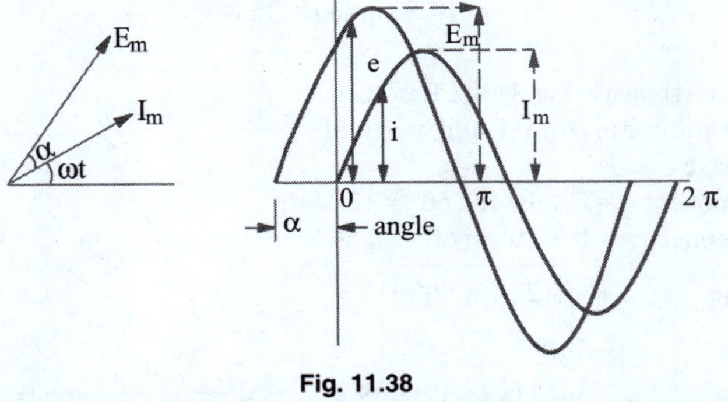

Fig. 11.38

Sine wave of different frequencies cannot be represented on the same vector diagram in a still picture because due to difference in speed of different vectors, the phase angles between them will be continuously changing.

11.26. Addition of Two Alternating Quantities

In Fig. 11.39 (*a*) are shown two rotating vectors representing the maximum values of two sinusoidal voltage waves represented by $e_1 = E_{m1} \sin \omega t$ and $e_2 = E_{m2} \sin (\omega t - \phi)$. It is seen that the sum of the two sine waves of the same frequency is another sine wave of the same frequency but of a different maximum value and phase. The value of the instantaneous resultant voltage e_r, at any instant is obtained by algebraically adding the projections of the two vectors on the *Y*-axis. If these projections are e_1 and e_2, then, $e_r = e_1 + e_2$ at that time. The resultant curve is drawn in this way by adding the ordinates. It is found that the resultant wave is a sine wave of the same frequency as the component waves but lagging behind E_{m1} by an angle α. The vector diagram of Fig. 11.39 (*a*) can be very easily drawn. Lay off E_{m2} lagging φ° behind E_{m1} and then complete the parallelogram to get E_r.

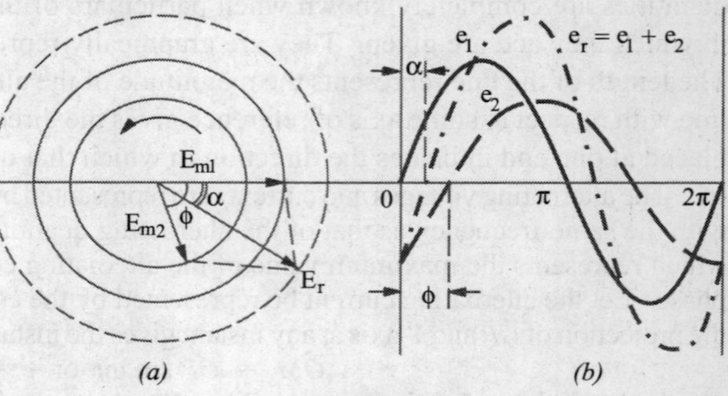

Fig. 11.39

Example 11.24. *Add the following currents as waves and as vectors.*

$$i_1 = 7 \sin \omega t \text{ and } i_2 = 10 \sin (\omega t + \pi/3)$$

Solution. As Waves

$$i_r = i_1 + i_2 = 7 \sin \omega t + 10 \sin (\omega t + 60^\circ)$$
$$= 7 \sin \omega t + 10 \sin \omega t \cos 60^\circ + 10 \cos \omega t \sin 60^\circ$$
$$= 12 \sin \omega t + 8.66 \cos \omega t$$

14.8
8.66
α

Fig. 11.40

Dividing both sides by $\sqrt{(12^2 + 8.66^2)} = 14.8$, we get

$$i_r = 14.8 \left(\frac{12}{14.8} \sin \omega t + \frac{8.66}{14.8} \cos \omega t \right)$$
$$= 14.8 (\cos \alpha \sin \omega t + \sin \alpha \cos \omega t)$$

where $\cos \alpha = 12/14.8$ and $\alpha = 8.66/14.8$ —as shown in Fig. 11.40

$\therefore \quad i_r = 14.8 \sin (\omega t + \alpha)$

where $\tan \alpha = 8.66/12$ or $\alpha = \tan^{-1}(8.66/12) = 35.8^\circ$

$\therefore \quad i_r = 14.8 \sin (\omega t + 35.8^\circ)$

As Vectors

Vector diagram is shown in Fig. 11.41. Resolving the vectors into their horizontal and vertical components, we have

X – component $= 7 + 10 \cos 60^\circ = 12$

Y – component $= 0 + 10 \sin 60^\circ = 8.66$

Resultant $= \sqrt{(12^2 + 8.66^2)}$

$= 14.8$ A

and $\alpha = \tan^{-1} (8.66/12) = 35.8^\circ$

Hence, the resultant equation can be written as

$$\boldsymbol{i_r = 14.8 \sin (\omega t + 35.8^\circ)}$$

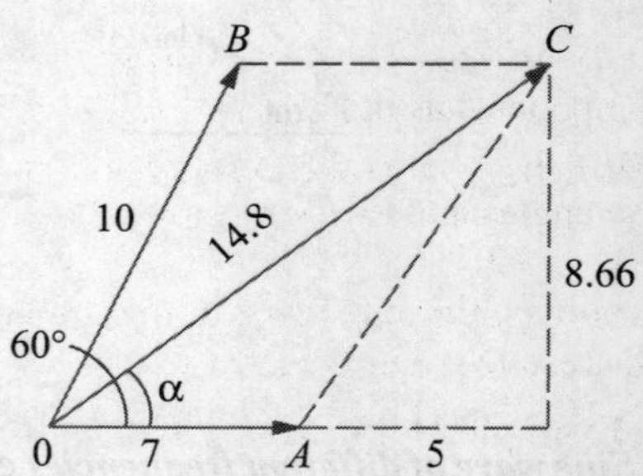

Fig. 11.41

11.27. Addition and Subtraction of Vectors

***(i)* Addition.** In a.c. circuit problems we may be concerned with a number of alternating voltages or currents of the same frequency but of different phases and it may be required to obtain the resultant voltage or current. As explained earlier (Art. 11.23) if the quantities are sinusoidal, they may be represented by a number of rotating vectors having a common axis of rotation and displaced from one another by fixed angles which are equal to the phase differences between the respective alternating quantities. The instantaneous value of the resultant voltage is given by the algebraic sum of the projections of the different vectors on *Y*-axis. The maximum value (or r.m.s. value if the vectors represent that value) is obtained by compounding the several vectors by using the parallelogram and polygon laws of vector addition.

However, another easier method is to resolve the various vectors into their *X*-and *Y*-components and then to add them up as shown in Example 11.30 and 31.

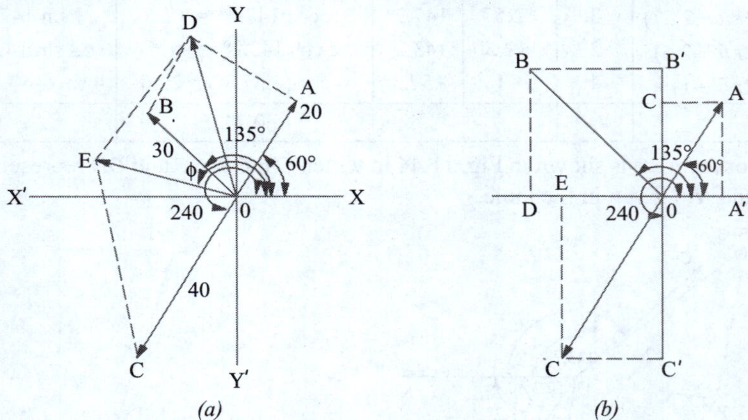

Fig. 11.42

Suppose we are given the following three alternating e.m.fs. and it is required to find the equation of the resultant e.m.f.

$$e_1 = 20 \sin (\omega t + \pi/3)$$
$$e_2 = 30 \sin (\omega t + 3\pi/4)$$
$$e_3 = 40 \sin (\omega t + 4\pi/3)$$

Then the vector diagram can be drawn as explained before and solved in any of the following three ways :

(i) By compounding according to parallelogram law as in Fig. 11.42 (*a*)

(ii) By resolving the various vectors into their *X*-and *Y*-components as in Fig. 11.42 (*b*).

(iii) By laying off various vectors end-on end at their proper phase angles and then measuring the closing vector as shown in Fig. 11.43.

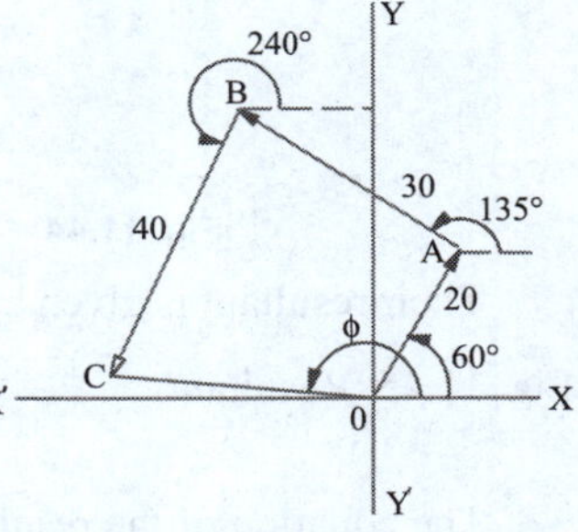

Fig. 11.43

Knowing the magnitude of the resultant vector and its inclination ϕ with *X* axis, the equation of the resultant e.m.f. can be written as $e = E_m \sin (\omega t + \phi)$.

Example 11.25. *Represent the following quantities by vectors :*
5 sin $(2\pi ft - 1)$ *; 3 cos* $(2\pi ft + 1)$ *; 2 sin* $(2\pi ft + 2.5)$ *and 4 sin* $(2\pi ft - 1)$
Add the vectors and express the result in the form : A sin $(2\pi ft \pm \Phi)$

Solution. It should be noted that all quantities have the same frequency f, hence they can be represented vertically on the same vector diagram and added as outlined in Art. 11.27. But before doing this, it would be helpful to express all the quantities as sine functions. Therefore, the second expression 3 cos $(2\pi ft + 1)$ can be written as

$$3 \sin\left(2\pi ft+1+\frac{\pi}{2}\right) = 3 \sin(2\pi ft + 1 + 1.57) = 3 \sin(2\pi ft + 2.57)$$

The maximum value of each quantity, its phase with respect to the quantity of reference *i.e.* $X \sin 2\pi f t$, its horizontal and vertical components are given in the table on next page :

Quantity	*Max value*	*Phase*		*Horizontal component*	*Vertical component*
		radians	*angles*		
(i) $5 \sin(2\pi ft - 1)$	5	−1	−57.3°	$5 \times \cos(-57.3°) = 2.7$	$5 \sin(-57.3°) = -4.21$
(ii) $3 \sin(2\pi ft + 2.57)$	3	+2.57	147.2°	$3 \times \cos 147.2° = -2.52$	$3 \sin 147.2° = 1.63$
(iii) $2 \sin(2\pi ft + 2.5)$	2	+ 2.5	143.2°	$2 \cos 143.2° = -1.6$	$2 \sin 143.2° = 1.2$
(iv) $4 \sin(2\pi ft - 1)$	4	− 1	− 57.3°	$4 \cos(-57.3°) = 2.16$	$4 \sin(-57.3°) = -3.07$
Total				0.74	− 4.75

The vector diagram is shown in Fig. 11.44 in which *OA, OB, OC* and *OD* represent quantities ***(i)***, ***(ii)***, ***(iii)*** and ***(iv)*** given in the table.

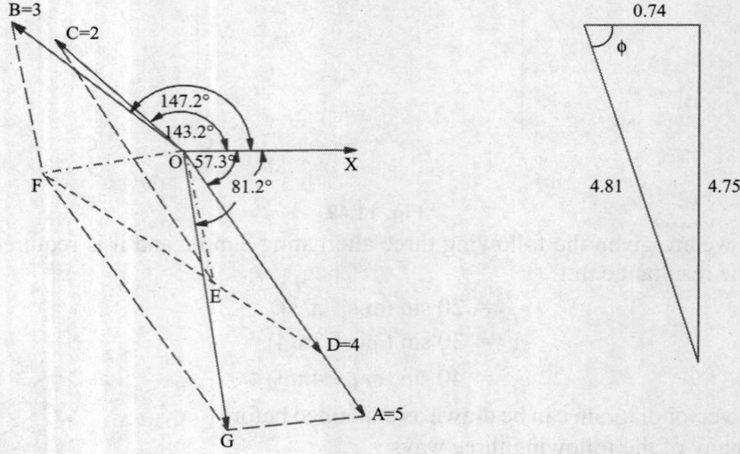

Fig. 11.44 **Fig. 11.45**

Their resultant is given by *OG* and the net horizontal and vertical components are shown in Fig. 11.45. Resultant $= \sqrt{[0.74^2 + (-4.75)^2]} = 4.81$ and $\tan\theta = -4.75 / 0.74$

$\therefore \quad \theta = \tan^{-1}(-4.75/0.74) = -81.2° = -1.43$ radians

The equation of the resultant quantity is **4.81 sin (2πft – 1.43)**

Example 11.26. *Two currents i_1 and i_2 are given by the expressions*

$$i_1 = 10 \sin(314t + \pi/4) \text{ amperes and } i_2 = 8 \sin(313\,t - \pi/3) \text{ amperes}$$

*Find **(a)** $i_1 + i_2$ and **(b)** $i_1 - i_2$. Express the answer in the form $i = I_m \sin(314\,t \pm \phi)$*

Solution. **(*a*)** The current vectors representing maximum values of the two currents are shown in Fig. 11.53 (*a*). Resolving the currents into their *X*-and *Y*-components, we get

X- component $= 10 \cos 45° + 8 \cos 60° = 10/\sqrt{2} + 8/2 = 11.07$ A

Y- component = 10 sin 45° – 8 sin 60 ° = 0.14 A

$\therefore \quad I_m = \sqrt{11.07^2 + 0.14^2} = 11.08$ A

tan φ = (0.14/11.07) = 0.01265 $\therefore$ φ = 44′

Hence, the equation for the resultant current is ***i* = 11.08 sin (314 *t* + 44′) amperes**

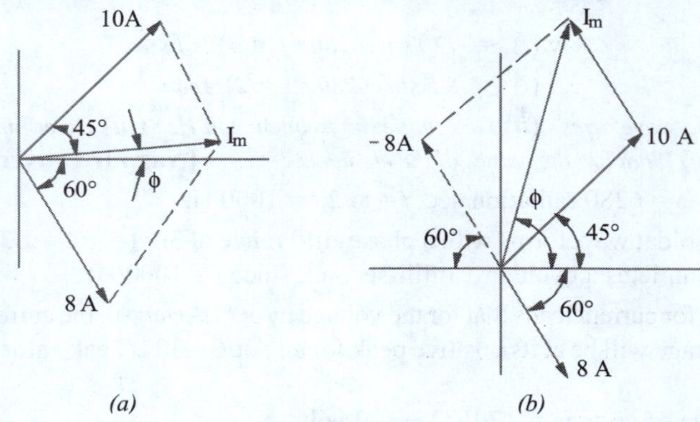

Fig. 11.46

(*b*) *X* – component = 10 cos 45° – 8 cos 60° = 3.07 A

Y – component = 10 sin 45° + 8 sin 60° = 14 A

$\therefore \quad I_m = \sqrt{3.07^2 + 14^2} = 14.33$ A ...Fig. 11.46 (*b*)

φ = tan–¹ (14/3.07) = 77° 38′

Hence, the equation of the resultant current is

***i* = 14.33 sin (314 + 77° 38′) amperes**

Example 11.27. *The maximum values of the alternating voltage and current are 400 V and 20 A respectively in a circuit connected to a 50 Hz supply. The instantaneous values of voltage and current are 283 V and 10 A respectively at time t = 0, both increasing positively.*

(*i*) *Write down the expression for voltage and current at time t.*

(*ii*) *Determine the power consumed in the circuit.*

Take the voltage and current to be sinusoidal **[Nagpur University]**

Solution. V_m = 400, I_m = 20, ω = 314 rad./sec

(*i*) Let the expressions be as follows :

$v(t) = V_m \sin(\omega t + \theta_1) = 400 \sin(314\, t + \theta_1)$

$I(t) = I_m \sin(\omega t + \theta_2) = 20 \sin(314\, t + \theta_2)$

where θ_1 and θ_2 indicate the concerned phase-shifts with respect to some reference. Substituting the given instantaneous values at *t* = 0,

$\theta_1 = 45°$ and $\theta_2 = 30°$

The required expressions are :

$V(t) = 400 \sin(314\, t + 45°)$

$i(t) = 20 \sin(314\, t + 30°)$

Thus, the voltage leads the current by 15°.

V = RMS voltage = 400/1.414 *t* = 283 V

I = RMS voltage = 20/1.414 = 14.14 A

Power-factor, cos ϕ = cos 15° = 0.966 lagging, since current lags behind the voltage.

(ii) Power = V I cos ϕ = 3865 watts

Additional Hint : Draw these two wave forms

Example 11.28. *Voltage and current for a circuit with two elements in series are experssed as follows :*

$$v(t) = 170 \sin(6280\,t + \pi/3) \text{ Volts}$$
$$i(t) = 8.5 \sin(6280\,t + \pi/2) \text{ Amps}$$

(i) *Plot the two waveforms.* ***(ii)*** *Determine the frequency in Hz.* ***(iii)*** *Determine the power factor stating its nature.* ***(iv)*** *What are the values of the elements ?* **[Nagpur University, April 1996]**

Solution. ***(ii)*** ω = 6280 radiation/sec, $f = \omega/2\pi$ = 1000 Hz

(i) Two sinusoidal waveforms with a phase-difference of 30° (= $\pi/2 - \pi/3$) are to be drawn. Each waveform completes a cycle in 1 milli-second, since f = 1000 Hz.

The waveform for current leads that for the voltage by 30°. At $\omega t = 0$, the current is at its positive peak, while the voltage will be at its positive peak for $\omega t = \pi/6 = 30°$. Peak value are 170 volts and 8.5 amp.

(iii) RMS value of voltage = $170/\sqrt{2} = 120$ volts

RMS value of current = $8.5/\sqrt{2} = 6$ amp.

Impedance = V/I = 120/6 = 20 ohms

Power factor = cos 30°, Leading = 0.866,2

Since the current leads the voltage, the two elements must be R and C.

$R = Z \cos\phi = 20 \times 0.866 = 17.32$ ohms

$X_c = Z \sin\phi = 20 \times 0.50 = 10$ ohms

$$C = 1/(\omega X_c) = \frac{1000 \times 1000}{6280 \times 10} = 15.92 \text{ mF}$$

Example 11.29. *Three sinusoidally alternating currents of rms values 5,7.5, and 10 A are having same frequency of 50 Hz, with phase angles of 30°, – 60° and 45°.*

(i) *Find their average values,* ***(ii)*** *Write equations for their instantaneous values,* ***(iii)*** *Draw waveforms and phasor diagrams taking first current as the reference,* ***(iv)*** *Find their instantaneous values at 100 mSec from the original reference.* **[Nagpur University, Nov. 1996]**

Solution. ***(i)*** Average value of alternating quantity in case of sinusoidal nature of variation = (RMS Value)/1.11

Average value of first current = 5/1.11 = 4.50 A

Average value of second current = 7.5/1.11 = 6.76 A

Average value of third current = 10/1.11 = 9.00 A

(ii) Instantaneous Values : $\omega = 2\pi \times 50 = 314$ rad/sec

$i_1(t) = 5\sqrt{2}\,\sin(314\,t + 30°)$

$i_2(t) = 7.5\sqrt{2}\,\sin(314\,t - 60°)$

$i_3(t) = 10\sqrt{2}\,\sin(314\,t + 45°)$

(iii) First current is to be taken as a reference, now. From the expressions, second current lags behind the first current by 90°. Third current leads the first current by 15°. Waveforms with this description are drawn in Fig. 11.47 (*a*) and the phasor diagrams, in Fig. 11.47 (*b*).

(iv) A 50 Hz a.c. quantity completes a cycle in 20 m sec. In 100 m sec, it completes five cycles. Original reference is the starting point required for this purpose. Hence, at 100 m sec from the reference.

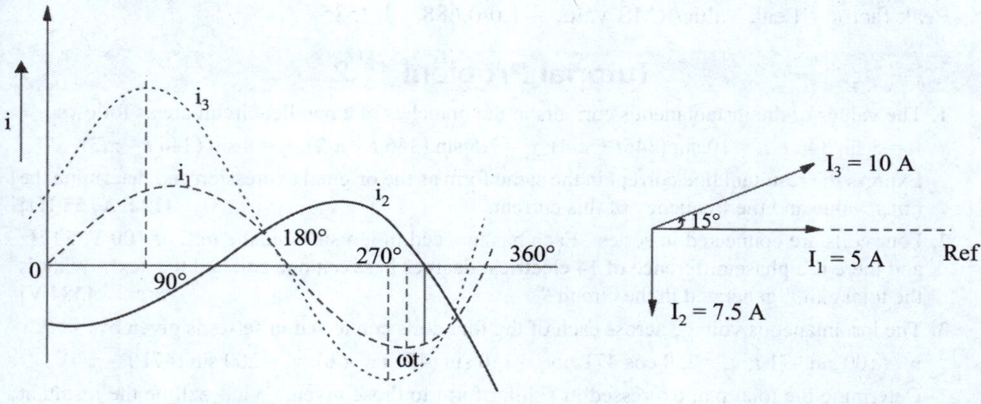

Fig. 11.47 (*a*) **Fig. 11.47 (*b*)**

(v) instantaneous value of $i_1(t) = 5\sqrt{2}\sin 30° = 3.53$ A

instantaneous value of $i_2(t) = 7.5\sqrt{2}\sin(-60°) = -9.816$ A

instantaneous value of $i_3(t) = 10\sqrt{2}\sin(45°) = 10$ A

Example 11.30. *Determine the form factor and peak factor for the unshaded waveform,in Fig. 11.55.* **[Bombay University, 2000]**

Solution. $v(\theta) = V_m \sin\theta$, except for the region between

$\theta = 60°$ to $\theta = 90°$, wherein $v = 0.866\, V_m$.

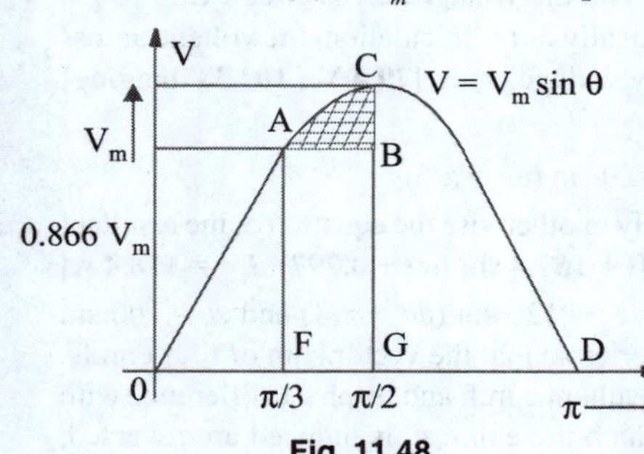

Fig. 11.48

Area under the curve will be worked out first, for calculating the average value.

Area OAF $= V_m \int_0^{\pi/3} \sin\theta d\theta = 0.5\, V_m$

Area FABG $= 0.866\, V_m(\pi/2 - \pi/3) = 0.4532\, V_m$

Area GCD $= V_m$. Total area under the curve $= V_m(1 + 0.432 + 0.5)$

Average value, $V_{av} = (1.9532\, V_m)/3.14 = 0.622\, V_m$

For evaluating rms value, the square of the function is to be taken, its mean value calculated and square-root of the mean value found out.

Area under the squared function :

$$\text{For Portion OF : } V_m^2 \int_0^{\pi/2} \sin^2\theta d\theta = V_m^2/2 \int_0^{\pi/2} (1 - \cos 2\theta)\, d\theta = 0.307\, V_m^2$$

$$\text{For Portion FG : } (0.866\, V_m)^2 \times \pi \times (1/2 - 1/3) = 0.3925\, V_m^2$$

$$\text{For Portion GD : } V_m^2 \int_0^{\pi/2} \sin^2\theta\, d\theta = 0.785\, V_m^2$$

Total area $= V_m^2\,[0.307 + 0.3925 + 0.785] = 1.4845\, V_m^2$

Let R.M.S. Value be V_c

$V_m^2\,\pi = 1.4845\,V_m^2$, or $V_c = 0.688 V_m$

Form factor = RMS Value/Average Value = 0.688/0.622 = 1.106

Peak factor = Peak Value/RMS Value = 1.0/0.688 = 1.4535

Tutorial Problem 11.2

1. The values of the instantaneous currents in the branches of a parallel circuit are as follows :

$i_1 = 5 \sin 346\, t$; $i_2 = 10 \sin (346t + \pi/4)$; $i_3 = 7.5 \sin (346\, t + \pi/2)$; $i_4 = 8 \sin (346\, t - \pi/3)$

Express the resultant line current in the same form as the original expression and determine the r.m.s. value and the frequency of this current. **[12.5 A; 55 Hz]**

2. Four coils are connected in series. Each has induced in it a sinusoidal e.m.f. of 100 V, 50 Hz and there is a phase difference of 14 electrical degrees between one coil and the next. What is the total e.m.f. generated in the circuit ? **[384 V]**

3. The instantaneous voltage across each of the four coils connected in series is given by

$v_1 = 100 \sin 471\, t$; $v_2 = 250 \cos 471\, t$; $v_3 = 150 \sin (471\, t + \pi/6)$; $v_4 = 200 \sin (471\, t - \pi/4)$

Determine the total p.d. expressed in similar form to those given. What will be the resultant p.d. If v_2 is reversed in sign ? **[v = 414 sin (471 *t* + 26.5°); v = 486 sin (471 *t* – 40°)]**

4. An alternating voltage of $v = 100 \sin 376.8\, t$ is applied to a circuit consisting of a coil having a resistance of 6Ω and an inductance of 21.22 mH.

(*a*) Express the current flowing in the circuit in the form $i = I_n \sin (376.8\, t \pm \phi)$

(*b*) If a moving-iron voltmeter, a wattmeter and a frequency meter are connected in the circuit, what would be the respective readings on the instruments ?

[i = 10 sin (376.8 *t* – 53.1°); 70.7 V; 300 W; 60 Hz]

5. Three circuits *A*, *B* and *C* are connected in series across a 200-V supply. The voltage across circuit *A* is 50 V lagging the supply voltage by 45° and the voltage across circuit *C* is 100 V leading the supply voltage by 30°. Determine graphically or by calculation, the voltage across circuit *B* and its phase displacement from the supply voltage. **[79.4 V ; 10° 38′ lagging]**

6. Three alternating currents are given by

$i_1 = 141 \sin (\omega t + \pi/4)$ $i_2 = 30 \sin (\omega t + \pi/2)$ $i_3 = 20 \sin (\omega t - \pi/6)$

and are fed into a common conductor. Find graphically or otherwise the equation of the resultant current and its r.m.s. value. **[i = 167.4 sin (ω*t* + 0.797), I_{rms} = 118.4 A]**

7. Four e.m.fs $e_1 = 100 \sin \omega t$, $e_2 = 80 \sin (\omega t - \pi/6)$, $e_3 = 120 \sin (\omega t + \pi/4)$ and $e_4 = 100 \sin (\omega t - 2\pi/3)$ are induced in four coils connected in series so that the vector sum of four e.m.fs. is obtained. Find graphically or by calculation the resultant e.m.f. and its phase difference with (*a*) e_1 and (*b*) e_2.If the connections to the coil in which the e.m.f. e_2 is induced are reversed, find the new resultant e.m.f.

[208 sin (ω*t* – 0.202) (*a*) 11°34′ lag (*b*) 18′26′ lead; 76 sin (ω*t* + 0.528)]

11.28. A.C. Through Resistance, Inductance and Capacitance

We will now consider the phase angle introduced between an alternating voltage and current when the circuit contains resistance only, inductance only and capacitance only. ***In each case, we will assume that we are given the alternating voltage of equation e = E_m sin ωt*** and will proceed to find the equation and the phase of the alternating current produced in each case.

11.29. A.C. Through Pure Ohmic Resistance Alone

The circuit is shown in Fig. 11.56. Let the applied voltage be given by the equation.

$$v = V_m \sin \theta = V_m \sin \omega t \qquad ...(i)$$

Let R = ohmic resistance ; i = instantaneous current

Obviously, the applied voltage has to supply ohmic voltage drop only. Hence for equilibrium

$$v = iR;$$

Putting the value of 'v' from above, we get $V_m \sin \omega t = iR$; $i = \frac{V_m}{R} \sin \omega t$ ...(ii)

Current 'i' is maximum when $\sin \omega t$ is unity $\therefore$ $I_m = V_m/R$ Hence, equation (ii) becomes, $i = I_m \sin \omega t$ ...(iii)

Comparing (*i*) and (*ii*), we find that the alternating voltage and current are in phase with each other as shown in Fig. 11.50. It is also shown vectorially by vectors V_R and I in Fig. 11.47.

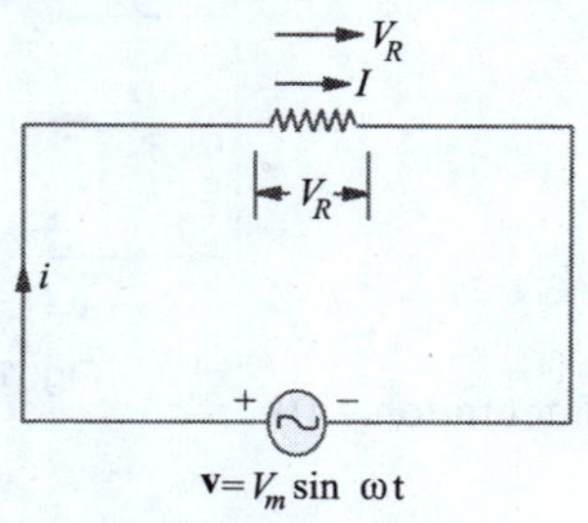

Fig. 11.49

Fig. 11.50

Power. Instantaneous power, $p = vi = V_m I_m \sin^2 \omega t$ ...(Fig. 11.58)

$$= \frac{V_m I_m}{2}(1 - \cos 2\omega t) = \frac{V_m I_m}{2} - \frac{V_m I_m}{2} \cos 2\,\omega t$$

Power consists of a constant part $\frac{V_m I_m}{2}$ and a fluctuating part $\frac{V_m I_m}{2} \cos 2\,\omega t$ of frequency double that of voltage and current waves. For a complete cycle. the average value of $\frac{V_m I_m}{2} \cos 2\omega t$ is zero.

Hence, power for the whole cycle is

$$P = \frac{V_m I_m}{2} = \frac{V_m}{\sqrt{2}} \times \frac{I_m}{\sqrt{2}}$$

or $P = V \times I$ watt

where V = r.m.s. value of applied voltage.

I = r.m.s. value of the current.

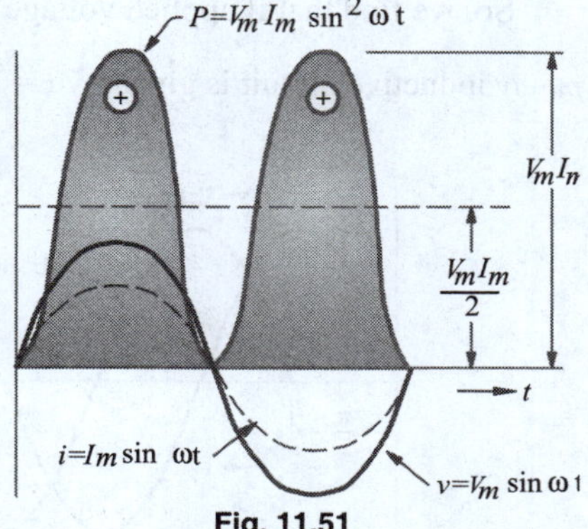

Fig. 11.51

It is seen from Fig. 11.51 that no part of the power cycle becomes negative at any time. In other words, in a purely resistive circuit, power is never zero. This is so because the instantaneous values of voltage and current are always either both positive or negative and hence the product is always positive.

Example 11.31. *A 60-Hz voltage of 115 V (r.m.s.) is impressed on a 100 ohm resistance :*

(i) *Write the time equations for the voltage and the resulting current. Let the zero point of the voltage wave be at t = 0* **(ii)** *Show the voltage and current on a time diagram.* **(iii)** *Show the voltage and current on a phasor diagram.*

[Elect Technology Hyderabad Univ. 1992, Similar Example, U.P. Technical Univ. 2001]

Solution. (*i*) $V_{max} = \sqrt{2}\ V = \sqrt{2} \times 115 = 163$ V

$I_{max} = V_{max}/R = 163/100 = 1.63$ A; $\phi = 0$; $\omega = 2\pi f = 2\pi \times 60 = 377$ rad/s

The required equations are : $v(t) = 1.63 \sin 377\, t$ and $i(t) = 1.63 \sin 377\, t$.

(ii) and **(iii)** These are similar to those shown in Fig. 11.49 and 11.50

11.30. A.C. Through Pure Inductance Alone

Whenever an alternating voltage is applied to a purely inductive coil*, a back e.m.f. is produced due to the self-inductance of the coil. The back e.m.f., at every step, opposes the rise or fall of current through the coil. As there is no ohmic voltage drop, the applied voltage has to overcome this self-induced e.m.f. only. So at every step

$$v = L\frac{di}{dt}$$

Now $\qquad v = V_m \sin \omega t$

$$\therefore \quad V_m \sin \omega t = L\frac{di}{dt} \quad \therefore \quad di = \frac{V_m}{L}\sin \omega t\, dt$$

Fig. 11.52

Integrating both sides, we get $i = \dfrac{V_m}{L}\int \sin \omega t\, dt$

$$= \frac{V_m}{\omega L}(-\cos \omega t) \qquad \text{...(constant of integration = 0)}$$

$$\therefore \quad = \frac{V_m}{\omega L}\sin\left(\omega t - \frac{\pi}{2}\right) = \frac{V_m}{X_L}\sin(\omega t - \pi/2) \qquad \text{...(ii)}$$

Max. value of i is $I_m = \dfrac{V_m}{\omega L}$ when $\sin\left(\omega t - \dfrac{\pi}{2}\right)$ is unity.

Hence, the equation of the current becomes $i = I_m \sin(\omega t - \pi/2)$.

So, we find that if applied voltage is represented by $v = V_m \sin \omega t$, then current flowing in a *purely* inductive circuit is given by $i = I_m \sin\left(\omega t - \dfrac{\pi}{2}\right)$

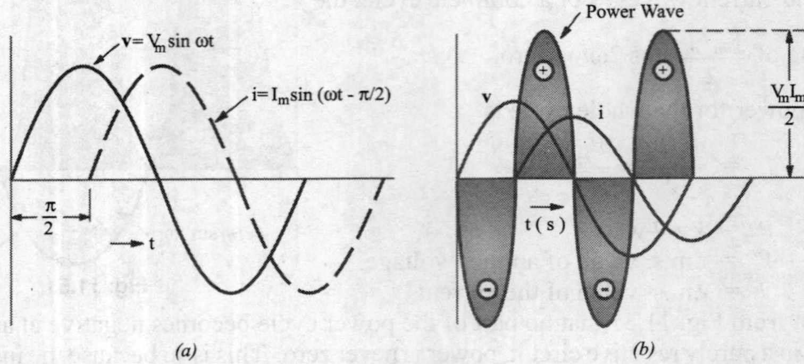

Fig. 11.53

Clearly, the current lags behind the applied voltage by a quarter cycle (Fig. 11.53) or the phase difference between the two is $\pi/2$ with voltage leading. Vectors are shown in Fig. 11.54 where voltage has been taken along the reference axis. We have seen that $I_m = V_m/\omega L = V_m/X_L$. Here '$\omega L$' plays the part of 'resistance'. It is called the (inductive) ***reactance*** X_L of the coil and is given in ohms if L is in henry and ω is in radian/second.

* By purely inductive coil is meant one that has no ohmic resistance and hence no I2R loss. Pure inductance is actually no attainable, though it is very nearly approached by a coil wound with such thick wire that its resistance is negligible. If it has some actual resistance, then it is represented by a separate equivalent resistance joined in series with it.

Now, $X_L = \omega L = 2\pi f L$ ohm. It is seen that X_L depends directly on frequency of the voltage. Higher the value of f, greater the reactance offered and ***vice-versa***.

Power

Instantaneous power $= v_i = V_m I_m \sin \omega t \sin\left(\omega t - \frac{\pi}{2}\right) = -V_m I_m \sin \omega t . \cos \omega t = -\frac{V_m I_m}{2} \sin 2\,\omega t$ *

Power for whole cycle is $P = -\frac{V_m I_m}{2} \int_0^{2\pi} \sin 2\omega\, t\, dt = 0$

It is also clear from Fig. 11.54 (*b*) that the average demand of power from the supply for a complete cycle is zero. Here again it is seen that power wave is a sine wave of frequency double that of the voltage and current waves. The maximum value of the instantaneous power is $V_m I_m/2$.

Example 11.32. *Through a coil of inductance 1 henry, a current of the wave-form shown in Fig. 11.61 (a) is flowing. Sketch the wave form of the voltage across the inductance and calculate the r.m.s. value of the voltage.* **(Elect. Technology, Indore Univ.)**

Solution. The instantaneous current $i\,(t)$ is given by

(i) $0 < t < 1$ second, here slope of the curve is $1/1 = 1$.

$\therefore \quad i = 1 \times t = t$ ampere

(ii) $1 < t < 3$ second, here slope is $\frac{1-(-1)}{2} = 1$

$\therefore \quad t = 1 - (1)(t-1) = 1 - (t-1) = (2-t)$ ampere

(iii) $3 < t < 4$ second, here slope is $\frac{1-0}{1} = -1$

(b) $i = -1 - (-1)(t-3) = (t-4)$ ampere

The corresponding voltage are (*i*) $v_1 = Ldi/dt = 1 \times 1 = 1$ V (*ii*) $v_2 = Ldi/dt = 1 \times \frac{d}{dt}(2-t) = -1$ V

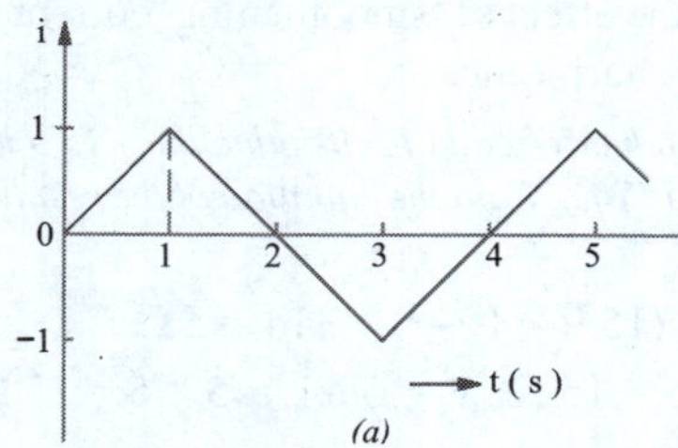

(a)

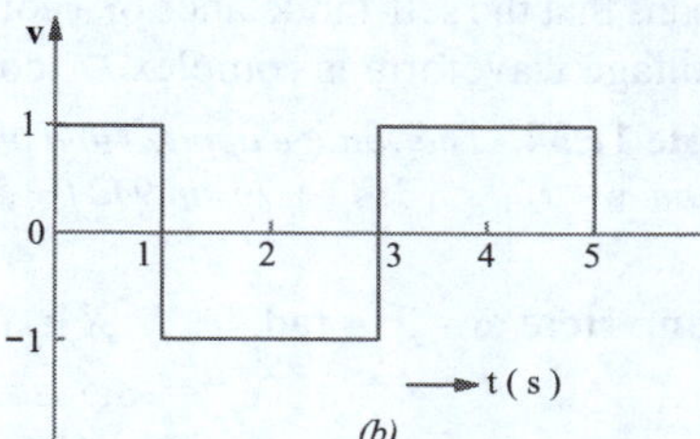

(b)

Fig. 11.54

(iii) $v_3 = Ldi/dt = 1 \times \frac{d}{dt}(t-4) = 1$ V

The voltage waveform is sketched in Fig. 11.54. Obviously, the r.m.s. value of the symmetrical square voltage waveform is **1 V.**

Example 11.33. *A 60-Hz voltage of 230-V effective value is impressed on an inductance of 0.265 H.* ***(i)*** *Write the time equation for the voltage and the resulting current. Let the zero axis of the voltage wave be at t = 0.* ***(ii)*** *Show the voltage and current on a phasor diagram.* ***(iii)*** *Find the maximum energy stored in the inductance.* **(Elect. Engineering, Bhagalpur Univ.)**

Solution. $V_{max} = \sqrt{2}\,V = \sqrt{2} \times 230\,V, f = 60\,Hz.$

---

* Or $p = \frac{1}{2} E_m I_m [\cos 90° - \cos(2\,\omega t - 90°)]$. The constant component $= \frac{1}{2} E_m I_m \cos 90° = 0$. The pulsating component is $-\frac{1}{2} E_m I_m \cos(2\omega t - 90°)$ whose average value over one complete cycle is zero.

$\omega = 2\pi f = 2\pi \times 60 = 377$ rad/s, $X_L = \omega L = 377 \times 0.265 = 100\ \Omega$

(*i*) The time equation for voltage is $v(t) = 230\sqrt{2} \sin 377\, t$;

$I_{max} = V_{max}/X_L = 230\sqrt{2}/100 = 2.3\sqrt{2}, \phi = 90°$ (lag)

$\therefore$ Current equation is $i(t) = 2.3\sqrt{2} \sin (377\, t - \pi/2)$ or $= 2.3\sqrt{2} \cos 377\, t$.

(*ii*) It is shown in Fig. 11.56. (*iii*) $E_{max} = \frac{1}{2} L I_{max}^2 = \frac{1}{2} \times 0.265 \times (2.3\sqrt{2})^2 = \mathbf{1.4\ J}$

11.31. Complex Voltage Applied to Pure Inductance

In Art. 11.30, the applied voltage was a pure sine wave (*i.e.* without harmonics) given by

$$v = V_m \sin \omega t.$$

The current was given by $i = I_m \sin (\omega t - \pi/2)$

Now, it is applied voltage has a complex form and is given by*

$$v = V_{1m} \sin \omega t + V_{3m} \sin 3\omega t + V_{5m} \sin 5\omega t$$

then the reactances offered to the fundamental voltage wave and the harmonics would be different.

For the fundamental wave, $X_1 = \omega L$. For 3rd harmonic ; $X_3 = 3\omega L$. For 5th harmonic ; $x_5 = 5\omega L$.

Hence, the current would be given by the equation.

$$i = \frac{V_{1m}}{\omega L} \sin\left(\omega t - \frac{\pi}{2}\right) + \frac{V_{3m}}{3\omega L} \sin\left(3\omega t - \frac{\pi}{2}\right) + \frac{V_{5m}}{5\omega L} \sin\left(5\omega t - \frac{\pi}{2}\right)$$

Obviously, the harmonics in the current wave are much smaller than in the voltage wave. For example, the 5th harmonic of the current wave is of only 1/5th of the harmonic in the voltage wave. It means that the self-inductance of a coil has the effect of 'smoothening' current waveform when the voltage waveform is complex *i.e.* contains harmonics.

Example 11.34. *The voltage applied to a purely inductive coil of self-inductance 15.9 mH is given by the equation, v = 100 sin 314 t + 75 sin 942 t + 50 sin 1570 t. Find the equation of the resulting current wave.*

Solution. Here $\omega = 314$ rad/s $\quad \therefore X_1 = \omega L = (15.9 \times 10^{-3}) \times 314 = 5\ \Omega$

$$X_3 = 3\omega L = 3 \times 5 = 15\ \Omega;\ X_5 = 5\omega L = 5 \times 5 = 25\ \Omega$$

Hence, the current equation is

$$i = (100/5) \sin (314\, t - \pi/2) + (75/15) \sin (942t - \pi/2) + (50/25) \sin (1570t - \pi/2)$$

or $i = \mathbf{20 \sin (314t - \pi/2) + 5 \sin (942t - \pi/2) + 2 \sin (1570\, t - \pi/2)}$

11.32. A.C. Through Pure Capacitance Alone

When an alternating voltage is applied to the plates of a capacitor, the capacitor is charged first in one direction and then in the opposite direction. When reference to Fig. 11.55, let

v = p.d. developed between plates at any instant

q = Charge on plates at that instant.

Then $\quad q = Cv$ ...where C is the capacitance

$\quad = C V_m \sin \omega t$ ...putting the value of v.

* It is assumed that the harmonics have no individual phase differences.

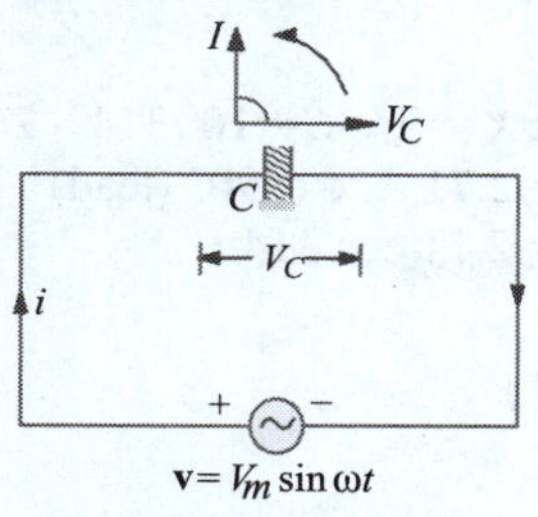

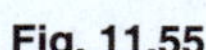

Fig. 11.55

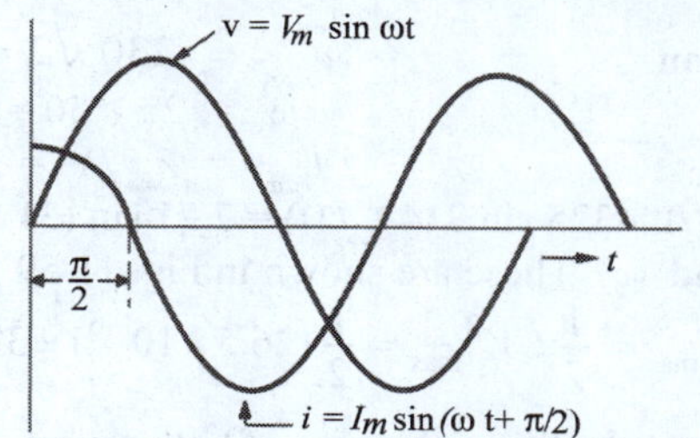

Fig. 11.56

Now, current i is given by the rate of flow of charge.

$$\therefore i=\frac{dq}{dt}=\frac{d}{dt}(CV_m \sin \omega t)=\omega CV_m \cos \omega t \text{ or } i=\frac{V_m}{1/\omega C}\cos \omega t=\frac{V_m}{1/\omega C}\sin\left(\omega t+\frac{\pi}{2}\right)$$

Obviously, $I_m=\frac{V_m}{1/\omega C}=\frac{V_m}{X_C}\quad \therefore\ i=I_m \sin\left(\omega t+\frac{\pi}{2}\right)$

The denominator $X_C = 1/\omega C$ is known as capacitive *reactance* and is in ohms if C is in farad and ω in radian/second. It is seen that if the applied voltage is given by $v = V_m \sin \omega t$, then the current is given by $i = I_m \sin(\omega t + \pi/2)$.

Hence, we find that the current in a pure capacitor leads its voltage by a quarter cycle as shown in Fig. 11.56 or phase difference between its voltage and current is π/2 with the current leading. Vector representation is given in Fig. 11.55. Note that V_c is taken along the reference axis.

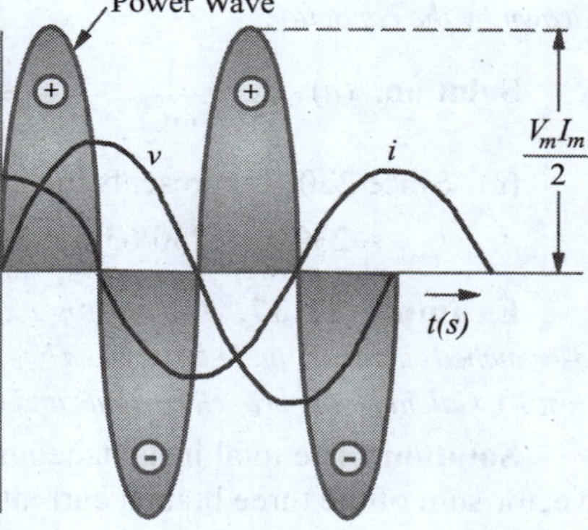

Fig. 11.57

Power. Instantaneous power

$$p = vi = V_m \sin \omega t.\ I_m \sin(\omega t + 90°)$$

$$= V_m I_m \sin \omega t \cos \omega t = \frac{1}{2} V_m I_m \sin 2\omega t\ ^*$$

Power for the whole cycle

$$= \frac{1}{2} V_m I_m \int_0^{2\pi} \sin 2\omega t\ dt = 0$$

This fact is graphically illustrated in Fig. 11.64. We find that in a purely capacitive circuit**, the average demand of power from supply is zero (as in a purely inductive circuit). Again, it is seen that power wave is a sine wave of frequency double that of the voltage and current waves. The maximum value of the instantaneous power is $V_m I_m/2$.

Example 11.35. *A 50-Hz voltage of 230 volts effective value is impressed on a capacitance of 26.5 μF.* **(a)** *Write the time equations for the voltage and the resulting current. Let the zero axis of the voltage wave be at t = 0.* **(b)** *Show the voltage and current on a time diagram.*

(c) *Show the voltage and current on a phasor diagram.* **(d)** *Find the maximum energy stored in the capacitance. Find the relative heating effects of two current waves of equal peak value, the one sinusoidal and the other rectangular in waveform.* **(Elect. Technology, Allahabad Univ. 1991)**

* Or power $p = \frac{1}{2} E_m I_m [\cos 90° - \cos(2\omega t - 90°)]$. The constant component is again zero. The pulsating component averaged over one complete cycle is zero.

** By pure capacitor is meant one that has neither resistance nor dielectric loss. If there is loss in a capacitor, then it may be represented by loss in (*a*) high resistance joined in parallel with the pure capacitor or (*b*) by a comparatively low resistance joined in series with the pure capacitor. But out of the two alternatives usually. (*a*) is chosen.

Solution.

$$V_{max} = 230\sqrt{2} = 325 \text{ V}$$
$$\omega = 2\pi \times 50 = 314 \text{ rad/s}; X_C = 1/\omega C = 10^6/314 \times 26.5 = 120\ \Omega$$
$$I_{max} = V_{max}/X_C = 325/120 = 2.71 \text{ A}, \phi = 90° \text{ (lead)}$$

(*a*) $v(t) = 325 \sin 314\, t;\ i(t) = 2.71 \sin(314t + \pi/2) = 2.71 \cos 314\, t$.

(*b*) and (*c*) These are shown in Fig. 11.59.

(*d*) $E_{max} = \frac{1}{2} CV^2{}_{max} = \frac{1}{2}(26.5 \times 10^{-6}) \times 325^2 = \mathbf{1.4\ J}$

(*e*) Let I_m be the peak value of both waves.

For sinusoidal wave : $H \propto I^2 R \propto (I_m/\sqrt{2})^2 R \propto I_m^2 R/2$. For rectangular wave : $H \propto I_m^2 R$.

$$\frac{H \text{ rectangular}}{H \text{ sinusoidal}} = \frac{I_m^2 R}{I_m^2 R/2} = \mathbf{2}$$

Example 11.36. *A 50-μF capacitor is connected across a 230-V, 50-Hz supply. Calculate* **(*a*)** *the reactance offered by the capacitor* **(*b*)** *the maximum current and* **(*c*)** *the r.m.s. value of the current drawn by the capacitor.*

Solution. (*a*) $X_C = \frac{1}{\omega C} = \frac{1}{2\pi f_C} = \frac{1}{2\pi \times 50 \times 50 \times 10^{-6}} = \mathbf{63.6\ \Omega}$

(*c*) Since 230 V represents the r.m.s. value,

$\therefore\ I_{r.m.s.} = 230/X_c = 230/63.6 = \mathbf{3.62\ A}$ **(*b*)** $I_m = I_{r.m.s.} \times \sqrt{2} = 3.62 \times \sqrt{2} = \mathbf{5.11\ A}$

Example 11.37. *The voltage applied across 3-branched circuit of Fig. 11.58 is given by v = 100 sin (5000t + π/4). Calculate the branch currents and total current.*

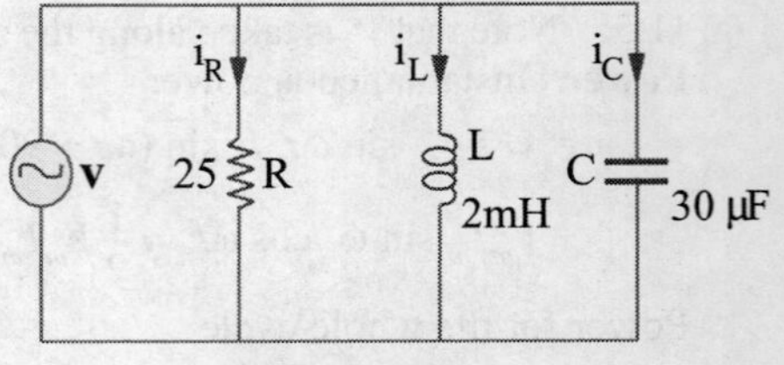

Fig. 11.58

Solution. The total instantaneous current is the vector sum of the three branch currents.

$$i_r = i_R + i_L + i_C.$$

Now $i_R = v/R = 100 \sin(5000\, t + \pi/4)/25$

$$= 4 \sin(5000\, t + \pi/4)$$

$$i_L = \frac{1}{L}\int v\, dt = \frac{10^3}{2}\int 100 \sin\left(5000t + \frac{\pi}{4}\right) dt$$

$$= \frac{10^3 \times 100}{2}\left[\frac{-\cos(5000\, t + \pi/4)}{5000}\right] = -10 \cos(5000\, t + \pi/4)$$

$$i_C = C\frac{dv}{dt} = C.\frac{d}{dt}[100 \sin(5000\, t + \pi/4)]$$

$$= 30 \times 10^{-6} \times 100 \times 5000 \times \cos(5000\, t + \pi/4) = 15 \cos(5000\, t + \pi/4)$$

$$i_r = 4 \sin(5000\, t + \pi/4) - 10 \cos(5000\, t + \pi/4) + 15 \cos(5000t + \pi/4)$$

$$= \mathbf{4 \sin(5000\, t + \pi/4) + 5 \cos(5000\, t + \pi/4)}$$

Tutorial Problem 11.3

1. Define r.m.s. and average value as applied to ac voltage, prove that in pure inductive circuit, current lags behind applied voltage at an angle 90°. (Down waveform)

(Nagpur University, Summer 2002)

2. Find the rms value, average value, form factor and peak factor for the waveform shown in figure.
(Nagpur University, Winter 2003)

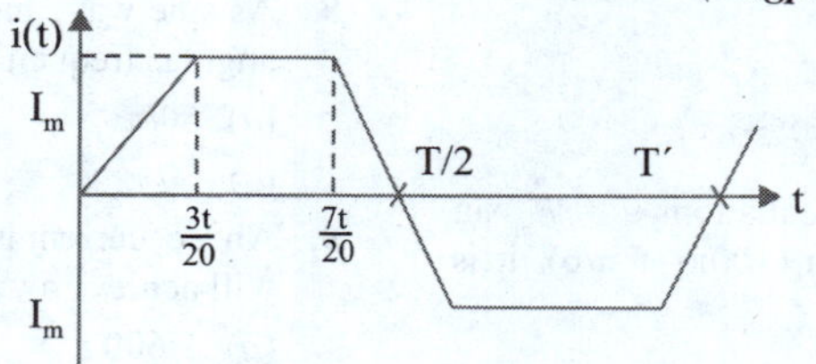

Fig. 11.59

3. Derive an expression for the instantaneous value of alternating sinusoidal e.m.f. in terms of its maximum value, angular freq. and time. *(Gujrat University, June/July 2003)*
4. Prove that average power consumption in pure inductor is zero when a.c. voltage is applied.
(Gujrat University, June/July 2003)
5. Define and explain the following :
(*i*) time period (*ii*) amplitude (*iii*) phase difference. *(Gujrat University, June/July 2003)*
6. What is the r.m.s value of an a.c. quantity? Obtain expression for the r.m.s. value of a sinusoidal current in terms of its maximum value.
(V.T.U., Belgaum Karnataka University, February 2002)
7. Deduce an expression for the average power in a single phase series R.L. circuit and therefrom explain the term power factor. *(V.T.U., Belgaum Karnataka University, February 2002)*
8. Derive an expression for the RMS value of a sine wave.
(V.T.U., Belgaum Karnataka University, Summer 2002)
9. With a neat sketch briefly explain how an alternating voltage is produced when a coil is rotated in a magnetic field. *(V.T.U., Belgaum Karnataka University, Summer 2003)*
10. Derive expressions for average value and RMS value of a sinusoidally varying AC voltage.
(V.T.U., Belgaum Karnataka University, Summer 2003)
11. A circuit having a resistance of 12Ω, an inductance of 0.15 H and a capacitance of 100μ*f* in series is connected across a 100V, 50Hz supply. Calculate the impedance, current, the phase difference between the current and supply voltage.*(V.T.U., Belgaum Karnataka University, Summer 2003)*
12. Two circuits with impedances of $Z_1 = 10 + j15\Omega$ and $Z_2 = 6 - j8\Omega$ are connected in parallel. If the supply current is 20A, what is the power dissipated in each branch?
(V.T.U., Belgaum Karnataka University, Summer 2003)
13. Show that the power consumed in a pure inductance is zero.
(U.P. TechnicalUniversity 2002) (RGPV Bhopal 2002)
14. What do you understand by the terms power factor, active power and reactive power?
(U.P. TechnicalUniversity 2002) (RGPV Bhopal 2002)
15. Current flowing through each line. *(RGPV Bhopal December 2002)*
16. Distinguish between (*i*) apparent power (*ii*) active power and (*iii*) reactive power in A.C. circuits.
(U.P. TechnicalUniversity 2002) (RGPV Bhopal June 2003)

OBJECTIVE TESTS – 11

1. An a.c. current given by $i = 14.14 \sin (\omega t + \pi/6)$ has an r.m.s value of — amperes.
(*a*) 10 (*b*) 14.14
(*c*) 1.96 (*d*) 7.07

and a phase of — degrees.
(*e*) 180 (*f*) 30
(*g*) – 30 (*h*) 210

2. If e_1 = A sin ωt and $e_2 = B \sin(\omega t - \phi)$, then
   (a) e_1 lags e_2 by θ
   (b) e_2 lags e_1 by θ
   (c) e_2 leads e_2 by θ
   (d) e_1 is in phase with e_2
3. From the two voltage equations $e_A = E_m \sin 100\pi t$ and $e_B = E_m \sin(100\pi t + \pi/6)$, it is obvious that
   (a) A leads B by 30°
   (b) B achieves its maximum value 1/600 second before A does.
   (c) B lags behind A
   (d) A achieves its zero value 1/600 second before B.
4. The r.m.s. value of a half-wave rectified current is 10A, its value for full-wave rectification would be — amperes.
   (a) 20 (b) 14.14
   (c) 20/π (d) 40/π
5. A resultant current is made of two components : a 10 A d.c. component and a sinusoidal component of maximum value 14.14 A. The average value of the resultant current is — amperes.
   (a) 0 (b) 24.14
   (c) 10 (d) 4.14
   and r.m.s. value is — amperes.
   (e) 10 (f) 14.14
   (g) 24.14 (h) 100
6. The r.m.s. value of sinusoidal a.c. current is equal to its value at an angle of — degree
   (a) 60 (b) 45
   (c) 30 (d) 90
7. Two sinusoidal currents are given by the equations : $i_2 = 10 \sin(\omega t + \pi/3)$ and $i_2 = 15 \sin(\omega t - \pi/4)$. The phase difference between them is — degrees.
   (a) 105 (b) 75
   (c) 15 (d) 60
8. As sine wave has a frequency of 50 Hz. Its angular frequency is — radian/second.
   (a) 50/π (b) 50/2 π
   (c) 50 π (d) 100 π
9. An a.c. current is given by i = 100 sin 100. It will achieve a value of 50 A after — second.
   (a) 1/600 (b) 1/300
   (c) 1/1800 (d) 1/900
10. The reactance offered by a capacitor to alternating current of frequency 50 Hz is 10Ω. If frequency is increased to 100 Hz reactance becomes—ohm.
    (a) 20 (b) 5
    (c) 2.5 (d) 40
11. A complex current wave is given by $i = 5 + 5 \sin 100\,\pi t$ ampere. Its average value is — ampere.
    (a) 10 (b) 0
    (c) $\sqrt{50}$ (d) 5

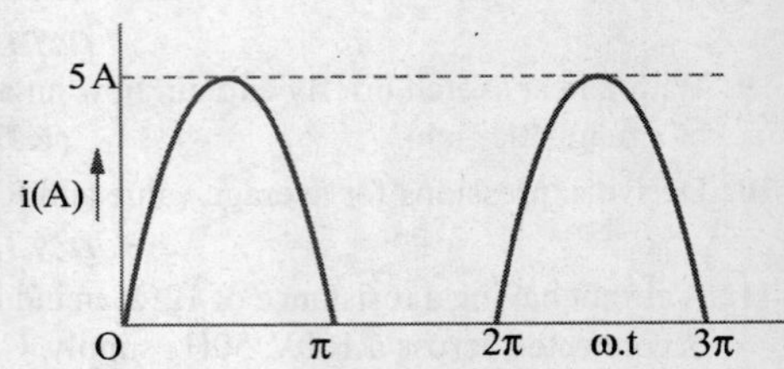

Fig. 11.60

12. The current through a resistor has a waveform as shown in Fig. 11.60. The reading shown by a moving coil ammeter will be—ampere.
    (a) $5/\sqrt{2}$ (b) $2.5/\sqrt{2}$
    (c) 5/π (d) 5

(Principles of Elect. Engg. Delhi Univ.)

ANSWERS

1. *a, f* **2.** *b* **3.** *b* **4.** *b* **5.** *c, f* **6.** *b* **7.** *a* **8.** *d* **9.** *a* **10.** *b* **11.** *d* **12.** *c*

CHAPTER

12

Complex Numbers

Chapter Outline

- ❖ Mathematical
- ❖ Representation of Vectors
- ❖ Symbolic Notation
- ❖ Significance of Operator j
- ❖ Conjugate Complex Numbers
- ❖ Polar Form of Vector Representation
- ❖ Addition and Subtraction of Vector Quantities
- ❖ Multiplication and Division of Vector Quantities
- ❖ Power and Root of Vectors
- ❖ The 120° Operator

12.1. Mathematical Representation of Vectors

There are various forms or methods of representing vector quantities, all of which enable those operations which are carried out graphically in a phasor diagram, to be performed analytically. The various methods are :

(i) Rectangular form. According to this method, a vector quantity is expressed algebraically in terms of its rectangular components. Hence, this form of representation is also known as Cartesian form of notation or representation.

(ii) Polar Form.

12.2. Symbolic Notation

A vector can be specified in terms of its *X*-component and *Y*-component. For example, the vector OE_1 (Fig. 12.1) may be completely described by stating that its horizontal component is a_1 and vertical component is b_1. But instead of stating this verbally, we may express symbolically

$$\mathbf{E}_1 = a_1 + jb_1$$

where symbol j, known as an operator, indicates that component b_1 is perpendicular to component a_1 and that the two terms are ***not*** to be treated like terms in any algebraic expression. The vector

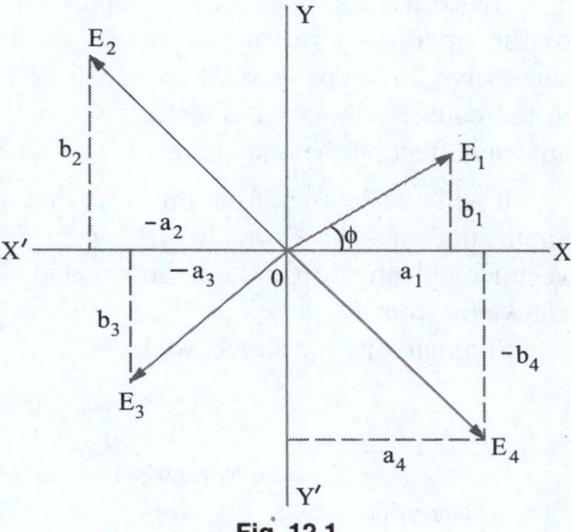

Fig. 12.1

written in this way is said to be written in '***complex form***'. In Mathematics, a_1 is known as real component and b_1 as imaginary component but in electrical engineering, these are known as ***in phase*** (or active) and ***quadrature*** (or reactive) components respectively.

The other vectors OE_2, OE_3 and OE_4 can similarly, be expressed in this form.

$$\mathbf{E}_2 = -a_2 + jb_2 \; ; \; \mathbf{E}_3 = -a_3 - jb_3 \; ; \; \mathbf{E}_4 = +a_4 - jb_4$$

It should be noted that in this book, a vector quantity would be represented by letters in heavy type and its numerical or scalar value by the same letter in ordinary type.* Other method adopted for indicating a vector quantity is to put an arrow about the letter such as $\vec{E}$.

The numerical value of vector E_1 is $\sqrt{a_1^2 + b_1^2}$. Its angle with X-axis is given by $\phi = \tan^{-1}(b_1/a_1)$.

12.3. Significance of Operator *j*

The letter j used in the above expression is a symbol of an operation. Just as symbols ×, + $\sqrt{\ }$, ∫ etc. are used with numbers for indicating certain operations to be performed on those numbers, similarly, symbol j is used to indicate the counter-clockwise rotation of a vector through 90°. It is assigned a value of $\sqrt{(-1)}$ **. The double operation of j on a vector rotates it counter-clockwise through 180° and hence reverses its sense because

$$jj = j^2 = \sqrt{(-1)}^2 = -1$$

When operator j is operated on vector **E**, we get the new vector j**E** which is displaced by 90° in counter-clockwise direction from **E** (Fig. 12.2). Further application of j will give $j^2\mathbf{E} = -\mathbf{E}$ as shown.

If the operator j is applied to the vector $j^2\mathbf{E}$, the result is $j^3\mathbf{E} = -j\mathbf{E}$. The vector $j^3\mathbf{E}$ is 270° counter-clockwise from the reference axis and is directly opposite to $j\mathbf{E}$. If the vector $j^3\mathbf{E}$ is, turn, operated on by j, the result will be

$$j^4\mathbf{E} = \left[\sqrt{(-1)}\right]^4 \mathbf{E} = \mathbf{E}$$

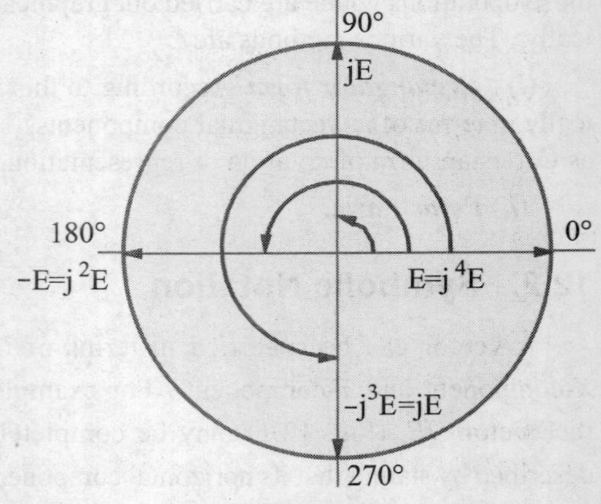

Fig. 12.2

Hence, it is seen that successive applications of the operator j to the vector **E** produce successive 90° steps of rotation of the vector in the counter-clockwise direction without in anyway affecting the magnitude of the vector.

It will also be seen from Fig. 12.2 that the application of $-j$ to **E** yields $-j\mathbf{E}$ which is a vector of identical magnitude but rotated 90° ***clockwise*** from **E**.

Summarising the above, we have

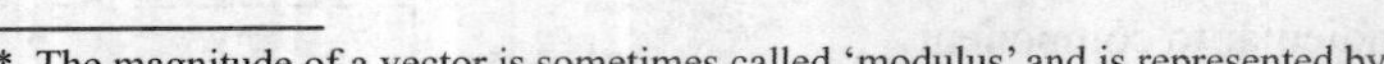

$$j = 90^\circ \text{ ccw rotation} = \sqrt{(-1)}$$

* The magnitude of a vector is sometimes called 'modulus' and is represented by |E| or E.

** In Mathematics, $\sqrt{(-1)}$. is denoted by i but in electrical engineering j is adopted because letter i is reserved for representing current. This helps to avoid confusion.

$$j^2 = 180^\circ \text{ ccw rotation} = [\sqrt{(-1)}]^2 = -1;$$

$$j^3 = 270^\circ \text{ ccw rotation} = [\sqrt{(-1)}]^3 = -\sqrt{(-1)} = -j$$

$$j^4 = 360^\circ \text{ ccw rotation} = [\sqrt{(-1)}]^4 = +1;$$

$$j^5 = 450^\circ \text{ ccw rotation} = [\sqrt{(-1)}]^5 = -\sqrt{(-1)} = j$$

It should also be noted that $\frac{1}{j} = \frac{j}{j^2} = \frac{j}{-1} = -j$

12.4. Conjugate Complex Numbers

Two complex numbers are said to be conjugate if they differ only in the algebraic sign of their quadrature components. Accordingly, the numbers $(a + jb)$ and $(a - jb)$ are conjugate. The sum of two conjugate numbers gives in-phase (or active) component and their difference gives quadrature (or reactive) component.

12.5. Polar Form of Vector Representation

The expression $E(\cos\theta + j\sin\theta)$ is written in the simplified form of $E\angle\theta$. In this expression, E represents the magnitude of the vector and θ its inclination (in ccw direction) with the X-axis. For angles in clockwise direction the expression becomes $E\angle-\theta$. In general, the expression is written as $E\angle\pm\theta$. ***It may be pointed out here that E $\angle\pm\theta$ is simply a short-hand or symbolic style of writing $Ee^{\pm j\theta}$***. Also, the form is purely conventional and does not possess the mathematical elegance of the various other forms of vector representation given above.

Summarizing, we have the following alternate ways of representing vector quantities

(*i*) Rectangular form (or complex form) $\mathbf{E} = a + jb$

(*ii*) Polar form (conventional) $\mathbf{E} = E\angle\pm\theta$.

Example 12.1. *Write the polar forms of vector 3 + j4. How will you illustrate the vector means of diagram ?*

Solution. With reference to Fig. 12.3., magnitude of the vector is $= \sqrt{3^2 + 4^2} = 5$. $\tan\theta = 4/3$.

$\therefore \quad \theta = \tan^{-1}(4/3) = 53.1^\circ$

The angle may also be expressed in radians.

Polar form = $5\angle 53.1^\circ$.

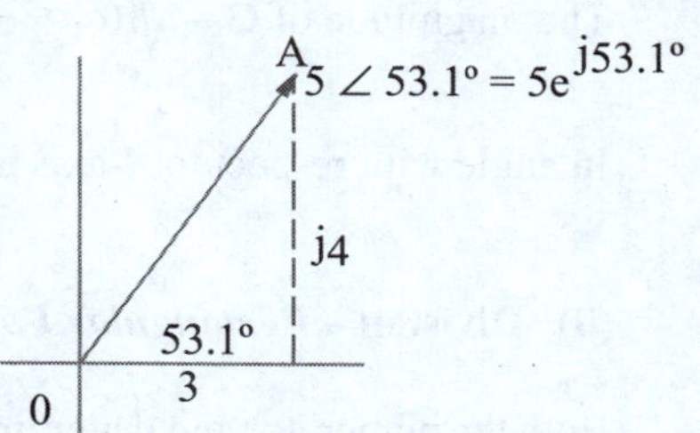

Fig. 12.3

Example 12.2. *A vector is represented by $20\,e^{-j2\pi/3}$. Write the various equivalent forms of the vector and illustrate by means of a vector diagram, the magnitude and position of the above vector.*

Solution. The vector is drawn in a direction making an angle of $2\pi/3 = 120^\circ$ in the clockwise direction (Fig. 12.4). The clockwise direction is taken because the angle is negative.

(*i*) **Rectangular Form** $\quad a = 20\cos(-120^\circ) = -10$;

$b = 20\sin(-120^\circ) = -17.32$

$\therefore \quad$ Expression is = $(-10 - j17.32)$

(*ii*) **Polar Form** is $20\angle-120^\circ$

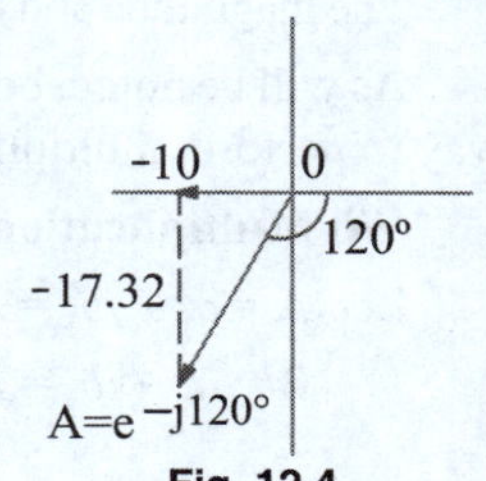

Fig. 12.4

12.6. Addition and Subtraction of Vector Quantities

Rectangular form is best suited for addition and subtraction of vector quantities. Suppose we are given two vector quantities $\mathbf{E}_1 = a_1 + jb_1$ and $\mathbf{E}_2 = a_2 + jb_2$ and it is required to find their sum and difference.

Addition. $\mathbf{E} = \mathbf{E}_1 + \mathbf{E}_2 = a_1 + jb_1 + a_2 + jb_2 = (a_1 + a_2) + j(b_1 + b_2)$

The magnitude of resultant vector **E** is $\sqrt{(a_1 + a_2)^2 + (b_1 + b_2)^2}$

The position of **E** with respect to *X*-axis is $\theta = \tan^{-1}\left(\dfrac{b_1 + b_2}{a_1 + a_2}\right)$

Subtraction. $\mathbf{E} = \mathbf{E}_1 - \mathbf{E}_2 = (a_1 + jb_1) - (a_2 + jb_2) = (a_1 - a_2) + j(b_1 - b_2)$

Magnitude of $\mathbf{E} = \sqrt{(a_1 - a_2)^2 + (b_1 - b_2)^2}$

Its position with respect to *x*-axis is given by the angle $\theta = \tan^{-1}\left(\dfrac{b_1 - b_2}{a_1 - a_2}\right)$.

12.7. Multiplication and Division of Vector Quantities

Multiplication and division of vectors becomes very simple and easy ***if they are represented in the polar or exponential form***. As will be shown below, the rectangular form of representation is not well-suited for this process.

(*i*) Multiplication – *Rectangular Form*

Let the two vectors be given by $\mathbf{A} = a_1 + jb_1$ and $\mathbf{B} = a_2 + jb_2$

$$\therefore \quad \mathbf{A} \times \mathbf{B} = \mathbf{C} = (a_1 + jb_1)(a_2 + jb_2) = a_1a_2 + j^2b_1b_2 + j(a_1b_2 + b_1a_2)$$

$$= (a_1a_2 - b_1b_2) + j(a_1b_2 + b_1a_2) \qquad (\because j^2 = -1)$$

The magnitude of $\mathbf{C} = \sqrt{[(a_1a_2 - b_1b_2)^2 + (a_1b_2 + b_1a_2)^2]}$

In angle with respect to *X*-axis is given by $\theta = \tan^{-1}\left(\dfrac{a_1b_2 + b_1a_2}{a_1a_2 - b_1b_2}\right)$

(ii) Division – *Rectangular Form* : $\dfrac{\mathbf{A}}{\mathbf{B}} = \dfrac{a_1 + jb_1}{a_2 + jb_2} = \dfrac{(a_1 + jb_1)(a_2 - jb_2)}{(a_2 + jb_2)(a_2 - jb_2)}$

Both the numerator and denominator have been multiplied by the conjugate of $(a_2 + jb_2)$ *i.e.* by $(a_2 - jb_2)$

$$\therefore \quad \frac{\mathbf{A}}{\mathbf{B}} = \frac{(a_1a_2 + b_1b_2) + j(b_1a_2 - a_1b_2)}{a_2^2 + b_2^2} = \frac{a_1a_2 + a_1b_2}{a_2^2 + b_2^2} + j\frac{b_1a_2 + a_1b_2}{a_2^2 + b_2^2}$$

The magnitude and the angle with respects *X*-axis can be found in the same way as given above.

As will be noted, both the results are somewhat awkward but unfo tunately, there is no easier way to perform multiplication in rectangular form.

(*iii*) Multiplication – *Polar Form*

Let $\mathbf{A} = a_1 + jb_1 = A\angle\alpha = A\,e^{j\alpha}$ where $\alpha = \tan^{-1}(b_1/a_1)$

$\mathbf{B} = a_2 + jb_2 = B\angle\beta = B\,e^{j\beta}$ where $\beta = \tan^{-1}(b_2/a_2)$

$\therefore$ $\mathbf{AB} = A\angle\alpha \times B\angle\beta = AB\angle(\alpha+\beta)$* or $AB = Ae^{j\alpha} \times Be^{j\beta} = ABe^{j(\alpha+\beta)}$

Hence, product of any two vector **A** and **B** is given by another vector equal in length to **A** × **B** and having a phase angle equal to the sum of the angles of **A** and **B**.

(iv)* Division – *Polar Form

$$\frac{\mathbf{A}}{\mathbf{B}} = \frac{A\angle\alpha}{B\angle\beta} = \frac{A}{B}\angle(\alpha-\beta)$$

Hence, the quotient **A** ÷ **B** is another vector having a magnitude of **A** ÷ **B** and phase angle equal to angle of **A** minus the angle of **B**.

Also $$\frac{\mathbf{A}}{\mathbf{B}} = \frac{Ae^{j\alpha}}{Be^{j\beta}} = \frac{A}{B}e^{j(\alpha-\beta)}$$

As seen, the division and multiplication become extremely simple if vectors are represented in their polar or exponential form.

Example 12.3. *Add the following vectors given in rectangular form and illustrate the process graphically.*

$$A = 16 + j\,12, \quad B = -6 + j\,10.4$$

Solution. $\mathbf{A}+\mathbf{B} = C = (16 + j\,12) + (-6 + j\,10.4) = 10 + j\,22.4$

$\therefore$ Magnitude of **C** $= \sqrt{(10^2 + 22.4^2)} = 25.5$ units

Slope of **C** $= \theta = \tan^{-1}\left(\frac{22.4}{10}\right) = 65.95°$

The vector addition is shown in Fig. 12.5.

$\alpha = \tan^{-1}(12/16) = 36.9°$

$\beta = \tan^{-1}(-10.4/6) = -240°$ or $120°$

The resultant vector is found by using parallelogram law of vectors (Fig. 12.5).

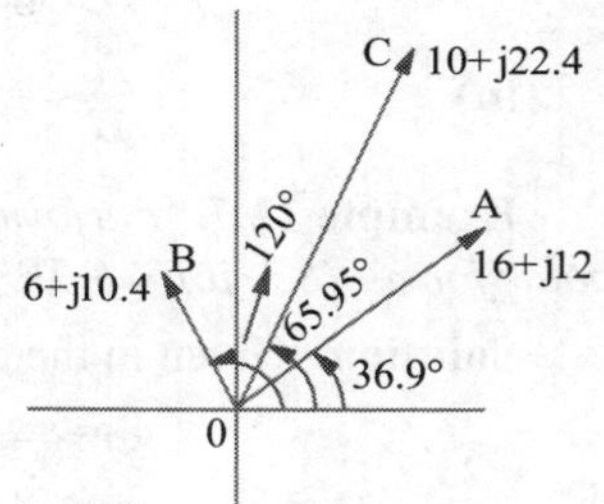

Fig. 12.5

Example 12.4. *Perform the following operation and express the final result in the polar form : 5 ∠ 30° + 8 ∠ – 30°.* **(Elect. Engg. & Electronics Bangalore Univ. 1989)**

Solution. $5\angle 30° = 5(\cos 30° + j\sin 30°) = 4.33 + j\,2.5$

$8\angle -30° = 8[\cos(-30°) + j\sin(-30°)] = 8(0.866 - 0.5) = 6.93 - j\,4$

$\therefore 5\angle 30° + 8\angle -30° = 4.33 + j2.5 + 6.93 - j4 = 11.26 - j1.5 = \sqrt{11.26^2 + 1.5^2}\,\angle\tan^{-1}(-1.5/11.26) = 11.35$

$\tan^{-1}(-0.1332) = 11.35\angle 7.6°$

Example 12.5. *Subtract the following given vectors from one another :*

$$A = 30 + j\,52 \text{ and } B = -39.5 - j\,14.36$$

Solution. $\mathbf{A} - \mathbf{B} = \mathbf{C} = (30 + j\,52) - (-39.5 - j\,14.36) = 69.5 + j\,66.36$

$\therefore$ Magnitude of $C = \sqrt{(66.5^2 + 66.36^2)} = 96$

---

* $\mathbf{A} = A(\cos\alpha + j\sin\alpha)$ and $\mathbf{B} = B(\cos\beta + j\sin\beta)$

$\therefore$ $\mathbf{AB} = AB(\cos\alpha\cos\beta + j\sin\alpha\cos\beta + j\cos\alpha\sin\beta + j^2\sin\alpha\sin\beta)$

$= AB[\cos\alpha\cos\beta - \sin\alpha\sin\beta) + j(\sin\alpha\cos\beta + \cos\alpha\sin\beta)]$

$= AB[\cos(\alpha+\beta) + j\sin(\alpha+\beta)] = AB\angle(\alpha+\beta)$

Slope of $C = \tan^{-1}(66.36/69.5) = 43.6°$ $\therefore$ $\mathbf{C} = 96 \angle 43.6°$.

Similarly $\mathbf{B} - \mathbf{A} = -69.5 - j\,66.36 = 96 \angle 223.6°$* or $= 96 \angle -136.4°$

Example 12.6. *Given the following two vectors :*

$$A = 20 \angle 60° \text{ and } B = 5 \angle 30°$$

*Perform the following indicated operations and illustrate graphically **(i)** A × B and **(ii)** A/B.*

Solution. (*i*) $\mathbf{A} \times \mathbf{B} = \mathbf{C} = 20 \angle 60° \times 5 \angle 30° = 100 \angle 90°$

Vectors are shown in Fig. 12.6.

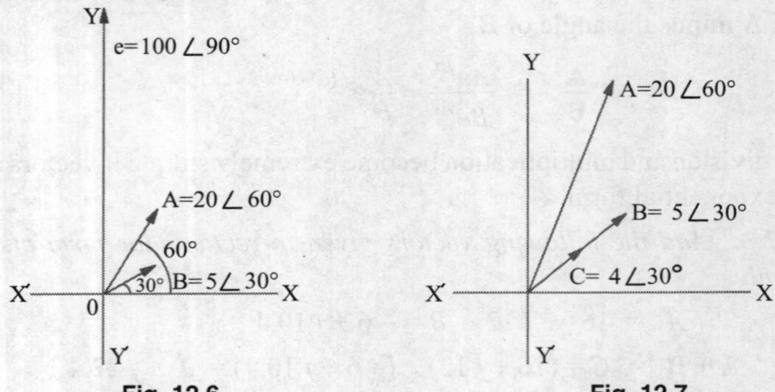

Fig. 12.6 **Fig. 12.7**

(*ii*) $\dfrac{\mathbf{A}}{\mathbf{B}} = \dfrac{20 \angle 60°}{5 \angle 30°} = 4 \angle 30°$ —Fig. 12.10

Example 12.7. *Perform the following operation and the final result may be given in the polar form : (8 + j6) × (–10 – j 7.5)* **(Elect. Engg. & Electronics Bangalore Univ. 1990)**

Solution. Given in the question, polar form: $(8 + j6) \times (-10 - j7.5)$

$\because$ $8 + j6 = 10\underline{|36.9°}$

$-10 - j7.5 = 12.5\ \underline{|216.9°}$

$\therefore$ $10\underline{|36.9°} \times 12.5\underline{|216.9°}$

$10 \times 12.5\ \underline{|36.9° + 216.9°}$

Answer: $125\underline{|253.8°}$

Example 12.8. *Given two current $i_1 = 10 \sin(\omega t + \pi/4)$ and $i_2 = 5 \cos(\omega t - \pi/2)$, find the r.m.s. value of $i_1 + i_2$ using the complex number representation.*

[Elect. Circuit Theory, Kerala Univ.]

Solution. The maximum value of first current is 10 A and it leads the reference quantity by 45°. The second current can be written as

$$i_2 = 5 \cos(\omega t - \pi/2) = 5 \sin[90 + (\omega t - \pi/2)] = 5 \sin \omega t$$

Hence, its maximum value is 5 A and is in phase with the reference quantity.

$\therefore$ $\mathbf{I}_{m1} = 10(\cos 45° + j \sin 45°) = (7.07 + j\,7.07)$

$\mathbf{I}_{m2} = 5(\cos 0° + j \sin 0°) = (5 + j\,0)$

* $\tan\theta = 66.36/69.5$ or $\theta = \tan{-1}(66.36/69.5) = 43.6°$. Since both components are negative, the vector lies in third quadrant. Hence, the angle measured from + ve direction of X-axis and in the CCW directions is = (180 + 43.6) = 223.6°.

The maximum value of resultant current is

$$\mathbf{I}_m = (7.07 + j\,7.07) + (5 + j\,0) = 12.07 + j\,7.07 = 14 \angle 30.4°$$

$\therefore$ R.M.S. value $= 14/\sqrt{2} = \mathbf{10\ A}$ **10 A**

12.8. The 120° Operator

In three-phase work where voltage vectors are displaced from one another by 120°, it is convenient to employ an operator which rotates a vector through 120° forward or backwards without changing its length. This operator is '*a*'. Any operator which is multiplied by '*a*' remains unchanged in magnitude but is rotated by 120° in the counter-clockwise (ccw) direction.

$$\therefore \quad a = 1 \angle 120°$$

This, when expressed in the cartesian form, becomes

$$a = \cos 120° + j \sin 120° = -0.5 + j\,0.866$$

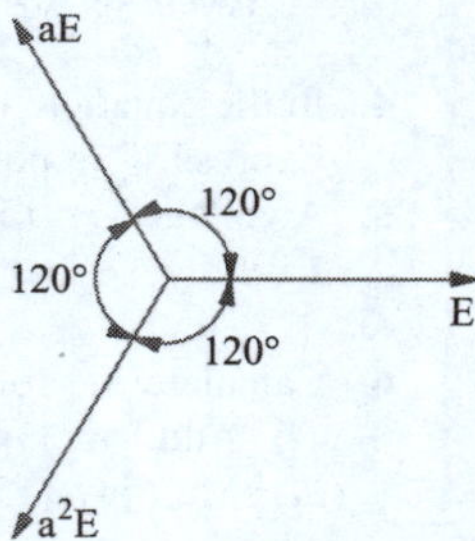

Fig. 12.8

Similarly, $a^2 = 1 \angle 120° \times 1 \angle 120° = 1 \angle 240° = \cos 240° + j \sin 240° = -0.5 - j\,0.866$

Hence, operator 'a^2' will rotate the vector in ccw by 240°. This is the same as rotating the vector in *clockwise* direction by 120°.

$\therefore$ $a^2 = 1 \angle -120°$. Similarly, $a^3 = 1 \angle 360° = 1$*

As shown in Fig. 12.8, the 3-phase voltage vectors with standard phase sequence may be represented as E, a^2E and aE or as E, $E\,(-0.5 - j\,0.866)$ and $E\,(-0.5 + j\,0.866)$

It is easy to prove that

(i) $a^2 + a = -1$ ***(ii)*** $a^2 + a + 1 = 0$ ***(iii)*** $a^3 + a^2 + a = 0$

Note. We have seen that operator $-j$ turns a vector through $-90°$ *i.e.* through 90° in *clockwise* direction. But it should be clearly noted that operator '$-a$' *does not* turn a vector through $-120°$. Rather '$-a$' turns a vector through $-60°$ as shown below.

Example 12.9. *Evaluate the following expressions in the polar form* ***(i)*** $a^2 - 1$, ***(ii)*** $1 - a - a^2$ ***(iii)*** $2a^2 + 3 + a$ ***(iv)*** ja.

[Elect. Meas and Meas. Inst., Madras Univ.]

Solution. ***(i)*** $a^2 = a \times a = 1 \angle 120° = 1 \angle 240° = 1 \angle -120° = -\frac{1}{2} - j\frac{\sqrt{3}}{2}$

$$\therefore \quad a^2 - 1 = -\frac{1}{2} - j\frac{\sqrt{3}}{2} - 1 - j\frac{\sqrt{3}}{2}\sqrt{3}\left(-\frac{\sqrt{3}}{2} - j\frac{1}{2}\right) = \sqrt{3} \angle 30° + 180° = \mathbf{\sqrt{3} \angle 210°}$$

(ii)
$$a = 1 \angle 120° = -\frac{1}{2} + j\frac{\sqrt{3}}{2};\; a^2 = 1 \angle 240° = -\frac{1}{2} - j\frac{\sqrt{3}}{2}$$

$$\therefore \quad 1 - a - a^2 = 1 + \frac{1}{2} - j\frac{\sqrt{3}}{2} + \frac{1}{2} + j\frac{\sqrt{3}}{2} = 2 + j\,0 = \mathbf{2 \angle 0°}$$

(iii)
$$2a^2 = 2\left(-\frac{1}{2} - j\frac{\sqrt{3}}{2}\right) = -1 - j\sqrt{3}$$

$$2a = 2\left(-\frac{1}{2} + j\frac{\sqrt{3}}{2}\right) = \mathbf{-1 + j\sqrt{3}}$$

* Numerically, *a* is equivalent to the cube root of unity.

$\therefore \quad 2a^2 + 3 + 2a = 3 - 1 - j\sqrt{3} - 1 + j\sqrt{3} = 1\angle 0°$

(iv) $\quad ja = j \times a = 1\angle 90° \times 1\angle 120° = 1\angle 210°$

Tutorial Problem 12.1

1. Perform the following indicated operations :
   (*a*) $(60 + j\,80) + (30 - j\,40)$ (*b*) $(12 - j\,6) - (40 - j\,20)$ (*c*) $(6 + j\,8)\,(3 - j\,4)$ (*d*) $16 + j8) \div (3 - j\,4)$
   [(*a*) (90 + *j* 40) (*b*) (– 28 + *j* 14) (*c*) (50 + *j* 0) (*d*) (– 0.56 + *j* 1.92)]
2. Two impedances $\mathbf{Z}_1 = 2 + j\,6\;\Omega$ and $\mathbf{Z}_2 = 6 - 12\;\Omega$ are connected in a circuit so that they are additive. Find the resultant impedance in the polar form. **[10 ∠ – 36.9°]**
3. Express in rectangular form and polar form a vector, the magnitude of which is 100 units and the phase of which with respect to reference axis is
   (*a*) + 30° (*b*) + 180° (*c*) – 60° (*d*) + 120° (*e*) – 120° (*f*) – 210°.
   [(*a*) 86.6 + *j*50 ∠30° (*b*) (–100 + *j*0), 100 ∠180° (*c*) 50 –*j*86.6, 100 ∠–60° (*d*) (–50 + *j*86.6), 100 ∠ –120° (*e*) (– 50 – *j* 86.6), 100 ∠ – 120° (*f*) (–50 + *j* 86.6), 100 ∠ – 210°]
4. In the equation $\mathbf{V}_m = V - \mathbf{ZI}$, $\mathbf{V} = 100\angle 0°$ volts, $\mathbf{Z} = 10\angle 60°\;\Omega$ and $\mathbf{I} = 8\angle -30°$ amperes. Express $\mathbf{V}_m$ in polar form. **[50.5 ∠ – 52°]**
5. A voltage $\mathbf{V} = 150 + j\,180$ is applied across an impedance and the current flowing is found to be $\mathbf{1} = 5 - j\,4$. Determine (*i*) scalar impedance (*ii*) resistance (*iii*) reactance (*iv*) power consumed.
   [(*i*) 3.73 Ω (*ii*) 0.75 Ω (*iii*) 36.6 Ω (*iv*) 30 W]
6. Calculate the following in polar form:
   (*i*) Add (40 + j20) to (20 + j120) (*ii*) Subtract (10 + j30) from (20 – j20)
   (*iii*) Multiply (15 + j20) with (20 + j30) (*iv*) Devide (6 + j7) by (5 + j3)
   (Gujrat University, June/July 2003)
7. Why is impedance represented by a complex number? ***(RGPV Bhopal December 2002)***

OBJECTIVE TESTS – 12

1. The symbol *j* represents counterclockwise rotation of a vector through—degrees.
   (*a*) 180 (*b*) 90
   (*c*) 360 (*d*) 270
2. The operator *j* has a value of
   (*a*) + 1 (*b*) – 1
   (*c*) $\sqrt{-1}$ (*d*) $\sqrt{+1}$
3. The vector $j^5 E$ is the same as vector
   (*a*) jE (*b*) $j^2 E$
   (*c*) $j^3 E$ (*d*) $j^4 E$
4. The conjugate of $(-a + jb)$ is
   (*a*) $(a - jb)$ (*b*) $(-a - jb)$
   (*c*) $(a + jub)$ (*d*) $(jb - a)$

ANSWERS

1. *b* **2.***c* **3.** *a* **4.** *b*

CHAPTER

13 A.C. Series Circuits

Chapter Outline

13.1. A.C. Through Resistance and Inductance

A pure resistance R and a pure inductive coil of inductance L are shown connected in series in Fig. 13.1.

Let V = r.m.s. value of the applied voltage, I = r.m.s. value of the resultant current

V_R = IR – voltage drop across R (in phase with I), $V_L = I \cdot X_L$ –voltage drop across coil (ahead of I by 90°)

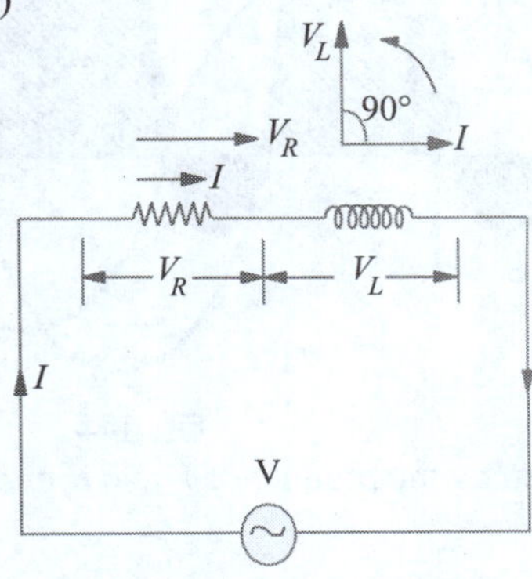

Fig. 13.1

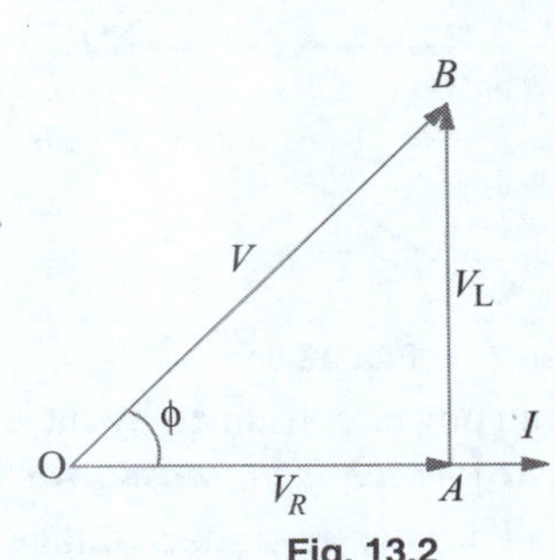

Fig. 13.2

These voltage drops are shown in voltage triangle OAB in Fig. 13.2. Vector OA represents ohmic drop V_R and AB represents inductive drop V_L. The applied voltage V is the vector sum of the two *i.e.* OB.

$$\therefore \quad V = \sqrt{(V_R^2 + V_L^2)} = \sqrt{[(IR)^2 + (I \,.\, X_L)^2]} = I\sqrt{R^2 + X_L^2}\,, \quad \frac{V}{\sqrt{(R^2 + X_L^2)}} = I$$

The quantity $\sqrt{(R^2 + X_L^2)}$ is known as the ***impedance*** (Z) of the circuit. As seen from the impedance triangle ABC (Fig. 13.3) $Z^2 = R^2 + X_L^2$.

i.e. **(*Impedance*)^2 = (*resistance*)^2 + (*reactance*)^2**

From Fig. 13.2, it is clear that the applied voltage V leads the current I by an angle ϕ such that

$$\tan\phi = \frac{V_L}{V_R} = \frac{I \,.\, X_L}{I \,.\, R} = \frac{X_L}{R} = \frac{\omega L}{R} = \frac{\text{reactance}}{\text{reactance}} \quad \therefore \ \phi = \tan^{-1}\frac{X_L}{R}$$

The same fact is illustrated graphically in Fig. 13.4.

In other words, current I lags behind the applied voltage V by an angle ϕ.

Hence, if applied voltage is given by $v = V_m \sin\omega t$, then current equation is

$$i = I_m \sin(\omega t - \phi) \text{ where } I_m = V_m / Z$$

Fig. 13.3

Fig. 13.4

In Fig. 13.5, I has been resolved into its two mutually perpendicular components, $I\cos\phi$ along the applied voltage V and $I\sin\phi$ in quadrature (*i.e.* perpendicular) with V.

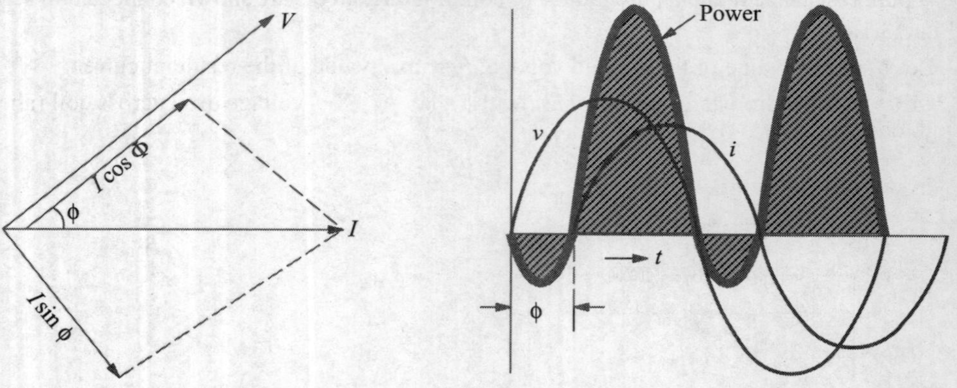

Fig. 13.5

Fig. 13.6

The mean power consumed by the circuit is given by the product of V and ***that component of the current I which is in phase with V.***

So $P = V \times I\cos\phi$ = r.m.s. voltage × r.m.s. current × $\cos\phi$

The term 'cos ϕ' is called the power factor of the circuit.

Remember that in an a.c. circuit, the product of r.m.s. volts and r.m.s. amperes gives volt-amperes (VA) and not true power in watts. True power (W) = volt-amperes (VA) × power factor.

or **Watts = VA × cos ϕ****

It should be noted that power consumed is due to ohmic resistance only because pure inductance does not consume any power.

Now $P = VI \cos \phi = VI \times (R/Z) = (V/Z) \times I.\ R = I^2 R$ ($\because \cos \phi = R/Z$) or $P = I^2 R$ watt

Graphical representation of the power consumed is shown in Fig. 13.6.

Let us calculate power in terms of instantaneous values.

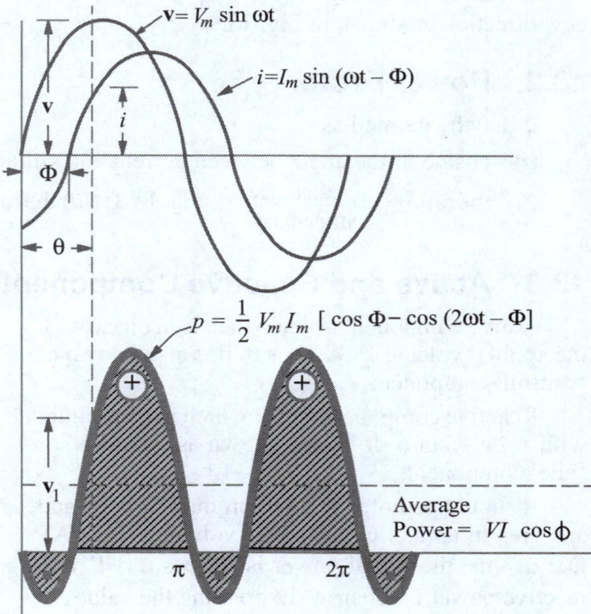

Fig. 13.7

Instantaneous power is $= vi = V_m \sin \omega t \times I_m \sin(\omega t - \phi) = V_m I_m \sin \omega t \sin (\omega t - \phi)$

$$= \frac{1}{2} ü_m \ _m[\cos \phi - \cos (2\omega \ - \Phi)]$$

Obviously, this power consists of two parts (Fig. 13.7).

(*i*) a constant part $\frac{1}{2} V_m I_m \cos \phi$ which contributes to real power.

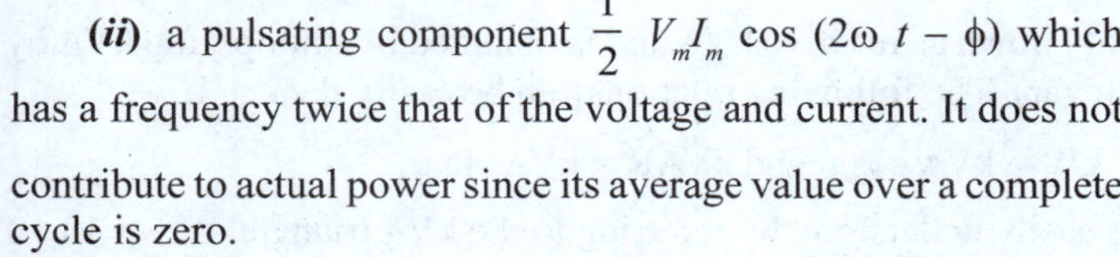

(*ii*) a pulsating component $\frac{1}{2} V_m I_m \cos (2\omega t - \phi)$ which has a frequency twice that of the voltage and current. It does not contribute to actual power since its average value over a complete cycle is zero.

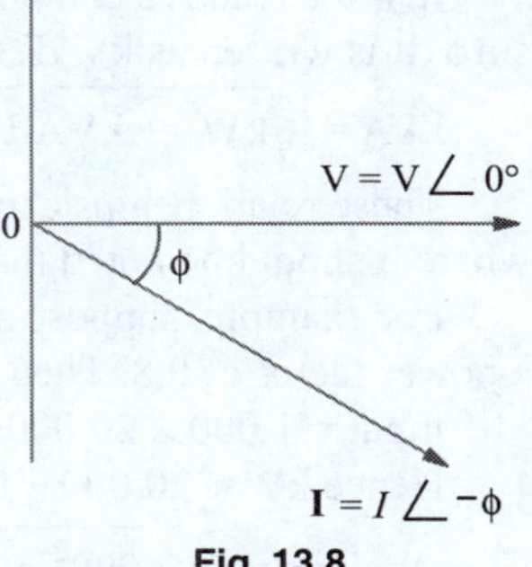

Fig. 13.8

Hence, average power consumed $= \frac{1}{2} V_m I_m \cos \phi = \frac{V_m}{\sqrt{2}} \frac{I_m}{\sqrt{2}} \cos \phi = VI \cos \phi$, where V and I represent the r.m.s values.

Symbolic Notation. $Z = R + jX_L$

Impedance vector has numerical value of $\sqrt{(R^2 + X_L^2)}$.

Its phase angle with the reference axis is $\phi = \tan^{-1} (X_L/R)$

It may also be expressed in the polar form as $\mathbf{Z} = \mathbf{Z} \angle \phi°$

(*i*) Assuming $\mathbf{V} = V \angle 0°$; $\mathbf{I} = \frac{V}{Z} = \frac{V \angle 0°}{Z \angle \phi°} = \frac{V}{Z} \angle -\phi°$ (Fig. 13.8)

It shows that current vector is lagging behind the voltage vector by $\phi°$. The numerical value of current is V/Z.

---

* While dealing with large supplies of electric power, it is convenient to use kilowatt as the unit kW = kVA × cos ϕ

(*ii*) However, if we assumed that

$\mathbf{I} = I \angle 0$, then

$\mathbf{V} = \mathbf{IZ} = I \angle 0° \times Z \angle \phi°$

$= IZ \angle \phi°$

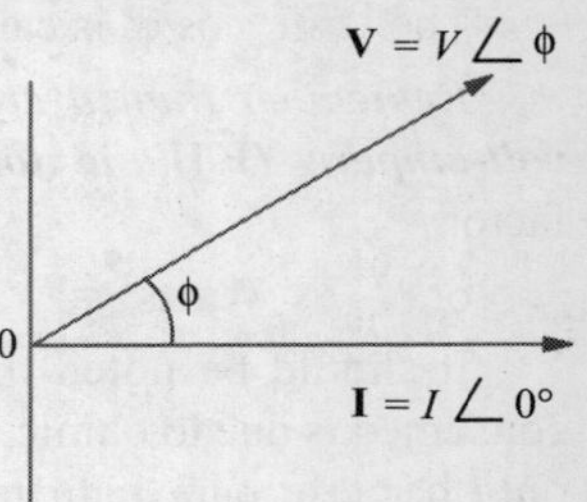

Fig. 13.9

It shows that voltage vector is φ° ahead of current vector in ccw direction as shown in Fig. 13.9.

13.2. Power Factor

It may be defined as

(*i*) cosine of the angle between current and voltage

(*ii*) the ratio $\frac{R}{Z} = \frac{\text{resistance}}{\text{impedance}}$ (...Fig. 13.3) (***iii***) the ratio $\frac{\text{true power}}{\text{apparent power}} = \frac{\text{watts}}{\text{volt amperes}} = \frac{W}{VA}$

13.3. Active and Reactive Components of Circuit Current I

Active component is that which is in phase with the applied voltage *V i.e. I* cos φ. It is also known as 'wattful' component.

Reactive component is that which in quadrature with *V i.e. I* sin φ. It is also known as 'wattless' or 'idle' component.

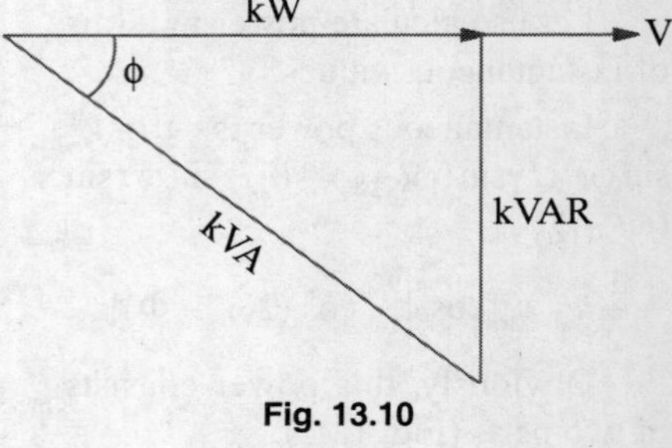

Fig. 13.10

It should be noted that the product of volts and amperes in an a.c. circuit gives voltamperes (VA). Out of this, the actual power is *VA* cos φ = *W* and reactive power is *VA* sin φ. Expressing the values in kVA, we find that it has two rectangular components :

(*i*) active component which is obtained by multiplying kVA by cos φ and this gives power in kW.

(*ii*) the reactive component known as reactive kVA and is obtained by multiplying kVA by sin φ. It is written as kVAR (kilovar). The following relations can be easily deduced.

$\text{kVA} = \sqrt{\text{kW}^2 + \text{kVAR}^2}$; kW = kVA cos φ and kVAR = kVA sin φ

These relationships can be easily understood by referring to the kVA triangle of Fig. 13.10 where it should be noted that lagging kVAR has been taken as negative.

For example, suppose a circuit draws a current of 1000 A at a voltage of 20,000 V and has a power factor of 0.8. Then

input = 1,000 × 20,000/1000 = 20,000 kVA; cos φ = 0.8; sin φ = 0.6

Hence kW = 20,000 × 0.8 = 16,000 ; kVAR = 20,000 × 0.6 = 12,000

Obviously, $\sqrt{16000^2 + 12000^2} = 20,000$ *i.e.* $\text{kVA} = \sqrt{\text{kW}^2 + \text{kVAR}^2}$

13.4. Active, Reactive and Apparent Power

Let a series *R-L* circuit draw a current of *I* when an alternating voltage of r.m.s. value *V* is applied to it. Suppose that current lags behind the applied voltage by φ. The three powers drawn by the circuit are as under :

(*i*) **apparent power (*S*)**

It is given by the product of r.m.s. values of applied voltage and circuit current.

$\therefore \quad S = VI = (IZ) \,.\, I = I^2 Z$ volt-amperes (*VA*)

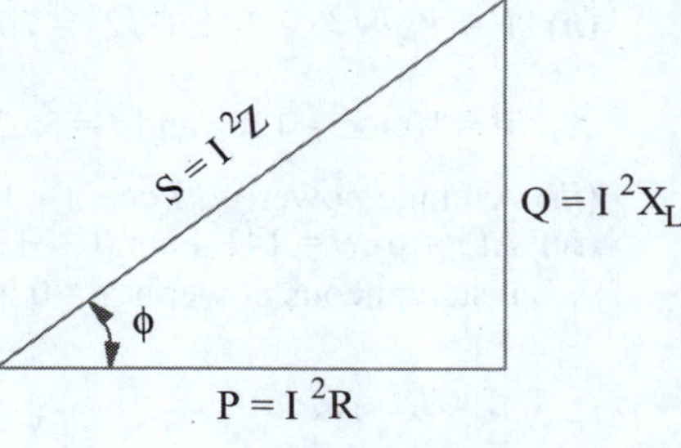

Fig. 13.11

***(ii)* active power (*P* or *W*)**

It is the power which is actually dissipated in the circuit resistance. $P = I^2R = VI \cos \phi$ watts

***(iii)* reactive power (Q)**

It is the power developed in the inductive reactance of the circuit.

$Q = I^2X_L = I^2 \, . \, \boldsymbol{Z} \sin \phi = I \, . \, (IZ). \sin \phi = VI \sin \phi$ volt-amperes-reactive (VAR)

These three powers are shown in the power triangle of Fig. 13.11 from where it can be seen that

$$S^2 = P^2 + Q^2$$

or $$S = \sqrt{P^2 + Q^2}$$

13.5. Q-factor of a Coil

Reciprocal of power factor is called the *Q*-factor of a coil or its figure of merit. It is also known as quality factor of the coil.

$$Q \text{ factor} = \frac{1}{\text{power factor}} = \frac{1}{\cos \phi} = \frac{Z}{R}$$

If *R* is small as compared to reactance, then *Q*-factor = $Z/R = \omega L/R$

Also, $$Q = 2\pi \frac{\text{maximum energy stored}}{\text{energy dissipated per cycle}} \quad \text{—in the coil}$$

Example 13.1. *In a series circuit containing pure resistance and a pure inductance, the current and the voltage are expressed as :*

$$i\,(t) = 5 \sin (314\, t + 2\, \pi/3) \text{ and } v\,(t) = 15 \sin (314\, t + 5\, \pi/6)$$

(*a*) *What is the impedance of the circuit ?* **(*b*)** *What is the value of the resistance ?* **(*c*)** *What is the inductance in henrys ?* **(*d*)** *What is the average power drawn by the circuit ?* **(*e*)** *What is the power factor ?* **[Elect. Technology, Indore Univ.]**

Solution. Phase angle of current = 2 π/3 = 2 180°/3 = 120° and phase angle of voltage = 5 π/6 = 5 × 180°/6 = 150°. Also, $Z = V_m/I_m = 3\ \Omega$.

Hence, current *lags* behind voltage by 30°. It means that it is an *R-L* circuit. Also 314 = 2π *f* or *f* = 50 Hz. Now, $R/Z = \cos 30° = 0.866$; $R = 2.6\ \Omega$; $X_L/Z = \sin 30° = 0.5$

$\therefore \quad X_L = 1.5\ \Omega \qquad 314\, L = 1.5, \qquad L = 4.78$ mH

(*a*) $Z = \mathbf{3}\ \Omega$ (*b*) *R* = **2.6 Ω** **(c)** *L* = **4.78 mH**

(*d*) $P = I^2 R = (5/\sqrt{2})^2 \times 2.6 =$ **32.5 W** (*e*) p.f. = cos 30° = **0.866 (lag).**

Example 13.2. *In a circuit the equations for instantaneous voltage and current are givn by*

$$v = 141.4 \sin \left(\omega t - \frac{2\pi}{3}\right) \text{ volt and } i = 7.07 \sin \left(\omega t - \frac{\pi}{2}\right) \text{ amp, where } \omega = 314 \text{ rad/sec.}$$

(*i*) *Sketch a neat phasor diagram for the circuit.* **(*ii*)** *Use polar notation to calculate impedance with phase angle.* **(*iii*)** *Calculate average power & power factor.* **(*iv*)** *Calculate the instantaneous power at the instant t* = 0. **(F.Y. Engg. Pune Univ.)**

Solution. (*i*) From the voltage equation, it is seen that the voltage lags behind the reference quantity by 2π/3 radian or 2 × 180/3 = 120°. Similarly, current lags behind the reference quantity by π/2 radian or 180/2 = 90°. Between themselves, voltage lags behind the current by (120 – 90) = 30° as shwon in Fig. 13.12 (*b*).

(ii) $V = V_m/\sqrt{2} = 141.4\sqrt{2} = 100$ V; $I = I_m/\sqrt{2} = 7.07/\sqrt{2}$ = **5 A.**

$\therefore \quad V = 100 \angle -120°$ and $I = 5 \angle -90°$ $\therefore$ $Z = \dfrac{100 \angle -120°}{5 \angle -90°} = 20 \angle -30°$

(iii) Average power = $VI \cos \phi = 100 \times 5 \times \cos 30°$ = **433 W**

(iv) At $t = 0$; $v = 141.4 \sin (0 - 120°) = -122.45$ V; $i = 7.07 \sin (0 - 90°) = -7.07$ A.

$\therefore$ instantaneous power at $t = 0$ is given by $vi = (-122.45) \times (-7.07)$ = **865.7 W.**

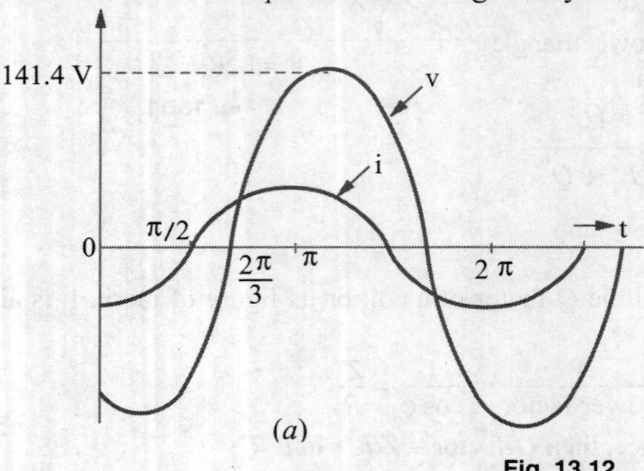

(a) (b)

Fig. 13.12

Example 13.3. *The potential difference measured across a coil is 4.5 V, when it carries a direct current of 9 A. The same coil when carries an alternating current of 9 A at 25 Hz, the potential difference is 24 V. Find the current, the power and the power factor when it is supplied by 50 V, 50 Hz supply.* **(F.Y. Pune Univ.)**

Solution. Let R be the *d.c.* resistance and L be the inductance of the coil.

$\therefore \quad R = V/I = 4.5/9 = 0.5\ \Omega$;

With a.c. current of 25 Hz, $Z = V/I = 24/9 = 2.66\ \Omega$.

$X_L = \sqrt{Z^2 - R^2} = \sqrt{2.66^2 - 0.5^2} = 2.62\ \Omega$. Now $X_L = 2\pi \times 25 \times L$; $L = 0.0167\ \Omega$

At 50 Hz

$X_L = 2.62 \times 2 = 5.24\ \Omega$; $Z = \sqrt{0.5^2 + 5.24^2} = 5.26\ \Omega$

Current $I = 50/5.26$ = **9.5 A** ; Power = $I^2 R = 9.5^2 \times 0.5$ = **45 W.**

Example 13.4. *A 60 Hz sinusoidal voltage v = 141 sin ω t is applied to a series R-L circuit. The values of the resistance and the inductance are 3 Ω and 0.0106 H respectively.*

(i) *Compute the r.m.s. value of the current in the circuit and its phase angle with respect to the voltage.*

(ii) *Write the expression for the instantaneous current in the circuit.*

(iii) *Compute the r.m.s. value and the phase of the voltages appearing across the resistance and the inductance.*

(iv) *Find the average power dissipated by the circuit.*

(v) *Calculate the p.f. of the circuit.* **(F.E. Pune Univ.)**

Solution. $V_m = 141$ V; $V = 141/\sqrt{2} = 100\ V$ $\quad\therefore\quad V = 100 + j0$

$X_L = 2\pi \times 60 \times 0.0106 = 4\ \Omega$. $Z = 3 + j4 = 5 \angle 53.1°$

(i) $I = V/Z = 100 \angle 0°/5 \angle 53.1° = 20 \angle -53.1°$

Since angle is minus, the current lags behind the voltage by 53.1

(*ii*) $I_m = \sqrt{2} \times 20 = 28.28$; $\therefore$ $i = 28.28 \sin(\omega t - 53.1°)$

(*iii*) $VR = IR = 20 \angle -53.1° \times 3 = 60 \angle -53.1°$ volt.

$V_L = jIX_L = 1 \angle 90° \times 20 \angle -53.1° \times 4 = 80 \angle 36.9°$

(*iv*) $P = VI \cos\phi = 100 \times 20 \times \cos 53.1° =$ **1200 W.**

(*v*) p.f. $= \cos\phi = \cos 53.1° =$ **0.6.**

Example 13.5. *In an alternating circuit, the impressed voltage is given by V = (100 – j50) volts and the current in the circuit is I = (3 – j4) A. Determine the real and reactive power in the circuit.* **(Electrical Engg., Calcutta Univ. 1991)**

Solution. Power will be found by the conjugate method. Using current conjugate, we have

$$P_{VA} = (100 - j\,50)(3 + j\,4) = 300 + j\,400 - j\,150 + 200 = 500 + j\,250$$

Hence, real power is **500 W** and reactive power of VAR is **250.** Since the second term in the above expression is positive, the reactive volt-amperes of 250 are inductive.*

Example 13.6. *In the circuit of Fig. 14.13, applied voltage V is given by (0 + j10) and the current is (0.8 + j 0.6) A. Determine the values of R and X and also indicate if X is inductive or capacitive.* **(Elect. Technology, Nagpur Univ. 1991)**

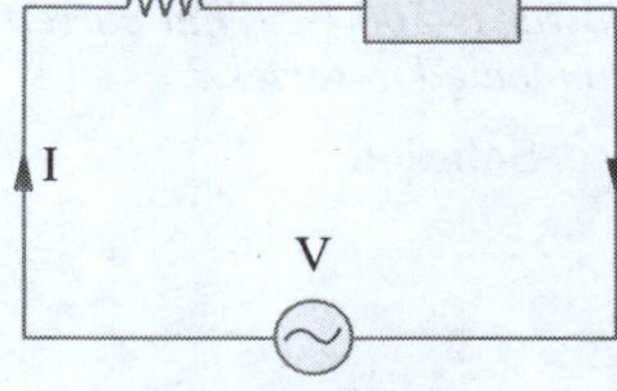

Fig. 13.13

Solution. $\mathbf{V} = 0 + j10 = \angle 90°$; $\mathbf{I} = 0.8 + j0.6 = 1\angle 36.9°$

As seen, V leads the reference quantity by 90° whereas I leads by 36.9°. In other words , *I lags* behind the applied voltage by (90° – 36.9°) = 53.1°

Hence, the circuit of Fig. 13.14 is an *R-L* circuit.

Now, $\mathbf{Z} = \mathbf{V/I} = 10 \angle 90°/1 \angle 36.9° = 10 \angle 53.1° = 6 + j8$

Hence, $R = \mathbf{6\ \Omega}$ and $X_L = \mathbf{8\ \Omega}$.

Example 13.7. *A two-element series circuit is connected across an a.c. source e =* $200\sqrt{2}$ *sin (ωt + 20°) V. The current in the circuit then is found to be i =* $10\sqrt{2}$ *cos (314 t – 25°) A. Determine the parameters of the circuit.* **(Electromechanic Allahabad Univ. 1991)**

Solution. The current can be written as $i = 10\sqrt{2} \sin(314\,t - 25° + 90°) = 10\sqrt{2} \sin(314\,t + 65°)$. It is seen that applied voltage leads by 20° and current leads by 65° with regards to the reference quantity, their mutual phase difference is = 65° – (20°) = 45°. Hence, p.f. = cos 45° = $\mathbf{1/\sqrt{2}}$ **(lead).**

Now, $V_m = 200\sqrt{2}$ and $I_m = 10\sqrt{2}$ $\therefore$ $Z = V_m/I_m = 200\sqrt{2}/10\sqrt{2} = 20\ \Omega$

$R = Z\cos\phi = 20/\sqrt{2}\ \Omega = \mathbf{14.1\ \Omega}$; $X_c = Z \sin\phi = 20/\sqrt{2} = \mathbf{14.1\ \Omega}$

Now, $f = 314/2\pi = 50$ Hz. Also, $X_c = 1/2\pi fC$ $\therefore$ $C = 1/2\pi \times 50 \times 14.1 =$ **226 μF**

Hence, the given circuit is an *R-C* circuit.

Example 13.8. *Transform the following currents to the time domain* : (*i*) 6 – *j* 8 (*ii*) – 6 + *j*8 (*iii*) – *j*5.

Solution. (*i*) Now, (6 – *j*8) when expressed in the polar form is $\sqrt{6^2 + 8^2}\ \angle -\tan^{-1} 8/6 = 10 \angle -53.1°$. The time domain representation of this current is $i(t) = 10 \sin(\omega t - 53.1°)$

(*ii*) $-6 + j8 = \sqrt{6^2 + 8^2}\ \angle \tan^{-1} 8/-6 = 10 \angle 126.9°$

$\therefore$ $i(t) = 10 \sin(\omega t + 126.9°)$

* If voltage conjugate is used, then capacitive VARs are positive and inductive VARs negative. If current conjugate is used, then capacitive VARs are negative and inductive VARs are positive.

(iii) $-j\,5 = 10 \angle -90°$ $\therefore$ $i(t) = 10 \sin(\omega t - 90°)$

Example 13.9. *An inductive circuit draws 10 A and 1 kW from a 200-V, 50 Hz a.c. supply. Determine :*

(i) *the impedance in cartesian from (a + jb)* ***(ii)*** *the impedance in polar from Z ∠ θ* ***(iii)*** *the power factor* ***(iv)*** *the reactive power* ***(v)*** *the apparent power.*

Solution. $Z = 200/10 = 20\ \Omega$; $P = I^2R$ or $1000 = 10^2 \times R$; $R = 10\ \Omega$; $X_L = \sqrt{20^2 - 10^2}$ $= 17.32\Omega$.

(i) Z $= 10 + j17.32$ **(ii)** $|Z| = \sqrt{(10^2 + 17.32^2)} = 20\ \Omega$; $\tan\phi = 17.32/10 = 1.732$; $\phi = \tan^{-1}(1.732)$
$= 60°$ $\therefore$ $\mathbf{Z = 20 \angle 60°}$. ***(iii)*** p.f. $= \cos\phi = \cos 60° =$ **0.5 lag** ***(iv)*** reactive power $= VI \sin\phi$ $= 200 \times 10 \times 0.866 =$ **1732 VAR** (*v*) apparent power $= VI = 200 \times 10 =$ **2000 VA.**

Example 13.10. *When a voltage of 100 V at 50 Hz is applied to a choking coil A, the current taken is 8 A and the power is 120 W. When applied to a coil B, the current is 10 A and the power is 500 W. What current and power will be taken when 100 V is applied to the two coils connected in series ?* **(Elements of Elect. Engg., Bangalore Univ.)**

Solution.

$$Z_1 = 100/8 = 12.5\ \Omega\ ;\ P = I^2 . R_1 \text{ or } 120 = 8^2 \times R_1\ ;\ R_1 = 15/8\ \Omega$$
$$X_1 = \sqrt{Z_1^2 - R_1^2} = \sqrt{12.5^2 - (15/8)^2} = 12.36\ \Omega$$
$$Z_2 = 100/10 = 10\ \Omega\ ;\ 500 = 10^2 \times R_2 \quad \text{or} \quad R_2 = 5\ \Omega$$
$$X_2 = \sqrt{10^2 - 5^2} = 8.66\ \Omega$$

With Joined in Series

$$R = R_1 + R_2 = (15/8) + 5 = 55/9\ \Omega;\ X = 12.36 + 8.66 = 21.02\ \Omega$$
$$Z = \sqrt{(55/8)^2 + (21.02)^2} = 22.1\ \Omega,\ I = 100/22.1 = 4.52\text{ A},\ P = I^2R = 4.52^2 \times 55/8 = \mathbf{140\ W}$$

Example 13.11. *A coil takes a current of 6 A when connected to a 24-V d.c. supply. To obtain the same current with a 50-Hz a.c. supply, the voltage required was 30 V.*

Calculate ***(i)*** *the inductance of the coil* ***(ii)*** *the power factor of the coil.*

(F.Y. Engg. Pune Univ.)

Solution. It should be kept in mind the coil offers only resistance to direct voltage whereas it offers impedance to an alternating voltage.

$\therefore$ $$R = 24/6 = 4\ \Omega;\ Z = 30/6 = 5\ \Omega$$

(*i*) $\therefore$ $$X_L = \sqrt{Z^2 - R^2} = \sqrt{5^2 - 4^2} = 3\ \Omega\ \textbf{(ii)}\ \text{p.f.} = \cos\phi = R/Z = 4/5 = \mathbf{0.8\ (lag)}$$

Example 13.12. *A resistance of 20 ohm, inductance of 0.2 H and capacitance of 150 μF are connected in series and are fed by a 230 V, 50 Hz supply. Find X_L, X_C, Z, Y, p.f., active power and reactive power.* **(Elect. Science-I, Allahabad Univ. 1992)**

Solution.

$$X_L = 2\pi fL = 2\pi \times 50 \times 0.2 = 62.8\Omega\ ;\ X_C = 1/2\pi fC$$
$$= 10^{-6}\ 2\pi \times 50 \times 150 = 21.2\ \Omega\ ;\ X = (X_L - X_C) = 41.6\ \Omega\ ;$$
$$Z = \sqrt{R^2 + X^2} = \sqrt{20^2 + 41.6^2} = 46.2\ \Omega\ ;\ I = V/Z = 230/46.2 = 4.98\ A$$

Also, $$Z = R + jX = 20 + j\,41.6 = 46.2 \angle 64.3° \text{ ohm}$$

$\therefore$ $$Y = 1/Z = 1/46.2 \angle 64.3° = 0.0216 \angle -64.3° \text{ siemens}$$

$$\text{p.f.} = \cos 64.3° = 0.4336 \text{ (lag)}$$

Active power $$= VI \cos\phi = 230 \times 4.98 \times 0.4336 = 497\text{ W}$$

Reactive power $$= VI \sin\phi = 230 \times 4.98 \times \sin 64.3° = 1031\text{ VAR}$$

Example 13.13. *Two coils A and B are connected in series across a 240-V, 50-Hz supply. The resistance of A is 5 Ω and the inductance of B is 0.015 H. If the input from the supply is 3 kW and 2 kVAR, find the inductance of A and the resistance of B. Calculate the voltage across each coil.* **(Elect. Technology Hyderabad Univ. 1991)**

Solution. The kVA triangle is shown in Fig. 13.14(*b*) and the circuit in Fig. 13.14(*a*). The circuit kVA is given by, kVA $= \sqrt{(3^2 + 2^2)} = 3.606$ or $VA = 3{,}606$ voltmeters

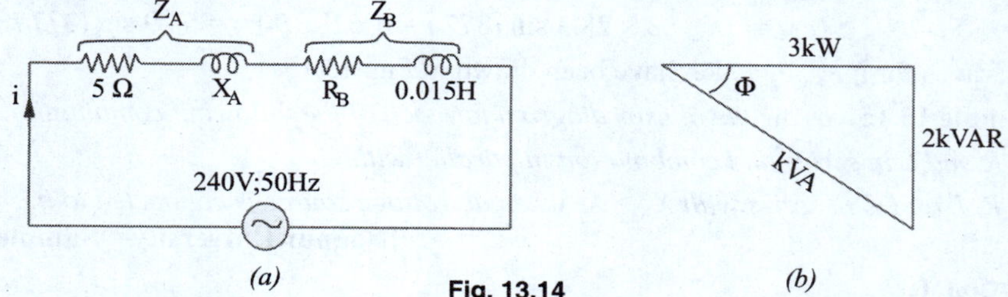

(*a*) (*b*)

Fig. 13.14

Circuit current = 3,606/240 = 15.03 A $\therefore$ $15.03^2 (R_A + R_B) = 3{,}000$

$\therefore$ $R_A + R_B = 3{,}000/15.03^2 = 13.3\ \Omega$ $\therefore$ $R_B = 13.3 - 5 =$ **8.3 Ω**

Now, impedance of the whole circuit is given by $Z = 240/15.03 = 15.97\ \Omega$

$$\therefore \quad X_A + X_B = \sqrt{(Z^2 - (R_A + R_B)^2} = \sqrt{15.97^2 - 13.3^3} = 8.84\ \Omega$$

Now $\quad X_B = 2\pi \times 50 \times 0.015 = 4.713\ \Omega \quad \therefore \quad X_A = 8.843 - 4.713 = 4.13\ \Omega$

or $\quad 2\pi \times 50 \times L_A = 4.13 \quad \therefore \quad L_A =$ **0.0132 H (approx)**

$$\text{Now} \quad Z_A = \sqrt{R_A^2 + X_A^2} = \sqrt{5^2 + 4.13^2} = 6.585\ \Omega$$

P.D. across coil $A = I \,.\, Z_A = 15.03 \times 6.485 =$ **97.5 V;** $Z_b = \sqrt{8.3^2 + 4.713^2} = 9.545\ \Omega$

$\therefore$ p.d. across coil $B = I \,.\, Z_B = 15.03 \times 9.545 =$ **143.5 V**

Example 13.14. *An e.m.f. $e_0 = 141.4 \sin (377\, t + 30°)$ is impressed on the impedance coil having a resistance of 4 Ω and an inductive reactance of 1.25 Ω, measured at 25 Hz. What is the equation of the current ? Sketch the waves for i, e_R, e_L and e_0.*

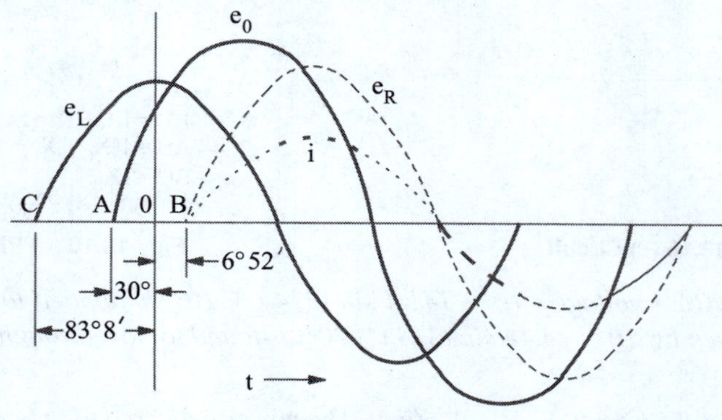

Fig. 13.15

Solution. The frequency of the applied voltage is $f = 377/2\pi = 60$ Hz

Since coil reactance is 1.25 Ω at 25 Hz, its value at 60 Hz = 1.25 × 60/25 = 3 Ω

Coil impedance, $Z = \sqrt{4^2 + 3^2}, = 5\ \Omega$; $\phi = \tan^{-1} (3/4) = 36°52'$

It means that circuit current lags behind the applied voltage by 36°52′. Hence, equation of the circuit current is

$$i = (141.4/5) \sin (377\,t + 30° - 36°52') = 28.3 \sin (377\,t - 6°52')$$

Since, resistance drop is in phase with current, its equation is $e_R = iR = 113.2 \sin (377\,t - 6°52')$

The inductive voltage drop ***leads the current*** by 90°, hence its equation is

$$e_L = iX_L = 3 \times 28.3 \sin (377\,t - 6°52' + 90°) = 54.9 \sin (377\,t + 83°8')$$

The waves for i, e_R, e_L and e_0 have been drawn in Fig. 13.15.

Example 13.15. *Draw the phasor diagram for each of the following combinations:*

(i) *R and L in series and combination in parallel with C.*

(ii) *R, L and C in series with $X_C > X_L$ when ac voltage source is connected to it.*

[Nagpur University—Summer 2000]

Solution. (*i*)

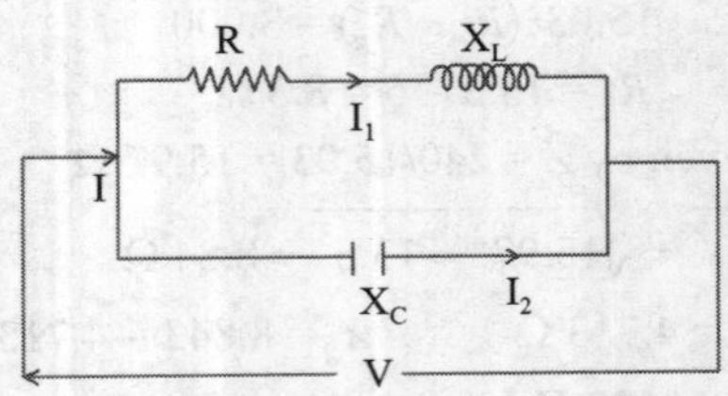

Fig. 13.16 (*a*) Circuit **Fig. 13.16 (*b*) Phasor diagram**

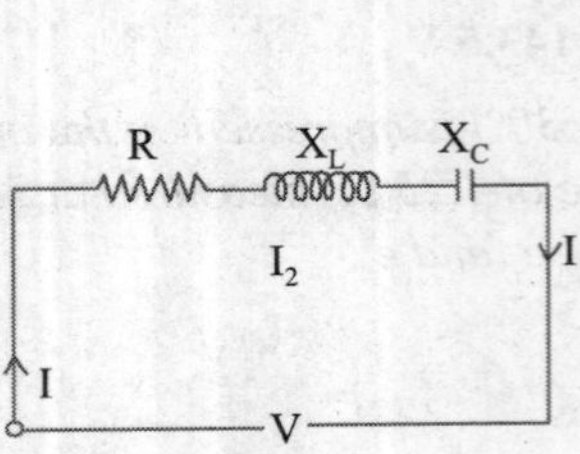

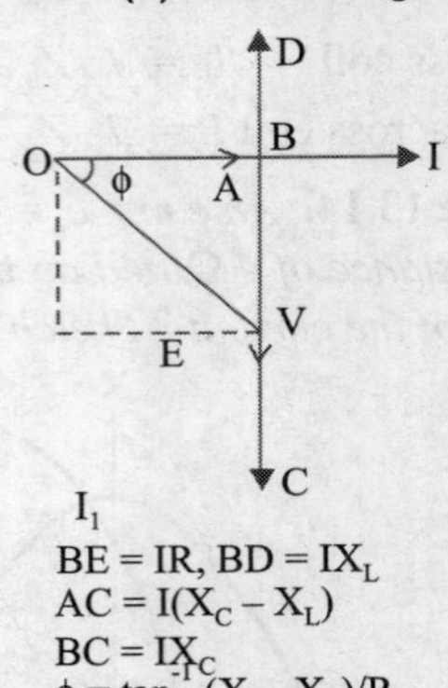

Fig. 13.16 (*c*) Circuit **Fig . 13.16 (*d*) Phasor diagram**

Example 13.16. *A voltage v (t) = 141.4 sin (314 t + 10°) is applied to a circuit and the steady current given by i(t) = 14.14 sin (314 t – 20°) is found to flow through it.*

Determine

(i) *The p.f. of the circuit* ***(ii)*** *The power delivered to the circuit*

(iii) *Draw the phasor diagram.* **[Nagpur University Summer 2000]**

Solution. $v\,(t) = 141.4 \sin (314\,t + 10°)$

This expression indicates a sinusoidally varying alternating voltage at a frequency

$\omega = 314$ rad/sec, $f = 50$ Hz

V = RMS voltage (Peak voltage)/ $\sqrt{2}$ = 100 volts

The expression for the current gives the following data:

I = RMS value = $14.14/\sqrt{2}$ = 10 amp

frequency = 50 Hz, naturally.

Phase shift between I and V = 30°, I lags behind V.

(i) Power factor of the circuit = cos 30° = 0.866 lag

(ii) $P = VI \cos \phi = 100 \times 10 \times 0.866 = 866$ watts

(iii) Phasor diagram as drawn below, in Fig. 13.17 (*a*).

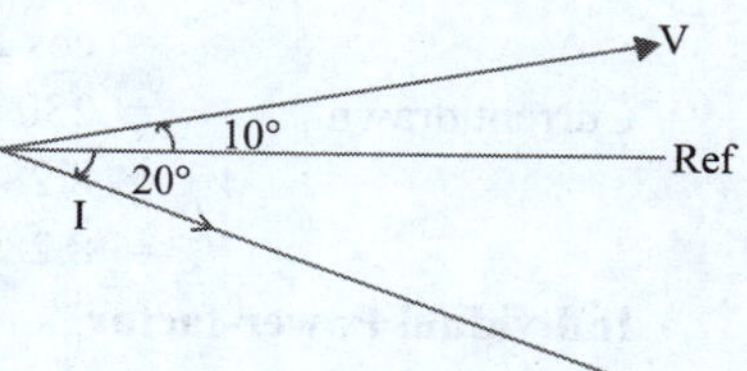

Fig. 13.17 (*a*) Phasor diagram

Example 13.17. *A coil of 0.8 p.f. is connected in series with 110 micro-farad capacitor. Supply frequency is 50 Hz. The potential difference across the coil is found to be equal to that across the capacitor. Calculate the resistance and the inductance of the coil. Calculate the net power factor.* **[Nagpur University, November 1997]**

Solution. $X_C = 1/(3.14 \times C) = 28.952$ ohms

∴ Coil Impedance, $Z = 28.952\ \Omega$

Coil resistance = $28.952 \times 0.8 = 23.162\ \Omega$

Coil reactance = 17.37 ohms

Coil-inductance = 17.37/314 = 55.32 milli-henrys

Total impedance, $Z_T = 23.16 + j\,17.37 - j\,28.952 = 23.162 - j\,11.582 = 25.9$ ohms

Net power-factor = 23.162/25.9 = 0.8943 leading

Example 13.18. *A coil is connected in series with a pure capacitor. The combination is fed from a 10 V supply of 10,000 Hz. It was observed that the maximum current of 2 Amp flows in the circuit when the capacitor is of value 1 microfarad. Find the parameters (R and L) of the coil.* **[Nagpur University April 1996]**

Solution. This is the situation of resonance in A.C. Series circuit, for which $X_L = X_C$

$$Z = R = V/I = 10/2 = 5 \text{ ohms}$$

If ω_0 is the angular frequency, at resonance, L and C are related by $\omega_0^2 = 1/(LC)$,

which gives $L = 1/(\omega_0^2 C) = 2.5 \times 10^{-4}$ H = 0.25 mH

Example 13.19. *Two impedances consist of (resistance of 15 ohms and series-connected inductance of 0.04 H) and (resistance of 10 ohms, inductance of 0.1 H and a capacitance of 100 μF, all in series) are connectd in series and are connected to a 230 V, 50 Hz a.c. source. Find :* ***(i)*** *Current drawn,* ***(ii)*** *Voltage across each impedance,* ***(iii)*** *Individual and total power factor. Draw the phasor diagram.* **[Nagpur University, Nov. 1996]**

Solution. Let suffix 1 be used for first impedance, and 2 for the second one. At 50 Hz,

$$Z_1 = 15 + j\,(314 \times 0.04) = 15 + j\,12.56 \text{ ohms}$$

$$Z_1 = \sqrt{15^2 + 12.56^2} = 19.56 \text{ ohms},$$

Impedance-angle, $\theta_1 = \cos^{-1}(15/19.56) = +40°$,

$$Z_2 = 10 + j\,(314 \times 0.1) - j\,\{1/(314 \times 100 \times 16^{-6})\}$$

$$= 10 + j\,31.4 - j\,31.85 = 10 - j\,0.45$$

$= 10.01$ ohms, Impedance angle, $\theta_2 = -2.56°$,

Total Impedance, $Z = Z_1 = Z_2 = 15 + j\,12.56 + 10 - j\,0.85$

$= 25 + j\,12.11 = 27.78 \underline{|25.85°}$

For this, Phase-angle of + 25.85°, the power-factor of the total impedance

$= \cos 25.85° = 0.90$, Lag.

Current drawn $= 230/27.78 = 8.28$ Amp, at 090 lagging p.f.

$V_1 = 8.28 \times 19.56 = 162$ Volts

$V_2 = 8.28 \times 10.01 = 82.9$ Volts

Individual Power-factor

$\cos \theta_1 = \cos 40° = 0.766$ Lagging

$\cos \theta_2 = \cos 2.56° = 0.999$ leading

Phasor diagram : In case of a series circuit, it is easier to treat the current as a reference. The phasor diagram is drawn as in Fig. 13.18.

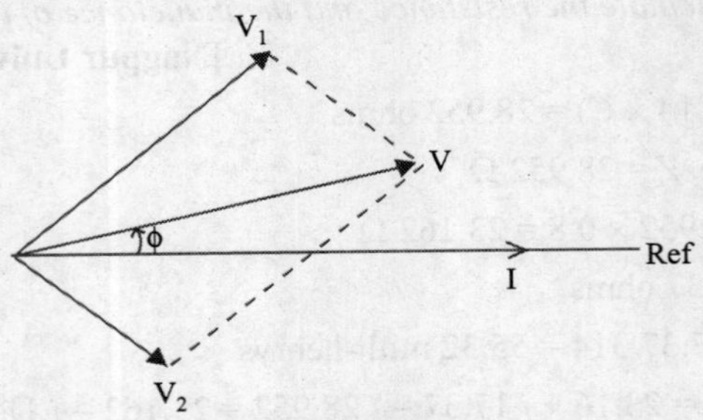

Fig. 13.18

13.6. Power in an Iron-cored Choking Coil

Total power, P taken by an iron-cored choking coil is used to supply

(*i*) power loss in ohmic resistance *i.e.* $I^2 R$. **(*ii*)** iron-loss in core, P_i

$\therefore \quad P = I^2 R + P_i$ or $\dfrac{P}{I^2} = R + \dfrac{P_1}{I^2}$ is known as the **effective resistance*** of the choke.

$\therefore$ effective resistance = true resistance = equivalent resistance $\dfrac{P_i}{I^2}$ $\therefore R_{eff} = \dfrac{P}{I^2} = R + \dfrac{P_i}{I^2}$

Example 13.20. *An iron-cored choking coil takes 5 A when connected to a 20-V d.c. supply and takes 5 A at 100 V a.c. and consumes 250 W. Determine* **(a)** *impedance* **(b)** *the power factor* **(c)** *the iron loss* **(d)** *inductance of the coil.* **(Elect. Engg. & M.A.S.I. June, 1991)**

Solution. (*a*) $Z = 100/5 = \mathbf{20\ \Omega}$

(*b*) $P = VI\cos\phi$ or $250 = 100 \times 5 \times \cos\phi \quad \therefore \quad \cos\phi = 250/500 = \mathbf{0.5}$

(*c*) Total loss = loss in resistance + iron loss $\therefore 250 = 20 \times 5 + P_i \therefore P_i = 250 - 100 = \mathbf{150\ W}$

(*d*) Effective resistance of the choke is $\dfrac{P}{I^2} = \dfrac{250}{25} = 10\ \Omega$

$$\therefore \quad X_L = \sqrt{(Z^2 - R^2)} = \sqrt{(400 - 100)} = \mathbf{17.32\ \Omega}$$

Example 13.21. *An iron-cored choking coil takes 5 A at a power factor of 0.6 when supplied at 100-V, 50 Hz. When the iron core is removed and the supply reduced to 15 V, the current rises to 6 A at power factor of 0.9.*

* At higher frequencies like radio frequencies, there is skin-effect loss also.

Determine ***(a)*** *the iron loss in the core* ***(b)*** *the copper loss at 5 A* ***(c)*** *the inductance of the choking coil with core when carrying a current of 5 A.*

Solution. When core is removed, then $Z = 15/6 = 2.5\ \Omega$

True resistance, $R = Z \cos \phi = 2.5 \times 0.9 = 2.25\ \Omega$

With Iron Core

Power input $= 100 \times 5 \times 0.6 = 300$ W

Power wasted in the true resistance of the choke when current is 5 A $= 5^2 \times 2.25 = 56.2$ W

(*a*) Iron loss $= 300 - 56.2 =$ **244 W (approx)** **(*b*)** Cu loss at 5 A $=$ **56.2 W**

(c) $Z = 100/5 = 20\ \Omega$; $X_L = Z \sin \phi = 20 \times 0.8 = 16\Omega$ $\therefore$ $2\pi \times 50 \times L = 16$ $\quad \therefore \quad L =$ **0.0509 N**

Tutorial Problem 13.1

1. The voltage applied to a coil having $R = 200\ \Omega$, $L = 638$ mH is represented by $e = 20 \sin 100\ \pi$ t. Find a corresponding expression for the current and calculate the average value of the power taken by the coil. **[$i = 0.707 \sin (100\ \pi t - \pi/4)$; 50 W] (*I.E.E. London*)**
2. The coil having a resistance of 10 Ω and an inductance of 0.2 *H* is connected to a 100-V, 50-Hz supply. Calculate (*a*) the impedance of the coil (*b*) the reactance of the coil (*c*) the current taken and (*d*) the phase difference between the current and the applied voltage. **[(*a*) 63.5 Ω (*b*) 62.8 Ω (*c*) 1.575 A (*d*) 80°57′]**
3. An inductive coil having a resistance of 15 Ω takes a current of 4 A when connected to a 100-V, 60 Hz supply. If the coil is connected to a 100 -V, 50 Hz supply, calculate (*a*) the current (*b*) the power (*c*) the power factor. Draw to scale the vector diagram for the 50-Hz conditions, showing the component voltages. **[(*a*) 4.46 A (*b*) 298 W (*c*) 0.669]**
4. When supplied with current at 240-V, single-phase at 50 Hz, a certain inductive coil takes 13.62 A. If the frequency of supply is changed to 40 Hz, the current increases to 16.12 A. Calculate the resistance and inductance of the coil. **[17.2 W, 0.05 H] (*London Univ.*)**
5. A voltage $v(t) = 141.4 \sin (314\,t + 10°)$ is applied to a circuit and a steady current given by $i(t) = 14.4 \sin (314\,t - 20°)$ is found to flow through it. Determine (*i*) the p.f. of the circuit and (*ii*) the power delivered to the circuit. **[0.866 (lag); 866 W]**
6. A circuit takes a current of 8 A at 100 V, the current lagging by 30° behind the applied voltage. Calculate the values of equivalent resistance and reactance of the circuit. **[10.81 Ω ; 6.25 Ω]**
7. Two inductive impedance *A* and *B* are connected in series. *A* has $R = 5\ \Omega$, $L = 0.01$ *H;* B has $R = 3\ \Omega$, $L = 0.02$ H. If a sinusoidal voltage of 230 V at 50 Hz is applied to the whole circuit calculate (*a*) the current (*b*) the power factor (*c*) the voltage drops. Draw a complete vector diagram for the circuit. **[(*a*) 18.6 (*b*) 0.648 (*c*) $V_A = 109.5$ V, $V_B = 129.5$ V] (*I.E.E. London*)**
8. A coil has an inductance of 0.1 H and a resistance of 30 Ω at 20°C. Calculate (*i*) the current and (*ii*) the power taken from 100-V, 50-Hz mains when the temperature of the coil is 60° C, assuming the temperature coefficient of resistance to be 0.4% per°C from a basic temperature of 20°C. **[(*i*) 2.13 A (*ii*) 158.5 W] (*London Univ.*)**
9. An air-cored choking coil takes a current of 2 A and dissipates 200 W when connected to a 200-V, 50-Hz mains. In other coil, the current taken is 3 A and the power 270 W under the same conditions. Calculate the current taken and the total power consumed when the coils are in series and connected to the same supply. **[1.2, 115 W] (*City and Guilds, London*)**
10. A circuit consists of a pure resistance and a coil in series. The power dissipated in the resistance is 500 W and the drop across it is 100 V. The power dissipated in the coil is 100 W and the drop across it is 50 V. Find the reactance and resistance of the coil and the supply voltage. **[9.168Ω; 4Ω; 128.5V]**

11. A choking coil carries a current of 15 A when supplied from a 50-Hz, 230-V supply. The power in the circuit is measured by a wattmeter and is found to be 1300 watt. Estimate the phase difference between the current and p.d. in the circuit. **[0.3768]** ***(I.E.E. London)***

12. An ohmic resistance is connected in series with a coil across 230-V, 50-Hz supply. The current is 1.8 A and p.ds. across the resistance and coil are 80 V and 170 V respectively. Calculate the resistance and inductance of the coil and the phase difference between the current and the supply voltage. **[61.1 Ω, 0.229 H, 34°20′]** ***(App. Elect. London Univ.)***

13. A coil takes a current of 4 A when 24 V d.c. are applied and for the same power on a 50-Hz a.c. supply, the applied voltage is 40. Explain the reason for the difference in the applied voltage. Determine (*a*) the reactance (*b*) the inductance (*c*) the angle between the applied p.d. and current (*d*) the power in watts. **[(*a*) 8 Ω (*b*) 0.0255 H (*c*) 53°7′ (*d*) 96 W]**

14. An inductive coil and a non-inductive resistance *R* ohms are connected in series across an a.c. supply. Derive expressions for the power taken by the coil and its power factor in terms of the voltage across the coil, the resistance and the supply respectively. If $R = 12\ \Omega$ and the three voltages are in order, 110 V, 180 V and 240 V, calculate the power and the power factor of the coil. **[546 W; 0.331]**

15. Two coils are connected in series. With 2 A d.c. through the circuit, the p.ds. across the coils are 20 and 30 V respectively. With 2 A a.c. at 40 Hz, the p.ds. across the coils are 140 and 100 V respectively. If the two coils in series are connected to a 230-V, 50-Hz supply, calculate (*a*) the current (*b*) the power (*c*) the power factor. **[(*a*) 1.55 A (*b*) 60 W (*c*) 0.1684]**

13.7. A.C. Through Resistance and Capacitance

The circuit is shown in Fig. 13.19 (*a*). Here $V_R = IR$ = drop across *R* in phase with *I*.

$V_C = IX_C$ = drop across capacitor which is lagging *I* by $\pi/2$

As capacitive reactance X_C is taken negative, V_C is shown along negative direction of *Y*-axis in the voltage triangle [Fig. 13.19 (*b*)]

The denominator is called the ***impedance*** of the circuit. So, $Z = \sqrt{R^2 + X_C^2}$

Impedance triangle is shown in Fig. 13.19 (*c*)

From Fig. 13.19 (*b*) it is found that *I* leads *V* by angle ϕ such that $\tan\phi = -X_C/R$

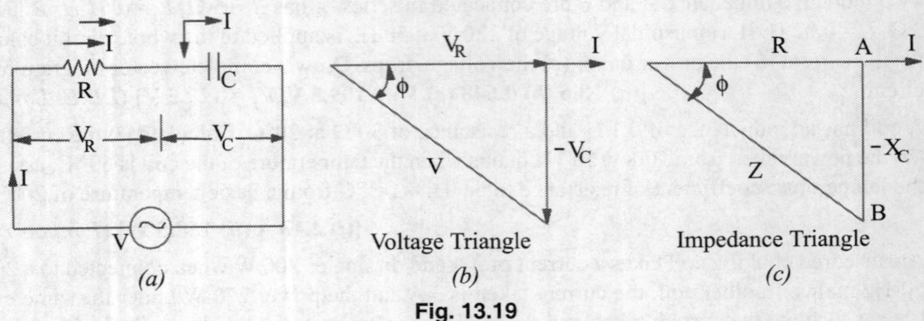

Fig. 13.19

$$\text{Now } V = \sqrt{V_R^2 + (-V_C)^2} = \sqrt{(IR)^2 + (-IX_C)^2} = I\sqrt{R^2 + X_C^2} \text{ or } I = \frac{V}{\sqrt{R^2 + X_C^2}} = \frac{V}{Z}$$

Hence, it means that if the equation of the applied alternating voltage is $v = V_m \sin \omega t$, the equation of the resultant current in the *R-C* circuit is $i = I_m \sin(\omega t + \phi)$ so that current ***leads*** the applied voltage by an angle ϕ. This fact is shown graphically in Fig. 13.19.

Example 13.22. *An a.c. voltage (80 + j 60) volts is applied to a circuit and the current flowing is (– 4 + j 10) amperes. Find* ***(i)*** *impedance of the circuit* ***(ii)*** *power consumed and* ***(iii)*** *phase angle.* **[Elect. Technology, Indore, Univ., Bombay Univ. 1999]**

Solution. $\mathbf{V} = (80 + j\,60) = 100 \angle 36.9°$;

$\mathbf{I} = -4 + j\,10 = 10.77 \angle \tan^{-1}(-2.5) = 10.77 \angle (180° - 68.2°) = 10.77 \angle 111.8°$

(i) $\mathbf{Z} = \mathbf{V/I} = 100 \angle 36.9°/10.77\angle 111.8°$

$= 9.28 \angle -74.9°$

$= 9.28 (\cos 74.9° - j \sin 74.9°) = 2.42 - j\,8.96\ \Omega$

Hence $R = 2.42\ \Omega$ and $X_C = 8.96\ \Omega$ capacitive

(ii) $P = I^2 R = 10.77^2 \times 2.42 = \mathbf{2.81\ W}$

(iii) Phase angle between voltage and current = 74.9° with current ***leading*** as shown in Fig. 13.21.

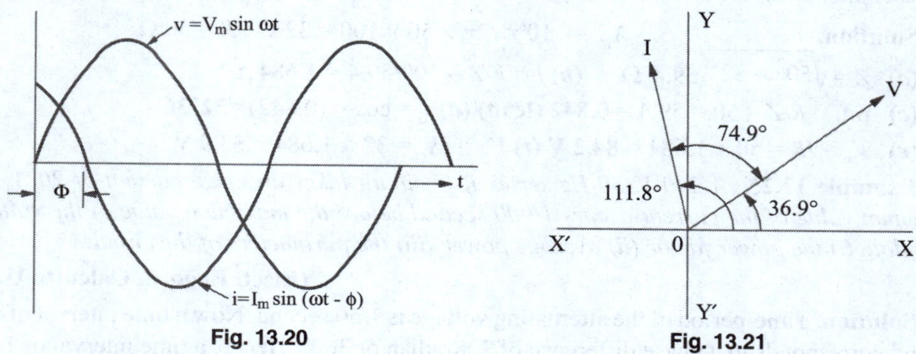

Fig. 13.20 Fig. 13.21

Alternative Method for Power

The method of conjugates will be used to determine the real power and reactive volt-ampere. It is a convenient way of calculating these quantities when both voltage and current are expressed in cartesian form. If the conjugate of *current* is multiplied by the voltage in cartesian form, the result is a complex quantity, the real part of which gives the real power *j* part of which gives the reactive volt-amperes (*VAR*). It should, however, be noted that real power as obtained by this method of conjugates is the same regardless of whether *V* or *I* is reversed although sign of voltamperes will depend on the choice of *V* or *I*.*

Using current conjugate, we get $P_{VA} = (80 + j\,60)(-4 - j\,10) = 280 - j\,1040$

∴ Power consumed = **280 W**

Example 13.23. *A tungsten filament bulb rated at 500-W, 100-V is to be connected to series with a capacitance across 200-V, 50-Hz supply. Calculate :*

(a) *the value of capacitor such that the voltage and power consumed by the bulb are according to the rating of the bulb.* ***(b)*** *the power factor of the current drawn from the supply.* ***(c)*** *draw the phasor diagram of the circuit.* **(Elect. Technology-1, Nagpur Univ. 1991)**

Solution. The rated values for bulb are : voltage = 100 *V* and current $I = W/V = 500/100 = 5$A. Obviously, the bulb has been treated as a pure resistance :

---

* If voltage conjugate is used, then capacitive VARs are positive and inductive VARs negative. If current conjugate is used, then capacitive VARs are negative and inductive VARs are positive.

(*a*) $V_C = \sqrt{220^2 - 100^2} = 196\ V$

Now, $IX_C = 196$ or $5\ X_C = 196, X_C = 39.2\ \Omega$

$\therefore\ I/\omega C = 39.2$ or $C = 1/314 \times 39.2 =$ **81 μF**

(*b*) p.f. $= \cos\phi = V_R/V = 100/200$

$= $ **0.455 (lead)**

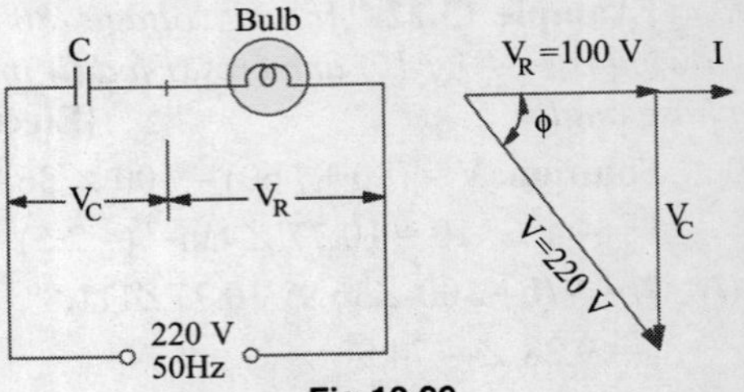

Bulb C V_C V_R 220 V 50Hz V_R=100 V I φ V=220 V V_C

Fig 13.22

(*c*) The phasor diagram is shown in Fig 13.22.

Example 13.24. *A pure resistance of 50 ohms is in series with a pure capacitance of 100 microfarads. The series combination is connected across 100-V, 50-Hz supply. Find* **(a)** *the impedance (b) current* **(c)** *power factor* **(d)** *phase angle* **(e)** *voltage across resistor* **(f)** *voltage across capacitor. Draw the vector diagram.* **(Elect. Engg.-1, JNT Univ. Warrangel)**

Solution. $X_C = 10^6\pi/2\pi \times 50 \times 100 = 32\ \Omega\ ;\ R = 50\ \Omega$

(*a*) $Z = \sqrt{50^2 + 32^2}$ **59.4 Ω** (*b*) $I = V/Z = 100/59.4 =$ **1.684 A**

(**c**) p.f. $= R/Z = 50 = 59.4 =$ **0.842 (lead) (*d*)** $\phi = \cos^{-1}(0.842) =$ **32°36′**

(*e*) $V_R = IR = 50 \times 1.684 =$ **84.2 V (*f*)** $V_C = IX_C = 32 \times 1.684 =$ **53.9 V**

Example 13.25. *A 240-V, 50-Hz series R-C circuit takes an r.m.s. current of 20 A. The maximum value of the current occurs 1/900 second before the maximum value of the voltage. Calculate* **(i)** *the power factor* **(ii)** *average power* **(iii)** *the parameters of the circuit.*

(Elect. Engg.-I, Calcutta Univ.)

Solution. Time-period of the alternating voltage is 1/50 second. Now a time interval of 1/50 second corresponds to a phase difference of 2 π radian or 360°. Hence, a time interval of 1/900 second corresponds to a phase difference of 360 × 50/900 = 20°.

Hence, current leads the voltage by 20º.

(*i*) power factor = cos 20° = **0.9397 (lead)**

(*ii*) averge power = 240 × 20 × 0.9397 = **4,510 W**

(*iii*) $Z = 240/20 = 12\ \Omega;\ R = Z\cos\Phi = 12 \times 0.9397 =$ **11.28 Ω**

$X_C = Z\sin\Phi = 12 \times \sin 20° = 12 \times 0.342 = 4.1\ \Omega$

$C = 10^6/2\pi \times 50 \times 4.1 =$ **775 μF**

Example 13.26. *A voltage v = 100 sin 314 t is applied to a circuit consisting of a 25 Ω resistor and an 80 μF capacitor in series. Determine :* **(a)** *an expression for the value of the current flowing at any instant* **(b)** *the power consumed* **(c)** *the p.d. across the capacitor at the instant when the current is one-half of its maximum value.*

Solution. $X_C = 1/(314 \times 80 \times 10^{-6}) = 39.8\ \Omega,\ Z = \sqrt{25^2 + 39.8^2} = 47\ \Omega$

$I_m = V_m/Z = 100/47 = 2.13$ A

$\phi = \tan^{-1}(39.8/25) = 57°52' = 1.01$ radian (lead)

(*a*) Hence, equation for the instantaneous current

$i =$ **2.13 sin (314 *t* + 1.01)** (*b*) Power $= I^2R = (2.13/\sqrt{2})^2 \times 25 =$ **56.7 W** (*c*) The voltage across the capacitor lags the circuit current by π/2 radians. Hence, its equation is given by

$$v_c = V_{cm}\sin\left(314\,t + 1.01 - \frac{\pi}{2}\right) \text{ where } V_{cm} = I_m \times X_C = 2.13 \times 39.8 = 84.8\ V$$

Now, when *i* is equal to half the maximum current (say, in the positive direction) then

$$i = 0.5 \times 2.13 \text{ A}$$

$\therefore$ $0.5 \times 2.13 = 2.13 \sin (314\,t + 1.01)$ or $314\,t + 1.01 = \sin^{-1} (0.5) = \frac{\pi}{6}$ or $\frac{5\pi}{6}$ radian

$\therefore$ $v_C = 84.8 \sin\left(\frac{\pi}{6} - \frac{\pi}{2}\right) = 84.8 \sin(-\pi/3) = -73.5$ V

or $v_c = 34.8 \sin\left(\frac{5\pi}{6} - \frac{\pi}{2}\right) = 84.4 \sin \pi/3 =$ **73.5 V**

Hence, p.d. across the capacitor is **73.5 V**

Example 13.27. *A capacitor and a non-inductive resistance are connected in series to a 200-V, single-phase supply. When a voltmeter having a non-inductive resistance of 13,500 Ω is connected across the resistor, it reads 132 V and the current then taken from the supply is 22.35 mA.*

Indicate on a vector diagram, the voltages across the two components and also the supply current **(a)** *when the voltmeter is connected and* **(b)** *when it is disconnected.*

Solution. The circuit and vector diagrams are shown in Fig. 13.23 (*a*) and (*b*) respectively.

(*a*) $V_C = \sqrt{200^2 - 132^2} =$ **150 V**

It is seen that $\phi = \tan^{-1}(150/132) = 49°$ in Fig. 13.23 (*b*). Hence

(*i*) Supply voltage lags behind the current by 49°. **(*ii*)** V_R leads supply voltage by 49° (*iii*) V_C lags behind the supply voltage by (90° – 49°) = 41°

The supply current is, as given equal to **22.35 mA.** The value of unknown resistance *R* can be found as follows:

Current through voltmeter = 132/13,500 = 9.78 mA

$\therefore$ Current through $R = 22.35 - 9.78 = 12.57$ mA $\therefore$ $R = 132/12.57 \times 10^{-3} = 10{,}500\ \Omega$;

$X_C = 150/22.35 \times 10^{-3} = 6.711\ \Omega$

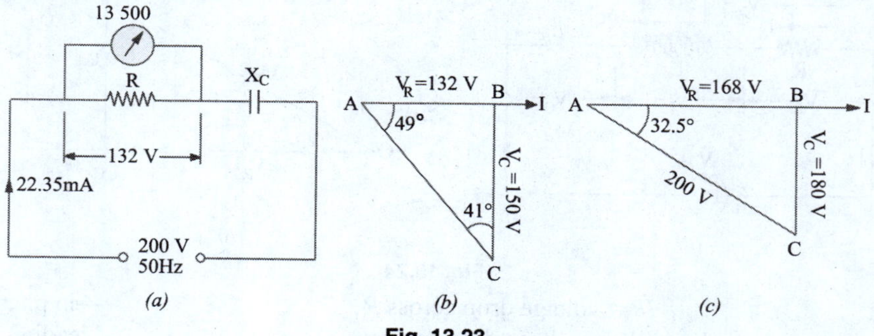

Fig. 13.23

(*b*) When voltmeter is disconnected, $Z = \sqrt{R^2 + X_L^2} = \sqrt{10{,}500^2 + 6{,}730^2} = 12{,}500\ \Omega$

Supply current = 200/12,461 = **16.0 mA**

In this case, $V_R = 16.0 \times 10^{-3} \times 10{,}500 =$ **168 V**

$V_C = 16.0 \times 10^{-3} \times 6711 =$ **107.4 V;** $\tan \phi = 107.4/168$

$\therefore$ $\phi = 32.5°$.

In this case, the supply voltage lags the circuit current by 32.5° as shown in Fig. 13.23 (*c*).

Tutorial Problem 13.2

1. A capacitor having a capacitance of 20 μF is connected in series with a non-inductive resistance of 120 Ω across a 100-V, 50-Hz supply, Calculate (*a*) voltage (*b*) the phase difference between the current and the supply voltage (*c*) the power. Also draw the vector diagram.

[(*a*) 0.501A (*b*) 52.9° (*c*) 30.2 W]

2. A capacitor and resistor are connected in series to an a.c. supply of 50 V and 50 Hz. The current is 2 A and the power dissipated in the circuit is 80 W. Calculate the resistance of the resistor and the capacitance of the capacitor. **[20 Ω ; 212 μF]**
3. A voltage of 125 *V* at 50 Hz is applied to a series combination of non-inductive resistor and a lossless capacitor of 50 μF. The current is 1.25 A. Find (*i*) the value of the resistor (*ii*) power drawn by the network (*iii*) the power factor of the network. Draw the phasor diagram for the network. **[(*i*) 77.3 Ω (*ii*) 121 W (*iii*) 0.773 (lead)] (*Electrical Technology-1, Osmania Univ.*)**
4. A black box contains a two-element series circuit. A voltage (40 – *j*30) drives a current of (40 – *j*3) A in the circuit. What are the values of the elements ? Supply frequency is 50 Hz. **[*R* = 1.05 ; *C* = 4750 μF] (*Elect. Engg. and Electronics Bangalore Univ.*)**
5. Following readings were obtained from a series circuit containing resistance and capacitance: $V = 150\ V$; $I = 2.5$ A; $P = 37.5\ W$, $f = 60$ Hz.
   Calculate (*i*) Power factor (*ii*) effective resistance (*iii*) capacitive reactance and (*iv*) capacitance. **[(*i*) 0.1 (*ii*) 6 Ω (*iii*) 59.7 Ω (*iv*) 44.4 μF]**
6. An alternating voltage of 10 volt at a frequency of 159 kHz is applied across a capacitor of 0.01 μF. Calculate the current in the capacitor. If the power dissipated within the dielectric is 100 μW, calculate (*a*) loss angle (*b*) the equivalent series resistance (*c*) the equivalent parallel resistance. **[0.A (*a*) 10^{-4} radian (*b*) 0.01 Ω (*c*) 1 MΩ]**

13.8. Resistance, Inductance and Capacitance in Series

The three are shown in Fig. 13.24 (*a*) joined in series across an a.c. supply of r.m.s. voltage *V*.

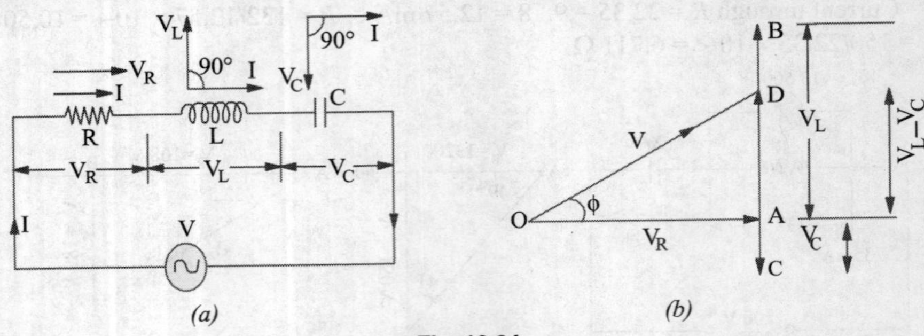

Fig. 13.24

Let $V_R = IR$ = voltage drop across R —in phase with I

$V_L = I.X_L$ = voltage drop across L —leading I by $\pi/2$

$V_C = I.X_C$ = voltage drop across C —lagging I by $\pi/2$

In voltage triangle of Fig. 13.24 (*b*), *OA* represents V_R, *AB* and *AC* represent the inductive and capacitive drops respectively. It will be seen that V_L and V_C are 180° out of phase with each other *i.e.* they are in direct opposition to each other.

Subtracting *BD* (= *AC*) from *AB*, we get the net reactive drop $AD = I(X_L - X_C)$

The applied voltage *V* is represented by *OD* and is the vector sum of *OA* and *AD*

$$\therefore\quad OD = \sqrt{OA^2 + AD^2} \quad \text{or} \quad V = \sqrt{(IR)^2 + (IX_L - IX_C)^2} = I\sqrt{R^2 + (X_L - X_C)^2}$$

$$\text{or} \quad I = \frac{V}{\sqrt{R^2 + (X_L - X_C)^2}} = \frac{V}{\sqrt{R^2 + X^2}} = \frac{V}{Z}$$

The term $\sqrt{R^2 + (X_L - X_C)^2}$ is known as the impedance of the circuit. Obviously,

***(impedance)*² = (*resistance*)² + (*net reactance*)²**

or $Z^2 = R^2 + (X_L - X_C)^2 = R^2 + X^2$

where X is the net reactance (Fig. 13.25 and 13.26).

Phase angle ϕ is given by $\tan \phi = (X_L - X_C)/R = X/R$ = net reactance/resistance

Power factor is $\cos \phi = \dfrac{R}{Z} = \dfrac{R}{\sqrt{R^2 + (X_L - X_C)^2}} = \dfrac{R}{\sqrt{R^2 + X^2}}$

Hence, it is seen that if the equation of the applied voltage is $v = V_m \sin \omega t$, then equation of the resulting current in an *R-L-C* circuit is given by $i = I_m \sin (\omega t \pm \phi)$

The + ve sign is to be used when current leads *i.e.* $X_C > X_L$.

The − ve sign is to be used when current lags *i.e.* when $X_L > X_C$.

In general, the current lags or leads the supply voltage by an angle ϕ such that $\tan \phi = X/R$

Using symbolic notation, we have (Fig. 13.26), $\mathbf{Z} = R + j(X_L - X_C)$

Numerical value of impedance

$$Z = \sqrt{R^2 + (X_L - X_C)^2}$$

Its phase angle is $\Phi = \tan^{-1} [X_L - X_C/R]$

$Z = Z \angle \tan^{-1} [(X_L - X_C)/R] = Z \angle \tan^{-1} (X/R)$

If $\mathbf{V} = V \angle 0$, then, $\mathbf{I} = \mathbf{V}/\mathbf{Z}$

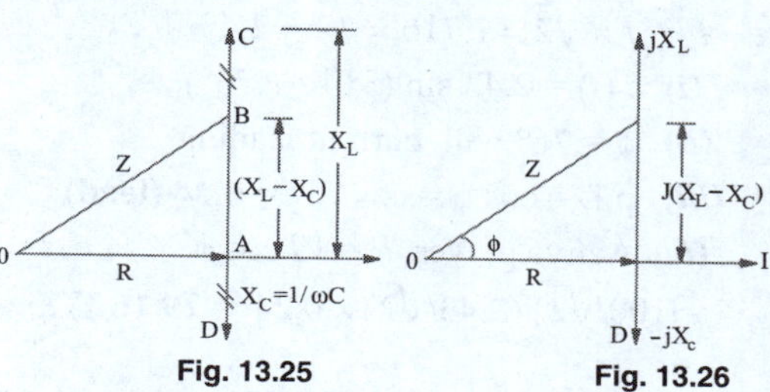

Fig. 13.25 Fig. 13.26

Summary of Results of Series AC Circuits

Type of Impedance	*Value of Impedance*	*Phase angle for current*	*Power factor*
Resistance only	R	0°	1
Inductance only	ωL	90° lag	0
Capacitance only	1/ω C	90° lead	0
Resistance and Inductance	$\sqrt{[R^2 + (\omega L)^2]}$	$0 < \phi < 90°$ lag	1 > p.f. > 0 lag
Resistance and Capacitance	$\sqrt{[R^2 + (-1/\omega C)^2]}$	$0 < \phi < 90°$ lead	1 > p.f. > 0 lead
R-L-C	$\sqrt{[R^2 + (\omega L \sim 1/\omega C)^2]}$	between 0° and 90° lag or lead	between 0 and unity lag or lead

Example 13.28. *A resistance of 20 Ω, an inductance of 0.2 H and a capacitance of 100 μF are connected in series across 220-V, 50-Hz mains. Determine the following* ***(a)*** *impedance* ***(b)*** *current* ***(c)*** *voltage across R, L and C* ***(d)*** *power in watts and VA* ***(e)*** *p.f. and angle of lag.*

(Elect. Engg. A.M.Ae S.I. 1992)

Solution. $X_C = 0.2 \times 314 = 63\ \Omega$, $C = 10\ \mu F = 100 \times 10^{-6} = 10^{-4}$ farad

$X_C = \dfrac{1}{\omega C} = \dfrac{1}{314 \times 10^{-4}} = 32\ \Omega$, $X = 63 - 32 =$ **31 Ω (*inductive*)**

(a) $Z = \sqrt{(20^2 + 31)^2} =$ **37 Ω** ***(b)*** $I = 220/37 =$ **6 A (approx)**

(c) $V_R = I \times R = 6 \times 20 =$ **120 V;** $V_L = 6 \times 63 =$ **278 V,** $V_C = 6 \times 32 =$ **192 V**

(d) Power in *VA* = 6 × 220 = **1320**

Power in watts = 6 × 220 × 0.54 = **713 W**

(e) p.f. = $\cos \phi = R/Z = 20/37 =$ **0.54;** $\phi = \cos^{-1} (0.54) =$ **57°18′**

Example 13.29. *A voltage e(t) = 100 sin 314 t is applied to series circuit consisting of 10 ohm resistance, 0.0318 henry inductance and a capacitor of 63.6 μF. Calculate* **(i)** *expression for i (t)* **(ii)** *phase angle between voltage and current* **(iii)** *power factor* **(iv)** *active power consumed* **(v)** *peak value of pulsating energy.* **(Elect. Technology, Indore Univ.)**

Solution. Obviously, $\omega = 314$ rad/s ; $X_l = \omega L = 314 \times 0.0318 = 10\ \Omega$

$X_C = 1/\omega C = 1/314 \times 63.6 \times 10^{-6} = 50\ \Omega$; $X = X_L - X_C = (10 - 50) = -40\ \Omega$ (capacitive)

$\mathbf{Z} = 10 - j\,40 = 41.2 \angle -76°$;

$$\mathbf{I} = \frac{\mathbf{V}}{\mathbf{Z}} = \frac{(100/\sqrt{2}) \angle 0°}{41.2 \angle -76°} = 1.716 \angle 76°$$

$I_m = I \times \sqrt{2} = 1.716 \times \sqrt{2} = 2.43\ A$

(*i*) $i\,(t) = 2.43 \sin (314\,t + 76°)$

(*ii*) ϕ = **76° with current leading**

(*iii*) p.f. = cos ϕ = cos 76° = **0.24 (lead)**

(*iv*) Active power, $P = VI \cos \phi$

$= (100/\sqrt{2})\,(2.43/\sqrt{2}) \times 0.24 =$ **29.16 W**

(*v*) As seen from Fig. 13.27, peak value of pulsating energy is $\frac{V_m I_m}{2} + \frac{V_m I_m}{2} \cos \phi$

$= \frac{V_m I_m}{2}(1 + \cos \phi) = \frac{100 \times 2.43}{2}(1 + 0.24) =$ **151 W**

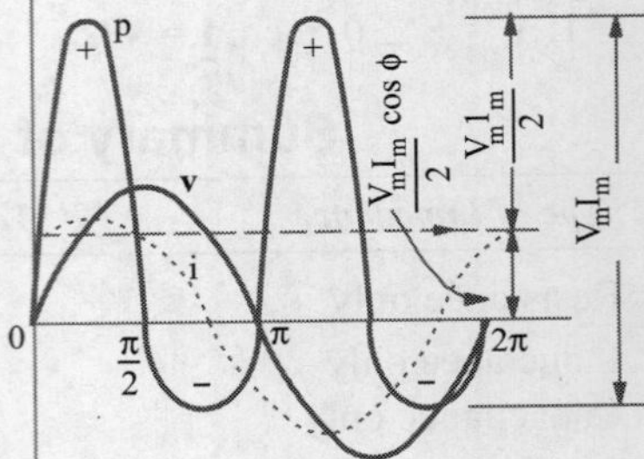

Fig. 13.27

Example 13.30. *A resistance R, an inductance L = 0.01 H and a capacitance C are connected in series. When a voltage v = 400 cos (300 t – 10°) votls is applied to the series combination, the current flowing is $10\sqrt{2}$ cos (3000 t – 55°) amperes. Find R and C.*

(Elect. Circuits Nagpur Univ. 1992)

Solution. The phase difference between the applied voltage and circuit current is (55° – 10°) = 45° with current lagging. The angular frequency is ω = 3000 radian/second. Since current lags, $X_L > X_C$.

Net reactance $X = (X_L - X_C)$. Also $X_L = \omega L = 3000 \times 0.01 = 30\ \Omega$

$\tan \phi = X/R$ or $\tan 45° = X/R$ $\therefore$ $X = R$ Now, $Z = \frac{V_m}{I_m} = \frac{400}{10\sqrt{2}} = 28.3\ \Omega$

$Z^2 = R^2 + X^2 = 2R^2$ $\therefore R = Z/\sqrt{2} = 28.3/\sqrt{2} = 20\ \Omega$; $X = X_L - X_C = 30 - X_C = 20$

$X_C = 10\ \Omega$ or $\frac{1}{\omega C} = 10$ or $\frac{1}{3000\,C} =$ or $C =$ **33 μF**

Example 13.31. *A coil is in series with a 20 μF capacitor across a 230-V, 50-Hz supply. The current taken by the circuit is 8 A and the power consumed is 200 W. Calculate the inductance of the coil if the power factor of the circuit is* **(i)** *leading* **(ii)** *lagging.*

Sketch a vector diagram for each condition and calculate the coil power factor in each case.

(Elect. Engg.-I Nagpur Univ. 1993)

Solution. (*i*) Since power factor is leading, net reactance $X = (X_C - X_L)$ as shown in Fig. 13.28 (*a*).

$I^2R = 200$ or $8^2 \times R = 200$; $\therefore$ $R = 200/64 = 25/8\ \Omega = 3.125\ \Omega$

$Z = V/I = 230/8 = 28.75\ \Omega$; $X_C = 10^6/2\pi \times 50 \times 20 = 159.15\ \Omega$

$R_2 + X_2 = 28.75^2$ $\therefore$ $X = 28.58\ \Omega$ $\therefore$ $(X_C - X_L) = 28.58$ or $159.15 - X_L = 28.58$

$\therefore\quad X_L = 130.57\ \Omega$ or $2\pi \times 50 \times L = 130.57 \quad \therefore \quad L = 0.416\ H$

If θ is the p.f. angle of the coil, then $\tan\theta = R/X_L = 3.125/130.57 = 0.024$; $\theta = 1.37°$, p.f. of the coil = 0.9997

(ii) When power factor is lagging, net reactance is $(X_L - X_C)$ as shown in Fig. 13.28 (*b*).

$\therefore\quad X_L - 159.15 = 28.58 \quad$ or $\quad X_L = 187.73\ \Omega \quad \therefore \quad 187.73 = 2\pi \times 50 \times L$ or $L = 0.597$ H.

In this case, $\tan\theta = 3.125/187.73 = 0.0167$; $\theta = 0.954° \ \therefore\ \cos\theta = 0.9998$.

The vector diagrams for the two conditions are shown in Fig. 13.28.

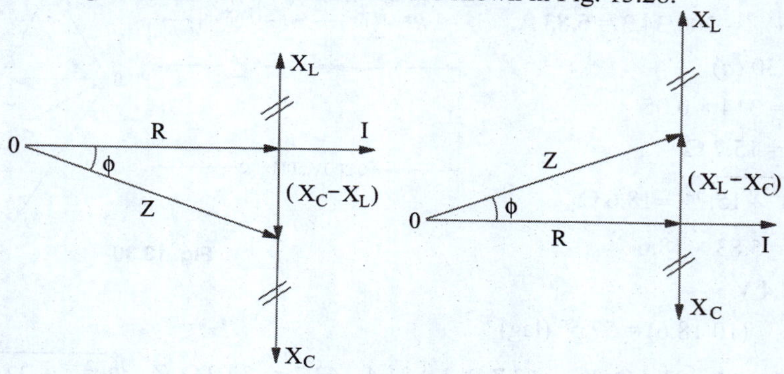

Fig. 13.28

Example 13.32. *In Fig. 13.29, calculate* **(i)** *current* **(ii)** *voltage drops V_1, V_2, and V_3 and* **(iii)** *power absorbed by each impedance and total power absorbed by the circuit. Take voltage vector along the reference axis.*

Solution. $\mathbf{Z}_1 = (4 + j3)\ \Omega$; $\mathbf{Z}_2 = (6 - j8)\ \Omega$; $\mathbf{Z}_3 = (4 + j0)\Omega$

$\mathbf{Z} = \mathbf{Z}_1 + \mathbf{Z}_2 + \mathbf{Z}_3 = (4 + j3) + (6 - j8) + (4 + j0) = (14 - j5)\Omega$

Taking $\mathbf{V} = V \angle 0° = 100 \angle 0° = (100 + j0)$

$$\therefore\ \mathbf{I} = \frac{\mathbf{V}}{\mathbf{Z}} = \frac{100}{(14 - j5)} = \frac{100\,(14 + j5)}{(14 - j5)\,(14 + j5)} = 6.34 + j2.26$$

(i) Magnitude of the current

$= \sqrt{(6.34^2 + 2.26^2)} = \mathbf{6.73\ A}$

(ii) $\mathbf{V}_1 = \mathbf{IZ}_1 = (6.34 + j\,2.26)\,(4 + j3) = 18.58 + j\,28.06$

$\mathbf{V}_2 = \mathbf{IZ}_2 = (6.34 + j\,2.26)\,(6 - j8) = 56.12 - j\,37.16$

$\mathbf{V}_3 = \mathbf{IZ}_3 = (6.34 + j\,2.26)\,(4 + j0) = 25.36 + j\,9.04$

$\mathbf{V} = 100 + j\,0$ (check)

(iii) $P_1 = 6.73^2 \times 4 = 181.13$ W.

$P_2 = 6.73^2 \times 6 = 271.74$ W, $P_3 = 6.73^2 \times 4$

$= 181.13$ W,

Total = 34 W

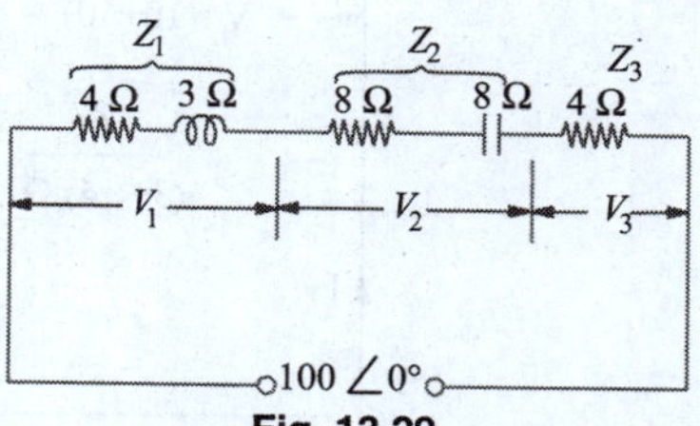

Fig. 13.29

Otherwise $\quad \mathbf{P}_{VA} = (100 + j\,0)\,(6.34 - j\,2.26)$

(using current conjugate)

$= 634 - j\,226$

real power = 634 W (as a check)

Example 13.33. *Draw a vector for the circuit shown in Fig. 13.30 indicating the resistance and reactance drops, the terminal voltages V_1 and V_2 and the current. Find the values of* **(*i*)** *the current I* **(*ii*)** *V_1 and V_2 and* **(*iii*)** *p.f.* **(Elements of Elect Engg-I, Bangalore Univ.)**

Solution. $L = 0.05 + 0.1 = 0.15$ H; $X_L = 314 \times 0.5 = 47.1\ \Omega$

$X_C = 10^6/314 \times 50 = 63.7\ \Omega$; $X = 47.1 - 63.7 = -16.6\ \Omega$, $R = 30\ \Omega$,

$Z = \sqrt{30^2 + (-16.6)^2} = 34.3\ \Omega$

(*i*) $I = V/Z = 200/34.3 =$ **5.83 A,**

from Fig. 13.30 (*a*)

(*ii*) $X_{L1} = 314 \times 0.05$

$= 15.7\ \Omega$

$Z_1 = \sqrt{10^2 + 15.7^2} = 18.6\ \Omega$

$V_1 = IZ_1 = 5.83 \times 18.6$

$=$ **108.4 V**

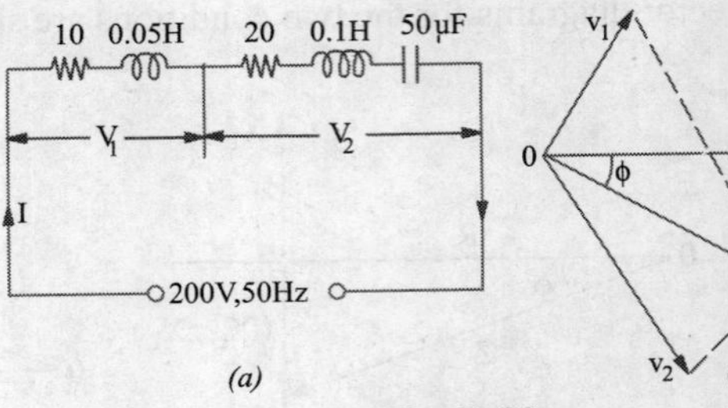

Fig. 13.30

$\phi_1 = \cos^{-1}(10/18.6) = 57.5°$ (lag)

$X_{L2} = 314 \times 0.1 = 31.4\ \Omega, X_c = -63.7\ \Omega, X = 31.4 - 63.7 = -32.2\ \Omega, Z_2\ \sqrt{20^2 + (-32.3)^2} =$ **221 V**

$\phi_2 = \cos^{-1}(20/38) = 58.2$ (lead)

(*iii*) Combined p.f. $= \cos\phi = R/Z = 30/34.3 =$ **0.875 (lead)**, from Fig. 13.30 (*b*).

Example 13.34. *In a circuit, the applied voltage is found to lag the current by 30°.*

(a) *Is the power factor lagging or leading ?* **(*b*)** *What is the value of the power factor ?* **(*c*)** *Is the circuit inductive or capacitive ?*

In the diagram of Fig 13.31, the voltage drop across Z_1 is (10 + j0) volts. Find out **(*i*)** *the current in the circuit* **(*ii*)** *the voltage drops across Z_2 and Z_3* **(*iii*)** *the voltage of the generator.*

(Elect. Engg.-I, Bombay Univ. 1991)

Solution. **(*a*)** Power factor is ***leading*** because current leads the voltage.

(*b*) p.f. = cos 30° = **0.86 (lead)** **(*c*)** The circuit is **capacitive.**

(*i*) Circuit current can be found by dividing voltage drop $\mathbf{V}_1$ by $\mathbf{Z}_1$

$$\mathbf{I} = \frac{10 + j\,0}{3 + j\,4} = \frac{10\angle 0°}{5\angle 53.1°} = 2\angle -53.1° = 2(\cos 53.1° - j\sin 53.1°)$$

$$= 2\,(0.6 - j\,0.8) = 1.2 - j\,1.6$$

V_1=(10+j0) V_2 V_3

Z_1 (3+ j4) Ω Z_2 (2+ j3.46) Ω Z_3 (1- j7.46) Ω

I

Fig. 13.31

$$\mathbf{Z}_2 = 2 + j\,3.46;\ \mathbf{V}_2 = \mathbf{IZ}_2 = (1.2 - j\,1.6)\,(2 + j3.46) = (7.936 + j\,0.952) \text{ volt}$$

$$\mathbf{V}_3 = (1.2 - j\,1.6)\,(1 - j\,7.46) = (-\,10.74 - j\,10.55) \text{ volt}$$

(ii)

$$\mathbf{V} = \mathbf{V}_1 + \mathbf{V}_2 + \mathbf{V}_3 = (10 + j\,0) + (7.936 + j\,0.952) + (-\,10.74 - j\,10.55)$$

$$= (7.2 - j\,9.6) = 12\angle -53.1°$$

Incidentally, it shows that current **I** and voltage **V** are in phase with each other.

Tutorial Problem 13.3

1. An e.m.f. represented by $e = 100 \sin 100\,\pi\,t$ is impressed across a circuit consisting of 40-Ω resistor in series with a 40-μF capacitor and a 0.25 H indicator.
   Determine (*i*) the r.m.s. value of the current (*ii*) the power supplied (*iii*) the power factor.
   [(*i*) 1.77 A (*ii*) 125 W (*iii*) 1.0] *(London Univ.)*
2. A series circuit with a resistor of 100 Ω capacitor of 25 μF and inductance of 0.15 H is connected across 220-V, 60-Hz supply. Calculate (*i*) current (*ii*) power and (*iii*) power factor in the circuit.
   [(*i*) 1.97 A; (*ii*) 390 W (*iii*) 0.9 (lead)] *(Elect. Engg. and Electronics Bangalore Univ.)*
3. A series circuit with R = 10 Ω, L = 50 mH and C = 100 μF is supplied with 200 V/50 Hz. Find (*i*) the impedance (*ii*) current (*iii*) power (*iv*) power factor.
   [(*i*) 18.94 Ω (*ii*) 18.55 A (*iii*) 1966 W (*iv*) 0.53 (leading)] *(Elect. Engg. & Electronics Bangalore Univ.)*
4. A coil of resistance 10 Ω and inductance 0.1 H is connected in series with a 150-μF capacitor across a 200-V, 50-Hz supply. Calculate (*a*) the inductive reactance, (*b*) the capacitive reactance, (*c*) the impedance (*d*) the current, (*e*) the power factor (*f*) the voltage across the coil and the capacitor respectively. **[(*a*) 31.4 Ω (*b*) 21.2 Ω (*c*) 14.3 Ω (*d*) 14 A (*e*) 0.7 lag (*f*) 460 V, 297 V]**
5. A circuit is made up of 10 Ω resistance, 12 mH inductance and 281.5 μF capacitance in series. The supply voltage is 100 V (constant). Calculate the value of the current when the supply frequency is (*a*) 50 Hz and (*b*) 150 Hz. **[8 A leading; 8 A lagging]**
6. A coil having a resistance of 10 Ω and an inductance of 0.2 H is connected in series with a capacitor of 59.7 μF. The circuit is connected across a 100-V, 50-Hz a.c. supply. Calculate (*a*) the current flowing (*b*) the voltage across the capacitor (*c*) the voltage across the coil. Draw a vector diagram to scale. **[(*a*) 10 A (*b*) 628 V (*c*) 635 V]**
7. A coil is in series with a 20 μF capacitor across a 230-V , 50-Hz supply. The current taken by the circuit is 8 A and the power consumed is 200 W. Calculate the inductance of the coil if the power factor of the circuit is (*a*) leading and (*b*) lagging.
   Sketch a vector diagram for each condition and calculate the coil power factor in each case.
   [0.415 H; 0.597 H; 0.0238 ; 0.0166]
8. A circuit takes a current of 3 A at a power factor of 0.6 lagging when connected to a 115-V, 50-Hz supply. Another circuit takes a current, of 5 A at a power factor of 0.707 leading when connected to the same supply. If the two circuits are connected in series across a 230-V, 50Hz supply, calculate
   (*a*) the current (*b*) the power consumed and (*c*) the power factor.
   [(*a*) 5.5 A (*b*) 1.188 kW (*c*) 0.939 lag]
9. A coil of insulated wire of resistance 8 ohms and inductance 0.03 H is connected to an a.c. supply at 240 V, 50-Hz. Calculate (*a*) the current, the power and power factor (*b*) the value of a capacitance which, when connected in series with the above coil, causes no change in the values of current and power taken from the supply. **[(*a*) 19.4 A, 3012 W, 0.65 lag (*b*) 168.7 μF] *(London Univ.)***
10. A series circuit, having a resistance of 10 Ω, an inductance of 0.025 H and a variable capacitance is connected to a 100-V, 25-Hz single-phase supply. Calculate the capacitance when the value of the current is 8 A. At this value of capacitance, also calculate (*a*) the circuit impedance (*b*) the circuit power factor and (*c*) the power consumed. **[556 μF (*a*) 1.5 Ω (*b*) 0.8 leading (*c*) 640 W]**

11. An alternating voltage is applied to a series circuit consisting of a resistor and iron-cored inductor and a capacitor. The current in the circuit is 0.5 A and the voltages measured are 30 V across the resistor, 48 V across the inductor, 60 V across the resistor and inductor and 90 V across the capacitor. Find (*a*) the combined copper and iron losses in the inductor (*b*) the applied voltage.

[(*a*) 3.3 W (*b*) 56 V] (*City & Guilds, London*)

12. When an inductive coil is connected across a 250-V, 50-Hz supply, the current is found to be 10 A and the power absorbed 1.25 kW. Calculate the impedance, the resistance and the inductance of the coil.

A capacitor which has a reactance twice that of the coil, is now connected in series with the coil across the same supply. Calculate the p.d. across the capacitor. **[25 Ω; 12.5 Ω; 68.7 mH; 433 V]**

13.9. Resonance in R-L-C Circuits

We have seen that net reactance in an *R-L-C* circuit of Fig. 13.32 (*a*) is

$$X = X_L - X_C \text{ and } Z = \sqrt{[R^2 + (X_L - X_C)^2]} = \sqrt{R^2 + X^2}$$

Let such a circuit be connected across an a.c. source of constant voltage V but of frequency varying from zero to infinity. There would be a certain frequency of the applied voltage which would make X_L equal to X_C in magnitude. In that case, $X = 0$ an $Z = R$ as shown in Fig. 13.32 (*c*). Under this condition, the circuit is said to be in electrical resonance.

As shown in Fig. 13.32 (*c*), $V_L = I.X_L$ and $V_C = I.X_C$ and the two are equal in magnitude but opposite in phase. Hence, they cancel each other out. The two reactances taken together act as a short-circuit since no voltage develops across them. Whole of the applied voltage drops across R so that $V = V_R$. The circuit impedance $Z = R$. The phasor diagram for series resonance is shown in Fig. 13.32 (*d*).

Calculation of Resonant Frequency

The frequency at which the net reactance of the series circuit is zero is called the resonant frequency f_0. Its value can be found as under : $X_L - X_C = 0$ or $X_L = X_C$ or $\omega_0 L = 1/\omega_0 C$

or $$\omega_0^2 = \frac{1}{LC} \text{ or } (2\pi f_0)^2 = \frac{1}{LC} \text{ or } f_0 = \frac{1}{2\pi\sqrt{LC}}$$

If L is in henry and C in farad, then f_0 is given in *Hz*.

When a series *R-L-C* circuit is in resonance, it possesses minimum impedance $Z = R$. Hence, circuit current is maximum, it being limited by value of R alone. The current $I_0 = V/R$ and is in phase with V.

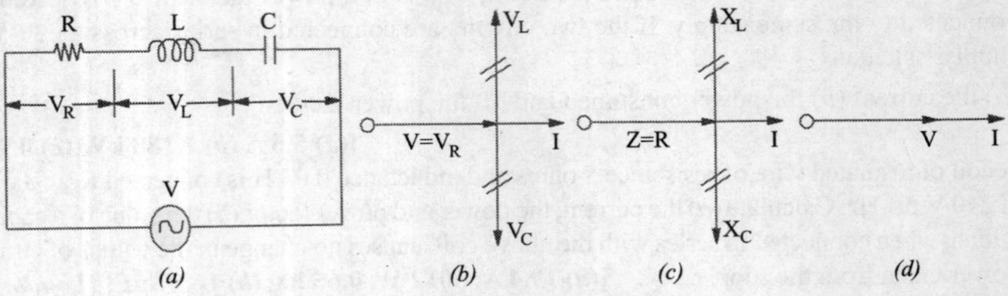

Fig. 13.32

Since circuit current is maximum, it produces large voltage drops across L and C. But these drops being equal and opposite, cancel each other out. Taken together, L an C from part of a circuit across which no voltage develops, however, large the current flowing. If it were not for the presence of R, such a resonant circuit would act like a short-circuit to currents of the frequency to which it resonates. Hence, a series resonant circuit is sometimes called ***acceptor*** circuit and the series resonance is often referred to as voltage resonance.

In fact, at resonance the series *RLC* circuit is reduced to a purely resistive circuit, as shown in Fig. 13.32.

Incidentally, it may be noted that if X_L and X_C are shown at any frequency *f*, that the value of the resonant frequency of such a circuit can be found by the relation $f_0 = f\sqrt{X_C/X_L}$.

Summary

When an *R-L-C* circuit is in resonance

1. net reactance of the circuit is zero *i.e.* $(X_L - X_C) = 0$. or $X = 0$.
2. circuit impedance is minimum *i.e.* $Z = R$. Consequently, circuit admittance is maximum.
3. circuit current is maximum and is given by $I_0 = V/Z_0 = V/R$.
4. power dissipated is maximum *i.e.* $P_0 = I_0^2 R = V^2/R$.
5. circuit power factor angle $\theta = 0$. Hence, power factor $\cos\theta = 1$.
6. although $V_L = V_C$ yet V_{coil} is greater than V_C because of its resistance.
7. at resonance, $\omega^2 LC = 1$
8. $Q = \tan\theta = \tan 0° = 0$*.

13.10. Graphical Representation of Resonance

Suppose an alternating voltage of constant magnitude, but of varying frequency is applied to an *R-L-C* circuit. The variations of resistance, inductive reactance X_L and capacitive reactance X_C with frequency are shown in Fig. 13.33 (*a*).

(i) Resistance : It is independent of *f*, hence, it is represented by a straight line.

(ii) Inductive Reactance : It is given by $X_L = \omega L = 2\pi fL$. As seen, X_L is directly proportional to *f i.e.* X_L increases linearly with *f*. Hence, its graph is a straight line passing through the origin.

(iii) Capacitive Reactance : It is given by $X_C = 1/\omega C = 1/2\pi fC$. Obviously, it is inversely proportional to *f*. Its graph is a rectangular hyperbola which is drawn in the fourth quadrant because X_C is regarded negative. It is asymptotic to the horizontal axis at high frequencies and to the vertical axis at low frequencies.

(iv) Net Reactance : It is given by $X = X_L \sim X_C$. Its graph is a hyperbola (not rectangular) and crosses the *X*-axis at point *A* which represents resonant frequency f_0.

(v) Circuit Impedance : It is given by $Z = \sqrt{[R^2 + (X_L \sim X_C)^2]} = \sqrt{R^2 + X^2}$

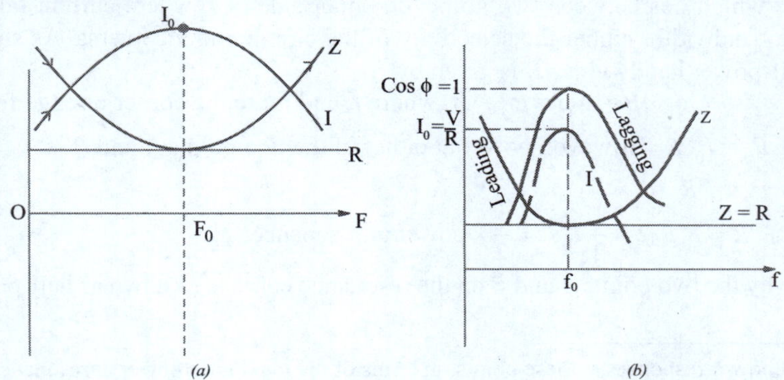

Fig. 13.33

* However, value of Q0 is as given in Art 13.5, 13.9 and 13.17.

At low frequencies Z is large because X_C is large. Since $X_C > X_L$, the net circuit reactance X is capacitive and the p.f. is leading [Fig. 13.33 (*b*)]. At high frequencies, Z is again large (because X_L is large) but is inductive because $X_L > X_C$. Circuit impedance has minimum values at f_0 given by $Z = R$ because $X = 0$.

(vi) Current I_0: It is the reciprocal of the circuit impedance. When Z is low, I_0 is high and vice versa. As seen, I_0 has low value on both sides of f_0 (because Z is large there) but has maximum value of $I_0 = V/R$ at resonance. Hence, maximum power is dissipated by the series circuit under resonant conditions. At frequencies below and above resonance, current decreases as shown in Fig. 13.33 (*b*). Now, $I_0 = V/R$ and $I = V/Z = V/\sqrt{(R^2 + X^2)}$. Hence $I/I_0 = R/Z = V/\sqrt{(R^2 + X^2)}$ where X is the net circuit reactance at any frequency f.

(vii) Power Factor

As pointed out earlier, X is capacitive below f_0. Hence, current leads the applied voltage. However, at frequencies above f_0, X is inductive. Hence, the current lags the applied voltage as shown in Fig. 13.33. The power factor has maximum value of unity at f_0.

13.11. Resonance Curve

The curve, between circuit current and the frequency of the applied voltage, is known as resonance curve. The shapes of such a curve, for different values of R are shown in Fig. 13.34. For smaller values of R, the resonance curve is sharply peaked and such a circuit is said to be sharply resonant or highly selective. However, for larger values of R, resonance curve is flat and is said to have poor selectivity. The ability of a resonant circuit to discriminate between one particular frequency and all others is called its selectivity. The selectivities of different resonant circuits are compared in terms of their half-power bandwidths (Art. 13.13).

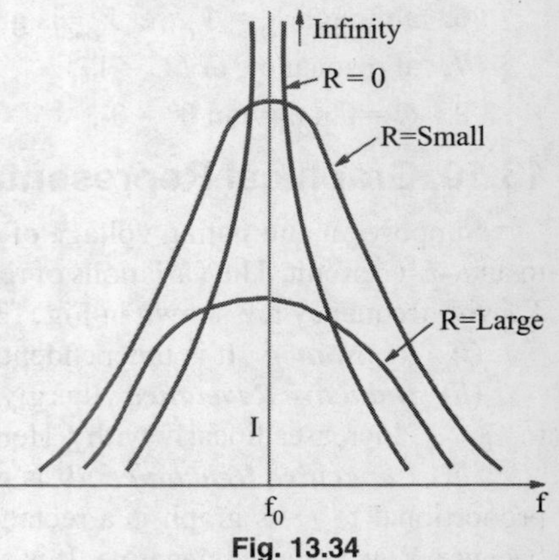

Fig. 13.34

13.12. Half-Power Bandwidth of a Resonant-Circuit

A discussed earlier, in an *R-L-C* circuit, the maximum current at resonance is solely determined by circuit resistance R ($\because X = 0$) but at off-resonance frequencies, the current amplitude depends on Z (where $X \neq 0$). The half-wave bandwidth of a circuit is given by the band of fequencies which lies between two points on either side of f_0 where current falls to $I/\sqrt{2}$. Narrower the bandwidth, higher the selectivity of the circuit and vice versa. As shown in Fig. 13.47 the half-power bandwidth AB is given by

$AB = \Delta f = f_2 - f_1$ or $AB = \Delta\omega = \omega_2 - \omega_1$ where f_1 and f_2 are the corner or edge frequencies.

As seen, $P_0 = I_0^2 R$. However, power at either of the two points A and B is

$$P_1 = P_2 = I^2R$$

$$= (I_0/\sqrt{2})^2 R = I_0^2 R/2 = \frac{1}{2} I_0^2 R = \frac{1}{2} \times \text{power at resonance}$$

That is why the two points A and B on the resonance curve are known as half-power points*

* The decibel power responses at these points, in terms of the maximum power at resonance, is

$$10 \log_{10} \frac{I_m^2 R/2}{I_m^2 R} = 10 \log_{10} \frac{1}{2} = -10 \log_{10} 2 = -3\ dB$$

Hence, the half-power points are also referred to as - 3 dB points.

and the corresponding value of the bandwidth is called half-power bandwidth B_{hp}. It is also called $-3dB$* bandwidth. The following points regarding half-power point A and B are worth noting. At these points,

1. current is $I_0/\sqrt{2}$
2. impedance is $\sqrt{2}.R$ or $\sqrt{2}.Z_0$
3. $P_1 = P_2 = P_0/2$
4. the circuit phase angle is $\theta = \pm 45°$
5. $Q = \tan\theta = \tan 45° = 1$
6. $B_{hp} = f_2 - f_1 = f_0/Q_0 = \sqrt{f_1 f_2}/Q_0 = R/2\pi L.$

It is interesting to note that B_{hp} is independent of the circuit capacitance.

13.13. Bandwidth B at any Off-resonance Frequency

It is found that the bandwidth of a given *R-L-C* circuit at any off-resonance frequencies f_1 and f_2 is given by

$B = f_0\, Q/Q_0 = \sqrt{f_1 f_2};.Q/Q_0 = f_2 - f_1$

where f_1 and f_2 are any frequencies (not necessarily half-power frequencies) below and above f_0.

Q = tangent of the circuit phase angle at the off-resonance frequencies f_1 and f_2.

Q_0 = quality factor at resonance $= \dfrac{\omega_0 L}{R} = \dfrac{1}{R}\sqrt{\dfrac{L}{C}}$

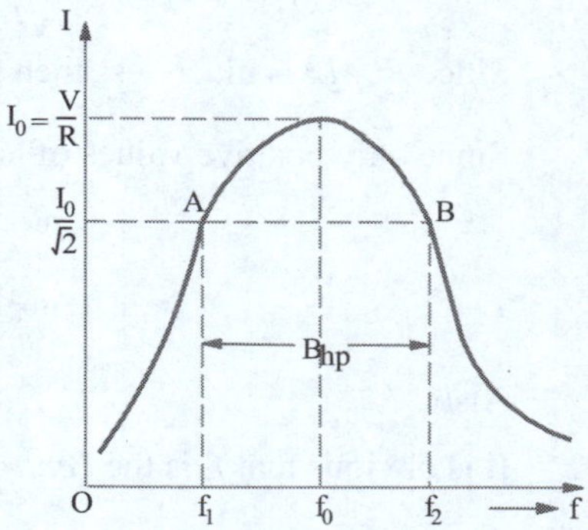

Fig. 13.35

13.14. Determination of Upper and Lower Half-Power Frequencies

As mentioned earlier, at lower half-power frequencies, $\omega_1 < \omega_0$ so that $\omega_1 L < 1/\omega_1 C$ and $\phi = -45°$

$$\therefore \quad \frac{1}{\omega_1 C} - \omega_1 L = R \quad \text{or} \quad \omega_1^2 + \frac{R}{L}\omega_1 - \frac{1}{LC} = 0$$

Putting $\dfrac{\omega_0}{Q_0} = \dfrac{R}{L}$ and $\omega_0^2 = \dfrac{1}{LC}$ in the above equation, we get $\omega_1^2 + \dfrac{\omega_0}{Q_0}\omega_1 - \omega_0^2 = 0$

The positive solution of the above equation is, $\omega_1 = \omega_0\left[\sqrt{\left(1 + \dfrac{1}{4Q_0^2}\right)} - \dfrac{1}{2Q_0}\right]$

Now at the upper half-power frequency, $\omega_2 > \omega_0$ so that $\omega_2 > 1/\omega_2 C$ and $\phi = +45°$

$$\therefore \quad \omega_2 L - \frac{1}{\omega_2 C} = R \quad \text{or} \quad \omega_2^2 - \frac{\omega_0}{Q_0}\omega_2 - \omega_0^2 = 0$$

The positive solution of the above equation is $\omega_2 = \omega_0\left[\sqrt{\left(1 + \dfrac{1}{4Q_0^2}\right)} + \dfrac{1}{2Q_0}\right]$

In case $Q_0 > 1$; then the term $1/4\,Q_0^2$ is negligible as compared to 1.

Hence, in that case $\qquad \omega_1 \cong \omega_0\left(1 - \dfrac{1}{2Q_0}\right)$ and $\omega_2 \cong \omega_0\left(1 + \dfrac{1}{2Q_0}\right)$

Incidentally, it may be noted from above that $\omega_2 - \omega_1 = \omega_0/Q_0$.

13.15. Values of Edge Frequencies

Let us find the values of ω_1 and ω_2, $I_0 = V/R$...at resonance

$$I = \frac{V}{[R^2 + (\omega L - 1/\omega C)^2]^{1/2}} \text{ ... at any frequency}$$

At points A and B, $I = \frac{I_0}{\sqrt{2}} = \frac{1}{\sqrt{2}} \cdot \frac{V}{R}$

$$\therefore \frac{1}{\sqrt{2}} \cdot \frac{V}{R} = \frac{V}{[(R^2 + (\omega L - 1/\omega C)^2]^{1/2}} \text{ or } R = \pm(\omega L - 1/\omega C) = \pm X$$

It shows that at half-power points, net reactance is equal to the resistance.

(Since resistance equals reactance, p.f. of the circuit at these points is $= 1/\sqrt{2}$ *i.e.* 0.707, though leading at point A and lagging at point B).

Hence $R^2 = (\omega L - 1/\omega C)^2 \quad \therefore \quad \omega = \pm\frac{R}{2L} \pm \sqrt{\frac{R^2}{4L^2} + \frac{1}{LC}} = \pm\alpha \pm \sqrt{\alpha^2 + \omega_0^2}$

where $\alpha = \frac{R}{2L}$ and $\omega_0 = \frac{1}{\sqrt{LC}}$

Since $R^2/4L^2$ is much less then $1/\sqrt{LC}$ $\therefore \omega = \pm\frac{R}{2L} \pm \frac{1}{\sqrt{LC}} = \pm\frac{R}{2L} \pm \omega_0$

Since only positive values of ω_0 are considered, $\omega = \omega_0 \pm R/2L = \omega_0 \pm \alpha$

$$\therefore \quad \omega_1 = \omega_0 - \frac{R}{2L} \text{ and } \omega_2 = \omega_0 + \frac{R}{2L}$$

$$\therefore \quad \Delta\omega = \omega_2 - \omega_1 = \frac{R}{L} \text{ rad/s and } \Delta f = f_2 - f_1 = \frac{R}{2\pi L} \text{ Hz} = \frac{f_0}{Q_0} \text{ Hz}$$

Also
$$f_1 = f_0 - \frac{R}{4\pi L} \text{ Hz and } f_2 = f_0 + \frac{R}{4\pi L} \text{ Hz}$$

It is obvious that f_0 is the *centre* frequency between f_1 and f_2.

Also, $\omega_1 = \omega_0 - \frac{1}{2}\Delta\omega$ and $\omega_2 = \omega_0 + \frac{1}{2}\Delta\omega$

As stated above, bandwidth is a measure of circuits selectivity. Narrower the bandwidth, higher the selectivity and vice versa.

13.16. Q-Factor of a Resonant Series Circuit

The Q-factor of an R-L-C series circuit can be defined in the following different ways.

(*i*) it is given by the voltage magnification produced in the circuit at resonance.

We have seen that at resonance, current has maximum value $I_0 = V/R$. Voltage across either coil or capacitor $= I_0X_{L0}$ or I_0X_{C0}, supply voltage $V = I_0 R$

$$\therefore \quad \text{Voltage magnification} = \frac{V_{L_0}}{V} = \frac{I_0 X_{L_0}}{I_0 R} = \frac{\text{reactive power}}{\text{active power}} = \frac{X_{L_0}}{R} = \frac{\omega_0 L}{R} = \frac{\text{reactance}}{\text{resistance}}$$

or
$$= \frac{V_{C0}}{V} = \frac{I_0 X_{C0}}{I_0 R} = \frac{\text{reactive power}}{\text{active power}} = \frac{X_{C0}}{R} = \frac{\text{reactance}}{\text{resistance}} = \frac{1}{\omega_0 CR}$$

$$\therefore \quad Q - \text{factor}, \quad Q_0 = \frac{\omega_0 L}{R} = \frac{2\pi f_0 L}{R} = \tan\phi \quad \text{...(i)}$$

where ϕ is power factor of the coil.

(*ii*) The Q-factor may also be defined as under.

$$Q\text{-factor} = 2\pi\frac{\text{maximum stored energy}}{\text{energy dissipated per cycle}} \quad \text{...in the circuit}$$

$$= 2\pi \frac{\frac{1}{2}LI_0^2}{I^2 R T_0} = 2\pi \frac{\frac{1}{2}L(\sqrt{2}I)^2}{I^2 R(1/f_0)} = \frac{I^2\, 2\pi f_0 L}{I^2 R} = \frac{\omega_0 L}{R} = \frac{1}{\omega_0 CR} \qquad ...(T_0 = 1/f_0)$$

In other words, $Q_0 = \dfrac{\text{energy stored}^*}{\text{energy lost}}$ **...in the circuit**

(iii) We have seen above that resonant frequency, $f_0 = \dfrac{1}{2\pi\sqrt{(LC)}}$ or $2\pi f_0 = \dfrac{1}{\sqrt{(LC)}}$

Substituting this value in Eq. (*i*) above, we get the *Q*-factor, $Q_0 = \dfrac{1}{R}\sqrt{\left(\dfrac{L}{C}\right)}$

(iv) In the case of series resonance, higher *Q*-factor means not only higher voltage magnification but also higher selectivity of the tuning coil. In fact, *Q*-factor of a resonant series circuit may be written as $Q_0 = \dfrac{\omega_0}{\text{bandwidth}} = \dfrac{\omega_0}{\Delta\omega} = \dfrac{\omega_0}{R/L} = \dfrac{\omega_0 L}{R} = \dfrac{L}{R\sqrt{LC}} = \dfrac{1}{R}\sqrt{\dfrac{L}{C}}$ **...as before**

Obviously, *Q*-factor can be increased by having a coil of large inductance but of small ohmic resistance.

(v) In summary, we can say that

$$Q_0 = \frac{\omega_0 L}{R} = \frac{1}{\omega_0 C R} = \frac{1}{R}\sqrt{\frac{L}{C}} = \sqrt{\frac{X_{L0}\, X_{C0}}{R}} = \frac{f_0}{B_{hp}} = \frac{\omega_0}{\omega_2 - \omega_1} = \frac{f_0}{f_2 - f_1}$$

13.17. Circuit Current at Frequencies Other Than Resonant Frequencies

At resonance, $I_0 = V/R$

At any other frequency above the resonant frequency, the current is given by I

$$\frac{V}{Z} = \frac{V}{\sqrt{R^2 + (\omega L - 1/\omega C)^2}}$$

This current lags behind the applied voltage by a certain angle ϕ

$$\frac{I}{I_0} = \frac{V}{\sqrt{R^2 + (\omega L - 1/\omega C)^2}} \times \frac{R}{V} = \frac{1}{\sqrt{1 + \dfrac{1}{R^2}(\omega L - 1/\omega C)^2}} = \frac{1}{\left[1 + \left(\dfrac{\omega_0 L}{R}\right)^2 \left(\dfrac{\omega}{\omega_0} - \dfrac{\omega_0}{\omega}\right)^2\right]^{1/2}}$$

Now, $\omega_0 L/R = Q_0$ and $\omega/\omega_0 = f/f_0$ hence, $\dfrac{I}{I_0} = \dfrac{1}{\left[1 + Q_0^2\left(\dfrac{f}{f_0} - \dfrac{f_0}{f}\right)^2\right]^{1/2}}$

13.18. Relation between Resonant Power P_0 and Off-resonant Power P

In a series *RLC* resonant circuit, current is maximum *i.e.* I_0 at the resonant frequency f_0. The maximum power P_0 is dissipated by the circuit at this frequency where X_L equals X_C. Hence, circuit impedance $Z_0 = R$.

---

* The author often jokingly tells students in his class that these days the quality of a person is also measured in terms of a quality factor given by

$$Q = \frac{\text{money earned}}{\text{money spent}}$$

Obviously, a person should try to have a high quality factor as possible by minimising the denominator and/or maximizing the numerator

$\therefore \quad P_0 = I_0^2 R = (V/R)^2 \times R = V^2/R$

At any other frequency either above or below f_0 the power is (Fig. 13.36).

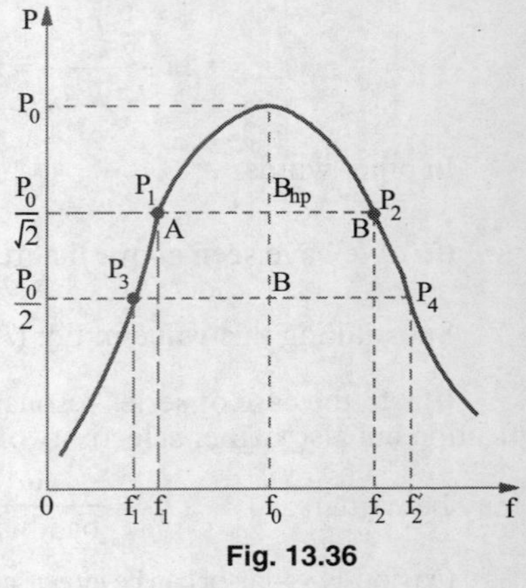

Fig. 13.36

$$P = I^2 R = \left(\frac{V}{Z}\right)^2 \times R = \frac{V^2 R}{Z^2} = \frac{V^2 R}{R^2 + X^2} = \frac{V^2 R}{R^2 + X^2 R^2/R^2}$$

$$= \frac{V^2 R}{R^2 + R^2 Q^2} = \frac{V^2 R}{R^2 (1+Q^2)} = \frac{V^2}{R(1+Q^2)} = \frac{P_0}{(1+Q^2)}$$

The above equation shows that any frequency other than f_0, the circuit power P is reduced by a factor of $(1 + Q^2)$ where Q is the tangent of the circuit phase angle (and not Q_0). At resonance, circuit phase angle $\theta = 0$, and $Q = \tan\theta = 0$. Hence, $P = P_0 = V^2/R$ (values of Q_0 are given in Art.)

Example 13.35. *For a series R.L.C circuit the inductor is variable. Source voltage is 200 $\sqrt{2}$ sin 100πt. Maximum current obtainable by varying the inductance is 0.314 A and the voltage across the capacitor then is 300 V. Find the circuit element values.*

(Circuit and Field Theory, A.M.I.E. Sec B, 1993)

Solution. Under resonant conditions, $I_m = V/R$ and $V_L = V_C$.

$\therefore \quad R = V/I_m = 200/0.314 = 637\ \Omega$, $V_C = I_m \times X_{CD} = I_m/\omega_0 C$

$\therefore \quad C = I_m/\omega_0 V_C = 0.314/100\,\pi \times 300 = 3.33\ \mu F.$

$V_L = I_m \times X_L = I_m\, \omega_0 L;\ L = V_l/\omega_0 I_m = 300/100\,\pi \times 0.314 = \mathbf{3.03\ H}$

Example 13.36. *A coil having an inductance of 50 mH and resistance 10 Ω is connected in series with a 25 µF capacitor across a 200 V ac supply. Calculate* **(a)** *resonance frequency of the circuit* **(b)** *current flowing at resonance and* **(c)** *value of Q_0 by using different data.*

(Elect. Engg. A.M.Ae. S.I, June 1991)

Solution. (a) $$f_0 = \frac{1}{2\pi\sqrt{LC}} = \frac{1}{2\pi\sqrt{50\times10^{-3}\times25\times10^{-6}}} = 142.3\text{ Hz}$$

(b) $I_0 = V/R = 200/10 = 20$ A

(c) $$Q_0 = \frac{\omega_0 L}{R} = \frac{2\pi\times142.3\times50\times10^{-3}}{10} = 4.47$$

$$Q_0 = \frac{1}{\omega_0 CR} = \frac{1}{2\pi\times142.3\times25\times10^{-6}\times10} = 4.47$$

$$Q_0 = \frac{1}{R}\sqrt{\frac{L}{C}} = \frac{1}{10}\sqrt{\frac{50\times10^{-3}}{25\times10^{-6}}} = 4.47$$

Example 13.37. *An R-L-C series circuit consists of a resistance of 1000 Ω, an inductance of 100 mH an a capacitance of 10 µµ F. If a voltage of 100 V is applied across the combination, find* **(i)** *the resonance frequency* **(ii)** *Q-factor of the circuit and* **(iii)** *the half-power points.*

(Elect. Circuit Analysis, Bombay Univ.)

Solution. (*i*) $f_0 = \dfrac{1}{2\pi\sqrt{10^{-1} \times 10^{-11}}} = \dfrac{10^6}{2\pi} =$ **159 kHz**

(*ii*) $Q = \dfrac{1}{R}\sqrt{\dfrac{L}{C}} = \dfrac{1}{1000} \times \sqrt{\dfrac{10^{-1}}{10^{-11}}} = \mathbf{100}$

(*iii*) $f_1 = f_0 - \dfrac{R}{4\pi L} = 159 \times 10^3 - \dfrac{1000}{4\pi \times 10^{-1}} =$ **158.2 kHz**

$f_2 = f_0 + \dfrac{R}{4\pi L} = 159 \times 10^3 + \dfrac{1000}{4\pi \times 10^{-1}} =$ **159.8 kHz**

Example 13.38. *A series R-L-C circuit consists of R = 1000 Ω, L = 100 mH and C = 10 picofarads. The applied voltage across the circuit is 100 V.*

(*i*) *Find the resonant frequency of the circuit.*
(*ii*) *Find the quality factor of the circuit at the resonant frequency.*
(*iii*) *At what angular frequencies do the half power points occur ?*
(*iv*) *Calculate the bandwidth of the circuit.*

(Networks-I, Delhi Univ. & U.P. Tech. Univ. 2001)

Solution. (*i*) $f_0 = \dfrac{1}{2\pi\sqrt{LC}} = \dfrac{1}{2\pi\sqrt{100 \times 10^{-3} \times 10 \times 10^{-12}}} = 159.15 \text{ kHz}$

(*ii*) $Q_0 = \dfrac{1}{R}\sqrt{\dfrac{L}{C}} = \dfrac{1}{1000} \times \sqrt{\dfrac{100 \times 10^{-3}}{10 \times 10^{-12}}} = 100$

(*iii*) $B_{hp} = \dfrac{R}{2\pi L} = \dfrac{1000}{2\pi \times 100 \times 10^{-3}} = 1591.5 \text{ Hz}$

Also. $B_{hp} = f_0/Q_0 = 159.15 \text{ kHz}/100 = 1.5915 \text{ kHz} = \mathbf{1591.5\ Hz}$ ...as above

(*iv*) $\omega_1 = \omega_0\left(1 - \dfrac{1}{2Q_0}\right) = 2\pi \times 159.15\left(1 - \dfrac{1}{2 \times 100}\right) = 994.969$ radia/sec.

$\omega_2 = \omega_0\left(1 + \dfrac{1}{2\,Q_0}\right) = 2\pi \times 159.15\left(1 + \dfrac{1}{2 \times 100}\right) = 1004.969$ rad/sec

(*v*) Band width $= (\omega_2 - \omega_1) = 1004.969 - 994.969 = 10.00$ rad/sec.

Example 13.39. *An R-L-C series resonant circuit has the following parameters* :
Resonance frequency = 5000/2π Hz; impedance at resonance = 56 Ω and Q-factor = 25.
Calculate the capacitance of the capacitor and the inductance of the inductor.
Assuming that these values are independent of the frequency, find the two frequencies at which the circuit impedance has a phase angle of π/4 radian.

Solution. Here $\omega_0 = 2\pi f_0 = 2\pi \times 5000/2\pi = 5000$ rad/s

Now, $Q = \dfrac{\omega_0 L}{R}$ or $25 = \dfrac{5000\,L}{56}$ or $L = 0.28$ H

Also at resonance $\omega_0 L = 1/\omega_0 C$ or $5000 \times 0.28 = 1/5000 \times C$ ∴ $C = 0.143\ \mu$F

The circuit impedance has a phase shift of 45° and the two half-power frequencies which can be found as follows :

$BW = \dfrac{f_0}{Q} = \dfrac{5000/2\pi}{25} =$ **31.83 Hz**

Therefore lower half-power frequency = $(f_0 - 31.83/2) = 5000/2\pi - 15.9 =$ **779.8 Hz.**

Upper half-power frequency = $(f_0 + 31.83/2) = 5000/2\pi + 15.9 =$ **811.7 Hz.**

Example 13.40. *An R-L-C series circuit is connected to a 20-V variable frequency supply. If R = 20 Ω, L = 20 mH and C = 0.5 μF, calculate the following :*

(a) *resonant frequency f_0* ***(b)*** *resonant circuit Q_0 using L/C ratio* ***(c)*** *half-power bandwidth using f_0 and Q_0* ***(d)*** *half-power bandwidth using the general formula for any bandwidth* ***(e)*** *half-power bandwidth using the given component values* ***(f)*** *maximum power dissipated at f_0.*

Solution. (*a*) $f_0 = 1/2\pi\sqrt{LC} = 1/2\pi\sqrt{(20\times10^{-3}\times0.5\times10^{-6})} = \textbf{159 Hz}$

(*b*) $Q_0 = \frac{1}{R}\cdot\sqrt{\frac{L}{C}} = \frac{1}{20}\cdot\sqrt{\frac{20\times10^{-3}}{0.5\times10^{-6}}} = \textbf{10}$

(*c*) $B_{hp} = f_0/Q_0 = 1591/10 = \textbf{159.1 Hz}$

(*d*) $B_{hp} = f_0\,Q/Q_0 = 1591\times\tan 45°/10 = \textbf{159.1 Hz}$

It is so because the power factor angle at half-power frequencies is ± 45°.

(*e*) $B_{hp} = R/2\pi L = 20/2\pi\times20\times10-^{3} = \textbf{159.1 Hz}$

(*f*) $f_0 = V^2/R = 20^2/20 = \textbf{20 W}$

Example 13.41. *An inductor having a resistance of 25 Ω and a Q_0 of 10 at a resonant frequency of 10 kHz is fed from a 100 ∠ 0° supply. Calculate*

(a) *Value of series capacitance required to produce resonance with the coil*

(b) *the inductance of the coil* ***(c)*** *Q_0 using the L/C ratio* ***(d)*** *voltage across the capacitor* ***(e)*** *voltage across the coil.*

Solution. (*a*) $X_{L0} = Q_0R = 10\times25 = 250\ \Omega.$

Now, $X_{C0} = X_{L0} = 250\ \Omega.$

Hence, $C = 1/2\pi f_0\times X_{C0} = 1/2\pi\times10^4\times250 = 63.67\times10-^{9}F = \textbf{63.67 μF}$

(*b*) $L = X_{L0}/2\pi f_0 = 250/2\pi\times10^4 = \textbf{3.98 mH}$

(*c*) $Q_0 = \frac{1}{R}\cdot\sqrt{\frac{L}{C}}$

Now, $\frac{L}{C} = \frac{3.98\times10^{-3}}{63.67\times10^{-9}} = 6.25\times10^4$

∴ $Q_0 = \frac{1}{25}\times\sqrt{6.25\times10^4} = \textbf{10}$ (verification)

(*d*) $V_{C0} = -jQ_0V = -j\,100\angle0°\times10 = -j\,1000\text{V} = \mathbf{-100\angle-90°\ V}$

(*e*) Since $V_{L0} = V_{C0}$ in magnitude, hence, $V_{L0} = +j\,1000$ V

$= 1000\angle90°\text{V}$; Also, $V_R = V = 100\angle0°$

Hence, $V_{coil} = V_R + V_{L0}$

$= 100 + 1000\angle90° = 100 + j\,1000 = \mathbf{1005\angle84.3°}$

Example 13.41(a). *A series L.C circuit has L = 100 pH, C = 2500 μF and Q = 70. Find* ***(a)*** *resonant frequency f_0* ***(b)*** *half-power points and* ***(c)*** *bandwidth.*

Solution. (*a*) $f_0 = \frac{1}{2\pi\sqrt{LC}} = \frac{10^9}{2\pi\sqrt{100\times2500}} = \textbf{318.3 kHz}$

(*b*) $f_2 - f_1 = \Delta f = f_0/Q = 318.3/70 = \textbf{4.55 kHz}$

(*c*) $f_1 f_2 = f_0^2 = 318.3^2; f_2 - f_1 = 4.55$ kHz

Solving for f_1 and f_2, we get, f_1 = **3104 kHz** and f_2 = **320.59 kHz**

Note. Since Q is very high, there would be negligible error in assuming that the half-power points are equidistant from the resonant frequency.

Example 13.42. *A large coil of inductance 1.405 H and resistance 40 ohms is connected in series with a capacitor of 20 microfarads. Calculate the frequency at which the circuit resonates. If a voltage of 100 V at the corresponding frequency is applied to the circuit, calculate the current drawn from the supply and the voltages across the coil and across the capacitor.*

[Nagpur University Nov. 1999]

Solution. $\omega_0 = \frac{1}{\sqrt{LC}} = \frac{1}{\sqrt{1.405 \times 20 \times 10^{-6}}} = 188.65$ radians/sec.

$$f_0 = \frac{188.65}{2\pi} = 30.04 \text{ Hz}$$

Reactance at 30.04 Hz have to be calculated for voltages across the coil and the capacitor

$$X_L = \omega_0 L = 188.65 \times 1.405 = 265 \ \Omega$$

$$X_C = \frac{1}{\omega_0 C} = \frac{1}{188.65 \times 20 \times 10^{-6}} = 265 \ \Omega$$

Coil Impedance $= \sqrt{40^2 + 265^2} = 268 \ \Omega$

Impedance of the total circuit

$$= 40 + j\,265 - j\,265 = 40 \ \Omega$$

Supply Current $= \frac{100}{40} = 2.5$ amp, at unity p.f.

Voltage across the coil = 2.5 × 268 = 670 V

Voltage across the capacitor = 2.5 × 265 = 662.5 V

The phasor diagram is drawn below; in Fig. 13.37(*a*) for the circuit in Fig. 13.37(*b*).

$$OB = V, BC = V_C$$

$$AB = IX_L$$

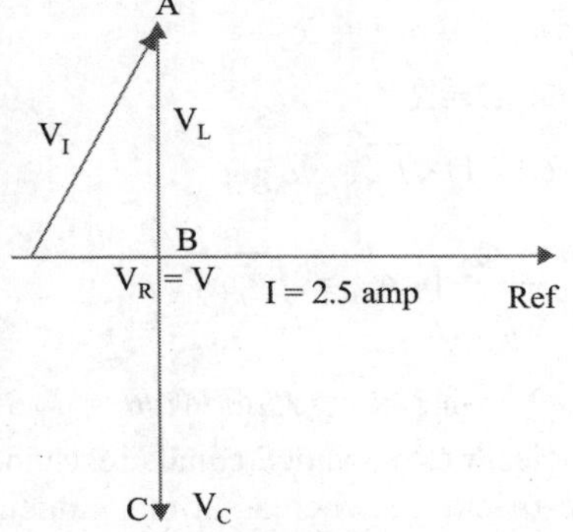

Fig. 13.37(*a*) Phasor diagram

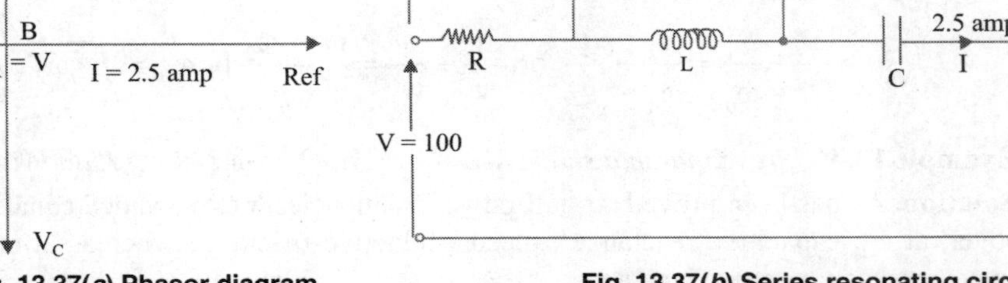

Fig. 13.37(*b*) Series resonating circuit

Example 13.43. *A coil of inductance 9 H and resistance 50 Ω in series with a capacitor is supplied at constant voltage from a variable frequency source. If the maximum current of 1A occurs at 75 Hz, find the frequency when the current is 0.5 A.*

(Principles of Elect. Engg. Delhi Univ.)

Solution. Here, $N = 1_0/I = I/0.5 = 2$; $Q = \omega_0 L/R = 2\pi \times 75 \times 9/50 = 84.8$

Let f_1 and f_2 be the frequencies at which current falls to half its maximum value at resonance frequency. Then, as seen from above

$$\frac{f_0}{f_1} - \frac{f_1}{f_0} = \frac{\sqrt{3}}{Q} \text{ or } \frac{75}{f_1} - \frac{f_1}{75} = \frac{\sqrt{3}}{84.4}$$

or $$(75^2 - f_1^2)/75 f_1 = 0.02 \text{ or } f_1^2 + 1.5\, f_1 - 5625 = 0 \text{ or } f_1 = \mathbf{74.25\ Hz}$$

Also $$\frac{f_2}{75} - \frac{75}{f_2} = \frac{\sqrt{3}}{84.4} \text{ or } f_2^2 - 1.5 f_2 - 5625 = 0 \text{ or } f_2 = \mathbf{75.75\ Hz.}$$

Example 13.44. *Find the following when the power drops to 4 W on either side of the maximum power at resonance.*

(a) *circuit Q (b) circuit phase angle ϕ* ***(c)*** *4-W bandwidth B*

(d) *lower freqency f_1* ***(e)*** *upper frequency f_2.*

(f) *resonance frequency using the value of f_1 and f_2.*

Solution. ***(a)*** $$P = \frac{P}{(1+Q_0^2)} \quad \therefore Q = \sqrt{(P_0/P_1) - 1} = \sqrt{(20/4) - 1} = \mathbf{2}$$

(b) $$\tan(\pm\theta) = 2\, j \pm \theta/\tan^{-1} 2 = \pm\, \mathbf{63.4°}$$

(c) $$B_{hp} = \frac{f_0 Q}{Q_0} = \frac{1591 \times 2}{10} = 318.2 \text{ Hz}$$

(d) $$f_1 = f_0 - B/2 = 1591 - (318.2/2) = 1431.9 \text{ Hz}$$

(e) $$f_2 = f_0 + B/2 = 1591 + (318.2/2) = 1750.1 \text{ Hz}$$

(f) $$f_0 = \sqrt{f_1 f_2} = \sqrt{1431.9 \times 1750.2} = \mathbf{1591\ Hz.}$$

It shows that regardless of the bandwidth magnitude, f_0 is always the geometric mean of f_1 and f_2.

Example 13.45. *Show that in R-L-C circuit, the resonant frequency ω_0 is the geometric mean of the lower and upper half-power frequencies ω_1 and ω_2 respectively.*

Solution. As stated earlier, at lower half-power radiant frequency ω_1; $X_C > X_L$ and at frequencies higher than half-power frequencies $X_L > X_C$. However, the difference between the two equals R.

$$\therefore \quad \text{at } \omega_1, X_C - X_L = R \text{ or } 1/\omega_1 C - \omega_1 L = R \quad \text{...(i)}$$

$$\text{At } \omega_2, X_L - X_C = R \text{ or } \omega_2 L - 1/\omega_2 C = R$$

Multiplying both sides of Eq. (*i*) by C and substitutin $\omega_0^2 = 1/\sqrt{LC}$, we get

$$\frac{1}{\omega_1} - \frac{\omega_1}{\omega_0^2} = \frac{\omega_2}{\omega_0^2} - \frac{1}{\omega_2} \text{ or } \frac{1}{\omega_1} + \frac{1}{\omega_2} = \frac{\omega_1 + \omega_2}{\omega_0^2} \text{ or } \omega_0 = \sqrt{\omega_1 \omega_2}$$

Example 13.46. *Prove that in a sereis R-L-C circuit, $Q_0 = \omega_0 L/R = f_0$/bandwidth $= f_0/BW$.*

Solution. As has been proved, at half power frequencies, net reactance equals resistance. Moreover, at ω_1, capacitive reactance exceeds inductive reactance whereas at ω_2, inductive reactance exceeds capacitive reactance.

$$\therefore \quad \frac{1}{2\pi f_1 C} - 2\pi\, f_1 L = R \text{ or } f_1 = \frac{\aleph R \quad \sqrt{R^2 \quad 4L/C}}{4\pi L}$$

Also $2\pi f_2 L - \dfrac{1}{2\pi f_2 C} =$ or $f_2 = \dfrac{R + \sqrt{R^2 + 4L/C}}{4\pi L}$

Now, $BW = f_2 - f_1 = R/2\pi L$. Hence, $Q_0 = f_0/BW = 2\pi f_0 L/R = \omega_0 L/R$.

Tutorial Problem 13.4

1. An a.c. series circuit has a resistance of 10 Ω, an inductance of 0.2 H and a capacitance of 60 μF. Calculate
   (*a*) the resonant frequency (*b*) the current and (*c*) the power at resonance.
   Give that the applied voltage is 200 V. **[46 Hz; 20 A; 4 kW]**
2. A circuit consists of an inductor which has a resistance of 10 Ω and an inductance of 0.3 H, in series with a capacitor of 30 μ*F* capacitance. Calculate
   (*a*) the impedance of the circuit to currents of 40 Hz (*b*) the resonant frequency (*c*) the peak value of stored energy in joules when the applied voltage is 200 V at the resonant frequency. **[58.31 Ω; 53 Hz; 120 J]**
3. A resistor and a capacitor are connected in series with a variable inductor. When the circuit is connected to a 240-V, 50-Hz supply, the maximum current given by varying the inductance is 0.5 A. At this current, the voltage across the capacitor is 250 V. Calculate the values of
   (*a*) the resistance (*b*) the capacitance (*c*) the inductance. **[480 Ω, 6.36 μF; 1.59 H]**
   Neglect the resistance of the inductor
4. A circuit consisting of a coil of resistance 12Ω and inductance 0.15 H in series with a capacitor of 12μF is connected to a variable frequency supply which has a constant voltage of 24 V. Calculate (*a*) the resonant frequency (*b*) the current in the circuit at resonance (*c*) the voltage across the capacitor and the coil at resonance. **[(*a*) 153 Hz (*b*) 2 A (*c*) 224 V]**
5. A resistance, a capacitor and a variable inductance are connected in series across a 200-V, 50-Hz supply. The maximum current which can be obtained by varying the inductance is 314 mA and the voltage across the capacitor is then 300 V. Calculate the capacitance of the capacitor and the values of the inductance and resistance. **[3.33 μF, 3.04 H, 637 Ω] (*I.E.E. London*)**
6. A 250-V circuit, consisting of a resistor, an inductor and a capacitor in series, resonates at 50 Hz. The current is then 1A and the p.d. across the capacitor is 500 V. Calculate (*i*) the resistance (*ii*) the inductance and (*iii*) the capacitance. Draw the vector diagram for this condition and sketch a graph showing how the current would vary in a circuit of this kind if the frequency were varied over a wide range, the applied voltage remaining constant.
   [(*i*) 250 Ω (*ii*) 0.798 H (*iii*) 12.72 μ F] (*City & Guilds, London*)
7. A resistance of 24 Ω, a capacitance of 150 μF and an inductance of 0.16 H are connected in series with each other. A supply at 240 V, 50 Hz is applied to the ends of the combination. Calculate (*a*) the current in the circuit (*b*) the potential differences across each element of the circuit (*c*) the frequency to which the supply would need to be changed so that the current would be at unity power-factor and find the current at this frequency.
   [(*a*) 6.37 A (*b*) V_R = 152.9 V, V_C = 320 V, V_C = 123.3 V (*c*) 32 Hz; 10 A] (*London Univ.*)
8. A series circuit consists of a resistance of 10 Ω, an inductance of 8 mH and a capacitance of 500 μμ*F*. A sinusoidal E.M.F. of constant amplitude 5 V is introduced into the circuit and its frequency varied over a range including the resonant frequency.
   At what frequencies will current be (*a*) a maximum (*b*) one-half the-maximum ?
   [(*a*) 79.6 kHz (*b*) 79.872 kHz, 79.528 kHz] (*App. Elect. London Univ.*)

9. A circuit consists of a resistance of 12 ohms, a capacitance of 320 μF and an inductance of 0.08 H, all in series. A supply of 240 V, 50 Hz is applied to the ends of the circuit. Calculate :

(*a*) the current in the coil.

(*b*) the potential differences across each element of the circuit.

(*c*) the frequency at which the current would have unity power-factor.

[(*a*) 12.4 A (*b*) 149 V, 311 V (*c*) 32 Hz] *(London Univ.)*

10. A pure capacitor is connected in series with practical inductor (coil). The voltage source is of 10 volts, 10,000 Hz. It was observed that the maximum current of 2 Amp. flows in the circuit when the value of capacitor is 1 micro-farad. Find the parameter (R & L) of the coil.

(Nagpur University, Winter 2002)

11. Draw the phasor diagram for each of the following combinations :

(*i*) Rand L in series and combination in parallel with C.

(*ii*) Rand C in series and combination in parallel with L.

(*iii*) R, Land C in series, with $X_L > X_C$, when ac source is connected to it.

(Nagpur University, Summer 2003)

12. Two choke coils are connected in series as shown in Fig. 13.38. :

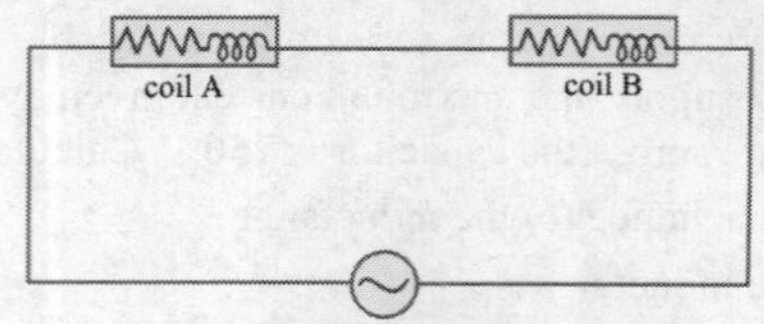

Fig. 13.38

Internal resistance and its inductive reactance of coil A is 4 Ω and 8 Ω resp. Supply voltage is 200 V. Total power consumed in the circuit is 2.2 kW and reactive power consumed is 1.5 kVAR. Find the internal resistance and inductive reactance of coil B. ***(Nagpur University, Summer 2004)***

13. A series circuit having a resistance of 10Ω, and inductance of (1/2π) H and a variable cap. is connected to 100 V, 50 Hz supply. Calculate the value of capacitor to form series resonance. Calculate resonant current, power and power factor. ***(Gujrat University, June/July 2003)***

14. Derive expressions for impedance, current and power factor for an R-L-C series circuit when applied with a.c. voltage. Draw also the vector diagram. ***(Gujrat University, June/July 2003)***

15. Explain the terms active power, reactive power and power factor. Also describe the series resonance of RLC circuit and list its important properties. ***(R.G.P.V. Bhopal University, June 2004)***

16. A 60 Hz sinusoidal voltage v = 100 sin ω t is applied to a series R-L circuit. Assuming R = 10 Ω, L = 0.01 H, find the steady state current and relative phase angle. Also compute the effective magnitude and phase of voltage drops across each circuit elements.

(R.G.P.V. Bhopal University, June 2004)

17. With reference to Fig 13.39, find the values of R and X so that $V_1 = 3V_2$ and V_1 and V_2 are in quadrature. Applied voltage across AB is 240V. ***(Belgaum Karnataka University, February 2002)***

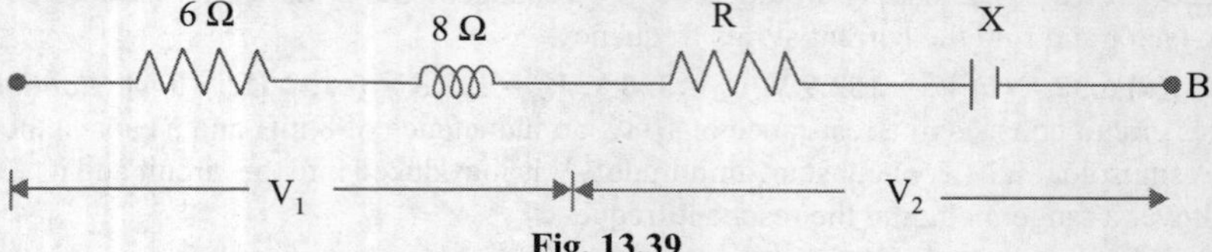

Fig. 13.39

18. A choke coil takes a current of 2 A lagging 60° behind the applied voltage of 220V at 50Hz. Calculate the inductance and resistance of the coil.

(V.T.U., Belgaum Karnataka University Winter 2003)

19. The instantaneous values ofthe voltage across a two element series circum and the current flowing through it are given by V = 100 sin $(314t - \pi/4)V$, $i = 20sin\ (314t - 90^0)A$. Find the frequency and the circuit elements. ***(V.T.U., Belgaum Karnataka University, Winter 2003)***

20. Show that the power consumed in a pure inductance is zero. ***(U.P. TechnicalUniversity 2003) (RGPV Bhopal 2002)***

21. What do you understand by the terms power factor, active power and reactive power? ***(Mumbai University 2003) (RGPV Bhopal 2002)***

22. Series R-L-C circuit. ***(Mumbai University 2003) (RGPV Bhopal 2002)***

23. Describe the properties of (*i*) Resistance (*ii*) Inductance and (*iii*) capacitance used in A.C. Circuit. ***(RGPV Bhopal June 2003)***

24. Define Apparant Power and Power factor in a.c. circuit. Describe parallel resonance and list its important properties. ***(Mumbai University 2003) (RGPV Bhopal December 2003)***

OBJECTIVE TESTS -13

1. In a series *R-L* circuit, V_L—V_R by—degrees.
(*a*) lags, 45 (*b*) lags, 90
(*c*) leads, 90 (*d*) leads, 45

2. The voltage applied across an *R-L* circuit is equal to—of V_R and V_L.
(*a*) arithmetic sum
(*b*) algebraic sum
(*c*) phasor sum
(*d*) sum of the squares.

3. The power in an a.c. circuit is given by
(*a*) $VI \cos \phi$ (*b*) $VI \sin \phi$
(*c*) I^2Z (*d*) I^2X_L

4. The p.f. of an *R-C* circuit is
(*a*) often zero
(*b*) between zero and 1
(*c*) always unity
(*d*) between zero and – 1.0

5. In a series RLC circuit at resonance, the magnitude of the voltage developed across the capacitor
(*a*) is always zero
(*b*) can never be greater than the input voltage.
(*c*) can be greater than the input voltage, however, it is 90⁰ out of phase with the input voltage
(*d*) can be greater than the input voltage,and in in phase with the input voltage.
(GATE 2001)

6. The total impedance Z (jω) of the circuit shown above is

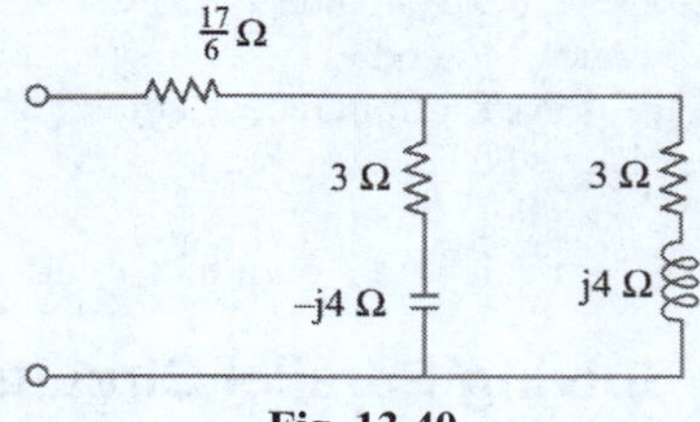

Fig. 13.40

(*a*) $(6 + j0)\ \Omega$ (*b*) $(7 + j0)\ \Omega$
(*c*) $(0 + j8)\ \Omega$ (*d*) $(6 + j8)\ \Omega$
(ESE 2003)

7. The impedance of a parallel RC network is $Z(s) = \dfrac{58}{s^2 + 0.5s + 100}$. Then the values of R, L and C are, respectively

(*a*) $10\Omega,\ \frac{1}{20}\text{H},\ \frac{1}{5}\text{F}$

(*b*) $1\Omega,\ \frac{1}{2}\text{H},\ \frac{1}{5}\text{F}$

(*c*) $10\Omega,\ \frac{1}{20}\text{H},\ \frac{1}{5}\text{F}$

(*d*) $2\Omega,\ \frac{1}{20}\text{H},\ \frac{1}{5}\text{F}$

(Engineering Services Exam. 2003)

ANSWERS

1. *c* **2.** *c* **3.** *a* **4.** *b*

CHAPTER

14

Parallel A.C. Circuits

Chapter Outline

- Solving Parallel Circuits
- Vector or Phasor Method
- Admittance Method
- Application of Admittance Method
- Complex or Phasor Algebra
- Series-Parallel Circuits
- Series Equivalent of a Parallel Circuit
- Parallel Equivalent of a Series Circuit
- Resonance in Parallel Circuits
- Graphic Representation of Parallel Resonance
- Points to Remember
- Bandwidth of a Parallel Resonant Circuit
- Q-factor of a Parallel Circuit

14.1. Solving Parallel Circuits

When impedances are joined in parallel, there are three methods available to solve such circuits:

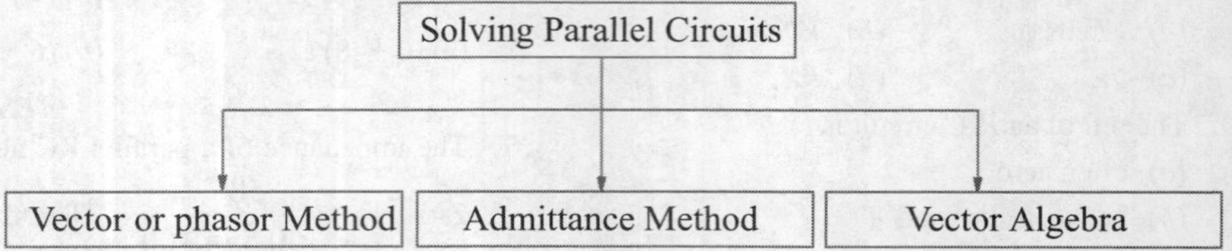

14.2. Vector or Phasor Method

Consider the circuits shown in Fig. 14.1. Here, two reactors *A* and *B* have been joined in parallel across an r.m.s. supply of *V* volts. The voltage across two parallel branches *A* and *B* is the same, but currents through them are different.

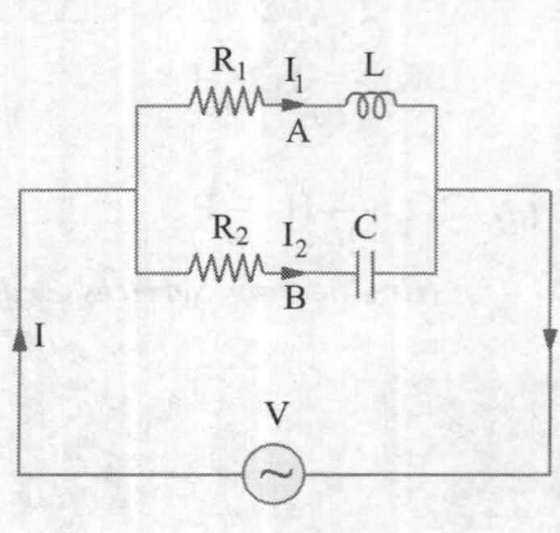

Fig. 14.1

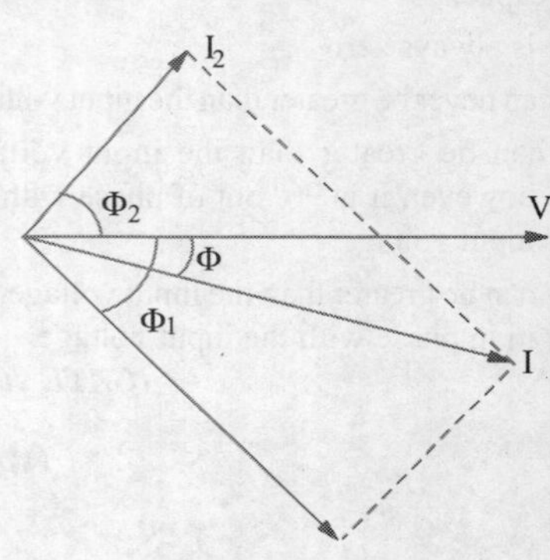

Fig. 14.2

For Branch A**,** $Z_1 = \sqrt{(R_1^2 + X_L^2)}$; $I_1 = V/Z_1$; $\cos\phi_1 = R_1/Z_1$ or $\phi_1 = \cos^{-1}(R_1/Z_1)$

Current I_1 lags behind the applied voltage by ϕ_1 (Fig. 14.2).

For Branch B, $Z_2 = \sqrt{(R_2^2 + X_c^2)}$; $I_2 = V/Z_2$; $\cos\phi_2 = R_2/Z_2$ or $\phi_2 = \cos^{-1}(R_2/Z_2)$

Current I_2 leads V by ϕ_2 (Fig. 14.2).

Resultant Current I

The resultant circuit current I is the vector sum of the branch currents I_1 and I_2 and can be found by **(*i*)** using parallelogram law of vectors, as shown in Fig. 14.2. or **(*ii*)** resolving I_2 into their X- and Y-components (or active and reactive components respectively) and then by combining these components, as shown in Fig. 14.3. Method **(*ii*)** is preferable, as it is quick and convenient.

With reference to Fig. 14.3. (*a*) we have

Sum of the active components of I_1 and I_2

$$= I_1\cos\phi_1 + I_2\cos\phi_2$$

Sum of the reactive components of I_1 and $I_2 = I_2\sin\phi_2 - I_1\sin\phi_1$

If I is the resultant current and ϕ its phase, then its active and reactive components must be equal to these X-and Y-components respectively [Fig. 14.3. (*b*)]

$$I\cos\phi = I_1\cos\phi_1 + I_2\cos\phi_2 \text{ and } I\sin\phi = I_2\sin\phi_2 - I_1\sin\phi_1$$

$$\therefore \quad I = \sqrt{[(I_1\cos\phi_1 + I_2\cos\phi_2)^2 + (I_2\sin\phi_2 - I_1\sin\phi_1)^2}$$

and

$$\tan\phi = \frac{I_2\sin\phi_2 - I_1\sin\phi_1}{I_1\cos\phi_1 + I_2\cos\phi_2} = \frac{Y\text{ component}}{X\text{ component}}$$

If $\tan\phi$ is positive, then current leads and if $\tan\phi$ is negative, then current lags behind the applied voltage V. Power factor for the whole circuit is given by

$$\cos\phi = \frac{I_1\cos\phi_1 + I_2\cos\phi_2}{I} = \frac{X\text{ comp.}}{I}$$

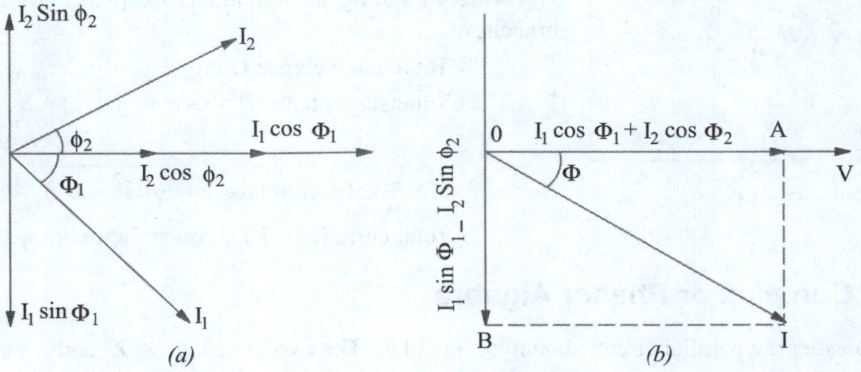

(*a*) (*b*)

Fig. 14.3

14.3. Admittance Method

Admittance of a circuit is defined as ***the reciprocal of its impedance***. Its symbol is Y.

$$\therefore \quad Y = \frac{1}{Z} = \frac{I}{V} \text{ or } Y = \frac{\text{r.m.s. amperes}}{\text{r.m.s. volts}}$$

Its unit is Siemens (S). A circuit having an impedance of one ohm has an admittance of one Siemens. The old unit was mho (ohm spelled backwards).

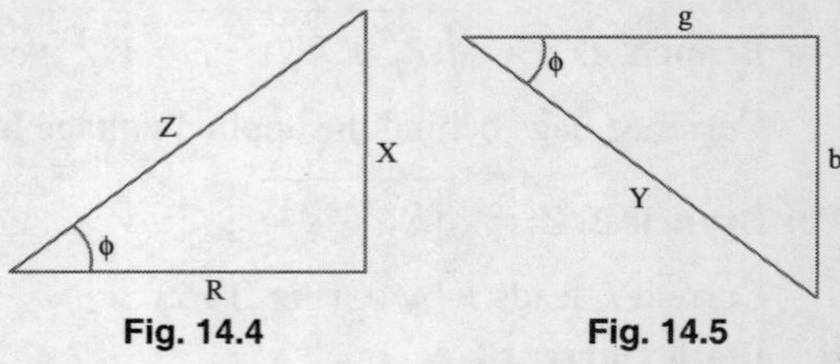

Fig. 14.4 Fig. 14.5

As the impedance Z of a circuit has two components X and R (Fig. 14.4.), similarly, admittance Y also has two components as shown in Fig. 14.5. The X-component is known as ***conductance*** and Y-component as ***susceptance.***

Obviously, conductance $g = Y \cos \phi$

or $g = \dfrac{1}{Z} \cdot \dfrac{R}{Z}$ (from Fig. 14.4)

$$\therefore \quad g^{*} = \frac{R}{Z^2} = \frac{R}{R^2 + X^2}$$

Similarly, susceptance $b = Y \sin \phi = \dfrac{1}{2}, \dfrac{X}{Z}$ $\quad \therefore \; b^{**} = X/Z^2 = X/(R^2 + X^2)$ (from Fig. 14.5)

The admittance $Y = \sqrt{(g^2 + b^2)}$ just as $Z = \sqrt{(R^2 + X^2)}$

The unit of g, b and Y is Siemens. We will regard the ***capacitive susceptance as positive and inductive susceptance as negative.***

14.4. Application of Admittance Method

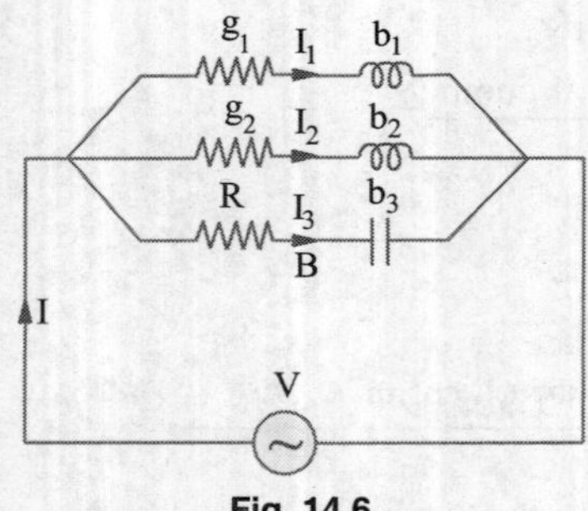

Fig. 14.6

Consider the 3-branched circuit of Fig. 14.6. Total conductance is found by merely adding the conductances of three branches. Similarly, total susceptance is found by ***algebraically*** adding the individual susceptances of different branches.

Total conductance $G = g_1 + g_2 + g_3$ ....

Total susceptance $B = (-b_1) + (-b_2) + b_3$ ......

(algebraic sum)

$\therefore$ Total admittance $Y = \sqrt{(G^2 + B^2)}$

Total current $I = VY$; Power factor $\cos \phi = G/Y$

14.5. Complex or Phasor Algebra

Consider the parallel circuit shown in Fig. 14.7. The two impedances, $\mathbf{Z}_1$ and $\mathbf{Z}_2$, being in parallel, have the same p.d. across them.

* In the special case when $X = 0$, then $g = 1/R$ *i.e.*, conductance becomes reciprocal of resistance, not otherwise.

** Similarly, in the special case when $R = 0$, $b = 1/X$ *i.e.*, susceptance becomes reciprocal of reactance, not otherwise.

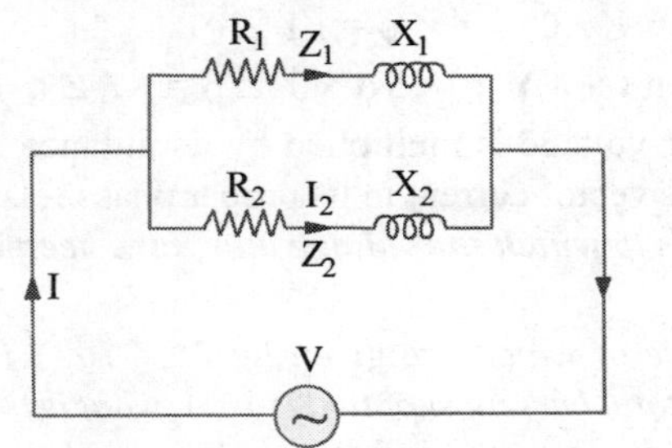

Fig. 14.7

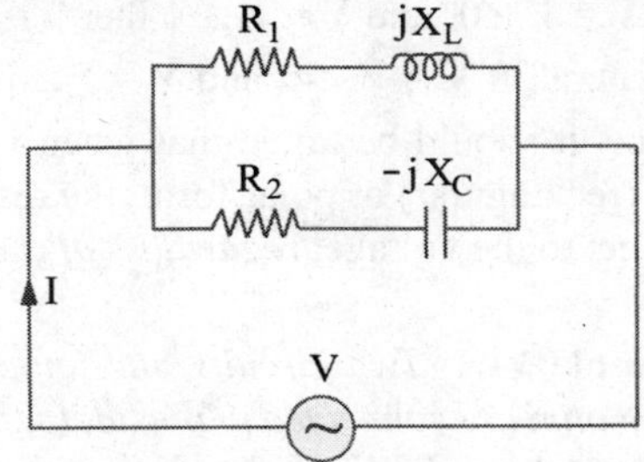

Fig. 14.8

Now $\mathbf{I_1} = \dfrac{\mathbf{V}}{\mathbf{Z_1}}$ and $\mathbf{I_2} = \dfrac{\mathbf{V}}{\mathbf{Z_2}}$

Total current $\mathbf{I} = \mathbf{I_1} + \mathbf{I_2} = \dfrac{\mathbf{V}}{\mathbf{Z_1}} + \dfrac{\mathbf{V}}{\mathbf{Z_2}} = \mathbf{V}\left(\dfrac{1}{\mathbf{Z_1}} + \dfrac{1}{\mathbf{Z_2}}\right) = \mathbf{V}\,(\mathbf{Y_1} + \mathbf{Y_2}) = \mathbf{VY}$

where $\mathbf{Y}$ = total admittance = $\mathbf{Y_1} + \mathbf{Y_2}$

It should be noted that ***admittances* are *added*** for parallel branches, whereas for branches in series, it is the *impedances* which are *added*. ***However, it is important to remember that since both admittances and impedances are complex quantities, all additions must be in complex form.*** Simple arithmetic additions must not be attempted.

Considering the two parallel branches of Fig. 14.8, we have

$$\mathbf{Y_1} = \frac{1}{\mathbf{Z_1}} = \frac{1}{R_1 + jX_L} = \frac{(R_1 - jX_L)}{(R_1 + jX_L)(R_1 - jX_L)}$$

$$= \frac{R_1 - jX_L}{R_1^2 + X_L^2} = \frac{R_1}{R_1^2 + X_L^2} - j\frac{X_L}{R_1^2 + X_L^2} = g_1 - jb_1$$

where $g_1 = \dfrac{R_1}{R_1^2 + X_L^2}$ – conductance of upper branch,

$b_1 = -\dfrac{X_L}{R_1^2 + X_L^2}$ – susceptance of upper branch

Similarly, $\mathbf{Y_2} = \dfrac{1}{Z_2} = \dfrac{1}{R_2 - jX_C}$

$$= \frac{R_2 + jX_C}{(R_2 - jX_C)(R_2 + jX_C)} = \frac{R_2 + jX_C}{R_2^2 + X_C^2} = \frac{R_2}{R_2^2 + X_C^2} + j\frac{X_C}{R_2^2 + X_C^2} = g_2 + jb_2$$

Total admittance $\mathbf{Y} = \mathbf{Y_1} + \mathbf{Y_2} = (g_1 - jb_1) + (g_2 + jb_2) = (g_1 + g_2) - j\,(b_1 - b_2) = G - jB$

$$\mathbf{Y} = \sqrt{[(g_1 + g_2)^2 + (b_1 - b_2)^2]}\;;\; \phi = \tan^{-1}\left(\frac{b_1 - b_2}{g_1 + g_2}\right)$$

The polar form for admittance is $\mathbf{Y} = \mathbf{Y} \angle \phi°$ where ϕ is as given above.

$$Y = \sqrt{G^2 + B^2}\ \angle \tan^{-1}(B/G)$$

Total current $\mathbf{I} = \mathbf{VY}$; $\mathbf{I_1} = \mathbf{VY_1}$ and $\mathbf{I_2} = \mathbf{VY_2}$

If $\mathbf{V} = V \angle 0°$ and $\mathbf{Y} = Y \angle \phi$ then $\mathbf{I} = \mathbf{VY} = V \angle 0° \times Y \angle \phi = VY \angle \phi$

In general, if $\mathbf{V} = V \angle \alpha$ and $\mathbf{Y} = Y \angle \beta$, then $\mathbf{I} = \mathbf{VY} = V \angle \alpha \times Y \angle \beta = VY \angle \alpha + \beta$

Hence, it should be noted that when vector voltage is multiplied by admittance either in complex (rectangular) or polar form, the result is vector current in its proper phase relationship with respect to the voltage, ***regardless of the axis to which the voltage may have been referred to.***

Example 14.1. *Two circuits, the impedance of which are given by $Z_1 = 10 + j\,15$ and $Z_2 = 6 - j8$ ohm are connected in parallel. If the total current supplied is 15 A, what is the power taken by each branch? Find also the p.f. of individual circuits and of combination. Draw vector diagram.* **(Elect. Technology, Vikram Univ, Ujjain)**

Solution. Let $\mathbf{I} = 15 \angle 0°\,;\ \mathbf{Z}_1 = 10 + j15 = 18 \angle 57°$

$$\mathbf{Z}_2 = 6 - j8 = 10 \angle -53.1°$$

Total impedance, $\mathbf{Z} = \dfrac{\mathbf{Z}_1\mathbf{Z}_2}{\mathbf{Z}_1 + \mathbf{Z}_2}$

$$= 9.67 - j\,3.6 = 10.3 \angle -20.4°$$

Applied voltage is given by

$$\mathbf{V} = \mathbf{IZ} = 15 \angle 0° \times 10.3 \angle -20.4° = 154.4 \angle -20.4°$$

$$\mathbf{I}_1 = \mathbf{V}/\mathbf{Z}_1 = 154.5 \angle -20.4°/18 \angle 57° = 8.58 \angle -77.4°$$

$$\mathbf{I}_2 = \mathbf{V}/\mathbf{Z}_2 = 154.5 \angle -20.4°/10 \angle -53.1°$$

$$= 15.45 \angle 32.7°$$

I₂ 32.7° I 20.4° 77.4° V I₁

Fig. 14.9

We could also find branch currents as under :

$$\mathbf{I}_1 = \mathbf{I}.\,\mathbf{Z}_2/(\mathbf{Z}_1 + \mathbf{Z}_2) \text{ and } \mathbf{I}_2 = \mathbf{I}.\mathbf{Z}_1/(\mathbf{Z}_1 + \mathbf{Z}_2)$$

It is seen from phasor diagram of Fig. 14.9 that $\mathbf{I}_1$ lags behind $\mathbf{V}$ by (77.4° – 20.4°) = 57° and $\mathbf{I}_2$ leads it by (32.7° + 20.4°) = 53.1°.

$\therefore\quad P_1 = I_1^2 R_1 = 8.58^2 \times 10 =$ **736 W; p.f.** = cos 57° = **0.544 (lag)**

$\quad\quad P_2 = I_2^2 R_2 = 15.45^2 \times 6 =$ **1432 W** ; p.f. = cos 53.1° = **0.6**

Combined p.f. = cos 20.4° = **0.937 (lead)**

Example 14.2. *Two impedance $Z_1 = (8 + j6)$ and $Z_2 = (3 - j4)$ are in parallel. If the total current of the combination is 25 A, find the current taken and power consumed by each impedance.* **(F.Y. Engg. Pune Univ.)**

Solution. $Z_1 = (8 + j6) = 10 \angle 36.87°\,;\ Z_2 = (3 - j4) = 5 \angle -53.1°$

$$Z = \frac{Z_1 Z_2}{Z_1 + Z_2} = \frac{(10 \angle 36.87°)(5 \angle -53.1°)}{(8 + j6) + (3 - j4)} = \frac{50 \angle -16.23°}{11 + j2} = \frac{50 \angle -16.23°}{11.18 \angle 10.3°}$$

$$= 4.47 \angle -26.53°$$

Let $I = 25 \angle 0°\,;\ V = I Z = 25 \angle 0° \times 4.47 \angle -26.53° = 111.75 \angle -26.53°$

$I_1 = V/Z_1 = 111.75 \angle -26.53°/10 \angle 36.87° = 11.175 \angle -63.4°$

$I_2 = 111.75 \angle -26.53/5 \angle -53.1° = 22.35 \angle 26.57°$

Now, the phase difference between V and I_1 is 63.4° – 26.53° = 36.87° with current lagging. Hence, $\cos \phi_1 = \cos 36.87° = 0.8$.

Power consumed in $\mathbf{Z}_1 = VI_1 \cos \phi = 11.175 \times 111.75 \times 0.8 =$ **990 W**

Similarly, $\phi_2 = 26.57 - (-26.53) = 53.1°\,;\ \cos 53.1° = 0.6$

Power consumed in $\mathbf{Z}_2 = VI_2 \cos \phi_2 = 111.75 \times 22.35 \times 0.6 =$ **1499 W**

Example 14.3. *Refer to the circuit of Fig. 14.10 (a) and determine the resistance and reactance of the lagging coil load and the power factor of the combination when the currents are as indicated.* **(Elect. Engg. A.M.Ae. S.I.)**

Solution. As seen from the $\Delta\, ABC$ of Fig. 14.10 (*b*).

$5.6^2 = 2^2 + 4.5^2 + 2 \times 2 \times 4.5 \times \cos\theta$, $\therefore$ $\cos\theta = 0.395$, $\sin\theta = 0.919$.

$Z = 300/4.5 = 66.67\ \Omega$

$R = Z\cos\theta = 66.67 \times 0.919 = 61.3\ \Omega$

p.f. $= \cos\phi = AC/AD = (2 + 4.5 \times 0.395)/5.6 =$ **0.67 (lag)**

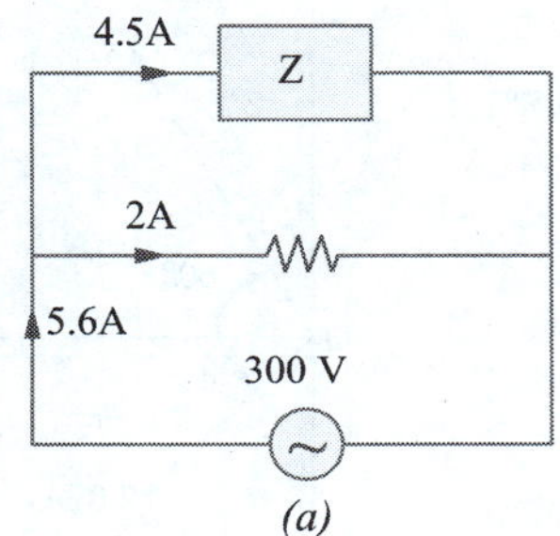

(*a*)

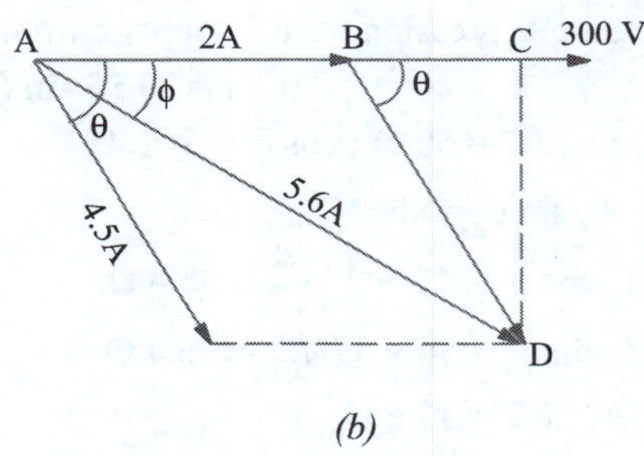

(*b*)

Fig. 14.10

Example 14.4. *A mercury vapour lamp unit consists of a 25µf condenser in parallel with a series circuit containing the resistive lamp and a reactor of negligible resistance. The whole unit takes 400 W at 240 V, 50 Hz at unity p.f. What is the voltage across the lamp ?* **(F.Y. Engg. Pune Univ.)**

Solution. $X_C = \dfrac{1}{2\pi fC} = \dfrac{1}{2\pi \times 50 \times (25 \times 10^{-6})} = 127.3\ \Omega,$

$\therefore \quad I_C = \dfrac{240}{127.3} = 1.885$ A

$W = VI\cos\phi = VI \quad \therefore \quad I = W/V = 400/240 = 1.667$ A

In the vector diagram of Fig. 14.10 (*b*) I_C leads V by 90° and current I_1 in the series circuit lags V by ϕ_1 where ϕ_1 is the power factor angle of the series circuit. The vector sum of I_C and ϕ_1 gives the total current I. As seen $\tan\phi_1 = I_C/I = 1.885/1.667 = 1.13077$. Hence, $\phi_1 = 48.5°$ lag. The applied voltage V is the vector sum of the drop across the resistive lamp which is in phase with I_1 and drop across the coil which leads I_1 by 90°.

Voltage across the lamp $= V\cos\Phi_1 = 340 \times \cos 48.5 = 240 \times 0.662 =$ **159 V.**

Example 14.5. *The currents in each branch of a two-branched parallel circuit are given by the expression* $i_a = 7.07 \sin(314t - \pi/4)$ *and* $i_b = 21.2 \sin(314\,t + \pi/3)$

The supply voltage is given by the expression $v = 354 \sin 314t$. *Derive a similar expression for the supply current and calculate the ohmic value of the component, assuming two pure components in each branch. State whether the reactive components are inductive or capacitive.*

(Elect. Engineering., Calcutta Univ.)

Solution. By inspection, we find that i_a lags the voltage by $\pi/4$ radian or 45° and i_b leads it by π / 3 radian or 60°. Hence, branch *A* consists of a resistance in series with a pure inductive reactance. Branch *B* consists of a resistance in series with pure capacitive reactance as shown in Fig. 14.11 (*a*).

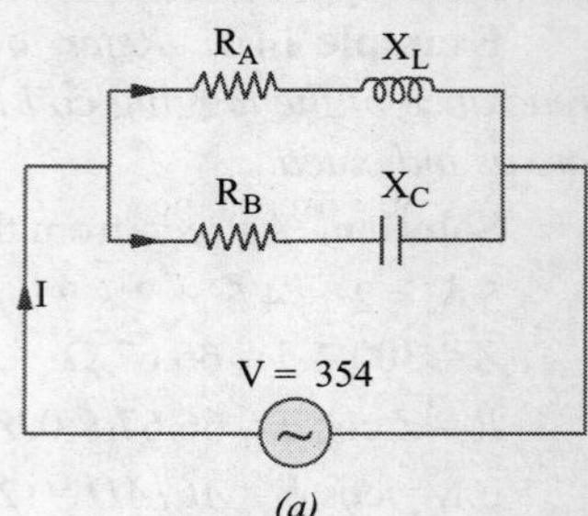

(a)

Maximum value of current in branch A is 7.07 A and in branch *B* is 21.2 A. The resultant current can be found vectorially. As seen from vector diagram.

X-comp = 21.2 cos 60° + 7.07 cos 45° = 15.6 A

Y-comp = 21.2 sin 60° – 7.07 sin 45° = 13.36 A

Maximum value of the resultant current is

$$= \sqrt{15.6^2 + 13.36^2} = 20.55 \text{ A}$$

$\phi = \tan^{-1}(13.36/15.6) = \tan^{-1}(0.856) = 40.5°$ (lead)

Hence, the expression for the supply current is

$$\mathbf{i = 20.55 \sin (314\, t + 40.5°)}$$

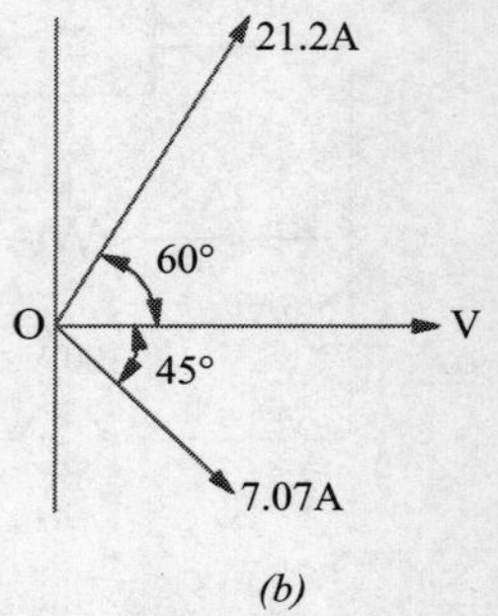

(b)

Fig. 14.11

$Z_A = 354/7.07 = 50\ \Omega$; $\cos \phi_A = \cos 45°$

$= 1/\sqrt{2}$. $\sin \phi_A = \sin 45° = 1/\sqrt{2}$

$R_A = Z_A \cos \phi_A = 50 \times 1/\sqrt{2} = \mathbf{35.4\ \Omega}$

$X_L = Z_A \sin \phi_A = 50 \times 1/\sqrt{2} = \mathbf{35.4\ \Omega}$

$Z_B = 354/20.2 = \mathbf{17.5\ \Omega}$

$R_B = 17.5 \times \cos 60° = \mathbf{8.75\ \Omega}$

$X_C = 17.5 \times \sin 60° = \mathbf{15.16\ \Omega}$

Example 14.6 (*a*). *A total current of 10 A flows through the parallel combination of three impedance : (2 – j5) Ω, (6 + j3) Ω and (3 + j4) Ω. Calculate the current flowing through each branch. Find also the p.f. of the combination.* **(Elect. Engg., -I Delhi Univ.)**

Solution. Let $\mathbf{Z_1} = (2 - j5)$, $\mathbf{Z_2} = (6 + j3)$, $\mathbf{Z_3} = (3 + j4)$

$\mathbf{Z_1Z_2} = (2 - j5)(6 + j3) = 27 - j24$. $\mathbf{Z_2\,Z_3} = (6 + j3)(3 + j4) = 6 + j33$

$\mathbf{Z_3Z_1} = (3 + j4)(2 - j5) = 26 - j7$; $\mathbf{Z_1\,Z_2 + Z_2\,Z_3 + Z_3\,Z_1} = 59 + j2$

With reference to Art, 1.25

$$\mathbf{I_1} = \mathbf{I} \cdot \frac{\mathbf{Z_2\,Z_3}}{\mathbf{Z_1\,Z_2 + Z_2\,Z_3 + Z_3\,Z_1}} = (10 + j0) \times \frac{6 + j\,33}{59 + j\,2} = 1.21 + j\,5.55$$

$$\mathbf{I_2} = \mathbf{I} \cdot \frac{\mathbf{Z_3\,Z_1}}{\Sigma \mathbf{Z_1\,Z_2}} = (10 + j0) \times \frac{26 \quad j7}{59 + j2} = 4.36 - j1.33$$

$$\mathbf{I_3} = \mathbf{I} \cdot \frac{\mathbf{Z_1\,Z_2}}{\Sigma \mathbf{Z_1\,Z_2}} = (10 + j0) \times \frac{27 - j24}{59 + j2} = 4.43 - j4.22$$

Now,
$$\mathbf{Z} = \frac{\mathbf{Z_1Z_2Z_3}}{\mathbf{Z_1Z_2 + Z_2Z_3 + Z_3Z_1}} = \frac{(2 - j5)(6 + j33)}{59 + j2} = 3.01 + j0.51$$

$$\mathbf{V} = 10\,\angle 0° \times 3.05\,\angle 9.6° = 30.5\,\angle 9.6°$$

Combination p.f. = cos 9.6° = **0.986 (lag)**

Example 14.7. *A 15-mH inductor is in series with a parallel combination of an 80 Ω resistor and 20 μF capacitor. If the angular frequency of the applied voltage is ω = 1000 rad/s, find the admittance of the network.* **(Basic Circuit Analysis Osmania Univ. Jan/Feb 1992)**

Solution. $X_L = \omega L = 1000 \times 15 \times 10^{-3} = 15\Omega$; $X_C = 1/\omega C = 10^6/1000 \times 20 = 50\ \Omega$

Impedance of the parallel combination is given by

$$Z_p = 80 || -J\,50 = -j4000/(80 - j50) = 22.5 - j36,$$

$$\text{Total impedance} = j15 + 22.5 - j36 = 22.5 - j21$$

$$\text{Admittance} \quad Y = \frac{1}{Z} = \frac{1}{22.5 - j21} = 0.0238 - j0.022 \text{ Siemens}$$

Example 14.8. *An impedance (6 + j 8) is connected across 200-V, 50-Hz mains in parallel with another circuit having an impedance of (8 – j6) Ω. Calculate* **(a)** *the admittance, the conductance, the susceptance of the combined circuit* **(b)** *the total current taken from the mains and its p.f.* **(Elect. Engg-AMIE, S.I. 1992)**

Solution. $\mathbf{Y}_1 = \dfrac{1}{6 + j8} = \dfrac{6 - j8}{6^2 + 8^2} = 0.06 - j0.08$ Siemens,

$$\mathbf{Y} = \frac{1}{8 - j6} = \frac{8 + j6}{100} = 0.08 + j0.06 \text{ Siemens}$$

(*a*) Combined admittance is $\mathbf{Y} = \mathbf{Y}_1 + \mathbf{Y}_2 = 0.14 - j0.02 =$ **0.1414 ∠– 88°′ Siemens**
Conductance, $G =$ **0.14 Siemens** ; Susceptance, $\boldsymbol{B} =$ **–0.02 Siemens** (inductive)

(*b*) Let $\mathbf{V} = 200\,\angle 0°$; $\mathbf{I} = \mathbf{VY} = 200 \times 0.1414\,\angle{-8°8'}$ $V = 28.3\,\angle{-8°8'}$

p.f. = cos 8°8′ = **0.99 (lag)**

Example 14.9. *If the voltmeter in Fig. 14.12 reads 60 V, find the reading of the ammeter.*

Solution. $I_2 = 60/4 = 15$ A. Taking it as reference quantity, we have $\mathbf{I}_2 = 15\,\angle 0$.

Obviously, the applied voltage is

$$\mathbf{V} = 15\,\angle 0° \times (4 - j4) = 84.8\,\angle{-45°}$$

$$\mathbf{I}_1 = 84.8\,\angle{-45°}/(6 + j\,3)$$
$$= 84.8\,\angle{-45} + 6.7\,\angle 26.6$$
$$= 12.6\,\angle{-71.6°} = (4 - j\,12)$$

$$\mathbf{I} = \mathbf{I}_1 + \mathbf{I}_2 = (15 + j\,0) + (4 - j\,12)$$
$$= 19 - j\,12 = 22.47\,\angle{-32.3°}$$

Hence, ammeter **reads 22.47.**

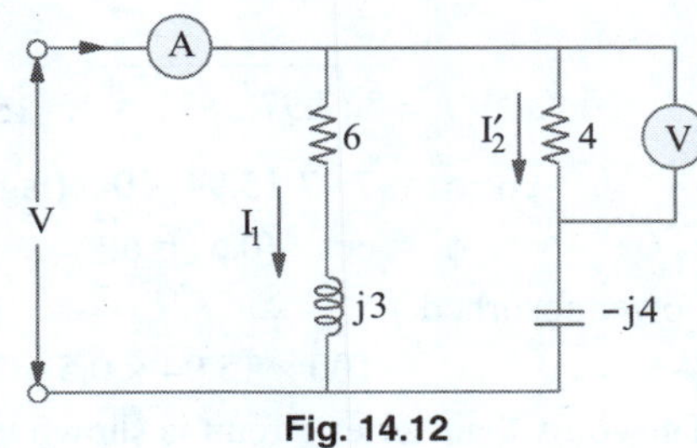

Fig. 14.12

Example 14.10. *Find the reading of the ammeter when the voltmeter across the 3 ohm resistor in the circuit of Fig. 14.13 reads 45 V.*

(Elect. Engg. & Electronics Bangalore Univ.)

Solution. Obviously $I_1 = 45/3 = 15\,A$. If we take it as reference quantity, $I_1 = 3\,\angle 0°$

Now, $Z_1 = 3 - j3 = 4.24\,\angle{-45°}$.

Hence, $V = I_1 Z_1 = 15\,\angle 0° \times 4.24\,\angle{-45°} = 63.6\,\angle{-45°}$

$$I_2 = \frac{V}{Z_2} = \frac{63.6\,\angle{-45°}}{5 + j2} = \frac{63.6\,\angle{-45°}}{5.4\,\angle 21.8°}$$

$$= 11.77\,\angle{-66.8°} = 4.64 - j\,10.8$$

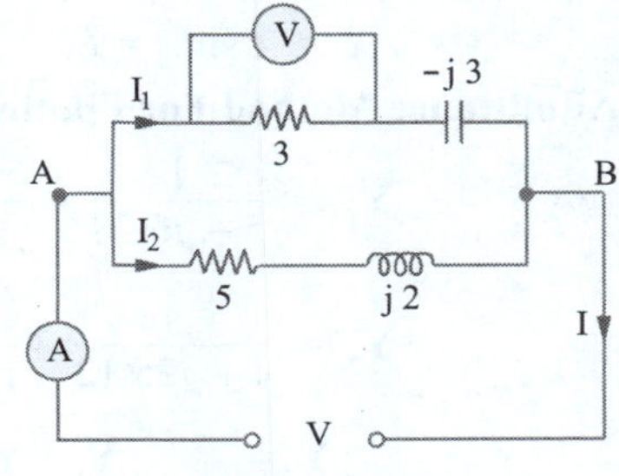

Fig. 14.13

$I = I_1 + I_2 = 19.64 - j10.8 =$ **22.4 ∠28.8°**

Example 14.11. *A coil having a resistance of 5 Ω and an inductance of 0.02 H is arranged in parallel with another coil having a resistance of 1 Ω and an inductance of 0.08 H. Calculate the current through the combination and the power absorbed when a voltage of 100 V at 50 Hz* *is applied. Estimate the resistance of a single coil which will take the same current at the same power factor.*

Solution. The circuit and its phasor diagram are shown in Fig. 14.14.

Branch No. 1

$X_1 = 314 \times 0.02 = 6.28\ \Omega$

$Z_1 = \sqrt{5^2 + 6.28^2} = 8\ \Omega$

$I_1 = 100/8 = 12.5$ A

$\cos\phi_1 = R_1/Z_1 = 5/8$

$\sin\phi_1 = 6.28/8$

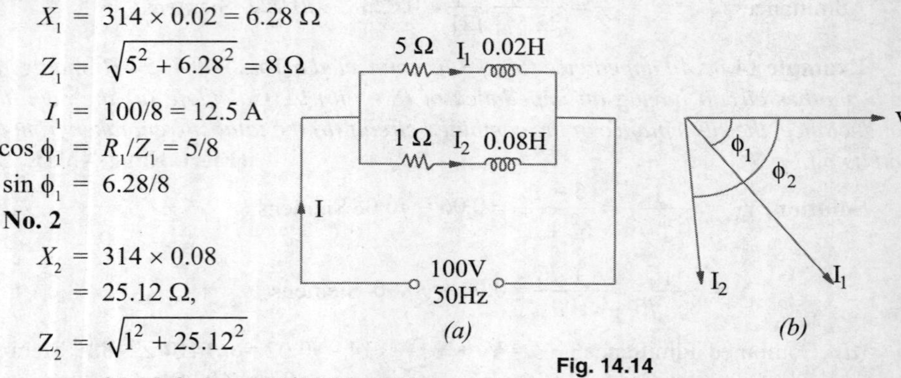

Fig. 14.14

Branch No. 2

$X_2 = 314 \times 0.08$

$= 25.12\ \Omega,$

$Z_2 = \sqrt{1^2 + 25.12^2}$

$= 25.14\ \Omega,\ I_2 = 100/25.14 = 4$ A

$\cos\phi_2 = 1/25.14$ and $\sin\phi_2 = 25.12/25.14$

X - components of I_1 and $I_2 = I_1 \cos\phi_1 + I_2 \cos\phi_2 = (12.5 \times 5/8) + (4 \times 1/25.14) = 7.97$ A

Y - components of I_1 and $I_2 = I_1 \sin\phi_1 + I_2 \sin\phi_2 = (12.5 \times 6.28/8) + (4 \times 25.12/25.14)$

$= 13.8$ A

$I = \sqrt{7.97^2 + 13.8^2} = \mathbf{15.94\ A}$

$\cos\phi = 7.97/15.94 = 0.5$ (lag)

$\phi = \cos^{-1}(0.5) = 60°$

Power absorbed

$= 100 \times 15.94 \times 0.5 = \mathbf{797\ W}$

The equivalent series circuit is shown in Fig. 14.15 (*a*).

$V = 100$ V ; $I = 15.94$ A ; $\phi = 60°$

$Z = 100/15.94 = 6.27\ \Omega$;

$R = Z\cos\phi = 6.27 \times \cos 60° = \mathbf{3.14\ \Omega}$

$X = Z\sin\phi = 6.27 \times \sin 60° = \mathbf{5.43\ \Omega}$

Fig. 14.15

Admittance Method For Finding Equivalent Circuit

$$\mathbf{Y_1} = \frac{1}{5 + j6.28} = \frac{5 - j6.28}{5^2 + 6.28^2} = 0.078 - j\,0.098\ \text{S},$$

$$\mathbf{Y_2} = \frac{1}{1 + j25.12} = \frac{1 - j25.12}{1^2 + j25.12^2} = 0.00158 - j\,0.0397\ \text{S},$$

$\mathbf{Y} = \mathbf{Y_1} + \mathbf{Y_2} = 0.0796 - j0.138 = 0.159\ \angle -60°$

Here $G = 0.0796$ S, $B = -0.138$ S, $Y = 0.159\ \Omega$

$\therefore R_{eq} = G/Y^2 = 0.0796/0.159^2 = \mathbf{3.14\ \Omega},$

$X_{eq} = B/Y^2 = 0.138/0.159^2 = \mathbf{5.56\ \Omega}$

Example 14.12. *Two circuits, the impedances of which are given by $Z_1 = 15 + j12$ ohms and $Z_2 = 8 - j5$ ohms are connected in parallel. If the potential difference across one of the impedance is $250 + j0$ V, calculate.*

(i) total current and branch currents

(ii) total power and power consumed in each branch

and (iii) overall power-factor and power-factor of each branch.

(Nagpur University, November 1998)

Solution. *(i)*

$$I_1 = (250 + j\,0)/(15 + j\,12) = 250\,\angle 0°/19.21\,\angle 38.6°$$
$$= 13\,\angle{-38.6°}\text{ amp} = 13\,(0.78 - j\,0.6247)$$
$$= 10.14 - j\,8.12\text{ amp}$$
$$I_2 = (250 + j\,0)/(8 - j\,5) = 250\,\angle 0°/9.434\,\angle{-32°}$$
$$= 26.5\,\angle{+32°} = 26.5\,(0.848 + j\,0.530)$$
$$= 22.47 + j14.05\text{ amp}$$
$$I = I_1 + I_2 = 32.61 + j\,5.93 = 33.15\,\angle{+10.36°}$$

(ii) Power in branch 1 = $13^2 \times 15$ = 2535 watts

Power in branch 2 = $26.5^2 \times 8$ = 5618 watts

Total power consumed = 2535 + 5618

= 8153 watts

(iii) Power factor of branch 1 = cos 38.60°

= 0.78 lag

Power factor of branch 2 = cos 32°

= 0.848 lead.

Overall power factor = cos 10.36°

= 0.984 lead.

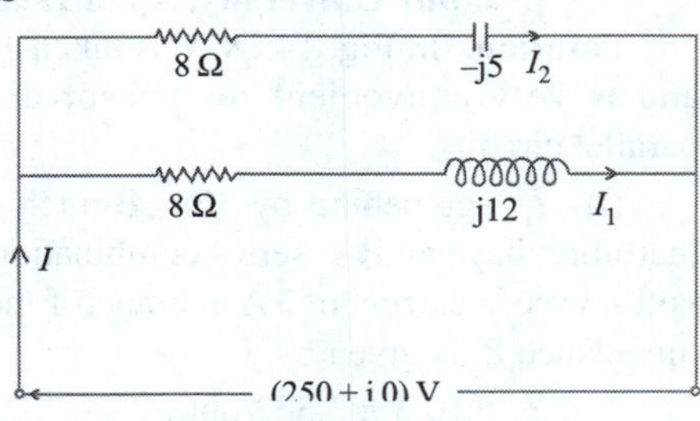

Fig. 14.16

Additional hint : Drawn phasor-diagram for these currents, in fig. 14.16, for the expressions written above,

Example 14.13. *An inductive circuit, in parallel with a resistive circuit of 20 ohms is connected across a 50 Hz supply. The inductive current is 4.3 A and the resistive current is 2.7 A. The total current is 5.8 A Find : **(a)** Power factor of the combined circuit. Also draw the phasor diagram.*

(Nagpur University, November 1997)

Solution. I_2 (= 2.7 A) is in phase with V which is 54 V in magnitude. The triangle for currents is drawn in the phasor diagram in fig. 14.17 (*b*)

Solving the triangle, $\phi_1 = 180° - \cos^{-1}[(2.7^2 = 4.3^2 - 5.8^2)/(2 \times 4.3 \times 2.7)] = 70.2°$

Further, $5.8 \sin\phi = 4.3 \sin\phi_1$, giving $\phi = 44.2°$.

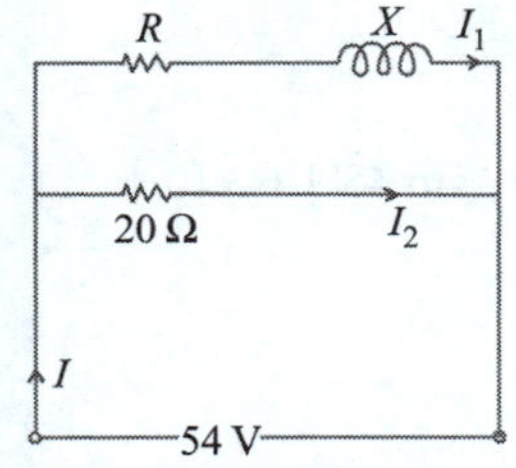

Fig. 14.17 (*a*)

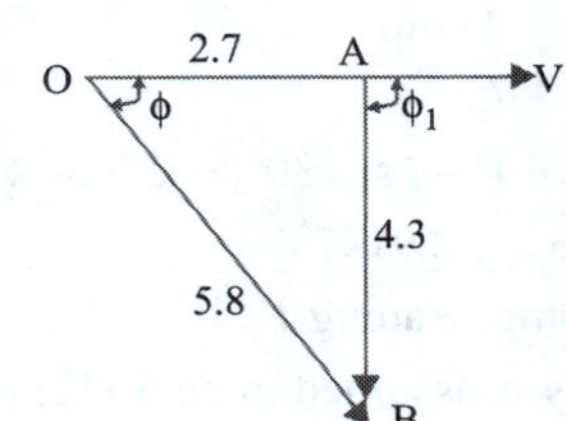

Fig. 14.17 (*b*)

OA = I_2 = 2.7 in phase with V

AB = I_1 = 4.3 lagging behind V by ϕ_1

OB = I = 5.8 lagging behind V by φ

$$|Z_1| = 54/4.3 = 12.56 \text{ ohms}$$
$$R = Z_1 \cos \phi_1 = 4.25 \text{ ohms}$$
$$X = Z_1 \sin \phi_1 = 11.82 \text{ ohms, since } \phi_1 \text{ is the lagging angle}$$

(a) Power absorbed by the Inductive branch

$$= 4.3^2 \times 4.25 = 78.6 \text{ watts}$$

(b) $L = 11.82/314 = 37.64$ mH

(c) P.f. of the combined circuit = $\cos \phi = 0.717$ lag

Check : Power consumed by 20 ohms resistor = $2.7^2 \times 20 = 145.8$ W

Total Power consumed in two branches = 78.6 + 145.8 = 224.4 W

This figure must be obtained by input power = VI $\cos \phi$

= 54 × 5.8 × cos 44.2° = 224.5 W. Hence checked.

Example 14.14. *In a particular A.C. circuit, three impedances are connected in parallel, currents as shown in fig. 14.18 are flowing through its parallel branches.*

(i) *Write the equations for the currents in terms of sinusoidal variations and draw the waveforms.*

Find the total current supplied by the source.

[Nagpur University, April 1998, 2008]

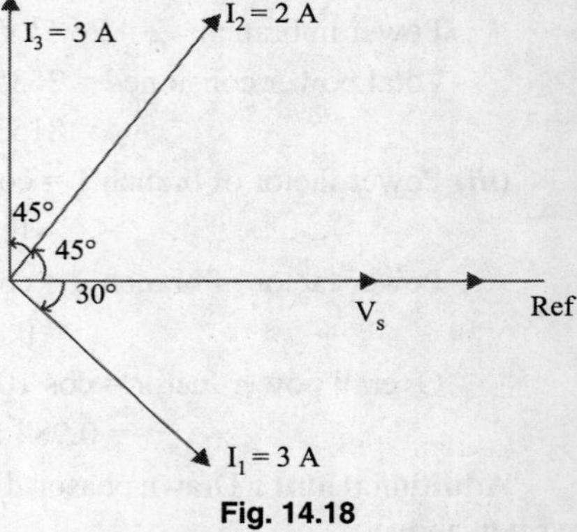

Fig. 14.18

Solution. In Fig. 14.18, V is taken as reference, and is very convenient for phasor diagrams for parallel circuits.

(i) I_1 lags behind by 30°. Branch no. 1 must, therefore, have an R-L series combination. With 10-volt source, a current of 3 A in branch 1 means that its impedance Z_1 is given by

$Z_1 = 10/3 = 3.333$ ohms

The phase-angle for I_1 is 30° lagging

$R_1 = 3.333 \cos 30° = 2.887$ ohms

$X_{L1} = 3.333 \sin 30° = 1.6665$ ohms

(ii) I_2 is 2 amp and it leads the voltage by 45°. Branch 2 must, therefore, have R-C series combination.

$$Z_2 = 10/2 = 5 \text{ ohms}$$
$$R_2 = 5 \cos 45° = 3.5355 \text{ ohms}$$
$$X_{c2} = 5 \sin 45° = 3.5355 \text{ ohms}$$

(iii) Third branch draws a current of 3 amp which leads the voltage by 90°. Hence, it can only have a capacitive reactance.

$$|Z_3| = X_{c3} = 10/3 = 3.333 \text{ ohms}$$

Total current supplied by the source = I amp

$$I = I_1 + I_2 + I_3$$
$$= 3 [\cos 30° - j \sin 30°] + 2 [\cos 45° + j \sin 45°] + 3 [0 + j\, 1]$$
$$= 4.0123 + j\, 2.9142$$
$$|I| = 4.96 \text{ amp, leading } V_s \text{, by } 36°.$$

Expressions for currents : Frequency is assumed to be 50 Hz

$$v_s = 10\sqrt{2} \sin (314\, t)$$

$$i_1 = 3\sqrt{2}\ \sin(314\,t - 30^\circ)$$
$$i_2 = 2\sqrt{2}\ \sin(314\,t + 45^\circ)$$
$$i_3 = 3\sqrt{2}\ \sin(314\,t + 90^\circ)$$
$$\text{Total current, } i\,(t) = 4.96\sqrt{2}\ \sin(314\,\text{t} + 36^\circ)$$

Total power consumed = Voltage × active (or in phase-) component of current

= 10 × 4.012 = 40.12 watts

Example 14.15. *An impedance $Z_1 = (8 - j5)\ \Omega$ is in parallel with an impedance $Z_2 = (3 + j7)\ \Omega$. If 100 V are impressed on the parallel combination, find the branch currents I_1, I_2 and the resultant current. Draw the corresponding phasor diagram showing each current and the voltage drop across each parameter. Calculate also the equivalent resistance, reactance and impedance of the whole circuit.* **(Elect. Techology-I, Gwalior Univ. 1998)**

Solution. Admittance Method

$\mathbf{Y_1} = 1/(8 - j5) = (0.0899 + j0.0562)\ S$

$\mathbf{Y_2} = 1/(3 + j7) = (0.0517 - j0.121)\ S$, $\mathbf{Y} = \mathbf{Y_1} + \mathbf{Y_2} = (0.1416 - j0.065)S$

Let $\mathbf{V} = (100 + j0)$; $\mathbf{I_1} = \mathbf{VY_1} = 100\ (0.0899 + j0.0562) = 8.99 + j5.62$

$\mathbf{I_2} = \mathbf{VY_2} = 100\ (0.0517 - j0.121) = 5.17 - j12.1;$

$\mathbf{I} = \mathbf{VY} = 100\ (0.416 - j0.056) = 14.16 - j6.5$

Now, $G = 0.1416\text{S}$, $B = -\ 0.065\ S$ (inductive;)

$$Y = \sqrt{G^2 + B^2} = \sqrt{0.1416^2 + 0.065^2} = 0.1558\ \text{S}$$

Equivalent series resistance, $R_{eq} = G/Y^2 = 0.1416/0.1558^2 = 5.38\ \Omega$

Equivalent series inductive reactance $X_{eq} = B/Y^2 = 0.065/0.1558^2 = 2.68\ \Omega$

Equivalent series impedance $\mathbf{Z} = 1/\mathbf{Y} = 1/0.1558 = \mathbf{6.42}\ \Omega$

Impedance Method

$$\mathbf{I_1} = \mathbf{V/Z_1} = (100 + j0) / (8 - j5) = 8.99 + j5.62$$
$$\mathbf{I_2} = \mathbf{V/Z_2} = 100/(3 + j7) = 5.17 - j12.1$$
$$\mathbf{Z} = \frac{\mathbf{Z_1\,Z_2}}{\mathbf{Z_1 + Z_2}} = \frac{(8 - j5)\,(3 + j7)}{(11 + j2)} = \frac{59 + j41}{(11 + j2)}$$
$$= 5.848 + j2.664 = 6.426 = \angle\ 24.5^\circ,$$
$$\mathbf{I} = 100/6.426\ \angle 24.5^\circ = 15.56\ \angle{-24.5^\circ} = 14.16 - j\,6.54$$

As seen from the expression for Z, equivalent series resistance is **5.848** Ω and inductive reactance is **2.664 ohm.**

Example 14.16. *The impedances $Z_1 = 6 + j\,8$, $Z_2 = 8 - j\,6$ and $Z_3 = 10 + j\,0$ ohms measured at 50 Hz, form three branches of a parallel circuit. This circuit is fed from a 100 volt. 50-Hz supply. A purely reactive (inductive or capacitive) circuit is added as the fourth parallel branch to the above three-branched parallel circuit so as to draw minimum current from the source. Determine the value of L or C to be used in the fourth branch and also find the minimum current.*

(Electrical Circuits, South Gujarat Univ.)

Solution. Total admittance of the 3-branched parallel circuit is

$$\mathbf{Y} = \frac{1}{6+j8} + \frac{1}{8-j6} + \frac{1}{10+j0} = 0.06 - j0.08 + 0.08 + j0.06 + 0.1 = 0.24 - j0.02$$

Current taken would be minimum when net susceptance is zero. Since combined susceptance is inductive, it means that we must add capacitive susceptance to neutralize it. Hence, we must connect a pure capacitor in parallel with the above circuit such that its susceptance equals $+j0.02$ S

$$\therefore \quad I/X_C = 0.02 \text{ or } 2\pi/C = 0.02\,;\ C = 0.2/314 = \mathbf{63.7\ \mu F}$$

Admittance of four parallel branches = $(0.24 - j0.02) + j0.02 = 0.24$ S

∴ Minimum current drawn by the circuit = 100 × 0.24 = **24 A**

Example 14.17. *The total effective current drawn by parallel circuit of Fig. 14.19 is 20 A. Calculate* ***(i)*** *VA* ***(ii)*** *VAR and* ***(iii)*** *watts drawn by the circuit.*

Solution. The combined impedance of the circuit is

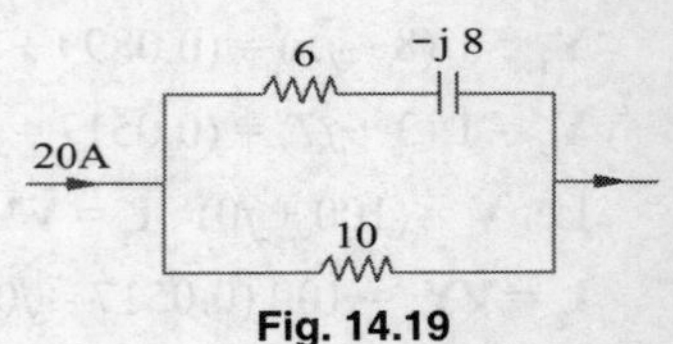

Fig. 14.19

$$\mathbf{Z} = \frac{\mathbf{Z_1 Z_2}}{\mathbf{Z_1 + Z_2}} = \frac{10\,(6-j8)}{(16-j\,8)} = (5 - j2.5) \text{ ohm}$$

(iii) Power = $I^2R = 20^2 \times 5 =$ **2000 W** *(ii)* $Q = I^2X = 20^2 \times 2.5 =$ **1000 VAR (leading)** ***(i)*** $S = P + j\,Q = 2000 + j1000 = 2236 \angle 27^\circ$; $S =$ **2236 VA**

Example 14.18. *Calculate (i) total current and (ii) equivalent impedance for the four-branched circuit of Fig. 14.20.*

Solution. $\mathbf{Y_1} = 1/20 = 0.05$ S, $\mathbf{Y_2} = 1/j10 = -j0.1$ S;

$\mathbf{Y_3} = 1/-j20 = j0.05$ S ;

$\mathbf{Y_4} = 1/5 - j8.66 = 1/10 \angle 60^\circ$

$= 0.1 \angle -60^\circ = (0.05 - j0.0866)$ S

$\mathbf{Y = Y_1 + Y_2 + Y_3 + Y_4}$

$= (0.1 - j\,0.1366)$ S

$= 0.169 \angle -53.8^\circ$ S

Fig. 14.20

(i) $\mathbf{I = VY} = 200 \angle 30^\circ \times 0.169 \angle -53.8^\circ = 33.8 \angle -23.8^\circ$ A

(ii) $\mathbf{Z = 1/Y} = 1/0.169 \angle -53.8^\circ = 5.9 \angle 53.8^\circ\ \Omega$

Example 14.19. *The power consumed by both branches of the circuit shown in Fig. 14.21 is 2200 W. Calculate power of each branch and the reading of the ammeter.*

Solution. $\mathbf{I_1 = V/Z_1}$

$= \mathbf{V}/(6 + j\,8) = \mathbf{V}/10 \angle 53.1^\circ,\ \mathbf{I_2 = V/Z_2 = V}/20$

$\therefore \quad I_1/I_2 = 20/10 = 2,\ P_1 = I_1^2 R_1 \text{ and } P_2 = I_2^2 R_2$

$$\therefore \quad \frac{P_1}{P_2} = \frac{I_1^2 R_1}{I_2^2 R_2} = 2^2 \times \left(\frac{6}{20}\right) = \frac{6}{5}$$

Fig. 14.21

Now, $\quad P = P_1 + P_2 \text{ or } \frac{P}{P_2} = \frac{P_1}{P_2} + 1 = \frac{6}{5} + 1 = \frac{11}{5}$

or $\quad P_2 = 2200 \times \frac{5}{11} = 1000\ W \quad \therefore \quad P_1 = 2200 - 1000 = 1200$ W

Since $P_1 = I_1^2 R_1$ or $1200 = I_1^2 \times 6$; $I_1 = 14.14$ A

If $\mathbf{V} = V\angle 0^\circ$, then $\mathbf{I}_1 = 14.14 \angle{-53.1^\circ} = 8.48 - j11.31$

Similarly, $P_2 = I_2^2 R_2$ or $1000 = I_2^2 \times 20$; $\mathbf{I}_2 = 7.07$ A or $\mathbf{I}_2 = \mathbf{7.07 \angle 0^\circ}$

Total current $\mathbf{I} = \mathbf{I}_1 + \mathbf{I}_2 = (8.48 - j11.31) + 7.07 = 15.55 - j11.31 = \mathbf{19.3 \angle{-36^\circ}}$

Hence, ammeter reads **19.3 A**

Example 14.20. *Consider an electric circuit shown in Figure 14.22 (a)*

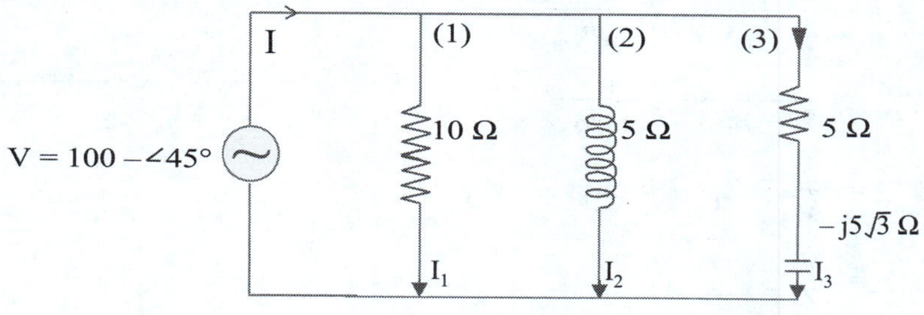

Fig. 14.22. (*a*)

Determine : ***(i)*** *the current and power consumed in each branch.*

(ii) *the supply current and power factor.* **(U.P. Technical University, 2001)**

Solution. Indicating branch numbers 1, 2, 3 as marked on the figure, and representing the source voltage by $100 \angle 45^\circ$,

$$Z_1 = 10 + j\,0 = 10\angle 0^\circ,\ I_1 = 100\angle 45^\circ/10\angle 0^\circ = 10\angle 45^\circ \text{ amp}$$

$$Z_2 = 5 + j5\sqrt{3} = 10\angle 60^\circ,\ I_2 = 100\angle 45^\circ/10\angle 60^\circ = 10\angle{-15^\circ} \text{ amp}$$

$$Z_3 = 5 - j5\sqrt{3} = 10\angle{-60^\circ},\ I_3 = 100\angle 45^\circ/10\angle{-60^\circ} = 10\angle 105^\circ \text{ amp}$$

Phasor addition of these three currents gives the supply current, I which comes out to be $I = 20 \angle 45^\circ$ amp.

This is in phase with the supply voltage.

(*i*) Power consumed by the branches :

Branch 1 : $10^2 \times 10 = 1000$ watts

Branch 2 : $10^2 \times 5 = 500$ watts

Branch 3 : $10^2 \times 5 = 500$ watts

Total power consumed = 2000 watts

(*ii*) Power factor = 1.0 since V and I are in phase

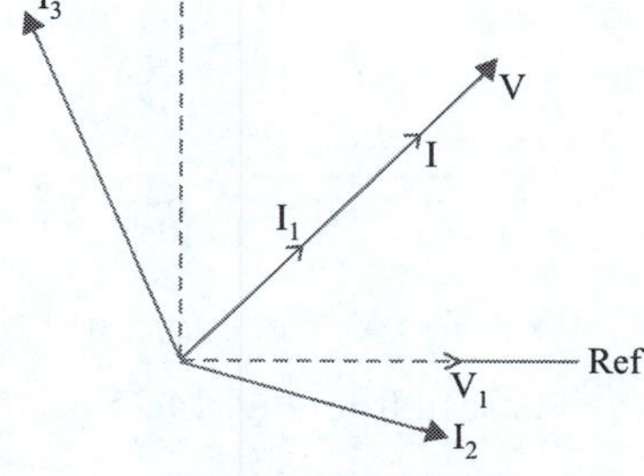

Fig. 14.22. (*b*)

14.6. Series-Parallel Circuits

(i) By Admittance Method

In such circuits, the parallel circuit is first reduced to an equivalent series circuit and then, as usual, combined with the rest of the circuit. For a parallel circuit,

Equivalent series resistance $R_{eq} = Z\cos\phi = \dfrac{1}{Y}\cdot\dfrac{G}{Y} = \dfrac{G}{Y^2}$

- Sec Ex. 14.14

Equivalent series reactance

$$X_{eq} = Z\sin\phi = \frac{1}{Y}\cdot\frac{B}{Y} = \frac{B}{Y^2}$$

(ii) *By Symbolic Method*

Consider the circuit of Fig. 14.23. First, equivalent impedance of parallel branches is calculated and it is then added to the series impedance to get the total circuit impedance. The circuit current can be easily found.

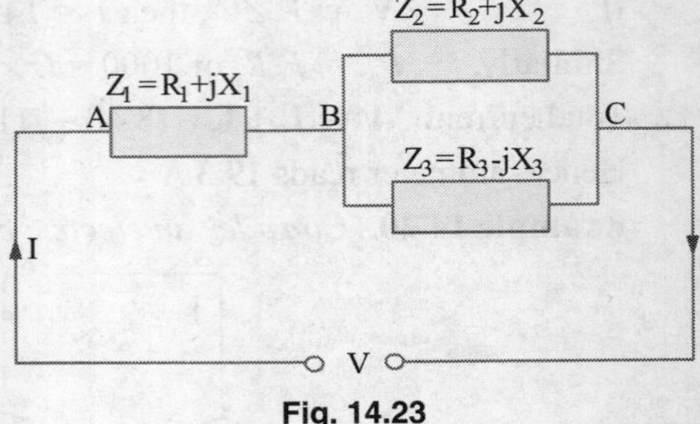

Fig. 14.23

$$\mathbf{Y}_2 = \frac{1}{R_2 + jX_2}; \quad \mathbf{Y}_3 = \frac{1}{R_3 - jX_3}$$

$$\therefore \quad \mathbf{Y}_{23} = \frac{1}{R_2 + jX_2} + \frac{1}{R_3 - jX_3}$$

$$\therefore \quad \mathbf{Z}_{23} = \frac{1}{\mathbf{Y}_{23}}; \; \mathbf{Z}_1 = R_1 + jX_1; \; \mathbf{Z} = \mathbf{Z}_{23} + \mathbf{Z}_1$$

$$\therefore \quad \mathbf{I} = \frac{\mathbf{V}}{\mathbf{Z}}$$

14.7. Series Equivalent of a Parallel Circuit

Consider the parallel circuit of Fig. 14.24 (*a*). As discussed in Art. 14.5

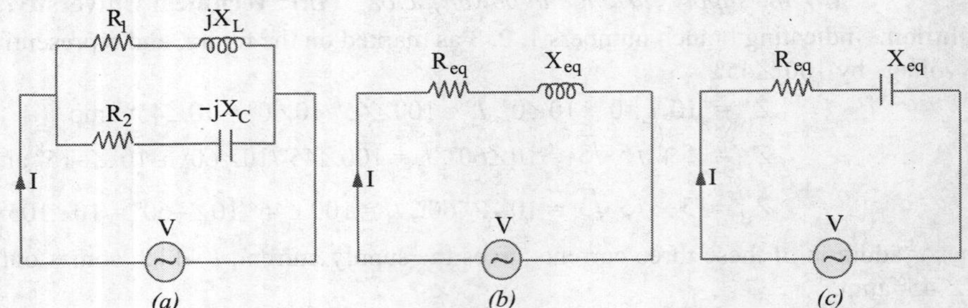

Fig. 14.24

$$\mathbf{Y}_1 = \frac{R_1}{R_1^2 + X_L^2} - j\frac{X_L}{R_1^2 + X_L^2} = g_1 - jb_1; \; \mathbf{Y}_2 = \frac{R_2}{R_2^2 + X_c^2} + j\frac{X_2}{R_2^2 + X_c^2} = g_2 + jb_2$$

$$\mathbf{Y} = \mathbf{Y}_1 + \mathbf{Y}_2 = g_1 - jb_1 + g_2 + jb_2 = (g_1 + g_2) + j(b_2 - b_1) = G + jB = \sqrt{(G^2 + B^2)} \angle \tan^{-1}(B/G)$$

As seen from Fig. 14.25.

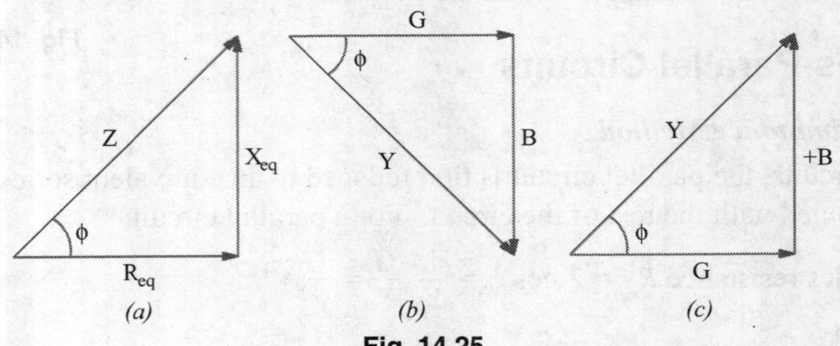

Fig. 14.25

$$R_{eq} = Z\cos\phi = \frac{1}{Y}\cdot\frac{G}{Y} = \frac{G}{Y^2}; \quad X_{eq} = Z\sin\phi = \frac{1}{Y}\cdot\frac{B}{Y} = \frac{B}{Y^2}$$

Hence, equivalent series circuit is as shown in Fig. 14.24 (*b*) or (*c*) depending on whether net susceptance *B* is negative (inductive) or positive (capacitive). If *B* is negative, then it is an *R-L* circuit of Fig. 14.24 (*b*) and if *B* is positive, then it is an *R-C* circuit of Fig. 14.24 (*c*).

14.8. Parallel Equivalent of a Series Circuit

The two circuits will be equivalent if **Y** of Fig. 14.26 (*a*) is equal to the **Y** of the circuit of Fig. 14.26. (*b*).

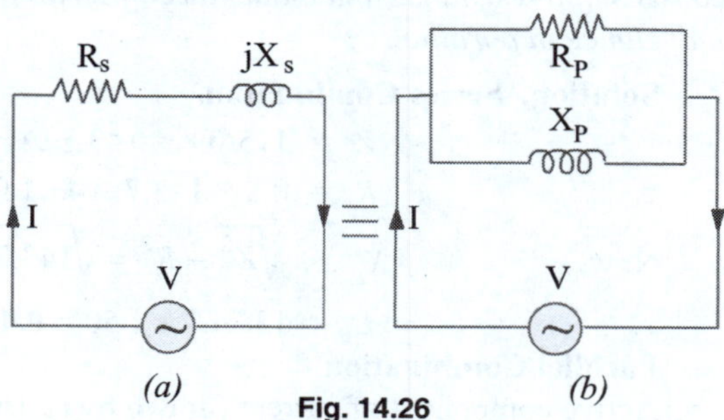

Fig. 14.26

Series Circuit

$$\mathbf{Y}_s = \frac{1}{R_s + jX_s}$$

$$= \frac{R_s - jX_s}{(R_s + jX_s)(R_s - jX_s)}$$

$$= \frac{R_s - jX_s}{R_s^2 + X_s^2} = \frac{R_s}{R_s^2 + X_s^2} - j\frac{X_s}{R_s^2 + X_s^2}$$

Parallel Circuit

$$\mathbf{Y}_p = \frac{1}{R_p + j0} + \frac{1}{0 + jX_p} = \frac{1}{R_p} + \frac{1}{jX_p} = \frac{1}{R_p} - \frac{j}{X_p}$$

$$\therefore \quad \frac{R_s}{R_s^2 + X_s^2} - j\frac{X_s}{R_s^2 + X_s^2} = \frac{1}{R_P} - \frac{j}{X_p}$$

$$\therefore \quad \frac{1}{R_p} = \frac{R_s}{R_s^2 + X_s^2} \text{ or } R_p = R_s + \frac{X_s^2}{R_s} = R_s\left(1 + \frac{X_s^2}{R_s^2}\right)$$

$$\text{Similarly } X_p = X_s + \frac{R_s^2}{X_s} = X_s\left(1 + \frac{R_s^2}{X_s^2}\right)$$

Example 14.21. *The admittance of a circuit is (0.03 – j 0.04) Siemens. Find the values of the resistance and inductive reactance of the circuit if they are joined* **(a)** *in series and (b) in parallel.*

Solution. (*a*) $\mathbf{Y} = 0.03 - j0.04$

$$\therefore \quad \mathbf{Z} = \frac{1}{\mathbf{Y}} = \frac{1}{0.03 - j0.04} = \frac{0.03 + j0.04}{0.03^2 + 0.04^2} = \frac{0.03 + j0.04}{0.0025} = 12 + j16$$

Hence, if the circuit consists of a resistance and inductive reactance in series, then resistance is **12 Ω** and inductive reactance is 16 Ω as shown in Fig. 14.27.

(***b***) Conductance = 0.03 mho ∴ Resistance = 1/0.03 = 33.3 Ω

Susceptance (inductive) = 0.04 S ∴ Inductive reactance = 1/0.04 = 25 Ω

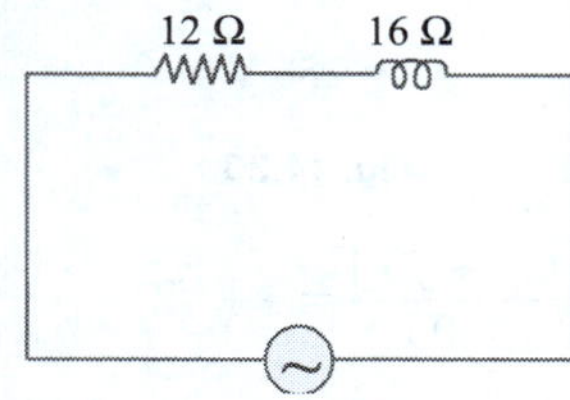

Fig. 14.27

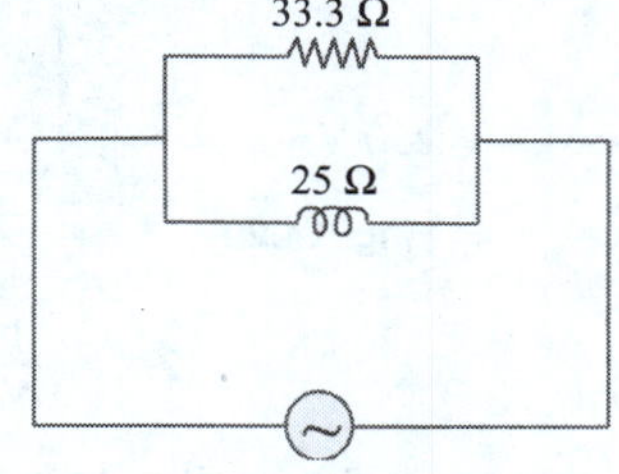

Fig. 14.28

Hence, if the circuit consists of a resistance connected in parallel with an inductive reactance, then resistance is **33.3 Ω** and inductive reactance is **25 Ω** as shown in Fig. 14.28.

Example 14.22. *A circuit connected to a 115-V, 50-Hz supply takes 0.8 A at a power factor of 0.3 lagging. Calculate the resistance and inductance of the circuit assuming* ***(a)*** *the circuit consists of a resistance and inductance in series and* ***(b)*** *the circuit consists of a resistance and inductance in paralllel.* **(Elect. Engg.-I, Sardar Patel Univ.)**

Solution. Series Combination

$$Z = 115/0.8 = 143.7\ \Omega\ ;\ \cos\phi = R/Z = 0.3$$

$$\therefore \quad R = 0.3 \times 143.7 = \mathbf{43.1\ \Omega}$$

Now $$X_L = \sqrt{Z^2 - R^2} = \sqrt{143.7^2 - 43.1^2} = 137.1\ \Omega$$

$$\therefore \quad L = 137.1/2\pi \times 50 = \mathbf{0.436\ H}$$

Parallel Combination

Active component of current (drawn by resistance)

$= \quad 0.8 \cos\phi = 0.8 \times 0.3 = 0.24$ A; $\quad R = 115/0.24 = \mathbf{479\ \Omega}$

Quadrature component of current (drawn by inductance) $= 0.8 \sin\phi = 0.8\sqrt{1 - 0.3^2}$

$= 0.763$ A

$\therefore \quad X_L = 115/0.763\ \Omega \quad \therefore \quad L = 115/0.763 \times 2\pi \times 50 = \mathbf{0.48\ H}$

Example 14.23. *The active and lagging reactive components of the current taken by an a.c. circuit from a 250-V supply are 50 A and 25 A respectively. Calculate the conductance, susceptance, admittance and power factor of the circuit. What resistance and reactance would an inductive coil have if it took the same current from the same mains at the same factor ?*

(Elect. Technology, Sumbal Univ.)

Solution. The circuit is shown in Fig. 14.29.

Resistance $= 250/50 = 5\ \Omega$; Reactance $= 250/25 = 10\ \Omega$

$\therefore$ Conductance $g = 1/5 = \mathbf{0.2\ S}$, Susceptance $b = -1/10 = \mathbf{-0.1\ S}$

Admittance $$Y = \sqrt{g^2 + b^2} = \sqrt{0.2^2 + (-0.1)^2} = \sqrt{0.05} = \mathbf{0.224\ S}$$

$\mathbf{Y} = 0.2 - j0.1 = 0.224\angle -26°34'$. Obviously, the total current lags the supply voltage by 26°34′, p.f. = cos 26°34′ = **0.894 (lag)**

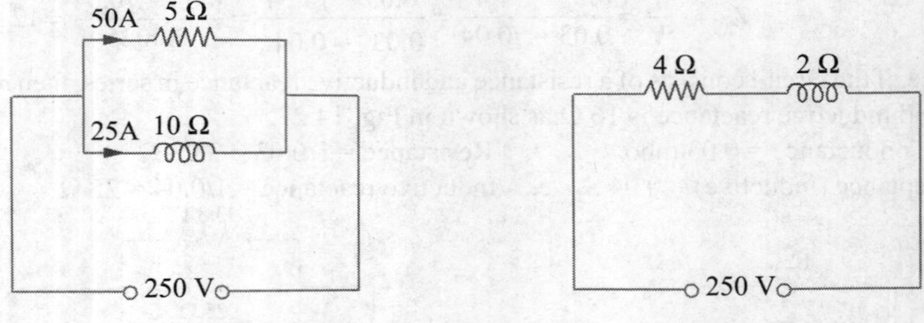

Fig. 14.29 Fig. 14.30

Now $$\mathbf{Z} = \frac{1}{\mathbf{Y}} = \frac{1}{0.2 - j0.1} = \frac{0.2 + j0.1}{0.05} = 4 + j2$$

Hence, resistance of the coil = **4 Ω** Reactance of the coil = **2 Ω** (Fig. 14.30)

Example 14.24. *The series and parallel circuits shown in Fig. 14.31 have the same impedance and the same power factor. If R = 3 Ω and X = 4 Ω, find the values of R_1 and X_1. Also, find the impedance and power factor.* **(Elect. Engg., Bombay Univ.)**

Solution. Series Circuit [Fig. 14.31 (*a*)]

$$\mathbf{Y}_S = \frac{1}{R+jX} = \frac{R-jX}{R^2+X^2} = \frac{R}{R^2+X^2} - j\frac{X}{R^2+X^2}$$

Parallel Circuit [Fig. 14.31 (*b*)]

$$\mathbf{Y}_P = \frac{1}{R_1+j0} = \frac{1}{0+jX_1} = \frac{1}{R_1} + \frac{1}{jX_1} = \frac{1}{R_1} - \frac{j}{X_1}$$

$$\therefore \quad \frac{R}{R^2+X^2} - j\frac{X}{R^2+X^2} = \frac{1}{R_1} - \frac{j}{X_1}$$

$\therefore \quad R_1 = R + X^2/R$ and $X_1 = X + R^2/X$

$\therefore \quad R_1 = 3 + (16/3) = \mathbf{8.33\ \Omega}$; $X_1 = 4 + (9/4) = \mathbf{6.25\ \Omega}$

Impedance $= 3 + j\,4 = 5\angle 53.1°$; Power factor $= \cos 53.1° = \mathbf{0.6\ (lag)}$

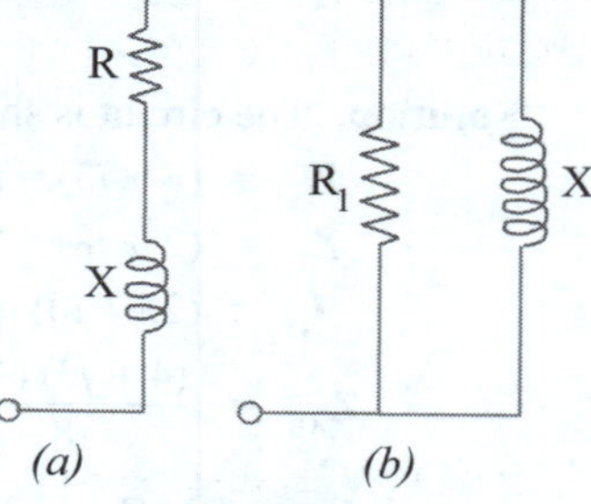

Fig. 14.31

Example 14.25. *Find the value of the resistance R and inductance L which when connected in parallel will take the same current at the same power factor from 400-V, 50-Hz mains as a coil of resistance R_1 = 8 Ω and an induction L_1 = 0.2 H from the same source of supply.*

Show that when the resistance R_1 of the coil is small as compared to its inductance L_1, then R and L are respectively equal to $\omega^2 L_1^2/R_1$ and L_1. **(Elect. Technology, Utkal Univ.)**

Solution. As seen from Art. 14.8 in Fig. 14.32.

$R = R_1 + X_1^2/R_1$ ...**(*i*)**

$X = X_1 + R_1^2/X_1$ ...**(*ii*)**

Now $R_1 = 8\ \Omega$, $X_1 = 2\pi \times 50 \times 0.2 = 62.8\ \Omega$

$\therefore \quad R = 8 + (62.8^2/8) = \mathbf{508\ \Omega}$

$X = 62.8 + (64/62.8) = \mathbf{63.82\ \Omega}$

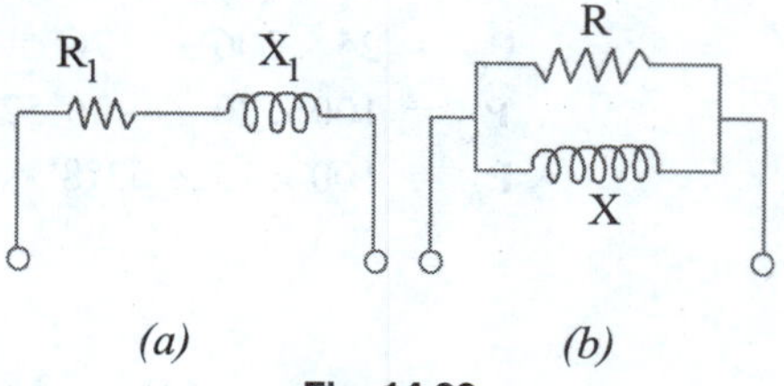

Fig. 14.32

From **(*i*)**, it is seen that if R_1 is negligible, then $R = X_1^2/R_1 = \omega^2 L_1^2/R_1$

Similarly, from **(*ii*)** we find that the term R_1^2/X_1 is negligible as compared to X_1,

$\therefore \quad X = X_1$ or $L = L_1$

Example 14.26. *For the circuit shown in Fig. 14.33 (a), find* **(i)** *total impedance* **(ii)** *total current* **(iii)** *total power absorbed and power-factor. Draw a vector diagram.*

(Elect. Tech. Osmania Univ. Jan/Feb 1992)

Solution. $Z_{BC} = (4 + j8) \parallel (5 - j8) = 9.33 + j0.89$

(*i*) $Z_{AC} = 3 + j6 + 9.33 + j0.89 = 12.33 + j6.89$

$= 14.13 \angle 29.2°$

(*ii*) $I = 100/14.13 \angle 29.2°$, as drawn in Fig. 14.33

$= 7.08 \angle -29.2°$

(*iii*) $\phi = 29.2°$; $\cos\phi = 0.873$; $P = VI\cos\phi$

$= 100 \times 7.08 \times 0.873 = 618$ W

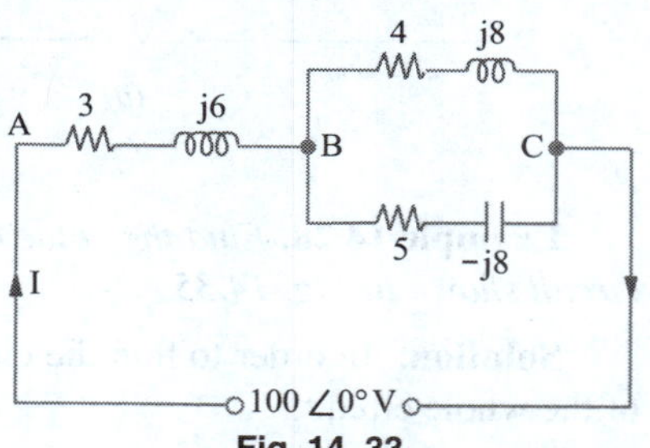

Fig. 14. 33

Example 14.27. *In a series-parallel circuit, the parallel branches A and B are in series with C. The* ***impedances*** *are : $Z_A = (4 + j\,3)$; $Z_B = (4 - j\,16/3)$; $Z_C = (2 + j8)$ ohm.*

If the current $I_C = (25 + j\,0)$, draw the complete phasor diagram determining the branch currents and voltages and the total voltage. Hence, calculate the complex power (the active and reactance powers) for each branch and the whole circuits.

(Basic Electricity, Bombay Univ.)

Solution. The circuit is shown in Fig. 14.34 (*a*)

$$\mathbf{Z_A} = (4 + j3) = 5 \angle 36°52' \text{ ; } \mathbf{Z_B} = (4 - j16/3) = 20/3 \angle -53°8' \text{ ;}$$

$$\mathbf{Z_C} = (2 + j8) = 8.25 \angle 76°$$

$$\mathbf{I_C} = (25 + j0) = 25 \angle 0° \text{ ; } \mathbf{V_C} = \mathbf{I_C Z_C} = 206 \angle 76°$$

$$\mathbf{Z_{AB}} = \frac{(4 + j3)\,(4 - j16/3)}{(8 - j7/3)} = \frac{(32 - j28/3)}{(8 - j7/3)} = 4 + j\,0 = 4 \angle 0°$$

$$\mathbf{V_{AB}} = \mathbf{I_C\, Z_{AB}} = 25 \angle 0° \times 4 \angle 0° = 100 \angle 0°$$

$$\mathbf{Z} = \mathbf{Z_C} + \mathbf{Z_{AB}} = (2 + j8) + (4 + j0) = (6 + j8) = 10 \angle 53°8 \text{ ;}$$

$$\mathbf{V} = \mathbf{I_C Z} = 25\angle 0° \times 10 \angle 53°8' = 250 \angle 53°8'$$

$$\mathbf{I_A} = \frac{\mathbf{V_{AB}}}{\mathbf{Z_A}} = \frac{100 \angle 0°}{5 \angle 36°52'} = 20 \angle -36°52' \text{ ; } \mathbf{I_B} = \frac{\mathbf{V_{AB}}}{\mathbf{Z_B}} = \frac{100 \angle 0°}{(20/3) \angle -53°8'} = 15 \angle 53°8'$$

Various voltages and currents are shown in Fig. 14.34 (*b*). Powers would be calculated by using voltage conjugates.

Power for whole circuit is $\mathbf{P} = \mathbf{VI_C} = 250 \angle\ 53°8' \times 25 \angle 0° = 6{,}250 \angle 53°8'$

$$= 6250 (\cos 53°8' - j \sin 53°8') = 3750 - j5000$$

$$\mathbf{P_C} = 25 \times 206 \angle -76° = 5150 (\cos 76° - j \sin 76°) = 1250 - j5000$$

$$\mathbf{P_A} = 100 \times 20 \angle -36°52' = 2000 \angle -36°52' = 1600 - j1200$$

$$\mathbf{P_B} = 100 \times 15 \angle 53°8' = (900 + j1200); \text{ Total} = 3{,}750 - j5000° \text{ (as a check)}$$

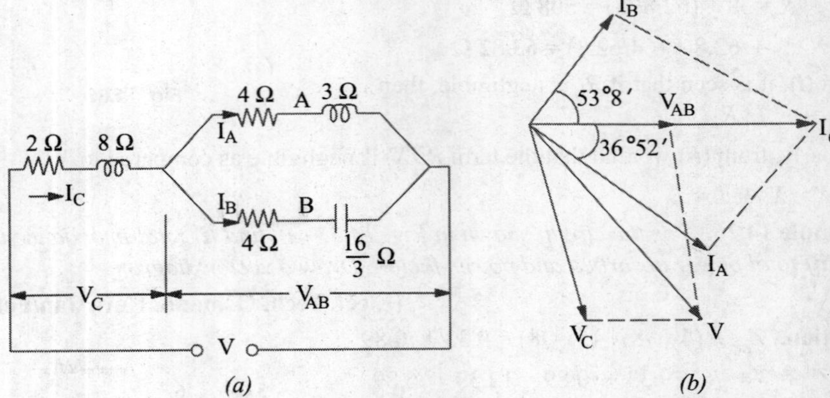

Fig. 14.34

Example 14.28. *Find the value of the power developed in each arm of the series-paralell circuit shown in Fig. 14.35.*

Solution. In order to find the circuit current, we must first find the equivalent impedance of the whole circuit.

$Z_{AB} = (5 + j12) \parallel (-j\,20)$

$$= \frac{(5 + j12)(-j20)}{5 + j12 - j20} = \frac{13 \angle 67.4° \times 20 \angle -90°}{9.43 \angle -58°}$$

$= 27.57 \angle 35.4° = (22.47 + j\,15.97)$

$Z_{AC} = (10 + j0) + (22.47 + j15.97) = (32.47 + j14.97)$

$= 36.2 \angle 26.2°$

$$I = \frac{V}{Z} = \frac{50 \angle 0°}{36.2 \angle 26.2°} = 1.38 \angle -26.2° A$$

Fig. 14.35

Power developed in 10 Ω resistor

$= I^2 R = 1.38^2 \times 10 =$ **19 W.**

Potential difference across 10 Ω resistor is

$IR \;=\; 1.38 \angle -26.2° \times 10 = 13.8 \angle -26.2° = (12.38 - j\,6.1)$

$V_{BC} \;=\;$ supply voltage – drop across 10 Ω resistor

$= (50 + j0) - (12.38 - j6.1) = (37.62 + j6.1) = 38.1 \angle 9.21°$

$$I_2 = \frac{V_{BC}}{(5 + j12)} = \frac{38.1 \angle 9.21°}{13 \angle 67.4°} = 2.93 \angle -58.2°$$

Power developed $= I_2^2 \times 5 = 2.93^2 \times 5 =$ **43 W**

No power is developed in the capacitor branch because it has no resistance.

Example 14.29. *In the circuit shown in Fig. 14.36 determine the voltage at a frequency of 50 Hz to be applied across AB in order that the current in the circuit is 10 A. Draw the phasor diagram.* **(Elect. Engg. & Electronics Bangalore Univ.)**

Solution. $X_{L1} = 2\,\pi \times 50 \times 0.05 = 15.71\ \Omega$; $X_{L2} = -\,2\,\pi \times 50 \times 0.02 = 6.28\ \Omega$,

$X_C = 1/2\pi \times 50 \times 400 \times 10^{-6} = 7.95\ \Omega$

$\mathbf{Z_1} = R_1 + jX_{L1} = 10 + j15.71 = 18.6 \angle 57°33'$

$\mathbf{Z_2} = R_2 + jX_{L2} = 5 + j6.28 = 8 \angle 51°30'$

$\mathbf{Z_3} = R_3 - jX_C = 10 - j7.95 = 12.77 \angle -38°30'$

$\mathbf{Z_{BC}} = Z_2 \,||\, Z_3 = (5 + j6.28) \,||\, (10 - j7.95) = 6.42 + j2.25 = 6.8 \angle 19°18'$

$Z = Z_1 + Z_{BC} = (10 + j15.71) + (6.42 + j2.25) = 16.42 + j17.96 = 24.36 \angle 47°36'$

Let $I = 10\angle 0°$; $\therefore$ $\mathbf{V} = IZ = 10\angle 0° \times 24.36\angle 47°36 = 243.6\angle 47°36'$

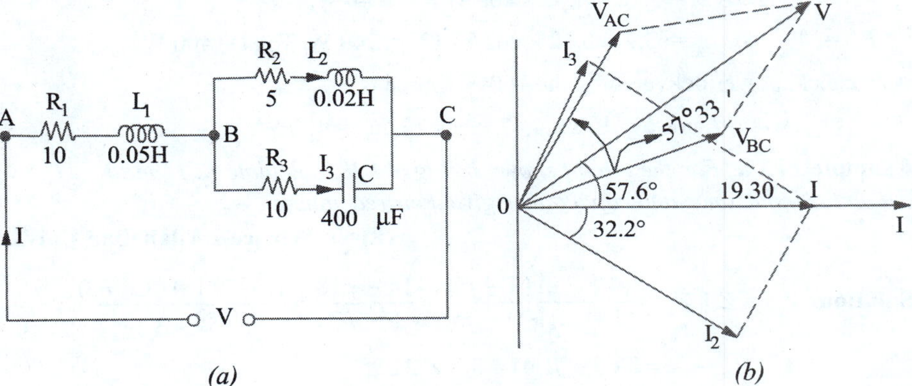

Fig. 14.36

$V_{BC} = IZ_{BC} = 10 \angle 0° \times 6.8 \angle 19°18' = 68 \angle 19°18'$;

$$I_2 = \frac{V_{BC}}{Z_2} = \frac{68 \angle 19° 18'}{8 \angle 51°31'} = 8.5 \angle -32°12'$$

$$I_3 = \frac{V_{BC}}{Z_3} = \frac{68 \angle 19° 18'}{12.77 \angle -38°30'} = 5.32 \angle 57°48' ;$$

$$V_{AC} = IZ_1 = 10 \angle 0° \times 18.6 \angle 57°33' = 186 \angle 57°33'$$

The phasor diagram is shown in Fig. 14.36 (*b*).

Example 14.30. *Determine the average power delivered to each of the three boxed networks in the circuit of Fig. 14.37.* **(Basic Circuit Analysis Osmania Univ. Jan/Feb 1992)**

Solution. $Z_1 = 6 - j8 = 10 \angle -53°13°$; $Z_2 = 2 + j14 = 14.14 \angle 81.87°$;

$$Z_3 = 6 - j8 = 10 \angle -53.13°$$

$$Z_{23} = \frac{Z_2 Z_3}{Z_2 + Z_3} = 14.14 \angle -8.13° = 14 - j2$$

Drop across two parallel impedances is given by

$$V_{23} = 100 \frac{14 - j2}{(6 - j8) + (14 - j2)} = 63.2 \angle 18.43° = 60 + j20$$

$$V_1 = 100 \frac{10 \angle -53.13°}{6 - j8 + (14 - j2)} = 47.7 \angle -26.57° = 40 - j20$$

$$I_1 = \frac{44.7 \angle -26.57°}{10 \angle -53.13°} = 4.47 \angle 26.56°$$

$$I_2 = \frac{63.2 \angle 18.43°}{14.14 \angle 81.87°} = 4.47 \angle -63.44°$$

$$I_3 = \frac{63.2 \angle 18.43°}{10 \angle -53.13°} = 6.32 \angle 71.56°$$

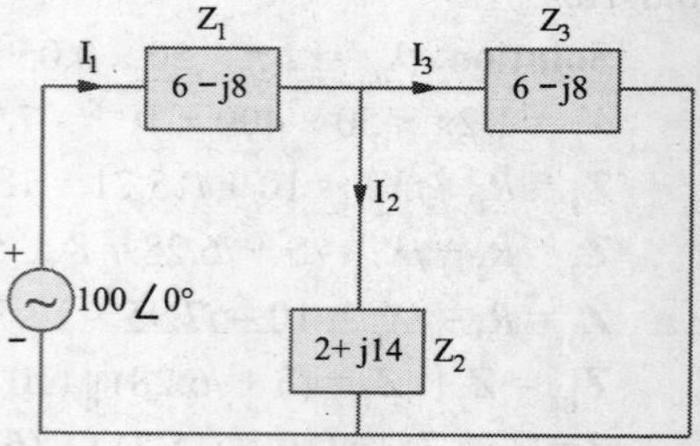

$$P_1 = V_1 I_1 \cos \phi_1 = 44.7 \times 4.47 \times \cos 53.13° = 120 \text{ W}$$

$$P_2 = V_2 I_2 \cos \phi_2 = 63.2 \times 4.47 \times \cos 81.87° = 40 \text{ W};$$

$P_3 = V_3 I_3 \cos \phi_3 = 63.2 \times 6.32 \times \cos 53.13° =$ **240 W,** Total = 400 W

Fig. 14.37

As a check, power delivered by the 100-V source is,

$$P = VI_1 \cos \phi = 100 \times 4.47 \times \cos 26.56° = 400 \text{ W}$$

Example 14.31. *For the circuit shown in Fig. 14.38, calculate I_1, I_2 and I_3. The values marked on the inductance and capacitance give their reactances.*

(Elect. Science-I Allahabad Univ. 1992)

Solution. $Z_{BC} = Z_2 \| Z_3 = \dfrac{(4 + j2)(1 - j5)}{(3 + j2) + (1 - j5)} = \dfrac{14 - j18}{5 - j3} = \dfrac{(14 - j18)(5 + j3)}{5^2 + 3^2}$

$$= 3.65 - j1.41 = 3.9 \angle 21.2°$$

$\mathbf{Z} = \mathbf{Z}_1 + \mathbf{Z}_{BC} = (2 + j3) + (3.65 - j1.41)$

$= 5.65 + j1.59 = 5.82 \angle 74.3°$

Let $V = 10 \angle 0°$; $I_1 = V/Z$

$= 10 \angle 0°/5.82 \angle 74.3° = \mathbf{1.72 \angle -74.3°}$

$V_{BC} = I_1 Z_{BC} = 1.72 \angle -74.3° \times 3.9 \angle 21.2°$

$= 6.7 \angle -53.1$

Now, $Z_2 = 4 + j2 = 4.47 \angle 63.4°$;

$Z_3 = 1 - j\,5 = 5.1 \angle -11.3°$

$I_2 = V_{BC}/Z_2 = 6.7 \angle -53.1°/4.47 \angle 63.4° = \mathbf{1.5 \angle 10.3°}$

$I_3 = V_{BC}/Z_3 = 6.7 \angle -53.1°/5.1 \angle -11.3° = \mathbf{1.3 \angle -41.8°}$

Fig. 14.38

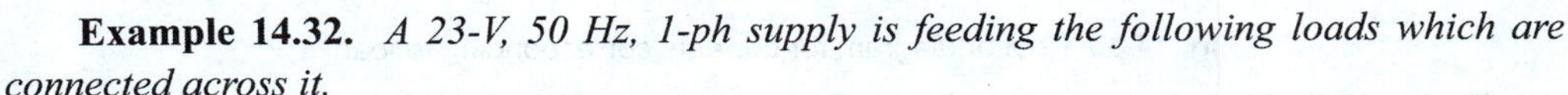

Example 14.32. *A 23-V, 50 Hz, 1-ph supply is feeding the following loads which are connected across it.*

(i) *A motor load of 4 kW, 0.8 lagging p.f.*

(ii) A rectifier of 3 kW at 0.6 leading p.f.

(ii) *A lighter-load of 10 kVA at unity p.f.*

(iv) A pure capacitive load of 8 kVA

Determine : Total kW, Total kVAR, Total kVA **(I BE Nagpur University Nov. 1999)**

Solution.

S. No.	Item	kW	P.f	kVA	kVAR	*I*	I_s	I_r
1	Motor	4	0.8 lag	5	3 – ve, Lag	21.74	17.4	13.04 Lag (–)
2	Rectifier	3	0.6 Lead	5	4 + ve, Lead	21.74	13.04	17.4 Lead (+)
3	Light-Load	10	1.0	10	zero	43.48	43.48	zero
4	Capacitive Load	Zero	0.0 Lead	8	8 + ve Lead	34.8	zero	34.8 Lead (+)
	Toal	17	-	Phasor Addition required	+ 9 + ve Lead	Phasor addition required	73.92	39.16 Lead (+)

Performing the calculations as per the tabular entries above, following answers are obtained

Total kW = 17

Total kVAR = + 9, leading

Total $kVA = \sqrt{17^2 + 9^2} = 19.2354$

Overall circuit p.f. $= \dfrac{\text{kW}}{\text{kVA}} = \dfrac{17}{19.2354} = 0.884$ leading

Overall Current $= \dfrac{19235.4}{230} = 83.63$ amp

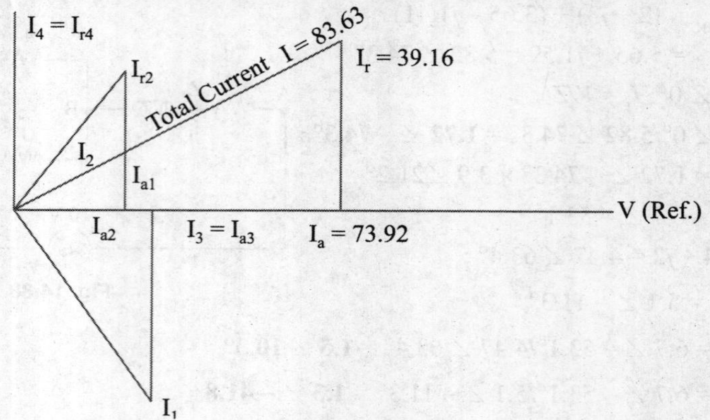

Fig. 14.39 Phasor diagram for currents corresponding to load

Example 14.33. *A three phase induction Motor delivers an output of 15 h.p. at 83 % efficiency. The motor is d (delta) connected and is supplied by 440 V, three phase, 50 Hz supply. Line current drawn by motor is 22.36 Amp. What is motor power factor ?*

It is now decided to improve the power factor to 0.95 lag by connecting three similar capacitors in delta across the supply terminals. Determine the value of the capacitance of each capacitor.

[Note : 1 h.p. = 745 watts] **(Bombay University, 2000)**

Solution. Power factor $= \dfrac{15 \times 745}{0.83 \times 1.732 \times 440 \times 22.36} = 0.79$, Lagging

$\phi = \cos^{-1} 0.79 = 37.8°$

$I_1 = I_{ph} = 22.36/1.732 = 12.91$ amp

Active Current $I_a = I_1 \cos \phi_1$

$= 12.91 \times 0.79$

$= 10.2$ amp

New Power-factor $\cos \phi_2 = 0.95, \phi_2 = 18.2°$

$I_2 = 10.2/0.95 = 10.74$ amp

Capacitive current per phase $= I_1 \sin \phi_1 - I_2 \sin \phi_2$

$= 4.563$

Capacitive reactance per phase = 440/4.563

= 96.43 ohms

Capacitance per phase = 33 μf

These have to be delta-connected

Example 14.34. *Draw admittance triangle between the terminals AB of Fig 14.40 (a) labelling its sides with appropriate values and units in case of :*

(i) *$X_L = 4$ and $X_C = 8$*

(ii) *$X_L = 10$ and $X_C = 5$* **[Bombay University 1999]**

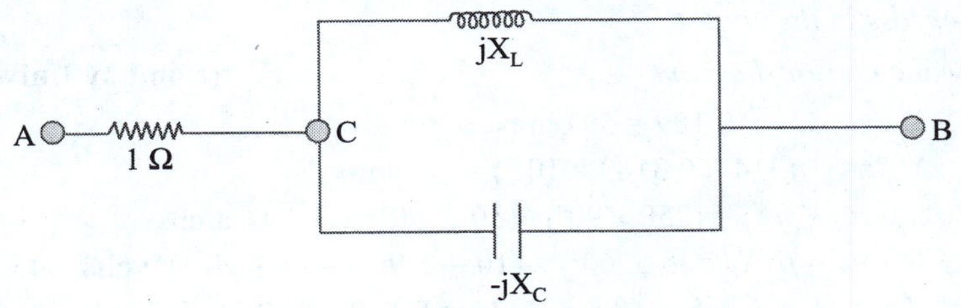

Fig. 14.40 (*a*)

Solution. (*i*)

$$X_L = 4\,\Omega, X_C = 8\,\Omega$$

$$Z_{CB} = \frac{jX_L(-jX_C)}{j(X_L - X_C)} = j8$$

$$Z_{AB} = 1 + j\,8 \text{ ohms}$$

$$Y_{AB} = 1/Z_{AB} = (1/65) - j\,(8/65) \text{ mho}$$

(*ii*)

$$X_L = 10\,\Omega, X_C = 5\,\Omega$$

$$Z_{CB} = \frac{j10 \times (-j5)}{j5} = -j10$$

$$Z_{AB} = 1 - j\,10 \text{ ohms}$$

$$Y_{AB} = (1/101) + j\,(10/101) \text{ mho}$$

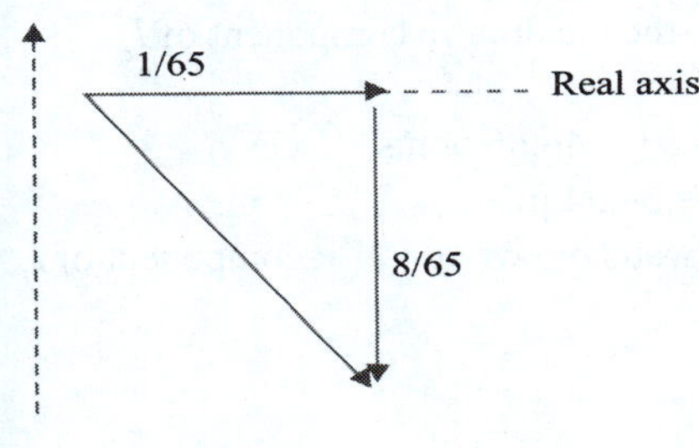

(*i*) Admittance triangle for first case

Fig. 14.40 (*b*)

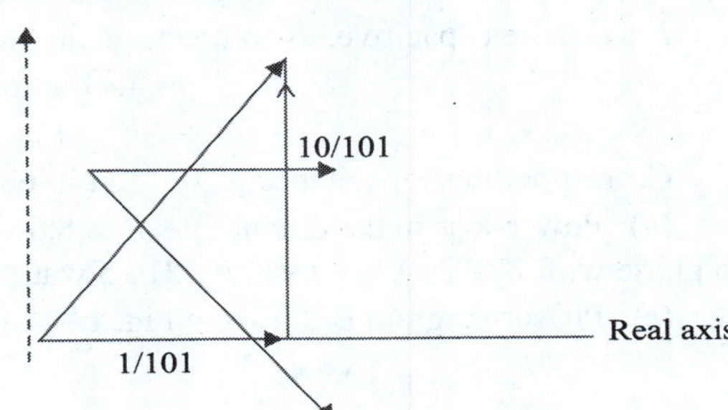

(*ii*) Admittance triangle for second case

Fig. 14.40 (*c*)

Example 14.35. *For the circuit in Fig. 14.41 (a), given that L = 0.159 H*

$$C = 0.3183 \text{ mf}$$

$$I_2 = 5 \angle 60° \text{ A}$$

$$V_1 = 250 \angle 90° \text{ volts.}$$

Find :-

(*i*) *Impedance Z_1 with its components.*

(*ii*) *Source voltage in the form of V_m cos $(\omega t + \phi)$.*

(*iii*) *Impedance Z_2 with its components so that source p.f. is unity, without adding to the circuit power loss.*

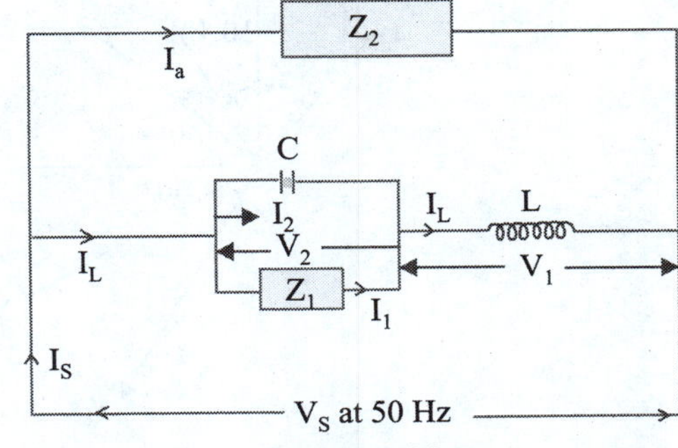

Fig. 14.41 (*a*)

(iv) *Power loss in the circuit*

(v) *Draw the phasor diagram.* **(Bombay University 1997)**

Solution. $X_L = 314 \times 0.159 = 50$ ohms

$X_C = 1/(314 \times 0.3183 \times 10^{-3}) = 10$ ohms

$I_L = V_1/jX_L = (250 \angle 90°) / (50 \angle 90°) = 5 \angle 0°$ amps

$V_2 = -jI_2X_C = (5 \angle 60°) \times (10 - \angle 90°) = 50 - \angle 30°$ volts $= 43.3 - j\,25$ volts

$I_1 = I_L - I_2 = 5 \angle 0° - 5 \angle 60° = 5 + j0 - 5\,(0.5 + j0.866)$

$= 2.5 - j4.33 = 5 \angle -60°$

(a) $Z_1 = V_2/I_1 = (50 \angle -30°) / 5 \angle -60° = 10 \angle +30°$

$= 10\,(\cos 30° + j \sin 30°) = 8.66 + j5$

(b) $V_s = V_1 + V_2 = 0\,j250 + 43.3 - j25 = 43.3 + j225$

$= 229.1 \angle 79.1°$ volts

V_s has a peak value of $(229.1 \times \sqrt{2} =)$ 324 volts

$V_s = 324 \cos (314\,t - 10.9°)$, taking V_1 as reference

or $V_s = 325 \cos (314\,t - 79\,.1°)$, taking I_L as reference.

(c) Source Current must be at unity P.f., with V_s

Component of I_L in phase with $V_s = 5 \cos 79.1° = 0.9455$ amp

Component of I_L in quadrature with V_S (and is lagging by 90°)

$= 5.00 \times \sin 79.1° = 4.91$ amp

Z_2 must carry I_a such that no power loss is there and I_S is at unity P.f. with V_s.

I_a has to be capacitive, to compensate, in magnitude, the quadrature component of I_L

$|I_a| = 4.91$ amp

$|Z_2| = V_s / |I_a| = 229.1/4.91 = 46.66$ ohms

Corresponding capacitance, $C_2 = 1/(46.66 \times 314) = 68.34\ \mu F$

(d) Power-loss in the circuit $= I_1^2 \times 8.66 = 216.5$ watts or power $= V_s \times$ component of I_L in phase with $V_s = 299.1 \times 0.9455 = 216.5$ watts

(e) Phasor diagram is drawn in Fig. 14.42 *(b)*

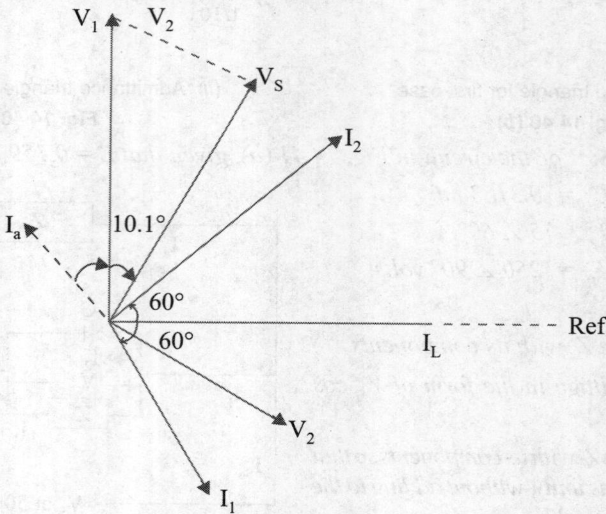

Fig. 14.42 *(b)* Phasor diagram

Tutorial Problem 14.1

1. A capacitor of 50 μF capacitance is connected in parallel with a reactor of 22 Ω resistance and 0.07 henry inductance across 200-V, 50-Hz mains. Calculate the total current taken. Draw the vector diagram in explanation. **[4.76 A lagging, 17°12′] (*City & Guilds, London*)**
2. A non-inductive resistor is connected in series with a capacitor of 100 μF capacitance across 200-V, 50-Hz mains. The p.d. measured across the resistor is 150 V. Find the value of resistance and the value of current taken from the mains if the resistor were connected in parallel-with the capacitor instead of in series. **[R = 36.1 Ω; 8.37 Ω] (*City & Guilds, London*)**
3. An impedance of $(10 + j15)$ Ω is connected in parallel with an impedance of $(6 - j8)$Ω. The total current is 15 A. Calculate the total power. **[2036 W] (*City & Guilds, London*)**
4. The load on a 250-V supply system is : 12 A at 0.8 power factor lagging ; 10 A at 0.5 power factor lagging ; 15 A at unity power factor ; 20 A at 0.6 power factor leading. Find (*i*) the total lead in kVA and (*ii*) its power factor. **[(*i*) 10.4 kVA (*ii*) 1.0] (*City & Guilds, London*)**
5. A voltage having frequency of 50 Hz and expressed by $\mathbf{V} = 200 + j100$ is applied to a circuit consisting of an impedance of $50\ \angle 30°$ Ω in parallel with a capacitance of 10 μF. Find (*a*) the reading on a ammeter connected in the supply circuit (*b*) the phase difference between the current and the voltage. **[(*a*) 4.52 (*b*) 26.6° lag] (*London University*)**
6. A voltage of $200°\ \angle 30°$ V is applied to two circuits *A* and *B* connected in parallel. The current in *A* is $20\ \angle 60°$ A and that in *B* is $40\ \angle{-30°}$ A. Find the kVA and kW in each branch circuit and the main circuit. Express the current in the main circuit in the form $A + jB$.(*City & Guilds, London*)
[$kVA_A = 4$, $kVA_B = 8$, $kV_A = 12$, $kW_A = 3.46$, $kW_B = 4$, $kW = 7.46$, $I = 44.64 - j2.68$]
(*City & Guilds, London*)
7. A coil having an impedance of $(8 + j6)$ Ω is connected across a 200-V supply. Express the current in the coil in (*i*) polar and (*ii*) rectangular co-ordinate forms.
If a capacitor having a susceptance of 0.1 S is placed in parallel with the coil, find (*iii*) the magnitude of the current taken from the supply. **[(*i*) 20 ∠ 36.8°A (*ii*) 16 – j12 A (*iii*) 17.9 A]**
(*City & Guilds, London*)
8. A coil-A of inductance 80 mH and resistance 120 Ω, is connected to a 230-V, 50 Hz single-phase supply. In parallel with it in a 16 μF capacitor in series with a 40 Ω non-inductive resistor *B*. Determine (*i*) the power factor of the combined circuit and (*ii*) the total power taken from the supply. **[(*i*) 0.945 lead (*ii*) 473 W] (*London University*)**
9. A choking coil of inductance 0.08 H and resistance 12 ohm, is connected in parallel with a capacitor of 120 μF. The combination is connected to a supply at 240 V, 50 Hz Determine the total current from the supply and its power factor. Illustrate your answers with a phasor diagram. **[3.94 A, 0.943 lag] (*London University*)**
10. A choking coil having a resistance of 20 Ω and an inductance of 0.07 henry is connected with a capacitor of 60 μF capacitance which is in series with a resistor of 50 Ω. Calculate the total current and the phase angle when this arrangement is connected to 200-V, 50 Hz mains.
[7.15 A, 24°39′ lag] (*City & Guilds, London*)
11. A coil of resistance 15 Ω and inductance 0.05 H is connected in parallel with a non-inductive resistance of 20 Ω. Find (*a*) the current in each branch (*b*) the total current (*c*) the phase angle of whole arrangement for an applied voltage of 200 V at 50 Hz. **[9.22 A ; 10A ; 22.1°]**
12. A sinusoidal 50-Hz voltage of 200 V (r.m.s) supplies the following three circuits which are in parallel : (*a*) a coil of inductance 0.03 H and resistance 3 Ω (*b*) a capacitor of 400 μF in series with a resistance of 100 Ω (*c*) a coil of inductance 0.02 H and resistance 7 Ω in series with a 300 μF capacitor. Find the total current supplied and draw a complete vector diagram.
[29.4 A] (*Sheffield Univ. U.K.*)
13. A 50-Hz, 250-V single-phase power line has the following loads placed across it in parallel : 4 kW at a p.f. of 0.8 lagging ; 6 kVA at a p.f. of 0.6 lagging; 5 kVA which includes 1.2 kVAR leading. Determine the overall p.f. of the system and the capacitance of the capacitor which, if connected across the mains would restore the power factor to unity. **[0.844 lag ; 336 μF]**

14. Define the terms admittance, conductance and susceptance with reference to alternating current circuits. Calculate their respective values for a circuit consisting of resistance of 20Ω, in series with an inductance of 0.07 H when the frequency is 50 Hz. **[0.336 S, 0.0226 S, 0.0248 S]** ***(City & Guilds, London)***
15. Explain the terms admittance, conductance, susceptance as applied to a.c. circuits. One branch *A*, of a parallel circuit consists of a coil, the resistance and inductance of which are 30 Ω and 0.1 *H* respectively. The other branch *B*, consists of a 100 μF capacitor in series with a 20 Ω resistor. If the combination is connected 240-V, Hz mains, calculate (*i*) the line current and (*ii*) the power. Draw to scale a vector diagram of the supply current and the branch-circuit currents. **[(*i*) 7.38 A (*ii*) 1740 W] *(City & Guilds, London)***
16. Find the value of capacitance which when placed in parallel with a coil of resistance 22 Ω and inductance of 0.07 H, will make it resonate on a 50-Hz circuit. **[72.33 μF] *(City & Guilds, London)***
17. A parallel circuit has two branches. Branch A consists of a coil of inductance 0.2 H and a resistance of 15 Ω ; branch B consists of a 30 mF capacitor in series with a 10 Ω resistor. The circuit so formed is connected to a 230-V, 50-Hz supply. Calculate (*a*) current in each branch (*b*) line current and its power factor (*c*) the constants of the simplest series circuit which will take the same current at the same power factor as taken by the two branches in parallel. **[3.57 A, 2.16 A ; 1.67 A, 0.616 lag, 8.48 Ω, 0.345 H]**
18. A coil having a resistance of 45 Ω and an inductance of 0.4 *H* is connected in parallel with a capacitor having a capacitance of 20 μ*F* across a 230-V, 50-Hz system. Calculate (*a*) the current taken from the supply (*b*) the power factor of the combination and (*c*) the total energy absorbed in 3 hours. **[(*a*) 0.615 (*b*) 0.951 (*c*) 0.402 kWh] *(London University)***
19. A series circuit consists of a resistance of 10 Ω and reactance of 5 Ω. Find the equivalent value of conductance and susceptance in parallel. **[0.08 S, 0.04 S]**
20. An alternating current passes through a non-inductive resistance *R* and an inductance *L* in series. Find the value of the non-inductive resistance which can be shunted across the inductance without altering the value of the main current. **[$\omega^2L^2/2R$] *(Elec. Meas. London Univ.)***
21. A coil (*A*) having an inductance of 0.2 *H* and resistance of 3.5 Ω is connected in parallel with another coil (*B*) having an inductance of 0.01 H and a resistance of 5 Ω. Calculate (*i*) the current and (*ii*) the power which these coils would take from a 100-V supply system having a frequency of 50-Hz. Calculate also (*iii*) the resistance and (*iv*) the inductance of a single coil which would take the same current and power. **[(*i*) 29.9 A (*ii*) 2116 W (*ii*) 2.365 Ω (*iv*) 0.00752 H] *(London Univ.)***
22. Two coils, one (*A*) having *R* = 5 Ω, *L* = 0.031 *H* and the other (*B*) having *R* = 7 Ω ; *L* = 0.023 *H*, are connected in parallel to an a.c. supply at 200 *V*, 50 Hz. Determine (*i*) the current taken by each coil and also (*ii*) the resistance and (*iii*) the inductance of a single coil which will take the same total current at the same power factor as the two coils in parallel. **[(*i*) I_A = 18.28 A, I_B = 19.9 A (*ii*) 3.12 Ω (*iii*) 0.0137 H] *(London Univ.)***
23. Two coils are connected in parallel across 200-V, 50-Hz mains. One coil takes 0.8 kW and 1.5 kVA and the other coil takes 1.0 kW and 0.6 kVAR. Calculate (*i*) the resistance and (*ii*) the reactance of a single coil which would take the same current and power as the original circuit. **[(*i*) 10.65 Ω (*ii*) 11.08 Ω] *(City & Guilds, London)***
24. An a.c. circuit consists of two parallel branches, one (*A*) consisting of a coil, for which *R* = 20 Ω and *L* = 0.1 H and the other (*B*) consisting of a 40-Ω non-inductive resistor in series with 60-μF capacitor. Calculate (*i*) the current in each branch (*ii*) the line current (*iii*) the power, when the circuit is connected to 230-V mains having a frequency of 50 Hz. Calculate also (*iv*) the resistance and (*b*) the inductance of a single coil which will take the same current and power from the supply. **[(*i*) 6.15 A, 3.46 A (*ii*) 5.89 (*iii*) 1235 W (*iv*) 35.7 Ω (*b*) 0.0509 H] *(London Univ.)***
25. One branch (*A*) of a parallel circuit, connected to 230-V, 50-Hz mains consists of an inductive coil (*L* = 0.15 H, *R* = 40 Ω) and the other branch (*B*) consists of a capacitor (*C* = 50 μF) in series with a 45 Ω resistor. Determine (*i*) the power taken (*ii*) the resistance and (*iii*) the reactance of the equivalent series circuit. **[(*i*) 946 W (*ii*) 55.4 Ω (*iii*) 4.6 Ω] *(London Univ.)***

14.9. Resonance in Parallel Circuits

We will consider the practical case of a coil in parallel with a capacitor, as shown in Fig. 14.43. Such a circuit is said to be in electrical resonance when the reactive (or wattless) component of line current becomes zero. The frequency at which this happens is known as ***resonant*** frequency.

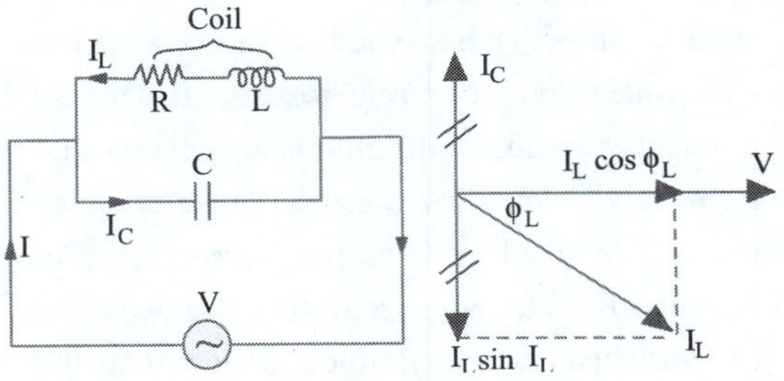

The vector diagram for this circuit is shown in Fig. 14.43 (*b*).

Fig. 14.43

Net reactive or wattless component

$= I_C - I_L \sin \phi_L$

As at resonance, its value is zero, hence $I_C - I_L \sin \phi_L = 0$ or $I_L \sin \phi_L = I_C$

Now, $I_L = V/Z_L$; $\sin \phi Z_L = X_L$ and $I_C = V/X_C$

Hence, condition for resonance becomes

$$\frac{V}{Z_L} \times \frac{X_L}{Z_L} = \frac{V}{X_C} \quad \text{or} \quad X_L \times X_C = Z_L$$

Now, $X_L = \omega L, X_C = \dfrac{1}{\omega C}$

$$\therefore \quad \frac{\omega L}{\omega C} = Z_L \quad \text{or} \quad \frac{L}{C} = Z_L \;. \qquad \text{.. (i)}$$

$$\text{or} \quad \frac{L}{C} = R^2 + X_L^2 = R^2 + (2\pi f_0 L)^2$$

$$\text{or} \quad (2\pi f_0 L)^2 = \frac{L}{C} - R^2$$

$$\text{or} \quad 2\pi f_0 = \sqrt{\frac{1}{LC} - \frac{R^2}{L^2}} \quad \text{or} \quad f_0 = \frac{1}{2\pi}\sqrt{\frac{1}{LC} - \frac{R^2}{L^2}}$$

This is the resonant frequency and is given in Hz, R is in ohm, L is the henry and C is the farad. If R is the negligible, then $f_0 = \dfrac{1}{2\pi\sqrt{(LC)}}$

... same as for series resonance

Current at Resonance

Since wattless component of the current is zero, the circuit current is $I = I_L \cos \phi_L = \dfrac{V}{Z_L} \cdot \dfrac{R}{Z_L}$

or $I = \dfrac{VR}{Z_L}$.

Putting the value of $Z_L = L/C$ from (*i*) above, we get $I = \dfrac{VR}{L/C} = \dfrac{V}{L/CR}$

The denominator L/CR is known as the ***equivalent*** or ***dynamic impedance*** of the parallel circuit at resonance. It should be noted that impedance is 'resistive' only. Since current is minimum at resonance, L/CR must, therefore, represent the maximum impedance of the circuit.

In fact, parallel resonance is a condition of maximum impedance or minimum admittance.

Current at resonance is minimum, hence such a circuit (when used in radio work) is sometimes known as ***rejector*** circuit because it rejects (or takes minimum current of) that frequency to which it resonates. This resonance is often referred to as current resonance also because the current circulating ***between*** the two branches is many times greater than the line current taken from the supply.

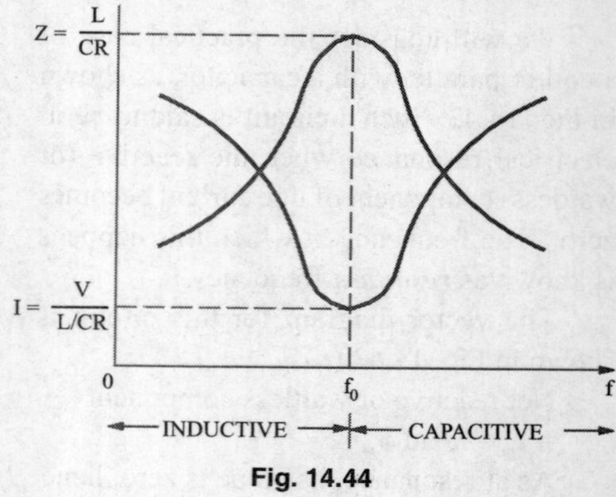

Fig. 14.44

The phenomenon of parallel resonance is of great practical importance because it forms the basis of tuned circuits in Electronics.

The variations of impedance and current with frequency are shown in Fig. 14.44. As seen, at resonant frequency, impedance is maximum and equals L/CR. Consequently, current at resonance is minimum and is $= V/(L/CR)$. At off-resonance frequencies, impedance decreases and, as a result, current increases as shown.

Alternative Treatment

$$\mathbf{Y}_1 = \frac{1}{R + jX_L} = \frac{R}{R^2 + X_L^2} - j\frac{X_L}{R^2 + X_L^2}; \quad \mathbf{Y}_2 = \frac{1}{-jX_C} = \frac{j}{X_C}$$

$$\mathbf{Y} = \frac{R}{R^2 + X_L^2} + j\left(\frac{1}{X_C} - \frac{X_L}{R^2 + X_L^2}\right)$$

Now, circuit would be in resonance when j-component of the complex admittance is zero

i.e. when $\dfrac{1}{X_C} - \dfrac{X_L}{R^2 + X_L^2} = 0$ or $\dfrac{X_L}{R^2 + X_L^2} = \dfrac{1}{X_C}$

or $X_L X_C = R^2 + X_L^2 = Z^2$ **—as before**

Talking in terms of susceptance, the above relations can be put as under :

Inductive susceptance $B_L = \dfrac{X_L}{R^2 + X_L^2}$; capacitive susceptance $B_C = \dfrac{1}{X_C}$

Net susceptance $B = (B_C - B_L)$ $\therefore$ $Y = G + j(B_C - B_L) = G + jB$.

The parallel circuit is said to be in resonance when $B = 0$.

$$\therefore \quad B_C - B_L = 0 \quad \text{or} \quad \frac{1}{X_C} = \frac{X_L}{R^2 + X_L^2}$$

The rest procedure is the same as above. It may be noted that at resonance, the admittance equals the conductance.

14.10. Graphic Representation of Parallel Resonance

We will now discuss the effect of variation of frequency on the susceptance of the two parallel branches. The variations are shown in Fig. 14.45.

***(i)* Inductive susceptance ;**

$B_L = -1/X_L = -1/2\pi f L$

It is inversely proportional to the frequency of the applied voltage. Hence, it is represented by a rectangular hyperbola drawn in the fourth quadrant ($\therefore$ it is assumed negative).

***(ii)* Capacitive susceptance ;**

$B_C = 1/X_C = \omega C = 2\pi f C$

It increases with increase in the frequency of the applied voltage. Hence, it is represented by a straight line drawn in the first quadrant (it is assumed positive).

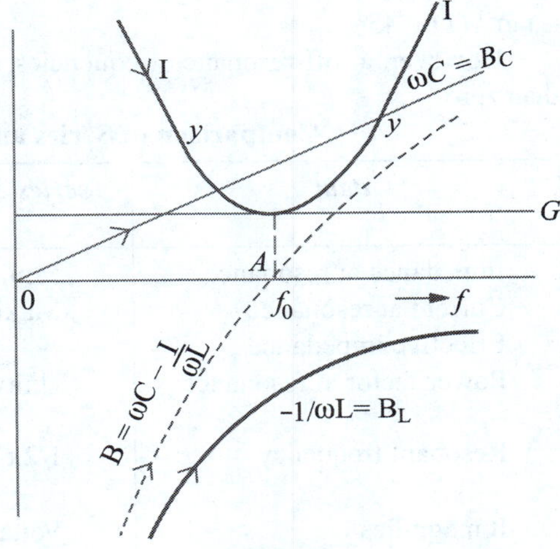

Fig. 14.45

***(iii)* Net Susceptance B**

It is the difference of the two susceptances and is represented by the dotted hyperbola. At point *A*, net susceptance is zero, hence admittance is minimum (and equal to *G*). So at point *A*, line current is minimum.

Obviously, below resonant frequency (corresponding to point *A*), inductive susceptance predominates, hence line current lags behind the applied voltage. But for frequencies above the resonant frequency, capacitive susceptance predominates, hence line current leads the applied voltage.

14.11. Points to Remember

Following points about parallel resonance should be noted and compared with those about series resonance. At resonance.

1. Net susceptance is zero *i.e.* $1/X_C = X_L/Z^2$ or $X_L \times X_C = Z^2$ or $L/C = Z^2$
2. The admittance equals conductance
3. Reactive or wattless component of line current is zero.
4. Dynamic impedance = L/CR ohm.
5. Line current at resonance is minimum and = $\dfrac{V}{L/CR}$ but is in phase with the applied voltage.
6. Power factor of the circuit is unity.

14.12. Bandwidth of a Parallel Resonant Circuit

The bandwidth of a parallel circuit is defined in the same way as that for a series circuit. This circuit also has upper and lower half-power frequencies where power dissipated is half of that at resonant frequency.

At bandwidth frequencies, the net susceptance *B* equals the conductance. Hence, at f_2,

$B = B_{C2} - B_{L2} = G$. At f_1, $B = B_{L1} - B_{C1} = G$. Hence, $Y = \sqrt{G^2 + B^2} = \sqrt{2}.G$ and $\phi = \tan^{-1}(B/G)$ $= \tan^{-1}(1) = 45°$.

However, at off-resonance frequencies, $Y > G$ and $B_C \neq B_L$ and the phase angle is greater than zero.

Comparison of Series and Parallel Resonant Circuits

item	***series circuit (R–L–C)***	***parallel circuit (R–L and C)***
Impedance at resonance	Minimum	Maximum
Current at resonance	Maximum = V/R	Minimum = $V/(L/CR)$
Effective impedance	R	L/CR
Power factor at resonance	Unity	Unity
Resonant frequency	$1/2\pi\sqrt{(LC)}$	$\frac{1}{2\pi}\sqrt{\left(\frac{1}{LC} - \frac{R^2}{L^2}\right)}$
It magnifies	Voltage	Current
Magnification is	$\omega L/R$	$\omega L/R$

14.13. Q-factor of a Parallel Circuit

It is defined as ***the ratio of the current circulating between its two branches to the line current drawn from the supply or simply, as the current magnification.*** As seen from Fig. 14.46, the circulating current between capacitor and coil branches is I_C.

Hence Q-factor $= I_C/I$

Now $\quad I_C = V/X_C = V/(1/\omega C) = \omega CV$

and $\quad I_C = V/(L/CR)$

$$\therefore Q-\text{factor} = \omega CV \div \frac{V}{L/CR} = \frac{\omega L}{R} = \frac{2\pi f_0 L}{R}$$

$= \tan\phi$ (same as for series circuit)

where ϕ is the power factor angle of the *coil*.

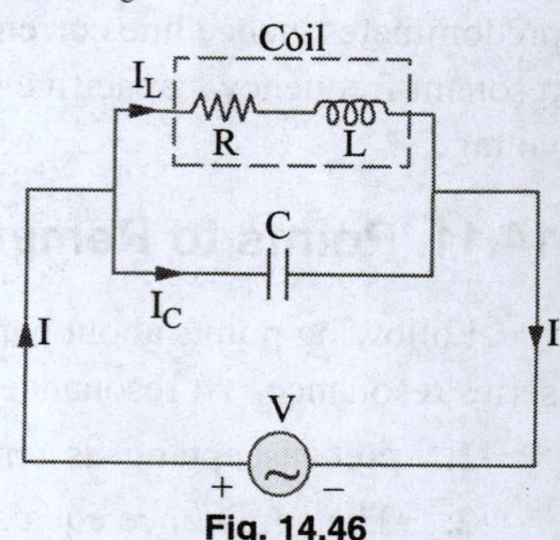

Fig. 14.46

Now, resonant frequency when R is negligible is,

$$f_0 = \frac{1}{2\pi\sqrt{(LC)}}$$

Putting this value above, we get, Q-factor $= \frac{2\pi f_0 L}{R} = \frac{1}{2\pi\sqrt{(LC)}} = \frac{1}{R}\sqrt{\frac{L}{C}}$

It should be noted that in series circuits, Q-factor gives the voltage magnification, whereas in parallel circuits, it gives the current magnification.

Again, $\quad Q = 2\pi \dfrac{\text{maximum stored energy}}{\text{energy dissipated/cycle}}$

Example 14.36. *A capacitor is connected in parallel with a coil having L = 5.52 mH and R = 10 Ω, to a 100-V, 50-Hz supply. Calculate the value of the capacitance for which the current taken from the supply is in phase with voltage.* **(Elect. Machines, A.M.I.E. Sec B, 1992)**

Solution. At resonance, $L/C = Z^2$ or $C = L/Z^2$

$X_L = 2\pi \times 50 \times 5.52 \times 10^{-3} = 1.734\ \Omega$, $Z^2 = 10^2 + 1.734^2$, $Z = 10.1\ \Omega$

$C = 5.52 \times 10^{-3}/10.1 =$ **54.6 μF**

Example 14.37. *Calculate the impedance of the parallel-turned circuit as shown in Fig. 14.47 at a frequency of 500 kHz and for bandwidth of operation equal to 20 kHz. The resistance of the coil is 5 Ω.* **(Circuit and Field Theory, A.M.I.E. Sec. B, 1993)**

Solution. At resonance, circuit impedance is L/CR. We have been given the value of R but that of L and C has to be found from the given data.

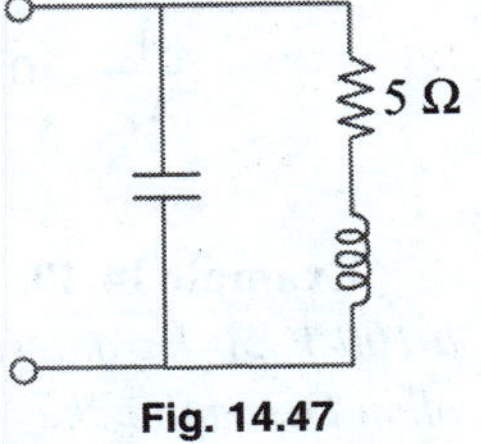

Fig. 14.47

$$BW = \frac{R}{2\pi L}, 20 \times 10^3 = \frac{5}{2\pi \times L} \quad \text{or} \quad L = 39\ \mu\text{H}$$

$$f_0 = \frac{1}{2\pi}\sqrt{\frac{1}{LC} - \frac{R^2}{L^2}} = \frac{1}{2\pi}\sqrt{\frac{1}{39 \times 10^{-6} C} - \frac{5^2}{(39 \times 10^{-6})^2}}$$

$\therefore \quad C = 2.6 \times 10^{-9}$ F, $Z = L/CR = 39 \times 10^{-6}/2.6 \times 10^{-9} \times 5 = \mathbf{3 \times 10^3\ \Omega}$

Example 14.38. *An inductive circuit of resistance 2 ohm and inductance 0.01 H is connected to a 250-V, 50-Hz supply. What capacitance placed in parallel will produce resonance ? Find the total current taken from the supply and the current in the branch circuits.*

(Elect. Engineering, Kerala Univ.)

Solution. As seen from Art. 14.9, at resonance $C = L/Z^2$

Now, $\quad R = 2\ \Omega, X_L = 314 \times 0.01 = 3.14\ \Omega\ ; Z = \sqrt{2^2 + 3.14^2} = 3.74\ \Omega$

$C = 0.01/3.74^2 = 714 \times 10^{-6}$ F $= \mathbf{714\ \mu F}$; $I_{RL} = 250/3.74 = \mathbf{66.83\ A}$

$\tan \phi_L = 3.14/2 = 1.57\ ; \phi_L = \tan^{-1}(1.57) = 57.5°$

Hence, current in R-L branch lags the applied voltage by 57.5°

$$\therefore \quad I_C = \frac{V}{X_C} = \frac{V}{1/\omega C} = \omega VC = 250 \times 314 \times 714 \times 10^{-6} = \mathbf{56.1\ A}$$

This current leads the applied voltage by 90°.
Total current taken from the supply under resonant condition is

$$I = I_{RL} \cos \phi_L = 66.83 \cos 57.5° = 66.83 \times 0.5373 = \mathbf{35.9\ A} \left(\text{or } \boldsymbol{I = \frac{V}{L/CR}}\right)$$

Example 14.39. *Find active and reactive components of the current taken by a series circuit consisting of a coil of inductance 0.1 henry and resistance 8 Ω and a capacitor of 120 μF connected to a 240-V, 50-Hz supply mains. Find the value of the capacitor that has to be connected in parallel with the above series circuit so that the p.f. of the entire circuit is unity.*

(Elect. Technology, Mysore Univ.)

Solution. $X_L = 2\pi \times 50 \times 0.1 = 31.4\ \Omega, X_C = 1/\omega C = 1/2\pi \times 50 \times 120 \times 10^{-6} = 26.5\ \Omega$

$X = X_L - X_C = 31.4 - 26.5 = 5\ \Omega,\ Z = \sqrt{(8^2 + 5^2)} = 9.43\ \Omega\ ; I = V/Z = 240/9.43 = 25.45$ A

$\cos \phi = R/Z = 8/9.43 = 0.848, \sin \phi = X/Z = 5/9.43 = 0.53$

active component of current $= I \cos \phi = 25.45 \times 0.848 = 21.58$ A

reactive component of current $= I \sin \phi = 25.45 \times 0.53 = 13.49$ A

Let a capacitor of capacitance C be joined in parallel across the circuit.

$$\mathbf{Z}_1 = R + jX = 8 + j5\ ; \mathbf{Z}_2 = -jX_C\ ;$$

$$\mathbf{Y} = \mathbf{Y}_1 + \mathbf{Y}_2 = \frac{1}{\mathbf{Z}_1} + \frac{1}{\mathbf{Z}_2} = \frac{1}{8 + j5} + \frac{1}{-jX_C}$$

$$= \frac{8-j5}{89} + \frac{j}{X_C} = 0.0899 - j0.056 + \frac{j}{X_C} = 0.0899 + j(1/X_C - 0.056)$$

For p.f. to be unit , the j-component of **Y** must be zero.

$$\therefore \quad \frac{1}{X_C} - 0.056 = 0 \quad \text{or} \quad 1/X_C = 0.056 \quad \text{or} \quad \omega C = 0.056 \quad \text{or} \quad 2\pi \times 50C = 0.056$$

$$\therefore \quad C = 0.056/100\pi = 180 \times 10^{-6}\ \text{F} = \mathbf{180\ \mu F}$$

Example 14.40. *Impedances Z_2 and Z_3 in parallel are in series with an impedance Z_1 across a 100-V, 50-Hz a.c. supply. $Z_1 = (6.25 + j\ 1.25)$ ohm ; $Z_2 = (5 + j0)$ ohm and $Z_3 = (5 - j\ X_C)$ ohm. Determine the value of capacitance of X_C such that the total current of the circuit will be in phase with the total voltage. When is then the circuit current and power ?*

(Elect. Engg-I, Nagpur Univ, 1992)

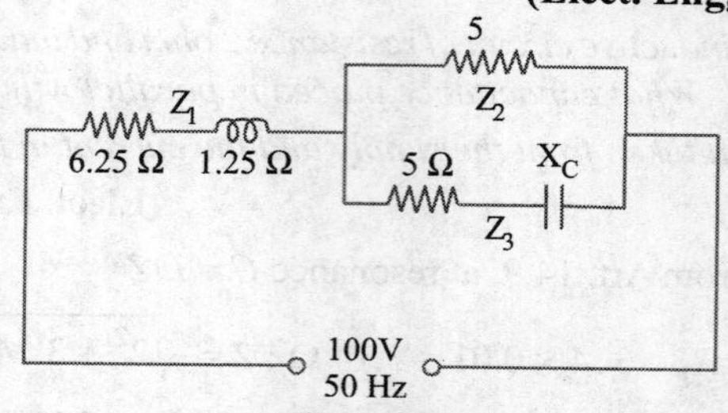

Fig. 14.48

Solution. $\mathbf{Z}_{23} = \frac{5(5 - jX_c)}{(10 - jX_C)}$, for the circuit in Fig. 14.48.

$$= \frac{25 - j5X_C}{(10 - jX_C)} \times \frac{10 + jX_C}{10 + jX_C} = \frac{250 + 5X_C^2}{100 + X_C^2} - j\frac{25X_C}{100 + X_C^2}$$

$$\mathbf{Z} = 6.25 + j\ 1.25 + \frac{250 + 5X_C^2}{100 + X_C^2} - j\frac{25X_C}{100 + X_C^2}$$

$$= \left(6.25 + \frac{250 + 5X_C^2}{100 + X_C^2}\right) - j\left(\frac{25X_C}{100 + X_C^2} - \frac{5}{4}\right)$$

Power factor will be unity or circuit current will be in phase with circuit voltage if the j term in the above equation is zero.

$$\therefore \quad \left(\frac{25X_C}{100 + X_C^2} - \frac{5}{4}\right) = 0 \text{ or } X_C = 10 \quad \therefore \quad 1/\omega C = 10 \text{ or } C = 1/314 \times 10 = \mathbf{318\ \mu F}$$

Substituting the value of $X_C = 10\ \Omega$ above, we get

$$Z = 10 - j0 = 10\angle 0° \text{ and } I = 100/10 = \mathbf{10\ A}\ ; \text{ Power} = I^2R = 10^2 \times 10 = \mathbf{1000\ W}$$

Example 14.41. *In the circuit given below, if the value of $R = \sqrt{L/C}$, then prove that the impedance of the entire circuit is equal to R only and is independent of the frequency of supply. Find the value of impedance for L = 0.02 H and C = 100 μF.*

(Communication System, Hyderabad Univ. 1991)

Solution. The impedance of the circuit of Fig. 14.49 is

$$\mathbf{Z} = \frac{(R + j\omega L)(R - j/\omega C)}{2R + j(\omega L - 1/\omega C)} = \frac{R^2 + (L/C) + jR(\omega L - 1/\omega C)}{2R + j(\omega L - 1/\omega C)}$$

If $R^2 = L/C$ or $R = \sqrt{L/C}$, then

$$\mathbf{Z} = \frac{R^2 + R^2 + jR(\omega L - 1/\omega C)}{2R + j(\omega L - 1/\omega C)}$$

$$= R\left[\frac{2R + j(\omega L - 1/\omega C)}{2R + j(\omega L - 1/\omega C)}\right] \quad \text{or} \quad \mathbf{Z = R}$$

Now, $R = \sqrt{\frac{L}{C}} = \sqrt{\frac{0.02}{100 \times 10^{-6}}} = \mathbf{14.14\ \Omega}$

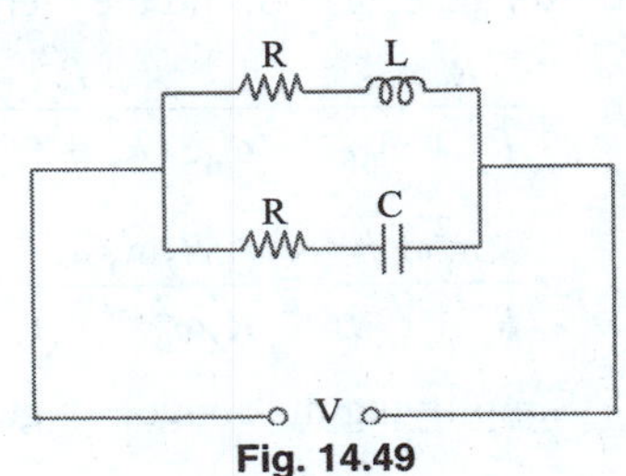

Fig. 14.49

Example 14.42. *Derive an expression for the resonant frequency of the parallel circuit shown in Fig. 14.50.* **(Electrical Circuit, Nagpur Univ, 1993)**

Solution. As stated in Art. 14.9 for resonance of a parallel circuit, total circuit susceptance should be zero. Susceptance of the *R-L* branch is

$$B_1 = -\frac{X_L}{R_1^2 + X_L^2}$$

Similarly, susceptance of the *R-C* branch is

$$B_2 = \frac{X_C}{R_2^2 + X_C^2}$$

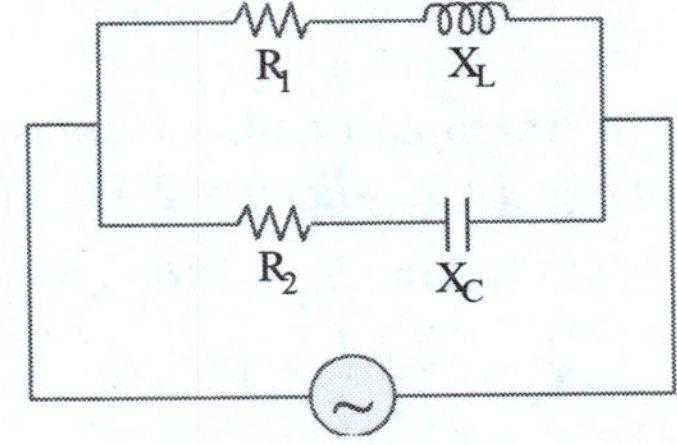

Fig. 14.50

Net susceptance is $B = -B_1 + B_2$

For resonance $B = 0$ or $0 = -B_1 + B_2 \quad \therefore \; B_1 = B_2$

$$\text{or} \quad \frac{X_L}{R_1^2 + X_L^2} = \frac{X_C}{R_2^2 + X_C^2} \quad \text{or} \quad X_L(R_2^2 + X_C^2) = X_C(R_1^2 + X_L^2)$$

$$2\pi f L\left[R_2^2 + \left(\frac{1}{2\pi f C}\right)^2\right] = \frac{1}{2\pi f C}[R_1^2 + (2\pi f L)^2];\; 4\pi^2 f^2 LC\left[R_2^2\left(\frac{1}{2\pi f C}\right)^2\right] = [R_1^2 + (2\pi f L)^2]$$

$$\therefore \quad 4\pi^2 f^2 LCR_2^2 + \frac{L}{C} = R_1^2 + 4\pi^2 f^2 L^2;\; 4\pi^2 f^2 [L(L - CR_2^2)] = \frac{L}{C} - R_1^2$$

$$\therefore f_0 = \frac{1}{2\pi}\sqrt{\left(\frac{L/C - R_1^2}{L(L - CR_2^2)}\right)} \quad \therefore \; f_0 = \frac{1}{2\pi}\sqrt{\left(\frac{L - CR_1^2}{LC(L - CR_2^2)}\right)};\; \omega_0 = \frac{1}{\sqrt{LC}}\sqrt{\left(\frac{L - CR_1^2}{L - CR_2^2}\right)}$$

Note. If both R_1 and R_2 are negligible, then $f_0 = \dfrac{1}{2\pi\sqrt{LC}}$

Example 14.43. *Calculate the resonant frequency of the network shown in Fig. 14.51.*

Solution. Total impedance of the network between terminals *A* and *B* is

$$Z_{AB} = (R_1 \| jX_L) + [R_2 \| (-jX_C)] = \frac{jR_1X_L}{R_1 + jX_L} + \frac{R_2(-jX_C)}{R_2 - jX_C} = \frac{jR_1\omega L}{R_1 + j\omega L} - \frac{jR_2/\omega C}{R_2 - j/\omega C}$$

$$= \frac{R_1\omega^2L^2}{R_1^2+\omega^2L^2} + \frac{R_2}{\omega C(R_2^2+1/\omega^2C^2)} + j\left[\frac{R_1^2\omega L}{R_1^2+\omega^2L^2} - \frac{R_2^2}{\omega C(R_2^2+1/\omega^2C^2)}\right]$$

At resonance, $\omega = \omega_0$ and the j term of Z_{AB} is zero.

$$\therefore \frac{R_1^2\omega_0 L}{R_1^2+\omega_0^2L^2} - \frac{R_2^2}{\omega_0 C(R_2^2+1/\omega_0^2C^2)} = 0$$

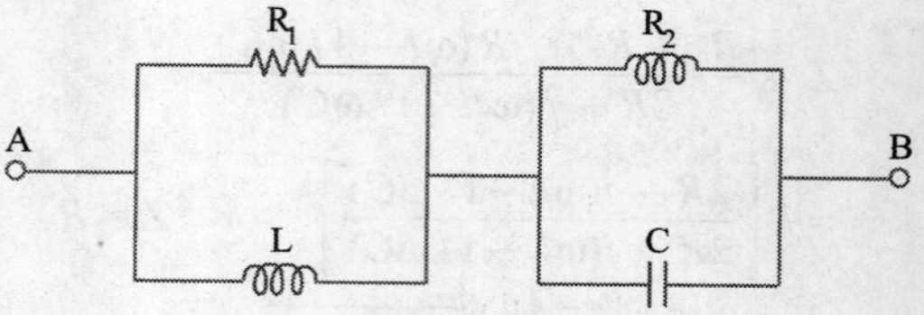

$$\text{or} \quad \frac{R_1^2\omega_0 L}{R_1^2+\omega_0^2L^2} = \frac{R_2^2\omega_0 C}{R_2^2\omega_0^2C^2+1}$$

Simplifying the above, we get

$$\omega_0^{\,2} = \frac{G_2^2 - C/L}{LC(G_1^2 - C/L)} \text{ where } G_1 = \frac{1}{R_1} \text{ and } G_2 = \frac{1}{R_2}$$

Fig. 14.51

The resonant frequency of the given network in Hz is

$$f_0 = \frac{\omega_0}{2\pi} = \frac{1}{2\pi}\sqrt{\left[\frac{G_2^2 - C/L}{LC\,(G_1^2 - C/L)}\right]}$$

Example 14.44. *Compute the value of C which results in resonance for the circuit shown in Fig. 14.52 when f = 2500/π Hz.*

Solution. $\mathbf{Y}_1 = 1/(6 + j8)$

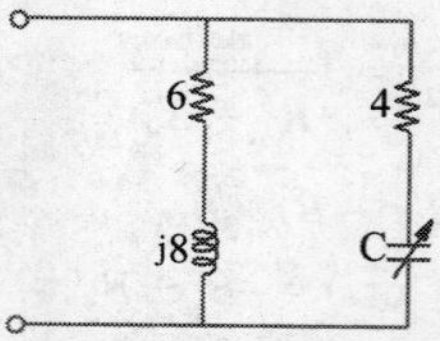

$\mathbf{Y}_2 = 1/(4 - jX_C)$

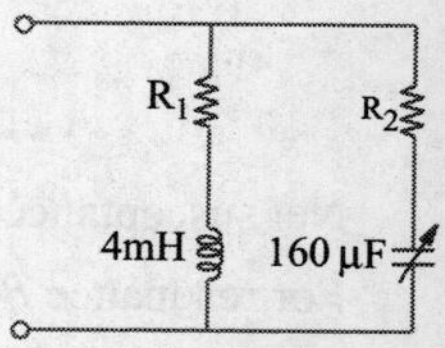

$$\mathbf{Y} = \mathbf{Y}_1 + \mathbf{Y}_2 = \frac{1}{6+j8} + \frac{1}{4 - jX_C}$$

$$= \left(0.06 + \frac{4}{16+X_C^2}\right) + j\left(\frac{X_C}{16+X_C^2} - 0.08\right)$$

Fig. 14.52 Fig. 14.53

For resonance, j part of admittance is zero, *i.e.* the complex admittance is real number.

$\therefore \quad X_C/(16 + X_C^2) - 0.08 = 0 \quad \text{or} \quad 0.08\,X_C^2 - X_C + 1.28 = 0$

$\therefore \quad X_C = 11.05 \text{ or } 1.45 \quad \therefore \quad 1/\omega C = 11.05 \text{ or } 1.45$

(*i*) $1/5000C = 11.05$ or $C =$ **18 μF** (*ii*) $1/5000\,C = 1.45$ or $C =$ **138 μF**

Example 14.45. *Find the values of R_1 and R_2 which will make the circuit of Fig. 14.53 resonate at all frequencies.*

Solution. As seen from Example 14.42, the resonant frequency of the given circuit is

$$\omega_0 = \frac{1}{\sqrt{LC}}\sqrt{\left(\frac{L - CR_1^2}{L - CR_2^2}\right)}$$

Now, ω_0 can assume any value provided $R_1^2 = R_2^2 = L/C$.

In the present case, $L/C = 4 \times 10^{-3}/60 \times 10^{-6} = 25$. Hence, $R_1 = R_2 = \sqrt{25} =$ **5 ohm.**

Tutorial Problem 14.2

1. A resistance of 20 W and a coil of inductance 31.8 mH and negligible resistance are connected in parallel across 230 V, 50 Hz supply. Find (*i*) The line current (*ii*) power factor and (*iii*) The power consumed by the circuit. **[(*i*) 25.73 A (*ii*) 0.44 T lag (*iii*) 246 W]** ***(F. E. Pune Univ.)***
2. Two impedances $Z_1 = (150 + j157)$ ohm and $Z_2 = (100 + j\ 110)$ ohm are connected in parallel across a 220-V, 50-Hz supply. Find the total current and its power factor. **[24 $\angle -47°$A ; 0.68 (lag)]** ***(Elect. Engg. & Electronics Bangalore Univ.)***
3. Two impedances $(14 + j5)\Omega$ and $(18 + j10)\ \Omega$ are connected in parallel across a 200-V, 50-Hz supply. Determine (*a*) the admittance of each branch and of the entire circuit ; (*b*) the total current, power, and power factor and (*c*) the capacitance which when connected in parallel with the original circuit will make the resultant power factor unity. **[(*a*) (0.0634 – j0.0226), (0.0424 – j0.023) (0.1058 – j0.0462 S) (*b*) 23.1 A, 4.232 kW, 0.915 (*c*) 147 μF]**
4. A parallel circuit consists of two branches *A* and *B*. Branch A has a resistance of 10 Ω and an inductance of 0.1 H in series. Branch *B* has a resistance of 20 Ω and a capacitance of 100 μF in series. The circuit is connected to a single-phase supply of 250 V, 50 Hz. Calculate the magnitude and the phase angle of the current taken from the supply. Verify your answer by measurement from a phasor diagram drawn to scale. **[6.05 $\angle -15.2°$]** ***(F. E. Pune Univ.)***
5. Two circuits, the impedances of which are given by $Z_1 = (10 + j15)\ \Omega$ and $Z_2 = (6 - j8)\ \Omega$ are connected in parallel. If the total current supplied is 15 A, what is the power taken by each branch ? **[737 W ; 1430 W]** ***(Elect. Engg. A.M.A.E. S.I.)***
6. A voltage of 240 V is applied to a pure resistor, a pure capacitor, and an inductor in parallel. The resultant current is 2.3 A, while the component currents are 1.5, 2.0 and 1.1 A respectively. Find the resultant power factor and the power factor of the inductor. **[0.88 ; 0.5]**
7. Two parallel circuits comprise respectively (*i*) a coil of resistance 20 Ω and inductance 0.07 H and (*ii*) a capacitance of 60 μF in series with a resistance of 50 Ω. Calculate the current in the mains and the power factor of the arrangement when connected across a 200-V, 50-Hz supply. **[7.05 A ; 0.907 lag]** ***(Elect. Engg. & Electronics, Bangalore Univ.)***
8. Two circuits having the same numerical ohmic impedances are joined in parallel. The power factor of one circuit is 0.8 lag and that of other 0.6 lag. Find the power factor of the whole circuit. **[0.707]** ***(Elect. Engg. Pune Univ.)***
9. How is a current of 10 A shared by three circuits in parallel, the impedances of which are $(2 - j5)\ \Omega$, $(6 + j3)\Omega$ and $(3 + j4)\ \Omega$. **[5.68 A ; 4.57 A, 6.12 A]**
10. A piece of equipment consumes 2,000 W when supplied with 110 V and takes a lagging current of 25 A. Determine the equivalent series resistance and reactance of the equipment. If a capacitor is connected in parallel with the equipment to make the power factor unity, find its capacitance. The supply frequency is 100 Hz. **[3.2 Ω, 3.02 Ω, 248 μF]** ***(Sheffield Univ. U.K.)***
11. A circuit, consisting of a capacitor in series with a resistance of 10 Ω, is connected in parallel with a coil having $L = 55.2$ mH and $R = 10\ \Omega$, to a 100-V, 50-Hz supply. Calculate the value of the capacitance for which the current take from the supply is in phase with the voltage. Show that for the particular values given, the supply current is independent of the frequency. **[153 μF]** ***(London Univ.)***
12. In a series-parallel circuit, the two parallel branches *A* and *B* are in series with *C*. The impedances are $Z_A = (10 - j8)\ \Omega$, $Z_B = (9 - j6)\ \Omega$ and $Z_C = (100 + j0)$. Find the currents I_A and I_B and the phase difference between them. Draw the phasor diagram. **[$I_A = 12.71 \angle -30°58'$ $I_B = 15\angle -35°56'$; 4°58']** ***(Elect. Engg. & Electronics Bangalore Univ.)***
13. Find the equivalent series circuits of the 4-branch parallel circuit shown in Fig. 14.54. **[(4.41 + j2.87) Ω]** [A resistor of 4.415 Ω in series with a 4.57 mH inductor]

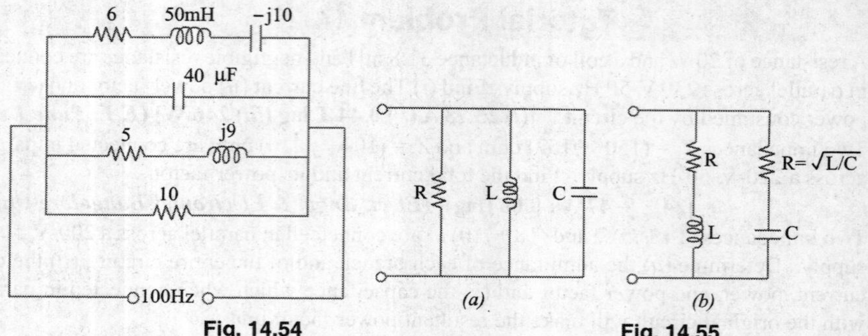

Fig. 14.54 Fig. 14.55

14. A coil of 20 Ω resistance has an inductance of 0.2 H and is connected in parallel with a 100-μF capacitor. Calculate the frequency at which the circuit will act as a non-inductive resistance of *R* ohms. Find also the value of *R*. **[31.8 Hz; 100 Ω]**

15. Calculate the resonant frequency, the impedance at resonance and the Q-factor at resonance for the two circuits shown in Fig. 14.55.

(a) $f_0 = \dfrac{1}{2\pi\sqrt{LC}}$; $Z_0 = R$; $Q_0 = \dfrac{R}{\sqrt{L/C}}$

(b) Circuit is resonant at all frequencies with a constant resistive impedance of $(\sqrt{L/C})$ ohm, Q = 0.]

16. Prove that the circuit shown in Fig. 14.56 exhibits both series and parallel resonances and calculate the frequencies at which two resonaces occur.

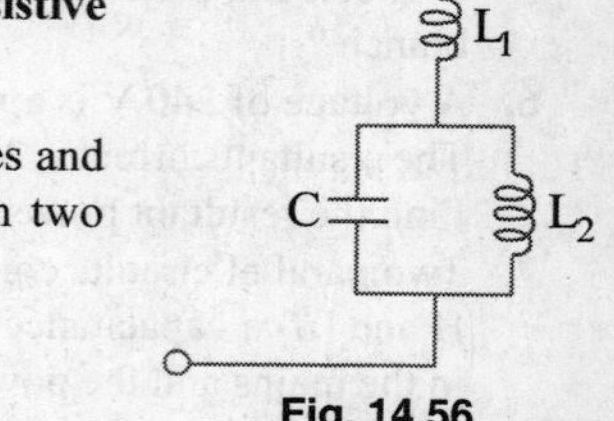

Parallel $f_0 = \dfrac{1}{2\pi}\sqrt{\dfrac{1}{(L_2C_2)}}$; **series** $f_0 = \dfrac{1}{2\pi}\sqrt{\dfrac{(L_1+L_2)}{L_1L_2C_2}}$

Fig. 14.56

17. A parallel *R-L-C* circuit is fed by a constant current source of variable frequency. The circuit resonates at 100 kHz and the *Q*-factor measured at this frequency is 5. Find the frequencies at which the amplitude of the voltage across the circuit falls to (*a*) 70.7% (*b*) 50% of the resonant frequency amplitude. **[(*a*) 90.5 kHz ; 110.5 kHz (*b*) 84.18 kHz ; 118.8 kHz]**

18. Two impedance Z_1 = (6 + j8) ohm and Z_2 = (8 - j6) ohm are connected in parallel across 100 V supply. Determine : (*i*) Current and power factor of each branch. (*ii*) Overall current and power factor (*iii*) Power consumed by each branch and total power.
(Nagpur University, Winter 2003)

19. The currents is each branch of a two branched parallel circuit is given as :

$$i_a = 8.07 \sin\left(314\,t - \frac{\pi}{4}\right)$$

$$i_b - 21.2 \sin\left(314\,t - \frac{\pi}{3}\right)$$

and supply voltage is v = 354 sin 314 t.

Calculate :

(*i*) Total current in the same form (*ii*) Calculate ohmic value of components in each branch.
(Nagpur University, Summer 2004)

20. Two coils are connected in parallel and a voltage of 200 V is applied between the terminals the total current taken by the circuit is 25 A and power dissipated in one of the coils is 1500 W. Calculate the resistance of each coil. ***(Gujrat University, June/July 2003)***

21. Compare the series and parallel resonance of R-L-C series and R-L-C parallel circuit.
(Gujrat University, June/July 2003)

22. Two circuits with impedances $Z_1 = (10 + r15)\ \Omega$ *and* $z_2 = (6 - r8_{\Omega})$ are connected in parallel. If the supply current is 20A, what is the power dissipated in each branch.
(V.T.U., Belgaum Karnataka University, Winter 2003)

23. Three impedances $z_1 = 8 + j6\Omega$, $z_2 = 2 - j1.5\Omega$ and $z_3 = 2\Omega$ are connected in parallel across a 50Hz supply. If the current through z_1 is $3 + j4$amp, calculate the current through the other impedances and also power absorbed by this parallel circuit.
(V.T.U. Belgaum Karnataka University, Winter 2004)

24. Show that the power consumed in a pure inductance is zero. *(RGPV Bhopal 2002)*

25. What do you understand by the terms power factor, active power and reactive power?
(RGPV Bhopal 2002)

26. Two circuits the impedances of which are given by Z_1 $(10 + j\,15)\Omega$ and $Z_2 = (6 - j\,8)\Omega$ are conected in parallel. If the total current supplied is 15 A. What is the power taken by each branch?
(RGPV Bhopal 2002)

27. Does an inductance draw instantaneous power as well as average power?
(RGPV Bhopal December 2002)

28. Describe the properties of (*i*) Resistance (*ii*) Inductance and (*iii*) capacitance used in A.C. Circuit. *(RGPV Bhopal June 2003)*

OBJECTIVE TYPES – 14

1. Fill in the blanks
   (*a*) unit of admittance is ....
   (*b*) unit of capacitive susceptance is ....
   (*c*) admittance equals the reciprocal of ....
   (*d*) admittance is given by the .... sum of conductance and susceptance.

2. An *R-L* circuit has **Z** = (6 + *j*8) ohm. Its susceptance is -Siemens.
   (*a*) 0.06 (*b*) 0.08 (*c*) 0.1 (*d*) –0.08

3. The impedances of two parallel branches of a circuit are (10 + *j*10) and (10 – *j*10) respectively. The impedance of the parallel combination is
   (*a*) 20 + *j*0 (*b*) 10 + *j*0
   (*c*) 5 – *j*5 (*d*) 0 – *j*20

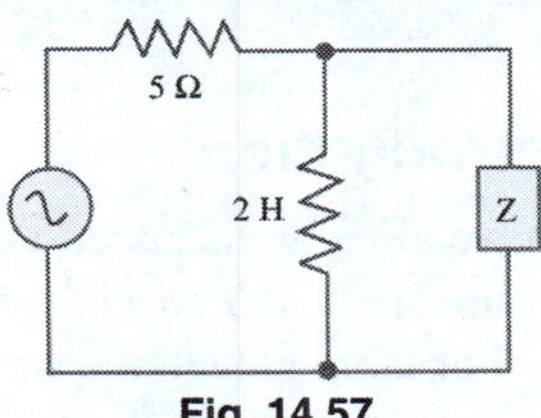

Fig. 14.57

4. The value of Z in Fig. 14.57 which is most appropriate to cause parallel resonance at 500 Hz is
   (*a*) 125.00 mH (*b*) 304.20 μF
   (*c*) 2.0 μF (*d*) 0.05 μF
   (GATE 2004)

CHAPTER

15

A.C. Network Analysis

Chapter Outline

- ❖ Introduction
- ❖ Kirchhoff's Laws
- ❖ Mesh Analysis
- ❖ Nodal Analysis
- ❖ Superposition Theorem
- ❖ Thevenin's Theorem
- ❖ Reciprocity Theorem
- ❖ Norton's Theorem
- ❖ Maximum Power Transfer Theorem-General Case
- ❖ Maximum Power Transfer Theorem
- ❖ Milliman's Theorem

15.1. Introduction

We have already discussed various d.c. network theorems in Chapter 2 of this book. The same laws are applicable to a.c. networks except that instead of resistances, we have impedances and instead of taking algebraic sum of voltages and currents we have to take the phasor sum.

15.2. Kirchhoff's Laws

The statements of Kirchhoff's laws are similar to those for d.c. networks except that instead of algebraic sum of currents and voltages, we take phasor or vector sums for a.c. networks.

1. ***Kirchhoff's Current Law.*** According to this law, in any electrical network, the phasor sum of the currents meeting at a junction is zero.

In other words, $\sum I = 0$ ...at a junction

Put in another way, it simply means that in any electrical circuit the phasor sum of the currents flowing towards a junction is equal to the phasor sum of the currents going away from that junction.

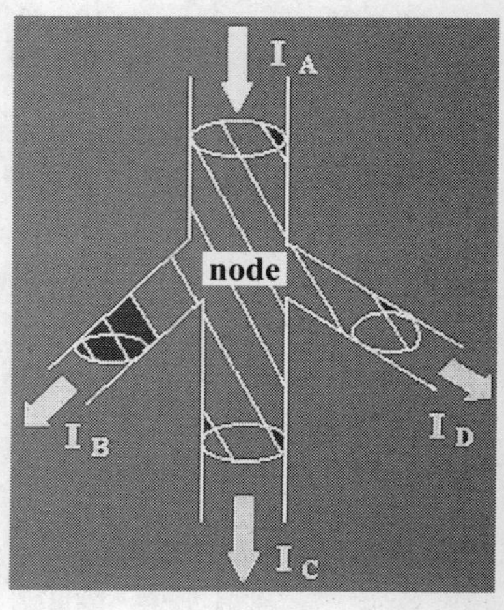

2. ***Kirchhoff's Voltage Law.*** According to this law, the phasor sum of the voltage drops across each of the conductors in any closed path (or mesh) in a network plus the phasor sum of the e.m.fs. connected in that path is zero.

In other words, $\sum IR + \sum$ e.m.f. $= 0$ ...round a mesh

Example 15.1. *Use Kirchhoff's laws to find the current flowing in each branch of the network shown in Fig. 15.1.*

Solution. Let the current distribution be as shown in Fig. 15.1 (*b*). Starting from point *A* and applying KVL to closed loop *ABEFA*, we get

$$-10(x+y) - 20\,x + 100 = 0 \quad \text{or} \quad 3x + y = 10 \qquad \text{...}(i)$$

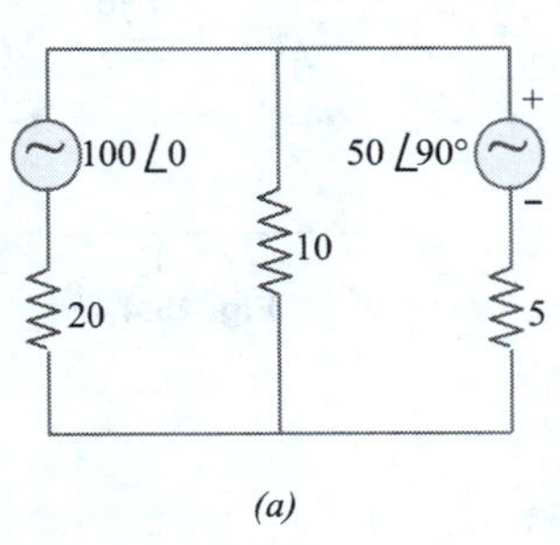

(*a*)

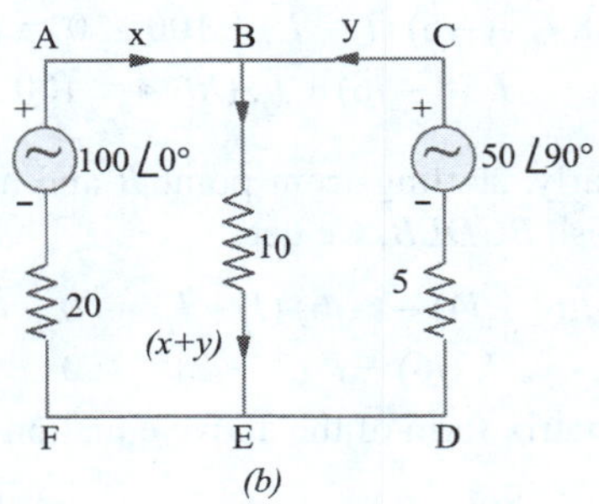

(*b*)

Fig. 15.1

Similarly, considering the closed loop *BCDEB* and starting from point *B*, we have

$$-50\,\angle 90^\circ + 5y + 10\,(x+y) = 0 \quad \text{or} \quad 2x + 3y = j10 \qquad \text{...}(ii)$$

Multiplying Eq. (*i*) by 3 and subtracting it from Eq. (*ii*), we get

$$7x = 30 - j10 \text{ or } x = 4.3 - j1.4 = 4.52\,\angle{-18^\circ}$$

Substituting this value of *x* in Eq. (*i*), we have

$$y = 10 - 3x = 5.95\,\angle\,\mathbf{119.15^\circ} = -\,2.9 + j5.2$$

$$\therefore \qquad x + y = 4.3 - j1.4 - 2.9 + j5.2 = 1.4 + j3.8$$

Tutorial Problem 15.1

1. Using Kirchhoff's Laws, calculate the current flowing through each branch of the circuit shown in Fig. 15.2 **[$I = 0.84\,\angle 47.15^\circ$ A; $I_1 = 0.7\,\angle -88.87^\circ$ A; $I_2 = 1.44\,\angle 67.12^\circ$A]**

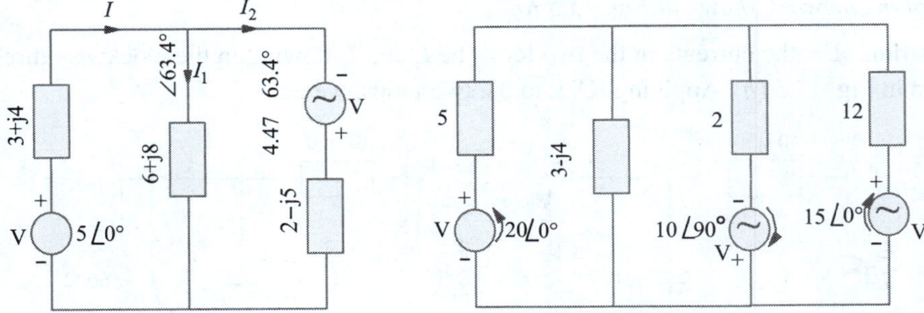

Fig. 15.2 **Fig. 15.3**

2. Use Kirchhoff's laws to find the current flowing in the capacitive branch of Fig. 15.3 **[5.87 A]**

15.3. Mesh Analysis

Sign convention regarding the voltage drops across various impedances and the e.m.f.s is the same. The circuits may be solved with the help of KVL or by use of determinants and Cramer's rule or with the help of impedance matrix [Z_m].

Example 15.2. *Find the power output of the voltage source in the circuit of Fig. 15.4. Prove that this power equals the power in the circuit resistors.*

Solution. Starting from point A in the clockwise direction and applying KVL to the mesh *ABEFA*, we get.

$$-8\,I_1 - (-j6)\,(I_1 - I_2) + 100 \angle 0^\circ = 0$$

or $\quad I_1\,(8 - j6) + I_2.\,(j6) \;=\; 100 \angle 0^\circ$ ...(*i*)

Fig. 15.4

Similarly, starting from point *B* and applying *KVL* to mesh *BCDEB*, we get

$$-I_2(3 + j4) - (-j6)\,(I_2 - I_1) = 0$$

or $\quad I_1\,(j6) + I_2\,(3 - j2) \;=\; 0$ ...(*ii*)

The matrix form of the above equation is

$$\begin{bmatrix} (8-j6) & j6 \\ j6 & (3-j2) \end{bmatrix} = \begin{bmatrix} I_1 \\ I_2 \end{bmatrix} = \begin{bmatrix} 100\angle 0^\circ \\ -0 \end{bmatrix}$$

$$\Delta = \begin{bmatrix} (8-j6) & j6 \\ j6 & (3-j2) \end{bmatrix} = (8 - j6)\,(3 - j2) - (j6)^2 = 62.5 \angle -39.8^\circ$$

$$\Delta_1 = \begin{vmatrix} 100\angle 0^\circ & j6 \\ 0 & (3-j2) \end{vmatrix} = (300 - j\,200) = 360 \angle -26.6^\circ$$

$$\Delta_2 = \begin{vmatrix} (8-j6) & 100\angle 0^\circ \\ j6 & 0 \end{vmatrix} = 600 \angle 90^\circ$$

$$I_1 = \frac{\Delta_1}{\Delta} = \frac{360\angle -26.6^\circ}{62.5\angle -39.8^\circ} = 5.76\angle 13.2^\circ\;;\; I_2 = \frac{\Delta_2}{\Delta} = \frac{600\angle 90^\circ}{62.5\angle -39.8^\circ} = 9.6\angle 129.8^\circ$$

Example 15.3. *Using Maxwell's loop current method, find the value of current in each branch of the network shown in Fig. 15.5 (a).*

Solution. Let the currents in the two loops be I_1 and I_2 flowing in the clockwise direction as shown in Fig. 15.5 (*b*). Applying KVL to the two loops, we get

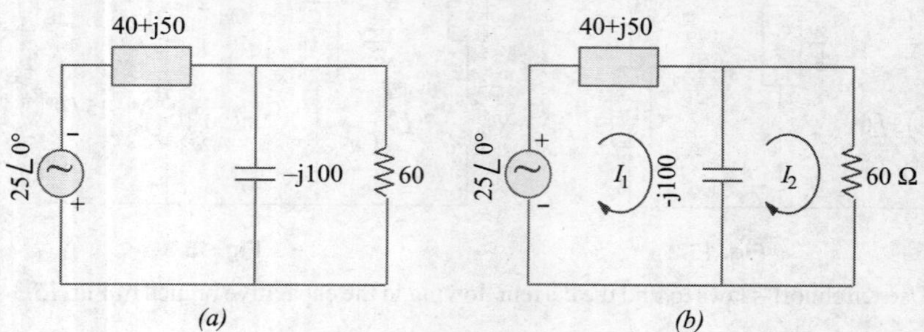

Fig. 15.5

Loop No. 1

$$25 - I_1\,(40 + j50) - (-j\,100)\,(I_1 - I_2) = 0$$

$$\therefore\; 25 - I_1\,(40 - j50) - j\,100\,I_2 = 0 \quad \text{...}(i)$$

Loop No. 2

$- 60\, I_2 - (-j100)\,(I_2 - I_1) = 0$

$\therefore \quad -j100\, I_1 - I_2\,(60 - j100) = 0$

$$\therefore \quad I_2 = \frac{-j100 I_1}{(60 - j100)} = \frac{100\angle -90^\circ I_1}{116.62\angle -59^\circ} = 0.8575\angle 31^\circ I_1 \qquad ...(ii)$$

Substituting this value of I_2 in (i) above, we get

$25 - I_1\,(40 - j50) - j100 \times 0.8575\ \angle 31^\circ\, I_1 = 0$

or $\quad 25 - 40\, I_1 + j50\, I_1 - 85.75\ \angle 59^\circ \quad I_1 = 0\ (j100 = 100\ \angle\ 90^\circ)$

or $\quad 25 - I_1\,(84.16 + j\,23.5) = 0.$

$$\therefore \quad I_1 = \frac{25}{(84.16 + j23.5)} = \frac{25}{87.38\angle 15.6^\circ}\ 0.286\ \angle -15.6^\circ \text{ A}$$

Also, $\quad I_2 = 0.8575\ \angle -31^\circ\, I_1 \times 0.286\ \angle -15.6^\circ = 0.2452\ \angle -46.6^\circ\text{A}$

Current through the capacitor $= (I_1 - I_2) = 0.286\ \angle -15.6^\circ - 0.2452\ \angle 46.6^\circ = 0.107 + j0.1013 = 0.1473\ \angle\ 43.43^\circ$ A.

Example 15.4. *Write the three mesh current equations for network shown in Fig. 15.6.*

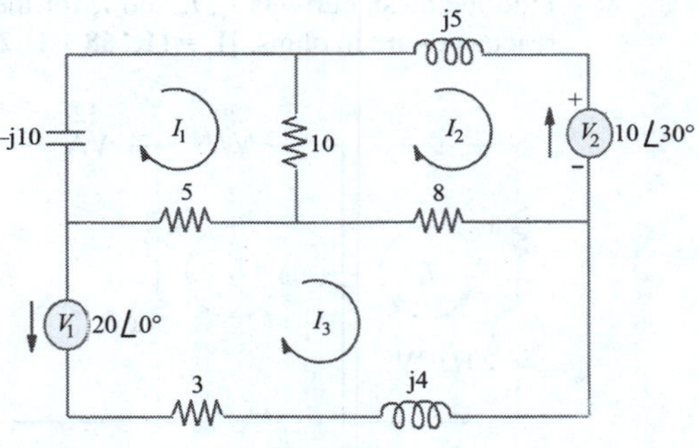

Fig. 15.6

Solution. While moving along I_1, if we apply KVL, we get

$-(-j10)\, I_1 - 10(I_1 - I_2) - 5\,(I_1 - I_3) = 0$

or $\quad I_1\,(15 - j10) - 10\, I_2 - 5 I_3 = 0 \qquad ...(i)$

In the second loop, current through the a.c. source is flowing upwards indicating that its upper end is positive and lower is negative. As we move along I_2, we go from the positive terminal of the voltage source to its negative terminal. Hence, we experience a decrease in voltage which as per Art. would be taken as negative.

$-j5\, I_2 - 10\ \angle 30^\circ - 8(I_2 - I_3) - 10\,(I_2 - I_1) = 0$

or $\quad 10\, I_1 - I_2\,(18 + j5) + 8\, I_3 = 10\ \angle\ 30^\circ \qquad ...(ii)$

Similarly, from third loop, we get

$-20\ \angle 0^\circ - 5(I_3 - I_1) - 8\,(I_3 - I_2) - I_3\,(3 + j4) = 0$

or $\quad 5\, I_1 + 8\, I_2 - I_3\,(16 + j4) = 20\ \angle 0^\circ \qquad ...(iii)$

The values of the three currents may be calculated with the help of Cramer's rule. However, the same values may be found with the help of mesh impedance $[Z_m]$ whose different items are as under:

$$Z_{11} = -j\,10 + 10 + 5 = (15 - j10);\ Z_{22} = (18 + j5)$$
$$Z_{33} = (16 + j5);\ Z_{12} = Z_{21} = -10;\ Z_{23} = Z_{32} = -8$$
$$Z_{13} = Z_{31} = -5;\ E_1 = 0;\ E_2 = -10\ \angle 30^\circ;\quad E_3 = -20\ \angle 0^\circ$$

Hence, the mesh equations for the three currents in the matrix form are as given below:

$$\begin{bmatrix} (15 - j10) & -10 & -5 \\ -10 & (18 + j5) & -8 \\ -5 & -8 & (16 + j5) \end{bmatrix} = \begin{bmatrix} I_1 \\ I_2 \\ I_3 \end{bmatrix} = \begin{bmatrix} 0 \\ -10\angle 30^\circ \\ -20\angle 0^\circ \end{bmatrix}$$

Tutorial Problem 15.2

1. Using mesh analysis, find current in the capacitor of Fig. 15.7. **[13.1 ∠70.12° A]**

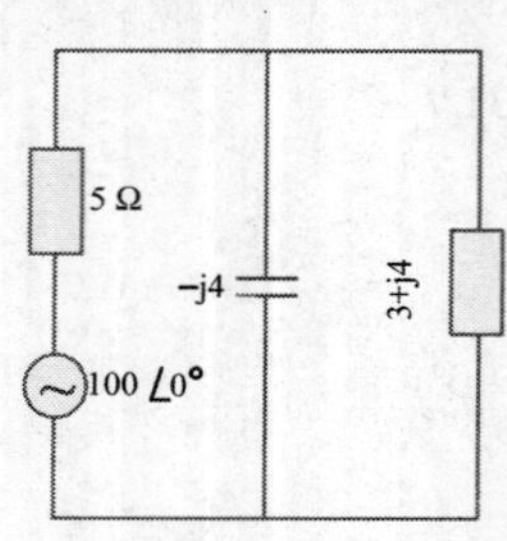

Fig. 15.7

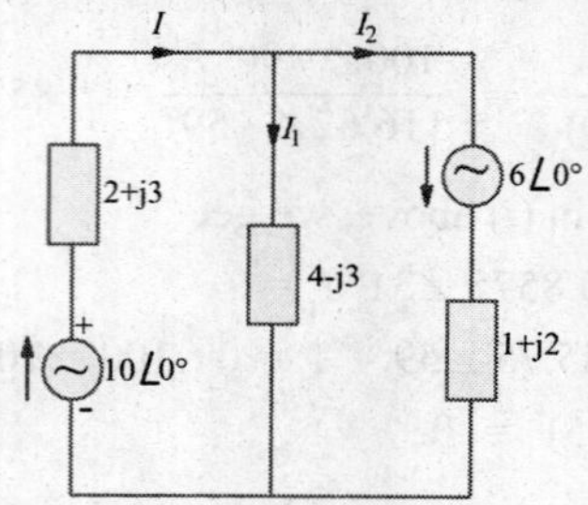

Fig. 15.8

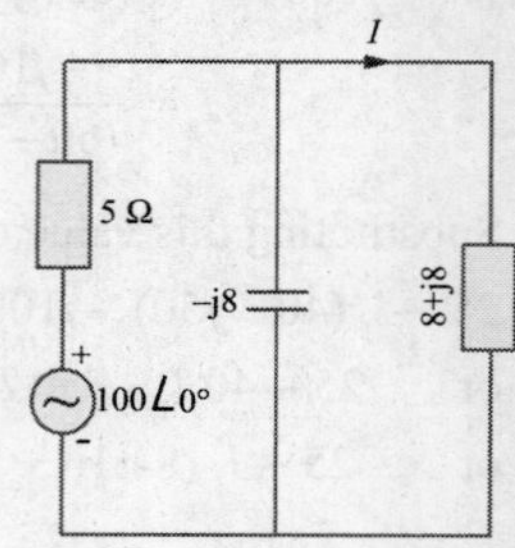

Fig. 15.9

2. Using mesh analysis or Kirchhoff's laws, determine the values of I, I_1 and I_2 (in Fig. 15.8) **[I = 2.7 ∠ – 58.8° A ; I_1 = 0.1 ∠97° A; I_2 = 2.8 ∠–59.6°A]**
3. Using mesh current analysis, find the value of current I and active power output of the voltage source in Fig. 15.9. **[7 ∠ – 50° A ; 645 W]**
4. Find the mesh currents I_1, I_2 and I_3 for the circuit shown in Fig. 15.10. All resistances and reactances are in ohms. **[I_1 = (1.168 + j1.281) ; I_2 = (0.527 – j0.135); I_3 = (0.718 + J0.412)]**

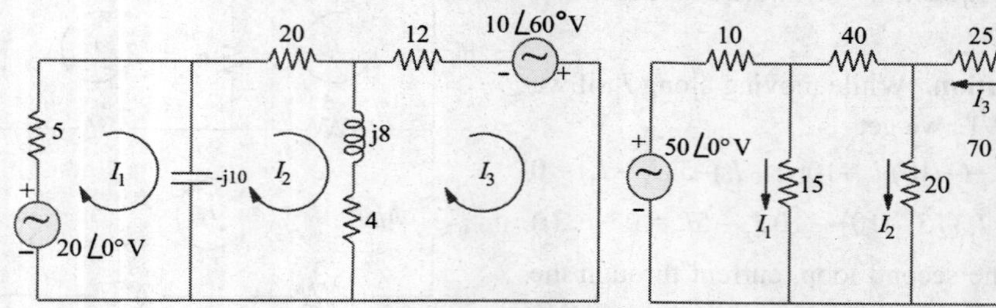

Fig.15.10

Fig. 15.11

5. Find the values of branch currents I_1, I_2 and I_3 in the circuit shown in Fig. 15.12 by using mesh analysis. All resistances are in ohms. **[I_1 = 2.008 ∠0° ; I_2 = 1.545 ∠0° ; I_3 = 1.564 ∠0°]**
6. Using mesh-current analysis, determine the current I_1, I_2 and I_3 flowing in the branches of the networks shown in Fig. 15.13. **[I_1 = 8.7 ∠–1.37° A; I_2 = 3 ∠–48.7° A ; I_3 = 7 ∠17.25° A]**

15.4. Nodal Analysis

This method has already been discussed in details Chapter 2. This technique is the same although we have to deal with circuit impedances rather than resistances and take phasor sum of voltages and currents rather than algebraic sum.

Example 15.5. *Use Nodal analysis to calculate the current flowing in each branch of the network shown in Fig. 15.12.*

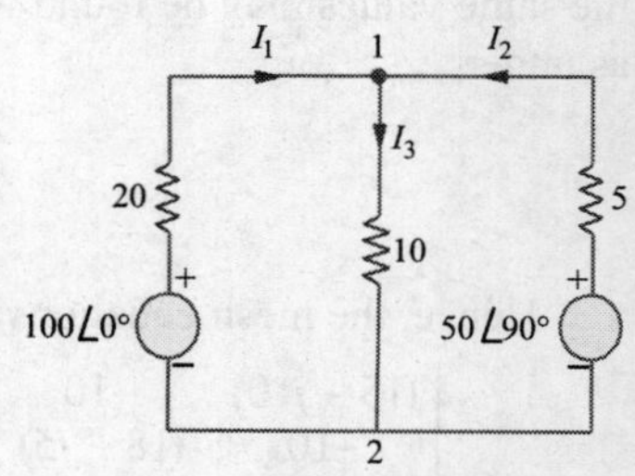

Fig. 15.12

Solution. As seen, there are only two principal nodes out of which node No. 2 has been taken as the reference node. The nodal equations are:

$$V_1\left(\frac{1}{20}+\frac{1}{10}+\frac{1}{5}\right)-\frac{100\angle 0^\circ}{20}-\frac{50\angle 90^\circ}{5}=0$$

$$\therefore \quad 0.35\,V_1 = 5 + j10;\; V_1 = \frac{5 + j10}{0.35}$$

$$= 14.3 + j28.6 = 32\angle 63.4^\circ$$

$$\therefore \quad I_1 = \frac{100\angle 0^\circ - V_1}{20} = \frac{100 - 14.3 - j28.6}{20}$$

$= 4.3 - j1.4 = 4.5 \angle - 18^\circ$ flowing towards node No. 1

(or $4.5 \angle - 18^\circ + 180^\circ = 45 \angle 162^\circ$ flowing away from node No. 1)

$$I_3 = \frac{V_1}{10} = \frac{32\angle 63.4^\circ}{10} \leq = 3.2\angle 63.4^\circ = 1.4 + j2.9$$ flowing from node No. 1 to node No. 2

$$I_2 = \frac{50\angle 90^\circ - V_1}{5} = \frac{j50 - 14.3 - j28.6}{5} = \frac{-14.3 + j21.4}{5} = -2.86 + j4.3 = 5.16\angle 123.6^\circ$$

flowing towards node No. 1

$\angle 123.6^\circ - 180^\circ = 5.16 \angle -56.4^\circ$ flowing away from node No. 1).

Example 15.6. *Find the current I in the j10 Ω branch of the given circuit shown in Fig. 15.13 using the Nodal Method.* **(Principles of Elect. Engg. Delhi Univ.)**

Solution. There are two principal nodes out of which node No. 2 has been taken as the reference node. As per Art.

$$V_1\left(\frac{1}{6 + j8} + \frac{1}{6 - j8} + \frac{1}{j10}\right) - \frac{100\angle 0^\circ}{6 + j8}\frac{100\angle -60^\circ}{6 - j8} = 0$$

$$V_1(0.06 - j0.08 + 0.06 + j0.08 - j0.1)$$

$$= 6 - j8 + 9.93 - j1.2 = 18.4 \angle -30^\circ$$

$$\therefore \quad V_1(0.12 - j0.1) = 18.4\angle 30^\circ \text{ or}$$

$$V_1 \times 0.156 \angle -85.6^\circ = 18.4 \angle - 30^\circ$$

Fig. 15.13

$$\therefore \quad V_1 = 18.4 \angle - 30^\circ / 0.156\angle -85.6^\circ = 118\angle 55.6^\circ\text{V}$$

$$\therefore \quad V = V_1/j10 = 118\angle 55.6^\circ/j10 = 11.8\angle -34.4^\circ\text{A}$$

Example 15.7. *Find the voltage V_{AB} in the circuit of Fig. 15.14 (a). What would be the value of V_1 if the polarity of the second voltage source is reversed as shown in Fig. 15.14 (b).*

Solution. In the given circuit, there are no principle nodes. However, if we take point *B* as the reference node and point *A* as node 1, then using nodal mathod, we get

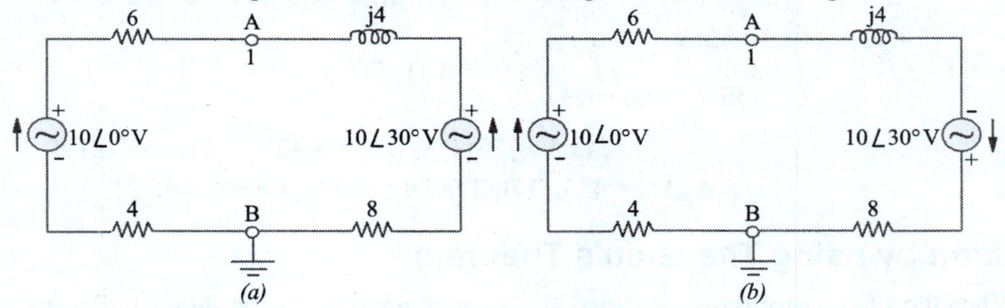

Fig. 15.14

$$V_1\left(\frac{1}{10}+\frac{1}{8+j4}\right) - \frac{10\angle 0^\circ}{10}-\frac{10\angle 30^\circ}{8+j4}=0$$

$$V_1 \times 0.2\angle -14.1^\circ = 1+1.116+j0.066 = 4.48\angle 1.78^\circ$$

$$\therefore \quad V_1 = 4.48\angle 1.78^\circ/0.2\angle -14.1^\circ = 22.4\angle 15.88^\circ$$

When source polarity is Reversed

$$V_1\left(\frac{1}{10}+\frac{1}{8+j4}\right)-\frac{10\angle 0^\circ}{10}+\frac{10\angle 30^\circ}{8+j4}=0 \text{ or } V_1 = 0.09\angle 223.7^\circ$$

Example 15.8. *Write the nodal equations for the network shown in Fig. 15.15.*

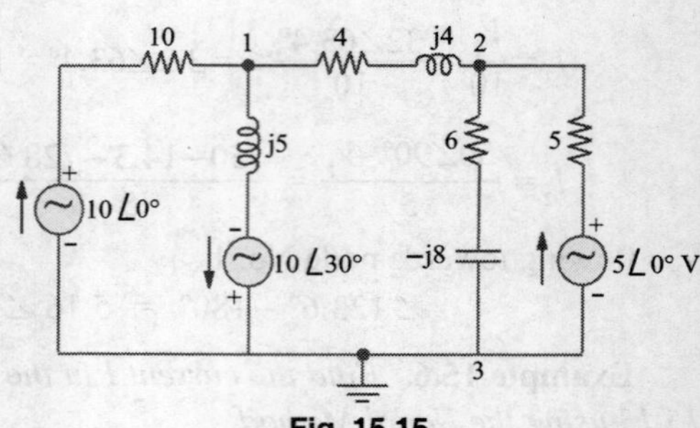

Fig. 15.15

Solution. Keeping in mind the guidance, it would be obvious that since current of the second voltage source is flowing away from node 1, it would be taken as negative. Hence, the term containing this source will become positive because it has been reversed twice. As seen, node 3 has been taken as the reference node. Considering node 1, we have

$$V_1\left(\frac{1}{10}+\frac{1}{4+j4}+\frac{1}{j5}\right)-\frac{V_2}{4+j4}-\frac{10\angle 0^\circ}{10}+\frac{10\angle 30^\circ}{j5}=0$$

Similarly, considering node 2, we have

$$V_2\left(\frac{1}{4+j4}+\frac{1}{5}+\frac{1}{6-j8}\right)-\frac{V_1}{4+j4}-\frac{5\angle 0^\circ}{5}=0$$

Solution by Using Mesh Impedance Matrix

The different items of the mesh resistance matrix $[R_m]$ are as under :

$$Z_{11} = (5+j2)\,;\, Z_{22} = 4\,;\, Z_{33} = (2-j2)\,;\, Z_{12} = Z_{21} = -j2;$$

$$Z_{23} = Z_{32} = j2\,;\, Z_{31} = Z_{13} = 0$$

The mesh equations in the matrix form are :

$$\begin{bmatrix} (5+j2) & -j2 & 0 \\ -j2 & 4 & j2 \\ 0 & j2 & (2-j2) \end{bmatrix} = \begin{bmatrix} I_1 \\ I_2 \\ I_3 \end{bmatrix} = \begin{bmatrix} 50\angle 30^\circ \\ 0 \\ -j50 \end{bmatrix}$$

$$\Delta = (5+j2)\,[4(2-j2) - (j2 \times j2)] - (-j2)\,[(-j2)\,(2-j2)] = 84 - j24 = 87.4\angle -15.9^\circ$$

$$\Delta_2 = \begin{bmatrix} (5+j2) & (43.3+j25) & -0 \\ -j2 & 0 & j2 \\ 0 & -j50 & (2-j2) \end{bmatrix} = (5+j2)\,[-j2\,(-j50)] +$$

$$j2\,[43.3+j25)\,(2-j2)] = -427 + j73 = 433\angle 170.3^\circ$$

$$\therefore \quad I_2\,\Delta_2/\Delta = 433\angle 170.3^\circ/87.4\angle -15.9^\circ = 4.95\angle 186.2^\circ$$

Solution by using Thevenin's Theorem

When the 4 Ω resistor is disconnected, the given figure becomes as shown in Fig. 15.16 (*a*). The voltage V_A is given by the drop across $j2$ reactance. Using the voltage-divider rule, we have

$$V_A = 50\angle 30° \times \frac{j2}{5+j2} = 18.57\angle 98.2° = -2.65 + j18.38$$

Similarly,

$$V_B = 50\angle 90° \frac{-j2}{2-j2} = 35.36\angle 45° = 25 + j25$$

$$\therefore \quad V_{th} = V_A - V_B = -2.65 + 18.38 - 25 - j25 = 28.43\angle 193.5°$$

$$R_{th} = 5\|j2 + 2\|(-j2) = \frac{j10}{5+j2} + \frac{-j4}{2-j2} + 1.689 + j0.72$$

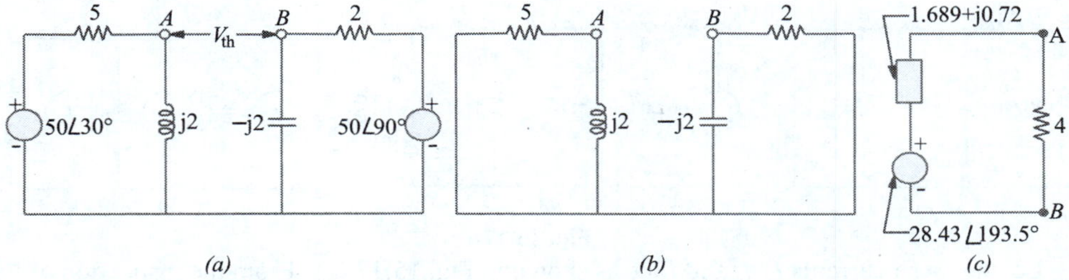

(a) (b) (c)

Fig. 15.16

The Thevenin's equivalent circuit consists of a voltage source of 28.43 ∠ 193.5° V and an impedance of $(1.689 + j0.72)\ \Omega$ as shown in Fig. 15.16 (*c*). Total resistance is $4 + (1.689 + j0.72) = 5.689 + j\,0.72 = 5.73\angle 7.2°$. Hence, current through the 4 Ω resistor is $28.43\angle 193.5°/5.73\angle 7.20° = 4.96\angle 186.3°$.

Note. The slight variations in the answers are due to the approximations made during calculations.

Example 15.9. *Using any suitable method, calculate the current through 4 ohm resistance of the network shown in Fig. 15.17.* **(Network Analysis AMIE Sec. B Summer 1990)**

Solution. We will solve this question with the help of **(*i*)** Kirchhoff's laws **(*ii*)** Mesh analysis and **(*iii*)** Nodal analysis.

(*i*) Solution by using Kirchhoff's Laws

Let the current distribution be as shown in Fig. 15.17 (*b*). Using the same sign convention as given in Art. we have

First Loop $\quad -10(I_1 + I_2 + I_3) - (-j5)\,I_1 + 100 = 0$

or $\quad I_1\,(10 - j5) + 10\,I_3 + 10I_2 = 100 \quad$ ...**(*i*)**

Second Loop $\quad -5(I_2 + I_3) - 4I_2 + (-j\,5)\,I_1 = 0$

or $\quad j5I_1 + 9\,I_2 + 5I_3 = 0 \quad$ ...**(*ii*)**

Third Loop $\quad -I_3\,(8 + j6) + 4I_2 = 0$

or $\quad 0I_1 + 4I_2 - I_3\,(8 + j6) = 0 \quad$ ...**(*iii*)**

The matrix form of the above three equations is

$$\begin{bmatrix} (10-j5) & -10 & 10 \\ j5 & 9 & 5 \\ 0 & 4 & -(8+j6) \end{bmatrix} \begin{bmatrix} I_1 \\ I_2 \\ I_3 \end{bmatrix} = \begin{bmatrix} 100 \\ 0 \\ 0 \end{bmatrix}$$

$$\Delta = (10 - j\,5)\,[-9\,(8 + j6) - 20] - j\,5\,[-10(8 + j\,6) - 40]$$
$$= -1490 + j\,520 = 1578\angle 160.8°$$

Since we are interested in finding I_2 only, we will calculate the value of Δ_2.

$$\Delta_2 = \begin{bmatrix} (10-j5) & 100 & 10 \\ j5 & 0 & 5 \\ 0 & 0 & -(8+j6) \end{bmatrix}$$

$$= -j5(-800 - j600) = -3000 + j4000 = 5000 \angle 126.9^\circ$$

$$I_2 = \frac{\Delta_2}{\Delta} = \frac{500.0\angle 126.9^\circ}{1578\angle 160.8^\circ} = 3.17 \angle -33.9^\circ \text{ A}$$

(*ii*) Solution by using Mesh Impedance Matrix

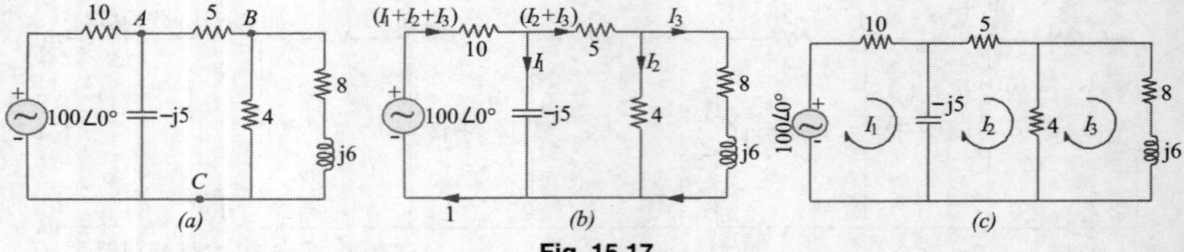

Fig. 15.17

Let the mesh currents I_1, I_2 and I_3 be as shown in Fig. 15.17 (*c*). From the inspection of Fig. 15.17 (*c*), the different items of the mesh impedance matrix $[Z_m]$ are as under :

$$Z_{11} = (10 - j5);\ Z_{22} = (9 - j5);\ Z_{33} = (12 + j6)$$

$$Z_{21} = Z_{12} = -(-j5) = j5;\ Z_{23} = Z_{32} = -4;\ Z_{31} = Z_{13} = 0$$

Hence, the mesh equations in the matrix form are :

$$\begin{bmatrix} (10-j5) & j5 & 0 \\ j5 & (9-j5) & -4 \\ 0 & -4 & (12+j6) \end{bmatrix} \begin{bmatrix} I_1 \\ I_2 \\ I_3 \end{bmatrix} = \begin{bmatrix} 100 \\ 0 \\ 0 \end{bmatrix}$$

$$\therefore \quad \Delta = (10 - j5)[(9 - j5)(12 + j6) - 16] - j5(j60 - 30)$$

$$= 1490 - j520 = 1578\angle 19.2^\circ$$

It should be noted that the current passing through 4 Ω resistance is the vector diference $(I_2 - I_3)$. Hence, we will find I_2 and I_3 only.

$$\Delta_2 = \begin{bmatrix} (10-j5) & 100 & 0 \\ j5 & 0 & -4 \\ 0 & 0 & (12+j6) \end{bmatrix} = j5(1200 + j600) = 3000 - j6000 = 6708 \angle -63.4^\circ$$

$$\Delta_2 = \begin{bmatrix} (10-j5) & j5 & 100 \\ j5 & (9-j5) & 0 \\ 0 & -4 & 0 \end{bmatrix} = -j5(400) = -j2000 = 2000 \angle -90^\circ$$

$$\therefore \quad I_2 = \frac{\Delta_2}{\Delta} = \frac{6708\angle -63.4^\circ}{1578\angle -19.2^\circ} = 4.25 \angle -44.2^\circ = 3.05 - j2.96$$

$$I_3 = \frac{\Delta_3}{\Delta} = \frac{2000\angle -90^\circ}{1578\angle -19.2^\circ} = 1.27 \angle -70.8^\circ = 0.42 - j1.2$$

Current $\quad (I_2 - I_3) = 2.63 - j1.76 = 3.17 \angle -33.9^\circ$

(*iii*) Solution by Nodal Analysis

The current passing through 4 Ω resistance can be found by finding the voltage V_B of node *B* with the help of Nodal analysis. For this purpose point *C* in Fig. 15.17 (*a*) has been taken as the reference node. Using the Nodal technique as explained in Art. we have

$$V_A\left(\frac{1}{10}+\frac{1}{5}+\frac{1}{-j5}\right)-\frac{V_B}{5}-\frac{100\angle 0^\circ}{10}=0 \qquad \textbf{...for node A}$$

$$V_A(3+j\,2)-2\,V_B = 100 \qquad ...(i)$$

Similarly, for node B, we have

$$V_B\left(\frac{1}{5}+\frac{1}{4}+\frac{1}{(8+j6)}\right)-\frac{V_A}{5}=0 \quad \text{or} \quad V_B(53-j6)-20\,V_A=0 \qquad ...(ii)$$

Estimating V_A from Eq. (*i*) and (*ii*), we have

$$V_B(131+j88) = 2000 \quad \text{or} \quad V_B = 12.67\angle{-33.9^\circ}$$

Current through 4 Ω resistor 12.67 ∠–33.9º/4 = 3.17 ∠–33.9º

Tutorial Problem 15.3

1. Apply nodal analysis to the network of Fig. 15.18 to determine the voltage at node A and the active power delivered by the voltage source. **[8∠3.7º V; 9.85 W]**
2. Using nodal analysis, determine the value of voltages at models 1 and 2 in Fig. 15.19. **[V_1 = 88.1 ∠33.88º A; V_2 = 58.7∠72.34ºA]**

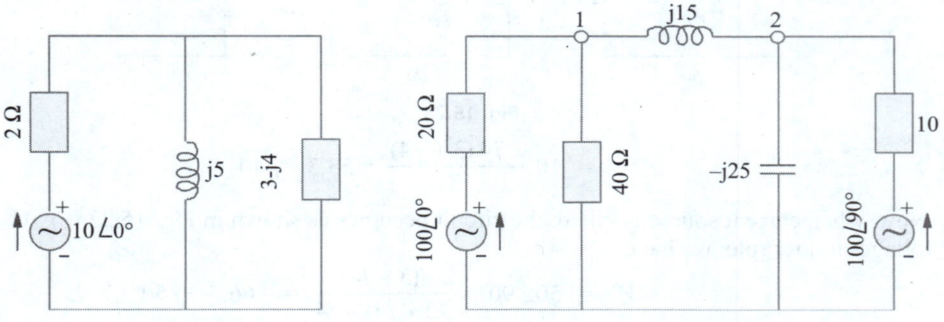

Fig. 15.18 Fig. 15.19

3. Using Nodal analysis, find the nodal voltages V_1 and V_2 in the circuit shown in Fig. 15.20. All resistances are given in terms of siemens **[V_1 = 1.64 V : V_2 = 0.38 V]**

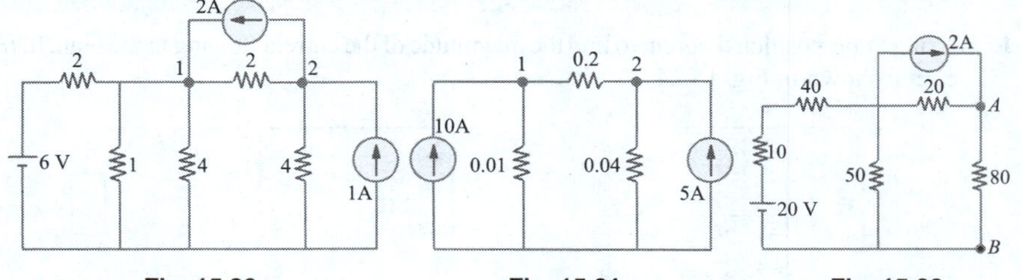

Fig. 15.20 Fig. 15.21 Fig. 15.22

4. Find the values of nodal voltages V_1 and V_2 in the circuit of Fig. 15.21. Hence, find the current going from node 1 to node 2. All resistances are given in siemens. **[V_1 = 327 V; V_2 = 293.35 V ; 6.73 A]**
5. Using Nodal analysis, find the voltage across points *A* and *B* in the circuit of Fig. 15.22: Check your answer by using mesh analysis. **[32 V]**

15.5. Superposition Theorem

As applicable to a.c. networks, it states as follows:

In any network made up of linear impedances and containing more than one source of e.m.f., the current flowing in any branch is the phasor sum of the currents that would flow in that branch if each source were considered separately, all other e.m.f. sources being replaced for the time being, by their respective internal impedances (if any).

Note. It may be noted that independent sources can be 'de-energised' *i.e.* removed leaving behind their internal impedances (if any) but dependent sources should not be de-energised.

Example 15.10. *Use Superposition theorem to find the voltage V in the network shown in Fig. 15.23.*

Solution. When the voltage source is killed, the circuit becomes as shown in the Fig. 15.23 (*b*) Using current-divider rule,

$$I = 10\angle 0^\circ \times \frac{-4}{(3+j4)-j4}, \text{ Now, } V' = I\,(3+j4)$$

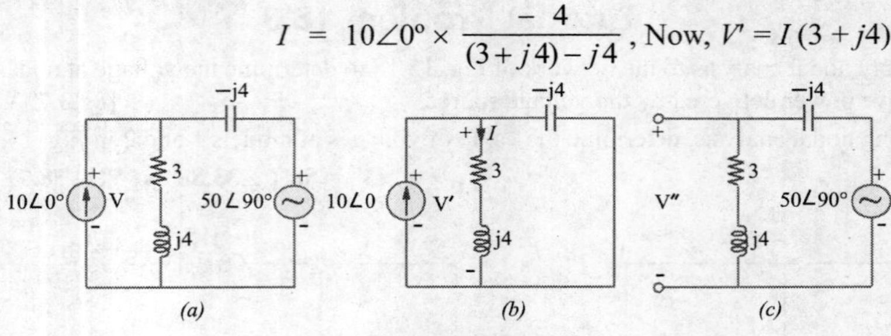

Fig. 15.23

$$\therefore \quad V' = 10\,\frac{-j4\,(3+j4)}{3} = 53.3 - j\,40$$

Now, when current source is killed, the circuit becomes as shown in Fig. 15.23 (*c*). Using the voltage-divider rule, we have

$$V'' = 50\angle 90^\circ \times \frac{(3+j4)}{(3+j4)-j4} = -66.7 + j\,50$$

$\therefore$ drop V = $V' + V''$ = 53.3 − j40 (−66.7 + j50) = − 13.4 + j10 = 16.7 ∠ 143.3° V

Tutorial Problem 15.4

1. Using Superposition theorem to find the magnitude of the current flowing in the branch *AB* of the circuit shown in Fig. 15.24.

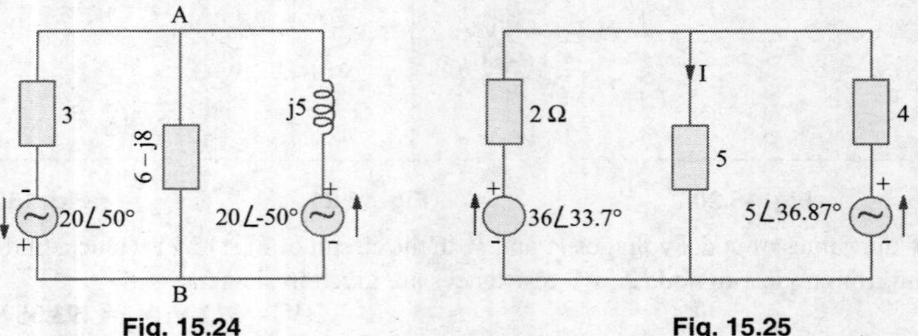

Fig. 15.24 Fig. 15.25

2. Apply Superposition theorem to determine the circuit *I* in the circuit of Fig. 15.25.

[0.53 ∠ 5.7° A]

15.6. Thevenin's Theorem

As applicable to a.c. networks, this theorem may be stated as follows:

The current through a load impedance Z_L connected across any two terminals A and B of a linear network is given by $V_{th}/(Z_{th} + Z_L)$ wher V_{th} is the open-circuit voltage across A and B and Z_{th} is the total impedance of the network as viewed from the open-circuited terminals A and B with all voltage sources replaced by their internal impedances (if any) and current sources by infinite impedance.

Example 15.11. *In the network shZown in Fig. 15.26.*

$Z_1 = (8 + j8)\ \Omega$; $Z_2 = (8 - j8)\ \Omega$; $Z_3 = (2 + j20)$; $V = 10 \angle 0^\circ$ *and* $Z_L = j\,10\ \Omega$

Find the current through the load Z_L *using Thevenin's theorem.*

Solution. When the load impedance Z_L is removed, the circuit becomes as shown in Fig. 15.26 (*b*). The open-circuit voltage which appears across terminals A and B represents the Thevenin voltage V_{th}. This voltage equals the drop across Z_2 because there is no current flow through Z_3.

Current flowing through Z_1 and Z_2 is

$$I = V(Z_1 + Z_2) = 10 \angle 0^\circ\ [(8 + j8) + (8 - j8)] = 10 \angle 0^\circ/16 = 0.625 \angle 0^\circ$$

$$V_{th} = IZ_2 = 0.625\ (8 - j\,8) = (5 - j\,5) = 7.07 \angle -45^\circ$$

The Thevenin impedance Z_{th} is equal to the impedance as viewed from open terminals A and B with voltage source shorted.

$$\therefore \quad Z_{th} = Z_3 + Z_1 \,||\, Z_2 = (2 + j\,20) + (8 + j\,8) \,||\, (8 - j\,8) = (10 + j\,20)$$

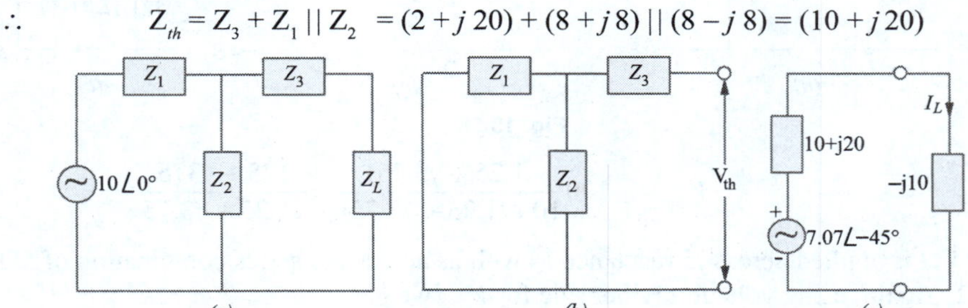

Fig. 15.26

The equivalent Thevenin circuit is shown in Fig. 15.26 (*c*) across which the load impedance has been reconnected. The load current is given by

$$\therefore \quad I_L = \frac{V_{th}}{Z_{th} + Z_L} = \frac{(5 - j5)}{(10 + j20) + (-j10)} = \frac{-j}{2}$$

Example 15.12 A. *Find the Thevenin equivalent circuit at terminals AB of the circuit given in Fig. 15.27 (a).*

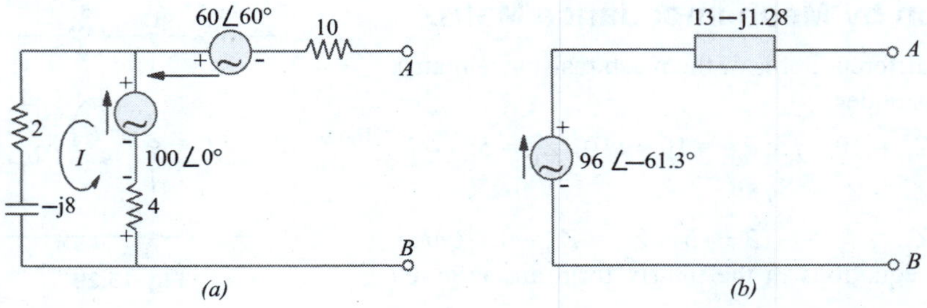

Fig. 15.27

Solution. For finding $V_{th} = V_{AB}$, we have to find the phasor sum of the voltages available on the way as we go from point *B* to point *A* because V_{AB} means voltage of point *A* with respect to that of point *B* (Art.). The value of current $I = 100\angle 0°/(6 - j8) = (6 + j8)A$.

Drop across 4 Ω resistor = $4(6 + j8) = (24 + j32)$

$$\therefore \quad V_{th} = V_{AB} = -(24 + j32) + (100 + j0) - 60(0.5 + j0.866)$$
$$= 46 - j84 = 96\angle -61.3°$$
$$Z_{AB} = Z_{th} = [10 + [4 \,||\, (2 - j8)] = (13 - j1.28)$$

The Thevenin equivalent circuit is shown in Fig. 15.27 (*b*).

Example 15.13. *For the network shown, determine using Thevenin's theorem, voltage across capacitor in. Fig. 15.28.* **(Elect. Network Analysis Nagpur Univ. 1993)**

$Z_{CD} = j5 \,||\, (10 + j5) = 1.25 + j3.75$. This impedance is in series with the 10Ω resistance. Using voltage divider rule, the drop over Z_{CD} is

Solution. When load of –j5 Ω is removed the circuit becomes as shown in Fig. 15.28 (*b*). Thevenin voltage is given by the voltage drop produced by 100-V source over $(5 + j5)$impedance. It can be calculated as under.

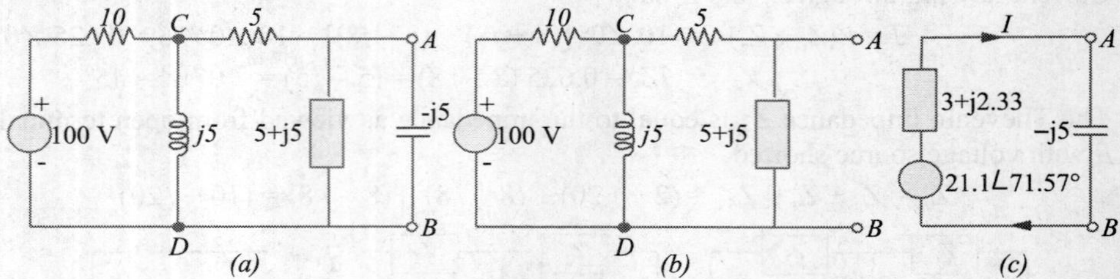

Fig. 15.28

$$V_{CD} = 100\frac{(1.25 + j3.75)}{10 + (1.25 + j3.75)} = \frac{125 + j375}{11.25 + j3.75}$$

This V_{CD} is applied across j5 reactance as well as across the series combination of 5Ω and $(5 + j5)$ Ω. Again, using voltage-divider rule for V_{CD}, we get

$$V_{AB} = V_{th} = V_{CD} \times \frac{5 + j5}{10 \times j5} = \frac{(125 + j375)}{11.25 + j3.75} \times \frac{5 + j5}{10 + j5} = 21.1\angle 71.57° = 6.67 + j20$$

As looked into terminals *A* and *B*, the equivalent impedance is given by

$$R_{AB} = R_{th} = (5 + j5) \,||\, (5 + 10 \,||\, j5) = (5 + j5) \,||\, (7 + j4) = 3 + j2.33$$

The equivalent Thevenin's source along with the load is shown in Fig. 15.28 (c).

$$\text{Total impedance} = 3 + j2.33 - j5 = 3 - j2.67 = 4.02\angle -41.67°$$
$$\therefore \quad I = 21.1\angle 71.57°/4.02\angle -41.67° = 5.25\angle 113.24°$$

Solution by Mesh Impedance Matrix

The different items of the mesh resistance matrix $[Z_m]$ are as under:

$Z_{11} = 10 + j5$; $Z_{22} = 10 + j10$; $Z_{33} = 5$;

$Z_{12} = Z_{21} = -j5$;

$Z_{23} = Z_{32} = -(5 + j5)$; $Z_{31} = Z_{13} = 0$. Hence, the mesh equations in the matrix form are as given below:

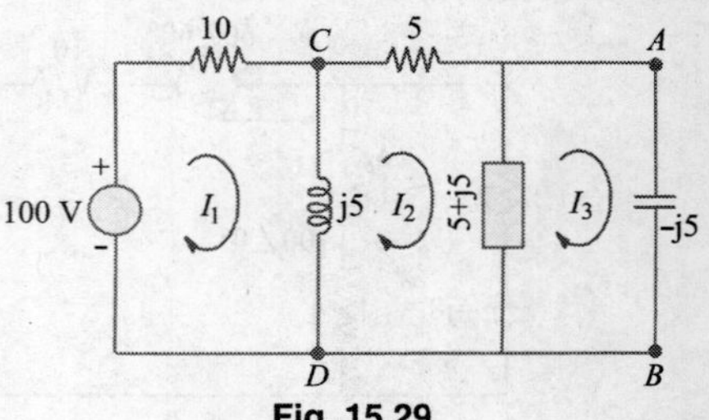

Fig. 15.29

$$\begin{bmatrix} (10+j5) & -j5 & 0 \\ -j5 & (10+j110) & -(5+j5) \\ 0 & -(5+j5) & 5 \end{bmatrix} = \begin{bmatrix} I_1 \\ I_2 \\ I_3 \end{bmatrix} = \begin{bmatrix} 100 \\ 0 \\ 0 \end{bmatrix}$$

$\therefore \quad \Delta = (10 + j5)\,[5(10 + j10) - (5 + j5)\,(5 + j5)] + j5\,(-j25) = 625 + j250 = 673\angle 21.8^\circ$

$$\begin{bmatrix} (10+j5) & -j5 & 100 \\ -j5 & (10+j10) & 0 \\ 0 & -(5+j5) & 0 \end{bmatrix} = j5\,(500 + j\,500) = 3535\angle 135^\circ$$

$$\therefore \qquad I_3 = \Delta_3/\Delta = 3535\angle 135^\circ/673\angle 21.8^\circ = 5.25\angle 113.2^\circ$$

Tutorial Problem 15.5

1. Determine the Thevenin's equivalent circuit with respect to terminals *AB* of the circuit shown in Fig. 15.30. **[$V_{th} = 14.3\angle 6.38^\circ$, $Z_m = (4 + j0.55)\ \Omega$]**

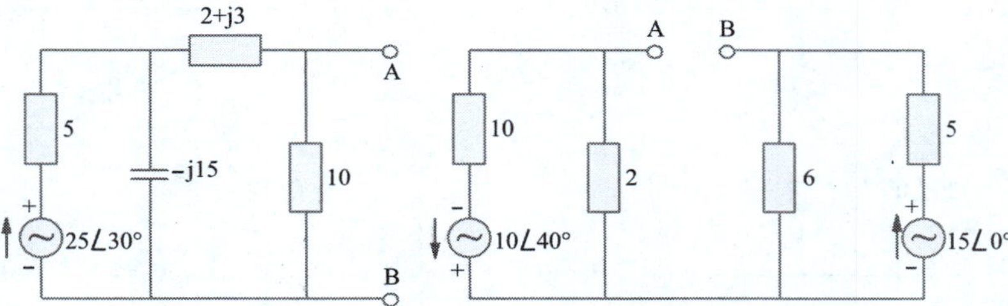

Fig. 15.30 **Fig. 15.31**

2. Determine Thevenin's equivalent circuit with respect to terminals *AB* in Fig. 15.31. **[$V_{th} = 9.5\angle 6.46^\circ$; $Z_{th} = 4.4\angle 0^\circ$]**
3. The e.m.fs. of two voltage source shown in Fig. 15.32 are in phase with each other. Using Thevenin's theorem, find the current which will flow in a 16 Ω resistor connected across terminals *A* and *B*. **[$V_{th} = 100$ V ; $Z_{th} = (48 + j32)$; $I = 1.44\angle -26.56^\circ$]**

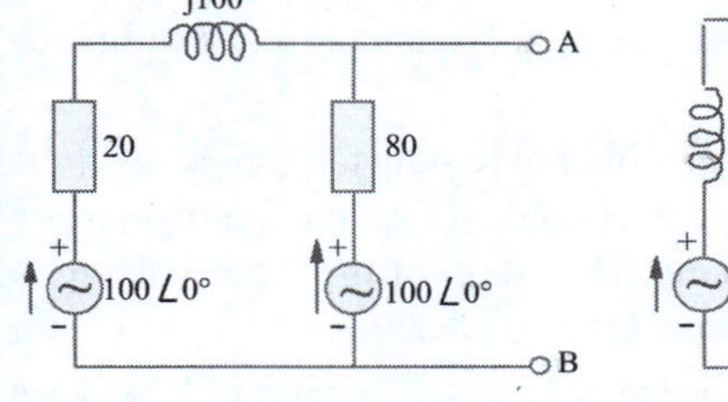

A, j4, 5−j4, 12∠0°, B

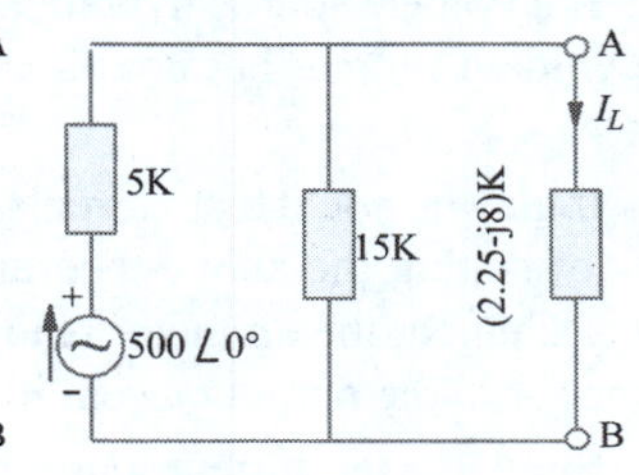

Fig. 15.32 **Fig. 15.33** **Fig. 15.34**

4. Find the Thevenin's equivalent circuit for terminals *AB* for the circuit shown in Fig. 15.34. **[$V_{th} = 15.37\angle -38.66^\circ$; $Z_{th} = (3.2 + j4)\ \Omega$]**
5. Using Thevenin's theorem, find the magnitude of the load current I_L passing through the load connected across terminals *AB* of the circuit shown in Fig. 15.34. **[37.5 mA]**
6. By using Thevenin's theorem, calculate the current flowing through the load connected across terminals *A* and *B* of circuit shown in Fig. 15.35. All resistances and reactances are in ohms. **[$V_{th} = 56.9\angle 50.15^\circ$; $3.11\angle 85.67^\circ$]**

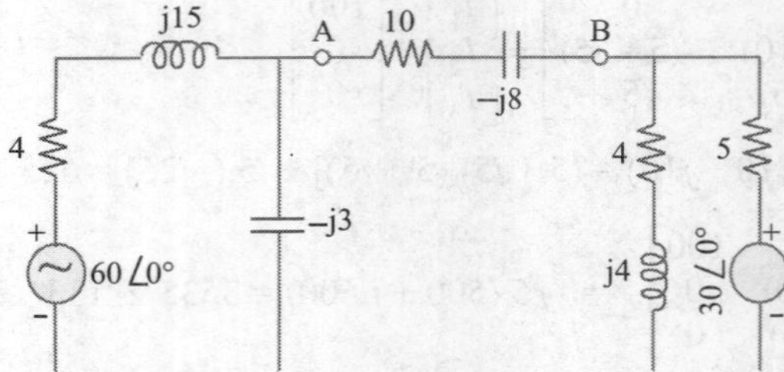

Fig. 15.35

7. Calculate the equivalent Thevenin's source with respect to the terminals *A* and *B* of the circuit shown in Fig. 15.36. **[V_{th} = (6.34 + j2.93) V ; Z_{th} = (3.17 – j5.07) Ω]**

8. What is the Thevenin's equivalent source with respect to the terminals *A* and *B* of the circuit shown in Fig. 15.37 ? **[V_{th} = (9.33 + j8) V ; Z_{th} = (8 – j11) Ω]**

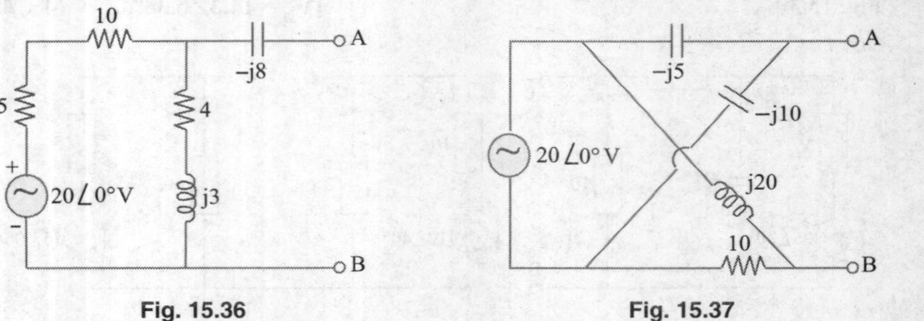

Fig. 15.36 Fig. 15.37

15.7. Reciprocity Theorem

This theorem applies to networks containing linear bilateral elements and a single voltage source or a single current source. This theorem may be stated as follows:

If a voltage source in branch *A* of a network causes a current of 1 in branch *B*, then shifting the voltage source (but not its impedance) to branch *B* will cause the same current *I* in branch A.

It may be noted that currents in other branches will generally not remain the same. A simple way of stating the above theorem is that if an ideal voltage source and an ideal ammeter are inter-changed, the ammeter reading would remain the same. The ratio of the input voltage in branch *A* to the output current in branch *B* is called the transfer impedance.

Similarly, if a current source between nodes 1 and 2 causes a potential difference of *V* between nodes 3 and 4, shifting the current source (but not its admittance) to nodes 3 and 4 causes the same voltage *V* between nodes 1 and 2.

In other words, the interchange of an ideal current source and an ideal voltmeter in any linear bilateral network does not change the voltmeter reading.

However, the voltages between other nodes would generally not remain the same. The ratio of the input current between one set of nodes to output voltage between another set of nodes is called the transfer admittance.

Example 15.14. *Verify Reciprocity theorem for V & I in the circuit shown in Fig. 15.38.*

[Elect. Network Analysis, Nagpur Univ. 1993]

Solution. We will find the value of the current I as read by the ammeter first by applying series parallel circuits technique and then by using mesh resistance matrix (Art.)

1. Series Parallel Circuit Technique

The total impedance as seen by the voltage source is

$$= 1 + [j\,1 \,||\, (2 - j\,1)] \;=\; 1 + \frac{j1(2 - j1)}{2} = 15 + j\,1$$

$$\therefore \qquad \text{total circuit current } i \;=\; \frac{5\angle 0^{\circ}}{1.5 + j1}$$

This current gets divided into two parts at point A, one part going through the ammeter and the other going along AB. By using current-divider rule. (Art), we have

$$I \;=\; \frac{5}{1.5 + j1} \times \frac{J1}{(2 + j1 - j1)} = \frac{j5}{3 + j2}$$

2. Mesh Impedance Matrix

In Fig. 15.48 (b), $Z_{11} = (1 + j1)$, $Z_{22} = (2 + j1 - j1) = 2$; $Z_{12} = Z_{21} = -j1$

$$\therefore \quad \begin{vmatrix} (1+j1) & -j1 \\ -j1 & 2 \end{vmatrix} \begin{vmatrix} I_1 \\ I_2 \end{vmatrix} \;=\; \begin{vmatrix} 5 \\ 10 \end{vmatrix} ; \Delta = 2(1+j1) - (-j1)(-j\,1) = 3 + j2$$

$$\Delta_2 = \begin{vmatrix} (1+j\,1) & 5 \\ -j1 & 0 \end{vmatrix} \;=\; 0 + j5 = j5\;;\; I_2 = I = \frac{\Delta_2}{\Delta} = \frac{j5}{(3 + j2)}$$

As shown in Fig. 15.38 (c), the voltage source has been interchanged with the ammeter. The polarity of the voltage source should be noted in particular. It looks as if the voltage source has been pushed along the wire in the counterclockwise direction to its new position, thus giving the voltage polarity as shown in the figure. We will find the value of I in the new position of the ammeter by using the same two techniques as above.

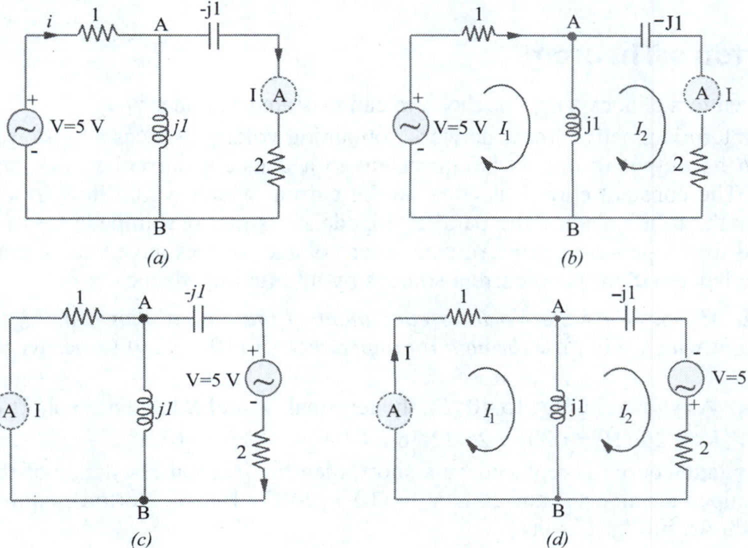

Fig. 15.38

1. Series Parallel Circuit Technique

As seen by the voltage source from its new position, the total circuit impedance is

$$= 2\,(2 - j\,1) + j\,1\,||\,1 = \frac{3 + j2}{1 + j1}$$

The total circuit current $\quad i = 5 \times \dfrac{1 + j1}{3 + j2}$

This current i gets divided into two parts at point B as per the current-divider rule.

$$I = \frac{5(1 + j1)}{3 + j2} \times j\frac{1}{1 + j1} = \frac{j5}{3 + j2}$$

2. Mesh Impedance Matrix

As seen from Fig. 15.48 (*d*).

$$\begin{vmatrix} (1 + j1) & -j1 \\ -j1 & 2 \end{vmatrix} \begin{vmatrix} I_1 \\ I_2 \end{vmatrix} = \begin{vmatrix} 0 \\ 5 \end{vmatrix};\ \Delta = 2(1 + j\,1) + 1 = 3 + j\,2$$

$$\Delta = \begin{vmatrix} 0 & -j1 \\ 5 & 2 \end{vmatrix} = j5;\ I = I_1 = \frac{\Delta_1}{\Delta} = \frac{j\,5}{3 + j\,2}$$

The reciprocity theorem stands verified from the above results.

Tutorial Problem 15.6

1. State reciprocity theorem. Verify for the circuit Fig. 15.39, with the help of any suitable current through any element.

(Elect. Network Analysis Nagpur Univ. 1993)

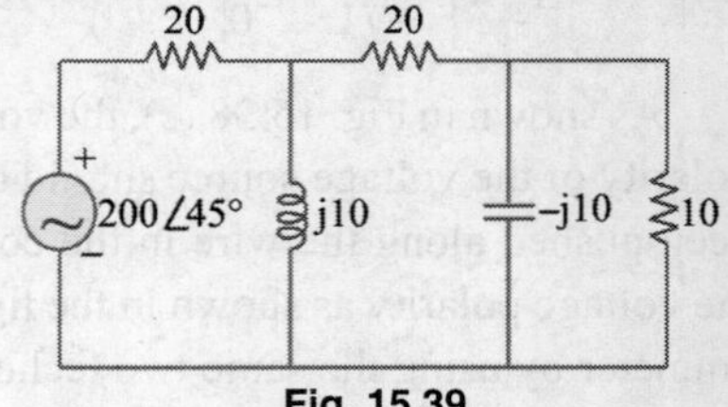

Fig. 15.39

15.8. Norton's Theorem

As applied to a.c. networks, this theorem can be stated as under :

Any two terminal active linear network containing voltage sources and impedances when viewed from its output terminals is equivalent to a constant current source and a parallel impedance. The constant current is equal to the current which would flow in a short-circuit placed across the terminals and the parallel impedance is the total impedance of the network when viewed from open-circuited terminals after voltage sources have been replaced by their internal impedances (if any) and current sources by infinite impedance.

Example 15.15. *Find the Norton's equivalent of the circuit shown in Fig. 15.40. Also find the current which will flow through an impedance of (10 – j 20) Ω across the terminals A and B.*

Solution. As shown in Fig. 15.40 (*b*), the terminals *A* and *B* have been short-cicuited.

$\therefore \quad I_{SC} = I_N = 25/(10 + j\,20) = 25/22.36\ \angle 63.4° = 1.118\ \angle{-63.4°}$

When voltage source is replaced by a short, then the internal resistance of the circuit, as viewed from open terminals *A* and *B*, is $R_N = (10 + j20)\Omega$. Hence, Norton's equivalent circuit becomes as shown in Fig. 15.40(*c*).

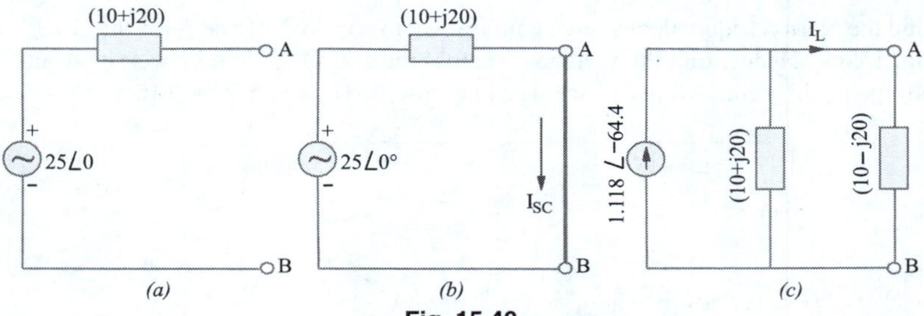

Fig. 15.40

When the load impedance of $(10 - j20)$ is applied across the terminals A and B, current through it can be found with the help of current-divider rule.

$$\therefore \qquad I_L = 1.118 \angle{-63.4^\circ} \times \frac{(10 + j20)}{(10 + j20) + (10 - j20)} = 1.25 \text{ A}$$

Example 15.16. *Use Norton's theorem to find current in the load connected across terminals A and B of the circuit shown in Fig. 15.41 (a).*

Solution. The first step is to short-circuit terminals A and B as shown in Fig. 15.41 (a)*. The short across A and B not only short-circuits the load but the $(10 + j10)$ impedance as well.

$$I_N = 100 \angle 0^\circ / (-j10) = j10 = 10 \angle 90^\circ$$

Since the impedance of the Norton and Thevenin equivalent circuits is the same, $Z_N = 10 - j10$.

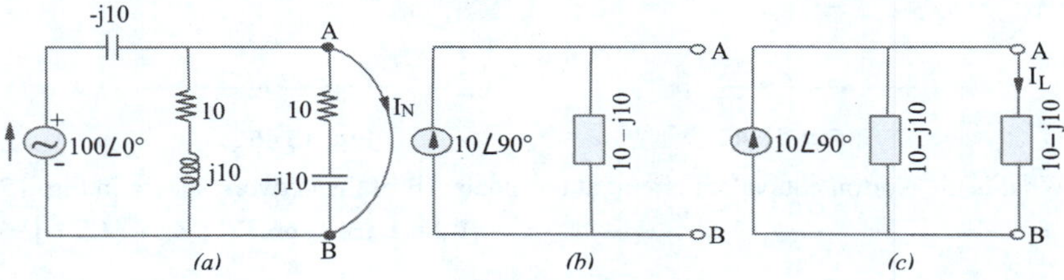

Fig. 15.41

The Norton's equivalent circuit is shown in Fig. 15.41 (b). In Fig. 15.41 (c), the load has been reconnected across the terminals A and B. Since the two impedances are equal, current through each is half of the total current *i.e.* $10\angle 90^\circ/2 = 5 \angle 90^\circ$.

Tutorial Problem 15.7

1. Find the Norton's equivalent source with respect to terminals A and B of the networks shown in Fig. 15.51 (a) (b). All resistances and reactances are expressed in siemens in Fig. 15.51 (a) and in ohms in Fig. 15.42.

[(*a*) I_N = - (2.1 - j3) A; $1/Z_N$ = (0.39 + j0.3)S (b) I_N = (6.87 + j0.5) A; $1/Z_N$ = (3.17 + j1.46S]

Fig. 15.42

Fig. 15.43

* For finding I_N, we may or may not remove the load from the terminals (because, in either case, it would be short-circuited) but for finding R_N, it has to be removed as in the case of Thevenin's theorem.

2. Find the Nortons equivalent source with respect to terminals A and B for the circuit shown in Fig. 15.44. Hence, find the voltage V_L across the 100 Ω load and check its result by using Millman's theorem. All resistances are in ohms. **[$I_N = 9A$; $Y_N = 0.15\ S$; $V_L = 56.25\ \angle 0°$]**

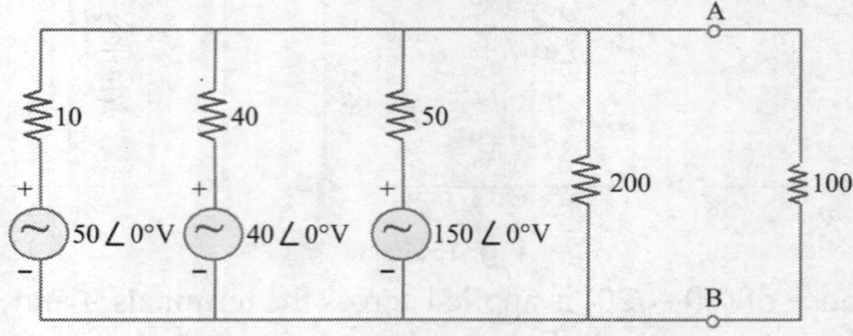

Fig. 15.44

3. Find the Norton's equivalent network at terminals AB of the circuit shown in Fig. 15.45.

[$I_{SC} = 2.17\ \angle{-44°}$ A; $Z_N = (2.4 + j1.47)\ \Omega$]

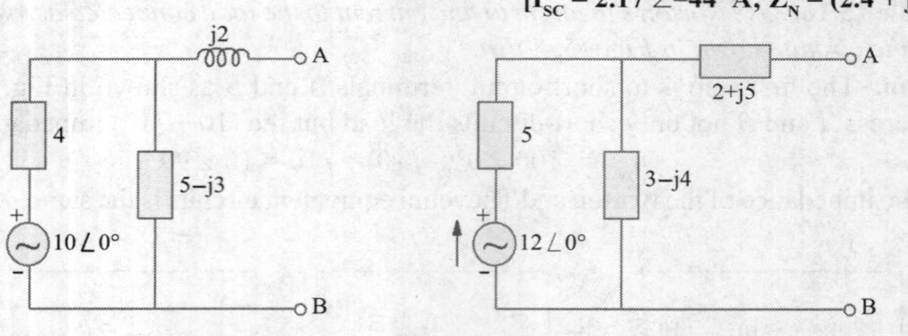

Fig. 15.45 **Fig. 15.46**

4. What is the Norton equivalent circuit at terminals AB of the network shown in Fig. 15.46

[$I_{SC} = 1.15\ \angle{-66.37°}$; $Z_N = (4.5 + j3.75)\ \Omega$]

15.9. Maximum Power Transfer Theorem

As explained earlier in Art. this theorem is particularly useful for analysing communication networks where the goals is transfer of maximum power between two circuits and not highest efficiency.

15.10. Maximum Power Transfer Theorems - General Case

We will consider the following maximum power transfer theorems when the source has a fixed complex impedance and delivers power to a load consisting of a variable resistance or a variable complex impedance.

Case 1. When load consists only for a variable resistance R_L [Fig. 15.47 (*a*)]. The circuit current is

$$I = \frac{V_g}{\sqrt{(R_g + R_L)^2 + X_g^2}}$$

Power delivered to R_L is P_L

$$= \frac{V_g^2 R_L}{(R_g + R_L)^2 + X_g^2}$$

To determine the value of R_L for maximum transfer of power, we should set the first derivative dP_L/dR_L to zero.

$$\frac{dP_L}{dR_L} = \frac{d}{dR_L}\left[\frac{V_g^2 R_L}{(R_g + R_L)^2 + X_g^2}\right] = V_g^2 \left\{\frac{[(R_g + R_L)^2 + X_g^2] - R_L(2)(Rg + R_L)}{[(R_g + R_L)^2 + X_g^2]^2}\right\} = 0$$

or $\quad R_g^2 + 2R_gR_L + R_L^2 + X_g^2 - 2R_LR_g - 2R_L^2 = 0$ and $R_g^2 + X_g^2 = R_L^2$

$$\therefore \qquad R_L = \sqrt{R_g^2 + X_g^2} = |Z_g|$$

It means that with a variable pure resistive load, maximum power is delivered across the terminals of an active network only when the load resistance is equal to the absolute value of the impedance of the active network. Such a match is called magnitude match.

Moreover, if X_g is zero, then for maximum power transfer $R_L = R_g$

Case 2. Load impedance having both variable resistance and variable reactance [Fig. 15.47 (*b*)].

The circuit current is $I = \dfrac{V_g}{\sqrt{(R_g + R_L)^2 + (X_g + X_L)^2}}$

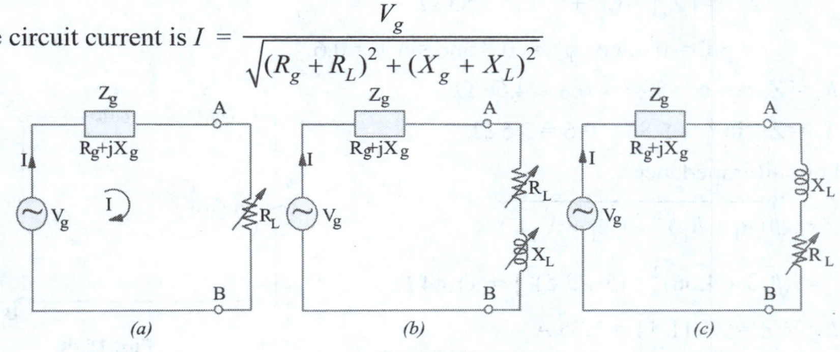

Fig. 15.47

The power delivered to the load is $= P_L = I^2R_L = \dfrac{V_g^2 R_L}{(R_g + R_L)^2 + (X_g + X_L)^2}$

Now, if R_L is held fixed, P_L is maximum when $X_g = -X_L$. In that case $P_{Lmax} = \dfrac{V_g^2 R_L}{(R_g + R_L)^2}$

If on the other hand, R_L is variable then, as in **Case 1** above, maximum power is delivered to the load when $R_L = R_g$. In that case if $R_L = R_g$ and $X_L = -X_g$, then $Z_L = Z_g$. Such a match is called conjugate match.

From the above, we come to the conclusion that in the case of a load impedance having both variable resistance and variable reactance, maximum power transfer across the terminals of the active network occurs when Z_L equals the complex conjugate of the network impedance Z_g *i.e.* the two impedances are conjugately matched.

Case 3. Z_L with variable resistance and fixed reactance [Fig. 15.47 (*c*)]. The equations for current *I* and power P_L are the same as in **Case** 2 above except that we will consider X_L to remain constant. When the first derivative of P_L with respect to R_L is set equal to zero, it is found that

$$R_L^2 = R_g^2 + (X_g + X_L)^2 \text{ and } R_L = |Z_g + jZ_L|$$

Since Z_g and X_L are both fixed quantities, these can be combined into a single impedance. Then with R_L variable, **Case** 3 is reduced to **Case** 1 and the maximum power transfer takes place when R_L equals the absolute value of the network impedance.

Summary

The above facts can be summarized as under :

1. When load is purely resistive and adjustable, MPT is achieved when

$R_L = |Z_g| = \sqrt{R_g^2 + X_g^2}$.

2. When both load and source impedances are purely resistive (*i.e.* $X_L = X_g = 0$), MPT is achieved when $R_L = R_g$.

3. When R_L and X_L are both independently adjustable, MPT is achieved when $X_L = -X_g$ and $R_L = R_g$.

4. When X_L is fixed and R_L is adjustable, MPT is achieved when $R_L = \sqrt{[R_g^2 + (X_g + X_L)^2]}$

Example 15.17. *In the circuit of Fig. 15.48, which load impedance of p.f. = 0.8 lagging when connected across terminals A and B will draw the maximum power from the source. Also find the power developed in the load and the power loss in the source.*

Solution. For maximum power transfer

$$|Z_L| = |Z_1| \sqrt{(3^2 + 5^2)} = 5.83\ \Omega.$$

For p.f. = 0.8, cos φ = 0.8 and sin φ = 0.6.

$\therefore$ $R_L = Z_L \cos\phi = 5.83 \times 0.8 = 4.66\ \Omega.$

$X_L = Z_L \sin\phi = 5.83 \times 0.6 = 3.5\ \Omega.$

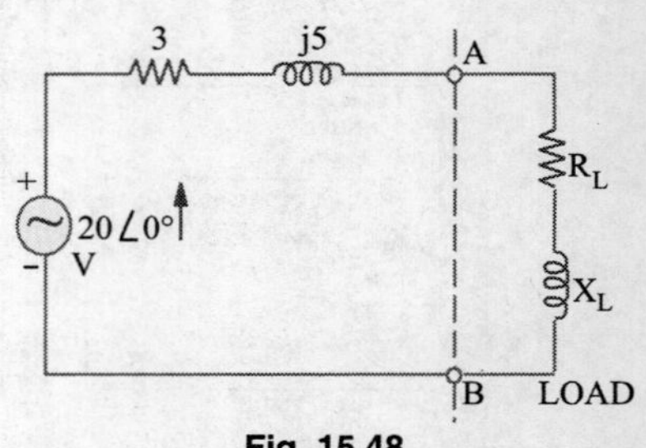

Fig. 15.48

Total circuit impedance

$$Z = \sqrt{[(R_1 + R_L)^2 + (X_1 + X_L)^2]}$$

$$= \sqrt{[(3 + 4.66)^2 + (5 + 3.5)^2]} = 11.44\ \Omega$$

$\therefore$ $I = V/Z = 20/11.44 = 1.75$ A.

Power in the load = $I^2R_L = 1.75^2 \times 4.66 = 14.3$ W

Power loss in the source = $1.75^2 \times 3 = 9.2$ W.

Example 15.18. *In the network shown in Fig. 15.49 find the value of load to be connected across terminals AB consisting of variable resistance R_L and capacitive reactance X_C which would result in maximum power transfer.* **(Network Analysis, Nagpur Univ. 1993)**

Solution. We will first find the Thevenin's equivalent circuit between terminals *A* and *B*. When the load is removed, the circuit become as shown in Fig. 15.49 (*b*).

$$V_{th} = \text{drop across } (2 + j10) = 50\angle 45^\circ \times \frac{2 + j\,10}{7 + j\,10}$$

$$= 41.8 \times 68.7^\circ = 15.2 + j\,38.9$$

$$R_{th} = 5\,||\,(2 + j\,10) = 4.1\angle 23.7^\circ = 3.7 + j\,1.6$$

(*a*) (*b*) (*c*)

Fig. 15.49

The Thevenin's equivalent source is shown in Fig. 15.49 (*c*)

Since for MPT, conjugate match is required hence, $X_C = 1.6\ \Omega$ and $R_L = 3.7\ \Omega$.

Tutorial Problem 15.8

1. In the circuit of Fig. 15.50 the load consists of a fixed inductance having a reactance of $j10$ Ω and a variable load resistor R_L. Find the value of R_L for MPT and the value of this power. **[58.3 Ω ; 46.2 W]**
2. In the circuit of Fig. 15.51, the source resistance R_g is variable between 5 Ω and 50 Ω but R_L has a fixed value of 25 Ω. Find the value of R_g for which maximum power is dissipated in the load and the value of this power. **[5 Ω ; 250 W]**

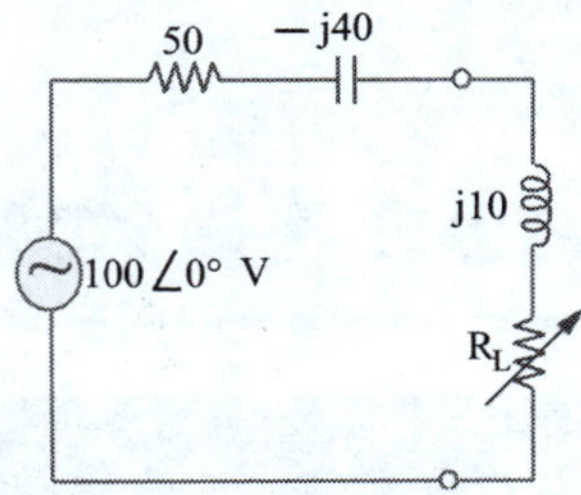

Fig. 15.50

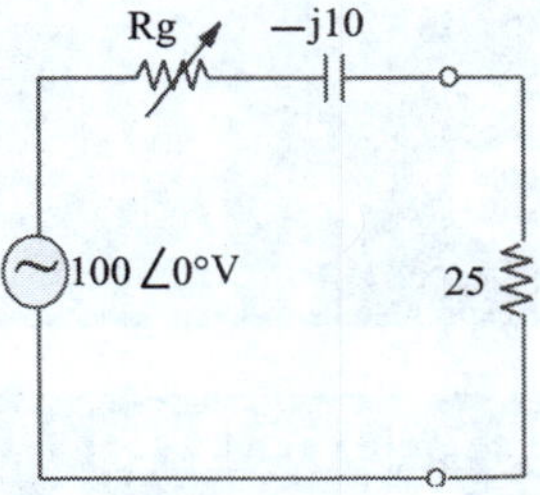

Fig. 15.51

OBJECTIVE TYPES – 15

1. The Thevenin's equivalent resistance R_{th} for the given network is

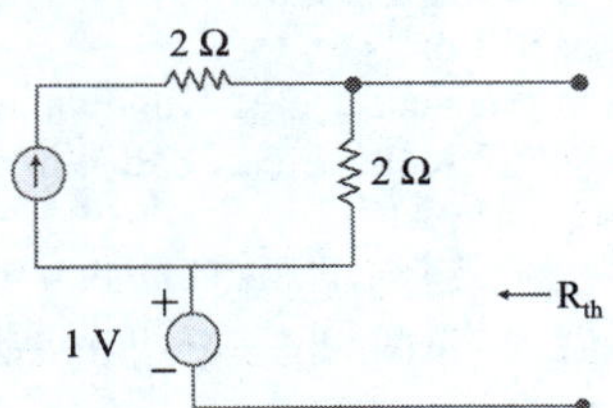

Fig. 15.52

(*a*) 1 Ω (*b*) 2 Ω
(*c*) 4 Ω (*d*) infinity

(ESE 2001)

2. The Norton's equivalent of circuit shown in Figure 15.53 is drawn in the circuit shown in Figure 15.54 The values of I_{SC} and R_{eq} in Figure II are respectively

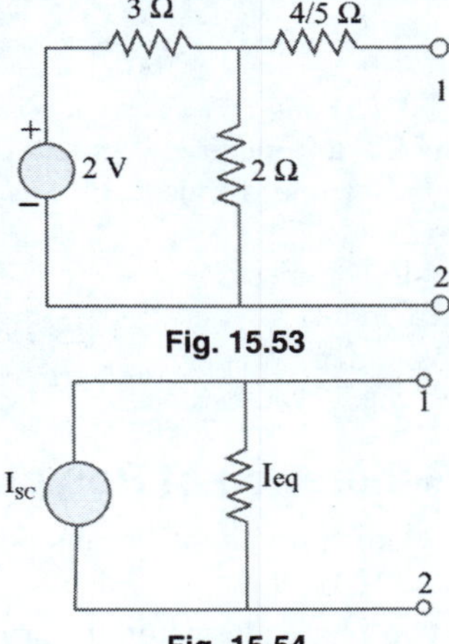

Fig. 15.54

(*a*) $\frac{5}{2}$ A and 2 Ω (*b*) $\frac{2}{5}$ A and 1 Ω
(*c*) $\frac{4}{5}$ A and $\frac{12}{5}$ Ω (*d*) $\frac{2}{5}$ A and 2 Ω

CHAPTER

16 Polyphase Circuits

Chapter Outline

- Generation of Polyphase Voltages
- Phase Sequence
- Interconnection of Three Phases
- Star or Wye (Y) Connection
- Voltages and Currents in Y-Connection
- Delta (D) or Mesh Connection
- Balanced Y/D and D/Y Conversions
- Star and Delta Connected Lighting Loads
- Power Factor Improvement
- Parallel Loads
- Power Measurement in 3-phase Circuits
- Three Wattmeter Method
- Two Wattmeter Method
- Balanced or Unbalanced load
- Variations in Wattmeter Readings
- Leading Power Factor
- Power Factor-Balanced Load
- Reactive Voltamperes with One Wattmeter
- One Wattmeter Method
- Double Subscript Notation
- Unbalanced Loads
- Four-wire Star-connected Unbalanced Load
- Unbalanced Y-connected Load Without Neutral
- Delta/Star and Star/Delta Conversions
- Unbalanced Star-connected Non-inductive Load
- Phase Sequence Indicators

16.1. Generation of Polyphase Voltage

The word 'polyphase' means poly (*i.e.* many or numerous) and phases (*i.e.* winding or circuit).

In a two-phase alternator, the armature windings are displaced 90 electrical degrees apart. A 3-phase alternator, as the name shows, has three independent armature windings which are 120 electrical degrees apart. Hence, the voltages induced in the three windings are 120° apart in time-phase. With the exception of two-phase windings, it can be stated that, in general, the electrical displacement between different phases is $360/n$ where n is the number of phases or windings.

In Fig. 16.1 is shown a two-pole, stationary-armature, rotating-field type three-phase alternator. It has three armature coils aa', bb' and cc' displaced 120° apart from one another. With the position and clockwise rotation of the poles as indicated in Fig. 16.1, it is found that the e.m.f. induced in conductor 'a' for coil aa' is maximum and its direction* is away

* The direction is found with the help of Fleming's Right-hand rule. But while applying this rule, it should be remembered that the relative motion of the conductor with respect to the field is anticlockwise although the motion of the field with respect to the conductor is clockwise as shown. Hence, thumb should point to the left.

from the reader. The e.m.f. in conductor '*b*' of coil *bb*′ would be maximum and away from the reader when the N-pole has turned through 120° *i.e.* when *N-S* axis lies along *bb*′. It is clear that the induced e.m.f. in conductor '*b*' reaches its maximum value 120° later than the maximum value in conductor '*a*'. In the like manner, the maximum e.m.f. induced (in the direction away from the reader) in conductor '*c*' would occur 120° later than that in '*b*' or 240° later than that in '*a*'.

Thus the three coils have three e.m.fs. induced in them which are similar in all respects except that they are 120° out of time phase with one another as pictured in Fig. 16.3. Each voltage wave is assumed to be sinusoidal and having maximum value of E_m.

In practice, the space on the armature is completely covered and there are many slots per phase per pole.

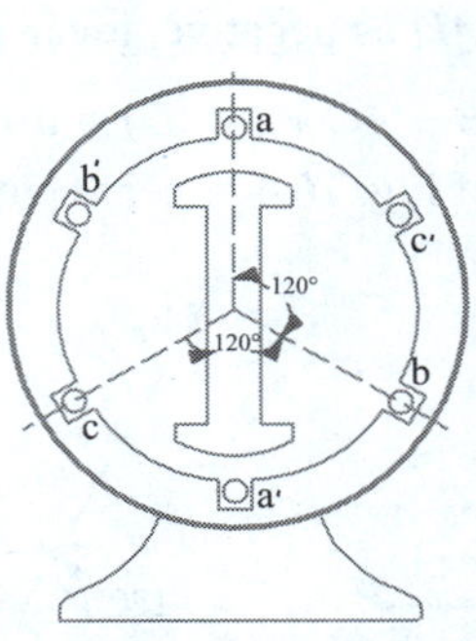

Fig. 16.1

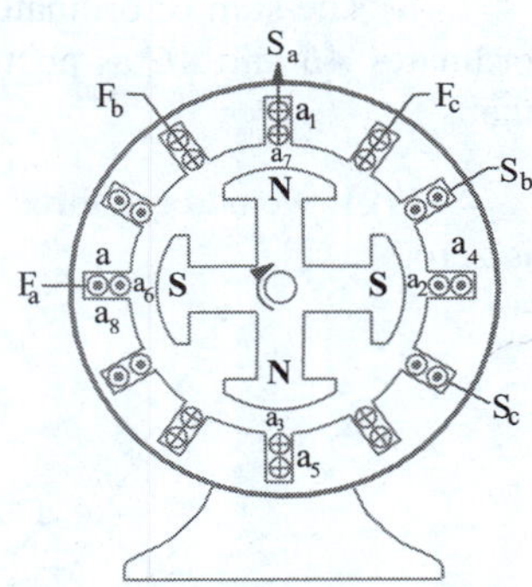

Fig. 16.2

Fig. 16.2 illustrates the relative positions of the windings of a 3-phase, 4-pole alternator. Assuming full-pitched winding and the direction of rotation as shown, phase '*a*' occupies the position under the centres of *N* and *S*-poles. It starts at S_a and ends or finishes at F_a.

The second phase '*b*' start at S_b which is 120 electrical degrees apart from the start of phase '*a*', progresses round the armature clockwise (as does '*a*') and finishes at F_b. Similarly, phase '*c*' starts at S_c, which is 120 electrical degrees away from S_b, progresses round the armature and finishes at F_c. As the three circuits are exactly similar but are 120 electrical degrees apart, the e.m.f. waves generated in them (when the field rotates) are displaced from each other by 120°. Assuming these waves to be sinusoidal and counting the time from the instant when the e.m.f. in phase '*a*' is zero, the instantaneous values of the three e.m.fs. will be given by curves of Fig. 16.3.

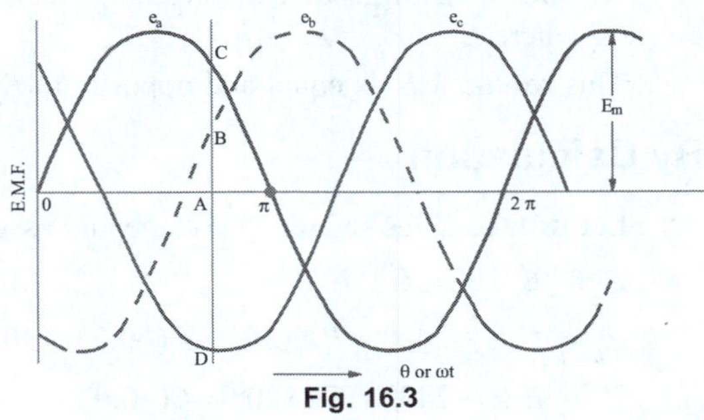

Fig. 16.3

Their equations are:

$$e_a = E_m \sin \omega t \qquad \text{... (i)}$$

$$e_b = E_m \sin(\omega t - 120°) \qquad \text{... (ii)}$$

$$e_c = E_m \sin(\omega t - 240°) \qquad \text{... (iii)}$$

The actual values of these voltages vary from peak positive to zero and to peak negative values in one revolution of the vectors. In Fig. 16.4 are shown the three vectors representing the r.m.s. voltages of the three phases E_a, E_b and E_c (in the present case $E_a = E_b = E_c = E$, say).

It can be shown that the sum of the three phase e.m.fs. is zero in the following three ways :

(i) The sum of the above three equations ***(i)***, ***(ii)*** and ***(iii)*** is zero as shown below:

Resultant instantaneous e.m.f. $= e_a + e_b + e_c$

$$= E_m \sin \omega t + E_m \sin(\omega t - 120°) + E_m (\omega t - 240)°$$

$$= E_m[\sin \omega t + 2\sin(\omega t - 180°)\cos 60°]$$

$$= E_m [\sin \omega t - 2\sin \omega t \cos 60°] = 0$$

(ii) The sum of ordinates of three e.m.f. curves of Fig. 16.3 is zero. For example, taking ordinates *AB* and *AC* as positive and *AD* as negative, it can be shown by actual measurement that

$$AB + AC + (-AD) = 0$$

(iii) If we add the three vectors of Fig. 16.4 either vectorially or by calculation, the result is zero.

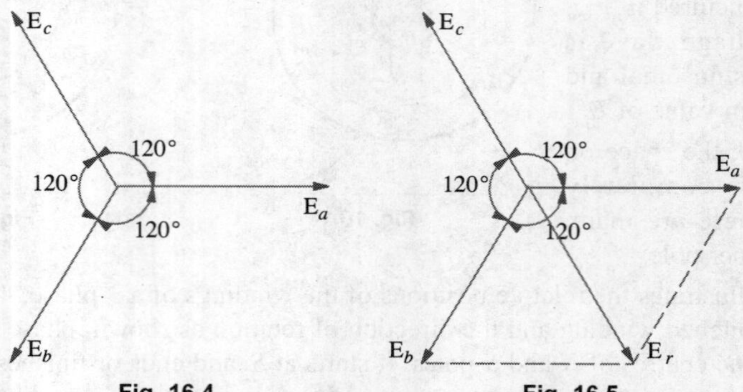

Fig. 16.4 **Fig. 16.5**

Vector Addition

As shown in Fig. 16.5, the resultant of E_a and E_b is E_r and its magnitude is 2*E* cos 60° = *E* where $E_a = E_b = E_c = E$.

This resultant E_r is equal and opposite to E_c. Hence, their resultant is zero.

By Calculation

Let us take E_a as reference voltage and assuming clockwise phase sequence

$\boldsymbol{E_a} = E\angle 0° = E + j0$

$\boldsymbol{E_b} = E\angle -240° = E\angle 120° = E(-0.05 + j0.866)$

$\boldsymbol{E_c} = E\angle -240° = E\angle 120° = E(-0.05 + j0.866)$

$\therefore\ \boldsymbol{E_a} + \boldsymbol{E_b} + \boldsymbol{E_c} = (E + j0) + E(-0.5 - 0.866) + E(-0.05 + j0.866) = 0$

16.2. Phase Sequence

By phase sequence is meant the order in which the three phases attain their peak or maximum values. In the development of the three-phase e.m.fs. in Fig. 16.6, clockwise rotation of the field system in Fig. 16.1 was assumed. This assumption made the e.m.fs. of phase '*b*' lag behind that of '*a*' by 120° and in a similar way, made that of '*c*' lag behind that of '*b*' by 120° (or that of '*a*' by 240°). Hence, the order in which the e.m.fs. of phases *a*, *b* and *c* attain their maximum values is *a b c*. ***It is called the phase order* or *phase sequence*** $a \rightarrow b \rightarrow c$ as illustrated in Fig. 16.6 (*a*).

If, now, the rotation of the field structure of Fig. 16.1 is reversed *i.e.* made anticlockwise, then the order in which the three phases would attain their corresponding maximum voltages would also be reversed. The phase sequence would become $a \rightarrow b \rightarrow c$. This means that e.m.f. of phase '*c*' would now lag behind that of phase '*a*' by 120° instead of 240° as in the previous case as shown in Fig. 16.6 (*b*). By repeating the letters, this phase sequence can be written as ***acbacba*** which is the same thing as ***cba***. Obviously, a three-phase system has only two possible sequences: ***abc*** and ***cba*** (*i.e.* ***abc*** read in the reverse direction).

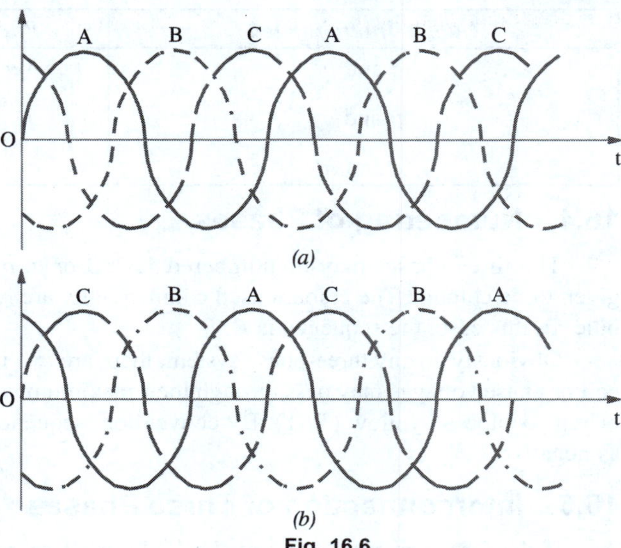

Fig. 16.6

16.3. Phase Sequence At Load

In general, the phase sequence of the voltages applied to load is determined by the order in which the 3-phase lines are connected. The phase sequence can be reversed by interchanging any pair of lines. In the case of an induction motor, reversal of sequence results in the reversed direction of motor rotation. In the case of 3-phase unbalanced loads, the effect of sequence reversal is, in general, to cause a completely different set of values of the currents. Hence, when working on such systems, it is essential that phase sequence be clearly specified otherwise unnecessary confusion will arise. Incidentally, reversing the phase sequence of a 3-phase generator which is to be paralleled with a similar generator can cause extensive damage to both the machines.

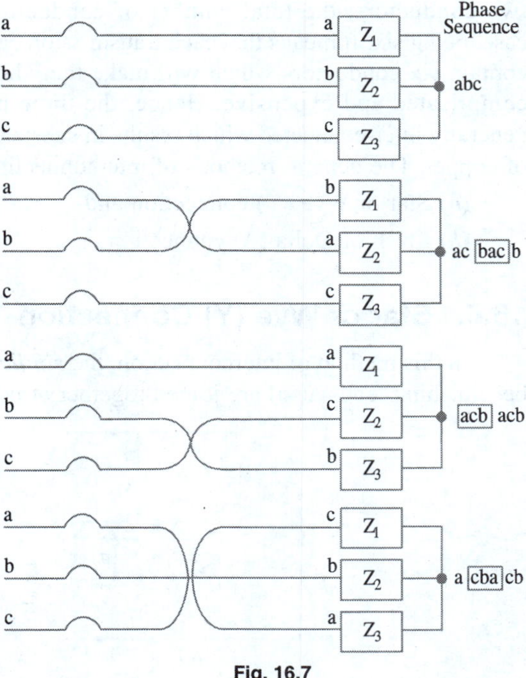

Fig. 16.7

Fig. 16.7 illustrates the fact that by interchanging any two of the three cables the phase sequence ***at the load*** can be reversed though sequence of 3-phase supply remains the same *i.e.* ***abc***. It is customary to define phase sequence at the load by reading repetitively from top to bottom. For example, load phase sequence in Fig. 16.7 (*a*) would be read as ***abcabcabc***– or simply ***abc***. The changes are as tabulated:

Cables Interchanged	*Phase Sequence*
a and *b*	*b a* *c b a c b a* *c* – or *c b a*
b and *c*	*a* *c b a c b a* *c b* – or *c b a*
c and *a*	*c* *b a c b a* *c b a* – or *c b a*

16.4. Numbering of Phases

The three phases may be numbered 1, 2, 3 or *a*, *b*, *c* or as is customary, they may be given three colours. The colours used commercially are red, yellow (or sometimes white) and blue. In this case, the sequence is *RYB*.

Obviously, in any three-phase system, there are two possible sequences in which the three coil or phase voltages may pass through their maximum values *i.e.* red → yellow → blue (*RYB*) or red → blue → yellow (*RBY*). By convention, sequence *RYB* is taken as positive and *RBY* as negative.

16.5. Interconnection of Three Phases

If the three armature coils of the 3-phase alternator (Fig. 16.7) are not interconnected but are kept separate, as shown in Fig. 16.8, then each phase or circuit would need two conductors, the total number of conductors, in that case, being six. It means that each transmission cable would contain six conductors which will make the whole system complicated and expensive. Hence, the three phases are generally inter-connected which results in substantial saving of copper. The general methods of interconnection are

(a) Star or Wye (Y) connection and

(b) Mesh or Delta (Δ) connection.

Fig. 16.8

16.6. Star or Wye (Y) Connection

In this method of interconnection, the ***similar****** ends say, 'star' ends of three coils (it could be 'finishing' ends also) are joined together at point *N* as shown in Fig. 16.9 (*a*).

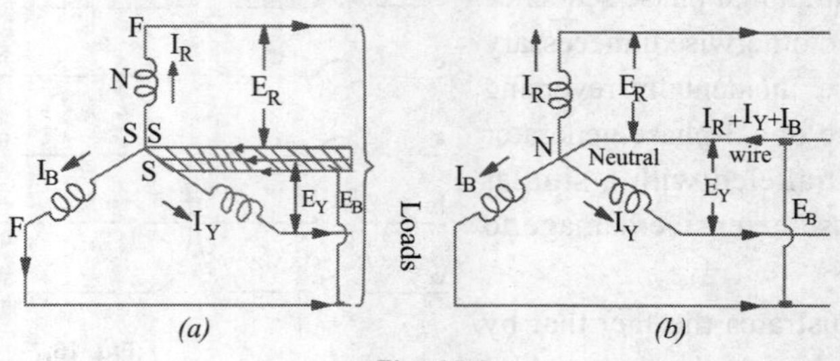

Fig. 16.9

The point *N* is known as ***star point*** or ***neutral point***. The three conductors meeting at point *N* are replaced by a single conductor known as ***neutral conductor*** as shown in

* As an aid to memory, remember that first letter *S* of Similar is the same as that of Star.

Fig. 16.9 (*b*). Such an interconnected system is known as four-wire, 3-phase system and is diagrammatically shown in Fig. 16.9 (*b*). If this three-phase voltage system is applied across a balanced symmetrical load, the neutral wire will be carrying three currents which are exactly equal in magnitude but are 120° out of phase with each other. Hence, their vector sum is zero.

i.e. $$\mathbf{I_R + I_Y + I_B = 0} \quad \text{... vectorially}$$

The neutral wire, in that case, may be omitted although its retention is useful for supplying lighting loads at low voltages. The p.d. between any terminal (or line) and neutral (or star) point gives the ***phase*** or ***star*** voltage. But the p.d. between any two lines gives the line-to-line voltage or simply line voltage.

16.7. Values of Phase Currents

When considering the distribution of current in a 3-phase system, it is extremely important to bear in mind that:

(i) the arrow placed alongside the currents I_R I_Y and I_B flowing in the three phases [Fig. 16.9 (*b*)] indicate the directions of currents when they are assumed to be ***positive*** and not the directions at a particular instant. It should be clearly understood ***that at no instant will all the three currents flow in the same direction either outwards or inwards.*** The three arrows indicate that first the current flows outwards in phase *R*, then after a phase-time of 120°, it will flow outwards from phase *Y* and after a further 120°, outwards from phase *B*.

(ii) the current flowing outwards in one or two conductors is always equal to that flowing inwards in the remaining conductor or conductors. In other words, ***each conductor in turn, provides a return path for the currents of the other conductors.***

In Fig. 16.10 are shown the three phase currents, having the same peak value of 20 A but displaced from each other by 120°. At instant '*a*', the currents in phases *R* and *B* are each + 10 A (*i.e.* flowing outwards) whereas the current in phase *Y* is – 20A (*i.e.* flowing inwards). In other words, at the instant '*a*', phase *Y* is acting as return path for the currents in phases *R* and *B*. At instant *b*, I_R = +15 A and I_Y = +5 A but I_B = –20A which means that now phase *B* is providing the return path.

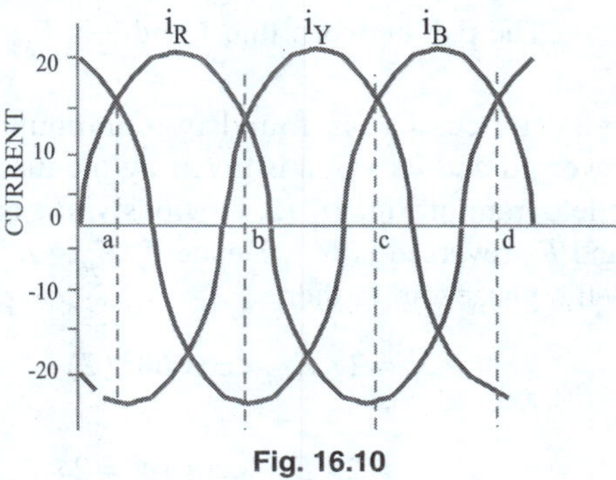

Fig. 16.10

At instant *c*, I_Y = +15 *A* and I_B = +5A and I_R = – 20A.

Hence, now phase *R* carries current inwards whereas *Y* and *B* carry current outwards. Similarly at point *d*, I_R = 0, I_B = 17.3 A and I_Y = – 17.3 A. In other words, current is flowing outwards from phase *B* and returning *via* phase *Y*.

In addition, it may be noted that although the distribution of currents between the three lines is continuously changing, yet at any instant the algebraic sum of the ***instantaneous*** values of the three currents is zero *i.e.*

$$i_R + i_Y + i_B = 0 \quad \text{– algebraically.}$$

16.8. Voltages and Currents in (Star) Y-Connection

The voltage induced in each winding is called the ***phase*** voltage V_{ph} and current in each winding is likewise known as ***phase*** current I_{ph}. However, the voltage available between any pair of terminals (or outers) is called ***line*** voltage (V_L) and the current flowing in each ***line*** is called ***line*** current (I_L).

As seen from Fig. 16.11 (*a*), in this form of interconnection, there are two phase windings between each pair of terminals but since their ***similar*** ends have been joined together, they are in opposition. Obviously, the ***instantaneous*** value of p.d. between any two terminals is the ***arithmetic difference*** of the two phase e.m.fs. concerned. However, the r.m.s. value of this p.d. is given by the ***vector difference*** of the two phase e.m.fs.

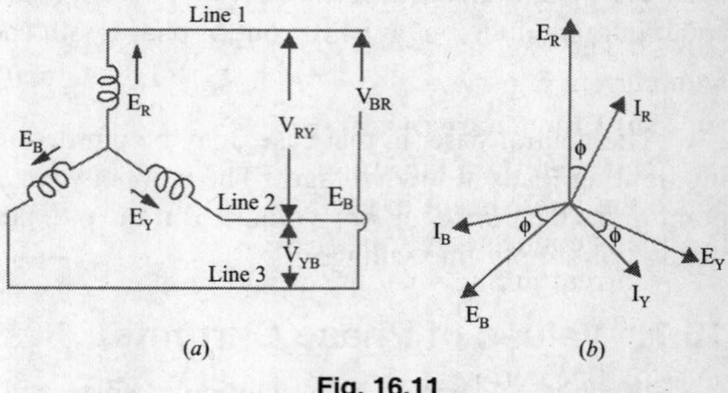

Fig. 16.11

The vector diagram for phase voltages and currents in a star connection is shown in Fig. 16.11.

(*b*) where a balanced system has been assumed.* It means that $E_R = E_Y = E_{ph}$ (phase e.m.f.).

Line voltage V_{RY} between line 1 and line 2 is the vector difference of E_R and E_Y.
Line voltage V_{YB} between line 2 and line 3 is the vector difference of E_Y and E_B.
Line voltage V_{BR} between line 3 and line 1 is the vector difference of E_B and E_R.

(*a*) Line Voltages and Phase Voltages

The p.d. between line 1 and 2 is $V_{RY} = E_R - E_Y$ ... vector difference.

Hence, V_{RY} is found by compounding E_R and E_Y reversed and its value is given by the diagonal of the parallelogram of Fig. 16.12. Obviously, the angle between E_R and E_Y reversed is 60°. Hence if $E_R = E_Y = E_B =$ say, E_{ph} – the phase e.m.f., then

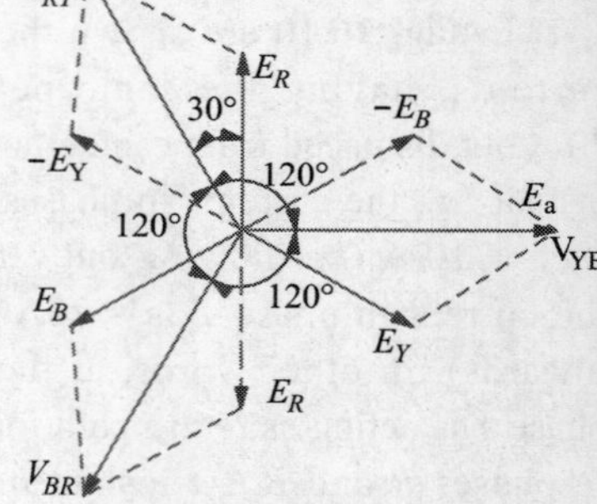

Fig. 16.12

$$V_{RY} = 2 \times E_{ph} \times \cos(60°/2)$$

$$= 2 \times E_{ph} \times \cos 30° = 2 \times E_{ph} \times \frac{\sqrt{3}}{2} = \sqrt{3}\, E_{ph}$$

Similarly, $V_{YB} = E_Y - E_B = \sqrt{3} \cdot E_{ph}$

...vector difference

and $V_{BR} = E_B - E_R = \sqrt{3} \cdot E_{ph}$

Now $V_{RY} = V_{YB} = Y_{BR} =$ line voltage, say V_L. Hence, in star connection $\boxed{V_L = \sqrt{3} \cdot E_{ph}}$

* A balanced system is one in which (*i*) the voltages in all phases are equal in magnitude and offer in phase from one another by equal angles, in this case, the angle = 360/3 = 120°, (*ii*) the currents in the three phases are equal in magnitude and also differ in phase from one another by equal angles. A 3-phase balanced load is that in which the loads connected across three phases are identical.

It will be noted from Fig. 16.13 that

1. Line voltages are 120° apart.

2. Line voltages are 30° ahead of their respective ***phase*** voltages.

3. The angle between the line currents and the corresponding line voltages is $(30 + \phi)$ with current lagging.

(*b*) Line Currents and Phase Currents

It is seen from Fig. 16.11 (*a*) that each line is in series with its individual phase winding, hence the line current in each line is the same as the current in the phase winding to which the line is connected.

Current in line 1 = I_R ; Current in line 2 = I_Y; Current in line 3 = I_B

Since $I_R = I_Y = I_B =$ say, I_{ph} – the phase current

$\therefore$ line current $\boxed{I_L = I_{ph}}$

(*c*) Power

The total active or true power in the circuit is the sum of the three phase powers. Hence, total active power = 3 × phase power or $P = 3 \times V_{ph} I_{ph} \cos\phi$

Now $V_{ph} = V_L / \sqrt{3}$ and $I_{ph} = I_L$

Hence, in terms of line values, the above expression becomes

$$P = 3 \times \frac{V_L}{\sqrt{3}} \times I_L \times \cos\phi \text{ or } \boxed{P = \sqrt{3}\, V_L I_L \cos\phi}$$

It should be particularly noted that ϕ is the angle between ***phase*** voltage and ***phase*** current and not between the line voltage and line current.

Similarly, total reactive power is given by $\boxed{Q = \sqrt{3}\, V_L I_L \sin\phi}$

By convention, reactive power of a coil is taken as positive and that of a capacitor as negative.

The total apparent power of the three phases is

$\boxed{S = \sqrt{3}\, V_L I_L}$ Obviously, $\boxed{S = \sqrt{P^2 + Q^2}}$

Example 16.1. *A balanced star-connected load of (8 + j6) Ω per phase is connected to a balanced 3-phase 400-V supply. Find the line current, power factor, power and total volt-amperes.* **(Elect. Engg., Bhagalpur Univ.)**

Solution. $Z_{ph} = \sqrt{8^2 + 6^2} = 10\,\Omega$

$V_{ph} = 400 / \sqrt{3} = 231$ V

$I_{ph} = V_{ph} / Z_{ph} = 231/10 = 23.1$ A

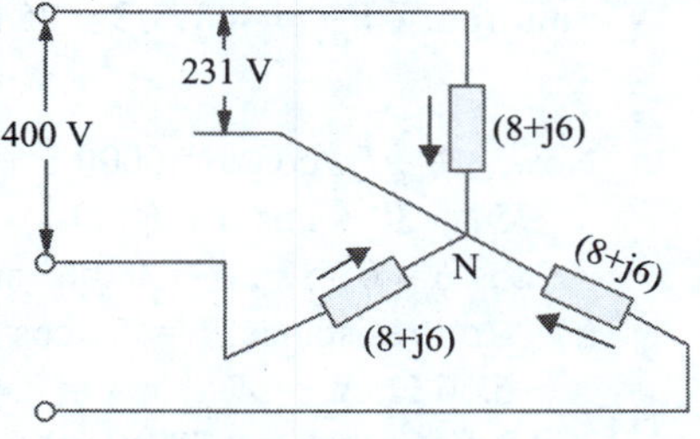

Fig. 16.13

(*i*) $I_L = I_{ph}$ = **23.1 A**

(*ii*) p.f. = $\cos\phi = R_{ph}/Z_{ph}$ = 8/10 = **0.8 (lag)**

(*iii*) Power $P = \sqrt{3}\, V_L I_L \cos\phi$

$= \sqrt{3} \times 400 \times 23.1 \times 0.8$ = **12,800 W**

[Also, $P = 3I_{ph}^2 \; R_{ph} = 3(23.1)^2 \times 8 = 12{,}800$ *W*]

(*iv*) Total volt-amperes, $S = \sqrt{3}\; V_L I_L = \sqrt{3} \times 400 \times 23.1$ = **16,000 VA**

Example 16.2. *Phase voltages of a star connected alternator are $E_R = 231 \angle 0°$ V; $E_Y = 231 \angle -120°$ V ; and $E_B = 231 \angle +120°$V. What is the phase sequence of the system ? Compute the line voltages E_{RY} and E_{YB}.* **(Elect. Mechines AMIE Sec. B Winter 1990)**

Solution. The phase voltage $E_B = 231 \angle -120°$ can be written as $E_B = 231 \angle -240°$. Hence, the three voltages are: $E_R = 231 \angle -0°$, $E_Y = 231 \angle -120°$ and $E_B = 231 \angle -240°$. It is seen that E_R is the reference voltage, E_Y lags behind it by 120° whereas E_B lags behind it by 240°. Hence, phase sequence is *RYB*. Moreover, it is a symmetrical 3-phase voltage system.

$$\therefore \quad E_{RY} = E_{YB} = \sqrt{3}\times 231 = \mathbf{400\ V}$$

Example 16.3 *Three equal star-connected inductors take 8 kW at a power factor 0.8 when connected across a 460 V, 3-phase, 3-phase, 3-wire supply. Find the circuit constants of the load per phase.* **(Elect. Machines AMIE Sec. B 1992)**

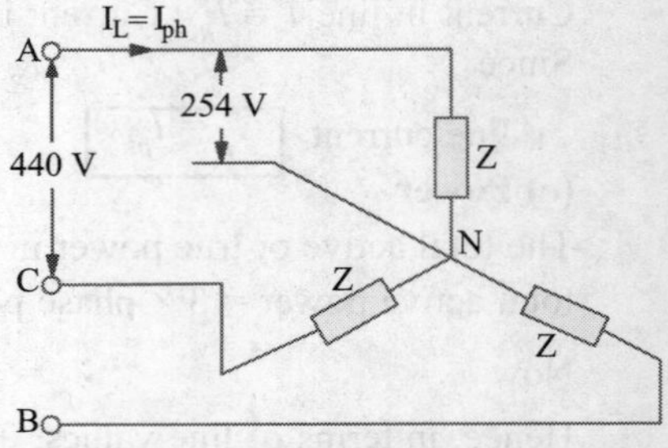

Solution. $P = \sqrt{3} V_L I_L \cos\phi$ or

$8000 = \sqrt{3}\times 460 \times I_L \times 0.8$

$\therefore \quad I_L = 12.55$ A $\therefore I_{ph} = 12.55$ *A*;

$V_{ph} = V_L / \sqrt{3} = 460/\sqrt{3} = 265 V$

$I_{ph} = V_{ph}/Z_{ph}; \therefore Z_{ph} = V_{ph}/I_{ph} = 265/12.55 = 21.1 \Omega$

$R_{ph} = Z_{ph}\cos\phi = 21.1\times 0.8 = 16.9\ \Omega$

$X_{ph} = Z_{ph}\sin\phi = 21.1\times 0.6 = 12.66 \Omega$;

The circuit is shown in Fig. 16.14.

Fig. 16.14

Example 16.4. *When the three identical star-connected coils are supplied with 440 V, 50Hz, 3-φ supply, the 1-φ wattmeter whose current coil is connected in line R and pressure coil across the phase R and neutral reads 6 kW and the ammeter connected in R-phase reads 30 Amp. Assuming RYB phase sequence find:*

(i) *resistance and reactance of the coil,*

(ii) *the power factor, of the load*

(iii) *reactive power of 3-φ load.*

(Elect. Engg.-I, Nagpur Univ. 1993)

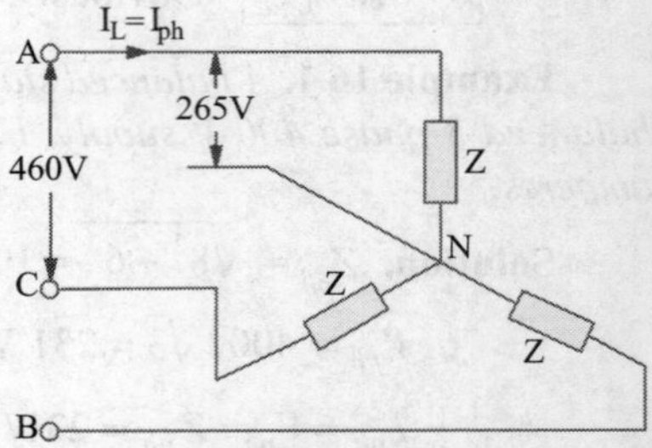

Solution. $V_{ph} = 440/\sqrt{3} = 254$ V; $I_{ph} = 30$A

(Fig. 16.15.)

Now, $V_{ph} I_{ph} \cos\phi = 6000$;

$254 \times 30 \times \cos\phi = 6000$

$\therefore \cos\phi = 0.787$; $\phi = 38.06°$ and $\sin\phi = 0.616°$; $Z_{ph} = V_{ph}/I_{ph} = 254/30 = 8.47\ \Omega$

(i) Coil resistance $R = Z_{ph}\cos\phi = 8.47 \times 0.787 =$ 6.66 Ω, $X_L = Z_{ph}\sin\phi = 8.47\times 0.616 =$ **5.22 Ω**

Fig. 16.15

(ii) p.f. = cos φ = **0.787 (lag)**

(iii) Reactive power = $\sqrt{3}\ V_L I_L \sin\phi = \sqrt{3}\times 440\times 30\times 0.616 = 14{,}083\text{VA} =$ **14.083 kVA**

Example 16.5. *Calculate the active and reactive components in each phase of Y-connected 10,000 V, 3-phase alternator supplying 5,000 kW at 0.8 p.f. If the total current remains the same when the load p.f is raised to 0.9, find the new output.*

(Elements of Elect. Engg.-I, Bangalore Univ.)

Solution. $5000 \times 10^3 = \sqrt{3} \times 10{,}000 \times I_L \times 0.8; I_L = I_{ph} = 361\text{A}$

active component = $I_L \cos\phi = 361 \times 0.8 =$ **288.8 A**
reactive component = $I_L \sin\phi = 361 \times 0.6 =$ **216.6 A**
New power $\quad P = \sqrt{3}V_L I_L \cos\phi = \sqrt{3} \times 10^4 \times 361 \times 0.9 =$ **5,625 kW**
[or new power = 5000 × 0.9/0.8 = 5625 kW]

Example 16.6. *Deduce the relationship between the phase and line voltages of a three-phase star-connected alternator. If the phase voltage of a 3-phase star-connected alternator be 200 V, what will be the line voltages **(a)** when the phases are correctly connected and **(b)** when the connections to one of the phases are reversed.*

Solution. (*a*) When phases are correctly connected, the vector diagram is as shown in Fig. 16.11. (*b*).

$$V_{RY} = V_{YB} V_{BR} = \sqrt{3}.E_{Ph}$$

Each line voltage = $\sqrt{3} \times 200 =$ **346 V**

(*b*) Suppose connections to *B*-phase have been reversed. Then voltage vector diagram for such a case is shown in Fig. 16.16. It should be noted that E_B has been drawn in the reversed direction, so that angles between the three-phase voltages are 60° (instead of the usual 120°)

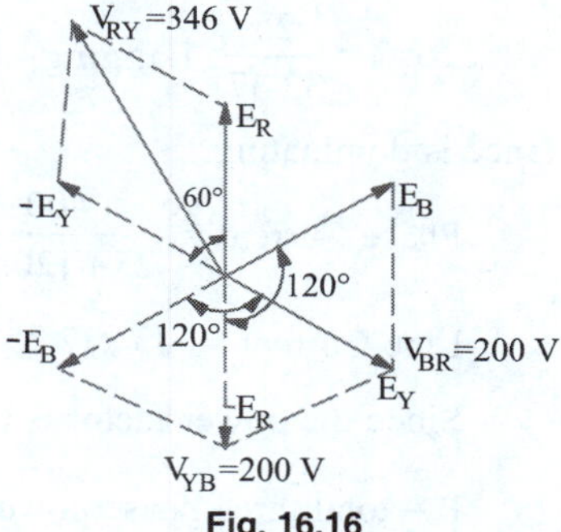

Fig. 16.16

$V_{RY} = E_R - E_Y$... vector difference

$= 2 \times E_{ph} \times \cos 30° = \sqrt{3} \times 200 =$ **346 V**

$V_{YB} = E_Y - E_B$... vector difference

$= 2 \times E_{ph} \times \cos 60° = 2 \times 200 \times \frac{1}{2} =$ **200V**

$V_{BR} = E_B - E_R$ ... vector difference = $2 \times E_{ph} \times \cos 60° = 2 \times 200 \times \frac{1}{2} =$ **200 V**

Example 16.7. *In a 4-wire, 3-phase system, two phases have currents of 10A and 6A at lagging power factors of 0.8 and 0.6 respectively while the third phase is open-circuited, Calculate the current in the neutral and sketch the vector diagram.*

Solution. The circuit is shown in Fig. 16.17 (*a*).

$\phi_1 = \cos^{-1}(0.8) = 36°54'; \phi_2 = \cos^{-1}(0.6) = 53°6'$

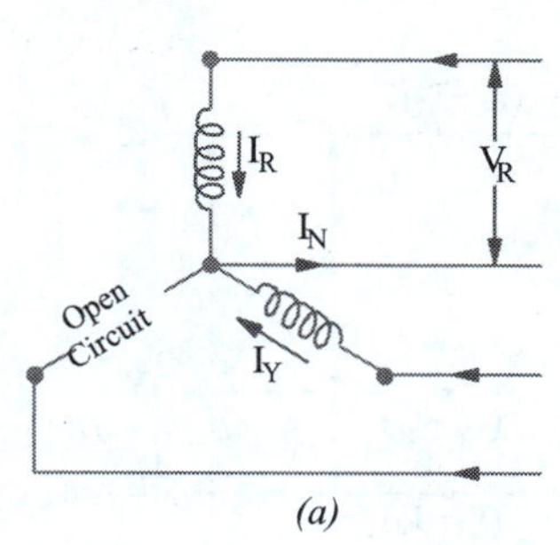

(*a*)

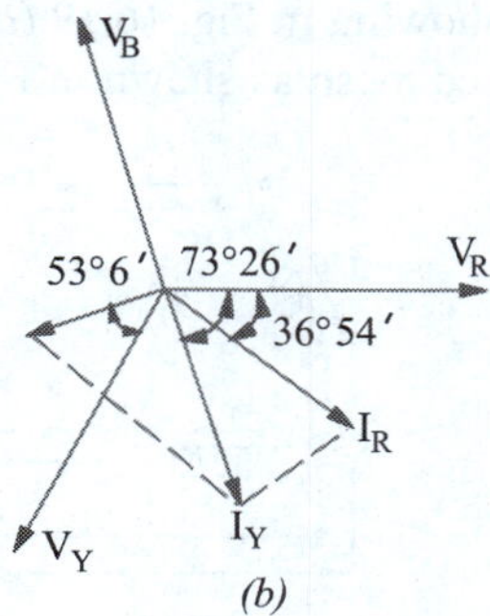

(*b*)

Fig. 16.17

Let V_R be taken as the reference vector. Then

$\mathbf{I}_R = 10\angle-36°54' = (8 - j6)$ $\mathbf{I}_y = 6\angle-173°6' = (-6 - j0.72)$

The neutral current $\mathbf{I}_N$, as shown in Fig. 16.16 (*b*), is the sum of these two currents.

$$\therefore \qquad \mathbf{I}_N = (8 - j6) + (-6 - j0.72) = 2 - j6.72 = 7\angle-73°26'$$

Example 16.8. *A three pahse 400-V, 50 Hz, a.c. supply is feeding a three phase delta-connected load with each phase having a resistance of 25 ohms, an inductance of 0.15 H, and a capacitor of 120 microfarads in series. Determine the line current, volt-amp, active power and reactive volt-amp.* **[Nagpur University, November 1999]**

Solution. Impedance per phase $r + jX_L - jX_C$

$$X_L = 2\pi\times50\times0.15 = 47.1\ \Omega$$

$$X_C = \frac{10^6}{32{\cdot}37} = 26.54\ \Omega$$

$$\cos\phi = \frac{25}{32.37}$$ Lagging, since inductive reactance is dominating.

Fig. 16.18

$$\text{Phase Current} = \frac{400}{25 + j20.56} = 12.357$$

$$\text{Line Current} = \sqrt{3}\times12.357 = 21.4 \text{ amp}$$

Since the power factor is 0.772 lagging,

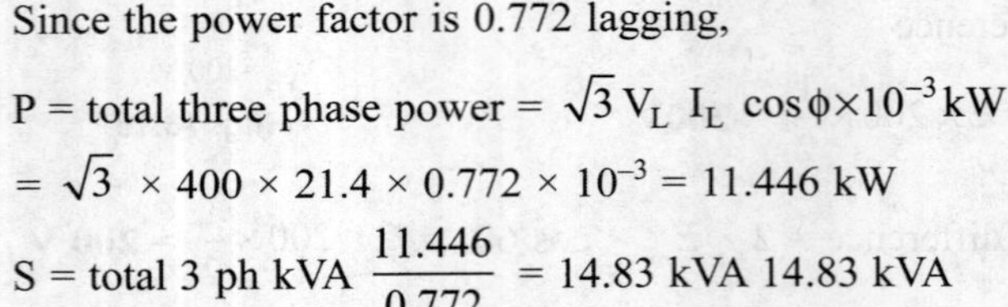

$$\text{P} = \text{total three phase power} = \sqrt{3}\,V_L\,I_L\,\cos\phi\times10^{-3}\,\text{kW}$$

$$= \sqrt{3}\times400\times21.4\times0.772\times10^{-3} = 11.446\ \text{kW}$$

$$\text{S} = \text{total 3 ph kVA } \frac{11.446}{0.772} = 14.83\ \text{kVA } 14.83\ \text{kVA}$$

Q = total 3 ph "reactive kilo-volt-amp" $\sqrt{3} = (S^2 - P^2)^{0.50} = 9.43$ kVAR lagging

16.9. Delta (Δ)* or Mesh Connection

In this form, of interconnection the ***dissimilar*** ends of the three phase winding are joined together *i.e.* the 'starting' end of one phase is joined to the 'finishing' end of the other phase and so on as showing in Fig. 16.19 (*a*). In other words, the three windings are joined in series to form a closed mesh as shown in Fig. 16.19 (*b*).

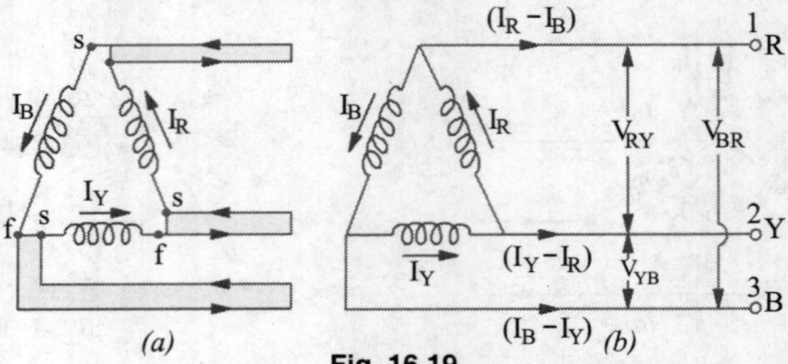

Fig. 16.19

* As an aid to memory, remember that first letter *D* of Dissimilar is the same as that of Delta.

Three leads are taken out from the three junctions as shown as outward directions are taken as positive.

It might look as if this sort of interconnection results in shortcircuiting the three windings. However, if the system is balanced then sum of the three voltages round the closed mesh is zero, hence no current of fundamental frequency can flow around the mesh when the terminals are open. It should be clearly understood that at any instant, the e.m.f. in one phase is equal and opposite to the resultant of those in the other two phases.

This type of connection is also referred to as 3-phase, 3-wire system.

(*i*) Line Voltages and Phase Voltages

It is seen from Fig. 16.19 (*b*) that there is only one phase winding completely included between any pair of terminals. Hence, in Δ-connection, the voltage between any pair of lines is equal to the phase voltage of the phase winding connected between the two lines considered. Since phase sequence is *R Y B*, the voltage having its positive direction from *R* to *Y* leads by 120° on that having its positive direction from *Y* to *B*. Calling the voltage between lines 1 and 2 as V_{RY} and that between lines 2 and 3 as V_{YB}, we find that V_{RY} lead V_{YB} by 120°. Similarly, V_{YB} leads V_{BR} by 120º as shown in Fig. 16.23. Let $V_{RY} = V_{YB} = V_{BR}$ = line voltage V_L. Then, it is seen that $V_L = V_{ph}$.

(*ii*) Line Currents and Phase Currents

It will be seen from Fig. 16.19 (*b*) that current in each line is the ***vector difference*** of the two phase currents flowing through that line. For example

$$\left.\begin{aligned}\text{Current in line 1 is } I_1 &= I_R - I_B \\ \text{Current in line 2 is } I_2 &= I_Y - I_R \\ \text{Current in line 3 is } I_3 &= I_B - I_Y\end{aligned}\right\} \text{ vector difference}$$

Current in line No. 1 is found by compounding I_R and I_B reversed and its value is given by the diagonal of the parallelogram of Fig. 16.20. The angle between I_R and I_B reversed (*i.e.* – I_B) is 60°. If $I_R = I_Y$ = phase current I_{ph} (say), then

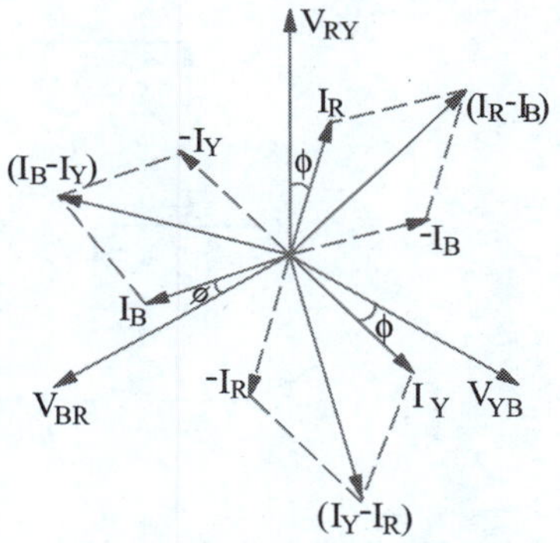

Fig. 16.20

Current in line No. 1 is

$I_1 = 2 \times I_{ph} \times \cos(60º/2) = 2 \times I_{ph}\ \sqrt{3}/2 = \sqrt{3}\, I_{ph}$

Current in line No. 2 is

$I_2 = I_B - I_Y$ ... vector difference = $\sqrt{3} I_{ph}$ and current in line No. 3 is $I_3 = I_B - I_Y$

∴ Vector difference = $\sqrt{3} \cdot I_{ph}$

Since all the line currents are equal in magnitude *i.e.* $I_1 = I_2 - I_3 = I_L$

∴ $I_L = \sqrt{3}\, I_{ph}$

With reference to Fig. 16.20, it should be noted that

1. line currents are 120º apart ;

2. line currents are 30º behind the respective phase currents ;

3. the angle between the line currents and the corresponding line voltages is (30 + ϕ) with the current lagging.

***(iii)* Power**

Power/phase = $V_{ph}I_{ph}$ cos φ; Total active or Real power

$= 3 \times V_{ph}I_{ph}$ cos φ. However, $V_{ph} = V_L$ and $I_{ph} = I_L/\sqrt{3}$

Hence, in terms of line values, the above expression for power becomes

$$P = 3 \times V_L \times \frac{I_L}{\sqrt{3}} \times \cos\phi = \sqrt{3}\, V_L I_L \cos\phi$$

where φ is the phase power factor angle.

Similarly total reactive power $Q = \sqrt{3}\ V_L I_L \sin\phi$

Total apparent power $S = \sqrt{3}\ V_L I_L$

16.10. Balanced Y/Δ and Δ/Y Conversion

In view of the above relationship between line and phase currents and voltages, any balanced *Y*-connected system may be completely replaced by an equivalent Δ-connected system. For example, a 3-phase, *Y*-connected system having the voltage of V_L and line current I_L may be replaced by a Δ-connected system in which phase voltage is V_L and phase current is $I_L/\sqrt{3}$.

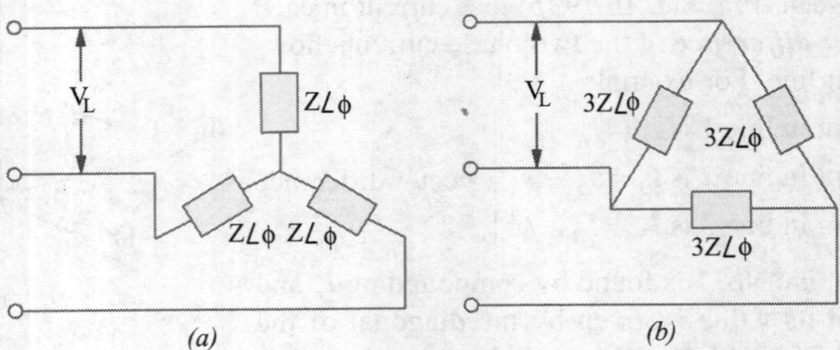

Fig. 16.21

Similarly, a balanced *Y*-connected load having equal branch impedances each of $Z\angle\phi$ may be replaced by an equivalent Δ-connected load whose each phase impedance is $3Z\angle\phi$. This equivalence is shown in Fig. 16.21.

For a balanced star-connected load, let

V_L = line voltage; I_L = line current ; Z_Y = impedance/phase

$\therefore\ V_{ph} = V_L/\sqrt{3}.\ I_{ph} = I_L;\ Z_Y = V_L/(\sqrt{3}\, I_L)$

Now, in the equivalent Δ-connected system, the line voltages and currents must have the same values as in the *Y*-connected system, hence we must have

$$V_{ph} = V_L,\quad I_{ph} = I_L/\sqrt{3}\ \ \therefore\ Z_\Delta = V_L/(I_L/\sqrt{3}) = \sqrt{3}\, V_L/I_L = 3Z_Y$$

$$\therefore \qquad Z_\Delta\angle\phi = 3Z_Y\angle\phi \qquad (\because V_L/I_L = \sqrt{3}Z_Y)$$

or $\quad Z_\Delta = 3Z_Y$ or $Z_Y = Z_\Delta/3$

The case of unbalanced load conversion is considered later.

Example 16.9. *A star-connected alternator supplies a delta connected load. The impedance of the load branch is (8 + j6) ohm/phase. The line voltage is 230 V. Determine* **(a)** *current in the load branch,* **(b)** *power consumed by the load,* **(c)** *power factor of load, (d) reactive power of the load.* **(Elect. Engg. A.M.Ae. S.I. June 1991)**

Solution. Considering the Δ-connected load, we have

$$Z_{ph} = \sqrt{8^2 + 6^2} = 10\Omega;\ V_{ph} = V_L = \mathbf{230\,V}$$

(*a*) $I_{ph} = V_{ph}/Z_{ph} = 230/10 = 23$ A

(*b*) $I_L = \sqrt{3}\,I_{ph} = \sqrt{3}\times 23 = 39.8\ A;\ \ P = \sqrt{3}\ V_L I_L\ \cos\phi = \sqrt{3}\times 230\times 39.8\times 0.8$

$= \mathbf{12{,}684\ W}$

(*c*) p.f. $\cos\phi = R \mid Z = 8/10 =$ **0.8 (lag)**

(*d*) Reactive power $Q = \sqrt{3}\ V_L I_L \sin\phi = \sqrt{3}\times 230\times 39.8\times 0.6 = \mathbf{9513\ W}$

Example 16.10. *A 220-V, 3-ϕ voltage is applied to a balanced delta-connected 3-ϕ load of phase impedance (15 + j20)Ω*

(*a*) *Find the phasor current in each line.* **(*b*)** *What is the power consumed per phase ?*

(*c*) *What is the phasor sum of the three line currents ? Why does it have this value ?*

(Elect. Circuits and Instruments, B.H.U.)

Solution. The circuit is shown in Fig. 16.27 (*a*).

$$V_{ph} = V_L = 220\ V;\ Z_{ph} = \sqrt{15^2 + 20^2} = 25\Omega,\ I_{ph} = V_{ph}/Z_{ph} = 220/25 = 8.8\,\text{A}$$

(*a*) $I_L = \sqrt{3}\ I_{ph} = \sqrt{3}\times 8.8 =$ **15.24 A** (*b*) $P = I_{ph}{}^2 R_{ph} = 8.8^2\times 15 =$ **462 W**

(*c*) Phasor sum would be zero because the three currents are equal in magnitude and have a mutual phase difference of 120°.

Solution by Symbolic Notation

Taking V_{RY} as the reference vector, we have [Fig. 16.22 (*b*)]

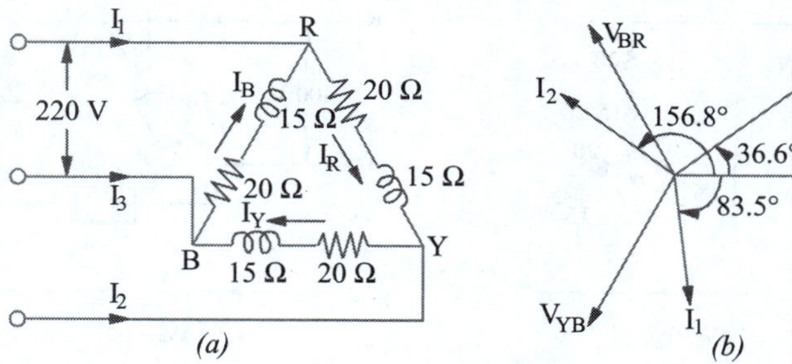

Fig. 16.22

$V_{RY} = 20\angle 0°$; $\qquad V_{YB} = 220\angle -120°$

$V_{BR} = 220°\angle 120°$; $\qquad \mathbf{Z} = 15 + j20 = 125\angle 53°8'$

$$I_R = \frac{V_{RY}}{Z} = \frac{220\angle 0°}{25\angle 53°8'} = 8.8\angle -53°8 = (5.28 - j7.04)\ A$$

$$I_Y = \frac{V_{YB}}{Z} = \frac{220\angle -120^\circ}{25\angle 53^\circ 8'} = 8.8\angle -173^\circ 8' = (-8.75 - j1.05)\ A$$

$$I_B = \frac{V_{BR}}{Z} = \frac{220\angle -120^\circ}{25\angle 53^\circ 8'} = 8.8\angle 66^\circ 55' = (3.56 + j8.1)$$

(*a*) Current in line No. 1 is

$\mathbf{I}_1 = \mathbf{I}_R - \mathbf{I}_B = (5.28 - j\,7.04) - (4.56 + j8.1) = (1.72 - j15.14) = 15.23 \angle -83.5^\circ$

$\mathbf{I}_2 = \mathbf{I}_Y - \mathbf{I}_R = (-8.75 - j\,1.05) - (5.28 - j7.04) = (-14.03 + j\,6.0) = 15.47 \angle -156.8^\circ$

$\mathbf{I}_3 = \mathbf{I}_B - \mathbf{I}_Y = (3.56 - j\,8.1) - (-8.75 - j1.05) = (12.31 + j\,9.15) = 15.26 \angle 36.8^\circ$

(*b*) Using conjugate of voltage, we get for *R*-phase

$\mathbf{P}_{VA} = \mathbf{V}_{RY} \cdot \mathbf{I}_R = (220 - j0)\,(5.28 - 7.04) = (1162 - j\,1550)$ voltampere

Real power per phase = **1162 W**

(*c*) Phasor sum of three line currents

$= \mathbf{I}_1 + \mathbf{I}_2 + \mathbf{I}_3 = (1.72 - j\,15.14) + (-14.03 + j\,6.0) + (12.31 + j\,9.15) = 0$

As expected, phasor sum of 3 line currents drawn by a balanced load is zero because these are equal in magnitude and have a phase difference of 120° amount themselves.

Example 16.11. *Three similar coils, each having a resistance of 20 ohms and an inductance of 0.05 H are connected in* **(i)** *star* **(ii)** *mesh to a 3-phase, 50-Hz supply with 400-V between lines. Calculate the total power absorbed and the line current in each case. Draw the vector diagram of current and voltages in each case.*

(Elect. Technology, Punjab Univ. 1990)

Solution. $X_L = 2\pi \times 50 \times 0.05 \cong 15\,\Omega,\ Z_{ph} = \sqrt{15^2 + 20^2} = \mathbf{25\ \Omega}$

(*i*) **Star Connection.** [Fig. 16.23 (*a*)]

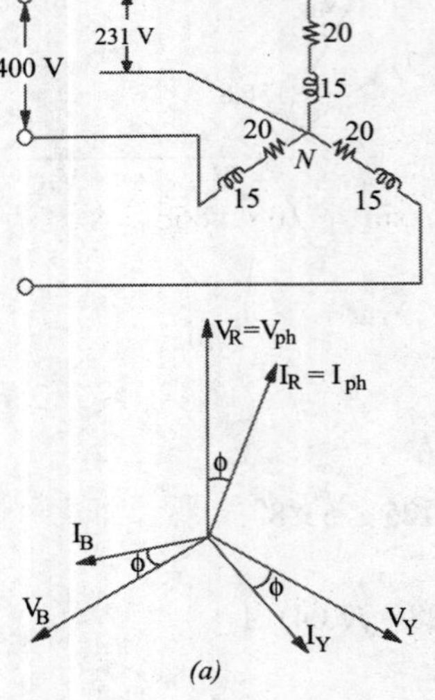

(*a*)

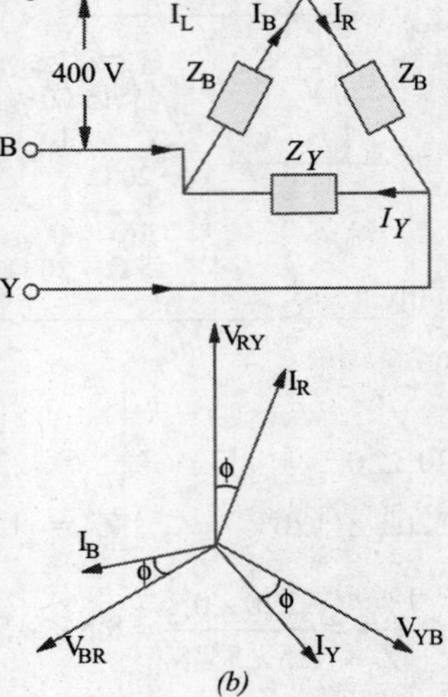

(*b*)

Fig. 16.23

$$V_{ph} = 400/\sqrt{3} = 231\text{ V};\ I_{ph} = V_{ph}/Z_{ph} = 231/25 = \mathbf{9.24\ \Omega}$$

$$I_L = I_{ph} = \mathbf{9.24\ A};\ P = \sqrt{3} \times 400 \times 9.24 \times (20/25) = \mathbf{5120\ W}$$

***(ii)* Delta Connection** [Fig. 16.24 (*b*)]

$$V_{ph} = V_L = 400\ V;\ I_{ph} = 400/25 = 16\ A;\ I_L = \sqrt{3}\, I_{ph} = \sqrt{3} \times 16 = \mathbf{27.7\ A}$$

$$P = \sqrt{3} \times 400 \times 27.7 \times (20/25) = \mathbf{15{,}360\ W}$$

Note. It may be noted that line current as well as power are three times the star values.

Example 16.12. *A Δ-connected balanced 3-phase load is supplied from a 3-phase, 400-V supply. The line current is 20 A and the power taken by the load is 10,000 W. Find (i) impedance in each branch (ii) the line current, power factor and power consumed if the same load is connected in star.* **(Electrical Machines, A.M.I.E. Sec. B. 1992)**

Solution. (*i*) Delta Connection.

$$V_{ph} = V_L = 400\ V;\ I_L = 20\ A;\ I_{ph} = 20/\sqrt{3}\ \text{A}$$

(*i*) $\therefore\ Z_{ph} = \dfrac{400}{20/\sqrt{3}} = 20\sqrt{3} = \mathbf{34.64\ \Omega}$

Now $P = \sqrt{3}\, V_L I_L \cos\phi$ $\therefore$ $\cos\phi = 10{,}000/\sqrt{3} \times 400 \times 20 = \mathbf{0.7217}$

***(ii)* Star Connection**

$$V_{ph} = \frac{400}{\sqrt{3}},\ I_{ph} = \frac{400/\sqrt{3}}{\sqrt{3}} = \frac{20}{3}\ A,\ I_L = I_{ph} - \frac{20}{3}\ \text{A}$$

Power factor remains the same since impedance is the same.

Power consumed $= \sqrt{3} \times 400 \times (20/3) + 0.7217 = \mathbf{3{,}330\ W}$

Note. The power consumed is 1/3 of its value of Δ-connection.

Example 16.13. *A balanced delta connected load, consisting of there coils, draws* $10\sqrt{3}$ *A at 0.5 power factor from 100 V, 3-phase supply. If the coils are re-connected in star across the same supply, find the line current and total power consumed.*

(Elect. Technology, Punjab Univ. Nov.)

Solution. Delta Connection

$$V_{ph} = V_L = 100V;\ I_L = 10\sqrt{3}\ A;\ I_{ph} = 10\sqrt{3}/\sqrt{3} = \mathbf{10\ A}$$

$$Z_{ph} = V_{ph}/I_{ph} = 100/10 = 10\Omega;\ \cos\phi = 0.5\ \text{(given)};\ \sin\phi = 0.866$$

$$\therefore\ R_{ph} = Z_{ph}\cos\phi = 10 \times 0.5 = 5\Omega;\ X_{ph} = Z_{ph}\sin\phi = 10 \times 0.866 = \mathbf{8.66\ \Omega}$$

Incidentally, total power consumed $= \sqrt{3}\, V_L I_L \cos\phi = \sqrt{3} \times 100 \times 10\sqrt{3} \times 0.5 = 1500\text{ W}$

Star Connection

$$V_{ph} = V_L/\sqrt{3} = 100/\sqrt{3};\ Z_{ph} = 10\Omega;\ I_{ph} = V_{ph}/Z_{ph} = 100/\sqrt{3} \times 10 = 10\sqrt{3}\ \text{A}$$

Total power absorbed $= \sqrt{3} \times 100 \times (10\sqrt{3}) \times 0.5 = \mathbf{500\ W}$

It would be noted that the line current as well as the power absorbed are one-third of that in the delta connection.

Example 16.14. *Three identical impedances are connected in delta to a 3 ϕ supply of 400 V. The line current is 35 A and the total power taken from the supply is 15 kW. Calculate the resistance and reactance values of each impedance.* **(Elect. Technology, Punjab Univ.,)**

Solution. $V_{ph} = V_L = 400$ V; $I_L = 35$ A $\therefore$ $I_{ph} = 35/\sqrt{3}$ A

$$Z_{ph} = V_{ph}/I_{ph} = 400 \times \sqrt{3}/35 = 19.8 \text{ A}$$

Now,

$$\text{Power } P = \sqrt{3}V_L I_L \cos\phi \quad \therefore \cos\phi = \frac{P}{\sqrt{3}V_L I_L} = \frac{15{,}000}{\sqrt{3} \times 400 \times 35} = 0.619; \text{ But } \sin\phi = 0.786$$

$\therefore R_{ph} = Z_{ph} \cos\phi = 19.8 \times 0.619 =$ **12.25 Ω**; $X_{ph} = Z_{ph} \sin\phi$ and $X_{ph} = 19.8 \times 0.786$ = **15.5 Ω**

Example 16.15. *Three 100Ω non-inductive resistances are connected in (a) star (b) delta across a 400-V, 50-Hz, 3-phase mains. Calculate the power taken from the supply system in each case. In the event of one of the three resistances getting open-circuited, what would be the value of total power taken from the mains in each of the two cases ?*

(Elect. Engg. A.M.Ae. S.I June, 1993)

Solution. (*i*) Star Connection [**Fig. 16.24 (*a*)**]

$V_{ph} = 400/\sqrt{3}$ V

$P = \sqrt{3}\, V_L I_L \cos\phi$

$= \sqrt{3} \times 400 \times 4 \times 1/\sqrt{3} =$ **1600 W**

(*ii*) Delta Connection Fig. 16.24 (*b*)

$V_{ph} = 400$ V; $R_{ph} = 100\,\Omega$

$I_{ph} = 400/100 = 4$ A

$I_L = 4 \times \sqrt{3}$ A

$P = \sqrt{3} \times 400 \times 4 \times \sqrt{3} \times 1 =$ **4800 W**

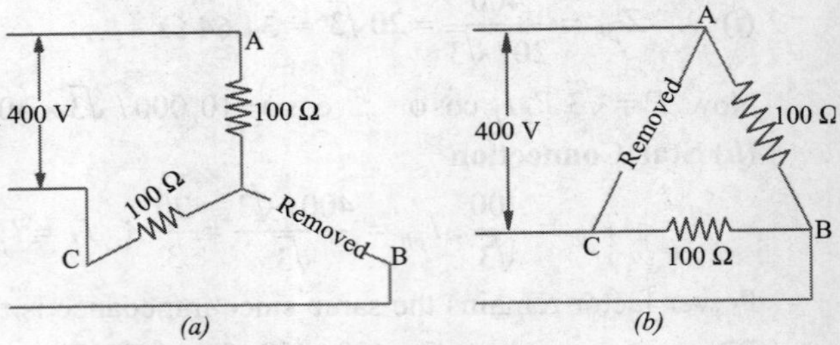

Fig. 16.24

When one of the resistors is disconnected

(*i*) Star Connection

The circuit no longer remains a 3-phase circuit but consists of two 100 Ω resistors in series across a 400-V supply. Current in lines *A* and *C* is = 400/200 = 2 A

Power absorbed in both = 400 × 2 = 800 W

Hence, by disconnecting one resistor, the power consumption is reduced by half.

(*ii*) Delta Connection [Fig. 16.28 (*b*)]

In this case, currents in *A* and *C* remain as usual 120° out of phase with each other.

Current in each phase = 400/100 = 4 A

Power consumption in both = 2 × 4^2 × 100 = **3200 W**

(or *P* = 2 × 4 × 400 = 3200 W)

In this case, when one resistor is disconnected, the power consumption is reduced by one-third.

Example 16.16. *The load connected to a 3-phase supply comprises three similar coils connected in star. The line currents are 25 A and the kVA and kW inputs are 20 and 11 respectively. Find the line and phase voltages, the kVAR input and the resistance and reactance of each coil.*

If the coils are now connected in delta to the same three-phase supply, calculate the line currents and the power taken.

Solution. Star Connection

$\cos\phi\ k$ = W/kVA = 11/20 $\quad I_L = 25$ A $\quad P = 11$ kW = 11,000 W

Now $\quad P = \sqrt{3}\ V_L I_L \cos\phi \quad \therefore\ 11{,}000 = \sqrt{3} \times V_L \times 25 \times 11/20$

$\therefore\ V_L =$ **462 V**; $\quad V_{ph} = 462/\sqrt{3} =$ **267 V**

$$\text{kVAR} = \sqrt{\text{kVA}^2 - \text{kW}^2} = \sqrt{20^2 - 11^2} = \mathbf{16.7};\ Z_{ph} = 267/2 = 10.68$$

$$\therefore\ R_{ph} = Z_{ph} \times \cos\phi = 10.68 \times 11/20 = \mathbf{5.87\ \Omega}$$

$$\therefore\ X_{ph} = Z_{ph} \times \sin\phi = 10.68 \times 0.838 = \mathbf{897\Omega}$$

Delta Connection

$$V_{ph} = V_L = 462\ V \text{ and } Z_{ph} = 10.68\Omega$$

$$\therefore\ I_{ph} = 462/10.68\,\text{A},\ I_L = \sqrt{3} \times 462/10.68 = 75\text{ A}$$

$$P = \sqrt{3} \times 462 \times 75 \times 11/20 = \mathbf{33{,}000\ W}$$

Example 16.17. *A 3-phase, star-connected system with 230 V between each phase and neutral has resistances of 4, 5 and 6 Ω respectively in the three phases. Estimate the current flowing in each phase and the neutral current. Find the total power absorbed.* **(I.E.E. London)**

Solution. Here, V_{ph} = 230 V [Fig. 16.25 (*a*)]

Current in 4-Ω resistor = 230/4 = 57.5 A

Current in 5-Ω resistor = 230/5 = 46 A

Current in 6-Ω resistor = 230/6 = 38.3 A

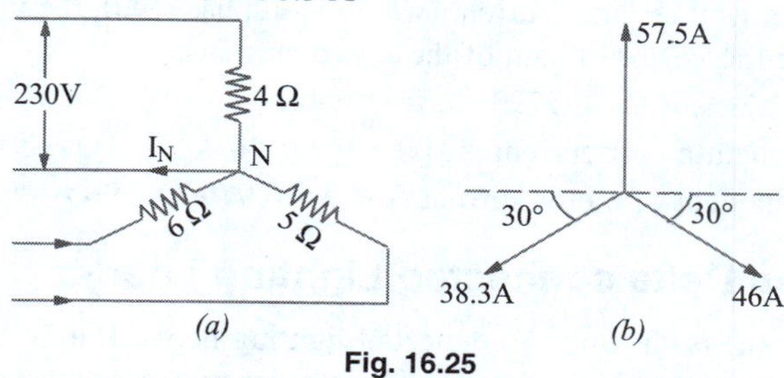

Fig. 16.25

These currents are mutually displaced by 120°. The neutral current I_N is the vector sum* of these three currents. I_N can be obtained by splitting up these three phase currents into their *X*-components and *Y*-components and then by combining them together, in diagram 16.32 (*b*).

X-component = 46 cos 30° – 38.3 cos 30° = **6.64 A**

Y-component = 57.5 – 46 sin 30° – 38.3 sin 30° = 15.3 A $\therefore\ I_N = \sqrt{6.64^2 + 15.3^2}$ = **16.71 A**

The power absorbed = 230 (57.5 + 46 + 38.3) = **32.610 W**

Example 16.18. *A 3-phase, 4-wire system supplies power at 400 V and lighting at 230 V. If the lamps in use require 70, 84 and 33 A in each of the three lines, what should be the current in the neutral wire ? If a 3-phase motor is now started, taking 200 A from the line at a power factor of 0.2, what would be the current in each line and the neutral current ? Find also the total power supplied to the lamps and the motor.* **(Elect. Technology, Aligarh Univ.)**

Solution. The lamp and motor connections are shown in Fig. 16.26.

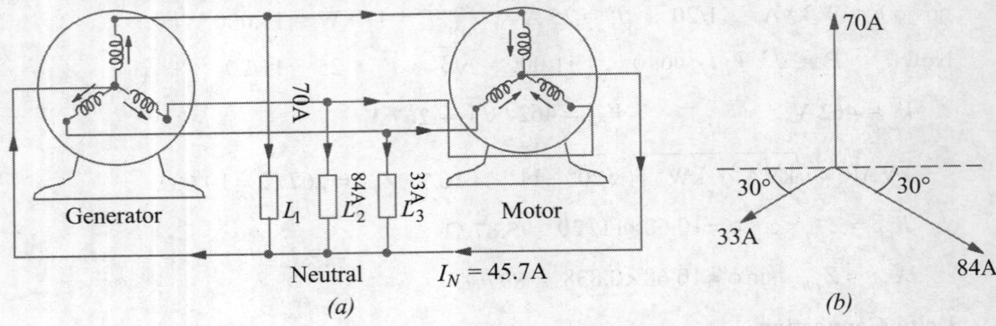

Fig. 16.26

When motor is not started

The neutral current is the vector sum of lamp currents. Again, splitting up the currents into their *X*- and *Y*-components, we get

X-component = 84 cos 30° – 33 cos 30° = 44.2 A

Y-component = 70 – 84 sin 30° – 33 sin 30° = 11.5 A

$$\therefore \quad I_N = \sqrt{44.2^2 + 11.5^2} = \mathbf{45.7\ A}$$

When motor is started

A 3-phase motor is a balanced load. Hence, when it is started, it will change the line currents but being a balanced load, it contributes nothing to the neutral current. Hence, ***the neutral current remains unchanged even after starting the motor.***

Now, the motor takes 200 A from the lines. It means that each line will carry motor current (which lags) as well as lamp current (which is in phase with the voltage). The current in each line would be the vector of sum of these two currents.

Motor p.f. = 0.2 ; sin ϕ = 0.9799 ... from tables

Active component motor current = 200 × 0.2 = 40 A

Reactive component of motor current = 200 × 0.9799 = 196 A

16.11. Star and Delta connected Lighting Loads

In Fig. 16.27 (*a*) is shown a *Y*-connected lighting network in a three storey house. For such a load, it is essential to have neutral wire in order to ensure uniform distribution of load among the three phases despite random switching on and off or burning of lamps. It is seen from Fig. 1934 (*a*), that network supplies two flats on each floor of the three storey residence and there is balanced distribution of lamp load among the three phases. There are house fuses at the cable entry into the building which protect the two mains against short-circuits in the main cable. At the flat entry, there are apartment (or flat) fuses in the single-phase supply which protect the two mains and other flats in the same building from short-circuits in a given building. There is no fuse (or switch) on the neutral wire of the mains because blowing of such a fuse (or disconnection of such a switch) would mean a break in the neutral wire. This would result in unequal voltages across different groups of lamps in case they have different power ratings or number. Consequently, filaments in one group would burn dim whereas in other groups they would burn too bright resulting in their early burn-out.

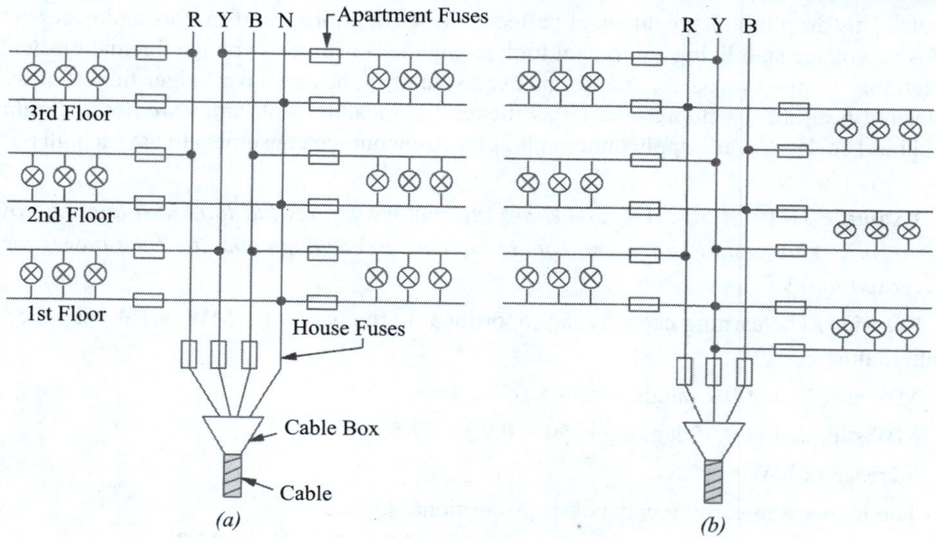

Fig. 16.27

The house-lighting wire circuit for Δ-connected lamps is shown in Fig. 16.27 (*b*).

16.12. Power Factor Improvement

The heating and lighting loads supplied from 3-phase supply have power factors, ranging from 0.95 to unity. But motor loads have usually low lagging power factors, ranging from 0.5 to 0.9. Single-phase motors may have as low power factor as 0.4 and electric wedding units have even lower power factors of 0.2 or 0.3.

The power factor is given by $\cos\phi = \dfrac{\text{kW}}{\text{kVA}}$ or $\text{kVA} = \dfrac{\text{kW}}{\cos\phi}$

In the case of single-phase supply, $\text{kVA} = \dfrac{\text{VI}}{1000}$ or $I = \dfrac{1000\,k\text{VA}}{\text{V}}$ $\therefore \quad I \propto \text{kVA}$

In the case of 3-phase supply $\text{kVA} = \dfrac{\sqrt{3}V_L I_L}{1000}$ or $I_L = \dfrac{1000\text{ kVA}}{\sqrt{3}\times V_L}$ $\therefore \quad I \propto \text{kVA}$

In each case, the kVA is directly proportional to current. The chief disadvantage of a low p.f. is that the current required for a given power, is very high. This fact leads to the following undesirable results.

(*i*) Large kVA for given amount of power

All electric machinery, like alternators, transformers, switchgears and cables are limited in their current-carrying capacity by the permissible temperature rise, which is proportional to I^2. Hence, they may all be fully loaded with respect to their rated kVA, without delivering their full power. Obviously, it is possible for an existing plant of a given kVA rating to increase its earning capacity (which is proportional to the power supplied in kW) if the overall power factor is improved *i.e.* raised.

(*ii*) Poor voltage regulation

When a load, having allow lagging power factor, is switched on, there is a large voltage drop in the supply voltage because of the increased voltage drop in the supply lines and transformers. This drop in voltage adversely affects the starting torques of motors and necessitates expensive voltage stabilizing equipment for keeping the consumer's voltage fluctuations within the statutory limits. Moreover, due to this excessive drop, heaters take longer time to provide the desired heat energy, fluorescent lights flicker and incandescent lamps are not as bright as they should be. Hence, all supply undertakings try to encourage consumers to have a high power factor.

Example 16.19. *A 50-MVA, 11-kV, 3-ϕ alternator supplies full load at a lagging power factor of 0.7. What would be the percentage increase in earning capacity if the power factor is increased to 0.95 ?*

Solution. The earning capacity is proportional to the power (in MW or kW) supplied by the alternator.

MW supplied at 0.7 lagging = 50 × 0.7 = 35

MW supplied at 0.95 lagging = 50 × 0.95 = 47.5

increase in MW = 12.5

The increase in earning capacity is proportional to 12.5

∴ Percentage increase in earning capacity = (12.5/35) × 100 = **35.7**

16.13. Power Correction Equipment

The following equipment is generally used for improving or correcting the power factor :

(*i*) Synchronous Motors (or capacitors)

These machines draw leading kVAR when they are over-excited and, especially, when they are running idle. They are employed for correcting the power factor in bulk and have the special advantage that the amount of correction can be varied by changing their excitation.

(*ii*) Static Capacitors

They are installed to improve the power factor of a group of a.c. motors and are practically loss-free (*i.e.* they draw a current leading in phase by 90°). Since their capacitances are not variable, they tend to over-compensate on light loads, unless arrangements for automatic switching off the capacitor bank are made.

(*iii*) Phase Advancers

They are fitted with individual machines.

However, it may be noted that the economical degree of correction to be applied in each case, depends upon the tariff arrangement between the consumers and the supply authorities.

Example 16.20. *A 3-phase, 37.3 kW, 440-V, 50-Hz induction motor operates on full load with an efficiency of 89% and at a power factor of 0.85 lagging. Calculate the total kVA rating of capacitors required to raise the full-load power factor at 0.95 lagging. What will be the capacitance per phase if the capacitors are* **(a)** *delta-connected and* **(b)** *star-connected ?*

Solution. It is helpful to approach such problems from the 'power triangle' rather than from vector diagram viewpoint.

Motor power input $P = 37.3/0.89 = 41.191$ kW

Power Factor 0.85 (lag)

$\cos\phi_1 = 0.85 : \phi_1 = \cos^{-1}(0.85) = 31.8° ; \tan\phi_1 = \tan 31.8° = 0.62$

Motor $\text{kVAR}_1 = P \tan\phi_1 = 41.91 \times 0.62 = 25.98$

Power Factor 0.95 (lag)

Motor power input $P = 41.91$ kW ... as before

It is the same as before because capacitors are loss-free *i.e.* they do not absorb any power.

$\cos\phi_2 = 0.95 \quad \therefore \phi_2 = 18.2°; \quad \tan 18.2° = 0.3288$

Motor $kVAR_2 = P \tan \phi_2 = 41.91 \times 0.3288 = 13.79$

The difference in the values of kVAR is due to the capacitors which supply ***leading*** kVAR to partially neutralize the ***lagging*** kVAR of the motor.

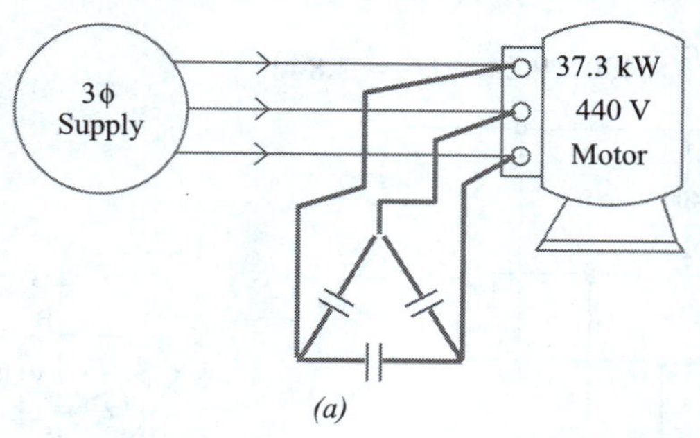

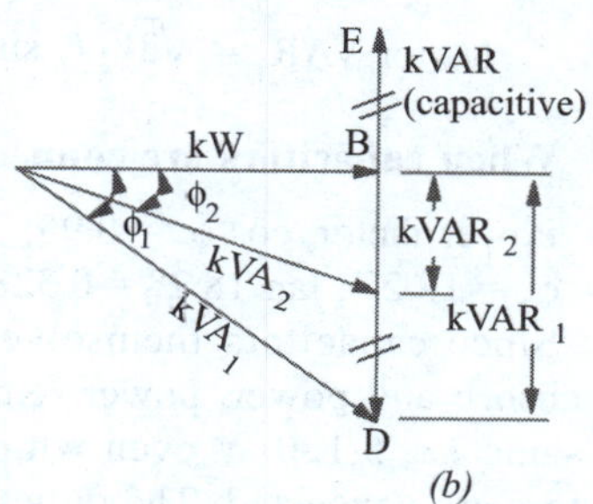

Fig. 16.28

$\therefore$ leading kVAR supplied by capacitors is

$= kVAR_1 - kVAR_2 = 25.98 - 13.79 =$ **12.19** ... *CD* in Fig. 16.28 (*b*)

Since capacitors are loss-free, their kVAR is the same as kVA

$\therefore$ kVA/capacitor = 12.19/3 = 4.063 $\quad \therefore$ VAR/capacitor = 4,063

(*a*) In Δ-connection, voltage across each capacitor is 440 V

Current drawn by each capacitor $I_C = 4063/440 = 9.23$ A

Now, $$I_c = \frac{V}{X_c} = \frac{V}{1/\omega C} = \omega VC$$

$\therefore$ $$C = I_c/\omega V = 9.23/2\pi\times 50\times 440 = 66.8\times 10^{-6}\text{ F} = \mathbf{66.8}$$

μF

(*b*) In star connection, voltage across each capacitor is = $440/\sqrt{3}$ volt

Current drawn by each capacitor, $I_c = \dfrac{4063}{440/\sqrt{3}} = 16.0$ A

$$I_c = \frac{V}{X_c} = \omega VC \quad \text{or } 16 = \frac{440}{\sqrt{3}}\times 2\pi\times 50\times C$$

$\therefore$ $$C = 200.4 \times 10^{-6}\text{F} = \mathbf{200.4\ \mu F}$$

Note. Star value is three times the delta value.

Example 16.21. *Three impedance coils, each having a resistance of 20 Ω and a reactance of 15 Ω, are connected in star to a 400-V, 3-ϕ, 50-Hz supply. Calculate* ***(i)*** *the line current* ***(ii)*** *power supplied and* ***(iii)*** *the power factor.*

If three capacitors, each of the same capacitance, are connected in delta to the same supply so as to form parallel circuit with the above impedance coils, calculate the capacitance of each capacitor to obtain a resultant power factor of 0.95 lagging.

Solution. $V_{ph} = 100/\sqrt{3}V, Z_{ph} = \sqrt{20^2 + 15^2} = 25\Omega$

$$\cos\phi_1 = R_{ph} / Z_{ph} = 20/25 = 0.8 \text{ lag}; \phi_1 = 0.6 \text{ lag}$$

where ϕ_1 is the power factor angle of the coils.

When capacitors are not connected

(i) $I_{ph} = 400/25\times\sqrt{3} = 9.24 \text{ A} \therefore I_L =$ **9.24 A**

(ii) $P = \sqrt{3}\, V_L I_L \cos\phi_1 = \sqrt{3}\times 400\times 9.24\times 0.8 =$ **5.120 W**

(iii) Power factor = **0.8 (lag)**

$\therefore$ Motor $VAR_1 = \sqrt{3} V_L I_L \sin\phi_1 = \sqrt{3}\times 400\times 9.24\times 0.6 =$ **3,840**

When capacitors are connected

Power factor, $\cos\phi_2 = 0.95$,
$\phi_2 = 18.2°$; $\tan 18.2° = 0.3288$

Since capacitors themselves do not absorb any power, power remains the same *i.e.* 5,120 W even whin capacitors are connected. The only thing that changes is the VAR.

Now $VAR_2 = P \tan\phi_2$

$= 5120 \times 0.3288 = 1684$

Leading VAR supplied by the three capacitors is

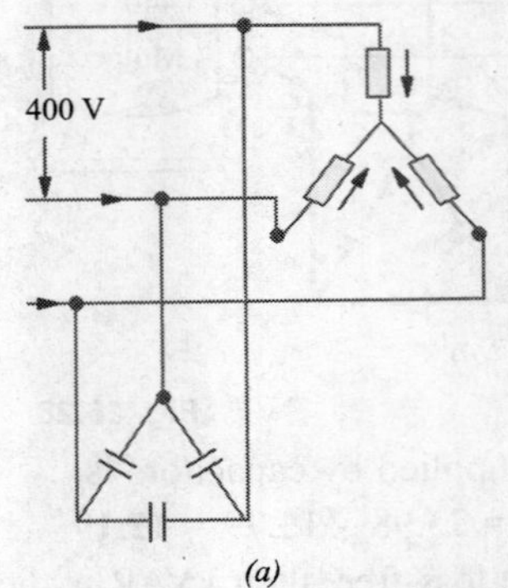

(a)

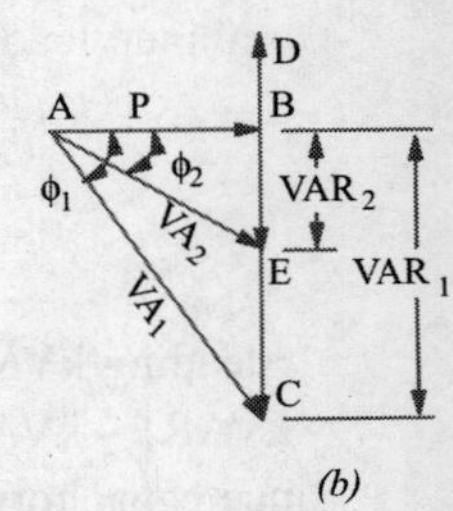

(b)

Fig. 16.29

$= VAR_1 - VAR_2 = 3840 - 1684 = 2156$ BD or CE in Fig 16.29 (*b*)

VAR/ Capacitor = 2156/3 = 719

For delta connection, voltage across each capacitor is 400 V $\therefore I_c = 719/400 = 1.798$ A

Also $I_c = \dfrac{V}{I/\omega C} = \omega VC \therefore C = 1.798/\pi\times 50\times 400 = 14.32\times 10^{-6}\text{F} =$ **14.32 μF**

16.14. Parallel Loads

A combination of balanced 3-phase loads connected in parallel may be solved by any one of the following three methods :

1. All the given loads may be converted into equivalent Δ-loads and then combined together according to the law governing parallel circuits.

2. All the given loads may be converted into equivalent *Y*-loads and treated as in **(1)** above.

3. The third method, which requires less work, is to work in terms of volt-amperes. The special advantage of this approach is that voltameters can be added regardless of the kind of connection involved. The real power of various loads can be added arithmetically and VARs may be added algebraically so that total voltamperes are given by

$$VA = \sqrt{W^2 + VAR^2} \quad \text{or} \quad S = \sqrt{P^2 + Q^2}$$

where *P* is the power in water and *Q* represents reactive voltamperes.

Example 16.22. *A symmetrical 3-phase, 3-wire supply with a line voltage of 173 V supplies two balanced 3-phase loads; one Y-connected with each branch impedance equal to (6 + j8) ohm and the other Δ-connected with each branch impedance equal to (18 + j24) ohm. Calculate*

(i) the magnitudes of branch currents taken by each 3-phase load

(ii) the magnitude of the total line current and

(iii) the power factor of the entire load circuit

Draw the phasor diagram of the voltages and currents for the two loads.

(Elect. Engineering-I, Bombay Univ.)

Solution. The equivalent *Y*-load of the given Δ-load is = $(18 + j24)/3 = (6 + j8)\,\Omega$.

With this, the problem now reduces to one of solving two equal *Y*-loads connected in parallel across the 3-phase supply as shown in Fig. 16.30 (*a*). Phasor diagram for the combined load for one phase only is given in Fig. 16.30 (*b*).

Combined load impedance

$$= (6 + j8)/2 = 3 + j4$$

$$= 5\angle 53.1^\circ \text{ ohm}$$

$$V_{ph} = 173/\sqrt{3} = 100 \text{ V}$$

Let $V_{ph} = 100\angle 0^\circ$

$$\therefore \quad I_{ph} = \frac{100\angle 0^\circ}{5\angle 53.1^\circ} = 20\angle -53.1^\circ$$

(*a*)

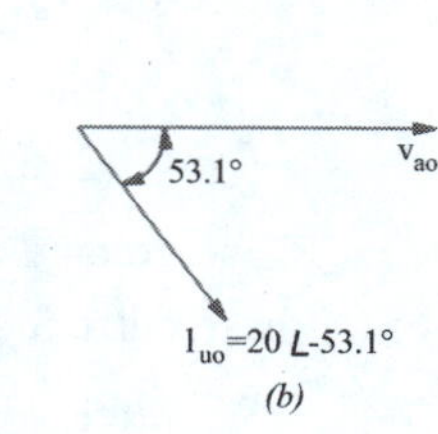

(*b*)

Fig. 16.30

Current in each load = $10\angle -53.1$ A

(*i*) branch current taken by each load is **10 A**; (*ii*) line current is **20 A**;

(*iii*) combined power factor = cos 53.1° = **0.6 (lag).**

Example 16.23. *Three identical impedances of 30 ∠30° ohms are connected in delta to a 3-phase, 3-wire, 208 V volt abc system by conductors which have impedances of (0.8 + j 0.63) ohm. Find the magnitude of the line voltage at the load end.*

(Elect. Engg. Punjab Univ. May 1990)

Solution. The equivalent Z_y, of the given Z_Δ is $30\angle 30/3 = 10\angle 30^\circ = (8.86 + j5)$. Hence, the load connections become as shown in Fig. 16.31.

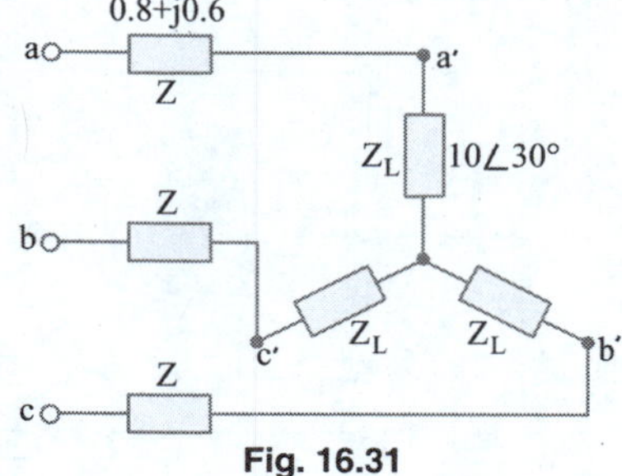

Fig. 16.31

$$Z_{an} = (0.8 + j0.6) + (8.86 + j5)$$

$$= 9.66 + j5.6 = 11.16\angle 30.1^\circ$$

$$V_{an} = V_{ph} = 208/\sqrt{3} = 120 \text{ V}$$

Let $V_{an} = 120\angle 0^\circ$

$$\therefore \quad I_{an} = 120\angle 0^\circ / 11.16\angle 30.1^\circ = 10.75\angle -30.1^\circ$$

Now, $Z_{aa'} = 0.8 + j0.6 = 1\angle 36.9^\circ$

Voltage drop on line conductors is

$$V'_{aa} = I_{an} Z'_{aa} = 10.75\angle -30.1^\circ \times 1\angle 36.9^\circ = 10.75\angle 6.8^\circ = 10.67 + j1.27$$

$$\therefore \quad V'_{an} = V_{an} - V'_{aa} = (120 + j0) - (10.67 + j1.27) = \mathbf{109.3 \angle 2.03^\circ}$$

Example 16.24. *A balanced delta-connected load having an impedance Z_L = (300 + j210) ohm in each phase is supplied from 400-V, 3-phase supply through a 3-phase line having an impedance of Z_s = (4 + j8) ohm in each phase. Find the total power supplied to the load as well as the current and voltage in each phase of the load.*

(Elect. Circuit Theory, Kerala Univ.)

Solution. The equivalent *Y*-load of the given Δ-load is

$$= (300 + j210)/3 = (100 + j70)\Omega$$

Hence, connections become as shown in Fig. 16.32

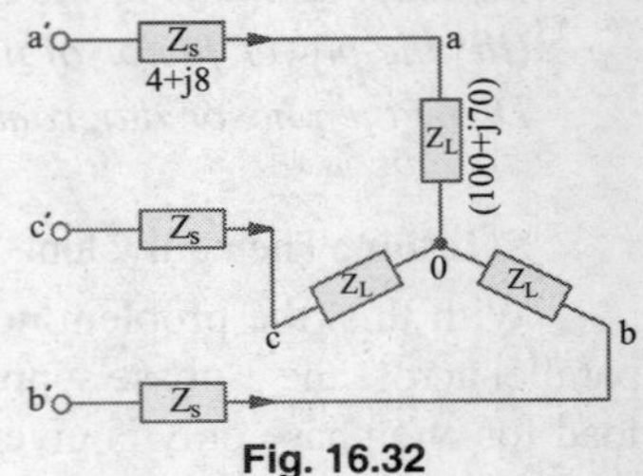

Fig. 16.32

$$\mathbf{Z'_{a0}} = (4 + j8) + (100 + j70) = 104 + j78 = 130\angle 36.9°$$

$$V'_{a0} = 400/\sqrt{3} = 231V,$$

$$I'_{a0} = 231\angle 0°/130\angle 36.9° = 1.78\angle -36.9°$$

Now, $Z'_{a0} = (4 + j8) = 8.94\angle 63.4°$

Line drop $\mathbf{V'_{aa}} = \mathbf{I'_{aa}Z'_{aa}} = 1.78\angle -36.9° \times 8.94\angle 63.4 = 15.9\angle 26.5° = 14.2 + j7.1$

$$\mathbf{V_{a0}} = \mathbf{V'_{a0}} - \mathbf{V'_{aa}} = (231 + j0) - (14.2 + j7.1)$$

$$= (216.8 - j7.1) = 216.9\angle -1°52'$$

Phase voltage at load end, V_{a0} = **216.9 V**

Phase current at load end, I_{a0} = **1.78 A**

Power supplied to load = 3 × 1.782 × 100 = **951 W**

Incidentally, line voltage at load end V_{ac} = 216.9 × $\sqrt{3}$ = **375.7 V***

Example 16.25. *Three non-inductive resistances, each of 100 ohms, are connected in star to a three-phase, 440-V supply. Three inductive coils, each of reactance 100 ohms connected in delta are also connected to the supply. Calculate: **(i)** Line-currents, and **(ii)** power factor of the system.* **[Nagpur University, November 1998]**

Solution. (*a*) Three resistances are connected in star. Each resistance is of 100 ohms and 254 – V appears across it. Hence, a current of 2.54 A flows through the resistors and the concerned power-factor is unity. Due to star-connection,

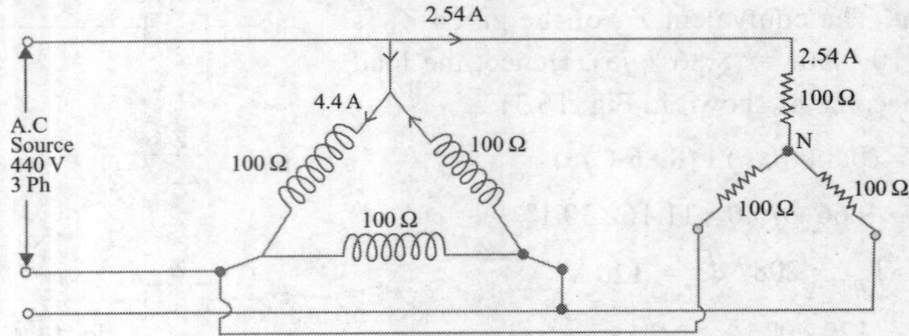

Fig. 16.33

Line-current = Phase-current = 2.54A

(*b*) Three inductive reactance are delta connected.

Line-Voltage = Phase – Voltage = 440 V

Phase Current = 440/100 = 4.4 A

* It should be noted that total line drop is not the numerical sum of the individual line drops because they are 120° out of phase with each other. By a laborious process $V_{ac} = V'_{ac} - V'_{aa} - V'_{ac}$.

Line current = 1.732 × 4.4 = 7.62 A

The current has a zero lagging power-factor.

Total Line Current = 2.54 - j 7.62 A

= 8.032 A, in each of the lines.

Power factor = 2.54/8.032 = 0.32 Lag.

Example 16.26. *The delta-connected generator of Fig 16.34 has the voltage;* $V_{RY} = 220 \angle 0°$, $V_{YB} = 220 \angle -120°$ *and* $V_{BR} = 220 \angle -240°$ *Volts.*

The load is balanced and delta-connected. Find:

(a) *Impedance per phase,* ***(b)*** *Current per phase,* ***(c)*** *Other line – currents* I_Y *and* I_B.

[Nagpur University, November 1997]

Solution. Draw phasors for voltages as mentioned in the data. V_{RY} naturally becomes a reference-phasor, along which the phasor I_R also must lie, as shown in Fig. 16.34 (*b*) & (*c*). I_R is the line voltage which is related to the phase-currents I_{RY} and $-I_{BR}$. In terms of magnitudes,

$|I_{RY}| = |I_{BR}| = |I_R|/\sqrt{3} = 10/\sqrt{3} = 5.8$ Amp

Thus, I_{RY} leads V_{RY} by 30°. This can take place only with a series combination of a resistor and a capacitor, as the simplest impedance in each phase.

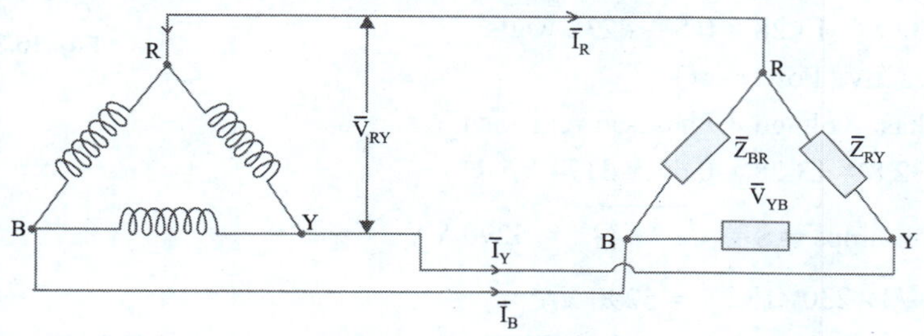

Fig 16.34 (*a*)

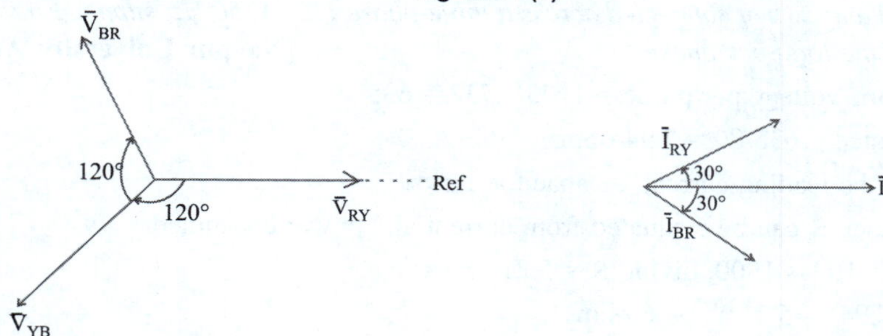

Fig. 16.34 (*b*) **Fig. 16.34 (*c*)**

(*a*) |Z| = 220/5.8 = 38.1 ohms

Resistance per phase = 38.1 × cos 30° = 33 ohms

Capacitive Reactance/phase = 38.1 × sin 30° = 16.05 ohms

(*b*) Current per phase = 5.8 amp, as calculated above.

(*c*) Otherline currents: Since a symmetrical three phase system is being dealt with, three currents have a mutual phase-difference of 120°. Hence

$I_R = 10 \angle 0°$ as given, $I_Y = 10 \angle -120°$ amp; $I_B = 10 \angle -240°$ amp.

Example 16.27. *A balanced 3-phase star-connected load of 8 + j 6 ohms per phase is connected to a three-phase 230 V supply. Find the line-current, power-factor, active power, reactive-power, and total volt-amperes.*

[Rajiv Gandhi Technical University, Bhopal, April 2001]

Solution. When a statement is made about three–phase voltage, when not mentioned otherwise, the voltage is the line-to-line voltage. Thus, 230 V is the line voltage, which means, in star-system, phase-voltage is 230/1.732, which comes to 132.8 V.

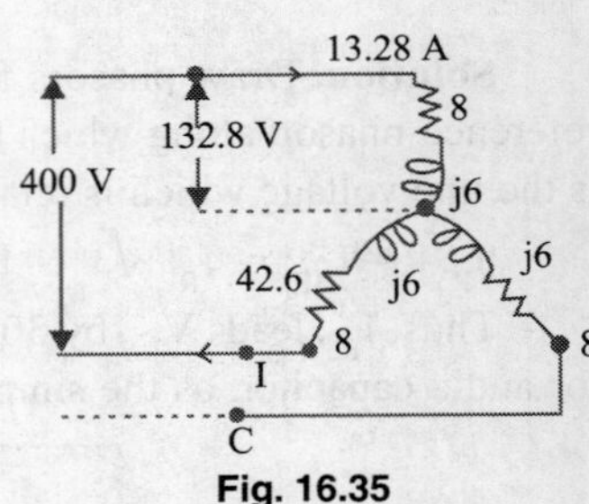

Fig. 16.35

$|Z| = \sqrt{8^2 + 6^2} = 10$ ohms

Line current = Phase current

= 132.8/10 = 13.28 amp

Power – factor = R/Z = 0.8, Lagging

Total Active Power = P = 1.732 × Line Voltage × Line Current × P.f.

Or = 3. Phase Voltage × Phase-current × P.f

= 3 × 132.8 × 13.28 × 0.8 = 4232 watts

Total Reactive Power = Q

= 3 × Phase-voltage × Phase-current × sin ϕ

= 3 × 132.8 × 13.28 × 0.60 = 3174 VAR

Total Volt-amps = S = $\sqrt{P^2 + Q^2}$ = 5290 VA

Or S = $\sqrt{3} \times 230 \times 13.28$ = 5290 VA

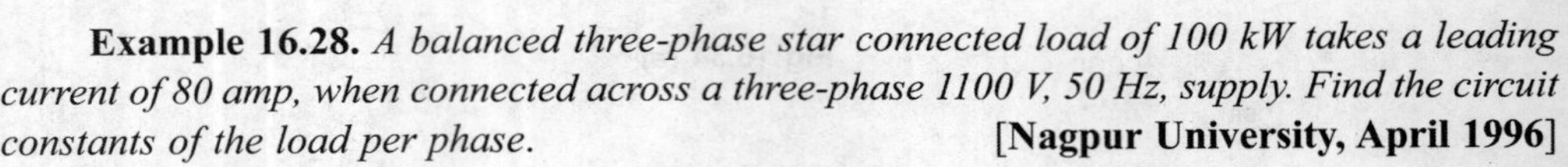

Example 16.28. *A balanced three-phase star connected load of 100 kW takes a leading current of 80 amp, when connected across a three-phase 1100 V, 50 Hz, supply. Find the circuit constants of the load per phase.* **[Nagpur University, April 1996]**

Solution. Voltage per phase = 1100/1.732 = 635 V

Impedance = 635/80 = 7.94 ohms.

Due to the leading current, a capacitor exists.

Resistance R can be evaluated from current and power consumed

$3\ I^2\ R = 100 \times 1000$, giving R = 5.21 ohms

$X_c = (7.94^2 - 5.21^2)^{0.5} = 6$ ohms

At 50 Hz, C = 1/(314 × 6) = 531 microfarads.

Tutorial Problem 16.1

1. Each phase of a delta-connected load comprises a resistor of 50 Ω and capacitor of 50 μF in series. Calculate (*a*) the line and phase currents (*b*) the total power and (*c*) the kilovoltamperes when the load is connected to a 440-V, 3-phase, 50-Hz supply.
   [(*a*) 9.46 A; 5.46 A (*b*) 4480 W (*c*) 7.24 kVA]
2. Three similar-coils, *A*, *B* and *C* are available. Each coil has 9 Ω resistance and 12 Ω reactance. They are connected in delta to a 3-phase, 440-V, 50-Hz supply. Calculate for this load:
   (*a*) the line current (*b*) the power factor
   (*c*) the total kilovolt-amperes (*d*) the total kilowatts

If the coils are reconnected in star, calculate for the new load the quantities named at (*a*), (*b*); (*c*) and (*d*) above. **[50.7 A; 0.6; 38.6 kVA; 23.16 kW; 16.9 A; 0.6; 12.867 kVA; 7.72 kW]**

3. Three similar choke coils are connected in star to a 3-phase supply. If the line currents are 15 A, the total power consumed is 11 kW and the volt-ampere input is 15 kVA, find the line and phase voltages, the VAR input and the reactance and resistance of each coil.

[577.3 V; 333.3 V; 10.2 kVAR; 15.1 Ω; 16.3 Ω]

4. The load in each branch of a delta-connected balanced 3-ϕ circuit consists of an inductance of 0.0318 H in series with a resistance of 10 Ω. The line voltage is 400 V at 50 Hz. Calculate (*i*) the line current and (*ii*) the total power in the circuit. **[(*i*) 49 A (*ii*) 24 kW] *(London Univ.)***

5. A 3-phase, delta-connected load, each phase of which has $R = 10$ Ω and $X = 8$ Ω, is supplied from a star-connected secondary winding of a 3-phase transformer each phase of which gives 230 V. Calculate

(*a*) the current in each phase of the load and in the secondary windings of the transformer

(*b*) the total power taken by the load

(*c*) the power factor of the load. **[(*a*) 31.1 A; 54 A (*b*) 29 kW (*c*) 0.78]**

6. A 3-phase load consists of three similar inductive coils, each of resistance 50 Ω and inductance 0.3 H. The supply is 415 V, 50 Hz, Calculate (*a*) the line current (*b*) the power factor and (*c*) the total power when the load is (*i*) star-connected and (*ii*) delta-connected.

[(*i*) 2.25 A, 0.47 lag, 762 W (*ii*) 6.75 A, 0.47 lag, 2280 W] (*London Univ.*)

7. A 415-V, 3-phase, 4-wire system supplies power to three non-inductive loads. The loads are 25 kW between red and neutral, 30 kW between yellow and neutral and 12 kW between blue and neutral.

Calculate (*a*) the current in each-line wire and (*b*) the current in the neutral conductor.

[(*a*) 104.2 A, 125 A, 50 A (*b*) 67 A] (*London Univ.*)

8. Non-inductive loads of 10, 6 and 4 kW are connected between the neutral and the red, yellow and blue phases respectively of a three-phase, four-wire system. The line voltage is 400 V. Find the current in each line conductor and in the neutral.

[(*a*) 43.3 A, 26A, 173. A, 22.9] (*App. Elect. London Univ.*)

9. A three-phase, star-connected alternator supplies a delta-connected load, each phase of which has a resistance of 20 Ω and a reactance of 10 Ω. Calculate (*a*) the current supplied by the alternator (*b*) the output of the alternator in kW and kVA, neglecting the losses in the lines between the alternator and the load. The line voltage is 400 V. **[(*a*) 30.95 A (b) 16.2 kW, 21.45 kVA]**

10. Three non-inductive resistances, each of 100 Ω, are connected in star to 3-phase, 440-V supply. Three equal choking coils each of reactance 100 Ω are also connected in delta to the same supply. Calculate:

(*a*) line current (*b*) p.f. of the system. **[(*a*) 8.04 A (*b*) 0.3156] (*I.E.E. London*)**

11. A 440-V, 50-Hz induction motor takes a line current of 45 A at a power factor of 0.8 (lagging). Three Δ-connected capacitors are installed to improve the power factor to 0.95 (lagging). Calculate the kVA of the capacitor bank and the capacitance of each capacitor.

[11.45 kVA, 62.7 μF] (*I.E.E. London*)

12. Three resistances, each of 500 Ω, are connected in star to a 400-V, 50-Hz, 3-phase supply. If three capacitors, when connected in delta to the same supply, take the same line currents, calculate the capacitance of each capacitor and the line current.

[2.123 μ F, 0.653 A] (*London Univ.*)

16.15. Power Measurement in 3-Phase Circuits

Following methods are available for measuring power in a 3-phase load.

(*a*) Three Wattmeter Method

In this method, three wattmeters are inserted one in each phase and the algebraic sum of their readings gives the total power consumed by the 3-phase load.

(b) Two Wattmeter Method

(*i*) This method gives true power in the 3-phase circuit without regard to balance or wave form provided in the case of *Y*-connected load. The neutral of the load is isolated from the neutral of the source of power. Or if there is a neutral connection, the neutral wire should not carry any current. This is possible only if the load is perfectly balanced and there are no harmonics present of triple frequency or any other multiples of that frequency.

(*ii*) This method can also be used for 3-phase, 4-wire system in which the neutral wire carries the neutral current. In this method, the current coils of the wattmeters are supplied from current transformers inserted in the principal line wires in order to get the correct magnitude and phase differences of the currents in the current coils of the wattmeter, because in the 3-phase, 4-wire system, the sum of the instantaneous currents in the principal line wires is not necessarily equal to zero as in 3-phase 3-wire system.

(c) One Wattmeter Method

In this method, a single wattmeter is used to obtain the two readings which are obtained by two wattmeters by the two-wattmeter method. This method can, however, be used only when the load is balanced.

16.16. Three Wattmeter Method

A wattmeter consists of (*i*) a low resistance current coil which is inserted in series with the line carrying the current and (*ii*) a high resistance pressure coil which is connected across the two points whose potential difference is to be measured.

A wattmeter shows a reading which is proportional to the product of the current through its current coil, the p.d. across its potential or pressure coil and cosine of the angle between this voltage and current.

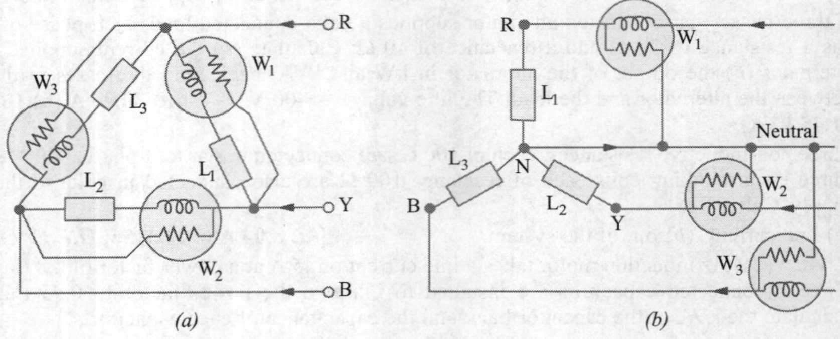

Fig. 16.36

As shown in Fig. 16.36 in this method three wattmeters are inserted in each of the three phases of the load whether Δ-connected or *Y*-connected. The current coil of each wattmeter carries the current of one phase only and the pressure coil measures the phase-voltage of this phase. Hence, each wattmeter measures the power in a single phase. The algebraic sum of the readings of three wattmeters must give the total power in the load.

The difficulty with this method is that under ordinary conditions it is not generally feasible to break into the phases of a delta-connected load nor is it always possible, in the case of a *Y*-connected load, to get at the neutral point which is required for connections as shown in Fig. 16.37 (*b*). However, it is not necessary to use three wattmeters to measure power, two wattmeters can be used for the purpose as shown below.

16.17. Two Wattmeter Method-Balanced or Unbalanced Load

As shown in Fig. 16.37, the current coils of the two wattmeters are inserted in ***any two*** lines and the potential coil of each joined to the third line. It can be proved that the sum of the instantaneous powers indicated by W_1 and W_2 gives the instantaneous power absorbed by the three loads L_1, L_2 and L_3. A star-connected load is considered in the following discussion although it can be equally applied to Δ-connected loads because a Δ-connected load can always be replaced by an equivalent *Y*-connected load.

Now, before we consider the currents through and p.d. across each wattmeter, it may be pointed out that ***it is important to take the direction of the voltage through the circuit the same as that taken for the current when establishing the readings of the two wattmeters.***

Instantaneous current through $W_1 = i_R$

p.d. across $W_1 = e_{RB} \qquad = e_R - e_B$

p.d. across power read by $W_1 = i_R\,(e_R - e_B)$

Instantaneous current through $W_2 = i_Y$

Instantaneous p.d. across $W_2 = e_{YB} = (e_Y - e_B)$

Instantaneous power read by $W_2 = i_Y\,(e_Y - e_B)$

$$\therefore \quad W_1 + W_2 = e_R(e_R - e_B) + i_Y(e_Y - e_B) = i_R e_R + i_Y e_Y - e_B(i_R + i_Y)$$

Now, $\qquad i_R + i_Y + i_B = 0 \qquad$ ... Kirchhoff's Current Law

$$\therefore \qquad i_R + i_Y = -i_B$$

or $\quad W_1 + W_2 = i_R . e_R + i_Y . e_y + i_B . e_B = p_1 + p_2 + p_3$

where p_1 is the power absorbed by load L_1, p_2 that absorbed by L_2 and p_3 that absorbed by L_3

$\therefore \quad W_1 + W_2$ = total power absorbed

The proof is true whether the load is balanced or unbalanced. If the load is *Y*-connected, it should have no neutral connection (*i.e.* 3 – ϕ, 3-wire connected) and if it has a neutral connection (*i.e.* 3 – ϕ, 4-wire connected) then it should be exactly balanced so that in each case there is no neutral current i_N otherwise Kirchoff's current law will give $i_N + i_R + i_y + i_B = 0$.

16.18. Two Wattmeter Method– *Balanced Load*

If the load is balanced, then power factor of the load can also be found from the two wattmeter readings. The *Y*-connected load in Fig. 16.37 (*b*) will be assumed inductive. The vector diagram for such a balanced *Y*-connected load is shown in Fig. 16.38. We will now consider the problem in terms of r.m.s. values instead of instantaneous values.

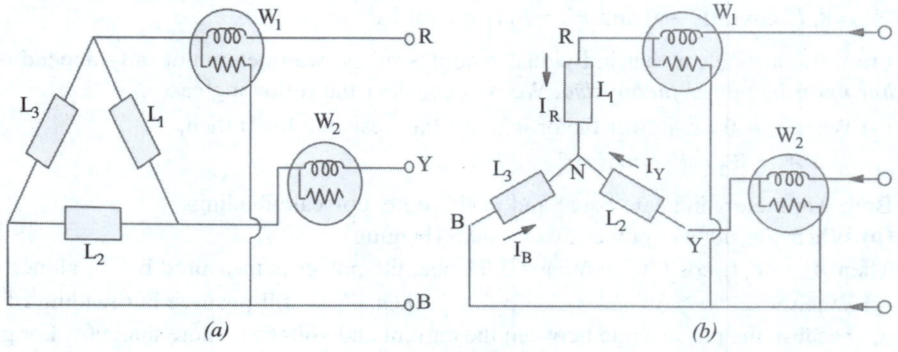

Fig. 16.37

Let V_R, V_Y and V_B be the r.m.s. values of the three phase voltages and I_R, I_Y and I_B the r.m.s. values of the currents. Since these voltages and currents are assumed sinusoidal, they can be represented by vectors, the currents lagging behind their respective phase voltages by ϕ.

Current through wattmeter W_1 [Fig. 16.37 (*b*)] is $= I_R$.

P.D. across voltage coil of W_1 is

$$V_{RB} = V_R - V_B \quad \text{... vectorially}$$

This V_{RB} is foundby compounding V_R and V_B reversed as shown in Fig. 16.33. It is seen that phase difference between V_{RB} and I_R $= (30° - \phi)$.

$\therefore$ Reading of $W_1 = I_R\, V_{RB} \cos(30° - \phi)$

Similarly, as seen from Fig. 16.47 (*b*).

Current through $W_2 = I_Y$

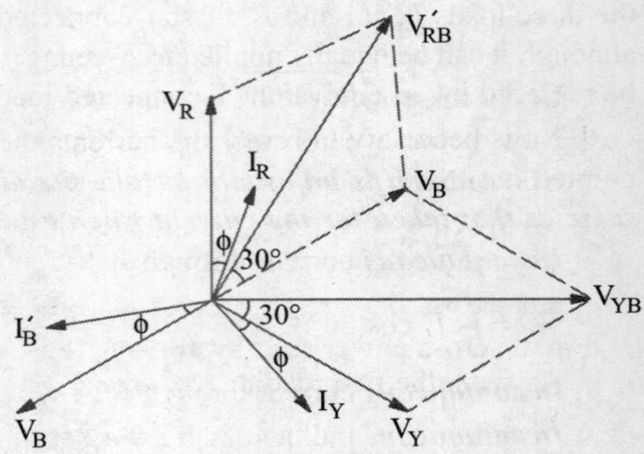

Fig. 16.38

P.D. across $W_2 = V_{YB} = V_Y - V_B$ ... vectorially

Again, V_{YB} is found by compounding V_Y and V_B reversed as shown in Fig. 16.38. The angle between I_Y and V_{YB} is $(30° + \phi)$. Reading of $W_2 = I_Y V_{YB} \cos(30° + \phi)$

Since load is balanced, $V_{RB} = V_{YB}$ = line voltage V_L; $I_Y = I_R$ = line current, I_L

$$\therefore \quad W_1 = V_L I_L \cos(30° - \Phi) \text{ and } W_2 = V_L I_L \cos(30° + \phi)$$

$$\therefore \quad W_1 + W_2 = V_L I_L \cos(30° - \phi) + V_L I_L (\cos(30° + \phi)$$

$$= V_L I_L [\cos 30° \cos\phi + \sin 30° \sin\phi + \cos 30° \cos\phi - \sin 30° \sin\phi]$$

$$W_1 + W_2 = V_L I_L (2\cos 30° \cos\phi) = \sqrt{3} V_L I_L \cos\phi = \text{total power in the 3-phase load}$$

Hence, the sum of the two wattmeter readings gives the total power consumption in the 3-phase load.

It should be noted that phase sequence of *RYB* has been assumed in the above discussion. Reversal of phase sequence will interchange the readings of the two wattmeters.

16.19. Variations in Wattmeter Readings

It has been shown above that for a lagging power factor

$$W_1 = V_L I_L \cos(30° - \varphi) \text{ and } W_2 = V_L I_L \cos(30° + \phi)$$

From this it is clear that individual readings of the wattmeters not only depend ***on the load but upon its power factor also.*** We will consider the following cases:

(*a*) When $\phi = 0$ *i.e.* power factor is unity (*i.e.* resistive load) then,

$$W_1 = W_2 = V_L I_L \cos 30°$$

Both wattmeters indicate equal and positive *i.e.* up-scale readings.

(*b*) When $\phi = 60°$ *i.e.* power factor = 0.5 (lagging)

Then $W_2 = V_L I_L \cos(30° + 60°) = 0$. Hence, the power is measured by W_1 alone.

(*c*) When $90° > \phi > 60°$ *i.e.* 0.5 > p.f. > 0, then W_1 is still positive but reading of W_2 is reversed because the phase angle between the current and voltage is more than 90°. For getting the total power, the reading of W_2 is to be subtracted from that of W_1.

ϕ	0°	60°	90°
$\cos\phi$	1	0.5	0
W_1	+ve	+ve	+ve
W_2	+ve $W_1 = W_2$	0	–ve $W_1 = W_2$

Under this condition, W_2 will read 'down scale' *i.e.* backwards. Hence, to obtain a reading on W_2 it is necessary to reverse either its pressure coil or current coil, usually the

All readings taken after reversal of pressure coil are to be taken as negative.

(*d*) When $\phi = 90°$ (*i.e.* pure inductive or capacitive load), then

$$W_1 = V_L I_L \cos(30° - 90°) = V_L I_L \sin 30°;$$

$$W_2 = V_L I_L \cos(30° + 90°) = -V_L I_L \sin 30°$$

As seen, the two readings are equal but of opposite sign.

$\therefore \quad W_1 + W_2 = 0$

The above facts have been summarised in the above table for a lagging power factor.

16.20. Leading Power Factor*

In the above discussion, lagging angles are taken positive. Now, we will see how wattmeter readings are changed if the power factor becomes leading. For $\phi = +60°$ (lag), W_2 is zero. But for $\phi = -60°$ (lead), W_1 is zero. So we find that for angles of lead, the reading of the two wattmeters are interchanged. Hence, for a ***leading*** power factor.

$$W_1 = V_L I_L \cos(30° + \phi) \text{ and } W_2 = V_L I_L \cos(30° - \phi)$$

16.21. Power Factor–*Balanced Load*

In case the load is balanced (and currents and voltages are sinusoidal) and for a *lagging* power factor:

$$W_1 + W_2 = V_L I_L \cos(30° - \phi) + V_L I_L \cos(30° + \phi) = \sqrt{3} V_L I_L \cos\phi \qquad \text{... }(i)$$

Similarly $W_1 - W_2 = V_L I_L \cos(30° - \phi) - V_L I_L \cos(30° + \phi)$

$$= V_L I_L (2 \times \sin\phi \times 1/2) = V_L I_L \sin\phi \qquad \text{... }(ii)$$

Dividing **(*ii*)** by **(*i*)**, we have $\tan\phi = \dfrac{\sqrt{3}(W_1 - W_2)}{(W_1 + W_2)}$ ** ... **(*iii*)**

Knowing tan ϕ and hence ϕ, the value of power factor cos ϕ can be found by consulting the trigonometrical tables. It should, however, be kept in mind that if W_2 reading has been taken after reversing the pressure coil *i.e.* if W_2 is negative, then the above relation becomes

$$\tan\phi = -\sqrt{3}\left(\frac{W_1 - W_2}{W_1 + W_2}\right)$$

* For a leading p.f., conditions are just the opposite of this. In that case, W_1 reads negative (Art. 16.22).

** For a leading power factor, this expression becomes

$$\tan\phi = -\sqrt{3}\left(\frac{W_1 - W_2}{W_1 + W_2}\right)$$

$$\tan\phi = \sqrt{3}\,\frac{W_1 - (-W_2)}{W_1 + (-W_2)} = \sqrt{3}\,\frac{W_1 + W_2}{W_1 - W_2}$$

Obviously, in this expression, only *numerical* values of W_1 and W_2 should be substituted. We may express power factor in terms of the ratio of the two wattmeters as under:

$$\text{Let}\quad \frac{\text{smaller reading}}{\text{larger reading}} = \frac{W_2}{W_1} = r$$

Then from equation **(*iii*)** above,

$$\tan\phi = \frac{\sqrt{3}[1-(W_2/W_1)]}{1+(W_2/W_1)} = \frac{\sqrt{3}(1-r)}{1+r}$$

Now $\sec^2\phi = 1 + \tan^2\phi$

$$\text{or}\quad \frac{1}{\cos^2\phi} = 1 + \tan^2\phi$$

$$\therefore\quad \cos\phi = \frac{1}{\sqrt{1+\tan^2\phi}}$$

$$= \frac{1}{\sqrt{1+3\left(\dfrac{1-r}{1+r}\right)^2}} = \frac{1+r}{2\sqrt{1-r+r^2}}$$

Fig. 16.39

If r is plotted against $\cos\phi$, then a curve called watt-ratio curve is obtained as shown in Fig. 16.39.

16.22. Balanced Load – *Leading Power Factor*

In this case, as seen from Fig. 16.40

$$W_1 = V_L I_L \cos(30 + \phi)$$

$$\text{and}\quad W_2 = V_L I_L \cos(30 - \phi)$$

$$\therefore\quad W_1 + W_2 = \sqrt{3} V_L I_L \cos\phi \quad \text{– as found above}$$

$$W_1 - W_2 = -V_L I_L \sin\phi$$

$$\therefore\quad \tan\phi = -\frac{\sqrt{3}(W_1 - W_2)}{(W_1 + W_2)}$$

Obviously, if $\phi > 60°$, then phase angle between V_{RB} and I_R becomes more than 90°. Hence, W_1 reads 'down-scale' *i.e.* it indicates negative reading. However, W_2 gives positive reading even in the extreme case when $\phi = 90°$.

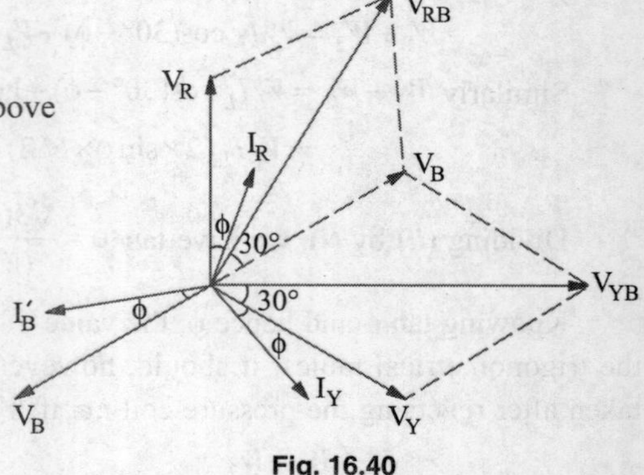

Fig. 16.40

16.23. Reactive Voltamperes with Two Wattmeters

We have seen that $\tan\phi = \dfrac{\sqrt{3}(W_1 - W_2)}{(W_1 + W_2)}$

Since the tangent of the angle of lag between phase current and phase voltage of a circuit is always equal to the ratio of the reactive power to the active power (in watts), it is clear that $\sqrt{3}\,(W_1 - W_2)$ represents the reactive power (Fig. 16.41). Hence, for a balanced load, the reactive power is given by $\sqrt{3}$ times the difference of the readings of the two wattmeters used to measure the power for a 3-phase circuit by the two wattmeter method. It may also be proved mathematically a follows:

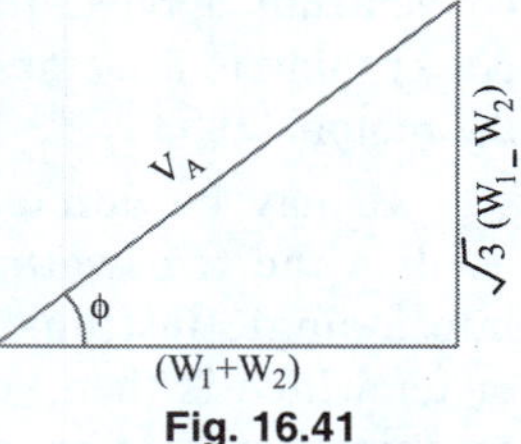

Fig. 16.41

$$= \sqrt{3}(W_1 - W_2) = \sqrt{3}[V_L I_L \cos(30° - \phi) - V_L I_L \cos\,(30° - \phi)]$$

$$= \sqrt{3}V_L I_L(\cos 30° \cos\phi + \sin 30° \sin\phi - \cos 30° \cos\phi + \sin 30° \sin\phi)$$

$$= \sqrt{3}V_L I_L \sin\phi$$

16.24. Reactive Voltamperes with One Wattmeter

For this purpose, the wattmeter is connected as shown in Fig. 16.42 (*a*) and (*b*). The pressure coil is connected across *Y* and *B* lines whereas the current coil is included in the *R* line. In Fig. 16.38 (*a*), the current coil is connected between terminals *A* and *B* whereas pressure coil is connected between terminals *C* and *D*. Obviously, current flowing through the wattmeter is I_R and p.d. is V_{YB}. The angle between the two, as seen from vector diagram of Fig. 16.38, is (30 + 30 + 30 – ϕ) = (90 – ϕ).

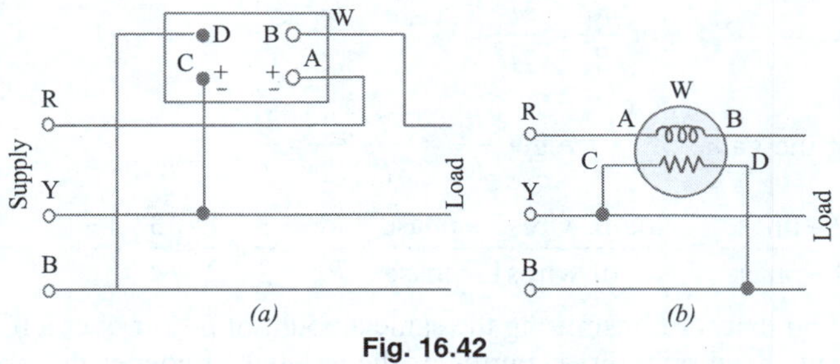

Fig. 16.42

Hence, reading of the wattmeter is $W = V_{YB} I_R \cos(90 - \phi) = V_{YB} I_R \sin\phi$

For a balanced load, V_{YB} equals the line voltage V_L and I_R equals the line current I_L, hence

$$W = V_L I_L \sin\phi$$

We know that the total reactive voltamperes of the load are $Q = \sqrt{3}V_L I_L \sin\phi$.

Hence, to obtain total *VARs*, the wattmeter reading must be multiplied by a factor of $\sqrt{3}$.

16.25. One Wattmeter Method

In this case, it is possible to apply two-wattmeter method by means of one wattmeter without breaking the circuit. The current coil is connected in any one line and the pressure coil is connected alternately between this and the other two lines (Fig. 16.43). The two readings so obtained, for a balanced load, correspond to those obtained by normal two wattmeter method. It should be kept in mind that this method is not of as much universal application as the two wattmeter method because it is restricted to fairly balanced loads only. However, it may be conveniently applied, for instance, when it is desired to find the power input to a factory motor in order to check the load upon the motor.

It may be pointed out here that the two wattmeters used in the two wattmeter method are usually combined into a single instrument in the case of switchboard wattmeter which is then known as a polyphase wat meter. The combination is affected by arranging the two sets of coils in such a way as to operate on a single moving system resulting in an indication of the total power on the scale.

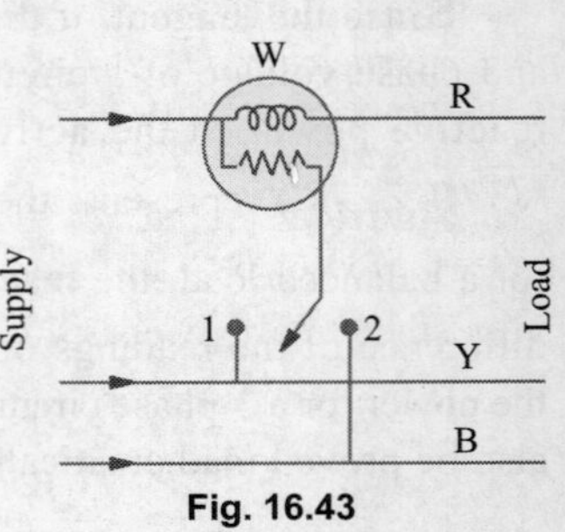

Fig. 16.43

16.26. Copper Required for Transmitting Power under Fixed Conditions

The comparison between 3-phase and single-phase systems will be done on the basis of a fixed amount of power transmitted to a fixed distance with the same amount of loss and at the same maximum voltage between conductors. In both cases, the weight of copper will be directly proportional to the number of wires (since the distance is fixed) and inversely proportional to the resistance of each wire. We will assume the same power factor and same voltage.

$$P_1 = VI_1 \cos\phi \text{ and } P_3 = \sqrt{3}VI_3 \cos\phi$$

where I_1 = r.m.s. value of current in 1-phase system

I_3 = r.m.s. value of line current in 3-phase system

$\because \quad P_1 = P_2 \therefore \; VI_1 \cos\phi = \sqrt{3}VI_3 \cos\phi \;\therefore\; I_1 = \sqrt{3}I_3$

also $I_1^2 R_1 \times 2 = I_2^3 R_3 \times 3$ or $\dfrac{R_1}{R_3} = \dfrac{3I_3^2}{2I_1^2}$

Substituting the value of I_1, we get $\dfrac{R_1}{R_3} = \dfrac{3I_3^2}{3I_3^2 \times 2} = \dfrac{1}{2}$

$$\therefore \quad \frac{\text{copper 3-phase}}{\text{copper 1-phase}} = \frac{\text{No. of wires 3-phase}}{\text{No. of wires 1-phase}} \times \frac{R_1}{R_3} = \frac{3}{2} \times \frac{1}{2} = \frac{3}{4}$$

Hence, we find that for transmitting the same amount of power over a fixed distance with a fixed line loss, we need only three-fourths of the amount of copper that would be required for a single phase or to put it in another way, one-third more copper is required for a 1-phase system than would be necessary for a three-phase system.

Example 16.29. *Phase voltage and current of a star-connected inductive load is 150 V and 25 A. Power factor of load is 0.707 (lag). Assuming that the system is 3-wire and power is measured using two wattmeters, find the readings of wattmeters.*

(Elect. Instrument & Measurements, Nagpur Univ. 1993)

Solution. $V_{ph} = 150\text{V}; V_L = 150 \times \sqrt{3}\text{V}; I_{ph} = I_L = 25$ A

Total power = $\sqrt{3}\, V_L I_L \cos\phi = \sqrt{3} \times 150 \times \sqrt{3} \times 25 \times 0.707$ = 7954 W

$\therefore \quad W_1 + W_2 = 7954\ W$ ... (*i*)

$\cos\phi = 0.707; \phi = \cos^{-1}(0.707) = 45°; \tan 45° = 1$

Now, for a lagging power factor, $\tan\phi = \sqrt{3}\,(W_1 - W_2)/(W_1 + W_2)$ or $1 = \sqrt{3}(W_1 - W_2)/7954$

$\therefore \quad (W_1 - W_2) = 4592\ W$ ... (*ii*)

From (*i*) and (*ii*) above, we get, W_1 = **6273 W; W$_2$ = 1681 W.**

Example 16.30. *In a balanced 3-phase 400-V circuit, the line current is 115.5 A. When power is measured by two wattmeter method, one meter reads 40 kW and the other zero. What is the power factor of the load? If the power factor were unity and the line current the same, what would be the reading of each wattmeter?*

Solution. Since $W_2 = 0$, the whole power is measured by W_1. In such a situation, p.f. = 0.5. However, it can be calculated as under.

Since total power is 40 kW, $\therefore$ $40{,}000 = \sqrt{3}\times 400\times 115.5\times\cos\phi;\ \cos\phi = 0.5$

If the power factor is unity with line currents remaining the same, we have

$$\tan\phi = \frac{\sqrt{3}(W_1 - W_2)}{(W_1 + W_2)} = 0 \text{ or } W_1 = W_2$$

Also, $(W_1 + W_2) = \sqrt{3}\times 400\times 115.5\times 1 = 80000\ \text{W} = 80\ \text{kW}$

At unity p.f., $W_1 = W_2$. Hence, each wattmeter reads = 80/2 = **40 kW.**

Example 16.31. *The input power to a three-phase motor was measured by two wattmeter method. The readings were 10.4 KW and – 3.4 KW and the voltage was 400 V. Calculate **(a)** the power factor **(b)** the line current.* **(Elect. Engg. A.M.Ae, S.I. June 1991)**

Solution. When W_2 reads negative, then we have

$\tan\phi = \sqrt{3}(W_1 + W_2)/(W_1 - W_2)$. Substituting numerical values of W_1 and W_2, we get

$\tan\phi = \sqrt{3}(10.4 + 3.4)/(10.4 - 3.4) = 1.97; \phi = \tan^{-1}(1.97) = 63.1°$

(*a*) p.f. = $\cos\phi = \cos 63.1° =$ **0.45 (lag)**

(*b*) W = 10.4 – 3.4 = 7 KW = 7,000 W

$7000 = \sqrt{3}I_L \times 400\times 0.45; I_L =$ **22.4 A**

Example 16.32. *A three-phase, three-wire, 100-V, ABC system supplies a balanced delta connected load with impedance of 20 ∠ 45° ohm.*

***(a)** Determine the phase and line currents and draw the phase or diagram **(b)** Find the wattmeter readings when the two wattmeter method is applied to the system.*

(Elect. Machines, A.M.I.E. Sec B.)

Solution. (*a*) The phasor diagram is shown in Fig. 16.44 (*b*)

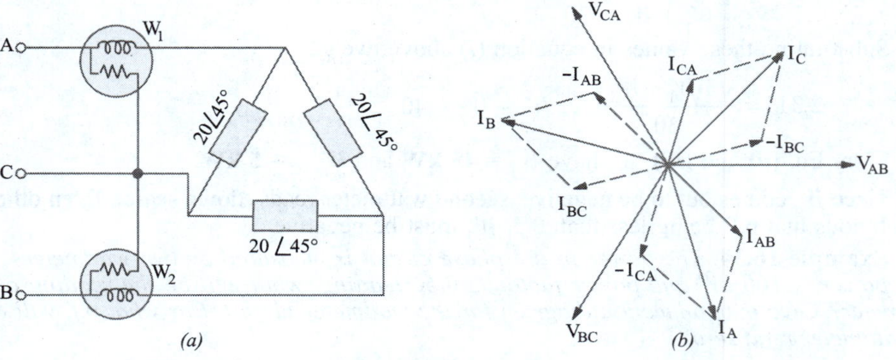

Fig. 16.44

Let $V_{AB} = 100\angle 0°$. Since phase sequence is *ABC*, $V_{BC} = 100\angle -120°$ and $V_{CA} = 100°$ $\angle 120°$

$$\text{Phase current } I_{AB} = \frac{V_{AB}}{Z_{AB}} = \frac{100\angle 0°}{20\angle 45°} = 5\angle -45°$$

$$I_{BC} = \frac{V_{BC}}{Z_{BC}} = \frac{100\angle -120°}{20\angle 45°} = 5\angle -165°, I_{CA} = \frac{V_{CA}}{Z_{CA}} = \frac{100\angle 120°}{20\angle 45°} = 5\angle 75°$$

Applying *KCL* to junction A, we have

$$I_A + I_{CA} - I_{AB} = 0 \text{ or } I_A = I_{AB} - I_{CA}$$

$\therefore$ Line current $I_A = 5\angle -45° - 5\angle 75° = 8.66\angle -75°$

Since the system is balanced, I_B will lag I_A by 120° and I_C will lag I_A by 240°.

$$\because \quad I = 8.66\angle(75° - 120°) = 8.66\angle -195°; I_C = 8.66\angle(-75° - 240°)$$
$$= 8.66\angle -315° = 8.66\angle 45°$$

(*b*) As shown in Fig. 16.44 (*b*), reading of wattmeter W_1 is $W_1 = V_{AC}\ I_C \cos \varphi$. Phasor V_{AC} is the reverse of phasor V_{CA}. Hence, V_{AC} is the reverse of phasor V_{CA}. Hence, V_{AC} lags the reference vector by 60° whereas I_A lags by 75°. Hence, phase difference between the two is (75° – 60°) = 15°.

$\therefore W_1 = 100 \times 8.66 \times \cos 15° = 836.5$ W

Similarly $W_2 = V_{BC} I_B \cos\phi = 100 \times 8.66 \times \cos 75° = 224.1$ W

$\therefore W_1 + W_2 = 836.5 + 224.1 = 1060.6$ W

Resistance of each delta branch = 20 cos 45° = 14.14 Ω

Total power consumed = $3\ I^2R = 3 \times 5^2 \times 14.14 = 1060.6$ W

Hence, it proves that the sum of the two wattmeter readings gives the total power consumed.

Example 16.33. *A 3-phase, 500-V motor load has a power factor of 0.4 Two wattmeters connected to measure the power show the input to be 30 kW. Find the reading on each instrument.* **(Electrical Meas., Nagpur Univ. 1991)**

Solution. As seen

$$\tan\phi = \frac{\sqrt{3}(W_1 - W_2)}{W_1 + W_2} \qquad \text{... }(i)$$

Now, $\cos\phi = 0.4; \phi = \cos^{-1}(0.4) = 66.6°;\ \tan 66.6° = 2.311$

$$W_1 + W_2 = 30 \qquad \text{... }(ii)$$

Substituting these values in equation (*i*) above, we get

$$2.311 = \frac{\sqrt{3}(W_1 - W_2)}{30} \quad \therefore W_1 - W_2 = 40 \qquad \text{... }(iii)$$

From Eq. (***ii***) and (***iii***), we have W_1 = **45 kW** and W_2 = **– 5 kW**

Since W_2 comes out to be negative, second wattmeter reads 'down scale'. Even otherwise it is obvious that p.f. being less than 0.5, W_2 must be negative

Example 16.34. *The power in a 3-phase circuit is measured by two wattmeters. If the total power is 100 kW and power factor is 0.66 leading, what will be the reading of each wattmeter? Give the connection diagram for the wattmeter circuit. For what p.f. will one of the wattmeter read zero?*

Solution. $\phi = \cos^{-1}(0.66) = 48.7°; \tan\phi = 1.1383$

Since p.f. is leading,

$$\therefore \tan\phi = -\frac{\sqrt{3}(W_1 - W_2)}{W_1 + W_2} \quad \therefore 1.1383 = -\sqrt{3}(W_1 - W_2)/100$$

$\therefore$ $W_1 - W_2 = -65.7$ and $W_1 + W_2 = 100$ $\therefore$ W_1 = **17.14 kW;** W_2 = **82.85 kW**

Connection diagram is similar to that shown in Fig. 16.37 (*b*). One of the wattmeters will read zero when p.f. = **0.5**

Example 16.35. *Two wattmeters are used for measuring the power input and the power factor of an over-excited synchronous motor. If the readings of the meters are (– 2.0 kW) and (+ 7.0 kW) respectively, calculate the input and power factor of the motor.*

(Elect. Technology, Punjab Univ., June, 1991)

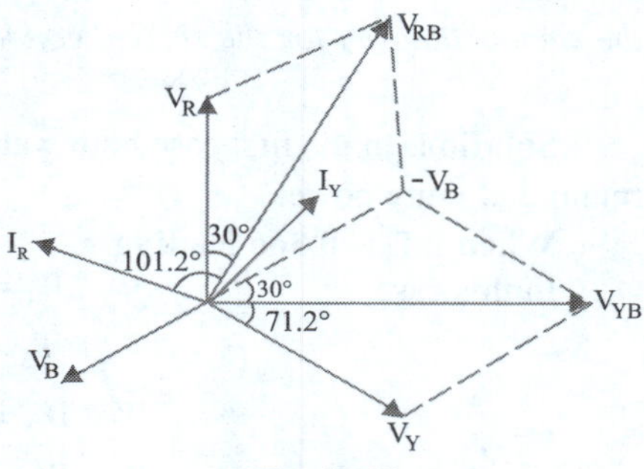

Fig. 16.45

Solution. Since an over-excited synchronous motor runs with a leading p.f..

$$\tan\phi = \frac{\sqrt{3}(W_1 - W_2)}{W_1 + W_2}$$

Moreover, as explained in the same article, it is W_1 that gives negative reading and not W_2.

Hence, $W_1 = -2\text{kW}$

$\therefore$ $$\tan\phi = -\frac{\sqrt{3}(-2-7)}{-2+7} = \sqrt{3}\times\frac{9}{5} = 3.1176$$

$\therefore$ $$\phi = \tan^{-1}(3.1176) = 71.2° \ \text{(lead)}$$

$\therefore$ $\cos\phi = \cos 71.2°$ = **0.3057 (lead)** and

Input = $W_1 + W_2 = -2 + 7$ = **5 kW**

Example 16.36. *A 440-V, 3-phase, delta-connected induction motor has an output of 14.92 kW at a p.f. of 0.82 and efficiency 85%. Calculate the readings on each of the two wattmeters connected to measure the input. Prove any formula used.*

If another star-connected load of 10 kW at 0.85 p.f. lagging is added in parallel to the motor, what will be the current draw from the line and the power taken from the line?

(Elect. Technology-I, Bombay Univ.)

Solution. Motor input = 14,920/0.85 = 17,600 W $\therefore$ $W_1 + W_2 = 17.6$ kW ... (*i*)

$\cos\phi = 0.82; \phi = 34.9°,\ \tan 34.9° = 0.6976;\ 0.6976 = \sqrt{3}\dfrac{W_1 - W_2}{17.6}$

$\therefore$ $W_1 - W_2 = 7.09$ kW ... (*ii*)

From (*i*) and (*ii*) above, we get W_1 = **12.35 kW** and W_2 = **5.26 kW**

$$\text{Motor } \textbf{kVA},\ S_m = \frac{\text{motor kW}}{\cos\phi_m} = \frac{17.6}{0.82} = 21.46$$

$\therefore$ $\mathbf{S_m} = 21.46\angle -34.9° = (17.6 - j\,12.28)$ kVA

Load p.f. = 0.85 $\therefore$ $\phi = \cos^{-1}(0.85) = 31.8°$; Load kVA, $S_Y = 10/0.85 = 11.76$

$\therefore$ $S_Y = 11.76\angle -31.8° = (10 - j6.2)$ kVA

Combined kVA, $\mathbf{S} = \mathbf{S_m} + \mathbf{S_Y} = (27.6 - j\,18.48) = 32.2\angle -33.8°$ kVA

$$I = \frac{S}{\sqrt{3}.V} = \frac{33.2\times10^3}{\sqrt{3}\times440} = \mathbf{43.56\ A}$$

Power taken = **27.6 kW**

Example 16.37. *The power input to a synchronous motor is measured by two wattmeters both of which indicate 50 kW. If the power factor of the motor be changed to 0.866 leading, determine the readings of the two wattmeters, the total input power remaining the same. Draw the vector diagram for the second condition of the load.*

(Elect. Technology, Nagpur Univ. 1992)

Solution. In the first case both wattmeters read equal and positive. Hence motor must be running at unity power.

When p.f. is 0.866 leading

In this case; $W_1 = V_L I_L \cos(30° + \phi)$;

$$W_2 = V_L I_L \cos(30° - \phi)$$

$$\therefore \quad W_1 + W_2 = \sqrt{3} V_L I_L \cos\phi$$

$$W_1 - W_2 = -V_L I_L \sin\phi$$

$$\therefore \quad \tan\phi = \frac{\sqrt{3}(W_1 - W_2)}{(W_1 + W_2)}$$

$$\phi = \cos^{-1}(0.866) = 30°$$

$$\tan\phi = 1/\sqrt{3}$$

$$\therefore \quad \frac{1}{\sqrt{3}} = \frac{-\sqrt{3}(W_1 - W_2)}{100}$$

$$\therefore \quad W_1 - W_2 = -100/3$$

and $\quad W_1 + W_2 = 100$

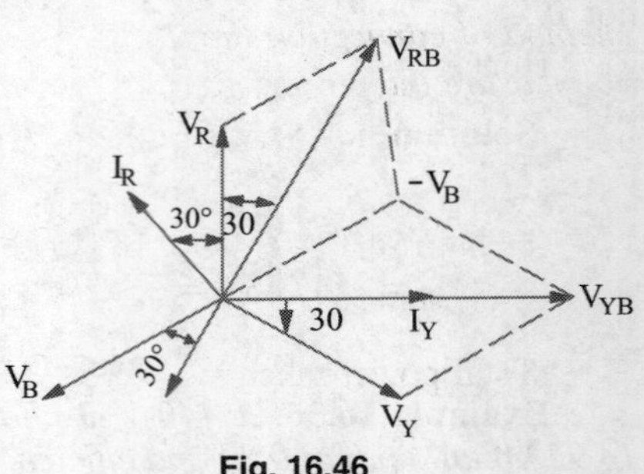

Fig. 16.46

$\therefore \quad 2W_1 = 200/3$; W_1 = **33.33 kW**; W_2 = **66.67 kW**

For connection diagram, please refer to Fig. 16.37. The vector or phasor diagram is shown in Fig. 16.56.

Example 16.38. (*a*). *A star-connected balanced load is supplied from a 3 – φ balanced supply with a line voltage of 416 volts at a frequency of 50 Hz. Each phase of the load consists of a resistance and a capacitor joined in series and the reading on two wattmeters connected to measure the total power supplied are 782 W and 1980 W, both positive. Calculate* **(*i*)** *power factor of circuit,* **(*ii*)** *the line current,* **(*iii*)** *the capacitance of each capacitor.*

(Elect. Engg. I, Nagpur Univ. 1993)

Solution. (*i*) $\tan\phi = -\frac{\sqrt{3}(W_1 - W_2)}{(W_1 + W_2)} = -\frac{\sqrt{3}(782 - 1980)}{(782 + 1980)} = 0.75;\ \phi = 36.9°,\ \cos\phi = 0.8$

(*ii*) $\sqrt{3} \times 416 \times I_L \times 0.8 = 2762, I_L = 4.8\,\text{A}$

(*iii*) $Z_{ph} = V_{ph}/I_{ph} = (416\sqrt{3})/4.8 = 50\Omega, X_C = Z_{ph}\sin\phi = 50 \times 0.6 = 30\Omega$

Now, $X_C = 1/2\pi fC = 1/2\phi \times 50 \times C = 106 \times 10^{-6}\,\text{F}$

Example 16.39. *Each phase of a 3-phase, Δ-connected load consists of an impedance Z = 20 ∠60° ohm. The line voltage is 440 V at 50 Hz. Compute the power consumed by each phase impedance and the total power. What will be the readings of the two wattmeters connected?* **(Elect. and Mech. Technology, Osmania Univ.)**

Solution. $Z_{ph} = 20\Omega; V_{ph} = V_L = 440\text{V}; I_{ph} = V_{ph} / Z_{ph} = 440/20 = 22\ A$

Since $\phi = 60°; \cos\phi = \cos 60° = 0.5°;\ R_{ph} = Z_{ph} \times \cos 60° = 20 \times 0.5 = 10\Omega$

$\therefore$ Power/phase = $I_{ph}^2 R_{ph} = 22^2 \times 10$ = **4,840 W**

Total power = 3 × 4,840 = **14,520 W** [or $P = \sqrt{3} \times 440 \times (\sqrt{3} \times 22) \times 0.5$ = 14,520 W]

Now, $W_1 + W_2 = 14{,}520.$

Also $\tan\phi = \sqrt{3}.\dfrac{W_1 - W_2}{W_1 + W_2}$ $\therefore$ $\tan 60° = \sqrt{3} = \sqrt{3}.\dfrac{W_1 - W_2}{14{,}520}$

$\therefore$ $W_1 - W_2$ = 14,520. Obviously, $W_2 = 0$

Even otherwise it is obvious that W_2 should be zero because p.f. = cos 60° = 0.5.

Example 16.40. *Three identical coils, each having a reactance of 20 Ω and resistance of 20 Ω are connected in* ***(a)*** *star* ***(b)*** *delta across a 440-V, 3-phase line. Calculate for each method of connection the line current and readings on each of the two wattmeters connected to measure the power.* **(Electro-mechanics, Allahabad Univ. 1992)**

Solution. (*a*) Star Connection

$$Z_{ph} = \sqrt{20^2 + 20^2} = 20\sqrt{2} = 28.3\Omega; V_{ph} = 440 / \sqrt{3} = 254\text{ V}$$

$$I_{ph} = 254 / 28.3 = 8.97 A; I_L = \textbf{8.97 A};\ \cos\phi = R_{ph} / Z_{ph} = 20/28.3 = 0.707$$

Total power taken = $\sqrt{3} V_L I_L \cos\phi = \sqrt{3} \times 440 \times 8.97 \times 0.707$ = 4830 W

If W_1 and W_2 are wattmeter readings, then $W_1 + W_2$ = 4830 W ... **(*i*)**

Now, $\tan\phi = 20/20 = \sqrt{3}(W_1 - W_2)/(W_1 + W_2);\ (W_1 - W_2); = 2790\text{ W}$ ... **(*ii*)**

From **(*i*)** and **(*ii*)** above, W_1 = **3810 W;** W_2 = **1020 W**

(*b*) Delta Connection

$$Z_{ph} = 28.3\Omega, V_{ph} = 440V, I_{ph} = 440/28.3 = 15.5A; I_L = 15.5 \times \sqrt{3} = 28.8\text{ A}$$

$P = \sqrt{3} \times 440 \times 28.8 \times 0.707$ = 14,490 W (it is 3 times the *Y*-power)

$\therefore$ $W_1 + W_2$ = 14,490 W ... **(*iii*)**

$\tan\phi = 20/20 = \sqrt{3}(W_1 - W_2)/14{,}490; W_1 - W_2 = 8370$ ... **(*iv*)**

From Eq. **(*iii*)** and **(*iv*)**, we get, W_1 = **11,430 W;** W_2 = **3060 W**

Note: These readings are 3-times the Y-readings.

Example 16.41. *Three identical coils are connected in star to a 200-V, three-phase supply and each takes 500 W. The power factor is 0.8 lagging. What will be the current and the total power if the same coils are connected in delta to the same supply? If the power is measured by two wattmeters, what will be their readings? Prove any formula used.*

(Elect. Engg. A.M. A. S.I. Dec. 1991)

Solution. When connected in star as shown in Fig. 16.47 (*a*), $V_{ph} = 200/\sqrt{3}$ = 115.5 V

Now, $V_{ph} I_{ph} \cos\phi$ = power per phase or 115.5 × I_{ph} × 0.8 = 500

$\therefore$ $I_{ph} = 5.41A; Z_{ph} = V_{ph} / I_{ph} = 115.5/5.41 = 21.34\Omega$

$R = Z_{ph} \cos\phi = 21.34 \times 0.8 = 17\Omega;\; X_L = Z_{ph} \sin\phi = 21.34 \times 0.6 = 12.8\Omega$

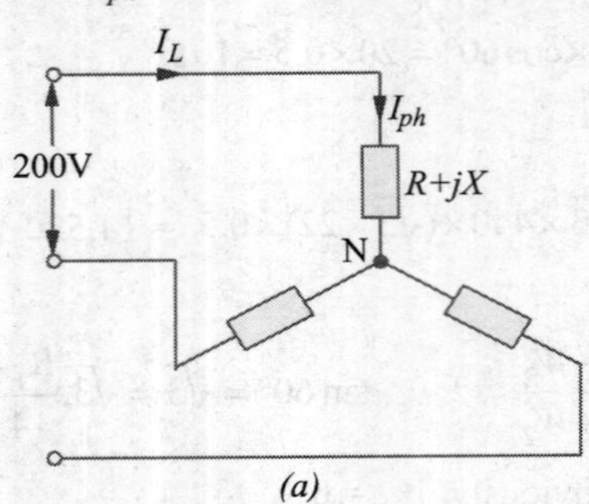

(a)

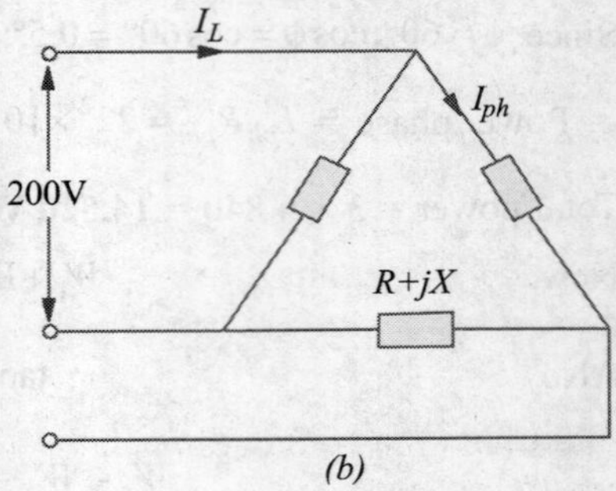

(b)

Fig. 16.47

The same three coils have been connected in delta in Fig. 16.47 (*b*). Here, $V_{ph} = V_L = 200$ V.

$$I_{ph} = 200/21.34 = 9.37\,A;\; I_L = \sqrt{3} I_{ph} = 9.37 \times 1.732 = \mathbf{16.23\ A}$$

Total power consumed $= \sqrt{3} \times 200 \times 16.23 \times 0.8 = \mathbf{4500\ W}$

It would be seen that when the same coils are connected in delta, they consume three times more power than when connected in star.

Wattmeter Readings

Now, $$W_1 + W_2 = 4500;\; \tan\phi = \frac{\sqrt{3}(W_1 - W_2)}{(W_1 + W_2)}$$

$$\phi = \cos^{-1}(0.8) = 36.87°;\; \tan\phi = 0.75$$

$$0.75 = \frac{\sqrt{3}(W_1 - W_2)}{4500} \quad \therefore\ (W_1 - W_2) = 950\ \text{W}$$

$\therefore$ $$W_1 = (4500 + 1950)/2 = 3225\ \text{W};\; \mathbf{W_2 = 1275\ W.}$$

Example 16.42. *A wattmeter reads 5.54 kW when its current coil is connected in R phase and its voltage coil is connected between the neutral and the R phase of a symmetrical 3-phase system supplying a balanced load of 30 A at 400 V. What will be the reading on the instrument if the connections to the current coil remain unchanged and the voltage coil be connected between B and Y phases? Take phase sequence RYB. Draw the corresponding phasor diagram.*

(Elect. Machines, A.M.I.E., Sec. B, 1992)

Fig. 16.47 (*b*)

Solution. As seen from Fig. 16.47 (*b*).

$$W_1 = V_R I_R \cos\phi \text{ or } 5.54 \times 10^3 = (400/\sqrt{3}) \times 30 \times \cos\phi;\quad \therefore\ \cos\phi = 0.8, \sin\phi = 0.6$$

In the second case (Fig. 16.57 (*b*))

$$W_2 = V_{YB} I_R \cos(90° - \phi) = 400 \times 30 \times \sin\phi = 400 \times 30 \times 0.6 = \mathbf{7.2\ kW}$$

Example 16.43. *A 3-phase, 3-wire balanced load with a lagging power factor is supplied at 400 V (between lines). A 1-phase wattmeter (scaled in kW) when connected with its current coil in the R-line and voltage coil between R and Y lines gives a reading of 6 kW. When the same terminals of the voltage coil are switched over to Y- and B-lines, the current-coil connections remaining the same, the reading of the wattmeter remains unchanged. Calculate the line current and power factor of the load. Phase sequence is R → Y → B.*

(Elect. Engg-1, Bombay Univ. 1985)

Solution. The current through the wattmeter is I_R and p.d. across its pressure coil is V_{RY}. As seen from the phasor diagram of Fig. 16.48, the angle between the two is (30° + ϕ).

$$\therefore \quad W_1 = V_{RY} I_R \cos(30° + \phi) = V_L I_L \cos(30° + \phi) \qquad \text{... }(i)$$

In the second case, current is I_R but voltage is V_{YB}. The angle between the two is (90° – ϕ)

$$\therefore \quad W_2 = V_{YB} I_R \cos(90° - \phi) = V_L I_L \cos(90° - \phi)$$

Since $W_1 = W_2$ we have

$$V_L I_L \cos(30° + \phi) = V_L I_L \cos(90° - \phi)$$

$\therefore$ 30° + ϕ = 90° – ϕ

or 2ϕ = 60° $\therefore$ ϕ = 30°

$\therefore$ load power factor = cos 30° = **0.866 (lag)**

Now $W_1 = W_2 = 6$ kW.

Hence, from (*i*) above, we get

$$6000 = 400 \times I_L \cos 60°; I_L = \mathbf{30\ A}$$

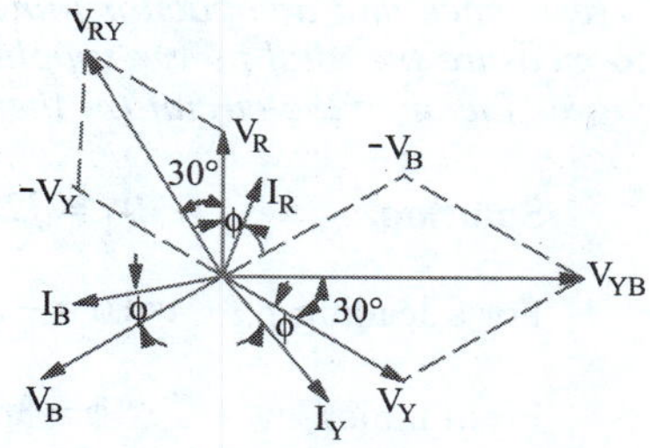

Fig. 16.48

Example 16.44. *In a balanced 3-phase system load I draws 60 kW and 80 leading kVAR whereas load 2 draws 160 kW and 120 lagging kVAR. If line voltage of the supply is 1000 V, find the line current supplied by the generator. (Fig. 16.49)*

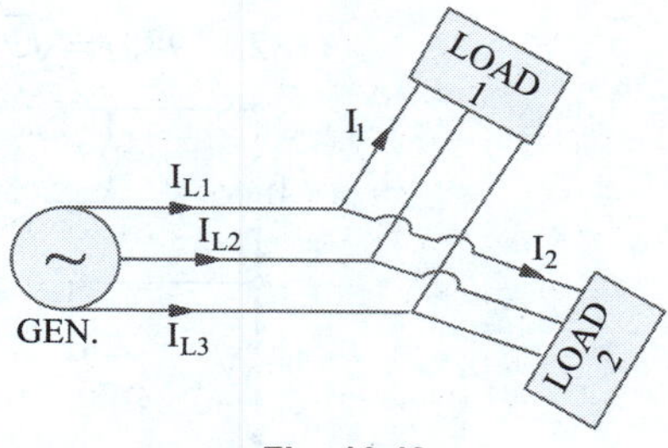

Fig. 16.49

Solution. For load 1 which is a leading load, tan $\phi_1 = Q_1/P_1 = 80/60 = -1.333$; $\phi_1 = 53.1°$, cos $\phi_1 = 0.6$. Hence, line current of this load is

$$I_1 = 60{,}000 / \sqrt{3} \times 1000 \times 0.6 = 57.8 \text{ A}$$

For load 2, tan $\phi_2 = 120/160 = 0.75$; $\phi_2 = 26.9°$, cos $\phi_2 = 0.8$. The line current drawn by this load is

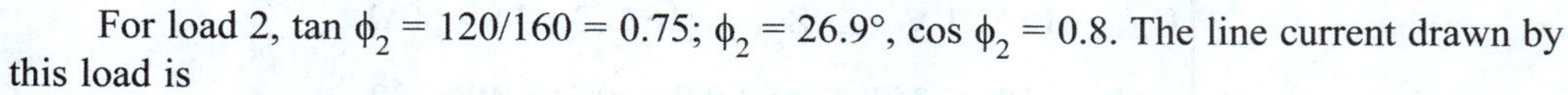

$$I_2 = 160{,}000 / \sqrt{3} \times 1000 \times 0.8 = 115.5 \text{ A}$$

If we take the phase voltage as the reference voltage *i.e.* $V_{ph} = (1000/\sqrt{3})\angle 0° = 578\angle 0°$;

then I_1 leads this voltage by 53.1° whereas I_2 lags it by 36.9°. Hence, $I_1 = 57.8 \angle 53.1°$ and $I_2 = 115.5\angle -36.9°$

$$\therefore \quad I_{L1} = I_1 + I_2 = 57.8\ \angle 53.1° + 115.5\ \angle -36.9° = 171.7\angle 42.3° \text{A}.$$

Example 16.45. *A synchronous motor absorbing 50 kW is connected in parallel with a factory load of 200 kW having a lagging power factor of 0.8. If the combination has a power factor of 0.9 lagging, find the kVAR supplied by the motor and its power factor.*

(Elect. Machines, A.M.I.E. Sec B)

Solution. Load kVA = 200/0.8 = 250

Load kVAR = 250 × 0.6 = 150 (lag) [cos ϕ = 0.8 sin ϕ = 0.6]

Total combined load = 50 + 200 = 250 kW

kVA of combined load = 250/0.9 = 277.8

Combined kVAR = 277.8 × 0.4356 = 121 (inductive) (combined cos ϕ = 0.9, sin ϕ = 0.4356)

Hence, leading kVAR supplied by synch, motor = 150 – 121 = 29 (capacitive)

kVA of motor alone = $\sqrt{(kW^2 + kVAR^2)} = \sqrt{50^2 + 29^2} = 57.8$

p.f. of motor = kW/kVA = 50/57.8 = 0.865 (leading)

Example 16.46. *A star-connected balanced load is supplied from a 3-phase balanced supply with a line voltage of 416 V at a frequency of 50 Hz. Each phase of load consists of a resistance and a capacitor joined in series and the readings on two wattmeters connected to measure the total power supplied are 782 W and 1980 W, both positive. Calculate **(a)** the power factor of the circuit **(b)** the line current and **(c)** the capacitance of each capacitor.*

(Elect. Machinery-I, Bombay Univ.)

Solution. $W_1 = 728 \text{ and } W_2 = 1980$

For a leading p.f. $\tan\phi = -\sqrt{3}\dfrac{W_1 - W_2}{W_1 + W_2}$ $\therefore$ $\tan\phi = -\sqrt{3} \times \dfrac{(782-1980)}{782+1980} = 0.75$

From tables, $\phi = 36°54'$

(*a*) $\therefore$ $\cos\phi = \cos 36°54'$ = **0.8 (leading)**

(*b*) power = $\sqrt{3}V_L I_L \cos\phi$ or $W_1 + W_2 = \sqrt{3}V_L I_L \cos\phi$

or $(782+1980) = \sqrt{3} \times 416 \times I_L \times 0.8$ $\therefore$ $I_L = I_{ph}$ = **4.8 A**

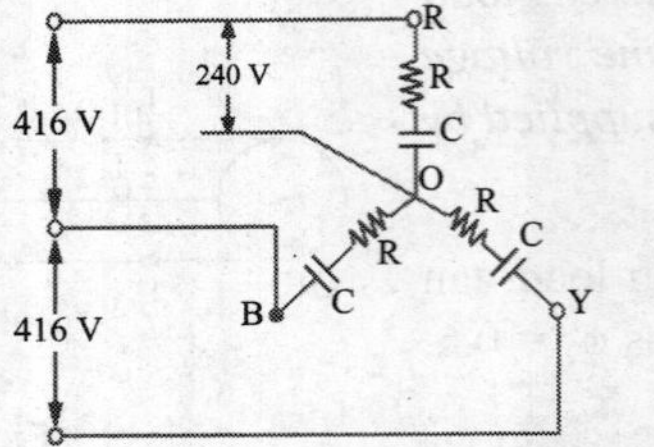

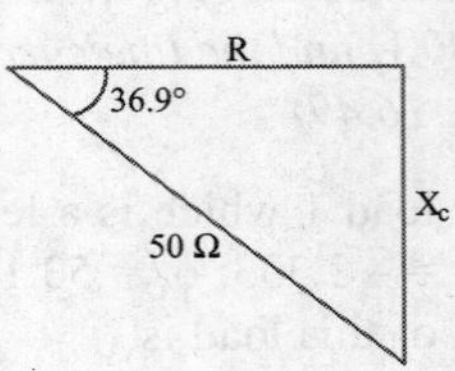

Fig. 16.50

(*c*) Now $V_{ph} = 416/\sqrt{3}$ V $\therefore$ $Z_{ph} = 416/\sqrt{3} \times 4.8 = 50\,\Omega$

$\therefore$ In Fig. 16.50, $Z_{ph} = 50\angle -36°54' = 50(0.8 - j0.6) = 40 - j30$

Capacitive reactance $X_C = 30$; or $\dfrac{1}{2\pi \times 50 \times C} = 30$ $\therefore$ C = **106 μF.**

Example 16.47. *The two wattmeters A and B, give readings as 5000 W and 1000 W respectively during the power measurement of 3-ϕ, 3-wire, balanced load system. **(a)** Calculate the power and power factor if **(i)** both meters read direct and **(ii)** one of them reads in reverse. **(b)** If the voltage of the circuit is 400 V, what is the value of capacitance which must be introduced in each phase to cause the whole of the power to appear on A. The frequency of supply is 50 Hz.* **(Elect. Engg-I, Nagpur Univ. 1992)**

Solution. **(*a*) (*i*)** Both Meters Read Direct

$W_1 = 5000\,\text{W}; W_2 = 1000\,\text{W}; \therefore W_1 + W_2 = 6000\,\text{W}; W_1 - W_2 = 4000\,\text{W}$

$\tan\phi = \sqrt{3}(W_1 - W_2)/(W_1 + W_2) = \sqrt{3}\times 4000/6000 = 1.1547$

$\therefore \quad \phi = \tan^{-1}(1.1547) = 49.1°$; p.f. = cos 49.1° = 0.655 (lag)

Total power = 5000 + 1000 = 6000 W

(*ii*) One Meter Reads in Reverse

In this case, $\tan\phi = \sqrt{3}(W_1 + W_2)/(W_1 - W_2) = \sqrt{3}\times 6000/4000 = 2.598$

$\therefore \quad \phi = \tan^{-1}(2.598) = 68.95°$; p.f. = cos 68.95° = 0.36 (lag)

Total power = $W_1 + W_2$ = 5000 – 1000 = 4000 W ...Art.

(*b*) The whole of power would be measured by wattmeter W_1 if the load power factor is 0.5 (lagging) or less. It means that in the present case p.f. of the load will have to be reduced from 0.655 to 0.5 In other words, capacitive reactance will have to be introduced in each phase of the load in order to partially neutralize the inductive-reactance.

Now, $\sqrt{3}V_L I_L \cos\phi = 6000$ or $\sqrt{3}\times 400\, I_L \times 0.655 = 6000$

$\therefore \quad I_L = 13.2\text{A}; \therefore I_{ph} = 13.2/\sqrt{3} = 7.63\text{ A}$

$Z_{ph} = V_{ph}/I_{ph} = 400/7.63 = 52.4\,\Omega$

$X_L = Z_{ph}\sin\phi = 52.4\times\sin 49.1° = 39.6\,\Omega$

When p.f. = 0.5

$\sqrt{3}\times 400\times I_L \times 0.5 = 6000; I_L = 17.32A; I_{ph} = 17.32/\sqrt{3} = 10A; Z_{ph} = 400/10 = 40\Omega$

$\cos\phi = 0.5; \phi = 60; \sin 60° = 0.886; X = Z_{ph}\sin\phi = 40\times 0.886 = 35.4\,\Omega$

$\therefore X = X_L - X_C = 35.4$ or $39.6 - X_C = 35.4; \quad \therefore X_C = 4.2\,\Omega.$

If C is the required capacitance, then $4.2 = 1/2\pi\times 50\times C; \therefore C = 758\,\mu\text{F}.$

Tutorial Problem 16.2

1. Two wattmeters connected to measure the input to a balanced three-phase circuit indicate 2500 W and 500 W respectively. Find the power factor of the circuit (*a*) when both readings are positive and (*b*) when the latter reading is obtained after reversing the connections to the current coil of one instrument.
[(*a*) 0.655 (*b*) 0.3591] (*City & Guilds, London*)
2. A 400-V, 3-phase induction motor load takes 900 kVA at a power factor of 0.707. Calculate the kVA rating of the capacitor bank to raise the resultant power factor of the installation of 0.866 lagging.

   Find also the resultant power factor when the capacitors are in circuit and the motor load has fallen to 300 kVA at 0.5 power factor. **[296 kVA, 0.998 leading] (*City & Guilds, London*)**
3. Two wattmeters measure the total power in three-phase circuits and are correctly connected. One reads 4,800 W while other reads backwards. On reversing the latter, it reads 400 W. What is the total power absorbed by the circuit and the power factor?**[4400 W; 0.49] (*Sheffield Univ. U.K.*)**
4. The power taken by a 3-phase, 400-V motor is measured by the two wattmeter method and the readings of the two wattmeters are 460 and 780 watts respectively. Estimate the power factor of the motor and the line current. **[0.913, 1.96 A] (*City & Guilds, London*)**

5. Two wattmeters, W_1 and W_2 connected to read the input to a three-phase induction motor running unloaded, indicate 3 kW and 1 kW respectively. On increasing the load, the reading on W_1 increases while that on W_2 decreases and eventually reverses.
   Explain the above phenomenon and find the unloaded power and power factor of the motor.
   [2 kW, 0.287 lag] ***(London Univ.)***

6. The power flowing in a 3-ϕ, 3-wire, balanced-load system is measured by the two wattmeter method. The reading on wattmeter *A* is 5,000 W and on wattmeter *B* is – 1,000 W
   (*a*) What is the power factor of the system?
   (*b*) If the voltage of the circuit is 440, what is the value of capacitance which must be introduced into each phase to cause the whole of the power measured to appear on wattmeter A?
   [0.359; 5.43 Ω] ***(Meters and Meas. Insts. A.M.I.E.E. London)***

7. Two wattmeters are connected to measure the input to a 400 V; 3-phase, connected motor outputting 24.4 kW at a power factor of 0.4 (lag) and 80% efficiency. Calculate the
   (*i*) resistance and reactance of motor per phase
   (*ii*) reading of each wattmeters.
   [(*i*) 2.55 Ω; 5.85 Ω; (*ii*) 34,915 W; – 4850 W] ***(Elect. Machines, A.M.I.E. Sec. B, 1993)***

8. The readings of the two instruments connected to a balanced three-phase load are 128 W and 56 W. When a resistor of about 25 Ω is added to each phase, the reading of the second instrument is reduced to zero. State, giving reasons, the power in the circuit before the resistors were added.
   [72 W] ***(London Univ.)***

9. A balanced star-connected load, each phase having a resistance of 10 Ω and inductive reactance of 30 Ω is connected to 400-*V*, 50-Hz supply. The phase rotation is red, yellow and blue. Wattmeters connected to read total power have their current coils in the red and blue lines respectively. Calculate the reading on each wattmeter and draw a vector diagram in explanation.
   [2190 W, – 583 W] ***(London Univ.)***

10. A 7.46 kW induction motor runs from a 3-phase, 400-V supply. On no-load, the motor takes a line current of 4. A at a power factor of 0.208 lagging. On full load, it operates at a power factor of 0.88 lagging and an efficiency of 89 per cent. Determine the readings on each of the two wattmeters connected to read the total power on (*a*) no load and (*b*) full load.
    [1070 W, – 494 W; 5500 W; 2890 W]

16.27. Double Subscript Notation

In symmetrically-arranged networks, it is comparatively easier and actually more advantageous, to use single-subscript notation. But for unbalanced 3-phase circuits, it is essential to use double subscript notation, in order to avoid unnecessary confusion which is likely to result in serious errors.

Suppose, we are given two coils are whose induced e.m.fs. are 60° out of phase with each other [Fig. 16.51 (*a*)]. Next, suppose that it is required to connect these coils in additive series *i.e.* in such a way that their e.m.fs. add at an angle of 60°. From the information given, it is impossible to know whether to connect terminal '*a*' to terminal '*c*' or to terminal '*d*'.

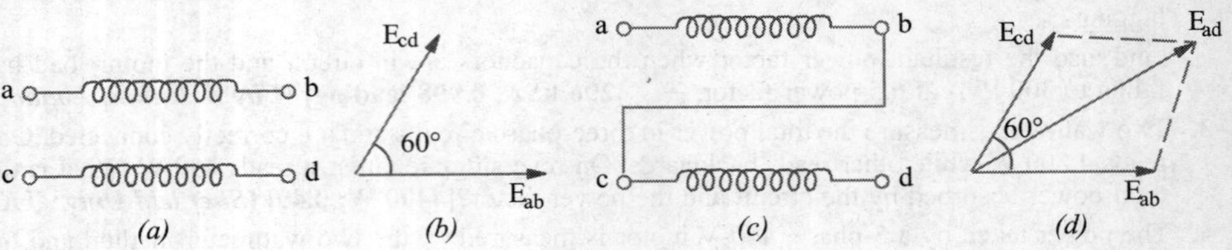

Fig. 16.51

But if additionally it were given that e.m.f. from terminal '*c*' to terminal '*d*' is 60° out of phase with that from terminal '*a*' to terminal '*b*', then the way to connect the coils is definitely fixed. The double-subscript notation is obviously very convenient in such cases. The order in which these subscripts are written indicates the direction along which the voltage acts (or current flows). For example the e.m.f. '*a*' to '*b*' [Fig. 16.59 (*a*)], may be written as E_{ab} and that from '*c*' to '*d*' as E_{cd}.... The e.m.f. between '*a*' and '*d*' is E_{ad} where $\mathbf{E}_{ad} = \mathbf{E}_{ab} + \mathbf{E}_{ad}$.

Example 16.48. *In Fig. 16.52 (a) with terminal 'b' connected to 'd', find* E_{ca}.

Solution. $\mathbf{E}_{ca} = \mathbf{E}_{cd} + \mathbf{E}_{ba} = \mathbf{E}_{cd} + (-\mathbf{E}_{ab})$

As shown in Fig. 16.53, vector $\mathbf{E}_{ab}$ is reversed and then combined with $\mathbf{E}_{cd}$ to get $\mathbf{E}_{ca}$. Magnitude of $\mathbf{E}_{ca}$ is given by 2 × 100 × cos 60° = 100 V but it leads $\mathbf{E}_{ab}$ by 120°.

$$\therefore \quad \mathbf{E}_{ca} = 100 \angle 120°$$

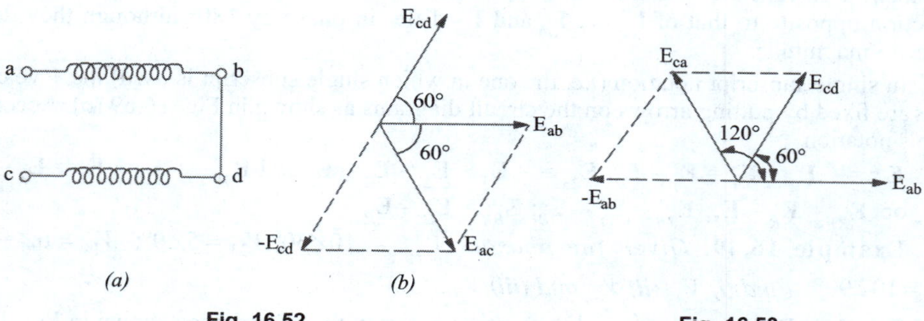

Fig. 16.52

Fig. 16.53

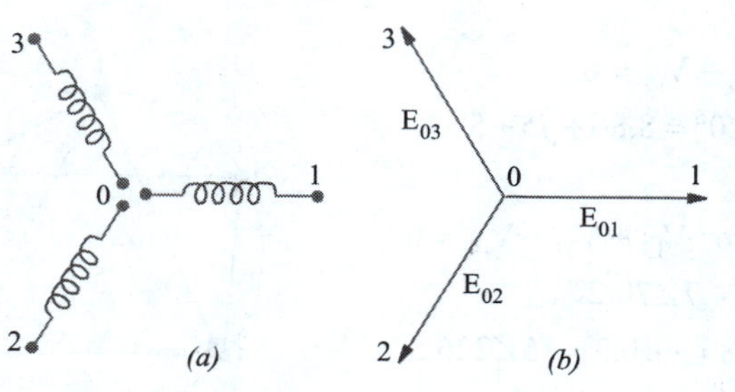

Fig. 16.54

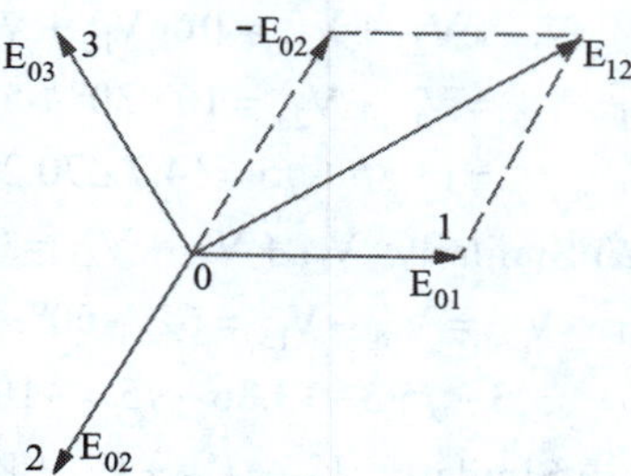

Fig. 16.55

In Fig. 16.54 (*b*) is shown the vector diagram of the e.m.fs induced in the three phases 1, 2, 3 (or *R*, *Y*, *B*) of a 3-phase alternator [Fig.16.54 (*a*)]. According to double subscript notation, each phase e.m.f. may be written as $\mathbf{E}_{01}$, $\mathbf{E}_{02}$ and $\mathbf{E}_{03}$, the order of the subscripts indicating the direction in which the e.m.fs. act. It is seen that while passing from phase 1 to phase 2 through the external circuit, we are in opposition to $\mathbf{E}_{02}$.

$$\mathbf{E}_{12} = \mathbf{E}_{20} + \mathbf{E}_{01} = (-\mathbf{E}_{02}) + \mathbf{E}_{01} = \mathbf{E}_{01} - \mathbf{E}_{02}$$

It means that for obtaining $\mathbf{E}_{12}$, $\mathbf{E}_{20}$ has to be reversed to obtain $-\mathbf{E}_{02}$ which is then combined with $\mathbf{E}_{01}$ to get $\mathbf{E}_{12}$ (Fig. 16.55). Similarly,

$$\mathbf{E}_{23} = \mathbf{E}_{30} + \mathbf{E}_{02} = (-\mathbf{E}_{03}) + \mathbf{E}_{02} = \mathbf{E}_{02} - \mathbf{E}_{03}$$

$$\mathbf{E}_{31} = \mathbf{E}_{10} + \mathbf{E}_{03} = (-\mathbf{E}_{01}) + \mathbf{E}_{03} = \mathbf{E}_{03} - \mathbf{E}_{01}$$

By now it should be clear that double-subscript notation is based on lettering every junction and terminal point of diagrams of connections and on the use of two subscripts with all vectors representing voltage or current. The subscripts on the vector diagram, taken from the diagram of connections, indicate that the positive direction of the current or voltage is from the first subscript to the second. For example, according to this notation $\mathbf{I}_{ab}$ represents a current whose + ve direction is from a to b in the branch ab of the circuit in the diagram of connections. In the like manner, $\mathbf{E}_{ab}$ represents the e.m.f. which produces this current. Further, $\mathbf{I}_{ba}$ will represent a current flowing from b to a, hence its vector will be drawn equal to but in a direction opposite to that of $\mathbf{I}_{ab}$ *i.e.* $\mathbf{I}_{ab}$ and $\mathbf{I}_{ba}$ differ in phase by 180° although they do not differ in magnitude.

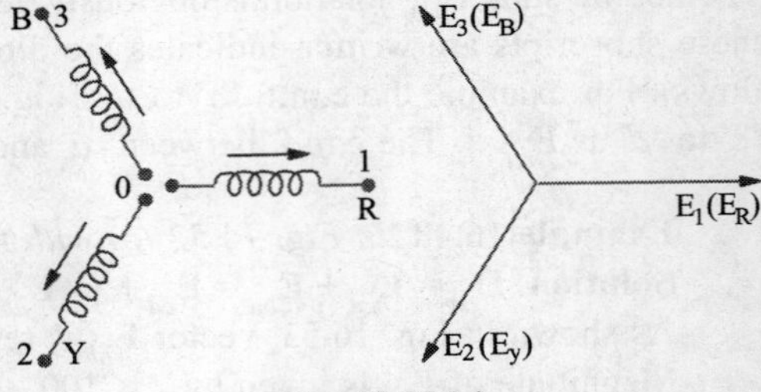

Fig. 16.55 (a)

In single subscript notation (*i.e.* the one in which single subscript is used) the + ve directions are fixed by putting arrows on the circuit diagrams as shown in Fig. 16.69 (*a*). According to this notation

$$\mathbf{E}_{12} = -\mathbf{E}_2 + \mathbf{E}_1 = \mathbf{E}_1 - \mathbf{E}_2;\ \mathbf{E}_{23} = -\mathbf{E}_3 + \mathbf{E}_2 = \mathbf{E}_2 - \mathbf{E}_3 \text{ and } \mathbf{E}_{31} = -\mathbf{E}_1 + \mathbf{E}_3 = \mathbf{E}_3 - \mathbf{E}_1$$

or $\mathbf{E}_{RY} = \mathbf{E}_R - \mathbf{E}_Y;\ \mathbf{E}_{YB} = \mathbf{E}_Y - \mathbf{E}_B;\ \mathbf{E}_{BR} = \mathbf{E}_B - \mathbf{E}_R$

Example 16.49. *Given the phasors* $V_{12} = 10\angle 30°; V_{23} = 5\angle 0°;\ V_{14} = 6\angle -60°;$ $V_{45} = 10\angle 90°$. *Find* **(i)** V_{13} **(ii)** V_{34} *and* **(iii)** V_{25}.

Solution. Different points and the voltage between them have been shown in Fig. 16.56.

(*i*) Using KVL, we have

$$V_{12} + V_{23} + V_{31} = 0 \text{ or } V_{12} + V_{23} - V_{13} = 0$$

or $V_{13} = V_{12} + V_{23} = 10\angle 30° + 5\angle 0° = 8.86 + j5 + 5$

$= 13.86 + j5 = 14.7\angle 70.2°$

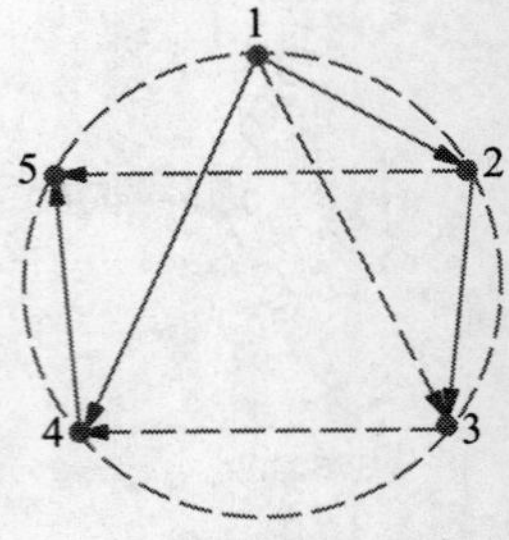

Fig. 16.56

(*ii*) Similarly, $V_{13} + V_{34} + V_{41} = 0$ or $V_{13} + V_{34} - V_{14} = 0$

or $V_{34} = V_{14} - V_{13} = 6\angle -60° - 14.7\angle 70.2°$

$= 3 - j5.3 - 13.86 - j5 = -10.86 - j10.3 = 15\angle 226.5°$

(*iii*) Similarly, $V_{23} + V_{34} + V_{45} + V_{52} = 0$

or $V_{23} + V_{34} + V_{45} - V_{52} = 0$

or $V_{25} = V_{23} + V_{34} + V_{45} = 5\angle 0° + 15\angle 226.5° + 10\angle 90°$

$= 5 - 10.86 - j10.3 + j10 = -5.86 - j0.3 = 5.86 - \angle 2.9°$.

Example 16.50. *In a balanced 3-phase Y-connected voltage source having phase sequence abc,* $V_{an} = 230\ \angle 30°$. *Calculate analytically* **(i)** V_{bn} **(ii)** V_{cn} **(iii)** V_{ab} **(iv)** V_{bc} *and* **(v)** V_{ca}. *Show the phase and line voltages on a phasor diagram.*

Solution. It should be noted that V_{an} stands for the voltage of terminal a with respect to the neutral point n. The usual positive direction of the phase voltages are as shown in Fig. 16.57 (*a*). Since the 3-phase system is balanced, the phase differences between the different phase voltages are 120°.

(*i*) $V_{bn} = \angle -120° = 230\ \angle(30° - 120°) = 230\angle -90°$

(*ii*) $V_{cn} = V_{an}\angle 120° = 230\ \angle(30° + 120°) = 230\angle 150°$ ... Fig. 16.57 (*b*)

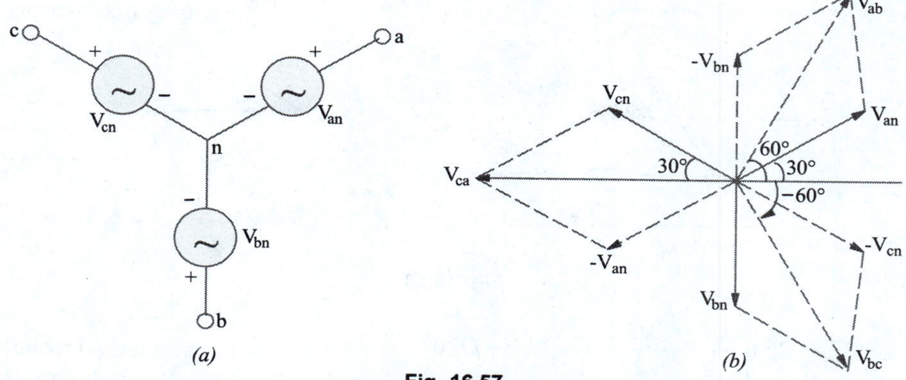

Fig. 16.57

(iii) It should be kept in mind that V_{ab} stands for the voltage of point a with respect to point *b*. For this purpose, we start from the reference point *b* in Fig.16.57 (*a*) and go to point *a* and find the sum of the voltages met on the way. As per sign convention as we go from *b* to *n*, there is a fall in voltage of by an amount equal to V_{bn}. Next as we go from *n* to *a*, there is increase of voltage given by V_{an}.

$$\therefore\ V_{ab} = -V_{bn} + V_{an} = V_{an} - V_{bn} = 230\angle 30° - 230\angle -90°$$

$$= 230(\cos 30° + j\sin 30°) - 230(0 - j\sin 90°)$$

$$= 230\left(\frac{\sqrt{3}}{2} + j\frac{1}{2}\right) + j230 = 230\left(\frac{\sqrt{3}}{2} + j\frac{3}{2}\right) = 230\sqrt{3}\left(\frac{1}{2} + j\frac{\sqrt{3}}{2}\right) = 400\angle 60°$$

(*iv*) $V_{bc} = V_{bn} - V_{cn} = 230\angle -90° - 230\angle 150° = -j230 - 230$

$$= \left(-\frac{\sqrt{3}}{2} + j\frac{1}{2}\right) = 230\sqrt{3}\left(\frac{1}{2} - j\frac{\sqrt{3}}{2}\right) = 400\angle -60°$$

(*v*) $V_{ca} = V_{cn} - V_{an} = 230\angle 150° - 230\angle 30°$

$$= \left(-\frac{\sqrt{3}}{2} + j\frac{1}{2}\right) - 230\left(\frac{\sqrt{3}}{2} + j\frac{1}{2}\right) = -400 = 400\angle 180°$$

These line voltages along with the phase voltages have been shown in the phasor diagram of Fig. 16.57 (*b*).

Example 16.51. *Three non-inductive resistances, each of 100 Ω are connected in star to a 3-phase, 440-V supply. Three equal choking coils are also connected in delta to the same supply; the resistance of one coil being equal to 100 Ω. Calculate **(a)** the line current and **(b)** the power factor of the system.* **(Elect. Technology-II, Sambal Univ.)**

Solution. The diagram of connections and the vector diagram of the *Y*-and Δ-connected impedances are shown in Fig. 16.58.

The voltage $\mathbf{E}_{10}$ between line 1 and neutral is taken along the *X*-axis. Since the load is balanced, it will suffice to determine the current in one line only. Applying Kirchhoff's Law to junction 1, we have

$$\mathbf{I}'_{11} = \mathbf{I}_{10} + \mathbf{I}_{12} + \mathbf{I}_{13}$$

Let us first get the vector expressions for $\mathbf{E}_{10}$, $\mathbf{E}_{20}$ and $\mathbf{E}_{30}$

$$\mathbf{E}_{10} = \frac{440}{\sqrt{3}}(1 + j0) = 254 + j0,\ ,\ \mathbf{E}_{20} = 254\left(-\frac{1}{2} - j\frac{\sqrt{3}}{2}\right) = -127 - j220$$

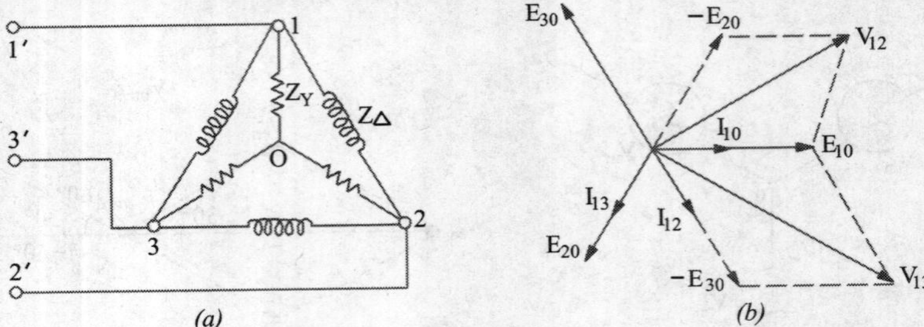

Fig. 16.58

$$\mathbf{E}_{30} = 254\left(-\frac{1}{2} + j\frac{\sqrt{3}}{2}\right) = -127 + j220$$

Let us now derive vector expressions for $\mathbf{V}_{12}$ and $\mathbf{V}_{31}$.

$$\mathbf{V}_{10} = \mathbf{E}_{10} + \mathbf{E}_{02} = \mathbf{E}_{10} - \mathbf{E}_{20} = (254 + j0) - (-127 - j220) = 381 + j220$$

$$\mathbf{V}_{13} = \mathbf{E}_{10} + \mathbf{E}_{03} = \mathbf{E}_{10} - \mathbf{E}_{30} = (254 - j0) - (-127 + j220) = 381 - j220$$

$$\mathbf{I}_{10} = \frac{\mathbf{E}_{10}}{\mathbf{Z}_y} = \frac{254 + j0}{100} = 2.54 + j0,\ \mathbf{I}_{12} = \frac{\mathbf{V}_{13}}{\mathbf{Z}_{\varnothing}} = \frac{381 + j220}{j100} = 2.2 - j3.81 = 4.4\angle -60^\circ$$

$$\mathbf{I}_{13} = \frac{V_{13}}{Z_\Delta} = \frac{381 - j220}{j100} = -2.2 - j3.81 = 4.4\angle -120^\circ$$

(*a*) $\mathbf{I}'_{11} = (2.54 + j0) + (2.2 - j3.81) + (-2.2 - j3.81) = (2.54 - j7.62) = 8.03 \angle -71.6^\circ$

(*b*) p.f. = cos 71.6° = **0.316 (lag)**

Alternative Method

This question may be easily solved by Δ/Y conversion. The star equivalent of the delta reactance is 100/3 Ω per phase.

As shown is Fig. 16.59, there are now two parallel circuits across each phase, one consisting of a resistance of 100 Ω and the other of a reactance of 100/3 Ω.

Taking E_{10} as the reference vector, we have

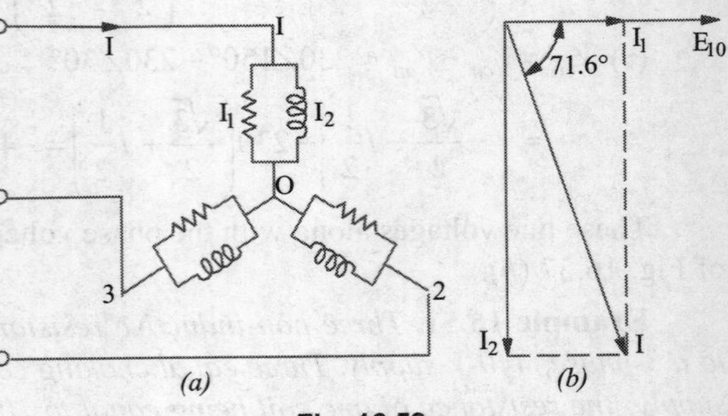

Fig. 16.59

$$\mathbf{E}_{10} = (254 + j0)$$

$$\mathrm{I}_1 = \frac{254 + j0}{100} = 2.54 + j0;\ \mathrm{I}_2 = \frac{254 + j0}{j100/3} = -j.7.62$$

Line current **I** = (2.54 + *j*0) + (–*j*7.62) = (2.54 – *j*7.62) = 8.03 ∠– 71.6° ... Fig. 16.59 (*b*)

16.28. Unbalanced Loads

Any polyphase load in which the impedances in one or more phases differ from the impedances of other phases is said to be an unbalanced load. We will now consider different methods to handle unbalanced star-connected and delta-connected loads.

16.29. Unbalanced Δ-Connected Load

Unlike unbalanced *Y*-connected load, the unbalanced Δ-connected load supplied from a balanced 3-phase supply does not present any new problems because the voltage across each load phase is fixed. It is independent of the nature of the load and is equal to line voltage. In fact, the problem resolves itself into three independent single-phase circuits supplied with voltages which are 120° apart in phase.

The different phase currents can be calculated in the usual manner and the three line currents are obtained by taking the vector difference of phase currents in pairs.

If the load consists of three different pure resistances, then trigonometrical method can be used with advantage, otherwise symbolic method may be used.

Example 16.52. *A 3-phase, 3-wire, 240 volt, CBA system supplies a delta-connected load in which* $Z_{AB} = 25\angle 90°$, $Z_{BC} = 15\angle 30°$, $Z_{CA} = 20\angle 0°$ *ohms. Find the line currents and total power.* **(Advanced Elect. Machines A.M.I.E. Sec. B, Summer 1991)**

Solution. A 3-phase system has only two possible sequences : *ABC* and *CBA*. In the *ABC* sequence, the voltage of phase *B* lags behind voltage of phase *A* by 120° and that of phase *C* lags behind phase *A* voltage by 240°. In the *CBA* phase which can be written as A → *C* → *B*, voltage of *C* lags behind voltage *A* by 120° and that of *B* lags behind voltage *A* by 240°. Hence, the phase voltage which can be written as

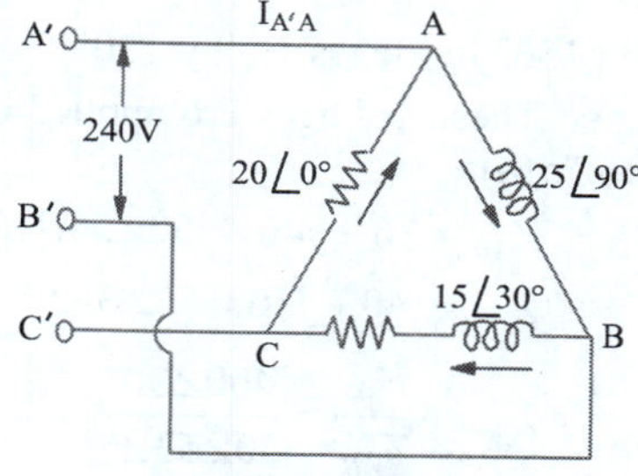

Fig. 16.60

$$E_{AB} = E\angle 0° \; ; \; E_{BC} = E\angle -120°$$

and $E_{CA} = E\angle -240°$ or $E_{CA} = \angle 120°$

$$\therefore \; I_{AB} = \frac{E_{AB}}{Z_{AB}} = \frac{240\angle 0°}{25\angle 90°} = 9.6\angle -90° = -j9.6 \text{ A}$$

$$I_{BC} = \frac{E_{BC}}{Z_{BC}} = \frac{240\angle 120°}{15\angle 30°} = 16\angle 90° = j16 \text{ A}$$

$$I_{CA} = \frac{E_{CA}}{Z_{CA}} = \frac{240\angle -120°}{20\angle 0°} = 12\angle -120° = 12(0.5 - j0.866) = (-6 - j10.4) \text{ A}$$

The circuit is shown in Fig. 16.60.

Line current $I_{A'A} = I_{AB} + I_{AC} = I_{AB} - I_{CA} = -j9.6 - (-6 - j10.4) = 6 + j0.08$

Line current $I_{B'B} = I_{BC} - I_{AB} = j16 - (-j9.6) = j25.6$ A

$$I_{C'C} = I_{CA} - I_{BC} = (-6 - j10.4) - j\,16 = (-5 - j26.4) \text{ A}$$

Now, $R_{AB} = 0$; $R_{BC} = 15 \cos 30 = 13\ \Omega$; $R_{CA} = 20\ \Omega$

Power

$W_{AB} = 0$; $W_{BC} = I_{BC}^2 R_{BC} = 16^2 \times 13 = 3328$ W; $W_{CA} = I_{CA}^2 \times R_{CA} = 27^2 \times 20 = 14{,}580$ W.

Total Power = 3328 + 14580 = 17,908 W.

Example 16.53. *In the network of Fig. 16.61,* $E_{na} = 230\angle 0°$ *and the phase sequence is abc. Find the line currents* I_a, I_b *and* I_c *as also the phase currents* I_{AB}, I_{BC} *and* I_{CA}.

E_{na}, E_{nb}, E_{nc} *is a balanced three-phase voltage system with phase sequence abc.*

(Network Theory, Nagpur Univ. 1993)

Solution. Since the phase sequence is *abc*, the generator phase voltages are:

$$E_{na} = 320\angle 0° ; E_{nb} = 230\angle -120° ; E_{nc} = 230\angle 120°$$

As seen from the phasor diagram of Fig. 16.61 (*b*), the line voltages are as under :-

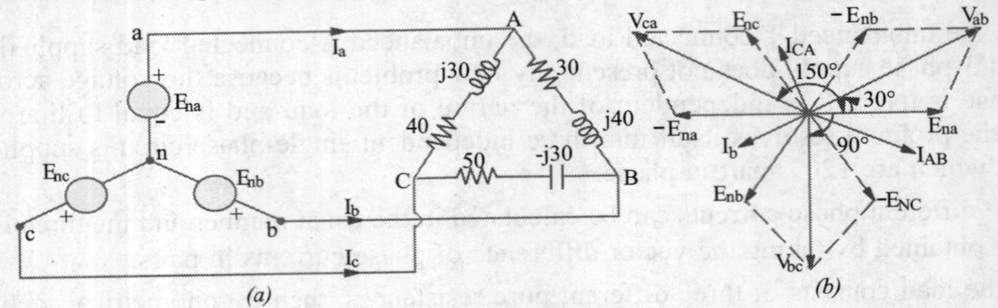

Fig. 16.61

$V_{ab} = E_{na} - E_{nb}$; $V_{bc} = E_{nb} - E_{nc}$; $V_{ca} = E_{nc} - E_{na}$

$\therefore \ V_{ab} = \sqrt{3}\times 230\angle 30° = 400\angle 30°$ *i.e* it is ahead of the reference generator phase voltage E_{na} by 30°.

$V_{bc} = \sqrt{3}\times 230\angle 90° = 400\angle -90°$. This voltage is 90° behind E_{na} but 120° behind V_{ab}.

$V_{ca} = \sqrt{3}\times 230\angle 150° = 400\angle 150°$ or $\angle -210°$. This voltage leads reference voltage E_{na} by 150° but leads V_{ab} by 120°.

These voltages are applied across the unbalanced Δ - connected lead as shown in Fig. 16.75(*a*).

$$Z_{AB} = 30 + j40 = 50\angle 53.1°; Z_{BC} = 50 - j30 = 58.3\angle -31°,$$

$$Z_{CA} = 40 + j30 = 50\angle 36.9°$$

$$I_{AB} = \frac{V_{ab}}{Z_{AB}} = \frac{400\angle 30°}{50\angle 53.1°} = 8\angle -23.1° = 7.36 - j3.14$$

$$I_{BC} = \frac{V_{bc}}{Z_{BC}} = \frac{400\angle -90°}{58.3\angle -31°} = 6.86\angle -59° = 3.53 - j5.88$$

$$I_{CA} = \frac{V_{ca}}{Z_{CA}} = \frac{400\angle 150°}{50\angle 36.9°} = 8\angle 113.1° = 3.14 + j7.36$$

$$I_a = I_{AB} - I_{CA} = 7.36 - j3.14 + 3.14 - j7.36 = 10.5 - j10.5 = 14.85\angle -45°$$

$$I_b = I_{BC} - I_{AB} = 3.53 - j5.88 - 7.36 + j3.14 = -3.83 - j2.74 = 4.71\angle -215.6°$$

$$I_c = I_{CA} - I_{BC} = -3.14 + j7.36 - 3.53 + j5.88 = -6.67 + j13.24 = 14.8\angle 116.7°$$

Example 16.54. *For the unbalanced Δ - connected load of Fig. 16.62 (a), find, the phase currents, line currents and the total power consumed by the load when phase sequence is* (**a**) *abc and* (**b**) *acb.*

Solution. (*a*) Phase sequence abc (Fig. 16.62).

Let $\mathbf{V}_{ab} = 100\angle 0° = 100 + j0$

$$\mathbf{V}_{bc} = 100\angle -120° = 100\left(-\frac{1}{2} j\frac{\sqrt{3}}{2}\right) = -50 - j86.6$$

$$\mathbf{V}_{ca} = 100\angle 102° = 100\left(-\frac{1}{2} + j\frac{\sqrt{3}}{2}\right) = -50 + j86.6$$

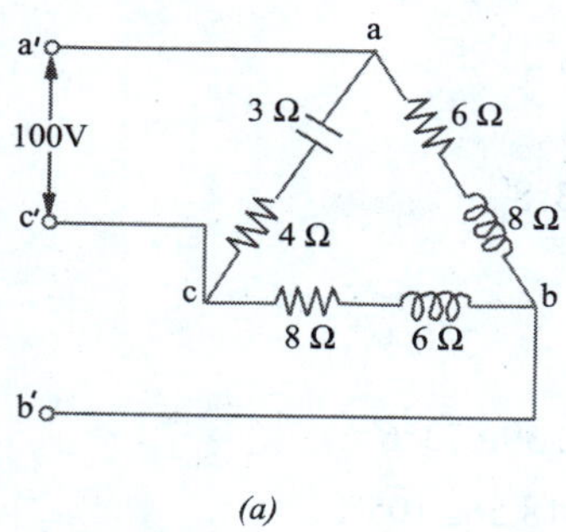

(a)

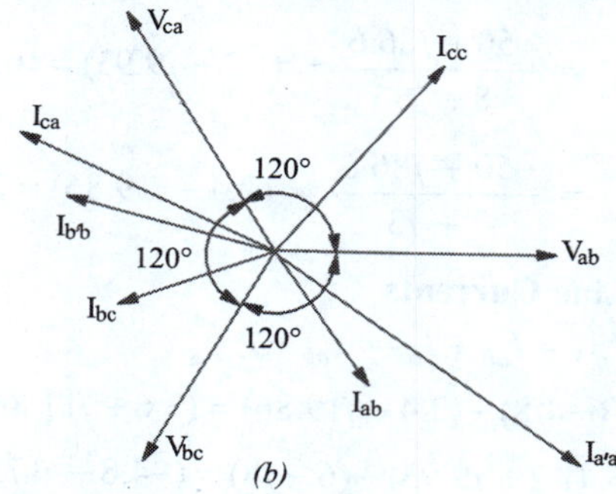

(b)

Fig. 16.62

(i) Phase currents

Phase current, $\mathbf{I} = \dfrac{\mathbf{V}_{ab}}{\mathbf{Z}_{ab}} = \dfrac{100 + j0}{6 + j8} = 6 - j8 = 10\angle -53°8'$

Similarly, $\mathbf{I}_{bc} = \dfrac{\mathbf{V}_{bc}}{\mathbf{Z}_{bc}} = \dfrac{-30 - j86.6}{8 + j6} = -9.2 - j3.93 = 10\angle -156°52'$

$$\mathbf{I}_{ca} = \frac{\mathbf{V}_{ca}}{\mathbf{Z}_{ca}} = \frac{-50 + j86.6}{4 - j3} = -18.39 + j7.86 = 20\angle 156°52'$$

(ii) Line Currents

Line Current $\mathbf{I}_{a'a} = \mathbf{I}_{ab} + \mathbf{I}_{ac} = \mathbf{I}_{ab} - \mathbf{I}_{ca} = (6 - j8) - (-18.39 + j7.86)$

$= 24.39 - j15.86 = 29.1 \angle -33°2'$

Similarly, $\mathbf{I}_{b'b} = \mathbf{I}_{bc} + \mathbf{I}_{ba} = \mathbf{I}_{bc} - \mathbf{I}_{ab}$

$= (-9.2 - j3.93) - (6 - j8) = -15.2 + j4.07 = 15.73\angle 165°30'$

$\mathbf{I}_{c'c} = \mathbf{I}_{ca} + \mathbf{I}_{cb} = \mathbf{I}_{ca} - \mathbf{I}_{bc} = (-18.39 + j7.86) - (-9.2 + j3.93)$

$= 9.19 + j11.79 = 14.94\angle 52°3'$

Check $\quad \Sigma\mathbf{I} = 0 + j0$

(iii) Power

$$W_{ab} = I_{ab}^2 R_{ab} = 10^2 \times 6 = 600 \text{ W}$$

$$W_{bc} = I_{bc}^2 R_{bc} = 10^2 \times 8 = 800 \text{ W}$$

$$W_{ca} = I_{ca}^2 R_{ca} = 20^2 \times 4 = 1600 \text{ W}$$

Total = 3000 W

(b) Phase sequence acb (Fig. 16.63)

Here, $V_{ab} = 100\angle 0° = 100 + j9$

$V_{bc} = 100\angle 120° = -50 + 86.6$

$V_{ca} = 100\angle -120° = -50 - j86.6$

(i) Phase Currents

$$I_{ab} = \frac{100}{6 + j8} = 6 - j8 = 10\angle -53°8'$$

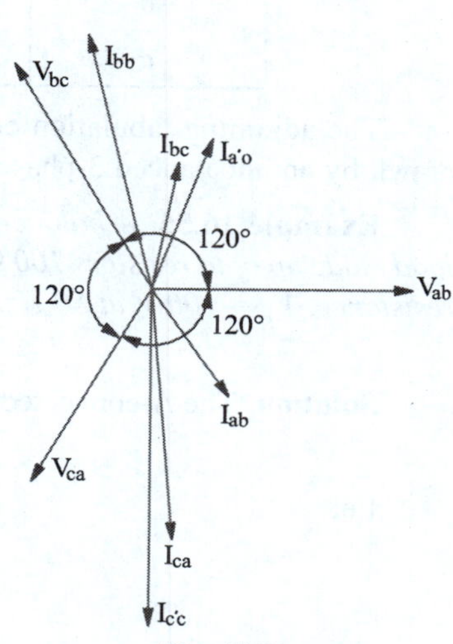

Fig. 16.63

$$I_{bc} = \frac{-50+j86.6}{8+j6} = (1.2+j9.93) = 10\angle 83°8'$$

$$I_{ca} = \frac{-50+j86.6}{7+j3} = (2.4-j19.86) = 20-83°8'$$

(ii) Line Currents

$$I_{a'a} = I_{ab} + I_{ac} = I_{ab} - I_{ca}$$

$$= (6-j8)-(2.4-j19.86) = (3.6+j11.86) = 12.39\angle 73°6'$$

$$I'_{bb} = (1.2+j9.93)-(6-j8) = (-4.8+j17.93) = 18.56\angle 105°$$

$$I'_{cc} = (2.4-j19.86)-(1.2-j9.93) = (1.2-j29.79) = 29.9\angle -87°42'$$

It is seen that $\Sigma \mathbf{I} = 0 + j0$

(iii) Power

$$W_{ab} = 10^2 \times 6 = 600\text{W}$$

$$W_{bc} = 10^2 \times 8 = 800\text{ W}$$

$$W_{ca} = 20^2 \times 4 = 1600\text{ W}$$

Total = 3000 W – as before

It will be seen that the effect of phase reversal on an unbalanced Δ-connected load is as under:

(*i*) phase currents change in angle only, their magnitudes remaining the same

(*ii*) consequently, phase powers remain unchanged

(*iii*) line currents change both in magnitude and angle.

Line	Ampere Sequence *a b c*	Sequence *c b a*
a	29.1 ∠ –33°2′	12.39 ∠ 73.1°
b	15.73 ∠ 165°	18.56 ∠105°
c	14.94 ∠ 52°3′	29.9 ∠ –87.7°

The adjoining tabulation emphasizes the effect of phase sequence on the line currents drawn by an unbalanced 3-phase load.

Example 16.55. *A balanced 3-phase supplies an unbalanced 3-phase delta-connected load made up of to resistors 100 Ω and a reactor having an inductance of 0.3 H with negligible resistance. V_L = 100 V at 50 Hz. Calculate (a) the total power in the system.*

(Elect. Engineering-I, Madras Univ.)

Solution. The Δ-connected load and its phasor diagram are shown in Fig. 16.64 (*a*).

$$X_L = \omega L = 314.2 \times 0.2 = 94.3\Omega$$

Let $V_{ab} = 100\angle 0° = 100 + j0$

$$V_{bc} = 100\angle -120°$$

$$= -50 - j86.6$$

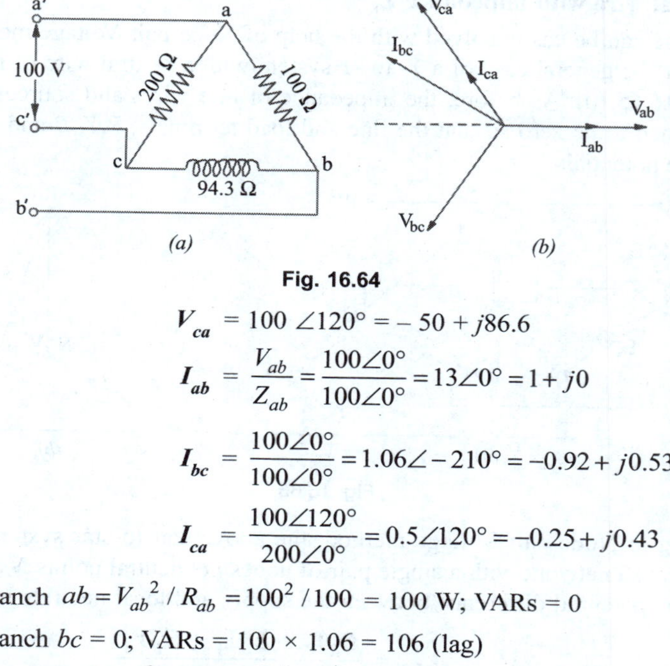

Fig. 16.64

$$V_{ca} = 100\angle 120° = -50 + j86.6$$

$$I_{ab} = \frac{V_{ab}}{Z_{ab}} = \frac{100\angle 0°}{100\angle 0°} = 13\angle 0° = 1 + j0$$

$$I_{bc} = \frac{100\angle 0°}{100\angle 0°} = 1.06\angle -210° = -0.92 + j0.53$$

$$I_{ca} = \frac{100\angle 120°}{200\angle 0°} = 0.5\angle 120° = -0.25 + j0.43$$

Watts in branch $ab = V_{ab}^2 / R_{ab} = 100^2/100$ = 100 W; VARs = 0

Watts in branch bc = 0; VARs = 100 × 1.06 = 106 (lag)

Watts in branch $ca = V_{ca}^2 / R_{ca} = 100^2/200$ = 50 W; VRAs = 0

(*a*) Total power = 100 + 50 = 150 W; VARs = **106 (lag)**

16.30. Four-Wire Star-Connected Unbalanced Load

It is the simplest case of an unbalanced load and may be treated as three separate single-phase systems with a common return wire. It will be assumed that impedance of the line wires and source phase windings is zero. Should such an assumption be unacceptable, these impedances can be added to the load impedances. Under these conditions, source and load line terminals are at the same potential.

Consider the following two cases:

(*i*) Neutral wire of zero impedance

Because of the presence of neutral wire (assumed to behaving zero impedance), the star points of the generator and load are tied together and are at the same potential. Hence, the voltages across the three load impedances are ***equalized*** and each is equal to the voltage of the corresponding phase of the generator. In other words, due to the provision of the neutral, each phase voltage is a ***forced*** voltage so that the three phase voltages are balanced when line voltages are balanced even though phase impedances are unbalanced. However, it is worth noting that a break or open ($Z_n = \infty$) in the neutral wire of a 3-phase, 2-wire system with ***unbalanced*** load always causes large (in most cases inadmissible) changes in currents and phase voltages. It is because of this reason that no fuses and circuit breakers are ever used in the neutral wire of such a 3-phase system.

The solution for currents follows a pattern similar to that for the unbalanced delta.

Obviously, the vector sum of the currents in the three lines is not zero but is equal to neutral current.

(ii) Neutral wire with impedance Z_N

Such a case can be easily solved with the help of Node-pair Voltage method as detailed below. Consider the general case of a *Y*- to -*Y* system with a neutral wire of impedance Z_n as shown in Fig. 16.65 (*a*). As before, the impedance of line wires and source phase windings would be assumed to be zero so that the line and load terminals, *R*,*Y*, *B* and *R*′,*Y*′, *B*′ are the same respective potentials.

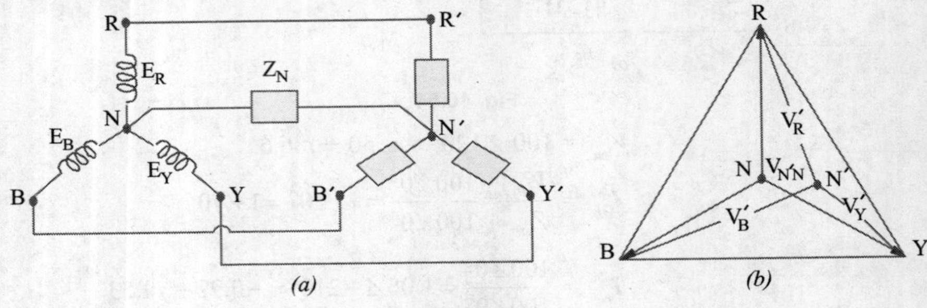

Fig. 16.65

According to Node-pair Voltage method, the above star-to-star system can be looked upon as multi-mesh network with a single pair of nodes *i.e.* neutral points *N* and *N*′. The node potential *i.e.* the potential difference between the supply and local neutrals is given by

$$V'_{NN} = \frac{E_R Y_R + E_Y Y_Y + E_B Y_B}{Y_R + Y_Y + Y_B + Y_N}$$

where Y_R, Y_Y and Y_B represent the ***load phase*** admittances. Obviously, the load neutral N' does not coincide with source neutral *N*. Hence, load phase voltages are no longer equal to one another even when phase voltages are as seen from Fig. 16.65 (*b*).

The load phase voltage are given by

$$V'_R = E_R - V'_{NN}; V'_Y = E_Y - V_{NN} \text{ and } V_{B'} = E_B - V'_{NN}$$

The phase currents are

$$I_R = V'_R Y_R,\ I'_Y = V'_Y Y_Y \text{ and } I_B = V'_B Y_B$$

The current in the neutral wire is $I_N = V_N Y_N$

Note. In the above calculations, $I_R = I'_R = I_{RR}$

Similarly, $I_Y = I'_Y = I'_{YY}$ and $I'_B = I_{BB}$.

Example 16.56. *A 3-phase, 4-wire system having a 254-V line-to-neutral has the following loads connected between the respective lines and neutral; $Z_R = 10\angle 0°$ ohm; $Z_Y = 10\angle 37°$ ohm and $Z_B = 10\angle -53°$ ohm. Calculate the current in the neutral wire and the power taken by each load when phase sequence is* **(i)** *RYB and* **(ii)** *RBY.*

Solution. (*i*) Phase sequence RYB (Fig. 16.66)

$V_{RN} = 254\ \angle 0°$; $V_{YN} = 254\ \angle -120°$; $V_{BN} = 254\ \angle 120°$

$$I_R = I_{RN} = \frac{V_{RN}}{R_R} = \frac{254\angle 0°}{10\angle 0°} = 25.4\angle 0\ ^*$$

* This method is similar to Millman's Theorem

$$\mathbf{I}_r = \mathbf{I}_{YN} = \frac{254\angle -120°}{10\angle 37°} = 25.4\angle -157° = 25.4(-0.9205 - j0.3907) = -23.38 - j9.95$$

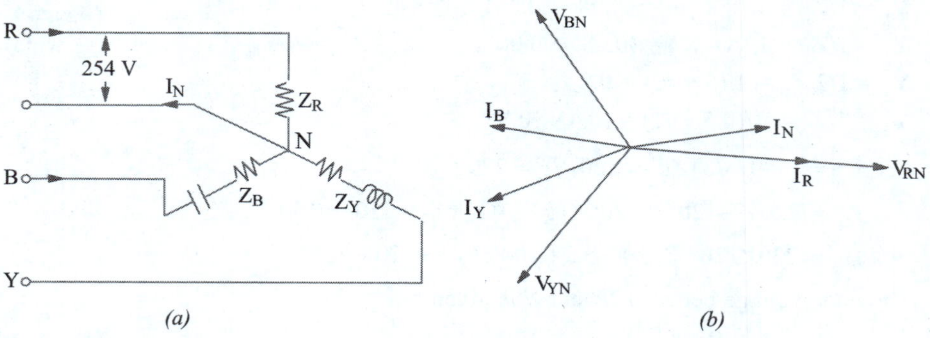

(a) (b)

Fig. 16.66

$$I_B = I_{BN} = \frac{254\angle 120°}{10\angle -53°} = 25.4\angle 173° = 25.4(-0.9925 + j0.1219) = -25.2 + j3.1$$

$I_N = -(I_R + I_Y + I_B) = -[25.4 + (-23.38 - j9.55) + (-25.21 + j3.1)] = 23.49 + j6.85$
$= \mathbf{24.46\ \angle 16°15'}$

Now $R_R = 10\,\Omega; R_Y = 10\cos 37° = 8\Omega; R_B = 10\cos 53° = 6\Omega$

$W_R = 25.4^2 \times 10 =$ **6.452 W**; $W_Y = 25.4^2 \times 8 =$ **5.162 W**
$W_B = 25.4^2 \times 6 =$ **3,871 W**

***(ii)* Phase sequence RBY** [Fig. 16.67]

$V_{RN} = 254\angle 0°; V_{YN} = 254\angle 120°$

$V_{BN} = 254\ \angle -120°$

$$I_R = \frac{254\angle 0°}{10\angle 0°} = 25.4\angle 0°$$

$$I_Y = \frac{254\angle 120°}{10\angle 37°} = 25.4\angle 83° = (3.1 + j25.2)I_B$$

$$I_B = \frac{254\angle -120°}{10\angle -53°} = 25.4\angle -67° = (9.95 - j23.4)$$

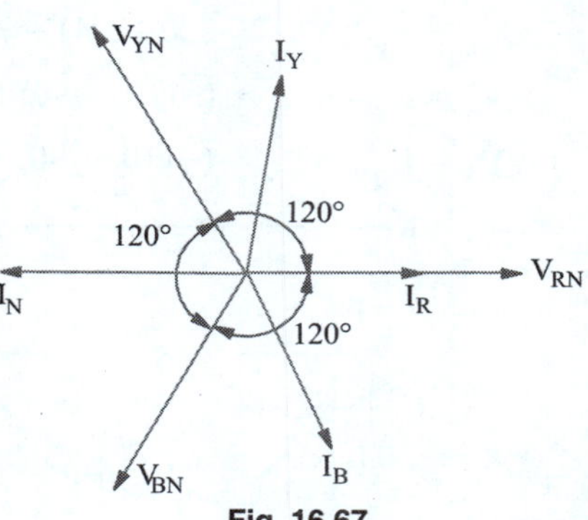

Fig. 16.67

$\mathbf{I}_N = -(\mathbf{I}_R + \mathbf{I}_Y + \mathbf{I}_B) = -(38.45 + j1.8)$
$= -38.45 - j1.8 = 38.5\angle -177.3°$.

Obviously, power would remain the same because magnitude of branch currents is unaltered. From the above, we conclude that phase reversal in the case of a 4-wire unbalanced load supplied from a balanced voltage system leads to the following changes:

(*i*) it changes the angles of phase currents but not their magnitudes.

(*ii*) however, power remains unchanged.

(*iii*) it changes the magnitude as well as angle of the neutral current $\mathbf{I}_N$.

Example 16.57. *A 3 – φ, 4-wire, 380-V supply is connected to an unbalanced load having phase impedances of: $Z_R = (8 + j6)\ \Omega$, $Z_y = (8 - j6)\ \Omega$ and $Z_B = 5\Omega$. Impedance of the neutral wire is $Z_N = (0.5 + j1)\Omega$.*

Ignoring the impedances of line wires and internal impedances of the e.m.f. sources, find the phase currents and voltages of the load.

Solution. This question will be solved by using Node-pair Voltage method. The admittances of the various branches connected between nodes N and N' in Fig. 16.68 (*a*).

$\mathbf{Y}_R = 1/\mathbf{Z}_R = 1/(8 + j6) = (0.08 - j0.06)$

$\mathbf{Y}_Y = 1/\mathbf{Z}_Y = 1/(8 - j6) = (0.08 + j0.06)$

$\mathbf{Y}_B = 1/2/\mathbf{Z}_\mathbf{B} = 1/(5 + j0) = 0.2$

$\mathbf{Y}_N = 1/\mathbf{Z}_N = 1/(0.5 + j1) = (0.4 - j0.8)$

Let $\mathbf{E}_R = (380/\sqrt{3})\angle 0° = 220\angle 0° = 220 + j0$

$\mathbf{E}_Y = 220\angle -120° = 220(-0.5 - j0.866) = -110 - j190$

$\mathbf{E}_B = 220\angle 120° = 220(-0.5 + j0.866) = -110 + j190$

The node voltage between N' and N is given by

$$V'_{NN} = \frac{E_R Y_R + E_Y Y_Y + E_B Y_B}{Y_R + Y_Y + Y_B + Y_N}$$

$$= \frac{200(0.08 - \text{j}0.06) + (-110 - j190)(0.08 + j0.06) + (-110 + j190)\times 0.2}{(0.08 - j0.06) + (0.08 + j0.06) + 0.2 + (0.4 - j0.8)}$$

$$= \frac{-1.8 + j3}{0.76 - j0.8} = -3.41 + j0.76$$

The three load phase voltages are as under:

$\mathbf{V}'_R = \mathbf{E}_R - \mathbf{V}'_{NN} = 220 + 3.41 - j0.76 = 223.41 - j0.76$

$\mathbf{V}'_Y = \mathbf{E}_Y - \mathbf{V}'_{NN} = (-110 - j90) + 3.41 - j0.76 = -106.59 - j190.76$

$\mathbf{V}'_B = \mathbf{E}_B - \mathbf{V}'_{NN} = (-110 - j90) + 3.41 - j0.76 = -106.59 + j190.76$

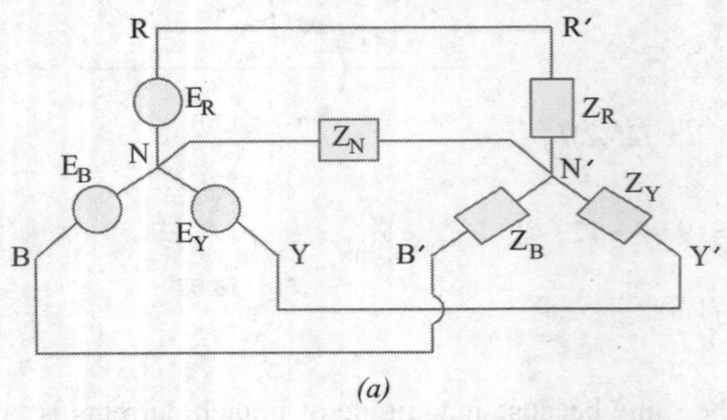

(a)

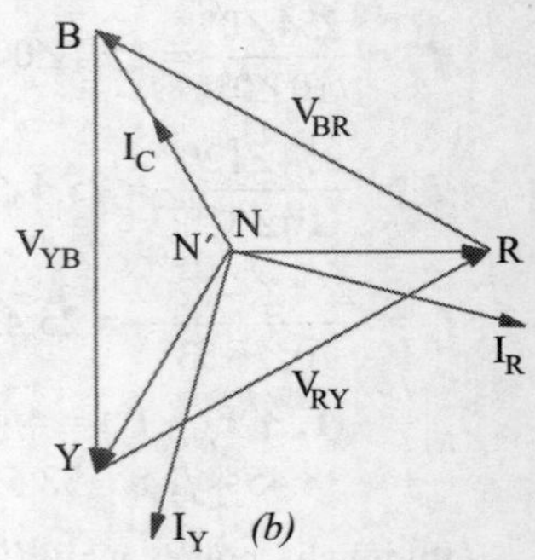

(b)

Fig 16.68

$\mathbf{I}'_R + \mathbf{V}'_R + \mathbf{Y}_R = (223.41 - \text{j}0.76)(0.08 - \text{j}0.06) = 17.83 - j13.1 = 22.1\angle -36.3°$ A

$\mathbf{I}'_Y + \mathbf{V}'_Y\mathbf{Y}_Y = (-106.59 + j190.76)(0.08 + j0.06) = 2.92 - j21.66 = 21.86\angle 82.4°$ A

$\mathbf{I}'_B + \mathbf{V}'_B\mathbf{Y}_B = (-106.59 + j190.76)\times 0.2 = -21.33 + j37.85 = 43.45\angle 119.4°$ A

$\mathbf{I}_N + \mathbf{V}_N\mathbf{Y}_N = (-3.41 + j0.76)(0.4 - j0.8) = -0.76 - j3.03 = 3.12\angle 104.1°$A

These voltage and currents are shown in the phasor diagram of Fig. 16.68 (b) where displacement of the neutral point has not been shown due to the low value of $\mathbf{V}'_{NN}$.

Note. It can be shown that $\boldsymbol{I}_N = \boldsymbol{I}'_R + \boldsymbol{I}'_Y + \boldsymbol{I}'_B$

16.31. Unbalanced Y-Connected Load Without Neutral

When a star-connected load is unbalanced and it has no neutral wire. Then its star point is isolated from the star-point of the generator. The potential of the load star-point is different from that of the generator star-point. The potential of the former is subject to variations according to the imbalance of the load and under certain conditions of loading, the potentials of the two star- point may differ considerably. Such an isolated load star-point or neutral point is called 'floating' neutral point because its potential is always changing and is not fixed.

All *Y*-connected unbalanced loads supplied from polyphase systems have floating neutral points without a neutral wire. Any unbalancing of the load causes variations not only of the potential of the star-point but also of the voltages across the different branches of the load. ***Hence, in that case, phase voltage of the load is not*** $1/\sqrt{3}$ ***of the line voltage.***

There are many methods to tackle such unbalanced Y-connected loads having isolated neutral points.

1. By converting the *Y*-connected load to an equivalent unbalanced Δ-connected load by using Y-Δ conversion theorem. The equivalent Δ-connection can be solved in Fig. 16.80. The line currents so calculated are equal in magnitude and phase to those taken by the original unbalanced *Y*-connected load.

2. By applying Kirchhoff's Laws.

3. By applying Millman's Theorem.

4. By using Maxwell's Mesh or Loop Current Method.

Example 16.58. *If in the unbalanced Y-connected load of Fig. 16.64,* $Z_a = (10 + j0)$, $Z_b = (3 + j4)$ *and* $Z_c = (0 - j10)$ *and the load is put across a 3-phase, 200-V circuit with balanced voltages, find the three line currents and voltages across each branch impedance. Assume phase sequence of* V_{ab}, V_{bc}, V_{ca}.

Solution. Take V_{ab} along the axis of reference. The vector expressions for the three voltages are

$$V_{ab} = 200 + j0$$

$$V_{bc} = 200\left(-\frac{1}{2} - j\frac{\sqrt{3}}{2}\right) = -100 - j173.2; \quad V_{ca} = 200\left(-\frac{1}{2} + j\frac{\sqrt{3}}{2}\right) = -100 + j173.2$$

From equation **(9)** given above

$$\mathbf{I}_{oa} = \frac{(-100 + j173.2)(3 + j4) - (200 + j0)(0 - j10)}{(0 - j10)(10 + j0) + (10 + j0)(3 + j4) + (3 + j4)(0 - j10)}$$

$$= \frac{-992.8 + j2119.6}{70 - j90} = -20.02 + j4.54$$

$$\mathbf{I}_{ob} = \frac{(200 + j0)(0 - j10) - (-100 - j173.2)(10 + j0)}{(10 + j0)(3 + j4) + (3 + j4)(0 - j10) + (0 - j10)(10 + j0)}$$

$$= \frac{1000 - j268}{70 - j90} = 7.24 + j5.48$$

Now, $\mathbf{I}_{oc}$ may also be calculated in the same way or it can be found easily from equation (4) of Art. 16.33.

$$\mathbf{I}_{oc} = \mathbf{I}_{ao} + \mathbf{I}_{bo} = -\mathbf{I}_{oa} - \mathbf{I}_{ob} = 20.02 - j4.54 - 7.24 - j5.48 = 12.78 - j10.02$$

Now
$$\mathbf{V}_{oa} = \mathbf{I}_{oa}\mathbf{Z}_a = (-20.02 + j4.54)(10 + j0) = 200.2 + j45.4$$

$$\mathbf{V}_{ob} = \mathbf{I}_{ob}\mathbf{Z}_b = (7.24 + j5.48)(3 + j4) = -0.2 + j45.4$$

$$\mathbf{V}_{oc} = \mathbf{I}_{oc}\mathbf{Z}_c = (12.78 - j10.02)(0 - j10) = -100.2 - j127.8$$

As a check, we may combine $\mathbf{V}_{oa}$, $\mathbf{V}_{ob}$ and $\mathbf{V}_{oc}$ to get the line voltages which should be equal to the applied line voltages. In passing from *a* to *b* through the circuit internally, we find that we are in opposition to $\mathbf{V}_{oa}$ but in the same direction as the positive direction of $\mathbf{V}_{ob}$.

$$\mathbf{V}_{ab} = \mathbf{V}_{ao} + \mathbf{V}_{ob} = -\mathbf{V}_{oa} + \mathbf{V}_{ob} = -(-200.2 + j45.4) + (-0.2 + j45.4) = 200 + j0$$

$$\mathbf{V}_{bc} = \mathbf{V}_{bo} + \mathbf{V}_{oc} = -\mathbf{V}_{ob} + \mathbf{V}_{oc} = -(-0.2 + j45.4) + (-100.2 - j127.8) = -100 - j173.2$$

$$\mathbf{V}_{ca} = \mathbf{V}_{co} + \mathbf{V}_{oa} = -\mathbf{V}_{oc} + \mathbf{V}_{oa} = -(-100.2 + j127.8) + (-200.2 + j45.4) = -100 + j173.2$$

16.32. Delta/Star and Star/Delta Conversions

Let us consider the unbalanced Δ-connected load of Fig. 16.69 (*a*) and *Y*-connected load of Fig. 16.69 (*b*). If the two systems are to be equivalent, then the impedances between corresponding pairs of terminals must be the same.

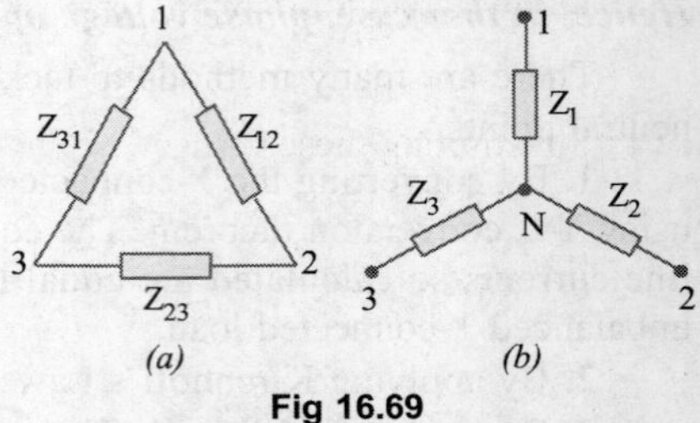

Fig 16.69

(*i*) Delta/Star Conversion

For *Y*-load, total impedance between terminals 1 and 2 is = $\mathbf{Z}_1 + \mathbf{Z}_2$ (it should be noted that double subscript notation of $\mathbf{Z}_{01}$ and $\mathbf{Z}_{02}$ has been purposely avoided).

Considering terminals 1 and 2 of Δ-load, we find that there are two parallel paths having impedances of $\mathbf{Z}_{12}$ and $(\mathbf{Z}_{31} + \mathbf{Z}_{23})$. Hence, the equivalent impedance between terminals 1 and 2 is given by

$$\frac{1}{\mathbf{Z}} = \frac{1}{\mathbf{Z}_{12}} + \frac{1}{\mathbf{Z}_{23} + \mathbf{Z}_{31}} \quad \text{or} \quad \mathbf{Z} = \frac{\mathbf{Z}_{12}(\mathbf{Z}_{23} + \mathbf{Z}_{31})}{\mathbf{Z}_{12} + \mathbf{Z}_{23} + \mathbf{Z}_{31}}$$

Therefore, for equivalence between the two systems $\mathbf{Z}_1 + \mathbf{Z}_2 = \dfrac{\mathbf{Z}_{12}(\mathbf{Z}_{23} + \mathbf{Z}_{31})}{\mathbf{Z}_{12} + \mathbf{Z}_{23} + \mathbf{Z}_{31}}$ ... **(1)**

Similarly $\mathbf{Z}_2 + \mathbf{Z}_3 = \dfrac{\mathbf{Z}_{23}(\mathbf{Z}_{31} + \mathbf{Z}_{12})}{\mathbf{Z}_{12} + \mathbf{Z}_{23} + \mathbf{Z}_{31}}$ ... **(2)** $\mathbf{Z}_3 + \mathbf{Z}_1 = \dfrac{\mathbf{Z}_{31}(\mathbf{Z}_{12} + \mathbf{Z}_{23})}{\mathbf{Z}_{12} + \mathbf{Z}_{23} + \mathbf{Z}_{31}}$ ... **(3)**

Adding equation **(3)** to **(1)** and subtracting equation **(2)**, we get

$$2\mathbf{Z}_1 = \frac{\mathbf{Z}_{12}(\mathbf{Z}_{23} + \mathbf{Z}_{31}) + \mathbf{Z}_{31}(\mathbf{Z}_{12} + \mathbf{Z}_{23}) - \mathbf{Z}_{23}(\mathbf{Z}_{31} + \mathbf{Z}_{12})}{\mathbf{Z}_{12} + \mathbf{Z}_{23} + \mathbf{Z}_{31}} = \frac{2\mathbf{Z}_{12}\mathbf{Z}_{31}}{\mathbf{Z}_{12} + \mathbf{Z}_{23} + \mathbf{Z}_{31}}$$

$$\therefore \quad \mathbf{Z}_1 = \frac{\mathbf{Z}_{12}\,\mathbf{Z}_{31}}{\mathbf{Z}_{12} + \mathbf{Z}_{23} + \mathbf{Z}_{31}} \quad \text{... (4)}$$

The other two results may be written down by changing the subscripts cyclically

$$\therefore \quad \mathbf{Z}_2 = \frac{\mathbf{Z}_{23}\mathbf{Z}_{12}}{\mathbf{Z}_{12} + \mathbf{Z}_{23} + \mathbf{Z}_{31}}; \quad \text{... (5)} \quad \mathbf{Z}_3 = \frac{\mathbf{Z}_{31}\mathbf{Z}_{23}}{\mathbf{Z}_{12} + \mathbf{Z}_{23} + \mathbf{Z}_{31}} \quad \text{... (6)}$$

The above expression can be easily obtained by remembering that

$$\text{Star } \mathbf{Z} = \frac{\text{Product of } \Delta\mathbf{Z}\text{'s connected to the same terminals}}{\text{Sum of } \Delta\mathbf{Z}\text{'s}}$$

In should be noted that all **Z'** are to be expressed in their complex form.

(iii) Star/Delta Conversion

The equations for this conversion can be obtained by rearranging equations **(4)**, **(5)** and **(6)**, Rewriting these equations, we get

$$\mathbf{Z_1(Z_{12} + Z_{23} + Z_{31}) = Z_{12}Z_{31}} \quad \text{... (7)}$$

$$\mathbf{Z_2(Z_{12} + Z_{23} + Z_{31}) = Z_{23}Z_{12}} \quad \text{... (8)}$$

$$\mathbf{Z_3(Z_{12} + Z_{23} + Z_{31}) = Z_{31}Z_{23}} \quad \text{... (9)}$$

Dividing equation **(7)** by **(9)**, we get $\mathbf{\frac{Z_1}{Z_3} = \frac{Z_{12}}{Z_{23}}} \quad \therefore \mathbf{Z_{23} = Z_{12}\frac{Z_3}{Z_1}}$

Dividing equation **(8)** by **(9)**, we get $\mathbf{\frac{Z_2}{Z_3} = \frac{Z_{12}}{Z_{31}}} \quad \therefore \mathbf{Z_{31} = Z_{12}\frac{Z_3}{Z_2}}$

Substituting these values in equation **(7)**, we have $\mathbf{Z_1\left(Z_{12} + Z_{12}\frac{Z_3}{Z_1} + Z_{31}\right) = Z_{12}.Z_{12}\frac{Z_3}{Z_2}}$

or $\mathbf{Z_1Z_{12}\left(1 + \frac{Z_3}{Z_1} + \frac{Z_{31}}{Z_{12}}\right) = Z_{12}Z_{12}\frac{Z_3}{Z_2}} \quad \therefore \mathbf{Z_1Z_2 + Z_2Z_3 + Z_3Z_1 = Z_{12} \times Z_3}$

$$\mathbf{Z_{12} = \frac{Z_1Z_2 + Z_2Z_3 + Z_3Z_1}{Z_3}} \quad \text{or} \quad \mathbf{Z_{12} = Z_1 + Z_2 + \frac{Z_1Z_2}{Z_3}}$$

Similarly, $\mathbf{Z_{23} = Z_2 + Z_3 + \frac{Z_2Z_3}{Z_1} = \frac{Z_1Z_2 + Z_2Z_3 + Z_3Z_1}{Z_1}}$

$$\mathbf{Z_{31} = Z_3 + Z_1 + \frac{Z_3Z_1}{Z_2} = \frac{Z_1Z_2 + Z_2Z_3 + Z_3Z_1}{Z_2}}$$

As in the previous case, it is to be noted that all impedances must be expressed in their complex form.

Another point for noting is that *the line currents of this equivalent delta are the currents in the phases of the Y-connected load.*

Example 16.59. *An unbalanced star-connected load has branch impedances of $Z_1 = 10 \angle 30°\ \Omega$, $Z_2 = 10 \angle -45°\ \Omega$, $Z_3 = 20 \angle 60°\ \Omega$ and is connected across a balanced 3-phase, 3-wire supply of 200 V. Find the line currents and the voltage across each impedance using Y/Δ conversion method.*

Solution. The unbalanced *Y*-connected load and its equivalent Δ-connected load are shown in Fig. 16.70.

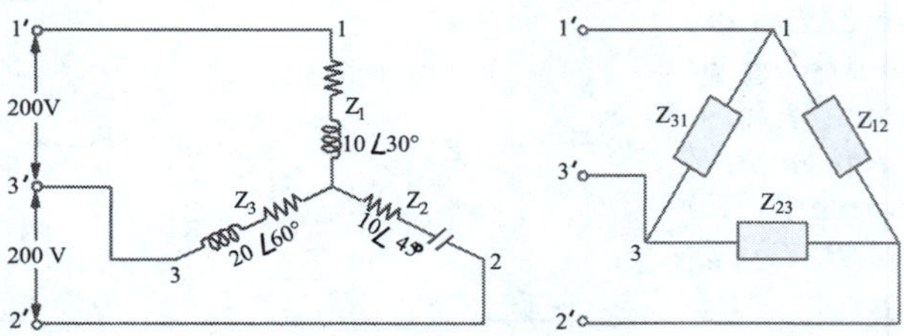

Fig 16.70

Now $\mathbf{Z_1Z_2 + Z_2Z_3 + Z_3Z_1}$ = (10 ∠30°) (10 ∠– 45°) + (10 ∠– 45°) (20 ∠60°) + (20 ∠60°) (10 ∠30°) = 100 ∠– 15° + 200 ∠15° + 200 ∠90°

Converting these into their cartesian form, we get

$= 100\,[\cos(-15°) - j\sin 15°] + 200\,(\cos 15° + j\sin 15°) + 200\,(\cos 90° + j\sin 90°)$

$= 96.6 - j25.9 + 193.2 + j51.8 + 0 + j200 = 289.8 + j225.9 = 368\angle 38°$

$$\mathbf{Z}_{12} = \frac{\mathbf{Z}_1\mathbf{Z}_2 + \mathbf{Z}_2\mathbf{Z}_3 + \mathbf{Z}_3\mathbf{Z}_1}{\mathbf{Z}_3} = \frac{368\angle 38°}{20\angle 60°} = 18.4\angle -22° = 17.0 - j6.9$$

$$\mathbf{Z}_{23} = \frac{368\angle 38°}{10\angle -30°} = 36.8\angle 8° = 36.4 + j5.1$$

$$\mathbf{Z}_{31} = \frac{368\angle 38°}{10\angle -45°} = 36.8\angle 83° = 4.49 + j36.5$$

Assuming clockwise phase sequence of voltages $\mathbf{V}_{12}$, $\mathbf{Z}_{23}$ and $\mathbf{V}_{31}$, we have

$\mathbf{V}_{12} = 200\angle 0°$, $\mathbf{V}_{23} = 200\angle -120°$, $\mathbf{V}_{31} = 200\angle 120°$

$$\mathbf{I}_{12} = \frac{\mathbf{V}_{12}}{\mathbf{Z}_{12}} = \frac{200\angle 0°}{18.4\angle -22°} = 10.86\angle 22° = 10.07 + j4.06$$

$$\mathbf{I}_{23} = \frac{\mathbf{V}_{23}}{\mathbf{V}_{23}} = \frac{200\angle -120°}{36.8\angle 8°} = 5.44\angle -128° = -3.35 - j4.29$$

$$\mathbf{I}_{31} = \frac{\mathbf{V}_{31}}{\mathbf{Z}_{31}} = \frac{200\angle 120°}{36.8\angle 83°} = 5.44\angle 37° = 4.34 + j3.2$$

Line current $= \mathbf{I'}_{11} = \mathbf{I}_{12} + \mathbf{I}_{13} = \mathbf{I}_{12} - \mathbf{I}_{31}$

$= (10.07 + j4.06) - (4.34 + j3.2) = 5.73 + j0.86 = 5.76\angle 8°32'$

$\mathbf{I'}_{22} = \mathbf{I}_{23} - \mathbf{I}_{12} = (-3.35 - j4.29) - (10.07 + j4.06) = -13.42 - j8.35 = 15.79\angle -148°6'$

$\mathbf{I'}_{33} = \mathbf{I}_{31} - \mathbf{I}_{23} = (4.34 - j3.2) - (-3.35 - j4.29) = 7.69 + j7.49 = 10.73\angle°44°16'$

These are currents in the *phases of the Y-connected unbalanced load.* Let us find voltage drop across each star-connected branch impedance.

Voltage drop across $\mathbf{Z}_1 = \mathbf{V}_{10} = \mathbf{I'}_{11}\mathbf{Z}_1 = 5.76\angle 8°32'.10\angle 30° = 57.6\angle 38°32'$

Voltage drop across $\mathbf{Z}_2 = \mathbf{V}_{20} = \mathbf{I'}_{22}\mathbf{Z}_2 = 15.79\angle -148°6.10\angle -45° = 157.9\angle -193°6'$

Voltage drop across $\mathbf{Z}_3 = \mathbf{V}_{30} = \mathbf{I'}_{33}\mathbf{Z}_3 = 10.73\angle 44°16'.20°\angle 60° = 214.6 * 104°16'$

Example 16.60. *A 300-V (line) 3-phase supply feeds! star-connected load consisting of non-inductive resistors of 15, 6 and 10 Ω connected to the R, Y and B lines respectively. The phase sequence is RYB. Calculate the voltage across each resistor.*

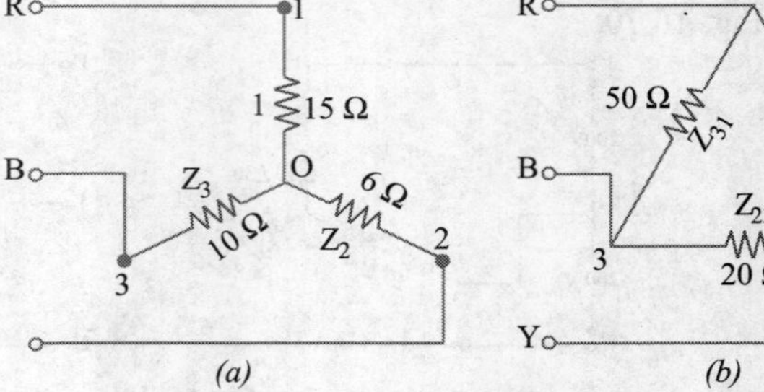

Fig 16.71

Solution. The *Y*-connected unbalanced load and its equivalent Δ-connected load are shown in Fig. 16.71. Using *Y*/Δ conversion method we have

$$\mathbf{Z}_{12} = \frac{\mathbf{Z}_1\mathbf{Z}_2 + \mathbf{Z}_2\mathbf{Z}_3 + \mathbf{Z}_3\mathbf{Z}_1}{\mathbf{Z}_3} = \frac{90 + 60 + 150}{10} = 30\Omega$$

$\mathbf{Z}_{23} = 300/15 = 20\Omega$

$\mathbf{Z}_{31} = 300/6 = 50\Omega$

Phase current $I_{RY} = V_{RY} / Z_{12} = 300/30 = 10\,\text{A}$

Similarly $I_{YB} = V_{YB} / Z_{23} = 300/20 = 15\,\text{A}$

$I_{BR} = V_{BR} / Z_{31} = 300/50 = 6\,\text{A}$

Each current is in phase with its own voltage because the load is purely resistive.

The line currents for the delta connection are obtained by compounding these phase currents in pairs, either trigonometrically or by phasor algebra. Using phasor algebra and choosing $\mathbf{V}_{RY}$ as the reference axis, we get

$$\text{I}_{RY} = 10 + j0;\ \text{I}_{YB} = 15(-\frac{1}{2} - j\sqrt{3}/2) = -7.5 - j13.0;\text{I}_{BR} = 6(-\frac{1}{2} + j\sqrt{3}/2) = -3.0 + j5.2$$

Line currents for delta-connection are

$\mathbf{I}_R = \mathbf{I}_{RY} + \mathbf{I}_{RB} = \mathbf{I}_{RY} - \mathbf{I}_{BR} = (10 + j0) - (-3 + j5.2) = 13 - j5.2$ or 14 A in magnitude

$\mathbf{I}_Y = \mathbf{I}_{YR} + \mathbf{I}_{YB} = \mathbf{I}_{YB} - \mathbf{I}_{RY} = (-7.5 - j13.0) - (10 + j0) = -17.5 - j13$ or 21.8 A in magnitude

$\mathbf{I}_B = \mathbf{I}_{BR} + \mathbf{I}_{BY} = \mathbf{I}_{BR} - \mathbf{I}_{YB} = (-3.0 + j5.2) - (-7.5 - j13.0) = 4.5 + j18.2 = 18.7$ A in magnitude

These line currents for Δ-connection are the phase currents for *Y*-connection. Voltage drop across each limb of *Y*-connected load is

$\mathbf{V}_{RN} = \mathbf{I}_R\,\mathbf{Z}_1 = (13 - j5.2)(15 + j0) = 195 - 78$ volt or 210 V

$\mathbf{V}_{YN} = \mathbf{I}_Y\mathbf{Z}_2 = (-17.5 - j13.0)(6 + j0) = -105 - j78$ volt or 131 V

$\mathbf{V}_{BN} = \mathbf{I}_B\mathbf{Z}_3 = (4.5 + j18.2)(10 + j0) = 45 + j182$ volt or 187 V

As a check, it may be verified that the difference of phase voltages taken in pairs should give the three line voltages. Going through the circuit internally, we have

$\mathbf{V}_{RY} = \mathbf{V}_{RN} + \mathbf{V}_{NY} = \mathbf{V}_{RN} - \mathbf{V}_{YN} = (195 - j78) - (105 - j78) = 300\angle 0°$

$\mathbf{V}_{YB} = \mathbf{V}_{YN} - \mathbf{V}_{BN} = (-105 - j78) - (45 + j182) = -150 - j260 = 300\angle -120°$

$\mathbf{V}_{BR} = \mathbf{V}_{BN} - \mathbf{V}_{RN} = (45 + j182) - (195 - j78) = -150 + j260 = 300\angle 120°$

This question could have been solved by direct geometrical methods.

Example 16.61. *A Y-connected load is supplied from a 400-V, 3-phase, 3-wire symmetrical system RYB. The branch circuit impedances are*

$$Z_R = 10\sqrt{3} + j10;\ Z_Y = 20 + j20\sqrt{3};\ Z_B = 0 - j10$$

Determine the current in each branch. Phase sequence is RYB.

(Network Analysis, Nagpur Univ. 1993)

Solution. The circuit is shown in Fig. 16.72. The problem will be solved by using all the four possible ways in which 3-wire unbalanced *Y* connected load can be handled.

Now, $\mathbf{Z}_R = 20\angle 30° = (17.32 + j10)$

$\mathbf{Z}_Y = 40\angle 60° = (20 + j34.64)$

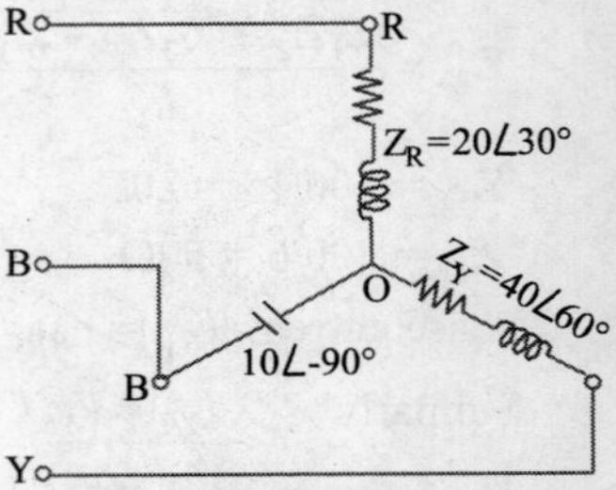

$$\mathbf{Z}_B = 10\angle-90° = -j10$$

Also, let
$$\mathbf{V}_{RY} = 400\angle0° = 400 + j0$$
$$\mathbf{V}_{RB} = 400\angle-120° = -200 - j346$$
$$\mathbf{V}_R = 400\angle120° = -200 + j346$$

R — R, $Z_R = 20\angle30°$, O, $Z_Y = 40\angle60°$, B, $10\angle-90°$, Y

Fig 16.72

(*a*) By applying Kirchhoff's Laws

It is seen that

$$I_{RO} = I_R = \frac{V_{RY}Z_B - V_{BR}Z_Y}{Z_RZ_Y + Z_YZ_B + Z_BZ_R};$$

$$I_{YO} = I_Y = \frac{V_{YB}Z_R - V_{RY}Z_B}{Z_RZ_Y + Z_YZ_B + Z_BZ_R}; \quad I_{BO} = I_B = \frac{V_{BR}Z_Y - V_{YB}Z_R}{Z_RZ_Y + Z_YZ_B + Z_BZ_R}$$

Now, $\mathbf{Z}_R\mathbf{Z}_Y + \mathbf{Z}_Y\mathbf{Z}_B + \mathbf{Z}_B\mathbf{Z}_R$

$$= 20\angle30°.40\angle60° + 40\angle60°.10\angle-90° + 10\angle-90°.20\angle30°$$
$$= 800\angle90° + 400\angle-30° + 200\angle-60° = 446 + j426 = 617\angle43.7°$$

$$\mathbf{V}_{RY}\mathbf{Z}_B - \mathbf{V}_{BR}\mathbf{Z}_Y = 400\times10\angle-90° - 400\angle120°.40\angle60°$$
$$= 16{,}000 - j4000 = 16{,}490\angle-14°3'$$

$$\therefore \quad \mathbf{I}_R = \frac{16{,}490\angle-14°3'}{617\angle43.7°} = 26.73\angle-57°45'$$

$$\mathbf{V}_{YB}\mathbf{Z}_R - \mathbf{V}_{RY}\mathbf{Z}_B = 400\angle-120°.20\angle30° - 400.10\angle-90° = -j4000 = 4000\angle-90°$$

$$\mathbf{I}_Y = \frac{4000\angle-90°}{617\angle43.7°} = 6.48\angle-133.7°$$

$$\mathbf{V}_{BR}\mathbf{Z}_Y - \mathbf{V}_{YB}\mathbf{Z}_R = 400\angle120°.40\angle60° - 400\angle-120°20.\angle30°$$
$$= -16{,}000 + j8{,}000 = 17{,}890\angle153°26'$$

$$\therefore \quad \mathbf{I}_B = \frac{17{,}890\angle153°26'}{617\angle43.7°} = 29\angle109°45'$$

(*b*) By Star/Delta Conversion (Fig. 16.73)

The given star may be converted into the equivalent delta with the help of equations given in Art. 16.34.

$$\mathbf{Z}_{RY} = \frac{\mathbf{Z}_R\mathbf{Z}_Y + \mathbf{Z}_Y\mathbf{Z}_B + \mathbf{Z}_B\mathbf{Z}_R}{\mathbf{Z}_B} = \frac{617\angle43.7°}{10\angle-90°} = 61.73\angle133.7°$$

$$\mathbf{Z}_{YB} = \frac{\Sigma\mathbf{Z}_R\mathbf{Z}_Y}{\mathbf{Z}_R} = \frac{617\angle43.7°}{20\angle30°} = 30.87\angle13.7°$$

$$\mathbf{Z}_{BR} = \frac{\Sigma\mathbf{Z}_R\mathbf{Z}_Y}{\mathbf{Z}_Y} = \frac{617\angle43.7°}{40\angle60°} = 15.43\angle-16.3°$$

$$\mathbf{I}_{RY} = \frac{\mathbf{V}_{RY}}{\mathbf{Z}_{RY}} = \frac{400}{61.73\angle133.7°} = 6.48\angle-133.7° = (-4.47 - j4.68)$$

$$\mathbf{I}_{YB} = \frac{\mathbf{V}_{YB}}{\mathbf{Z}_{YB}} = \frac{400\angle-120°}{30.87\angle13.7°} = 12.95\angle-133.7° = (-8.95 - j9.35)$$

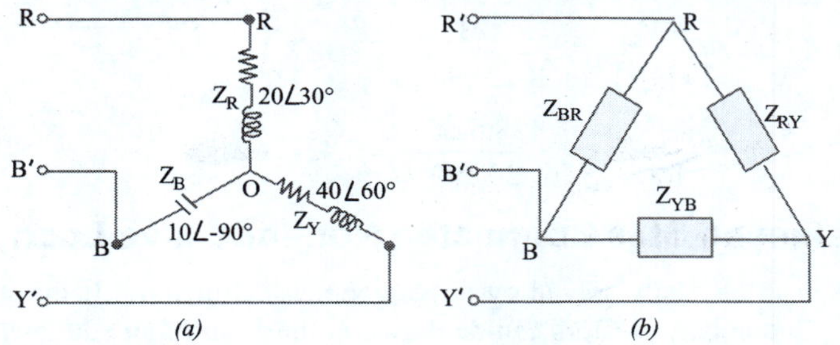

Fig. 16.73

$$\mathbf{I}_{BR} = \frac{\mathbf{V}_{BR}}{\mathbf{Z}_{BR}} = \frac{400\angle 120°}{15.43\angle -16.3°} = 25.9\angle 136.3° = (-18.7 + j17.9)$$

$$\mathbf{I}'_{RR} = \mathbf{I}_{RY} - \mathbf{I}_{BR} = 14.23 - j22.58 = 26.7\angle -57°48'$$

$$\mathbf{I}'_{YY} = \mathbf{I}_{YB} - \mathbf{I}_{RY} = -4.48 - j4.67 = 6.47\angle -134°6'$$

$$\mathbf{I}'_{YB} = \mathbf{I}_{BR} - \mathbf{I}_{YB} = -9.85 + j27.25 = 29\angle 109°48'$$

$\Sigma\mathbf{I} = (0 + j0)$ –as a check

These line currents of the equivalent delta represent the phase currents of the star-connected load of Fig. 16.73 (*a*).

Note. Minor differences are due to accumulated errors.

Example 16.62. *Three impedances, Z_R, Z_Y and Z_B are connected in star across a 440-V, 3-phase supply. If the voltage of star-point relative to the supply neutral is 200 ∠150° volt and Y and B line currents are 10 ∠– 90° A and 20 ∠90° A respectively, all with respect to the voltage between the supply neutral and the R line, calculate the values of Z_R, Z_Y and Z_B.*

(Elect Circuit; Nagpur Univ. 1991)

Solution. Let *O* and *O*′ be the supply and load neutrals respectively. Also, let,

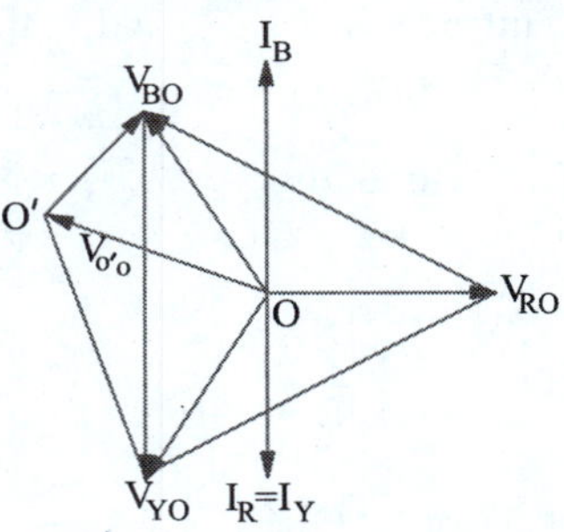

Fig. 16.74

$$\mathbf{V}_{RO} = \frac{440}{\sqrt{3}}\angle 0° = 254\angle 0° = 254 + j0$$

$$\mathbf{Y}_{YO} = 254\angle -120° = -127 - j220$$

$$\mathbf{V}_{BO} = 254\angle 120° = -127 + j220$$

$$\mathbf{I}_Y = 10\angle -90° = -j10; \ \mathbf{I}_B = 20\angle 90° = j20$$

$$\mathbf{I}_R = -(\mathbf{I}_Y + \mathbf{I}_B) = -j10$$

Also, $\mathbf{V}'_{OO} = 200\angle 150° = -173 + j100$

$$\mathbf{V}_{RO} - \mathbf{V}'_{OO} = 254 - (-173 + j100) = 427 - j100 = 438.5\angle -13.2°$$

$$\mathbf{V}_{YO} - \mathbf{V}'_{OO} = (-127 - j220) - (-173 + j100) = 46 - j320 = 323\angle -81.6°$$

$$\mathbf{V}_{BO} - \mathbf{V}'_{OO} = (-127 + j220) - (-173 + j100) = 46 + j120 = 128.6\angle 69°$$

As seen from Art. 16.32.

$$\mathbf{Z}_R = \frac{\mathbf{V}_{RO} - \mathbf{V}'_{00}}{\mathbf{I}_R} = \frac{438.5\angle -13.2°}{10\angle -90°} = 43.85\angle 76.8°$$

$$\mathbf{Z}_Y = \frac{\mathbf{V}_{YO} - \mathbf{V}'_{OO}}{\mathbf{I}_Y} = \frac{323}{10\angle -90°} = 32.3\angle 8.4°$$

$$\mathbf{Z}_B = \frac{\mathbf{V}_{BO} - \mathbf{V}'_{OO}}{\mathbf{I}_B} = \frac{128.6\angle 69°}{20\angle 90°} = 6.43\angle -21°$$

16.33. Unbalanced Star-Connected Non-Inductive Load

Such a case can be easily solved by direct geometrical method. If the supply system is symmetrical, the line voltage vectors can be drawn in the form of an equilateral triangle *RYB* (Fig. 16.75). As the load is an unbalanced one, its neutral point will not, obviously, coincide with the centre of the gravity or centroid of the triangle. Let it lie at any other point like *N*. If point *N* represents the potential of the neutral point if the unbalanced load, then vectors drawn from *N* to points, *R*, *Y* and *B* represent the voltages across the branches of the load. These voltages can be represented in their rectangular co-ordinates with respect to the rectangular axis drawn through *N*. It is seen that taking co-ordinates of *N* as (0, 0), the co-ordinates of point *R* are $[(V/2 - x), - y]$ of point *Y* are $[- (V/2 + x), - y]$ and of point *B* are $[-x, (\sqrt{3}V/2 - y)]$

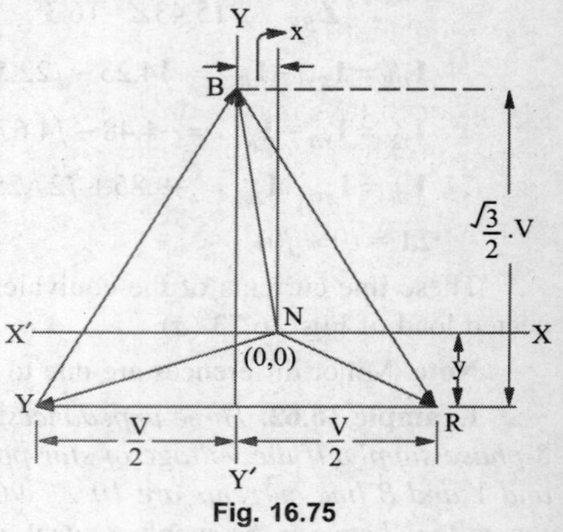

Fig. 16.75

$$V_{RN} = \left(\frac{V}{2} - x\right) - jy;\; V_{YN} = -\left(\frac{V}{2} + x\right) - iy$$

$$V_{BN} = -x + j\left(\frac{\sqrt{3}V}{2} - y\right)$$

Let R_1, R_2 and R_3 be the respective branch impedances, Y_1, Y_2 and Y_3 the respective admittances and $\mathbf{I}_R$, $\mathbf{I}_Y$ and $\mathbf{I}_B$ the respective currents in them.

Then $\mathbf{I}_R = \mathbf{V}_{RN}/\mathbf{R}_1 = \mathbf{V}_{RN}\,\mathbf{Y}_1$

Similarly, $\mathbf{I}_Y = \mathbf{V}_{YN}\mathbf{Y}_2$ and $\mathbf{I}_B = \mathbf{V}_{BN}\mathbf{Y}_3$. Since $\mathbf{I}_R + \mathbf{I}_Y + \mathbf{I}_B = 0$

$\therefore \quad \mathbf{V}_{RN}\mathbf{Y}_1 + \mathbf{V}_{YN}\mathbf{Y}_2 + \mathbf{V}_{BN}\mathbf{Y}_3 = 0$

$$\text{or } Y_1\left[\left(\frac{V}{2} - x\right) - jy\right] + Y_2\left[-\left(\frac{V}{2} + x\right) - jy\right] + Y_3\left[-x + j\left(\frac{\sqrt{3}V}{2} - y\right)\right] = 0$$

$$\text{or } -x(Y_1 + Y_2 + Y_3) + \frac{V}{2}(Y_1 - Y_2) + j\left[Y_3\frac{\sqrt{3}V}{2} - y(Y_1 + Y_2 + Y_3)\right] = 0$$

$$\therefore \quad -x(Y_1 + Y_2 + Y_3) + \frac{V}{2}(Y_1 - Y_2) = 0 \quad \therefore \; x = \frac{V(Y_1 - Y_2)}{2Y_1 + Y_2 + Y_3}$$

$$\text{Also } Y_3\frac{\sqrt{3}V}{2} - y(Y_1 + Y_2 + Y_3) = 0 \quad \therefore \; y = \frac{\sqrt{3}V}{2}\frac{Y_3}{(Y_1 + Y_2 + Y_3)}$$

Knowing the values of *x*, the values of $\mathbf{V}_{RN}$, $\mathbf{V}_{YN}$ and $\mathbf{V}_{BN}$ and hence, of $\mathbf{I}_R$, $\mathbf{I}_Y$ and $\mathbf{I}_B$ can be found.

Example 16.63. *Three non-inductive resistances of 5, 10 and 15 Ω are connected in star and supplied from a 230-V symmetrical 3-phase system. Calculate the line currents (magnitudes).* **(Principles of Elect. Engg. Jadavpur Univ.)**

Solution.

(*a*) Star/Delta Conversion Method

The *Y*-connected unbalanced load and its equivalent Δ-connected load are shown in Fig. 16.76 (*a*) and (*b*) respectively. Using *Y*/Δ conversion, we have

$$\mathbf{Z}_{12} = \frac{\mathbf{Z}_1\mathbf{Z}_2 + \mathbf{Z}_2\mathbf{Z}_3 + \mathbf{Z}_3\mathbf{Z}_1}{\mathbf{Z}_3} = \frac{50+150+75}{15} = \frac{55}{3}\Omega$$

$$Z_{23} = 275/5 = 55\Omega \text{ and } Z_{31} = 275/10 = 27.5\Omega$$

Phase current $\mathbf{I}_{RY} = \mathbf{V}_{RY}/\mathbf{Z}_{12} = 230/(55/3) = 12.56$ A

Similarly, $\mathbf{I}_{YB} = \mathbf{V}_{YB}/\mathbf{Z}_{23} = 230/55 = 4.18$ A; $\mathbf{I}_{BR} = \mathbf{V}_{BR}/\mathbf{Z}_{31} = 230/27.5 = 8.36$ A

The line currents for Δ-connection are obtained by compounding the above phase currents trigonometrically or vectorially. Choosing vector addition and taking $\mathbf{V}_{RY}$ as the reference vector, we get;

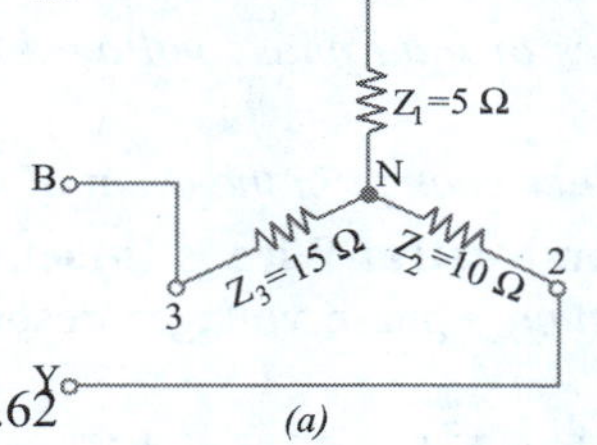

(*a*)

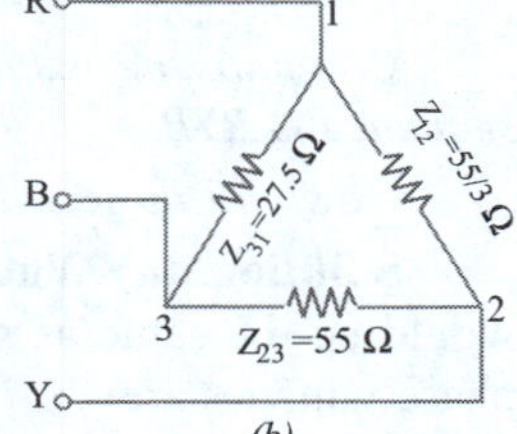

(*b*)

Fig. 16.76

$$\mathbf{I}_{RY} = (12.56 + j0)$$

$$\mathbf{I}_{YB} = 4.18\left(-\frac{1}{2} j\frac{\sqrt{3}}{2}\right) = -2.09 - j3.62$$

$$\mathbf{I}_{BR} = 8.36\left(-\frac{1}{2} + j\frac{\sqrt{3}}{2}\right) = -4.18 + j7.24$$

Hence, line currents for Δ-connection of Fig. 16.76 (*b*) are

$$\mathbf{I}_R = \mathbf{I}_{RY} + \mathbf{I}_{RB} = \mathbf{I}_{RY} - \mathbf{I}_{BR}$$

$$= (12.56 + j0) - (-4.18 + j7.24) = 16.74 - j7.24 \text{ or } 18.25 \text{ A – in magnitude}$$

$$\mathbf{I}_Y = \mathbf{I}_{YR} + \mathbf{I}_{YB} = \mathbf{I}_{YB} - \mathbf{I}_{RY}$$

$$= (-2.09 - j3.62) - (12.56 + j0) = -14.65 - j3.62 \text{ or } 15.08 \text{ A – in magnitude}$$

$$\mathbf{I}_B = \mathbf{I}_{BR} + \mathbf{I}_{BY} = \mathbf{I}_{BR} - \mathbf{I}_{YB}$$

$$= (-4.18 + j7.24) - (-2.09 - j3.62) = -2.09 + j10.86 \text{ or } 11.06 \text{ A –in magnitude}$$

(*b*) Geometrical Method

Here, $R_1 = 5\Omega$, $R_2 = 10\Omega$; and $R_3 = 15\Omega$; $Y_1 = 1/5$S; $Y_2 = 1/10$S; $Y_3 = 1/15$S

$$x = \frac{V}{2}(Y_1 - Y_2)/(Y_1 + Y_2 + Y_3)$$

$$= \frac{230}{2}\left(\frac{1}{5} - \frac{1}{10}\right)\Big/\left(\frac{1}{5} + \frac{1}{10} + \frac{1}{15}\right) = 31.4$$

$$y = \left(\frac{\sqrt{3}V}{2}.Y_3\right)\Big/(Y_1 + Y_2 + Y_3) = (\sqrt{3} \times 115 \times 1/15)/(11/30) = 36.2$$

$$V_{RN} = \left(\frac{V}{2} - x\right) - jy = (115 - 31.4) - j36.2 = 83.6 - j36.2$$

$$V_{YN} = -\left(\frac{V}{2}+x\right) - jy = -146.4 - j36.2$$

$$V_{BN} = -x + j\left(\frac{\sqrt{3}V}{2} - y\right) = -31.4 + j163$$

$$\mathbf{I}_R = \mathbf{V}_{RN}\,\mathbf{Y}_1 = (83.6 - j36.2) \times 1/5 = 16.72 - j7.24$$

$$\mathbf{I}_Y = \mathbf{V}_{YN}\,\mathbf{Y}_2 = (-146.4 - j36.2) \times 1/10 = -14.64 - j3.62$$

$$\mathbf{I}_B = \mathbf{V}_{BN}\,\mathbf{Y}_3 = (-31.4 + j163) \times 1/15 = -2.1 \times j10.9$$

These are the same currents as found before.

Example 16.64. *A 3 – ϕ, 4-wire, 400-V symmetrical system supplies a Y-connected load having following branch impedances:*

$$Z_R = 100\Omega, Z_Y = j10\Omega \text{ and } Z_B = -j10\Omega$$

Compute the values of load phase voltages and currents and neutral current. Phase sequence is RYB.

How will these values change in the event of an open in the neutral wire?

Solution. (*a*) When Neutral Wire is Intact. [Fig 16.98 (*a*)]. The *load* phase voltages would be the same as *supply* phase voltages despite imbalance in the load. The three load phase voltages are:

$$\mathbf{V}_R = 231\angle 0°, \mathbf{V}_Y = 231\angle -120° \text{ and } \mathbf{V}_B = 231\angle 120°$$

$$\mathbf{I}_R = 231\angle 0° / 100\angle 0° = 2.31\angle 0° = 2.31 + j0$$

$$\mathbf{I}_Y = 231\angle 120° / 10\angle 90° = 2.31\angle -210° = -20 + j11.5$$

$$\mathbf{I}_B = 231\angle 120° / 10\angle -90° = 23.1\angle 210° = -20 - j11.5$$

$$\mathbf{I}_N = -(\mathbf{I}_R + \mathbf{I}_Y + \mathbf{I}_B) = -(2.31 - 20 + j11.5 - 20 - j11.5) = 37.7\text{ A}$$

(*b*) When Neutral is Open [Fig.16.77 (*b*)]

In this case, the load phase voltages will be no longer equal. The node pair voltage method will be used to solve the question. Let the supply phase voltages be given by

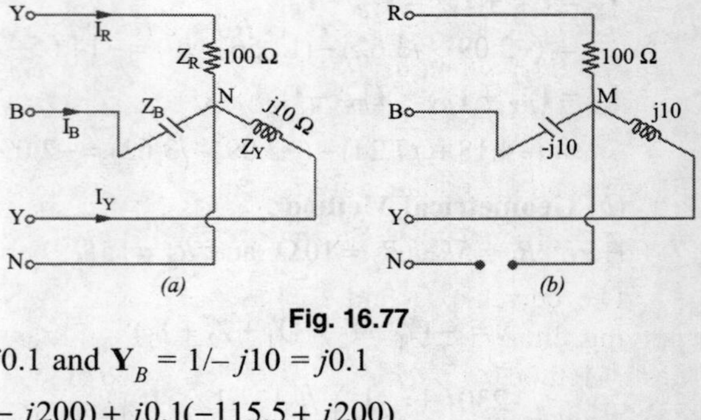

$$\mathbf{E}_R = 231\angle 0°, \mathbf{E}_Y = 231\angle -120° = -115.5 - j200$$

$$\mathbf{E}_B = 231\angle 120° = -115{\cdot}5 + j200$$

Fig. 16.77

$$\mathbf{Y}_R = 1/100 = 0.01;\ \mathbf{Y}_Y = 1/j10 = -j0.1 \text{ and } \mathbf{Y}_B = 1/-j10 = j0.1$$

$$\mathbf{V}'_{NN} = \frac{231 \times 0.01 + (-j0.1)(-115.5 - j200) + j0.1(-115.5 + j200)}{0.01 + (-j0.1) + j0.1} = -3769 + j0$$

The load phase voltages are given by

$$V'_R = E_R - V'_{NN} = (231 + j0) - (-3769 + j0) = 4000\text{ V}$$

$$\mathbf{V}'_Y = \mathbf{E}_Y - \mathbf{V}'_{NN} = -115.5 - j200 - (-3769 + j0) = (3653.5 - j200)$$

$$\mathbf{V}_B' = \mathbf{E}_B - \mathbf{V}'_{NN} = -115.5 + j200 - (-3769 + j0) = (3653.5 + j0)$$

$$\mathbf{I}_R = \mathbf{V}_R{}'\mathbf{Y}_R = 4000 \times 0.01 = -40\text{ A}$$

$$\mathbf{I}_Y = (-j0.1)(3653.5 - j200) = (20 - j3653.5)$$

$$\mathbf{I}_B = (j0.1)(3653.5 + j200) = -20 + j3653.5$$

Obviously, the neutral current will just not exist.

Note. The load phase voltages and currents become abnormally high.

Example 16.65. *For the circuit shown in Fig. 16.78 find the readings on the two wattmeters W_a and W_c.*

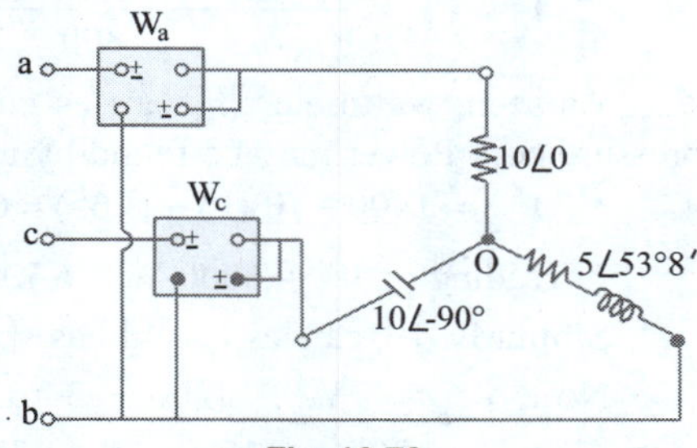

Fig. 16.78

Solution. The three line currents for this problem have already been determined.

$$\mathbf{I}_{aa} = 20.02 - j\,4.54$$

$$\mathbf{I}_{bo} = -7.24 - j\,5.48$$

$$\mathbf{I}_{co} = -12.78 + j\,10.12$$

The line voltages are given by

$$\mathbf{V}_{ab} = 200 + j\,0$$

$$\mathbf{V}_{bc} = -\,100 - j\,173.2$$

$$\mathbf{V}_{ca} = -\,100 + j\,173.2$$

Wattmeter W_a carries a current of using $\mathbf{I}_{ao} = 20.02 - j\,4.54$ and has voltage $\mathbf{V}_{ab}$ impressed across its pressure coil. Power can be found by using current conjugate.

$$\mathbf{P}_{VA} = (200 + j\,0)\,(20.02 + j\,4.54) = (200)\,(20.02) + j\,(200)(4.54)$$

Actual power = 200 × 20.02 = 4004 W ∴ W_a = **4004 W**

The other wattmeter W_c carries current of $\mathbf{I}_{co} = -\,12.78 + j\,10.02$ and has a voltage $\mathbf{V}_{cb} = -\,\mathbf{V}_{bc} = 100 + j\,173.2$ impressed across it. By the same method, wattmeter reading is

$$W_c = (100 \times -12.78) + (173.2 \times 10.02) = -1278 + 1735.5 = \mathbf{457.5\ W}$$

Example 16.66. *Three resistors 10, 20 and 20 Ω are connected in star to the terminals A, B and C of a 3 – φ, 3 wire supply through two single-phase wattmeters for measurement of total power with current coils in lines A and C and pressure coils between A and B and C and B. Calculate **(i)** the line currents **(ii)** the readings of each wattmeter.*

The line voltage is 400-V. **(Electrical Engineering-I, Bombay Univ.)**

Solution. Let $\mathbf{V}_{AB} = 400\ \angle 0°$; $\mathbf{V}_{BC} = 400\ \angle{-120°}$; and $\mathbf{V}_{CA} = 400\ \angle 120°$

As shown in Fig. 16.79, current through wattmeter W_1 is $\mathbf{I}_{AO}$ or $\mathbf{I}_A$ and that through W_2 is $\mathbf{I}_{CO}$ or $\mathbf{I}_C$ and the voltages are $\mathbf{V}_{AB}$ and $\mathbf{V}_{CB}$ respectively. Obviously,

$$\mathbf{Z}_A = 10\angle 0°;\ \mathbf{Z}_B = 20\angle 0°,\ \mathbf{Z}_C = 20\angle 0°$$

The currents $\mathbf{I}_A$ and $\mathbf{I}_C$ may be found by applying either Kirchhoff's laws or Maxwell's Mesh Method. Both methods will be used for illustration.

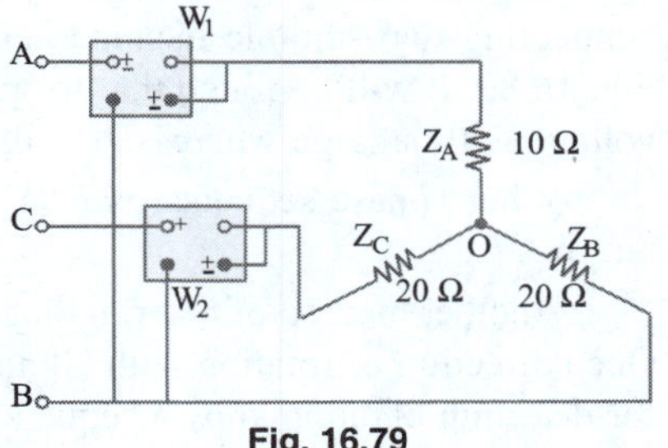

Fig. 16.79

(a) From Eq. (10), (11) and (12) of Art. 16.33, we have

$$\mathbf{I}_A = \frac{400 \times 20 - 20(-200 + j346)}{(10 \times 20) + (20 \times 20) + (20 \times 10)}$$

$$= \frac{12{,}000 - j6{,}920}{800} = 15 - j8.65\text{ A}$$

$$\mathbf{I}_C = \frac{20(-200+j346)-10(-200-j346)}{800} = \frac{-2000+j10,380}{800} = -2.5+j13$$

(*b*) From Eq. (*i*) and (*ii*) of solved example 17.48 (*c*) we get

$$\mathbf{I}_A = \mathbf{I}_1 = \frac{400\times 40+20(-200-j346)}{30\times 40-20^2} = 15-j8.65\text{ A}$$

$$\mathbf{I}_C = -\mathbf{I}_2 = \frac{30\times(-200-j346)+400\times 20}{800} = -25+j13$$

As seen, wattmeter W_1 carries current $\mathbf{I}_A$ and has a voltage $\mathbf{V}_{AB}$ impressed across its pressure coil. Power may be found by using voltage conjugate.

$$\mathbf{P}_{VA} = (400-j0)(15-j8.65) = 6000-j3,460$$

$\therefore$ reading of W_1 = 6000 W = **6 kW**

Similarly, W_2 carries $\mathbf{I}_C$ and has voltage $\mathbf{V}_{CB}$ impressed across its pressure coil.

Now, $\mathbf{V}_{CB} = -\mathbf{V}_{BC}$ = (200 + *j*346). Using voltage conjugate, we get

$$\mathbf{P}_{VA} = (200-j346)(-2.5+j13)$$

Real power = (200 × – 2.5) + (13 × 346) = 4000 W

$\therefore$ reading of W_2 = **4kW;** Total power = 10 kW

16.34. Phase Sequence Indicators

In unbalanced 3-wire star-connected loads, phase voltages change considerably if the phase sequence of the supply is reversed. One or the other load phase voltage becomes dangerously large which may result in damage to the equipment. Some phase voltage becomes too small which is equally deterimental to some types of electrical equipment. Since phase voltage depends on phase sequence, this fact has been made the basis of several types of phase sequence indicators. *A simple phase sequence indicator may be made by connecting two suitable incandescent lamps and a capacitor in a *Y*-connection as shown in Fig. 16.80. It will be found that for phase sequence *RYB*, lamp L_1 will glow because its phase voltage will be large whereas L_2 will not glow because of low voltage across it.

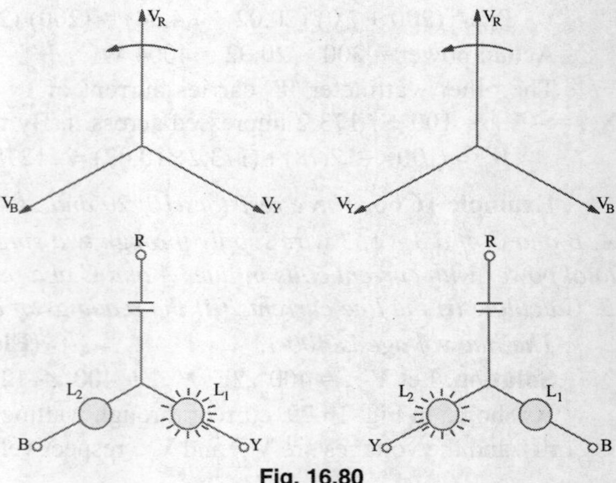

Fig. 16.80

When, phase sequence is *RBY*, opposite conditions develop so that this time L_2 glows but not L_1.

Another method of determining the phase sequence is by means of a small 3-phase motor. Once direction of rotation with a known sequence is found, the motor may be used thereafter for determining an unknown sequence.

* It may, however, be noted that phase sequence of currents in an unbalanced load is not necessarily the same as the voltage phase sequence. Unless indicated otherwise, voltage phase sequence is implied.

Tutorial Problem 16.3

1. Determine the line currents in an unbalanced Y connected load supplied from a symmetrical 3 – ϕ, 440-V, 3-wire system. The branch impedances of the load are : $\mathbf{Z}_1 = 5\ \angle 30°$ ohm, $\mathbf{Z}_2 = 10\ \angle 45°$ ohm and $\mathbf{Z}_3 = 10\ \angle 45°$ ohm and $\mathbf{Z}_4 = 10\ \angle 60°$ ohm. The sequence is *RYB*.
[35.7 A, 32.8 A; 27.7 A]

2. A 3 – ϕ, *Y*-connected alternator supplies an unbalanced load consisting of three impedances (10 + *j*20), (10 – *j*20) and 10Ω respectively, connected in star. There is no neutral connection. Calculate the voltage between the star point of the alternator and that of the load. The phase voltage of the alternator is 230 V. **[–245.2 V]**

3. Non-reactive resistors of 10, 20 and 25 Ω are star-connected to the *R*, *Y* are *B* phases of a 400-V, symmetrical system. Determine the current and power in each resistor and the voltage between star point and neutral. Phase sequence, *RYB*.
[16.5 A, 2.72 kW ; 13.1 A, 3.43 kW; 11.2 A, 3.14 kW ; 68 V]

4. Determine the line current in an unbalanced, star-connected load supplied from a symmetrical 3-phase, 440-V system. The branch impedance of the load are $\mathbf{Z}_R = 5\ \angle 30°\ \Omega$, $Z_Y = 10\ \angle 45°\ \Omega$ and $Z_B = 10\ \angle 60°\ \Omega$. The phase sequence is *RYB*. **[35.7 A, 32.8 A, 27.7 A]**

5. Three non-reactive resistors of 3, 4 and 5-Ω respectively are star-connected to a 3-phase, 400-V symmetrical system, phase sequence *RYB*. Find (*a*) the current in each resistor (*b*) the power dissipated in each resistor (*c*) the phase angles between the currents and the corresponding line voltages (*d*) the star-point potential. Draw to scale the complete vector diagram.
[(*a*) 66.5 A, 59.5 A, 51.8 A (*b*) 13.2, 14.15, 13.4 kW (*c*) 26°24′, 38°10′, 25°20′ (*d*) 34 V]

6. An unbalanced *Y*-connected load is supplied from a 400-V, 3 – ϕ, 3-wire symmetrical system. The branch circuit impedances and their connection are (2 + *j*2) Ω, *R* to *N* ; (3 – *j*3) Ω, *Y* to *N* and (4 + *j*1) Ω, *B* to *N* of the load. Calculate (*i*) the value of the voltage between lines *Y* and *N* and (*ii*) the phase of this voltage relative to the voltage between line *R* and *Y*. Phase sequence *RYB*.
[(*i*) (–216–j 135.2) or 225.5 V (*ii*) 2° or –178°]

7. Three non-reactive resistance of 5, 10 and 5 ohm are star-connected across the three lines of a 230-V 3-phase, 3-wire supply. Calculate the line currents.
[(18.1 + J21.1) A ; (– 10.9–j 10.45) A ; (– 7.3 + j8.4) A]

8. A 3 – ϕ, 400-V symmetrical supply feeds a star-connected load consisting of non-reactive resistors of 3, 4 and 5 Ω connected to the *R*, *Y* and *B* lines respectively. The phase sequence is *RYB*. Calculate (*i*) the load star point potential (*ii*) current in each resistor and power dissipated in each resistor. **[(*i*) 34.5 V (*ii*) 66.4 A, 59.7 A, 51.8 A (*iii*) 13.22 kW, 14.21 kW, 13.42 kW]**

9. A 20-Ω resistor is connected between lines *R* and *Y*, a 50-Ω resistor between lines *Y* and *B* and a 10-Ω resistor between lines *B* and *R* of a 415-V, 3-phase supply. Calculate the current in each line and the reading on each of the two wattmeters connected to measure the total power, the respective current coils of which a connected in lines *R* and *Y*.
[(25.9 – j9); (– 24.9 – j7.2); (– 1.04 + j16.2); 8.6 kW; 7.75 kW]

10. A three-phase supply, giving sinusoidal voltage of 400 V at 50 Hz is connected to three terminals marked *R*, *Y* and *B*. Between *R* and *Y* is connected a resistance of 100 Ω, between *Y* and *B* an inductance of 318 mH and negligible resistance and between *B* and *R* a capacitor of 31.8 μF. Determine (*i*) the current flowing in each line and (*ii*) the total power supplied. Determine (*iii*) the resistance of each phase of a balanced star-connected, non-reactive load, which will take the same total power when connected across the same supply.
[(*i*) 7.73 A, 7.73 A, (*ii*) 1,600 W (*iii*) 100 Ω (*London Univ.*)]

11. An unbalanced, star-connected load is fed from a symmetrical 3-phase system. The phase voltages across two of the arms of the load are $V_B = 295\ \angle 97°\ 30'$ and $V_R = 206\ \angle -25°$. Calculate the voltage between the star-point of the load and the supply neutral. **[52.2 ∠–49.54']**

12. A symmetrical 440-V, 3-phase system supplies a star-connected load with the following branch impedances: $Z_R = 100\ \Omega$, $Z_Y = j\,5\ \Omega$, $Z_B = -j5\ \Omega$. Calculate the voltage drop across each branch and the potential of the neutral point to earth. The phase sequence is *RYB*. Draw the vector diagram. [$8800\ \angle -30°$, $8415\angle -31.5°$, $8420 - \angle -28.5°$, $8545\angle 150°$]

13. Three star-connected impedances, $Z_1 = (20 + j37.7)\ \Omega$ per phase are in parallel with three delta-connected impedances, $Z_2 = (30 - j159.3)\ \Omega$ per phase. The line voltage is 398 V. Find the line current, power factor, power and reactive volt-amperes taken by the combination.

[3.37 ∠10.4°; 0.984 lag; 2295 lag; 2295 W; 420 VAR.]

14. A 415 V, 50 Hz, 3-phase supply of phase sequence *RYB* is connected to a delta connected load in which branch *RY* consists of $R_1 = 100\ \Omega$, branch Y_B consists of $R_2 = 20\ \Omega$ in series with $X_2 = 60\ \Omega$ and branch *BR* consists of a capacitor $C = 30\ \Omega$. Take V_{RY} as the reference and calculate the line currents. Draw the complete phasor diagrams.

(Elect. Machines, A.M.I.E. Sec. B, 1989)

15. Three resistances of 5, 10 and 15 Ω are connected in delta across a 3-phase supply. Find the values of the three resistors, which if connected in star across the same supply, would take the same line currents.

If this star-connected load is supplied from a 4-wire, 3-phase system with 260 V between lines, calculate the current in the neutral. **[2.5 Ω, 1.67 Ω, 5 Ω; 52 A] *(London Univ.)***

16. Show that the power consumed by three identical phase loads connected in delta is equal to three times the power consumed when the phase loads are connected in star.

(Nagpur University, Summer 2002)

17. Prove, that the power consumed in balanced three- phase Delta-connected load is three times the power consumed in starconnected load. ***(Nagpur University, Winter 2002)***

18. A three-phase 230 volts systems supplies a total load of 2000 watts at a line current of 6 Amp when three identical impedances are in star-connection across the line terminals of the systems. Determine the resistive and reactive components of each impedance.

(Nagpur University, Winter 2002)

19. Three similar coils each of impedance z = (8 + j10) ohms, are connected in star and supplied from 3-phase 400V, 50 Hz supply. Find the line current, power factor, power and total volt amperes.

(Nagpur University, Summer 2003)

20. Three similar each having a resistance of 20 ohm and an inductance of 0.05 H are connected in star to a 3-phase 50 Hz supply with 400 V between lines. Calculate power factor, total power absorb and line current. If the same coil are reconnected in delta across the same supply what will be the power factor, total power absorbed and line current?

(Pune University 2003) (Nagpur University, Winter 2003)

21. A 3 φ star connected load when supplied from 440 V, 50 Hz source takes a line current of 12 amp lagging w.r.t. line voltage by 70^0. Calculate : (*i*) limpedance parameters (*ii*) Power factor and its nature (*iii*) Draw phasor diagram indicating all voltages and currents.

(Nagpur University, Summer 2004)

22. Derive the relationship between line current and phase current for Delta connected 3 phase load when supplied from 3 phase balanced supply. ***(Nagpur University, Summer 2004)***

23. Derive the relationship between line voltage, phase voltage, line current and phase current in a 3 phase star connected and delta connected circuit. ***(Gujrat University, June/July 2003)***

24. Show that power input to a 3 phase circuit can be measured by two wattmeters connected properly in the circuit. Draw vector diagram. ***(Gujrat University, June/July 2003)***

25. A balanced 3 phase star connected load of 100 kW takes a leading current of 100 A when connected across a 3 phase, 1100 V, 50 Hz supply. Calculate the circuit constants of the load per phase.

(Mumbai University 2003) (Gujrat University, June/July 2003)

26. Establish relationship between line and phase voltages and currents in a balanced 3-phase star connection. Draw complete phasor diagram for voltages and currents.

(R.G.P.V. Bhopal University, June 2004)

27. A delta connected load has the following impedances :
$Z_{RY} = j\ 10\ \Omega$, $Z_{YB} = 10 \angle 0^o\ \Omega$ and $Z_{BR} = -j\ 10\ \Omega$. If the load is connected across 100 volt balanced 3-phase supply, obtain the line currents. *(R.G.P.V. Bhopal University, June 2004)*

28. Two wattmeters ω_1 and ω_2 are used to measure power in a 3 phase balanced circuit. Mention the conditions under which (*i*) $\omega_1 = \omega_2$ (*ii*) $\omega_2 = 0$ (*iii*) $\omega_1 = 2\omega_2$.

(V.T.U. Belgaum Karnataka University, February 2002)

29. Three coils each of impedance $20 \mid 60^0\Omega$ are connected across a 400V, 3 phase supply. Find the reading of each of the two wattmeters connected to measure the power when the coils are connected in (*i*) star (*ii*) Delta. *(V.T.U. Belgaum Karnataka University, February 2002)*

30. The power input to a 3 phase circuit was measured by two wattmeter method and the readings were 3400 and - 1200 watts respectively. Calculate the total power and powerfactor.

(V.T.U. Belgaum Karnataka University, July/August 2002)

31. Obtain the relationship between line and phase values of current in a three phase, balanced, delta connected system. *(V.T.U. Belgaum Karnataka University, January/February 2003)*

32. Show that in a three phase, balanced circuit, two wattmeters are sufficient to measure the total three phase power and power factor of the circuit.

(V.T.U. Belgaum Karnataka University, January/February 2003)

33. Each of the two wattmeters connected to measure the input to a three phase circuit, reads 20kW. What does each instrument reads, when the load p.f. is 0.866 lagging with the total three phase power remaining unchanged in the altered condition?

(V.T.U. Belgaum Karnataka University, January/February 2003)

34. Two wattmeters connected to measure power in a 3 phase circuit read 5KW and 1KW, the latter reading being obtained after reversing current coil connections. Calculate power factor of the load and the total power consumed.

(V.T.U. Belgaum Karnataka University, January/February 2003)

35. Derive the relationship between phase and line values of voltages in a connected load.

(V.T.U. Belgaum Karnataka University, January/February 2003)

36. Three coils each of impedance $20\angle 60^0\Omega$ are connected in delta across a 400, 3 phase, 50Hz, 50Hz Ac supply. Calculate line current and total power.

(V.T.U. Belgaum Karnataka University, January/February 2003)

37. What are the advantages of a three phase system over a single phase system?

(V.T.U. Belgaum Karnataka University, July/August 2003)

38. With a neat circuit diagram and a vector diagram prove that two wattmeters are sufficient to measure total power in a 3 phase system.

(V.T.U. Belgaum Karnataka University, July/August 2003)

39. Two watt meters are used to measure the power delivered to a balance 3 phase load of power factor 0.281. One watt meter reads 5.2kW. Determine the reading of the second watt meter. What is the line current if the line voltage is 415 wolt?

(V.T.U. Belgaum Karnataka University, January/February 2004)

40. Write the equations for wattmeter reading W_1 and W_2 in 3 phase power measurement and therefrom for power factor. *(Anna University, October/November 2002)*

41. Show that the wattmeters will read equal in two wattmeter method under unity power factor loading condition. *(Anna University, November/December 2003)*

42. A star connected 3-phase load has a resistance of 6Ω and an inductive reactance of 8Ω in each brance. Line voltage is 220 volts. Write the phasor expressions for voltage across each branch, line voltages and line currents. Calculate the total power.

(Anna University, November/December 2003)

43. Two wattmeters connected to measure the total power in a 3-phase balanced circuit. One measures 4,800 W, while the other reads backwards. On reversing the latter it is found to read 400 W. What is the total power and power and power factor? Draw the connection diagram and phasor diagram of the circuit. ***(Mumbai University 2003) (RGPV Bhopal 2001)***

44. Explain with diagram the measurement of 3-phase power by two-wattmeter method. ***(RGPV Bhopal 2002)***

45. Show that the power taken by a 3-phase circuit can measured by two wattmeters connected properly in the circuit. ***(RGPV Bhopal)***

46. With the aid of star-delta connection diagram, state the basic equation from which star-delta conversionequation canbe derived. ***(Pune University, 2003) (RGPV Bhopal 2001)***

47. Star-delta connections in a 3-phase supply and their inter-relationship. ***(RGPV Bhopal 2001)***

48. Measurement of power in three-phase circuit in a balanced condition. ***(RGPV Bhopal 2001)***

49. Measurement of reactive power in three-phase circuit. ***(RGPV Bhopal 2001)***

50. Differentiate between balanced and unbalancedthree-phase supply and balanced and unbalanced three-phase load. ***(RGPV Bhopal June 2002)***

51. A 3-phase 3 wire supply feeds a load consisting of three equal resistors. By how much is the load reduced if one of the resistors be removed ? ***(RGPV Bhopal June 2002)***

52. Establish relationship between line and phase voltages and currents in a balanced delta connection. Draw complete phasor diagram of voltages and currents. ***(RGPV Bhopal December 2003)***

OBJECTIVE TESTS – 16

1. The minimum number of wattmeter (s) required to measure 3-phase, 3-wire balanced or unbalanced power is
(*a*) 1 (*b*) 2
(*c*) 3 (*d*) 4 ***(GATE 2001)***

2. A wattmeter reads 400 W when its current coil is connected in the R phase and its pressure coil is connected between this phase and the neutral of a symmetrical 3-phase system supplying a balanced star connected 0.8 p.f. inductive load. The phase sequence is RYB. What will be the reading of this wattmeter if its pressure coil alone is reconnected between the B and Y phases, all other connections remaining as before?
(*a*) 400.0 (*b*) 516.6
(*c*) 300.0 (*d*) 692.8
(GATE 2003)

3. Total instantaneous power supplied by a 3-phase ac supply to a balanced R-L load is
(*a*) zero
(*b*) constant
(*c*) pulsating with zero average
(*d*) pulsating with non-zero average
(GATE 2004)

4. A balanced 3-phase, 3-wire supply feeds balanced star connected resistors. If one of the resistors is disconnected, then the percentage reduction in the load will be
(*a*) $33\frac{1}{3}$ (*b*) 50
(*c*) $66\frac{2}{3}$ (*d*) 75 ***(GATE)***

5. The voltages induced in the three windings of a 3-phase alternator are----degree apart in time phase.
(*a*) 120 (*b*) 60
(*c*) 90 (*d*) 30

6. If positive phase sequence of a 3-phase load is *abc*, the negative sequence would be
(*a*) *bac* (*b*) *cba*
(*c*) *acb* (d) all of the above

7. In the balanced 3-phase voltages system generated by a *Y*-connected alternator, V_{YB} lags E_R by ________ electrical degrees.
(*a*) 90 (*b*) 120
(*c*) 60 (*d*) 30

8. The power taken by a 3-ϕ load is given by the expression
(*a*) $3V_L I_L \cos\phi$ (*b*) $\sqrt{3}\, V_L I_L \cos\phi$
(*c*) $3V_L I_L \sin\phi$ (*d*) $\sqrt{3}\, V_L I_L \sin\phi$

9. In a balanced 3-phase voltage generator, the different phase voltages reach their maximum values __________ degree apart.

(*a*) 120 (*b*) 60

(*c*) 240 (*d*) 30

10. If the B-phase of a 3-phase, *Y*-connected alternator become reverse connected by mistake, it will not affect

(*a*) V_{YB} (*b*) V_{RY}

(*c*) V_{BR} (*d*) V_{BY}

11. Three equal impedances are first connected in star across a balanced 3-phase supply. If connected in delta across the same supply.

(*a*) phase current will be tripled

(*b*) phase current will be doubled

(*c*) line current will become on-third

(*d*) power consumed will increase three-fold.

12. A 3-phase, 4-wire, 230/440-V system is supplying lamp load at 230 V. If a 3-phase motor is now switched on across the same supply, then

(*a*) neutral current will increase

(*b*) all line currents will decrease

(*c*) neutral current will remain unchanged

(*d*) power factor will be improved.

13. Power factor improvement

(*a*) does not affect the performance characteristics of the original load

(*b*) employs series resonance

(*c*) increase the active power drawn by the load

(*d*) increases the reactive power taken by the load

14. The chief disadvantage of a low power factor is that

(*a*) more power is consumed by the load

(*b*) current required for a given load power is higher

(*c*) active power developed by a generator exceeds its rated output capacity

(*d*) heat generated is more than the desired amount

15. In the 2-wattmeter method of measuring 3-phase power, the two wattmeters indicate equal and opposite readings when load power factor angle is ________ degrees lagging.

(*a*) 60 (*b*) 0

(*c*) 30 (*d*) 90

16. When phase sequence at the 3-phase load is reversed

(*a*) phase powers are changed

(*b*) phase currents are changed

(*c*) phase currents change in angle but not in magnitude

(*d*) total power consumed is changed.

17. Phase reversal of a 4-wire unbalanced load supplied from a balanced 3-phase supply changes.

(*a*) magnitude of phase currents

(*b*) magnitudes as well as phase angle of neutral current

(*c*) the power consumed

(*d*) only the magnitude of neutral current.

CHAPTER

17 Harmonics

Chapter Outline

- Fundamental Wave and Harmonics
- General Equation of a Complex Wave
- R.M.S. Value of a Complex Wave
- Form Factor of a Copmplex Wave
- Power Supplied by a Complex Wave
- Harmonics in Single-phase A.C. Circuits
- Selective Resonance Due to Harmonics
- Effect of Harmonics on Measurement of Inductance and Capacitance
- Harmonics in Single and 3-Phase Transformers

17.1. Fundamental Wave and Harmonics

Upto this stage, while dealing with alternating voltages and currents, it has been assumed that they have sinusoidal waveform or shape. Such a waveform is an ideal one and much sought after by the manufacturers and designers of alternators. But it is nearly impossible to realize such a waveform in practice. All the alternating waveforms deviate, to a greater or lesser degree, from this ideal sinusoidal shape. Such waveforms are referred to as ***non-sinusoidal*** or ***distorted*** or ***complex waveforms***.

Complex waveforms are produced due to the superposition of sinusoidal waves of different frequencies. Such waves occur in speech, music, TV, rectifier outputs and many other applications of electronics. On analysis, it is found that a complex wave essentially consists of

(*a*) ***a fundamental*** wave – it has the lowest frequency, say 'f'

(*b*) a number of other sinusoidal waves whose frequencies are an integral multiple of the fundamental or basic frequency like $2f$, $3f$ and $4f$ etc.

The fundamental and its higher multiples form a ***harmonic*** series.

As shown in Fig. 17.1, fundamental wave itself is called the ***first*** harmonic. The ***second*** harmonic has frequency ***twice*** that of the fundamental, the ***third*** harmonic has frequency ***thrice*** that of the fundamental and so on.

Waves having frequencies of $2f$, $4f$ and $6f$ etc. are called ***even harmonics*** and those having frequencies of $3f$, $5f$ and $7f$ etc. are called *odd* harmonics. Expressing the above is angular frequencies, we may say that successive odd harmonics have frequencies of 3ω, 5ω and 7ω etc. and even harmonics have frequencies of 2ω, 4ω and 6ω etc.

As mentioned earlier, harmonics are introduced in the output voltage of an alternator due to many reason such as the irregularities of the flux distribution in it. Considerations of

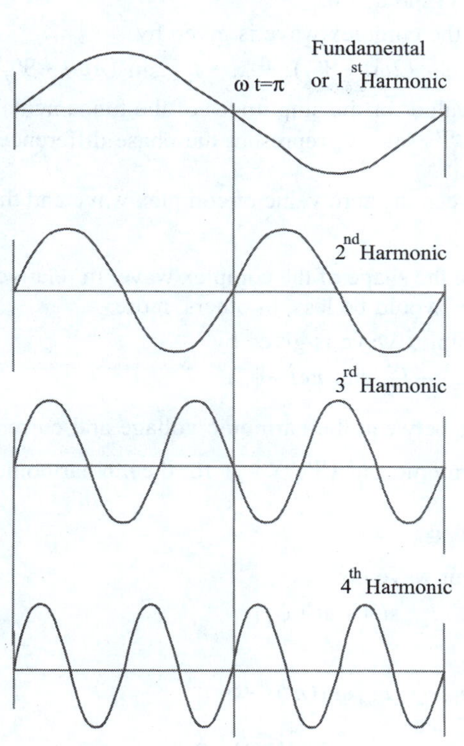

Fig. 17.1

Complex Wave
Fundamental
2nd Harmoni

Fig. 17.2

waveform and form factor are very important in the transmission of a. c. power but they are of much greater importance in radio work where the intelligibility of a signal is critically dependent on the faithful transmission of the harmonic structure of sound waves. In fact, it is only the rich harmonic content of the consonants and lesser at still plentiful harmonic content of vowels which helps the ear to distinguish a well regulated speech from a more rhythmical succession of musical sounds.

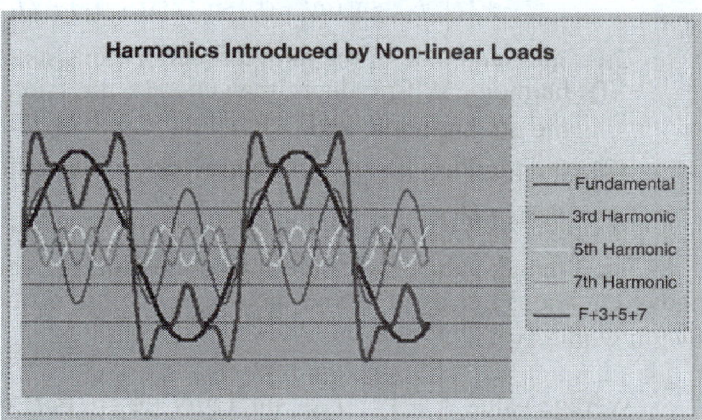

17.2. General Equation of a Complex Wave

Consider a complex wave which is built up of the fundamental and a few harmonics, each of which has its own peak value of phase angle. The fundamental may be represented by

$$e_1 = E_{1m} \sin(\omega t + \Psi_1)$$

the second harmonic by $e_2 = E_{2m} \sin(2\omega t + \Psi_2)$

the third harmonic by $e_3 = E_{3m} \sin(3\omega t + \Psi_3)$ and so on.

The equation for the instantaneous value of the complex wave is given by

$$e = e_1 + e_2 + ... e_n = E_{1m} \sin(\omega t + \Psi_1) \times E_{2m} \sin(2\omega t + \Psi_2) + ... + I_{nm} \sin(n\omega t + \Psi_n)$$

when E_{1m}, E_{2m} and E_{nm} etc. denote the maximum values or the amplitudes of the fundamental, second harmonic and nth harmonic etc. and Ψ_1, Ψ_2 and Ψ_n represent the phase differences

with respect to the complex wave* (*i.e.* angle between the zero value of complex wave and the corresponding zero value of the harmonic).

The number of terms in the series depends on the shape of the complex wave. In relatively simple waves, the number of terms in the series would be less, in others, more.

Similarly, the instantaneous value of the complex wave is given by

$$i = I_{1m} \sin(\omega t + \phi_1) + I_{2m} \sin(2\omega t + \phi_2) + ... + I_{nm} \sin(n\omega t + \phi_n)$$

Obviously $(\Psi_1 - \phi_1)$ is the phase difference between the harmonic voltage and current for the fundamental, $(\Psi_2 - \phi_2)$ for the second harmonic and $(\Psi_n - \Phi_2)$ for the nth harmonic.

17.3. R.M.S. Value of a Complex Wave

Let the equation of the given complex current wave be

$$i = I_{1m} \sin(\omega t + \phi_1) + I_{2m} \sin(2\omega t + \phi_2) + ...I_{nm} \sin(n\omega t + \phi_n)$$

Its r.m.s. value is given by

Now $i^2 = [I_{1m} \sin(\omega t + \phi_1) + I_{2m} \sin(2\omega t + \phi_1) + ...I_{nm} \sin(n\omega t + \phi_n)]^2$

$$= I_{1m}{}^2 \sin^2(\omega t + \phi_1) + I_{2m}{}^2 \sin^2(2\omega t + \phi_2) + ...I_{nm}{}^2 \sin^2(n\omega t + \phi_n)]^2$$

$$+2I_{1m}I_{2m} \sin(\omega t + \phi_1)\sin(2\omega t + \phi_2) + 2I_{1m}I_{3m} \sin(\omega t + \Phi_1)\sin(\omega t + \Phi_2) + ...$$

The right-hand side of the above equation consists of two types of terms

(*i*) harmonic self-products, the general expression for which is $\mathbf{I}_{pm}{}^2 \sin^2(p\omega t + \phi_p)$ for the pth harmonic and

(*ii*) the products of different harmonics of the general form $2I_{pm}I_{qm} \sin(p\omega t + \phi_n) \sin(q\omega t + \phi_q)$

The average value of i^2 is the sum of the average values of these individual terms in the above equation. Let us now find the average value of the general term $I_{pm}{}^2 \sin^2(p\omega t + \phi_p)$ over a whole cycle.

$$\text{Average value} = \frac{1}{\pi}\int_0^{2\pi} I_{pm}{}^2 \sin^2(p\omega t + \phi_p)\, d(\omega t) = \frac{I_{pm}{}^2}{2\pi}\int_0^{2\pi} \sin^2(p\theta + \phi_p)\, d\theta$$

$$= \frac{I_{pm}{}^2}{2\pi}\int_0^{2\pi}\left(\frac{1 - \cos 2(p\theta + \phi_p)}{2}\right) d\theta = \frac{I_{pm}{}^2}{2\pi}\left|\theta - \sin^2(p\theta + \phi_p)\right|_0^{2\pi}$$

$$= \frac{I_{pm}{}^2}{4\pi} \times 2\pi = \frac{I_{pm}{}^2}{2}$$

* We could also express these phase angles with respect to the fundamentals wave instead of the complex wave.

From this result, we can generalize that

Average value of $I_{1m}{}^2 \sin^2(\omega t + \phi_1) = \dfrac{I_{1m}{}^2}{2}$

Average value of $I_{2m}{}^2 \sin^2(2\omega t + \phi_2) = \dfrac{I_{2m}{}^2}{2}$

Average value of $I_{nm}{}^2 \sin^2(n\omega t + \phi_n) = \dfrac{I_{nm}{}^2}{2}$ and so on.

Now, the average value of the product terms is

$$= \frac{1}{2\pi}\int_0^{2\pi} I_{pm} I_{qm} \sin(p\omega t + \phi_p)\sin(q\omega t + \phi_q)\, d(\omega t)$$

$$= \frac{I_{pm} I_{qm}}{5\pi}\int_0^{2\pi} \sin(p\theta + \phi_p)\sin(q\theta + \phi_q)\, d\theta = 0$$

$\therefore$ Average value of $i^2 = \dfrac{I_{1m}{}^2}{2} + \dfrac{I_{2m}{}^2}{2} + ... + \dfrac{I_{nm}{}^2}{2}$

$\therefore$ r.m.s. value, $I = \sqrt{\text{average value of } i^2} = \sqrt{\dfrac{I_{1m}{}^2}{2} + \dfrac{I_{2m}{}^2}{2} + ... \dfrac{I_{nm}{}^2}{2}}$ ... **(*i*)**

$$= 0.707\sqrt{I_{1m}{}^2 + I_{2m}{}^2 + ... + I_{nm}{}^2}$$

Equation **(*i*)** above may also be put in the form

$$I = \sqrt{\left(\frac{I_{1m}}{\sqrt{2}}\right)^2 + \left(\frac{I_{2m}}{\sqrt{2}}\right)^2 + ... + \left(\frac{I_{nm}}{\sqrt{2}}\right)^2} = \sqrt{I_1^2 + I_2^2 + ...I_n^2}$$

where $I_1 = I_{1m}/\sqrt{2}$ – r.m.s. value of fundamental

$I_2 = I_{2m}/\sqrt{2}$ – r.m.s. value of 2nd harmonic

$I_n = I_{nm}/\sqrt{2}$ – r.m.s. value of *n*th harmonic

Similarly, the r.m.s. value of a complex voltage wave is

$$E = 0.707\sqrt{E_{1m}{}^2 + E_{2m}{}^2 + ...E_{nm}{}^2} = \sqrt{E_1{}^2 + E_2{}^2 + ...E_n{}^2}$$

Hence, the **rule *is that the r.m.s. value of the complex current (or voltage) wave is given by the square-root of the sum of the squares of the r.m.s. values of its individual components.***

Note. If complex current wave contains a d.c. component of constant value I_D then its equation is given by

$$i = I_D + I_{1m}\sin(\omega t + \phi_1) + I_{2m}\sin(2\omega t + \phi_2) + ...I_{nm}\sin(n\omega t + \phi_n)$$

r.m.s. value, $I = \sqrt{I_D{}^2 + (I_{1m}/\sqrt{2})^2 + (I_{2m}/\sqrt{2})^2 .....(I_{nm}/\sqrt{2})^2} = I_D{}^2 + I_1{}^2 + I_2{}^2 + ...I_n{}^2$

17.4. Form Factor of a Complex Wave

In general, it may be defined as $k_f = \dfrac{\text{R.M.S. value}}{\text{average value}}$

A general expression for form factor in some simple cases may be found as under :

(*i*) Sine Series. Suppose the equation of a complex voltage wave is

$$v = V_{1m}\sin\omega t \pm V_{3m}\sin 3\omega t \pm V_{5m}\sin 5\omega t$$

$$= V_{1m}\sin\theta \pm V_{3m}\sin 3\theta \pm V_{5m}\sin 5\theta \quad \text{where} \quad \omega = 2\pi/T.$$

Obviously, zeros occurs at $t = 0$ or at $\theta = 0°$ and $\theta = 180°$ or $t = T/2$.
Mean value over half-cycle is

$$V_{av} = \frac{1}{\pi}\int_0^{\pi} v d\theta$$

$$= \frac{1}{\pi}\left[V_{1m}\int_0^{\pi}\sin\theta.d\theta \pm V_{3m}\int_0^{\pi}\sin 3\theta.d\theta \pm V_{5m}\int_0^{\pi}\sin 5\theta.d\theta\right] = \frac{2}{\pi}\left(\frac{V_{1m}}{1} \pm \frac{V_{3m}}{3} \pm \frac{V_{5m}}{5}\right)$$

As found in Art. 17.3,

$$V = (V_1^2 + V_3^2 + V_5^2)^{1/2} = \frac{1}{\sqrt{2}}(V_{1m}^{\ 2} + V_{3m}^{\ 2} + V_{5m}^{\ 2})^{1/2}$$

$$\therefore \quad k_f = \frac{(1/\sqrt{2})(V_{1m}^{\ 2} + V_{3m}^{\ 2} + V_{5m}^{\ 2})^{1/2}}{\left(\frac{2}{\pi}\right)\left(V_{1m} \pm \frac{V_{3m}}{3} \pm \frac{V_{5m}}{5}\right)}$$

(*ii*) **Cosine Series.** Consider the following cosine series:

$$v = V_{1m}\cos\omega t \pm V_{3m}\cos 3\omega t \pm V_{5m}\cos 5\omega t$$

$$= V_{1m}\cos\theta \pm V_{3m}\cos 3\theta \pm V_{5m}\cos 5\theta$$

Obviously, in this case, zeros occur at $\theta = \pm\pi/2$ or 90°. Moreover, positive and negative half-cycles are symmetrical.

$$\therefore \quad V_{av} = \frac{1}{\pi}\int_{-\pi/2}^{\pi/2}(V_{1m}\cos\theta \pm V_{3m}\cos 3\theta \pm V_{5m}\cos 5\theta)\,d\theta = \frac{2}{\pi}\left(V_{1m} \pm \frac{V_{3m}}{3} \pm \frac{V_{5m}}{5}\right)$$

$$\therefore \quad k_f = \frac{(1/\sqrt{2})(V_{1m}^{\ 2} + V_{3m}^{\ 2} + V_{5m}^{\ 2})^{1/2}}{\left(\frac{2}{\pi}\right)\left(V_{1m} \pm \frac{V_{3m}}{3} \pm \frac{V_{5m}}{5}\right)}$$

Example 17.1. *Draw one complete cycle of the following wave*

$$i = 100 \sin \omega t + 40 \sin 5\,\omega t$$

Determine the average value, the r.m.s. value and form factor of the wave.

(Elect. Engineering, Osmania Univ.)

Solution. $I_{av} = \frac{2}{\pi}\left(\frac{I_{2m}}{1} + \frac{I_{5m}}{5}\right) = \frac{2}{\pi}\left(\frac{100}{1} + \frac{40}{5}\right)$ = **68.7 A**

$$I = \sqrt{\frac{I_{1m}^{\ 2}}{2} + \frac{I_{5m}^{\ 2}}{2}} = \frac{1}{\sqrt{2}}(100^2 + 40^2)^{1/2} = \textbf{76.2 A}$$

Form factor $= \frac{I}{I_{av}} = \frac{76.2}{68.7}$ = **1.109**

17.5. Power Supplied by a Complex Wave

Let the complex voltage be represented by the equation

$$e = E_{1m}\sin\omega t + E_{2m}\sin 2\omega t + ...E_{nm}\sin n\omega t$$

be applied to a circuit. Let the equation of the resultant current wave be

$$i = I_{1m}\sin(\omega t + \phi_1) + I_{2m}\sin(2\omega t + \phi_2) + \ldots\ldots I_{nm}\sin(n\omega t + \phi_n)$$

The instantaneous value of the power in the circuit is $p = ei$ watt

For obtaining the value of this product, we will have to multiply every term of the voltage wave, in turn, by every term in the current wave. The average power supplied during a cycle would be equal to the sum of the average values over one cycle of each individual product term. However, as proved in Art. 17.3 earlier, the average value of all product terms involving harmonics of different frequencies will be zero over one cycle, so that we need consider only the products of current and voltage harmonics of the same frequency.

Let us consider a general term of this nature *i.e.* E_{nm} sin $n\omega t \times I_{nm}$ sin $(n\omega t - \phi_n)$ and find its average value over one cycle of the fundamental.

$$\text{Average value of power} = \frac{1}{2\pi}\int_0^{2\pi} E_{nm}I_{nm}\sin n\omega t\sin(n\omega t - \phi_n)\,d(\omega t)$$

$$= \frac{E_{nm}I_{nm}}{2\pi}\int_0^{2\pi} \sin n\theta\sin(n\theta - \phi_n)\,d\theta$$

$$= \frac{E_{nm}I_{nm}}{2\pi}\int_0^{2\pi} \frac{\cos\phi_n - \cos(2n\theta - \phi_n)}{2}\,d\theta$$

$$= \frac{E_{nm}I_{nm}\cos\phi}{2} = \frac{E_{nm}}{\sqrt{2}}\cdot\frac{I_{nm}}{\sqrt{2}}\cdot\cos\phi_n = E_nI_n\cos\phi_n$$

where E_n and I_n are the r.m.s. values of the voltage and current respectively. Hence, total average power supplied by a complex wave is ***the sum of the average power supplied by each harmonic component acting independently***.

∴ Total power is $P = E_1I_1\cos\phi_1 + E_2I_2\cos\phi_2 + \ldots\ldots E_nI_n\cos\phi_n$

The overall power factor is given by

$$\text{pf.}^* = \frac{\text{total watts}}{\text{total voltamperes}} = \frac{E_1I_1\cos\phi_1 + E_2I_2\cos\phi_2 + \ldots\ldots}{E \times I}$$

when E = r.m.s. value of the complex voltage wave

I = r.m.s. value of the complex current wave

Example 17.2. *A single-phase voltage source 'e' is given by*

$$e = 141\sin\omega t + 42.3\sin 3\omega t + 28.8\sin 5\omega t$$

The corresponding current in the load circuit is given by

$$i = 16.5\sin(\omega t + 54.5^\circ) + 8.43\sin(3\omega t - 38^\circ) + 4.65\sin(5\omega t - 34.3^\circ)$$

Find the power supplied by the source. **(Electrical Circuits, Nagpur Univ. 1991)**

Solution. In problems of such type, it is best to deal with each harmonic separately

Power at fundamental

$$= E_1I_1\cos\phi_1 = \frac{E_{1m}}{\sqrt{2}}\cdot\frac{I_{1m}}{\sqrt{2}}\cos\phi_1 = \frac{E_{1m}I_{1m}}{2}\cos\phi_1 = \frac{141 \times 16.5}{2}\cos 54.5^\circ = 675.5\text{ W}$$

$$\text{Power at 3rd harmonic} = \frac{E_{3m}I_{3m}}{2}\cos\phi_3 = \frac{42.3 \times 8.43}{2}\cos 38^\circ = 140.5\text{ W}$$

---

* When harmonics are present, it is obvious that the overall p.f. of the circuit cannot be stated lagging or leading. It is simply the ratio of power in watts of voltamperes.

Power at 5th harmonic $= \dfrac{28.8 \times 4.65}{2} \cos 34.3° = 55.5\text{ W}$

Total power supplied = 675.5 + 140.5 + 55.5 = **871.5 W**

Example 17.3. *A complex voltage is given by e = 60 sin ωt + 24 sin (3ωt + ω/6) + 12 sin (5ωt +π/3) is applied across a certain circuit and the resulting current is given by*

$$i = 0.6\ \sin(\omega t - 2\pi/10) + 0.12 \sin(\omega t - 2\pi/24) + 0.1 \sin(5\pi - 3\pi/4)$$

*Find **(i)** r.m.s value of current and voltage **(ii)** total power supplied and **(iii)** the overall power factor.*

Solution. In such problems where harmonics are involved, it is best to deal with each harmonic separately.

$$\text{Power at fundamental} = E_1 I_1 \cos\phi_1 = \frac{E_{1m} I_{1m}}{2} \cos\phi_1 = \frac{60 \times 0.6}{2} \times \cos 36° = 14.56\text{ W}$$

$$\text{Power at 3rd harmonic} = \frac{E_{3m} I_{3m}}{2} \cos 45° = \frac{24 \times 0.12}{2} \times 0.707 = 1.02\text{ W}$$

$$\text{Power at 5th harmonic} = \frac{E_{5m} I_{5m}}{2} \cos 75° = \frac{12 \times 0.1}{2} \times 0.2588 = 0.16\text{ W}$$

$$\textit{(i)}\ \text{R.M.S. current}\quad I = \sqrt{I_1^2 + I_3^2 + I_5^2} = \sqrt{\left(\frac{I_{1m}}{\sqrt{2}}\right)^2 + \left(\frac{I_{3m}}{\sqrt{2}}\right)^2 + \left(\frac{I_{5m}}{\sqrt{2}}\right)^2}$$

$$= \sqrt{\frac{0.6^2}{2} + \frac{0.12^2}{2} + \frac{0.1^2}{2}} = \mathbf{0.438\ A}$$

$$\text{R.M.S. volts,}\ E = \sqrt{\frac{60^2}{2} + \frac{24^2}{2} + \frac{12^2}{2}} = \mathbf{46.5\ V}$$

(ii) Total power = 14.56 + 1.02 + 0.16 = **15.74 W**

$$\textit{(iii)}\ \text{Overall p.f.} = \frac{\text{watts}}{\text{voltamperes}} = \frac{15.74}{46.5 \times 0.438} = \mathbf{0.773}$$

17.6. Harmonics in Single-Phase A.C. Circuits

If an alternating voltage, containing various harmonics, is applied to a single-phase circuit containing linear circuit elements, then the current so produced also contains harmonics. Each harmonic voltage will produce its own current independent of others. By the principle of superposition, the combined current can be found. We will now consider some of the well-known elements like pure resistance, pure inductance and pure capacitance and then various combinations of these. In each case, we will assume that the applied complex voltage is represented by

$$e = E_{1m} \sin \omega t + E_{2m} \sin 2\omega t + \ldots + E_{nm} \sin n\omega t$$

(a) Pure Resistance

Let the circuit have a resistance of R which is independent of frequency.

The instantaneous current i_1 due to fundamental voltage is

$$i_1 = \frac{E_{1m} \sin \omega t}{R}$$

Similarly, $\quad i_2 = \dfrac{E_{2m} \sin 2\omega t}{R}$ for 2nd harmonic and $$i_n = \frac{E_{nm} \sin n\omega t}{R} \quad \text{... for } n\text{th harmonic}$$

total current $$i = i_1 + i_2 + ... + i_n$$

$$= \frac{E_{1m} \sin \omega t}{R} + \frac{E_{2m} \sin 2\omega t}{R} + ... + \frac{E_{nm} \sin n\omega t}{R}$$

$$= I_{1m} \sin \omega t + I_{2m} \sin 2\omega t + ... + I_{nm} \sin n\omega t$$

It shows that

(i) the waveform of the resulting current is similar to that of the applied voltage *i.e.* the two waves are identical.

(ii) the percentage of harmonic content in the current wave is the same as in the applied voltage.

(*b*) Pure Inductance

Let the inductance of the circuit be L henry whose reactance varies directly as the frequency of the applied voltage. Its reactance for the fundamental would be $X_1 = \omega L$ for the second harmonic, $X_2 = 2\omega L$ for the third harmonic, $X_3 = 2\omega L$ and for the nth harmonic $X_n = 2\omega L$ However, for every harmonic term, the current will lag behind the voltage by 90°.

Current due to fundamental, $i_1 = \frac{E_{1m}}{\omega L} \sin(\omega t - \pi / 2)$

Current due to 2nd harmonic, $i_2 = \frac{E_{2m}}{2\omega L} \sin(2\omega t - \pi / 2)$

Current due to 3rd harmonic, $i_3 = \frac{E_{3m}}{3\omega L} \sin(3\omega t - \pi / 2)$

Current due to nth harmonic, $i_n = \frac{E_{nm}}{n\omega L} \sin(n\omega t - \pi / 2)$

$\therefore$ Total current $i = i_1 + i_2 + ...i_n$

$$= \frac{E_{1m}}{\omega L} \sin(\omega t - \pi / 2) + \frac{E_{2m}}{2\omega L} \sin(2\omega t - \pi / 2) + ... + \frac{E_{nm}}{n\omega L} \sin(n\omega t - \pi / 2)$$

It can be seen from the above equation that

(i) the waveform of the current differs from that of the applied voltage.

(ii) for the nth harmonic, the percentage harmonic content in the current-wave is $1/n$ of the corresponding harmonic content in the voltage wave. It means that in an inductive circuit, the current waveform shows less distortion that the voltage waveform. In this case, current more nearly approaches a sine wave than it does in a circuit containing resistance.

(*c*) Pure Capacitance

In this case,

$$X_1 = \frac{1}{\omega C} \text{ – for fundamental ; } X_2 = \frac{1}{2\omega C} \text{ – for 2nd harmonic}$$

$$X_3 = \frac{1}{3\omega C} \text{ – for 3rd harmonic ; } X_n = \frac{1}{n\omega C} \text{ – for } n\text{th harmonic}$$

$$i_1 = \frac{E_{1m}}{1/\omega C}\sin(\omega t + \pi/2) = \omega CE_{1n}\sin(\omega t + \pi/2)$$

$$i_2 = \frac{E_{2m}}{1/2\omega C}\sin(2\omega t + \pi/2) = 2\omega CE_{2m}\sin(2\omega t + \pi/2)$$

$$i_n = \frac{E_{nm}}{1/n\omega C}\sin(n\omega t + \pi/2) = n\omega CE_{nm}\sin(n\omega t + \pi/2)$$

For every harmonic term, the current will lead the voltage by 90°.

Now $i = i_1 + i_2 + \ldots\ldots + i_n$

$$= \omega C E_{1m}\sin(\omega t + \pi/2) + 2\omega C E_{2m}\sin(2\omega t + \pi/2) + \ldots + n\omega CE_{nm}\sin(n\omega t + \pi/2)$$

This equation shows that

(i) the current and voltage waveforms are dissimilar.

(ii) percentage harmonic content of the current is larger than that of the applied voltage wave. For example, for *n*th harmonic, it would be *n* time larger.

(iii) as a result, the current wave is more distorted than the voltage wave.

(iv) effect of capacitor on distortion is just the reverse of that of inductance.

Example 17.4. *A complex wave of r.m.s. value 240 V has 20% 3rd harmonic content, 5% 5th harmonic content and 2% 7th harmonic content. Find the r.m.s. value of the fundamental and of each harmonic.* **(Elect. Circuits, Gujarat Univ.)**

Solution. Let V_1, V_3, V_5 and V_7 be the r.m.s. values of the fundamental and harmonic voltages. Then

$V_3 = 0.2\ V_1$; $V_5 = 0.05\ V_1$ and $V_7 = 0.02\ V_1$

$240 = (V_1^2 + V_3^2 + V_5^2 + V_7^2)^{1/2}$

$\therefore\ 240 = [V_1^2 + (0.2\ V_1)^2 + (0.05\ V_1)^2 + (0.02\ V_1)^2]^{1/2}$

$\therefore\ V_1 = \mathbf{235\ V}$; $V_3 = 0.2 \times 235 = \mathbf{47\ V}$

$V_5 = 0.05 \times 235 = \mathbf{11.75\ V}$; $V_7 = 0.02 \times 235 = \mathbf{4.7\ V}$

Example 17.5. *Derive an expression for the power, power factor and r.m.s. value for a complex wave.*

A voltage e = 250 sin ωt sin $(3\omega t + \pi/3)$ *+ 2 sin* $(\omega t + 5\pi/6)$ *is applied to a series circuit of resistance* 20 Ω *and inductance 0.05 H. Derive* ***(a)*** *an expression for the current* ***(b)*** *the r.m.s. value of the current and for the voltage* ***(c)*** *the total power supplied and (d) the power factor. Take ω = 314 rad/s.* **(Electrical Circuits, Nagpur Univ. 1991)**

Solution. For Fundamental

$$X_1 = \omega L = 314 \times 0.05 = 15.7\ \Omega;\ Z_1 = 20 + j15.7 = 25.4\angle 38.1°\ \Omega$$

For Third Harmonic

$$X_3 = 3\omega L = 3 \times 15.7 = 47.1\Omega;\ Z_3 = 20 + j47.1 = 51.2\angle 67°\ \Omega$$

For Fifth Harmonic

$$X_5 = 5\omega L = 5 \times 15.7 = 78.5\Omega;\ Z_5 = 20 + j78.5 = 81\angle 75.7°\ \Omega$$

(a) Expression for the current is

$$i = \frac{250}{25.4}\sin(\omega t - 38.1°) + \frac{50}{51.2}\sin(3\omega t + 60° - 67°) + \frac{20}{81}\sin(5\omega t + 150° - 75.7°)$$

$$\therefore\ i = 9.84\sin(\omega t - 38.1°) + 0.9\sin(3\omega t - 7°) + 0.25\sin(5\omega t + 74.3°)$$

(*b*) *R.M.S. current* $I = \sqrt{\frac{I_{1m}^2}{2} + \frac{I_{3m}^2}{2} + \frac{I_{5m}^2}{2}}$

$$I^2 = \frac{9.84^2}{2} + \frac{0.9^2}{2} + \frac{0.25^2}{2} = 48.92$$

$$\therefore \quad I = \sqrt{48.92} = \mathbf{6.99\ A}$$

R.M.S. voltage $V = \sqrt{\frac{250^2}{2} + \frac{50^2}{2} + \frac{20^2}{2}} = \mathbf{180.8\ V}$

(*c*) Total power $= I^2R = 48.92 \times 20 = \mathbf{978\ W}$

(*d*) Power factor $= \frac{\text{Watts}}{\text{VI}} = \frac{978}{180.8 \times 6.99} = \mathbf{0.773}$

Example 17.6. *An r.m.s. current of 5 A, which has a third harmonic content, is passed through a coil having a resistance of 1 Ω and an inductance of 10 mH. The r.m.s. voltage across the coil is 20 V. Calculate the magnitude of the fundamental and harmonic components of current if the fundamental frequency is 300/2Ω Hz. Also, find the power dissipated.*

Solution. (*i*) Fundamental Frequency

$\omega = 300$ rad/s ; $X_L = 300 \times 10^{-2} = 3\ \Omega$ ∴ $\mathbf{Z}_1 = 1 + j\,3 = 3.16\ \angle\ 71.6°$ ohm

If V_1 is the r.m.s. value of the fundamental voltage across the coil, then

$$V_1 = I_1Z_1 = 3.16\ I_1$$

(*ii*) Third Harmonic

$X_3 = 3 \times 3 = 9\ \Omega$; $\mathbf{Z}_3 = 1 + j\,9 = 9.05\ \angle 83.7°$ ohm ; $V_3 = I_3Z_3 = 9.05\ I_3$

Since r.m.s. current of the complex wave is 5 A and r.m.s. voltage drop 20 V

$$5 = \sqrt{I_1^2 + I_3^2} \quad \text{and} \quad 20 = \sqrt{V_1^2 + V_3^2}$$

Substituting the values of V_1 and V_3, we get, $20 = [(3.16\ I_1)^2 + (9.05\ I_3)^2]^{1/2}$

Solving for I_1 and I_3, we have I_1 = **4.8 A** and I_3 = **1.44 A**

Power dissipated = $I^2R = 5^2 \times 1$ = **25 W**

Example 17.7. *An e.m.f. represented by the equation e = 150 sin 314 t + 50 sin 942 t is applied to a capacitor having a capacitance 20 μF. What is the r.m.s. value of the charging current ?*

Solution. For Fundamental

$$X_{C1} = 1/\omega C = 10^6/20 \times 314 = 159\Omega;\ I_{1m} = E_{1m}/X_{C1} = 150/159 = 0.943\ \text{A}$$

For Third Harmonic

$X_{C1} = 1/3\ \omega C = 159/3 = 53\ \Omega$ ∴ $I_{3m} = E_{3m}/X_{C1} = 50/53 = 0.943$ A

r.m.s. value of charging current,

$$I = \sqrt{\frac{I_{1m}^2}{2} + \frac{I_{3m}^2}{2}} = \sqrt{\frac{0.943^2}{2} + \frac{0.943^2}{2}}$$

or I = **0.943 A**

Example 17.8. *The voltage given by v = 100 cos 314 t + 50 sin (1570t – 30°) is applied to a circuit consisting of a 10 Ω resistance, a 0.02 H inductance and a 50 μF capacitor. Determine the instantaneous current through the circuit. Also find the r.m.s. value of the voltage and current.*

Solution. For Fundamental

$\omega = 314$ rad/s; $X_L = 314 \times 0.02 = 6.28\ \Omega$

$X_C = 10^6/314 \times 50 = 63.8\ \Omega$; $X = X_L - X_C = 6.28 - 63.8 = -57.32\ \Omega$

$$Z = \sqrt{10^2 + (-57.32)^2} = 58.3\ \Omega\ ;\ I_{1m} = 100/58.3 = 1.71\ \text{A}$$

$\phi_1 = \tan^{-1}(-57.32/10) = -80.2°$ (lead) ; $i_1 = 1.71 \cos(314t + 80.2°)$

For Fifth Harmonic

Inductive reactance $= 5\ X_L = 5 \times 6.28 = 31.4\ \Omega$

Capacitive reactance $= X_C/5 = 63.8/5 = 12.76\ \Omega$

Net reactance $= 31.4 - 12.76 = 18.64\ \Omega$

$$Z = \sqrt{10^2 + 18.64^2} = 21.2\Omega$$

$I_{5m} = 50/21.2 = 2.36$ A; $\phi_5 = \tan^{-1}(18.64/10) = 61.8°$ (lag)

$i_5 = 2.36 \sin(1570\ t - 30° - 61.8°) = 2.36 \sin(1570t - 91.8°)$

Hence, total instantaneous current is

$i = i_1 + i_5 = 1.71 \cos(314\ t + 80.2°) + 2.36 \sin(1570t - 91.8°)$

$$\textit{R.M.S. voltage} = \sqrt{\frac{100^2}{2} + \frac{50^2}{5}} = \mathbf{79.2\ V}$$

$$\textit{R.M.S. current} = \sqrt{\frac{1.71^2}{2} + \frac{2.36^2}{2}} = \mathbf{2.06\ A}$$

17.7. Selective Resonance due to Harmonics

When a complex voltage is applied across a circuit containing both inductance and capacitance, it may happen that the circuit resonates at one of the harmonic frequencies of the applied voltage. This phenomenon is known as ***selective resonance.***

If it is a series circuit, then large currents would be produced at resonance, even though the applied voltage due to this harmonic may be small. Consequently, it would result in large harmonic voltage appearing across both the capacitor and the inductance.

If it is a parallel circuit, then at resonant frequency, the resultant current drawn from the supply would be minimum.

It is because of the possibility of such selective resonance happening that every effort is made to eliminate harmonics in supply voltage.

However, the phenomenon of selective resonance has been usefully employed in some wave analyses for determining the harmonic content of alternating waveforms. For this purpose, a variable inductance, a variable capacitor, a variable non-inductive resistor and a fixed non-inductive resistance or shunt for an oscillograph are connected in series and connected to show the wave-form of the voltage across the fixed non-inductive resistance. The values of inductance and capacitance are adjusted successively to give resonance for the first, third, first and seventh harmonics and a record of the waveform is obtained by the oscillograph. A quick inspection of the shape of the waveform helps to detect the presence or absence of a particular harmonic.

Example 17.9. *An e.m.f. $e = 200 \sin \omega t + 40 \sin 3\omega t + 10 \sin 5\omega t$ is impressed on a circuit comprising of a resistance of 10 Ω a variable inductor and a capacitance of 30 μF, all connected in series. Find the value of the inductance which will give resonance with triple frequency component of the pressure and estimate the effective p.f. of the circuit, $\omega = 300$ radian/second.* **(Elect. Engg. I, Bombay Univ.)**

Solution. For resonance at third harmonic

$3\omega L = 1/3\omega C \quad \therefore L = 1/9\omega^2 C = 10^6/9 \times 300^2 \times 30 =$ **0.041 H**

$$\mathbf{Z}_1 = 10 + j\left(300 \times 0.041 - \frac{10^6}{300 \times 30}\right) = 10 + j\,(12.3 - 111.1) = 10 - j98.8 = 99.3\ \angle{-84.2^\circ}$$

$$Z_3 = 10 + j\left(3\omega L - \frac{1}{3\omega C}\right) = 10 + j(36.9 - 37.0) = 10\angle 0^\circ$$

$$Z_5 = 10 + j\left(5\omega L - \frac{1}{5\omega L}\right) = 10 + j(61.5 - 22.2) = 10 + j39.3 = 40.56\angle 75.7^\circ$$

$I_{1m} = 200/99.3 = 2.015$ A ; $I_{3m} = 40/10 = 4$A ; $I_{5m} = 10/40.56 = 0.246$ A

$$I = \sqrt{\frac{2.015^2}{2} + \frac{4^2}{2} + \frac{0.246^2}{2}} = \sqrt{10.06} = 3.172 \text{ A}$$

$$V = \sqrt{\frac{200^2}{2} + \frac{40^2}{2} + \frac{10^2}{2}} = 144.5 \text{ V} \text{ ; Power} = I^2R = 10.06 \times 10 = 100.6 \text{ W}$$

Volt-amperes $VI = 144.5 \times 3.172 = 458$ VA ; Power factor = 100.6/458 = **0.22**

Example 17.10. *A coil having R = 100 Ω and L = 0.1 H is connected in series with a capacitor across a supply, the voltage of which is given by e = 200 sin 314t + 5 sin 3454t. What capacitance would be required to produce resonance with the 11th harmonic. Find* **(a)** *the equation of the current and* **(b)** *the r.m.s. value of the current, if this capacitance is in circuit.*

Solution. For series resonance, $X_L = X_C$

Since resonance is required for 11th harmonic whose frequency is 3454 rad/s, hence

$$3454\,L = \frac{1}{3454\,C};\ C = \frac{1}{3454^2 \times 0.1} \text{ farad} = \mathbf{0.838} \quad \text{F}$$

(*a*) For Fundamental

Inductive reactance = $\omega L = 314 \times 0.1 = 31.4\ \Omega$

Capacitive reactance = $1/\omega C = 10^6 / 0.836 \times 314 = 3796\Omega$

∴ Net reactance = 3796 – 31.4 = 3765 Ω; Resistance = 100 Ω

∴ $Z_1 = \sqrt{100^2 + 3765^2} = 3767\ \Omega$; $\tan\phi_1 = 3765/100 = 37.65$

∴ $\phi_1 = 88^\circ 28'$ (leading) = 1.546 radian

Now $E_{1m} = 200\ V$; $Z_1 = 3767\ \Omega \quad \therefore I_{1m} = 200/3767 = 0.0531$ A

Eleventh Harmonic

New reactance = 0; Impedance $Z_{11} = 100\ \Omega$

∴ Current $I_{11m} = 5/100 = 0.05$ A ; $\phi_{11} = 0$ ... at resonance

Hence, the equation of the current is

$$i = \frac{200}{3767}(\sin 314t + 1.546) + \frac{5}{100}\sin(3454t + 0)$$

$$\mathbf{i = 0.0531 \sin (314\ t + 1.546) + 0.05 \sin 3454\ t}$$

(*b*) $I = \sqrt{(0.0531)^2/2 + (0.05)^2/2} = \mathbf{0.052\ A}$

17.8. Effect of Harmonics on Measurement of Inductance and Capacitance

Generally, with the help of ammeter and voltmeter readings, the value of impedance, inductance and capacitance of a circuit can be calculated. But while dealing with complex voltages, the use of instrument readings does not, in general, give correct values of inductance and capacitance except in the case of a circuit containing only pure resistance. It is so because, in the case of resistance, the voltage and current waveforms are similar and hence the values of r.m.s. volts and r.m.s amperes (as read by the voltmeter and ammeter respectively) would be the same whether they ware sinusoidal or non-sinusoidal (*i.e.* complex).

(*i*) Effect on Inductances

Let L be the inductance of a circuit and E and I the r.m.s. values of the applied voltage and current as read by the instruments connected in the circuit. For a complex voltage

$$E = 0.707\sqrt{(E_{1m}{}^2 + E_{3n}{}^2 + E_{5m}{}^2 + ...)}$$

Hence $$I = 0.707\sqrt{\left[\left(\frac{E_{1m}}{\omega L}\right)^2 + \left(\frac{E_{3m}}{3\omega L}\right)^2 + \left(\frac{E_{5m}}{5\omega L}\right)^2 + ...\right]}$$

$$= \frac{0.707}{\omega L}\sqrt{\left(E_{1m}{}^2 + \frac{1}{9}E_{3m}{}^2 + \frac{1}{25}E_{5m}{}^2 + ...\right)}$$

$$\therefore \quad L = \frac{0.707}{\omega I}\sqrt{\left(E_{1m}{}^2 + \frac{1}{9}E_{3m}{}^2 + \frac{1}{25}E_{5m}{}^2 + ...\right)}$$

For calculating the value of L from the above expression, it is necessary to known the absolute value of the amplitudes of several harmonic voltages. But, in practices, it is more convenient to deal with relative values than with absolute values. For this purpose, let us multiply and divide the right-hand side of the above expression by E but write the E ***in the denominator*** in its form 0.707 $\sqrt{(E_{1m}{}^2 + E_{3m}{}^2 + E_{5m}{}^2 + ...)}$

$$\therefore \quad L = \left[\frac{0.707}{\omega I}\sqrt{E_{1m}{}^2 + 1/9.E_{3m}{}^2 + 1/25.E_{5m}{}^2 + ...}\right] \times \left[\frac{E}{0.707\sqrt{(E_{1m}{}^2 + E_{3m}{}^2 + E_{5m}{}^2 + ...)}}\right]$$

or $$L = \frac{E}{\omega I}\sqrt{\left[\frac{E_{1m}{}^2 + 1/9 \cdot E_{3m}{}^2 + 1/25 \cdot E_{5m}{}^2 + ...}{(E_{1m}{}^2 + E_{3m}{}^2 + E_{5m}{}^2 + ...)}\right]} = \frac{E}{\omega I}\sqrt{\left[\frac{1 + 1/9 \cdot (E_{3m}/E_{1m})^2 + 1/25 \cdot (E_{5m}/E_{1m})^2 + ...}{1 + (E_{3m}/E_{1m})^2 + (E_{5m}/E_1)^2 + ...}\right]}$$

If the effect of harmonics were to be neglected, then the value of the inductance would appear to be E/ω I but the true or actual value is less than this. The apparent value has to be multiplied by the quantity under the radical to get the true value of inductance when harmonics are present.

The quantity under the radical is called the correction factor *i.e.*

True inductance (L) = Apparent inductance (L') × correction factor

(*ii*) Effect on Capacitance

Let the capacitance of the circuit be C farads and E and I the instrument readings for voltage and current. Since the instruments read r.m.s. values, hence, as before,

$$E = 0.707\sqrt{(E_{1m}{}^2 + E_{3m}{}^2 + E_{5m}{}^2 + ...)}$$

$$\text{Hence } I = 0.707\sqrt{\left[\left(\frac{E_{1m}}{1/\omega C}\right)^2 + \left(\frac{E_{3m}}{1/3\omega C}\right)^2 + \left(\frac{E_{5m}}{1/5\omega C}\right) + ...\right]}$$

$$= 0.707\sqrt{(\omega C E_{1m})^2 + (3\omega C E_{3m})^2 + (5\omega C\, E_{5m})^2 + ...}$$

$$= 0.707\ \omega C\sqrt{E_{1m}{}^2 + 9E_{3m}0^2 + 25E_{sm}{}^2 + ...}$$

$$\therefore\ C = \frac{I}{0.707\omega\sqrt{(E_{1\ddot{u}}{}^2 + 9E_3{}^2 + 25E_5{}^2 + ...)}}$$

Again, we will multiply and divide the right-hand side E but in this case, we will write E in the *numerator* in its form $[0.707\sqrt{(E_{1m}{}^2 + E_{3m}{}^2 + E_{5m}{}^2 + ...)}]$

$$\therefore\ C = \left[\frac{1}{0.707\,\omega E\sqrt{(E_{1m}{}^2 + 9E_{3m}{}^2 + 25\,E_{5m}{}^2 + ...)}}\right] \times \left[0.707\sqrt{(E_{1m}{}^2 + E_{3m}{}^2 + E_{5m}{}^2 + ...)}\right]$$

$$= \frac{1}{\omega E}\sqrt{\left[\frac{E_{1m}{}^2 + E_{3m}{}^2 + E_{5m}{}^2 .....}{E_{1m}{}^2 + 9E_{3m}{}^2 + 25E_{5m}{}^2 + ...}\right]} = \frac{1}{\omega E}\sqrt{\left[\frac{1 + (E_{3m}/E_{1m})^2 + (E_{5m}/E_{1m})^2 .....}{1 + 9(E_{3m}/E_{1m})^2 + 25(E_{5m}/E_{1m})^2 + ...}\right]}$$

Again, if the effects of harmonics were neglected, the value of capacitance would appear to be I/ω E but its true value is less than this. For getting the true value, this apparent value will have to be multiplied by the quantity under the radical (which, therefore, is referred to as correction factor).*

∴ True capacitance (C) = Apparent capacitance (C') × correction factor

Example 17.11. *A current of 50-Hz containing first, third and fifth harmonics of maximum values 100, 15 and 12 A respectively, is sent through an ammeter and an inductive coil of negligibly small resistance. A voltmeter connected to the terminals shows 75 V. What would be the current indicated by the ammeter and what is the exact value of the inductance of the coil in henrys?*

Solution. The r.m.s. current is

$$I = 0.707\sqrt{I_{1m}{}^2 + I_{3m}{}^2 + I_{5m}{}^2} = 0.707\sqrt{(100^2 + 15^2 + 12^2)} = 72 \text{ A}$$

Hence, current indicated by the ammeter is **72 A**

Now $\quad E = 0.707\sqrt{(E_{1m}{}^2 + E_{3m}{}^2 + E_{5m}{}^2)}$

Also $\quad I_{1m} = \frac{E_{1m}}{\omega L};\ I_{3m} = \frac{E_{3m}}{3\omega L};\ I_{5m} = \frac{E_{5m}}{5\omega L}$

* It may be noted that this correction factor is different from that in the case of pure inductance.

$$\therefore \quad E_{1m} = I_{1m} \cdot \omega L; E_{3m} = I_{3m} 3\omega L; E_{5m} = I_{5m} \cdot 5\omega L$$

$$\therefore \; E = 0.707\sqrt{(I_{1m}\omega L)^2 + (I_{3m} 3\,\omega L)^2 + (I_{5m} 5\omega L)^2} = 0.707\omega L\sqrt{I_{1m}{}^2 + 9I_{3m}{}^2 + 25I_{5m}{}^2}$$

$$\therefore \quad 75 = 0.707\, L \times 2\pi \times 50\sqrt{100^2 + 9\times 15^2 + 25\times 12^2}$$

$$\therefore \; L = \mathbf{0.0027\ H}$$

Note. Apparent inductance $L' = \dfrac{E}{\omega I} = \dfrac{75}{2\pi \times 50 \times 72} = 0.00331$ H

Example 17.12. *The capacitance of a 20 μF capacitor is checked by direct connection to an alternating voltage which is supposed to be sinusoidal, an electrostatic voltmeter and a dynamometer ammeter being used for measurement. If the voltage actually follows the law,*

$$e = 100 \sin 250\, t + 20 \sin (500 - \phi) + 10 \sin (750\, t - \phi)$$

Calculate the value of capacitance as obtained from the direct ratio of the instrument readings.

Solution. True value, $C = 20\ \mu F$

Apparent value C' = value read by the instruments

Now, $C = C' \times$ correction factor.

Let us find the value of correction factor.

Here $E_{1m} = 100$; $E_{2m} = 20$ and $E_{3m} = 10$

$$\therefore \text{ Correction factor } = \sqrt{\left[\frac{E_{\text{üm}}{}^{\text{ü}} + E_{m} + E_{m}}{E_{1m} + 4E_{2m}{}^2 + 9E_{5m}{}^2}\right]} = \sqrt{\left[\frac{100^{\text{ü}} + 20 + 10}{100^2 + 4\times 20^2 + 9\times 10^2}\right]} = 0.9166$$

$$20 = C' \times 0.9166 \qquad \therefore \; C' = 20/0.9166 = \mathbf{21.82\ \mu F.}$$

17.9. Harmonics in Single and 3-Phase Transformers

The flux density in transformer course is usually maintained at a fairly high value in order to keep the required volume of iron to the minimum. However, due to the non-linearity of magnetisation curve, some third harmonic distortion is always produced. Also, there is usually a small percentage of fifth harmonic. The magnetisation current drawn by the primary contains mainly third harmonic whose proportion depends on the size of the primary applied voltage. Hence, the flux is sinusoidal.

In the case of three-phase transformers, the production of harmonics will be affected by the method of connection and the type of construction employed.

***(a)* Primary Windings Δ-CONNECTED**

Each primary phase can be considered as separately connected across the sinusoidal supply.

(i) The core flux will be sinusoidal which means that magnetizing current will contain 3rd harmonic component in addition to relatively small amounts of other harmonics of higher order.

(ii) In each phase, these third harmonic currents will be in phase and so produce a circulating current round the mesh with the result that there will be no third harmonic component in the line current.

(*b*) Primary Winding Connected in 4-wire Star

Each phase of the primary can again be considered as separately connected across a sinusoidal supply.

(*i*) The flux in the transformer core would be sinusoidal and so would be the output voltage.

(*ii*) The magnetizing current will contain 3rd harmonic component. This component being in phase in each winding will, therefore, return through the neutral wire.

(*c*) Primary Windings Connected 3-wire Star

Since there is no neutral wire, there will be no return path for the 3rd harmonic component of the magnetizing current. Hence, there will exist a condition of forced magnetization so that core flux must contain third harmonic component which is in phase in each limb of the transformer core. Although there will be a magnetic path for these fluxes in the case of shell type 3-phase transformer, yet in the case of three-limb core type transformer, the third harmonic component of the flux must return *via* the air. Because of the high reluctance magnetic path in such transformers, the third harmonic flux is reduced to a very small value. However, if the secondary of the transformer is delta-connected, then a third harmonic circulating current would be produced. This current would be in accordance with Lenz's law tend to oppose the very cause producing it *i.e.* it would tend to minimize the third harmonic component of the flux.

Should the third and fifth secondary be *Y*-connected, then provision of an additional Δ-connected winding, in which this current can flow, becomes necessary. This tertiary winding additionally served the purpose of preserving magnetic equilibrium of the transformer in the case of unbalanced loads. In this way, the output voltage from the secondary can be kept reasonably sinusoidal.

Example 17.13. *Determine whether the following two waves are of the same shape*

$e = 10 \sin(\omega t + 30^\circ) - 50 \sin(3\omega t - 60^\circ) + 25 \sin(5\omega t + 40^\circ)$

$i = 1.0 \sin(\omega t - 60^\circ) + 5 \sin(3\omega t - 150^\circ) + 2.5 \cos(5\omega t - 140^\circ)$

(Principles of Elect. Engg-II Jadavpur Univ.)

Solution. Two waves possess the same waveshape

(*i*) if they contain the same harmonics

(*ii*) if the ratio of the corresponding harmonics to their respective fundamentals is the same

(*iii*) if the harmonics are similarly spaced with respect to their fundamentals.

In other words,

(*a*) the ratio of the magnitudes of corresponding harmonics must be constant and

(*b*) with fundamentals in phase, the corresponding harmonics of the two waves must be in phase.

The test is applied first by checking the ratio of the corresponding harmonics and then coinciding the fundamentals by shifting one wave. If the phase angles of the corresponding harmonics are the same, then the two waves have the same shape.

In the present case, condition (*i*) is fulfilled because the voltage and current waves contain the same harmonics, *i.e.* third and fifth.

Secondly, the ratio of the magnitude of corresponding current and voltage harmonics is the same *i.e.* 1/10.

Now, let the fundamental of the current wave be shifted ahead by 90° so that it is brought in phase with the fundamental of the voltage wave. It may be noted that the third and fifth harmonics of the current wave will be shifted by 3 × 90° = 270° and 5 × 90° = 450° respectively. Hence, the current wave becomes

$$i' = 1.0\sin(\omega t - 60° + 90°) + 5\sin(3\omega t - 150° + 270°) + 2.5\cos(5\omega t - 140° + 450°)$$

$$= 1.0\sin(\omega t + 30°) + 5\sin(3\omega t + 120°) + 2.5\cos(5\omega t + 310°)$$

$$= 1.0\sin(\omega t + 30°) - 5\sin(3\omega t - 60°) + 2.5\sin(5\omega t + 40°)$$

It is seen that now the corresponding harmonics of the voltage and current waves are in phase. Since all conditions are fulfilled, the two waves are of the same waveshape.

Tutorial Problem 17.1

1. A series circuit consist of a coil of inductance 0.1 H and resistance 25 Ω and a variable capacitor. Across this circuit is applied a voltage whose instantaneous value is given by

$$v = 100\sin\omega t + 20\sin(3\omega t - 45) + 5\sin(5\omega t - 30) \text{ where } \omega = 314 \text{ rad/s}$$

Determine the value of C which will produce response at third harmonic frequency and with this value of C, find

(*a*) an expression for the current in the circuit (*b*) the r.m.s. value of this current (*c*) the total power absorbed. **[11.25μF, (*a*) i = 0.398 sin (ω*t* + 84.3) + 0.8 sin (3ω*t* + 45) + 0.485 sin (5ω*t* + 106) (*b*) 0.633 A (*c*) 10W]**

2. A voltage given by v = 200 sin 314 t + 520 sin (942 t + 45°) is applied to a circuit consisting of a resistance of 20 Ω, and inductance of 20 mH and a capacitance of 56.3 μF all connected in series.

Calculate the r.m.s. values of the applied voltage and current. Find also the total power absorbed by the circuit. **[146 V ; 3.16 A 200 W]**

3. A certain coil has a resistance of 20 Ω and an inductance of 0.04 H. If the instantaneous current flowing in it is represented by i = 5 sin 300 t + 0.8 sin 900 t amperes, derive an expression for the instantaneous value of the voltage applied across the ends of the coil and calculate the r.m.s. value of that voltage. **[V = 117 sin (300 t + 0.541) + 33 sin (900 t + 1.06) ; 0.86 V]**

4. A voltage given by the equation $v = \sqrt{2}100\sin 2\pi \times 50t + \sqrt{2.20}\sin 2\pi.\ 150t$ is applied to the terminals of a circuit made up of a resistance of 5 Ω, an inductance of 0.0318 H and a capacitor of 12.5 μF all in series. Calculate the effective current and the power supplied to the circuit. **[0.547 A ; 1.5 W]**

5. An alternating voltage given by the expression v = 1,000 sin 314t + 100 sin 942t is applied to a circuit having a resistance of 100 Ω and an inductance of 0.5 H. Calculate r.m.s. value of the current and p.f. **[3.81 A ; 0.535]**

6. The current in a series circuit consisting of a 159 μF capacitor, a reactor with a resistance of 10Ω and an inductances of 0.0254 H is given by $i = \sqrt{2}(8\sin\omega t + 2\sin 3\omega t)$ amperes. Calculate the power input and the power factor. Given $\omega = 100\pi$ radian/second. **[680 W ; 0.63]**

7. If the terminal voltage of a circuit is 100 sin $\omega t + 50\sin(3\omega t + \pi/4)$ and the current is 10 sin$(\omega t + \pi 3)$ +5 sin 3ωt, calculate the power consumption. **[522.6 W]**

8. Explain what is meant by harmonic resonance in a.c. circuits.

A current having an instantaneous value of 2 $(\sin\omega t + \sin 3\omega t)$ amperes is passed through a circuit which consists of a coil of resistance R and inductance L in series with a capacitor C. Derive an expression for the value of ω at which the r.m.s. circuit voltage is a minimum. Determine the voltage if the coil has inductance 0.1 H and resistance 150 Ω and the capacitance is 10 μ F. Determine also the circuit voltage at the fundamental resonant frequency.

[$\omega = 1/\sqrt{(LC)}$; 378 V ; 482 V]

9. An r.m.s. current of 5 A which has a third-harmonic content, is passed through a coil having a resistance of 1Ω and an inductance of 10 mH. The r.m.s. voltage across the coil is 20 V. Calculate the magnitudes of the fundamental and harmonic components of current if the fundamental frequency is $300/2\pi$ Hz Also, find the power dissipated. **[4.8 A ; 1.44 A ; 25 W]**

10. Derive a general expression for the form factor of a complex wave containing only odd-order harmonics. Hence, calculate the form factor of the alternating current represented by
i = 2.5 sin 157 t + 0.7 sin 471 t + 0.4 sin 785 t **[1.038]**

OBJECTIVE TESTS – 17

1. Non-sinusoidal waveforms are made up of
(*a*) different sinusoidal waveforms
(*b*) fundamental and even harmonics
(*c*) fundamental and odd harmonics
(*d*) even and odd harmonics only.

2. The positive and negative halves of a complex wave are symmetrical when
(*a*) it contains even harmonics
(*b*) phase difference between even harmonics and fundamental is 0 or π
(*c*) it contains odd harmonics
(*d*) phase difference between even harmonics and fundamental is either $\pi/2$ or $3\pi/2$.

3. The r.m.s. value of the complex voltage given by $v = 16\sqrt{2}\sin\omega t + 12\sqrt{2}\sin 3\omega t$ is
(*a*) $20\sqrt{2}$ (*b*) 20
(*c*) $28\sqrt{2}$ (*d*) 192

4. In a 3-phase system, ___th harmonic has negative phase sequence of RBY.
(a) 9 (*b*) 13
(*c*) 5 (*d*) 15

5. A complex current wave is given by the equation i = 14 sin $\omega t + 2\sin 5\omega t$. The r.m.s. value of the current is ___ ampere.
(*a*) 16 (*b*) 12
(*c*) 10 (*d*) 8

6. When a pure inductive coil is fed by a complex voltage wave, its current wave
(*a*) has larger harmonic content
(*b*) is more distorted
(*c*) is identical with voltage wave
(*d*) shows less distortion.

7. A complex voltage wave is applied across a pure capacitor. As compared to the fundamental voltage, the reactance offered by the capacitor to the third harmonic voltage would be
(*a*) nine time (*b*) three times
(*b*) one-third (*d*) one-ninth

8. Which of the following harmonic voltage components in a 3-phase system would be in phase with each other ?
(*a*) 3rd, 9th, 15th etc.
(*b*) 7th, 13th, 19th etc.
(*c*) 5th, 11th, 17th etc.
(*d*) 2nd, 4th, 6th etc.

ANSWERS

1. *(a)* **2.** *(c)* **3.** *(b)* **4.** *(c)* **5.** *(c)* **6.** *(d)* **7.** *(c)* **8.** *(d)*

CHAPTER

18 Fourier Series

Chapter Outline

- ❖ Harmonic Analysis
- ❖ Periodic Functions
- ❖ Trigonometric Fourier Series
- ❖ Alternate Forms of Trigonometric Fourier Series
- ❖ Certain Useful Integral Calculus Theorems
- ❖ Evalulation of Fourier Constants
- ❖ Different Types of Functional Symmetries
- ❖ Line or Frequency Spectrum
- ❖ Procedure for Finding the Fourier Series of a Given Function

18.1. Harmonic Analysis

By harmonic analysis is meant the process of determining the magnitude, order and phase of the several harmonics present in a complex periodic wave.

For carrying out this analysis, the following methods are available which are all based on Fourier theorem:

(i) Analytical Method–the standard Fourier Analysis

(ii) Graphical Method– ***(a)*** by Superposition Method (Wedgemore' Method) ***(b)*** Twenty four Ordinate Method

(iii) Electronic Method–by using a special instrument called 'harmonic analyser'

We will consider the first and third methods only.

18.2. Periodic Functions

A function $f(t)$ is said to be periodic if $f(t + T) = f(t)$ for all values of t where T is some positive number. This T is the interval between two successive repetitions and is called the period of $f(t)$. A sine wave having a period of $T = 2\pi/\omega$ is a common example of periodic function.

18.3. Trigonometric Fourier Series

Suppose that a given function $f(t)$ satisfies the following conditions (known as Dirichlet conditions):

1. $f(t)$ is periodic having a period of T.

2. $f(t)$ is single-valued everywhere.

3. In case it is discontinuous, $f(t)$ has a finite number of discontinuities in any one period.

4. $f(t)$ has a finite number of maxima and minima in any one period.

The function $f(t)$ may represent either a voltage or current waveform. According to Fourier theorem, this function $f(t)$ may be represented in the trigonometric form by the infinite series.

$$f(t) = a_0 + a_1 \cos\omega_0 t + a_2 \cos 2\omega_0 t + a_3 \cos 3\omega_0 t + ... + a_n \cos n\omega_0 t$$
$$+ \; b_1 \sin\omega_0 t + b_2 \sin 2\omega_0 t + b_3 \sin 3\omega_0 t + ... + b_n \sin n\omega_0 t$$

$$= a_0 + \sum_{n=1}^{\infty}(a_n \cos n\omega_0 t + b_n \sin n\omega_0 t) \qquad ... (i)$$

Putting $\omega_0 t = \theta$, we can write the above equation as under

$$f(\theta) = a_0 + a_1 \cos\theta + a_2 \cos 2\theta + a_3 \cos 3\theta + ... + a_n \cos n\theta + b_1 \sin\theta + b_2 \sin 2\theta +$$
$$+ b_3 \sin 3\theta + ... + b_n \sin n\theta$$

$$= a_0 + \sum_{n=1}^{\infty}(a_n \cos n\theta + b_n \sin n\theta) \qquad ... (ii)$$

Since $\omega_0 = 2\pi / T$, Eq. (*i*) above can be written as

$$f(t) = a_0 + \sum_{n=1}^{\infty}\left(a_n \cos\frac{2\pi n}{T} t + b_n \sin\frac{2\pi n}{T} t\right) \qquad ... (iii)$$

where ω_0 is the fundamental angular frequency, T is the period and a_0, a_n and b_n are constants which depend on n and $f(t)$. The process of determining the values of the constants a_0, a_n and b_n is called Fourier Analysis. Also, $\omega_0 = 2\pi / T = 2\pi f_0$ where f_0 is the fundamental frequency.

It is seen from the above Fourier Series that the periodic function consists of sinusoidal components of frequency 0, ω_0, $2\omega_0 .... n\omega_0$. This representation of the function $f(t)$ is in the frequency domain. The first component a_0 with zero frequency is called the *dc* component. The sine and cosine terms represent the harmonics. The number n represents the order of the harmonics.

When $n = 1$, the component $(a_1 \cos\omega_0 t + b_1 \sin\omega_0 t)$ is called the first harmonic or the fundamental component of the waveform.

When $n = 2$, the component $(a_2 \cos 2\omega_0 t + b_2 \sin 2\omega_0 t)$ is called the second harmonic of the waveform.

The *n*th harmonic of the waveform is represented by $(a_n \cos n\omega_0 t + b_n \sin n\omega_0 t)$. It has a frequency of $n\omega_0$ *i.e.* n times the frequency of the fundamental component.

18.4. Alternate Forms of Trigonometric Fourier Series

Eq. (*i*) given above can be written as

$$f(t) = a_0 + (a_1 \cos\omega_0 t + b_1 \sin\omega_0 t) + (a_2 \cos 2\omega_0 t + b_2 \sin 2\omega_0 t) +$$
$$... + (a_n \cos n\omega_0 t + b_n \sin n\omega_0 t)$$

Let, $a_n \cos n\omega_0 t + b_n \sin n\omega_0 t = A_n \cos(n\omega_0 t - \phi_n)$

$= A_n \cos n\omega_0 t \cos\phi_n + A_n \sin n\omega_0 t \sin\phi_n$

$\therefore \quad a_n = A_n \cos\phi_n$ and $b_n = A_n \sin\phi_n$

$\therefore A_n = \sqrt{a_n^2 + b_n^2}$ and $\phi_n = \tan^{-1} b_n / a_n$

Similarly, let $(a_n \cos n\omega_0 t + b_n \sin n\omega_0 t = A_n \sin(n\omega_0 t + \Psi_n))$

$= A_n \sin n\omega_0 t \cos\Psi_n + A_n \cos n\omega_0 t \sin\phi_n$

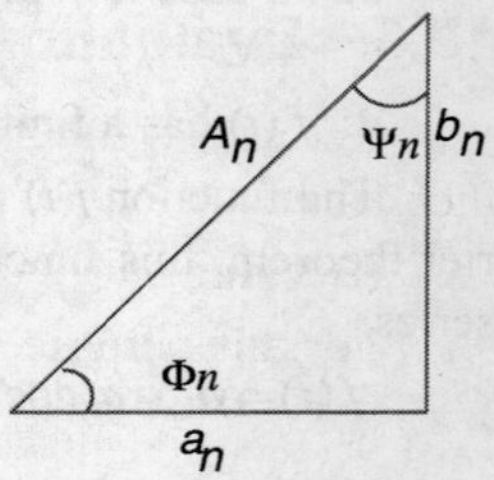

As seen from Fig. 18.1, $b_n = A_n \cos\psi_n$ and $a_n = A_n \sin\psi_n$

$\therefore \quad A_n = \sqrt{a_n^2 + b_n^2}$ and $\psi = \tan^{-1} a_n / b_n$

The two angles ϕ_n and ψ_n are complementary angles.

Fig. 18.1

Hence, the Fourier series given in Art. 18.2 may be put in the following two alternate forms

$$f(t) = A_0 + \sum_{n=1}^{\infty} A_n \cos(n\omega_0 t - \phi_n)$$

or $$f(t) = A_0 + \sum_{n=1}^{\infty} A_n \sin(n\omega_0 t + \psi_n)$$

18.5. Certain Useful Integral Calculus Theorems

The Fourier coefficients or constants $a_0, a_1, a_2 ... a_n$ and $b_1, b_2 ..... b_n$ can be evaluated by integration process for which purpose the following theorems will be used.

(*i*) $\int_0^{2\pi} \sin n\theta d\theta = \frac{1}{b} |\cos n\theta|_0^{2\pi} = \frac{1}{n}(1-1) = 0$

(*ii*) $\int_0^{2\pi} \sin n\theta d\theta = -\frac{1}{n} |\cos n\theta|_0^{2\pi} = -\frac{1}{n}(0-0) = 0$

(*iii*) $\int_0^{2\pi} \sin^2 n\theta d\theta = \frac{1}{2}\int_0^{2\pi} (1 - \cos 2n\theta) d\theta = \frac{1}{2} \left| \theta - \frac{1}{2n} \sin 2n\theta \right|_0^{2\pi} = \pi$

(*iv*) $\int_0^{2\pi} \cos^2 n\theta d\theta = \frac{1}{2}\int_0^{2\pi} (\cos 2n\theta + 1) d\theta = \frac{1}{2} \left| \frac{1}{2n} \sin 2n\theta + \theta \right|_0^{2\pi} = \pi$

(*v*) $\int_0^{2\pi} \sin m\theta \cos n\theta d\theta = \frac{1}{2}\int_0^{2\pi} \{\sin(m+n)\theta + \sin(m-n)\theta\} d\theta$

$= \frac{1}{2} \left| \frac{1}{m+n} \cos(m+n)\theta - \frac{1}{m-n} \cos(m-n)\theta \right|_0^{2\pi} = 0$

(*vi*) $\int_0^{2\pi} \cos m\theta \cos n\theta d\theta = \frac{1}{2}\int_0^{2\pi} \{\cos(m+n)\theta + \cos(m-n)\theta\} d\theta$

$= \frac{1}{2} \left| \frac{1}{m+n} \sin(m+n)\theta + \frac{1}{m-n} \sin(m-n)\theta \right|_0^{2\pi} = 0 ... \text{for } n \neq m$

(*vii*) $\int_0^{2\pi} \sin m\theta \sin n\theta d\theta = \frac{1}{2}\int_0^{2\pi} \{\cos(m-n)\theta - \cos(m+n)\theta\} d\theta$

$= \frac{1}{2} \left| \frac{1}{m-n} \sin(m+n)\theta + \frac{1}{m+n} \sin(m-n)\theta \right|_0^{2\pi} = 0 ... \text{for } n \neq m$

where m and n are any positive integers.

18.6. Evaluation of Fourier Constants

Let us now evaluate the constants a_0, a_n and b_n by using the above integral calculus theorems

***(i)* Value of a_0**

For this purpose we will integrate both sides of the series given below over one period *i.e.* for $\theta = 0$ to $\theta = 2\pi$.

$$f(\theta) = a_0 + a_1\cos\theta + a_2\cos 2\theta + ... + a_n\cos n\theta \; + \; b_1\sin\theta + b_2\sin 2\theta + ... + b_n\sin n\theta$$

$$\therefore \int_0^{2\pi} f(\theta)d\theta = \int_0^{2\pi} a_0 d\theta + a_1\int_0^{2\pi}\cos\theta d\theta + a_2\int_0^{2\pi}\cos 2\theta d\theta + ... + a_n\int_0^{2\pi}\cos n\theta d\theta$$

$$+ \; b_1\int_0^{2\pi}\sin\theta d\theta + b_2\int_0^{2\pi}\sin 2\theta d\theta + ... + b_n\int_0^{2\pi}\sin n\theta d\theta$$

$$= a_0 \,|\, \theta \,|_0^{2\pi} + 0 + 0 + ...0 + 0 + 0 + ... + 0 = 2\pi a_0$$

$$\therefore \qquad a_0 = \frac{1}{2\pi}\int_0^{2\pi} f(\theta)d\theta \text{ or } = \frac{1}{2\pi}\int_{-\pi}^{\pi} f(\theta)d\theta$$

= mean value of $f(\theta)$ between the limits 0 to 2π *i.e.* over one cycle or period.

$$\text{Also,} \quad a_0 = \frac{1}{2\pi}[\text{net area}]_0^{2\pi}$$

If we take the periodic function as $f(t)$ and integrate over period T (which corresponds to 2π),

$$\text{we get } a_0 = \frac{1}{T}\int_0^T f(t)dt = \frac{1}{T}\int_{-T/2}^{T/2} f(t)dt = \frac{1}{T}\int_{t_1}^{t_1+T} f(t)dt$$

where t_1 can have any value.

***(ii)* Value of a_n**

For finding the value of a_n, multiply both sides of the Fourier Series by cos $n\theta$ and integrate between the limits $\theta = 0$ to 2π

$$\therefore \int_0^{2\pi} f(\theta)\cos n\theta d\theta = a_0\int_0^{2\pi}\cos n\theta d\theta + a_1\int_0^{2\pi}\cos\theta\cos n\theta d\theta + a_2\int_0^{2\pi}\cos 2\theta\cos n\theta d\theta$$

$$+ a_n\int_0^{2\pi}\cos^2 n\theta d\theta + \; b_1\int_0^{2\pi}\cos\theta\cos n\theta d\theta + b_2\int_0^{2\pi}\sin 2\theta\cos n\theta d\theta$$

$$+ ... + b_n\int_0^{2\pi}\sin n\theta\cos n\theta d\theta$$

$$= 0 + 0 + 0 + ... + a_n\int_0^{2\pi}\cos^2 n\theta d\theta + 0 + 0 + ... + 0 = a_n\int_0^{2\pi}\cos^2 n\theta d\theta = \pi a_n$$

$$\therefore \; a_n = \frac{1}{\pi}\int_0^{2\pi} f(\theta)\cos n\theta d\theta = 2\times\frac{1}{2\pi}\int_0^{2\pi} f(\theta)\cos n\theta \, d\theta$$

= 2 × average value of $f(\theta)$ cos $n\theta$ over one cycle of the fundamental.

$$\text{Also,} \; a_n = \frac{1}{\pi}\int_{-\pi}^{\pi} f(\theta)\cos n\theta \, d\theta = 2\times\frac{1}{2\pi}\int_{-\pi}^{\pi} f(\theta)\cos n\theta d\theta$$

If we take periodic function as $f(t)$, then different expressions for a_n are as under.

$$a_n = \frac{2}{T}\int_0^T f(t)\cos\frac{2\pi n}{T}t \, dt = \frac{2}{T}\int_{-T/2}^{T/2} f(t)\cos\frac{2\pi n}{T}t \, dt$$

Giving different numerical values to n, we get

$a_1 = 2 \times$ average of $f(\theta)\cos\theta$ over one cycle ....$n = 1$

$a_2 = 2 \times$ average value of $f(\theta)\cos 2\theta$ over one cycle etc. .... $n = 2$

(iii) Value of b_n

For finding its value, multiply both sides of the Fourier Series of Eq. (*i*) by $\sin n\theta$ and integrate between limits $\theta = 0$ to $\theta = 2\pi$

$$\therefore \int_0^{2\pi} f(\theta)\sin n\theta d\theta = a_0\int_0^{2\pi}\sin n\theta d\theta + a_1\int_0^{2\pi}\cos\theta\sin n\theta d\theta$$

$$+ a_2\int_0^{2\pi}\cos 2\theta\sin n\theta d\theta + ... + a_n\int_0^{2\pi}\cos n\theta\sin n\theta d\theta$$

$$+ b_1\int_0^{2\pi}\sin\theta\sin n\theta d\theta + b_2\int_0^{2\pi}\sin 2\theta\sin n\theta d\theta + ... + b_n\int_0^{2\pi}\sin^2 n\theta d\theta$$

$$= 0+0+0+...+0+0+...+b_n\int_0^{2\pi}\sin^2 n\theta d\theta = b_n\int_0^{2\pi}\sin^2 n\theta\, d\theta = b_n\pi$$

$$\therefore \int_0^{2\pi} f(\theta)\sin\theta d\theta = b_n \times \pi$$

$$\therefore b_n = \frac{1}{\pi}\int_0^{2\pi} f(\theta)\sin n\theta d\theta = 2\times\frac{1}{2\pi}\int_0^{2\pi} f(\theta)\sin n\theta d\theta$$

$= 2 \times$ average value of $f(\theta)\sin n\theta$ over one cycle of the fundamental.

$\therefore b_1 = 2 \times$ average value of $f(\theta)\sin 2\theta$ over one cycle ... $n = 1$

$b_2 = 2 \times$ average value of $f(\theta)\sin 2\theta$ over one cycle ... $n = 2$

$$\text{Also, } b_n = \frac{2}{t}\int_0^T f(t)\sin\frac{2\pi n}{T}t\,dt + \frac{2}{T}\int_{T/2}^{T/2} f(t)\sin\frac{2\pi n}{T}t\,dt$$

$$= \frac{2}{T}\int_0^T f(t)\sin n\omega_0 t\,dt = \frac{2}{T}\int_{T/2}^{T/2} f(t)\sin n\omega_0 tdt$$

Hence, for Fourier analysis of a periodic function, the following procedure should be adopted:

(*i*) Find the term a_0 by integrating both sides of the equation representing the periodic function between limits 0 to 2π or 0 to T or $-T/2$ to $T/2$ or t_1 to $(t_1 + T)$.

$$\therefore a_0 = \frac{1}{2\pi}\int_0^{2\pi} f(\theta)d\theta = \frac{1}{T}\int_0^T f(t)dt = \frac{1}{T}\int_{-T/2}^{T/2} f(t)dt = \frac{1}{T}\int_{t_1}^{t_1+T} f(t)dt$$

= average value of the function over one cycle.

(*ii*) Find the value of a_n by multiplying both sides of the expression for Fourier series by $\cos n\theta$ and then integrating it between limits 0 to 2π or 0 to T or $-T/2$ to $T/2$ or t_1 to $(t_1 + T)$.

$$\therefore \quad a_n = \frac{1}{\pi}\int_0^{2\pi} f(\theta)\cos n\theta d\theta = 2\times\frac{1}{2\pi}\int_0^{2\pi} f(\theta)\cos n\theta d\theta$$

Since $\pi = T/2$, we have

$$a_n = \frac{2}{T}\int_0^T f(t)\cos\frac{2\pi n}{T}t\,dt = \frac{2}{T}\int_{-T/2}^{T/2} f(t)\cos\frac{2\pi n}{T}t\,dt = \frac{2}{T}\int_{t_1}^{t_1+T} f(t)\cos\frac{2\pi n}{T}t\,dt$$

$$= \frac{2}{T}\int_0^T f(t)\cos n\omega_0 tdt = \frac{2}{T}\int_{-T/2}^{T/2} f(t)\cos n\omega_0 tdt = \frac{2}{T}\int_{t_1}^{t_1+T} f(t)\cos n\omega_0 tdt$$

$= 2 \times$ average value of $f(\theta)\cos n\theta$ over one cycle of the fundamental.

Values of a_1, a_2, a_3 etc. can be found from above by putting $n = 1, 2, 3$ etc.

(iii) Similarly, find the value of b_n by multiplying both sides of Fourier series by sin $n\theta$ and integrating it between the limits 0 to 2π or 0 to T or $-T/2$ to $T/2$ or t_1 to $(t_1 + T)$.

$$\therefore \quad b_n = \frac{1}{\pi}\int_0^{2\pi} f(\theta)\sin n\theta = 2\times\frac{1}{2\pi}\int_0^{2\pi} f(\theta)\sin n\theta d\theta$$

$$= \frac{2}{T}\int_0^{T} f(t)\sin\frac{2\pi n}{T}t\,dt = \frac{2}{T}\int_{-T/2}^{T/2} f(t)\sin\frac{2\pi n}{T}t\,dt = \frac{2}{T}\int_{t_1}^{t_1+T} f(t)\sin\frac{2\pi n}{t}t\,dt$$

$$= \frac{2}{T}\int_0^{T} f(t)\sin n\omega_0 t dt = \frac{2}{T}\int_{-T/2}^{T/2} f(t)\sin n\omega_0 t\,dt = \frac{2}{T}\int_{t_1}^{t_1+T} f(t)\sin n\omega_0 t\,dt$$

$= 2 \times$ average value of $f(\theta)$ sin of $f(t)$ sin $\frac{2\pi n}{T}t$ or $f(t) f$ sin $n\omega_0 t$ over one cycle of the fundamental.

Values of b_1, b_2, b_3 etc. can be found from above by putting $n = 1, 2, 3$ etc.

18.7. Different Types of Functional Symmetries

A non-sinusoidal wave can have the following types of symmetry:

1. Even Symmetry

The function $f(t)$ is said to possess even symmetry if $f(t) = f(-t)$.

It means that as we travel equal amounts in time to the left and right of the origin (*i.e.* along the $+X$-axis and $-X$-axis), we find the function to have the same value. For example in Fig. 18.2 (*a*), points *A* and *B* are equidistant from point *O*. Here the two function values are equal and positive. At points *C* and *D*, the two values of the function are again equal, though negative. Such a function is symmetric with respect to the vertical axis. Examples of even function are: t^2, cos $3t$, $\sin^2 5t\,(2 + t^2 + t^4)$ and a constant *A* because the replacement of *t* by $(-t)$ does not change the value of any of these functions. For example, $\cos\omega t = \cos(-\omega t)$.

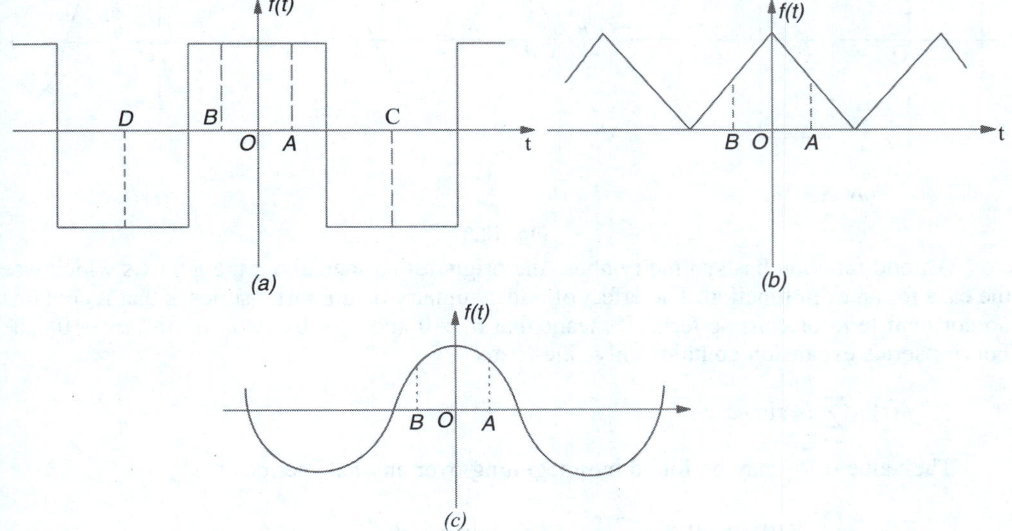

Fig. 18.2

This type of symmetry can be easily recognised graphically because mirror symmetry exists about the vertical or $f(t)$ axis. The function shown in Fig. 18.2 has even symmetry because when folded along vertical axis, the portions of the graph of the function for positive and negative time fit exactly, one on top of the other.

The effect of the even symmetry on Fourier series is that the constant $b_n = 0$ *i.e.* the wave has no sine terms. In general, $b_1, b_2, b_3 \ldots b_n = 0$. The Fourier series of an even function contains only a constant term and cosine terms *i.e.*

$$f(t) = a_0 + \sum_{n=1}^{\infty} a_n \cos n\omega_0 t = a_0 + \sum_{n=1}^{\infty} a_n \cos \frac{2\pi n}{T} t$$

The value of a_n may be found by integrating over any half-period.

$$\therefore \quad a_n = \frac{2}{\pi}\int_0^{\pi} f(\theta)\cos n\theta d\theta = \frac{4}{T}\int_0^{T/2} f(t)\cos n\omega t\, dt$$

2. Odd Symmetry

A function $f(t)$ is said to possess odd symmetry if $f(-t) = -f(t)$.

It means that as we travel an equal amount in time to the left or right from the origin, we find the function to be the same except for a reversal in sign. For example, in Fig. 18.3 the two points *A* and *B* are equidistant from point *O*. The two function values at *A* and *B* are equal in magnitude but opposite in sign. In other words, if we replace *t* by (– *t*), we obtained the negative of the given function. The *X*-axis divides an odd function into two halves with equal areas above and below the *X*-axis. Hence, $a_0 = 0$.

Examples of odd functions are: *t*, sin *t*, *t* cos 50 $t(t+t^3+t^5)$ and $t\sqrt{(1+t^2)}$

A sine function is an odd function because sin $(-\omega t) = -\sin\omega t$.

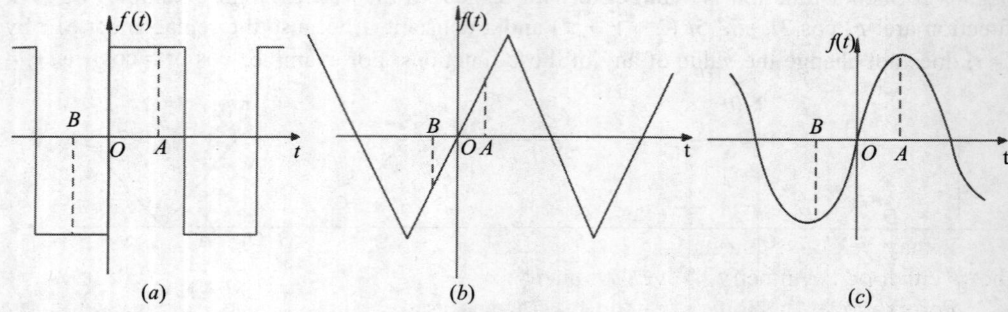

Fig. 18.3

An odd function has symmetry about the origin rather than about the $f(t)$ axis which was the case for an even function. The effect of odd symmetry on a Fourier series is that it contains no constant term or consine term. It means that $a_0 = 0$ and $a_n = 0$ *i.e.* $a_1, a_2, a_3 \ldots. a_n = 0$. The Fourier series expansion contains only sine terms.

$$\therefore \ f(t) = \sum_{n=1}^{\infty} b_n \sin n\omega_0 t$$

The value of b_n may be found by integrating over any half-period.

$$\therefore \ b_n = \frac{2}{\pi}\int_0^{\pi} f(\theta)\sin n\theta d\theta = \frac{4}{T}\int_0^{T/2} f(t)\sin n\omega t\, dt$$

3. Half-wave Symmetry or Mirror-Symmetry or Rotational Symmetry

A function $f(t)$ is said to possess half-wave symmetry if $f(t) = -f(t \pm T/2)$ or $-f(t) = f(t \pm T/2)$.

It means that the function remains the same if it is shifted to the left or right by half a period and then flipped over (*i.e.* multiplied by – 1) in respect to the *t*-axis or horizontal axis. It is called mirror symmetry because the negative portion of the wave is the mirror image of the positive portion of the wave displaced horizontally a distance $T/2$.

In other words, a waveform possesses half symmetry only when we invert its negative half-cycle and get an exact duplicate of its positive half-cycle. For example, in Fig. 18.4 (*a*) if we invert the negative half-cycle, we get the dashed *ABC* half-cycle which is exact duplicate of the positive half-cycle. Same is the case with the waveforms of Fig. 18.4 (*b*) and Fig. 18.4 (*c*). In case of doubt, it is helpful to shift the inverted half-cycle by a half-period to the left and see if it super-imposes the positive half-cycle. If it does so, there exists half-wave symmetry otherwise not. It is seen that the waveform of Fig. 18.4 (*d*) does not possess half-wave symmetry. It is so because when its negative half-cycle is inverted and shifted by half a period to the left it does not superimpose the positive half-cycle.

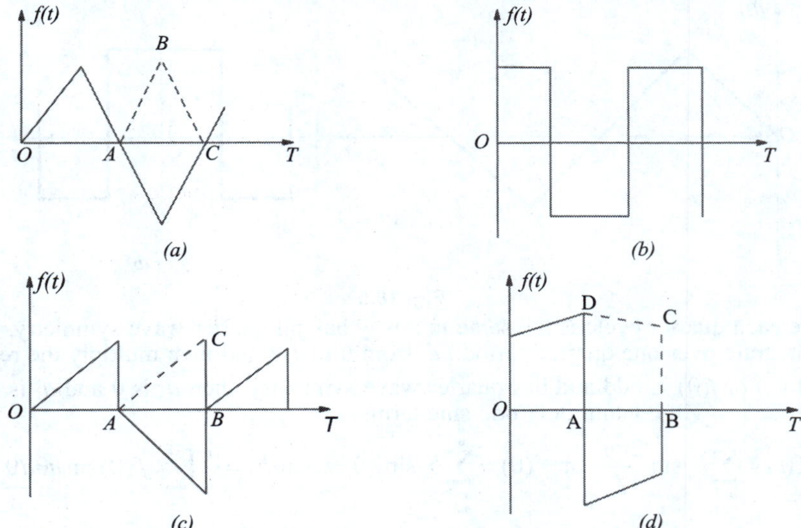

Fig. 18.4

It may be noted that half-wave symmetry may be present in a waveform which also shows either odd symmetry or even symmetry:

For example, the square waveform shown in Fig. 18.4 (*a*) possesses even symmetry whereas the triangular waveform of Fig. 18.4 (*b*) has odd symmetry. All cosine and sine waves possess half-wave symmetry because

$$\cos\frac{2\pi}{T}\left(t \pm \frac{T}{2}\right) = \cos\left(\frac{2\pi}{T}t \pm \pi\right) = -\cos\frac{2\pi}{T}t; \sin\frac{2\pi}{T}\left(t \pm \frac{T}{2}\right) = \sin\left(\frac{2\pi}{T}t \pm \pi\right) = -\sin\frac{2\pi}{T}t$$

It is worth noting that the Fourier series of any function which possesses half-wave symmetry has zero average value and contains only odd harmonics and is given by

$$f(t) = \sum_{\substack{n=1 \\ odd}}^{\infty}\left(a_n \cos\frac{2\pi n}{T}t + b_n \sin\frac{2\pi n}{T}t\right) = \sum_{\substack{n=1 \\ odd}}^{\infty}(a_n \cos n\omega_0 t + b_n \sin n\omega_0 t)$$

$$= \sum_{\substack{n=1 \\ odd}}^{\infty}(a_n \cos n\theta + b_n \sin n\theta)$$

where, $a_n = \frac{4}{T}\int_0^{T/2} f(t)\cos\frac{2\pi n}{T}dt = \frac{2}{\pi}\int_0^{\pi} f(\theta)\cos n\theta d\theta$ ... n odd

$b_n = \frac{4}{T}\int_0^{T/2} f(t)\sin\frac{2\pi n}{T}dt = \frac{2}{\pi}\int_0^{\pi} f(\theta)\sin n\theta d\theta$ ... n odd

4. Quarter-wave Symmetry

An odd or even function with rotational symmetry is said to possess quarter-wave symmetry.

Fig. 18.5 (*a*) possesses half-wave symmetry as well as odd symmetry. The wave shown in Fig. 18.5 (*b*) has both half-wave symmetry and even symmetry.

The mathematical test for quarter-wave symmetry is as under:

Odd quarter-wave $f(t) = -f(t+T/2)$ and $f(-t) = -f(t)$

Even quarter-wave $f(t) = -f(t+T/2)$ and $f(t) = f(-t)$

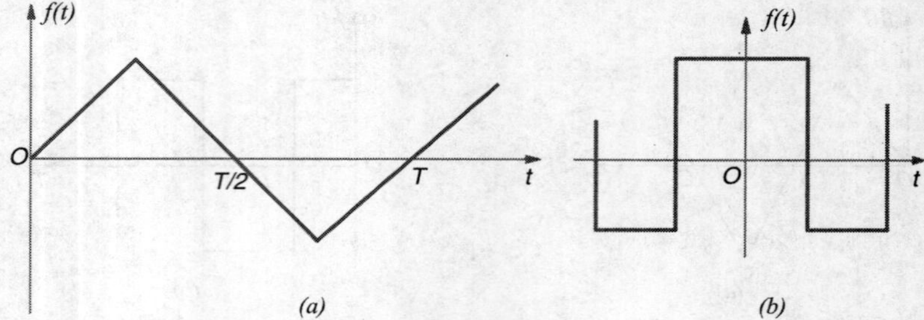

Fig. 18.5

Since each quarter cycle is the same in a way having quarter-wave symmetry, it is sufficient to integrate over one quarter period *i.e.* from 0 to $T/4$ and then multiply the result by 4.

(i) If $f(t)$ or $f(\theta)$ is odd and has quarter-wave symmetry, then a_0 is 0 and a_n is 0. Hence, the Fourier series will contain only odd sine terms.

$$\therefore\ f(t) = \sum_{\substack{n=1\\ odd}}^{\infty} b_n \sin\frac{2\pi nt}{T} \text{ or } f(\theta) = \sum_{\substack{n=1\\ odd}}^{\infty} b_n \sin n\theta, \text{ where } b_n = \frac{1}{\pi}\int_0^{2\pi} f(\theta)\sin n\theta d\theta$$

It may be noted that in the case of odd quarter-wave symmetry, the integration may be carried over a quarter cycle.

$\therefore\ a_n = \frac{4}{\pi}\int_0^{\pi/2} f(\theta)\cos n\theta d\theta$ ... n odd

$= \frac{8}{T}\int_0^{T/4} f(t)\sin n\omega t dt$ ... n odd

(ii) If $f(t)$ or $f(\theta)$ is even and, additionally, has quarter-wave symmetry, then a_0 is 0 and b_n is 0. Hence, the Fourier series will contain only odd cosine terms.

$$\therefore\ f(\theta) = \sum_{\substack{n=1\\ odd}}^{\infty} a_n \cos n\theta d\theta = \sum_{\substack{n=1\\ odd}}^{\infty} a_n \cos n\omega_0 t dt; \text{ where } a_n = \frac{1}{\pi}\int_0^{2\pi} f(\theta)\cos n\theta d\theta$$

In this case a_n may be found by integrating over any quarter period.

$a_n = \frac{4}{\pi}\int_0^{\pi/2} f(\theta)\sin n\theta d\theta$ ... n odd

$= \frac{8}{T}\int_0^{T/4} f(t)\cos n\omega t\, dt$ ... n odd

18.8. Line or Frequency Spectrum

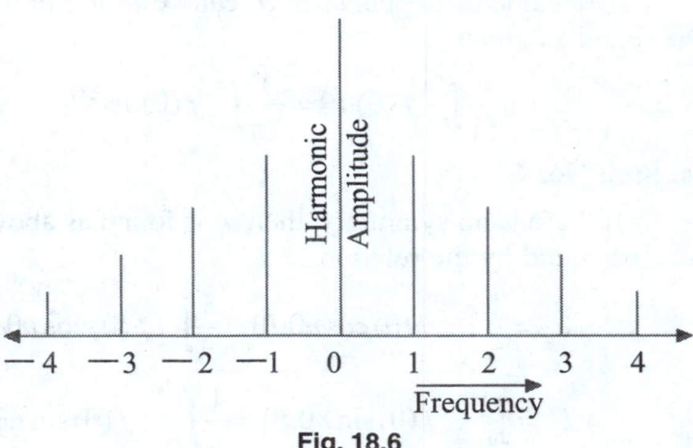

Fig. 18.6

A plot which shows the amplitude of each frequency component in a complex waveform is called the line spectrum or frequency spectrum (Fig. 18.6). The amplitude of each frequency component is indicated by the length of the vertical line located at the corresponding frequency. Since the spectrum represents frequencies of the harmonics as discrete lines of appropriate height, it is also called a discrete spectrum. The lines decrease rapidly for waves having convergent series. Waves with discontinuities such as the sawtooth and square waves have spectra whose amplitudes decrease slowly because their series have strong high harmonics. On the other hand, the line spectra of waveforms without discontinuities and with a smooth appearance have lines which decrease in height very rapidly.

The harmonic content and the line spectrum of a wave represent the basic nature of that wave and never change irrespective of the method of analysis. Shifting the zero axis changes the symmetry of a given wave and gives its trigonometric series a completely different appearance but the same harmonics always appear in the series and their amplitude remains constant.

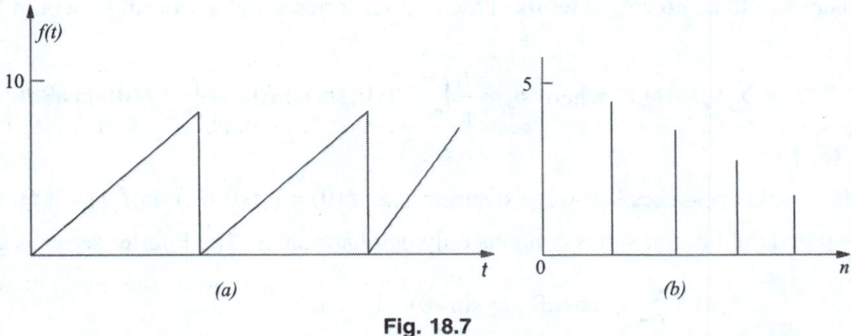

Fig. 18.7

Fig. 18.7 shows a smooth wave alongwith its line spectrum. Since there are only sine terms in its trigonometric series (apart from $a_0 = \pi$), the harmonic amplitudes are given by b_n.

18.9. Procedure for Finding the Fourier Series of a Given Function

It is advisable to follow the following steps:

1. Step No. 1

If the function is defined by a set of equations, sketch it approximately and examine for symmetry.

2. Step No. 2

Whatever be the period of the given function, take it as 2π and find the Fourier series in the form

$$f(\theta) = a_0 + a_1 \cos\theta + a_2 \cos 2\theta + ... + a_n \cos n\theta + b_1 \sin\theta + b_2 \sin 2\theta + ... + b_n \sin n\theta$$

3. Step No. 3

The value of the constant a_0 can be found in most cases by inspection. Otherwise it can be found as under:

$$a_0 = \frac{1}{2\pi}\int_0^{2\pi} f(\theta)d\theta = \frac{1}{2\pi}\int_{-\pi}^{\pi} f(\theta)d\theta$$

4. Step No. 4

If there is no symmetry, then a_0 is found as above whereas the other two fourier constants can be found by the relation.

$$a_n = \frac{1}{\pi}\int_0^{2\pi} f(\theta)\cos n\theta d\theta = \frac{1}{\pi}\int_{-\pi}^{\pi} f(\theta)\cos n\theta d\theta$$

$$b_n = \int_0^{2\pi} f(\theta)\sin n\theta d\theta = \frac{1}{\pi}\int_{-\pi}^{\pi} f(\theta)\sin n\theta d\theta$$

5. Step No. 5

If the function has even symmetry *i.e.* $f(\theta) = f(-\theta)$, then $b_n = 0$ so that the Fourier series will have no sine terms. The series would be given by

$$f(\theta) = a_0 + \sum_{n=1}^{\infty} a_n \cos n\theta d\theta \text{ where } a_n = \frac{1}{\pi}\int_0^{2\pi} f(\theta)\cos n\theta = \frac{2}{\pi}\int_0^{\pi} f(\theta)\cos n\theta d\theta$$

6. Step No. 6

If the given function has odd symmetry *i.e.* $f(-\theta) = -f(\theta)$ then $a_0 = 0$ and $a_n = 0$. Hence, there would be no cosine terms in the Fourier series which accordingly would be given by

$$f(\theta) = \sum_{n=1}^{\infty} b_n \sin \omega_0 t\text{ ; where } b_n = \frac{1}{\pi}\int_0^{2\pi} f(\theta)\sin n\theta d\theta = \frac{2}{\pi}\int_0^{\pi} f(\theta)\sin n\theta d\theta$$

7. Step No. 7

It the function possesses half-wave symmetry *i.e.* $f(\theta) = -f(\theta \pm \pi)$ or $f(t) = -f(t \pm T/2)$, then a_0 is 0 and the Fourier series contains only odd harmonics. The Fourier series is given by

$$f(\theta) = \sum_{\substack{n=1\\ odd}}^{\infty} a_n(\cos n\theta + b_n \sin n\theta)$$

where $\quad a_n = \dfrac{1}{\pi}\displaystyle\int_0^{2\pi} f(\theta)\cos n\theta d\theta ... n$ odd, $b_n = \dfrac{2}{\pi}\displaystyle\int_0^{\pi} f(\theta)\sin n\theta d\theta \quad$ ... n odd

8. Step No. 8

If the function has even quarter-wave symmetry then $a_0 = 0$ and $b_n = 0$. It means the Fourier series will contain no sine terms but only odd cosine terms. It would be given by

$$f(\theta) = \sum_{\substack{n=1\\ odd}}^{\infty} a_n \cos n\theta ;\text{ where } a_n = \frac{1}{\pi}\int_0^{2\pi} f(\theta)\cos n\theta d\theta = \frac{2}{\pi}\int_0^{\pi} f(\theta)\cos n\theta d\theta$$

$$= \frac{4}{\pi}\int_0^{\pi/2} f(\theta)\cos n\theta d\theta f \quad \text{... } n \text{ odd}$$

9. Step No. 9

If the function has odd quarter-wave symmetry, then $a_0 = 0$ and $a_n = 0$. The Fourier series will contain only odd sine terms (but no cosine terms).

$$\therefore \quad f(\theta) = \sum_{\substack{n=1 \\ odd}}^{\infty} b_n \sin n\theta; \text{ where } b_n = \frac{1}{\pi}\int_0^{2\pi} f(\theta)\sin n\theta d\theta = \frac{2}{\pi}\int_0^{\pi} f(\theta)\sin n\theta d\theta$$

$$= \frac{1}{4}\int_0^{\pi/2} f(\theta)\sin n\theta d\theta \quad \text{... } n \text{ odd}$$

10. Step No. 10

Having found the coefficients, the Fourier series as given in step No. 2 can be written down.

11. Step No. 11

The different harmonic amplitudes can be found by combining similar sine and cosine terms *i.e.* $A_n = \sqrt{a_n^2 + b_n^2}$

where A_n is the amplitude of the n_{th} harmonic.

Table No. 18.1

Wave form	*Appearance*	*Equation*
A. Sine wave	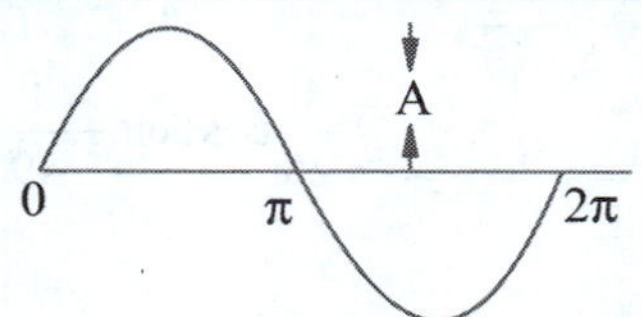	$f(t) = A = A \sin \omega t$
B. Half-wave rectified sin wave	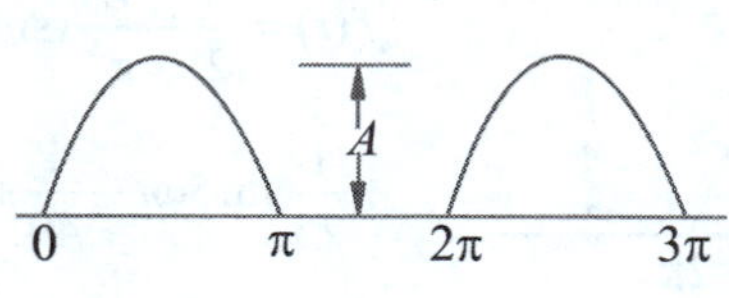	$f(t) = A\left(\frac{1}{\pi} + \frac{1}{2}\sin \omega t - \frac{2}{3\pi}\cos 2\omega t\right.$ $- \frac{2}{15\pi}\cos 4\omega t - \frac{2}{35\pi}\cos 6\,\omega t$ $\left.- \frac{2}{63\pi}\cos 8\omega t .....\right)$
C. Full-wave rectified sin wave		$f(t) = \frac{2A}{\pi}(1 - \frac{2}{3}\cos 2\omega t$ $- \frac{2}{15}\cos 4\omega t - \frac{2}{35}\cos 6\omega t .....)$
D. Rectangular or square wave	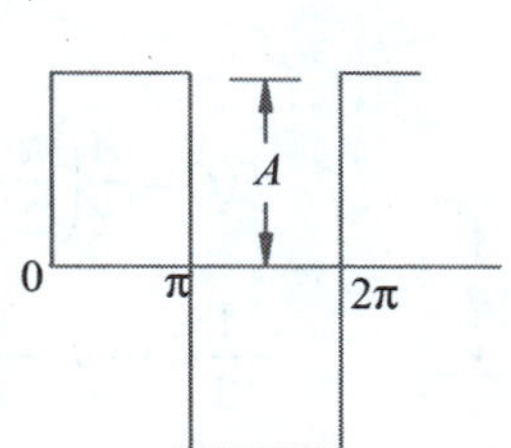	$f(t) = \frac{44}{\pi}(\sin \omega t + \frac{1}{3}\sin 3\omega t .....)$ $+ \frac{1}{5}\sin 5\omega t + \frac{1}{7}\sin 7\omega t + ...)$ $f(t) = \frac{4A}{\pi}(\cos \omega t - \frac{1}{3}\cos 3\omega t +$

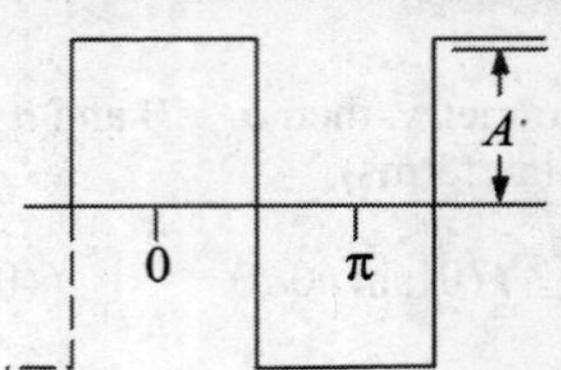

$\frac{1}{5}\cos 5\omega t........)$

E. Rectangular or square wave pulse

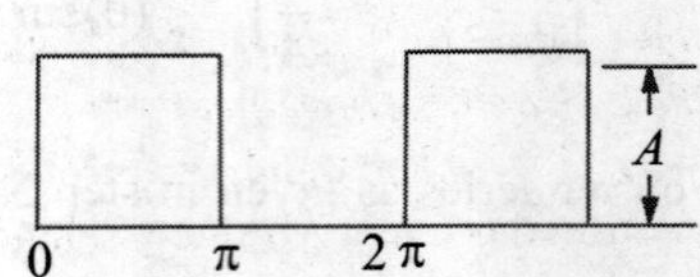

$f(t) = \frac{A}{2} + \frac{2A}{\pi}(\sin \omega t + \frac{1}{3}\sin 3\omega t + \frac{1}{5}\sin 5\omega t + \frac{1}{7}\sin 7\omega t + ...)$

F. Triangular wave

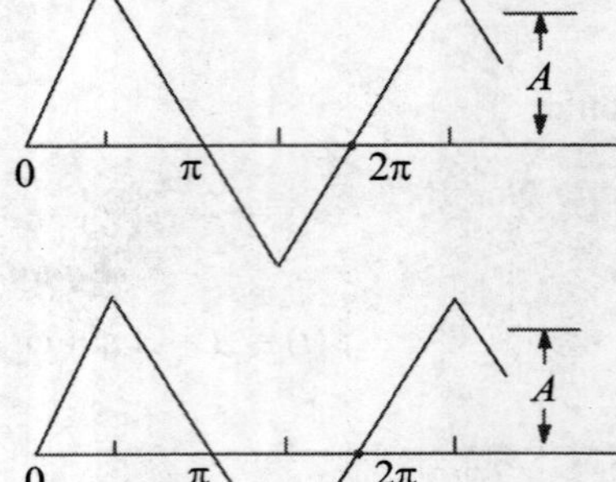

$f(t) = \frac{8A}{\pi^2}(\sin \omega t - \frac{1}{9}\sin 3\omega t + \frac{1}{25}\sin 5\omega t - \frac{1}{49}\sin 7\omega t + ...)$

$f(t) = \frac{8A}{\pi^2}(\cos \omega t + \frac{1}{9}\cos 3\omega t + \frac{1}{25}\cos 5\omega t + \frac{1}{49}\cos 7\omega t + ...)$

G. Triangular pulse

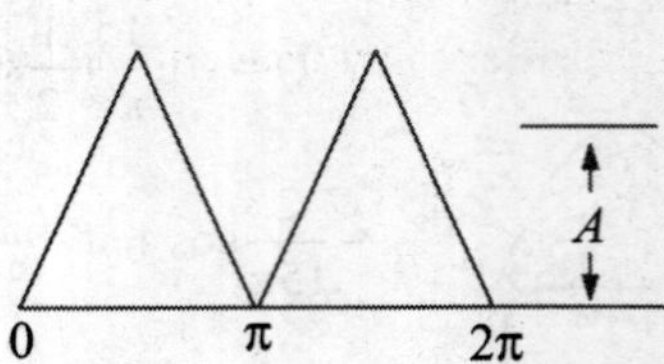

$f(t) = \frac{A}{2} + \frac{4A}{\pi^2}(\sin \omega t - \frac{1}{9}\sin 3\omega t + \frac{1}{25}\sin 5\omega t - \frac{1}{49}\sin 7\omega t + ...)$

H. Sawtooth wave

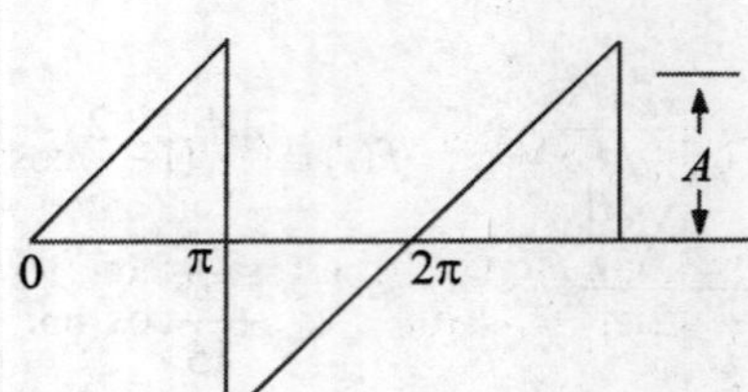

$f(t) = \frac{2A}{\pi}(\sin \omega t - \frac{1}{2}\sin 2\omega t + \frac{1}{3}\sin 3\omega t - \frac{1}{4}\sin 4\omega t + ...)$

I. Sawtooth pulse

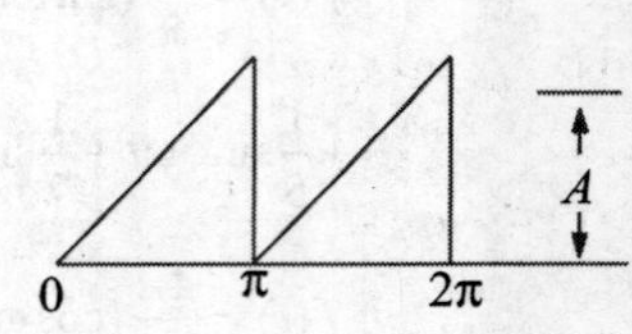

$f(t) = \frac{A}{\pi}\left(\frac{\pi}{2} - \sin \omega\ t - \frac{1}{2}\sin 2\omega t - \frac{1}{3}\sin 3\omega t - \frac{1}{4}\sin 4\omega t\right)$

J. Trapezoidal wave

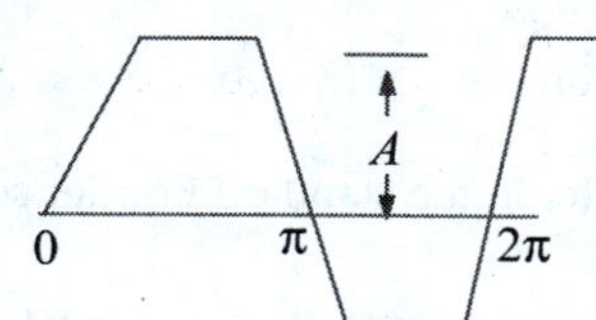

$$f(t)=\frac{3\sqrt{3}A}{\pi^2}\left(\sin\omega t-\frac{1}{25}\sin 5\omega t+\frac{1}{49}\sin 7\omega t......\right)$$

Example 18.1. *Find the Fourier series of the "half sinusoidal" voltage waveform which represents the output of a half-wave rectifier. Sketch its line spectrum.*

Solution. As seen from Fig. 18.8 (*a*), T = 0.2 second, $f_0 = 1/T = 1/0.2 = 5$ Hz and $\omega_0 = 2\pi\ f_0 = 10\pi$ rad/s. Moreover, the function has even symmetry. Hence, the Fourier series will contain no sine terms because $b_n = 0$.

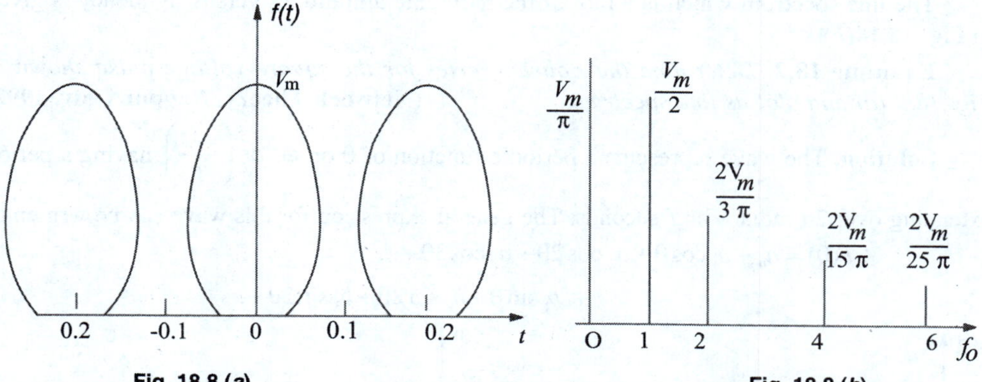

Fig. 18.8 (*a*) **Fig. 18.8 (*b*)**

The limits of integration would not be taken from $t = 0$ to $t = 0.2$ second, but from $t = -0.5$ to $t = 0.15$ second in order to get fewer equations and hence fewer integrals. The function can be written as

$$f(t)=V_m\cos 10\pi t \qquad -0.05<t<0.05$$

$$=0 \qquad 0.05<t<0.15$$

$$a_0=\frac{1}{T}\int_{-0.05}^{0.15}f(t)dt=\frac{1}{0.2}\left[\int_{-0.05}^{0.05}V_m\cos 10\pi t dt+\int_{0.05}^{0.15}0.dt\right]$$

$$=\frac{V_m}{0.2}\left|\frac{\sin 10\pi t}{10\pi}\right|_{-0.05}^{0.05}=\frac{V_m}{\pi}a_n=\frac{2V_m}{0.2}\int_{-0.05}^{0.05}\cos 10\pi t.\cos 10\pi nt\,dt$$

The expression we obtain after integration cannot be solved when $n = 1$ although it can be solved when n is other than unity. For $n = 1$, we have

$$a_1=10V_m\int_{-0.05}^{0.05}\cos^2 10\pi t\,dt=\frac{V_m}{2}$$

When $n\neq 1$, then $a_n=10V_m\int_{-0.05}^{0.05}\cos 10\pi t.\cos 10\pi n t dt$

$$=\frac{10V_m}{2}\int_{-0.05}^{0.05}[\cos 10\pi(1+n)t\cos 10\pi(1-n)t]dt=\frac{2V_m}{\pi}.\frac{\cos(\pi n/2)}{(1-n^2)} \quad ...n\neq 1$$

$$a_2=\frac{2V_m}{\pi}.\frac{\cos\pi}{1-4}=\frac{2V_m}{\pi}.\frac{-1}{-3}=\frac{2V_m}{2\pi};a_3=\frac{2V_m}{\pi}.\frac{\cos 3\pi/2}{1-3^2}=0;a_4=\frac{2V_m}{\pi}.\frac{\cos 2\pi}{1-4^2}=-\frac{2V_m}{15\pi}$$

$$a_5 = 0; a_6 = \frac{2V_m}{\pi}.\frac{\cos 3\pi}{1-6^2} = \frac{2V_m}{35\pi} \text{ and so on}$$

Substituting the values of a_0, a_1, a_2, a_4 etc. in the standard Fourier series expression given in Art. 20.3. we have

$$f(t) = a_0 + a_1 \cos 2\omega_0 t + a_2 \cos_2 \omega_0 t + a_4 \cos 4\omega_0 t + a_6 \cos 6\omega_0 t + ...$$

$$= \frac{V_m}{\pi} + \frac{V_m}{2}\cos 10\pi t + \frac{2V_m}{3\pi}\cos 20\pi t - \frac{2V_m}{15\pi}\cos 40\pi t + \frac{2V_m}{35}\cos 60\pi t....$$

$$= V_m\left(\frac{1}{\pi} + \frac{1}{2}\cos\omega_0 t + \frac{2}{3\pi}\cos 2\omega_0 t\pi - \frac{2}{15\pi}\cos 4\omega_0 t + \frac{2}{35\pi}\cos 6\omega_0 t...\right)$$

The line spectrum which is a plot of the harmonic amplitudes versus frequency is given in Fig. 18.14(*b*).

Example 18.2. *Determine the Fourier series for the square voltage pulse shown in Fig. 18.9 (a) and plot its line spectrum.* **(Network Theory, Nagpur Univ. 1992)**

Solution. The wave represents a periodic function of θ or ωt or $\left(\frac{2\pi t}{T}\right)$ having a period extending over 2π radians or T seconds. The general expression for this wave can be written as

$$f(\theta) = a_0 + a_1 \cos\theta + a_2 \cos 2\theta + a_3 \cos 3\theta + ...$$

$$...+ b_1 \sin\theta + b_2 \sin 2\theta + b_3 \sin 3\theta + ...$$

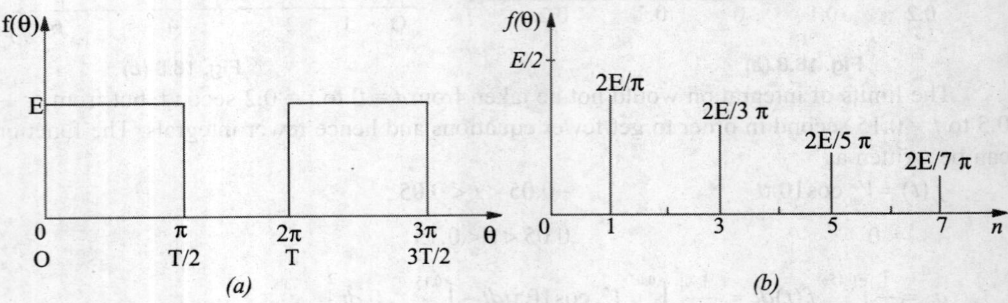

Fig. 18.9

(*i*) Now, $f(\theta) = V;\ \theta = 0 \text{ to } \theta = \pi;\ f(\theta) = 0, \text{ from } \theta = \pi \text{ to } \theta = 2\pi$

$$\therefore \quad a_0 = \frac{1}{2\pi}\int_0^{2\pi}(\theta)d\theta = \frac{1}{2\pi}\left\{\int_0^{\pi} f(\theta)d\theta + \int_{\pi}^{2\pi} f(\theta)d\theta\right\}$$

$$\text{or} \quad a_0 = \frac{1}{2\pi}\left\{\int_0^{\pi} Vd\theta + \int_{\pi}^{2\pi} 0d\theta\right\} = \frac{V}{2\pi}|\theta|_0^{\pi} + 0 = V_{2\pi} \times \pi = \frac{V}{2}$$

$$(ii)\ a_n = \frac{1}{\pi}\int_0^{2\pi} f(\theta)\cos n\theta d\theta = \frac{1}{\pi}\left\{\int_0^{\pi} V\cos n\theta d\theta + \int_{\pi}^{2\pi} 0 \times \cos n\theta d\theta\right\}$$

$$= \frac{V}{\pi}\int_0^{\pi}\cos n\theta d\theta + 0 = \frac{V}{n\pi}|\sin n\theta|_0^{\pi} = 0 \text{ ... whether } n \text{ is odd or even}$$

$$(iii)\ b_n = \frac{1}{\pi}\int_0^{2\pi} f(\theta)\sin n\theta d\theta = \frac{1}{\pi}\left\{\int_0^{\pi} V\sin n\theta d\theta + \int_{\pi}^{2\pi} 0 \times \sin n\theta d\theta\right\}$$

$$= \frac{V}{\pi}\int_0^{\pi} \sin n\theta d\theta + 0 = \frac{V}{\pi}\left|\frac{-\cos n\theta}{n}\right|_0^{\pi} = \frac{V}{n\pi}(-\cos n\pi + 1)$$

Now, when n is odd, $(1 - \sin n\pi) = 2$ but when n is even, $(1 - \cos n\pi) = 0$.

$$\therefore\ b_1 = \frac{2V}{\pi} \ ...n = 1;\ b_2 = \frac{V}{2\pi}\times 0 = 0\ ...\ n = 2;\ b_3 = \frac{V}{3\pi}\times 2 = \frac{2V}{3\pi}\ ...n = 3 \text{ and so on.}$$

Hence, substituting the values of a_0, a_1, a_2 etc. and b_1, b_2 etc. in the above given Fourier series, we get

$$f(\theta) = \frac{V}{2} + \frac{2V}{\pi}\sin\theta + \frac{2V}{3\pi}\sin 3\theta + \frac{2V}{5\pi}\sin 5\theta + ...$$

$$= \frac{E}{2} + \frac{2V}{\pi}\left(\sin\omega_0 t + \frac{1}{3}\sin 3\omega_0 t + \frac{1}{5}\sin 5\omega_0 t + ...\right)$$

It is seen that the Fourier series contains a constant term $V/2$ and odd harmonic components whose amplitudes are as under:

Amplitude of fundamental or first harmonic $= \dfrac{2V}{\pi}$

Amplitude of second harmonic $= \dfrac{2V}{3\pi}$

Amplitude of third harmonic $= \dfrac{2V}{5\pi}$ and so on.

The plot of harmonic amplitude versus the harmonic frequencies (called line spectrum) is shown in Fig. 18.15 (*b*).

Example 18.3. *Obtain the Fourier series for the square wave pulse train indicated in Fig. 18.16.* **(Network Theory and Design, AMIETE June, 1990)**

Solution. Here $T = 2$ second, $\omega_0 = 2\pi/T = \pi$ rad/s. The given function is defined by

$$f(t) = 1 \quad 0 < t < 1 = 0 \text{ and}$$

$$= 0; \quad 1 < t < 2$$

$$a_0 = \frac{1}{T}\int_0^T f(t)dt = \frac{1}{2}\int_0^2 1.dt = \frac{1}{2}\left[\int_0^1 1.dt + \int_1^2 0.dt\right] = \frac{1}{2}$$

Even otherwise by inspection $a_0 = (1 + 0)/2 = 1/2$

$$a_n = \frac{2}{T}\int_0^T f(t)\cos n\omega_0 t dt = \frac{2}{2}\int_0^2 1.\cos n\pi t dt = \left[\int_0^1 1.\cos n\pi t dt + \int_1^2 (0).\cos n\pi t dt\right]$$

$$= \int_0^1 \cos n\pi t dt = \left|\frac{\sin n\pi t}{n\pi}\right|_0^1 = 0$$

$$b_n = \frac{2}{T}\int_0^T f(t)\sin n\omega_0 t dt = [\int_0^1 1.\sin n\pi t\, dt] + \int_1^2 (0)\sin n\pi t dt] = \int_0^1 \sin n\pi t\, dt$$

$$= \left|\frac{-\cos n\pi t}{n\pi}\right|_0^1 = \frac{1 - \cos n\pi}{n\pi}$$

$\therefore \quad b_n = 2/n\pi$ ..... when n is odd; = 0 ... when n is even

$$\therefore \quad f(t) = a_0 + \frac{2}{\pi}\sum_{\substack{n=1\\odd}}^{\infty}\frac{1}{n}\sin 3n\pi t = \frac{1}{2} + \frac{2}{\pi}\left(\sin \pi t + \frac{1}{3}\sin 3\pi t + \frac{1}{5}\sin 5\pi\, t,\ etc.\right)$$

Example 18.4. *Find the trigonometric Fourier series for the square voltage waveform shown in Fig. 18.10(a) and sketch the line spectrum.*

Solution. The function shown in Fig. 18.10 (*a*) is an odd function because at any time $f(-t) = -f(t)$. Hence, its Fourier series will contain only sine terms *i.e.* $a_n = 0$. The function also possesses half-wave symmetry, hence, it will contain only odd harmonics.

The Fourier series for the above wave is given by

$$f(\theta) = \sum_{\substack{n=1\\odd}}^{\infty} b_n \sin n\theta t \quad \text{where } b_n = \frac{1}{\pi}\int_0^{2\pi} f(\theta)\sin n\theta d\theta$$

$$= \frac{1}{\pi}\left\{\int_0^{\pi} V\sin n\theta d\theta + \int_{\pi}^{2\pi} -V\sin n\theta d\theta\right\} = \frac{V}{\pi n}\left|-\cos n\theta\right|_0^{\pi} + \frac{V}{\pi n}\left|\cos n\theta\right|_{\pi}^{2\pi}$$

$$= \frac{V}{\pi n}\{(-\cos n\pi + \cos 0) - \cos \pi n\}$$

$$= \frac{V}{\pi n}\{(1-\cos n\pi) + (1-\cos n\pi)\} = \frac{2V}{\pi n}(1-\cos n\pi)$$

Now, $1 - \cos n\pi = 2$ when n is odd

and $= 0$ when n is even

$$\therefore \quad b_1 = \frac{2V}{\pi}\times 2 = \frac{4V}{\pi} \text{ ... putting } n = 1;\ b_2 = 0 \text{ ... putting } n = 2$$

$$b_3 = \frac{2V}{\pi 3}\times 2 = \frac{4V}{3\pi} \text{ ... putting } n = 3;\ b_4 = 0\text{... putting } n = 4$$

$$b_5 = \frac{2V}{\pi.5}\times 2 = \frac{4V}{5\pi} \text{ .. putting } n = 5 \text{ and so on.}$$

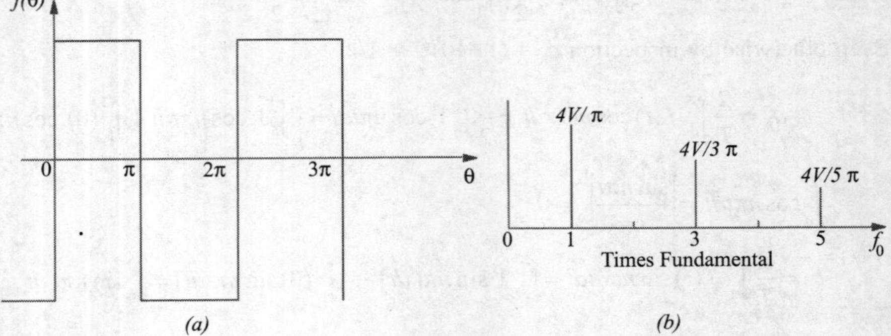

Fig. 18.10

Hence, the Fourier series for the given waveform is

$$f(\theta) = \frac{4V}{\pi}\sin\theta + \frac{4V}{3\pi}\sin 3\theta + \frac{4V}{5\pi}\sin 5\theta + ...$$

$$= \frac{4V}{\pi}\left(\sin\omega_0 t + \frac{1}{3}3\omega_0 t + \frac{1}{5}\sin 5\omega_0 t + ...\right) = \frac{4V}{\pi}\left(\sin\frac{2\pi}{T}t + \frac{1}{3}\sin\frac{6\pi}{T}t + \frac{1}{5}\sin\frac{10\pi}{T}t + ...\right)$$

The line spectrum of the function is shown in Fig. 18.10 (*b*). It would be seen that the harmonic amplitudes decrease as $1/n$, that is, the third harmonic amplitude is 1/3 as large as the fundamental, the fifth harmonic is 1/5 as large and so on.

Example 18.5. *Determine the Fourier series for the square voltage waveform shown in Fig. 18.10 (a). Plot its line spectrum.*

Solution. This is the same question as given in Ex. 18.6 but has been repeated to illustrated a singhtly different technique. As seen from Fig. 18.10(a) $T = 2\pi$, hence, $\omega_0 = 2\pi f_0 = 2\pi / T = 2\pi / \pi = 1$. Over one period the function can be defined as

$$f(t) = V \;\; 0 < t < 5$$

$$= -V, \; \pi < t < 2\pi$$

$$b_n = \frac{1}{\pi}\int_0^{2\pi} f(t)\sin n\omega_0 t dt + \frac{1}{\pi}\int_0^{2\pi} f(t)\sin ntdt$$

$$= \frac{1}{\pi}\left[\int_0^{\pi} f(t)\sin ntdt + \frac{1}{\pi}\int f(t)\sin ntdt\right]$$

$$= \frac{1}{\pi}\int_0^{\pi} V\sin ntdt + \frac{1}{\pi}\int_{\pi}^{2\pi}(-V)\sin ntdt = \frac{V}{\pi}\left|\frac{-\cos nt}{n}\right|_0^{\pi} + \frac{V}{\pi}\left|\frac{\cos nt}{n}\right|_{\pi}^{2\pi}$$

$$= -\frac{V}{n\pi}(\cos n\pi - \cos 0) + \frac{V}{n\pi}(\cos 2n\pi - \cos n\pi)$$

Since cos 0 is 1 and cos $2n\pi = 1 \quad \therefore \; b_n = \frac{2V}{n\pi}(1 - \cos n\pi)$

When n is even, $\cos n\pi = 1 \quad \therefore \quad b_n = 0$

When n is odd, cos $n\pi$ = – 1 $\therefore \; b_n = \frac{2V}{n\pi}(1+1) = \frac{4V}{n\pi}, n = 1, 3, 5...$

Substituting the value of b_n, the Fourier series become

$$f(t) = \sum_{\substack{n=1\\ odd}}^{\infty}\frac{4V}{n\pi}\sin nt = \frac{4V}{\pi}\sum_{\substack{n=1\\ odd}}^{\infty}\frac{1}{n}\sin nt = \frac{4V}{\pi}\left(\sin t + \frac{1}{3}\sin 3t + \frac{1}{5}\sin 5t + ...\right)$$

Since $\omega_0 = 1$, the above expression in general terms becomes

$$f(t) = \frac{4V}{\pi}\left(\sin\omega_0 t + \frac{1}{3}\sin 3\omega_0 t + \frac{1}{5}\sin 5\omega_0 t + ...\right)$$

The line spectrum is as shown in Fig. 18.10 (*b*).

Example 18.6. *Determine the Fourier series of the square voltage waveform shown in Fig. 18.11.*

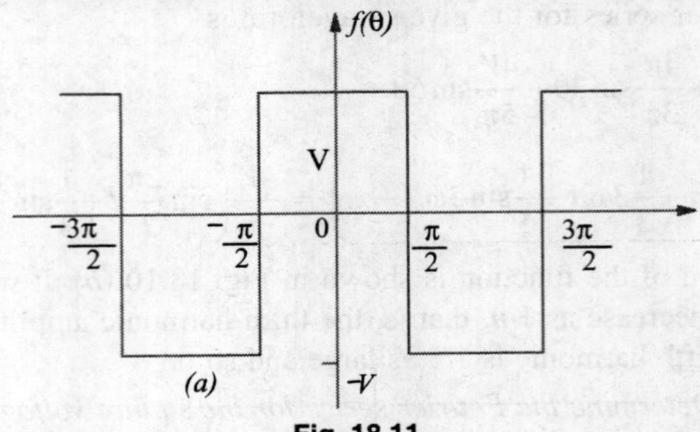

Fig. 18.11

Solution. As compared to Fig. 18.10 (*a*) given above, the vertical axis of figure has been shifted by $\pi/2$ radians. Replacing *t* by $(t+\pi/2)$ in the above equation, the Fourier series of the waveform shown in Fig. 18.11 becomes

$$f(t)=\frac{4V}{\pi}\left[\sin\left(t+\frac{\pi}{2}\right)+\frac{1}{3}\sin 3\left(t+\frac{\pi}{2}\right)+\frac{1}{5}\sin 5\left(t+\frac{\pi}{2}\right)+...\right]$$

$$=\frac{4V}{\pi}\left(\cos t-\frac{1}{3}\cos 3t+\frac{1}{5}\cos 5t...\right)$$

Example 18.7. *Determine the trigonometric Fourier series for the half-wave rectified sine wave form shown in Fig. 18.12 (a) and sketches line spectrum.*

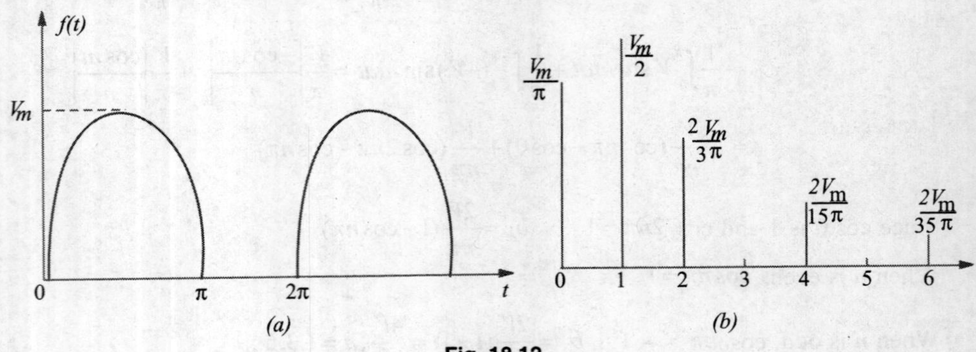

Fig. 18.12

Solution. The given waveform shows no symmetry, hence its series would contain both sine and cosine terms. Moreover, its average value is not obtainable by inpsection, hence it will have to be found by integration.

Here, $T=2\pi\omega\ =2\pi/T=1$. Hence, equation of the given waveform is $V=V_m\sin\omega t=V_m\sin t$.

The given waveform is defined by $f(t)=V_m\sin t, 0<t<\pi=0, \pi<t<2\pi$

$$a_0=\frac{1}{2\pi}\int_0^{2\pi}f(t)dt=\frac{1}{2\pi}\left[\int_0^{\pi}V_m\sin tdt+\int_{\pi}^{2\pi}dt=\frac{1}{2\pi}\int_0^{\pi}V_m\sin tdt\right]$$

$$=\frac{V_m}{2\pi}|-\cos t|_0^{\pi}=-\frac{V_m}{2\pi}(\cos\pi-\cos 0)=\frac{V_m}{\pi}$$

$$a_n = \frac{1}{\pi}\int_0^{2\pi} f(t)\cos ntdt = \frac{1}{\pi}\int_0^{\pi} V_m \sin t\cos ntdt$$

$$= \frac{V_m}{2\pi}\int_0^{\pi}[\sin(n+1)t - \sin(n-1)t]dt = \frac{V_m}{2\pi}\left|\frac{-\cos(n+1)t}{n+1} + \frac{\cos(n-1)t}{n-1}\right|_0^{\pi}$$

$$= \frac{V_m}{2\pi}\left[-\frac{\cos(n+1)\pi}{n+1} + \frac{\cos(n-1)\pi}{n-1} + \frac{1}{n+1} - \frac{1}{n-1}\right]$$

when n is even, $\cos(n+1)\pi = -1$ and $\cos(n-1)\pi = -1$

$$\therefore \quad a_n = \frac{V_m}{2\pi}\left[\frac{1}{n+1} - \frac{1}{n-1} + \frac{1}{n+1} - \frac{1}{n-1}\right] = -\frac{2V_m}{\pi(n^2-1)} \qquad \text{... } n = 2, 4, 6 \text{ etc.}$$

when n is odd and $\neq 1$, $\cos(n+1)\pi = 1$ and $\cos(n-1)\pi = 1$

$$\therefore \quad a_n = \frac{V_m}{2\pi}\left(-\frac{1}{n+1} + \frac{1}{n-1} + \frac{1}{n+1} - \frac{1}{n-1}\right) = 0 \qquad \text{... } n = 3, 5, 7 \text{ etc.}$$

when $n = 1$, $a_1 = \frac{1}{\pi}\int_0^{\pi} V_m \sin t.\cos tdt = \frac{V_m}{\pi}\int_0^{\pi}\sin t\cos tdt = \frac{V_m}{2\pi}\int_0^{\pi}\sin 2tdt = 0$

Hence, $a_n = 0$ ...$n = 1, 3, 5$...

$$= b_n = \frac{1}{\pi}\int_0^{2\pi} f(t)\sin ntdt = \frac{1}{\pi}\left[\int_0^{\pi} V_m \sin t\sin ntdt + \int_{\pi}^{2\pi}(0)\sin ntdt\right]$$

$$= \frac{V_m}{\pi}\int_0^{\pi}\sin t\sin ntdt = 0 \text{ for } n = 2, 3, 4, 5 \text{ etc.}$$

However, the expression indeterminate for $n = 1$ so that b_1 has to be evaluated separately.

$$b_1 = \frac{1}{\pi}\int_0^{\pi} V_m \sin t.\sin tdt = \frac{V_m}{\pi}\int_0^{\pi}\sin^2 tdt = \frac{V_m}{2}$$

The required Fourier series for the half-wave rectified voltage waveform is

$$f(t) = \frac{V_m}{\pi} + \frac{V_m}{2}\sin t - \frac{2V_m}{\pi}\sum_{\substack{n=1\\ even}}^{\infty}\left(\frac{\cos nt}{n^2-1}\right)$$

$$= \frac{V_m}{\pi}\left(1 + \frac{\pi}{2}\sin t - \frac{2}{3}\cos 2t - \frac{2}{15}\cos 4t - \frac{2}{35}\cos 6t\ldots\right)$$

$$= \frac{V_m}{\pi}\left(1 + \frac{\pi}{2}\sin\omega_0 t - \frac{2}{3}\cos 2\omega_0 t - \frac{2}{15}\cos\omega_0 t - \frac{2}{35}\cos 6\omega_0 t - \ldots\right)$$

$$= \frac{V_m}{T}\left(1 + \frac{\pi}{2}\sin\theta - \frac{2}{3}\cos 2\theta - \frac{2}{15}\cos 4\theta - \frac{2}{35}\cos 6\theta - \ldots\right)$$

The line spectrum is shown in Fig. 18.12 (*b*) which has a strong fundamental term with rapidly decreasing amplitudes of the higher harmonics.

Example 18.8. *Find the trigonometrical Fourier series for the full wave rectified voltage sine wave shown in Fig. 18.13.*

Solution. Since the given function has even symmetry, $b_n = 0$ *i.e.* it will contain no sine terms in its series.

The equation of the sinusoidal sine wave given by $V = V_m \sin\theta$. In other words, $f(\theta) = V_m = \sin\theta$.

$$a_0 = \frac{1}{2\pi}\int_0^{2\pi} f(\theta)d\theta = \frac{1}{2\pi}\int_0^{2\pi} V_m \sin\theta d\theta = \frac{2V_m}{2\pi}\int_0^{\pi} \sin\theta d\theta$$

It is so because the two parts $0 - \pi$ and $\pi - 2$ are identical.

$$\therefore \quad a_0 = \frac{V_m}{\pi}|-\cos\theta|_0^{\pi} = \frac{2V_m}{\pi}$$

$$a_n = \frac{1}{\pi}\int_0^{2\pi} f(\theta)\cos n\theta d\theta = \frac{2V_m}{\pi}\int_0^{\pi} \sin\theta\cos n\theta d\theta$$

Now, $\sin A \cos B = \frac{1}{2}[\sin(A+B)+\sin(A-B)]$

$$\therefore \quad a_n = \frac{2V_m}{\pi}\int_0^{\pi}[\sin(1+n)\theta + \sin(1-n)\theta]d\theta$$

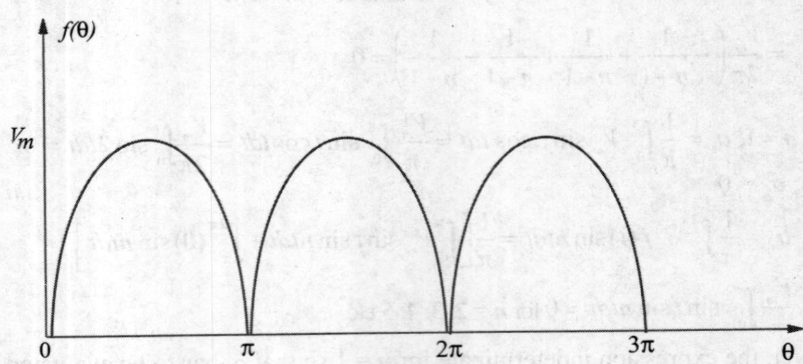

Fig. 18.13

$$= \frac{V_m}{\pi}\frac{\cos(1+n)\theta}{(1+n)} + \frac{\cos(1-n)\theta}{(1-n)}$$

$$= 0 = \frac{-V}{\pi}\left[\frac{\cos(1+n)\pi}{1+n} + \frac{\cos(1-n)\pi}{1-n} - \frac{1}{1+n} - \frac{1}{1-n}\right] \text{ ... when } n \text{ is odd}$$

However, when n is even, then

$$a_n = \frac{-V_m}{\pi}\left[\frac{1}{1+n} - \frac{1}{1-n} - \frac{1}{1+n} - \frac{1}{1-n}\right] = \frac{2V_m}{\pi}\left[\frac{1}{1+n} + \frac{1}{1-n}\right] = \frac{-4V_m}{\pi(n^2-1)}$$

$$\therefore \quad f(\theta) = a_0 - \frac{4V_m}{\pi}\sum_{\substack{n=2\\ even}}^{\infty}\frac{\cos 2\theta}{(n^2-1)}$$

$$\therefore \quad f(\theta) = \frac{2V_m}{\pi} - \frac{4V_m}{\pi}\left(\frac{1}{3}\cos^2\theta - \frac{1}{15}\cos 4\theta + \frac{1}{35}\cos 6\theta + \ldots\right)$$

Example 18.9. *Determine the Fourier series for the waveform shown in Fig. 18.14 (a) and sketch its line spectrum.*

Solution. Its is seen from Fig. 18.14 (*a*) that the waveform equation is $f(\theta) = (V_m / \pi)\theta$.

The given function $f(\theta)$ is defined by

$$f(\theta) = \left(\frac{V_m}{\pi}\right)\theta, \quad 0 < \theta < \pi$$

$$= 0, \; \pi < \theta < 2\pi$$

Since the function possesses neither even nor odd symmetry, it will contain both sine and cosine terms.

Average value of the wave over one cycle is $V_m/4$ or $a_0 = V_m/4$. It is so because the average value over the first half cycle is $V_m/2$ and over the second half cycle is 0 hence, the average value for full cycle is $= \frac{(V_m/2)+0}{2} = \frac{V_m}{4}$

$$a_n = \frac{1}{\pi}\int_0^{2\pi} f(\theta)n\theta d\theta = \frac{1}{\pi}\left[\int_0^{\pi}(V_m/\pi)\theta\cos n\theta d\theta + \int_{\pi}^{2\pi}(0)\cos n\theta d\theta\right]$$

$$= \frac{V_m}{\pi^2}\int_0^{\pi}\theta\cos n\theta d\theta = \frac{V_m}{\pi^2}\left|\frac{\cos n\theta}{n^2} + \frac{\theta}{n}\sin n\theta\right|_0^{\pi} = \frac{V_m}{\pi^2 n^2}(\cos n\pi - 1)$$

$\therefore \quad a_n = 0$ when n is odd because $\cos n\pi - 1 = 0$

$= -2V_m/\pi^2 n^2$ when n is even

$$b_n = \frac{1}{\pi}\int_0^{2\pi} f(\theta)\sin\theta d\theta = \frac{1}{\pi}\int_0^{\pi}(V_m/\pi)\theta\sin n\theta d\theta + \int_{\pi}^{2\pi}(0)\sin n\theta d\theta$$

$$= \frac{1}{\pi}\int_0^{\pi}\left(\frac{V_m}{\pi}\right)\theta\sin n\theta d\theta = \frac{V_m}{\pi^2}\left|\frac{\sin n\theta}{n^2} - \frac{\theta}{n}\cos n\theta\right|_0^{\pi} = \frac{-V_m}{\pi n}\cos n\pi$$

$\therefore \quad b_n = -V_m \mid \pi n$ when n is even $b_n = +V_m \mid \pi_n$ when n is odd

Substituting the values of various constants in the general expression for Fourier series, we get

$$f(\theta) = \frac{V_m}{4} - \frac{2V_m}{\pi^2}\cos\theta - \frac{2V_m}{(3\pi)^2}\cos 3\theta - \frac{2V_m}{(5\pi)^2}\cos 5\theta ... + \frac{V_m}{\pi}\sin\theta - \frac{V_m}{2\pi}\sin 2\theta + \frac{V_m}{3\pi}\sin 3\theta ....$$

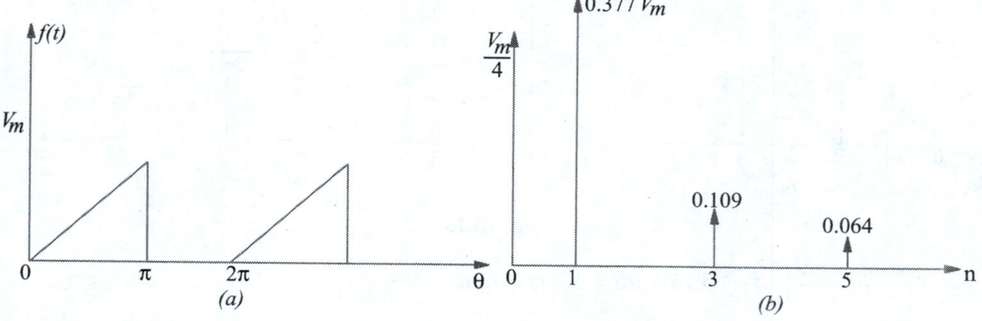

Fig. 18.14

The amplitudes of even harmonics are given directly by b_n but amplitudes of odd harmonics are given by $A_n = \sqrt{a_n^2 + b_n^2}$

For example,

$$A_1 = \sqrt{(2V_m/\pi^2)^2 + (V_m/\pi)^2} = 0.377\ V_m$$

$$A_3 = \sqrt{\left(\frac{2V_m}{(3\pi)^2}\right)^2 + \left(\frac{V_m}{2\pi}\right)^2} = 0.109\ V_m$$

$$A_5 = \left(\frac{2V_m}{(5\pi)^2}\right)^2 + \left(\frac{V_m}{5\pi}\right)^2 = 0.064\, V_m \text{ and so on.}$$

The line spectrum is as shown in Fig. 18.14 (*b*).

Example 18.10. *Find the Fourier series for the sawtooth waveform shown in Fig. 18.15 (a). Sketch its line spectrum.*

Solution. Using by the relation $y = mx$, the equation of the function becomes $f(t) = 1$, t or $f(t) = t$.

$$T = 2, \omega_0 = 2\pi / t = 2\pi / 2 = \pi$$

By inspection it is clear that $a_0 = 2/2 = 1$, $a_n = \frac{2}{T}\int_0^T f(t)\cos n\omega_0 t = \int_0^2 t.\cos n\pi t.dt$

Since we have to find the integral of two functions, we use the technique of integration by parts *i.e.*

$$\int uvdx = u\int vdx - \int\left(\frac{du}{dx}\int vdx\right)dx$$

$$\therefore \quad a_n = t.\int_0^2 \cos n\pi t dt - \int_0^2\left(1.\int_0^2 \cos n\pi t dt\right)dt$$

$$= \left|\frac{t}{n\pi}\sin n\pi t\right|_0^2 + \left|\frac{\cos n\pi t}{(n\pi)^2}\right|_0^2 = 0 + \frac{1}{(n\pi)^2}(\cos 2n\pi - \cos 0)$$

Since $\cos 2n\pi = \cos 0$ for all values of n, hence $a_n = 0$

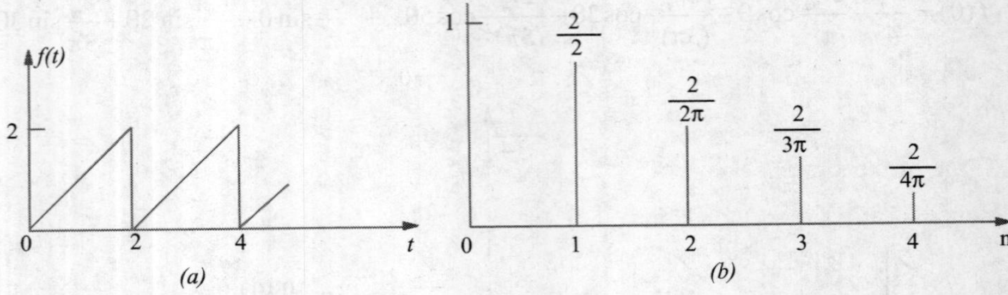

Fig. 18.15

$$b_n = \frac{2}{T}\int_0^T f(t)\sin n\omega_0 t dt = \int_0^T t\sin n\pi t dt$$

Employing integration by parts, we get,

$$b_n = t\int_0^2 \sin n\pi t\, dt - \int_0^2\left(1.\int_0^2 \sin n\pi t dt\right)dt$$

$$= \left|t.\frac{-\cos n\pi t}{n\pi}\right|_0^2 dt - \int_0^2 \frac{-\cos n\pi t}{n\pi} = \left|\frac{-t}{n\pi}.\cos n\pi t\right|_0^2 + \left|\frac{\sin n\pi t}{(n\pi)^2}\right|_0^2 = \frac{\sin 2n\pi}{(n\pi)^2} - \frac{2}{n\pi}\cos 2n\pi$$

The sine term is 0 for all values of n because sign of any multiple of 2π is 0. Since value of cosine term is 1 for any multiple of 2π, we have, $b_n = -2/n\pi$.

$$\therefore \quad f(t) = a_0 + \sum_{n=1}^{\infty} b_n \sin n\omega_0 t = a_0 - \frac{2}{\pi}\sum_{n=1}^{\infty} \frac{1}{n}\sin n\pi t$$

$$= 1 - \frac{2}{\pi}\left(\sin \pi t + \frac{1}{2}\sin 2\pi t + \frac{1}{3}\sin 3\pi t + ...\right)$$

The line spectrum showing the amplitudes of various harmonics is shown in Fig. 18.15 (*b*).

Example 18.11. *Determine the trigonometric Fourier series of the triangular waveform shown in Fig. 18.16.*

Solution. Since the waveform possesses odd symmetry, hence $a_0 = 0$ and $a_n = 0$ *i.e.* there would be n_o cosine terms in the series. Moreover, the waveform has half-wave symmetry. Hence, series will have only odd harmonics. In the present case, there would be only odd sine terms. Since the waveform possesses quarter-wave symmetry, it is necessary to integrate over only one quarter period of finding the Fourier coefficients.

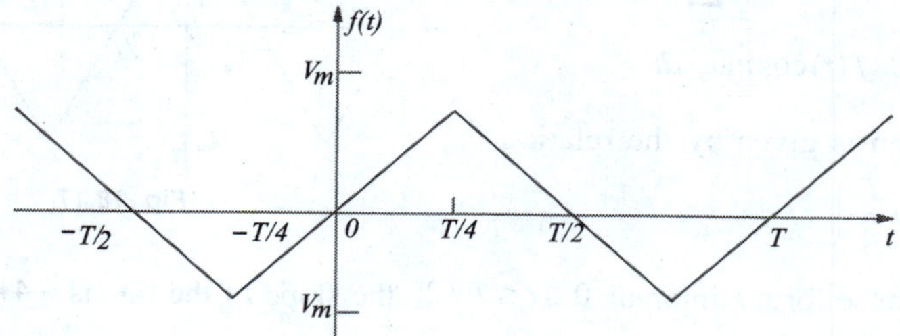

Fig. 18.16

$$f(t) = \sum_{\substack{n=1 \\ odd}}^{\infty} b_n \sin n\omega_0 t$$

$$\text{where} \quad b_n = \frac{8}{T}\int_0^{t/4} f(t) \sin n\omega_0 t$$

The quarter-wave of the given waveform can be represented by equation of a straight line. Slope of the straight line is $V_m/(T/40) = 4V_m/T$.

Hence, using $Y = mx$, we have

$$f(t) = \left(\frac{4V_m}{T}\right) t 0 < t < T/4 \quad \therefore \; b_n = \frac{8}{T}\int_0^{T/4}\left(\frac{4V_m}{T}\right) t.\sin n\omega_0 t = \frac{32V_m}{T^2}\int_0^{T/4} t.\sin n\omega_0 t dt$$

Using the theorem of integration by parts, we have

$$b_n = \frac{32V_m}{T^2}\left[t\int_0^{t/4} \sin n\omega_0 t \, dt - \int_0^{T/4}\left(1.\int_0^{T/4} \sin n\omega_0 t \, dt\right)dt\right]$$

$$= \frac{32V_m}{T^2}\left[\left(\frac{t}{4}\cdot\frac{-\cos n\omega_0 T/4}{n\omega_0}\right) + \left|\frac{\sin n\omega_0 t}{(n\omega_0)^2}\right|^{T/4} = \frac{32V_m}{T^2}\left[\left(\frac{T}{4}\frac{\cos n\omega_0 T/4}{n\omega_0}\right) + \left(\frac{\sin n\omega_0 T/4}{(n\omega_0)^2}\right)\right]\right.$$

$$\text{Now,} \quad \omega_0 = 2\pi/T \text{ or } \omega_0 T = 2\pi \quad \therefore \quad n\omega_0 T/4 = n\pi/2$$

$\therefore \quad \cos n\omega_0 T/4 = \cos n\pi/2 = 0$ when n is odd

$$\therefore \quad b_n = \frac{32V_m}{n^2\omega_0^2T^2}\sin n\omega_0\frac{T}{4} = \frac{32V_m}{n^2(2\pi)^2}\sin\frac{n\pi}{2} = \frac{8V_m}{n^2\pi^2}\sin\frac{n\pi}{2}$$

$$\therefore \quad b_n = \frac{8V_m}{n^2\pi^2} \ldots n = 1, 5, 9, 13 \ldots b_n = \frac{-8V_m}{n^2\pi^2} \ldots n = 3, 7, 11, 15, \ldots$$

Substituting this value of b_n, the Fourier series for the given waveform becomes

$$f(t) = \frac{8V_m}{\pi^2}\left(\sin\omega_0 t - \frac{1}{3^2}3\omega_0 t + \frac{1}{5^2}\sin 5\omega_0 t - \frac{1}{7^2}\sin 7\omega_0 t + \ldots\right)$$

Example 18.12. *Determine the Fourier series of the triangular waveform shown in Fig. 18.17.*

Solution. Since the function has even symmetry, $b_n = 0$. Moreover, it also has half-wave symmerty, hence, $a_0 = 0$. The Fourier series can be written as $f(t) = \sum_{n=1}^{\infty} a_n \cos n\omega_0 t$

where $a_n = \frac{2}{T}\int_0^T f(t)\cos n\omega_0 t\, dt.$

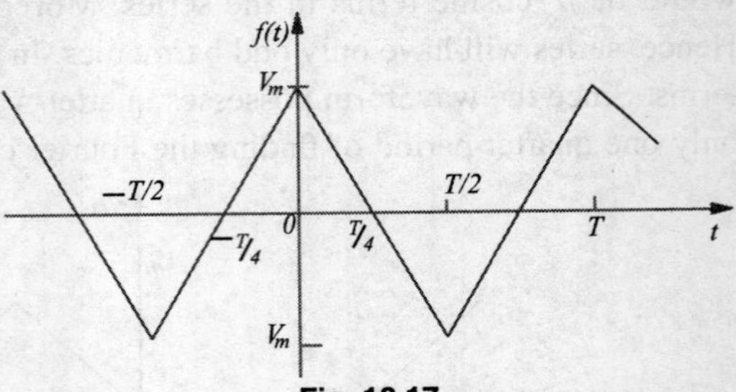

Fig. 18.17

The function is given by the relation

$$f(t) = \frac{-4V_m}{T}\left(t - \frac{T}{4}\right)$$

It is so because for the interval $0 \le t \le T/2,$ the slope of the line is $-4V_m/T.$

$$\therefore \quad a_n = \frac{2}{T}\int_0^T f(t)\cos n\omega_0 t\, dt = \frac{4}{T}\int_0^{T/4} f(t)\cos n\omega_0 t\, dt$$

$$= \frac{-16V_m}{T^2}\int_0^{T/2}\left(t - \frac{T}{4}\right)\cos n\omega_0 t\, dt = \frac{-16V_m}{T^2}\int_0^{T/2} t\cos n\omega_0 t\, dt + \frac{4V_m}{T}\int_0^{T/2} \cos n\omega_0 t\, dt$$

$$= \frac{-16V_m}{T^2}\left|\frac{1}{4n^2\omega_0^2}\cdot\cos n\omega_0 t + \frac{t}{n\omega_0}\sin n\omega_0 t\right|_0^{T/2} + \frac{4V}{T}\left|\frac{\sin n\omega_0 t}{n\omega_0}\right|_0^{T/2}$$

Substituting $\omega = 2\pi/T,$ we get

$$a_n = \frac{-16V_m}{T^2}\left[\frac{T^2}{4\pi^2 n^2}(\cos n\pi - 1) + \frac{T^2}{4\pi n}(\sin n\pi)\right] + \frac{2V}{\pi n}\cdot\sin n\pi$$

Now, $\sin n\pi = 0$ for all values of n, $\cos n\pi = 1$ when n is even add -1 when n is odd.

$$\therefore \quad a_n = \frac{8V_m}{\pi^2 n^2}$$

$$\therefore \quad f(t) = \sum_{\substack{n=0\\odd}}^{\infty}\frac{8V_m}{\pi^2 n^2}\cos n\omega_0 t = \frac{8V_m}{\pi^2}\sum_{\substack{n=1\\odd}}^{\infty}\frac{\cos n\omega_0 t}{n^2}$$

$$= \frac{8V_m}{\pi^2}\left(\cos\omega_0 t + \frac{1}{9}\cos 3\omega_0 t + \frac{1}{25}.\cos 5\omega_0 t + \frac{1}{49}\cos 7\omega_0 t + \ldots\right)$$

Alternative Solution

We can deduce the Fourier series from Fig. of Ex. 18.9 by shifting the vertical axis by π/2 radians to the right. Replacing t by $(t+\pi/2)$ in the Fourier series of Ex. 18.9, we get

$$f(t) = \frac{8V}{\pi^2}\left[\sin\omega_0\left(t+\frac{\pi}{2}\right)-\frac{1}{3^2}\sin 3\omega_0\left(t+\frac{\pi}{2}\right)+\frac{1}{5^2}\sin 5\omega_0\left(t+\frac{\pi}{2}\right)\right.$$

$$\left.-\frac{1}{7^2}\sin 7\omega_0\left(t+\frac{n}{2}\right)+...\right]$$

$$= \frac{8V_m}{\pi^2}\left(\cos\omega_0 t+\frac{1}{9}\cos 3\omega_0 t+\frac{1}{25}\cos 5\omega_0 t+\frac{1}{49}\cos\omega_0 t+...\right)$$

Example 18.13. *Obtain the Fourier series representation of the sawtooth waveform shown in Fig. 18.18 (a) and plot its spectrum.*

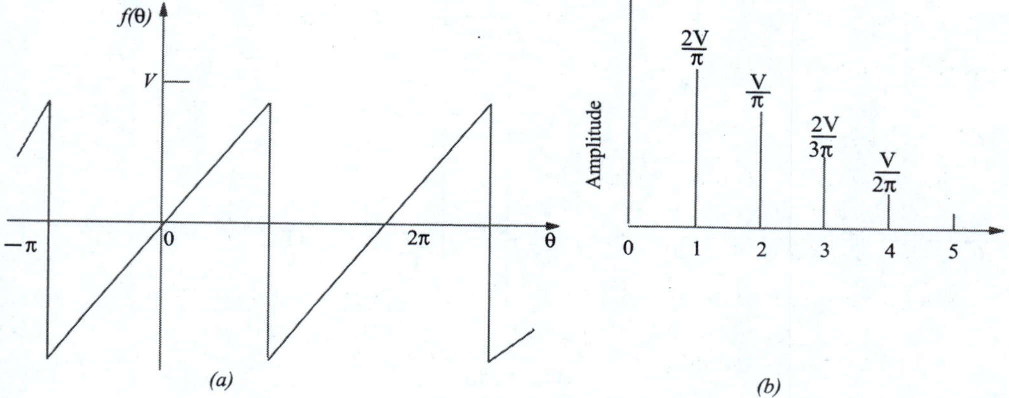

Fig. 18.18

Solution. By inspection, we know that the average value of the wave is zero over a cycle because the height of the curve below and above the X-axis is the same hence, $a_0 = 0$. Moreover, it has odd symmetry so that $a_n = 0$ *i.e.* there would be no cosine terms. The series will contain only sine terms.

$$\therefore\ f(\theta) = \sum_{n=1}^{\infty} b_n \sin n\theta \text{ where } b_n = \frac{1}{\pi}\int_0^{2\pi} f(\theta)\sin n\theta\, d\theta$$

The slope of curve is $m = V/\pi$

$\therefore$ we get, $f(\theta) = (V/\pi)\,\theta$.

If we are the limit of integration form $-\pi$ to $+\pi$ then

$$b_n \frac{1}{\pi}\int_{-\pi}^{\pi}\left(\frac{V}{\pi}\right)\theta\sin n\theta\, d\theta = \frac{V}{\pi^2}\left|\frac{1}{n^2}\sin n\theta-\frac{\theta}{n}\cos n\theta\right|_{-\pi}^{+\pi} = \frac{-2V}{n\pi}\cos n\pi$$

The above result has been obtained by making use of integration by parts as explained earlier. $\cos n\pi$ is positive when n is even and is negative when n is odd and thus the signs of the coefficients alternate. The required Fourier series is

$$f(\theta) = \frac{2V}{\pi}\left(\sin\theta-\frac{1}{2}\sin 2\theta+\frac{1}{3}\sin 3\theta-\frac{1}{4}\sin 4\theta+...\right)$$

or $f(t) = \frac{2V}{\pi}\left(\sin \omega_0 t - \frac{1}{2}\sin 2\omega_0 t + \frac{1}{3}\sin 3\omega_0 t - \frac{1}{4}\sin 4\omega_0 t + ...\right)$

As seen, the coefficients decrease as $1/n$ so that the series converges slowly as shown by the line spectrum of Fig. 18.18 (*b*).

The amplitudes of the fundamental of first harmonic, second harmonic, third harmonic and fourth harmonic are $(2/\pi), (2V/2\pi), (2V/3\pi)$ and $(2V/4\pi)$ respectively.

Tutorial Problem 18.1

1. Determine the Fourier series for the triangular waveform shown in Fig. 18.19 (*a*)

(Network Theory and Design, AMIETE June 1990)

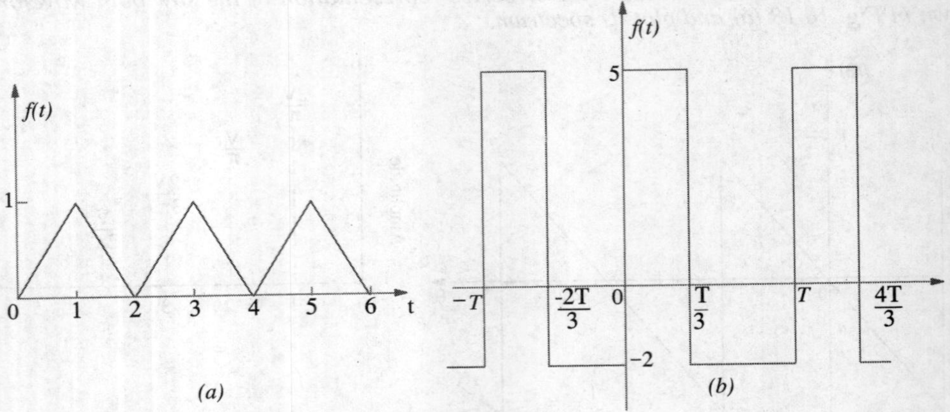

Fig. 18.19

$$\left[\frac{1}{2} - \frac{4}{\pi^2}\cos\omega_0 t + \frac{1}{3^2}\cos 3\omega_0 t + \frac{1}{5^2}\cos 5\omega_0 t + ...\right]$$

2. Find the values of the Fourier coefficients a_0, a_n and b_n for the function given in fig. 18.26 (*b*).

$$\left[\mathbf{a_0} = \frac{2}{3}; \mathbf{a_n} = \frac{7}{n\pi}\sin\frac{2\pi n}{3}; \mathbf{b_n} = \frac{7}{n\pi}\left(1 - \cos\frac{2\pi n}{3}\right)\right]$$

3. Determine the trigonometric series of the triangular waveform shown in Fig. 18.20. Sketch its line spectrum.

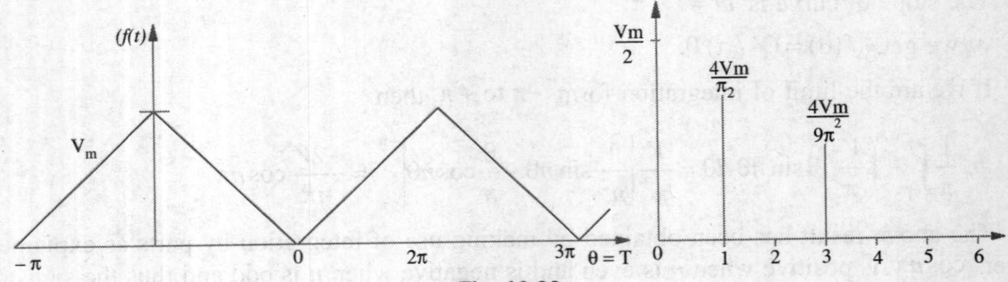

Fig. 18.20

$$\left[\frac{V_m}{2} + \frac{4V_m}{\pi^2}\left(\cos\omega_0 t + \frac{1}{3^2}\cos 3\omega_0 t + \frac{1}{5^2}\cos 5\omega_0 t + ...\right)\right]$$

4. Determine the Fourier series for the sawtooth waveform shown in Fig. 18.21.

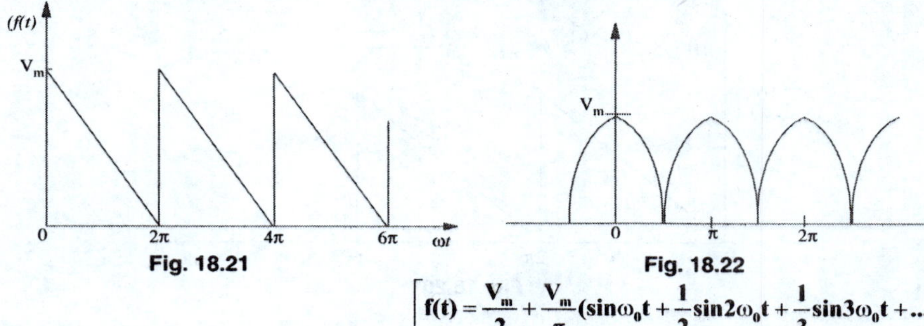

Fig. 18.21 Fig. 18.22

$$\left[f(t)=\frac{V_m}{2}+\frac{V_m}{\pi}(\sin\omega_0 t+\frac{1}{2}\sin 2\omega_0 t+\frac{1}{3}\sin 3\omega_0 t+....\right]$$

5. Represent the full-wave rectified voltage sine waveform shown in Fig. 18.22 by a Fourier series.

$$\left[f(t)=\frac{2V_m}{\pi}+\left(1+\frac{2}{3}\cos 2\omega t-\frac{2}{15}\cos 4\omega t+\frac{2}{35}\cos 6\omega t...\right)\right]$$

6. Obtain trigonometric Fourier series for the wave form shown in Fig. 18.23.

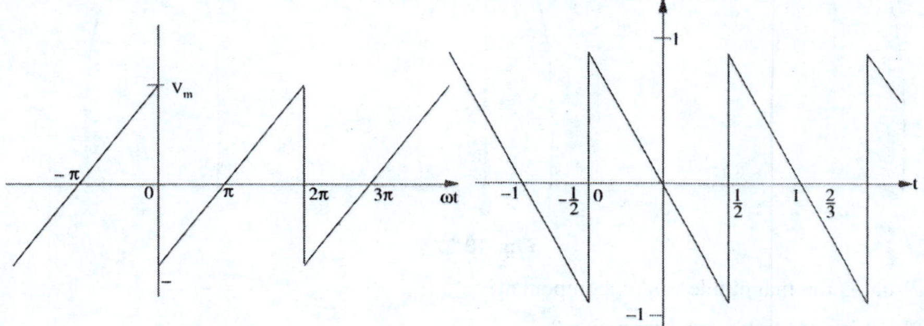

Fig. 18.23 Fig. 18.24

$$\left[f(t)=-\frac{2V_m}{\pi}\left(\sin\omega_0 t+\frac{1}{2}\sin 2\omega_0 t+\frac{1}{3}\sin 3\omega_0 t+\frac{1}{4}\sin 4\omega_0 t...\right)\right]$$

7. Find the Fourier series for the sawtooth waveform shown in Fig. 18.24.

$$\left[f(t)=\frac{2}{\pi}\left(-\sin 2\pi t+\frac{1}{2}\sin 4\pi t-\frac{1}{3}\sin 6\pi t+...\right)\right]$$

8. For the waveform of Fig. 18.25, find the Fourier series terms up to the 5th harmonic.

(Network Theory Nagpur Univ. 1993)

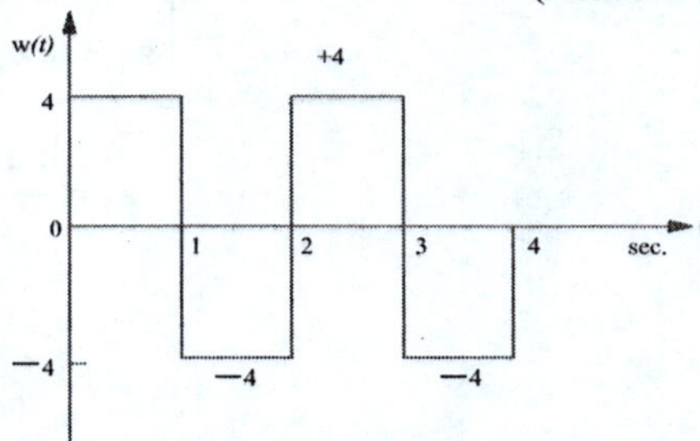

Fig. 18.25

9. Determine Fourier series of a repetitive triangular wave as shown in Fig. 18.26.

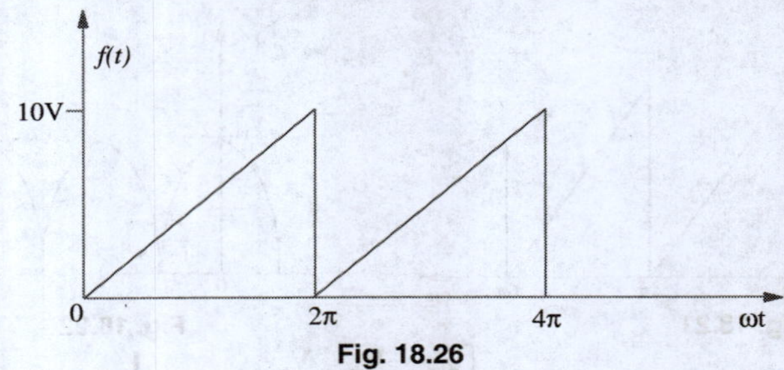

Fig. 18.26

ω(t)

V(t) Vm

0.4 0.2 0 0.2 0.4 t

Fig. 18.27

(a) What is the magnitude of d.c. component?

(b) What is the fundamental frequency?

(c) What is the magnitude of the fundamental?

(d) Obtain its frequency spectrurm. **(*Network Theory Nagpur Univ.1993*)**

[(*a*) 5 V (*b*) 1 Hz (*c*) $10/\pi$ volt]

10. Determine the Fourier series of voltage responses obtained at the o/p of a half wave rectifier shown in Fig. 18.27. Plot the discrete spectrum of the waveform.

(*Elect. Network Analysis Nagpur Univ. 1993*)

11. Determine the Fourier coefficients and plot amplitude and phase spectral.

(*Network Analysis Nagpur Univ. 1993*)

$$\left[\mathbf{a_3 = 0, b_n = 0, a_1 = \frac{4V_m}{\pi}, a_2 = 0} \right.$$

$$\left. \mathbf{a_3 = \frac{-4V_m}{3\pi}, a_4 = 0, a_5 = \frac{4V_m}{\pi}} \right]$$

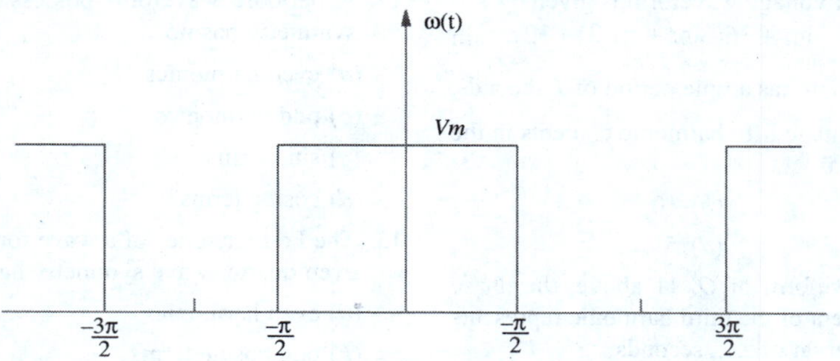

Fig. 18.28

OBJECTIVE TESTS – 18

1. A given function $f(t)$ can be represented by a Fourier series if it
   (a) is periodic
   (b) is single valued
   (c) has a finite number of maxima and minima in any one period
   (d) all of the above.
2. In a Fourier series expansion of a periodic function, the coefficient a_0 represents its
   (a) net area per cycle
   (b) d.c value
   (c) average value over half cycle
   (d) average a.c value per cycle
3. If in the Fourier series of a periodic function, the coefficient a_0 is zero, it means that the function has
   (a) odd symmetry
   (b) even quarter-wave symmetry
   (c) odd quarter-wave symmetry
   (d) any of the above.
4. A periodic function $f(t)$ is said to possess odd quarter-wave symmetry if
   (a) $f(t) = f(-t)$
   (b) $f(-t) = -f(t)$
   (c) $f(t) = -f(t + T/2)$
   (a) both (b) and (c).
5. If the average value of a periodic function over one period is zero and it consists of only odd harmonics then it must be possessing _______ symmetry.
   (a) half-wave
   (b) even quarter-wave
   (c) odd quarter-wave
   (d) odd.
6. If in the Fourier series of a periodic function, the coefficient $a_0 = 0$ and $a_n = 0$, then it must be having _______ symmetry.
   (a) odd
   (b) odd quarter-wave
   (c) even
   (d) either (a) or (b).
7. In the case of a periodic function possessing half-wave symmetry, which Fourier coefficient is zero?
   (a) a_n (b) b_n
   (c) a_0 (d) none of above.
8. A periodic function has zero average value over a cycle and its Fourier series consists of only odd cosine terms. What is the symmetry possessed by this function.
   (a) even (b) odd
   (c) even quarter-wave
   (d) odd quarter-wave
9. Which of the following periodic function possesses even symmetry ?
   (a) cos 3t (b) sin t
   (c) t . cos 50 t (d) $(t + t^2 + t^5)$.
10. If the Fourier coefficient b_n of a periodic function is zero, then it must possess _______ symmetry.
    (a) even
    (b) even quarter-wave
    (c) odd
    (d) either (a) and (b).

11. A complex voltage waveform is given by $V = 120 \sin \omega t + 36(3\omega t + \pi/2) + 12 \sin (5\omega t + \pi)$. It has a time period of T seconds.

The percentage fifth harmonic contents in the waveform is

(*a*) 12 (*b*) 10
(*c*) 36 (*d*) 5

12. In the waveform of *Q*. 11 above, the phase displacement of the third harmonic represents a time interval of ____ seconds.

(*a*) T/12 (*b*) T/3
(*c*) 3T (*d*) T/36

13. When the negative half-cycle of a complex waveform is reversed, it becomes identical to its positive half-cycle. This feature indicates that the complex waveform is composed of

(*a*) fundamental
(*b*) odd harmonics (*c*) even harmonics
(*d*) both (*a*) and (*b*) (*e*) both (*a*) and (*c*)

14. A periodic waveform possessing half-wave symmetry has no

(*a*) even harmonics
(*b*) odd harmonics
(*c*) sine terms
(*d*) cosine terms

15. The Fourier series of a wave form possessing even quarter-wave symmetry has only

(*a*) even harmonics
(*b*) odd cosine terms
(*c*) odd sine terms
(*d*) both (*b*) and (*c*).

16. The Fourier series of a waveform possessing odd quarter-wave symmetry contains only

(*a*) even harmonics
(*b*) odd cosine terms
(*c*) odd cosine terms
(*d*) none of above

ANSWERS

1. (*d*)	**2.** (*b*)	**3.** (*d*)	**4.** (*d*)	**5.** (*a*)	**6.** (*d*)	**7.** (*c*)	**8.** (*c*)
9. (*a*)	**10.** (*d*)	**11.** (*b*)	**12.** (*a*)	**13.** (*d*)	**14.** (*a*)	**15.** (*b*)	**16.** (*c*)

CHAPTER

19

Transients

Chapter Outline

- ❖ Introduction
- ❖ Types of Transients
- ❖ Important Differential Equations
- ❖ Transients in R-L Circuits (D.C.),
- ❖ Short Circuit Current
- ❖ Time Constant
- ❖ Transients in R-L Circuits (A.C.)
- ❖ Transients in R-C Series Circuits (D.C.)
- ❖ Transients in R-C Series Circuits (A.C)
- ❖ Double Energy Transients

19.1. Introduction

It is quite an easy job to calculate the steady current, which flows in a circuit, when it is connected to a d.c. generator or a battery. Similarly, the alternating current which flows in a circuit when connected to an alternator can also be calculated by the various method discussed in Chapters 13 and 14. These currents are known as ***steady*** currents because in such cases, it is assumed that **(*i*)** the circuit components are constant and **(*ii*)** the circuit has been connected to the generator long enough for any disturbance produced on initial switching, to resolve itself.

In general, transients disturbances are produced whenever

(*a*) an apparatus or circuit is ***suddenly*** connected to or disconnected from the supply,

(*b*) a circuit is shorted and

(*c*) there is a ***sudden*** change in the applied voltage from one finite value to another.

We will now discuss the transients produced whenever different circuits are suddenly switched on or off from the supply voltage. In each case, we will assume that the resultant current consists of two parts **(*i*)** a final steady-stage or normal current and **(*ii*)** a transient current superimposed on the steady-stage current.

It is essential to remember that the transient currents are not driven by any part of the applied voltage but are entirely associated with the changes in the stored energy in inductors and capacitors. Since there is no stored energy in resistors, ***there are no transients in pure resistive circuits.***

19.2. Types of Transients

There are ***single-energy*** transients and ***double-energy*** transients. Single-energy transients are those in which only one form of energy, either electromagnetic or electrostatic is involved as in *R-L* and *R-C* circuits.

However, double-energy transients are those in which both electromagnetic or electrostatic is involved as in *R-L-C* circuits Transient disturbances may be further classified as follows :

(*a*) Initiation Transients : These are produced when a circuit, which is originally dead, is energised.

(*b*) Subsidence Transients : These are produced when an energised circuit is rapidly de-energised and reaches an eventual steady-stage of zero current or voltage, as in the case of short-circuiting an *R-L* or *R-C* circuit suddenly.

(*c*) Transition Transients : These are due to sudden but energetic changes from one steady state to another.

(*d*) Complex Transients : These are produced in a circuit which is simultaneously subjected to two transients due to two independent disturbances or when the disturbing force producing the transients is itself variable.

(*e*) Relaxation Transients : In these transients, the transition occurs cyclically towards states, which when reached, become unstable themselves.

A distinction may also be made between free and forced transients which are produced due to the applied voltage being itself transient.

19.3. Important Differential Equations

Some of the important differential equations, used in the treatment of single and double energy transients, are given below. We will consider both first-order and second-order differential equations.

1. First Order Equations

(*i*) Let $\dfrac{dy}{dx} + ay = 0$ where *a* is a constant.

Its solution is $y = k\,e^{-ax}$ where *k* is the constant of integration whose value can be found from the boundary conditions *i.e.* conditions prevalent at the instant when the voltage to a circuit is applied or excluded.

(*ii*) If $\dfrac{dy}{dx} + ay = b$ where *a* and *b* are constants, then solution is $y = \dfrac{b}{a} + k\,e^{-ax}$

The value of *k* can again be found from boundary conditions.

(*iii*) If $\dfrac{dy}{dx} + Ay = B$

where *A* and *B* are not constants but are functions of *x*, then the solution is given by

$$y = e^{-\int A\cdot dx} \int e^{\int A\cdot dx} B dx + k e^{-\int A\cdot dx}$$

If $A = a =$ constant, then the above equation simplifies to

$$y = e^{-ax} \int e^{ax}\, B\, dx + ke^{-ax}$$

2. Second Order Equations

(*i*) Suppose $\dfrac{d^2y}{dx^2} + \text{a}\dfrac{dy}{dx} + by = 0$ where *a* and *b* are constants, then the solution is

$$y = k_1 e^{\lambda_2 x} + k_2 e^{\lambda_2 x}$$

where $_1$ and λ_2 are constants of integration and whose values are,

$$\lambda_1 = -\frac{a}{2} - \sqrt{\left(\frac{a^2}{4} - b\right)} \text{ and } \lambda_2 = -\frac{a}{2} + \sqrt{\left(\frac{a^2}{4} - b\right)}$$

(a) If $a^2/4 > b$, the roots are real and the above solution can be applied without any difficulty.

(b) If $a^2/4 < b$, the radicals contain a negative quantity. In that-case, the solution is given by

$$y = e^{-\frac{1}{2}ax}(k_3 \sin \lambda_0 x + k_4 \cos \lambda_0 x)$$

where k_3 and k_4 are the new constants of integration and

$$\lambda_0 = \sqrt{b - \frac{a^2}{4}}$$

(c) If $a^2/4 = b$, then both roots are equal and each is $= -a/2$.

Hence, in this case, the solution becomes $y = k_5 e^{\lambda t} + k_6 t.e^{\lambda t}$

(ii) Let $\dfrac{d^2y}{dx^2} + a\dfrac{dy}{dx} + by = c$

where *a*, *b* and *c* are constant. In this case also, the solution will again depend on the root as discussed above.

$$y = k_1 e^{\lambda} 1^x + k_2 e_{\lambda 2} x + c/b$$

(iii) (a) Let the differential equation be given by

$$\frac{d^2y}{dx^2} + a\frac{dy}{dx} + by = u$$

where *a*, *b* are *c* are constants but *u* is a particular function of the variable *x*. The solution of such an equation consists of a ***particular integral*** and a complementary function.

(b) Let *y* be a sinusoidal function of *x*, then

$$\frac{d^2y}{dx^2} + a\frac{dy}{dx} + by = c \sin \omega t$$

In this case, particular integral is

$$y_1 = \frac{-c}{\sqrt{a^2\omega^2 + (\omega^2 - b)^2}} \cos\left[\omega x - \tan^{-1}\left(\frac{\omega^2 - b}{a\omega}\right)\right]$$

The complementary function is given by

$$y_2 = k_1 e^{\lambda_1 x} + k_2 e^{\lambda_2 x} \text{ where } \lambda_1 = -\frac{a}{2} - \sqrt{\left(\frac{a^2}{4} - b\right)} \text{ and } \lambda_2 = -\frac{a}{2} + \sqrt{\left(\frac{a^2}{4} - b\right)}$$

The complete solution for the above equation is $y = y_1 + y_2$

Further treatment is the same as for case 2 (*i*) above.

19.4. Transients in R-L Circuits (D.C.)

If i, I_s and i_t be the resultant current, steady-state current and transient current respectively in $R–L$ circuit of Fig. 19.1 (*a*), then by superimposition, the equation for the resultant current, for the duration of initiation transient, is

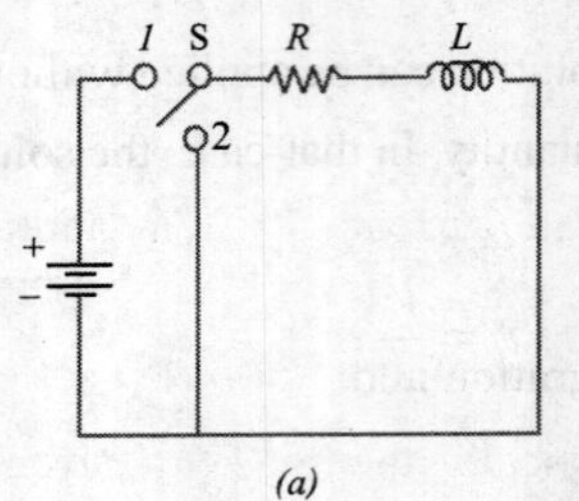

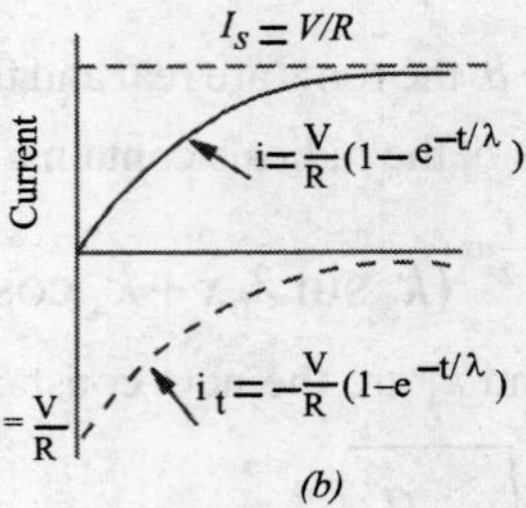

Fig. 19.1

$$i = I_s + i_t \qquad \text{... (i)}$$

Since the applied voltage V drives the steady-state current, hence

$$I_s = V/R$$

Since the transient current i_t is not associated with any voltage,

$$\because \quad i_t R + L\frac{di_t}{dt} = 0 \qquad \text{... (ii)}$$

or

$$\frac{di_t}{dt} = -\frac{R}{L}dt \qquad \text{... (iii)}$$

or

$$\int \frac{di_t}{i_t} = -\frac{R}{L}\int dt \quad \because \quad \text{logh } i_t = -\frac{R}{L}t + K \,* \qquad \text{... (iv)}$$

where K is the constant of integration whose value may be found from the initial conditions.

Now when $t = 0$, $i_t = I_0$ (say). Then from Eq. **(*iv*)** above we get, logh $I_0 = 0 + K$

Putting this value of K in Eq. **(*iv*)**, we have

$$\text{logh } i_t - \text{logh } I_0 = \frac{R}{L}t \text{ or logh } i_t / I_0 = -\frac{R}{L}t = \frac{-t}{\lambda}$$

$$\therefore \quad i_t = I_0 e^{-t/\lambda} \qquad \text{... (v)}$$

where $\lambda = L/R$ is called the ***time-constant*** of the circuit. Its reciprocal R/L is called the ***damping coefficient*** of the circuit. The current decreases exponentially as shown in Fig. 19.1 (*b*). From Eq. **(*i*)** and **(*v*)**, we have

$$i = I_s + I_0 e^{-t/\lambda} \qquad \text{... (vi)}$$

If the time is reckoned when the voltage V is applied, so that when $t = 0$, $i = 0$, then from equation **(*vi*)**, we get

$$0 = I_s + i_0 e^{-0} = I_s + I_0 \quad \therefore \quad I_0 = -I_s = -\frac{V}{R}$$

* $\log_e x$ is written as logh x. Obviously, logh $x = 2.3 \log_{10} x$.

In that case, Eq. (*vi*) becomes

$$i = \frac{V}{R} - \frac{V}{R}e^{-t/\lambda} \qquad \text{... (vii)}$$

$$= \frac{V}{R}(1 - e^{-t/\lambda}) \qquad \text{... (viii)}$$

Curves for I_s and i_t have been plotted in Fig. 19.1 (*b*). The curve for resultant current has been obtained by the superposition of steady-state current I_s (= V/R) and transient current

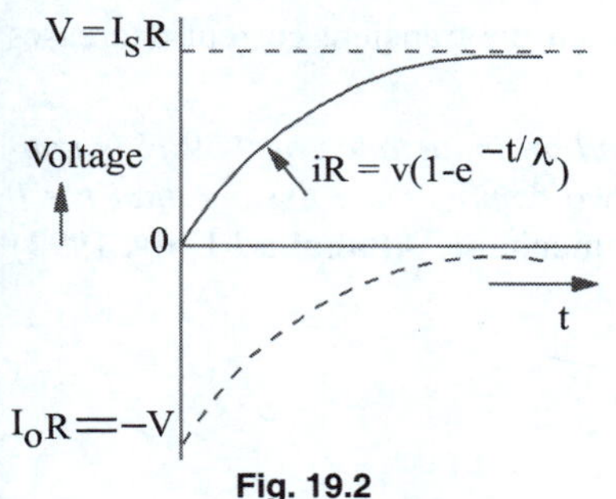

Fig. 19.2

$$i_t = \frac{V}{R}e^{-t/\lambda}$$

Theoretically, the transient current i_t takes infinite time to die off but, in practice, it disappears in a very short time.

The values of resultant, steady-state and transient voltages across the resistor can be found by multiplying Eq. (*vii*) by R and are shown in Fig. 19.2. The e.m.f. of self-induction–Ldi/dt is only transient in nature and equals $i_t R$ as seen from Eq. (*ii*) above.

19.5. Short Circuit Current

After some time, the transient current would disappear and the only current flowing in the circuit would be the steady-state current $I_s = V/R$. Let the *R-L* circuit be closed upon itself *i.e.* be short-circuited by shifting the switch [Fig. 19.1 (*a*)] to position 2. Since the voltage V has been excluded from the circuit, the trapped current I_s will immediately cease to be a steady-state current, but on the other hand, will become the initial value I_0 of a new subsidence transient current i_t. If time is measured at the instant of short-circuit, so that when $t = 0$, the current is $I_s = V/R$, then Eq. (*v*) becomes.

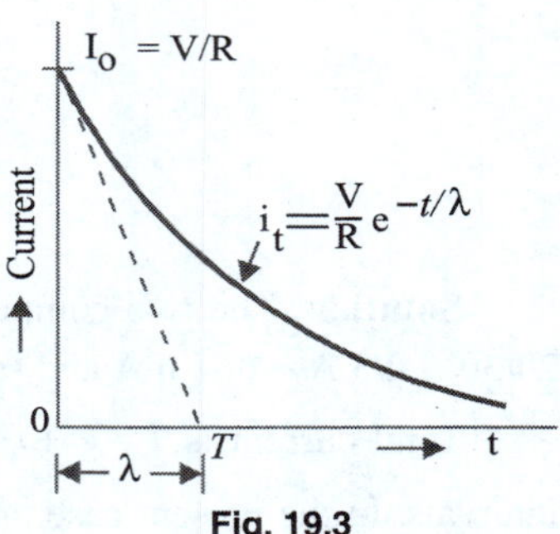

Fig. 19.3

$$i_t = \frac{V}{R}e^{-t/\lambda} \qquad \text{... (ix)}$$

This equation has been plotted in Fig. 19.3. The only voltage acting in the circuit is that due to self-induction *i.e.* $-L\,di/dt$ which equals $i_t R$.

19.6. Time Constant

The time constant of a circuit is defined as the time it would take for the transient current to decrease to zero, if the decrease were linear instead of being exponential.

In other words, it is the time during which the transient current would have decreased to zero, had it maintained its initial rate of decrease.

The initial rate of decrease can be found by differentiating Eq. (*vi*) and putting $t = 0$

$$\therefore \quad \frac{di_t}{dt} = \frac{I_0}{\lambda}e^{-t/\lambda} \qquad \therefore \quad \left(\frac{di_t}{dt}\right)_{t=0} = -\frac{I_0}{\lambda}$$

If the rate of decrease were constant throughout and equal to $-I_0/\lambda$, then the straight line showing the relation between i_t and t would be given by

$$i_t = -\frac{I_0}{\lambda}t$$

The time-period would be equal to the sub-tangent OT drawn to the exponential curve of Fig. 19.3 at $i_t = I_0$ *i.e.* at the beginning of the curve.

If we put $t = \lambda$ in Eq. (*v*), then $i_t = I_0\, e^{-1} = I_0/e = I_0/2.718 = 0.37\, I_0$

Hence, time period of a circuit is the time during which the transient current decreases to 0.37 of its initial value.

Example 19.1. *A coil having a resistance of 30 ω and an inductance of 0.09 H is connected across a battery of 20 V. Plot the current and its two components. Assume that t = 0 when the circuit is completed.* **(Electromechanic Allahabad Univ. 1992)**

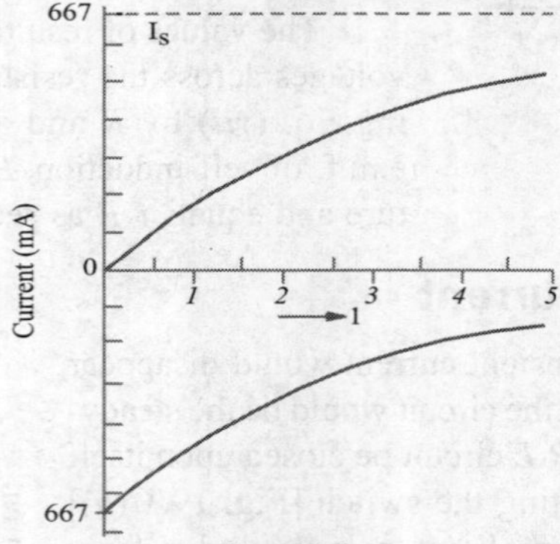

Fig. 19.4

Solution. The two components of the circuit current are **(*i*)** steady current $I_s = V/R$ = 20/30 = 2/3 A = 667 mA and **(*ii*)** Transient current $i_t = -(V/R)e^{-t/\lambda}$.

Total current is $i = I_s + i_t$. Let us find the value of transient current after various time intervals. In the present case, $\lambda = L/R = 0.09/30 = 0.003$ second = 3 millisecond.

The values of i_t and i at various times are tabulated below. Value of $i = I_s + i_t$

t (ms)	$e^{-t/\lambda}$	i_t (mA)	i	t (ms)	$e^{-t/\lambda}$	i_t (mA)	i
0.0	1.000	–667	0	2.5	0.435	–290	377
0.5	0.847	–565	102	3.0	0.368	–244	423
1.0	0.716	–477	190	3.5	0.311	–208	459
1.5	0.606	–405	262	4.0	0.264	–176	491
2.0	0.514	-344	323	4.5	0.223	–148	519

It is seen that whereas transient current decreases exponentially, total circuit current increases exponentially as expected (Fig. 19.4).

Example 19.2. *A circuit of resistance 10 Ω and inductance 0.1 H in series has a direct voltage of 200 V suddenly applied to it. Find the voltage drop across the inductance at the instant of switching on and at 0.01 second. Find also the flux-linkages at these instants.*

(Basic Electricity, Bombay Univ.)

Solution. (*i*) Switching instant

At the instant of switching on, $i = 0$, so that $iR = 0$ hence all applied voltage must drop across the inductance only. Therefore, voltage drop across inductance = **200 V**.

Since at this instant $i = 0$, there are no flux-linkages of the coil.

(*ii*) When t = 0.01 second

As time passes, current grows so that the applied voltage is partly dropped across the resistance and partly across the coil. Let us first find *iR* drop for which purpose, we need the value of *i* at $t = 0.01$ second.

Now, time period of the circuit is $\lambda = L/R = 0.1/10 = 0.01$ second. Since the given time happens to be equal to time constant,

$\therefore \quad i = (200/10) \times 0.632 = 12.64$ A; $iR = 152.64 \times 10 =$ **126.4 V**

Drop across inductance = $\sqrt{200^2 - 126.4^2}$ = **155 V**

Now, $\quad L = N\,\Phi / i \quad$ or $\quad N\Phi = Li$

$\therefore$ Flux-linkages $Li = 0.1 \times 12.64 =$ **1.264 Wb-turns.**

Example 19.3. *A coil of 10 H inductance and 5 Ω resistance is connected in parallel with a 20 Ω resistor across a 100-V d.c. supply which is suddenly disconnected. Find*

(a) *the initial rate of change of current after switching.*

(b) *the voltage across the 20 Ω resistor initially and after 0.3 s.*

(c) *the voltage across the switch contacts at the instant of separation and*

(d) *the rate at which the coil is losing stored energy 0.3 second after switching.*

Solution. (*a*) Since the steady-state current is zero, $i = I_0\, e^{-t/\lambda}$

Now, when $t = 0$, current is = 100/5 = 20 A. It means the current flowing through the coil immediately before opening the switch is 20 A.

$$\therefore \quad I_0 = 20 \text{ A}$$

Hence, the above equation becomes $i = 20e^{-t/\lambda}$

Now $\quad \lambda = L/R = 10/25 = 1/2.5 \quad \therefore \quad i = 20e^{-2.5t}$

$$\left(\frac{di}{dt}\right)_{t=0} = (-20 \times 2.5e^{-2.5t})_{t=0} = \mathbf{-50\ A/s}$$

The negative sign merely shows that the current is decreasing.

(*b*) After the supply has been disconnected, the current through the 20 – Ω resistor is *i* since it is in series with the coil.

Initial p.d. across the 20 Ω resistor = (current at $t = 0$) × 20 = 20 × 20 = **400 V**

Current through the resistor after 0.3 second = $20e^{-2.5\times 0.3} = 9.45$ A

$\therefore$ Voltage across the resistor after 0.3 second

= (current at $t = 0.3$ second) × 20 = 9.45 × 20 = **189 V**

(*c*) The e.m.f. induced in the coil at break tends to maintain the current through it in the original direction. Hence, the direction of the current through 20 Ω resistor is upwards so that the p.d. across the switch contacts will be the *sum* of supply voltage and the voltage across 20 Ω resistor.

$\therefore$ Initial voltage across switch contacts = 400 + 100 = **500 V** (*d*) The rate of loss of energy = power = induced e.m.f. in coil × current (after 0.3 s)

$$= L\left(\frac{di}{dt}\right)_{t=0.3} \times \text{(after 0.3 second)}$$

Now, after 0.3 second, $i = 9.45$ A

Value of di/dt after 0.3 second = $-\ 20 \times 2.5 \times e^{-0.75} = -\ 23.6$ A/second

∴ Rate of loss of energy = $-\ 10 \times 23.6 \times 9.45 = -\ 2{,}230$ joule/second

19.7. Transients in R-L Circuits (A.C.)

Let a voltage given by $v = V_m \sin(\omega t + \Psi)$ be ***suddenly*** applied across an *R-L* circuit [Fig. 19.5 (*a*)] at a time when $t = 0$. It means that the voltage is applied when it is passing through the value $V_m \sin \Psi$ Since the contact may be closed at any point of the cycle, angle Ψ may have any value lying between zero and Ψ radians. The resultant current, as before, is given by

$$i = i_s + i_t$$

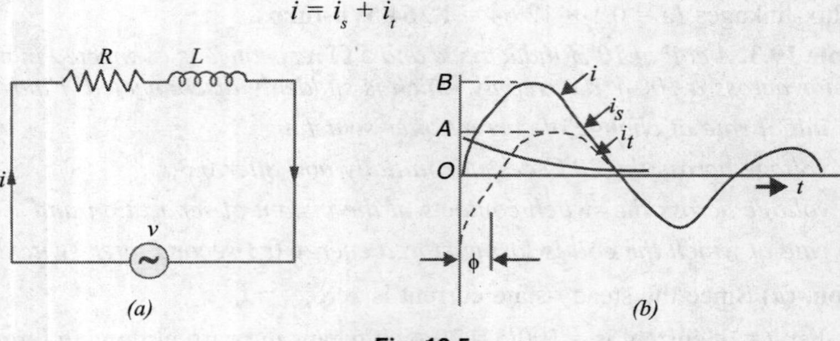

Fig. 19.5

The value of steady-state current is found by the normal circuit theory. The peak steady-state current is given by

$$I_m = \frac{V_m}{\sqrt{R^2 + X_L^{\ 2}}} = \frac{V_m}{Z}$$

where $\sqrt{R^2 + X_L^{\ 2}}$ is the impedance of the circuit. This current lags behind the applied voltage by an angle ϕ such that $\tan \phi = X_L/R$ or $\phi = \tan^{-1}(X_L/R)$

Hence, the equation for the instantaneous value of the steady-state current becomes

$$i_s = I_m \sin(\omega t + \Psi - \phi)$$

As before, the transient current is given by

$$i_t = I_0\, e^{-t/\lambda} \qquad \therefore \qquad i = I_m \sin(\omega t + \Psi - \phi) + I_0\, e^{-t/\lambda} \qquad \text{... (i)}$$

Now, when $t = 0$, $i = 0$, hence putting these values in Eq. (*i*) above, we get

$$0 = I_m \sin(\Psi - \phi) + I_0 \qquad \therefore \qquad I_0 = -I_m \sin(\Psi - \phi)$$

Hence Eq. (*i*) can be written as

$$i = I_m \sin(\omega t + \Psi - \phi) - I_m \sin(\Psi - \phi)\, e^{-t/\lambda} \qquad \text{... (ii)}$$

From the above, it is seen that the value of I_0 and hence the size of the transient current depends on angle Ψ *i.e.* it depends on the instant in the cycle at which the circuit is closed. We will consider the following three cases:

Case 1

When $t = 0$, let the voltage pass through its zero value and become positive *i.e.* let $\Psi = 0$. In that case, putting this value of Ψ in Eq. (*ii*), we get

$$i = I_m \sin(\omega t - \phi) - I_m \sin(-\phi) e^{-t/\lambda} = I_m[\sin(\omega t - \phi) + \sin\phi e^{-t/\lambda}]$$

This is shown in Fig. 19.5 (*b*). It is seen that maximum instantaneous peak current *OB* is larger than the normal peak current *OA*.

Case 2

Let $t = 0$ when voltage is passing through its value V_m sin (ϕ) so that $\Psi = \phi$ or $\Psi - \phi = 0$

In that, $I_0 = 0$, there is no transient current at the time of switching on (*i.e.* $i_t = 0$). It corresponds to the contacts closing at the instant when the steady state current itself is zero.

Case 3

When $t = 0$, let the voltage be passing through

$$V_m \sin\left(\phi \pm \frac{\pi}{2}\right) \text{ i.e. } \Psi = \phi \pm \frac{\pi}{2} \text{ and } \Psi - \phi = \pm\pi/2$$

In this case, the transient [as found from Eq. (*ii*)] would be given by

$$i_t = -I_m \sin\left(\pm\frac{\pi}{2}\right) . e^{-t/\lambda} = \mp I_m e^{-t/\lambda}$$

Under these conditions, the transient would have its maximum possible initial value.

Example 19.4. *A 1.0 H choke has a resistance of 50 Ω. This choke is supplied with an a.c. voltage given by e = 141 sin 314 t. Find the expression for the transient component of the current flowing through the choke after the voltage is suddenly switched on.*

(Principles of Elect. Engg-II, Jadavpur Univ.)

Solution. The equation of the transient component of the current is

$$i_t = I_m \sin\phi e^{-t/\lambda}$$

Here, $\lambda = L/R = 1/50 = 0.02$ second ; $\mathbf{Z} = 50 + j\,314 = 318\angle 80.95^\circ$

$I_m = V_m/Z = 141/318 = 0.443$ A; $\sin 80.95^\circ = 0.9875$

$\therefore \quad i_t = 0.443 \times 0.9875\, e^{-t/0.02} = 0.4376e^{-t/0.02}$

Example 19.5. *A 50-Hz sinusoidal voltage of maximum value of 400 V is applied to a series circuit of resistance 10 Ω and inductance 0.1 H. Find tan expression for the value of the current at any instant after the voltage is applied, assuming that voltage is zero at the instant of application. Calculate its value 0.02 second after switching on.*

(Electric Circuit, Punjab Univ. 1990)

Solution. In such cases, the current consists of a steady-state component and a transient component. The equation of the resultant current is

$$i = \underbrace{I_m \sin(\omega t - \phi)}_{\text{steady-state current}} + \underbrace{I_m \sin\phi e^{t/\lambda}}_{\text{transient current}}$$

where $I_m = V_m / Z$; $\phi = \tan^{-1} (X_L / R)$; $\lambda = L / R$ second

$$R = 10\Omega; X_L = 314 \times 0.1 = 31.4\Omega; Z = 10 + j31.4 = 33\angle 72.3^\circ$$

$I_m = 400 / 33 = 12.1$ A ; $\phi = 72.3^\circ = 1.26$ rad.

$\sin\phi = \sin 72.3^\circ = 0.9527$; $\lambda = 0.1/10 = 1/100$ second

$i = 121\ \{\sin(314\,t - 1.262) + 0.9527\, e^{100t}\}$

Substituting $t = 0.02$ second, we get

$$i = 12.1 (\sin (314 \times 0.02 - 1.262) + 0.9527 e^{-2}\}$$
$$= 12.1 (\sin 5.02 + 0.9527 e^{-2}) = 12.1 (\sin 288° + 0.9527 e^{-2})$$
$$= 12.1 (-\sin 72° + 0.9527 \times 0.1353) = 12.1 (-0.9511 + 0.1289) = \mathbf{-9.95\ A}$$

Example 19.6. *An alternating voltage $v = 400 \sin (314\, t + \Psi)$ is suddenly applied across a coil of resistance 0.2 Ω and inductance 6.36 mH. Determine the first peak value of the resultant current when the transient current has maximum value.*

Solution. Obviously, $\omega = 314$ rad/s

$$X_L = \omega L = 314 \times 6.36 \times 10^{-3} = 2\ \Omega$$

Coil impedance $\mathbf{Z} = 0.2 + j2 \cong 2 \angle 84.3°$

Max. value of steady-state current = 400/2 = 200 A

The maximum value of transient current will occur when

$\Psi = \phi \pm \pi/2$ where $\phi = 84.3°$ *i.e.* the phase angle of the current w.r.t. voltage

$\therefore \quad \Psi = 84.3° - 90° = -5.7°$

$\therefore \quad$ resultant current, $i = 400 \sin (314\, t - 90°) + I_0 e^{-31.4t}$

Now, at $t = 0$, $i = 0$ $\qquad \therefore \quad 0 = 400 \sin (-90°) + I_0$

$\therefore \qquad I_0 = 400$ A

Hence, the above equation becomes

$$i = 400 \sin (\omega t - 90°) + 400\, e^{-31.4t}$$

The procedure for determining an exact solution for the first peak of the resultant current is first to differentiate the above expression, next to equate the result to zero and then to solve the resulting expression graphically for *t*. However, sufficiently accurate result can be obtained by determining the instant at which steady-state current reaches its first positive peak value and then to add to it the value of the transient current at this instant. The first peak value of steady-state current occurs when

$(314\, t - 90°) = \pi / 2$ rad ; *i.e.* when $t = \pi / 314 = 0.01$ second

At this time, $\qquad i_t = 400\, e^{-0.314} = 292$ A

$\therefore \quad$ Resultant current *i* at this time = 200 + 292 = **492 A**

19.8. Transients in R-C Series Circuits (D.C.)

When a d.c. voltage *V* is ***suddenly*** applied to an *R-C* series circuit (Fig. 19.6), the voltage v_c across the capacitor rises from zero value to the steady-state value *V*. If v_c is the voltage across capacitor, V_{ct} the transient voltage, then

$$v_c = V + v_{ct} \qquad \text{... (i)}$$

The charging current is maximum at the beginning but then is reduced to zero so that there is no steady-state current but a transient one.

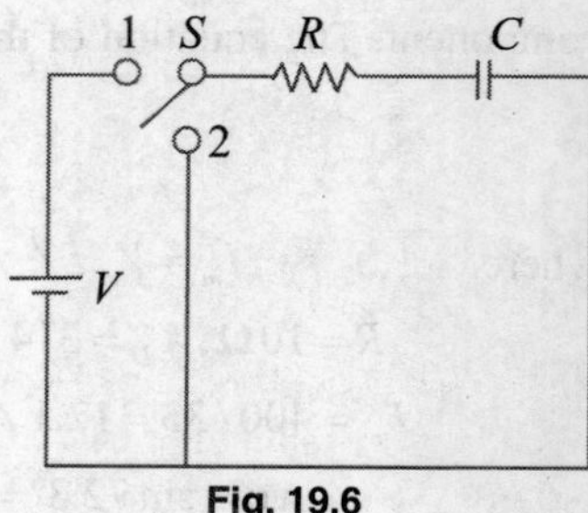

Fig. 19.6

Since the transient current is not associated with any applied voltage, hence

$$i_t R + v_{ct} = 0 \qquad \text{.... (ii)}$$

Now, capacitor voltage $v_{ct} = q_1/C$

Hence, Eq. (*ii*) becomes

$$i_t R + \frac{q_t}{C} = 0$$

or $R\cdot\frac{di_t}{dt}+\frac{1}{C}\cdot\frac{dq_t}{dt}=0$ or $\frac{di_t}{dt}=-\frac{1}{CR}\frac{dq_t}{dt}=-\frac{1}{CR}i_t\ (\because dq_t/dt=i_t)$

$\therefore\quad \frac{di_t}{i_t}=-\frac{d_t}{CR};$ As before $i_t=I_0e^{-t/CR}=I_0e^{-t/\lambda}$

where $CR=\lambda$ = time constant. The reciprocal I/CR is known as damping coefficient.

(i) Charging Current

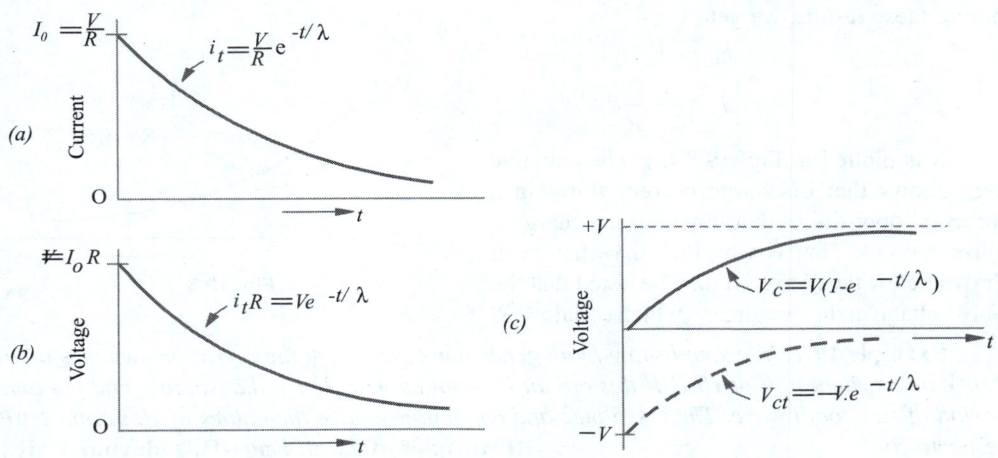

Fig. 19.7

When $t=0$, transient current $i_t=I_0$, so that from Eq. (*ii*) $v_{ct}=-I_0R$. Moreover, when $t=0$, $v_c=0$, hence from Eq. (*i*), $v_{ct}=-V$

Combining these results, we get

$$I_0 = V/R$$

$$\therefore\quad i_t = {}_0e^{-t/\lambda}=\frac{V}{R}e^{-t/\lambda}$$

This is plotted in Fig. 19.7 (*a*)

The transient voltage across the resistor R is given by

$$i_tR = \frac{V}{R}e^{-t/\lambda}\times R$$

$$= Ve^{-t/\lambda} \quad \text{... Fig. 19.7 (b)}$$

From Eq. (***ii***) the value of transient voltage across the capacitor is $v_a=-i_tR$

Hence, Eq. (***i***) becomes

$$v_c = V-i_tR=V-Ve^{-t/\lambda}$$

or

$$v_c = v_c=V(1-e^{-t/\lambda}) \quad \text{... (iii)}$$

The voltage across the capacitor v_c which is the sum of the transient voltage v_{ct} and steady-state V has been plotted in Fig. 19.7 (*c*).

The charge across the capacitor is given by

$$q=v_c=CV(1-e^{-t/\lambda}) \text{ or } q=Q(1-e^{-t/\lambda})\quad (\therefore Q=CV)$$

(ii) Discharge Current

When the capacitor has become fully charged so that charging current has ceased, then the R-C circuit is short-circuited by shifting the switch S from position 1 to position 2 (Fig. 19.6). On doing so, a transient discharge current will start flowing immediately. If time is reckoned from the instant of short-circuit, then when $t = 0$, $i_t = I_0$, hence from Eq. (*ii*) above $v_{ct} = -I_0, R$. Moreover, when $t = 0$, $v_c = V$. However, since there is no steady-state voltage across the capacitor, from Eq. (*i*), we get $v_c = v_{ct}$. Combining these results, we get

$$I_0 = V/R$$

$$\therefore \quad i_t = -\frac{V}{R}e^{-t/\lambda}$$

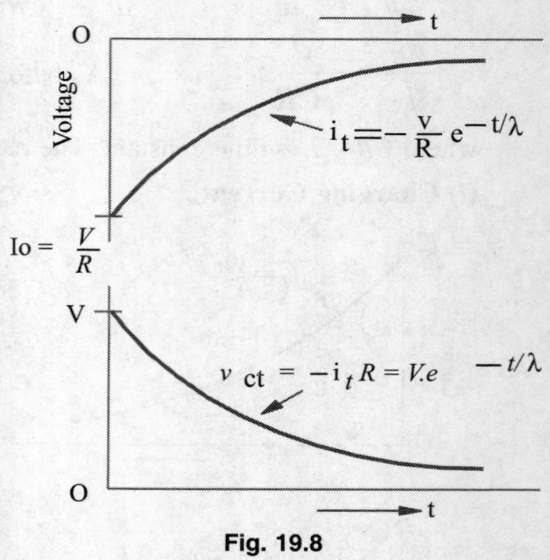

Fig. 19.8

It is plotted in Fig. 19.8 (*a*). The negative sign shows that discharge current flows in a direction opposite to that in which the charging current flows. That is why the curve has been drawn below the X-axis. It may be noted that the only voltage in the circuit is v_{ct} which equals $-i_t R$.

Example 19.7. *In a simple saw-tooth generator circuit with the thyratron switches on at 150 V and switches off at 10 V. If this circuit is supplied with 250 V d.c. source; find the time period of saw-tooth wave. The resistance and capacitance have the values of 10 Ω and 1 μF respectively.* **(Principles of Elect. Engg-II, Jadavpur Univ.)**

Solution. With reference to Fig. 19.9, let

V = applied voltage

v_{cl} = switching-off voltage of the thyratron = 10 V

v_{c2} = switching-on voltage of the thyratron = 150 V

Now, $\quad v = V(1 - e^{-t/\lambda})$

$$\therefore \quad v_{c1} = V(1 - e^{-t/\lambda}) \qquad \text{... } (i)$$

$$V(1 - e^{(-t_1+T)}/\lambda \qquad \text{... } (ii)$$

v_{c2} = 150 V

v_{c1} = 10 V

t_1 $\quad$ t_1+T

Fig. 19.9

where T is the time-period of the saw-tooth wave. From Eq. (*i*) and (*ii*), we get

$$T = \lambda \log_e (V - v_{c1})/(V - v_{c2})$$

Now $\quad \lambda = CR = 10^4 \times 10^{-6} = 10^{-2}$ second

$$V - v_{c1} = 250 - 10 = 240\text{ V}; V - v_{c2} = 250 - 150 = 100\text{ V}$$

$$\therefore \quad \frac{V - v_{c1}}{V - v_{c2}} = \frac{240}{100} = 2.4 \quad \therefore \quad T = 10^{-2}\log_e^{2.4} = \mathbf{0.00875\ second}$$

Example 19.8. *A simple neon-tube time base for a cathode-ray oscillorgraphy employs a 300 kΩ and a 0.016 μF capacitor. The striking and extinction voltages of the neon-tube are 170 V and 140 V respectively. Calculate the frequency of the time base if the supply voltage is 200 V.*

Solution. The voltage across the capacitor increases according to the equation

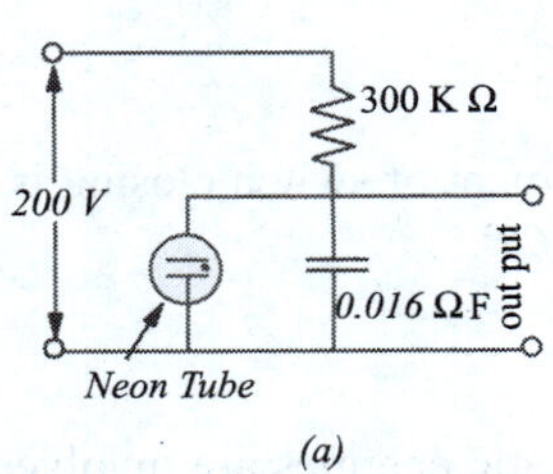

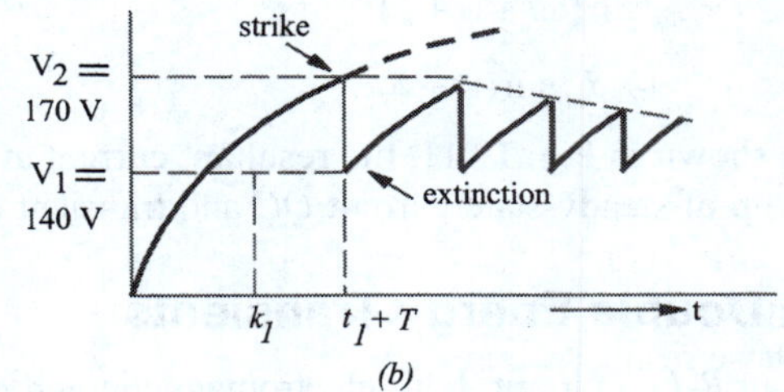

Fig. 19.10

$$v_c = V(1-e^{-t/CR})$$

It is shown in Fig. 19.10 (*b*)

$$\therefore \qquad v_{c1} = V(1-e^{-t_1/\lambda}) \qquad \text{... (i)}$$

$$v_{c2} = V(1-e^{(-t_1+T)/\lambda}) \qquad \text{... (ii)}$$

From eq. **(*i*)** and **(*ii*)**, we get

$$T = \lambda \text{logh}(V - v_{C1})/(V - v_{C2})$$

Now $\qquad \lambda = CR = 0.016\times10^{-6}\times300\times10^{3} = 4.8\times10^{-3}$ second

$$V - v_{c1} = 200-140 = 60 \text{ V and } V - v_{c2} = 200-170 = 30 \text{ V}$$

$$\therefore \qquad T = 4.8\times10^{-3}\log 60/30 = 1/300 \text{ second}$$

$\therefore$ Frequency of time base $= 1/T =$ **300 Hz**

19.9. Transients in R-C Series Circuits (A.C.)

In this case, the resultant currents can be determined in the same way as for an *R-L* circuit. It is given by

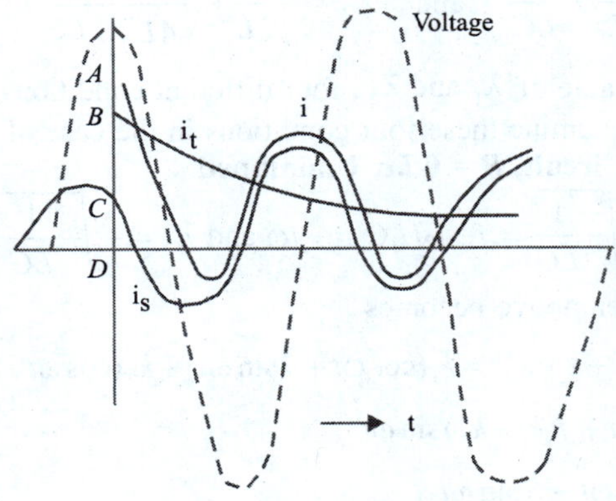

Fig. 19.11

$$i = i_s + i_t = I_m \sin(\omega t + \psi + \phi) + I_0 e^{-t/\lambda} \quad \text{where} \quad I_m = V_m / \sqrt{R^2 + X_C{}^2}$$

$$\text{and} \quad v = V_m \sin(\omega t + \psi)$$

The value of I_0 as found from initial known conditions ($t = 0$, $i = 0$) is given by $I_0 = -I_m \sin(\Psi + \phi)$

Hence, the resultant current becomes

$$i = I_m \sin(\omega t + \psi + \phi)$$

$$- I_m \sin(\psi + \phi) e^{-t/\lambda}$$

As shown in Fig. 19.11, the resultant current at the moment of switch closing is *OA* and is made up of steady-state current *OC* and transient current *OB*.

19.10. Double Energy Transients

In an *R-L-C* circuit, both electromagnetic and electrostatic energies are involved, hence any sudden change in the conditions of the circuit involves the redistribution of these two forms of energy. The transient currents produced due to this redistribution are known as double-energy transients. The transient current produced may be unidirectional or a decaying oscillatory current.

In an *R-L-C* circuit, the transient voltages across the three circuit parameters are $i_t, R, (L di_t / dt)$ and q_i / C. Hence, the equation of the transient voltage is

$$i_t R + L\frac{di_t}{dt} + \frac{q_t}{C} = 0 \quad \text{... } (i)$$

Differentiating the above equation and putting i_t for dq_t/dt, we get

$$\frac{d^2 i_t}{dt^2} + \frac{R}{L}\frac{di_t}{dt} + \frac{1}{LC} i_t = 0 \quad \text{... } (ii)$$

This is a linear differential equation of the second order with constant coefficient like 2 (*i*). Its solution is given by

$$i_t = k_1 e^{\lambda_1 t} + k_2 e^{\lambda_1 t} \quad \text{... } (iii)$$

where k_1 and k_2 are constants whose values are found from the boundary conditions. The values of λ_1 and λ_2 are given by

$$\lambda_1 = -\frac{R}{2L} - \sqrt{\frac{R^2}{4L^2} - \frac{1}{LC}} \quad \text{and} \quad \lambda_2 = -\frac{R}{2L} + \sqrt{\frac{R^2}{4L^2} - \frac{1}{LC}}$$

Depending on the value of λ_1 and λ_2, four different conditions of the circuit are distinguishable. We will now examine these four conditions in the case of an *R-L-C* circuit.

Case 1. Loss-free Circuit, *R* = 0 *i.e.* Undamped

In this case, $\lambda_1 = \sqrt{-\frac{1}{LC}} = j - 1\sqrt{LC} = -j\omega$ and $\lambda_2 = -\sqrt{-\frac{1}{LC}} = -j - 1\sqrt{LC} = -j\omega$

Hence, Eq. (***iii***) given above becomes

$$i_t = k_1 e^{j\omega t} + k_2 e^{-j\omega t} = k_1(\cos\omega t + j\sin\omega t) + k_2(\cos\omega t - j\sin\omega t)$$

$$= (k_1 + k_2)\cos\omega t + j(k_1 - k_2)\sin\omega t$$

or $$i_t = A\cos\omega t + B\sin\omega t \quad \text{... } (iv)$$

where $A = k_1 + k_2$ and $B = j(k_1 - k_2)$

Eq. (***iv***) can be still further simplified to

$$i_t = I_m \sin(\omega t + \phi) \quad \text{... } (v)$$

where $I_m = \sqrt{A^2 + B^2}$ and $\phi = \tan^{-1}(A/B)$

As seen from Eq. (**v**), the transient current in this case is sinusoidal wave of constant peak value and frequency $f = 1/2\pi\sqrt{LC}$ as shown in Fig. 19.12 (*a*). The values of two constant terms I_m and ϕ can be determined from any *two* known initial circuit conditions which are (*i*) the initial current in the inductance and (*ii*) the initial voltage across the capacitor.

Case 2. Low-loss Circuit: $\frac{R^2}{4L^2} < \frac{1}{LC}$ ***i.e.* Under-damped**

In this case, λ_1 and λ_2 would be conjugate complex numbers because the term under the square root sign in each case would be negative.

$$\therefore \quad \lambda_1 = \frac{R}{2L} + j\sqrt{\frac{1}{LC} - \frac{R^2}{4L^2}}$$

If $a = \frac{R}{2L}$ and $\omega = \sqrt{\frac{1}{LC} - \frac{R^2}{4L^2}}$ then $\lambda_1 = -a + j\omega$ and $\lambda_2 = -a - j\omega$

Putting these values in equation (*v*), we get

$$i_t = k_1 e^{(-a+j\omega)t} + k_2 e^{(-a-j\omega)t} = e^{-at}(k_1 e^{j\omega t} + k_2 e^{-j\omega t})$$

This equation can be reduced, as before, to the form

$$i_t = I_m e^{-at} \sin(\omega t + \phi) \qquad \text{... } (iv)$$

where I_m and ϕ are constants as before. Equation (***vi***) represents damped transient oscillatory current as shown in Fig. 19.12 (*b*).

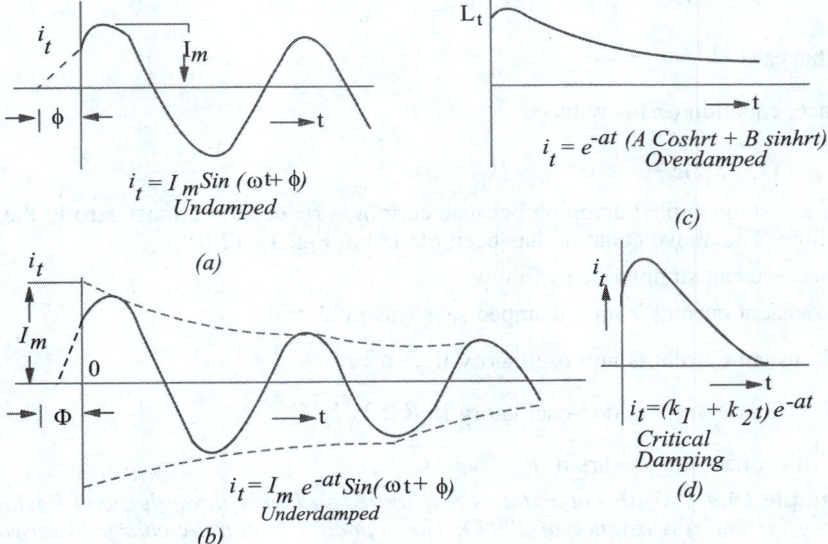

Fig. 19.12

The exponential term e^{-at} which accounts for the decay of oscillations, is called the decay or damping factor or merely ***decrement.*** It makes each current peak a definite fraction less than that preceding it. The logarithm to the Naperian base 'e' of the ratio of peaks one cycle apart in time is $a/f = R/2fL$ and is referred to as ***logarithmic decrement.*** The frequency of damped oscillations is given by

$f = \sqrt{\frac{1}{LC} - \frac{R^2}{4L^2}}$ and is called the natural frequency of the circuit

If $\frac{R^2}{4L^2} < \frac{1}{LC}$, then $f = \frac{1}{2\pi\sqrt{LC}}$

Case 3. High-loss Circuit: $\frac{R^2}{4L^2} > \frac{1}{LC}$ *i.e.* **overdamped**

In this case, λ_1 and λ_2 will be pure numbers.

$$\lambda_1 - \frac{R}{2L} + \sqrt{\frac{R^2}{4L^2} - \frac{1}{LC}} = -a + \gamma \text{ and } \gamma_2 = -a - \gamma$$

$$\therefore\ i_t = k_1 e^{(-a+\gamma)t} + k_2 e^{(-a-\gamma)t} = e^{-at}(k_1 e^{\gamma t} + k_2 e^{-\gamma t})$$

Now $\quad e^{\gamma t} = \sinh \gamma t + \cosh \gamma t$

and $\quad e^{-\gamma t} = \cosh \gamma t - \sinh \gamma t$

$\therefore \quad i_t = e^{-at}\{(k_1 + k_2)\cosh \gamma t + (k_1 - k_2)\sinh \gamma t\}$

or $\quad i_t = e^{-at}(A \cosh \gamma t + B \sinh \gamma t)$

A typical curve of this equation is shown in Fig. 19.12(*c*)

Case 4. $\frac{\mathbf{R^2}}{\mathbf{4L^2}} = \frac{\mathbf{1}}{\mathbf{LC}}$ ***i.e.* Critical Damping**

In this case, $\lambda_1 = \lambda_2 = -\frac{R}{2L}$

Hence, equation (***iii***) is reduced to

$$i_t = (k_1 + k_2 t)e^{-\frac{R}{2L}t} \quad \text{or} \quad i_t = (k_1 + k_2 t)e^{-at}$$

It is a case of critical damping because current is reduced to almost zero in the shortest possible time. The above equation has been plotted in Fig. 19.12 (*d*).

Hence, we can summarize as follows:

1. Transient current is an undamped sine wave if $R = 0$

2. Transient current is non-oscillatory if $R < 2\sqrt{L/C}$

3. Transient current is non-oscillatory if $R \geq 2\sqrt{L/C}$

4. Critical damping occurs if $R = 2\sqrt{L/C}$

Example 19.9. *A 5-μF capacitor is discharged suddenly through a coil having an inductance of 2H and a resistance of 200 Ω. The capacitor is initially charged to a voltage of 10 V. Find*

(a) *an expression for the current*

(b) *the additional resistance required to give critical damping.*

Solution. Since there is no battery or generator in the circuit (Fig. 19.13), the steady-state current must be zero. It means that resultant current is simply the transient current.

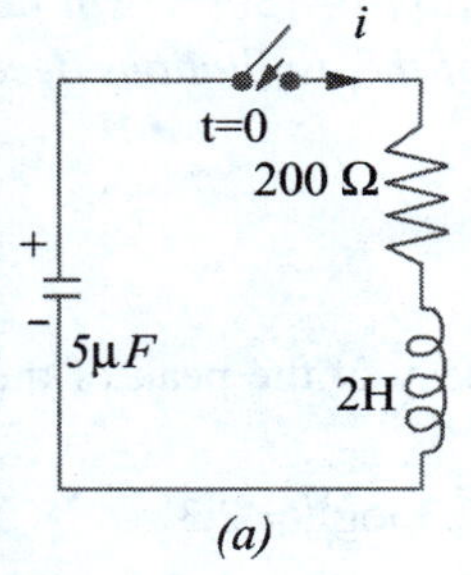

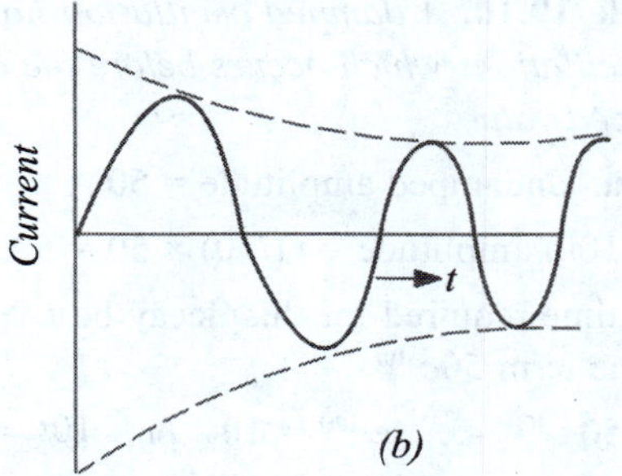

Fig. 19.13

Value of $2\sqrt{L/C} = 2\sqrt{2/5\times10^{-6}}$

$= 1265\ \Omega$

Since $R < 2\sqrt{L/C}$, the circuit is originally oscillatory.

(a) the expression for the transient current, therefore, is

$$i_t = I_m e^{-at}\sin(\omega t+\phi)$$

where $$\omega = \sqrt{\frac{1}{LC}-\frac{R^2}{4L^2}} = \sqrt{100{,}000-2500}$$

$$i_t = I_m e^{-50t}\sin(312.3t+\phi) = i = 312.3 \text{ rad/s}$$

$$\therefore \quad i_t = I_m e^{-50}\ \sin(312.3t+\phi) = i \qquad \text{... (i)}$$

Two initial conditions are known from which I_m and ϕ can be found (a) at $t = 0$; $i = 0$ and **(b)** at $t = 0$; $v_c = 10$ V. Applying condition **(a)** to Eq. (i), we get

$$0 = I_m \sin\phi, \text{ hence } \phi = 0 \quad \therefore \quad i = I_m e^{-50t}\sin 312.3t \qquad \text{... (ii)}$$

Now, at $t = 0$, the voltage across the inductance must be 10 V because the current in the resistance is zero.

i.e. $$(Ldi/dt)_{t=0} = 10 \text{ V} \therefore (di/dt)_{t=0} = 10/L = 5 \text{ A/s} \qquad \text{... (iii)}$$

Now, from equation **(ii)**, we have

$$\frac{di}{dt} = -50I_m e^{-50t}\sin 312.3t + 312.3I_m e^{-50t}\cos 312.3t \qquad \text{... (iv)}$$

Putting $t = 0$, it becomes $(di/dt)_{t=0} = 312.3I_m$

From equation **(iii)**, we have

$312.3I_m = 5 \qquad \therefore \quad I_m = 5/312.3 = 0.016$ A

Hence, the general expression for the current becomes

$$\boldsymbol{i = 0.016\ e^{-50t}\sin 312.3t}$$

It is roughly plotted (the first few cycles only) in Fig. 19.13 (b).

(b) Critical damping is achieved when $R = 2\sqrt{L/C}$

$\therefore \quad R = 2\sqrt{2/5\times10^{-6}} = 1265\ \Omega$

$\therefore$ Additional resistance reqd. = 1265 − 200 = **1065 Ω**

Example 19.10. *A damped oscillation has the equation* $i = 50e^{-10t} \sin 628t$. *Find the number of oscillations which occurs before the amplitude of the oscillations decays to 1/10th of its undamped value.*

Solution. Undamped amplitude = 50 A

1/10th amplitude = (1/10) × 50 = 5A

Let the time required for this decay be *t*. Now, the decay of the peak of the oscillations is given by the term $50e^{-10t}$

$$\therefore\ 5 = 50e_1^{10t} \quad \therefore \quad e_1^{10t1} = 10 \quad \text{or} \quad 10t_1 = \log h^{10} = 2.3 \log_{10}^{10} = 2.3$$

$$\therefore \qquad t_1 = 0.23 \text{ second}$$

Frequency of oscillations = 628 / 2π = 100 Hz.

Hence, the number of oscillations which occur before the amplitude falls to 1/10th of its undamped value is = 0.23 × 100 = **23**

Example 19.11. *If, in Fig. 19.14, a break occurs at a point marked X, what would be the voltage across the break? It may be assumed that prior to the break, steady conditions existed in the circuit.*

Solution. Steady-state current through the inductance = 120/60 = 2A

Energy stored in the inductor prior to the break

$$= \frac{1}{2}LI^2 = \frac{1}{2} \times 12 \times 10^{-3} \times 4 = 24 \times 10^{-3} \text{ J}$$

Energy initially stored in the capacitor

$$= \frac{1}{2}CV^2 = \frac{1}{2} \times 10^{-8} \times 120^2 = 72 \times 10^{-6} \text{ J} = \text{0-practically}$$

Fig. 19.14

When the break occurs, the energy stored in the inductor is transferred to the capacitor. If loss of energy during first transfer is neglected, then maximum energy stored in the capacitor is

$$= 20 \times 10^{-3} \text{ J} \quad \therefore \quad \frac{1}{2}CV_m^2 = 24 \times 10^{-3}$$

$$\therefore \qquad V_m = \sqrt{2 \times 24 \times 10^{-3} \times 10^8} = 2{,}190 \text{ V}$$

Maximum voltage across the break is = 2190 + 120 = 2310 V

The voltage would be oscillatory because the energy alternates between the inductor and capacitor.

Frequency of voltage oscillation is

$$f = 1/2\pi\sqrt{LC} = 10^5 / 2\pi \times \sqrt{1.2} = 14{,}530 \text{ Hz}$$

Decay or damping factor = e^{-at}

Here, $a = R/2L$ = 60/2 × 12 × 10^{-3} = 2500 $\quad \therefore$ damping factor = e^{2500t}

Hence, voltage across the break is

$$= 120 + 2190e^{-2500t} \sin 2\pi \times 14{,}500t = \mathbf{120 + 2190e^{-2500t} \sin 91{,}290t}$$

Tutorial Problem 19.1

1. Deduce an expression for the growth of current in an inductive circuit.
   A 15-H inductance coil of 10 Ω resistance is suddenly connected to a 20 V d.c. supply. Calculate:
   (*a*) the initial rate of change of current
   (*b*) the current after 2 second
   (*c*) the rate of change of current after 2 second
   (*d*) the energy stored in the magnetic field in this time
   (*e*) the energy lost as heat in this time
   (*f*) the time constant.
   [(*a*) 1.33 A/s (*b*) 1.47 A (*c*) 0.352 A/s (*d*) 16.3 joules (*e*) 19.5 joules (*f*) 1.5 s]
2. A circuit consisting of a 20 Ω resistor in series with a 0.2 H inductor is supplied from 200 V (r.m.s.) 50 Hz a.c. mains. Deduce equations showing how the current varies with time if the supply is suddenly switched on (*a*) at the instant when the voltage is zero (*b*) at the instant when the voltage is a maximum.
   [(*a*) 4.11 e^{-100} + 4.32 sin (314 t – 70°16') A (*b*) – 1.245 e^{-100t} + 4.32 cos (314t – 72°16') A]
3. A circuit consisting of a 20 Ω resistor, 20 mH inductor and a 100 μF capacitor in series is connected to a 200 V, d.c. supply. The capacitor is initially uncharged. Determine the equation relating the instantaneous current to the time and find the maximum instantaneous current.
   [(20 e^{-500t} sin 500 t) A; 6.44 A]
4. Find an expression for the value of current at any instant after a sinusoidal voltage of amplitude 600 V at 50 Hz is applied to a series circuit of resistance 10 Ω and inductance 0.1 H, assuming that voltage is zero at the instant of switching. Also, find the value of transient current at $t = 0.02$ second. **[– 15.14 A; 2.17 A] (*Electric Circuits and Fields, Gujarat Univ.*)**
5. A 40 Ω resistor and a 50 μ F capacitor are connected in series and supplied with an alternating voltage $v = 283 \sin 314\, t$. The supply is switched on at the instant when the voltage is zero. Determine the expression for the instantaneous current at time *t*.
   [– 3.18e^{-500t} + 3.76 sin (314 t + 57°50')]
6. A d.c. voltage of 100 V is suddenly applied to a circuit consisting of a 100 Ω resistor, a 0.1 H inductor and a 100 μ F capacitor in series. The capacitor is initially uncharged. Obtain the equation which shows how the capacitor voltage varies with time.
   [100 – 115.3 e^{-500t} sin (866 t + π/3 V)]
7. The voltage $v = 200 \sin 314t$ is suddenly applied at $t = 0$ to a circuit consisting of a 10 Ω resistor in series with a 0.1 H inductor. Deduce an equation showing how the current varies with time.
   [5.78 e^{-100t} + 6.06 sin (314 t – 72°20')]
8. A 20 Ω resistor, a 0.01 H inductor and a 100 Ω capacitor are connected in series. A d.c. voltage of 100 V is suddenly applied to the circuit. Obtain the equation showing how the current through the circuit varies with time. Find the maximum current and the time at which it occurs.
   [10^4 e^{-100t}; 3.67 A; 0.001 second]
9. A μ F capacitor is initially charged to 300 V. It is discharged through a 100 mH inductance and a resistor in series:
   (*a*) find the frequency of the discharge if the resistance is zero.
   (*b*) how many cycles at the above frequency will occur before the discharge oscillation decays to 1/10 of its initialy value if the resistance is 1Ω.
   (*c*) find the value of the resistance which would just prevent oscillations.
   [(*a*) 796 Hz (*b*) 36.6 (*c*) 100 Ω]

OBJECTIVE TESTS – 19

1. Transient disturbance is produced in a circuit whenever
(*a*) it is suddenly connected or disconnected from the supply
(*b*) it is shorted
(*c*) its applied voltage is changed suddenly
(*d*) all of the above.

2. There are no transients in pure resistive circuits because they
(*a*) offer high resistance
(*b*) obey Ohm's law
(*c*) have no stored energy
(*d*) are linear circuits.

3. Transient currents in electrical circuit are associated with
(*a*) inductors (*b*) capacitors
(*c*) resisters (*d*) both (*a*) and (*b*).

4. The transients which are produced due to sudden but energetic changes from one steady state of a circuit to another are called .... transients.
(*a*) initiation (*b*) transition
(*c*) relaxation (*d*) subsidence

5. In an *R-L* circuit connected to an alternating sinusoidal voltage, size of transient current primarily depends on
(*a*) the instant in the voltage cycle at which circuit is closed
(*b*) the peak value of steady-state current
(*c*) the circuit impedance
(*d*) the voltage frequency.

6. Double-energy transients are produced in circuits consisting of
(*a*) two or more resistors
(*b*) resistance and inductance
(*c*) resistance and capacitance
(*d*) resistance, inductance and capacitance.

7. The transient current in a loss-free *L-C* circuit when excited from an ac source is a/an ..... sine wave.
(*a*) over damped
(*b*) undamped
(*c*) under damped
(*d*) critically damped.

8. Transient current in an *R-L-C* circuit is oscillatory when
(*a*) $R = 0$
(*b*) $R > 2\sqrt{L/C}$
(*c*) $R < 2\sqrt{L/C}$
(*d*) $R = 2\sqrt{L/C}$

ANSWERS

1. (*d*) **2.** (*c*) **3.** (*d*) **4.** (*b*) **5.** (*a*) **6.** (*d*) **7.** (*b*) **8.** (*c*)

CHAPTER 20

Introduction to Electrical Energy Generation

Chapter Outline

- ❖ Importance of Electrical Energy
- ❖ Comparison of Sources of Power
- ❖ Sources for Generation of Electricity
- ❖ Brief Aspects of Electrical Energy Systems
- ❖ Utility and Consumers
- ❖ Why is the Three-phase a.c. System Most Popular?
- ❖ Cost of Generation
- ❖ Staggering of Loads during peak-demand Hours
- ❖ Classifications of Power Transmission
- ❖ Selecting A.C. Transmission Voltage for a Particular Case
- ❖ Conventional Sources of Electrical Energy
- ❖ Steam Power Stations (Coal-fired)
- ❖ Nuclear Power Stations
- ❖ Advantages of Nuclear Generation
- ❖ Disadvantages
- ❖ Hydroelectric Generation
- ❖ Non-Conventional Energy Sources
- ❖ Photo Voltaic Cells (P.V. Cells or SOLAR Cells)
- ❖ Fuel Cells
- ❖ Principle of Operation
- ❖ Chemical Process (with Acidic Electrolyte)
- ❖ Schematic Diagram
- ❖ Array for Large outputs
- ❖ High Lights
- ❖ Wind Power
- ❖ Introduction
- ❖ Basic Scheme

20.1. Importance of Electrical Energy

Energy is vital for all living-beings on earth. Modern life-style has further increased its importance, since a faster life means faster transport, faster communication, and faster manufacturing processes. All these lead to an increase in energy required for all those modern systems.

Arising out of comparison of status of nations, the progress is related in terms of per capita consumption of electrical energy (i.e. kWH consumed per person per year). At present, this parameter for India is about 300, for UK it is 12 to 15 times more, and for USA, it is about 30 times more.

It simply means that Electrical energy is the most popular form of energy, whether we require it in the usable thermal form (= heating applications), in mechanical form (= electrical motor-applications in Industries), for lighting purposes (= illumination systems), or for transportation systems.

Following are the main reasons for its popularity.

1. Cleaner environments for user
2. Higher efficiency
3. Better controllability
4. Easier bulk-power, long-distance transportation of power using overhead transmission or underground cables
5. Most versatile devices of energy conversions from Electrical to other forms are available for different purposes, such as thermal, illumination, mechanical, sound, chemical, etc.

20.2. Comparison of Sources of Power

While selecting a method of generating electricity, following factors are taken into account for purposes of comparison:

(*a*) Initial cost: For a given rating of a unit (in the minds of planners), investment must be known. Naturally, lower the initial cost, better it is.

(*b*) Running Cost:- To produce a given amount of electrical energy, the cost of conversion process (including proportional cost of maintenance/repairs of the system) has to be known.

(*c*) Limitations:- Whether a particular resource is available, whether a unit size of required rating is available from a single unit or from an array of large number of units, and whether a particular method of generation is techno-economically viable and is time-proven, are typical queries related to the limitations of the concerned method.

(*d*) **(1)** perpetuity, **(2)** efficiency, **(3)** reliability, **(4)** cleanliness and **(5)** simplicity.

It is naturally desirable that the source must have perpetuity (= be of endless duration), high conversion efficiency, and reliability (in terms of availability in appropriate quantity). The energy conversion must be through a cleaner process (specially from the view- points of toxicity, pollution or any other hazardous side effects). Further, a simpler overall system is always preferred with regards to maintenance/repairs problems and is supposed to be more reliable.

20.3. Sources for Generation of Electricity

Following types of resources are available for generating electrical energy (No doubt, this list can be extended to include some more up-coming resources. The following list, however, gives the popular and potential resources).

(*a*) Conventional methods

(*a*) **Thermal:** Thermal energy (from fossil fuels) or Nuclear Energy used for producing steam for turbines which drive the alternators (= rotating a.c. generators).

(*b*) **Hydro-electric:** Potential of water stored at higher altitudes is utilized as it is passed through water-turbines which drive the alternators.

(*b*) Non-conventional methods

(*c*) **Wind power:** High velocities of wind (in some areas) are utilized in driving wind turbines coupled to alternators. Wind power has a main advantage of having zero production cost. The cost of the equipment and the limit of generating-unit-rating is suitable for a particular location (= geographically) are the important constraints. This method has exclusive advantages of being pollution free and renewable. It is available in plentiful quantity, at certain places. It suffers from the disadvantages of its availability being uncertain (since dependent on nature) and the control being complex (since wind-velocity has wide range of variation, as an input, and the output required is at constant voltage and constant frequency). Single large-power units cannot be planned due to techno-economic considerations.

(*d*) **Fuel cells:** These are devices which enable direct conversion of energy, chemically, into electrical form. This is an up-coming technology and has a special merit of being pollution-free and noise-free. It is yet to become popular for bulk-power generation.

(*e*) **Photo voltaic cells:** These directly convert solar energy into electrical energy through a chemical action taking place in solar cells. These operate based on the photo-voltaic effect, which develops an emf on absorption of ionizing radiation from Sun.

Power Scenario In India:

Following approximate statistical data give an idea about some aspects in this regard.

Total installed capacity	:	150,000 MW
Hydropower	:	50,000 MW
Nuclear	:	10,000 MW
Thermal (fossil fuels)	:	80,000 MW
Other methods	:	10,000 MW

Other methods include partly exploited Potential such as Wind, Solar, Co-generation, Methods using Bio-fuels, etc.

20.4. Brief Aspects of Electrical Energy Systems

20.4.1. Utility and Consumers

At generating stations, power is generated at the best locations. Load-centres are generally away from these. Generation-units and Loads are connected by transmission systems. Thus, the energy system is divided into two main parts.

(*a*) Utility (including sources and transmission network) and,

(*b*) Consumers (who utilize the electrical energy)

20.4.2. Why is the three- phase a. c. system most popular?

(*a*) It is well known that a.c. generation is simpler (than d.c. generation through electrical machines because of absence of commutators in a.c. machines). Further, mechanical commutation system in d.c. machines sets an upper limit of their size, while the rating of the individual generators in ***modern power stations*** is too large, say about 1500 times the rating of a single largest feasible d.c. machine. A.C. further facilitates in stepping down or stepping-up of a voltage to suit a particular requirement, with the help of a simple device, the well-known transformer.

(*b*) Changing over from a.c. to d.c. is very easy these days due to the rectifiers of sufficiently high power ratings, so that a wide range of d.c. – applications can now be catered to.

(*c*) As the number of phases goes on increasing the power-output (from a device using a given quantity of active material, namely, that used for core and for winding) increases, but the number of circuits (*i.e.* connecting lines/ wires, switches, etc) also increases.

These two points are contradictory. A choice will be in favour of such a number (of phases) which will be high enough from power-output point of view but low enough from viewpoints of complexity of connecting/controlling large number of phases. There is a Golden compromise. It has resulted into popularity of the three-phase a. c. systems over the entire time period. Now it has almost become a standard practice for all purposes.

20.4.3. Cost of Generation

Cost of generation for one unit of electrical energy depends on the method of generation, formulae worked to assess its running cost under the specified conditions, and the cost of transmission line loss to transport power upto the load. These days, a modern utility (= electricity board) has a large number of generators sharing the responsibility of supplying power to all the customers connected to the Grid (= common supply-network). Then, for an increase in load-demand, at a known location, the most economical generating unit is to be identified and that unit should be monitored to meet the increased demand.

20.4.4. Staggering of Loads during Peak-demand Hours

Incentives to consumers (by way of supplying at reduced rates during light load hours e.g. Late night 'hours, after noon hours) help in even demand throughout the day.

20.4.5. Classifications of Power Transmission

(*a*) Using underground cables or using overhead transmission lines.

(*b*) Extra High Voltage A.C.-versus-Extra High Voltage D.C. transmission systems.

20.4.6. Selecting A.C. Transmission Voltage for a Particular Case

In general, for transporting a given power of V I watts, either V can be high or V can be low. Accordingly, the current can be either low or high respectively. Higher voltage means higher cost of insulation, and larger clearances. Higher current means larger cross section of conductors. Considering these together, the most economical voltage has to be found out for a particular requirement.

20.5. Conventional Sources of Electrical Energy

Thermal (coal, gas, nuclear) and hydro-generations are the main conventional methods of generation of Electrical Energy. These enjoy the advantages of reaching perfections in technologies for these processes. Further, single units rated at large power-outputs can be manufactured along with main components, auxiliaries and switch- gear due to vast experiences during the past century. These are efficient and economical.

These suffer from the disadvantages listed below:

1. The fuels are likely to be depleted in near future, forcing us to conserve them and find alternative resources.
2. Toxic, hazardous fumes and residues pollute the environment.
3. Overall conversion efficiency is poor.
4. Generally, these are located at remote places with respect to main load centres, increasing the transmission costs and reducing the system efficiency.
5. Maintenance costs are high.

Out of these, only two such types will be dealt here, which have a steam turbine working as the prime mover. While remaining two use Internal Combustion Engines (I.C. engines) or Gas turbine as the prime mover, and these will *not* be dealt with, in this Introductory treatment.

The steam-turbine driven systems are briefly discussed below.

20.5.1. Steam Power Stations (Coal-fired)

India has rich stocks of coal as a natural resource. Chemical energy stored within the coal is finally transformed into Electrical energy through the process of these stations. Heat released by the combustion of coal produces steam in a boiler at elevated temperatures and pressures. It is then passed through steam turbines, which drive the alternator, the output of which is the electrical energy.

Figure 20.1 shows a simple schematic diagram of a modern coal-fired thermal station.

In India, coal is generally of low grade containing ash upto 40 %. It poses two problems. ***(i)*** Calorific value is low and hence system efficiency can be increased only by additional processes like pulverizing the coal and using oil firing to start with, ***(ii)*** large volumes of ash have to be handled after ensuring that ash is extracted to the maximum possible extent (upto 99 %) by using electrostatic precipitators, before flue gases from boiler are finally passed on to the atmosphere.

Coal is burnt in the boiler. This heat converts water into steam when passed through the boiler tubes. Modern plants have super heaters to raise the temperature and pressure of steam so that plant efficiency is increased. Condenser and cooling tower deal with steam coming out of turbine. Here, maximum heat is extracted from steam (which then takes the form of water) to pre-heat the incoming water and also to recycle the water for its best utilization.

Steam-turbine receives controlled steam from boiler and converts its energy into mechanical energy which drives the 3-ph a. c. generator (=alternator). The alternator delivers electrical energy, at its rated voltage (which may be between 11 to 30 kV). Through a circuit breaker, the step up transformer is supplied. Considering the bulk-power to be transmitted over long distances, the secondary rating of this transformer may be 220 or 400 kV (as per figures for India). Through Extra-High Voltage Switch gear, the Bus is energized and the EHV primary transmission line can transport power to the Load centres connected to it.

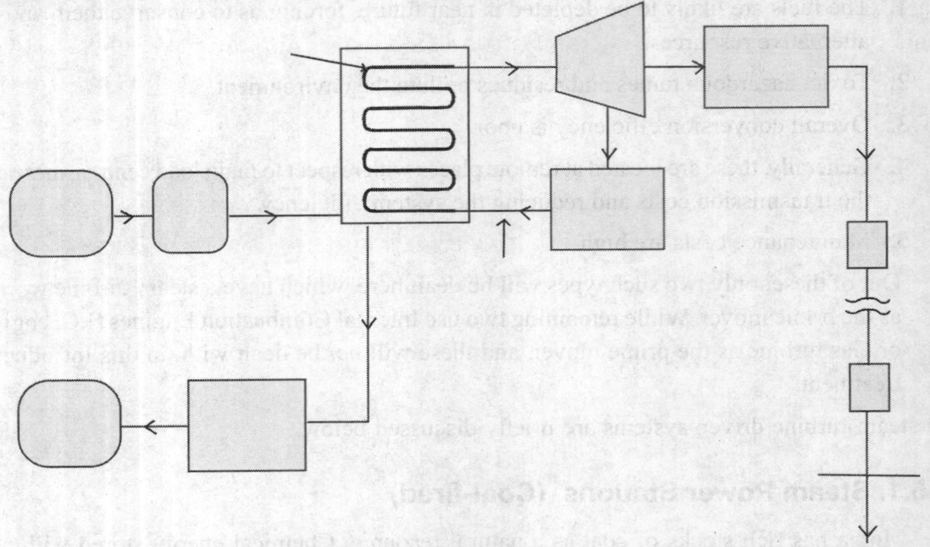

Fig. 20.1. Schematic Diagram of a Coal-fired Thermal Station

A modern coal-fired thermal power station consumes about 10 % of its power for supplying to the Auxiliaries. These are mainly as follows.

(*a*) Main-exciter for alternator.

(*b*) Water pumps.

(*c*) Fans: Forced draught and Induced draught fans for Pre-heaters and Chimney.

(*d*) Coal handling plant including pulverising mill.

(*e*) Ash handling plant including Electrolytic Precipitator.

Naturally, whenever such a station is to be brought into operation (either at commissioning or after repairs/maintenance schedule) the power required for the auxiliaries has to be supplied by the grid. Once the system is energized fully, it will look after supplying power to its own auxiliaries.

Merits of Coal-fired thermal stations

1. Fuel (=coal) is cheap. **2.** Less initial cost is required. **3.** It requires less space.
4. As a combination of all above points, the cost of generating unit of electrical energy is less.

Demerits

1. Atmospheric pollution is considerable.

2. Coal may have to be transported over long distances, in some cases, after some years, and then the energy cost may be quite high.

20.5.2. Nuclear Power Stations

Nuclear energy is available as a result of fission reaction. In a typical system, Uranium 235 is bombarded with neutrons and Heat energy is released. In chain-reaction, these release more neutrons, since more Uranium 235 atoms are fissioned. Speeds of Neutrons must be reduced to critical speeds for the chain reaction to take place. Moderators (= speed-reducing agents like graphite, heavy water, etc). are used for this purpose. Nuclear fuel rods (of Uranium 235) must be embedded in speed-reducing agents. Further, control rods (made of cadmium) are required since they are strong neutron-absorbers and help in finely regulating this reaction so that power control of the generator is precisely obtainable. When control rods are pulled out and are away from fuel rods, intensity of chain reaction increases, which increases the power output of the system. While if they are pushed in and closer to the fuel rods, the power-output decreases. Thus, the electrical load demand on the generator decides (automatically) the control-rod positions through a very sophisticated control system.

Fig. 20.2 shows a basic scheme of such a Nuclear power-station.

Advantages of Nuclear Generation

1. Quantity of fuel required is small for generating a given amount of electrical energy, compared to that with other fuels.
2. It is more reliable, cheaper for running cost, and is efficient when operated at rated capacity.

Disadvantages

1. Fuel is expensive and not abundantly available everywhere.
2. It has high capital cost.
3. Maintenance charges are high.
4. Nuclear waste disposal is a problem.

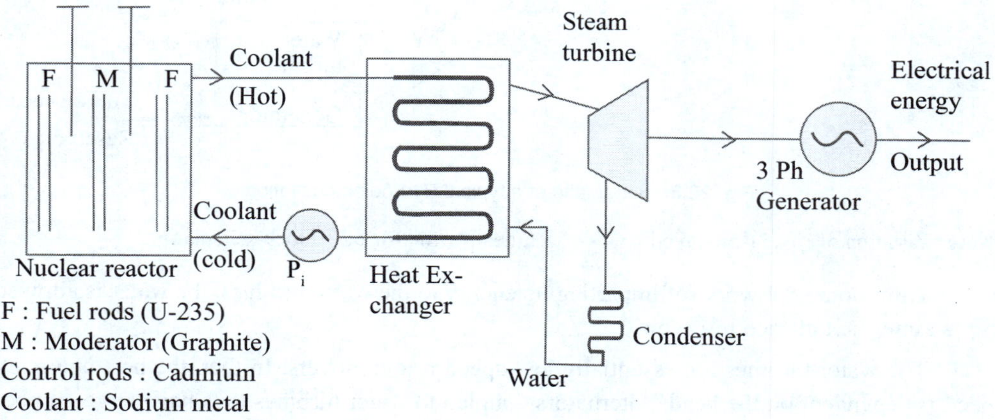

Fig. 20.2. Schematic Diagram of a Nuclear Power Station indicating Main Components

20.5.3. Hydroelectric Generation

Water-reservoir at higher altitudes is a pre-requisite for this purpose. Power-house is located at a lower level. The difference in these two levels is known as "Head."

Based on the "Heads", the Hydroelectric stations are categorized below:

1. Low head up to 60 metres.

2. Medium head between 60 and 300 metres.
3. High heads above 300 metres.

In this method of generation, water from higher height is passed through penstock as controlled in the valve-house, into the water turbine. Thus, potential energy of water stored at higher altitudes is first converted into Kinetic energy. As the water reaches the turbine, it gains speed after losing the Potential energy. Kinetic energy of this speedy water drives the water turbine, which converts this into mechanical output. It drives the coupled generator, which gives Electrical energy output.

A schematic diagram of such a system is shown in fig.

The valve house has a controlling valve (=main sluice valve) and a protecting valve (= an automatic, isolating, "butterfly" type valve). As is obvious, power control is done by the main sluice valve, while "butterfly" valve comes into action if water flows in opposite direction as a result of a sudden drop in load on the generator. Otherwise, the penstock is subjected to extreme strains and it has a tendency to burst due to pressure of water as a result of sudden load reduction.

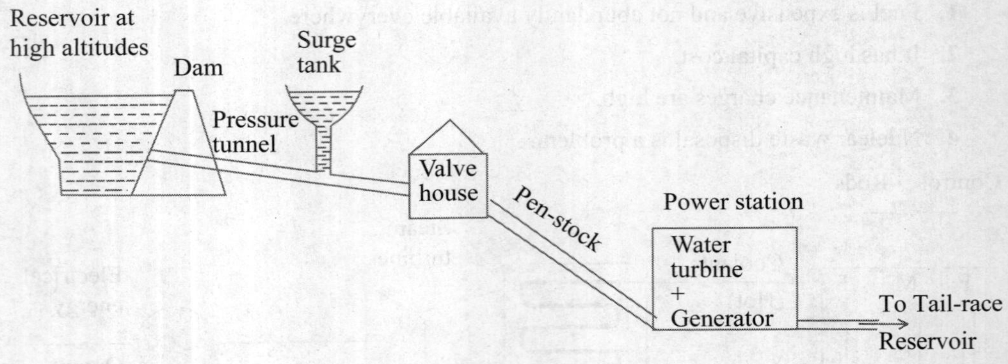

Fig. 20.3. Schematic of a typical Hydroelectric station

Note: Variation of Head (low-, medium-, or high-head) will affect the block-schematic

After doing the work (of imparting its energy to the water turbine), the water is allowed to pass into the tail-race reservoir.

The water turbines are essentially low-speed prime movers. In that, the best operating speed is dependent on the head. Alternators coupled to water turbines thus have large number of poles (since P= 120 f/N). Such alternators have the Salient-Pole type rotor.

There are different types of water turbines suitable for different cases (i.e. Heads, Power rating, Load-variation curve, etc). Since, this is only an elementary treatment, these aspects will not be discussed here.

20.6. Non-Conventional Energy Sources

Considering the previously discussed thermal methods of conventional energy-generations, it is necessary to understand the non-conventional energy sources, since they have two points in their favour.

(*a*) Non-polluting processes are used.

(*b*) Perpetuity and renewability of the main source (which is a natural atmospheric resource) generally exists.

The non-conventional energy sources are further advantageous due to virtually zero running cost, since wind energy or solar energy is the input-source of power.

However, they are disadvantageous due to high initial cost (per MW of installed capacity) and due to uncertainty resulting out of weather changes. For example, dense clouds (or night hours) lead to non-availability of solar energy. Similarly, "still-air" condition means no possibility of wind power generation, and during stormy weathers, wind turbines cannot be kept in operation (due to dangerously high speeds they would attain if kept in operation).

20.6.1. Photo Voltaic Cells (P.V. Cells or SOLAR Cells)

When ionized solar radiation is incident on a semi-conductor diode, energy conversion can take place with a voltage of 0.5 to 1 volt (d.c.) and a current density of 20-40 mA/cm^2, depending on the materials used and the conditions of Sunlight. Area of these solar cells decides the current output. An array of large number of such diodes (i.e. Solar cells) results into higher d.c. output voltage.

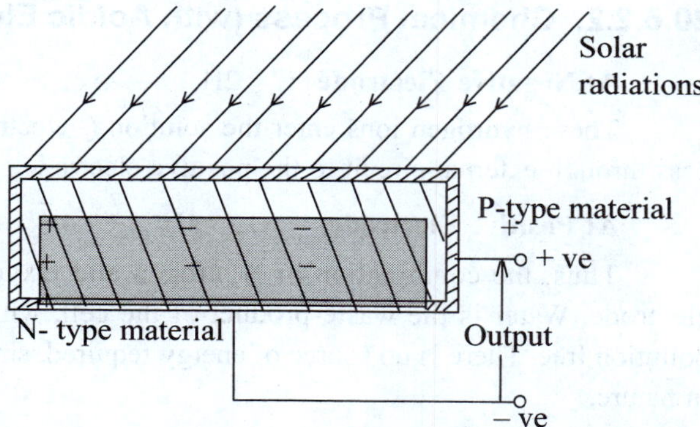

Fig. 20.4. Photo Voltaic (or Solar Cell)

Since, the final form of electrical energy required is generally an alternating current, it is realized from d.c. using inverters.

At quite a few locations in India, for realizing few hundred kilowatts of power-rating, huge arrays are accommodated in horizontal as well as vertical stacks, so that land area required is not too vast. Electrically, they are connected in series and in parallel combinations of cells so that rated voltage and current are realized.

Just to understand the principle of operation of solar cells, let a semi-conductor diode receive ionized radiation from Sun, as in Fig. 20.4.

Typical materials used for these cells are: material doped with boron, cadmium sulphide, gallium-arsenide, etc. Their choice is mainly decided by conversion efficiency. Best material may lead to the efficiency being typically 15%. Since solar energy is available free of cost, this low-efficiency does not matter.

This method suffers from the disadvantages of having high initial cost and uncertainty (since dependent on weather conditions) including non operative night periods. Main advantages are: **(*i*)** no running cost (however, replacements of components may be a botheration), **(*ii*)** no pollution, **(*iii*)** location can be near the load (hence transportation of power is not required over long distances). **(*iv*)** since natural source is involved, it is perpetual.

Individual stations using solar cells are in operation with ratings of the order of 250-1000 kW. With manufacturing costs of semi-conductor devices going down and with the advent of better and better quality of cells which will be available in future, this method of generation has bright prospects.

20.6.2. Fuel Cells

20.6.2.1. Principle of Operation

In Fuel cells, negative porous electrode is fed by hydrogen and the positive porous electrode is fed by oxygen. Both the electrodes are immersed in an electrolyte. The porous electrodes are made of such a conducting material that both the fuel (oxygen and hydrogen) and the electrolyte can pass through them. Such a material for electrodes is nickel. The electrolyte is a solution of sulphuric acid or potassium hydroxide. The electrodes have a catalyst (= platinum or sintered nickel) which break the fuel compound into more reactive atoms.

20.6.2.2. Chemical Process (with Acidic Electrolyte)

At Negative Electrode : $2H_2 \longrightarrow 4H^+ + 4\overline{e}$

These hydrogen ions enter the solution (=electrolyte) leaving behind electrons which pass through external circuit to the positive electrode.

At Positive Electrode: $O_2 + 4H^+ + 4\overline{e} \rightarrow 2H_2O$

Thus, the combination of Hydrogen and Oxygen results into water at the positive electrode. Water is the waste-product of the cell, which is harmless. The process is, thus, pollution free. There is no source of energy required, since the process is basically **"chemical"** in nature.

20.6.2.3. Schematic Diagram. **It is given in fig 20.5**

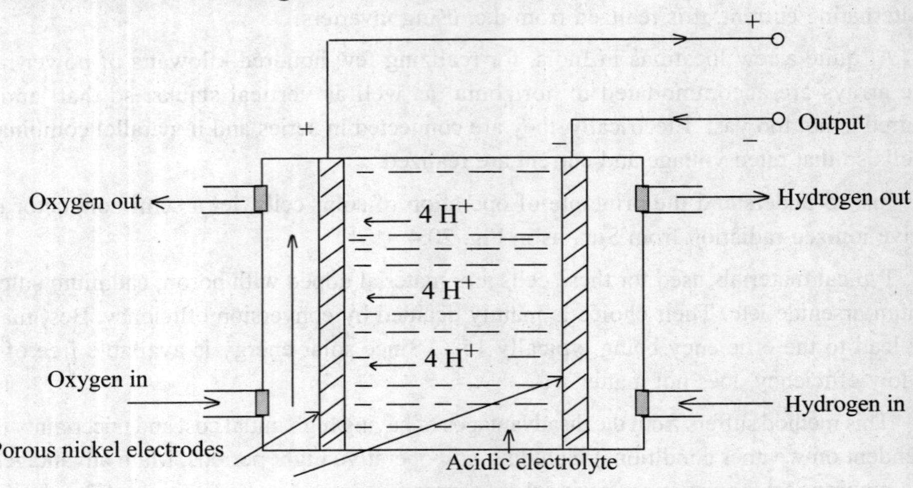

Fig. 20.5. Fuel Cell

20.6.2.4. Array for Large outputs

A fuel cell has a d. c. output voltage typically of 1.23 volts at normal atmospheric pressure and temperature. Raising pressure and temperature increases this voltage. To realize large output parameters (= voltage, current, and hence power), an array of a large number of fuel cells (connected in series as well as in parallel) is made. Voltage levels of 100 to 1000 V and power levels in kilowatts can be realized.

20.6.2.5. High Lights

1. Pollution-free, noiseless.
2. No outside source of energy is required.
3. Efficient.
4. No restriction on location

   (*a*) High initial cost. (*b*) Working life is short.

Note: Solar energy can also be used for generating electrical energy through an intermediate stage of producing steam, which is used later for driving an alternator. However, this method is not discussed here.

20.6.3. Wind Power

20.6.3.1. Introduction

Wind power has been in use for serving the mankind, since centuries through what has been popularly known as "Wind-mills." There is no "electrical" stage of energy in old-styled uses where wind-velocity is directly used for performing the jobs such as wheat-grinding, pumping water for irrigation, sailing vessels, etc. It enjoys the advantages of being plentiful, inexhaustible, renewable and non-polluting, over and above being cheap for running costs. It suffers from disadvantages of being unreliable, and being economically un-viable for large power generation. In India, a large number of such units with small and medium power ratings (up to 100 kW) are already in operation mainly in coastal or hilly areas. With the modern system, it is now preferred to have suitable power-control circuits on the output side of wind-generators so that these can pump energy into low voltage lines of the grid over a wide range of variation of wind speeds.

20.6.3.2. Basic Scheme

A large variety of wind-turbines naturally exist arising out of large variation in wind-pattern all over and out of different manufacturers producing systems with different designs. Since the aim here is to understand the basic system, only one type of system is presented here.

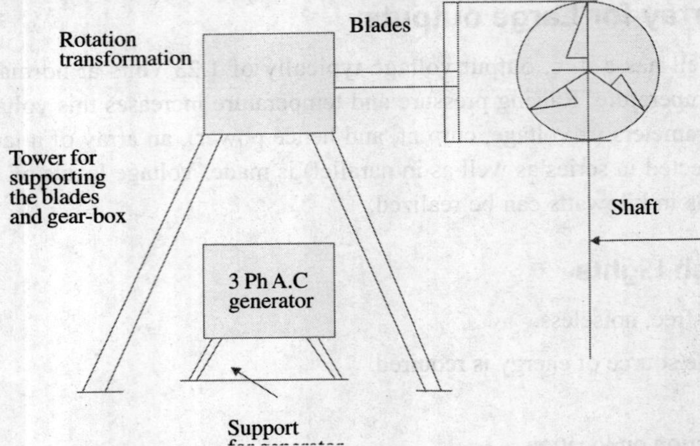

Fig. 20.6 (*a*) . Wind-generation, a Schematic view **Fig. 20.6 (*b*)**. Part-side-view to show a typical three bladed wind turbine

In Fig. 20.6 (*a*) an arrangement wherein, a horizontal three-bladed system is shown mounted on a tower. Through rotation-transformation using gears to step up the speed and to link the horizontal axis of turbine with vertical axis of generator. The speed of wind varies, as such turbine speed also varies so that output frequency and voltage of three-phase alternator vary over a wide range. Further, its waveform is also a distorted one. To increase its utility, it is necessary to modulate (through proper power-control) to derive line-frequency constant voltage output and hook-up to local grid for pumping the available wind-energy into it. This is schematically represented in Fig 20.7.

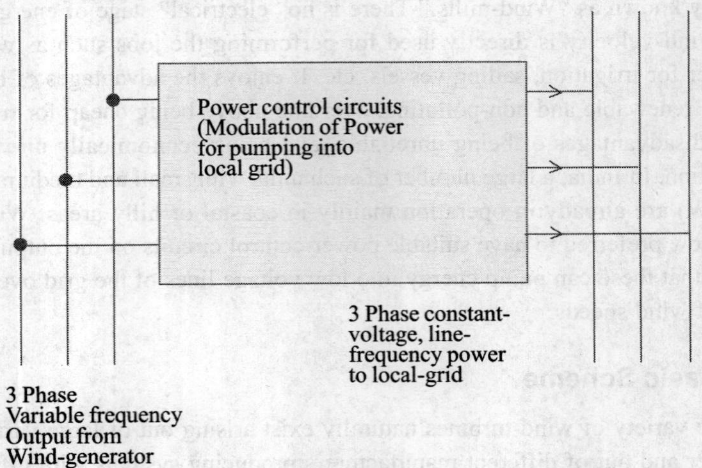

Fig. 20.7